# About the Cover

The graph on the cover of this book is reproduced below with more contextual information, so you can tell what the data *mean*. Each point represents an operating roller coaster in the United States. The horizontal coordinate of a point indicates the height of the roller coaster, while its vertical coordinate indicates the coaster's maximum speed. The color indicates what roller coaster aficionados call its "scale"—Kiddie, Family, Thrill, or Extreme.

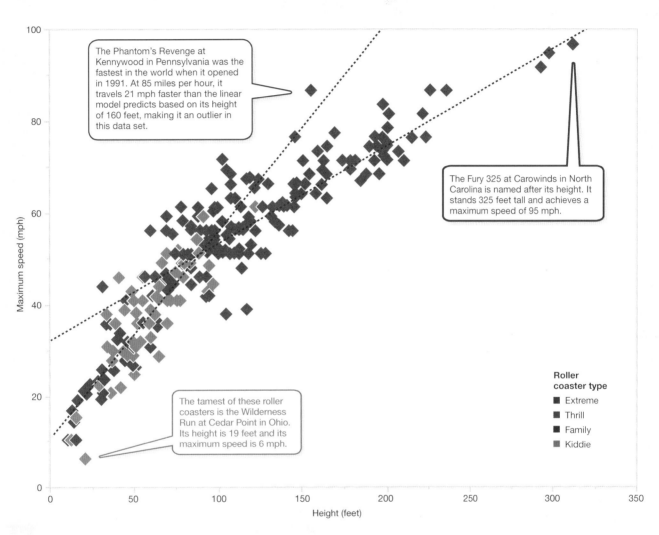

The graph shows that—not surprisingly!—taller roller coasters tend to go faster. But it shows more than just that. We can see how strong the association is—very, indicating that height is a good predictor of roller coaster speed. More subtly, we can see that the relationship between Height and Speed is not quite linear, but slightly curved. Additional height tends to confer additional speed, but less so for roller coasters that are already among the tallest and fastest. For that reason, we fit two separate regression lines to these data points. The less steep of the two lines models the relationship between Speed and Height for only the Extreme rides; the steeper line models that relationship for all the other rides.

In Chapter 7 of this text you'll learn to interpret regression analyses, such as the computer output shown below, which gives information about the relationship between Speed and Height for the 244 Extreme roller coasters depicted in red. For example, the line's slope is about 0.20, indicating that these coasters are predicted to be about one mile per hour faster for each additional five feet of height. We also learn from the R-square value of about 0.78 that nearly 80% of the variability in speeds can be explained by this model based on only height.

**Linear Fit is extreme==1**
Speed = 31.864511 + 0.1985705*Height
**Summary of Fit**
| | |
|---|---|
| RSquire | 0.778813 |
| RSquire Adj | 0.777899 |
| Root Mean Square Error | 6.197679 |
| Mean of Response | 57.55082 |
| Observations (or Sum Wgts) | 244 |

**Parameter Estimates**
| Terms | Estimate | Std Error | t Ratio | Prob>|t| |
|---|---|---|---|---|
| Intercept | 31.864511 | 0.965263 | 33.01 | <.0001* |
| Height | 0.1985705 | 0.006803 | 29.19 | <.0001* |

*We're grateful to the owners of rcdb.com (roller coaster database) for sharing their data with the world.*

**SIXTH EDITION**

# Stats
## Modeling The World

**David E. Bock**
Ithaca High School (Retired)

**Floyd Bullard**
North Carolina School of Science and Mathematics

**Paul F. Velleman**
Cornell University

**Richard D. De Veaux**
Williams College

*with contributions by*

**Corey Andreasen**
The American School of The Hague

and

**Jared Derksen**
Rancho Cucamonga High School

**Content Management:** *Dawn Murrin, Laura Briskman, Monique Bettencourt*
**Content Production:** *Robert Carroll, Jean Choe, Rachel S. Reeve, Prerna Rochlani*
**Product Management:** *Karen Montgomery*
**Product Marketing:** *Demetrius Hall, Jordan Longoria*
**Rights and Permissions:** *Tanvi Bhatia, Rimpy Sharma*

Please contact https://support.pearson.com/getsupport/s/ with any queries on this content

Cover Image by Apomares/E+/Getty Images, Elenamiv/Shutterstock

Microsoft and/or its respective suppliers make no representations about the suitability of the information contained in the documents and related graphics published as part of the services for any purpose. All such documents and related graphics are provided "as is" without warranty of any kind. Microsoft and/or its respective suppliers hereby disclaim all warranties and conditions with regard to this information, including all warranties and conditions of merchantability, whether express, implied or statutory, fitness for a particular purpose, title and non-infringement. In no event shall Microsoft and/or its respective suppliers be liable for any special, indirect or consequential damages or any damages whatsoever resulting from loss of use, data or profits, whether in an action of contract, negligence or other tortious action, arising out of or in connection with the use or performance of information available from the services.

The documents and related graphics contained herein could include technical inaccuracies or typographical errors. Changes are periodically added to the information herein. Microsoft and/or its respective suppliers may make improvements and/or changes in the product(s) and/or the program(s) described herein at any time. Partial screen shots may be viewed in full within the software version specified.

Microsoft® and Windows® are registered trademarks of the Microsoft Corporation in the U.S.A. and other countries. This book is not sponsored or endorsed by or affiliated with the Microsoft Corporation.

Copyright © 2023, 2019, 2015 by Pearson Education, Inc. or its affiliates, 221 River Street, Hoboken, NJ 07030. All Rights Reserved. Manufactured in the United States of America. This publication is protected by copyright, and permission should be obtained from the publisher prior to any prohibited reproduction, storage in a retrieval system, or transmission in any form or by any means, electronic, mechanical, photocopying, recording, or otherwise. For information regarding permissions, request forms, and the appropriate contacts within the Pearson Education Global Rights and Permissions department, please visit www.pearsoned.com/permissions/.

Acknowledgments of third-party content appear on the appropriate page on page A-79, which constitutes an extension of this copyright page.

PEARSON and MYLAB are exclusive trademarks owned by Pearson Education, Inc. or its affiliates in the U.S. and/or other countries.

Unless otherwise indicated herein, any third-party trademarks, logos, or icons that may appear in this work are the property of their respective owners, and any references to third-party trademarks, logos, icons, or other trade dress are for demonstrative or descriptive purposes only. Such references are not intended to imply any sponsorship, endorsement, authorization, or promotion of Pearson's products by the owners of such marks, or any relationship between the owner and Pearson Education, Inc., or its affiliates, authors, licensees, or distributors.

**Library of Congress Cataloging-in-Publication Data On File**

2 2022

Print Offer
ISBN-10: 0-13-768535-1
ISBN-13: 978-0-13-768535-6
Rental
ISBN-10: 0-13-768539-4
ISBN-13: 978-0-13-768539-4

# Pearson's Commitment to Diversity, Equity, and Inclusion

**Pearson is dedicated to creating bias-free content that reflects the diversity, depth, and breadth of all learners' lived experiences.**

We embrace the many dimensions of diversity, including but not limited to race, ethnicity, gender, sex, sexual orientation, socioeconomic status, ability, age, and religious or political beliefs.

Education is a powerful force for equity and change in our world. It has the potential to deliver opportunities that improve lives and enable economic mobility. As we work with authors to create content for every product and service, we acknowledge our responsibility to demonstrate inclusivity and incorporate diverse scholarship so that everyone can achieve their potential through learning. As the world's leading learning company, we have a duty to help drive change and live up to our purpose to help more people create a better life for themselves and to create a better world.

**Our ambition is to purposefully contribute to a world where:**

- Everyone has an equitable and lifelong opportunity to succeed through learning.
- Our educational content accurately reflects the histories and lived experiences of the learners we serve.
- Our educational products and services are inclusive and represent the rich diversity of learners.
- Our educational content prompts deeper discussions with students and motivates them to expand their own learning (and worldview).

### Accessibility

We are also committed to providing products that are fully accessible to all learners. As per Pearson's guidelines for accessible educational Web media, we test and retest the capabilities of our products against the highest standards for every release, following the WCAG guidelines in developing new products for copyright year 2022 and beyond.

 You can learn more about Pearson's commitment to accessibility at
**https://www.pearson.com/us/accessibility.html**

### Contact Us

While we work hard to present unbiased, fully accessible content, we want to hear from you about any concerns or needs with this Pearson product so that we can investigate and address them.

 Please contact us with concerns about any potential bias at
**https://www.pearson.com/report-bias.html**

 For accessibility-related issues, such as using assistive technology with Pearson products, alternative text requests, or accessibility documentation, email the Pearson Disability Support team at **disability.support@pearson.com**

*To Greg and Becca, great fun as kids and great friends as adults,
and especially to my wife and best friend, Joanna,
for her understanding, encouragement, and love*

—*Dave*

*To my husband Rick, who is always supportive and caring and
loving, and to whom I am so grateful for everything*

—*Floyd*

*To my sons, David and Zev, from whom I've learned so much,
and to my wife, Sue, for taking a chance on me*

—*Paul*

*To Sylvia, who has helped me in more ways than she'll ever know,
and to Nicholas, Scyrine, Frederick, and Alexandra,
who make me so proud in everything that they are and do*

—*Dick*

# MEET THE AUTHORS

**David E. Bock** taught mathematics at Ithaca High School for 35 years. He has taught Statistics at Ithaca High School, Tompkins-Cortland Community College, Ithaca College, and Cornell University. Dave has won numerous teaching awards, including the MAA's Edyth May Sliffe Award for Distinguished High School Mathematics Teaching (twice), Cornell University's Outstanding Educator Award (three times), and has been a finalist for New York State Teacher of the Year.

Dave holds degrees from the University at Albany in Mathematics (B.A.) and Statistics/Education (M.S.). Dave has been a reader and table leader for the AP Statistics exam and a Statistics consultant to the College Board, leading workshops and institutes for AP Statistics teachers. His understanding of how students learn informs much of this book's approach.

**Floyd Bullard** first taught high school math as a Peace Corps volunteer in Benin, West Africa, when he was 23 years old. Today he teaches at the North Carolina School of Science and Mathematics in Durham, North Carolina, where he has been since 1999. Floyd has served on the AP Statistics test development committee and presents regularly at workshops and conferences for Statistics teachers.

Floyd's academic degrees are from the Johns Hopkins University (B.S., Applied Mathematics, 1991), the University of North Carolina at Chapel Hill (M.S., Statistics, 1997), and Duke University (Ph.D., Statistics, 2009). He plays Dungeons & Dragons regularly and also enjoys playing the piano.

**Paul F. Velleman** has an international reputation for innovative Statistics education. He is the author and designer of the multimedia Statistics program *ActivStats*, for which he was awarded the EDUCOM Medal for innovative uses of computers in teaching statistics, and the ICTCM Award for Innovation in Using Technology in College Mathematics. He also developed the award-winning statistics program, *Data Desk*, the Internet site Data and Story Library (DASL) (DASL.datadesk.com), which provides data sets for teaching Statistics (and is one source for the datasets used in this text), and the tools referenced in the text for simulation and bootstrapping. Paul's understanding of using and teaching with technology informs much of this book's approach.

Paul taught Statistics at Cornell University, where he was awarded the MacIntyre Award for Exemplary Teaching. He is Emeritus Professor of Statistical Science from Cornell and lives in Maine with his wife, Sue Michlovitz. He holds an A.B. from Dartmouth College in Mathematics and Social Science, and M.S. and Ph.D. degrees in Statistics from Princeton University, where he studied with John Tukey. His research often deals with statistical graphics and data analysis methods. Paul co-authored (with David Hoaglin) *ABCs of Exploratory Data Analysis*. Paul is a Fellow of the American Statistical Association and of the American Association for the Advancement of Science. Paul is the father of two boys. In his spare time he sings with the *a capella* group VoXX and studies tai chi.

**Richard D. De Veaux** is an internationally known educator and consultant. He has taught at the Wharton School and the Princeton University School of Engineering, where he won a "Lifetime Award for Dedication and Excellence in Teaching." He is the C. Carlisle and M. Tippit Professor of Statistics at Williams College, where he has taught since 1994. Dick has won both the Wilcoxon and Shewell awards from the American Society for Quality. He is a fellow of the American Statistical Association (ASA) and an elected member of the International Statistical Institute (ISI). In 2008, he was named Statistician of the Year by the Boston Chapter of the ASA and was the 2018–2021 Vice-President of the ASA. Dick is also well known in industry, where for more than 30 years he has consulted for such Fortune 500 companies as American Express, Hewlett-Packard, Alcoa, DuPont, Pillsbury, General Electric, and Chemical Bank. Because he consulted with Mickey Hart on his book *Planet Drum*, he has also sometimes been called the "Official Statistician for the Grateful Dead." His real-world experiences and anecdotes illustrate many of this book's chapters.

Dick holds degrees from Princeton University in Civil Engineering (B.S.E.) and Mathematics (A.B.) and from Stanford University in Dance Education (M.A.) and Statistics (Ph.D.), where he studied dance with Inga Weiss and Statistics with Persi Diaconis. His research focuses on the analysis of large data sets and data mining in science and industry.

In his spare time, he is an avid cyclist and swimmer. He also is the founder of the "Diminished Faculty," an *a cappella* Doo-Wop quartet at Williams College and sings bass in the college concert choir and with the Choeur Vittoria of Paris. Dick is the father of four children.

# TABLE OF CONTENTS

Preface xi

## PART I Exploring and Understanding Data

### 1 Stats Starts Here 1
So, What Is (Are?) Statistics? ◆ Statistics in a Word ◆ But What *Are* Data? ◆ *Who* and *What* ◆ How the Data Are Collected ◆ More About Variables (*What*?)

### 2 Displaying and Describing Categorical Data 14
The Three Rules of Data Analysis ◆ The Area Principle ◆ Frequency Tables: Making Piles ◆ Bar Charts ◆ Pie Charts ◆ Contingency Tables: Children and First-Class Ticket Holders First? ◆ Conditional Distributions ◆ Other Types of Graphs ◆ Is the Difference Meaningful?

### 3 Displaying and Summarizing Quantitative Data 44
Histograms ◆ Stem-and-Leaf Displays (aka Stemplots) ◆ Dotplots ◆ Cumulative Distribution Graphs ◆ Think Before You Draw, Again ◆ The Shape of a Distribution ◆ A Measure of Center: The Median ◆ Spread: Home on the Range ◆ Spread: The Interquartile Range ◆ 5-Number Summary ◆ Boxplots ◆ Summarizing Symmetric Distributions: The Mean ◆ Mean or Median? ◆ What About Spread? The Standard Deviation ◆ What to *Tell* About a Quantitative Variable

### 4 Understanding and Comparing Distributions 86
The Big Picture ◆ Comparing Groups with Histograms ◆ Comparing Groups with Boxplots ◆ Outliers ◆ Timeplots: Order, Please! ◆ Looking into the Future ◆ *Re-expressing Data: A First Look

### 5 The Standard Deviation as a Ruler and the Normal Model 110
The Standard Deviation as a Ruler ◆ Standardizing with $z$-Scores ◆ Shifting Data: Move the Center ◆ Rescaling Data: Adjust the Scale ◆ Back to $z$-Scores ◆ When Is a $z$-Score BIG? ◆ The 68–95–99.7 Rule ◆ The First Three Rules for Working with Normal Models ◆ Determining Normal Percentiles ◆ From Percentiles to Scores: $z$ in Reverse ◆ *Are You Normal? Find Out with a Normal Probability Plot ◆ The Truth About Assessing Normality

Review of Part I Exploring and Understanding Data 143

## PART II Exploring Relationships Between Variables

### 6 Scatterplots, Association, and Correlation 155
Looking at Scatterplots; Describing Associations ◆ Roles for Variables ◆ Correlation ◆ Correlation Conditions ◆ Correlation Properties ◆ Accidental Correlation ◆ Warning: Correlation ≠ Causation

### 7 Linear Regression 179

Residuals ◆ Squared Residuals as a Measure of Fit ◆ Correlation and the Line ◆ How Big Can Predicted Values Get? ◆ The Regression Line in Real Units ◆ Residuals Revisited ◆ The Typical Residual Size (aka Standard Error) ◆ Different Samples, Different Regression Lines ◆ $R^2$—The Variation Accounted For ◆ How Big Should $R^2$ Be? ◆ More About Regression Assumptions and Conditions ◆ A Tale of Two Regressions ◆ Reality Check: Is the Regression Reasonable?

### 8 Regression Wisdom 215

Getting the "Bends": When the Residuals Aren't Straight ◆ Extrapolation: Reaching Beyond the Data ◆ Outliers, Leverage, and Influence ◆ Lurking Variables and Causation ◆ Working with Summary Values

### 9 Re-expressing Data: Get It Straight! 239

Straight to the Point ◆ Goals of Re-expression ◆ The Ladder of Powers ◆ Plan B: Attack of the Logarithms ◆ Why Not Just Use a Curve?

**Review of Part II Exploring Relationships Between Variables** 264

## PART III Gathering Data

### 10 Understanding Randomness 276

It's Not Easy Being Random ◆ Let's Simulate!

### 11 Sample Surveys 290

Idea 1: Examine a Part of the Whole ◆ Bias ◆ Idea 2: Randomize ◆ Idea 3: It's the Sample Size ◆ Does a Census Make Sense? ◆ Populations and Parameters ◆ Simple Random Samples ◆ Stratified Sampling ◆ Stratified Sampling: A Simulation ◆ Cluster and Multistage Sampling ◆ Systematic Samples ◆ The Valid Survey ◆ Lots Can Go Wrong: How to Sample Badly

### 12 Experiments and Observational Studies 315

Observational Studies ◆ Randomized, Comparative Experiments ◆ The Four Principles of Experimental Design ◆ Diagrams ◆ Does the Difference Make a Difference? ◆ Experiments and Samples ◆ Control Treatments ◆ Blinding ◆ Placebos ◆ Blocking ◆ Adding More Factors ◆ Confounding ◆ Lurking or Confounding? ◆ Statistical Significance, Revisited

**Review of Part III Gathering Data** 343

## PART IV Randomness and Probability

### 13 From Randomness to Probability 354

Random Phenomena ◆ The Law of Large Numbers ◆ The Nonexistent "Law of Averages" ◆ Modeling Probability ◆ Subjective Probability ◆ The First Three Rules for Working with Probability ◆ Formal Probability ◆ Checking for Independence

## 14 Probability Rules! 374
The General Addition Rule ◆ It Depends . . . ◆ Independence ◆ Independent ≠ Disjoint ◆ Tables, Venn Diagrams, and Probability ◆ The General Multiplication Rule ◆ Drawing Without Replacement ◆ Tree Diagrams ◆ Reversing the Conditioning

## 15 Random Variables 402
Expected Value: Center ◆ First Center, Now Spread . . . ◆ Representations of Random Variables ◆ More About Means and Variances ◆ Continuous Random Variables

## 16 Probability Models 429
Searching for Simone: Bernoulli Trials ◆ The Geometric Model: Waiting for Success ◆ The Binomial Model: Counting Successes ◆ The Normal Model to the Rescue! ◆ Should I Be Surprised? A First Look at Statistical Significance

**Review of Part IV Randomness and Probability 450**

# PART V — From the Data at Hand to the World at Large

## 17 Sampling Distribution Models 460
The Sampling Distribution of a Proportion ◆ Can We Always Use a Normal Model? ◆ A Sampling Distribution Model for a Proportion ◆ The Sampling Distributions of Other Statistics ◆ Simulating the Sampling Distribution of a Mean ◆ The Fundamental Theorem of Statistics ◆ Assumptions and Conditions ◆ But Which Normal? ◆ The CLT When the Population Is Very Skewed ◆ About Variation ◆ The Real World vs. the Model World ◆ Sampling Distribution Models

## 18 Confidence Intervals for Proportions 488
A Confidence Interval ◆ What Does "95% Confidence" Really Mean? ◆ Margin of Error: Degree of Certainty vs. Precision ◆ Critical Values ◆ Assumptions and Conditions ◆ Choosing Your Sample Size ◆ The Proof is in the Pudding

## 19 Testing Hypotheses About Proportions 511
Hypotheses ◆ A Jury Trial as a Hypothesis Test ◆ P-Values: Are We Surprised? ◆ Alternative Alternatives ◆ The Reasoning of Hypothesis Testing ◆ P-Values and Decisions: What to Tell About a Hypothesis Test ◆ Hypothesis Testing by Simulation

## 20 More About Tests and Intervals 535
Zero In on the Null ◆ How to Think About P-Values ◆ Alpha Levels ◆ Practical vs. Statistical Significance ◆ Confidence Intervals and Hypothesis Tests ◆ *A 95% Confidence Interval for Small Samples ◆ Making Errors ◆ Power

## 21 Comparing Two Proportions 562
Another Ruler ◆ The Standard Deviation of the Difference Between Two Proportions ◆ Assumptions and Conditions ◆ A Confidence Interval ◆ Screen Time Before Bed: A Cultural Difference? ◆ Everyone into the Pool ◆ The Two-Proportion z-Test ◆ *A Permutation Test for Two Proportions

**Review of Part V From the Data at Hand to the World at Large 583**

## PART VI  Learning About The World

### 22  Inferences About Means  593

Getting Started: The Central Limit Theorem (Again) ◆ The Story of Gosset's *t* ◆ A Confidence Interval for Means ◆ Assumptions and Conditions ◆ Using Table T to Find *t*-Values ◆ Be Careful When Interpreting Confidence Intervals ◆ A Hypothesis Test for the Mean ◆ Intervals and Tests ◆ Choosing the Sample Size

### 23  Comparing Means  624

Plot the Data ◆ Comparing Two Means ◆ Assumptions and Conditions ◆ Another One Just Like the Other Ones? ◆ A Test for the Difference Between Two Means ◆ Back into the Pool? ◆ *The Pooled *t*-Test (In Case You're Curious) ◆ Is the Pool All Wet? ◆ *A Permutation Test for Two Means

### 24  Paired Samples and Blocks  652

Paired Data ◆ Assumptions and Conditions ◆ Confidence Intervals for Matched Pairs ◆ Effect Size ◆ Blocking

Review of Part VI Learning About The World  675

## Part VII  Inference When Variables Are Related

### 25  Comparing Counts  689

Goodness-of-Fit Tests ◆ Homogeneity: Comparing Observed Distributions ◆ Chi-Square Test of Independence ◆ Chi-Square and Causation

### 26  Inferences for Regression  720

The Model and the Data ◆ Assumptions and Conditions ◆ Intuition About Regression Inference ◆ Standard Error for the Slope ◆ Regression Inference ◆ Another Example: Breaking Up Is Hard to Predict ◆ Different Samples, Different Slopes

Review of Part VII Inference When Variables Are Related  753

### Appendixes

A Selected Formulas  A-1 ◆ B Guide to Statistical Software  A-3 ◆ C Answers  A-29 ◆ D Photo and Text Acknowledgments  A-79 ◆ E Index  A-87 ◆ F Tables  A-97

# PREFACE

## About the Book

Yes, a preface is supposed to be "about this book"—and we'll get there—but first we want to talk about the bigger picture: the ongoing growth of interest in Statistics. These days it seems Statistics is everywhere, from Major League Baseball's innovative StatsCast analytics to the challenges of predicting election outcomes to *Wall Street Journal* and *New York Times* articles touting the explosion of job opportunities for graduates with degrees in Statistics. Public awareness of the widespread applicability, power, and importance of statistical analysis has never been higher. Each year, more students sign up for Stats courses and discover what drew us to this field: it's interesting, stimulating, and even fun. Statistics helps students develop key tools and critical thinking skills needed to become well-informed consumers, parents, and citizens. We think Statistics isn't as much a math course as a civics course, and we're delighted that our books can play a role in preparing a generation for life in the Information Age.

## New to the Sixth Edition

This new edition of *Stats: Modeling the World* extends the series of innovations pioneered in our books, teaching Statistics and statistical thinking as it is practiced today. We've made some important revisions and additions, each with the goal of making it even easier for students to put the concepts of Statistics together into a coherent whole.

- ◆ *Full inclusion of all new topics in the AP Statistics syllabus.* In the Fall of 2020 the College Board updated the AP Statistics syllabus to better reflect the way the course is taught in colleges around the United States. With this edition of *Stats: Modeling the World* we incorporate all the changes so that teachers and students can prepare for the exam without needing to supplement the text. The new topics range from new graphical tools like the cumulative distribution plot to stricter guidelines on how to carry out some hypothesis tests. Every change is reflected in the exercises as well as the text, to give students opportunities to practice what they've learned.

- ◆ *Updated examples, exercises, and data.* We've updated our innovative *Think/Show/Tell Step-by-Step* examples with new contexts and data. We've added hundreds of new exercises and updated continuing exercises with the most recent data. Whenever possible, we've provided those data, available through MyLab Statistics or at www.pearsonhighered.com/mathstatsresources. Many of the examples and exercises are based on recent news stories, research articles, and other real-world sources. We've listed many of those sources so students can explore them further.

## Our Goal: Read This Book!

The best text in the world is of little value if students don't read it. Starting with the first edition, our goal has been to create a book that students would willingly read, easily learn from, and even like. We've been thrilled with the glowing feedback we've received from instructors and students using the first five editions of *Stats: Modeling the World*. Our conversational style, our interesting anecdotes and examples, and even our humor[1] engage

---

[1] And, yes, those footnotes!

students' interest as they learn statistical thinking. We hear from grateful instructors that their students actually do read this book (sometimes even voluntarily reading ahead of the assignments). And we hear from (often amazed) students that they actually enjoyed their textbook.

Here are some of the ways we have made *Stats: Modeling the World*, Sixth Edition, engaging:

- *Readability.* You'll see immediately that this book doesn't read like other Statistics texts. The style is both colloquial and informative, enticing students to actually read the book to see what it says.

- *Informality.* Our informal style doesn't mean that the subject matter is covered superficially. Not only have we tried to be precise, but wherever possible we offer deeper explanations and justifications than those found in most introductory texts.

- *Focused lessons.* The chapters are shorter than in most other texts, making it easier for both instructors and students to focus on one topic at a time.

- *Consistency.* We've worked hard to demonstrate how to do Statistics well. From the very start and throughout the book we model the importance of plotting data, of checking assumptions and conditions, and of writing conclusions that are clear, complete, concise, and in context.

- *The need to read.* Because the important concepts, definitions, and sample solutions aren't set in boxes, students won't find it easy to just to skim this book. We intend that it be read, so we've tried to make the experience enjoyable.

# Continuing Features

Along with the improvements we've made, you'll still find the many engaging, innovative, and pedagogically effective features responsible for the success of our earlier editions.

- *Chapter 1 (and beyond).* Chapter 1 gets down to business immediately, looking at data. And throughout the book chapters lead with new up-to-the-minute motivating examples and follow through with analyses of the data, and real-world examples provide a basis for sample problems and exercises.

- *Think, Show, Tell.* The worked examples repeat the mantra of *Think, Show,* and *Tell* in every chapter. They emphasize the importance of thinking about a Statistics question (What do we know? What do we hope to learn? Are the assumptions and conditions satisfied?) and reporting our findings (the *Tell* step). The *Show* step contains the mechanics of calculating results and conveys our belief that it is only one part of the process.

- *Step-by-Step examples* guide students through the process of analyzing a problem by showing the general explanation on the left and the worked-out solution on the right. The result: better understanding of the concept, not just number crunching.

- *For Example.* In every chapter, an interconnected series of *For Example* elements present a continuing discussion, recapping a story and moving it forward to illustrate how to apply each new concept or skill.

- *Just Checking.* At key points in each chapter, we ask students to pause and think with questions designed to be a quick check that they understand the material they've just read. Answers are at the end of the exercise sets in each chapter so students can easily check themselves.

- *TI Tips.* Each chapter's easy-to-read "TI Tips" show students how to use TI-84 Plus CE Statistics functions with the StatWizard operating system. (Help using a TI-Nspire appears in Appendix B, and help with a TI-89 is available through MyLab Statistics or at www.pearsonhighered.com/mathstatsresources. As we strive for a sound understanding of formulas and methods, we want students to use technology for actual calculations. We do emphasize that calculators are just for "Show"—they cannot Think about what to do or Tell what it all means.

- *Math Boxes.* In many chapters we present the mathematical underpinnings of the statistical methods and concepts. By setting these proofs, derivations, and justifications apart from the narrative, we allow students to continue to follow the logical development of the topic at hand, yet also explore the underlying mathematics for greater depth.

- *TI-Nspire Activities.* Margin pointers identify demonstrations and investigations for TI-Nspire handhelds to enhance each chapter. They're available through MyLab Statistics or at www.pearsonhighered.com/mathstatsresources.

- *What Can Go Wrong?* Each chapter still contains our innovative *What Can Go Wrong?* sections that highlight the most common errors people make and the misconceptions they have about Statistics. Our goals are to help students avoid these pitfalls and to arm them with the tools to detect statistical errors and to debunk misuses of statistics, whether intentional or not.

- *What Have We Learned?* Chapter-ending study guides help students review key concepts and terms.

- *Exercises.* We've maintained the pairing of examples so that each odd-numbered exercise (with an answer in the back of the book) is followed by an even-numbered exercise illustrating the same concept. Exercises are ordered by approximate level of complexity.

- *Practice Exams.* At the end of each of the book's seven parts you'll find a practice exam, consisting of both multiple choice and free response questions. These cumulative exams encourage students to keep important concepts and skills in mind throughout the course while helping them synthesize their understanding as they build connections among the various topics.

- *Reality Check.* We regularly remind students that Statistics is about understanding the world with data. Results that make no sense are probably wrong, no matter how carefully we think we did the calculations. Mistakes are often easy to spot with a little thought, so we ask students to stop for a reality check before interpreting their result.

- *Notation Alerts.* Clear communication is essential in Statistics, and proper notation is part of the vocabulary students need to learn. We've found that it helps to call attention to the letters and symbols statisticians use to mean very specific things.

- *On the Computer.* Because real-world data analysis is done on computers, at the end of each chapter we summarize what students can find in most Statistics software, usually with an annotated example.

## Our Approach

We've been guided in the choice of topics and emphasis on clear communication by the requirements of the Advanced Placement Statistics course. In our order of presentation, we have tried to ensure that each new topic fits logically into the growing structure of understanding that we hope students will build.

## GAISE Guidelines

We have worked to provide materials to help each class, in its own way, follow the guidelines of the GAISE (Guidelines for Assessment and Instruction in Statistics Education) project sponsored by the American Statistical Association. That report urges that Statistics education should

1. emphasize statistical literacy and develop statistical thinking,
2. use real data,
3. stress conceptual understanding rather than mere knowledge of procedures,
4. foster active learning,
5. use technology for developing concepts and analyzing data, and
6. make assessment a part of the learning process.

## Mathematics

Mathematics traditionally appears in Statistics texts in several roles:

1. It can provide a concise, clear statement of important concepts.
2. It can embody proofs of fundamental results.
3. It can describe calculations to be performed with data.

Of these, we emphasize the first. Mathematics can make discussions of Statistics concepts, probability, and inference clear and concise. We have tried to be sensitive to those who are discouraged by equations by also providing verbal descriptions and numerical examples.

This book is not concerned with proving theorems about Statistics. Some of these theorems are quite interesting, and many are important. Often, though, their proofs are not enlightening to Introductory Statistics students and can distract the audience from the concepts we want them to understand. However, we have not shied away from the mathematics where we believed that it helped clarify without intimidating. You will find some important proofs, derivations, and justifications in the Math Boxes that accompany the development of many topics.

Nor do we concentrate on calculations. Although statistics calculations are generally straightforward, they are also usually tedious. And, more to the point, they are often unnecessary. Today, virtually all statistics are calculated with technology, so there is little need for students to work by hand. The equations we use have been selected for their focus on understanding concepts and methods, and we point out to students when they should use technology to do the computational heavy lifting.

# Technology and Data

To experience the real world of Statistics, it's best to explore real data sets using modern technology. This fact permeates *Stats: Modeling the World*, Sixth Edition, where we use real data for the book's examples and exercises. Technology lets us focus on teaching statistical thinking rather than getting bogged down in calculations. The questions that motivate each of our hundreds of examples are not "How do you find the answer?" but "How do you think about the answer?"

**Technology.** We assume that students are using some form of technology in this Statistics course. That could include a graphing calculator along with a Statistics package or spreadsheet. Rather than adopt any particular software, we discuss generic computer output. "TI-Tips"—included in most chapters—show students how to use Statistics features of the TI-84 Plus series. In Appendix B, we offer general guidance (by chapter) to help students get started on common software platforms (StatCrunch, Excel, MINITAB, Data Desk, and JMP) and a TI-Nspire. MyLab Statistics and the resource site (www.pearsonhighered.com/mathstatsresources) include additional guidance for students using a TI-89. Applets that let students explore key concepts are also included.

**Data.** Because we use technology for computing, we don't limit ourselves to small, artificial data sets. In addition to including some small data sets, we have built examples and exercises on real data with a moderate number of cases—usually more than you would want to enter by hand into a program or calculator. These data are included in MyLab Statistics and at www.pearsonhighered.com/mathstatsresources.

# MyLab Statistics Resources for Success

MyLab Statistics is available to accompany Pearson's market-leading text options, including *Stats: Modeling the World,* **6th Edition** (access code required).

MyLab™ is the teaching and learning platform that empowers you to reach every student. MyLab Statistics combines trusted author content—including full eText and assessment with immediate feedback—with digital tools and a flexible platform to personalize the learning experience and improve results for each student.

MyLab Statistics supports all learners, regardless of their ability and background, to provide an equal opportunity for success. Accessible resources support learners for a more equitable experience no matter their abilities. And options to personalize learning and address individual gaps helps to provide each learner with the specific resources they need to achieve success.

## Student Resources

**Motivate Your Students** - Students are motivated to succeed when they're engaged in the learning experience and understand the relevance and power of data and statistics.

- **NEW Applets** designed to help students understand a wide range of topics covered in introductory statistics by presenting the opportunity to further explore data in interactive versions of the distribution tables and test the randomization inference methods described in their textbook.

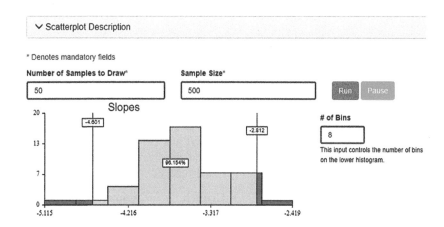

- **Real-World Data Examples and Data Sets** - Statistical concepts are applied to everyday life through the extensive current, real-world data examples and exercises provided throughout the text. Data sets from the book are available in file formats to apply in most statistical analysis software, including TI, Excel, and StatCrunch.

- **Example Video Program** Updated Step-by-Step Example videos guide students through the process of analyzing a problem using the "Think, Show, and Tell" strategy from the textbook.

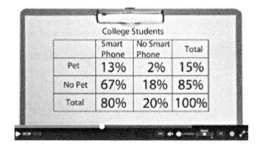

- Integrated directly into MyLab Statistics, **StatCrunch**® is a powerful web-based statistical software that allows users to perform complex analyses, share data sets, and generate compelling reports of their data. The vibrant online community offers tens of thousands shared data sets for students to analyze.

  - **Collect.** Users can upload their own data to StatCrunch or search a large library of publicly shared data sets, spanning almost any topic of interest. Data sets from the text and from online homework exercises can also be accessed and analyzed in StatCrunch. An online survey tool allows users to quickly collect data via web-based surveys.

  - **Crunch.** A full range of numerical and graphical methods allows users to analyze and gain insights from any data set. Interactive graphics help users understand statistical concepts, and are available for export to enrich reports with visual representations of data.

  - **Communicate.** Reporting options help users create a wide variety of visually appealing representations of their data.

StatCrunch is also available by itself to qualified adopters. It can be accessed on your laptop, smartphone, or tablet when you visit the StatCrunch website from your device's browser.

- **Technology Tutorial Videos** offer step-by-step guidance with popular technologies, including Excel, JMP, Minitab, SPSS, and R. These are also available within the homework learning aids for every exercise where technology is expected to perform statistical procedures.

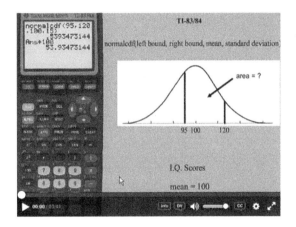

- **New! Personal Inventory Assessments** are a collection of online exercises designed to promote self-reflection and metacognition in students. These 33 assessments include topics such as a Stress Management Assessment, Diagnosing Poor Performance and Enhancing Motivation, and Time Management Assessment.

**Address Under Preparedness** - Each student learns at a different pace. Personalized learning pinpoints the precise areas where each student needs practice, giving all students the support they need—when and where they need it—to be successful.

**Getting Ready questions** can be used to help students prepare for key statistical procedures with a full understanding of prerequisite mathematic skills and concepts.

**Personalized Homework** - With Personalized Homework, students take a quiz or test and receive a subsequent homework assignment that is personalized based on their performance. This way, students can focus on just the topics they have not yet mastered.

Other student resources include:

- **Pearson eText** - The eText is "reflowable" to adapt to use on tablets and smartphones. You can insert your own highlights, notes, and bookmarks. It is also fully accessible using screen-readers. Download the Pearson+ app to access your eText on your smartphone or tablet anytime—even offline.

- **Study Slides** - PowerPoint slides featuring key ideas and examples are available for students within the Video & Resource Library. These slides are compatible with screen readers.

- **Graphing Calculator Manual**, is organized to follow the sequence of topics in the text, and is an easy-to-follow, step-by-step guide on how to use the TI-84 Plus, TI-Nspire, and Casio graphing calculators. It provides worked-out examples to help students fully understand and use the graphing calculator.

- **StatTalk Videos** follow fun-loving statistician Andrew Vickers as he takes to the streets of Brooklyn, NY to demonstrate important statistical concepts through interesting stories and real-life events. This series of 24 videos will actually help you understand statistics.

## Instructor Resources

Your course is unique. So whether you'd like to build your own assignments, teach multiple sections, or set prerequisites, MyLab gives you the flexibility to easily create your course to fit your needs.

**Pre-Built Assignments** are designed to maximize students' performance. All assignments are *fully editable* to make your course your own.

**MyLab Statistics Question Library** is correlated to the exercises in the text, reflecting each author's approach and learning style. They regenerate algorithmically to give students unlimited opportunity for practice and mastery. Below are a few exercise types available to assign:

▼ **New! Applet Exercises** use the enhanced applets to solve key statistical questions as they explore real-world, modern data through simulations.

- **Updated!** Real data exercises with jittering applied where necessary to ensure appropriate variation.

- **T exercises** identify problems that require the use of technology to perform the statistical procedures.

- **Getting Ready questions.** 450 questions that help prepare students with the mathematical foundations to explore statistical processes and procedures.

- **Conceptual Questions.** 1,000 questions that go beyond the textbook problems to assess students conceptual understanding of critical statistical topics.

**Learning Catalytics** - With Learning Catalytics, you'll hear from every student when it matters most. You pose a variety of questions in class (choosing from pre-loaded questions or your own) that help students recall ideas, apply concepts, and develop critical-thinking skills. Your students respond using their own smartphones, tablets, or laptops.

**Performance Analytics** enable instructors to see and analyze student performance across multiple courses. Based on their current course progress, individuals' performance is identified above, at, or below expectations through a variety of graphs and visualizations.

Now included with Performance Analytics, **Early Alerts** use predictive analytics to identify struggling students—even if their assignment scores are not a cause for concern. In both Performance Analytics and Early Alerts, instructors can email students individually or by group to provide feedback.

**Accessibility** - Pearson works continuously to ensure our products are as accessible as possible to all students. Currently we work toward achieving WCAG 2.1 AA for our existing products and Section 508 standards, as expressed in the Pearson Guidelines for Accessible Educational Web Media (https://www.pearson.com/accessibility-guidelines.html).

Other instructor resources include:

- **Instructor's Edition eText** A page-for-page eText of the Instructor's Edition is available within the Instructor Resources section of MyLab Math.

- **Instructional Lessons** provide daily topical lessons, activities, instructional tips and resources to guide students through each unit.

- **Instructor Solution Manual** - The Instructor's Solutions Manual provides complete solutions for all end-of-section exercises.

- **PowerPoint Lecture Slides** feature editable presentations written and designed specifically for this text, including figures and examples from the text.

- **Online test bank and resource guide** contains chapter-by-chapter comments on major concepts; tips on presenting topics (and what to avoid); teaching examples; suggested assignments; Web links and lists of other resources; additional sets of chapter quizzes, unit tests, and investigative tasks; TI-Nspire activities; and suggestions for projects.

- **TestGen** enables instructors to build, edit, print, and administer tests using a computerized bank of questions developed to cover all the objectives of the text. TestGen is algorithmically based, allowing instructors to create multiple but equivalent versions of the same questions or test with the click of a button. Instructors can also modify test bank questions or add new questions. The software and test bank are available for download at pearson.com.

Learn more at pearson.com/mylab/statistics

# ACKNOWLEDGMENTS

Many people have contributed to this book in all five of its editions. This edition would have never seen the light of day without the assistance of the incredible team at Pearson. Deirdre Lynch, was central to the genesis, development, and realization of the book from day one. Suzy Bainbridge and Laura Briskman, Content Strategy Managers; Karen Montgomery, Product Manager; and Dawn Murrin, Director, Content Strategy, provided much needed support. Rachel Reeve, Senior Content Producer; Prerna Rochlani, Assistant Content Producer; and Monique Bettencourt, Associate Analyst, were essential in managing all of the behind-the-scenes work that needed to be done for both the print book and the eText. Jean Choe, Producer, and Bob Carroll, Manager Content Development, put together a top-notch media package for this book. Demetrius Hall, Product Marketer, and Jordan Longoria, Marketing Manager, Savvas Learning Company, are spreading the word about *Stats: Modeling the World*. Carol Melville, Manufacturing Buyer, LSC Communications, worked miracles to get the print book in your hands. Special thanks go out to Straive for the wonderful work they did on this book, and in particular to Rose Kernan, Project Manager, for her close attention to detail, for keeping the cogs from getting into the wheels where they often wanted to wander, and for keeping the authors on task and on schedule . . . no easy feat!

We would like to thank Corey Andreasen of The American School of The Hague and Jared Derksen of Rancho Cucamonga High School for their invaluable help with updating the exercises, answers, and data sets. We'd also like to thank our accuracy checkers whose monumental task was to make sure we said what we thought we were saying. They are Bill Craine, Lansing High School, Lansing, NY and Adam Yankay, Highland School, Warrenton VA.

We extend our sincere thanks for the suggestions and contributions made by the following reviewers, focus group participants, and class-testers:

John Arko
*Glenbrook South High School, IL*

Kathleen Arthur
*Shaker High School, NY*

Allen Back
*Cornell University, NY*

Beverly Beemer
*Ruben S. Ayala High School, CA*

Judy Bevington
*Santa Maria High School, CA*

Susan Blackwell
*First Flight High School, NC*

Gail Brooks
*McLennan Community College, TX*

Walter Brown
*Brackenridge High School, TX*

Darin Clifft
*Memphis University School, TN*

Bill Craine
*Lansing High School, NY*

Sybil Coley
*Woodward Academy, GA*

Kevin Crowther
*Lake Orion High School, MI*

Caroline DiTullio
*Summit High School, NJ*

Jared Derksen
*Rancho Cucamonga High School, CA*

Sam Erickson
*North High School, WI*

Laura Estersohn
*Scarsdale High School, NY*

Laura Favata
*Niskayuna High School, NY*

David Ferris
*Noblesville High School, IN*

Linda Gann
*Sandra Day O'Connor High School, TX*

Landy Godbold
*The Westminster Schools, GA*

Randall Groth
*Illinois State University, IL*

Donnie Hallstone
*Green River Community College, WA*

Howard W. Hand
*St. Marks School of Texas, TX*

Bill Hayes
*Foothill High School, CA*

Miles Hercamp
*New Palestine High School, IN*

Michelle Hipke
*Glen Burnie Senior High School, MD*

Carol Huss
*Independence High School, NC*

Sam Jovell
*Niskayuna High School, NY*

Peter Kaczmar
*Lower Merion High School, PA*

John Kotmel
*Lansing High School, NY*

Beth Lazerick
*St. Andrews School, FL*

Michael Legacy
*Greenhill School, TX*

Guillermo Leon
*Coral Reef High School, FL*

John Lieb
*The Roxbury Latin School, MA*

Mark Littlefield
*Newburyport High School, MA*

Martha Lowther
*The Tatnall School, DE*

John Maceli
*Ithaca College, NY*

Jim Miller
*Alta High School, UT*

Timothy E. Mitchell
*King Philip Regional High School, MA*

Maxine Nesbitt
*Carmel High School, IN*

Jordan Pare
*St. Francis High School CA*

Elizabeth Ann Przybysz
*Dr. Phillips High School, FL*

Diana Podhrasky
*Hillcrest High School, TX*

Rochelle Robert
*Nassau Community College, NY*

Karl Ronning
*Davis Senior High School, CA*

Bruce Saathoff
*Centennial High School, CA*

Agatha Shaw
*Valencia Community College, FL*

Murray Siegel
*Sam Houston State University, TX*

Chris Sollars
*Alamo Heights High School, TX*

Darren Starnes
*The Webb Schools, CA*

Jane Viau
*The Frederick Douglass Academy, NY*

And we'd be especially remiss if we did not applaud the outstanding team of teachers whose creativity, insight, knowledge, and dedication create the many valuable resources so helpful for Statistics students and their instructors:

Corey Andreasen
*The American School of The Hague*

Anne Carroll
*Kennett High School, PA*

Ruth Carver
*Germantown Academy in Fort Washington, PA*

William B. Craine
*Lansing High School, NY*

Jared Derksen
*Rancho Cucamonga High School, CA*

John Diehl
*Hindsdale Central High School, IL*

Lee Kucera
*Capistrano Valley High School, CA*

John Mahoney
*Benjamin Banneker Academic High School, Washington, D.C.*

Susan Peters
*University of Louisville, KY*

Janice Ricks
*Marple Newtown High School, PA*

Jane Viau
*The Frederick Douglass Academy, NY*

Webster West
*Creator of StatCrunch*

Adam Yankay
*Highland School, Warrenton VA*

*David Bock*
*Floyd Bullard*
*Paul Velleman*
*Richard De Veaux*

# Stats Starts Here[1]

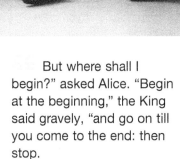

"But where shall I begin?" asked Alice. "Begin at the beginning," the King said gravely, "and go on till you come to the end: then stop.

—Lewis Carroll, Alice's Adventures in Wonderland

Statistics gets no respect. People say things like "You can prove anything with Statistics." People will write off a claim based on data as "just a statistical trick." And a Statistics course may not be your friends' first choice for a fun elective.

But Statistics *is* fun! That's probably not what you heard on the street, but it's true. Statistics is about how to think clearly with data. We'll talk about data in more detail soon, but for now, think of **data** as any collection of numbers, characters, images, or other items that provide information about something. Whenever there are data and a need for understanding the world, you'll find Statistics. A little practice thinking statistically is all it takes to start seeing the world more clearly and accurately.

## So, What Is (Are?) Statistics?

How does it feel to live near the dawn of the Information Age? More data are collected today (a lot of it on *us*) than ever before in human history. Consider these examples of how data are used today:

◆ If you have a Facebook account, you have probably noticed that the ads you see online tend to match your interests and activities. Coincidence? Hardly. According to *Wired* magazine,[2] much of your personal information has probably been sold to marketing or tracking companies. Why would Facebook give you a free account and let you

---

[1] We could have called this chapter "Introduction," but nobody reads the introduction, and we wanted you to read this. We feel safe admitting this here, in the footnote, because nobody reads footnotes either.

[2] http://www.wired.com/story/wired-guide-personal-data-collection/

> Q: What is Statistics?
>
> A: Statistics is a way of reasoning, along with a collection of tools and methods, designed to help us understand the world.
>
> Q: What are statistics?
>
> A: Statistics (plural) are particular calculations made from data.
>
> Q: So what is data?
>
> A: You mean, "What *are* data?" Data is the plural form. The singular is datum.
>
> Q: OK, OK, so what *are* data?
>
> A: Data are values along with their context.

upload as much as you want to its site? Because your data are valuable! Using your Facebook profile, a company might build a profile of your interests and activities: what movies and sports you like; your age, sex, education level, and hobbies; where you live; and, of course, who your friends are and what *they* like.

- Americans spend an average of 4.9 hours per day on their smartphones. About 9.4 trillion text messages are sent each year.[3] Some of these messages are sent or read while the sender or the receiver is driving. How dangerous is texting while driving?

  How can we study the effect of texting while driving? One way is to measure reaction times of drivers faced with an unexpected event while driving and texting. Researchers at the University of Utah tested drivers on simulators that could present emergency situations. They compared reaction times of sober drivers, drunk drivers, and texting drivers.[4] The results were striking. The texting drivers actually responded more slowly and were more dangerous than drivers who were above the legal limit for alcohol.

- Many athletes use data-gathering devices while they exercise to motivate themselves and to share their routines on social media. In early 2018 the New York Times reported[5] that some of this publicly available data could potentially compromise U.S. national security and even the lives of some of its military service members. Some of the data revealed the times and routes that soldiers were running on military bases worldwide, and even routes they traveled between bases when they were not exercising.

With this text you'll learn how to design good studies and discern messages in data, should you become a researcher yourself. More important—especially to those who, like most of us, *don't* become researchers—you'll learn to judge with a more skilled and critical eye the conclusions drawn from data by *others*. With so much data everywhere, you need such judgment just to be an informed and responsible citizen.

> **ARE YOU A STATISTIC?**
> The ads say, "Don't drink and drive; you don't want to be a statistic." But you can't be a statistic.
>   We say: "Don't be a datum."

# Statistics in a Word

It can be fun, and sometimes useful, to summarize a discipline in only a few words. So,

> Economics is about . . . *Money (and why it is good)*.
> Psychology: *Why we think what (we think) we think*.
> Biology: *Life*.
> Anthropology: *Who?*
> History: *What, where, and when?*
> Philosophy: *Why?*
> Engineering: *How?*
> Accounting: *How much?*

In such a caricature, Statistics is about . . . **Variation.**

Some of the reasons why we act and think differently from one another can be explained—perhaps by our education or our upbringing or how our friends act and think. But some of that variation among us will always remain unexplained.[6] Statistics is largely about trying to find explanations for why data vary while acknowledging that some amount of the variation will always remain a mystery.

---

[3] https://www.textrequest.com/blog/texting-statistics-answer-questions/
[4] "Text Messaging During Simulated Driving," Drews, F. A. et al. Human Factors: hfs.sagepub.com/content/51/5/762
[5] "Strava Fitness App Can Reveal Military Sites, Analysts Say," *New York Times*, January 21, 2018; https://www.nytimes.com/2018/01/29/world/middleeast/strava-heat-map.html
[6] And that's a good thing!

# But What *Are* Data?

Amazon.com opened for business in July 1995, billing itself as "Earth's Biggest Bookstore." By 1997, Amazon had a catalog of more than 2.5 million book titles and had sold books to more than 1.5 million customers in 150 countries. In 2016, the company's sales reached $136 billion (more than a 27% increase from the previous year). Amazon has sold a wide variety of merchandise, including a $400,000 necklace, yak cheese from Tibet, and the largest book in the world. How did Amazon become so successful and how can it keep track of so many customers and such a wide variety of products? The answer to both questions is *data*.

But what are data? Think about it for a minute. What exactly *do* we mean by "data"? You might think that data have to be numbers, but data can be text, pictures, web pages, and even audio and video. If you can sense it, you can measure it. The amount of data collected in the world is growing exponentially.

Let's look at some hypothetical data values that Amazon might collect:

| | | | | | | | | | |
|---|---|---|---|---|---|---|---|---|---|
| B0000010AA | 0.99 | Chris G. | 902 | 105-2686834-3759466 | 1.99 | 0.99 | Illinois |
| Los Angeles | Samuel R. | Ohio | N | B000068ZVQ | Amsterdam | New York, New York, | Katherine H. |
| Will S. | 002-1663369-6638649 | Beverly Hills | N | N | 103-2628345-9238664 | 0.99 | Massachusetts |
| 312 | Monique D. | 105-9318443-4200264 | 413 | B0000015Y6 | 440 | B000002BK9 | 0.99 |
| Canada | Detroit | 567 | 105-1372500-0198646 | N | B002MXA7Q0 | Ohio | Y |

Try to guess what they represent. Why is that hard? Because there is no *context*. If we don't know what values are measured and what is measured about them, the values are meaningless. We can make the meaning clear if we organize the values into a **data table** such as this one:

| Order Number | Name | State/Country | Price | Area Code | Download | Gift? | ASIN | Artist |
|---|---|---|---|---|---|---|---|---|
| 105-2686834-3759466 | Katherine H. | Ohio | 0.99 | 440 | Amsterdam | N | B0000015Y6 | Cold Play |
| 105-9318443-4200264 | Samuel R | Illinois | 1.99 | 312 | Detroit | Y | B000002BK9 | Red Hot, Chili Peppers |
| 105-1372500-0198646 | Chris G. | Massachusetts | 0.99 | 413 | New York, New York | N | B000068ZVQ | Frank Sinatra |
| 103-2628345-9238664 | Monique D. | Canada | 0.99 | 902 | Los Angeles | N | B0000010AA | Blink 182 |
| 002-1663369-6638649 | Will S. | Ohio | 0.99 | 567 | Beverly Hills | N | B002MXA7Q0 | Weezer |

```
THE W'S:
WHO
WHAT    and in what units
WHEN
WHERE
WHY
HOW
```

Now we can see that these are purchase records for album download orders from Amazon. The column titles tell what information has been recorded. Each row is about a particular purchase.

What information would provide a **context**? Newspaper journalists know that the lead paragraph of a good story should establish the "Five W's": *who, what, when, where,* and (if possible) *why*. Often, we add *how* to the list as well. The answers to the first two questions are essential. If we don't know *what* values are measured and *who* those values are measured on, the values are meaningless.

## Who and What

In general, the rows of a data table correspond to individual **cases** about *Who*m (or about which—if they're not people) we record some characteristics. Cases go by different names, depending on the situation.

- Individuals who answer a survey are called **respondents.**
- People on whom we experiment are **subjects** or **participants.**
- Animals, plants, websites, and other inanimate subjects are often called **experimental units.**
- Often we simply call cases what they are: for example, *customers*, *hurricanes*, or *counties*.
- In a database, rows are called **records**—in the hypothetical Amazon data, purchase records. Perhaps the most generic term is *cases*, but in any event the rows represent the *who* of the data.[7]

The characteristics recorded about each individual are called **variables**. These are usually shown as the columns of a data table, and they should have a name that identifies *What* has been measured. The data table of Amazon purchases shows that some of the variables that Amazon collected data for were *Name*, *Price*, and whether the purchase was a *Gift*.

Often, the cases are a **sample** of cases selected from some larger **population** that we'd like to understand. Amazon certainly cares about its customers, but also wants to know how to attract all those other Internet users who may never have made a purchase from Amazon's site. To be able to generalize from the sample of cases to the larger population, we'll want the sample to be *representative* of that population—a kind of snapshot image of the larger world.

We must know *who* and *what* to analyze data. Without knowing these two, we don't have enough information to start. Of course, we'd always like to know more. The more we know about the data, the more we'll understand about the world. If possible, we'd like to know the *when* and *where* of data as well. Values recorded in 1803 may mean something different than similar values recorded last year. Values measured in Tanzania may differ in meaning from similar measurements made in Mexico. And knowing *why* the data were collected can tell us much about its reliability and quality.

### FOR EXAMPLE
#### Identifying the *Who*

*Consumer Reports* included an evaluation of 126 tablets from a variety of manufacturers.

QUESTION: Describe the population of interest, the sample, and the *Who* of the study.

ANSWER: The population of interest is all tablets currently available. The sample (and the Who of the study) is the particular collection of 126 tablets that the consumer organization purchased and studied.

## How the Data Are Collected

*How* the data are collected can make the difference between insight and nonsense. For example, as we'll see later, data that come from a voluntary survey on the Internet are almost always worthless. One primary concern of Statistics is the design of sound methods

---

[7] If you're going through the *W*'s in your head and your data are measurements of objects, "*Who*" should still remind you to ask, "What objects were measured?"

for collecting data.⁸ Throughout this book, whenever we introduce data, we'll provide a margin note listing the W's (and H) of the data. Identifying the W's is a habit we recommend.

The first step of any data analysis is to know what you are trying to accomplish and what you want to know. To help you use Statistics to understand the world and make decisions, we'll lead you through the entire process of *thinking* about the problem, *showing* what you've found, and *telling* others what you've learned. Every guided example in this book is broken into these three steps: *Think*, *Show,* and *Tell*. Identifying the problem and the *who* and *what* of the data is a key part of the *Think* step of any analysis. Make sure you know these before you proceed to *Show* or *Tell* anything about the data.

# More About Variables (*What*?)

The Amazon data table displays information about several variables: *Order Number*, *Name*, *State/Country*, *Price*, and so on. These identify *what* we know about each individual. Variables such as these can play different roles, depending on how we plan to use them. While some are merely identifiers, others may be categorical or quantitative. Making that distinction is an important step in our analysis.

### Identifiers

Amazon wants to know who you are when you sign in again and it doesn't want to confuse you with some other customer. So it assigns you a unique identifying number.⁹ Amazon also wants to send you the right product, so it assigns a unique Amazon Standard Identification Number (ASIN) to each item it carries. Both of these numbers are useful to Amazon, but they aren't measurements of anything. They're generated by Amazon and *assigned* uniquely to customers and products. Data like these values are called **identifier variables**. Other examples are student ID numbers and Social Security numbers. Identifier variables are typically used to identify single cases, not to look for patterns in collections of data, so they are seldom used in data analysis.

### Categorical Variables

Some variables just tell us what group or category each individual belongs to. Are you fluent in Spanish? Do you have any piercings? What's your favorite music genre? We call variables like these **categorical variables**.¹⁰ Some variables are clearly categorical, like the variable *State/Country*. Its values are text and those values tell us what category the particular case falls into. Descriptive responses to questions are often categories. For example, the responses to the questions "Who is your cell phone provider?" or "What is your marital status?" yield categorical values. But numerals are sometimes used to label categories, so categorical data can also be numerals. ZIP codes, for example, are numerals but they convey geographic information, which is categorical.

### Quantitative Variables

When a variable contains *measured* numerical values, we call it a **quantitative variable**. Quantitative variables typically have **units**, such as centimeters or years, but they may also be *counts* of something associated with each case, such as the number of siblings a person

> **PRIVACY AND THE INTERNET**
>
> You have many identifiers: a Social Security number, a student ID number, possibly a passport number, a health insurance number, and probably a Facebook account name. Privacy experts are worried that Internet thieves may match your identity in these different areas of your life, allowing, for example, your health, education, and financial records to be merged. Even online companies such as Facebook and Google are able to link your online behavior to some of these identifiers, which carries with it both advantages and dangers. The National Strategy for Trusted Identities in Cyberspace (www.wired.com/images_blogs/threatlevel/2011/04/NSTICstrategy_041511.pdf) proposes ways that we may address this challenge in the near future.

---

⁸Coming attractions: to be discussed in Part III. We sense your excitement.
⁹Or sometimes a code containing numerals and letters.
¹⁰You may also see them called *qualitative* variables.

has, or how many languages they speak. Whenever you look at a quantitative variable, be sure you know what its units are. Without units, the values of a measured variable have no meaning. It does little good to be promised a raise of 5000 a year if you don't know whether it will be paid in euros, dollars, pennies, yen, or bronze knuts.[11]

## Either/Or?

Some variables with numeric values can be treated as either categorical or quantitative depending on what we want to know. Amazon could record your *Age* in years. That seems quantitative, and it would be if the company wanted to know the average age of those customers who visit their site after 3 A.M. But suppose Amazon wants to decide which album to feature on its site when you visit. Then thinking of your age in one of the categories Child, Teen, Adult, or Senior might be more useful. So, sometimes whether a variable is treated as categorical or quantitative is more about the question we want to ask than an intrinsic property of the variable itself.

## Ordinal Variables

Suppose a course evaluation survey asks, "How valuable do you think this course will be to you?" 1 = Worthless; 2 = Slightly; 3 = Somewhat; 4 = Reasonably; 5 = Invaluable. Is *Educational Value* categorical or quantitative? A teacher might just count the number of students who gave each response for her course, treating *Educational Value* as a categorical variable. Or if she wants to see whether the course is improving, she might treat the responses as the *amount* of perceived value—in effect, treating the variable as quantitative.

But what are the units? There is certainly an *order* of perceived worth: Higher numbers indicate higher perceived worth. A course that averages 4.2 seems more valuable than one that averages 2.1. But is it *twice* as valuable? Does that even mean anything? Variables like this that have a natural order but no units are often called *ordinal variables*. Other examples are college class (freshman, sophomore, junior, or senior) and hurricane level (1, 2, 3, 4, or 5). Ordinal variables can be a little tricky to analyze and for the most part they are not considered in this text.

### FOR EXAMPLE
#### Identifying the *What* and *Why* of Tablets

**RECAP:** A *Consumer Reports* article about 126 tablet computers lists each tablet's manufacturer, cost, battery life (hours), operating system (Android, IOS, or Windows) and overall performance score (0–100).

**QUESTION:** Are these variables categorical or quantitative? Include units where appropriate, and describe the *Why* of this investigation.

**ANSWER:** The variables are:

- manufacturer (categorical)
- cost (quantitative, $)
- battery life (quantitative, hours)
- operating system (categorical)
- performance score (quantitative, with no units—essentially an ordinal variable)

*Why?* The magazine hopes to provide consumers with information to help them choose a tablet to meet their needs.

---

[11] Wizarding money. Just seeing who's paying attention.

## JUST CHECKING

In the 2004 Tour de France, Lance Armstrong made history by winning the race for an unprecedented sixth time. In 2005, he became the only 7-time winner and set a new record for the fastest average speed—41.65 kilometers per hour (about 26 mph)—that stands to this day. Then in 2012, Armstrong was banned for life for doping offenses, stripped of all his titles and his records expunged. Here are the first three and last nine lines of a data table of all Tour de France races. Keep in mind that the dataset has over 100 entries.

1. List as many of the W's as you can for this dataset.
2. Classify each variable as categorical or quantitative; if quantitative, identify the units.

| Year | Winner | Country of Origin | Age | Team | Total Time (h/min/s) | Avg. Speed (km/h) | Stages | Total Distance Ridden (km) | Starting Riders | Finishing Riders |
|---|---|---|---|---|---|---|---|---|---|---|
| 1903 | Maurice Garin | France | 32 | La Française | 94.33.00 | 25.7 | 6 | 2428 | 60 | 21 |
| 1904 | Henri Cornet | France | 20 | Cycles JC | 96.05.00 | 25.3 | 6 | 2428 | 88 | 23 |
| 1905 | Louis Trousseller | France | 24 | Peugeot | 112.18.09 | 27.1 | 11 | 2994 | 60 | 24 |
| ... | | | | | | | | | | |
| 2012 | Bradley Wiggins | Great Britain | 32 | Sky | 87.34.47 | 39.83 | 20 | 3488 | 198 | 153 |
| 2013 | Christopher Froome | Great Britain | 28 | Sky | 83.56.40 | 40.55 | 21 | 3404 | 198 | 169 |
| 2014 | Vincenzo Nibali | Italy | 29 | Astana | 89.56.06 | 40.74 | 21 | 3663.5 | 198 | 164 |
| 2015 | Christopher Froome | Great Britain | 30 | Sky | 84.46.14 | 39.64 | 21 | 3660.3 | 198 | 160 |
| 2016 | Christopher Froome | Great Britain | 31 | Sky | 89.04.48 | 39.62 | 21 | 3529 | 198 | 174 |
| 2017 | Christopher Froome | Great Britain | 32 | Sky | 86.34 | 40.997 | 21 | 3540 | 198 | 167 |
| 2018 | Geraint Thomas | Great Britain | 32 | Sky | 83.28 | 40.210 | 21 | 3349 | 176 | 145 |
| 2019 | Egan Bernal | Colombia | 22 | INEOS | 82.57.00 | 40.576 | 21 | 3365.8 | 176 | 155 |
| 2020 | Tadej Pogacar | Slovenia | 21 | UAE Team Emirates | 87.20.05 | 39.872 | 21 | 3482.2 | 176 | 146 |

### THERE'S A WORLD OF DATA ON THE INTERNET

These days, one of the richest sources of data is the Internet. With a bit of practice, you can learn to find data on almost any subject. Many of the datasets we use in this book were found in this way. The Internet has both advantages and disadvantages as a source of data. Among the advantages is the fact that often you'll be able to find even more current data than those we present. The disadvantages include the fact that references to Internet addresses can "break" as sites evolve, move, and die.

Our solution to these challenges is to offer the best advice we can to help you search for the data, wherever they may be residing. We usually point you to a website. We'll sometimes suggest search terms and offer other guidance.

Some words of caution, though: Data found on Internet sites may not be formatted in the best way for use in statistics software. Although you may see a data table in standard form, an attempt to copy the data may leave you with a single column of values. You may have to work in your favorite statistics or spreadsheet program to reformat the data into variables. You will also probably want to remove commas from large numbers and extra symbols such as money indicators ($, ¥, £); few statistics packages can handle these.

## WHAT CAN GO WRONG?

- **Don't label a variable as categorical or quantitative without thinking about the question you want it to answer.** The same variable can sometimes take on different roles.
- **Just because your variable's values are numbers, don't assume that it's quantitative.** Categories are often given numerical labels. Don't let that fool you into thinking they have quantitative meaning. Look at the context.
- **Always be skeptical.** One reason to analyze data is to discover the truth. Even when you are told a context for the data, it may turn out that the truth is a bit (or even a lot) different. The context colors our interpretation of the data, so those who want to influence what you think may slant the context. A survey that seems to be about all students may in fact report just the opinions of those who visited a fan website. The question that respondents answered may be posed in a way that influences responses.

## TI TIPS

### Working with Data

You'll need to be able to enter and edit data in your calculator. Here's how:

**TO ENTER DATA:** Hit the STAT button, and choose EDIT from the menu. You'll see a set of columns labeled L1, L2, and so on. Here is where you can enter, change, or delete a set of data.
 Let's enter the heights (in inches) of the five starting players on a basketball team: 71, 75, 75, 76, and 80. Move the cursor to the space under L1, type in 71, and hit ENTER (or the down arrow). There's the first player. Now enter the data for the rest of the team.

**TO CHANGE A DATUM:** Suppose the 76″ player grew since last season; his height should be listed as 78″. Use the arrow keys to move the cursor onto the 76, then change the value and ENTER the correction.

**TO ADD MORE DATA:** We want to include the sixth man, 73″ tall. It would be easy to simply add this new datum to the end of the list. However, sometimes the order of the data matters, so let's place this datum in numerical order. Move the cursor to the desired position (atop the first 75). Hit 2ND INS (for "insert"), then ENTER the 73 in the new space.

**TO DELETE A DATUM:** The 78″ player just quit the team. Move the cursor there. Hit DEL. Bye.

**TO CLEAR THE DATALIST:** Finished playing basketball? Move the cursor atop the L1. Hit CLEAR, then ENTER (or down arrow). You should now have a blank datalist, ready for you to enter your next set of values.

**LOST A DATALIST?** Oops! Is L1 now missing entirely? Did you delete L1 by mistake, instead of just *clearing* it? Easy problem to fix: buy a new calculator. No? OK, then simply go to the STAT EDIT menu, and run SetUpEditor to return lists L1 through L6 to the STAT EDIT screen.

# WHAT HAVE WE LEARNED?

We've learned that data are information in a context.
- The W's help nail down the context *Who*, *What*, *When*, *Why*, *Where*, and *hoW*.
- We must know at least the *Who*, *What*, and *hoW* to be able to say anything useful based on the data. The *Who* are the cases. The *What* are the variables—the measurements made on each case. The *hoW*—how the data were collected—helps us evaluate the trustworthiness of the data, as does sometimes the *Why*.

We usually treat variables in one of two basic ways: as *categorical* or *quantitative*.
- Categorical variables identify a category for each case. (An exception is an identifier variable that just names each case.)
- Quantitative variables record measurements or amounts of something; they must have *units*.
- Sometimes we treat a variable as categorical or quantitative depending on what we want to learn from it.

## TERMS

| | |
|---|---|
| **Data** | Systematically recorded information, whether numbers or labels, together with its context. (p. 1) |
| **Data table** | An arrangement of data in which each row represents a case and each column represents a variable. (p. 3) |
| **Context** | The context ideally tells *Who* was measured, *What* was measured, *hoW* the data were collected, *Where* the data were collected, and *When* and *Why* the study was performed. (p. 3) |
| **Case** | A case is an individual about whom or which we have data. (*Who*). (p. 4) |
| **Respondent** | Someone who answers, or responds to, a survey. (p. 4) |
| **Subject** | A human experimental unit. Also called a participant. (p. 4) |
| **Participant** | A human experimental unit. Also called a subject. (p. 4) |
| **Experimental unit** | An individual in a study for whom or for which data values are recorded. Human experimental units are usually called subjects or participants. (p. 4) |
| **Record** | Information about an individual in a database. (p. 4) |
| **Variable** | A variable holds information about the same characteristic for many cases. (*What*). (p. 4) |
| **Sample** | The cases we actually examine in seeking to understand the larger population. (p. 4) |
| **Population** | All the cases we wish we knew about. (p. 4) |
| **Identifier variable** | A categorical variable that assigns a unique value for each case, used to name or identify it. (p. 5) |
| **Categorical variable** | A variable that names categories (whether with words or numerals) is called *categorical*. (p. 5) |
| **Quantitative variable** | A variable in which *measured* numbers act as numerical values is called *quantitative*. Quantitative variables always either have units or are counts of something associated with the cases. (p. 5) |
| **Units** | A quantity or amount adopted as a standard of measurement, such as dollars, hours, or grams. (p. 5) |

## ON THE COMPUTER

### Data

> Computers are useless; they can only give you answers.
> —Pablo Picasso

Most often we find statistics on a computer using a program, or *package*, designed for that purpose. There are many different statistics packages, but they all do essentially the same things. If you understand what the computer needs to know to do what you want and what it needs to show you in return, you can figure out the specific details of most packages pretty easily.

For example, to get your data into a computer statistics package, you need to tell the computer:

▶ Where to find the data. This usually means directing the computer to a file stored on your computer's disk or to data on a database. Or it might just mean that you have copied the data from a spreadsheet program or Internet site and it is currently on your computer's clipboard. Usually, the data should be in the form of a data table. Most computer statistics packages prefer the *delimiter* that marks the division between elements of a data table to be a *tab* character and the delimiter that marks the end of a case to be a *return* character.

▶ Where to put the data. (Usually this is handled automatically.)

▶ What to call the variables. Some data tables have variable names as the first row of the data, and often statistics packages can take the variable names from the first row automatically.

## EXERCISES

1. **Grocery shopping** Many grocery store chains offer customers a card they can scan when they check out, and offer discounts to people who do so. To get the card, customers must give information, including a mailing address and e-mail address. The actual purpose is not to reward loyal customers but to gather data. What data do these cards allow stores to gather, and why would they want that data?

2. **Online shopping** Online retailers such as Amazon.com keep data on products that customers buy, and even products they look at. What does Amazon hope to gain from such information?

3. **Parking lots** Sensors in parking lots are able to detect and communicate when spaces are filled in a large covered parking garage next to an urban shopping mall. How might the owners of the parking garage use this information both to attract customers and to help the store owners in the mall make business plans?

4. **Satellites and global climate change** Satellites send back nearly continuous data on the earth's land masses, oceans, and atmosphere from space. How might researchers use this information in both the short and long term to help study changes in the earth's climate?

5. **Super Bowl** Sports announcers love to quote statistics. During the Super Bowl, they particularly love to announce when a record has been broken. They might have a list of all Super Bowl games, along with the scores of each team, total scores for the two teams, margin of victory, passing yards for the quarterbacks, and many more bits of information. Identify the *Who* in this list.

6. **Nobel laureates** The website www.nobelprize.org allows you to look up all the Nobel prizes awarded in Any year. The data are not listed in a table. Rather you drag a slider to the year and see a list of the awardees for that year. Describe the *Who* in this scenario.

7. **Health records** The National Center for Health Statistics (NCHS) conducts an extensive survey consisting of an interview and medical examination with a representative sample of about 5000 people a year. The interview includes demographic, socioeconomic, dietary, and other health-related questions. The examination "consists of medical, dental, and physiological measurements, as well as laboratory tests administered by highly trained medical personnel" (www.cdc.gov/nchs/nhanes/about_nhanes.htm). Describe the *Who* and the *What* of this study.

8. **Facebook** Facebook uploads more than 350 million photos every day onto its servers. For this collection, describe the *Who* and give a plausible *What*.

9. **Gay marriage** A May 2020 Gallup poll asked, "Do you think marriages between same-sex couples should or should not be recognized by the law as valid, with the same rights as traditional marriages?" Sixty-seven percent of respondents said they should be valid—a new high in approval. If the choices were "Should be valid" and "Should not be valid," what kind of variable is the response?

10. **Gay marriage by party** The May 2020 Gallup poll cited in Exercise 9 also differentiated respondents according to the political party they identified with. Gallup reports that 83% of Democrats responded that gay marriage should be recognized, but only 49% of those self-identifying as Republican did so. What kind of variables were recorded for the responses to the poll?

11. **Medicine** A pharmaceutical company conducts an experiment in which a subject takes 100 mg of a substance orally. The researchers measure how many minutes it takes for half of the substance to exit the bloodstream. What kind of variable is the company studying?

12. **Stress** A medical researcher measures the increase in heart rate of patients who are taking a stress test. What kind of variable is the researcher studying?

13. **The news** Find a newspaper or magazine article in which some data are reported. For the data discussed in the article, identify as many of the W's as you can. Include a copy of the article with your report.

14. **Popular media** Find a non-news item, such as a blog or twitter post, that reports on a study or otherwise summarizes some data. For the data that are discussed, identify *Who* and *What* were investigated and the population of interest. Identify the source of the item and include a web address if the article is on-line.

(*Exercises 15–20*) *Each of these exercises describes some data. For each exercise, identify Who and What are described by the data, and what the population of interest is.*

15. **Bicycle and pedestrian safety** The New York City Department of Transportation determined that between 2010 and 2014, left-turning vehicles were the cause of death or injury for a considerably greater percentage (19%) of killed or injured pedestrians and bicyclists than were right-turning vehicles (6%).

16. **Investments** Some companies offer 401(k) retirement plans to employees, permitting them to shift part of their before-tax salaries into investments such as mutual funds. Employers typically match 50% of the employees' contribution up to about 6% of salary. One company, concerned with what it believed was a low employee participation rate in its 401(k) plan, sampled 30 other companies with similar plans and asked for their 401(k) participation rates.

17. **Fake news** Researchers at Stanford University studied middle school, high school, and college students in 12 states to determine their ability to evaluate the quality of information found in different online resources. Among other things, they found that middle school students had difficulty differentiating between news sources and advertisements, high school students could not tell the difference between real and fake news sources on Facebook, and college students did not suspect potential bias in a tweet from an activist group. (www.npr.org/sections/thetwo-way/2016/11/23/503129818/study-finds-students-have-dismaying-inability-to-tell-fake-news-from-real)

18. **Not-so-diet soda** A look at 474 participants in the San Antonio Longitudinal Study of Aging found that participants who drank two or more diet sodas a day "experienced waist size increases six times greater than those of people who didn't drink diet soda." (*J Am Geriatr Soc.* 2015 Apr; 63(4):708–15. doi: 10.1111/jgs.13376. Epub 2015 Mar 17.)

19. **Blindness** A study begun in 2011 examines the use of stem cells in treating two forms of blindness: Stargardt's disease and dry age-related macular degeneration. Each of the 24 patients entered one of two separate trials in which embryonic stem cells were to be used to treat the condition.

20. **Molten iron** The Cleveland Casting Plant is a large, highly automated producer of iron automotive castings for Ford Motor Company. The company is interested in keeping the pouring temperature of the molten iron (in degrees Fahrenheit) close to the specified value of 2550 degrees. Cleveland Casting measured the pouring temperature for 10 randomly selected castings.

(*Exercises 21–34*) *For each description of data, identify as many of the W's as you can from what's given, identify what a case is, name the variables, specify for each variable whether its use indicates that it should be treated as categorical or quantitative, and, for any quantitative variable, identify the units in which it was measured (or note that they were not provided).*

21. **Weighing bears** Because of the difficulty of weighing a bear in the woods, researchers caught and measured 54 bears, recording their weight, neck size, length, and sex. They hoped to find a way to estimate weight from the other, more easily determined quantities.

22. **Schools** The State Education Department requires local school districts to keep these records on all students: age, race or ethnicity, days absent, current grade level, standardized test scores in reading and mathematics, and any disabilities or special educational needs.

23. **Arby's menu** A listing posted by the Arby's restaurant chain gives, for each of the sandwiches it sells, the type of meat in the sandwich, the number of calories, and the serving size in ounces. The data might be used to assess the nutritional value of the different sandwiches.

24. **E-bikes** In May of 2020 *Bicycling* magazine reviewed electric bicycles. For each bike reviewed, they reported the motor size (in watts), maximum speed (mph), wheel base (mm), brand name, and whether the battery can be removed for security. They also provided pictures of each bike.

25. **Babies** Medical researchers at a large city hospital investigating the impact of prenatal care on newborn health collected data from 882 births during 1998–2000. They kept track of the mother's age, the number of weeks the pregnancy lasted, the type of birth (cesarean, induced, natural), the level of prenatal care the mother had (none, minimal, adequate), the birth weight and sex of the baby, and whether the baby exhibited health problems (none, minor, major).

26. **Flowers** In a study appearing in the journal *Science*, a research team reports that plants in southern England are flowering earlier in the spring. Records of the first flowering dates for 385 species over a period of 47 years show that flowering has

advanced an average of 15 days per decade, an indication of climate warming, according to the authors.

**27. Herbal medicine** Scientists at a major pharmaceutical firm conducted an experiment to study the effectiveness of an herbal compound to treat the common cold. They exposed each patient to a cold virus, then gave them either the herbal compound or a sugar solution known to have no effect on colds. Several days later they assessed each patient's condition, using a cold severity scale ranging from 0 to 5. (They found no evidence of the benefits of the compound.)

**28. Vineyards** Business analysts hoping to provide information helpful to American grape growers compiled these data about vineyards: size (acres), number of years in existence, state, varieties of grapes grown, average case price, gross sales, and percent profit.

**29. Streams** In performing research for an ecology class, students at a college in upstate New York collect data on local streams each year. They record a number of biological, chemical, and physical variables, including the stream name, the substrate of the stream (limestone, shale, or mixed), the acidity of the water (pH), the temperature (°C), and the BCI (a numerical measure of biological diversity).

**30. Fuel economy** The Environmental Protection Agency (EPA) tracks fuel economy of automobile models based on information from the manufacturers (Ford, Toyota, etc.). Among the data the agency collects are the manufacturer, vehicle type (car, SUV, etc.), weight, horsepower, and gas mileage (mpg) for city and highway driving.

**31. Dogs detecting coronavirus** In 2020, researchers at the University of Helsinki studied whether dogs can be trained to identify coronavirus (COVID-19) by smell. Researchers exposed dogs who had previously learned to identify cancer by scent to urine samples from individuals who had either tested positive or negative for COVID-19. They reported that the dogs were quick to learn the new scent. (Source: https://www.helsinki.fi/en/news/health-news/the-finnish-covid-dogs-nose-knows)

**32. Walking in circles** People who get lost in the desert, mountains, or woods often seem to wander in circles rather than walk in straight lines. To see whether people naturally walk in circles in the absence of visual clues, researcher Andrea Axtell tested 32 people on a football field. One at a time, they stood at the center of one goal line, were blindfolded, and then tried to walk to the other goal line. She recorded each individual's sex, height, handedness, the number of yards each was able to walk before going out of bounds, and whether each wandered off course to the left or the right. (No one made it all the way to the far end of the field without crossing one of the sidelines.) (*STATS* No. 39, Winter 2004)

| Year | Winner | Jockey | Trainer | Owner | Time |
|---|---|---|---|---|---|
| 1875 | Aristides | O. Lewis | A. Williams | H. P. McGrath | 2:37.75 |
| 1876 | Vagrant | R. Swim | J. Williams | William Astor | 2:38.25 |
| 1877 | Baden Baden | W. Walker | E. Brown | Daniel Swigert | 2:38 |
| 1878 | Day Star | J. Carter | L. Paul | T. J. Nichols | 2:37.25 |
| ... | | | | | |
| 2013 | Orb | J. Rosario | S. McGaughey | Stuart Janney & Phipps Stable | 2:02.89 |
| 2014 | California Chrome | Victor Espinoza | Art Sherman | California Chrome, LLC | 2:03.66 |
| 2015 | American Pharoah | Victor Espinoza | Bob Baffert | Zayat Stables, LLC | 2:03.03 |
| 2016 | Nyquist | M. Gutierrez | Doug F. O'Neill | Reddam Racing LLC | 2:01.31 |
| 2017 | Always Dreaming | J. Velazquez | Todd Pletcher | Meb Racing Stables | 2:03.59 |
| 2018 | Justify | M. Smith | Bob Baffert | China Horse Club | 2:04:.20 |
| 2019 | Country House | Flavien Prat | Bill Mott | J.O. Shields et al, | 2:03.93 |
| 2020 | Authentic | John Velazquez | Bob Baffert | Spendthrift Farm and others | 2:00.61 |

*Source:* Excerpt from HorseHats.com. Published by Thoroughbred Promotions.

**33. Kentucky Derby 2020** The Kentucky Derby is a horse race that has been run every year since 1875 at Churchill Downs in Louisville, Kentucky. The race started as a 1.5-mile race, but in 1896, it was shortened to 1.25 miles because experts felt that 3-year-old horses shouldn't run such a long race that early in the season. (It has been run in May every year but two—1901—when it took place on April 29, and 2020, when it was held on September 5.) The accompanying table shows the data for the first four and eight recent races.

**34. Indy 500 2020** The 2.5-mile Indianapolis Motor Speedway has been the home to a race on Memorial Day nearly every year since 1911. Even during the first race, there were controversies. Ralph Mulford was given the checkered flag first but took three extra laps just to make sure he'd completed 500 miles. When he finished, another driver, Ray Harroun, was being presented with the winner's trophy, and Mulford's protests were ignored. Harroun averaged 74.6 mph for the 500 miles. In 2013, the winner, Tony Kanaan, averaged over 187 mph, beating the previous record by over 17 mph!

Here are the data for the first five races and five recent Indianapolis 500 races.

| Year | Driver | Time (hr:min:sec) | Speed (mph) |
|---|---|---|---|
| 1911 | Ray Harroun | 6:42:08 | 74.602 |
| 1912 | Joe Dawson | 6:21:06 | 78.719 |
| 1913 | Jules Goux | 6:35:05 | 75.933 |
| 1914 | René Thomas | 6:03:45 | 82.474 |
| 1915 | Ralph DePalma | 5:33:55.51 | 89.840 |
| ... | | | |
| 2015 | Juan Pablo Montoya | 3:05:56.5286 | 161.341 |
| 2016 | Alexander Rossi | 3:00:02.0872 | 166.634 |
| 2017 | Takuma Sato | 3:13:3.3584 | 115.395 |
| 2018 | Will Power | 2:59:42.6365 | 166.935 |
| 2019 | Simon Pagenaud | 2:50:39.27948 | 175.79362 |
| 2020 | Takuma Sato | 3:10:05.0880 | 157.824 |

## JUST CHECKING

Answers

1. *Who*—All Tour de France races from 1903 to 2020. *What*—year, winner, country of origin, age, team, total time, average speed, stages, total distance ridden, starting riders, finishing riders; *hoW*— official statistics at race; *Where*—France (for the most part); *When*—1903 to 2020; *Why*—not specified.

2. 
| Variable | Type | Units |
|---|---|---|
| Year | Quantitative (or identifier) | Years |
| Winner | Categorical | |
| Country of Origin | Categorical | |
| Age | Quantitative | Years |
| Team | Categorical | |
| Total Time | Quantitative | Hours/minutes/seconds |
| Average Speed | Quantitative | Kilometers per hour |
| Stages | Quantitative | Counts (stages) |
| Total Distance | Quantitative | Kilometers |
| Starting Riders | Quantitative | Counts (riders) |
| Finishing Riders | Quantitative | Counts (riders) |

# 2 Displaying and Describing Categorical Data

| WHO | People on the *Titanic* |
|---|---|
| WHAT | Name, survival status, age, age category, sex, price paid, ticket class |
| WHEN | April 14, 1912 |
| WHERE | North Atlantic |
| HOW | A variety of sources and Internet sites |
| WHY | Historical interest |

Table 2.1
Part of a data table showing seven variables for eleven people aboard the *Titanic*

What happened on the *Titanic* at 11:40 on the night of April 14, 1912, is well known. Frederick Fleet's cry of "Iceberg, right ahead" and the three accompanying pulls of the crow's nest bell signaled the beginning of a nightmare that has become legend. By 2:15 A.M., the *Titanic*, thought by many to be unsinkable, had sunk. Only 712 of the 2208 people on board survived. The others (nearly 1500) met their icy fate in the cold waters of the North Atlantic.

The table below shows some data about the passengers and crew aboard the *Titanic*. Each case (row) of the data table represents a person on board the ship. The variables are the person's *Name*, *Survival* status (Lost or Saved), *Age* (in years), *Age Category* (Adult or Child), *Sex* (Male or Female), *Price* paid (in British pounds, £), and ticket *Class* (First, Second, Third, or Crew). Some of these, such as *Age* and *Price*, are quantitative variables. Others, like *Survival* and *Class*, place each case in a single category, so are categorical variables.

| Name | Survived | Age | Gender | Price | Class |
|---|---|---|---|---|---|
| ABBING, Mr Anthony | LOST | 41 | Male | 7.55 | 3rd |
| ABBOTT, Mr Eugene Joseph | LOST | 13 | Male | 20.25 | 3rd |
| ABBOTT, Mr Rossmore Edward | LOST | 16 | Male | 20.25 | 3rd |
| ABBOTT, Mrs Rhoda Mary "Rosa" | SAVED | 39 | Female | 20.25 | 3rd |
| ABELSETH, Miss Kalle (Karen) Marie Kristiane | SAVED | 16 | Female | 7.65 | 3rd |
| ABELSETH, Mr Olaus Jørgensen | SAVED | 25 | Male | 7.65 | 3rd |
| ABELSON, Mr Samuel | LOST | 30 | Male | 24 | 2nd |
| ABELSON, Mrs Anna | SAVED | 28 | Female | 24 | 2nd |
| ABĪ SA'B, Mr Jirjis Yūsuf | LOST | 45 | Male | 7.23 | 3rd |
| ABĪ SA'B, Mrs Sha'nīnah | SAVED | 38 | Female | 7.23 | 3rd |
| ABĪ SHADĪD, Mr Dāhir | LOST | 19 | Male | 7.23 | 3rd |

# CHAPTER 2 Displaying and Describing Categorical Data

The problem with a data table like this—and in fact with all data tables—is that you can't *see* what's going on. And seeing is just what we want to do. We need ways to show the data so that we can see patterns, relationships, trends, and exceptions.

# The Three Rules of Data Analysis

There are three things you should always do first with data:

1. **Make a picture.** A display of your data will reveal things you are not likely to see in a table of numbers and will help you to *Think* clearly about the patterns and relationships that may be hiding in your data.
2. **Make a picture.** A well-designed display will *Show* the important features and patterns in your data. A picture will also show you the things you did not expect to see: the extraordinary (possibly wrong) data values or unexpected patterns.
3. **Make a picture.** The best way to *Tell* others about your data is with a well-constructed picture.

We make graphs for two primary reasons: to understand data better ourselves and to show others what we have learned and want them to understand. For either purpose, a graph should be very easy to understand and should represent the facts of the data honestly. Among other things, this requires that if a graph has axes, each should be clearly labeled with the name of the variable it displays and the variable's units, if it has any. Any numeric scales should be easy to read. If colors or special symbols are used in the graph, then there should be a clear "key" identifying what they mean.[1] And all graphs should carry a title or caption that summarizes what the graph displays.

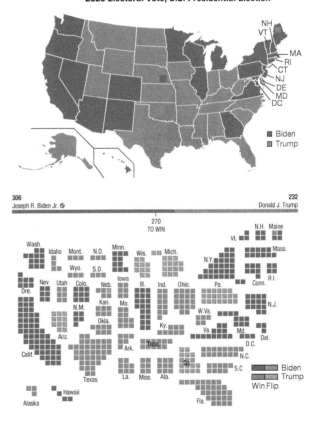

**Figure 2.1**

**Same data, different purposes** In November 2020 Joe Biden was elected to be the forty-sixth president of the Unites States, winning 306 electoral votes compared to 232 won by his Republican opponent, Donald Trump. Both of the maps to the right accurately show where the electoral votes were won, although their purposes are different. The upper map accurately displays the shapes and relative sizes of the 50 states, but it can leave the impression that big, sparsely populated states like Montana have more electoral clout than they really do, and that small, densely populated states like New Jersey have less. The lower map sacrifices an accurate representation of states' shapes and relative geographic sizes in favor of visually communicating an accurate sense of electoral clout, by making each equally sized square represent one electoral vote

The squares that are pale blue represent electoral votes in states that "flipped" between the elections of 2016 and 2020—their electoral votes went to the Republican candidate (Trump) in 2016 but to the Democratic candidate (Biden) in 2020.

---

[1] Colors and odd symbols should usually *not* be used if they don't mean anything at all. Using them risks puzzling the reader and/or distracting attention from the actual information in the graph.

# The Area Principle

A bad picture can distort our understanding rather than help it. What impression do you get from Figure 2.2 about who was on board the *Titanic*?

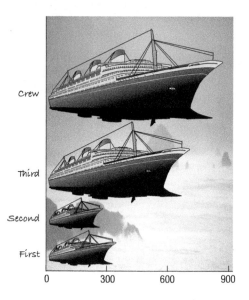

**Figure 2.2**
**How many people were in each class on the *Titanic*?** From this display, it looks as though the service must have been great, since most aboard were crew members. Although the length of each ship here corresponds to the correct number, the impression is all wrong. In fact, only about 40% were crew.

The *Titanic* was certainly a luxurious ship, especially for those in first class, but Figure 2.2 gives the mistaken impression that most of the people on the *Titanic* were crew members, with a few passengers along for the ride. The problem is that the graphic uses the ships' *lengths* to indicate counts of people, but our brains tend to be more impressed by the visual *area* that they take up on the page. There were about three times as many crew members as second-class passengers, and that's accurately represented by those two ships' lengths—but the ship image for the crew takes up about *nine* times as much space on the page as does the one for the second-class passengers. Graphs shouldn't leave people with incorrect impressions—especially easily avoidable ones.

The best graphs observe a fundamental principle called the **area principle**, which says that the area occupied by a part of a graph should be proportional to the magnitude of the value it represents.

# Frequency Tables: Making Piles

Categorical variables are easy to summarize in a **frequency table** that lists the categories and how many cases belong to each one. For ticket *Class*, the categories are First, Second, Third, and Crew:

| Class | Count |
|---|---|
| First | 325 |
| Second | 285 |
| Third | 706 |
| Crew | 885 |

**Table 2.2**
A frequency table of the *Titanic* passengers

| Class | % |
|---|---|
| First | 14.77 |
| Second | 12.95 |
| Third | 32.08 |
| Crew | 40.21 |

**Table 2.3**
A relative frequency table for the same data

A **relative frequency table** like Table 2.3 displays *percentages* (or *proportions*) rather than the counts in each category. A little information (actual counts) is lost, but the easy interpretability of percentages is gained. (The percentages should total 100%, but the sum can seem a little too high or too low if the individual category percentages have been rounded.) Tables like these show what is called the **distribution** of a categorical variable, because they show how the cases are distributed among the categories.

# Bar Charts

> **MIND THE GAP**
> A bar graph should always have gaps separating the bars, a characteristic that distinguishes it from its close cousin the histogram—which we'll introduce in the next chapter.

Although not as visually entertaining as the ships in Figure 2.2, the **bar charts** in Figure 2.3 and Figure 2.4 are *accurate* visual representations of the distribution of ticket *Class* because they obey the area principle.

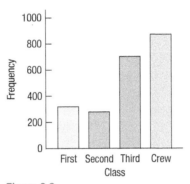

**Figure 2.3**
**People on the *Titanic* by Ticket Class** With the area principle satisfied, we can see the true distribution more clearly.

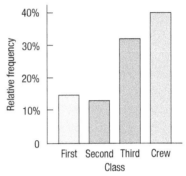

**Figure 2.4**
The relative frequency bar chart looks the same as the bar chart (Figure 2.3) but shows the proportion of people in each category rather than the counts.

> **WHAT A BAR CHART IS, AND ISN'T**
> For some reason, some computer programs give the name "bar chart" to any graph that uses bars. And others use different names according to whether the bars are horizontal or vertical. Don't be misled. "Bar chart" is the term for a *display of counts or proportions of a categorical variable* using bars.

Now it's easy to see that the crew did not make up a majority of the people on board. We can also easily see that there were about three times as many crew members as second-class passengers. Whether the bar chart showing counts or the one showing percentages is a better display depends on what you want to communicate. Because the *Titanic* captivates so many people with its human stories, this is a case where we prefer the display of counts, because they more easily make you think of the people behind the data.

## FOR EXAMPLE

### What Do You Think of Congress?

In June 2019 the Gallup poll asked a representative sample of U.S. adults how much confidence they had in various institutions in American society. The response about Congress was: (https://news.gallup.com/poll/1600/congress-public.aspx)

| | |
|---|---|
| Great deal | 4% |
| Quite a lot | 7% |
| Some | 36% |
| Very little | 48% |
| None | 4% |

**QUESTION:** What kind of table is this? What would be an appropriate display?

ANSWER: This is a relative frequency table because the numbers displayed are percentages, not counts. A bar chart would be appropriate:

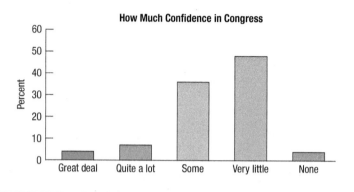

## FOR EXAMPLE
### Which Gadgets Do You Use?

The Pew Research Organization asked 1005 U.S. adults which of the following electronic items they use: cell phone, smartphone, computer, handheld e-book reader (e.g., Kindle), or tablet. The results were as follows:

| Device | Percentage (%) using the device |
|---|---|
| Cell phone | 86.8 |
| Smartphone | 54.0 |
| Computer | 77.5 |
| E-book reader | 32.2 |
| Tablet | 41.9 |

QUESTION: Is this a frequency table, a relative frequency table, or neither? How could you display these data graphically?

ANSWER: This is not a frequency table because the numbers displayed are not counts. Although the numbers are percentages, they do not sum to 100%. A person can use more than one device, so this is not a relative frequency table either. A bar chart might still be appropriate, but the numbers do not sum to 100%.

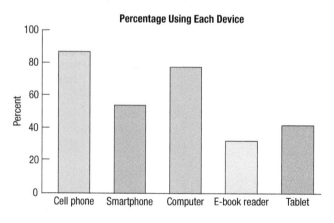

## Pie Charts

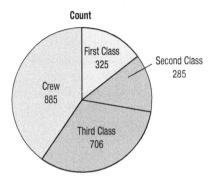

Figure 2.5
**Ticket Class Distribution** Number of *Titanic* passengers in each ticket class

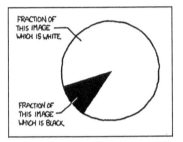

© 2013 Randall Munroe. Reprinted with permission. All rights reserved.

Another common display that shows how a whole group breaks into several categories is a pie chart. **Pie charts** show the whole group of cases as a circle. They slice the circle into pieces whose sizes are proportional to the fraction of the whole in each category.

Pie charts give a quick impression of how a whole group is partitioned into smaller groups. Because we're used to cutting up pies into 2, 4, or 8 pieces, pie charts are good for seeing relative frequencies near 1/2, 1/4, or 1/8. For example, you may be able to tell that the pink slice, representing the second-class passengers, is very close to 1/8 of the total. It's harder to see that there were about twice as many third-class as first-class passengers. Which category had the most passengers? Were there more crew or more third-class passengers? Comparisons such as these are easier in a bar chart.

### Think Before You Draw

Our first rule of data analysis is *Make a picture*. But what kind of picture? We don't have a lot of options—yet. There's more to statistics than pie charts and bar charts, and knowing when to use each type of graph is a critical first step in data analysis. That decision depends in part on what type of data we have.

It's important to check that the data are appropriate for whatever method of analysis you choose. Before you make a bar chart or a pie chart, always check the **Categorical Data Condition**: The data are counts or percentages of individuals in non-overlapping categories.

## Contingency Tables: Children and First-Class Ticket Holders First?

Only 32% of those aboard the *Titanic* survived. Was that survival rate the same for men and women? For children and adults? For all ticket classes? Another way to phrase questions like these is to ask whether two variables are related. For example, was there a relationship between the kind of ticket a passenger held and the passenger's chances of making it into a lifeboat?

To answer that question we can arrange the counts for the two categorical variables, *Survival* and ticket *Class*, in a table. Table 2.4 shows each person aboard the *Titanic* classified according to both their ticket *Class* and their *Survival*. The columns of this table show how many people were saved and how many were lost, contingent upon their ticket *Class*. And the rows show how many were in each ticket *Class*, contingent upon their *Survival* status. A table like this is called a **contingency table**.

Table 2.4
**Contingency table of ticket *Class* and *Survival*** The bottom line of "Totals" is the same as the previous frequency table.

|  |  | Class | | | | |
|---|---|---|---|---|---|---|
|  |  | First | Second | Third | Crew | Total |
| Survival | Saved | 203 | 118 | 178 | 212 | **711** |
|  | Lost | 122 | 167 | 528 | 673 | **1490** |
|  | Total | **325** | **285** | **706** | **885** | **2201** |

Each *cell* of the table gives the count for a combination of values of the two variables. The margins of the table, both on the right and at the bottom, give totals. The bottom line of the table is just the frequency distribution of ticket *Class*. The right column of the table is the frequency distribution of the variable *Survival*. When presented like this, in the margins of a contingency table, the frequency distribution of one of the variables is called its **marginal distribution**. The marginal distribution can be expressed either as counts or percentages.

If you look down the column for second-class passengers to the first row, you'll find the cell containing the 118 second-class passengers who were saved. Looking at the cell to its right we see that more third-class passengers (178) were saved. But, does that mean that third-class passengers were more *likely* to survive? It's true that *more* third-class passengers were saved, but there were many more third-class passengers on board the ship. To compare the two numbers fairly, we need to express them as percentages—but as a percentage of what?

For any cell, there are three choices of percentage. We could express the 118 second-class survivors as 5.4% of all the passengers on the *Titanic* (the *overall percent*), as 16.6% of all the survivors (the *row percent*), or as 41.4% of all second-class passengers (the *column percent*). Each of these percentages is potentially interesting.

Statistics programs offer all three. Unfortunately, they often put them all together in each cell of the table. The resulting table holds lots of information, but it can be hard to understand:

### Table 2.5
**Another contingency table of ticket *Class*** This time we see not only the counts for each combination of *Class* and *Survival* (in bold) but the percentages these counts represent. For each count, there are three choices for the percentage: by row, by column, and by table total. There's probably too much information here for this table to be useful.

|  |  |  | Class |  |  |  | |
|---|---|---|---|---|---|---|---|
|  |  |  | First | Second | Third | Crew | Total |
| Survival | Saved | Count | **203** | **118** | **178** | **212** | **711** |
|  |  | % of Row | 28.6% | 16.6% | 25.0% | 29.8% | 100% |
|  |  | % of Column | 62.5% | 41.4% | 25.2% | 24.0% | 32.3% |
|  |  | % of Table | 9.2% | 5.4% | 8.1% | 9.6% | 32.3% |
|  | Lost | Count | **122** | **167** | **528** | **673** | **1490** |
|  |  | % of Row | 8.2% | 11.2% | 35.4% | 45.2% | 100% |
|  |  | % of Column | 37.5% | 58.6% | 74.8% | 76.0% | 67.7% |
|  |  | % of Table | 5.6% | 7.6% | 24.0% | 30.6% | 67.7% |
|  | Total | Count | **325** | **285** | **706** | **885** | **2201** |
|  |  | % of Row | 14.8% | 12.9% | 32.1% | 40.2% | 100% |
|  |  | % of Column | 100% | 100% | 100% | 100% | 100% |
|  |  | % of Table | 14.8% | 12.9% | 32.1% | 40.2% | 100% |

To simplify the table, let's first pull out the % of table values:

|  |  | Class |  |  |  | |
|---|---|---|---|---|---|---|
|  |  | First | Second | Third | Crew | Total |
| Survival | Saved | 9.2% | 5.4% | 8.1% | 9.6% | 32.3% |
|  | Lost | 5.6% | 7.6% | 24.0% | 30.6% | 67.7% |
|  | Total | 14.8% | 12.9% | 32.1% | 40.2% | 100% |

### Table 2.6
**A contingency table of *Class* by *Survival* with only the table percentages**

These percentages tell us what fraction of *all* passengers belong to each combination of column and row category. For example, we see that although 8.1% of the people aboard the *Titanic* were surviving third-class ticket holders, only 5.4% were surviving second-class ticket holders. Is this fact useful? Comparing these percentages, you might think that the chances of surviving were better in third class than in second. But be careful. There were many more third-class than second-class passengers on the *Titanic*, which might be the only reason there were more third-class survivors. That group is a larger percentage of the passengers, but that's not really what we want to know. Overall percentages don't answer questions like this.

A bell-shaped artifact from the *Titanic*

## PERCENT OF WHAT?

The English language can be tricky when we talk about percentages. If you're asked, "What percent *of the survivors* were in second class?" it's pretty clear that we're interested only in survivors. It's as if we're restricting the *Who* in the question to the survivors, so we should look at the number of second-class passengers among all the survivors—in other words, the row percent.

But if you're asked, "What percent were second-class passengers who survived?" you have a different question. Be careful; here, the *Who* is everyone on board, so 2201 should be the denominator, and the answer is the table percent.

And if you're asked, "What percent of the second-class passengers survived?" you have a third question. Now the *Who* is the second-class passengers, so the denominator is the 285 second-class passengers, and the answer is the column percent.

Always be sure to ask "percent of what?" That will help you to know the *Who* and whether we want *row, column,* or *table* percentages.

### FOR EXAMPLE: Finding Marginal Distributions

A recent Gallup poll asked 1008 Americans age 18 and over whether they planned to watch the upcoming Super Bowl. The pollster also asked those who planned to watch whether they were looking forward more to seeing the football game or the commercials. The results are summarized in the table.

| | | Sex | | |
|---|---|---|---|---|
| | | Male | Female | Total |
| Response | Game | 279 | 200 | **479** |
| | Commercials | 81 | 156 | **237** |
| | Won't watch | 132 | 160 | **292** |
| | Total | **492** | **516** | **1008** |

QUESTION: What's the marginal distribution of the responses?

ANSWER: To determine the percentages for the three responses, divide the count for each response by the total number of people polled:

$$\frac{479}{1008} = 47.5\% \quad \frac{237}{1008} = 23.5\% \quad \frac{292}{1008} = 29.0\%$$

According to the poll, 47.5% of American adults were looking forward to watching the Super Bowl game, 23.5% were looking forward to watching the commercials, and 29% didn't plan to watch at all.

# Conditional Distributions

Rather than look at the overall percentages, it's more interesting to ask whether the chance of surviving the *Titanic* sinking *depended* on ticket class. We can look at this question in two ways. First, we could ask how the distribution of ticket *Class* changes between survivors and nonsurvivors. To do that, we look at the *row percentages*:

Table 2.7
**The conditional distribution of ticket *Class* conditioned on each value of *Survival*: Saved and Lost**

| | | Class | | | | |
|---|---|---|---|---|---|---|
| | | First | Second | Third | Crew | Total |
| Survival | Saved | 203 | 118 | 178 | 212 | **711** |
| | | 28.6% | 16.6% | 25.0% | 29.8% | **100%** |
| | Lost | 122 | 167 | 528 | 673 | **1490** |
| | | 8.2% | 11.2% | 35.4% | 45.2% | **100%** |

By focusing on each row separately, we can see the *conditional distribution* of class under the condition of surviving or not. (A **conditional distribution** shows the distribution on one variable for a subgroup of individuals that satisfy a condition on the other variable.) The bar charts in Figure 2.6 and Figure 2.7 show two different ways to contrast the distribution of class among those who were saved and those who were lost.

Figure 2.6
**The distribution of ticket *Class* among those who were saved, next to that of those who were lost**

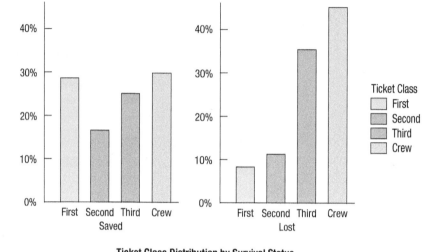

Figure 2.7
**Segmented Bar Chart** The same bars, stacked up neatly. Segmented bar charts allow clear comparisons of distributions of categorical data.

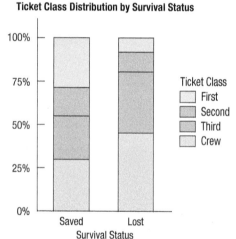

Figure 2.6 shows how the distribution of ticket *Class* among the survivors compares with the distribution of ticket *Class* among the victims. Because both distributions have been given the same scale (percentage of their group), it makes it easier to compare the distributions. For example, it's pretty easy to see that first- and second-class passengers together constitute about 45% of the survivors, but only about 20% of those who were lost. Figure 2.7 is just like Figure 2.6, only the bars have been stacked on top of one another, making the differences more obvious. A bar plot like this is called a **segmented bar chart**, and is often a good way to compare conditional distributions.

## FOR EXAMPLE

### Finding Conditional Distributions

RECAP The table shows results of a poll asking adults whether they were looking forward to the Super Bowl game, looking forward to the commercials, or didn't plan to watch.

CHAPTER 2  Displaying and Describing Categorical Data

|  | | Sex | | |
|---|---|---|---|---|
| | | Male | Female | Total |
| Response | Game | 279 | 200 | 479 |
| | Commercials | 81 | 156 | 237 |
| | Won't watch | 132 | 160 | 292 |
| | Total | 492 | 516 | 1008 |

QUESTION: Construct a graph that makes it easy to contrast the gender distributions across the three response groups. Describe how the distributions differ.

ANSWER: Determine the percentages of men and women among those who are most interested in the game. Then do the same for the other two responses.

Game:         Men: $\frac{279}{479} \approx 58.2\%$      Women: $\frac{200}{479} \approx 41.8\%$

Commercials: Men: $\frac{81}{237} \approx 34.2\%$      Women: $\frac{156}{237} \approx 65.8\%$

Won't watch: Men: $\frac{132}{292} \approx 45.2\%$      Women: $\frac{160}{292} \approx 54.8\%$

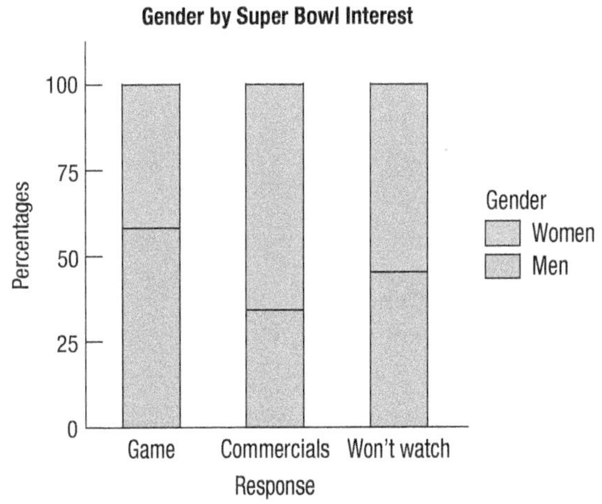

**Gender by Super Bowl Interest**

The graph shows that a majority of those most interested in the game are men, that an even greater majority of those most interested in the ads are women, and that among those who don't plan to watch the game, there are slightly more women than men.

We can also turn the question around. We can look at the distribution of *Survival* for each category of ticket *Class*. To do this, we look at the *column percentages*. Those show us whether the chance of being saved was roughly the same *for each of the four classes*. Now the percentages in each *column* add to 100%, because we've restricted the *Who*, in turn, to each of the four ticket classes:

**Table 2.8**

**A contingency table of *Class* by *Survival* with only counts and column percentages** Each column represents the conditional distribution of *Survival* for a given category of ticket *Class*.

| | | | Class | | | |
|---|---|---|---|---|---|---|
| | | | First | Second | Third | Crew | Total |
| Survival | Saved | Count | 203 | 118 | 178 | 212 | 711 |
| | | % of Column | 62.5% | 41.4% | 25.2% | 24.0% | 32.3% |
| | Lost | Count | 122 | 167 | 528 | 673 | 1490 |
| | | % of Column | 37.5% | 58.6% | 74.8% | 76.0% | 67.7% |
| | Total | Count | 325 | 285 | 706 | 885 | 2201 |
| | | | 100% | 100% | 100% | 100% | 100% |

Looking at how the percentages change across each row, it sure looks like ticket class mattered in whether a passenger was saved. To make it more vivid, we could display the percentages saved and not for each *Class* in side-by-side segmented bar charts:

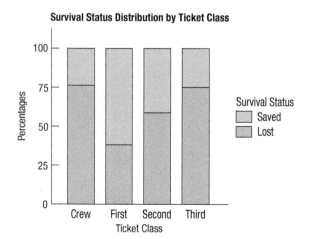

#### Figure 2.8
**Survival rates of each ticket class**
When a categorical variable like *Survival* status has only two possible data values, the graph wouldn't lose any information by showing only one of those values. (After all, anyone who wasn't lost must have been saved.) But this segmented bar chart isn't cluttered much by the inclusion of both, so in this case it seemed like a good graphing choice to include colored regions for both Saved and Lost.

> TI-*nspire*
> **Conditional distributions and association.** Explore the *Titanic* data to see which passengers were most likely to survive.

Figures 2.6 and 2.7 show that the distribution of ticket *Class* is not the same among those who were saved and those who were lost in the *Titanic* disaster. Figure 2.8 shows something similar (though not identical)—that the survival rates are not the same among those in the different ticket classes.

If the survival rates were about the same for all four classes, then we would say that ticket *Class* and *Survival* status were *independent* of one another. But they aren't. The first-class passengers survived at a greater rate than the other three classes, and the third-class passengers and crew survived at a much lower rate. Consequently, we say that ticket *Class* and *Survival* status are related, or associated, or dependent.

The most interesting things that data reveal are often associations between variables. (Case in point: The fate of the third-class passengers, trapped behind locked gates below decks, was one of the most frightening and memorable story elements in the blockbuster 1997 movie *Titanic*.) But associations are not always simple, because variables can be associated in many different ways and to different degrees. Something simpler and easier to think about is whether two variables are *not* associated with one another. In a contingency table, when the distribution of one variable is the same for all categories of another variable, we say that the variables are **independent**. That tells us there's *no* association between these variables.

## FOR EXAMPLE
### Looking for Associations Between Variables

**RECAP** The table shows results of a poll asking adults whether they were looking forward to the Super Bowl game, looking forward to the commercials, or didn't plan to watch.

|          | Sex Male | Sex Female | Total |
|----------|------|--------|-------|
| Game     | 279  | 200    | 479   |
| Commercials | 81 | 156   | 237   |
| Won't watch | 132 | 160  | 292   |
| Total    | 492  | 516    | 1008  |

**QUESTION:** Does it seem that there's an association between interest in Super Bowl TV coverage and a person's sex?

ANSWER: First find the distribution of the three responses for the men (the column percentages):

$$\frac{279}{492} = 56.7\% \quad \frac{81}{492} = 16.5\% \quad \frac{132}{492} = 26.8\%$$

Then do the same for the women who were polled, and display the two distributions with a side-by-side segmented bar chart.[2]

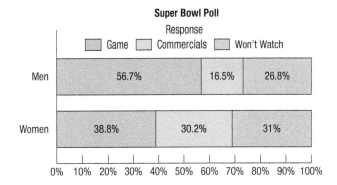

Based on this poll it appears that women were only slightly less interested than men in watching the Super Bowl telecast: 31% of the women said they didn't plan to watch, compared to just under 27% of men. Among those who planned to watch, however, there appears to be an association between the viewer's sex and what the viewer is most looking forward to. While more women are interested in the game (39%) than the commercials (30%), the margin among men is much wider: 57% of men said they were looking forward to seeing the game, compared to only 16.5% who cited the commercials.

## JUST CHECKING

A statistics class reports the following data on *Sex* and *Eye Color* for students in the class:

|  |  | Eye Color | | | |
|---|---|---|---|---|---|
|  |  | Blue | Brown | Green/Hazel /Other | Total |
| Sex | Males | 6 | 20 | 6 | 32 |
|  | Females | 4 | 16 | 12 | 32 |
|  | Total | 10 | 36 | 18 | 64 |

1. What percent of females are brown-eyed?
2. What percent of brown-eyed students are female?
3. What percent of students are brown-eyed females?
4. What's the distribution of *Eye Color*?
5. What's the conditional distribution of *Eye Color* for the males?
6. Compare the percent who are female among the blue-eyed students to the percent of all students who are female.
7. Does it seem that *Eye Color* and *Sex* are independent? Explain.

---

[2]Segmented bar charts can run vertically or horizontally. We were getting tired of all the vertical ones.

## STEP-BY-STEP EXAMPLE

### Examining Contingency Tables

Medical researchers followed 6272 Swedish men for 30 years to see if there was any association between the amount of fish in their diet and prostate cancer ("Fatty Fish Consumption and Risk of Prostate Cancer," *Lancet*, June 2001). Their results are summarized in this table:

|  | Prostate Cancer | |
|---|---|---|
| Fish Consumption | No | Yes |
| Never/seldom | 110 | 14 |
| Small part of diet | 2420 | 201 |
| Moderate part | 2769 | 209 |
| Large part | 507 | 42 |

We asked for a picture of a man eating fish. This is what we got.

**QUESTION:** Is there an association between fish consumption and prostate cancer?

---

**THINK**

**PLAN** Be sure to state what the problem is about.

**VARIABLES** Identify the variables and report the W's.

Be sure to check the appropriate condition.

I want to know if there is an association between fish consumption and prostate cancer.

The individuals are 6272 Swedish men followed by medical researchers for 30 years. The variables record their fish consumption and whether or not they were diagnosed with prostate cancer.

✓ **Categorical Data Condition:** I have counts for both fish consumption and cancer diagnosis. The categories of diet do not overlap, and the diagnoses do not overlap. It's okay to draw pie charts or bar charts.

---

**SHOW**

**MECHANICS** It's a good idea to check the marginal distributions first before looking at the two variables together.

|  | Prostate Cancer | | |
|---|---|---|---|
| Fish Consumption | No | Yes | Total |
| Never/seldom | 110 | 14 | 124 (2.0%) |
| Small part of diet | 2420 | 201 | 2621 (41.8%) |
| Moderate part | 2769 | 209 | 2978 (47.5%) |
| Large part | 507 | 42 | 549 (8.8%) |
| Total | 5806 (92.6%) | 466 (7.4%) | 6272 (100%) |

Two categories of the diet are quite small, with only 2.0% Never/Seldom eating fish and 8.8% in the "Large part" category. Overall, 7.4% of the men in this study had prostate cancer.

Then, make appropriate displays to see whether there is a difference in the relative proportions. These pie charts compare fish consumption for men who have prostate cancer to fish consumption for men who don't.

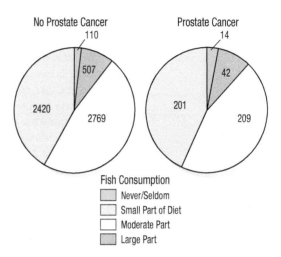

It's hard to see much difference in the pie charts. So, I made a display of the row percentages. Because there are only two alternatives, I chose to display the risk of prostate cancer for each group:

Both pie charts and bar charts can be used to compare conditional distributions. Here we compare prostate cancer rates based on differences in fish consumption.

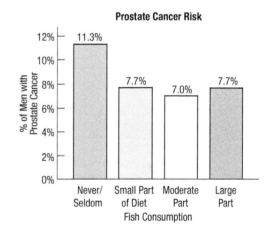

**TELL** CONCLUSION Interpret the patterns in the table and displays in context. If you can, discuss possible real-world consequences. Be careful not to overstate what you see. The results may not generalize to other situations.

Overall, there is a 7.4% rate of prostate cancer among men in this study. Most of the men (89.3%) ate fish either as a moderate or small part of their diet. From the pie charts, it's hard to see a difference in cancer rates among the groups. But in the bar chart, it looks like the cancer rate for those who never/seldom ate fish may be somewhat higher.

However, only 124 of the 6272 men in the study fell into this category, and only 14 of them developed prostate cancer. More study would probably be needed before we would recommend that men change their diets.[3]

---

[3]The original study actually used pairs of twins, which enabled the researchers to discern that the risk of cancer for those who never ate fish actually *was* substantially greater. Using pairs is a special way of gathering data. We'll discuss such study design issues and how to analyze the data in later chapters.

# Other Types of Graphs

We've shown two different ways we can represent conditional distributions graphically to make it easy to contrast them across groups. Figure 2.6 represented the distributions of ticket *Class of Titanic* passengers in separate bar charts—one for those who were saved and one for those who were lost. Figure 2.7 showed these two distributions as a single segmented bar chart, which may be easier to read.[4] There are many other ways to represent conditional distributions graphically. We'll show you two of the more common ones.

Figure 2.9 shows what the two bar charts in Figure 2.6 would look like if you "meshed" them together, putting the bar for those saved side-by-side with the bar for those lost in each of the four groups. All eight bars are the same heights as those in Figure 2.6; we've only changed their order and their colors, making it easier for the eye to separate the two bars in a pair.

Figure 2.9

**Side-by-side bar chart** This graph contrasts the distribution of ticket *class* among those who were saved to the one for those who were lost. It can be tricky to read correctly—although the bars have been grouped by ticket class, the vertical scale is percentages out of those saved or out of those lost. We think there are usually better options than a side-by-side bar graph for contrasting distributions.

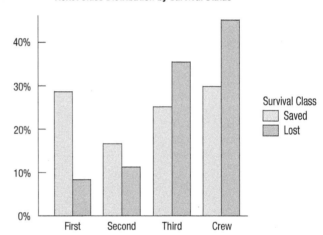

A graph that weaves together two or more conditional distributions like this is called a **side-by-side bar chart**. These are pretty popular, but we recommend extra care when reading a side-by-side bar graph, because it can be easy to mistakenly think that each grouping of bars is a conditional distribution—but it's not. For example, in Figure 2.9, the first pair of bars corresponds to first-class passengers, and the bar for Saved appears to be about three times taller than the one for Lost. Does that mean that three times as many of the first-class passengers were saved as were lost? Or does it mean that first-class passengers made up about three times as great a proportion of those who were saved than of those who were lost? Do you have to think about it very hard? That's because the graph seems to invite either interpretation. The correct interpretation is the second of these. But because it's easy to read this type of graph incorrectly, we recommend using one to display conditional distributions only after having considered other options.

If you want to contrast conditional distributions across different groups using a graph, it's critical that every distribution be scaled to relative frequencies *within that group*. (For example, in a segmented bar chart, each bar should reach to exactly 100%.) Otherwise the different sizes of the groups get in the way of making meaningful visual comparisons. Unfortunately, one piece of information is lost when you rescale like this: the relative sizes of the different groups. We can see from Figure 2.8 that the proportion of the second-class passenger class who were saved was greater than that of the third-class passengers. But the graph can't tell us which ticket class had the most *people* survive.

There is a clever type of graph that includes all the desirable characteristics of a segmented bar chart while also preserving the information about how large the different groups are. Figure 2.10 is called a **mosaic plot**. It is identical to the segmented bar chart shown in Figure 2.8, except that the *widths* of the bars—which were previously conveying no information—are now proportional to the sizes of the different groups. (And the gaps have also been removed

---

[4]Or it may not. Which is easier to read is a matter of opinion, and it also depends on the data. Just to give one reason, segmented bar charts grow more difficult to read when there are more categories.

from between the bars.) A segmented bar chart obeys the area principle within each bar (as it should!) but a mosaic plot does that and also obeys the area principle across the entire dataset!

Unlike in Figure 2.8, we can see in Figure 2.10 that although second-class passengers were saved at a greater *rate* than were third-class passengers, the actual roster of those saved included more third-class passengers, because there were so many more of them on the *Titanic* to start with. This is indicated in the graph by the fact that the mosaic plot's "tile" corresponding to third-class passengers who were saved has a greater area than does the tile corresponding to second-class passengers who were saved.

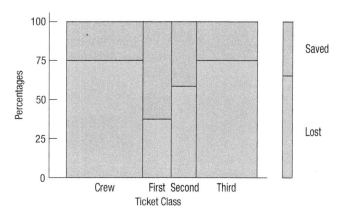

**Figure 2.10**
**Mosaic plot of survival by ticket class** This graph's vertical axis shows percentages within ticket *class* groups. The widths of the bars show the marginal distribution of ticket *Class*. To the right of the graph is a vertical strip showing the marginal distribution of the variable *Survival*. (Not all mosaic plots include this strip.)

## Is the Difference Meaningful?

A question we'll be asking a lot throughout this text is whether an apparent phenomenon in a dataset is "real" or just a "quirk" of random chance. For example, Figure 2.8 shows that the survival rates were different for *Titanic* passengers in different ticket classes. But were first-class passengers really less likely to perish than others? Or did they just happen to have the greatest survival rate by chance? After all, even if class had nothing to do with survival, it would be highly unlikely for all four ticket classes to have *exactly* the same survival rate, so *some* class had to come out on top.

One way to address this question is to **simulate** survival data for *Titanic* passengers under the assumption that ticket *Class* has *nothing to do with Survival status*. If the actual survival data look pretty similar to what might happen at random, then the different survival rates could easily just be due to chance.

We used a computer to *randomly* select 711 survivors from among everyone on board the Titanic, *without any regard to ticket Class*. We made a graph similar to Figure 2.8 with our simulated data, and then we did that again four more times. The five graphs we obtained are shown in Figure 2.11, with the actual data also shown for contrast.

**Figure 2.11**
**Simulated data, real data** The first five of these graphs show the sorts of differences in survival rates that could arise just by chance if ticket *Class* and *Survival* status were unrelated. The actual survival rates among the *Titanic*'s ticket classes clearly differed by much more than that.

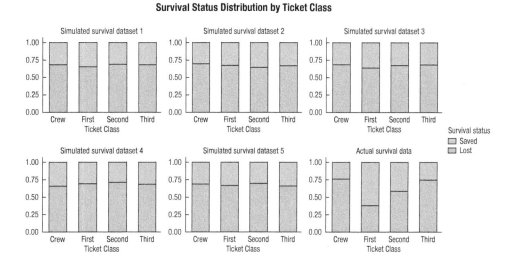

This collection of graphs shows us that even if survival had nothing to do with ticket class, it would still be reasonable to expect the survival rates across the four ticket classes to be a bit different from one another—but only a *little* bit different. The graph of the actual data stands out from the other five graphs like a sore thumb, because the differences in survival rates among *Titanic*'s ticket classes were so much greater than we would have expected had class and survival been independent.

This kind of reasoning recurs throughout Statistics. It can be summarized pretty succinctly like this: "If *A* were true, I'd have expected my data to look like *B*. But my data don't look anything like *B*. So I don't think *A* is true." The whole second half of this text is devoted to many different situations in which we'll draw conclusions from data by using exactly this reasoning process.

# WHAT CAN GO WRONG?

- **Don't violate the area principle.** This is probably the most common mistake in a graphical display. It is often made in the cause of artistic presentation. Here, for example, are two displays of the pie chart of the *Titanic* passengers by class:

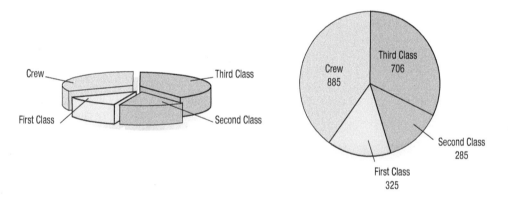

The one on the left looks pretty, doesn't it? But showing the pie on a slant violates the area principle and makes it much more difficult to compare fractions of the whole made up of each class—the principal feature that a pie chart ought to show.

- **Keep it honest.** Here's a pie chart that displays data on the percentage of high school students who engage in specified dangerous behaviors as reported by the Centers for Disease Control. What's wrong with this plot?

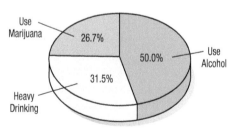

Try adding up the percentages. Or look at the 50% slice. Does it look right? Then think: What are these percentages of? Is there a "whole" that has been sliced up? In a pie chart, the proportions shown by each slice of the pie must add up to 100% and each individual must fall into exactly one category. Of course, showing the pie on a slant makes it even harder to detect the error.

The following chart shows the average number of texts in various time periods by American cell phone customers in the period 2006 to 2011.

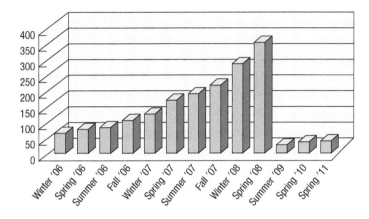

It may look as though text messaging decreased suddenly sometime around 2010, but that really doesn't make sense. In fact, this chart has several problems. First, it's not a bar chart. Bar charts display counts of categories. This bar chart is a plot of a quantitative variable (average number of texts) against time—although to make it worse, some of the time periods are missing. Even though these flaws are already fatal, the worst mistake is one that can't be seen from the plot. In 2010, the company reporting the data switched from reporting the average number of texts per year (reported each quarter) to average number of texts per month. So, the numbers in the last three quarters should be multiplied by 12 to make them comparable to the rest.

- **Don't confuse similar-sounding percentages.** These percentages sound similar but are different:
    - The percentage of the passengers who were both in first class and survived: This would be 203/2201, or 9.2%.
    - The percentage of the first-class passengers who survived: This is 203/325, or 62.5%.
    - The percentage of the survivors who were in first class: This is 203/711, or 28.6%.

In each instance, pay attention to the *Who* implicitly defined by the phrase. Often there is a restriction to a smaller group (all aboard the *Titanic*, those in first class, and those who survived, respectively) before a percentage is found. Your discussion of results must make these differences clear.

|  |  | Class | | | | |
|---|---|---|---|---|---|---|
|  |  | First | Second | Third | Crew | Total |
| Survival | Saved | 203 | 118 | 178 | 212 | **711** |
|  | Lost | 122 | 167 | 528 | 673 | **1490** |
|  | Total | **325** | **285** | **706** | **885** | **2201** |

- **Don't forget to look at the variables separately, too.** When you make a contingency table or display a conditional distribution, be sure you also examine the marginal distributions. It's important to know how many cases are in each category.

- **Be sure to use enough individuals.** When you consider percentages, take care that they are based on a large enough number of individuals. Take care not to make a report such as this one:

*We found that 66.67% of the rats improved their performance with training. The other rat died.*

- **Don't overstate your case.** Independence is an important concept, but it is rare for two variables to be *entirely* independent. We can't conclude that one variable has no effect whatsoever on another. Usually, all we know is that little effect was observed in our study. Other studies of other groups under other circumstances could find different results.

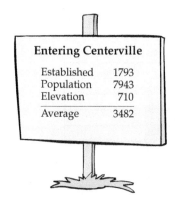

Table 2.9
**On-time flights by *Time of Day* and *Pilot*** Look at the percentages within each *Time of Day* category. Who has a better on-time record during the day? At night? Who is better overall?

## Simpson's Paradox

◆ **Don't use unfair or silly averages.** Sometimes averages can be misleading. Sometimes they just don't make sense at all. Be careful when averaging different variables that the quantities you're averaging are comparable. The Centerville sign says it all.

When using averages of proportions across several different groups, it's important to make sure that the groups really are comparable.

It's easy to make up an example showing that averaging across very different values or groups can give absurd results. Here's how that might work: Suppose there are two pilots, Moe and Jill. Moe argues that he's the better pilot of the two, since he managed to land 83% of his last 120 flights on time compared with Jill's 78%. But let's look at the data a little more closely. Here are the results for each of their last 120 flights, broken down by the time of day they flew:

|  |  | Time of Day | | |
|---|---|---|---|---|
|  |  | Day | Night | Overall |
| Pilot | Moe | 90 out of 100<br>90% | 10 out of 20<br>50% | 100 out of 120<br>83% |
|  | Jill | 19 out of 20<br>95% | 75 out of 100<br>75% | 94 out of 120<br>78% |

Look at the daytime and nighttime flights separately. For day flights, Jill had a 95% on-time rate and Moe only a 90% rate. At night, Jill was on time 75% of the time and Moe only 50%. So Moe is better "overall," but Jill is better both during the day and at night. How can this be?

What's going on here is a problem known as **Simpson's paradox**, named for the statistician who discovered it in the 1950s. It comes up rarely in real life, but there have been several well-publicized cases. As we can see from the pilot example, the problem is *unfair averaging* over different groups. Jill has mostly night flights, which are more difficult, so her *overall average* is heavily influenced by her nighttime average. Moe, on the other hand, benefits from flying mostly during the day, with its higher on-time percentage. With their very different patterns of flying conditions, taking an overall average is misleading. It's not a fair comparison.

The moral of Simpson's paradox is to be careful when you average across different levels of a second variable. It's always better to compare percentages or other averages *within* each level of the other variable. The overall average may be misleading.

### Simpson's Paradox

One famous example of Simpson's paradox arose during an investigation of admission rates for men and women at the University of California at Berkeley's graduate schools. As reported in an article in *Science*, about 45% of male applicants were admitted, but only about 30% of female applicants got in. It looked like a clear case of discrimination. However, when the data were broken down by school (Engineering, Law, Medicine, etc.), it turned out that, within each school, the women were admitted at nearly the same or, in some cases, much *higher* rates than the men. How could this be? Women applied in large numbers to schools with very low admission rates (Law and Medicine, for example, admitted fewer than 10%). Men tended to apply to Engineering and Science. Those schools have admission rates above 50%. When the *average* was taken, the women had a much lower *overall* rate, but the average didn't really make sense.

# WHAT HAVE WE LEARNED?

We've learned to analyze categorical variables.
- The methods in this chapter apply to categorical variables only. We always check the Categorical Variable Condition before proceeding.
- We summarize categorical data by counting the number of cases in each category, sometimes expressing the resulting distribution as percents.
- We display the distributions in a pie chart, a bar chart, or a segmented bar chart.

When we want to see how two categorical variables are related, we put the counts (and/or percentages) in a contingency table.
- We look at the marginal distribution of each variable.
- We also look at the conditional distribution of a variable within each category of the other variable.
- We compare these marginal and conditional distributions by using pie charts, bar charts, segmented bar charts, or mosaic plots.
- We examine the association between categorical variables by comparing conditional and marginal distributions. If the conditional distributions of one variable are roughly the same for each category of the other, we say the variables are independent.

## TERMS

**Area principle** — In a statistical display, each data value should be represented by the same amount of area. (p. 16)

**Frequency table** — A frequency table lists the categories of a categorical variable and gives the number of observations of each category. (p. 16)

**Relative frequency table** — A relative frequency table lists the categories of a categorical variable and gives the fraction or percent of observations of each category (p. 17)

**Distribution** — The distribution of a variable gives
- the possible values of the variable and
- the relative frequency of each value. (p. 17)

**Bar chart** — Bar charts show a bar whose area represents the count of observations for each category of a categorical variable. (p. 17)

**Pie chart** — Pie charts show how a "whole" divides into categories by showing a wedge of a circle whose area corresponds to the proportion in each category. (p. 19)

**Categorical Data Condition** — The methods in this chapter are appropriate for displaying and describing categorical data. Be careful not to use them with quantitative data. (p. 19)

**Contingency table** — A contingency table displays counts and, sometimes, percentages of individuals falling into named categories on two or more variables. The table categorizes the individuals on all variables at once to reveal possible patterns in one variable that may be contingent on the category of the other. (p. 19)

**Marginal distribution** — In a contingency table, the distribution of either variable alone is called the *marginal distribution*. The counts or percentages are the totals found in the margins (last row or column) of the table. (p. 20)

**Conditional distribution** — The distribution of a variable restricting the *Who* to consider only a smaller group of individuals is called a conditional distribution. (p. 22)

| | |
|---|---|
| **Segmented bar chart** | A segmented bar chart is a bar chart whose bars are stacked on top of one another in a vertical graph, or lined up side-by-side in a horizontal graph. A segmented bar chart usually shows relative frequencies so that the distribution of the categorical variable can be more easily compared between different groups. (p. 22) |
| **Independence** | Variables are said to be independent if the conditional distribution of one variable is the same for each category of the other. If the variables are not independent, we say there is an *association*. (p. 24) |
| **Side-by-side bar chart** | A side-by-side bar chart interweaves the bars of two or more conditional distributions to facilitate contrasting the distributions. Be careful interpreting them, and be cautious using them. (p. 28) |
| **Mosaic plot** | A mosaic plot is a special kind of segmented bar chart whose bars' widths display the marginal distribution of the variable represented by the bars. (p. 28) |
| **Simulation** | A random re-enactment of data collection under one or more assumptions. If real data look very different from simulated data, then the assumptions are called into question. (p. 29) |
| **Simpson's paradox** | When averages are taken across different groups, they can appear to contradict the overall averages. This is known as "Simpson's paradox." (p. 32) |

# ON THE COMPUTER

## Displaying Categorical Data

Although every package makes a slightly different bar chart, they all have similar features:

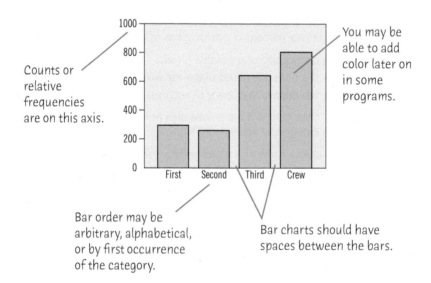

Counts or relative frequencies are on this axis.

You may be able to add color later on in some programs.

Bar order may be arbitrary, alphabetical, or by first occurrence of the category.

Bar charts should have spaces between the bars.

Sometimes the count or a percentage is printed above or on top of each bar to give some additional information. You may find that your statistics package sorts category names in annoying orders by default. For example, many packages sort categories alphabetically or by the order the categories are seen in the dataset. Often, neither of these is the best choice.

# EXERCISES

1. **Graphs in the news** Find a bar graph of categorical data from a newspaper, a magazine, or the Internet.
   a) Is the graph clearly labeled?
   b) Does it violate the area principle?
   c) Does the accompanying article tell the W's of the variable?
   d) Do you think the article correctly interprets the data? Explain.

2. **Graphs in the news II** Find a pie chart of categorical data from a newspaper, a magazine, or the Internet.
   a) Is the graph clearly labeled?
   b) Does it violate the area principle?
   c) Does the accompanying article tell the W's of the variable?
   d) Do you think the article correctly interprets the data? Explain.

3. **Tables in the news** Find a frequency table of categorical data from a newspaper, a magazine, or the Internet.
   a) Is it clearly labeled?
   b) Does it display percentages or counts?
   c) Does the accompanying article tell the W's of the variable?
   d) Do you think the article correctly interprets the data? Explain.

4. **Tables in the news II** Find a contingency table of categorical data from a newspaper, a magazine, or the Internet.
   a) Is it clearly labeled?
   b) Does it display percentages or counts?
   c) Does the accompanying article tell the W's of the variables?
   d) Do you think the article correctly interprets the data? Explain.

5. **Movie genres** The pie chart summarizes the genres of the 1300 movies released between 2006 and 2018.

   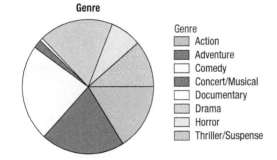

   a) Is this an appropriate display for the genres? Why or why not?
   b) Which category was least common?

6. **Movie ratings** The Motion Picture Association of America (MPAA) rates each film to designate the appropriate audience. The ratings are G, NC-17, PG, PG-13, and R. The pie chart shows the MPAA ratings of 891 movies.

   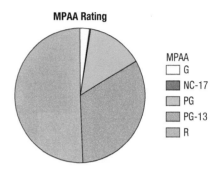

   a) Is this an appropriate display for the ratings? Why or why not?
   b) Which was the most common rating?

7. **Movie Genres, again** Write a few sentences describing the pie graph of movies genres seen in Exercise 5.

8. **Movie ratings, again** Write a few sentences describing the pie graph of movies ratings seen in Exercise 6.

9. **Diet and politics** A survey of 299 undergraduate students asked about respondents' diet preference (Carnivore, Omnivore, Vegetarian) and political alignment (Liberal, Moderate, Conservative). Here are side-by-side stacked bar charts of the 285 responses:

   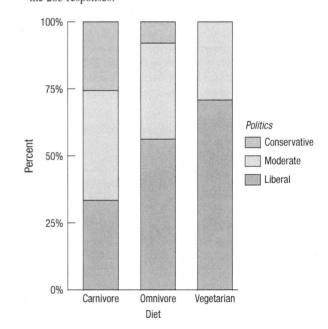

   a) Describe what this plot shows using the concept of a conditional distribution.
   b) Do you think the differences here are real? Explain.

**10. Diet and politics revisited** Here are the same data as in Exercise 9 but displayed differently:

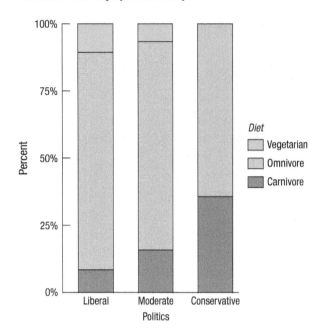

a) Describe what this plot shows using the concept of a conditional distribution.
b) Do you think the differences here are real? Explain.

**11. Magnet schools** An article in *Chance* magazine reported on the Houston Independent School District's magnet schools programs. Of the 1755 qualified applicants, 931 were accepted, 300 were wait-listed, and 524 were turned away for lack of space. Find the relative frequency distribution of the decisions made, and write a sentence describing it.

**12. Magnet schools again** The *Chance* article about the Houston magnet schools program described in Exercise 11 also indicated that 517 applicants were black or Hispanic, 292 Asian, and 946 white. Summarize the relative frequency distribution of ethnicity with a sentence or two (in the proper context, of course).

**13. Causes of death 2017** The Centers for Disease Control and Prevention lists causes of death in the United States during 2017:

| Cause of Death | Percent |
| --- | --- |
| Heart disease | 24.2% |
| Cancer | 21.9 |
| Accidents | 7.6 |
| Lung diseases | 5.2 |
| Stroke | 4.3 |

Source: https://www.cdc.gov/healthequity/lcod/men/2017/all-races-origins/index.htm

a) Is it reasonable to conclude that heart or lung diseases were the cause of approximately 29.4% of U.S. deaths in 2017?
b) What percentage of deaths were from causes not listed here?
c) Create an appropriate display for these data.

**14. Plane crashes** An investigator compiled information about recent nonmilitary plane crashes. The causes, to the extent that they could be determined, are summarized in the table.

| Causes of Fatal Accidents | |
| --- | --- |
| Cause | Percent |
| Pilot error | 46% |
| Other human error | 8 |
| Weather | 9 |
| Mechanical failure | 28 |
| Sabotage | 9 |
| Other causes | 1 |

Source: www.planecrashinfo.com/cause.htm

a) Is it reasonable to conclude that the weather or mechanical failures caused only about 37% of recent plane crashes?
b) Why do the numbers in the table add to 101%?
c) Create an appropriate display for these data.

**15. Oil spills 2016** Data from the International Tanker Owners Pollution Federation Limited (www.itopf.com) give the cause of spillage for 460 large oil tanker accidents from 1970–2016. Here are two displays.

a) Write a brief report interpreting what the displays show.

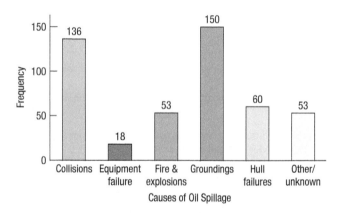

b) Is a pie chart an appropriate display for these data? Why or why not?

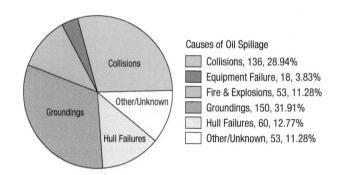

**16. Summer Olympics 2016** Fifty-nine countries won gold medals in the 2016 Summer Olympics. The table lists them, along with the total number of gold medals each won.

| Country | Medals | Country | Medals |
|---|---|---|---|
| United States | 46 | South Africa | 2 |
| Great Britain | 27 | Ukraine | 2 |
| China | 26 | Serbia | 2 |
| Russia | 19 | Poland | 2 |
| Germany | 17 | North Korea | 2 |
| Japan | 12 | Belgium | 2 |
| France | 10 | Thailand | 2 |
| South Korea | 9 | Slovakia | 2 |
| Italy | 8 | Georgia | 2 |
| Australia | 8 | Azerbaijan | 1 |
| Netherlands | 8 | Belarus | 1 |
| Hungary | 8 | Turkey | 1 |
| Brazil | 7 | Armenia | 1 |
| Spain | 7 | Czech Republic | 1 |
| Kenya | 6 | Ethiopia | 1 |
| Jamaica | 6 | Slovenia | 1 |
| Croatia | 5 | Indonesia | 1 |
| Cuba | 5 | Romania | 1 |
| New Zealand | 4 | Bahrain | 1 |
| Canada | 4 | Vietnam | 1 |
| Uzbekistan | 4 | Chinese Taipei | 1 |
| Kazakhstan | 3 | Bahamas | 1 |
| Colombia | 3 | Côte d'Ivoire | 1 |
| Switzerland | 3 | Fiji | 1 |
| Iran | 3 | Jordan | 1 |
| Greece | 3 | Kosovo | 1 |
| Argentina | 3 | Puerto Rico | 1 |
| Denmark | 2 | Singapore | 1 |
| Sweden | 2 | Tajikistan | 1 |

a) Try to make a display of these data. What problems do you encounter?
b) Organize the data so that the graph is more successful.

**17. Global warming** The Yale Program on Climate Change Communication surveyed 911 American adults in May of 2020 and asked them about their attitudes on global climate change. Here's a display of the percentages of respondents choosing each of the major alternatives offered to account for global warming. List the errors in this display.

**Causes of Global Warming**

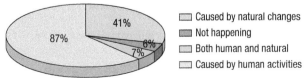

**18. Modalities** A survey of athletic trainers asked what modalities (treatment methods such as ice, whirlpool, ultrasound, or exercise) they commonly use to treat injuries. Respondents were each asked to list three modalities. The article included the following figure reporting the modalities used:

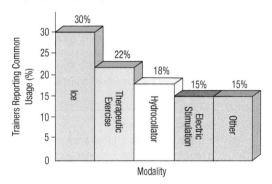

*Source:* Scott F. Nadler, Michael Prybicien, Gerard A. Malanga, and Dan Sicher, "Complications from Therapeutic Modalities: Results of a National Survey of Athletic Trainers." *Archives of Physical Medical Rehabilitation* 84 (June 2003).

a) What problems do you see with the graph?
b) Consider the percentages for the named modalities. Do you see anything odd about them?

**19. Teen smokers** As part of an effort to reduce smoking among teenagers, the organization Monitoring the Future (www.monitoringthefuture.org) asked 2048 eighth graders who said they smoked cigarettes what brands they preferred. The table below shows brand preferences for two regions of the country. Write a few sentences describing the similarities and differences in brand preferences among eighth graders in the two regions listed.

| Brand Preference | South | West |
|---|---|---|
| Marlboro | 58.4% | 58.0% |
| Newport | 22.5% | 10.1% |
| Camel | 3.3% | 9.5% |
| Other (over 20 brands) | 9.1% | 9.5% |
| No usual brand | 6.7% | 12.9% |

**20. Driving Enjoyment** In July 2018 the Gallup Organization (news.Gallup.com) polled a sample of 1503 adults in the United States and asked them, "How much do you personally enjoy driving—a great deal, a moderate amount, not much or not at all?" The table below shows separately the distributions of responses from men and women in the sample. (Percentage totals add up to less than 100 because of rounding.) Write a few sentences comparing the two distributions.

| Response | Men | Women |
|---|---|---|
| A great deal | 41 | 27 |
| A moderate amount | 40 | 48 |
| Not much or not at all | 18 | 24 |

21. **College and financial well-being** In 2019, Pew Research Center conducted a survey of 6878 American adults to ask "How would you describe your household's financial situation?" Participants were asked to choose one of the following options for their response: "live comfortably," "meet basic expenses with a little left over for extras," "just meet basic expenses," or "don't even have enough to meet basic expenses." Some participants said they didn't know or chose not to respond. The table below summarizes the responses by whether or not the respondent completed college. Respondents who gave no answer have been removed from the table.

| Completed college? | Can't meet basic expenses | Just meet basic expenses | Meet basic expenses+ | Live comfortably |
|---|---|---|---|---|
| No | 293 | 886 | 1526 | 932 |
| Yes | 76 | 380 | 1226 | 1537 |

Source: pewsocialtrends.org/2019/12/11/most-americans-say-thecurrent-economy-is-helping-the-rich-hurting-the-poor-and-middle-class/

a) What percent of respondents who could not meet basic expenses did not complete college?
b) What percent of respondents who did not complete college could not meet basic expenses?
c) What are the conditional distributions of household financial situation by college completion status?
d) Do you think that household financial situation is independent of college completion? Use a suitable visualization to support your answer.

22. **College and financial situation improvement** The American adults surveyed in Exercise 21 were also asked to rate the financial situations of most Americans along the same categories. The table below summarizes the respondents' perceived own household financial situation relative to that of most Americans by whether or not the respondent completed college. Respondents who gave no answer have been removed from the table.

| Completed college? | Own household financial situation (relative to most American households) | | |
|---|---|---|---|
| | Worse | Same as | Better |
| No | 577 | 1239 | 1800 |
| Yes | 214 | 680 | 2307 |

a) What percent of respondents who felt their own household financial situation was better than that of most American households did not complete college?
b) What percent of respondents who did not complete college felt their own household financial situation was better than that of most American households?
c) What are the conditional distributions of relative household financial situation by college completion status?
d) Do you think that the perceived household financial situation relative to most American households is independent of college completion? Use a suitable visualization to support your answer.

23. **Seniors** Prior to graduation, a high school class was surveyed about its plans. The following table displays the results for white and minority students (the "Minority" group included African-American, Asian, Hispanic, and Native American students):

| | | Seniors | |
|---|---|---|---|
| | | White | Minority |
| Plans | 4-year college | 198 | 44 |
| | 2-year college | 36 | 6 |
| | Military | 4 | 1 |
| | Employment | 14 | 3 |
| | Other | 16 | 3 |

a) What percent of the seniors are white?
b) What percent of the seniors are planning to attend a 2-year college?
c) What percent of the seniors are white and planning to attend a 2-year college?
d) What percent of the white seniors are planning to attend a 2-year college?
e) What percent of the seniors planning to attend a 2-year college are white?

24. **Politics** Students in an Intro Stats course were asked to describe their politics as "Liberal," "Moderate," or "Conservative." Here are the results:

| | | Politics | | | |
|---|---|---|---|---|---|
| | | L | M | C | Total |
| Sex | Female | 35 | 36 | 6 | 77 |
| | Male | 50 | 44 | 21 | 115 |
| | Total | 85 | 80 | 27 | 192 |

a) What percent of the class is male?
b) What percent of the class considers themselves to be "Conservative"?
c) What percent of the males in the class consider themselves to be "Conservative"?
d) What percent of all students in the class are males who consider themselves to be "Conservative"?

25. **More about seniors** Look again at the table of post-graduation plans for the senior class in Exercise 23.

a) Find the conditional distributions (percentages) of plans for the white students.
b) Find the conditional distributions (percentages) of plans for the minority students.
c) Create a graph comparing the plans of white and minority students.
d) Do you see any important differences in the post-graduation plans of white and minority students? Write a brief summary of what these data show, including comparisons of conditional distributions.

26. **Politics revisited** Look again at the table of political views for the Intro Stats students in Exercise 24.

a) Find the conditional distributions (percentages) of political views for the females.

b) Find the conditional distributions (percentages) of political views for the males.
c) Make a graphical display that compares the two distributions.
d) Do the variables *Politics* and *Sex* appear to be independent? Explain.

**27. Magnet schools revisited** The *Chance* magazine article described in Exercise 11 further examined the impact of an applicant's ethnicity on the likelihood of admission to the Houston Independent School District's magnet schools programs. Those data are summarized in the table below:

|  | Admission Decision | | | |
|---|---|---|---|---|
| Ethnicity | Accepted | Wait-listed | Turned away | Total |
| Black/Hispanic | 485 | 0 | 32 | 517 |
| Asian | 110 | 49 | 133 | 292 |
| White | 336 | 251 | 359 | 946 |
| Total | 931 | 300 | 524 | 1755 |

a) What percent of all applicants were Asian?
b) What percent of the students accepted were Asian?
c) What percent of Asians were accepted?
d) What percent of all students were accepted?

**28. More politics** Look once more at the table summarizing the political views of Intro Stats students in Exercise 24.
a) Produce a graphical display comparing the conditional distributions of males and females among the three categories of politics.
b) Comment briefly on what you see from the display in part (a).

**29. Back to school** Examine the table about ethnicity and acceptance for the Houston Independent School District's magnet schools program, shown in Exercise 27. Does it appear that the admissions decisions are made independent of the applicant's ethnicity? Explain.

**30. Cars** A survey of autos parked in student and staff lots at a large university classified the brands by country of origin, as seen in the table.

|  | Driver | |
|---|---|---|
| Origin | Student | Staff |
| American | 107 | 105 |
| European | 33 | 12 |
| Asian | 55 | 47 |

a) What percent of all the cars surveyed were foreign?
b) What percent of the American cars were owned by students?
c) What percent of the students owned American cars?
d) What is the marginal distribution of origin?
e) What are the conditional distributions of origin by driver classification?
f) Do you think that the origin of the car is independent of the type of driver? Explain.

**31. Diet and politics III** Are the patterns seen in Exercises 9 and 10 relating diet to political opinion the same for men and women? Here are two contingency tables:

**Men**

|  | Carnivore | Omnivore | Vegetarian |
|---|---|---|---|
| Liberal | 9 | 74 | 5 |
| Moderate | 12 | 54 | 1 |
| Conservative | 9 | 14 | 0 |

**Women**

|  | Carnivore | Omnivore | Vegetarian |
|---|---|---|---|
| Liberal | 4 | 53 | 12 |
| Moderate | 4 | 27 | 6 |
| Conservative | 1 | 4 | 0 |

a) Are women or men more likely to be conservative carnivores?
b) Are liberal vegetarians more likely to be women or men?

**32. COVID and blood type** Public health scientists confronted with a new disease must search for variables that may predict infection or complications. Some of the apparent associations may, in time, turn out to be false leads, but without any other information they are followed. Scientists observed that the prognosis of SARS-19, the virus that causes COVID-19, might be associated with the blood type of patients. Here is a table of data from one hospital in Wuhan, China. (https://doi.org/10.1101/2020.03.11.20031096)

|  | | Blood Type | | | |
|---|---|---|---|---|---|
|  | Total | A | B | AB | O |
| No disease | 3694 | 1188 | 920 | 336 | 1250 |
| Hospital patients | 1775 | 670 | 469 | 178 | 458 |
| Deaths | 206 | 85 | 50 | 19 | 52 |

a) Would column percentages or row percentages be most useful in examining whether the proportion of hospitalization or death was different according to blood type?
b) Calculate the percentages for (part a).
c) Does it appear that blood type might help to predict the course of the disease? Briefly explain.

**33. Super students** Since 2010, the American Statistical Association has encouraged participation by students in the "Census at School" survey. One question in the survey asks students which of five super powers they would most like to have: *Fly, Freeze Time, Invisibility, Super Strength*, or *Telepathy*. From the very large database of past responses, we selected a random sample of 100 students and plotted the distribution of *Super Power* choice in the graph shown below.

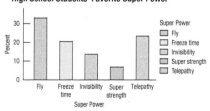

a) Describe the distribution of *Super Power* among students in this sample. Include at least three distinct characteristics of the distribution.

We wondered whether the apparent characteristics of the *Super Power* distribution that we saw in our sample of 100 students reflected similar characteristics of the entire population, or if they were just chance "quirks" of this particular sample. So we selected three more different random samples of 100 students and did the same thing. The graphs below show the distributions resulting from these samples.

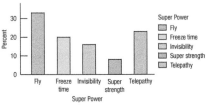

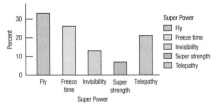

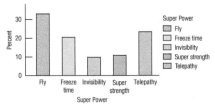

b) For each of the characteristics you described in part a, use these new graphs to decide whether the characteristic is likely to be descriptive of the whole population, or if it may easily just have been a chance "quirk" of the first sample.

**34. Super students II** Using the same large database described in Exercise 33, we selected 100 survey respondents at random and constructed side-by-side stacked bar graphs of *Super Power*, separately for Females and Males. The graph is shown below.

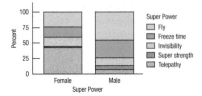

a) Describe at least three ways that the distribution of *Super Power* differs between Females and Males in this sample.

We wanted to see whether any apparent associations between *Sex* and *Super Power* in this sample were likely to be "real"— that is, to reflect associations between *Sex* and *Super Power* in the whole population. (Alternatively, they could just be chance "quirks" of the particular random sample we selected.) So we selected three more different random samples of 100 survey respondents from the same database and produced the graphs below.

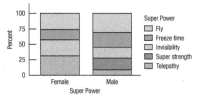

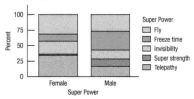

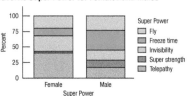

b) For each of the three differences between Females and Males that you described in part a, use the three new graphs to decide whether the difference appears "real" (it likely exists in the population) or a chance "quirk" of that sample (it may or may not exist in the population).

**35. Fish and prostate cancer revisited** Here is a mosaic plot of the data on *Fish consumption* and *Prostate cancer* from the Step-by-Step Example on page 26.

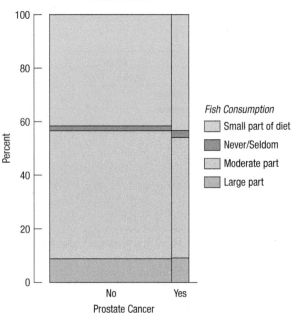

a) From the mosaic plot, about what percent of all men in this survey were diagnosed with prostate cancer?
b) Are there more men who had cancer and never/seldom ate fish, or more who didn't have cancer and never/seldom ate fish?

**36.** College value? The Pew Research Center asked 2143 U.S. adults and 1055 college presidents to "rate the job the higher education system is doing in providing value for the money spent by students and their families" as Excellent, Good, Only Fair, or Poor. The data table follows, as well as a mosaic plot of these data.

|  | Poor | Only Fair | Good | Excellent | DK/NA |
|---|---|---|---|---|---|
| U.S. Adults | 321 | 900 | 750 | 107 | 64 |
| Presidents | 32 | 222 | 622 | 179 | 0 |

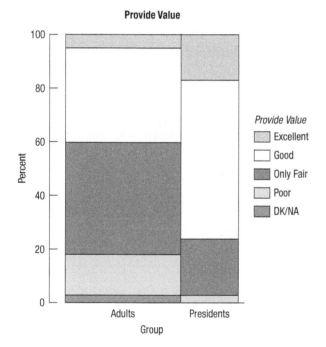

a) From the mosaic plot, about what percent of the respondents were college presidents?
b) From the mosaic plot, is it easy to see if there are more U.S. adults in the sample who said college provides excellent value or more presidents who said college provides excellent value? Explain briefly.
c) From the mosaic plot, is it easy to see how the percentage of college presidents who said college provides excellent value compares with the percentage of U.S. adults in the sample who said that? Explain briefly
d) From the mosaic plot, do you think that there is an association between the distribution of responses and whether the respondent is a college president? Explain briefly.

**37.** Diet and politics IV Here is a mosaic plot of the data on *Diet* and *Politics* from Exercise 9 combined with data on *Gender*.

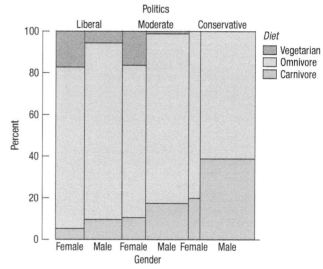

a) Are there more men or women in the survey? Explain briefly.
b) Does there appear to be an association between *Politics* and *Gender*? Explain briefly.
c) Does there appear to be an association between *Politics* and *Diet*? Explain briefly.
d) Does the association between *Politics* and *Diet* seem to differ between men and women? Explain briefly.
e) Are there more females who are liberal and vegetarian than males who are liberal and vegetarian? Or does it appear that the number of people in those two categories is about the same?
f) Is there a higher percentage of females who are liberal and vegetarian than males?

**38.** Being successful In a random sample of U.S. adults surveyed in December 2011, Pew Research asked how important it is "to you personally" to be successful in a high-paying career or profession. Here are a table reporting the responses and a mosaic plot. (Percentages may not add to 100% due to rounding.) (Data from www.pewsocialtrends.org/files/2012/04/Women-in-the-Workplace.pdf)

|  | Women | | Men | |
|---|---|---|---|---|
| Age | 18–34 | 35–64 | 18–34 | 35–64 |
| One of the most important things | 18% | 7% | 11% | 9% |
| Very important, but not the most | 48% | 35% | 47% | 34% |
| Somewhat important | 26% | 34% | 31% | 37% |
| Not important | 8% | 24% | 10% | 20% |
|  | 100% | 100% | 100% | 100% |

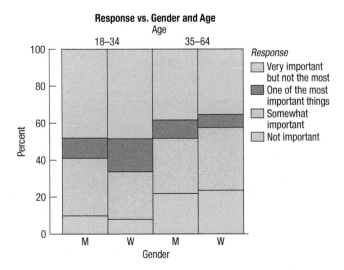

a) Are the differences in sample sizes in the four groups very large? Explain briefly.
b) Which factor seems more important in determining how someone responded: *Age* or *Gender*? Explain briefly.
c) Judging by the top two categories of importance, which of the four groups thinks being successful is most important?
d) Is the number of male 18- to 34-year-olds who feel this is not important about the same as the number of females who feel this is not important?
e) Is the percentage of male 18- to 34-year-olds who feel this is not important about the same as the percentage of females?

**39. Blood pressure** A company held a blood pressure screening clinic for its employees. The results are summarized in the table below by age group and blood pressure level:

|  |  | Age | | |
|---|---|---|---|---|
|  |  | Under 30 | 30–49 | Over 50 |
| Blood Pressure | Low | 27 | 37 | 31 |
|  | Normal | 48 | 91 | 93 |
|  | High | 23 | 51 | 73 |

a) Find the marginal distribution of blood pressure level.
b) Find the conditional distribution of blood pressure level within each age group.
c) Compare these distributions with a segmented bar graph.
d) Write a brief description of the association between age and blood pressure among these employees.
e) Does this prove that people's blood pressure increases as they age? Explain.

**40. Obesity and exercise** The Centers for Disease Control and Prevention (CDC) has estimated that 19.8% of Americans over 15 years old are obese. The CDC conducts a survey on obesity and various behaviors. Here is a table on self-reported exercise classified by body mass index (BMI):

|  | Body Mass Index | | |
|---|---|---|---|
|  | Normal (%) | Overweight (%) | Obese (%) |
| Inactive | 23.8 | 26.0 | 35.6 |
| Irregularly active | 27.8 | 28.7 | 28.1 |
| Regular, not intense | 31.6 | 31.1 | 27.2 |
| Regular, intense | 16.8 | 14.2 | 9.1 |

a) Are these percentages column percentages, row percentages, or table percentages?
b) Use graphical displays to show different percentages of physical activities for the three BMI groups.
c) Do these data prove that lack of exercise causes obesity? Explain.

**41. Anorexia** Hearing anecdotal reports that some patients undergoing treatment for the eating disorder anorexia seemed to be responding positively to the antidepressant Prozac, medical researchers conducted an experiment to investigate. They found 93 women being treated for anorexia who volunteered to participate. For one year, 49 randomly selected patients were treated with Prozac and the other 44 were given an inert substance called a placebo. At the end of the year, patients were diagnosed as healthy or relapsed, as summarized in the table:

|  | Prozac | Placebo | Total |
|---|---|---|---|
| Healthy | 35 | 32 | 67 |
| Relapse | 14 | 12 | 26 |
| Total | 49 | 44 | 93 |

Do these results provide evidence that Prozac might be helpful in treating anorexia? Explain.

**42. Antidepressants and bone fractures** For a period of five years, physicians at McGill University Health Center followed more than 5000 adults over the age of 50. The researchers were investigating whether people taking a certain class of antidepressants (SSRIs) might be at greater risk of bone fractures. Their observations are summarized in the table:

|  | Taking SSRI | No SSRI | Total |
|---|---|---|---|
| Experienced fractures | 14 | 244 | 258 |
| No fractures | 123 | 4627 | 4750 |
| Total | 137 | 4871 | 5008 |

Do these results suggest there's an association between taking SSRI antidepressants and experiencing bone fractures? Explain.

**43. Drivers' licenses 2014** The table below shows the number of licensed U.S. drivers (in millions) by age and by sex in 2014. (www.fhwa.dot.gov/policyinformation/statistics.cfm)

| Age | Male Drivers (Millions) | Female Drivers (Millions) | Total (Millions) |
|---|---|---|---|
| ≤19 | 4.3 | 4.2 | 8.5 |
| 20–24 | 8.9 | 8.7 | 17.6 |
| 25–29 | 9.3 | 9.4 | 18.7 |
| 30–34 | 9.2 | 9.4 | 18.7 |
| 35–39 | 8.7 | 8.8 | 17.5 |
| 40–44 | 9.1 | 9.2 | 18.4 |
| 45–49 | 9.4 | 9.5 | 18.9 |
| 50–54 | 10.2 | 10.4 | 20.6 |
| 55–59 | 9.7 | 10.0 | 19.8 |
| 60–64 | 8.4 | 8.7 | 17.1 |
| 65–69 | 6.9 | 7.1 | 14.0 |
| 70–74 | 4.8 | 5.0 | 9.8 |
| 75–79 | 3.2 | 3.4 | 6.6 |
| 80–84 | 2.1 | 2.3 | 4.4 |
| ≥85 | 1.6 | 1.9 | 3.6 |
| Total | 105.9 | 108.2 | 214.1 |

a) What percent of total drivers are under 20?
b) What percent of total drivers are male?
c) Write a few sentences comparing the number of male and female licensed drivers in each age group.
d) Do a driver's age and sex appear to be independent? Explain?

**44. Tattoos** A study by the University of Texas Southwestern Medical Center examined 626 people to see if an increased risk of contracting hepatitis C was associated with having a tattoo. If the subject had a tattoo, researchers asked whether it had been done in a commercial tattoo parlor or elsewhere. Write a brief description of the association between tattooing and hepatitis C, including an appropriate graphical display.

|  | Tattoo Done in Commercial Parlor | Tattoo Done Elsewhere | No Tattoo |
|---|---|---|---|
| Has Hepatitis C | 17 | 8 | 18 |
| No Hepatitis C | 35 | 53 | 495 |

**45. Hospitals** Most patients who undergo surgery make routine recoveries and are discharged as planned. Others suffer excessive bleeding, infection, or other postsurgical complications and have their discharges from the hospital delayed. Suppose your city has a large hospital and a small hospital, each performing major and minor surgeries. You collect data to see how many surgical patients have their discharges delayed by postsurgical complications, and you find the results shown in the following table:

|  | Discharge Delayed | |
|---|---|---|
|  | Large Hospital | Small Hospital |
| Major Surgery | 120 of 800 | 10 of 50 |
| Minor Surgery | 10 of 200 | 20 of 250 |

a) Overall, for what percent of patients was discharge delayed?
b) Were the percentages different for major and minor surgery?
c) Overall, what were the discharge delay rates at each hospital?
d) What were the delay rates at each hospital for each kind of surgery?
e) The small hospital advertises that it has a lower rate of postsurgical complications. Do you agree?
f) Explain, in your own words, why this confusion occurs.

**46. Delivery service** A company must decide which of two delivery services it will contract with. During a recent trial period, the company shipped numerous packages with each service and kept track of how often deliveries did not arrive on time. Here are the results:

| Delivery Service | Type of Service | Number of Deliveries | Number of Late Packages |
|---|---|---|---|
| Pack Rats | Regular | 400 | 12 |
|  | Overnight | 100 | 16 |
| Boxes R Us | Regular | 100 | 2 |
|  | Overnight | 400 | 28 |

a) Compare the two services' overall percentage of late deliveries.
b) On the basis of the results in part a, the company has decided to hire Pack Rats. Do you agree that Pack Rats delivers on time more often? Explain.
c) The results here are an instance of what phenomenon?

**47. Graduate admissions** A 1975 article in the magazine *Science* examined the graduate admissions process at Berkeley for evidence of sex discrimination. The table below shows the number of applicants accepted to each of four graduate programs:

|  |  | Males Accepted (of applicants) | Females Accepted (of applicants) |
|---|---|---|---|
| Program | 1 | 511 of 825 | 89 of 108 |
|  | 2 | 352 of 560 | 17 of 25 |
|  | 3 | 137 of 407 | 132 of 375 |
|  | 4 | 22 of 373 | 24 of 341 |
|  | Total | 1022 of 2165 | 262 of 849 |

a) What percent of total applicants were admitted?
b) Overall, was a higher percentage of males or females admitted?
c) Compare the percentage of males and females admitted in each program.
d) Which of the comparisons you made do you consider to be the most valid? Why?

**48. Be a Simpson** Can you design a Simpson's paradox? Two companies are vying for a city's "Best Local Employer" award, to be given to the company most committed to hiring local residents. Although both employers hired 300 new people in the past year, Company A brags that it deserves the award because 70% of its new jobs went to local residents, compared to only 60% for Company B. Company B concedes that those percentages are correct, but points out that most of its new jobs were full-time, while most of Company A's were part-time. Not only that, says Company B, but a higher percentage of its full-time jobs went to local residents than did Company A's, and the same was true for part-time jobs. Thus, Company B argues, it's a better local employer than Company A.

Show how it's possible for Company B to fill a higher percentage of both full-time and part-time jobs with local residents, even though Company A hired more local residents overall.

## JUST CHECKING

Answers

1. 50.0%.
2. 44.4%.
3. 25.0%.
4. 15.6% Blue, 56.3% Brown, 28.1% Green/Hazel/Other.
5. 18.8% Blue, 62.5% Brown, 18.8% Green/Hazel/Other.
6. 40% of the blue-eyed students are female, while 50% of all students are female.
7. Since blue-eyed students appear less likely to be female, it seems that *Sex* and *Eye Color* may not be independent. (But the numbers are small.)

# 3
# Displaying and Summarizing Quantitative Data

On March 11, 2011, the most powerful earthquake ever recorded in Japan created a wall of water that devastated the northeast coast of Japan and left 20,000 people dead or missing. Tsunamis like this are most often caused by earthquakes beneath the sea that shift the earth's crust, displacing a large mass of water. The 2011 tsunami in Japan was caused by a 9.1 magnitude earthquake and brought the Fukushima Daiichi nuclear power plant perilously close to a complete meltdown.

As disastrous as it was, the Japan tsunami was not nearly as deadly as the tsunami of December 26, 2004, off the west coast of Sumatra that killed an estimated 297,248 people, making it the most lethal tsunami on record. The earthquake that caused it was also a magnitude 9.1 earthquake. Were these earthquakes truly extraordinary, or did they just happen at unlucky times and places? The U.S. National Geophysical Data Center[1] has data on more than 5000 earthquakes dating back to 2150 B.C.E., and we have estimates of the magnitudes of the underlying earthquakes for the 1017 that were known to cause tsunamis. What can we learn from these data?

## Histograms

Let's start with a picture. For categorical variables, it is easy to draw the distribution because each category is a natural "pile." But for quantitative variables, there's no obvious way to choose piles. So, usually, we slice up all the possible values into equal-width "bins." We then count the number of cases that fall into each bin. The bins, together with these counts, give the **distribution** of the quantitative variable and provide the building blocks for a graph called a *histogram*. By representing the counts as heights of bars and plotting them against the bin values, the **histogram** displays the distribution at a glance.

[1] www.ngdc.noaa.gov

For example, here are the *Magnitudes* (on the Richter scale) of the 1017 earthquakes in the NGDC data:

| WHO | 1017 earthquakes known to have caused tsunamis for which we have data or good estimates |
|---|---|
| WHAT | Magnitude (Richter scale), depth (m), date, location, and other variables |
| WHEN | From 2150 B.C.E. to the present |
| WHERE | All over the earth |

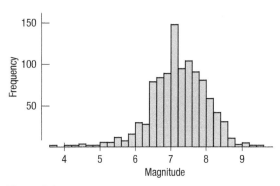

Figure 3.1
A histogram of earthquake magnitudes shows the number of earthquakes with magnitudes (in Richter scale units) in each bin.

## Designing Your Histogram

Different features of the distribution may appear more obvious at different bin width choices. When you use technology, it's usually easy to vary the bin width interactively so you can make sure that a feature you think you see isn't just a consequence of a certain bin width choice.

Like a bar chart, a histogram plots the bin counts as the heights of bars. In this histogram of earthquake magnitudes, each bin has a width of 0.2, so, for example, the height of the tallest bar says that there were about 150 earthquakes with magnitudes between 7.0 and 7.2. Does the distribution look as you expected? It is often a good idea to *imagine* what the distribution might look like before you make the display. That way you'll be less likely to be fooled by errors.

From the histogram, we can see a typical earthquake among these has a magnitude of around 7.2. Most are between 6 and 9, and some are as small as 4 and as big as 9.6. Now we can answer the question about the Japan and Sumatra tsunamis. An earthquake of magnitude 9.1 is clearly very powerful—among the largest on record.

The bar charts of categorical variables we saw in Chapter 2 had spaces between the bars to separate the counts of different categories. But in a histogram, the bins slice up *all the possible values* of the quantitative variable, so any spaces in a histogram are actual **gaps** in the data, indicating an interval where there are no values.

Sometimes it is useful to make a **relative frequency histogram**, replacing the counts on the vertical axis with the *percentage* of the total number of cases falling in each bin. The shape of the histogram is exactly the same; only the vertical scale is different.

### WHY SO MANY 7'S?

One surprising feature of the earthquake magnitudes is the spike around magnitude 7.0. These values include historical data for which the magnitudes were estimated by experts and not measured by modern seismographs. Perhaps the experts thought 7 was a typical and reasonable value for a tsunami-causing earthquake when they lacked detailed information. That would explain the overabundance of magnitudes right at 7.0 rather than spread out near that value.

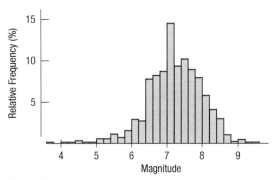

Figure 3.2
A relative frequency histogram looks just like a frequency histogram except for the scale on the *y*-axis, which now shows the percentage of earthquakes in each bin.

## TI TIPS

### Making a Histogram

Your calculator can create histograms. First you need some data. For an agility test, fourth-grade children jump from side to side across a set of parallel lines, counting the number of lines they clear in 30 seconds. Here are their scores:

22, 17, 18, 29, 22, 22, 23, 24, 23, 17, 21, 25, 20,
12, 19, 28, 24, 22, 21, 25, 26, 25, 16, 27, 22

Enter these data into L1.
Now set up the calculator's plot:

- Go to 2nd STATPLOT, choose Plot1, then ENTER.

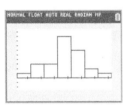

- In the Plot1 screen choose On, select the little histogram icon, then specify Xlist:L1 and Freq:1.
- Be sure to turn off any other graphs the calculator may be set up for. Just hit the Y= button, and deactivate any functions seen there.

All set? To create your preliminary plot go to ZOOM, select 9:ZoomStat, and then ENTER.

You now see the calculator's initial attempt to create a histogram of these data. Not bad. We can see that the distribution is roughly symmetric. But without looking at the scale we don't know what the bin intervals are yet, and the weird positions of the tick marks on the horizontal axis relative to the bars suggest that we can probably make this histogram easier to understand. Let's fix it up a bit.

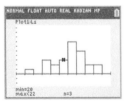

- Under WINDOW, let's reset the bins to convenient, sensible values. Try Xmin = 10, Xmax = 32, and Xscl = 2. These boundaries will extend the window a little beyond the data values, making it clearer that no part of the graph has "slopped over" out of view. Setting "Xscl" equal to 2 defines the widths of the bins.
- Hit GRAPH (not ZoomStat—this time we want control of the scale!).

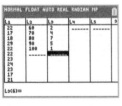

Note that you can now find out exactly what the bars indicate by activating TRACE and then moving across the histogram using the arrow keys. For each bar the calculator will indicate the interval of values and the number of data values in that bin. We see that 3 kids had agility scores of 20 or 21.

Play around with the WINDOW settings. A different Ymax will make the bars appear shorter or taller. What happens if you set the bar width (Xscl) smaller? Or larger? You don't want to lump lots of values into just a few bins or make so many bins that the overall shape of the histogram is not clear. Choosing a good bin width takes practice.

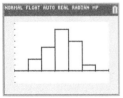

Finally, suppose the data are given as a frequency table. Consider a set of test scores, with two grades in the 60s, four in the 70s, seven in the 80s, five in the 90s, and one 100. Enter the group cutoffs 60, 70, 80, 90, 100 in L2 and the corresponding frequencies 2, 4, 7, 5, 1 in L3. When you set up the histogram STATPLOT, specify Xlist:L2 and Freq:L3. Can you specify the WINDOW settings to make this histogram look the way you want it? (By the way, if you get a DIM MISMATCH error, it means you can't count. Look at L2 and L3; you'll see the two lists don't have the same number of entries. Fix the problem by correcting the data you entered.)

# Stem-and-Leaf Displays (aka Stemplots)

Histograms provide an easy-to-understand summary of the distribution of a quantitative variable, but they don't show the data values themselves. Here's a histogram of the pulse rates of 24 women, taken by a researcher at a health clinic:

Figure 3.3
The pulse rates of 24 women at a health clinic

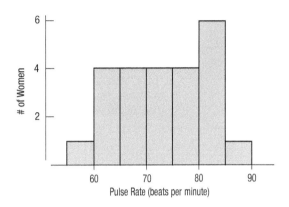

The story seems pretty clear. We can see the entire span of the data and can easily see what a typical pulse rate might be. But is that all there is to these data?

A **stem-and-leaf display** is like a histogram, but it shows the individual values. It's also easier to make by hand. Here's a stem-and-leaf display of the same data:

```
8 | 8
8 | 000044
7 | 6666
7 | 2222
6 | 8888
6 | 0444
5 | 6
```
Pulse Rate
(8 | 8 means 88 beats/min)

### WHAT'S IN A NAME?

The stem-and-leaf display was devised by John W. Tukey, one of the greatest statisticians of the 20th century. It is called a "Stemplot" in some texts and computer programs, but we prefer Tukey's original name for it.

Turn the stem-and-leaf on its side (or turn your head to the right) and squint at it. It should look roughly like the histogram of the same data. Does it? Well, it's backwards because now the higher values are on the left, but other than that, it has the same shape.[2]

Look at the top row of the display, where it says 8 | 8. That stands for a pulse of 88 beats per minute (bpm). We've taken the tens place of the number and made that the "stem." Then we sliced off the ones place and made it a "leaf." The next line down is 8 | 000044. That shows that there were four pulse rates of 80 and two of 84 bpm.

Stem-and-leaf displays are especially useful when you make them by hand for batches of fewer than a few hundred data values. They are a quick way to display—and even to record—numbers. Because the leaves show the individual values, we can sometimes see even more in the data than the distribution's shape. Take another look at all the

---

[2]You could make the stem-and-leaf with the higher values on the bottom. Usually, though, higher on the top makes more sense.

leaves of the pulse data. See anything unusual? At a glance you can see that they are all even. With a bit more thought you can see that they are all multiples of 4—something you couldn't possibly see from a histogram. How do you think the researcher took these pulses? Counting beats for a full minute or counting for only 15 seconds and multiplying by 4?

### How Do Stem-and-Leaf Displays Work?

Stem-and-leaf displays work like histograms, but they show more information. They use part of the number itself (called the stem) to name the bins. To make the "bars," they use the next digit of the number. For example, if we had a test score of 83, we could write it 8|3, where 8 serves as the stem and 3 as the leaf. Then, to display the scores 83, 76, and 88 together, we would write

```
8 | 3 8
7 | 6
```

For the pulse data, we have

```
8 | 0000448
7 | 22226666
6 | 04448888
5 | 6
Pulse Rate
(5|6 means 56 beats/min)
```

This display is OK, but a little crowded. A histogram might split each line into two bars. With a stem-and-leaf, we can do the same by putting the leaves 0–4 on one line and 5–9 on another, as we saw above:[3]

```
8 | 8
8 | 000044
7 | 6666
7 | 2222
6 | 8888
6 | 0444
5 | 6
Pulse Rate
(8|8 means 88 beats/min)
```

For numbers with three or more digits, you'll often decide to truncate (or perhaps round) the number to two places, using the first digit as the stem and the second

---

[3] You may occasionally see a stem-and-leaf plot, usually for a large dataset, in which the first-digit groupings are split into *five* groups representing *two* next-digits each, instead of *two* groups representing *five* next-digits each. (At the start of Chapter 5 we display a dataset that way.) That's fine, but don't ever construct a stem-and-leaf plot with groupings of any other number of digits than two or five. If you do, the rows won't all represent the same number of possible values, and the display will leave a distorted impression of the data.

as the leaf. So, if you had 432, 540, 571, and 638, you might display them as shown below with an indication that 6|3 means 630–639. (We truncate; it's easier.)

```
6 | 3
5 | 4 7
4 | 3
```

When you make a stem-and-leaf by hand, make sure to give each digit the same width, in order to preserve the area principle. (That can lead to some fat 1's and thin 8's—but it makes the display honest.)

# Dotplots

A **dotplot** is a simple display. It just places a dot along an axis for each case in the data. It's like a stem-and-leaf display, but with dots instead of digits for all the leaves. Dotplots are a great way to display a small dataset if you really don't need to see every individual data value. Here's a dotplot of the time (in seconds) that the winning horse took to win the Kentucky Derby in each race between the first Derby in 1875 and the 2020 Derby.

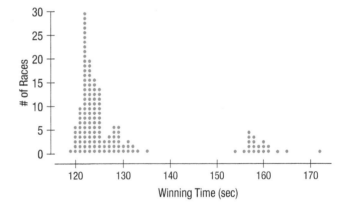

Figure 3.4
A dotplot of Kentucky Derby winning times shows a bimodal distribution begging for an explanation.

Dotplots show basic facts about the distribution. We can find the slowest and quickest races by finding times for the topmost and bottommost dots. It's also clear that there are two clusters of points, one just below 160 seconds and the other at about 122 seconds. Once we know to look for an explanation, we can find out that in 1896 the distance of the Derby race was changed from 1.5 miles to the current 1.25 miles. That explains the two clusters of winning times.

Some variations in dotplots are fairly common. For example, the data axis can run horizontal, like a histogram, or vertical, like a stem-and-leaf plot. Newspapers sometimes offer dotplots with the dots made up of little pictures. (There should be a key to tell you how many observations each picture represents.)

# Cumulative Distribution Graphs

Histograms, stem-and-leaf plots, and dotplots are all good tools for understanding the distribution of a quantitative variable. They differ in the level of detail they reveal—stem-and-leaf plots reveal the most, histograms the least—but they have much in common with one another. Each obeys the area principle, and each reveals intervals where data values are common—by tall bars, tall stacks of dots, or long rows of digits. As we'll discuss shortly, they all reveal the *shape* of a distribution.

They also all share a shortcoming. They show where data are dense, but they don't make it very easy to tell with much precision just how many data values, or what proportion of all the data, lie in any particular interval. For example, consider the histogram in Figure 3.5 that shows the distribution of the ages of the 100 U.S. senators at the start of the 117th Congress.[4] We can see from the histogram that a lot of the 117th Congressional senators are in their 60's and early 70's. But we can't tell with very much precision just how many senators that is.

**Figure 3.5**

A histogram showing the ages of the 100 U.S. senators at the start of the 117th Congress.

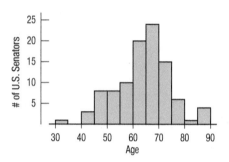

Here's how we might construct a different kind of graph that can help us estimate with greater precision how many senators are between 60 and 75 years old—or any other age range. We'll order all the ages from youngest to oldest, and plot the ranks against the ages. (A senator's rank is what position they hold in the ordering from 1 to 100; the youngest has rank 1 and the oldest has rank 100.) Figure 3.6 shows what that looks like:

**Figure 3.6**

A cumulative distribution plot of senators' ages. The horizontal axis shows the senators' ages; the vertical axis shows their age *rank*—that is, how many senators are as old as, or younger than, a given age.

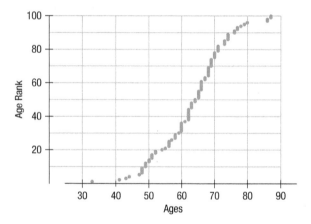

We were interested in learning how many senators were between 60 and 75 years old. This graph can help us do that with pretty good precision. We see that several of the senators are all 60 years old. The vertical axis indicates that the 60-year old senators are the 31st through the 36th when all the senators are ordered by age. That means that there are 30 senators younger than 60 years old.

There appear to be no senators who are exactly 75 years old, but a 76-year-old senator ranks 90th in age, as indicated by the vertical axis. Therefore there are 89 senators younger than 75.

Because there are 30 senators younger than 60 years old and 89 senators younger than 75, there must be 59 senators who are between 60 and 75 years old, inclusive. By ordering the ages from youngest to oldest and then simply plotting the ranks against the ages, we've made it so that *vertical* distances between points on the graph indicate how many senators fall between the corresponding ages.

[4]https://fiscalnote.com/blog/how-old-is-the-117th-congress

If we replace the ranks on the vertical axis with percentages, the resulting graph is called a **cumulative distribution plot**, or an **ogive**. (In the very special case of the U.S. senators, the ranks are equal to the percentages because there are exactly 100 senators!) Using percentages makes it easier to compare distributions of datasets that aren't the same size. Figure 3.7 shows two cumulative distribution plots on the same axes, one showing ages of U.S. senators and the other showing ages of U.S. representatives, also at the start of the 117th Congress.

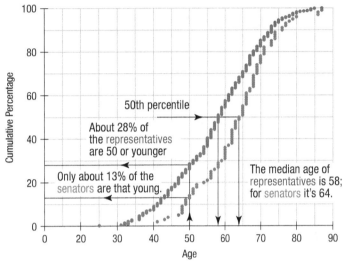

Figure 3.7
Plots showing the cumulative distributions of ages of the U.S. senators of the 117th Congress and of the U.S. representatives of the same Congress. Ogives can appear somewhat jagged (like these) or smooth, but they all are read the same way: the vertical axis shows what percentage of the data values lie *at or below* any particular value on the horizontal axis.

- Madison Cawthorn, a Republican representative from North Carolina, is the youngest member of the 117th Congress. He was 25 when he was sworn in.
- Don Young, a Republican representative from Alaska, is the oldest member of the 117th Congress. He was 87 when he was sworn in. (And older than Senator Dianne Feinstein of California by 13 days!)

A careful examination of these two overlaid cumulative distribution graphs can help us to compare the age distributions of the senators and the representatives. For example, we can see that about 28% of the representatives are 50 years old or younger, while only about 13% of the senators are that young. Reading the ages at the 50% point on both graphs we see that about half the senators are 64 or younger, while about half the representatives are 58 or younger. The big picture here is that in the 117th U.S. Congress, the representatives tend to be somewhat younger than the senators.

Cumulative distribution graphs are trickier to interpret than histograms, dotplots, and stem-and-leaf plots. But they are occasionally useful when you want to communicate quantiles precisely (like percentiles) for a distribution of data.

# Think Before You Draw, Again

Suddenly, we face a lot more options when it's time to invoke our first rule of data analysis and make a picture. You'll need to *Think* carefully to decide which type of graph to make. In the previous chapter you learned to check the Categorical Data Condition before making a pie chart or a bar chart. Now, before making a stem-and-leaf display, a histogram, or a dotplot, you need to check the

**Quantitative Data Condition**: The data are values of a quantitative variable whose units are known.

Although a bar chart and a histogram may look somewhat similar, they're not the same display. You can't display categorical data in a histogram or quantitative data in a bar chart. Always check the condition that confirms what type of data you have before proceeding with your display.

# The Shape of a Distribution

## What Is the Mode?

The *mode* is sometimes defined as the single value that appears most often. That definition is fine if the data include only a few possible values or if measurements are rounded so much that similar values are recorded as the same number. But if the measurements are *very* precise, then similar data may be spread out over many slightly different values, none of them standing out by itself. For this reason, it's better to think of a mode not as a single, precise number that shows up most often, but more generally as a peak in a histogram. In this sense, the important feature of the Kentucky Derby races is that there are two distinct modes, representing the two different versions of the race and warning us to consider those two versions separately.

Step back from a histogram, dotplot, or stem-and-leaf display. What can you say about the distribution? When you describe a distribution, you should always tell about three things: its **shape**, **center**, and **spread**. Let's start with shape:

1. *Does the histogram have a single, central hump of data or several separated humps?* These humps are called **modes**.[5] The earthquake magnitudes have a single mode at just about 7. A histogram with one peak, such as the earthquake magnitudes, is dubbed **unimodal**; histograms with two peaks, such as the winning Kentucky Derby times, are **bimodal**, and those with three or more are called multimodal.[6]

   Look back at the dotplot of winning Kentucky Derby times. We described it as bimodal, but perhaps you would describe it as multimodal. Does the large peak of faster times appear *itself* to have two or even more "mini-peaks"? If these "mini-peaks" are the result of real causes, then we might properly describe the distribution as multimodal. But if instead they're just the result of random chance (data tend to be messy!), then it's better not to call them out as modes. It takes practice to make judgment calls like this. Or sometimes it takes a little bit of research. We did a little research and found that during each Kentucky Derby race, the racetrack condition (like "Dusty" or "Muddy") is recorded for posterity. The dotplot below shows the winning Derby times color-coded by the racetrack conditions.

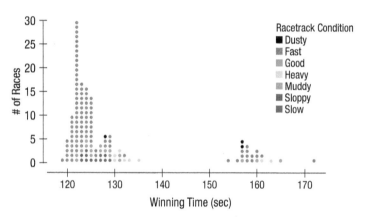

Figure 3.8
The winning Kentucky Derby times, color-coded by racetrack conditions

The color-coded dots suggest that the apparent multimodality *isn't* just a chance occurrence. The lowest winning times all took place when the racetrack condition was described (fittingly) as "Fast," while higher winning times mostly took place during other conditions. Without this additional information it wouldn't have been clear whether the apparent multimodality was a real phenomenon or just a "quirk" of chance. We'd have opted—as we initially did—for describing the overall distribution as bimodal.

A histogram that doesn't appear to have any obvious mode and in which all the bars are approximately the same height is called **uniform**.

---

[5] Well, technically, it's the value on the horizontal axis of the histogram that is the mode, but anyone asked to point to the mode would point to the hump.
[6] Apparently, statisticians don't like to count past two.

Figure 3.9
In a uniform histogram, the bars are all about the same height. The histogram doesn't appear to have a mode.

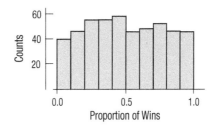

2. *Is the histogram **symmetric**?* Can you fold it along a vertical line through the middle and have the edges match pretty closely, or are more of the values on one side?

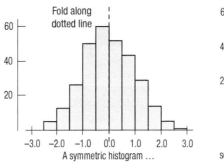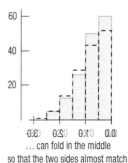

A symmetric histogram … … can fold in the middle so that the two sides almost match.

The (usually) thinner ends of a distribution are called the **tails**. If one tail stretches out farther than the other, the histogram is said to be **skewed** to the side of the longer tail.

**PIE À LA MODE?**
You've heard of pie à la mode. Is there a connection between pie and the mode of a distribution? Actually, there is! The mode of a distribution is a *popular* value near which a lot of the data values gather. And "à la mode" means "in style"— *not* "with ice cream." That just happened to be a *popular* way to have pie in Paris around 1900.

Figure 3.10
Two skewed histograms showing data on two variables for all female heart attack patients in New York state in one year. The blue histogram (age in years) is skewed to the left, or negatively skewed. The purple one (charges in $) is skewed to the right, or positively skewed.

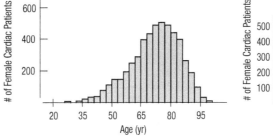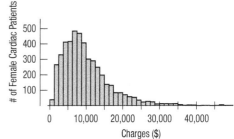

3. *Do any unusual features stick out?* Often such features tell us something interesting or exciting about the data. You should always mention any stragglers, or **outliers**, that stand off away from the body of the distribution. If you're collecting data on nose lengths and Pinocchio is in the group, you'd probably notice him, and you'd certainly want to mention it.

Outliers can affect almost every data analysis method we discuss in this course. So we'll always be on the lookout for them. An outlier can be the most informative part of your data. Or it might just be an error. But don't throw it away without comment. Treat it specially and discuss it when you tell about your data. Or find the error and fix it if you can. Be sure to look for outliers. Always. The histogram on the next page indicates that three cities have a substantially lower average number of people per housing unit than do all of the other cities. Whether your purpose is simply to describe the distribution or to do further analysis, you should identify those cities and decide whether they really belong to the same population as the others.

Figure 3.11
**A histogram with outliers.** There are three cities in the leftmost bar.

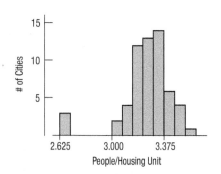

Soon you'll learn a handy rule-of-thumb for deciding when a data value might be considered an outlier.

### FOR EXAMPLE
### Describing Histograms

A credit card company wants to see how much customers in a particular segment of their market use their credit card. They have provided you with data[7] on the amount spent by 500 selected customers during a 3-month period and have asked you to summarize the expenditures. Of course, you begin by making a histogram.

**QUESTION:** Describe the shape of this distribution.

**ANSWER:** The distribution of expenditures is unimodal and positively skewed. There is an extraordinarily large value at about $7000, and some of the expenditures are negative. (Refunds to a credit card count as negative expenditures.)

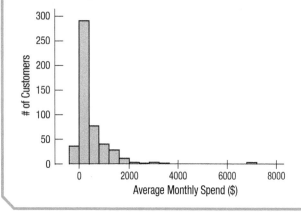

### Toto, I've a Feeling We're Not in Math Class Anymore . . .

When Dorothy and her dog Toto land in Oz, everything is more vivid and colorful, but also more dangerous and exciting. Dorothy has new choices to make. She can't always rely on the old definitions, and the yellow brick road has many branches. You may be coming to a similar realization about Statistics.

When we summarize data, our goal is usually more than just developing a detailed knowledge of the data we have at hand. We want to know what the data say about the world, so we'd like to know whether the patterns we see in histograms and summary statistics generalize to other individuals and situations. Scientists generally don't care about the particular guinea pigs in their experiment, but rather about what their reactions to different treatments say about how other animals (and, perhaps, humans) might respond.

---

[7] These data are real, but cannot be further identified for obvious privacy reasons.

Because we want to see broad patterns, rather than focus on the details of the dataset we're looking at, many of the most important concepts in Statistics are not precisely defined. Whether a histogram is symmetric or skewed, whether it has one or more modes, whether a case is far enough from the rest of the data to be considered an outlier—these are all somewhat vague concepts. There are some rules-of-thumb for helping to make judgments like these. (On page 59 you'll learn one for deciding whether a data value should count as an outlier, for example.) But they're still just rules-of-thumb, not meant to replace thinking carefully about data and using good judgment.

You may be used to finding a single correct and precise answer, but in Statistics, there may be more than one interpretation. That may make you a little uncomfortable at first, but soon you'll see that leaving room for judgment brings you both power and responsibility. It means that your own knowledge about the world and your judgment matter. You'll use them, along with the statistical evidence, to draw conclusions and make decisions about the world.

## JUST CHECKING

It's often a good idea to think about what the distribution of a dataset might look like before we collect the data. What do you think the distribution of each of the following datasets will look like? Be sure to discuss its shape. Where do you think the center might be? How spread out do you think the values will be?

1. Number of miles run by Saturday morning joggers at a park
2. Hours spent by U.S. adults watching football on Thanksgiving Day
3. Amount of winnings of all people playing a particular state's lottery last week
4. Ages of the faculty members at your school
5. Last digit of phone numbers on your campus

# A Measure of Center: The Median

Let's return to the tsunami earthquakes. But this time, let's look at recent data: the 200 earthquakes that occurred since 1 January, 2000. (See Figure 3.12.) These should be more accurately measured than prehistoric quakes because seismographs were in wide use. Try to put your finger on the histogram at the value you think is typical. (Read the value from the horizontal axis and remember it.)

Figure 3.12
**Tsunami-causing earthquakes (2000–2020)** The median of the earthquake magnitudes is actually 7.2—about half that many magnitudes are lower and about half are higher.

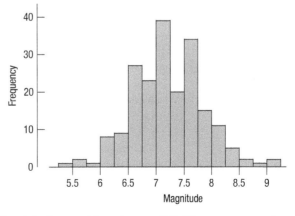

You probably picked something close to 7.2. There are several reasons you might argue that it's typical. One is that the histogram makes it look like about half the earthquakes have magnitudes less than 7.2 and about half have magnitudes greater than 7.2.

The middle value that divides a histogram into two equal areas is called the **median**. There are 200 earthquakes represented in the histogram. If you order the magnitudes from lowest to highest, the 100th and 101st magnitudes are both 7.2, so that's the median. (Had they been different values, the median would have been their average.) Half the earthquakes were less powerful than 7.2, half were more powerful.

> **NOTATION ALERT**
> We always use $n$ to indicate the number of values. Some people even say, "How big is the $n$?" when they mean the number of data values.

### How Do Medians Work?

Finding the median of a batch of $n$ numbers is easy as long as you remember to order the values first. If $n$ is odd, the median is the middle value. Counting in from the ends, we find this value in the $\frac{n+1}{2}$ position.

When $n$ is even, there are two middle values. Then, the median is the average of the two values in positions $\frac{n}{2}$ and $\frac{n}{2} + 1$.

Here are two examples:
Suppose the batch has these values: 14.1, 3.2, 25.3, 2.8, −17.5, 13.9, 45.8.
First we order the values: −17.5, 2.8, 3.2, 13.9, 14.1, 25.3, 45.8.
Since there are 7 values, the median is the $(7 + 1)/2 = $ 4th value, counting from the top or bottom: 13.9. Notice that 3 values are lower, 3 higher.
Suppose we had the same batch with another value at 35.7. Then the ordered values are −17.5, 2.8, 3.2, 13.9, 14.1, 25.3, 35.7, 45.8.
The median is the average of the 8/2 or 4th, and the $(8/2) + 1$, or 5th, values. So the median is $(13.9 + 14.1)/2 = 14.0$. Four data values are lower, and four higher.
Usually you don't have to determine the median yourself—software will do it for you. You have to know what it *means*.

The median is one way to find the center of the data. But there are others. We'll look at an even more important measure later in this chapter.

# Spread: Home on the Range

> **What We Don't Know**
> Statistics pays close attention to what we *don't* know as well as what we do know. Understanding how spread out the data are is a first step in understanding what a summary *cannot* tell us about the data. It's the beginning of telling us what we don't know.

If every earthquake that caused a tsunami registered 7.2 on the Richter scale, then knowing the median would tell us everything about the distribution of earthquake magnitudes. The more the data vary, however, the less the median alone can tell us. So we need to measure how much the data values vary around the center. In other words, how spread out are they? When we describe a distribution numerically, we always report a measure of its spread along with its center. After all, Statistics is about variation—remember?

How should we measure the spread? We could simply look at the extent of the data. How far apart are the two extremes? The **range** of the data is defined as the *difference* between the maximum and minimum values:

$$Range = max - min.$$

Notice that the range is a *single number*, not an interval of values, as you might think from everyday speech. The maximum magnitude of these earthquakes is 9.1 and the minimum is 5.3, so the *range* is $9.1 - 5.3 = 3.8$.

The range has the disadvantage that a single extreme value can make it very large, giving a value that doesn't really represent the data overall.

# Spread: The Interquartile Range

A better way to describe the spread of a variable might be to ignore the extremes and concentrate on the middle of the data. We could, for example, find the range of just the middle

## Percentiles

The lower and upper quartiles are also known as the 25th and 75th **percentiles** of the data, respectively, since the lower quartile falls above 25% of the data and the upper quartile falls above 75% of the data. If we count this way, the median is the 50th percentile. We could, of course, define and calculate any percentile that we want. For example, the 10th percentile would be the number that falls above the lowest 10% of the data values. Percentiles are more meaningful with large datasets than with small ones.

half of the data. What do we mean by the middle half? Divide the data in half at the median. Now divide both halves in half again, cutting the data into four quarters. We call these new dividing points **quartiles**. One quarter of the data lies below (or equal to) the **lower quartile**, and one quarter of the data lies above the **upper quartile**, so half the data lies between them. The quartiles border the middle half of the data. We just pointed out that the range of a dataset is a number, not an interval. The same is true of quartiles. The "upper quartile" may sound like an interval, but it's actually a single number.

### How Do Quartiles Work?

A simple way to find the quartiles is to start by splitting the batch into two halves at the median. (When *n* is odd, some statisticians include the median in both halves; others omit it.) The lower quartile is the median of the lower half, and the upper quartile is the median of the upper half.

Here are our two examples again:

The ordered values of the first batch were −17.5, 2.8, 3.2, 13.9, 14.1, 25.3, and 45.8, with a median of 13.9. Excluding the median, the two halves of the list are −17.5, 2.8, 3.2 and 14.1, 25.3, 45.8.

Each half has 3 values, so the median of each is the middle one. The lower quartile is 2.8, and the upper quartile is 25.3.

The second batch of data had the ordered values −17.5, 2.8, 3.2, 13.9, 14.1, 25.3, 35.7, and 45.8.

Here *n* is even, so the two halves of 4 values are −17.5, 2.8, 3.2, 13.9 and 14.1, 25.3, 35.7, 45.8.

Now the lower quartile is $(2.8 + 3.2)/2 = 3.0$, and the upper quartile is $(25.3 + 35.7)/2 = 30.5$.

The difference between the quartiles tells us how much territory the middle half of the data covers and is called the **interquartile range**. It's commonly abbreviated IQR:

$$IQR = upper\ quartile - lower\ quartile.$$

For the earthquakes, there are 100 values below the median and 100 values above the median. The middle value among the lower 100 turns out to be 6.8, and the middle value among the upper 100 happens to be 7.6. The *difference* between the quartiles gives the IQR:

$$IQR = 7.6 - 6.8 = 0.8.$$

Now we know that the middle half of the earthquake magnitudes extends across a (interquartile) range of 0.8 Richter scale units. This seems like a reasonable summary of the spread of the distribution, as we can see from this histogram:

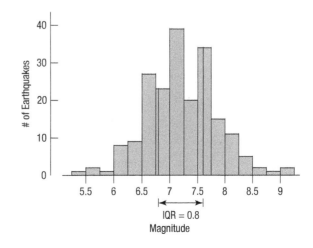

**Figure 3.13**
The quartiles are the boundaries of the middle 50% of the values of the distribution. This gives a visual indication of the spread of the data. Here we see that the IQR is 0.8 Richter scale units.

The IQR is almost always a reasonable summary of the spread of a distribution. Even if the distribution itself is skewed or has some outliers, the IQR should provide useful information. One exception is when the data are strongly bimodal. For example, remember the dotplot of winning times in the Kentucky Derby (page 52)? Because the race distance changed, we really have data on two different races, and they shouldn't be summarized together.

> ### So, What Is a Quartile Anyway?
> Finding the quartiles sounds easy, but surprisingly, the quartiles are not well-defined. It's not always clear how to find a value such that exactly one quarter of the data lies above or below that value. We offered a simple rule for finding quartiles in the box on page 57: Find the median of each half of the data split by the median. When $n$ is odd, we (and your TI calculator) omit the median from each of the halves. Some other texts include the median in both halves before finding the quartiles. Both methods are commonly used. If you are willing to do a bit more calculating, there are several other methods that locate a quartile somewhere between adjacent data values. We know of at least six different rules for finding quartiles. Remarkably, each one is in use in some software package or calculator.
>
> So don't worry too much about getting the "exact" value for a quartile. All of the methods agree pretty closely when the dataset is large. When the dataset is small, different rules will disagree more, but in that case there's little need to summarize the data anyway.
>
> Remember, Statistics is about understanding the world, not about calculating the right number. The "answer" to a statistical question is a sentence about the issue raised in the question.

# 5-Number Summary

The **5-number summary** of a distribution reports its median, quartiles, and extremes (maximum and minimum). The 5-number summary for the recent tsunami earthquake *Magnitudes* looks like this:

> *NOTATION ALERT*
> We always use Q1 to label the lower (25%) quartile and Q3 to label the upper (75%) quartile. We skip the number 2 because the median would, by this system, naturally be labeled Q2—but we don't usually call it that.

| Max | 9.1 |
|---|---|
| Q3 | 7.6 |
| Median | 7.2 |
| Q1 | 6.8 |
| Min | 5.3 |

It's good idea to report the number of data values and the identity of the cases (the *Who*). Here there are 200 recent tsunami-causing earthquakes.

The 5-number summary provides a good overview of the distribution of magnitudes of these tsunami-causing earthquakes. For a start, we can see that the median magnitude is 7.2. Because the IQR is only $7.7 - 6.8 = 0.8$, we see that many quakes are close to the median magnitude. Indeed, the quartiles show us that the middle half of these earthquakes had magnitudes between 6.8 and 7.6.

# Boxplots

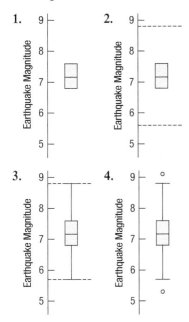

Once we have a 5-number summary of a (quantitative) variable, we can display that information in a **boxplot**. To make a boxplot of the earthquake magnitudes, follow these steps:

1. Draw a single vertical axis spanning the extent of the data.[8] Draw short horizontal lines at the lower and upper quartiles and at the median. Then connect them with vertical lines to form a box. The box can have any width that looks OK.
2. To construct the boxplot, erect temporary "fences" around the main part of the data. Place the upper fence 1.5 IQRs above the upper quartile and the lower fence 1.5 IQRs below the lower quartile. For the earthquake magnitude data, we compute

$$Upper\ fence = Q3 + 1.5\ IQR = 7.6 + 1.5 \times 0.8 = 8.8$$

and

$$Lower\ fence = Q1 - 1.5\ IQR = 6.8 - 1.5 \times 0.8 = 5.6.$$

The fences are just for construction and are not part of the display. We show them here with dotted lines for illustration. You should never include them in your boxplot.

3. We use the fences to grow "whiskers." Draw lines from the ends of the box up and down to *the most extreme data values found within the fences.* If a data value falls outside one of the fences, we do *not* connect it with a whisker. Data falling outside the fences are considered outliers.
4. Finally, we add the outliers by displaying any data values beyond the fences with dots. (Since there were two earthquakes whose magnitudes were both 9.1, they overlap with one another in the boxplot, which can be misleading. Statistical software often indicates overlapping data with special symbols or side-by-side dots.)

A boxplot highlights several features of the distribution. The central box shows the middle half of the data, between the quartiles. The height of the box is equal to the IQR. The whiskers show skewness if they are not roughly the same length. Any outliers are displayed individually to encourage you to give them special attention. They may be mistakes, or they may be the most interesting cases in your data.

> **TI-nspire**
> **Boxplots and dotplots.** Drag data points around to explore what a boxplot shows (and doesn't).

> **WHY 1.5 IQRS?**
> One of the authors asked the prominent statistician John W. Tukey, the originator of the boxplot, why the outlier nomination rule cut at 1.5 IQRs beyond each quartile. He answered that the reason was that 1 IQR would be too small and 2 IQRs would be too large. That works for us.

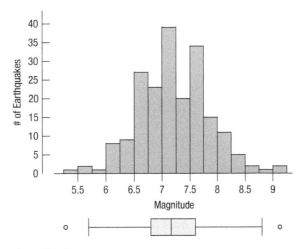

Figure 3.14
By turning the boxplot and putting it on the same scale as the histogram, we can compare both displays of the earthquake magnitudes and see how each represents the distribution.

---

[8]The axis could also run horizontally.

## STEP-BY-STEP EXAMPLE

### Shape, Center, and Spread: Flight Cancellations

The U.S. Bureau of Transportation Statistics (www.bts.gov) reports data on airline flights. Let's look at data giving the percentage of flights cancelled each month from January 1994 through September 2019.

QUESTION: How often are flights cancelled?

| WHO | Months |
|---|---|
| WHAT | Percentage of flights cancelled at U.S. airports |
| WHEN | January 1994 through September 2019 |
| WHERE | United States |

 **THINK**   **VARIABLE** Identify the *variable,* and decide how you wish to display it.

To identify a variable, report the W's.

Select an appropriate display based on the nature of the data and what you want to know.

I want to learn about the monthly percentage of flight cancellations at U.S airports.

I have data from the U.S. Bureau of Transportation Statistics giving the percentage of flights cancelled at U.S. airports each month from January 1994 through September 2019.[9]

✓ **Quantitative Data Condition**: Percentages are quantitative. A histogram and numerical summaries would be appropriate.

**SHOW**   **MECHANICS** We usually make histograms and boxplots with a computer or graphing calculator.

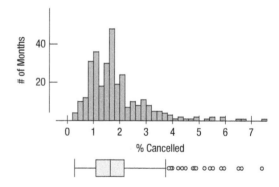

The histogram shows a unimodal (possibly bimodal?) distribution that is also right-skewed with numerous outliers—months with a relatively large number of cancelled flights.

In most months, only about 1% or 2% of flights are cancelled. That seems reasonable.

| Count | 308 |
|---|---|
| Max | 7.41 |
| Q3 | 2.19 |
| Median | 1.65 |
| Q1 | 1.10 |
| Min | 0.29 |
| IQR | 1.09 |

**REALITY CHECK** It's always a good idea to think about what you expect to see so that you can check whether the histogram looks like what you expected.

With 308 cases, we probably have more data than you'd choose to work with by hand. The results given here are from technology.

---

[9] These data include cancelled flights for all months between January 1994 and September 2019, with one exception. The month of September 2001 is excluded because it is an explainable outlier. Over 20% of flights at U.S. airports were cancelled that month following the tragedy of 9/11.

**TELL** **INTERPRETATION** Describe the shape, center, and spread of the distribution. Report on the symmetry, number of modes, and any gaps or outliers. You should also mention any concerns you may have about the data.

The possible bimodality invites the question: Are there really two modes? You can't always figure this out, but we looked deeper into the data and found that the months of June and July typically have more cancellations than other months, and they account for many of the months whose cancellation rates were close to 1.75%. (This is an example of a "seasonal effect.")

The distribution of flight cancellation percentages is unimodal, with about 1% to 2% of flights cancelled during most months. The distribution is also right-skewed, with several months that had quite a few more cancelled flights than is typical.

The median percentage of cancelled flights is 1.65% and the IQR is 1.09%. The middle 50% of all months had flight cancellation rates that were within slightly more than 1% of one another.

### The Three Rules of Data Analysis

So, what should we do with data like these? There are three things you should always do first with data:

1. **Make a picture.** A display of your data will reveal things you are not likely to see in a table of numbers and will help you to *Think* clearly about the patterns and relationships that may be hiding in your data.
2. **Make a picture.** A well-designed display will *Show* the important features and patterns in your data. A picture will also show you the things you did not expect to see: the extraordinary (possibly wrong) data values or unexpected patterns.
3. **Make a picture.** The best way to *Tell* others about your data is with a well-chosen picture.

These are the three rules of data analysis. There are pictures of data throughout the book, and new kinds keep showing up. These days, technology makes drawing pictures of data easy, so there is no reason not to follow the three rules.

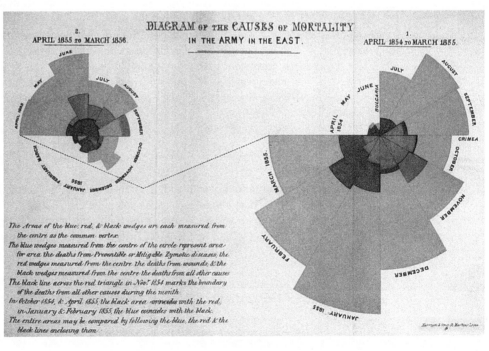

*Florence Nightingale (1820–1910), a founder of modern nursing, was also a pioneer in health management, statistics, and epidemiology. She was the first female member of the British Statistical Society and was granted honorary membership in the newly formed American Statistical Association.*

*To argue forcefully for better hospital conditions for soldiers, she and her colleague, Dr. William Farr, invented this display, which showed that in the Crimean War, far more soldiers died of illness and infection than of battle wounds. Her campaign succeeded in improving hospital conditions and nursing for soldiers.*

*Florence Nightingale went on to apply statistical methods to a variety of important health issues and published more than 200 books, reports, and pamphlets during her long and illustrious career.*

## TI TIPS

### Making a Boxplot (With a Bonus)

Last time we made a histogram to display the 4th-grade agility test data. It's just as easy to create a boxplot. You should still have those data in L1. (If not, enter them again—see page 46.)

Ready? First set up the plot:

- Go to 2nd STATPLOT, choose Plot1, and hit ENTER.
- Turn the plot On.
- Select the first boxplot icon (you always want your plot to indicate outliers).
- Specify Xlist:L1 and Freq:1.
- Select the Mark you want the calculator to use for displaying any outliers.

All set. ZoomStat will display the boxplot. (TI calculators construct horizontal boxplots instead of vertical ones.)

You can now TRACE the plot to see the statistics in the 5-number summary. Try it!

# Summarizing Symmetric Distributions: The Mean

The median is a natural statistic to choose when you want to summarize quantitative data, particularly if the data show skew or outliers. Even so, statisticians often prefer to summarize quantitative data with the mean (average) rather than the median, mainly because there are simple but powerful statistical tools that work with means but not medians. We'll see several such tools later on. For now, it's enough to know that if a set of numerical data shows strong skew or outliers primarily on one side, then the median is probably better for describing the center of the data. If the data are unimodal and somewhat symmetric, then the mean and the median should be about the same, and either would be a reasonable way to summarize the center of the data.

The earthquake magnitudes are pretty close to symmetric, so we can also summarize their center with a mean. The mean tsunami earthquake magnitude is about 7.197, which is consistent with what we might expect from the histogram. You already know how to average values, but this is a good place to introduce notation that we'll use throughout the book. We use the Greek capital letter sigma, $\Sigma$, to mean "sum" (sigma is "S" in Greek), and we'll write:

$$\bar{y} = \frac{Total}{n} = \frac{\Sigma y}{n}.$$

The formula says to add up all the values of the variable and divide that sum by the number of data values, $n$—just as you've always done.[10]

Once we've averaged the data, you'd expect the result to be called the *average*, but that would be too easy. Informally, we speak of the "average person" but we don't

---

[10] You may also see the variable called $x$ and the equation written $\bar{x} = \frac{Total}{n} = \frac{\Sigma x}{n}$. Don't let that throw you.

You are free to name the variable anything you want, but we'll generally use $y$ for variables like this that we want to summarize, model, or predict. (Later we'll talk about variables that are used to explain, model, or predict $y$. We'll call them $x$.)

add up people and divide by the number of people. A median is also a kind of average. To make this distinction, the value we calculated is called the mean, $\bar{y}$, and pronounced "y-bar."

> NOTATION ALERT
>
> In algebra you used letters to represent values in a problem, but it didn't matter what letter you picked. You could call the width of a rectangle $x$ or you could call it $w$ (or *Fred*, for that matter). But in Statistics, the notation is part of the vocabulary. For example, in Statistics $n$ is always the number of data values. Always.
>
> We have already begun to point out such special notation conventions: $n$, Q1, and Q3. Think of them as part of the terminology you need to learn in this course.
>
> Here's another one: Whenever we put a bar over a symbol, it means "find the mean."

The **mean** feels like the center because it is the point where the histogram balances:

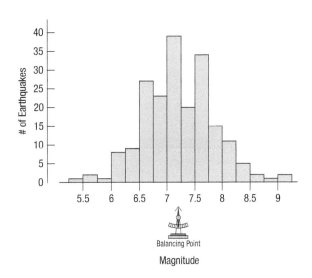

Figure 3.15
The mean is located at the balancing point of the histogram.

> AVERAGE, OR MEAN?
>
> In everyday language, sometimes "average" *does* mean what we want it to mean. We don't talk about your grade point mean or a baseball player's batting mean or the Dow Jones Industrial mean. So we'll continue to say "average" when that seems most natural. When we do, though, you may assume that what we mean is the mean.

## Mean or Median?

**APPLET**
Explore the effect of outliers

Although the mean and the median both summarize the center of a distribution of data, they do it in different ways. Both can be useful.

The median is less influenced by skewness or outliers than the mean. This makes it a good choice for summarizing skewed distributions such as income or company size. For example, the median ticket price on the *Titanic* might be the better summary for the typical price paid for the voyage. The median price was £14.4.

The mean pays attention to each value in the data. That's good news and bad. Its sensitivity makes the mean appropriate for overall summaries that need to take all of the data into account. But this sensitivity means that outliers and skewed distributions can pull the mean off to one side. For example, although the median ticket price is £14.4, the mean is £33.0 because it has been pulled to the right by a number of very high prices paid in first class and that one very large value of £512.3. (Nearly $90,000 in today's dollars!)

**Figure 3.16**
The distribution of ticket prices is skewed to the right. The mean (£33.0) is substantially higher than the median (£14.4). The higher prices at the right have pulled the mean toward them and away from the median.

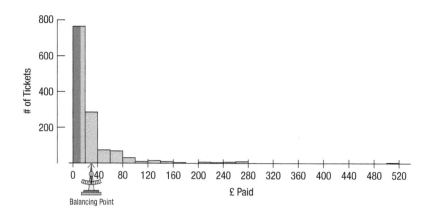

**TI-nspire**

**Mean, median, and outliers.** Drag data points around to explore how outliers affect the mean and median.

Nevertheless, the mean can be a useful summary even for a skewed distribution. For example, the White Star Line, which owned the *Titanic*, might have wanted to know the mean ticket price to get an idea of the revenue per passenger.

So, which summary should you use? Because technology makes it easy to calculate them, it is often a good idea to look at both. If the mean and median differ, you should think about the reasons for that difference. Are there outliers? If so, you should investigate them. Correct them if they are errors; set them aside if they really don't belong with your data. The fact that outliers do not affect the median doesn't get you "off the hook"—don't just use the median and think you've dealt with your outliers. If the mean and median differ because your data distribution is skewed, then you should consider what you want to know about your data. You may end up preferring one or the other—or you may decide to report both summaries. Some application areas may have a standard practice. For example, economists usually summarize income distributions with medians.

## FOR EXAMPLE
### Describing Center

**RECAP:** You want to summarize the expenditures of 500 credit card company customers and have looked at a histogram.

**QUESTION:** You have found the mean expenditure to be $478.19 and the median to be $216.28. Which is the more appropriate measure of center, and why?

**ANSWER:** Because the distribution of expenditures is skewed, the median is the more appropriate measure of center. Unlike the mean, it's not affected by the large outlying value or by the skewness. Half of these credit card customers had average monthly expenditures less than $216.28 and half had average monthly expenditures that were greater than that.

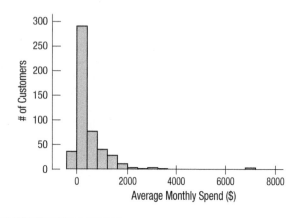

### When to Expect Skewness

Even without making a histogram, we can expect some variables to be skewed. When values of a quantitative variable are bounded on one side but not the other, the distribution may be skewed. For example, incomes and waiting times can't be less than zero, so they are often skewed to the right. Amounts of things (dollars, employees) are often skewed to the right for the same reason. If a test is too easy, the distribution of scores will be skewed to the left because many scores will bump against 100%.

# What About Spread? The Standard Deviation

The IQR is always a reasonable summary of spread, but because it uses only the two quartiles of the data, it ignores much of the information about how individual values vary. A different approach uses the *standard deviation*, which takes into account how far each value is from the mean. Like the mean, the standard deviation is most appropriate for symmetric data.

One way to think about spread is to examine how far each data value is from the mean. This difference is called a *deviation*. We could just average the deviations, but the positive and negative differences always cancel each other out. So the average deviation is always zero—not very helpful.

To keep them from canceling out, we *square* each deviation.[11] After squaring there are no negative values, so the sum won't be zero. That's great. Squaring also emphasizes larger differences, which is not so great.

When we add up these squared deviations and find their average (almost), we call the result the **variance**:

$$s^2 = \frac{\sum(y - \bar{y})^2}{n - 1}.$$

> NOTATION ALERT
> $s^2$ always means the variance of a set of data, and $s$ always denotes the standard deviation.

Why almost? It *would* be a mean if we divided the sum by $n$. Instead, we divide by $n - 1$. One of the most common questions statistics teachers get asked is "Why do we divide by $n - 1$ instead of $n$?" We're tempted to tell our students "to drive you crazy," because the real reasons are very technical. But there are good reasons . . . we promise.

The variance will play an important role later in this book, but it has a problem as a measure of spread. Whatever the units of the original data are, the variance is in *squared* units. We want measures of spread to have the same units as the data. And we probably don't want to talk about squared dollars or $mpg^2$. So, to get back to the original units, we take the square root of $s^2$. The result, $s$, is the **standard deviation**.

**APPLET**
Explore what standard deviation describes

Putting it all together, the standard deviation of the data is found by the following formula:

$$s = \sqrt{\frac{\sum(y - \bar{y})^2}{n - 1}}.$$

You will almost always rely on a calculator or computer to do the calculating.

Understanding what the standard deviation really means will take some time, and we'll revisit the concept in later chapters. For now, have a look at this histogram of resting pulse rates. The distribution is roughly symmetric, so it's okay to choose the mean and standard deviation as our summaries of center and spread. The mean pulse rate is 72.7 beats per minute, and we can see that's a typical heart rate. We also see that some heart rates are higher and some lower—but how much? Well, the standard deviation of 6.5 beats per minute indicates that, on average, we might expect people's heart rates to differ from

| WHO | 52 adults |
| WHAT | Resting heart rates |
| UNITS | Beats per minute |

---

[11] There are technical reasons why we square the deviations instead of taking their absolute values—although the Mean Absolute Deviation (or MAD) *is* an occasionally-used summary of spread in numeric data.

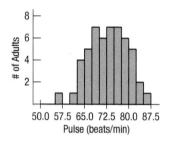

the mean rate by about 6.5 beats per minute. Looking at the histogram, we can see that 6.5 beats above or below the mean appears to be a typical deviation.

Measures of spread tell how well other summaries describe the data. That's why we always (always!) report a spread along with any summary of the center.

> **WAITING IN LINE**
> Why do banks favor a single line that feeds several teller windows rather than separate lines for each teller? The average waiting time is the same. But the time you can expect to wait is less variable when there is a single line, and people prefer consistency.

TI-*nspire*

**Standard deviation, IQR, and outliers**. Drag data points around to explore how outliers affect measures of spread.

### FOR EXAMPLE: Describing Spread

**RECAP:** The histogram has shown that the distribution of credit card expenditures is skewed, and you have used the median to describe the center. The quartiles are $73.84 and $624.80.

**QUESTION:** What is the IQR and why is it a suitable measure of spread?

**ANSWER:** For these data, the interquartile range (IQR) is $624.80 - $73.84 = $550.96. Like the median, the IQR is not affected by the outlying value or by the skewness of the distribution, so it is an appropriate measure of spread for the given expenditures.

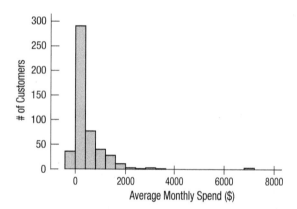

### JUST CHECKING

6. The U.S. Census Bureau reports the median family income in its summary of census data. Why do you suppose it uses the median instead of the mean? What might be the disadvantages of reporting the mean?

7. You've just bought a new car that claims to get a highway fuel efficiency of 31 miles per gallon. Of course, your mileage will "vary." If you had to guess, would you expect the IQR of gas mileage attained by all cars like yours to be 30 mpg, 3 mpg, or 0.3 mpg? Why?

8. A company selling a new smartphone advertises that the phone's battery has a mean lifetime of 5 years. If you were in charge of quality control at the factory, would you prefer that the standard deviation of life spans of the batteries be 2 years or 2 months? Why?

# What to *Tell* About a Quantitative Variable

What should you *Tell* about a quantitative variable?

- Start by making a histogram, dotplot, or stem-and-leaf display, and discuss the shape of the distribution.
- Next, discuss the center *and* spread.
  - Always pair the median with the IQR and the mean with the standard deviation. It's not useful to report one without the other. Reporting a center without a spread is dangerous. You may think you know more than you do about the distribution. Reporting only the spread leaves us wondering where we are.
  - If the shape is skewed, report the median and IQR. You may want to include the mean and standard deviation as well, but you should point out why the mean and median differ.
  - If the shape is symmetric, report the mean and standard deviation and possibly the median and IQR as well. For unimodal symmetric data, the IQR is usually a bit larger than the standard deviation.
- Also, discuss any unusual features, such as multiple modes, gaps in the data, or outliers.
  - If there are multiple modes, try to understand why. If you can identify a reason for separate modes (for example, women and men typically have heart attacks at different ages), it may be a good idea to split the data into separate groups.
  - If there are any clear outliers, you should point them out. If you are reporting the mean and standard deviation, report them with the outliers present and with the outliers removed. The differences may be revealing. (Of course, the median and IQR won't be affected very much by the outliers.)

### How "Accurate" Should We Be?

Don't think you should report means and standard deviations to a zillion decimal places; such implied accuracy is really meaningless. Although there is no ironclad rule, statisticians commonly report summary statistics to one or two decimal places more than the original data have.

### FOR EXAMPLE
#### Choosing Summary Statistics

**RECAP:** You have provided the credit card company's board of directors with a histogram of customer expenditures, and you have summarized the center and spread with the median and IQR. Knowing a little Statistics, the directors now insist on having the mean and standard deviation as summaries of the spending data.

**QUESTION:** Although you know that the mean is $478.19 and the standard deviation is $741.87, you need to explain to them why these are not suitable summary statistics for these expenditures data. What would you give as reasons?

**ANSWER:** The high outlier at $7000 pulls the mean up substantially and inflates the standard deviation. Locating the mean value on the histogram shows that it is not a typical value at all, and the standard deviation suggests that expenditures vary much more than they do. The median and IQR are more *resistant* to the presence of skewness and outliers, giving more realistic descriptions of center and spread.

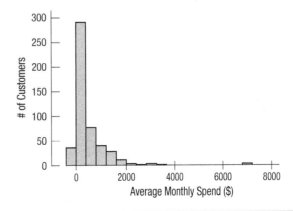

# STEP-BY-STEP EXAMPLE

## Summarizing a Distribution

One of the authors owned a Nissan Maxima for 8 years. Being a statistician, he recorded the car's fuel efficiency (in mpg) each time he filled the tank. He wanted to know what fuel efficiency to expect as "ordinary" for his car. (Hey, he's a statistician. What would you expect?[12]) Knowing this, he was able to predict when he'd need to fill the tank again and to notice if the fuel efficiency suddenly got worse, which could be a sign of trouble.

QUESTION: How would you describe the distribution of *Fuel efficiency* for this car?

**THINK**

**PLAN** State what you want to find out.

**VARIABLES** Identify the variable and report the W's.

Be sure to check the appropriate condition.

I want to summarize the distribution of Nissan Maxima fuel efficiency.

The data are the fuel efficiency values in miles per gallon for the first 100 fill-ups of a particular Nissan Maxima.

✓ **Quantitative Data Condition:** The fuel efficiencies are quantitative with units of miles per gallon. Histograms and boxplots are appropriate displays for displaying the distribution. Numerical summaries are appropriate as well.

**SHOW**

**MECHANICS** Make a histogram. Based on the shape, choose appropriate numerical summaries.

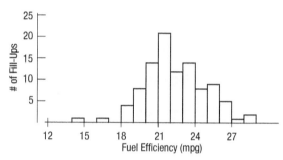

A histogram of the data shows a fairly symmetric distribution with a low outlier.

**REALITY CHECK** A value of 22 mpg seems reasonable for such a car. The spread is reasonable, although the range looks a bit large.

| Count | 100 |
|---|---|
| Mean | 22.4 mpg |
| StdDev | 2.45 |
| Q1 | 20.8 |
| Median | 22.0 |
| Q3 | 24.0 |
| IQR | 3.2 |

The mean and median are close, so the outlier doesn't seem to be a problem. I can use the mean and standard deviation.

**TELL**

**CONCLUSION** Summarize and interpret your findings in context. Be sure to discuss the distribution's shape, center, spread, and unusual features (if any).

The distribution of mileage is unimodal and roughly symmetric with a mean of 22.4 mpg. There is a low outlier that should be investigated, but it does not influence the mean very much. The standard deviation suggests that from tankful to tankful, I can expect the car's fuel economy to differ from the mean by an average of about 2.45 mpg.

---

[12] He also recorded the time of day, temperature, price of gas, and phase of the moon. (OK, maybe not phase of the moon.)

### I Got a Different Answer: Did I Mess Up?

When you calculate a mean, the computation is clear: You sum all the values and divide by the sample size. You may round your answer less or more than someone else (we recommend one more decimal place than the data), but all books and technologies agree on how to find the mean. Some statistics, however, are more problematic. For example, we've already pointed out that methods of finding quartiles differ.

Differences in numeric results can also arise from decisions in the middle of calculations. For example, if you round off your value for the mean before you calculate the sum of squared deviations, your standard deviation probably won't agree with a computer program that calculates using many decimal places. (We do recommend that you do calculations using as many digits as you can to minimize this effect.)

Don't be overly concerned with these discrepancies, especially if the differences are small. They don't mean that your answer is "wrong," and they usually won't change any conclusion you might draw about the data. Sometimes (in footnotes and in the answers in the back of the book) we'll note alternative results, but we could never list all the possible values, so we'll rely on your common sense to focus on the meaning rather than on the digits. Remember: Answers are sentences—not single numbers!

## TI TIPS

### Calculating the Statistics

Your calculator can easily find all the numerical summaries of data. To try it out, you simply need a set of values in one of your datalists. We'll illustrate using the agility test results from this chapter's earlier TI Tips (still in L1), but you can use any data currently stored in your calculator.

- Under the STAT CALC menu, select 1-Var Stats and hit ENTER.

- Specify List:L1, leave Freqlist: blank, then go to Calculate and hit ENTER
(OR on an older calculator, specify the location of your data by creating a command like 1-VarStats L1 and hit ENTER).

Voilà! Everything you wanted to know, and more. Among all of the information shown, you are primarily interested in these statistics: $\bar{x}$ (the mean), Sx (the standard deviation), n (the count), and—scrolling down—minX (the smallest datum), $Q_1$ (the first quartile), Med (the median), $Q_3$ (the third quartile), and maxX (the largest datum).

Sorry, but the TI doesn't explicitly tell you the range or the IQR. Just subtract:
IQR = $Q_3 - Q_1$ = 25 - 19.5 = 5.5. (What's the range?)

By the way, if the data come as a frequency table with the values stored in, say, L4 and the corresponding frequencies in L5, all you have to do is select 1-VarStats and specify List:L4, FreqList:L5, then go to Calculate and hit ENTER (OR on an older calculator ask for 1-VarStats L4, L5).

## WHAT CAN GO WRONG?

A data display should tell a story about the data. To do that, it must speak in a clear language, making plain what variable is displayed, what any axis shows, and what the values of the data are. And it must be consistent in those decisions.

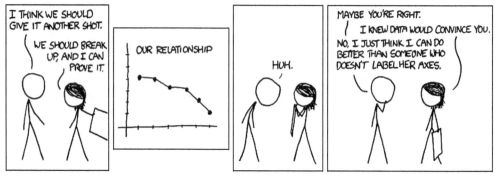

© 2013 Randall Munroe. Reprinted with permission. All rights reserved.

A display of quantitative data can go wrong in many ways. The most common failures arise from only a few basic errors:

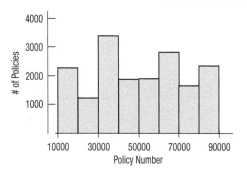

Figure 3.17
It's not appropriate to display these data with a histogram.

- ◆ **Don't make a histogram of a categorical variable.** Just because the variable contains numbers doesn't mean that it's quantitative. Here's a histogram of the insurance policy numbers of some workers. It's not very informative because the policy numbers are just labels. A histogram or stem-and-leaf display of a categorical variable makes no sense. A bar chart or pie chart would be more appropriate.
- ◆ **Don't look for shape, center, and spread of a bar chart.** A bar chart showing the sizes of the piles displays the distribution of a categorical variable, but the bars could be arranged in any order left to right. Concepts like symmetry, center, and spread make sense only for quantitative variables.
- ◆ **Don't use bars in every display—save them for histograms and bar charts.** In a bar chart, the bars indicate how many cases of a categorical variable are piled in each category. Bars in a histogram indicate the number of cases piled in each interval of a quantitative variable. In both bar charts and histograms, the bars represent counts of data values. Some people create other displays that use bars to represent individual data values. Beware: Such graphs are neither bar charts nor histograms. For example, a student was asked to make a histogram from data showing the number of juvenile bald eagles seen during each of the 13 weeks in the winter of 2003–2004 at a site in Rock Island, IL. Instead, he made this plot:

Figure 3.18
This isn't a histogram or a bar chart. It's an ill-conceived graph that uses bars to represent individual data values (numbers of juvenile eagles sighted) week by week.

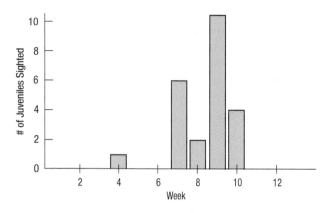

Look carefully. That's not a histogram. A histogram shows *What* we've measured along the horizontal axis and counts of the associated *Who*'s represented as bar heights. This student has it backwards: He used bars to show counts of birds for each week. We need counts of weeks. A correct histogram should have a tall bar at "0" to show there were many weeks when no eagles were seen, like this:

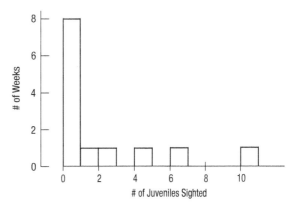

**Figure 3.19**
A histogram of the eagle-sighting data shows the number of weeks in which different counts of eagles occurred. This display shows the distribution of juvenile-eagle sightings.

- **Choose a bin width appropriate to the data.** Computer programs usually do a pretty good job of choosing histogram bin widths. Often there's an easy way to adjust the width, sometimes interactively. Both of the histograms below show the tsunami earthquake data, but the one on the left uses bins that are too wide, which makes some potentially interesting details invisible. The one on the right uses bins that are too narrow, creating the impression of meaningful details, like the spikes, that are probably really just meaningless chance "quirks."[13]

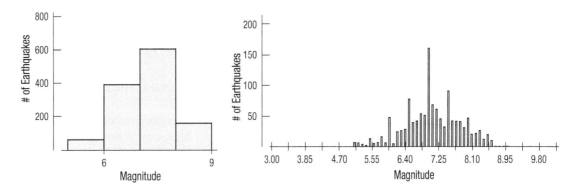

The task of summarizing a quantitative variable is relatively simple, and there is a simple path to follow. However, you need to watch out for certain features of the data that make summarizing them with a number dangerous. Here's some advice:

- **Don't forget to do a reality check.** Don't let the computer or calculator do your thinking for you. Make sure the calculated summaries make sense. For example, does the mean look like it is in the center of the histogram? Think about the spread: An IQR of 50 mpg would clearly be wrong for gas mileage. And no measure of spread can be negative. The standard deviation can take the value 0, but only in the very unusual case that all the data values equal the same number. If you see an IQR or standard deviation equal to 0, it's probably a sign that something's wrong with the data.

- **Don't forget to sort the values before finding the median or percentiles.** It seems obvious, but when you work by hand, it's easy to forget to sort the data first before counting in to find medians, quartiles, or other percentiles. Don't report that the median of the five values 194, 5, 1, 17, and 893 is 1 just because 1 is the middle number.

---

[13]TMI!!!

| Gold Card Customers—<br>Regions National Banks | | |
|---|---|---|
| Month | April 2007 | May 2007 |
| Average Zip Code | 45,034.34 | 38,743.34 |

- **Don't worry about small differences when using different methods.** Finding the 10th percentile or the lower quartile in a dataset sounds easy enough. But it turns out that the definitions are not exactly clear. If you compare different statistics packages or calculators, you may find that they give slightly different answers for the same data. These differences, though, are unlikely to be important in interpreting the data, the quartiles, or the IQR, so don't let them worry you.

- **Don't compute numerical summaries of a categorical variable.** Neither the mean zip code nor the standard deviation of social security numbers is meaningful. If the variable is categorical, you should instead report summaries such as percentages of individuals in each category. It is easy to make this mistake when using technology to do the summaries for you. After all, the computer doesn't care what the numbers mean.

- **Don't report too many decimal places.** Statistical programs and calculators often report a ridiculous number of digits. A general rule for numerical summaries is to report one or two more digits than the number of digits in the data. For example, earlier we saw a dotplot of the Kentucky Derby race times. The mean and standard deviation of those times could be reported as:

$$\bar{y} = 130.63401639344262 \text{ sec}; s = 13.66448201942662 \text{ sec}.$$

But we knew the race times only to the nearest quarter second, so the extra digits are meaningless.

- **Don't round in the middle of a calculation.** Don't *report* too many decimal places, but it's best not to do any rounding until the end of your calculations. Even though you might report the mean of the earthquakes as 7.2, it's really 7.1965. Use the more precise number in your calculations if you're finding the standard deviation by hand—or be prepared to see small differences in your final result.

- **Watch out for multiple modes.** The summaries of the Kentucky Derby times are meaningless for another reason. As we saw in the dotplot, the Derby was initially a longer race. It would make much more sense to report that the old 1.5 mile Derby had a mean time of 159.6 seconds, while the current Derby has a mean time of 124.6 seconds. If the distribution has multiple modes, consider separating the data into different groups and summarizing each group separately.

- **Beware of outliers.** The median and IQR are resistant to outliers, but the mean and standard deviation are not. To help spot outliers . . .

- **Don't forget to: Make a picture (make a picture, make a picture).** The sensitivity of the mean and standard deviation to outliers is one reason you should always make a picture of the data. Summarizing a variable with its mean and standard deviation when you have not looked at a histogram or dotplot to check for outliers or skewness invites disaster. You may find yourself drawing absurd or dangerously wrong conclusions about the data. And, of course, you should demand no less of others. Don't accept a mean and standard deviation blindly without some evidence that the variable they summarize is unimodal, symmetric, and free of outliers.

# WHAT HAVE WE LEARNED?

We've learned to *Think* about summarizing quantitative variables.

- All the methods of this chapter assume that the data are quantitative. The *Quantitative Data Condition* serves as a check that the data are, in fact, quantitative. One good way to be sure is to know the measurement units. You'll want those as part of the *Think* step of your answers.
- The median divides the data so that half of the data values are below the median and half are above.
- The mean is the point at which the histogram balances.
- The standard deviation summarizes how spread out all the data are around the mean.

- The median and IQR resist the effects of outliers, while the mean and standard deviation do not.
- In a skewed distribution, the mean is pulled in the direction of the skewness (toward the longer tail) relative to the median.
- We'll report the median and IQR when the distribution is skewed. If it's symmetric, we'll summarize the distribution with the mean and standard deviation (and possibly the median and IQR as well).

We've learned how to make a picture of quantitative data to help us see the story the data have to *Tell*.

- We can display the distribution of quantitative data with a *histogram*, a *stem-and-leaf* display, a *dotplot*, or a *boxplot*.
- We can display display quantitative data with a *cumulative distribution plot*, which shows precisely what fraction of the data lie in any given interval.

We've learned how to summarize distributions of quantitative variables numerically.

- Measures of center for a distribution include the median and the mean.
- Measures of spread include the range, IQR, and standard deviation.
- A 5-number summary includes the minimum and maximum values, the quartiles, and the median.
- We always pair the median with the IOR and the mean with the standard deviation.

We *Tell* what we see about the distribution by talking about *shape, center, spread*, and any *unusual features*.

## TERMS

**Distribution** — The distribution of a quantitative variable slices up all the possible values of the variable into equal-width bins and gives the number of values (or counts) falling into each bin. (p. 44)

**Histogram (relative frequency histogram)** — A histogram uses adjacent bars to show the distribution of a quantitative variable. Each bar represents the frequency (or relative frequency) of values falling in each bin. (p. 44, 45)

**Gap** — A region of the distribution where there are no values. (p. 45)

**Stem-and-leaf display** — A stem-and-leaf display shows quantitative data values in a way that sketches the distribution of the data. It's best described in detail by example. (p. 47)

**Dotplot** — A dotplot graphs a dot for each case against a single axis. (p. 49)

**Cumulative distribution plot** — A cumulative distribution plot displays the fraction of a dataset that lies at or below any given data value. (p. 51)

**Quantitative Data Condition** — The data are values of a quantitative variable whose units are known. (p. 51)

**Shape** — To describe the shape of a distribution, look for
- single vs. multiple modes,
- symmetry vs. skewness, and
- outliers and gaps. (p. 52)

**Center** — The place in the distribution of a variable that you'd point to if you wanted to attempt the impossible by summarizing the entire distribution with a single number. Measures of center include the mean and median. (p. 52)

**Spread** — A numerical summary of how tightly the values are clustered around the center. Measures of spread include the IQR and standard deviation. (p. 52)

**Mode** — A hump or local high point in the shape of the distribution of a variable. The apparent location of modes can change as the scale of a histogram is changed. (p. 52)

| | |
|---|---|
| **Unimodal (Bimodal)** | Having one mode. This is a useful term for describing the shape of a histogram when it's generally mound-shaped. Distributions with two modes are called **bimodal**. Those with more than two are **multimodal.** (p. 52) |
| **Uniform** | A distribution that's roughly flat is said to be uniform. (p. 52) |
| **Symmetric** | A distribution is symmetric if the two halves on either side of the center look approximately like mirror images of each other. (p. 53) |
| **Tails** | The tails of a distribution are the parts that typically trail off on either side. Distributions can be characterized as having long tails (if they straggle off for some distance) or short tails (if they don't). (p. 53) |
| **Skewed** | A distribution is skewed if it's not symmetric and one tail stretches out farther than the other. Distributions are said to be **skewed left** when the longer tail stretches to the left, and **skewed right** when it goes to the right. (p. 53) |
| **Outliers** | Outliers are extreme values that don't appear to belong with the rest of the data. They may be unusual values that deserve further investigation, or they may be just mistakes; there's no obvious way to tell. Don't delete outliers automatically—you have to think about them. Outliers can affect many statistical analyses, so you should always be alert for them. (p. 53) |
| **Median** | The median is the middle value, with half of the data above and half below it. If $n$ is even, it is the average of the two middle values. It is usually paired with the IQR. (p. 56) |
| **Range** | The difference between the lowest and highest values in a dataset. $Range = max - min$. (p. 56) |
| **Quartile** | The **lower quartile** (Q1) is the value with a quarter of the data below (or equal to) it. The **upper quartile** (Q3) has three quarters of the data below (or equal to) it. The median and quartiles divide data into four parts with equal numbers of data values. (p. 57) |
| **Percentile** | The $i$th percentile is the number that falls above $i$% of the data. (p. 57) |
| **Interquartile range (IQR)** | The IQR is the difference between the first and third quartiles. $IQR = Q3 - Q1$. It is usually reported along with the median. (p. 57) |
| **5-Number Summary** | The 5-number summary of a distribution reports the minimum value, Q1, the median, Q3, and the maximum value. (p. 58) |
| **Boxplot** | A boxplot displays the 5-number summary as a central box, whiskers that extend to the nonoutlying data values, and any outliers shown. (p. 59) |
| **Mean** | The mean is found by summing all the data values and dividing by the count: $$\bar{y} = \frac{Total}{n} = \frac{\sum y}{n}.$$ It is usually paired with the standard deviation. (p. 63) |
| **Variance** | The variance is the sum of squared deviations from the mean, divided by the count minus 1: $$s^2 = \frac{\sum(y - \bar{y})^2}{n - 1}.$$ It is useful in calculations later in the book. (p. 65) |
| **Standard deviation** | The standard deviation is the square root of the variance: $$s = \sqrt{\frac{\sum(y - \bar{y})^2}{n - 1}}.$$ It is usually reported along with the mean. (p. 65) |

## Use Technology!

It doesn't hurt to know these formulas, but in practice you'll almost always use technology to compute the mean, variance, or standard deviation of a dataset.

## ON THE COMPUTER
### Displaying and Summarizing Quantitative Variables

Almost any program that displays data can make a histogram, but some will do a better job of determining where the bars should start and how they should partition the span of the data.

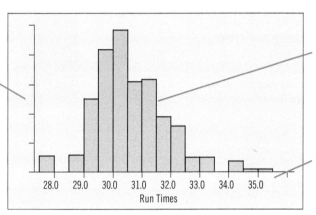

The vertical scale may be counts or proportions. Sometimes it isn't clear which. But the shape of the histogram is the same either way.

Most packages choose the number of bars for you automatically. Often you can adjust that choice.

The axis should be clearly labeled so you can tell what "pile" each bar represents. You should be able to tell the lower and upper bounds of each bar.

Many statistics packages offer a prepackaged collection of summary measures. The result might look like this:

```
Variable: Weight
N = 234
Mean = 143.3      Median = 139
St. Dev = 11.1    IQR = 14
```

Alternatively, a package might make a table for several variables and summary measures:

| Variable | N | mean | median | stdev | IQR |
|---|---|---|---|---|---|
| Weight | 234 | 143.3 | 139 | 11.1 | 14 |
| Height | 234 | 68.3 | 68.1 | 4.3 | 5 |
| Score | 234 | 86 | 88 | 9 | 5 |

It is usually easy to read the results and identify each computed summary. You should be able to read the summary statistics produced by any computer package.

Packages often provide many more summary statistics than you need. Of course, some of these may not be appropriate when the data are skewed or have outliers. It is your responsibility to check a histogram or stem-and-leaf display and decide which summary statistics to use.

It is common for packages to report summary statistics to many decimal places of "accuracy." Of course, it is rare data that have such accuracy in the original measurements. The ability to calculate to six or seven digits beyond the decimal point doesn't mean that those digits have any meaning. Generally it's a good idea to round these values, allowing perhaps one more digit of precision than was given in the original data.

Displays and summaries of quantitative variables are among the simplest things you can do in most statistics packages.

## EXERCISES

**1. Histogram** Find a histogram that shows the distribution of a variable in a newspaper, a magazine, or the Internet.
   a) Does the article identify the W's?
   b) Discuss whether the display is appropriate.
   c) Discuss what the display reveals about the variable and its distribution.
   d) Does the article accurately describe and interpret the data? Explain.

**2. Not a histogram** Find a graph other than a histogram that shows the distribution of a quantitative variable in a newspaper, a magazine, or the Internet.
   a) Does the article identify the W's?
   b) Discuss whether the display is appropriate for the data.
   c) Discuss what the display reveals about the variable and its distribution.
   d) Does the article accurately describe and interpret the data? Explain.

3. **Centers in the news** Find an article in a newspaper, a magazine, or the Internet that discusses an "average."
   a) Does the article discuss the W's for the data?
   b) What are the units of the variable?
   c) Is the average referring to the median or the mean? How can you tell?
   d) Is the choice of median or mean appropriate for the situation? Explain.

4. **Spreads in the news** Find an article in a newspaper, a magazine, or the Internet that discusses a measure of spread.
   a) Does the article discuss the W's for the data?
   b) What are the units of the variable?
   c) Does the article use the range, IQR, or standard deviation?
   d) Is the choice of measure of spread appropriate for the situation? Explain.

5. **Thinking about shape** Would you expect distributions of these variables to be uniform, unimodal, or bimodal? Symmetric or skewed? Explain why.
   a) The number of speeding tickets each student in the senior class of a college has ever had.
   b) Players' scores (number of strokes) at the U.S. Open golf tournament in a given year.
   c) Weights of female babies born in a particular hospital over the course of a year.
   d) The length of hair on the heads of students in a large class.

6. **More shapes** Would you expect distributions of these variables to be uniform, unimodal, or bimodal? Symmetric or skewed? Explain why.
   a) Ages of people at a Little League game.
   b) Number of siblings of people in your class.
   c) Pulse rates of college-age males.
   d) Number of times each face of a die shows in 100 tosses.

7. **Cereals** The dotplot shows the carbohydrate content of 77 breakfast cereals (in grams).

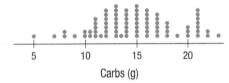

   a) Describe the distribution.
   b) Given what you know about breakfast cereals, what additional variable do you think would help describe the shape you see on this graph?

8. **Singers** The display shows the heights of some of the signers in a chorus, collected so that the signers could be positioned on stage with shorter ones in front and taller ones in back.

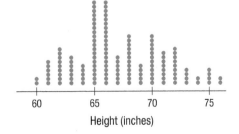

   a) Describe the distribution.
   b) Can you account for the features you see here?

9. **Vineyards** The histogram shows the sizes (in acres) of 36 vineyards in the Finger Lakes region of New York.

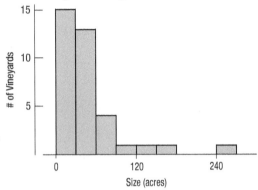

   a) Approximately what percentage of these vineyards are under 60 acres?
   b) Write a brief description of this distribution (shape, center, spread, unusual features).

10. **Run times** One of the authors collected the times (in minutes) it took him to run 4 miles on various courses during a 10-year period. Here is a histogram of the times.

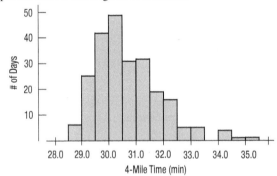

   Describe the distribution and summarize the important features. What is it about running that might account for the shape you see?

11. **Heart attack stays** The histogram shows the lengths of hospital stays (in days) for all the female patients admitted to hospitals in New York during one year with a primary diagnosis of acute myocardial infarction (heart attack).

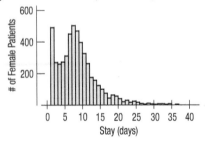

   a) From the histogram, would you expect the mean or median to be larger? Explain.
   b) Write a few sentences describing this distribution (shape, center, spread, unusual features).
   c) Which summary statistics would you choose to summarize the center and spread in these data? Why?

**12. Emails** A university teacher saved every emails received from students in a large Introductory Statistics class during an entire term. He then counted, for each student who had sent him at least one email, how many emails each student had sent.

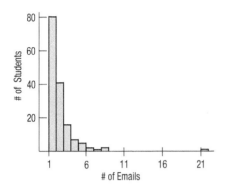

a) From the histogram, would you expect the mean or the median to be larger? Explain.
b) Write a few sentences describing this distribution (shape, center, spread, unusual features).
c) Which summary statistics would you choose to summarize the center and spread in these data? Why?

**13. Super Bowl points 2020** How many points do football teams score in the Super Bowl? Here are the total numbers of points scored by both teams in each of the first 54 Super Bowl games.

45, 47, 23, 30, 29, 27, 21, 31, 22, 38, 46, 37, 66, 50,
37, 47, 44, 47, 54, 56, 59, 52, 36, 65, 39, 61, 69, 43,
75, 44, 56, 55, 53, 39, 41, 37, 69, 61, 45, 31, 46, 31,
50, 48, 56, 38, 65, 51, 52, 34, 62, 74, 16, 51

a) Find the median.
b) Find the quartiles.
c) Make a histogram and write a brief description of the distribution.

**14. Super Bowl edge 2020** In the Super Bowl, by how many points does the winning team outscore the losers? Here are the winning margins for the first 54 Super Bowl games.

25, 19, 9, 16, 3, 21, 7, 17, 10, 4, 18, 17, 4, 12, 17, 5, 10, 29,
22, 36, 19, 32, 4, 45, 1, 13, 35, 17, 23, 10, 14, 7, 15, 7, 27,
3, 27, 3, 3, 11, 12, 3, 4, 14, 6, 4, 3, 35, 4, 14, 6, 8, 10, 11

a) Find the median.
b) Find the quartiles.
c) Make a histogram of the distribution.
d) Make a boxplot of the distribution.
e) Write a brief description of the distribution.

**15. Details** In the next column are histograms for four manufactured sets of numbers. The histograms look rather different, but all four sets have the same 5-number summary, so the boxplots for all four sets are identical to the one shown.

a) Using these plots as examples, identify some features of a distribution that a boxplot may not show.
b) What does this tell you about the limitations of using a boxplot to assess the shape of a distribution?

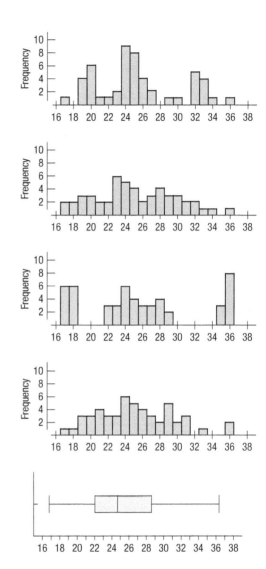

**16. Opposites** In a way, boxplots are the opposite of histograms. A histogram divides the number line into equal intervals and displays the number of data values in each interval. A boxplot divides the data into equal parts and displays the portion of the number line each part covers. These two plots display the number of incarcerated prisoners in each state as of June 2006.

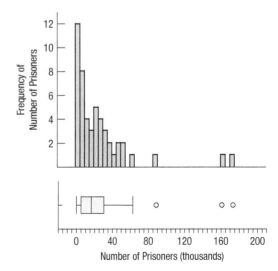

a) Explain how you could tell, by looking at a boxplot, where the tallest bars on the histogram would be located.
b) Explain how both the boxplot and the histogram can indicate a skewed distribution.
c) Identify one feature of the distribution that the histogram shows but the boxplot does not.
d) Identify one feature of the distribution that the boxplot shows but the histogram does not.

**17. Adoptions** The U.S. Census Bureau keeps track of the number of adoptions in each state and Washington, DC. (www.census.gov). Here are the histogram and boxplot of the distribution of adoptions.

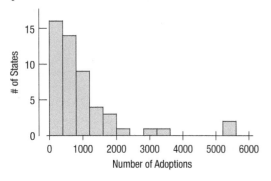

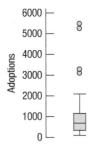

a) What features of the distribution can you see in both the histogram and the boxplot?
b) What features of the distribution can you see in the histogram that you could not see in the boxplot?
c) What summary statistic would you choose to summarize the center of this distribution? Why?
d) What summary statistic would you choose to summarize the spread of this distribution? Why?

**18. Adoptions again** Rather than look at the number of adoptions in each state given in Exercise 17, the histogram and boxplot below look at the number of adoptions per 100,000 residents of each state.

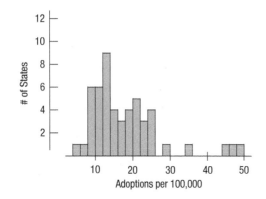

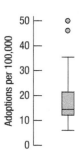

a) What features of the distribution can you see in both the histogram and the boxplot?
b) What features of the distribution can you see in the histogram that you could not see in the boxplot?
c) What summary statistic would you choose to summarize the center of this distribution? Why?
d) What summary statistic would you choose to summarize the spread of this distribution? Why?

**19. Camp sites** Shown below are the histogram and summary statistics for the number of camp sites at public parks in Vermont.

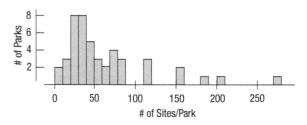

| Count | 46 |
|---|---|
| Mean | 62.8 sites |
| Median | 43.5 |
| StdDev | 56.2 |
| Min | 0 |
| Max | 275 |
| Q1 | 28 |
| Q3 | 78 |

a) Which statistics would you use to identify the center and spread of this distribution? Why?
b) How many parks would you classify as outliers? Explain.
c) Create a boxplot for these data. The five biggest parks had the following number of campsites: 150, 159, 181, 200, 273.
d) Write a few sentences describing the distribution.

**20. Outliers** The 5-number summary for the run times in minutes of the 150 highest grossing movies of a recent year looks like this:

| Min | QI | Med | Q3 | Max |
|---|---|---|---|---|
| 43 | 98 | 104.5 | 116 | 160 |

a) Are there any outliers in these data? How can you tell?
b) Construct a boxplot. Based on your plot, say what you can about the shape of the distribution. To assist your work, here are the four shortest and the four longest movies: 43, 48, 59, 75, 135, 149, 155, 160.

**21. Cereal variation** In Exercise 7 you described the distribution of cereal carbohydrates. The standard deviation was calculated to be 3.8 grams.

a) Interpret the standard deviation, in context.
b) What concerns do you have in using standard deviation as a measure of variation for these data? Explain.

**22. Singers short and tall** In Exercise 8 you described the distribution of the heights of the singers in a choir. The standard deviation of this dataset is 3.9 inches.

a) Interpret the standard deviation, in context.
b) What concerns do you have about using standard deviation as a measure of variation for these data? Explain.

**23. Pizza prices** The histogram shows the distribution of the prices of plain pizza slices (in $) for 156 weeks in Dallas, TX.

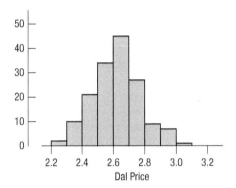

Which summary statistics would you choose to summarize the center and spread in these data? Why?

**24. Neck size** The histogram shows the neck sizes (in inches) of 250 men recruited for a health study in Utah.

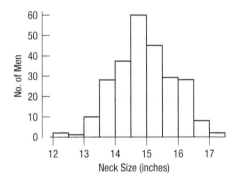

Which summary statistics would you choose to summarize the center and spread in these data? Why?

**25. Pizza prices again** Look again at the histogram of the pizza prices in Exercise 23.

a) Is the mean closer to $2.40, $2.60, or $2.80? Why?
b) Is the standard deviation closer to $0.15, $0.50, or $1.00? Explain.

**26. Neck sizes again** Look again at the histogram of men's neck sizes in Exercise 24.

a) Is the mean closer to 14, 15, or 16 inches? Why?
b) Is the standard deviation closer to 1 inch, 3 inches, or 5 inches? Explain.

**27. Movie lengths** The histogram shows the running times in minutes of 150 top grossing films of a recent year.

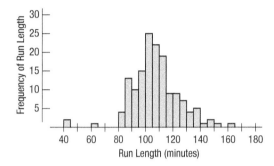

a) You plan to see a movie this weekend. Based on these movies, how long do you expect a typical movie to run?
b) Would you be surprised to find that your movie ran for 2.5 hours (150 minutes)?
c) Which would you expect to be higher: the mean or the median run time for all movies? Why?

**28. Golf Drives 2021** The display show the average drive distance (in yards) for 217 professional golfers during the first week of May on the PGA tour in 2021.

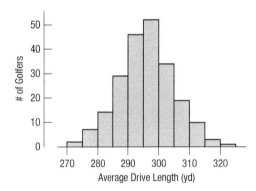

a) Describe the distribution.
b) Your friend states that the range for this distribution is 55 yards. Do you agree? Explain.
c) Estimate the median by examining this histogram.
d) Do you expect the mean to be smaller than, approximately equal to, or larger than the median? Why?

**29. Movie lengths II** Exercise 27 looked at the running times of movies released in a recent year. The standard deviation of these running times is $s = 17.3$ minutes, and the quartiles are $Q_1 = 97$ minutes and $Q_3 = 115$ minutes.

a) Write a sentence or two describing the spread in running times based on
   i) the quartiles.
   ii) the standard deviation.
b) Do you have any concerns about using either of these descriptions of spread? Explain.

**30. Golf drives II** Exercise 28 looked at distances PGA golfers can hit the ball. The standard deviation of these average drive distances is 8.8 yards, and the quartiles are $Q_1 = 290.5$ yards and $Q_3 = 301$ yards.
a) Write a sentence or two describing the spread in distances based on
   i) the quartiles.
   ii) the standard deviation.
b) Do you have any concerns about using either of these descriptions of spread? Explain.

**31. Fuel economy** The boxplot shows the fuel economy ratings for 67 subcompact cars. Some summary statistics are also provided. The extreme outlier is the Mitsubishi i-MiEV, an electric car whose electricity usage is *equivalent* to 112 miles per gallon.

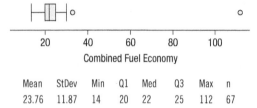

| Mean  | StDev | Min | Q1 | Med | Q3 | Max | n  |
|-------|-------|-----|----|-----|----|-----|----|
| 23.76 | 11.87 | 14  | 20 | 22  | 25 | 112 | 67 |

a) If the electric car was removed from the dataset, would the standard deviation change? Would the IQR change? Justify your answer.
b) Again, removing the electric car, would the mean change? The median? Justify your answer.

**32. Test scores correction** After entering the test scores from her Statistics class of 25 students, the instructor calculated some statistics of the scores. Upon checking, she discovered that she had entered the top score as 46, but it should have been 56.
a) When she corrects this score, how will the median and mean be affected? Justify your answer.
b) What effect will correcting the error have on the IQR and standard deviation? Justify your answer.

**33. Mistake** A clerk entering salary data into a company spreadsheet accidentally put an extra "0" in the boss's salary, listing it as $2,000,000 instead of $200,000. Explain how this error will affect these summary statistics for the company payroll:
a) Measures of center: median and mean.
b) Measures of spread: range, IQR, and standard deviation.

**34. Cold weather** A meteorologist preparing a talk about global warming compiled a list of weekly low temperatures (in degrees Fahrenheit) he observed at his southern Florida home last year. The coldest temperature for any week was 36°F, but he inadvertently recorded the Celsius value of 2°. Assuming that he correctly listed all the other temperatures, explain how this error will affect these summary statistics:
a) Measures of center: mean and median.
b) Measures of spread: range, IQR, and standard deviation.

**35. Movie earnings 2015** The histogram shows total gross earnings (in millions of dollars) of the top 200 major release movies in 2015.

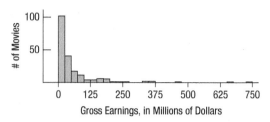

An industry publication reports that the typical movie makes $23.2 million, but a watchdog group concerned with rising ticket prices says that the average earnings are closer to $55.7 million. What statistic do you think each group is using? Explain.

**36. Sick days** During contract negotiations, a company seeks to change the number of sick days employees may take, saying that the annual "average" is 7 days of absence per employee. The union negotiators counter that the "average" employee misses only 3 days of work each year. Explain how both sides might be correct, identifying the measure of center you think each side is using and why the difference might exist.

**37. Payroll** A small warehouse employs a supervisor at $1200 a week, an inventory manager at $700 a week, six stock boys at $400 a week, and four drivers at $500 a week.
a) Find the mean and median wage.
b) How many employees earn more than the mean wage?
c) Which measure of center best describes a typical wage at this company: the mean or the median?
d) Which measure of spread would best describe the payroll: the range, the IQR, or the standard deviation? Why?

**38. Singers** The frequency table shows the heights (in inches) of 130 members of a choir.

| Height | Count | Height | Count |
|--------|-------|--------|-------|
| 60     | 2     | 69     | 5     |
| 61     | 6     | 70     | 11    |
| 62     | 9     | 71     | 8     |
| 63     | 7     | 72     | 9     |
| 64     | 5     | 73     | 4     |
| 65     | 20    | 74     | 2     |
| 66     | 18    | 75     | 4     |
| 67     | 7     | 76     | 1     |
| 68     | 12    |        |       |

a) Find the median and IQR.
b) Find the mean and standard deviation.
c) Display these data with a histogram.
d) Write a few sentences describing the distribution.

**39. Gasoline** Sixteen gas stations in eastern Wisconsin posted these prices for a gallon of regular gasoline:

| 2.19 | 2.32 | 2.31 | 2.33 |
| 2.28 | 2.35 | 2.22 | 2.29 |
| 2.23 | 2.30 | 2.25 | 2.29 |
| 2.25 | 2.21 | 2.45 | 2.25 |

a) Make a stem-and-leaf display of these gas prices. Use split stems; for example, use two 2.3 stems—one for prices between $2.30 and $2.34 and the other for prices from $2.35 to $2.39.

b) Describe the shape, center, and spread of this distribution.
c) What unusual feature do you see?

**40. The Great One** During his 20 seasons in the NHL, Wayne Gretzky scored 50% more points than anyone who ever played professional hockey. He accomplished this amazing feat while playing in 280 fewer games than Gordie Howe, the previous record holder. Here are the number of games Gretzky played during each season:

79, 80, 80, 80, 74, 80, 80, 79, 64, 78,
73, 78, 74, 45, 81, 48, 80, 82, 82, 70

a) Create a stem-and-leaf display for these data, using split stems.
b) Describe the shape of the distribution.
c) Describe the center and spread of this distribution.
d) What unusual feature do you see? What might explain this?

**41. States** The stem-and-leaf display shows populations of the 50 states, in millions of people, according to the 2010 census.

```
Population
 3 | 7
 3 |
 2 | 5
 2 |
 1 | 99
 1 | 000233
 0 | 555556666667789
 0 | 111111111111222233333334444
(2|5 means 25)
```

a) What measures of center and spread are most appropriate?
b) Without doing any calculations, which must be larger: the median or the mean? Explain how you know.
c) From the stem-and-leaf display, find the median and the interquartile range.
d) Write a few sentences describing this distribution.

**42. Wayne Gretzky** In Exercise 40, you examined the number of games played by hockey great Wayne Gretzky during his 20-year career in the NHL.
a) Would you use the median or the mean to describe the center of this distribution? Why?
b) Find the median.
c) Without actually finding the mean, would you expect it to be higher or lower than the median? Explain.

**43. A-Rod** Alex Rodriguez (known to fans as A-Rod) was the youngest player ever to hit 500 home runs. Here is a stem-and-leaf display of the number of home runs hit by A-Rod during the 1994–2016 seasons (www.baseball-reference.com/players/r/rodrial01.shtml). Describe the distribution, mentioning its shape and any unusual features.

```
Home Runs
 5 | 247
 4 | 12278
 3 | 0035566
 2 | 3
 1 | 68
 0 | 0579
(5|2 means 52)
```

**44. Bird species 2013** The Cornell Lab of Ornithology holds an annual Christmas Bird Count (www.birdsource.org), in which bird watchers at various locations around the country see how many different species of birds they can spot. Here are the numbers of species counted from the 20 sites with the most species in 2013:

184  98  101  126  150  166  82  136  124  118
133  83  86  101  105  97  88  131  128  106

a) Create a stem-and-leaf display of these data.
b) Write a description of the distribution.

**45. Major hurricanes 2013** The data below give the number of hurricanes classified as major hurricanes in the Atlantic Ocean each year from 1944 through 2013, as reported by *NOAA* (www.nhc.noaa.gov):

3, 3, 1, 2, 4, 3, 8, 5, 3, 4, 2, 6, 2, 2, 5, 2, 2, 7, 1, 2, 6, 1, 3,
1, 0, 5, 2, 1, 0, 1, 2, 3, 2, 1, 2, 2, 2, 3, 1, 1, 1, 3, 0, 1, 3, 2,
1, 2, 1, 1, 0, 5, 6, 1, 3, 5, 3, 4, 2, 3, 6, 7, 2, 2, 5, 2, 5, 4, 2, 0

a) Create a dotplot of these data.
b) Describe the distribution.

**46. Horsepower** Create a stem-and-leaf display for these horsepowers of autos reviewed by *Consumer Reports* one year, and describe the distribution:

| 155 | 103 | 130 | 80  | 65  |
| 142 | 125 | 129 | 71  | 69  |
| 125 | 115 | 138 | 68  | 78  |
| 150 | 133 | 135 | 90  | 97  |
| 68  | 105 | 88  | 115 | 110 |
| 95  | 85  | 109 | 115 | 71  |
| 97  | 110 | 65  | 90  |     |
| 75  | 120 | 80  | 70  |     |

**47. A-Rod again** Students were asked to make a histogram of the number of home runs Alex Rodriguez hit from 1994 to 2016 (see Exercise 43). One student submitted the following display:

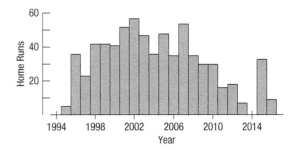

a) Critique this graph.
b) Create your own histogram of the data.

**48. Return of the birds 2013** Students were given the assignment to make a histogram of the data on bird counts reported in Exercise 44. One student submitted the following display:

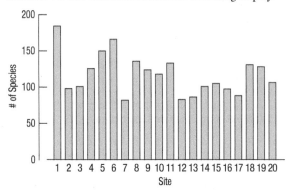

a) Critique this graph.
b) Create your own histogram of the data.

**49. Acid rain** Two researchers measured the pH (a scale on which a value of 7 is neutral and values below 7 are acidic) of water collected from rain and snow over a 6-month period in Allegheny County, PA.

4.57  5.62  4.12  5.29  4.64  4.31  4.30  4.39  4.45
5.67  4.39  4.52  4.26  4.26  4.40  5.78  4.73  4.56
5.08  4.41  4.12  5.51  4.82  4.63  4.29  4.60

Describe their data with a graph and a few sentences.

**50. Marijuana 2015** In 2015 the Council of Europe published a report entitled *The European School Survey Project on Alcohol and Other Drugs* (www.espad.org). Among other issues, the survey investigated the percentages of 16-year-olds who had used marijuana. Shown below are the results for 38 European countries. Create an appropriate graph of these data, and describe the distribution.

| Country | Cannabis | Country | Cannabis |
|---|---|---|---|
| Albania | 4 | Latvia | 24 |
| Belgium | 24 | Liechtenstein | 21 |
| Bosnia and Herz. | 4 | Lithuania | 20 |
| Bulgaria | 24 | Malta | 10 |
| Croatia | 18 | Moldova | 5 |
| Cyprus | 7 | Monaco | 37 |
| Czech Republic | 42 | Montenegro | 5 |
| Denmark | 18 | Netherlands | 27 |
| Estonia | 24 | Norway | 5 |
| Faroe Islands | 5 | Poland | 23 |
| Finland | 11 | Portugal | 16 |
| France | 39 | Romania | 7 |
| Germany | 19 | Russian Fed. | 15 |
| Greece | 8 | Serbia | 7 |
| Hungary | 19 | Slovak Republic | 27 |
| Iceland | 10 | Slovenia | 23 |
| Ireland | 18 | Sweden | 9 |
| Italy | 21 | Ukraine | 11 |
| Kosovo | 2 | United Kingdom | 25 |

**51. Final grades** A professor (of something other than Statistics!) distributed the following histogram to show the distribution of grades on his 200-point final exam. Comment on the display.

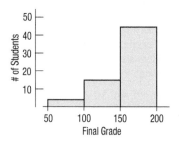

**52. Final grades revisited** After receiving many complaints about his final-grade histogram from students currently taking a Statistics course, the professor from Exercise 51 distributed the following revised histogram:

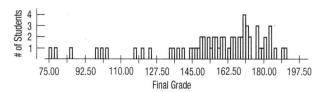

a) Comment on this display.
b) Describe the distribution of grades.

**53. Used Cars 2020** A random sample of 1500 cars was taken from truecar.com in the fall of 2020, within a 50 mile radius of Rancho Cucamonga, CA. Make, model, year, and asking price were collected. Below is a histogram of asking price, in U.S. dollars.

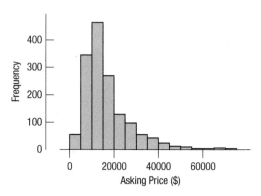

a) Describe the distribution.
b) Estimate the median asking price.
c) Approximately what percentage of cars have an asking price above $40,000?
d) Is the mean asking price more than, less than, or about the same as the median price? Explain.

**54. Poop** A Statistics teacher had his students perform a quick experiment. Each student was given two large copies of the infamous Poop emoji. Students were timed as they traced each emoji, once with their dominant hand and also with their nondominant hand. The class was instructed to investigate the difference in speed (in seconds) between the different hands, subtracting times (nondominant − dominant). This teacher collected the students' times for several years. Following is a histogram of these differences for 367 students.

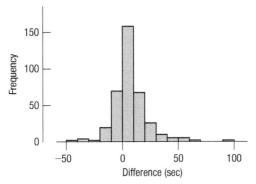

a) Describe the distribution of the differences, in context.
b) Are you certain that the largest observation is an outlier? Why or why not? Justify your answer.
c) Approximately what percentage of students were slower with their nondominant hand? You don't need to be precise, but use the graph to give an estimate and show your reasoning.
d) Given your answer to part (c), do you think it is reasonable to conclude that students, on average, trace slower with their nondominant hand? Do you have any questions about how the experiment was conducted? What additional information would help you determine if this was true?

**55. Math scores 2019** The National Center for Education Statistics reported average mathematics achievement scores for eighth graders in all 50 states, the District of Columbia, and Puerto Rico (nces.ed.gov/nationsreportcard/):

| State | Score | State | Score |
|---|---|---|---|
| Alabama | 269 | Montana | 284 |
| Alaska | 274 | Nebraska | 285 |
| Arizona | 280 | Nevada | 274 |
| Arkansas | 274 | New Hampshire | 287 |
| California | 276 | New Jersey | 292 |
| Colorado | 285 | New Mexico | 269 |
| Connecticut | 286 | New York | 280 |
| Delaware | 277 | North Carolina | 284 |
| District of Columbia | 269 | North Dakota | 286 |
| Florida | 279 | Ohio | 286 |
| Georgia | 279 | Oklahoma | 276 |
| Hawaii | 275 | Oregon | 280 |
| Idaho | 286 | Pennsylvania | 285 |
| Illinois | 283 | Puerto Rico | 222 |
| Indiana | 286 | Rhode Island | 276 |
| Iowa | 282 | South Carolina | 276 |
| Kansas | 282 | South Dakota | 287 |
| Kentucky | 278 | Tennessee | 280 |
| Louisiana | 272 | Texas | 280 |
| Maine | 282 | Utah | 285 |
| Maryland | 280 | Vermont | 287 |
| Massachusetts | 294 | Virginia | 287 |
| Michigan | 280 | Washington | 286 |
| Minnesota | 291 | West Virginia | 272 |
| Mississippi | 274 | Wisconsin | 289 |
| Missouri | 281 | Wyoming | 286 |

a) Using technology and the provided data file, find the median, IQR, mean, and standard deviation of these state averages.
b) Which summary statistics would you report for these data? Why?
c) Write a brief summary of the performance of eighth graders nationwide.

**56. Boomtowns 2020** In 2020, the website NewGeography.com (http://www.newgeography.com/content/005974-all-cities-rankings-2018-best-cities-job-growth) listed its ranking of the best cities for job growth in the United States. The magazine's top 20 large cities, along with their weighted job rating indices, are given in the table. The full dataset contains 71 cities.

a) Using technology and the provided data file, make a suitable display of the weighted growth indices for all 71 cities.
b) Summarize the typical growth index among these cities with a median and mean.
c) Summarize the spread of the growth index distribution with a standard deviation and with an IQR.
d) Suppose we subtract from each of the preceding growth rates the average U.S. large-city growth index of 62.56%, so that we can look at how much these indices exceed the U.S. rate. How would this change the values of the summary statistics you calculated above? (*Hint:* You need not recompute any of the summary statistics from scratch.)
e) If we were to omit Austin-Round Rock, TX, from the data, how would you expect the mean, median, standard deviation, and IQR to change? Explain your expectations for each.
f) Write a brief report about all of these growth indices.

| | |
|---|---|
| Dallas-Plano-Irving, TX Metro Div | 95.4 |
| Austin-Round Rock, TX | 94.3 |
| Nashville-Davidson–Murfreesboro–Franklin, TN | 94.3 |
| San Jose-Sunnyvale-Santa Clara, CA | 94.1 |
| Charlotte-Concord-Gastonia, NC-SC | 93.2 |
| Orlando-Kissimmee-Sanford, FL | 92.7 |
| Raleigh, NC | 92.4 |
| San Francisco-Redwood City-South San Francisco | 92.4 |
| Seattle-Bellevue-Everett, WA Metro Div | 92.4 |
| Riverside-San Bernardino-Ontario, CA | 89.5 |
| Salt Lake City, UT | 89.3 |
| Denver-Aurora-Lakewood, CO | 88.7 |
| San Antonio-New Braunfels, TX | 85.5 |
| Jacksonville, FL | 85.5 |
| Portland-Vancouver-Hillsboro, OR-WA | 84.5 |
| Phoenix-Mesa-Scottsdale, AZ | 79.6 |
| Las Vegas-Henderson-Paradise, NV | 79.4 |
| Atlanta-Sandy Springs-Roswell, GA | 79 |
| Fort Worth-Arlington, TX Metro Div | 77.9 |
| Tampa-St. Petersburg-Clearwater, FL | 77.3 |

**57. Population growth 2010** The following data show the percentage change in population for the 50 states and the District of Columbia from the 2000 census to the 2010 census. Using appropriate graphical displays and summary statistics, write a report on the percentage change in population by state.

| State | Percent Increase | State | Percent Increase |
|---|---|---|---|
| Alabama | 7.5 | Montana | 9.7 |
| Alaska | 13.3 | Nebraska | 6.7 |
| Arizona | 24.6 | Nevada | 35.1 |
| Arkansas | 9.1 | New Hampshire | 6.5 |
| California | 10.0 | New Jersey | 4.5 |
| Colorado | 16.9 | New Mexico | 13.2 |
| Connecticut | 4.9 | New York | 2.1 |
| Delaware | 14.6 | North Carolina | 18.5 |
| District of Columbia | 5.2 | North Dakota | 4.7 |
| Florida | 17.6 | Ohio | 1.6 |
| Georgia | 18.3 | Oklahoma | 8.7 |
| Hawaii | 12.3 | Oregon | 12.0 |
| Idaho | 21.1 | Pennsylvania | 3.4 |
| Illinois | 3.3 | Rhode Island | 0.4 |
| Indiana | 6.6 | South Carolina | 15.3 |
| Iowa | 4.1 | South Dakota | 7.9 |
| Kansas | 6.1 | Tennessee | 11.5 |
| Kentucky | 7.4 | Texas | 20.6 |
| Louisiana | 1.4 | Utah | 23.8 |
| Maine | 4.2 | Vermont | 2.8 |
| Maryland | 9.0 | Virginia | 13.0 |
| Massachusetts | 3.1 | Washington | 14.1 |
| Michigan | −0.6 | West Virginia | 2.5 |
| Minnesota | 7.8 | Wisconsin | 6.0 |
| Mississippi | 4.3 | Wyoming | 14.1 |
| Missouri | 7.0 | | |

Source: www.census.gov/compendia/statab/rankings.html

**58. Coasters 2021** The dataset **Coasters 2021** contains data on 257 roller coasters. The variables measured include:
- Track (steel or wood)
- Speed (mph)
- Height (feet)
- Drop (feet)
- Length (feet)
- Duration (total seconds)
- Inversions (the number of times the rider is turned upside-down)

Using technology and appropriate graphical displays and summary statistics, write a report on several of the variables.

**59. Derby speeds 2020** How fast do horses run? Kentucky Derby winners run well over 30 miles per hour, as shown in this cumulative frequency plot. Note that few have won running less than 33 miles per hour, but about 86% of the winning horses have run less than 37 miles per hour.

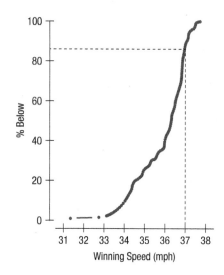

Suppose you had only the ogive (and no access to the data).
a) Estimate the median winning speed.
b) Estimate the quartiles.
c) Estimate the range and the IQR.
d) Create a boxplot of these speeds.
e) Write a few sentences about the speeds of the Kentucky Derby winners.

**60. Cholesterol** The Framingham Heart Study recorded the cholesterol levels of more than 1400 participants. Here is a cumulative frequency plot of the distribution of these cholesterol measures. Construct a boxplot for these data, and write a few sentences describing the distribution.

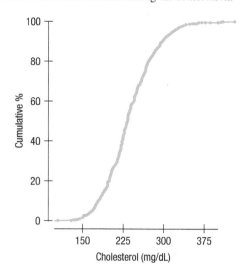

## JUST CHECKING

**Answers**

(Thoughts will vary.)

1. Slightly skewed to the right. Center around 3 miles? Few over 10 miles.

2. Probably bimodal or perhaps even multimodal, with a mode of zero hours for those who watch no football, a mode of perhaps 2—3 hours for people who watch one game, and possibly other modes corresponding to people who watch more than one game.

3. Strongly skewed to the right, with almost everyone at $0; a few small prizes, with the winner an outlier.

4. Fairly symmetric, somewhat uniform, perhaps slightly skewed to the right. Center in the 40s? Few ages below 25 or above 70.

5. Uniform, symmetric. Center near 5. Roughly equal counts for each digit 0–9.

6. Incomes are probably skewed to the right and not symmetric, making the median the more appropriate measure of center. The mean will be influenced by the high end of family incomes and not reflect the "typical" family income as well as the median would. It will give the impression that the typical income is higher than it is.

7. An IQR of 30 mpg would mean that only 50% of the cars get gas mileages in an interval 30 mpg wide. Fuel economy doesn't vary that much. 3 mpg is reasonable. It seems plausible that 50% of the cars will be within about 3 mpg of each other. An IQR of 0.3 mpg would mean that the gas mileage of half the cars varies little from the estimate. It's unlikely that cars, drivers, and driving conditions are that consistent.

8. We'd prefer a standard deviation of 2 months. Making a consistent product is important for quality. Customers want to be able to count on the phone's battery lasting somewhere close to 5 years, and a standard deviation of 2 years would mean that lifespans were highly variable.

# 4

# Understanding and Comparing Distributions

| WHO | Days during 2016 |
| --- | --- |
| WHAT | Average daily wind speed (mph), Average barometric pressure (mb), Average daily temperature (deg Celsius) |
| WHEN | 2016 |
| WHERE | Hopkins Memorial Forest, in Massachusetts, New York, and Vermont |
| WHY | Long-term observations to study ecology and climate |

The Hopkins Memorial Forest is a 2500-acre reserve in Massachusetts, New York, and Vermont managed by the Williams College Center for Environmental Studies (CES). As part of its mission, CES monitors forest resources and conditions over the long term.[1]

One of the variables measured in the forest is wind speed. Three remote sensors record the minimum, maximum, and average wind speed (in mph) for each day.

Wind is caused as air flows from areas of high pressure to areas of low pressure. Centers of low pressure often accompany storms, so both high winds and low pressure are associated with some of the fiercest storms. Wind speeds can vary greatly during a day and from day to day, but if we step back a bit farther, we can see patterns. By modeling these patterns, we can understand things about *Average Wind Speed* that we may not have known.

In Chapter 2 we looked at the association between two categorical variables using contingency tables and displays. Here we'll explore different ways of examining the relationship between two variables when one is quantitative and the other indicates groups to compare. We are given wind speed averages for each day of 2016. But we can collect the days together into different size groups and compare the wind speeds among them. If we partition *Time* in different ways, we'll gain enormous flexibility for our analysis. We'll discover new insights as we change from viewing the whole year's data at once, to comparing seasons, to looking for patterns across months, and, finally, to looking at the data day by day.

---

[1] www.williams.edu/CES/hopkins.htm

# The Big Picture

Let's start with the "big picture." Here's a histogram, 5-number summary, and boxplot of the *Average Wind Speed* for every day in 2016. Because of the skewness, we'll report the median and IQR. We can see that the distribution of *Average Wind Speed* is unimodal and skewed to the right. Median daily wind speed is about 2.1 mph, and on half of the days, the average wind speed is between about 1.4 and 2.9 mph. We also see some outliers, including a rather windy 6.17-mph day. Were those unusual weather events, or just the windiest days of the year? To answer that, we'll need to work with the summaries a bit more.

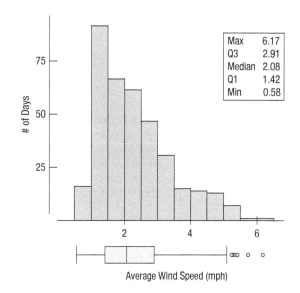

Figure 4.1
**A histogram of daily *Average Wind Speed* for 2016** It is unimodal and skewed to the right. The boxplot below the histogram suggests several possible outliers that may deserve our attention.

# Comparing Groups with Histograms

It is almost always more interesting to compare groups. Is it windier in the winter or the summer? Are any months particularly windy? Let's start by splitting the year into the four seasons. The histograms below show how the distributions of *Average Wind Speed* differ by season.

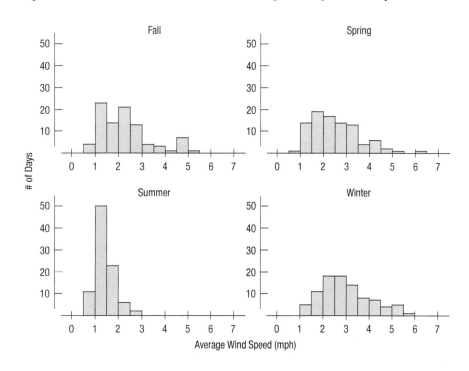

Figure 4.2
**Histograms of *Average Wind Speed* grouped by season** Graphs that are grouped like these should have a horizontal scale common to all graphs in a column, and a vertical scale common to all graphs in a row, to facilitate comparison.

Three of these distributions look pretty similar, but one—Summer—is clearly different. If you're not careful, you might see the tall spike in Summer's histogram and think it represents very windy days. But not so fast! It's the horizontal scale, not the vertical scale, that indicates wind speed! The tall bar in Summer's histogram indicates that there were a lot of days in the summer with wind speeds between 1.0 and 1.5 miles per hour, which are very calm. The windiest day of 2016 was actually in the spring, on a day when the average wind speed was over 6 miles per hour.

### FOR EXAMPLE
#### Comparing Groups with Stem-and-Leaf Displays

The Nest Egg Index, devised by the investment firm of A.G. Edwards, is a measure of saving and investment performance for each of the 50 states, based on 12 economic factors, including participation in retirement savings plans, personal debt levels, and home ownership. The average index is 100 and the numbers indicate the percentage above or below the average. There are only 50 values, so a back-to-back stem-and-leaf plot is an effective display. Here's one comparing the Nest Egg Index in the Northeast and Midwest states to those in the South and West. In this display, the stems run down the middle of the plot, with the leaves for the two regions to the left or right. Be careful when you read the values on the left: 5|8 means a Nest Egg Index of 85% for a southern or western state.

| South and West | | Northeast and Midwest |
|---:|:---:|:---|
| 5778 | 8 | 8 |
| 12344 | 9 | 03 |
| 6667778899 | 9 | 67 |
| 02334 | 10 | 012233334 |
| 56 | 10 | 6779 |
| | 11 | 122444 |

(4|9|3 means 94% for a South/West state
and 93% for a Northeast/Midwest state)

QUESTION: How do Nest Egg Indices compare for these regions?

ANSWER: The distribution of Nest Egg Indices is nearly symmetric for the South and West, but skewed to the left for the Northeast and Midwest.[2] Indices were generally higher and more variable in the Northeast/Midwest region than in the South/West. Nine northeastern and midwestern states had higher indices than any states in the South or West. The big picture is that people in the Northeast and Midwest tend to be better prepared for retirement than are people in the South and West.

## Comparing Groups with Boxplots

Are some months windier than others? Even residents may not have a good idea of which parts of the year are the most windy. (Do you know for your hometown?) We're not interested just in the centers, but also in the spreads. Are wind speeds equally variable from month to month, or do some months show more variation?

Histograms or stem-and-leaf displays are a fine way to look at a small number of distributions. But it would be hard to see patterns by comparing 12 histograms. Boxplots, however, ingeniously show all their information in just one dimension, so you can line them up side-by-side in the other dimension and make it easy to compare many groups.

---

[2] Even if the orientation of a graph is vertical, we still say "skewed left" or sometimes "negatively skewed"—definitely not "skewed up" (or down).

**TI-nspire**

**Histograms and boxplots.** See that the shape of a distribution is not always evident in a boxplot.

(In fact, that's exactly what they were invented for. If you want to show just one or two distributions, histograms or stem-and-leaf plots are usually better because they don't hide as much information about *shape* as boxplots do.)

Here are boxplots of the *Average Wind Speed* by month:

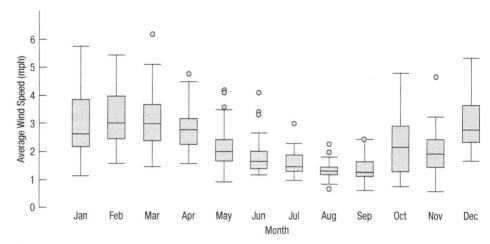

**Figure 4.3**
**Boxplots of the *Average Wind Speed* plotted for each *Month* show seasonal patterns in both the centers and spreads** New outliers appear because they are now judged relative to the month in which they occurred.

Here we see that wind speeds tend to decrease in the summer. The months in which the winds are both strongest and most variable are October through March.

When we looked at a boxplot of wind speeds for the entire year, there were only 6 outliers. But the monthly boxplots show different outliers than before because some days that seemed ordinary when placed against the entire year's data looked like outliers for the month that they're in. That windy day in July certainly wouldn't stand out in January or February, but for July it was remarkable.

### FOR EXAMPLE
#### Comparing Distributions

Roller coaster riders want a coaster that goes fast.[3] There are two main types of roller coasters: those with wooden tracks and those with steel tracks. Do they typically run at different speeds? Here are boxplots.

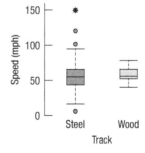

QUESTION: Compare the speeds of wood and steel roller coasters.

ANSWER: The median speed for the wood track coasters may be just a bit higher than for the steel coasters, but overall their typical speeds are about the same, around 50 mph. However, the speeds of the steel track coasters vary much more and include both high and low outliers.

---

[3] See the Roller Coaster Data Base at www.rcdb.com.

## STEP-BY-STEP EXAMPLE

### Comparing Groups

Most scientific studies compare two or more groups. It is almost always a good idea to start an analysis of data from such studies by comparing boxplots for the groups. Here's an example:

For her class project, a student compared the efficiency of various coffee containers. For her study, she decided to try 4 different containers and to test each of them 8 different times. Each time, she heated water to 180°F, poured it into a container, and sealed it. (We'll learn the details of how to set up experiments in Chapter 12.) After 30 minutes, she measured the temperature again and recorded the difference in temperature. Because these are temperature differences, smaller differences mean that the liquid stayed hot—just what we would want in a coffee mug.

QUESTION: What can we say about the effectiveness of these four mugs?

---

**THINK**    **PLAN** State what you want to find out.

**VARIABLES** Identify the *variables* and report the W's.

Be sure to check the appropriate condition.

I want to compare the effectiveness of the different mugs in maintaining temperature. I have 8 measurements of *Temperature Change* for each of the mugs.

✓ **Quantitative Data Condition:** The *Temperature Changes* are quantitative, with units of °F. Boxplots are appropriate displays for comparing the groups. Numerical summaries of each group are appropriate as well.

---

**SHOW**    **MECHANICS** Report the 5-number summaries of the four groups. Including the IQR is a good idea as well.

|            | Min  | Q1    | Median | Q3    | Max   | IQR   |
|------------|------|-------|--------|-------|-------|-------|
| CUPPS      | 6°F  | 6     | 8.25   | 14.25 | 18.50 | 8.25  |
| Nissan     | 0    | 1     | 2      | 4.50  | 7     | 3.50  |
| SIGG       | 9    | 11.50 | 14.25  | 21.75 | 24.50 | 10.25 |
| Starbucks  | 6    | 6.50  | 8.50   | 14.25 | 17.50 | 7.75  |

Make a picture. Because we want to compare the distributions for four groups, boxplots are an appropriate choice.

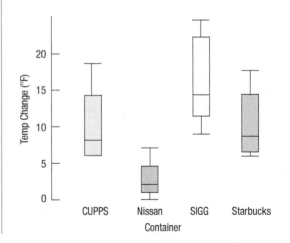

**TELL** ▶ **CONCLUSION** Interpret what the boxplots and summaries say about the ability of these mugs to retain heat. Compare the shapes, centers, and spreads, and note any outliers.

The individual distributions of temperature changes are all slightly skewed to the high end. The Nissan cup does the best job of keeping liquids hot, with a median loss of only 2°F, and the SIGG cup does the worst, typically losing 14°F. The difference is large enough to be important: A coffee drinker would be likely to notice a 14° drop in temperature. And the mugs are clearly different: 75% of the Nissan tests showed less heat loss than any of the other mugs in the study. The IQR of results for the Nissan cup is also the smallest of these test cups, indicating that it is a consistent performer.

## JUST CHECKING

The Bureau of Transportation Statistics of the U.S. Department of Transportation collects and publishes statistics on airline travel (www.transtats.bts.gov). Here are three displays of the monthly percentages of flights arriving on time between January 1994 and September 2019.

1. Describe what the histogram says about on-time arrivals.
2. What does the boxplot of on-time arrivals suggest that you can't see in the histogram?
3. Describe the patterns shown in the graph showing boxplots by month. At what time of year are flights most likely to be late? Can you suggest reasons for this pattern?

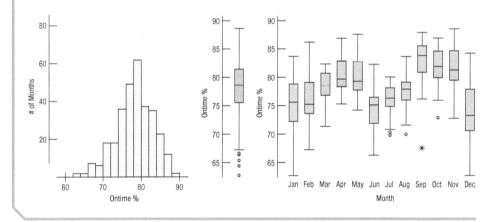

## TI TIPS

### Comparing Groups with Boxplots

Earlier in this chapter we compared the top speeds of some steel-track and wooden-track roller coasters. Now let's make comparative boxplots for their drop heights. (That's the height of the initial drop at the beginning of a roller coaster ride.) The drop heights below are given in feet. Enter the data in L1 (steel-track) and L2 (wooden-track).

> Drop heights (feet) of 24 roller coasters:
> Steel-track: 300, 155, 105, 146, 95, 64, 95, 190, 84, 109, 226, 206
> Wooden-track: 150, 124, 115, 100, 76, 90, 70, 52, 78, 75, 106, 58

Set up STATPLOT's Plot1 to make a boxplots of the steel-track coasters' data:

♦ Turn the plot On.
♦ Always choose the boxplot icon that shows outliers.
♦ Specify Xlist:L1 and Freq:1, and select an outlier Mark.

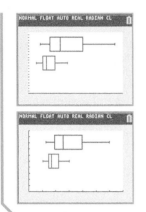

Use `ZoomStat` to display the boxplot for the steel-track coasters' drop heights.

As you did for the steel-track coasters, set up `Plot2` to display the data for the wooden-track coasters. This time when you use `ZoomStat` with both plots turned on, the display shows the parallel boxplots.

Sometimes when comparing two distributions it's helpful to see the magnitudes more clearly by including the value zero in the graph. Here we've changed the horizontal scale to go from zero feet to 350 feet, with tick marks at every 50 feet.

This is a great opportunity to use your "Tell" skills. Compare the drop heights of the steel-track and the wooden-track roller coasters in this dataset.

# Outliers

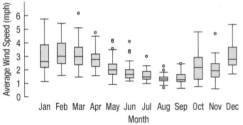

In the boxplots for the *Average Wind Speed* by *Month*, several days are nominated by the boxplots as possible outliers. Cases that stand out almost always deserve attention. An outlier is a value that doesn't fit with the rest of the data, but exactly how different it should be to receive special treatment is a judgment call. Boxplots provide a rule of thumb to highlight these unusual cases, but that rule is only a guide, and it doesn't tell you what to do with them. So, what *should* we do with outliers?

Outliers arise for many reasons. They *may* be the most important values in the dataset, pointing out an exceptional case or illuminating a pattern by being the exception to the rule. They may be values that just happen to lie above the limits suggested by the boxplot rule. In practice, they often represent incorrectly entered data. (In this text, we'll never give you incorrect data, though!)

The boxplots of *Average Wind Speed* by Month show outliers in 8 of the months. With one exception, these indicate days that were unusually windy for *that month*. (The exception is an outlier in August that indicates an unusually calm day.) For example, there was a day in July that was quite windy for July, but it would have been typically windy for February.

There are two things you should *never* do with outliers. You should not leave an outlier in place and proceed as if nothing were unusual. Analyses of data with outliers are very likely to be wrong. Nor should you omit an outlier from the analysis without comment. If you want to exclude an outlier, you must announce your decision and, to the extent you can, justify it. Finally, keep in mind that a case lying just over the fence suggested by the boxplot may just be the largest (or smallest) value at the end of a stretched-out tail. A histogram is often a better way to examine how the outlier fits in (or doesn't) by seeing how large the gap is between it and the rest of the data.

In the aftermath of Hurricane Irene, the Hoosic River in Western Massachusetts rose more than 10 feet over its banks, swallowing portions of Williams College, including the soccer and baseball fields.

## FOR EXAMPLE

### Checking Out the Outliers

**RECAP:** We've looked at the speeds of roller coasters and found a difference between steel- and wooden-track coasters. We also noticed an extraordinarily high value.

**QUESTION:** The fastest coaster in this collection turns out to be "Formula Rossa" at Ferrari World in Abu Dhabi. What might make this roller coaster unusual? You'll have to do some research, but that's often what happens with outliers.

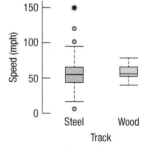

**ANSWER:** Formula Rossa is easy to find in an Internet search. We learn that it is a "hydraulic launch" coaster. That is, it doesn't get its remarkable speed just from gravity, but rather from a kick-start by a hydraulic piston. That could make it different from the other roller coasters. It accelerates to 149 miles per hour (240 km/h) in 4.8 seconds. Riders must wear safety goggles.

# Timeplots: Order, Please!

The Hopkins Memorial Forest wind speeds are reported as daily averages. Previously, we grouped the days into months or seasons, but we could look at the wind speed values day by day. Whenever we have data measured over time, it is a good idea to look for patterns by plotting the data in time order. Here are the daily *Average Wind Speeds* plotted over time:

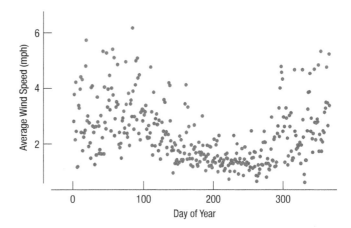

Figure 4.4

**A timeplot of *Average Wind Speed***
The timeplot shows the overall pattern and changes in variation.

A display of individual values against time is sometimes called a **timeplot**. This timeplot reflects the pattern that we saw when we plotted the wind speeds by month. But without the arbitrary divisions between months, we can see a calm period during the summer, starting around day 150 (the beginning of June), when the wind is relatively mild and doesn't vary greatly from day to day. We can also see that the wind becomes both more variable and stronger during the early and late parts of the year.

In Chapter 2 we wondered whether it might have been just coincidence that a relatively large fraction of first-class passengers aboard the *Titanic* survived. We simulated what the survival rates of the different ticket classes might have looked like if survival had been completely random, unrelated to ticket class. We learned that if survival had been unrelated to ticket class, we should not have seen differences in survival rates nearly so great as what occurred on the *Titanic*.

Let's do something similar with the timeplot shown in Figure 4.4. It may seem obvious that the description of summer months as relatively calm was describing a real phenomenon. But could it have been just a coincidence—a random "quirk" in the data? Maybe any old ordering of these wind speeds would show something similar.

Figure 4.5 shows the same wind speeds that actually were recorded in Hopkins Memorial Forest in 2016, but the days of the year have been re-arranged into a random order three different times.

Figure 4.5

When the wind speeds are put in a random order, the calm summer phenomenon is consistently absent—indicating that it is a real phenomenon and not a random "quirk."

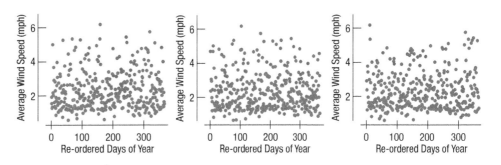

The pattern of the calm summer disappears when the days are put in a random order. That means that 2016's summer months in Hopkins Memorial Forest really are associated with calmer days; it isn't just chance that such a long calm spell exists in the data. (On the other hand, you might easily *imagine* you see patterns of other sorts in the *randomized* timeplots, even though they can't possibly represent real phenomena. It's human nature to look for patterns, and studies confirm that we humans are prone to imagining patterns in randomness. Simulations are a useful tool for helping determine whether an apparent phenomenon is real or a random quirk.)

# Looking into the Future

It is always tempting to try to extend what we see in a timeplot into the future. Sometimes that makes sense. Most likely, the Hopkins Memorial Forest climate follows regular seasonal patterns. It's probably safe to predict a less windy June next year and a windier November.

Other patterns are riskier to extend into the future. If a stock has been rising, will it continue to go up? No stock has ever increased in value indefinitely, and no stock analyst has consistently been able to forecast when a stock's value will turn around. Stock prices, unemployment rates, and other economic, social, or psychological concepts are much harder to predict than physical quantities. The path a ball will follow when thrown from a certain height at a given speed and direction is well understood. The path interest rates will take is much less clear. Unless we have strong (nonstatistical) reasons for doing otherwise, we should resist the temptation to think that any trend we see will continue, even into the near future.

Statistical models often tempt those who use them to think beyond the data. We'll pay close attention later in this book to understanding when, how, and how much we can justify doing that.

# *Re-expressing Data: A First Look

## Re-expressing to Improve Symmetry

When data are skewed, it can be hard to summarize them simply with a center and spread, and hard to decide whether the most extreme values are outliers or just part of the stretched-out tail. How can we say anything useful about such data? The secret is to *re-express* the data by applying a simple function to each value.

Many relationships and "laws" in the sciences and social sciences include functions such as logarithms, square roots, and reciprocals. Similar relationships often show up in data.

In 1980, large companies' chief executive officers (CEOs) made, on average, about 42 times what workers earned. In the next two decades, CEO compensation soared compared to the average worker. By 2008, that multiple had jumped to 344. What does the

Figure 4.6
CEOs' compensation for the *Forbes* 500 companies in 2018

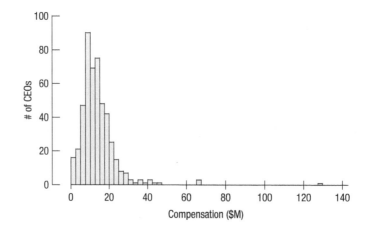
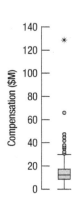

distribution of the compensation of *Forbes* 500 companies' CEOs look like? Look back at the histogram and boxplot for 2018 compensation on the previous page.[4]

We have data for 480 CEOs and nearly 60 possible histogram bins, about half of which are empty—and several of which contain only one or two CEOs. The boxplot indicates that some CEOs received extraordinarily high pay, while the majority received relatively "little." But look at the values of the bins. The first bin covers compensations of $0 to $2,500,000. Imagine receiving a salary survey with these categories:

What is your income?

a) $0 to $2,500,000
b) $2,500,001 to $5,000,000
c) $5,000,001 to $7,500,000
d) More than $7,500,000

What we *can* see from this histogram and boxplot is that this distribution is highly skewed to the right.

It can be hard to decide what we mean by the "center" of a skewed distribution, so it's hard to pick a typical value to summarize the distribution. What would you say was a typical CEO total compensation? The mean value is $14,110,520, while the median is "only" $12,623,906. Each tells us something different about the data.

One approach is to **re-express**, or **transform**, the data by applying a simple function to make the skewed distribution more symmetric. For example, we could take the square root or logarithm of each pay value. Taking logs works pretty well for the CEO compensations, as you can see:[5]

According to the Associated Press annual report on CEO salaries, Lisa Su, CEO of Advanced Micro Devices, was the highest-paid female CEO in 2019—she made $58.5 million.

Figure 4.7
The logarithms of 2018 CEO compensations are much more nearly symmetric.

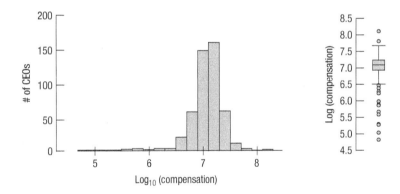

The histogram of the logs of the total CEO compensations is much more nearly symmetric, and we can see that the log compensations are between 4.8, which corresponds to about $64,000, and 8.1, corresponding to about $129,000,000. And it's easier to talk about a typical value for the logs. Both the mean and median log compensations are about 7.0. (That's about $10,000,000, but who's counting?)

Against the background of a generally symmetric main body of data, it's easier to decide whether the largest or smallest pay values are outliers. In fact, even after this re-expression the four most highly compensated CEOs are still identified as outliers by the boxplot rule of thumb. It's impressive to be an outlier CEO in annual compensation. It's even more impressive to be an outlier on the log scale!

Variables that are skewed to the right often benefit from a re-expression by square roots, logs, or reciprocals. Those skewed to the left may benefit from squaring the data. Because computers and calculators can do the calculations, re-expressing data is quite easy. Consider re-expression as a helpful tool whenever you have skewed data.

---

[4] https://aflcio.org/paywatch/company-pay-ratios
[5] We've set aside Larry Page (Google) and Jack Dorsey (Twitter), who had official compensations of $1. David Hess of Arconic, Inc. is then the low outlier with compensation of $67,418.

### Dealing with Logarithms

You have probably learned about logs in math courses and seen them in psychology or science classes. In this book, we use them only for making data behave better. Base 10 logs are the easiest to understand, but natural logs are often used as well. (Either one is fine.) You can think of base 10 logs as roughly one less than the number of digits you need to write the number. So 100, which is the smallest number to require three digits, has a $\log_{10}$ of 2. And 1000 has a $\log_{10}$ of 3. The $\log_{10}$ of 500 is between 2 and 3, but you'd need a calculator to find that it's approximately 2.7. All salaries of "six figures" have $\log_{10}$ between 5 and 6. Logs are incredibly useful for making skewed data more symmetric. But don't worry—nobody does logs without technology and neither should you. Often, remaking a histogram or other display of the re-expressed data is as easy as pushing another button.

## Re-expressing to Equalize Spread Across Groups

Researchers measured the concentration (nanograms per milliliter) of cotinine in the blood of three groups of people: nonsmokers who have not been exposed to smoke, nonsmokers who have been exposed to smoke (ETS), and smokers. Cotinine is left in the blood when the body metabolizes nicotine, so its value is a direct measurement of the effect of passive smoke exposure. The boxplots of the cotinine levels of the three groups tell us that the smokers have higher cotinine levels, but if we want to compare the levels of the passive smokers to those of the nonsmokers, we're in trouble, because on this scale, the cotinine levels for both nonsmoking groups are too low to be seen.

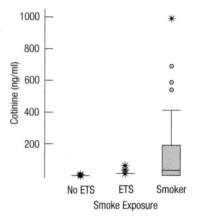

Figure 4.8

**Cotinine levels (nanograms per milliliter) for three groups with different exposures to tobacco smoke** Can you compare the ETS (exposed to smoke) and No ETS groups?

Re-expressing can help alleviate the problem of comparing groups that have very different spreads. For measurements like the cotinine data, whose values can't be negative and whose distributions are skewed to the high end, a good first guess at a re-expression is the logarithm.

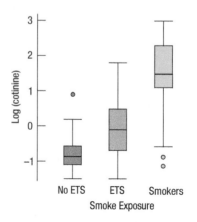

Figure 4.9

**Blood cotinine levels after taking logs** What a difference a log makes!

After taking logs, it is easy to see that the nonsmokers exposed to environmental smoke (the ETS group) do show increased levels of (log) cotinine, although not the high levels found in the blood of smokers.

Notice that the same re-expression has also improved the symmetry of the cotinine distribution for smokers and pulled in most of the apparent outliers in all of the groups. It is not unusual for a re-expression that improves one aspect of data to improve others as well. We'll talk about other ways to re-express data as the need arises throughout the book.

## WHAT CAN GO WRONG?

- **Avoid inconsistent scales.** Parts of displays should be mutually consistent—no fair changing scales in the middle or plotting two variables on different scales but on the same display. When comparing two groups, be sure to compare them on the same scale.
- **Label clearly.** Variables should be identified clearly and axes labeled so a reader knows what the plot displays. Here's a remarkable example of a plot gone wrong. It illustrated a news story about rising college costs. It uses timeplots, but it gives a misleading impression. First think about the story you're being told by this display. Then try to figure out what has gone wrong. What Can Go Wrong?

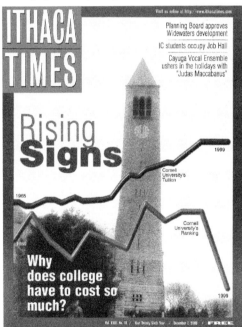

What's wrong? Just about everything.

- The horizontal scales are inconsistent. Both lines show trends over time, but exactly for what years? The tuition sequence starts in 1965, but rankings are graphed from 1989. Plotting them on the same (invisible) scale makes it seem that they're for the same years.
- The vertical axis isn't labeled. That hides the fact that it's inconsistent. Does it graph dollars (of tuition) or ranking (of Cornell University)?

This display violates three of the rules. And it's even worse than that: It violates a rule that we didn't even bother to mention.

- The two inconsistent scales for the vertical axis don't point in the same direction! The line for Cornell's rank shows that it has "plummeted" from 15th place to 6th place in academic rank. Most of us think that's an *improvement*, but that's not the message of this graph.

- **Beware of outliers.** If the data have outliers that are clearly wrong or impossible, you should remove them and report on them. Otherwise, consider summarizing the data both with and without the outliers.

# WHAT HAVE WE LEARNED?

- We've learned the value of comparing groups and looking for patterns among groups and over time.
- We've seen that histograms or stem-and-leaf plots can compare two distributions well, if drawn on the same scale. Boxplots are more effective for comparing several groups.
- We've learned that, when we compare groups, we must compare their shape, center, spreads, and any unusual features.
- We've experienced the value of identifying and investigating outliers. And we've seen that when we group data in different ways, it can allow different cases to emerge as possible outliers.
- We've graphed data that have been measured over time against a time axis and looked for long-term trends.

© 2013 Randall Munroe. Reprinted with permission. All rights reserved.

## TERMS

**Comparing distributions**  When comparing the distributions of several groups using histograms or stem-and-leaf displays, compare their:
- Shape
- Center
- Spread (p. 87)

**Comparing boxplots**  When comparing groups with boxplots:
- Compare the shapes. Do the boxes look symmetric or skewed? Are there differences between groups?
- Compare the medians. Which group has the higher center? Is there any pattern to the medians?
- Compare the IQRs. Which group is more spread out? Is there any pattern to how the IQRs change?
- Using the IQRs as a background measure of variation, do the medians seem to be different, or do they just vary much as you'd expect from the overall variation?
- Check for possible outliers. Identify them if you can and discuss why they might be unusual. Of course, correct them if you find that they are errors. (p. 88)

**Timeplot**  A timeplot displays data that change over time to show long-term patterns and trends. (p. 93)

**Re-express (Transform)**  Applying a simple function (such as a logarithm or square root) to the data can make a skewed distribution more symmetric or equalize spread across groups. (p. 94)

# ON THE COMPUTER

## Comparing Distributions

Most programs for displaying and analyzing data can display plots to compare the distributions of different groups. Typically these are boxplots displayed side-by-side.

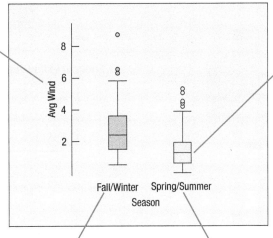

Side-by-side boxplots should be on the same y-axis scale so they can be compared.

Some programs offer a graphical way to assess how much the medians differ by drawing a band around the median or by "notching" the boxes.

Boxes are typically labeled with a group name. Often they are placed in alphabetical order by group name—not the most useful order.

# EXERCISES

**1. In the news** Find an article in a newspaper, magazine, or the Internet that compares two or more groups of data.
   a) Does the article discuss the W's?
   b) Is the chosen display appropriate? Explain.
   c) Discuss what the display reveals about the groups.
   d) Does the article accurately describe and interpret the data? Explain.

**2. In the news II** Find an article in a newspaper, magazine, or the Internet that shows a timeplot.
   a) Does the article discuss the W's?
   b) Is the timeplot appropriate for the data? Explain.
   c) Discuss what the timeplot reveals about the variable.
   d) Does the article accurately describe and interpret the data? Explain.

**3. Time on the Internet** Find data on the Internet (or elsewhere) that give results recorded over time. Make an appropriate display and discuss what it shows.

**4. Groups on the Internet** Find data on the Internet (or elsewhere) for two or more groups. Make appropriate displays to compare the groups, and interpret what you find.

**5. Pizza prices** A company that sells frozen pizza to stores in four markets in the United States (Denver, Baltimore, Dallas, and Chicago) wants to examine the prices that the stores charge for pizza slices. Here are boxplots comparing data from a sample of stores in each market:

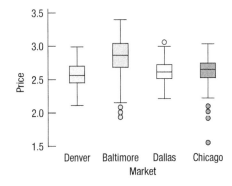

   a) Do prices appear to be the same in the four markets? Explain.
   b) Does the presence of any outliers affect your overall conclusions about prices in the four markets?

6. **Load factors 2019** The Research and Innovative Technology Administration of the Bureau of Transportation Statistics (www.TranStats.bts.gov/Data_Elements.aspx?Data=2) reports load factors (passenger-miles as a percentage of available seat-miles) for commercial airlines for every month from 2002. Here are histograms and summary statistics for the domestic and international load factors for 2002 through September 2019. Compare and contrast the distributions.

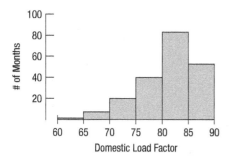

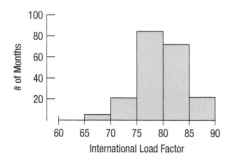

|  | Mean | Median | StdDev | IQR | n |
|---|---|---|---|---|---|
| Domestic Load Factor | 81.0610 | 82.1150 | 5.19554 | 6.89500 | 204 |
| International Load Factor | 79.4891 | 79.6950 | 4.23966 | 5.67500 | 204 |

7. **Hopkins** Below and at the top of the next column are histograms and the five-number summaries for the average wind speeds in the Hopkins Memorial Forest for the year 2011 and the year 2016. Compare these distributions, and be sure to address shape (including outliers if there are any), center, and spread.

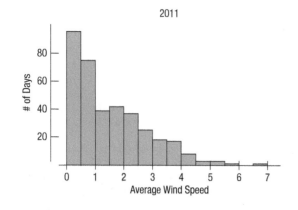

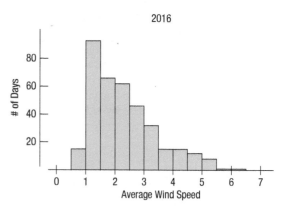

**Summary statistics:**

| Year | Min | Q1 | Median | Q3 | Max |
|---|---|---|---|---|---|
| 2016 | 0.58 | 1.42 | 2.075 | 2.91 | 6.17 |
| 2011 | 0 | 0.46 | 1.12 | 2.28 | 6.73 |

8. **Camping** Here are summary statistics and histograms for the number of campsites at public parks in Wisconsin and Vermont. Write a few sentences comparing the numbers of campsites in these two states' public parks. Be sure to talk about shape (including outliers), center, and spread.

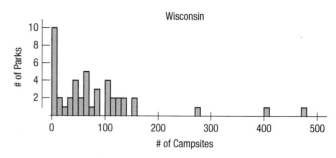

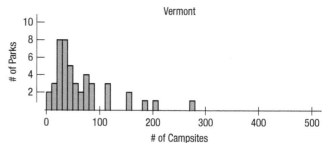

|  | Wisconsin | Vermont |
|---|---|---|
| Count | 45 | 46 |
| Mean | 81.9 | 62.8 |
| Median | 60 | 43.5 |
| StdDev | 96.9 | 56.2 |
| Min | 0 | 0 |
| Max | 472 | 275 |
| Q1 | 14 | 28 |
| Q3 | 108 | 78 |

**9. Cereals** Sugar is a major ingredient in many breakfast cereals. The histogram displays the sugar content as a percentage of weight for 45 brands of cereal. The boxplot compares sugar content for adults' and children's cereals.

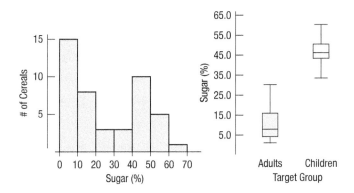

a) What is the range of the sugar contents of these cereals?
b) Describe the shape of the distribution.
c) What aspect of breakfast cereals might account for this shape?
d) Are all children's cereals higher in sugar than adults' cereals?
e) Which group of cereals varies more in sugar content? Explain.

**10. Tendon transfers** People with spinal cord injuries may lose function in some, but not all, of their muscles. The ability to push oneself up is particularly important for shifting position when seated and for transferring into and out of wheelchairs. Surgeons compared two operations to restore the ability to push up in children. The histogram shows scores rating pushing strength two years after surgery and boxplots compare results for the two surgical methods. (Mulcahey, Lutz, Kozen, Betz, "Prospective Evaluation of Biceps to Triceps and Deltoid to Triceps for Elbow Extension in Tetraplegia," *Journal of Hand Surgery*, 28, 6, 2003)

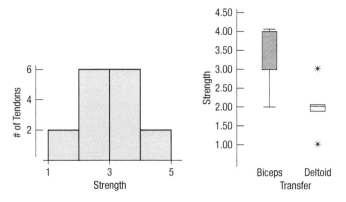

a) Describe the shape of the strength distribution.
b) What is the range of the strength scores?
c) What fact about results of the two procedures is hidden in the histogram?
d) Which method had the higher (better) median score?
e) Was that method always best?
f) Which method produced the most consistent results? Explain.

**11. Population growth, by region** This "back-to-back" stem-and-leaf plot displays two datasets at once—one going to the left, one to the right. The plot compares the percent change in population for two regions of the United States (based on census figures for 2000 and 2010). The fastest growing state was Nevada at 35%. To show the distributions better, this display breaks each stem into two lines, putting leaves 0–4 on one stem and leaves 5–9 on the other.

```
   NE/MW States  |   | S/W States
     44333322 00 | 0 | 1 3 4
    987777666555 | 0 | 7 7 9 9
                2| 1 | 0 0 2 3 3 4 4
                5| 1 | 5 7 8 8 9
                 | 2 | 1
                 | 2 | 5
                 | 3 | 4
                 | 3 | 5
```
Population Growth Rate
(2|1|0 means 12% for a NE/MW state and 10% for a S/W state)

a) Use the data displayed in the stem-and-leaf display to construct comparative boxplots.
b) Write a few sentences describing the difference in growth rates for the two regions of the United States.

**12. Strike outs 2019** Here is a "back-to-back" stemplot that shows two datasets at once—one going to the left and one to the right. The display compares the number of strikeouts (when a player fails to get a hit against the pitcher) for Major League Baseball teams in the National League and the American League in the 2019 season. The number of strikeouts has been rounded to the nearest ten.

| Team Strikeouts | | |
|---:|:---:|:---|
| National League | | American League |
|  | 11 | 7 |
| 1 | 12 | 8 |
| 8661 | 13 | 3348 |
| 7765442 | 14 | 1449 |
| 860 | 15 | 1588 |
|  | 16 | 0 |

Key: 0 | 15 | 1 means 1500 strikeouts for a team in the National League and 1510 strikeouts for a team in the American League.

a) Use the data in the stemplot to create comparative boxplots.
b) Compare the strikeout distribution between the two leagues.

**13. Hospital stays** The U.S. National Center for Health Statistics compiles data on the length of stay by patients in short-term hospitals and publishes its findings in *Vital and Health Statistics*. Data from a sample of 39 male patients and 35 female patients on length of stay (in days) are displayed in the histograms below.

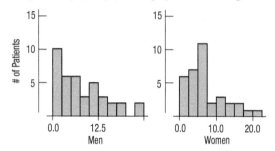

a) What would you suggest be changed about these histograms to make them easier to compare?
b) Describe these distributions by writing a few sentences comparing the duration of hospitalization for men and women.
c) Can you suggest a reason for the peak in women's length of stay?

**14. Deaths 2014** A National Vital Statistics Report (www.cdc.gov/nchs/) provides information on deaths by age, sex, and race. Below are displays of the distributions of ages at death for White and Black males:

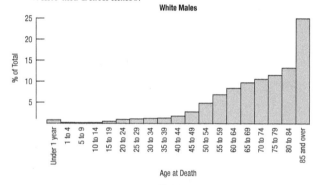

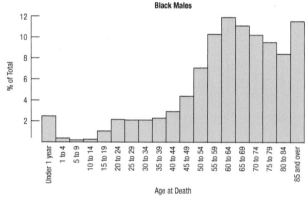

a) Describe the overall shapes of these distributions.
b) How do the distributions differ?
c) Look carefully at the bar definitions. Where do these plots violate the rules for statistical graphs?

**15. Women's basketball** In the next column are boxplots of the points scored during the first 10 games of the season for both Scyrine and Alexandra:

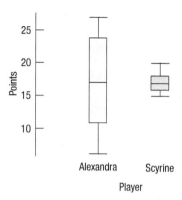

a) Summarize the similarities and differences in their performance so far.
b) The coach can take only one player to the state championship. Which one should she take? Why?

**16. Gas prices 2019** Here are boxplots of weekly gas prices for regular gas in the United States as reported by the U.S. Energy Information Administration for 2009 through October 2019:

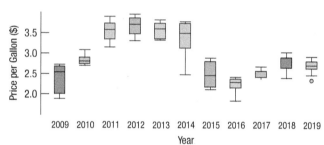

a) Compare the distribution of prices over the eleven years.
b) Compare the stability of prices over the eleven years.

**17. Cost of living 2020, selected countries** To help travelers know what to expect, researchers collected the prices of commodities in 101 countries throughout the world. Here are boxplots comparing the average prices of a bottle of water, a dozen eggs, and a loaf of bread in the 101 countries (prices are all in US$ as of June 2020).

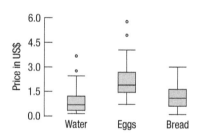

a) In general, which commodity is the most expensive?
b) These are prices for the same countries. But can you tell whether eggs are more expensive than water and bread in all countries? Explain.

**18. Cost of living 2020—cities** Here are the same three prices as in Exercise 17 but for 449 cities around the world. (Prices are all in US$ as of June 2020.)

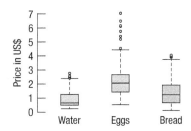

a) In general, which commodity is the most expensive?
b) Is a carton of eggs always more expensive than a bottle of water? Explain.
c) Is a carton of eggs ever more expensive than a loaf of bread? Explain.

**T 19. Fuel economy** The U.S. Department of Energy (www.fueleconomy.gov/feg/download.shtml) provides fuel economy and pollution information on 170 2021 car models. Here are boxplots of *Combined Fuel Economy* (using an average of driving conditions) for the 144 car models whose Class is either compact car, standard pickup truck, or SUV. Summarize what you see about the fuel economies of these three vehicle classes.

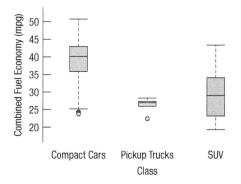

**T 20. Ozone** Ozone levels (in parts per billion, ppb) were recorded at sites in New Jersey monthly between 1926 and 1971. Here are boxplots of the data for each month (over the 46 years), lined up in order (January = 1):

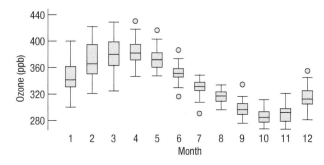

a) In what month was the highest ozone level ever recorded?
b) Which month has the largest IQR?
c) Which month has the smallest range?
d) Write a brief comparison of the ozone levels in January and June.
e) Write a report on the annual patterns you see in the ozone levels.

**21. Load factors by month** The Research and Innovative Technology Administration of the Bureau of Transportation Statistics (www.TranStats.bts.gov) reports load factors (passenger-miles as a percentage of available seat-miles) for commercial airlines for every month from 2000 through 2011. Here is a display of the load factors for international flights by month for the period from 2000 to 2011. Describe the patterns you see.

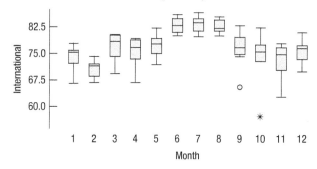

**22. Load factors by year** Here is a display of the load factors (passenger-miles as a percentage of available seat-miles) for domestic airline flights by year. Describe the patterns you see.

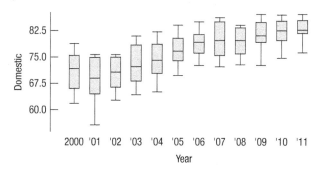

**23. Test scores** Three Statistics classes all took the same test. Histograms and boxplots of the scores for each class are shown below. Match each class with the corresponding boxplot.

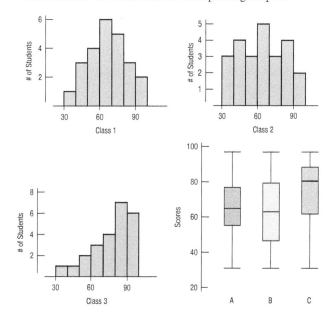

**24. Eye and hair color** A survey of 1021 school-age children was conducted by randomly selecting children from several large urban elementary schools. Two of the questions concerned eye and hair color. In the survey, the following codes were used:

| Hair Color | Eye Color |
|---|---|
| 1 = Blond | 1 = Blue |
| 2 = Brown | 2 = Green |
| 3 = Black | 3 = Brown |
| 4 = Red | 4 = Grey |
| 5 = Other | 5 = Other |

The Statistics students analyzing the data were asked to study the relationship between eye and hair color. They produced this plot:

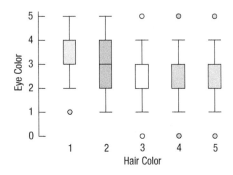

Is their graph appropriate? If so, summarize the findings. If not, explain why not.

**25. Graduation?** A survey of major universities asked what percentage of incoming freshmen usually graduate "on time" in 4 years. Use the summary statistics given to answer the questions that follow.

| | % on Time |
|---|---|
| Count | 48 |
| Mean | 68.35 |
| Median | 69.90 |
| StdDev | 10.20 |
| Min | 43.20 |
| Max | 87.40 |
| Range | 44.20 |
| 25th %tile | 59.15 |
| 75th %tile | 74.75 |

a) Would you describe this distribution as symmetric or skewed? Explain.
b) Are there any outliers? Explain.
c) Create a boxplot of these data.
d) Write a few sentences about the graduation rates.

**26. Vineyards** Here are summary statistics for the sizes (in acres) of Finger Lakes vineyards:

| | |
|---|---|
| Count | 36 |
| Mean | 46.50 |
| StdDev | 47.76 |
| Median | 33.50 |
| IQR | 36.50 |
| Min | 6 |
| Q1 | 18.50 |
| Q3 | 55 |
| Max | 250 |

a) Would you describe this distribution as symmetric or skewed? Explain.
b) Are there any outliers? (The four largest vineyards are: 100, 130, 150, 250.) Explain.
c) Create a boxplot of these data.
d) Write a few sentences about the sizes of the vineyards.

**27. Caffeine** A student study of the effects of caffeine asked volunteers to take a memory test 2 hours after drinking soda. Some drank caffeine-free cola, some drank regular cola (with caffeine), and others drank a mixture of the two (getting a half-dose of caffeine). Here are the 5-number summaries for each group's scores (number of items recalled correctly) on the memory test:

| | n | Min | Q1 | Median | Q3 | Max |
|---|---|---|---|---|---|---|
| No Caffeine | 15 | 16 | 20 | 21 | 24 | 26 |
| Low Caffeine | 15 | 16 | 18 | 21 | 24 | 27 |
| High Caffeine | 15 | 12 | 17 | 19 | 22 | 24 |

a) Describe the W's for these data.
b) Name the variables and classify each as categorical or quantitative.
c) Create parallel boxplots to display these results as best you can with this information.
d) Write a few sentences comparing the performances of the three groups.

**28. SAT scores** Here are the summary statistics for Verbal SAT scores for a high school graduating class:

| | n | Mean | Median | SD | Min | Max | Q1 | Q3 |
|---|---|---|---|---|---|---|---|---|
| Male | 80 | 590 | 600 | 97.2 | 310 | 800 | 515 | 650 |
| Female | 82 | 602 | 625 | 102.0 | 360 | 770 | 530 | 680 |

a) Create parallel boxplots comparing the scores of boys and girls as best you can from the information given.
b) Write a brief report on these results. Be sure to discuss the shape, center, and spread of the scores.
c) There are many variables that could be studied that affect SAT scores. What are some other variables that a researcher might find of interest to explain differences in scores?

**29. Coaster height** Let's revisit the Coaster 2021 data from earlier in this chapter. Here are summary statistics and boxplots for the height of the coasters in feet. The analysis divides the coasters into two groups: steel coasters vs. wooden coasters.

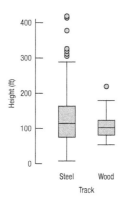

| Track | n | Mean | Std | Median | Min | Max | Q1 | Q3 |
|---|---|---|---|---|---|---|---|---|
| Steel | 225 | 128.2 | 71.6 | 116.5 | 8 | 420 | 75 | 163 |
| Wood | 32 | 106.9 | 36.9 | 100.75 | 55 | 218 | 80 | 123.5 |

a) Compare the distributions.
b) Find the upper outlier fence for both graphs. Do the outliers for the highest coasters appear to be correctly graphed?
c) Explain why the difference between the mean and median is fairly large for both distributions.

**30. Coaster speed** In Exercise 29 we examined the height of roller coasters. Now we shift our focus to the fastest speed of the ride, measured in miles per hour. Separating the steel and the wooden again, here are the results.

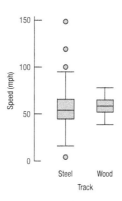

| Track | n | Mean | Std | Median | Min | Max | Q1 | Q3 |
|---|---|---|---|---|---|---|---|---|
| Steel | 225 | 55.6 | 19.6 | 55 | 4.5 | 149.1 | 44.7 | 66 |
| Wood | 32 | 58.2 | 9.1 | 55.95 | 40 | 78.4 | 52.5 | 65 |

a) Compare the distributions.
b) Verify that the steel outliers are correctly drawn.
c) Because steel is stronger than wood, it makes sense that the fastest coasters are made of steel. However, there are other variables that help explain the speed of a roller coaster. What other variables might you consider to help explain speed?

**31. Reading scores** A class of fourth graders takes a diagnostic reading test, and the scores are reported by reading grade level. The 5-number summaries for the 14 boys and 11 girls are shown:

| 5-Number Summaries | | | | | |
|---|---|---|---|---|---|
| Group | Min | Q1 | Median | Q3 | Max |
| Boys | 2.0 | 3.9 | 4.3 | 4.9 | 6.0 |
| Girls | 2.8 | 3.8 | 4.5 | 5.2 | 5.9 |

a) Which group had the highest score?
b) Which group had the greater range?
c) Which group had the greater interquartile range?
d) Which group's scores appear to be more skewed? Explain.
e) Which group generally did better on the test? Explain.
f) If the mean reading level for boys was 4.2 and for girls was 4.6, what is the overall mean for the class?

**32. Rainmakers?** In an experiment to determine whether seeding clouds with silver iodide increases rainfall, 52 clouds were randomly assigned to be seeded or not. The amount of rain they generated was then measured (in acre-feet). Here are the summary statistics:

| | n | Mean | Median | SD | IQR | Q1 | Q3 |
|---|---|---|---|---|---|---|---|
| Unseeded | 26 | 164.59 | 44.20 | 278.43 | 138.60 | 24.40 | 163 |
| Seeded | 26 | 441.98 | 221.60 | 650.79 | 337.60 | 92.40 | 430 |

a) Which of the summary statistics are most appropriate for describing these distributions? Why?
b) Do you see any evidence that seeding clouds may be effective? Explain.

**33. Industrial experiment** Engineers at a computer production plant tested two methods for accuracy in drilling holes into a PC board. They tested how fast they could set the drilling machine by running 10 boards at each of two different speeds. To assess the results, they measured the distance (in inches) from the center of a target on the board to the center of the hole. The data and summary statistics are shown in the table:

| Distance (in.) | Speed | Distance (in.) | Speed |
|---|---|---|---|
| 0.000101 | Fast | 0.000098 | Slow |
| 0.000102 | Fast | 0.000096 | Slow |
| 0.000100 | Fast | 0.000097 | Slow |
| 0.000102 | Fast | 0.000095 | Slow |
| 0.000101 | Fast | 0.000094 | Slow |
| 0.000103 | Fast | 0.000098 | Slow |
| 0.000104 | Fast | 0.000096 | Slow |
| 0.000102 | Fast | 0.975600 | Slow |
| 0.000102 | Fast | 0.000097 | Slow |
| 0.000100 | Fast | 0.000096 | Slow |
| Mean | 0.000102 | Mean | 0.097647 |
| StdDev | 0.000001 | StdDev | 0.308481 |

Write a report summarizing the findings of the experiment. Include appropriate visual and verbal displays of the distributions, and make a recommendation to the engineers if they are most interested in the accuracy of the method.

**34. Cholesterol** A study examining the health risks of smoking measured the cholesterol levels of people who had smoked for at least 25 years and people of similar ages who had smoked for no more than 5 years and then stopped. Create appropriate graphical displays for both groups, and write a brief report comparing their cholesterol levels. Here are the data:

| Smokers | | | | Ex-Smokers | | |
|---|---|---|---|---|---|---|
| 225 | 211 | 209 | 284 | 250 | 134 | 300 |
| 258 | 216 | 196 | 288 | 249 | 213 | 310 |
| 250 | 200 | 209 | 280 | 175 | 174 | 328 |
| 225 | 256 | 243 | 200 | 160 | 188 | 321 |
| 213 | 246 | 225 | 237 | 213 | 257 | 292 |
| 232 | 267 | 232 | 216 | 200 | 271 | 227 |
| 216 | 243 | 200 | 155 | 238 | 163 | 263 |
| 216 | 271 | 230 | 309 | 192 | 242 | 249 |
| 183 | 280 | 217 | 305 | 242 | 267 | 243 |
| 287 | 217 | 246 | 351 | 217 | 267 | 218 |
| 200 | 280 | 209 | | 217 | 183 | 228 |

**35. MPG** A consumer organization wants to compare gas mileage figures for several models of cars made in the United States with autos manufactured in other countries. The data for a random sample of cars classified as "midsize" are shown in the table.

| Gas Mileage (mpg) | Country | Gas Mileage (mpg) | Country |
|---|---|---|---|
| 22 | U.S. | 17 | Other |
| 39 | U.S. | 26 | Other |
| 39 | U.S. | 18 | Other |
| 22 | U.S. | 20 | Other |
| 22 | U.S. | 24 | Other |
| 21 | U.S. | 22 | Other |
| 29 | U.S. | 28 | Other |
| 21 | U.S. | 23 | Other |
| 21 | U.S. | 30 | Other |
| 24 | U.S. | 19 | Other |
| 23 | U.S. | 27 | Other |
| 17 | U.S. | 21 | Other |
| 30 | U.S. | 22 | Other |
| 19 | U.S. | 29 | Other |
| 23 | U.S. | 29 | Other |
| 21 | U.S. | 28 | Other |
| 24 | U.S. | 26 | Other |
| | | 21 | Other |
| | | 20 | Other |
| | | 21 | Other |
| | | 50 | Other |
| | | 24 | Other |
| | | 35 | Other |

a) Create graphical displays for these two groups.
b) Write a few sentences comparing the distributions.

**36. Baseball 2019** American League baseball teams play their games with the designated hitter rule, meaning that pitchers do not bat. The league believes that replacing the pitcher, typically a weak hitter, with another player in the batting order produces more runs and generates more interest among fans. Following are the average number of runs scored by each team in the 2019 season:

| American League | | National League | |
|---|---|---|---|
| Team | Runs per Game | Team | Runs per Game |
| NY Yankees | 5.82 | Los Angeles Dodgers | 5.47 |
| Minnesota Twins | 5.80 | Washington Nationals | 5 39 |
| Houston Astros | 5.68 | Atlanta Braves | 5.28 |
| Boston Red Sox | 5.56 | Colorado Rockies | 5.15 |
| Oakland Athletics | 5.22 | Arizona Diamondbacks | 5.02 |
| Texas Rangers | 5.00 | Chicago Cubs | 5 02 |
| Cleveland Indians | 4.75 | NY Mets | 4.88 |
| Los Angeles Angels | 4.75 | Philadelphia Phillies | 4.78 |
| Tampa Bay Rays | 4.75 | Milwaukee Brewers | 4.75 |
| Seattle Mariners | 4.68 | St Louis Cardinals | 4.72 |
| Baltimore Orioles | 4.50 | Pittsburgh Pirates | 4.68 |
| Toronto Blue Jays | 4.48 | Cincinnati Reds | 4.33 |
| Chicago White Sox | 4.40 | San Diego Padres | 4.21 |
| Kansas City Royals | 4.27 | San Francisco Giants | 4.19 |
| Detroit Tigers | 3.61 | Miami Marlins | 3.80 |

a) Create an appropriate graphical display of these data.
b) Write a few sentences comparing the average number of runs scored per game in the two leagues. (Remember: shape, center, spread, unusual features!)
c) The runs per game leaders were the Yankees and the Dodgers in the American and National Leagues, respectively. Did either of those teams score an unusually large number of runs per game? Explain briefly.

**37. Time travel** Using the **Census at School** database, 50 students were randomly sampled. Below is a boxplot showing the time it took the 24 boys and the 26 girls to travel to school.

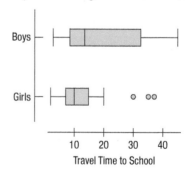

a) Compare the distribution for boys to the distribution for girls.

One could wonder if the differences in the distributions are a real difference, or simply a difference that is chance variation in the random sample we selected. To investigate this question, we selected four new random samples, each of size 50. Graphs of those data are on the next page.

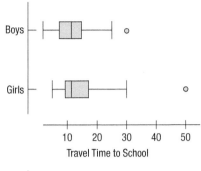

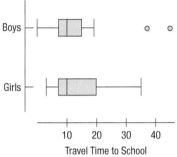

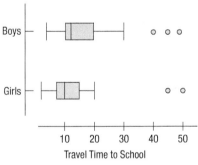

b) What do you observe about differences in travel time to school between boys and girls? Do you think there are real differences between the groups?

38. React! As part of the **Census at Schools** database, students complete a test of reaction time. Students click a button as soon as they see a square change color. As you can see from the dotplots below, most students click the button just under one second.

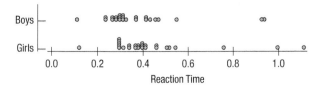

a) Compare the distributions for boys and girls.

As usual, we are curious if the differences between boys and girls are real, or if they simply occurred in the random sample we chose. Here are data displays from four more random samples;

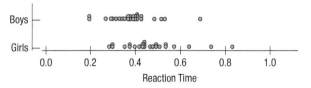

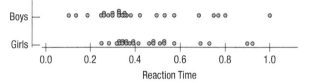

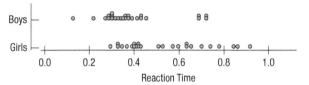

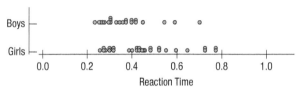

b) Comparing the five random samples, do you think there is a real difference in reaction times between boys and girls? (If there is a difference, it may be a difference between boys' and girls' inclinations to report the truth, and not between actual reaction times. The times were measured by a computer but entered in the database by the students, on the "honor system".)

39. Fruit flies Researchers tracked a population of 1,203,646 fruit flies, counting how many died each day for 171 days. Here are three timeplots offering different views of these data. One shows the number of flies alive on each day, one the number who died that day, and the third the mortality rate—the fraction of the number alive who died. On the last day studied, the last 2 flies died, for a mortality rate of 1.0.

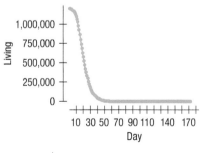

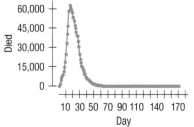

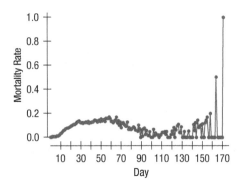

a) On approximately what day did the most flies die?
b) On what day during the first 100 days did the largest *proportion* of flies die?
c) When did the number of fruit flies alive stop changing very much from day to day?

T 40. Drunk driving 2019 Accidents involving drunk drivers account for about 30–40% of all deaths on the nation's highways. The table below tracks the number of alcohol-related fatalities for 38 years. (www.iii.org)

| Year | Deaths (thousands) | Year | Deaths (thousands) |
| --- | --- | --- | --- |
| 1982 | 26.2 | 2001 | 17.4 |
| 1983 | 24.6 | 2002 | 17.5 |
| 1984 | 24.8 | 2003 | 17.1 |
| 1985 | 23.2 | 2004 | 16.9 |
| 1986 | 25 | 2005 | 16.9 |
| 1987 | 24.1 | 2006 | 15.8 |
| 1988 | 23.8 | 2007 | 15.4 |
| 1989 | 22.4 | 2008 | 13.8 |
| 1990 | 22.6 | 2009 | 10.8 |
| 1991 | 20.2 | 2010 | 10.1 |
| 1992 | 18.3 | 2011 | 9.9 |
| 1993 | 17.9 | 2012 | 10.3 |
| 1994 | 17.3 | 2013 | 10.1 |
| 1995 | 17.7 | 2014 | 9.9 |
| 1996 | 17.7 | 2015 | 10.3 |
| 1997 | 16.7 | 2016 | 11 |
| 1998 | 16.7 | 2017 | 10.9 |
| 1999 | 16.6 | 2018 | 10.7 |
| 2000 | 17.4 | 2019 | 10.1 |

a) Create a stem-and-leaf display or a histogram of these data.
b) Create a timeplot.
c) Using features apparent in the stem-and-leaf display (or histogram) and the timeplot, write a few sentences about deaths caused by drunk driving.

T 41. Assets Here is a histogram of the assets (in millions of dollars) of 79 companies chosen from the *Forbes* list of the nation's top corporations:

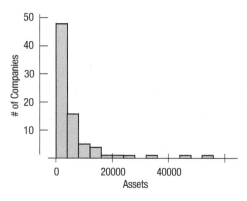

a) What aspect of this distribution makes it difficult to summarize, or to discuss, center and spread?
b) What would you suggest doing with these data if we want to understand them better?

T 42. Music library Students were asked how many songs they had in their digital music libraries. Here's a display of the responses:

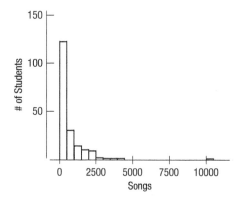

a) What aspect of this distribution makes it difficult to summarize, or to discuss, center and spread?
b) What would you suggest doing with these data if we want to understand them better?

T 43. Assets again Here are the same data you saw in Exercise 41 after re-expressions as the square root of assets and the logarithm of assets:

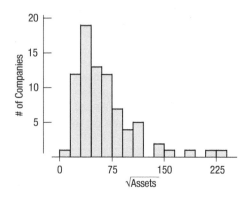

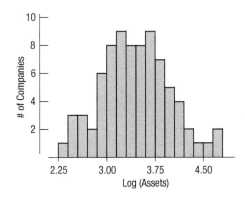

a) Which re-expression do you prefer? Why?
b) In the square root re-expression, what does the value 50 actually indicate about the company's assets?
c) In the logarithm re-expression, what does the value 3 actually indicate about the company's assets?

**44. Rainmakers again** The table lists the amount of rainfall (in acre-feet) from the 26 clouds seeded with silver iodide discussed in Exercise 32:

| Rainfall (acre-ft) | | | | | | |
|---|---|---|---|---|---|---|
| 2745 | 703 | 302 | 242 | 119 | 40 | 7 |
| 1697 | 489 | 274 | 200 | 118 | 32 | 4 |
| 1656 | 430 | 274 | 198 | 115 | 31 | |
| 978 | 334 | 255 | 129 | 92 | 17 | |

a) Why is acre-feet a good way to measure the amount of precipitation produced by cloud seeding?
b) Plot these data, and describe the distribution.
c) Create a re-expression of these data that produces a more advantageous distribution.
d) Explain what your re-expressed scale means.

**45. Stereograms** Stereograms appear to be composed entirely of random dots. However, they contain separate images that a viewer can "fuse" into a three-dimensional (3D) image by staring at the dots while defocusing the eyes. An experiment was performed to determine whether knowledge of the embedded image affected the time required for subjects to fuse the images. One group of subjects (group NV) received no information or just verbal information about the shape of the embedded object. A second group (group VV) received both verbal information and visual information (specifically, a drawing of the object). The experimenters measured how many seconds it took for the subject to report that he or she saw the 3D image.

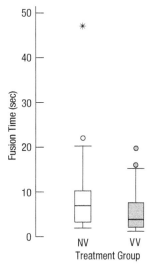

a) What two variables are discussed in this description?
b) For each variable, is it quantitative or categorical? If quantitative, what are the units?
c) The boxplots compare the fusion times for the two treatment groups. Write a few sentences comparing these distributions. What does the experiment show?

**46. Stereograms, revisited** Because of the skewness of the distributions of fusion times described in Exercise 45, we might consider a re-expression. Here are the boxplots of the *log* of fusion times. Is it better to analyze the original fusion times or the log fusion times? Explain.

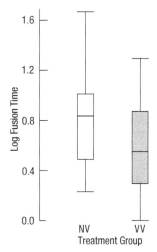

### JUST CHECKING

Answers

1. The percentages of on-time arrivals have a unimodal, symmetric distribution centered at about 79%. In most months, between 70% and 86% of the flights arrived on time.

2. The boxplot of percentage on-time arrivals nominates several months as low outliers.

3. The boxplots by month show a strong seasonal pattern. Flights are more likely to be on time in the spring and fall and less likely to be on time in the winter and summer. One possible reason for the pattern is snowstorms in the winter and thunderstorms in the summer.

# 5

# The Standard Deviation as a Ruler and the Normal Model

**GRADING ON A CURVE**
If you score 79% on an exam, what grade should you get? One teaching philosophy looks only at the raw percentage, 79, and bases the grade on that alone. Another looks at your *relative* performance and bases the grade on how you did compared with the rest of the class. Teachers and students often debate which method is better.

The women's heptathlon in the Olympics consists of seven track-and-field events: the 200 m and 800 m runs, 100 m high hurdles, shot put, javelin, high jump, and long jump. In the 2016 Olympics, Nafissatou Thiam of Belgium won the long jump with a jump of 6.58 meters, and Katarina Johnson-Thompson of Great Britain won the 200 m run with a time of 23.26 seconds.

Each contestant is awarded points for each event based on her performance. So, which performance deserves more points? It's not clear how to compare them. They aren't measured in the same units, or even in the same *direction* (longer jumps are better but shorter times are better).

To see which value is more extraordinary, we need a way to judge them against the background of data they come from. Chapter 3 discussed the standard deviation, but really didn't tell you what it's for. Statisticians use the standard deviation as a ruler, to judge how far a value is from the center of the distribution. Knowing that a value is 5 *units* higher than the mean doesn't tell us much unless we know the context, but knowing it's 5 *standard deviations* above the mean tells us that it's truly extraordinary.

## The Standard Deviation as a Ruler

To compare the two Olympic events, let's start with a picture. We'll use stem-and-leaf displays because they show individual values, and because it is easy to orient them either high-to-low or low-to-high so the best performances can be at the top of both displays.

### Figure 5.1

Stem-and-leaf displays for the 200 m race and the long jump in the 2016 Olympic heptathlon. Katarina Johnson-Thompson (green scores) won the 200 m, and Nafissatou Thiam (red scores) won the long jump. The stems for the 200 m race run from faster to slower and the stems for the long jump from longer to shorter so that the best scores are at the top of each display. Note: The stem-and-leaf display on the left uses two stems for each ten values (leaves 0–4 on one and 5–9 on the other). The other display uses five stems for each meter (0/1 labeled "O," 2/3 labeled "T," 4/5 "F," 6/7 "S," and "*" for the last two). These alternatives, along with the standard form, offer three alternatives for stem-and-leaf displays.

```
    200 m Run              Long Jump
Stem    Leaf          Stem    Leaf
 23  | 24              6F  | 44555
 23  | 799             6T  | 222233333
 24  | 011123334       6O  | 0000111111
 24  | 66677999        5*  | 889
 25  | 012334          5S  | 7
 25  |                 5F  | 5
 26  | 3
 23|2 = 23.2 seconds   6F|5 = 6.5 meters
```

Which of the two winning scores is the better one? Thiam's winning 6.58 m long jump was 1.66 SDs better (longer) than the mean. Johnson-Thompson's winning time of 23.26 sec was 2.02 SDs better (faster) than the mean time. That's more impressive. (Be patient: We'll show you the details of these calculations on the next page.)

We could turn each of the seven events' results into a score in this way, and add them up for each athlete to determine the winner. Olympic judges actually use a point system based on similar calculations (but based on performances from many competitions and a larger pool of contestants).

## Standardizing with z-Scores

Expressing a distance from the mean in standard deviations *standardizes* the performances. To **standardize** a value, we subtract the mean and then divide this difference by the standard deviation:

$$z = \frac{y - \bar{y}}{s}.$$

The values are called **standardized values** and are commonly denoted with the letter $z$. Usually we just call them **z-scores**.

*NOTATION ALERT*
We always use the letter $z$ to denote values that have been standardized with the mean and standard deviation.

z-scores measure the distance of a value from the mean in standard deviations. A z-score of 2 says that a data value is 2 standard deviations above the mean. It doesn't matter whether the original variable was measured in fathoms, dollars, or carats; those units don't apply to z-scores. Data values below the mean have negative z-scores, so a z-score of $-1.6$ means that the data value was 1.6 standard deviations below the mean. Of course, regardless of the direction, the farther a data value is from the mean, the more unusual it is, so a z-score of $-1.3$ is more extraordinary than a z-score of 1.2.

### FOR EXAMPLE
#### Standardizing Skiing Times

The men's super combined skiing event consists of two races: a downhill and a slalom. Thirty seven skiers finished both races. Times for the two events are added together, and the skier with the lowest total time wins. There is ongoing debate whether this format favors the technical slalom skiers or the downhill specialists. At the 2018 Winter Olympics in PyeongChang, the mean slalom time was 50.13 seconds with a standard deviation of 3.58 seconds. The mean downhill time was 81.72 seconds with a standard deviation of 1.46 seconds. Marcel Hirscher of Austria, who won the gold medal with a combined time of 126.52 seconds, skied the slalom in 45.96 seconds and the downhill in 80.56 seconds.

**QUESTION:** On which race did he do better compared to the competition?

**ANSWER:** $z_{slalom} = \dfrac{y - \bar{y}}{s} = \dfrac{45.96 - 50.13}{3.58} = -1.16$

$z_{downhill} = \dfrac{y - \bar{y}}{s} = \dfrac{81.84 - 81.72}{1.56} = -0.82$

Keeping in mind that better times are below the mean, Hirscher's slalom time of 1.16 SDs below the mean is more remarkable than his downhill time, which was 0.82 SDs below the mean.

Here are the calculations for the women's heptathlon. For each event, first subtract the mean of all participants from the individual's score. Then divide by the standard deviation to get the z-score. For example, Nafissatou's 6.58 m long jump is compared to the mean of 6.17 m for all 29 heptathletes: $(6.58 - 6.17 = 0.41 \text{ m})$. Dividing this difference by the standard deviation gives her a z-score of $0.41/0.247 = 1.66$.

|  |  | Event | |
|---|---|---|---|
|  |  | Long Jump | 200 m Run |
|  | Mean | 6.17 m | 24.58 s |
|  | SD | 0.247 m | 0.654 s |
| Thiam | Performance | 6.58 m | 25.10 s |
|  | z-score | $(6.58 - 6.17)/0.247 = 1.66$ | $(25.10 - 24.58)/0.654 = 0.795$ |
|  | Total for two events | \multicolumn{2}{c}{$1.66 - 0.795 = \mathbf{0.865}$} |
| Johnson-Thompson | Performance | 6.51 m | 23.26 s |
|  | z-score | $(6.51 - 6.17)/0.247 = 1.38$ | $(23.26 - 24.58)/0.654 = -2.02$ |
|  | Total for two events | \multicolumn{2}{c}{$1.38 + 2.02 = \mathbf{3.40}$} |

When we combine the two events, we change the sign of the run because faster is *better*. After two events, Thiam's two scores total only 0.865 compared with Johnson-Thompson's 3.40. But Thiam went on to take the gold once all seven events were counted, placing first in the high jump and the shot put as well.

### FOR EXAMPLE
### Combining z-Scores

Before winning the gold in the men's super combined in PyeongChang, Marcel Hirscher had earlier earned a gold medal in the super-G slalom. Alexis Pinturault, the winner of the bronze medal in the super-G in 2018 and 2014, was just over half a second behind him in the combined slalom, with a time of 46.67. But he beat Hirscher in the downhill by almost a half a second at 80.28. His combined time was 0.23 seconds slower and he finished in second with the silver medal. But the standard deviation in the downhill is much smaller than the SD of the slalom, so that half second may be more remarkable than the half second he lost in the slalom.

**QUESTION:** Would the placement have changed if each event had been treated equally by standardizing each and adding the standardized scores?

**ANSWER:** We've seen that Hirscher's z-scores for slalom and downhill were $-1.16$ and $-0.82$, respectively. That's a total of $-1.98$. Pinturault's z-scores were:

$z_{slalom} = \dfrac{y - \bar{y}}{s} = \dfrac{46.47 - 50.13}{3.58} = -1.021$

$z_{downhill} = \dfrac{y - \bar{y}}{s} = \dfrac{80.28 - 81.72}{1.56} = -1.00$

So his total z-score was $-1.021 - 1.00 = -2.0$. That's slightly better than Hirscher. Using the standardized scores would have given Pinturault the gold for his relatively faster downhill. (Do you think this supports the view of those who say that the super combined favors slalom specialists?)

### JUST CHECKING

1. Your Statistics teacher has announced that the lower of your two tests will be dropped. You got a 90 on test 1 and an 80 on test 2. You're all set to drop the 80 until she announces that she grades "on a curve." She standardized the scores in order to decide which is the lower one. The mean on the first test was 88 with a standard deviation of 4 and the mean on the second was 75 with a standard deviation of 5.
   a) Which one will be dropped?
   b) Does this seem "fair"?

2. A distribution of incomes shows a strongly right skewed distribution. Will the maximum or the minimum income be closer to the mean? Explain.

When we standardize data to get a z-score, we do two things. First, we shift the data by subtracting the mean. Then, we rescale the values by dividing by their standard deviation. We often shift and rescale data. What happens to a grade distribution if *everyone* gets a five-point bonus? If we switch from feet to meters, what happens to the distribution of heights of students in your class? Even though your intuition probably tells you the answers to these questions, we need to look at exactly how shifting and rescaling work.

## Shifting Data: Move the Center

Since the 1960s, the Centers for Disease Control's National Center for Health Statistics has been collecting health and nutritional information on people of all ages and backgrounds. The National Health and Nutrition Examination Survey (NHANES) 2001–2002[1] measured a wide variety of variables, including body measurements, cardiovascular fitness, blood chemistry, and demographic information on more than 11,000 individuals.

Included in this group were 80 men between 19 and 24 years old of average height (between 5'8" and 5'10" tall). Here are a histogram and boxplot of their weights:

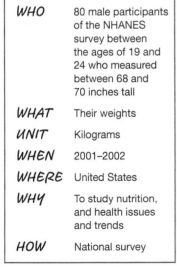

| WHO | 80 male participants of the NHANES survey between the ages of 19 and 24 who measured between 68 and 70 inches tall |
| WHAT | Their weights |
| UNIT | Kilograms |
| WHEN | 2001–2002 |
| WHERE | United States |
| WHY | To study nutrition, and health issues and trends |
| HOW | National survey |

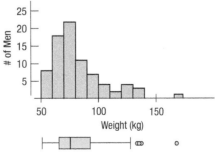

Figure 5.2
Histogram and boxplot for the men's weights. The shape is skewed to the right with several high outliers.

[1] www.cdc.gov/nchs/nhanes.htm

Their mean weight is 82.36 kg. For this age and height group, the National Institutes of Health recommends a maximum healthy weight of 74 kg, but we can see that some of the men are heavier than the recommended weight. To compare their weights to the recommended maximum, we could subtract 74 kg from each of their weights. What would that do to the center, shape, and spread of the histogram? Here's the picture:

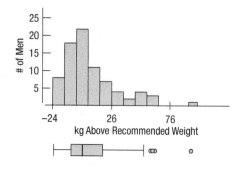

Figure 5.3
Subtracting 74 kilograms shifts the entire histogram down but leaves the spread and the shape exactly the same.

On average, they weigh 82.36 kg, so on average they're 8.36 kg overweight. And, after subtracting 74 from each weight, the mean of the new distribution is $82.36 - 74 = 8.36$ kg. In fact, when we **shift** the data by adding (or subtracting) a constant to each value, all measures of position (center, percentiles, min, max) will increase (or decrease) by the same constant.

What about the spread? What does adding or subtracting a constant value do to the spread of the distribution? Look at the two histograms again. Adding or subtracting a constant changes each data value equally, so the entire distribution just shifts. Its shape doesn't change and neither does the spread. None of the measures of spread we've discussed—not the range, not the IQR, not the standard deviation—changes.

*Adding (or subtracting) a constant to every data value adds (or subtracts) the same constant to measures of position, but leaves measures of spread unchanged.*

> **SHIFTING HEIGHTS**
>
> Doctors' height and weight charts sometimes give ideal weights for various heights that include 2-inch heels. If the mean height of adult women is 66 inches including 2-inch heels, what is the mean height of women without shoes? Each woman is shorter by 2 inches when barefoot, so the mean is decreased by 2 inches, to 64 inches.

# Rescaling Data: Adjust the Scale

Not everyone thinks naturally in metric units. Suppose we want to look at the weights in pounds instead. We'd have to **rescale** the data. Because there are about 2.2 pounds in every kilogram, we'd convert the weights by multiplying each value by 2.2. Multiplying or dividing each value by a constant changes the measurement units. Here are histograms of the two weight distributions, plotted on the same numeric scale, so you can see the effect of multiplying:

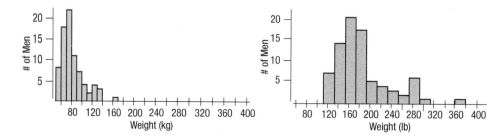

Figure 5.4
**Men's weights in both kilograms and pounds.** How do the distributions and numerical summaries change?

What happens to the shape of the distribution? Although the histograms don't look exactly alike, we see that the shape really hasn't changed: Both are unimodal and skewed to the right. (The apparent minor differences occur because the bins in the two histograms don't correspond to exactly the same intervals of weights.)

What happens to the mean? Not too surprisingly, it gets multiplied by 2.2 as well. The men weigh 82.36 kg on average, which is 181.19 pounds. As the boxplots and 5-number summaries show, all measures of position act the same way. They all get multiplied by this same constant.

What happens to the spread? Take a look at the boxplots. The spread in pounds (on the right) is larger. How much larger? If you guessed 2.2 times, you've figured out how measures of spread get rescaled.

**Figure 5.5**
The boxplots (plotted using the same axis) show the weights measured in kilograms (on the left) and pounds (on the right). Because 1 kg is 2.2 lb, all the points in the right box are 2.2 times larger than the corresponding points in the left box. So each measure of position and spread is 2.2 times as large when measured in pounds rather than kilograms.

|  | Weight (kg) | Weight (lb) |
|---|---|---|
| Min | 54.3 | 119.46 |
| Q1 | 67.3 | 148.06 |
| Median | 76.85 | 169.07 |
| Q3 | 92.3 | 203.06 |
| Max | 161.5 | 355.30 |
| IQR | 25 | 55 |
| SD | 22.27 | 48.99 |

When we multiply (or divide) all the data values by any constant, all measures of position (such as the mean, median, and percentiles) and measures of spread (such as the range, the IQR, and the standard deviation) are multiplied (or divided) by that same constant.

### FOR EXAMPLE

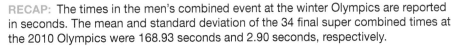

#### Rescaling the Men's Combined Times

**RECAP:** The times in the men's combined event at the winter Olympics are reported in seconds. The mean and standard deviation of the 34 final super combined times at the 2010 Olympics were 168.93 seconds and 2.90 seconds, respectively.

**QUESTION:** Suppose instead that we had reported the times in minutes—that is, that each individual time was divided by 60. What would the resulting mean and standard deviation be?

**ANSWER:** Dividing all the times by 60 would divide both the mean and the standard deviation by 60:

Mean = 168.93/60 = 2.816 minutes;   SD = 2.90/60 = 0.048 minute.

### JUST CHECKING

3. In 1995 the Educational Testing Service (ETS) adjusted the scores of SAT tests. Before ETS recentered the SAT Verbal test to 500, the mean of all test scores was 450.

   a) How would adding 50 points to each score affect the mean?
   b) The standard deviation was 100 points. What would the standard deviation be after adding 50 points?
   c) Suppose we drew boxplots of test takers' scores a year before and a year after the recentering. How would the boxplots of the two years differ?

4. A company manufactures wheels for in-line skates. The diameters of the wheels have a mean of 3 inches and a standard deviation of 0.1 inches. Because so many of their customers use the metric system, the company decided to report their production statistics in millimeters (1 inch = 25.4 mm). They report that the standard deviation is now 2.54 mm. A corporate executive is worried about this increase in variation. Should he be concerned? Explain.

# Back to z-Scores

Standardizing data into z-scores is just shifting them by the mean and rescaling them by the standard deviation. Now we can see how standardizing affects the distribution. When we subtract the mean of the data from every data value, we shift the mean to zero. As we have seen, such a shift doesn't change the standard deviation.

When we *divide* each of these shifted values by *s*, however, the standard deviation should be divided by *s* as well. Because the standard deviation was *s* to start with, the new standard deviation becomes 1.

How, then, does standardizing affect the distribution of a variable? Let's consider the three aspects of a distribution: the shape, center, and spread.

> *Standardizing into z-scores does not change the **shape** of the distribution of a variable.*
> *Standardizing into z-scores changes the **center** by making the mean 0.*
> *Standardizing into z-scores changes the **spread** by making the standard deviation 1.*

**z-Scores**
z-scores have mean 0 and standard deviation 1.

## STEP-BY-STEP EXAMPLE
### Working with Standardized Variables

Many colleges and universities require applicants to submit scores on standardized tests such as the SAT Evidence-Based Reading and Writing (EBRW) and Math tests. The college your little sister wants to apply to says that while there is no minimum score required, the middle 50% of its students have combined SAT scores between 1100 and 1350. You'd feel confident if you knew her score was in the college's top 25%, but unfortunately she took the ACT test, an alternative standardized test.

**QUESTION:** How high does her ACT need to be to make it into the top quarter of equivalent SAT scores?

To answer that question you'll have to standardize all the scores, so you'll need to know the mean and standard deviations of scores for some group on both tests. The college doesn't report the mean or standard deviation for its applicants on either test, so we'll use the group of all test takers nationally. For college-bound seniors, the average combined SAT score is about 1083 and the standard deviation is about 191 points. For the same group, the ACT average is 20.8 with a standard deviation of 4.8.

---

**THINK** **PLAN** State what you want to find out.

**VARIABLES** Identify the variables and report the W's (if known).

Check the appropriate conditions.

I want to know what ACT score corresponds to the upper-quartile SAT score. I know the mean and standard deviation for both the SAT and ACT scores based on all test takers, but I have no individual data values.

✓ **Quantitative Data Condition:** Scores for both tests are quantitative but have no meaningful units other than points.

---

**SHOW** **MECHANICS** Standardize the variables.

The middle 50% of SAT scores at this college fall between 1100 and 1350 points. To be in the top quarter, my sister would have to have a score of at least 1350. That's a z-score of

$$z = \frac{(1350 - 1083)}{191} = 1.40$$

So an SAT score of 1350 is 1.40 standard deviations above the mean of all test takers.

The *y*-value we seek is *z* standard deviations above the mean.

For the ACT, 1.40 standard deviations above the mean is $20.8 + 1.40(4.8) = 27.52$.

| TELL | CONCLUSION Interpret your results in context. | Normal models estimate that to be in the top quarter of applicants in terms of combined SAT score, she'd need to have an ACT score of at least 27.52. |

# When Is a *z*-Score BIG?

> **IS NORMAL NORMAL?**
> Don't be misled. The name "Normal" doesn't mean that these are the *usual* shapes for histograms. The name follows a tradition of positive thinking in Mathematics and Statistics in which functions, equations, and relationships that are easy to work with or have other nice properties are called "normal," "common," "regular," "natural," or similar terms. It's as if by calling them ordinary, we could make them actually occur more often and simplify our lives.

Champion runners turn in exceptionally low times, and fishermen brag about "the big one."[2] Not only do things that are unusually large or unusually small catch our interest, they can provide especially important information about whatever we are investigating. Being far from typical, they have *z*-scores far from 0. How far from 0 does a *z*-score have to be to be interesting or unusual? There is no universal standard, but the larger the score is (negative or positive), the more unusual it is. We know that 50% of the data lie between the quartiles. For symmetric data, the standard deviation is usually a bit smaller than the IQR, and it's not uncommon for most of the data to have z-scores between −1 and 1. But no matter what the shape of the distribution, a *z*-score of 3 (plus or minus) or more is rare, and a *z*-score of 6 or 7 shouts out for attention.

To say more about how big we expect a *z*-score to be, we need to *model* the data's distribution. A model will let us say much more precisely how often we'd be likely to see *z*-scores of different sizes. Of course, like all models of the real world, the model will be wrong—wrong in the sense that it can't match reality exactly. But it can still be useful. Like a physical model, it's something we can look at and manipulate in order to learn more about the real world.

Models help our understanding in many ways. Just as a model of an airplane in a wind tunnel can give insights even though it doesn't show every rivet,[3] models of data give us summaries that we can learn from and use, even though they don't fit each data value exactly. It's important to remember that they're only *models* of reality and not reality itself. But without models, what we can learn about the world at large is limited to only what we can say about the data we have at hand.

> **NOTATION ALERT**
> $N(\mu, \sigma)$ always denotes a Normal model. The $\mu$, pronounced "myoo," is the Greek letter for "m" and always represents the mean in a model. The $\sigma$, sigma, is the lowercase Greek letter for "s" and always represents the standard deviation in a model.

There is no universal standard for *z*-scores, but there is a model that shows up over and over in Statistics. You may have heard of "bell-shaped curves." Statisticians call them Normal models. **Normal models** are appropriate for many distributions whose shapes are unimodal and roughly symmetric. For these distributions, they provide a measure of how extreme a *z*-score is. Fortunately, there is a Normal model for every possible combination of mean and standard deviation. We write $N(\mu, \sigma)$ to represent a Normal model with a mean of $\mu$ and a standard deviation of $\sigma$. Why the Greek? Well, *this* mean and standard deviation are not numerical summaries of data. They are part of the model. They don't come from the data. Rather, they are numbers that we choose to help specify the model. Such numbers are called **parameters** of the model.

We don't want to confuse the parameters with summaries of the data such as $\bar{y}$ and $s$, so we use special symbols. In Statistics, we almost always use Greek letters for parameters. By contrast, summaries of data are called **statistics** and are usually written with Latin letters.

If we model data with a Normal model and standardize them using the corresponding $\mu$ and $\sigma$, we still call the standardized value a **z-score,** and we write

$$z = \frac{y - \mu}{\sigma}.$$

> **Is the Standard Normal a Standard?**
> Yes. We call it the "Standard Normal" because it models standardized values. It is also a "standard" because this is the particular Normal model that we almost always use.

Usually it's easier to standardize data first (using its mean and standard deviation). Then we need only the model $N(0, 1)$. The Normal model with mean 0 and standard deviation 1 is called the **standard Normal model** (or the **standard Normal distribution**).

---
[2] Especially if it got away.
[3] In fact, the model is useful *because* it doesn't have every rivet. It is because models offer a simpler view of reality that they are so useful as we try to understand reality.

But be careful. You shouldn't use a Normal model for just any dataset. Remember that standardizing won't change the shape of the distribution. If the distribution is not unimodal and symmetric to begin with, standardizing won't make it Normal.

When we use the Normal model, we assume that the distribution of the data is, well, Normal. Practically speaking, there's no way to check whether this **Normality Assumption** is true. In fact, it almost certainly is not true. Real data don't behave like mathematical models. Models are idealized; real data are real. The good news, however, is that to use a Normal model, it's sufficient to check the following condition:

> **Nearly Normal Condition**. The shape of the data's distribution is unimodal and symmetric. Check this by making a histogram or a dotplot (or a Normal probability plot, which we'll explain later). It's better not to check this condition with a boxplot because you won't see the shape of the distribution, and Normal is about shape. (The bell-curve shape.)

Don't model data with a Normal model without checking whether the condition is satisfied.

All models make *assumptions*. Whenever we model—and we'll do that often—we'll be careful to point out the assumptions that we're making. And, what's even more important, we'll check the associated *conditions* in the data to make sure that those assumptions are reasonable.

> " All models are wrong—but some are useful. "
>
> —George Box, famous statistician

## The 68–95–99.7 Rule

> **ONE IN A MILLION**
>
> These magic 68, 95, 99.7 values come from the Normal model. As a model, it can give us corresponding values for any z-score. For example, it tells us that fewer than 1 out of a million values have z-scores smaller than −5.0 or larger than +5.0. So if someone tells you you're "one in a million," they must really admire your z-score.

Normal models give us an idea of how extreme a value is by telling us how likely it is to find one that far from the mean. We'll soon show how to find these numbers precisely—but one simple rule is often all we need.

It turns out that in a Normal model, about 68% of the values fall within 1 standard deviation of the mean, about 95% of the values fall within 2 standard deviations of the mean, and about 99.7%—almost all—of the values fall within 3 standard deviations of the mean. These facts are summarized in a rule that we call (let's see . . .) the **68–95–99.7 Rule**.[4]

> TI-*nspire*
>
> **The 68–95–99.7 Rule.** See it work for yourself.

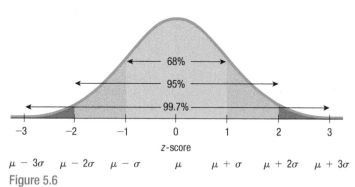

Figure 5.6
Reaching out one, two, and three standard deviations (or z-scores) on a Normal model gives the 68–95–99.7 Rule, seen as proportions of the area under the curve. The top line of numbers shows the z-scores. The bottom line shows the same distances in terms of the mean and standard deviation.

---

[4] This rule is also called the "Empirical Rule" because it originally came from observation. The rule was first published by Abraham de Moivre in 1733, 75 years before the Normal model was discovered. Maybe it should be called "de Moivre's Rule," but that wouldn't help us remember the important numbers, 68, 95, and 99.7.

## FOR EXAMPLE
### Using the 68–95–99.7 Rule

A study of men's health measured 14 body characteristics of 250 men, one of which was wrist circumference (in cm). The mean wrist circumference was 18.22 cm and the standard deviation was 0.91 cm. A histogram of all 250 values looks like this:

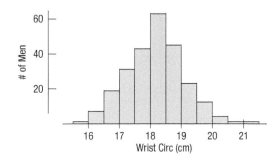

QUESTION: Will the 68–95–99.7 Rule provide useful percentages from these data? About how many of these men does the rule predict to have a wrist circumference larger than 20.04 cm?

ANSWER: Yes, the distribution is unimodal and symmetric and so the 68–95–99.7 Rule should be applicable to these data. 20.04 cm is about 2 standard deviations above the mean. We expect 5% of the data to be either above or below two standard deviations, so 2.5% should be above. That's about 6 men (which, it turns out, is the number of men in the dataset who have wrists larger than 20.04 cm).

## FOR EXAMPLE
### Using the 68–95–99.7 Rule II

Since 2017, public companies have been required to disclose the ratio of CEO pay to median worker pay. The Glassdoor Economic Research Blog has published the data for 2018. Here are some of those data: the total compensation (in $M) of CEOs of 480 top public companies.

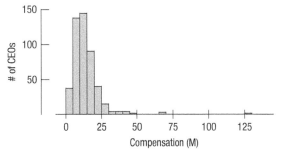

QUESTION: Will the 68–95–99.7 Rule provide useful percentages from these data? Explain briefly.
 Using the rule, about how many CEOs should have compensations within one SD of the mean (between $4.12M and $23.77M?

ANSWER: No, the distribution is strongly skewed to the right. We shouldn't expect the 68–95–99.7 Rule to be useful. According to the rule, about 68% of the CEO compensations should be within 1 standard deviation of the mean. In fact, from the dataset, 420 CEOs (about 88%) had compensation in that interval. The 68–95–99.7% Rule is not appropriate for strongly skewed data such as these.

## JUST CHECKING

5. As a group, the Dutch are among the tallest people in the world. The average Dutch man is 184 cm tall—just over 6 feet (and the average Dutch woman is 170.8 cm tall—just over 5′7″). If a Normal model is appropriate and the standard deviation for men is about 8 cm, what percentage of all Dutch men should be over 2 meters (6′6″) tall?

6. Suppose it takes you 20 minutes, on average, to drive to school, with a standard deviation of 2 minutes. Suppose a Normal model is appropriate for the distributions of driving times.
   a) How often should you arrive at school in less than 22 minutes?
   b) How often should it take you more than 24 minutes?
   c) Do you think the distribution of your driving times is unimodal and symmetric?
   d) What does this say about the accuracy of your predictions? Explain.

# The First Three Rules for Working with Normal Models

1. Make a picture.
2. Make a picture.
3. Make a picture.

Although we're thinking about models, not histograms of data, the three rules don't change. To help you think clearly, a simple hand-drawn sketch is all you need. Expert statisticians sketch pictures to help them think about Normal models. You should too.

Of course, when we have data, we'll also need to make a histogram to check the **Nearly Normal Condition** to be sure we can use the Normal model to model the data's distribution. Other times, we may be told that a Normal model is appropriate based on prior knowledge of the situation or on theoretical considerations.

TI-*nspire*

**Normal models.** Watch the Normal model react as you change the mean and standard deviation.

### How to Sketch a Normal Curve That Looks Normal

To sketch a good Normal curve, you need to remember only three things:
- The Normal curve is bell-shaped and symmetric around its mean. Start at the middle, and sketch to the right and left from there.
- Even though the Normal model extends forever on either side, you need to draw it only for 3 standard deviations. After that, there's so little left that it isn't worth sketching.
- The place where the bell shape changes from curving downward to curving back up—the *inflection point*—is exactly one standard deviation away from the mean.

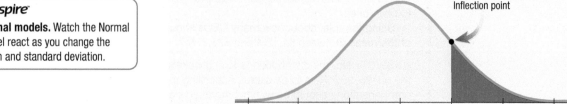

## STEP-BY-STEP EXAMPLE

### Working with the 68–95–99.7 Rule

The SAT test has two parts: Evidence-Based Reading and Writing (EBRW) and Math. Each part has a distribution of scores that is roughly unimodal and symmetric and is designed to have an overall mean of about 500 and a standard deviation of 100 for all test takers. In any one year, the mean and standard deviation may differ from these target values by a small amount, but they are a good overall approximation.

QUESTION: Suppose you earned a 600 on one part of your SAT. Where do you stand among all students who took that test?

You could calculate your *z*-score and find out that it's $z = (600 - 500)/100 = 1.0$, but what does that tell you about your percentile? You'll need the Normal model and the 68–95–99.7 Rule to answer that question.

| | | |
|---|---|---|
| **THINK** | **PLAN** State what you want to know. | I want to see how my SAT score compares with the scores of all other students. To do that, I'll need to model the distribution. |
| | **VARIABLES** Identify the variable and report the W's. | Let $y =$ my SAT score. Scores are quantitative but have no meaningful units other than points. |
| | Be sure to check the appropriate conditions. | ✓ **Nearly Normal Condition:** If I had data, I would check the histogram. I have no data, but I am told that the SAT scores are roughly unimodal and symmetric. |
| | Specify the parameters of your model. | I will model SAT score with a $N(500, 100)$ model. |
| **SHOW** | **MECHANICS** Make a picture of this Normal model. (A simple sketch is all you need.) | 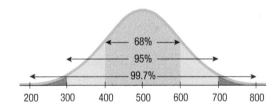 |
| | Locate your score. | My score of 600 is 1 standard deviation above the mean. That corresponds to one of the points of the 68–95–99.7 Rule. |
| **TELL** | **CONCLUSION** Interpret your result in context. | The Normal model predicts that about 68% of those taking the test would have scores that fall no more than 1 standard deviation from the mean, so $100\% - 68\% = 32\%$ of all students would have scores more than 1 standard deviation away. Only half of those would be on the high side, so about 16% (half of 32%) of the test scores would be better than mine. My score of 600 should be higher than about 84% of all scores on this test, placing me at the 84th percentile. |

The bounds of SAT scoring at 200 and 800 can also be explained by the 68–95–99.7 Rule. Because 200 and 800 are three standard deviations from 500, it hardly pays to extend the scoring any farther on either side. We'd get more information only on $100 - 99.7 = 0.3\%$ of students.

Remember: when a distribution is not Nearly Normal, then the 68–95–99.7 Rule does not apply. But data values more than 3 standard deviations from the mean are still unusual, whether the distribution is Normal or not. They may not be among the most unusual 0.3% of values, as the 68–95–99.7 Rule would imply, but they're still unusual. And data values more than 4 or 5 standard deviations from the mean are highly unusual—again, regardless of the shape of the distribution.

# Determining Normal Percentiles

An SAT score of 600 is easy to assess, because we can think of it as one standard deviation above the mean. If your score was 680, though, where do you stand among the rest of the people tested? Your $z$-score is 1.80, so you're somewhere between 1 and 2 standard deviations above the mean. We figured out that no more than 16% of people score better than 600. By the same logic, no more than 2.5% of people score better than 700. Can we be more specific than "between 16% and 2.5%"?

When the value doesn't fall exactly 1, 2, or 3 standard deviations from the mean, we can look it up in a table of **Normal percentiles** or use technology.[5] Either way, we first convert our data to $z$-scores. Your SAT score of 680 has a $z$-score of $(680 - 500)/100 = 1.80$.

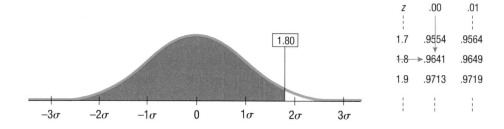

Figure 5.7
A table of Normal percentiles (Table Z in Appendix G) lets us determine the percentage of individuals in a Standard Normal distribution falling below any specified $z$-score value.

In the piece of the table shown, we find your $z$-score by looking down the left column for the first two digits, 1.8, and across the top row for the third digit, 0. The table gives the percentile as 0.9641. That means that 96.4% of the $z$-scores are less than or equal to 1.80. Only 3.6% of people, then, scored at least 680 on the SAT.

Most of the time, though, you'll do this with your calculator.

> TI-*nspire*
>
> **Normal percentiles.** Explore the relationship between $z$-scores and areas in a Normal model.

## TI TIPS

### Finding Normal Percentages

Your calculator knows the Normal model. Have a look under `2nd DISTR`. There you will see three "norm" functions, `normalpdf(`, `normalcdf(`, and `invNorm(`. Let's play with the first two.

- `normalpdf(` calculates $y$-values for graphing a Normal curve. You probably won't use this very often, if at all. If you want to try it, graph `Y1 = normalpdf(X, 0, 1)` in a graphing `WINDOW` with `Xmin = -4`, `Xmax = 4`, `Ymin = -0.1`, and `Ymax = 0.5`. The 0 and the 1 in the parentheses represent, respectively, the mean and the standard deviation. So this will produce a graph of the Standard Normal distribution.

- `normalcdf(` determines the proportion of area under the curve between two $z$-score cut points, by specifying `normalcdf(zLeft,zRight)`. *Do* make friends with this function; you will use it often!

---

[5] See Table Z in Appendix G, if you're curious. But your calculator (and any statistics computer package) does this, too—and more easily!

### CHAPTER 5  The Standard Deviation as a Ruler and the Normal Model

**EXAMPLE 1** The Normal model shown shades the region between $z = -0.5$ and $z = 1.0$.

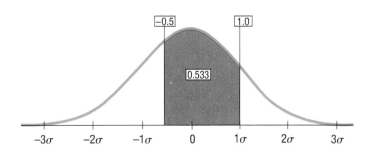

To determine the shaded area:

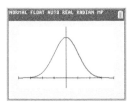

- Under `2nd DISTR` select `normalcdf(` and hit `ENTER`.
- Specify `lower: -.5, upper:1, `$\mu$`:0, sigma:1`, and then go to `Paste` and hit `ENTER` twice (OR on an older calculator, just enter `normalcdf( -.5,1)`).

There's the area. Approximately 53% of a Normal model lies between half a standard deviation below and one standard deviation above the mean.

**EXAMPLE 2** In the example in the text we used Table Z to determine the fraction of SAT scores above your score of 680. Now let's do it again, this time using your TI.

First we need $z$-scores for the cut points:

- Because 680 is 1.8 standard deviations above the mean, your $z$-score is 1.8; that's the left cut point.
- Theoretically the standard Normal model extends rightward forever, but you can't tell the calculator to use infinity as the right cut point. Recall that for a Normal model almost all the area lies within $\pm 3$ standard deviations of the mean, so any upper cut point beyond, say, $z = 5$ does not cut off anything very important. We suggest you always use 99 (or –99) when you really want infinity as your cut point—it's easy to remember and way beyond any meaningful area.

Now you're ready. Use the command `normalcdf(` as above with `lower:1.8, upper:99`.

There you are! The Normal model estimates that approximately 3.6% of SAT scores are higher than 680.

---

### STEP-BY-STEP EXAMPLE

## Working with Normal Models Part I

Below is a histogram of the heights of all of Major League Baseball's pitchers at the start of the 2017 season.

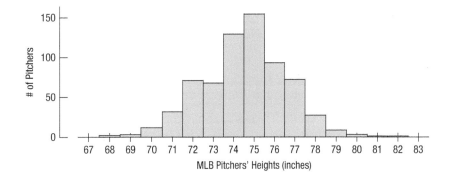

The mean height of the pitchers is about 74.5 inches and the standard deviation is about 2.1 inches.

QUESTION: What proportion of baseball pitchers should we expect to be at least 78 inches tall?

| | |
|---|---|
| **THINK**   **PLAN** State the problem. | I want to know the proportion of baseball pitchers we'd expect to be at least 78 inches tall. |
| **VARIABLES** Name the variable. <br><br> Check the appropriate conditions and state which Normal model to use. | Let $y$ = height. <br><br> ✓ **Nearly Normal Condition**: The histogram of pitchers' heights is unimodal and roughly symmetric, so the Nearly Normal Condition is met. <br><br> I'll model baseball pitchers' heights with a $N(74.5, 2.1)$ model, using the mean and standard deviation given. |
| **SHOW**   **MECHANICS** Make a picture of this Normal model. Locate the desired values and shade the region of interest. <br><br><br><br><br> Determine the $z$-score corresponding to the boundary of the region of interest. Use technology to find the area under the curve. (We just showed you how to do this, under TI Tips.) <br><br> (Using a table is okay too—but Table Z shows areas to the *left* of each $z$-score, and we're interested in the area to the *right* of $z = 1.67$. So we need to read 0.9525 from the table and then subtract from 1.) | 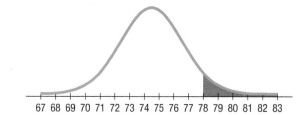 <br><br> The standardized $z$-score for a 78-inch-tall baseball player is: $$z = \frac{y - \mu}{\sigma} = \frac{78 - 74.5}{2.1} \approx 1.67.$$ Using technology, I find that the area to the right of 1.67 under the standard Normal curve is about 0.047. |
| **TELL**   **CONCLUSION** Interpret your conclusion in context. | Using the Normal model, I estimate that about 4.7% of Major League pitchers should be at least 78 inches tall. |

# From Percentiles to Scores: *z* in Reverse

**APPLET**
Play with Normal model areas and cut points

Determining areas from $z$-scores is the simplest way to work with the Normal model. But sometimes we start with areas and are asked to work backward to find the corresponding $z$-score or even the original data value. For instance, what $z$-score cuts off the top 10% in a Normal model?

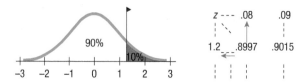

Make a picture like the one shown, shading the rightmost 10% of the area. Notice that this is the 90th percentile. If you're determined to do it the hard way, look in Table Z for an area of 0.900. The exact area is not there, but 0.8997 is pretty close. That shows up in the table with 1.2 in the left margin and .08 in the top margin. The *z*-score for the 90th percentile, then, is approximately $z = 1.28$.

Computers and calculators will determine the cut point[6] more precisely (and more easily).

## TI TIPS

### Finding Normal Cut Points

Let's find the *z*-score at the 25th percentile. This time we'll use the third of the built-in "norm(al)" functions: `invNorm(`, which stands for "inverse Normal."

- Select the `invNorm(` function from the distributions menu.
- In the dialog box, enter 0.25 for the area. Leave the default values of 0 and 1 for the mean and standard deviation.
- For the "Tail," choose LEFT, because we're interested in the *lowest* 25% of the distribution.
- Paste it into the home screen and hit Enter. The calculator says that the *z*-score that "cuts off" the lowest 25% of the standard Normal distribution is about –0.674.

We'll let you figure out how to find the *z*-score that cuts off the *highest* fraction of a Normal distribution. Try to use your calculator to show that the *z*-score that cuts off the upper 1% of the standard Normal distribution is about 2.33.

(We're going to get to the "CENTER" part of this dialog box later; you might be able to figure out what it's for just by experimenting.)

This TI Tip describes how to use `invNorm(` if your calculator's operating system uses stats dialog boxes. If yours doesn't, you can still determine cut points with the `invNorm(` function, directly on your home screen. But like Table Z, the TI function always requires you to input the area to the *left* of your cutoff. The *z*-score that cuts off the *upper* 1% of the standard Normal distribution is actually the 99th percentile of the distribution, so you would enter: `invNorm(0.99)`.

## STEP-BY-STEP EXAMPLE

### Working with Normal Models Part II

**QUESTION:** Suppose a college says it admits only people with SAT Verbal test scores among the top 10%. How high a score does it take to be eligible?

| THINK | |
|---|---|
| **PLAN** State the problem. | How high an SAT Verbal score do I need to be in the top 10% of all test takers? |
| **VARIABLE** Define the variable. | Let $y$ = my SAT score. |
| Check to see if a Normal model is appropriate, and specify which Normal model to use. | ✓ **Nearly Normal Condition:** I am told that SAT scores are nearly Normal. I'll model them with $N(500, 100)$. |

---

[6] We've called the boundary of a region of interest a "cut point" because we're "cutting off" some interval of interest. "Cut points" are actually called *quantiles*, which is the more generic term for "quartiles" and "percentiles."

| | | |
|---|---|---|
| **SHOW** MECHANICS Make a picture of this Normal model. Locate the desired percentile approximately by shading the rightmost 10% of the area. | 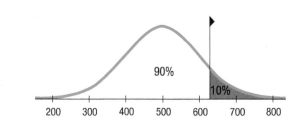 | |
| The college takes the top 10%, so its cutoff score is the 90th percentile. Find the corresponding z-score using your calculator as shown in the TI Tips. (OR: Use Table Z as shown on p. 124.) | The cut point is $z = 1.28$. | |
| Convert the z-score back to the original units. | A z-score of 1.28 is 1.28 standard deviations above the mean. Because the SD is 100, that's 128 SAT points. The cutoff is 128 points above the mean of 500, or 628. | |
| **TELL** CONCLUSION Interpret your results in the proper context. | Because the school wants SAT Verbal scores in the top 10%, the Normal model estimates that the cutoff is 628. (Actually, because SAT scores are reported only in multiples of 10, I'd have to score at least a 630.) | |

## STEP-BY-STEP EXAMPLE

### More Working with Normal Models

Working with Normal percentiles can be a little tricky, depending on how the problem is stated. Here are a few more worked examples of the kind you're likely to see.

A cereal manufacturer has a machine that fills the boxes. Boxes are labeled "16 ounces," so the company wants to have that much cereal in each box, but because no packaging process is perfect, there will be minor variations. If the machine is set at exactly 16 ounces and the Normal model applies (or at least the distribution is roughly symmetric), then about half of the boxes will be underweight, making consumers unhappy and exposing the company to bad publicity and possible lawsuits. To prevent underweight boxes, the manufacturer has to set the mean a little higher than 16.0 ounces.

Based on its experience with the packaging machine, the company believes that the amount of cereal in the boxes fits a Normal model with a standard deviation of 0.2 ounces. The manufacturer decides to set the machine to put an average of 16.3 ounces in each box. Let's use that model to answer a series of questions about these cereal boxes.

QUESTION 1: What fraction of the boxes will be underweight?

| | |
|---|---|
| **THINK** PLAN State the problem. | What proportion of boxes weigh less than 16 ounces? |
| VARIABLE Name the variable. | Let y = weight of cereal in a box. |
| Check to see if a Normal model is appropriate. | ✓ **Nearly Normal Condition**: I have no data, so I cannot make a histogram, but I am told that the company believes the distribution of weights from the machine is Normal. |
| Specify which Normal model to use. | I'll use a $N(16.3, 0.2)$ model. |

| | |
|---|---|
| **SHOW**    **MECHANICS** Make a picture of this Normal model. Locate the value you're interested in on the picture, label it, and shade the appropriate region.<br><br>**REALITY CHECK** Estimate from the picture the percentage of boxes that are underweight. (This will be useful later to check that your answer makes sense.) It looks like a low percentage. Less than 20% for sure.<br><br>Convert your cutoff value into a *z*-score.<br><br>Find the area with your calculator (or use the Normal table). | 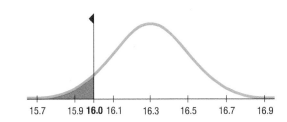<br><br>I want to know what fraction of the boxes will weigh less than 16 ounces.<br><br>$$z = \frac{y - \mu}{\sigma} = \frac{16 - 16.3}{0.2} = -1.50$$<br><br>$Area(y < 16) = Area(z < -1.50) = 0.0668$ |
| **TELL**    **CONCLUSION** State your conclusion, and check that it's consistent with your earlier guess. It's below 20%—seems okay. | I estimate that approximately 6.7% of the boxes will contain less than 16 ounces of cereal. |

**QUESTION 2:** The company's lawyers say that 6.7% is unacceptably high. They insist that no more than 4% of the boxes can be underweight. So the company needs to set the machine to put a little more cereal in each box. What mean setting does it need?

| | |
|---|---|
| **THINK**    **PLAN** State the problem.<br><br>**VARIABLE** Name the variable.<br><br>Check to see if a Normal model is appropriate.<br><br>Specify which Normal model to use. This time you are not given a value for the mean!<br><br>**REALITY CHECK** We found out earlier that setting the machine to $\mu = 16.3$ ounces made 6.7% of the boxes too light. We'll need to raise the mean a bit to reduce this fraction. | What mean weight will reduce the proportion of underweight boxes to 4%?<br><br>Let $y$ = weight of cereal in a box.<br><br>✓ **Nearly Normal Condition**: I am told that a Normal model applies.<br><br>I don't know $\mu$, the mean amount of cereal. The standard deviation for this machine is 0.2 ounces. The model is $N(\mu, 0.2)$.<br><br>No more than 4% of the boxes can be below 16 ounces. |
| **SHOW**    **MECHANICS** Make a picture of this Normal model. Center it at $\mu$ (because you don't know the mean), and shade the region below 16 ounces.<br><br><br>Using your calculator (or the Normal table), find the *z*-score that cuts off the lowest 4%.<br><br>Use this information to find $\mu$. It's located 1.75 standard deviations to the right of 16. Because $\sigma$ is 0.2, that's $1.75 \times 0.2$, or 0.35 ounces more than 16. | 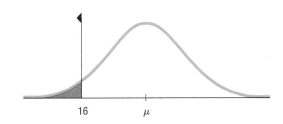<br><br>The *z*-score that has 0.04 area to the left of it is $z = -1.75$.<br><br>For 16 to be 1.75 standard deviations below the mean, the mean must be<br><br>$$16 + 1.75(0.2) = 16.35 \text{ ounces.}$$ |

| | |
|---|---|
| **TELL** **CONCLUSION** Interpret your result in context.<br>(This makes sense; we knew it would have to be just a bit higher than 16.3.) | Based on a Normal model, I estimate that the company must set the machine to average 16.35 ounces of cereal per box. |

**QUESTION 3:** The company president vetoes that plan, saying the company should give away less free cereal, not more. Her goal is to set the machine no higher than 16.2 ounces and still have only 4% underweight boxes. The only way to accomplish this is to reduce the standard deviation. What standard deviation must the company achieve, and what does that mean about the machine?

| | |
|---|---|
| **THINK** **PLAN** State the problem.<br><br>**VARIABLE** Name the variable.<br><br>Check conditions to be sure that a Normal model is appropriate.<br><br>Specify which Normal model to use. This time you don't know $\sigma$.<br><br>**REALITY CHECK** We know the new standard deviation must be less than 0.2 ounces. | What standard deviation will allow the mean to be 16.2 ounces and still have only 4% of boxes underweight?<br><br>Let $y$ = weight of cereal in a box.<br><br>✓ **Nearly Normal Condition:** The company believes that the weights are described by a Normal model.<br><br>I know the mean, but not the standard deviation, so my model is $N(16.2, \sigma)$. |
| **SHOW** **MECHANICS** Make a picture of this Normal model. Center it at 16.2, and shade the area you're interested in. We want 4% of the area to the left of 16 ounces.<br><br><br><br>Find the z-score that cuts off the lowest 4%.<br><br><br><br>Solve for $\sigma$. (We need 16 to be $1.75\sigma$ below 16.2, so $1.75\sigma$ must be 0.2 ounces. You could just start with that equation.) | 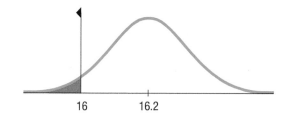<br><br>I know that the z-score with 4% below it is $z = -1.75$.<br><br>$$z = \frac{y - \mu}{\sigma}$$<br>$$-1.75 = \frac{16 - 16.2}{\sigma}$$<br>$$1.75\sigma = 0.2$$<br>$$\sigma = 0.114$$ |
| **TELL** **CONCLUSION** Interpret your result in context.<br>As we expected, the standard deviation is lower than before—actually, quite a bit lower. | According to the Normal model, the company must get the machine to box cereal with a standard deviation of only 0.114 ounces. This means the machine must be more consistent (by nearly a factor of 2) in filling the boxes. |

# *Are You Normal? Find Out with a Normal Probability Plot

In the examples we've worked through, we've assumed that the underlying data distribution was roughly unimodal and symmetric, so that using a Normal model makes sense. When you have data, you must *check* to see whether a Normal model is reasonable. How? Make a picture, of course! Drawing a histogram of the data and looking at the shape is one good way to see if a Normal model might be reasonable.

> **TI-*nspire***
> **Normal probability plots and histograms.** See how a Normal probability plot responds as you change the shape of a distribution.

There's a more specialized graphical display that can help you to decide whether the Normal model is appropriate: the **Normal probability plot**. If the distribution of the data is roughly Normal, the plot is roughly a diagonal straight line. Deviations from a straight line indicate that the distribution is not Normal. This plot shows deviations from Normality more clearly than the corresponding histogram, but it's usually easier to understand *how* a distribution fails to be Normal by looking at its histogram.

Some data on a car's fuel efficiency provide an example of data that are nearly Normal. The overall pattern of the Normal probability plot is straight. The two trailing low values correspond to the values in the histogram that trail off the low end. They're not quite in line with the rest of the dataset. The Normal probability plot shows us that they're a bit lower than we'd expect of the lowest two values in a Normal model.

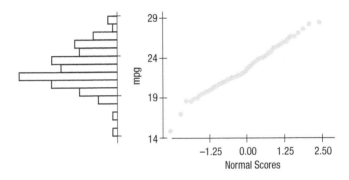

**Figure 5.8**
Histogram and Normal probability plot for gas mileage (mpg) recorded by one of the authors over the 8 years he owned a 1989 Nissan Maxima. The vertical axes are the same, so each dot on the probability plot would fall into the bar on the histogram immediately to its left. (The points are denser in the middle of the normal probability plot than at its ends, but because so many points overlap in the graph that feature is impossible to see.)

By contrast, the Normal probability plot of the men's *Weight*s from the NHANES study is far from straight. The weights are skewed to the high end, and the plot is curved. Conclusions from these data based on a Normal model would not be very accurate.

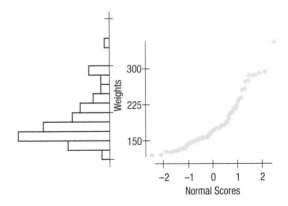

**Figure 5.9**
Histogram and Normal probability plot for men's weights. Note how a skewed distribution corresponds to a bent probability plot.

A Normal probability plot essentially works by comparing the data values' actual z-scores with those we'd expect to find in a dataset of the same size. When they match up well, the line is straight and the data satisfy the Nearly Normal Condition. We don't worry much if one or two points don't line up. But a plot that bends is a warning that the distribution is skewed (or strange in some other way) and we should not use a Normal model.

## TI TIPS

### *Creating a Normal Probability Plot

Let's make a Normal probability plot with the calculator. Here are the drop heights in feet of the twelve steel-track roller coasters that we looked at in Chapter 4; enter them in L1:

300, 155, 105, 146, 95, 64, 95, 190, 84, 109, 226, 206

Now you can create the plot:

- Turn a STATPLOT On.
- Tell it to make a Normal probability plot by choosing the last of the icons.
- Specify your datalist and which axis you want the data on. (We'll use Y so the plot looks like the others we showed you.)
- Specify the Mark you want the plot to use.
- Now ZoomStat does the rest.

The plot doesn't look very straight. The Normal model isn't very reasonable for these data.

(Not that it matters in making this decision, but that vertical line is the y-axis. Points to the left have negative z-scores and points to the right have positive z-scores.)

# The Truth About Assessing Normality

We've talked about using the Normal model for sets of data. And you now have tools for looking at data to see whether the Normal model is reasonable, including the histogram (unimodal and symmetric?) and the Normal probability plot (fairly straight?). Before we conclude this chapter, though, we need to be honest about something that we've been glossing over a little bit. Usually when we have a dataset, it's a sample of data from a larger population. And although we can construct histograms and Normal probability plots using the sample data, it's really the shape of the *population* distribution that we care about. So the question we want to answer isn't "Is a Normal model reasonable for this dataset?" Rather, it's this: *"Given this set of sample data, is a Normal model reasonable for the population that the data came from?"*

Uh, oh. We can't see the data for the *whole* population (unless we conduct a census). All we can see is a set of sample data. How can we assess the reasonableness of the Normal model for a population that we can't see? For better or worse, the answer is that we just do the best we can using the sample data, with the understanding that we have limited information and therefore might be able to draw only limited conclusions.

Let's see how this works using simulations. We'll suppose that the weights of newborn babies have a Normal distribution with a mean of 7.5 pounds and a standard deviation of 1.2 pounds (which is approximately true, incidentally). We'll simulate 25 random baby weights from this distribution. Then we'll construct a dotplot for those weights.

*"People can assess normality if they want. I think it's overrated."*

Here's one result from such a simulation.

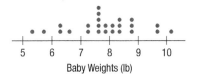

The dotplot is a *little* bit bumpy, but remember: we simulated these babies' weights as a random sample from a population model that *does* have a Normal distribution. So we know that at least a little bit of *apparent* non-Normality in a sample of data can show up even if the population really is Normal.

Let's do this a few more times with new simulated random samples. That way we can get an idea of just *how much* apparent non-Normality we might typically see in a sample of data that really do come from a Normal population model.

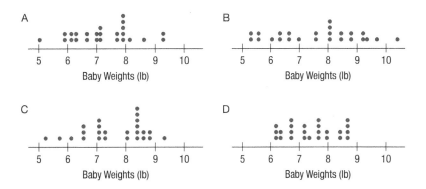

Each of these four datasets has characteristics that might lead you to think they came from a population that is not Normally distributed. For example, dataset A looks a little bit left-skewed, dataset C looks a little bit bimodal, and dataset D looks too "bunched up" in the middle. Yet all four are simulated samples from a Normal population model, so we know that these characteristics really are just *quirks*—chance variations—and not actual features of the population.

The upshot of this is that when you have a sample of data from a population, and you want to decide whether a Normal model is appropriate for the population, you *should* look at a dotplot or a histogram or a Normal probability plot of your data. But you should *not* demand perfect unimodality, perfect symmetry, or a perfectly straight Normal probability plot. Some wiggle room is allowed. And your conclusion should never be that the population "is Normal" or "is not Normal." Rather, your conclusion should be that the Normal model for the population "is reasonable" or "is not reasonable." Or you can say that the data "are consistent" with a Normal population model or "are inconsistent" with a Normal population model.

We've shown a few simulations using samples of 25 babies' weights. When the sample size is smaller, even more wiggle room should be allowed. Indeed, if your sample size is very small—say, four or five data points—then practically any dataset will be reasonably consistent with a Normal population model, so long as there isn't an extreme outlier among them.

It takes practice and experience to know when sample data are or are not reasonably consistent with a Normal population model. You'll get a chance to practice this more later on in the text. For now just remember that sample data will never look exactly like the population they were drawn from, and that we therefore can see some signs of non-Normality in a dataset yet still use a Normal model for the population, so long as those signs of non-Normality aren't extreme.

## WHAT CAN GO WRONG?

- **Don't use a Normal model when the distribution is not unimodal and symmetric.** Normal models are so easy and useful that it is tempting to use them even when they don't describe the data very well. That can lead to wrong conclusions. Don't use a Normal model without first checking the **Nearly Normal Condition**. Look at the distribution of the data to check that it is roughly unimodal and symmetric. A histogram, or a Normal probability plot, can help you tell whether a Normal model is appropriate.

  The CEOs (Example 5.5) had a mean total compensation of $13.94M and a standard deviation of $9.825M. Using the Normal model rule, we should expect about 95% of the CEOs to have compensations between $0 and $33.2M. In fact, more than 97% of the CEOs have annual compensations in this range. What went wrong? The distribution is skewed, not symmetric.

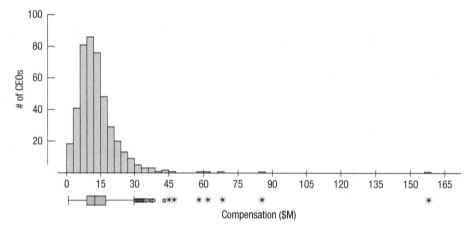

- **Don't use the mean and standard deviation when outliers are present.** Both means and standard deviations can be distorted by outliers, and no model based on distorted values will do a good job. A $z$-score calculated from a distribution with outliers may be misleading. It's always a good idea to check for outliers. How? Make a picture.

- **Don't round your results in the middle of a calculation.** We reported the mean of the heptathletes' 200 m run as 24.58 seconds. More precisely, it was 24.582069 seconds.

  You should use all the precision available in the data for all the intermediate steps of a calculation. Using the more precise value for the mean (and also carrying 8 digits for the SD), the $z$-score calculation for Thiam's run comes out to

  $$z = \frac{25.10 - 24.582069}{0.65449751} = 0.79134149.$$

  We'd likely report that as 0.791, as opposed to the rounded-off value of 0.795 we got earlier.

- **Do what we say, not what we do.** When we showed the $z$-score calculations for Thiam, we rounded the mean to 25.10 s and the SD to 0.654 s. Then to make the story clearer we used *those values* in the displayed calculation.

  We'll continue to show simplified calculations in the book to make the story simpler. When you calculate with full precision, your results may differ slightly from ours. So, we also advise . . .

- **Don't worry about minor differences in results.** Because various calculators and programs may carry different precision in calculations, your answers may differ slightly from those we show in the text and in the Step-by-Steps, or even from the values given in the answers in the back of the book. Those differences aren't anything to worry about. They're not the main story Statistics tries to tell.

# WHAT HAVE WE LEARNED?

We've learned that the story data can tell may be easier to understand after shifting or rescaling the data.
- Shifting data by adding or subtracting the same amount from each value affects measures of center and position but not measures of spread.
- Rescaling data by multiplying or dividing every value by a constant changes all the summary statistics—center, position, and spread.

We've learned the power of standardizing data.
- Standardizing uses the standard deviation as a ruler to measure distance from the mean, creating z-scores.
- Using these z-scores, we can compare apples and oranges. Because standardizing eliminates units, standardized values can be compared and combined even if the original variables had different units and magnitudes.
- And a z-score can identify unusual or surprising values among data.

We've learned that the 68–95–99.7 Rule can be a useful rule of thumb for understanding distributions.
- For data that are unimodal and symmetric, about 68% fall within 1 SD of the mean, 95% fall within 2 SDs of the mean, and 99.7% fall within 3 SDs of the mean.

Again we've seen the importance of *Thinking* about whether a method will work.
- Data can't be exactly Normal, so we check the Nearly Normal Condition by making a histogram (is it unimodal, symmetric, and free of outliers?) or a Normal probability plot (is it straight enough?).

## TERMS

**Standardized value**    A value found by subtracting the mean and dividing by the standard deviation. (p. 111)

**z-score**    A z-score tells how many standard deviations a value is from the mean, and in which direction; z-scores have a mean of 0 and a standard deviation of 1. When working with data, use the statistics $\bar{y}$ and $s$:

$$z = \frac{y - \bar{y}}{s}. \text{ (p.111)}$$

When working with models, use the parameters $\mu$ and $\sigma$:

$$z = \frac{y - \mu}{\sigma}. \text{ (p.117)}$$

**Shifting**    Adding a constant to each data value adds the same constant to the measures of position (mean, median, and quartiles), but does not change the measures of spread (standard deviation or IQR). (p. 114)

**Rescale**    Multiplying each data value by a constant multiplies both the measures of position (mean, median, and quartiles) and the measures of spread (standard deviation and IQR) by that constant. (p. 114)

**Normal model**    A useful family of models for unimodal, symmetric distributions. (p. 117)

**Parameter**    A numerically-valued attribute of a model. For example, the values of $\mu$ and $\sigma$ in a $N(\mu, \sigma)$ model are parameters. (p. 117)

**Statistic**    A value calculated from data to summarize aspects of the data. For example, the mean, $\bar{y}$, and standard deviation, $s$, are statistics. (p. 117)

**Standard Normal model**    A Normal model, $N(\mu, \sigma)$ with mean $\mu = 0$ and standard deviation $\sigma = 1$. Also called the **standard Normal distribution**. (p. 117)

| | |
|---|---|
| **Normality Assumption** | We must have a reason to believe a variable's distribution is approximately Normal before applying a Normal model. (p. 118) |
| **68–95–99.7 Rule** | In a Normal model, about 68% of values fall within 1 standard deviation of the mean, about 95% fall within 2 standard deviations of the mean, and about 99.7% fall within 3 standard deviations of the mean. (p. 118) |
| **Nearly Normal Condition** | A distribution is nearly Normal if it is roughly unimodal and symmetric. We can check by looking at a histogram (or a Normal probability plot). (p. 120) |
| **Normal percentile** | The Normal percentile corresponding to a $z$-score gives the percentage of values in a standard Normal distribution found at that $z$-score or below. (p. 122) |
| **\*Normal probability plot** | A display to help assess whether a distribution of data is approximately Normal. If the plot is nearly straight, the data satisfy the Nearly Normal Condition. (p. 129) |

# ON THE COMPUTER

## The Normal Model

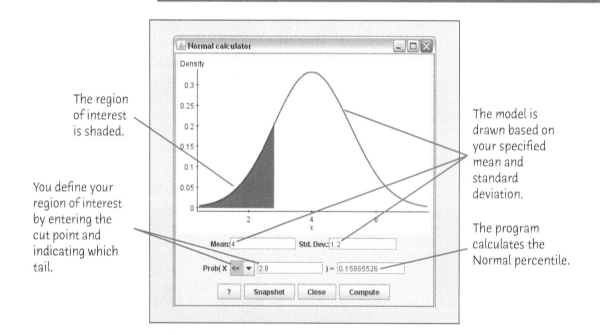

The best way to tell whether your data can be modeled well by a Normal model is to make a picture or two. We've already talked about making histograms. Normal probability plots are almost never made by hand because the values of the Normal scores are tricky to find. But most statistics software make Normal plots, though various packages call the same plot by different names and array the information differently.

© 2013 Randall Munroe. Reprinted with permission. All rights reserved.

# EXERCISES

1. **Stats test, part 1** Nicole's score on the Stats midterm was 80 points. The class average was 75 and the standard deviation was 5 points. What was her z-score?

2. **Horsepower** Cars currently sold in the United States have an average of 135 horsepower, with a standard deviation of 40 horsepower. What's the z-score for a car with 195 horsepower?

3. **Home runs** In 1998, Mark McGwire hit 70 home runs, to break the single season home run record of 61 home runs set by Roger Maris in 1961. McGwire later admitted to using performance enhancing steroids that season. Even so, how did McGwire's performance really compare to Maris's? The league was very different in 1998 than in 1961. The table below shows the mean and standard deviations for the number of home runs scored by all players with at least 502 plate appearances in their respective seasons. Use these to determine whose home run feat was more impressive.

|  | 1961 | 1998 |
|---|---|---|
| Mean | 18.8 | 20.7 |
| Standard deviation | 13.37 | 12.74 |

*source:* mlb. mlb.com

4. **Small dogs** Some people love very small dogs. Perlita is a very small Chihuahua that weighs only 2.3 lb. She has a playmate at the dog park who is a Yorkshire Terrier named Cocoa. Cocoa weighs only 2.1 lb. Given the information below about the distribution of the weights, which dog would be considered smaller for its breed?

|  | Yorkshire Terrier | Chihuahua |
|---|---|---|
| Mean | 5.5 lb | 5 lb |
| Standard deviation | 0.75 lb | 0.5 lb |

5. **SAT or ACT?** Each year thousands of high school students take either the SAT or the ACT, standardized tests used in the college admissions process. Combined SAT Math and EBRW scores go as high as 1600, while the maximum ACT composite score is 36. Because the two exams use very different scales, comparisons of performance are difficult. A convenient rule of thumb is SAT = 40 × ACT + 150; that is, multiply an ACT score by 40 and add 150 points to estimate the equivalent SAT score. An admissions officer reported the following statistics about the ACT scores of 2355 students who applied to her college one year. Find the summaries of equivalent SAT scores.

Lowest score = 19   Mean = 27   Standard deviation = 3
Q3 = 30   Median = 28   IQR = 6

6. **Cold U?** A high school senior uses the Internet to get information on February temperatures in the town where he'll be going to college. He finds a website with some statistics, but they are given in degrees Celsius. The conversion formula is °F = (9/5)°C + 32. Determine the Fahrenheit equivalents for the summary information below.

Maximum temperature = 11°C   Range = 33°
Mean = 1°                    Standard deviation = 7°
Median = 2°                  IQR = 16°

7. **Stats test, part II** Suppose your Statistics professor reports test grades as z-scores, and you got a score of 2.20 on an exam. Write a sentence explaining what that means.

8. **Checkup** One of the authors has an adopted grandson whose birth family members are very short. After examining him at his 2-year checkup, the boy's pediatrician said that the z-score for his height relative to American 2-year-olds was −1.88. Write a sentence explaining what that means.

9. **Stats test, part III** The mean score on the Stats exam was 75 points with a standard deviation of 5 points, and Gregor's z-score was −2. How many points did he score?

10. **Mensa** People with z-scores above 2.5 on an IQ test are sometimes classified as geniuses. If IQ scores have a mean of 100 and a standard deviation of 15 points, what IQ score do you need to be considered a genius?

11. **Temperatures** A town's January high temperatures average 36°F with a standard deviation of 10°, while in July the mean high temperature is 74° and the standard deviation is 8°. In which month is it more unusual to have a day with a high temperature of 55°? Explain.

12. **Placement exams** An incoming freshman took her college's placement exams in French and mathematics. In French, she scored 82 and in math 86. The overall results on the French exam had a mean of 72 and a standard deviation of 8, while the mean math score was 68, with a standard deviation of 12. On which exam did she do better compared with the other freshmen?

13. **Combining test scores** The first Stats exam had a mean of 65 and a standard deviation of 10 points; the second had a mean of 80 and a standard deviation of 5 points. Derrick scored an 80 on both tests. Julie scored a 70 on the first test and a 90 on the second. They both totaled 160 points on the two exams, but Julie claims that her total is better. How might Julie justify this claim?

14. **Combining scores again** The first Stats exam had a mean of 80 and a standard deviation of 4 points; the second had a mean of 70 and a standard deviation of 15 points. Reginald scored an 80 on the first test and an 85 on the second. Sara scored an 88 on the first but only a 65 on the second. Although Reginald's total score is higher, Sara feels she should get the higher grade. Explain her point of view.

**15. Final exams** Anna, a language major, took final exams in both French and Spanish and scored 83 on each. Her roommate Megan, also taking both courses, scored 77 on the French exam and 95 on the Spanish exam. Overall, student scores on the French exam had a mean of 81 and a standard deviation of 5, and the Spanish scores had a mean of 74 and a standard deviation of 15.

a) To qualify for language honors, a major must maintain at least an 85 average for all language courses taken. Based on that criterion, which student qualifies?
b) Which student's overall performance was better?

**16. MP3s** Two companies market new batteries targeted at owners of personal music players. DuraTunes claims a mean battery life of 11 hours, while RockReady advertises 12 hours.

a) Explain why you would also like to know the standard deviations of the battery life spans before deciding which brand to buy.
b) Suppose those standard deviations are 2 hours for DuraTunes and 1.5 hours for RockReady. You are headed for 8 hours at the beach. Which battery is most likely to last all day? Explain.
c) If your beach trip is all weekend, and you probably will have the music on for 16 hours, which battery is most likely to last? Explain.

**17. Quokka** Thanks to social media, the happiest creature on earth, an always smiling Australian marsupial called a quokka, has become well known. Suppose that weights of quokkas can be described by a Normal model with a mean of 8 pounds and a standard deviation of 1.5 pounds.

a) How many standard deviations from the mean would a quokka weighing 6 pounds be?
b) Which would be more unusual, a quokka weighing 6 pounds or one weighing 9 pounds?

**18. Car speeds** John Beale of Stanford, CA, recorded the speeds of cars driving past his house, where the speed limit was 20 mph. The mean of 100 readings was 23.84 mph, with a standard deviation of 3.56 mph. (He actually recorded every car for a two-month period. These are 100 representative readings.)

a) How many standard deviations from the mean would a car going under the speed limit be?
b) Which would be more unusual, a car traveling 34 mph or one going 10 mph?

**19. More marsupials** Recall that the quokka described in Exercise 17 had a mean weight of 8 pounds, with a standard deviation of 1.5 pounds.

a) A zoo keeper keeps a record of the weights of quokka in her zoo. If she subtracts the mean of 8 pounds from all the weights, what is the mean and standard deviation of the remainders?
b) Suppose such transporting small animals costs $0.50/pound. What is the mean and standard deviation of the cost of transporting a quokka?

**20. Car speeds again** For the car speed data of Exercise 18, recall that the mean speed recorded was 23.84 mph, with a standard deviation of 3.56 mph. To see how many cars are speeding, John subtracts 20 mph from all speeds.

a) What is the mean speed now? What is the new standard deviation?
b) His friend in Berlin wants to study the speeds, so John converts all the original miles-per-hour readings to kilometers per hour by multiplying all speeds by 1.609 (km per mile). What is the mean now? What is the new standard deviation?

**21. Quokka, part III** The quokkas in Exercise 19 are placed in 30 ounce boxes while they are moved from one location to another. There are 16 ounces in a pound, so a 9-pound quokka in a box would weigh $16(9) + 30 = 174$ ounces. In one group of quokkas the minimum weight was 5 pounds, the median 7.5 pounds, standard deviation 1.5 pounds, and IQR 2 pounds. Find the minimum, median, standard deviation, and IQR of these quokkas in ounces when they're in boxes for transport.

**22. Caught speeding** Suppose police set up radar surveillance on the Stanford street described in Exercise 18. For the many speeders who got tickets the mean was 28 mph, with a standard deviation of 2.4 mph, a maximum of 33 mph, and an IQR of 3.2 mph. Local law prescribes fines of $100, plus $10 per mile per hour over the 20 mph speed limit. For example, a driver convicted of going 25 mph would be fined $100 + 10(5) = \$150$. Find the mean, standard deviation, maximum, and IQR of all the potential fines.

**23. Professors** A friend tells you about a recent study dealing with the number of years of teaching experience among current college professors. He remembers the mean but can't recall whether the standard deviation was 6 months, 6 years, or 16 years. Tell him which one it must have been, and why.

**24. Rock concerts** A popular band on tour played a series of concerts in large venues. They always drew a large crowd, averaging 21,359 fans. While the band did not announce (and probably never calculated) the standard deviation, which of these values do you think is most likely to be correct: 20, 200, 2000, or 20,000 fans? Explain your choice.

**25. Guzzlers?** Environmental Protection Agency (EPA) fuel economy estimates for automobile models tested recently predicted a mean of 24.8 mpg and a standard deviation of 6.2 mpg for highway driving. Assume that a Normal model can be applied.

a) Draw the model for auto fuel economy. Clearly label it, showing what the 68–95–99.7 Rule predicts.
b) In what interval would you expect the central 68% of autos to be found?
c) About what percent of autos should get more than 31 mpg?
d) About what percent of cars should get between 31 and 37.2 mpg?
e) Describe the gas mileage of the worst 2.5% of all cars.

**26. IQ** Some IQ tests are standardized to a Normal model, with a mean of 100 and a standard deviation of 15.
 a) Draw the model for these IQ scores. Clearly label it, showing what the 68–95–99.7 Rule predicts.
 b) In what interval would you expect the central 95% of IQ scores to be found?
 c) About what percent of people should have IQ scores above 115?
 d) About what percent of people should have IQ scores between 70 and 85?
 e) About what percent of people should have IQ scores above 130?

**27. Small quokka** In Exercise 17 we suggested the model $N(8, 1.5)$ for weights in pounds of the adorable quokka. What weight would you consider to be unusually low for such an animal? Explain.

**28. High IQ** Exercise 26 proposes modeling IQ scores with $N(100, 15)$. What IQ would you consider to be unusually high? Explain.

**29. Trees** A forester measured 27 of the trees in a large woods that is up for sale. He found a mean diameter of 15.4 inches and a standard deviation of 4.7 inches. He believes that these trees provide an accurate description of the whole forest and that a Normal model applies.
 a) Draw the Normal model for tree diameters.
 b) What size would you expect the central 95% of all trees to be?
 c) About what percentage of the trees should be less than six inches in diameter?
 d) About what percentage of the trees should be between 10.7 and 15.4 inches in diameter?
 e) About what percentage of the trees should be over 20.1 inches in diameter?

**30. Rivets** The "shear strength" of a rivet is a measure of its resistance to certain types of damage. A company that manufactures rivets believes their shear strength (in pounds) is modeled by $N(800, 50)$.
 a) Draw and label the Normal model.
 b) Would it be safe to use these rivets in a situation requiring a shear strength of 750 pounds? Explain.
 c) About what percent of these rivets would you expect to fall below 900 pounds?
 d) Suppose you needed rivets to fail at a rate of one in a million or less. What is the maximum shear strength for which you would feel comfortable approving this company's rivets?

**31. Trees revisited** The forester in Exercise 29 measured the diameters of 27 trees in a different woods, and from these made projections about the whole forest based on a Normal model. The histogram displays his data. Do you think his analysis was justified? Explain, citing some specific concerns.

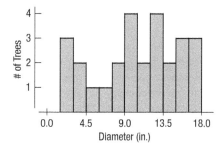

**32. Car speeds, the picture** For the car speed data of Exercise 18, here are the histogram, boxplot, and Normal probability plot of the 100 readings. Do you think it is appropriate to apply a Normal model here? Explain.

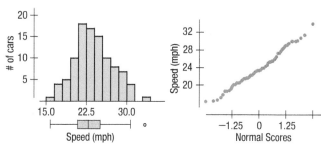

**33. Wisconsin ACT 2015** The histogram shows the distribution of mean ACT composite scores for all Wisconsin public schools in 2015. The vertical lines show the mean and one standard deviation above and below the mean. 80.1% of the data points are between the two outer lines.
 a) Give two reasons that a Normal model is not appropriate here.

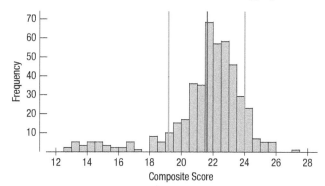

 b) The lower cluster, that is, the scores less than about 17, is almost entirely made up of schools in the Milwaukee school district. If those scores are removed, what would be the shape of the new distribution? What would happen to the mean and standard deviation?

**34. Wisconsin ACT 2015 II** This plot shows the mean ACT scores for Wisconsin schools with the Milwaukee schools and a couple alternative schools removed. The vertical lines show the mean, mean $- 1s$, and mean $+ 1s$.

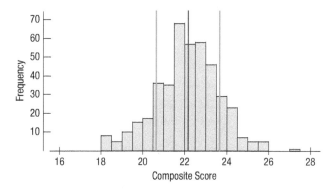

 a) Describe the shape of the distribution.
 b) Does a Normal model seem appropriate for this distribution?
 c) 65.7% of the school average scores are between the two outer lines. Does that support or refute your answer to part b?

d) Below, a Normal model with mean 23.64 and standard deviation 2.42 is drawn over the histogram. Explain how this demonstrates George Box's statement, "*All models are wrong—but some are useful.*"

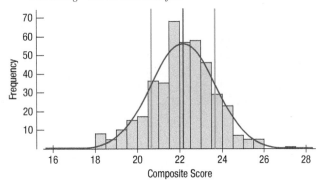

**35. Winter Olympics 2018 downhill** Fifty-three men completed the men's alpine downhill part of the super combined. The gold medal winner finished in 100.25 seconds. Here are the times (in seconds) for all competitors.

| | | | | |
|---|---|---|---|---|
| 100.25 | 101.75 | 102.84 | 103.88 | 105.86 |
| 100.37 | 101.76 | 102.96 | 104.02 | 106.42 |
| 100.43 | 101.86 | 102.96 | 104.37 | 106.6 |
| 100.79 | 101.89 | 102.98 | 104.48 | 107.87 |
| 101.03 | 102.18 | 102.99 | 104.65 | 107.99 |
| 101.08 | 102.22 | 103.03 | 104.79 | 108.57 |
| 101.19 | 102.23 | 103.19 | 105.01 | 108.81 |
| 101.39 | 102.39 | 103.61 | 105.21 | 109.5 |
| 101.46 | 102.53 | 103.72 | 105.36 | 109.98 |
| 101.62 | 102.59 | 103.78 | 105.42 | 111.72 |
| | 102.82 | 103.8 | 105.61 | |

a) The mean time was 103.88 seconds, with a standard deviation of 2.64 seconds. If the Normal model is appropriate, what percent of times will be greater than 108.52 seconds?
b) What is the actual percent of times greater than 108.52 seconds?
c) Why do you think the two percentages don't agree?
d) Make a histogram of these times. What do you see?

**36. Check the model** The mean of the 100 car speeds in Exercise 20 was 23.84 mph, with a standard deviation of 3.56 mph.

a) Using a Normal model, what values should border the middle 95% of all car speeds?
b) Here are some summary statistics.

| Percentile | | Speed |
|---|---|---|
| 100% | Max | 34.060 |
| 97.5% | | 30.976 |
| 90.0% | | 28.978 |
| 75.0% | Q3 | 25.785 |
| 50.0% | Median | 23.525 |
| 25.0% | Q1 | 21.547 |
| 10.0% | | 19.163 |
| 2.5% | | 16.638 |
| 0.0% | Min | 16.270 |

From your answer in part a, how well does the model do in predicting those percentiles? Are you surprised? Explain.

**37. Receivers 2019** NFL data from the 2019 football season reported the number of yards gained by each of the league's 508 receivers:

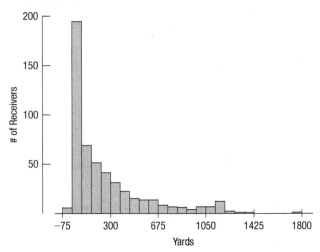

The mean is 253.76 yards, with a standard deviation of 312.36 yards.

a) According to the Normal model, what percent of receivers would you expect to gain more yards than 2 standard deviations above the mean number of yards?
b) For these data, what does that mean?
c) Explain the problem with using a Normal model here.

**38. Home Runs 2019** Here are the numbers of home runs hit by all 604 Major League Baseball players in the 2019 season (https://baseballsavant.mlb.com). As you can see, many players hit fewer than 5 home runs, while the best players hit more than 30.

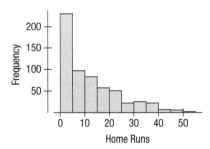

| Count | 604 |
|---|---|
| Mean | 11.38 |
| Median | 8 |
| StdDev | 11.25 |
| IQR | 16 |
| Q1 | 2 |
| Q3 | 18 |

a) Which would be the better summary of the number of home runs hit, the mean or the median? Why?
b) Which is a better description of the spread of the number of home runs hit, the standard deviation or the IQR? Explain.
c) Using a Normal model, what percentage of players should be within one standard deviation of the mean?
d) Using rough estimates, what percentage of players are within one standard deviation of the mean?
e) Explain the discrepancy between parts c and d.

**39. Normal quokka** Using $N(8, 1.5)$, the Normal model for weights of quokkas in Exercise 17, what percent of quokkas weigh

a) over 10.3 pounds?
b) under 9.1 pounds?
c) between 9.1 and 10.3 pounds?

**40. IQs revisited** Based on the Normal model $N(100, 15)$ describing IQ scores, what percent of people's IQs would you expect to be

a) over 80?
b) under 90?
c) between 112 and 132?

**41. More quokka** Based on the model $N(8, 1.5)$ describing quokka weights, what are the cutoff values for

a) the highest 10% of the weights?
b) the lowest 20% of the weights?
c) the middle 40% of the weights?

**42. More IQs** In the Normal model $N(100, 15)$, what cutoff value bounds

a) the highest 5% of all IQs?
b) the lowest 30% of the IQs?
c) the middle 80% of the IQs?

**43. Quokka, finis** Consider the quokka weights model $N(8, 1.5)$ one last time.

a) What weight represents the 40th percentile?
b) What weight represents the 99th percentile?
c) What's the IQR of the weights of these quokkas?

**44. IQ, finis** Consider the IQ model $N(100, 15)$ one last time.

a) What IQ represents the 15th percentile?
b) What IQ represents the 98th percentile?
c) What's the IQR of the IQs?

**45. Cholesterol** Assume the cholesterol levels of adult American women can be described by a Normal model with a mean of 188 mg/dL and a standard deviation of 24.

a) Draw and label the Normal model.
b) What percent of adult women do you expect to have cholesterol levels over 200 mg/dL?
c) What percent of adult women do you expect to have cholesterol levels between 150 and 170 mg/dL?
d) Estimate the IQR of the cholesterol levels.
e) Above what value are the highest 15% of women's cholesterol levels?

**46. Tires** A tire manufacturer believes that the treadlife of its snow tires can be described by a Normal model with a mean of 32,000 miles and standard deviation of 2500 miles.

a) If you buy a set of these tires, would it be reasonable for you to hope they'll last 40,000 miles? Explain.
b) Approximately what fraction of these tires can be expected to last less than 30,000 miles?
c) Approximately what fraction of these tires can be expected to last between 30,000 and 35,000 miles?
d) Estimate the IQR of the treadlives.
e) In planning a marketing strategy, a local tire dealer wants to offer a refund to any customer whose tires fail to last a certain number of miles. However, the dealer does not want to take too big a risk. If the dealer is willing to give refunds to no more than 1 of every 25 customers, for what mileage can he guarantee these tires to last?

**47. Kindergarten** Companies that design furniture for elementary school classrooms produce a variety of sizes for kids of different ages. Suppose the heights of kindergarten children can be described by a Normal model with a mean of 38.2 inches and standard deviation of 1.8 inches.

a) What fraction of kindergarten kids should the company expect to be less than 3 feet tall?
b) In what height interval should the company expect to find the middle 80% of kindergarteners?
c) At least how tall are the biggest 10% of kindergarteners?

**48. Body temperatures** Most people think that the "normal" adult body temperature is 98.6°F. That figure, based on a 19th-century study, has been challenged. In a 1992 article in the *Journal of the American Medical Association*, researchers reported that a more accurate figure may be 98.2°F. Furthermore, the standard deviation appeared to be around 0.7°F. Assume that a Normal model is appropriate.

a) In what interval would you expect most people's body temperatures to be? Explain.
b) What fraction of people would be expected to have body temperatures above 98.6°F?
c) Below what body temperature are the coolest 20% of all people?

**49. Eggs** Hens usually begin laying eggs when they are about 6 months old. Young hens tend to lay smaller eggs, often weighing less than the desired minimum weight of 54 grams.

a) The average weight of the eggs produced by the young hens is 50.9 grams, and only 28% of their eggs exceed the desired minimum weight. If a Normal model is appropriate, what would the standard deviation of the egg weights be?
b) By the time these hens have reached the age of 1 year, the eggs they produce average 67.1 grams, and 98% of them are above the minimum weight. What is the standard deviation for the appropriate Normal model for these older hens?
c) Are egg sizes more consistent for the younger hens or the older ones? Explain.

**50. Tomatoes** Agricultural scientists are working on developing an improved variety of Roma tomatoes. Marketing research indicates that customers are likely to bypass Romas that weigh less than 70 grams. The current variety of Roma plants produces fruit that averages 74 grams, but 11% of the tomatoes are too small. It is reasonable to assume that a Normal model applies.

a) What is the standard deviation of the weights of Romas now being grown?
b) Scientists hope to reduce the frequency of undersized tomatoes to no more than 4%. One way to accomplish this is to raise the average size of the fruit. If the standard deviation remains the same, what target mean should they have as a goal?
c) The researchers produce a new variety with a mean weight of 75 grams, which meets the 4% goal. What is the standard deviation of the weights of these new Romas?
d) Based on their standard deviations, compare the tomatoes produced by the two varieties.

**51. Wisconsin ACT 2015 finale** The histograms in Exercises 33 and 34 show the distribution of mean ACT composite scores for all Wisconsin public schools in 2015. 80.1% of the data points fall between one standard deviation below the mean and one standard deviation above the mean.

The Normal probability plot on the left shows the distribution of these scores. The plot on the right shows the same data with the Milwaukee area schools (mostly in the low mode) removed. What do these plots tell you about the shape of the distributions?

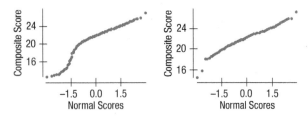

**52. Wisconsin math** The dotplot and Normal probability plot below show the average ACT Mathematics score for a random sample of 20 Wisconsin schools from the 2015–2016 school year.

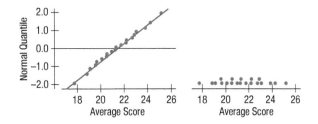

a) Does the distribution of *sample data* suggests that a Normal model would be reasonable for the ACT Mathematics scores for the *population* of all Wisconsin schools?
b) The mean of these 20 average scores is 21.4 and the standard deviation is 1.94. Can you use the 68–95–99.7 Rule to give an interval that you would expect to contain approximately the middle 95% of average scores for *all* Wisconsin schools? If so, what is that interval? If not, explain why not.

**53. Health study** The dotplot and Normal probability plot below show the ages of a random sample of 20 men from the NHANES health study.

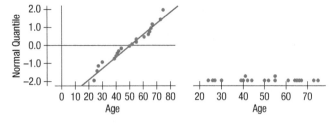

a) Does the distribution of *sample data* suggest that a Normal model would be reasonable for the ages of the *population* of all men in the NHANES study?
b) The mean of these 20 ages is 49.8 years and the standard deviation is 16.02 years. Can you use the 68–95–99.7 Rule to give an interval that you would expect to contain approximately the middle 95% of ages for *all* men in the study? If so, what is that interval? If not, explain why not.

**54. Life expectancy** The dotplot and Normal probability plot below show the life expectancies at birth for a random sample of 20 countries from Europe and Africa.

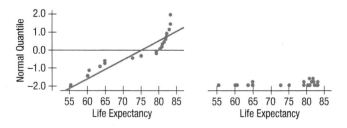

a) Does the distribution of *sample data* suggest that a Normal model would be reasonable for the life expectancies of the *population* of all countries in Europe and Africa?
b) The mean of these 20 life expectancies is 74.8 years and the standard deviation is 9.40 years. Can you use the 68–95–99.7 Rule to give an interval that you would expect to contain approximately the middle 95% of life expectancies for *all* countries in Europe and Africa? If so, what is that interval? If not, explain why not.

**55. CEO compensation 2019 sampled** The Glassdoor Economic Research Blog published the compensation (in millions of dollars) for the CEOs of large companies. The distribution looks like this:

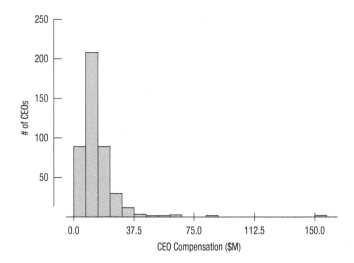

The mean CEO compensation is $14.1M and the standard deviation is $11.32M.

a) According to the Normal model, what percent of CEOs would you expect to earn more than 2 standard deviations above the mean compensation?
b) Is that percentage appropriate for these data?

Suppose we draw samples from the data and calculate the mean of each sample. How would we expect the *means* to vary? (Turn the page to find out.)

Here is a histogram of the means of 1000 samples of 30 drawn from the CEOs:

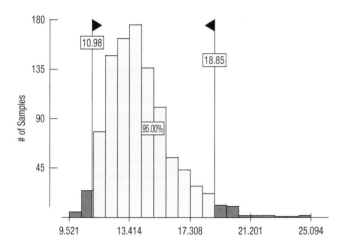

c) The standard deviation of these means is 2.0. The middle 95% of the means is colored in the histogram. Do you think the 68–95–99.7 Rule applies?

Suppose we draw samples of 100 instead. Now the histogram of means looks like this:

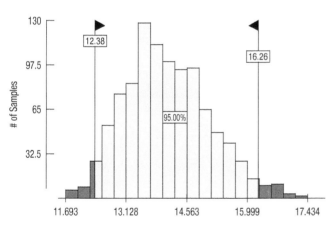

d) The standard deviation of these means is 1.0. Do you think the 68–95–99.7 Rule applies?

Once more, but this time with samples of 200:

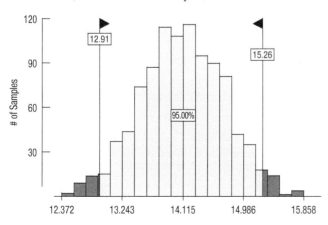

e) The standard deviation of these means is 0.60. Does the 68–95–99.7 Rule give a good idea of the middle 95% of the distribution?

**56. CEO compensation 2019 logged and sampled** Suppose we take logarithms of the CEO compensations in Exercise 55. The histogram of log Compensation looks like this:

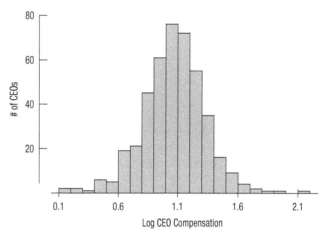

with a mean of 1.07 and a standard deviation of 0.26.

a) According to the Normal model, what percent of CEOs would you expect to earn more than 2 standard deviations above the mean compensation?
b) Is that percentage appropriate for these data?

Now let's draw samples of 30 CEOs from the logged data. We drew 1000 samples and found their means. The distribution of means looks like this:

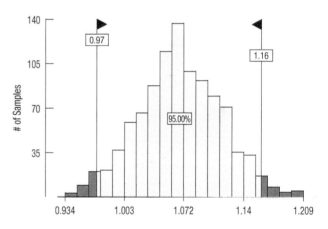

with a standard deviation of 0.05.

c) Do you think the 68–95–99.7 Rule applies to these means?

## JUST CHECKING

### Answers

1. a) On the first test, the mean is 88 and the SD is 4, so $z = (90 - 88)/4 = 0.5$. On the second test, the mean is 75 and the SD is 5, so $z = (80 - 75)/5 = 1.0$. The first test has the lower z-score, so it is the one that will be dropped.
   b) Yes. The second test is 1 standard deviation above the mean, farther away than the first test, so it's the better score relative to the class.

2. If incomes have a right-skewed distribution, then the higher-than-average incomes must be stretched out over a long interval and lower-than-average incomes must be bunched up over a short interval. So we'd expect the minimum income to be closer to the mean than the maximum income.

3. a) The mean would increase to 500.
   b) The standard deviation is still 100 points.
   c) The two boxplots would look nearly identical (the shape of the distribution would remain the same), but the later one would be shifted 50 points higher.

4. No, he should not be concerned. The standard deviation is now 2.54 millimeters, which is the same as 0.1 inches. Nothing has changed. The standard deviation has "increased" only because we're reporting it in millimeters now, not inches.

5. The mean is 184 centimeters, with a standard deviation of 8 centimeters. 2 meters is 200 centimeters, which is 2 standard deviations above the mean. We expect 5% of the men to be more than 2 standard deviations below or above the mean, so half of those, 2.5%, are likely to be above 2 meters.

6. a) We know that 68% of the time we'd expect to be within 1 standard deviation (2 min) of 20. So 32% of the time we'd expect to arrive in less than 18 or more than 22 minutes. Half of those times (16%) should be greater than 22 minutes, so 84% should be less than 22 minutes.
   b) 24 minutes is 2 standard deviations above the mean. Because of the 95% rule, we expect approximately 2.5% of the times will be more than 24 minutes.
   c) Traffic incidents may occasionally increase the time it takes to get to school, so the driving times may be skewed to the right, and there may be outliers.
   d) If so, the Normal model would not be appropriate and the percentages we predict would not be accurate.

# Review of Part I

## EXPLORING AND UNDERSTANDING DATA

### Quick Review

It's time to put it all together. Real data don't come tagged with instructions for use. So let's step back and look at how the key concepts and skills we've seen work together. This brief list and the review exercises that follow should help you check your understanding of Statistics so far.

- We treat most data in one of two ways: as either categorical or as quantitative.
- To describe categorical data:
  - Make a picture. Bar graphs work well for comparing counts in categories.
  - Summarize the distribution with a table of counts or relative frequencies (percents) in each category.
  - Pie charts and segmented bar charts display divisions of a whole.
  - Compare distributions with plots side by side using a common scale (like relative frequency).
  - Look for associations between variables by comparing marginal and conditional distributions. A mosaic plot can also help to reveal associations.
- To describe quantitative data:
  - Make a picture. Use histograms, boxplots, stem-and-leaf displays, or dotplots. Stem-and-leafs are great when working by hand and good for small datasets. Histograms are a good way to see the distribution. Boxplots are often best for comparing several distributions.
  - A cumulative distribution plot can show quantiles of a dataset precisely (e.g., quartiles, the median).
  - Describe distributions in terms of their shape, center, and spread, and note any unusual features such as gaps or outliers.
  - Regarding shape, most distributions you see will be unimodal. For unimodal distributions, address symmetry. For bimodal or multimodal distributions, describe the locations and spreads of the apparent subgroups, and think about whether they might represent different populations.
- A 5-number summary makes a good numerical description of a unimodal distribution: min, Q1, median, Q3, and max.
- If a distribution is skewed, be sure to include the median and interquartile range (IQR) when you describe its center and spread.
- A distribution that is severely skewed may benefit from re-expressing the data. If it is skewed to the high end, taking logs often works well.
- If the distribution is unimodal and symmetric, describe its center and spread with the mean and standard deviation.
- Use the standard deviation as a ruler to tell how unusual an observed value may be, or to compare or combine measurements made on different scales.
- Shifting a distribution by adding or subtracting a constant affects measures of position but not measures of spread. Rescaling by multiplying or dividing by a constant affects both.
- When a distribution is roughly unimodal and symmetric, a Normal model may be useful. For Normal models, the 68–95–99.7 Rule is a good rule of thumb.
- If the Normal model fits well (check a histogram or Normal probability plot), then Normal percentile tables or functions found in most statistics technology can provide more precise values.

Need more help with some of this? It never hurts to reread sections of the chapters! And in the following pages we offer you more opportunities[1] to review these concepts and skills.

The exercises that follow use the concepts and skills you've learned in the first five chapters. To be more realistic and more useful for your review, they don't tell you which of the concepts or methods you need. Making those decisions yourself is part of statistical thinking.

---
[1] If you doubted that we are teachers, this should convince you. Only a teacher would call additional homework exercises "opportunities."

## REVIEW EXERCISES

**1. Bananas** Here are the prices (in cents per pound) of bananas reported from 15 markets surveyed by the U.S. Department of Agriculture.

| 51 | 52 | 45 |
| 48 | 53 | 52 |
| 50 | 49 | 52 |
| 48 | 43 | 46 |
| 45 | 42 | 50 |

a) Display these data with an appropriate graph.
b) Report appropriate summary statistics.
c) Write a few sentences about this distribution.

**2. Prenatal care** Results of a 1996 American Medical Association report about the infant mortality rate for twins carried for the full term of a normal pregnancy are shown on the next page, broken down by the level of prenatal care the mother had received.

| Full-Term Pregnancies, Level of Prenatal Care | Infant Mortality Rate Among Twins (deaths per thousand live births) |
|---|---|
| Intensive | 5.4 |
| Adequate | 3.9 |
| Inadequate | 6.1 |
| **Overall** | **5.1** |

a) Is the overall rate the average of the other three rates? Should it be? Explain.
b) Do these results indicate that adequate prenatal care is important for pregnant women? Explain.
c) Do these results suggest that a woman pregnant with twins should be wary of seeking too much medical care? Explain.

**3. Truecars newer?** The boxplots shown display the prices (in dollars) of the 1500 cars in Chapter 3. However, now the asking price has been divided into two groups: cars that were made in 2011 or newer vs. cars made in 2010 or older.

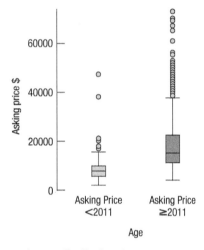

a) Compare the two distributions in context.
b) Clearly the age of a car affects its asking price. But there are other variables that have an influence on the price. What are other variables that a buyer would consider?
c) To convert these prices to Canadian dollars, you would multiply by a conversion rate (at the time of this writing, 1.21). How would this affect the graphs? The summary statstics?

**4. Dialysis** In a study of dialysis, researchers found that "of the three patients who were currently on dialysis, 67% had developed blindness and 33% had their toes amputated." What kind of display might be appropriate for these data? Explain.

**5. Beanstalks** Beanstalk Clubs are social clubs for very tall people. To join, a man must be over 6'2" tall, and a woman over 5'10". The National Health Survey suggests that heights of adults may be Normally distributed, with mean heights of 69.1" for men and 64.0" for women. The respective standard deviations are 2.8" and 2.5".

a) You are probably not surprised to learn that men are generally taller than women, but what does the greater standard deviation for men's heights indicate?
b) Who are more likely to qualify for Beanstalk membership, men or women? Explain.

**6. Bread** Clarksburg Bakery is trying to predict how many loaves its bakers need to bake. In the last 100 days, they have sold between 95 and 140 loaves per day. At the top of the next column is a histogram of the number of loaves they sold for the last 100 days.

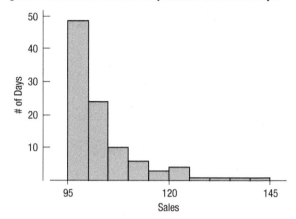

a) Describe the distribution.
b) Which should be larger, the mean number of sales or the median? Explain.
c) Here are the summary statistics for Clarksburg Bakery's bread sales. Use these statistics and the histogram above to create a boxplot. To assist you, here are the data from eight busiest days: 115, 121, 122, 124, 128, 130, 135, 140.

| Summary of Sales | |
|---|---|
| Median | 100 |
| Min | 95 |
| Max | 140 |
| 25th %tile | 97 |
| 75th %tile | 105.5 |

d) For these data, the mean was 103 loaves sold per day, with a standard deviation of 9 loaves. Do these statistics suggest that Clarksburg Bakery should expect to sell between 94 and 112 loaves on about 68% of the days? Explain.
e) A mistake was found in the data. The day with 135 sales was actually 107 sales. After this adjustment is made, which of the summary statistics will change? Which will stay the same? Explain.

**7. State University** Public relations staff at State U. phoned 850 local residents. After identifying themselves, the callers asked the survey participants their ages, whether they had attended college, and whether they had a favorable opinion of the university. The official report to the university's directors claimed that, in general, people had very favorable opinions about their university.

a) Identify the W's of these data.
b) Identify the variables, classify each as categorical or quantitative, and specify units if relevant.
c) Are you confident about the report's conclusion? Explain.

**8. Acid rain** Based on long-term investigation, researchers have suggested that the acidity (pH) of rainfall in the Shenandoah Mountains can be described by the Normal model $N(4.9, 0.6)$.

a) Draw and carefully label the model.
b) What percent of storms produce rainfall with pH over 6?
c) What percent of storms produce rainfall with pH under 4?
d) The lower the pH, the more acidic the rain. What is the pH level for the most acidic 20% of all storms? Reminder: the higher the pH, the lower the acidity.
e) What is the pH level for the least acidic 5% of all storms?
f) What is the IQR for the pH of rainfall?

9. **Fraud detection** A credit card bank is investigating the incidence of fraudulent card use. The bank suspects that the type of product bought may provide clues to the fraud. To examine this situation, the bank looks at the Standard Industrial Code (SIC) of the business related to the transaction. This is a code that was used by the U.S. Census Bureau and Statistics Canada to identify the type of every registered business in North America.[2] For example, 1011 designates Meat and Meat Products (except Poultry), 1012 is Poultry Products, 1021 is Fish Products, 1031 is Canned and Preserved Fruits and Vegetables, and 1032 is Frozen Fruits and Vegetables.

A company intern produces the following histogram of the SIC codes for 1536 transactions:

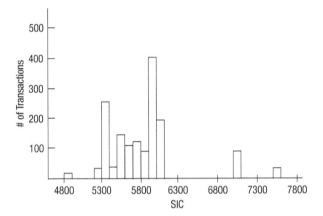

He also reports that the mean SIC is 5823.13 with a standard deviation of 488.17.

a) Comment on any problems you see with the use of the mean and standard deviation as summary statistics.
b) How well do you think the Normal model will work on these data? Explain.

10. **Streams** As part of the course work, a class at an upstate New York college collects data on streams each year. Students record a number of biological, chemical, and physical variables, including the stream name, the substrate of the stream (*limestone, shale,* or *mixed*), the pH, the temperature (°C), and the BCI, a measure of biological diversity.

| Group | Count | % |
|---|---|---|
| Limestone | 77 | 44.8 |
| Mixed | 26 | 15.1 |
| Shale | 69 | 40.1 |

a) Name each variable, indicating whether it is categorical or quantitative, and giving the units if available.
b) These streams have been classified according to their substrate—the composition of soil and rock over which they flow—as summarized in the table. What kind of graph might be used to display these data?
c) Should you describe the center, shape, and spread of this graph? Explain.

11. **Cramming** One Thursday, researchers gave students enrolled in a section of basic Spanish a set of 50 new vocabulary words to memorize. On Friday the students took a vocabulary test. When they returned to class the following Monday, they were retested—without advance warning. Both sets of test scores for the 25 students are shown below.

| Fri | Mon | Fri | Mon |
|---|---|---|---|
| 42 | 36 | 50 | 47 |
| 44 | 44 | 34 | 34 |
| 45 | 46 | 38 | 31 |
| 48 | 38 | 43 | 40 |
| 44 | 40 | 39 | 41 |
| 43 | 38 | 46 | 32 |
| 41 | 37 | 37 | 36 |
| 35 | 31 | 40 | 31 |
| 43 | 32 | 41 | 32 |
| 48 | 37 | 48 | 39 |
| 43 | 41 | 37 | 31 |
| 45 | 32 | 36 | 41 |
| 47 | 44 | | |

a) Create a graphical display to compare the two distributions of scores.
b) Write a few sentences about the scores reported on Friday and Monday.
c) Create a graphical display showing the distribution of the *changes* in student scores.
d) Describe the distribution of changes.

12. **e-Books** A study by the Pew Internet & American Life Project found that 72% of U.S. Adults say they have read a book in the past 12 months. They also found that 25% had read an e-book during that period and that 20% listened to an audiobook. This totals 116%, proving that an error has been made. (https://www.pewresearch.org/fact-tank/2019/09/25/onein-five-americans-now-listen-to-audiobooks/) Do you agree? Explain.

13. **Community service** You and a classmate are active in a community service organization where volunteers bring meals to people in need who lack transportation. To fulfill the requirements of your honor society, you keep a log of your hours and a description of the people you help with a final report.

a) For the people you help, you record the variable *age*. Is this variable categorical or quantitative?
b) Instead of recording a precise *age*, your classmate groups the people he helps into groups: Under 30, 30–50, and Over 50. Is this variable categorical or quantitative?
c) What graph might you make to show the distribution of *age*? What about your classmate?
d) Will you provide summary statistics for *age* in your report? Will your classmate? Explain.

---

[2] Since 1997, the SIC has been replaced by the North American Industry Classification System (NAICS), a code of six letters.

**14. Accidents** Progressive Insurance asked customers who had been involved in auto accidents how far they were from home when the accident happened. The data are summarized in the table.

| Miles from Home | % of Accidents |
|---|---|
| Less than 1 | 23 |
| 1 to 5 | 29 |
| 6 to 10 | 17 |
| 11 to 15 | 8 |
| 16 to 20 | 6 |
| Over 20 | 17 |

a) Create an appropriate graph of these data. Describe the graph.
b) Do these data indicate that driving near home is particularly dangerous? Explain.

**15. Hard water** In an investigation of environmental causes of disease, data were collected on the annual mortality rate (deaths per 100,000) for males in 61 large towns in England and Wales. In addition, the water hardness was recorded as the calcium concentration (parts per million, ppm) in the drinking water.

a) What are the variables in this study? For each, indicate whether it is quantitative or categorical and what the units are.
b) Here are histograms of calcium concentration and mortality. Describe the distributions of the two variables.

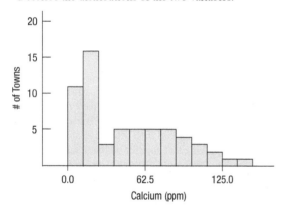

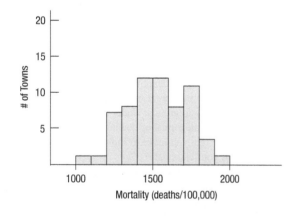

**16. Hard water II** The dataset from England and Wales also notes for each town whether it was south or north of Derby. Here are some summary statistics and a comparative boxplot for the two regions.

| Summary of Mortality | | | | |
|---|---|---|---|---|
| Group | Count | Mean | Median | StdDev |
| North | 34 | 1631.59 | 1631 | 138.470 |
| South | 27 | 1388.85 | 1369 | 151.114 |

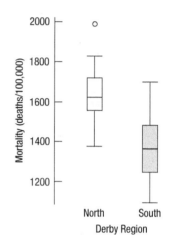

a) What is the overall mean mortality rate for the two regions?
b) Do you see evidence of a difference in mortality rates? Explain.

**17. Seasons** Average daily temperatures in January and July for 60 large U.S. cities are graphed in the following histograms.

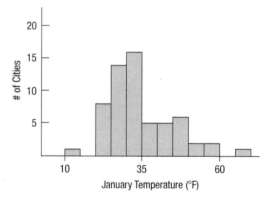

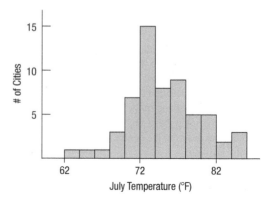

a) What aspect of these histograms makes it difficult to compare the distributions?
b) What differences do you see between the distributions of January and July average temperatures?

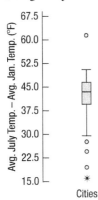

c) Differences in temperatures (July–January) for each of the cities are displayed in the boxplot above. Write a few sentences describing what you see.

**18. Old Faithful** It is a common belief that Yellowstone's most famous geyser erupts once an hour at very predictable intervals. The histogram below shows the time gaps (in minutes) between 222 successive eruptions. Describe this distribution.

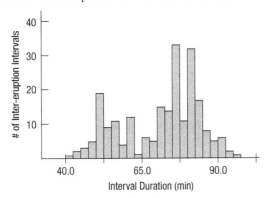

**19. Old Faithful?** Does the duration of an eruption have an effect on the length of time that elapses before the next eruption?
a) The histogram below shows the duration (in minutes) of those 222 eruptions. Describe this distribution.

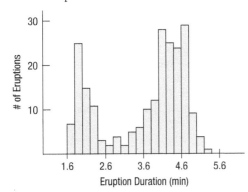

b) Explain why it is not appropriate to find summary statistics for this distribution.
c) Let's classify the eruptions as "long" or "short," depending upon whether or not they last at least 3 minutes. Describe what you see in the comparative boxplots.

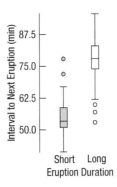

**20. Broadband for everyone?** In a survey conducted by Pew Research Center, respondents were sorted into three groups: rural, suburban, or urban. The respondents were asked if they have broadband Internet in their home. While 79% of suburban residents said that they had broadband, only 72% of rural residents had this service. (https://www.pewresearch.org/internet/2021/06/03/mobile-technology-and-home-broadband-2021/)

a) Use this difference to explain the concept of independence.
b) Is geographic location the only explanation of this difference? Can you think of another variable that might explain this discrepancy?

**21. Liberty's nose** Is the Statue of Liberty's nose too long? Her nose measures, 4′6″, but she is a large statue, after all. Her arm is 42 feet long. That means her arm is $42/4.5 = 9.3$ times as long as her nose. Is that a reasonable ratio? Shown in the table are arm and nose lengths of 18 girls in a Statistics class, and the ratio of arm-to-nose length for each.

| Arm (cm) | Nose (cm) | Arm/Nose Ratio |
|---|---|---|
| 73.8 | 5.0 | 14.8 |
| 74.0 | 4.5 | 16.4 |
| 69.5 | 4.5 | 15.4 |
| 62.5 | 4.7 | 13.3 |
| 68.6 | 4.4 | 15.6 |
| 64.5 | 4.8 | 13.4 |
| 68.2 | 4.8 | 14.2 |
| 63.5 | 4.4 | 14.4 |
| 63.5 | 5.4 | 11.8 |
| 67.0 | 4.6 | 14.6 |
| 67.4 | 4.4 | 15.3 |
| 70.7 | 4.3 | 16.4 |
| 69.4 | 4.1 | 16.9 |
| 71.7 | 4.5 | 15.9 |
| 69.0 | 4.4 | 15.7 |
| 69.8 | 4.5 | 15.5 |
| 71.0 | 4.8 | 14.8 |
| 71.3 | 4.7 | 15.2 |

a) Make an appropriate plot and describe the distribution of the ratios.
b) Summarize the ratios numerically, choosing appropriate measures of center and spread.
c) Is the ratio of 9.3 for the Statue of Liberty unrealistically low? Explain.

**22. Winter Olympics 2014 speed skating** The top 34 women's 500-meter speed skating times are listed in the table below. Times are the total time for two races.

| Name | Nation | Total Time |
|---|---|---|
| Lee Sang-hwa | South Korea | 74.7 |
| Olga Fatkulina | Russia | 75.06 |
| Margot Boer | Netherlands | 75.48 |
| Zhang Hong | China | 75.58 |
| Nao Kodaira | Japan | 75.61 |
| Jenny Wolf | Germany | 75.67 |
| Wang Beixing | China | 75.68 |
| Heather Richardson | United States | 75.75 |
| Maki Tsuji | Japan | 76.84 |
| Karolína Erbanová | Czech Republic | 76.86 |
| Laurine van Riessen | Netherlands | 76.99 |
| Christine Nesbitt | Canada | 77.15 |
| Brittany Bowe | United States | 77.19 |
| Miyako Sumiyoshi | Japan | 77.26 |
| Lauren Cholewinski | United States | 77.35 |
| Lotte van Beek | Netherlands | 77.4 |
| Yekaterina Malysheva | Russia | 77.55 |
| Angelina Golikova | Russia | 77.68 |
| Marrit Leenstra | Netherlands | 77.74 |
| Lee Bo-ra | South Korea | 77.75 |
| Denise Roth | Germany | 77.78 |
| Yekaterina Aydova | Kazakhstan | 77.85 |
| Qi Shuai | China | 77.89 |
| Kim Hyun-yung | South Korea | 78.23 |
| Yekaterina Lobysheva | Russia | 78.24 |
| Park Seung-ju | South Korea | 78.31 |
| Vanessa Bittner | Austria | 78.5 |
| Anastasia Bucsis | Canada | 78.52 |
| Sugar Todd | United States | 78.53 |
| Yvonne Daldossi | Italy | 78.64 |
| Zhang Shuang | China | 78.65 |
| Marsha Hudey | Canada | 79.22 |
| Danielle Wotherspoon-Gregg | Canada | 79.32 |
| Gabriele Hirschbichler | Germany | 79.51 |

a) Graph Total Time and find the summary statistics.
b) Describe the distribution.
c) The mean finishing time was 77.37 seconds, with a standard deviation of 1.28 seconds. If a Normal model were appropriate, what percentage of the times should be within 2 seconds of the mean?
d) What percentage of the times actually fall within this interval?
e) Explain the discrepancy between parts c and d.

**23. Sample** A vet collects data on the last 100 dogs she cared for. Here are the summary statistics for the weights of dogs, in pounds.

| Summary of Weights | |
|---|---|
| Median | 29 |
| Min | 3.5 |
| Max | 102 |
| 25th %tile | 21 |
| 75th %tile | 55 |

a) Draw a boxplot for weight.
b) Describe the distribution.
c) An error was found in the data. An 80-pound dog was incorrectly recorded as weighing 60 pounds. When the error is corrected, how will the summary statistics change?

**24. Sluggers** Babe Ruth was the first great "slugger" in baseball. His record of 60 home runs in one season held for 34 years until Roger Maris hit 61 in 1961. Mark McGwire (with the aid of steroids) set a new standard of 70 in 1998. Listed below are the home run totals for each season McGwire played. Also listed are Babe Ruth's home run totals.

**McGwire:** 3*, 49, 32, 33, 39, 22, 42, 9*, 9*, 39, 52, 58, 70, 65, 32*, 29*

**Ruth:** 54, 59, 35, 41, 46, 25, 47, 60, 54, 46, 49, 46, 41, 34, 22

a) Find the 5-number summary for McGwire's career.
b) Do any of his seasons appear to be outliers? Explain.
c) McGwire played in only 18 games at the end of his first big league season, and missed major portions of some other seasons because of injuries to his back and knees. Those seasons might not be representative of his abilities. They are marked with asterisks in the list above. Omit these values and make parallel boxplots comparing McGwire's career to Babe Ruth's.
d) Write a few sentences comparing the two sluggers.
e) Create a side-by-side stem-and-leaf display comparing the careers of the two players.
f) What aspects of the distributions are apparent in the stem-and-leaf displays that did not clearly show in the boxplots?

**25. Be quick!** Avoiding an accident when driving can depend on reaction time. That time, measured from the moment the driver first sees the danger until they step on the brake pedal, is thought to follow a Normal model with a mean of 1.5 seconds and a standard deviation of 0.18 seconds.

a) Use the 68–95–99.7 Rule to draw the Normal model.
b) Write a few sentences describing driver reaction times.
c) What percent of drivers have a reaction time less than 1.25 seconds?
d) What percent of drivers have reaction times between 1.6 and 1.8 seconds?
e) What is the interquartile range of reaction times?
f) Describe the reaction times of the slowest 1/3 of all drivers.

**26. Music and memory** Is it a good idea to listen to music when studying for a big test? In a study conducted by some Statistics students, 62 people were randomly assigned to listen to rap music, Mozart, or no music while attempting to memorize objects pictured on a page. They were then asked to list all the objects they could remember. Here are the 5-number summaries for each group:

| | n | Min | Q1 | Median | Q3 | Max |
|---|---|---|---|---|---|---|
| Rap | 29 | 5 | 8 | 10 | 12 | 25 |
| Mozart | 20 | 4 | 7 | 10 | 12 | 27 |
| None | 13 | 8 | 9.5 | 13 | 17 | 24 |

a) Describe the W's for these data: *Who, What, Where, Why, When, How.*
b) Name the variables and classify each as categorical or quantitative.
c) Create parallel boxplots as best you can from these summary statistics to display these results.
d) Write a few sentences comparing the performances of the three groups.

**27. Mail** Here are the number of pieces of mail received at a school office for 36 days.

| 123 | 70 | 90 | 151 | 115 | 97 |
|---|---|---|---|---|---|
| 80 | 78 | 72 | 100 | 128 | 130 |
| 52 | 103 | 138 | 66 | 135 | 76 |
| 112 | 92 | 93 | 143 | 100 | 88 |
| 118 | 118 | 106 | 110 | 75 | 60 |
| 95 | 131 | 59 | 115 | 105 | 85 |

a) Plot these data.
b) Find appropriate summary statistics.
c) Write a brief description of the school's mail deliveries.
d) What percent of the days actually lie within one standard deviation of the mean? Comment.

**28. Birth order** Is your birth order related to your choice of major? A Statistics professor at a large university polled his students to find out what their majors were and what position they held in the family birth order. The results are summarized in the table.

a) What percent of these students are oldest or only children?
b) What percent of Humanities majors are oldest children?
c) What percent of oldest children are Humanities students?
d) What percent of the students are oldest children majoring in the Humanities?

| | | Birth Order* | | | |
|---|---|---|---|---|---|
| | | 1 | 2 | 3 | 4+ | Total |
| **Major** | Math/Science | 34 | 14 | 6 | 3 | 57 |
| | Agriculture | 52 | 27 | 5 | 9 | 93 |
| | Humanities | 15 | 17 | 8 | 3 | 43 |
| | Other | 12 | 11 | 1 | 6 | 30 |
| | Total | 113 | 69 | 20 | 21 | 223 |

\* 1 = oldest or only child

**29. Herbal medicine** Researchers for the Herbal Medicine Council collected information on people's experiences with a new herbal remedy for colds. They went to a store that sold natural health products. There they asked 100 customers whether they had taken the cold remedy and, if so, to rate its effectiveness (on a scale from 1 to 10) in curing their symptoms. The Council concluded that this product was highly effective in treating the common cold.

a) Identify the W's of these data.
b) Identify the variables, classify each as categorical or quantitative, and specify units if relevant.
c) Are you confident about the Council's conclusion? Explain.

**30. Birth order revisited** Consider again the data on birth order and college majors in Exercise 28.

a) What is the marginal distribution of majors?
b) What is the conditional distribution of majors for the oldest children?
c) What is the conditional distribution of majors for the children born second?
d) Do you think that college major appears to be independent of birth order? Explain.

**31. Engines** One measure of the size of an automobile engine is its "displacement," the total volume (in liters or cubic inches) of its cylinders. Summary statistics for several models of new cars are shown. These displacements were measured in cubic inches.

| Summary of Displacement | |
|---|---|
| Count | 38 |
| Mean | 177.29 |
| Median | 148.5 |
| StdDev | 88.88 |
| Range | 275 |
| 25th %tile | 105 |
| 75th %tile | 231 |

a) How many cars were measured?
b) Why might the mean be so much larger than the median?
c) Describe the center and spread of this distribution with appropriate statistics.
d) Your neighbor is bragging about the 227-cubic-inch engine he bought in his new car. Is that engine unusually large? Explain.
e) Are there any engines in this dataset that you would consider to be outliers? Explain.
f) Is it reasonable to expect that about 68% of car engines measure between 88 and 266 cubic inches? (That's $177.289 \pm 88.8767$.) Explain.
g) We can convert all the data from cubic inches to cubic centimeters (cc) by multiplying by 16.4. For example, a 200-cubic-inch engine has a displacement of 3280 cc. How would such a conversion affect each of the summary statistics?

**32. Engines, again** Horsepower is another measure commonly used to describe auto engines. Here are the summary statistics and histogram displaying horsepowers of the same group of 38 cars discussed in Exercise 31.

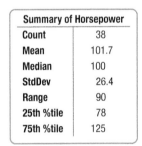

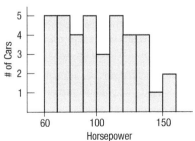

a) Describe the shape, center, and spread of this distribution.
b) What is the interquartile range?
c) Are any of these engines outliers in terms of horsepower? Explain.
d) Do you think the 68–95–99.7 Rule applies to the horsepower of auto engines? Explain.
e) From the histogram, make a rough estimate of the percentage of these engines whose horsepower is within one standard deviation of the mean.
f) A fuel additive boasts in its advertising that it can "add 10 horsepower to any car." Assuming that is true, what would happen to each of these summary statistics if this additive were used in all the cars?

**33. Age and party 2016** The Pew Research Center conducts surveys regularly asking respondents which political party they identify with or lean toward. Among their results is the following table relating preferred political party and age. (http://people-press.org/)

|  |  | Party |  |  |  |
|---|---|---|---|---|---|
|  |  | Republican/ Lean Rep. | Democrat/ Lean Dem. | Neither | Total |
| Generation | Millennial (18–35) | 574 | 909 | 112 | 1595 |
|  | Gen X (36–51) | 783 | 895 | 168 | 1846 |
|  | Baby-Boomer (52–70) | 1623 | 1491 | 199 | 3313 |
|  | Silent (71–88) | 623 | 470 | 83 | 1176 |
|  | Total | 3603 | 3765 | 562 | 7930 |

a) What percent of people surveyed were Republicans or leaned Republican?
b) Do you think this might be a reasonable estimate of the percentage of all voters who are Republicans or lean Republican? Explain.
c) What percent of people surveyed were under 36 or over 70?
d) What percent of people were classified as "Neither" and under the age of 36?
e) What percent of the people classified as "Neither" were under 36?
f) What percent of people under 36 were classified as "Neither"?

**34. Pay** According to the Bureau of Labor Statistics, the mean hourly wage for Chief Executives in 2009 was $80.43 and the median hourly wage was $77.27. By contrast, for General and Operations Managers, the mean hourly wage was $53.15 and the median was $44.55.

a) Are these wage distributions likely to be symmetric, skewed left, or skewed right? Explain.
b) An office has 3 executives that are paid an average of $85/hour and 20 managers who are paid $55/hour. What is the median hourly rate? What is the average hourly rate?

**35. Age and party 2016 II** Consider again the Pew Research Center results on age and political party in Exercise 33.

a) What is the marginal distribution of party affiliation?
b) Create segmented bar graphs displaying the conditional distribution of party affiliation for each age group.

c) Summarize these poll results in a few sentences that might appear in a newspaper article about party affiliation in the United States.
d) Do you think party affiliation is independent of the voter's age? Explain.

**36. Bike safety 2015** The Bicycle Helmet Safety Institute website includes a report on the number of bicycle fatalities per year in the United States. The table below shows the counts for the years 1994–2015.

| Year | Bicycle Fatalities |
|---|---|
| 1994 | 802 |
| 1995 | 833 |
| 1996 | 765 |
| 1997 | 814 |
| 1998 | 760 |
| 1999 | 754 |
| 2000 | 693 |
| 2001 | 732 |
| 2002 | 665 |
| 2003 | 629 |
| 2004 | 727 |
| 2005 | 784 |
| 2006 | 769 |
| 2007 | 699 |
| 2008 | 716 |
| 2009 | 628 |
| 2010 | 616 |
| 2011 | 675 |
| 2012 | 726 |
| 2013 | 743 |
| 2014 | 726 |
| 2015 | 811 |

a) What are the W's for these data?
b) Display the data in a stem-and-leaf display.
c) Display the data in a timeplot.
d) What is apparent in the stem-and-leaf display that is hard to see in the timeplot?
e) What is apparent in the timeplot that is hard to see in the stem-and-leaf display?
f) Write a few sentences about bicycle fatalities in the United States.

**37. Some assembly required** A company that markets build-it-yourself furniture sells a computer desk that is advertised with the claim "less than an hour to assemble." However, through postpurchase surveys the company has learned that only 25% of its customers succeeded in building the desk in under an hour. The mean time was 1.29 hours. The company assumes that consumer assembly time follows a Normal model.

a) Find the standard deviation of the assembly time model.
b) One way the company could solve this problem would be to change the advertising claim. What assembly time should the company quote in order that 60% of customers succeed in finishing the desk by then?

c) Wishing to maintain the "less than an hour" claim, the company hopes that revising the instructions and labeling the parts more clearly can improve the 1-hour success rate to 60%. If the standard deviation stays the same, what new lower mean time does the company need to achieve?

d) Months later, another postpurchase survey shows that new instructions and part labeling did lower the mean assembly time, but only to 55 minutes. Nonetheless, the company did achieve the 60%-in-an-hour goal, too. How was that possible?

**38.** *Global 500 2014* Here is a stem-and-leaf display showing profits (in $M) for 30 of the 500 largest global corporations (as measured by revenue). The stems are split; each stem represents a span of 5000 ($M), from a profit of 43,000 ($M) to a loss of 7000 ($M). Use the stem-and-leaf to answer the questions.

| Stem | Leaf | Count |
|---|---|---|
| 4 | 3 | 1 |
| 3 | 67 | 2 |
| 3 | 3 | 1 |
| 2 | 7 | 1 |
| 2 | 13 | 2 |
| 1 | 66899 | 5 |
| 1 | 1123 | 4 |
| 0 | 55677899 | 8 |
| 0 | 13334 | 5 |
| −0 | | |
| −0 | 7 | 1 |

**Profits ($M)**
−0 | 7 means −7000 ($M)

a) Find the 5-number summary.
b) Draw a boxplot for these data.
c) Find the mean and standard deviation.
d) Describe the distribution of profits for these corporations.

**39.** *Hopkins Forest investigation* The **Hopkins Forest** dataset includes all 24 weather variables reported by the researchers. Many of the variables (e.g., temperature, relative humidity, solar radiation, wind) are reported as daily averages, minima, and maxima. Using any of these variables, compare the distributions of the daily minima and maxima in a few sentences. Use summary statistics and appropriate graphical displays.

**40.** *Titanic investigation* The **Titanic** dataset includes more variables than just those discussed in Chapter 2. Others include the crew's job, and where each person boarded the ship. Stories, biographies, and pictures can be found on the site: www.encyclopedia-titanica.org/. Using the dataset, investigate some of the variables and their associations. Write a short report on what you discover. Be sure to include summary statistics, tables, and graphical displays.

# PRACTICE EXAM

## I. MULTIPLE CHOICE

1. Below are summary statistics for infant mortality rates for Wisconsin counties in 2011. The numbers represent infant deaths per 1000 residents.

   | Mean | 0.05958 |
   |---|---|
   | Median | 0.04492 |
   | Min | 0 |
   | Q1 | 0 |
   | Q3 | 0.0877265 |
   | Max | 0.324366 |

   Which of the following statements is true?

   A) About half the counties had more than 0.05958 infant deaths per 1000 residents.
   B) Because the distribution is skewed right, more than half the counties had more than 0.04492 infant deaths per 1000 residents.
   C) Because the distribution is skewed right, more than half the counties had less than 0.04492 infant deaths per 1000 residents.
   D) At least one fourth of the counties had no infant deaths.
   E) About half the counties had less than 0.08773 infant deaths per 1000 residents.

2. At a certain school 60 of the 100 boys and 60 of the 80 girls signed up for the senior trip. Is there an association between going on the trip and gender?

   A) We can't tell, because the class doesn't have the same number of boys and girls.
   B) Yes, because the same number of boys and girls signed up.
   C) Yes, because a lower percentage of boys signed up than of girls.
   D) No, because the people on the trip were 50% boys and 50% girls.
   E) No, because the sign-up rate was higher among girls than among boys.

3. During the 2019–20 NBA season, James Harden had an average of 34.3 points per game. The mean and standard deviation for the league were 11.98 points per game and 6.16 points per game, respectively. That same season the WNBA was led by Brittney Griner, who averaged 20.7 points per game. The mean and standard deviation for the entire WNBA that year were 6.97 points per game and 4.81 points per game, respectively. Which is the more remarkable performance compared to the rest of their league?

   A) James Harden had the more remarkable performance because he scored more points than Brittney Griner did.
   B) Brittney Griner had the more remarkable performance because the means show that it's harder to score points in the WNBA.
   C) Brittney Griner had the more remarkable performance because the standard deviations show that there's less variability in scoring in the WNBA.
   D) James Harden had the more remarkable performance because his average was more standard deviations above the mean than Brittney Griner's average.
   E) You cannot compare these performances because the two leagues are so different.

4. Below is a histogram of the run lengths of the 150 top-grossing films from 2011. Also shown is an incorrectly drawn boxplot for the data.

   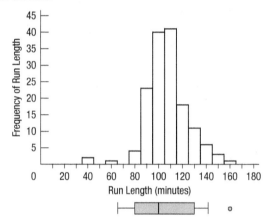

   Which of these is *not* a reason the boxplot is incorrect?

   A) The median should be between 105 and 115, and the boxplot shows a different value.
   B) The mean should be between 105 and 115, and the boxplot shows a different value.
   C) There should be more outliers.
   D) The box contains more than 50% of the data.
   E) The IQR shown in the boxplot is too large.

5. The fuel economy in miles per gallon for 2012 model cars in the U.S. is summarized in the table below.

   | Mean | s | Q1 | Median | Q3 | IQR |
   |---|---|---|---|---|---|
   | 22.0 | 6.69 | 18 | 21 | 25 | 7 |

   In order to compare these U.S. models to European cars, these statistics will be converted to km per liter by multiplying the numerical mpg rating for each car by 0.425. Which of the following statements is true?

   A) All of these summary statistics will be multiplied by 0.425.
   B) All of these summary statistics will remain unchanged.
   C) The mean and median will be multiplied by 0.425, but the other summary statistics will remain unchanged.
   D) The standard deviation and IQR will be multiplied by 0.425, but the other statistics will remain unchanged.
   E) The standard deviation and IQR will remain unchanged, but the other statistics will be multiplied by 0.425.

6. Relatively minor repetitive impacts to the head such as those experienced by elite male soccer players when hitting a soccer ball with their head may cause brain damage, according to *JAMA*, the *Journal of the American Medical Association*. One indicator of mild brain injury is called radial diffusivity. The radial diffusivity measurements of twelve right-handed male soccer players were compared to those of eleven competitive swimmers (who are not likely to have repeated head impacts). Which type of plot would *not* be appropriate for assessing the difference between the two groups? (www.medpagetoday.com/Neurology)

   A) Stacked dotplots, one for each group, on the same axis.
   B) Side-by-side boxplots.
   C) Stacked bar graphs.
   D) A back-to-back stem-and-leaf plot.
   E) A pair of histograms, one above the other, on the same axis.

7. American automobiles produced in 2012 and classified as "large" had a mean fuel economy of 19.6 miles per gallon with a standard deviation of 3.36 miles per gallon. A particular model on this list was rated at 23 miles per gallon, giving it a $z$-score of about 1.01. Which statement is true based on this information?

   A) Because the standard deviation is small compared to the mean, a Normal model is appropriate and we can say that about 84.4% of "large" automobiles have a fuel economy of 23 miles per gallon or less.
   B) Because a $z$-score was calculated, it is appropriate to use a Normal model to say that about 84.4% of "large" automobiles have a fuel economy of 23 miles per gallon or less.
   C) Because 23 miles per gallon is greater than the mean of 19.6 miles per gallon, the distribution is skewed to the right. This means the $z$-score cannot be used to calculate a proportion.
   D) Because no information was given about the shape of the distribution, it is not appropriate to use the $z$-score to calculate the proportion of automobiles with a fuel economy of 23 miles per gallon or less.
   E) Because no information was given about the shape of the distribution, it is not appropriate to calculate a $z$-score, so the $z$-score has no meaning in this situation.

8. The losing teams in all college basketball games for 2011 had scores that are approximately normally distributed with mean 64 points and standard deviation about 11.7 points. Based on the Normal model, we'd expect that the middle 90% of losing teams' scores would be between about

   A) 29 and 99 points.
   B) 41 and 87 points.
   C) 45 and 83 points.
   D) 49 and 79 points.
   E) 52 and 76 points.

9. A survey of students at a Wisconsin high school asked the following question:

   *Whom do you most often text during class?*

   _____ *family members*     _____ *girlfriend/boyfriend*
   _____ *friends inside school*     _____ *friends outside school*

   The results, sorted by grade, are summarized in this table:

   | Grade | Family Members | Girlfriend/ Boyfriend | Friends Inside School | Friends Outside School | Total |
   |---|---|---|---|---|---|
   | 9th | 19 | 10 | 8 | 5 | 42 |
   | 10th | 17 | 12 | 5 | 1 | 35 |
   | 11th | 13 | 9 | 11 | 5 | 38 |
   | 12th | 8 | 21 | 10 | 6 | 45 |
   | Total | 57 | 52 | 34 | 17 | 160 |

   Which statement about these results is correct?

   A) The proportion of 9th graders who said they text family members most is 19/57.
   B) Among those who said they text girlfriend/boyfriend the most, the proportion who are seniors is 21/45.
   C) The proportion of 10th graders who said they text friends inside school the most is greater than the proportion of *all* students who said that.
   D) A student who texts friends inside school is more likely to be a senior than is someone who texts friends outside school.
   E) Of all the grade levels, 11th graders are least likely to text a girlfriend or boyfriend.

10. The pie charts below show the percentages of users of Instagram and users of Twitter that fall into various age groups.

    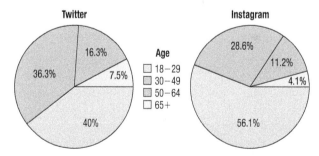

    *Source:* https://www.classy.org/blog/infographic-social-media-demographics-numbers/

    Which of the following cannot be concluded from the pie charts?

    A) Twitter has a larger proportion of users in the 65+ age range than Instagram.
    B) The smallest age group for both Facebook and Twitter users is the 65+ age group.
    C) There are about the same number of Instagram users as Twitter users.
    D) Both Twitter and Instagram have more people in the 18 to 29 age group than any other age group.
    E) A smaller proportion of Twitter users than Instagram users are in the 18 to 29 age group.

## II. FREE RESPONSE

1. The boxplots below show the average number of points scored per game in the 2019 season for both the NBA and the WNBA.

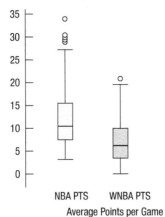

   a) Compare the distributions in context.
   b) The top two WNBA players were Brittney Griner (20.7 pts/game) and Elena Della Donne (19.5 pts/game). The first and third quartiles in the league were 3.3 and 10, respectively. Verify that these two players are correctly positioned on the boxplot. Show your work and justify your answer.
   c) For the 157 players in the WNBA, the average was 6.97 pts/game, while the 259 NBA players had an average of 11.998 pts/game. Find the average scoring of the two leagues combined.
   d) Explain why the NBA average is larger than the NBA median.

2. Many electronic devices use disposable batteries, which eventually have to be replaced. Assume that for one particular brand and type of battery, the distribution of the hours of useful power can be approximated well by a Normal model. The mean is 140 hours, and the standard deviation is 4 hours.
   a) Using the 68–95–99.7 Rule, sketch the appropriate model.
   b) The marketing department wants to write a guarantee for battery life. What life span should they quote so that they can expect 98% of the batteries to meet the guarantee?
   c) The company's research department has been given the task of improving the batteries, with a goal that at least 90% of the new batteries should last longer than the average of the current type. Initial research suggests that a mean of 143 hours is attainable. What other improvement must they make to achieve their goal?
   d) Explain what the improvement you suggest in part c would mean about the new batteries.

# Scatterplots, Association, and Correlation

| | |
|---|---|
| WHO | Years 1970–2018 |
| WHAT | Year and mean error in the position of Atlantic hurricanes as predicted 72 hours ahead by the NHC |
| UNITS | Years and nautical miles |
| WHEN | 1970–2018 |
| WHERE | Atlantic Ocean, the Gulf of Mexico, and the Caribbean |
| WHY | NHC wants to improve prediction models |

If you think that major hurricanes (and their cousins in the Pacific, typhoons) are getting stronger and more frequent, it may not be your imagination. There is growing evidence that although the number of hurricanes has been fairly constant, the frequency of the biggest and most destructive storms is increasing. In 2017, Hurricane Maria devastated the Caribbean. For the U.S. Virgin Islands and Puerto Rico, Hurricane Maria is considered the worst natural disaster in recorded history. Puerto Rico suffered catastrophic damage and a major humanitarian crisis due to extensive flooding and a lack of resources, compounded by a slow relief process. The official death toll from the storm was 3059, of which an estimated 2975 were killed and 60 missing on Puerto Rico alone, making it the deadliest hurricane there since 1899.

The first warnings of the storm were issued on September 16, and Maria made landfall on Dominica on September 18 and on Puerto Rico two days later.[1]

Where will a hurricane go? The impact of a hurricane depends on both its path and strength, so the National Hurricane Center (NHC) of the National Oceanic and Atmospheric Administration (NOAA) tries to predict the path each hurricane will take. But hurricanes tend to wander around aimlessly and are pushed by fronts and other weather phenomena in their area, so they are notoriously difficult to predict. Even relatively small changes in a hurricane's track can make big differences in the damage it causes.

To improve hurricane prediction, NOAA[2] relies on sophisticated computer models and has been working for decades to improve them. How well are they doing? Have predictions improved in recent years? Has the improvement been consistent? Figure 6.1 shows the mean error in nautical miles of the NHC's 72-hour predictions of Atlantic hurricanes for each year plotted against the year. NOAA refers to these errors as the Forecast error or the Prediction error and reports annual results.

---

[1]"Recent increases in tropical cyclone intensification rates," https://www.nature.com/articles/s41467-019-08471-z
[2]www.nhc.noaa.gov

Figure 6.1

A scatterplot of the average tracking error in nautical miles of the predicted position of Atlantic hurricanes for predictions made by the National Hurricane Center of NOAA, plotted against the *Year* in which the predictions were made.

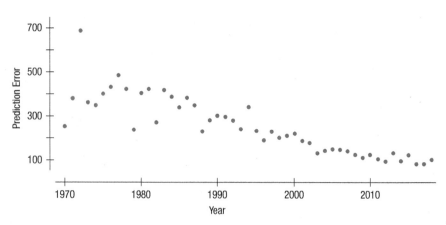

Figure 6.1 is an example of a general kind of display called a **scatterplot**. Because they show the relationship between two quantitative variables, scatterplots may be the most common displays for data. By just looking at them, you can see patterns, trends, relationships, and even the occasional extraordinary value sitting apart from the others. As the great philosopher Yogi Berra[3] once said, "You can observe a lot just by watching."[4] Scatterplots are the best way to start observing the relationship between two *quantitative* variables. When the *x*-variable is *Time,* as it is here, the plot is often referred to as a **timeplot**. (We saw these in Chapter 4.)

From Figure 6.1, it's clear that predictions have improved. The plot shows a fairly steady decline in the average tracking error, from almost 500 nautical miles in the late 1970s to less than 100 nautical miles in 2014. We can also see a few years when predictions were unusually good and that 1972 was a really bad year for predicting hurricane tracks.

Relationships between variables are often at the heart of what we'd like to learn from data:

◆ Are grades higher now than they used to be?
◆ Do people tend to reach puberty at a younger age than in previous generations?
◆ Does applying magnets to parts of the body relieve pain? If so, are stronger magnets more effective?
◆ Do students learn better when they get more sleep?

Questions such as these relate two quantitative variables and ask whether there is an **association** between them. Scatterplots are the ideal way to *picture* such associations.

### Where Did the Origin Go?

Scatterplots usually don't—and shouldn't—show the origin (the place where both *x* and *y* are zero) because often neither variable has values near 0. The display should focus on the part of the coordinate plane that actually contains the data. In our example about hurricanes, none of the prediction errors or years were anywhere near 0, so the computer drew the scatterplot with axes that don't meet.

## Looking at Scatterplots; Describing Associations

How would you describe the association of hurricane *Prediction Error* and *Year*? Everyone looks at scatterplots. But, if asked, many people would find it hard to say what to look for in a scatterplot. What do *you* see? Try to describe the scatterplot of *Prediction Error* against *Year*.

### Look for Direction:

What's my sign—positive, negative, or neither?

You might say that the **direction** of the association is important. Over time, the NHC's prediction errors have decreased. A pattern like this that runs from the upper left to the lower right shows what we call a **negative association**. A pattern running the other way shows a **positive association**.

A second thing to look for in a scatterplot is its **form**. If there is a straight line relationship, it will appear as a cloud or swarm of points stretched out in a generally consistent, straight form. For example, the scatterplot of *Prediction Error* vs. *Year* has such an underlying **linear** form, although some points stray away from it.

### Look for Form:

Is it straight, curved, something exotic, or no pattern?

---
[3]Hall of Fame catcher, outfielder, and manager of the New York Mets and Yankees.
[4]But then he also said, "I really didn't say everything I said." So we can't really be sure.

Scatterplots can reveal many kinds of patterns. Often they will not be straight, but straight line patterns are both the most common and the most useful for statistics.

If the relationship isn't straight, but curves gently, while still increasing or decreasing steadily, , we can often find ways to make it more nearly straight. But if it curves sharply—up and then down, for example —there is much less we can say about it with the methods of this book.

A third feature to look for in a scatterplot is the **strength** of the relationship. At one extreme, do the points appear tightly clustered in a single stream (whether straight, curved, or bending all over the place)? Or, at the other extreme, does the swarm of points seem to form a vague cloud through which we can barely discern any trend or pattern? The *Prediction Error* vs. *Year* plot shows moderate scatter around a generally straight form. This indicates that the linear trend of improving prediction is pretty consistent and moderately strong.

## Look for Strength:
How much scatter is there?

Finally, always look for the unexpected. Often the most interesting thing to see in a scatterplot is something you never thought to look for. One example of such a surprise is an **outlier** standing away from the overall pattern of the scatterplot. Such a point is almost always interesting and always deserves special attention. In the scatterplot of prediction errors, the year 1972 stands out as a year with very high prediction errors. An Internet search shows that it was a relatively quiet hurricane season. However, it included the very unusual—and deadly—Hurricane Agnes, which combined with another low-pressure center to ravage the northeastern United States, killing 122 and causing 1.3 billion 1972 dollars in damage. Possibly, Agnes was also unusually difficult to predict.

## Look for Unusual Features:
Are there outliers or subgroups?

You should also look for clusters or subgroups that stand away from the rest of the plot or that show a trend in a different direction. Deviating groups should raise questions about why they are different. They may be a clue that you should split the data into subgroups instead of looking at them all together.

### FOR EXAMPLE
#### Comparing Prices Worldwide

If you travel overseas, you know that what's really important is not the amount in your wallet but the amount it can buy. UBS (one of the largest banks in the world) prepared a report comparing prices, wages, and other economic conditions in cities around the world for its international clients. Some of the variables it measured in 73 cities are *Cost of Living, Food Costs, Average Hourly Wage*, average number of *Working Hours* per Year, average number of *Vacation Days*, hours of work (at the average wage) needed to buy an *iPhone*, minutes of work needed to buy a *Big Mac*, and *Women's Clothing Costs*.[5] For your burger fix, you might want to live in Tokyo, where it takes only about 9 minutes of work to afford a Big Mac. In Nairobi, you'd have to work almost an hour and a half.

---

[5]Detail of the methodology can be found in the report *Prices and Earning: A comparison of purchasing power around the globe/2012 edition*. https://www.economist.com/finance-and-economics/2012/06/09/burgernomics-to-go

Of course, these variables are associated, but do they consistently reflect costs of living? Plotting pairs of variables can reveal how and even if they are associated. The variety of these associations illustrates different directions and kinds of association patterns you might see in other scatterplots.

QUESTION: Describe the association shown in each of these scatterplots.

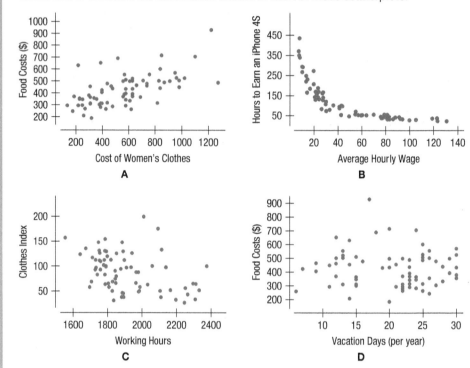

ANSWER: In Plot A, the association between *Food Costs* and *Cost of Women's Clothes* is positive and straight with a few high outliers, and moderately strong. In Plot B, the association between *Hours to Earn an iPhone 4S* and *Average Hourly Wage* is negative and very strong, but the form is not straight. In Plot C, the association between the *Clothes Index* and *Working Hours* is weak, negative, and generally straight with one or two possible high outliers. In Plot D, there does not appear to be any association between *Food Costs* and *Vacation Days*.

# Roles for Variables

*NOTATION ALERT*
In Statistics, the assignment of variables to the $x$- and $y$-axes (and the choice of notation for them in formulas) often conveys information about their roles as predictor or response variable. So $x$ and $y$ are reserved letters as well, but not just for labeling the axes of a scatterplot.

Which variable should go on the $x$-axis and which on the $y$-axis? What we want to know about the relationship can tell us how to make the plot. We often have questions such as:

♦ Do baseball teams that score more runs sell more tickets to their games?
♦ Do older houses sell for less than newer ones of comparable size and quality?
♦ Do students who score higher on their SAT or ACT tests have higher grade point averages in college?
♦ Can we estimate a person's percent body fat more simply by just measuring waist or wrist size?

In these examples, the two variables play different roles. We'll call the variable of interest the **response variable** and the other the **explanatory** or predictor variable.[6]

---
[6]The $x$- and $y$-variables have sometimes been referred to as the *independent* and *dependent* variables, respectively. The idea was that the $y$-variable depended on the $x$-variable and the $x$-variable acted independently to make $y$ respond. These names, however, conflict with other uses of the same terms in Statistics.

We'll continue our practice of naming the variable of interest *y*. Naturally we'll plot it on the *y*-axis and place the explanatory variable on the *x*-axis. Sometimes, we'll call them the **x-** and **y-variables**. When you make a scatterplot, you can assume that those who view it will think this way, so choose which variables to assign to which axes carefully.

The roles that we choose for variables are more about how we *think* about them than about the variables themselves. Just placing a variable on the *x*-axis doesn't necessarily mean that it explains or predicts *anything.* And the variable on the *y*-axis may not respond to it in any way. We plotted *Prediction Error* on the *y*-axis against *Year* on the *x*-axis because the National Hurricane Center is interested in how their predictions have changed over time. Could we have plotted them the other way? In this case, it's hard to imagine reversing the roles—knowing the prediction error and wanting to guess in what year it happened. But for some scatterplots, it can make sense to use either choice, so you have to think about how the choice of role helps to answer the question you have.

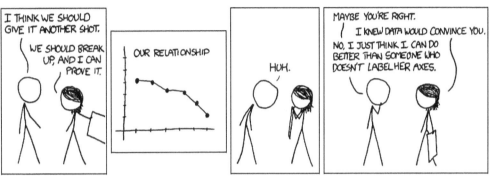

© 2013 Randall Munroe. Reprinted with permission. All rights reserved.

## TI TIPS

### Creating a Scatterplot

Let's use your calculator to make a scatterplot. For data, we'll use the average annual levels of carbon dioxide ($CO_2$) in the Earth's atmosphere for selected years, shown in the table below. $CO_2$ levels are measured in parts per million, or "ppm."

| Year | 1981 | 1986 | 1991 | 1996 | 2001 | 2006 | 2011 | 2016 |
|---|---|---|---|---|---|---|---|---|
| $CO_2$ | 340 | 347 | 355 | 362 | 370 | 381 | 390 | 403 |

#### NAMING THE LISTS

◆ Your calculator is programmed to have six empty lists ready to go, named L1 through L6. We could use those, but it's better practice to give the lists names that indicate what variables they represent, so let's get fancy. Go into STAT Edit, place the cursor on one of the list names (L1, say), and use the arrow key to move to the right across all the lists until you encounter a blank column.

◆ Type YEAR to name the first variable; then hit ENTER.

◆ Hit ENTER again until the slot beneath the list name is selected; then enter the eight selected years from 1981 to 2016 in the slots beneath the list name YEAR.

◆ Now go the next blank column, name this variable CO2, and enter the eight data values.

**160** PART II Exploring Relationships Between Variables

### MAKING THE SCATTERPLOT

- Set up the `Plot1 STATPLOT` by choosing the scatterplot icon, the first option. (Be sure that `Plot2` and `Plot3` are turned off.)
- Identify which lists you want as `Xlist` and `Ylist`. If the data are in `L1` and `L2`, that's easy to do—but your data are stored in lists with special names. To specify your `XList`, go to `2nd LIST NAMES`, scroll down the list of variables until you find `YEAR`, and then hit `ENTER`.
- Use `LIST NAMES` again to specify `Ylist: CO2`.
- Pick a symbol and color for displaying the points.
- Now `ZoomStat` to see your scatterplot. (Didn't work? If you see the message `ERR: DIM MISMATCH`, that means that the dim(ensions), or lengths, of your two lists don't match. You don't have the same number of $x$'s and $y$'s. Look carefully at your lists and you'll find what needs to be fixed.)
- Notice that if you `TRACE` the scatterplot, the calculator will tell you the $x$- and $y$-values at each point.

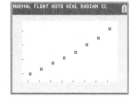

Here's a good chance to practice your skills describing an association based on a scatterplot. Before reading our description below, take a moment to write a description yourself of the association you see in the scatterplot.

Our description: There is a very strong positive association between $CO_2$ levels in the Earth's atmosphere and the years since 1981. The association appears to be approximately linear, but there may be some curvature, which would indicate an acceleration of the rise in $CO_2$ levels. (Did your description include direction, form, and strength? Did it give context by identifying the variables?)

## Correlation

| | |
|---|---|
| WHO | Roller coasters |
| WHAT | Length (feet), duration (seconds) |
| WHERE | Amusement parks everywhere |
| WHY | Data for fans of roller coasters |
| HOW | Data gathered from internet |

The two graphs below show, for 40 roller coasters around the world, how the lengths of the coasters' tracks are related to the rides' durations. The graph on the left measures lengths in feet and durations in seconds. The graph on the right uses meters and minutes.

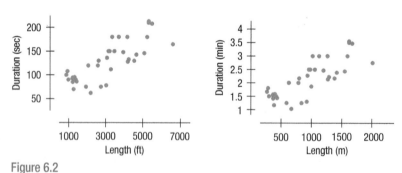

**Figure 6.2**

**Duration versus Length.** Plotting Duration versus Length in different units doesn't change the shape (or the strength) of the pattern.

In our dataset, the longest-lasting ride is the Intimidator roller coaster, located in Charlotte, North Carolina's Carowinds park. It lasts just over three and a half minutes.

It's no great surprise that there's a positive association between the two variables: coasters that are longer in length tend to give longer-lasting rides. The form of the scatterplot is fairly linear, and there are no outliers or clusters of points that seem to be following a different pattern from the others.

Let's talk about the strength of the association. The points don't follow a line (or any curve) perfectly, but the association still appears pretty strong. If you had to put a number on the strength of the association (say, between 0 and 1), what would it be? Whatever measure you use shouldn't depend on the choice of units for the variables. We can see by comparing the two graphs above that changing the units doesn't change the direction, form, or strength of the association, so it shouldn't change the number.

To ensure the units don't matter when we measure the strength, we can remove them by standardizing each variable. Now, for each point, instead of the values $(x, y)$ we'll have the standardized coordinates $(z_x, z_y)$. Remember that to standardize values, we subtract the mean of each variable and then divide by its standard deviation:

$$(z_x, z_y) = \left( \frac{x - \bar{x}}{s_x}, \frac{y - \bar{y}}{s_y} \right).$$

Because standardizing makes the means of both variables 0, the center of the new scatterplot is at the origin. (Take a look at the plot in Figure 6.3.) The scales on both axes are now standard deviation units.

We've color-coded the points in this standardized plot. For the green points (in the first and third quadrants) the coordinates $z_x$ and $z_y$ are either both positive or both negative. Either way, the product $z_x z_y$ is positive, and these points are evidence of a positive association. Coordinates of the square red points (in the second and fourth quadrants) have opposite signs, so the product $z_x z_y$ is negative, evidence of a negative association. Points with larger z-scores will have larger products and offer greater evidence. (Our graph has no points that lie exactly on either axis, although several are close. If there were any such points, they would offer no information on the direction of the association. Because they would have one or both z-scores equal to zero, they would also have the product $z_x z_y = 0$.)

Now we add up the $z_x z_y$ products for every point in the scatterplot: $\Sigma z_x z_y$. This summarizes the direction *and* strength of the association for all the points. But the sum gets bigger the more data we have. To adjust for this, the natural (for statisticians anyway) thing to do is to divide the sum by $n - 1$.[7] The result is the famous **correlation coefficient:**

$$r = \frac{\Sigma z_x z_y}{n - 1}.$$

For the roller coasters' lengths and durations, the correlation turns out to be about 0.794. Because it is based on z-scores, which have no units, correlation has no units either. It will stay the same regardless of how you measure length and duration.

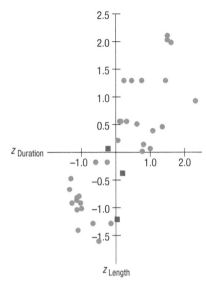

Figure 6.3
In this scatterplot of z-scores, points are colored according to how they affect the association: green for positive and red for negative.

*NOTATION ALERT*

The letter *r* is always used for correlation, so you can't use it for anything else in Statistics. Whenever you see an *r*, it's safe to assume it's a correlation.

## Correlation Conditions

**Correlation** measures the direction and strength of the *linear* association between two *quantitative* variables. Before you use correlation, you must check several *conditions*:

- **Quantitative Variables Condition** Don't make the common error of calling an association involving a categorical variable a correlation. Correlation is only about quantitative variables. (You can't find a z-score for Ticket Class on the Titanic.)
- **Straight Enough Condition** The best check for the assumption that the variables are truly linearly related is to look at the scatterplot to see whether it looks reasonably straight. That's a judgment call, but not a difficult one.
- **No Outliers Condition** Outliers can distort the correlation dramatically, making a weak association look strong or a strong one look weak. Outliers can even change the sign of the correlation. But it's easy to see outliers in the scatterplot, so to check this condition, just look.

Each of these conditions is easy to check with a scatterplot. Many correlations are reported without supporting data or plots. Nevertheless, you should still think about the conditions. And you should be cautious in interpreting (or accepting others' interpretations of) the correlation when you can't check the conditions for yourself.

---
[7]Yes, the same $n - 1$ as in the standard deviation calculation.

## FOR EXAMPLE

### Correlations for Scatterplot Patterns

Look back at the scatterplots of the economic variables in cities around the world (p. 158). The correlations for those plots are (A to D) 0.614, −0.791, −0.388, and −0.040, respectively.

QUESTION: Check the conditions for using correlation. If you feel they are satisfied, interpret the correlation.

ANSWER: All of the variables examined are quantitative and none of the plots shows an outlier. However, the relationship between *Hours to Earn an iPhone* and *Average Wage* is not straight, so the correlation coefficient isn't an appropriate summary. For the others:

A correlation of 0.614 between *Food Costs* and *Women's Clothing Costs* indicates a moderately strong positive association.

A correlation of −0.388 between *Clothes Index* and *Working Hours* indicates a moderately weak negative association.

The small correlation value of −0.040 between *Food Costs* and *Vacation Days* suggests that there may be no linear association between them.

## JUST CHECKING

Your Statistics teacher tells you that the correlation between the scores (points out of 50) on Exam 1 and Exam 2 was 0.75.

1. Before answering any questions about the correlation, what would you like to see? Why?
2. If she adds 10 points to each Exam 1 score, how will this change the correlation?
3. If she standardizes scores on each exam, how will this affect the correlation?
4. In general, if someone did poorly on Exam 1, are they likely to have done poorly or well on Exam 2? Explain.
5. If someone did poorly on Exam 1, can you be sure that they did poorly on Exam 2 as well? Explain.

## STEP-BY-STEP EXAMPLE

### Looking at Association

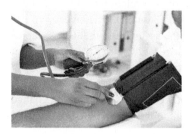

When your blood pressure is measured, it is reported as two values: systolic blood pressure and diastolic blood pressure.

QUESTION: How are these variables related to each other? Do they tend to be both high or both low? How strongly associated are they?

| THINK | |
|---|---|
| **PLAN** State what you are trying to investigate. | I'll examine the relationship between two measures of blood pressure. |
| **VARIABLES** Identify the two quantitative variables whose relationship we wish to examine. Report the W's, and be sure both variables are recorded for the same individuals. | |

**PLOT** Make the scatterplot. Use a computer program or graphing calculator if you can.

The variables are systolic and diastolic blood pressure (*SBP* and *DBP*), recorded in millimeters of mercury (mm Hg) for each of 1406 participants in the Framingham Heart Study, a famous health study in Framingham, MA.[8]

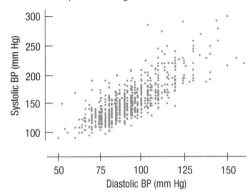

Check the conditions.

✓ **Quantitative Variables Condition**: Both *SBP* and *DBP* are quantitative and measured in mm Hg.

✓ **Straight Enough Condition**: The scatterplot looks straight.

✓ **Outlier Condition**: There are a few straggling points, but none far enough from the body of the data to be called outliers.

**REALITY CHECK** Looks like a strong positive linear association. We shouldn't be surprised if the correlation coefficient is positive and fairly large.

I have two quantitative variables that satisfy the conditions, so correlation is a suitable measure of association.

**SHOW**

**MECHANICS** We usually calculate correlations with technology. Here we have 1406 cases, so we'd never try it by hand.

The correlation coefficient is $r = 0.792$.

**TELL**

**CONCLUSION** Describe the direction, form, and strength you see in the plot, along with any unusual points or features. Be sure to state your interpretations in the proper context.

The scatterplot shows a positive direction, with higher *SBP* going with higher *DBP*. The plot is generally straight, with a moderate amount of scatter. The correlation of 0.792 is consistent with what I saw in the scatterplot.

## TI TIPS

### Finding the Correlation

Now let's use the calculator to find a correlation. Unfortunately, TI calculators fresh off the factory floor come with statistical "diagnostics" (which include correlations) turned off. To turn them on, once and for all:

◆ Hit `2nd CATALOG` (on the zero key). You now see a list of everything the calculator can do.

◆ Scroll down until you find `DiagnosticOn`. Hit `ENTER` once to select that function from the list, and a second time to "execute" the function on the home screen. It should say `Done`.

Now and forevermore your calculator will compute and display correlations.

---

[8]www.framinghamheartstudy.org/

**FINDING THE CORRELATION** We'll need some data—some bivariate,[9] quantitative data to be precise—for our correlation. Let's use the data shown in the table below, which are the average "unit value" (dollars per pound) of almonds and pistachios that were exported from the United States from 2010 to 2021. The table shows the years because that's part of the dataset and can be interesting, but we're only going to use the two other variables.

| Year | 2010 | 2011 | 2012 | 2013 | 2014 | 2015 | 2016 | 2017 | 2018 | 2019 | 2020 | 2021 |
|---|---|---|---|---|---|---|---|---|---|---|---|---|
| **Almonds** (dollars per pound) | 1.77 | 1.97 | 2.08 | 2.49 | 3.04 | 3.50 | 3.20 | 2.56 | 2.64 | 2.67 | 2.64 | 1.94 |
| **Pistachios** (dollars per pound) | 2.75 | 2.79 | 2.99 | 3.32 | 3.81 | 4.10 | 4.24 | 3.32 | 3.51 | 3.56 | 3.88 | 3.58 |

- The first thing we should do with data is (of course) *make a picture*. We showed you in an earlier TI Tip how to name lists on your calculator and make a scatterplot. TI List names can be at most five characters, so we're naming our lists ALMND and PSTCH.

- *Always* check the conditions before computing a correlation. The nut prices *are* both quantitative; the scatterplot *does* show an association that's pretty linear; and there *are no* outliers that could distort the meaning of the correlation. So we're good to go.

- Under the STAT CALC menu, select 8:LinReg(a+bx) and hit ENTER.

- Fill in the blanks in the dialog box with the data for Xlist and Ylist. You can leave the remaining items in the dialog box blank. Scroll down to Calculate and hit ENTER. (Or, on an older calculator, create the command LinReg(a+bx) LALMND, LPSTCH.)

- The calculator reports back four numbers. Later we'll see what the other three are, but for now we're just looking at the correlation. It's r, the last number among these in the output: about 0.863.

Now you know how to compute a correlation. How do you interpret it? What does it mean that almond and pistachio prices have a correlation of about 0.863?

# Correlation Properties

Here's a useful list of facts about the correlation coefficient:

- The sign of a correlation coefficient gives the direction of the association.
- Correlation is always between −1 and +1. Correlation *can* be exactly equal to −1.0 or +1.0, but these values are unusual in real data because they mean that all the data points fall *exactly* on a single straight line.
- Correlation treats *x* and *y* symmetrically. The correlation of *x* with *y* is the same as the correlation of *y* with *x*.

---

[9] *Bivariate* is, as you can probably tell, just a fancy way of saying "two variables."

## CHAPTER 6  Scatterplots, Association, and Correlation

- Correlation has no units. Correlation is sometimes given as a percentage, but you probably shouldn't do that because without units there's nothing for *r* to be a percentage *of*. We'll learn more in the next chapter, but for now you should just think of it as a measure of the strength of a linear association between two quantitative variables.
- Correlation is not affected by changes in the center or scale of either variable. Changing the units or baseline of either variable has no effect on the correlation coefficient. Correlation depends only on the $z$-scores, and they are unaffected by changes in center or scale.
- Correlation measures the strength of the *linear* association between the two variables. Variables can be strongly associated but still have a small (but meaningless) correlation if the association isn't linear.
- Correlation is sensitive to outliers. A single outlying value can make a small correlation large or make a large one small.

TI-*nspire*

**Correlation and Scatterplots.** See how the correlation changes as you drag data points around in a scatterplot.

**APPLET**
Explore Correlation

### HOW STRONG IS STRONG?

You'll often see correlations characterized as "weak," "moderate," or "strong," but be careful. There's no agreement on what those terms mean. The same numerical correlation might be strong in one context and weak in another. You might be thrilled to discover a correlation of 0.7 between the new summary of the economy you've come up with and stock market prices, but you'd consider it a design failure if you found a correlation of "only" 0.7 between two tests intended to measure the same skill. Deliberately vague terms like "weak," "moderate," or "strong" that describe a linear association can be useful additions to the numerical summary that correlation provides. But be sure to include the correlation, show a scatterplot, and mention the context so others can judge for themselves.

### FOR EXAMPLE  Changing Scales

**RECAP:** Several measures of prices and wages in cities around the world show a variety of relationships, some of which we can summarize with correlations.

**QUESTION:** Suppose that, instead of measuring prices in U.S. dollars and recording work time in hours, we had used euros and minutes. How would those changes affect the conditions, the values of correlation, or our interpretation of the relationships involving those variables?

**ANSWER:** Not at all. Correlation is based on standardized values ($z$-scores), so the conditions, value of *r*, and interpretation are all unaffected by changes in units.

# Accidental Correlation

In a recent baseball season, the Cincinnati Reds team had 43 players on its roster. The scatterplot below shows the association between their heights and ages.

Figure 6.4
The correlation between the Reds' ages and heights is about 0.221. Is that big enough to be meaningful?

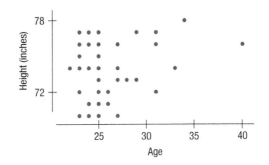

The correlation between age and height for these baseball players turns out to be about $r = 0.221$. That's not very strong, but it isn't zero—so there seems to be a weak association between the players' ages and their heights. You might think at first that makes sense, because people grow taller as they grow older. But hold on! That's true only for children and teenagers, not for grown adults like professional baseball players! What's going on here?

To understand what this correlation means (if anything) let's first recognize that any two variables, no matter how unrelated they actually are to one another, are almost certainly not going to have a calculated value of $r$ that's *exactly* zero. Random chance alone will produce a correlation that's at least a little bit positive or negative. So a reasonable question is: Just how far from zero should $r$ be before we believe that it reflects a *meaningful* association and not just random noise?

Let's use a simulation to address this question. From the entire population of 1,246 Major League Baseball players, we'll have a computer randomly select 43 of them and record their ages. Then we'll pair those ages with 43 *different* randomly selected players' heights. This way we can be sure that the ages and heights are realistic samples from the population of interest, but also that they're completely unrelated to one another. These resulting 43 "players" should help us understand better just how much *apparent* association between age and height can actually occur just by accident.

The results of the first four of our computer "teams" are shown below.

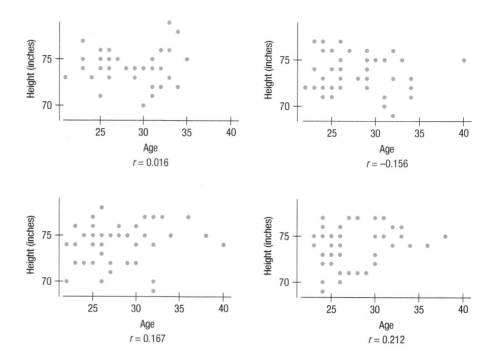

Figure 6.5

**Four scatterplots resulting from pairing random baseball players' ages with other random players' heights** Even though there's no meaningful association between these variables, the correlation will still almost never be exactly zero.

We've simulated only four made-up "teams," yet already we see one with a correlation between age and height of 0.212, very close to the actual value of $r = 0.221$ for the Reds. That suggests that maybe the Reds' apparent correlation is really meaningless, just a result of random chance.

To see how often a correlation as big as 0.221 might show up just by chance, we had the computer generate 10,000 more such made-up "teams" and keep track of the correlation between the ages and heights for each one. The histogram on the next page shows the distribution of all those correlations.

Wow! Even when we know that age and height are totally unrelated (because they were randomly paired together) we still can see occasional correlations of $r = 0.250$ or higher. In fact, in approximately 8% of the simulated "teams," the correlation between age

### Figure 6.6
Among 10,000 simulated "teams" with randomly paired ages and heights, 793 of them (just under 8%) had correlations that were at least 0.221, the actual correlation for the Reds.

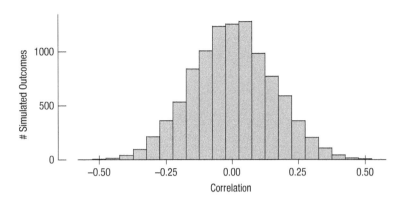

and height was at least as big as what we saw for the Reds. Eight percent is a little unusual, but not strikingly so. Without any better explanation for why the Reds' ages and heights would have a correlation this high, it seems reasonable to believe that it could have just happened by chance.

In fact, let's put the Reds' age/height correlation in a larger context. The dotplot below shows the actual correlations of players' ages and heights for all 30 Major League Baseball teams.

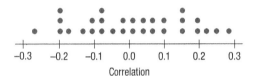

### Figure 6.7
The correlations between age and height for the 30 Major League Baseball teams have a distribution that's consistent with the one we saw in Figure 6.6, which represented made-up "teams" for which we knew age and height were unrelated.

The distribution of age/height correlations for the *actual* 30 Major League Baseball teams (Figure 6.7) looks very similar to the distribution of age/height correlations that we simulated *completely* at *random* (Figure 6.6). That's further evidence that the Reds' age/height correlation of 0.221, while perhaps a bit unusual for a single team, is consistent with there actually being no meaningful association between age and height at all. *Whew!* It was going to be difficult to explain why these adult, grown men were still growing taller as they aged—but now we see that no explanation is needed. Random chance alone could easily produce a team with an age/height correlation that large. (Indeed, two teams—the San Diego Padres and the New York Yankees—have even greater age/height correlations, also "accidental.")

We've used simulations in earlier chapters, but this may still have been a difficult section to understand. If so, don't worry: we're going to discuss simulations in more detail in a later chapter. For now, the only essential take-away message is that a nonzero correlation doesn't necessarily imply any meaningful association, strong or weak, between two variables. Sometimes random chance alone can explain a correlation. It takes practice to develop a reliable sense of when a correlation signifies something meaningful or not. There are also statistical tools that can help distinguish meaningful phenomena from random noise, and you'll learn some of them later in this text.

## Warning: Correlation ≠ Causation

Whenever we have a strong correlation, it's tempting to try to explain it by imagining that the predictor variable has *caused* the response to change. Humans are like that; we tend to see causes and effects in everything.

Sometimes this tendency can be amusing. The scatterplot on the next page shows, for 7 years in the 1930s, the human population ($y$) of Oldenburg, Germany, plotted against the number of storks nesting in the town ($x$).

**Figure 6.8**
**The number of storks in Oldenburg, Germany, plotted against the population of the town for 7 years in the 1930s.** The association is clear. How about the causation? (*Ornithologishe Monatsberichte*, 44, no. 2)

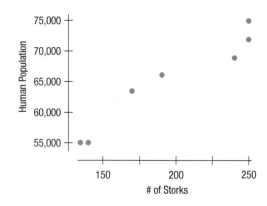

In 2009, the movie *Up* was preceded in theaters by the short film *Partly Cloudy* about an unfortunate stork tasked with delivering baby sharks, crocodiles, porcupines, and the like. The film was having fun with the old story told to children that storks deliver babies. But doesn't the scatterplot above support that notion? The two variables are obviously related to one another. The correlation is 0.97, and unlike that of the Cincinnati Reds' heights and ages, it's too big to be "explained away" as just random noise.

It turns out that storks nest on house chimneys. More people means more houses, more nesting sites, and so more storks. The causation is actually in the *opposite* direction, but you can't tell from the scatterplot or correlation. You need additional information—not just the data—to determine the real mechanism.

A scatterplot of the damage (in dollars) caused to a house by fire would show a strong correlation with the number of firefighters at the scene. Surely the damage doesn't cause firefighters. And firefighters do seem to cause damage, spraying water all around and chopping holes. Does that mean we shouldn't call the fire department? Of course not. There is an underlying variable that leads to both more damage and more firefighters: the size of the blaze.

A hidden variable that stands behind a relationship and determines it by simultaneously affecting the other two variables is called a **lurking variable**. You can often debunk claims made about data by finding a lurking variable behind the scenes.

Scatterplots and correlation coefficients *never* prove causation. That's one reason it took so long for the U.S. Surgeon General to get warning labels on cigarettes. Although there was plenty of evidence that increased smoking was *associated* with increased levels of lung cancer, it took years to provide compelling evidence that smoking actually *causes* lung cancer.

> ### DOES CANCER CAUSE SMOKING?
> Even if the correlation of two variables is due to a causal relationship, the correlation itself cannot tell us what causes what.
>
> Sir Ronald Aylmer Fisher (1890–1962) was one of the most influential statisticians of the 20th century. Fisher testified in court (in testimony paid for by the tobacco companies!) that a causal relationship might underlie the correlation of smoking and cancer:
>
> > "Is it possible, then, that lung cancer . . . is one of the causes of smoking cigarettes? I don't think it can be excluded . . . the pre-cancerous condition is one involving a certain amount of slight chronic inflammation . . . ."
>
> > "A slight cause of irritation . . . is commonly accompanied by pulling out a cigarette, and getting a little compensation for life's minor ills in that way. And . . . is not unlikely to be associated with smoking more frequently."
>
> Ironically, the proof that smoking indeed is the cause of many cancers came from experiments conducted following the principles of experimental design and analysis that Fisher himself developed—and that we'll see in Chapter 12.

# WHAT CAN GO WRONG?

- **Don't say "correlation" when you mean "association."** How often have you heard the word "correlation"? Chances are pretty good that when you've heard the term, it's been misused. When people want to sound scientific, they often say "correlation" when talking about the relationship between two variables. It's one of the most widely misused Statistics terms, and given how often statistics are misused, that's saying a lot. One of the problems is that many people use the specific term *correlation* when they really mean the more general term *association*. "Association" is a deliberately vague term describing a relationship between two variables.

  "Correlation" is a precise term that measures the strength and direction of the linear relationship between quantitative variables.

- **Don't correlate categorical variables.** People who misuse the term "correlation" to mean "association" often fail to notice whether the variables they discuss are quantitative. Be sure to check the Quantitative Variables Condition.

> **NOT CORRELATION**
>
> Did you know that there's a strong correlation between playing an instrument and drinking coffee? No? One reason might be that the statement doesn't make sense. Correlation is a statistic that's valid only for *quantitative* variables. If people who play an instrument include disproportionately many coffee-drinkers compared with people who don't play an instrument, then we should say that there is an *association* between playing an instrument and drinking coffee.

- **Don't confuse correlation with causation.** One of the most common mistakes people make in interpreting statistics occurs when they observe a high correlation between two variables and jump to the perhaps tempting conclusion that one thing must be causing the other. Scatterplots and correlations *never* demonstrate causation. At best, these statistical tools can only reveal an association between variables, and that's a far cry from establishing cause and effect. While it's true that some associations may be causal, the nature and direction of the causation can be very hard to establish, and there's always the risk of overlooking lurking variables.

- **Make sure the association is linear.** Not all associations between quantitative variables are linear. Correlation can miss even a strong nonlinear association. A student project evaluating the quality of brownies baked at different temperatures reports a correlation of −0.05 between judges' scores and baking temperature. That seems to say there is no relationship—until we look at the scatterplot:

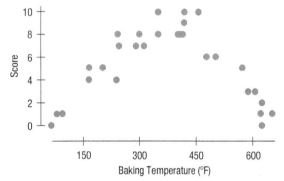

Figure 6.9
The relationship between brownie taste *Score* and *Baking Temperature* is strong, but not at all linear.

There is a strong association, but the relationship is not linear. Don't forget to check the Straight Enough Condition.

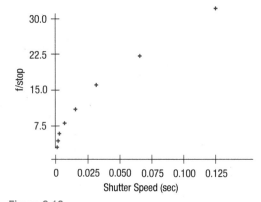

**Figure 6.10**
A scatterplot of *f/stop* vs. *Shutter Speed* shows a bent relationship even though the correlation is $r = 0.979$.

- **Don't assume the relationship is linear just because the correlation coefficient is high.** Some cameras lenses have an adjustable aperture (the hole that lets light in), whose size is expressed in a number called the f/stop. Changing the size of the hole requires also changing the shutter speed to get a good picture, as shown in the scatterplot. The correlation between f/stops and shutter speeds is 0.979 and yet the relationship is clearly not straight. Although a relationship must be straight for correlation to be an appropriate measure, a high correlation is no guarantee of straightness. Don't use correlation to judge the degree of straightness. It's always essential to look at the scatterplot.

- **Beware of outliers.** You can't interpret a correlation coefficient safely without a background check for outliers. Here's a silly example:

  The relationship between IQ and shoe size among comedians shows a surprisingly strong positive correlation of 0.50. To check assumptions, we look at the scatterplot:

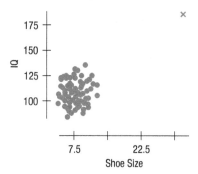

**Figure 6.11**

**A scatterplot of *IQ* vs. *Shoe Size*.** From this "study," what is the relationship between the two? The correlation is 0.50. Who does that point (the green x) in the upper right-hand corner belong to?

The outlier is Bozo the Clown, known for his large shoes, and widely acknowledged to be a comic "genius." Without Bozo, the correlation is near zero.

Even a single outlier can dominate the correlation value. That's why you need to check the Outlier Condition.

## WHAT HAVE WE LEARNED?

In recent chapters we learned how to listen to the story told by data from a single variable. Now we've turned our attention to the more complicated (and more interesting) story we can discover in the association between two quantitative variables.

We've learned to begin our investigation by looking at a scatterplot. We're interested in the *direction* of the association, the *form* it takes, and its *strength*.

We've learned that, although not every relationship is linear, when the scatterplot is straight enough, the *correlation coefficient* is a useful numerical summary.

- The sign of the correlation tells us the direction of the association.
- The magnitude of the correlation tells us the *strength* of a linear association. Strong associations have correlations near $-1$ or $+1$ and very weak associations near 0.
- Correlation has no units, so shifting or scaling the data, standardizing, or even swapping the variables has no effect on the numerical value.

Once again we've learned that doing Statistics right means we have to *Think* about whether our choice of methods is appropriate.

- The correlation coefficient is appropriate only if the underlying relationship is linear.
- We'll check the Straight Enough Condition by looking at a scatterplot.
- And, as always, we'll watch out for outliers!

Finally, we've learned not to make the mistake of assuming that a high correlation or strong association is evidence of a cause-and-effect relationship. Beware of lurking variables!

## TERMS

**Scatterplots**  A scatterplot shows the relationship between two quantitative variables measured on the same cases. In a **timeplot** the horizontal axis is time. (p. 156)

**Association**
- **Direction:** A **positive** direction or association means that, in general, as one variable increases, so does the other. When increases in one variable generally correspond to decreases in the other, the association is **negative**.
- **Form:** The form we care about most is **linear (straight)**, but you should certainly describe other patterns you see in scatterplots.
- **Strength:** A scatterplot is said to show a strong association if there is little scatter around the underlying relationship. (p. 156)

**Outlier**  A point that does not fit the overall pattern seen in the scatterplot. (p. 157)

**Response variable, Explanatory variable, $x$-variable, $y$-variable**  In a scatterplot, you must choose a role for each variable. Assign to the $y$-axis the response variable that you hope to predict or explain. Assign to the $x$-axis the explanatory or predictor variable that accounts for, explains, predicts, or is otherwise associated with the $y$-variable. (pp. 158, 159)

**Correlation coefficient**  The correlation coefficient is a numerical measure of the direction and strength of a linear association. (p. 161)

$$r = \frac{\sum z_x z_y}{n - 1}$$

**Lurking variable**  A variable other than $x$ and $y$ that simultaneously affects both variables, accounting for the association between the two. (p. 168)

© 2013 Randall Munroe. Reprinted with permission. All rights reserved.

## ON THE COMPUTER

### Scatterplots and Correlation

Statistics packages generally make it easy to look at a scatterplot to check whether the correlation is appropriate. Some packages make this easier than others.

Many packages allow you to modify or enhance a scatterplot, altering the axis labels, the axis numbering, the plot symbols, or the colors used. Some options, such as color and symbol choice, can be used to display additional information on the scatterplot.

# EXERCISES

1. **Association** Suppose you were to collect data for each pair of variables. You want to make a scatterplot. Which variable would you use as the explanatory variable and which as the response variable? Why? What would you expect to see in the scatterplot? Discuss the likely direction, form, and strength.

   a) Apples: weight in grams, weight in ounces
   b) For each week: ice cream cone sales, air-conditioner sales
   c) College freshmen: shoe size, grade point average
   d) Gasoline: number of miles you drove since filling up, gallons remaining in your tank

2. **Association II** Suppose you were to collect data for each pair of variables. You want to make a scatterplot. Which variable would you use as the explanatory variable and which as the response variable? Why? What would you expect to see in the scatterplot? Discuss the likely direction, form, and strength.

   a) Cell phone data plans: file size, cost
   b) Lightning strikes: distance from lightning, time delay of the thunder
   c) A streetlight: its apparent brightness, your distance from it
   d) Cars: weight of car, age of owner

3. **Scatterplots** Which of the four scatterplots show

   a) little or no association?
   b) a negative association?
   c) a linear association?
   d) a moderately strong association?
   e) a very strong association?

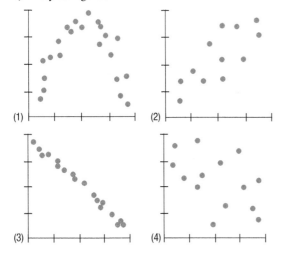

4. **Scatterplots II** Which of the scatterplots in the next column show

   a) little or no association?
   b) a negative association?
   c) a linear association?
   d) a moderately strong association?
   e) a very strong association?

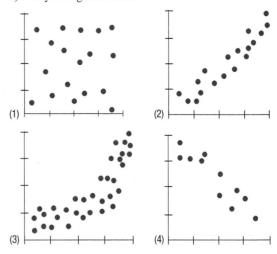

5. **Bookstore sales** Consider the following data from a small bookstore:

   | Number of Salespeople Working | Sales (in $1000) |
   |---|---|
   | 2 | 10 |
   | 3 | 11 |
   | 7 | 13 |
   | 9 | 14 |
   | 10 | 18 |
   | 10 | 20 |
   | 12 | 20 |
   | 15 | 22 |
   | 16 | 22 |
   | 20 | 26 |
   | $\bar{x} = 10.4$ | $\bar{y} = 17.6$ |
   | $SD(x) = 5.64$ | $SD(y) = 5.34$ |

   a) Prepare a scatterplot of *Sales* against *Number of Salespeople Working*.
   b) What can you say about the direction of the association?
   c) What can you say about the form of the relationship?
   d) What can you say about the strength of the relationship?
   e) Does the scatterplot show any outliers?

**6. Disk drives 2020** Disk drives have been getting larger. Their capacity is now often given in *terabytes* (TB), where 1 TB = 1000 gigabytes, or about a trillion bytes. A search of prices for external disk drives on Amazon.com in mid-2020 found the following data:

| Capacity (in TB) | Price (in $) |
|---|---|
| 0.25 | 24.99 |
| 0.5 | 36.79 |
| 1.0 | 47.99 |
| 2.0 | 59.99 |
| 3.0 | 79.99 |
| 4.0 | 97.99 |
| 5.0 | 104.99 |
| 6.0 | 124.99 |
| 8.0 | 218.80 |
| 10.0 | 269.97 |
| 12.0 | 319.99 |
| 16.0 | 519.99 |
| 32.0 | 1399.95 |

a) Prepare a scatterplot of *Price* against *Capacity*.
b) What can you say about the direction of the association?
c) What can you say about the form of the relationship?
d) What can you say about the strength of the relationship?
e) Does the scatterplot show any outliers?

**7. Performance IQ scores vs. brain size** A study examined brain size (measured as pixels counted in a digitized magnetic resonance image [MRI] of a cross section of the brain) and IQ (4 Performance scales of the Weschler IQ test) for college students. The scatterplot shows the Performance IQ scores vs. the brain size. Comment on the association between brain size and IQ.

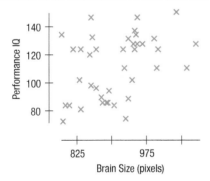

**8. Kentucky Derby 2020** The fastest horse in Kentucky Derby history was Secretariat in 1973. The scatterplot shows speed (in miles per hour) of the winning horses each year.

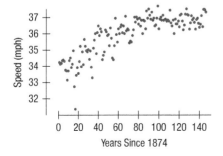

What do you see? In most sporting events, performances have improved and continue to improve, so surely we anticipate a positive direction. But what of the form? Has the performance increased at the same rate throughout the past 145 years?

**9. Correlation facts** If we assume that the conditions for correlation are met, which of the following are true? If false, explain briefly.

a) A correlation of $-0.98$ indicates a strong, negative association.
b) Multiplying every value of $x$ by 2 will double the correlation.
c) The units of the correlation are the same as the units of $y$.

**10. Correlation facts II** If we assume that the conditions for correlation are met, which of the following are true? If false, explain briefly.

a) A correlation of 0.02 indicates a strong positive association.
b) Standardizing the variables will make the correlation 0.
c) Adding an outlier can dramatically change the correlation.

**11. Bookstore sales again** A larger firm is considering acquiring the bookstore of Exercise 5. An analyst for the firm, noting the relationship seen in Exercise 5, suggests that when they acquire the store they should hire more people because that will drive higher sales. Is his conclusion justified? What alternative explanations can you offer? Use appropriate statistics terminology.

**12. Blizzards** A study finds that during blizzards, online sales are highly associated with the number of snow plows on the road; the more plows, the more online purchases. The director of an association of online merchants suggests that the organization should encourage municipalities to send out more plows whenever it snows because, he says, that will increase business. Comment.

**13. Firing pottery** A ceramics factory can fire eight large batches of pottery a day. Sometimes a few of the pieces break in the process. In order to understand the problem better, the factory records the number of broken pieces in each batch for 3 days and then creates the scatterplot shown.

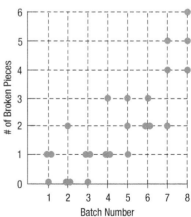

a) Make a histogram showing the distribution of the number of broken pieces in the 24 batches of pottery examined.
b) Describe the distribution as shown in the histogram. What feature of the problem is more apparent in the histogram than in the scatterplot?
c) What aspect of the company's problem is more apparent in the scatterplot?

14. **Coffee sales** Owners of a new coffee shop tracked sales for the first 20 days and displayed the data in a scatterplot (by day).

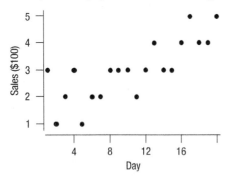

a) Make a histogram of the daily sales since the shop has been in business.
b) State one fact that is obvious from the scatterplot, but not from the histogram.
c) State one fact that is obvious from the histogram, but not from the scatterplot.

15. **Matching** Here are several scatterplots. The calculated correlations are −0.923, −0.487, 0.006, and 0.777. Which is which?

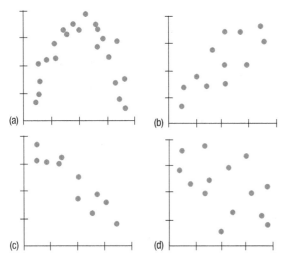

16. **Matching** Below are several scatterplots. The calculated correlations are −0.977, −0.021, 0.736, and 0.951. Which is which?

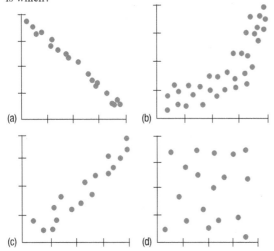

17. **Politics** A candidate for office claims that "there is a correlation between television watching and crime." Criticize this statement on statistical grounds.

18. **Car thefts** The National Insurance Crime Bureau reports that Honda Accords, Honda Civics, and Toyota Camrys are the cars most frequently reported stolen, while Ford Tauruses, Pontiac Vibes, and Buick LeSabres are stolen least often. Is it reasonable to say that there's a correlation between the type of car you own and the risk that it will be stolen?

19. **Bookstore sales final chapter** Let's examine the relationship for Exercise 5 one layer deeper.
    a) Find the correlation coefficient for the given data.
    b) Describe how this value confirms the relationship we described in Exercise 5. Think carefully about *all* the things you know from this value!

20. **Antidepressants** A study compared the effectiveness of several antidepressants by examining the experiments in which they had passed the FDA requirements. Each of those experiments compared the active drug with a placebo, an inert pill given to some of the subjects. In each experiment some patients treated with the placebo had improved, a phenomenon called the *placebo effect*. Patients' depression levels were evaluated on the Hamilton Depression Rating Scale, where larger numbers indicate greater improvement. (The Hamilton scale is a widely accepted standard that was used in each of the independently run studies.) The scatterplot below compares mean improvement levels for the antidepressants and placebos for several experiments.

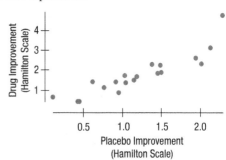

a) Is it appropriate to calculate the correlation? Explain.
b) The correlation is 0.898. Explain what we have learned about the results of these experiments.

21. **Streams and hard water** In a study of streams in the Adirondack Mountains, the following relationship was found between the water's pH and its hardness (measured in grains):

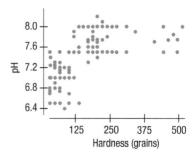

Is it appropriate to summarize the strength of association with a correlation? Explain.

**22. Traffic headaches** A study of traffic delays in 68 U.S. cities found the following relationship between total delays (in total hours lost) and mean highway speed:

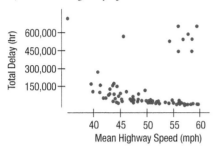

Is it appropriate to summarize the strength of association with a correlation? Explain.

**23. Cold nights** Is there an association between time of year and the nighttime temperature in North Dakota? A researcher assigned the numbers 1–365 to the days January 1–December 31 and recorded the temperature at 2:00 a.m. for each. What might you expect the correlation between *Day Number* and *Temperature* to be? Explain.

**24. Association** A researcher investigating the association between two variables collected some data and was surprised when he calculated the correlation. He had expected to find a fairly strong association, yet the correlation was near 0. Discouraged, he didn't bother making a scatterplot. Explain to him how the scatterplot could still reveal the strong association he anticipated.

**25. Prediction units** The errors in predicting hurricane tracks (examined in this chapter) were given in nautical miles. An ordinary mile is 0.86898 nautical miles. Most people living on the Gulf Coast of the United States would prefer to know the prediction errors in miles rather than nautical miles. Explain why converting the errors to miles would not change the correlation between *Prediction Error* and *Year*.

**26. More predictions** When Hurricane Katrina devastated New Orleans in 2005, hurricane force winds extended 120 miles from its center. Katrina was a big storm, and that affects how we think about the prediction errors. Suppose we add 120 miles to each error to get an idea of how far from the predicted track we might still find damaging winds. Explain what would happen to the correlation between *Prediction Error* and *Year*, and why.

**27. Correlation errors** Your Economics instructor assigns your class to investigate factors associated with the gross domestic product (*GDP*) of nations. Each student examines a different factor (such as *Life Expectancy, Literacy Rate*, etc.) for a few countries and reports to the class. Apparently, some of your classmates do not understand Statistics very well because you know several of their conclusions are incorrect. Explain the mistakes in their statements below.
a) "My very low correlation of $-0.772$ shows that there is almost no association between *GDP* and *Infant Mortality Rate*."
b) "There was a correlation of 0.44 between *GDP* and *Continent*."

**28. More correlation errors** Students in the Economics class discussed in Exercise 27 also wrote these conclusions. Explain the mistakes they made.
a) "There was a very strong correlation of 1.22 between *Life Expectancy* and *GDP*."
b) "The correlation between *Literacy Rate* and *GDP* was 0.83. This shows that countries wanting to increase their standard of living should invest heavily in education."

**29. Height and reading** A researcher studies children in elementary school and finds a strong positive linear association between height and reading scores.
a) Does this mean that taller children are generally better readers?
b) What might explain the strong correlation?

**30. Smart phones and life expectancy** A survey of the world's nations in 2014 shows a strong positive correlation between percentage of the country using smart phones and life expectancy in years at birth.
a) Does this mean that smart phones are good for your health?
b) What might explain the strong correlation?

**31. Correlation conclusions I** The correlation between *Age* and *Income* as measured on 100 people is $r = 0.75$. Explain whether or not each of these possible conclusions is justified:
a) When *Age* increases, *Income* increases as well.
b) The form of the relationship between *Age* and *Income* is straight.
c) There are no outliers in the scatterplot of *Income* vs. *Age*.
d) Whether we measure *Age* in years or months, the correlation will still be 0.75.

**32. Correlation conclusions II** The correlation between *Fuel Efficiency* (as measured by miles per gallon) and *Price* of 150 cars at a large dealership is $r = -0.34$. Explain whether or not each of these possible conclusions is justified:
a) The more you pay, the lower the fuel efficiency of your car will be.
b) The form of the relationship between *Fuel Efficiency* and *Price* is moderately straight.
c) There are several outliers that explain the low correlation.
d) If we measure *Fuel Efficiency* in kilometers per liter instead of miles per gallon, the correlation will increase.

**33. Eating out** Researchers interviewed 35,084 adults aged 20 years or older about their dietary habits. In follow-up studies, they found that the death rate among those who had reported eating away from home two or more times a day was significantly higher than for those who had reported eating away from home once a week or less. A story reporting on this study was titled, "This Eating Habit Could Lead To Premature Death, Study Finds," with a picture of two women eating at a restaurant. What unfounded assumption is the author of the article making?

**34. Coffee and cancer** In a study conducted between 2005 and 2018, 1171 patients with colorectal cancer who were being treated with chemotherapy were given a survey about their diet and lifestyle. One observation of the data showed that patients who drank more coffee had higher survival rates overall. Each additional cup of coffee (with caffeine or without) per day was associated with a higher survival rate. What headline might a journalist write about this discovery if they were assuming a causal link? What would be a more accurate headline?

**35. Baldness and heart disease** Medical researchers followed 1435 middle-aged men for a period of 5 years, measuring the amount of *Baldness* present (none $= 1$, little $= 2$, some $= 3$, much $= 4$, extreme $= 5$) and presence of *Heart Disease* (No $= 0$, Yes $= 1$). They found a correlation of 0.089 between the two variables. Comment on their conclusion that this shows that baldness is not a possible cause of heart disease.

**36. Sample survey** A polling organization is checking its database to see if the two data sources it used sampled the same ZIP codes. The variable *Datasource* = 1 if the data source is MetroMedia, 2 if the data source is DataQwest, and 3 if it's RollingPoll. The organization finds that the correlation between five-digit ZIP code and *Datasource* is −0.0229. It concludes that the correlation is low enough to state that there is no dependency between *ZIP Code* and *Source of Data*. Comment.

**37. Income and housing** The Office of Federal Housing Enterprise Oversight (www.fhfa.gov) collects data on various aspects of housing costs around the United States. Here is a scatterplot of the *Housing Cost Index* versus the *Median Family Income* for each of the 50 states. The correlation is 0.65.

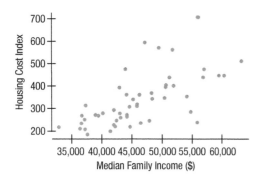

a) Describe the relationship between the *Housing Cost Index* and the *Median Family Income* by state.
b) If we standardized both variables, what would the correlation coefficient between the standardized variables be?
c) If we had measured *Median Family Income* in thousands of dollars instead of dollars, how would the correlation change?
d) Washington, DC, has a housing cost index of 548 and a median income of about $45,000. If we were to include DC in the dataset, how would that affect the correlation coefficient?
e) Do these data provide proof that by raising the median income in a state, the Housing Cost Index will rise as a result? Explain.

**38. Interest rates and mortgages 2020** Since 1985, average mortgage interest rates have fluctuated from a low of nearly 3% to a high of over 14%. Is there a relationship between the amount of money people borrow and the interest rate that's offered? Here is a scatterplot of *Mortgage Loan Amount* in the United States (in trillions of dollars) versus yearly *Interest Rate* since 1985. The correlation is −0.87.

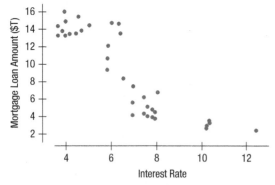

a) Describe the relationship between *Mortgage Loan Amount* and *Interest Rate*.
b) If we standardized both variables, what would the correlation coefficient between the standardized variables be?
c) If we were to measure *Mortgage Loan Amount* in billions of dollars instead of trillions of dollars, how would the correlation coefficient change?
d) Suppose that next year, interest rates were 11% and mortgages totaled $16 trillion. How would including that year with these data affect the correlation coefficient?
e) Do these data provide proof that if mortgage rates are lowered, people will take out larger mortgages? Explain.

**39. Fuel economy 2019** The scatterplot below shows engine size (displacement, in liters) and gas mileage (estimated combined city and highway) for a random sample of 35 2016 model cars (taken from **Fuel Economy 2019**).

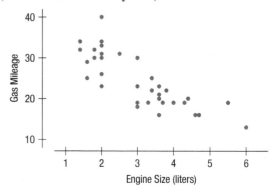

a) Describe the direction, form, and strength of the plot in context.
b) The correlation is −0.797. Write a few sentences to explain how this is related to your answer to part a.

**40. Drug abuse** A survey was conducted in the United States and 10 countries of Western Europe to determine the percentage of teenagers who had used marijuana and other drugs. The results are summarized in the table.

| Country | Percent Who Have Used | |
|---|---|---|
| | Marijuana | Other Drugs |
| Czech Rep. | 22 | 4 |
| Denmark | 17 | 3 |
| England | 40 | 21 |
| Finland | 5 | 1 |
| Ireland | 37 | 16 |
| Italy | 19 | 8 |
| No. Ireland | 23 | 14 |
| Norway | 6 | 3 |
| Portugal | 7 | 3 |
| Scotland | 53 | 31 |
| USA | 34 | 24 |

a) Create a scatterplot.
b) What is the correlation between the percent of teens who have used marijuana and the percent who have used other drugs?
c) Write a brief description of the association.
d) Do these results confirm that marijuana is a "gateway drug," that is, that marijuana use leads to the use of other drugs? Explain.

**41. Burgers** Fast food is often considered unhealthy because much of it is high in both fat and sodium. But are the two related? Here are the fat and sodium contents of several brands of burgers.

| Fat (g)     | 19  | 31   | 34   | 35  | 39   | 39  | 43   |
|-------------|-----|------|------|-----|------|-----|------|
| Sodium (mg) | 920 | 1500 | 1310 | 860 | 1180 | 940 | 1260 |

Analyze the association between fat content and sodium using correlation and scatterplots.

**42. Burgers II** In the previous exercise you analyzed the association between the amounts of fat and sodium in fast food hamburgers. What about fat and calories? Here are data for the same burgers:

| Fat (g)  | 19  | 31  | 34  | 35  | 39  | 39  | 43  |
|----------|-----|-----|-----|-----|-----|-----|-----|
| Calories | 410 | 580 | 590 | 570 | 640 | 680 | 660 |

Analyze the association between fat content and sodium using correlation and scatterplots.

**43. Attendance 2019** American League baseball games are played under the designated hitter rule, meaning that pitchers, often weak hitters, do not come to bat. Baseball owners believe that the designated hitter rule means more runs scored, which in turn means higher attendance. Is there evidence that more fans attend games if the teams score more runs? Data collected from American League games during the 2019 season indicate a correlation of 0.700 between runs scored and the average number of people at the home games. (www.espn.com/mlb/attendance)

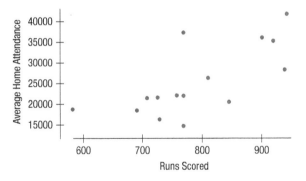

a) Does the scatterplot indicate that it's appropriate to calculate a correlation? Explain.
b) Describe the association between attendance and runs scored.
c) Does this association prove that the owners are right and that more fans will come to games if the teams score more runs?

**44. Vehicle weights** The Minnesota Department of Transportation hoped that they could measure the weights of big trucks without actually stopping the vehicles by using a newly developed "weight-in-motion" scale. To see if the new device was accurate, they conducted a calibration test. They weighed several stopped trucks (static weight) and assumed that this weight was correct. Then they weighed the trucks again while they were moving to see how well the new scale could estimate the actual weight. A scatterplot of their data is shown in the next column. The correlation is 0.965.

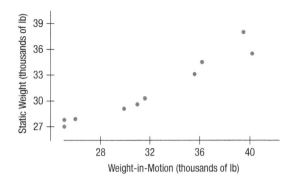

a) Describe the direction, form, and strength of the plot in context.
b) Write a few sentences telling what the plot says about the data. (*Note*: The sentences should be about weighing trucks, not about scatterplots.)
c) If the trucks were weighed in kilograms, how would this change the correlation? (1 kilogram = 2.2 pounds)
d) Do any points deviate from the overall pattern? What does the plot say about a possible recalibration of the weight-in-motion scale?

**45. Planets (more or less)** On August 24, 2006, the International Astronomical Union voted that Pluto is not a planet. Some members of the public have been reluctant to accept that decision. Let's look at some of the data. Is there any pattern to the locations of the planets? The table and scatterplot below show the average distance of each of the traditional nine planets from the sun.

| Planet  | Position Number | Distance from Sun (million miles) |
|---------|-----------------|-----------------------------------|
| Mercury | 1               | 36                                |
| Venus   | 2               | 67                                |
| Earth   | 3               | 93                                |
| Mars    | 4               | 142                               |
| Jupiter | 5               | 484                               |
| Saturn  | 6               | 887                               |
| Uranus  | 7               | 1784                              |
| Neptune | 8               | 2795                              |
| Pluto   | 9               | 3675                              |

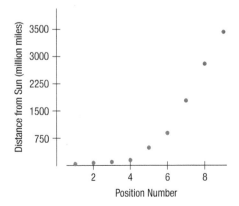

a) Describe the association.
(Remember: direction, form, and strength in context!)
b) Why would you not want to talk about the correlation between a planet's *Position Number* and *Distance* from the sun?

**46. Flights 2019** Here are the numbers of domestic flights flown in each year from 2000 to 2019 (www.transtats.bts.gov/homepage.asp):

| Year | Flights |
|------|---------|
| 2000 | 7,905,617 |
| 2001 | 7,626,312 |
| 2002 | 8,085,083 |
| 2003 | 9,458,818 |
| 2004 | 9,968,047 |
| 2005 | 10,038,373 |
| 2006 | 9,712,750 |
| 2007 | 9,839,578 |
| 2008 | 9,378,227 |
| 2009 | 8,768,938 |
| 2010 | 8,842,312 |
| 2011 | 8,803,572 |
| 2012 | 8,595,866 |
| 2013 | 8,462,141 |
| 2014 | 8,256,478 |
| 2015 | 8,198,906 |
| 2016 | 8,313,188 |
| 2017 | 8,309,843 |
| 2018 | 8,538,251 |
| 2019 | 8,723,835 |

a) Find the correlation of *Flights* with *Year*.
b) Make a scatterplot and describe the trend.
c) Why is the correlation you found in part a not a suitable summary of the strength of the association?

## JUST CHECKING

Answers

1. We know the scores are quantitative. We should check to see if the Straight Enough Condition and the Outlier Condition are satisfied by looking at a scatterplot of the two scores.
2. It won't change.
3. It won't change.
4. They are likely to have done poorly. The positive correlation means that low scores on Exam 1 are associated with low scores on Exam 2 (and similarly for high scores).
5. No. The general association is positive, but individual performances may vary.

# Linear Regression

| | |
|---|---|
| **WHO** | Operational roller coasters in the U.S. |
| **WHAT** | Height and top speed |
| **UNITS** | Feet (*Height*) and miles per hour (*Speed*) |
| **HOW** | Data downloaded from the roller coaster database (rcdb.com) |

Two of the things that make roller coaster rides thrilling are seeing the world from a towering height and traveling in open air really fast. Because roller coasters aren't powered (that's why they're called *coasters!*), getting a fast speed requires dropping from a great height, so coaster enthusiasts always get both together.

In this chapter we'll look at the relationship between a roller coaster's height and its top speed. Our dataset includes, for a recent year, all the operational roller coasters in the U.S. for which height and top speed were available on rcdb.com, a rich resource for roller coaster enthusiasts. (For the rest of this chapter we'll refer to these variables as "Height" and "Speed.") The scatterplot below shows the *Speed* (in miles per hour, mph) plotted against the *Height* (in feet, ft) for these 388 coasters. The shape and color of each point indicates which of four types of roller coasters it is: "Kiddie," "Family," "Thrill," or "Extreme."

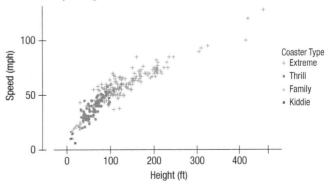

Figure 7.1
U.S. roller coasters' heights and top speeds

The association between these roller coasters' speeds and heights appears to be positive (not surprisingly!) and fairly strong. It also appears to be somewhat nonlinear—points for the coasters taller than 100 feet or so seem to cluster around one line, and the other points seem to cluster around a different, slightly steeper line. Sometimes two variables are simply related in a nonlinear way. But apparent nonlinearity can also result from mixing different populations, just as bimodality can appear in the distribution of a variable measured on a mixed population. (Remember the Kentucky Derby winners' times?)

Roller coasters aimed at different audiences, like small children and thrill-seekers, do not employ the same design elements, so we're not surprised that the different coaster types don't show quite the same association between their top speeds and their heights. Let's consider a smaller population—just the 91 of these that are classified as "Thrill" rides. The relationship between their top speeds and heights is shown in the scatterplot below.

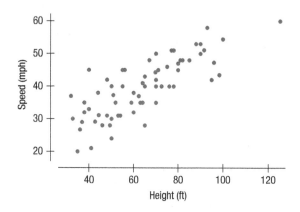

The association between top speed and height for just the "Thrill" rides looks positive, fairly strong, and also fairly linear. Because it's linear, it's okay to compute the correlation between *Speed* and *Height*, and use it to describe the strength of the association. It turns out that $r = 0.804$, which is consistent with a strong positive linear association.

But saying that *Speed* and *Height* have "a" strong positive linear association kind of leaves you hanging, doesn't it? Because it doesn't say *what the line is*. Looking at the scatterplot, it's clear that no straight line can include all those points. But you can imagine a line that would **model** the essence of the relationship, running from the lower left corner of the graph to the upper right. If we had an equation for such a line (called a **linear model**), we could use it to predict or estimate the top speed of a roller coaster given its height. Our goals in this chapter are to learn how to determine a good linear model and to understand the model better using several associated statistical tools.

> Statisticians, like artists, have the bad habit of falling in love with their models.
> —George Box, famous statistician

# Residuals

**Residual = Observed Value − Predicted Value**

A *negative* residual means the predicted value is too big—an overestimate. And a *positive* residual shows that the model makes an underestimate.

Not only can't we draw a line through all the points, the best line might not even hit *any* of the points. Then how can it be the "best" line? We want to find the line that somehow comes *closer* to all the points than any other line. Some of the points will be above the line and some below. For example, the line might suggest that the 78-foot-tall Blue Streak roller coaster at Cedar Point Park in Ohio would reach a top speed of 44 mph when, in fact, it reaches only 40 mph. We call the estimate made from a model the **predicted value**, and write it as $\hat{y}$ (called *y-hat*) to distinguish it from the true value *y* (called, uh, *y*). The difference between the observed value and its associated predicted value is called the **residual**. The residual value tells us how far off the model's prediction is at that point. The Blue Streak's residual would be $y - \hat{y} = 40 - 44 = -4$ miles per hour.

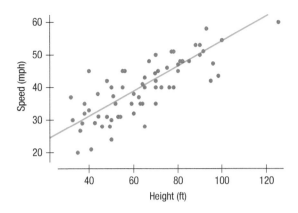

To find the residuals, we always subtract the predicted value from the observed one. The negative residual tells us that the actual speed of the Blue Streak is about 4 mph *less* than the model predicts for a typical 78-foot-tall roller coaster. (The Blue Streak is a wooden roller coaster built in 1964. Sometimes additional variables like these can help explain why a model over- or underestimates a *y*-value.)

Our challenge now is how to find a good line.

## Squared Residuals as a Measure of Fit

**APPLET**
Exploring regression lines

The size of the residuals tells us how well the line fits our data; a line that fits well will have very small residuals. But we can't assess how well the line fits by adding up the residuals—the positive and negative ones would just cancel each other out. We faced the same issue when we calculated a standard deviation to measure spread. And we deal with that issue the same way here: by squaring the residuals. Squaring makes them all positive (or zero). Now we can add them up. Squaring also emphasizes the large residuals. After all, points near the line are consistent with the model, but we're more concerned about points far from the line. When we add all the squared residuals together, that sum indicates how well the line we drew fits the data—the smaller the sum, the better the fit. A different line will produce a different sum, maybe bigger, maybe smaller. For each dataset, one unique line makes the sum of the squared residuals as small as it can be. That line is called the **least squares line**, or the least squares regression line, (sometimes LSRL for short), or the **line of best fit**.[1]

> TI-*nspire*
>
> **Least squares.** Try to minimize the sum of areas of residual squares as you drag a line across a scatterplot.

Thankfully, determining the equation of the least squares line doesn't involve trying out different possible lines. For over 200 years formulas have been known that compute the slope and intercept of the least squares line directly from the data. And these days we don't even use those formulas by hand—technology does the computations for us.

## Correlation and the Line

If you suspect that what we know about correlation can lead us to the equation of the linear model, you're headed in the right direction. It turns out that it's not a very big step. In Chapter 6 we learned a lot about how correlation worked by looking at a scatterplot of the standardized variables. Here's a scatterplot of $z_y$ (standardized *Speed*) vs. $z_x$ (standardized *Height*).

---

[1] We think "line of best fit" is somewhat misleading, because it implies a universal criterion for evaluating fit, when in fact there are other good ways to fit a line to data. We'll stick with the more descriptive "least squares line," or sometimes simply LSRL.

Figure 7.2

**The roller coaster scatterplot in z-scores**

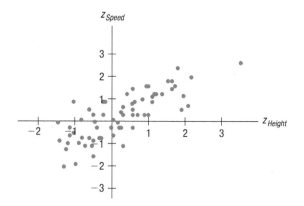

### NOTATION ALERT

"Putting a hat on it" is standard Statistics notation to indicate that something has been predicted by a model. Whenever you see a hat over a variable name or symbol, you can assume it is the predicted value of that variable or symbol (and look around for the model).

What line would you choose to model the relationship of the standardized values? Let's first consider a typical roller coaster. If there were a coaster whose height was the average of all their heights, $\bar{x}$, then without any more information about it, we'd expect its speed to be the average of all the roller coasters' speeds, $\bar{y}$. So a good line should go through the point $(\bar{x}, \bar{y})$. The z-scores of those averages are both zero, so in the plot of z-scores, the line passes through the origin $(0, 0)$.

The equation for a line that passes through the origin can be written with just a slope and no intercept:

$$y = mx.$$

The coordinates of our standardized points aren't written $(x, y)$; their coordinates are z-scores: $(z_x, z_y)$. We'll need to change our equation to show that. And we'll need to indicate that the point on the line corresponding to a particular $z_x$ is $\hat{z}_y$, the model's estimate of the actual value of $z_y$. So our equation becomes

$$\hat{z}_y = mz_x.$$

Many lines with different slopes pass through the origin. Which one fits our data the best? That is, which slope determines the line that minimizes the sum of the squared residuals? It turns out that the best choice for $m$ is the correlation coefficient itself, $r$! Wow! This line has an equation that's about as simple as we could possibly hope for:

$$\hat{z}_y = rz_x.$$

Great. It's simple, but what does it tell us? It says that in moving 1 standard deviation from the mean in $x$, we can expect to move about $r$ standard deviations away from the mean in $y$. Let's be more specific. For the roller coasters, the correlation is about 0.80. If we standardize both height and speed, we can write

$$\hat{z}_{Speed} = 0.80 z_{Height}.$$

Figure 7.3

**Standardized *Speed* vs. standardized *Height* with the regression line** Each 1 standard deviation in *Height* results in a predicted change of *r* standard deviations in *Speed*.

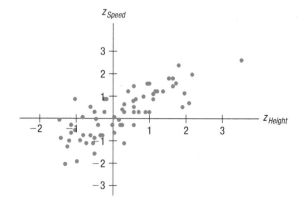

This model tells us that for every standard deviation above (or below) the mean a roller coaster is in height, we'll predict that its speed is 0.80 standard deviations above (or below) the mean in speed. For example, the Blue Streak at Cedar Point Park in Ohio is 78 feet tall, which is about 1 standard deviation taller than the mean ($\bar{x} + s_x = 59.4 + 18.7 = 78.1$). Based on our linear model, we'd expect its speed to be about 0.80 standard deviations faster than the mean, which is about 44.5 miles per hour ($\bar{y} + 0.80 \times s_y = 37.6 + (0.80) \times 8.6 = 44.48$).

In Chapter 6 we mentioned that you sometimes hear the correlation $r$ described as a percentage, which implies it's a fraction *of something*. Well here's what it's a fraction of: $r$ tells you what fraction of a $y$-standard deviation you expect to be from the mean $y$-value if you're one $x$-standard deviation from the mean $x$-value. We still shy away from saying that the correlation is "80 percent," though, because nobody would understand what it's a percent of.[2]

If $r = 0$, there's no linear relationship. The line is horizontal, and no matter how many standard deviations you move in $x$, the predicted value for $y$ doesn't change. On the other hand, if $r = 1.0$ or $-1.0$, there's a perfect linear association. In that case, moving any number of standard deviations in $x$ moves exactly the same number of standard deviations in $y$. In general, moving any number of standard deviations in $x$ moves $r$ times that number of standard deviations in $y$.

# How Big Can Predicted Values Get?

Suppose you were told that a new male student was about to join the class, and you were asked to guess his height in inches. What would be your guess? A safe guess would be the mean height of male students. Now suppose you are also told that this student has a grade point average (*GPA*) of 3.9—about 2 SDs above the mean *GPA*. Would that change your guess? Probably not. The correlation between *GPA* and *Height* is near 0, so knowing the *GPA* value doesn't tell you anything and doesn't move your guess. (And the equation tells us that as well, because it says that we should move $0 \times 2$ SDs from the mean.)

On the other hand, suppose you were told that, measured in centimeters, the student's height was 2 SDs above the mean. There's a perfect correlation of $r = 1$ between *height in inches* and *height in centimeters*, so you'd know he's 2 SDs above mean height in inches as well. (The equation would tell us to move $1.0 \times 2$ SDs from the mean.)

What if you're told that the student is 2 SDs above the mean in *shoe size*? Would you still guess that he's of average *height*? You might guess taller than average, because there's a positive correlation between *height* and *shoe size*. But would you guess that he's 2 SDs above the mean? When there was no correlation, we didn't move away from the mean at all. With a perfect correlation, we moved our guess the full 2 SDs. Any correlation between these extremes should lead us to move somewhere between 0 and 2 SDs above the mean. (To be exact, the equation tells us to move $r \times 2$ standard deviations away from the mean.)

Notice that if $x$ is 2 SDs above its mean, we won't ever guess more than 2 SDs away for $y$ because $r$ can't be bigger than 1.0. So, each predicted $y$ tends to be closer to its mean (in standard deviations) than its corresponding $x$ was. This property of linear models is called **regression to the mean**; that's why the least squares line is also called the **regression line**.

> **WHY IS CORRELATION "R"?**
>
> Did you wonder why we use the letter $r$ to mean correlation? You're going to think we're making this up, but we're not. It turns out that we should not go out on a limb and predict a 1 SD change in $y$ for each 1 SD change in $x$. We need to be more cautious. How cautious? We should back off a bit, predicting a change in $y$ of only $r$ SDs. Another word for backing off: *regression*. That's what the $r$ is for!

---

[2] Okay, you and a very few other people would understand.

### BOOMERANG CEOs

Sometimes when a company isn't doing well, it decides to replace its chief executive officer (CEO) with a former CEO who ran the company during good times. But so-called "boomerang CEOs" seldom do as well as hoped.[3] The explanation may be nothing more than regression to the mean. A company doing well is probably benefiting from a skillful CEO as well as some good luck. It can bring back the CEO later and benefit again from her skill, but it can't bring back the good luck that was also contributing to the company's earlier success. Or to put it another way: the success of a company under a CEO's first stint and its success under her second stint are likely to have a positive correlation, because she's contributing something meaningful to the company's success. But she's hardly the *only* thing contributing, so the correlation *r* will be less than 1. If the company did better than average under the CEO's first stint by some amount, then you'd expect it also to do better than average under her second stint—but by only a fraction (*r*) of that amount.

### JUST CHECKING

A scatterplot of house *Price* (in dollars) vs. house *Size* (in square feet) for houses sold recently in Saratoga, New York, shows a relationship that is straight, with only moderate scatter and no outliers. The correlation between house *Price* and house *Size* is 0.76. The least squares line is

$$\widehat{Price} = 8400 + 88.67\ Size.$$

1. You go to an open house and find that the house is 1 SD above the mean *Size*. How many SDs above the mean *Price* would you predict it to cost?

2. You read an ad for a house that's 2 SDs below the mean *Size*. What would you guess about its price?

3. A friend tells you about a house whose *Size* in square meters (he's European) is 1.5 SDs above the mean. What would you guess about its *Size* in square feet?

4. Suppose the standard deviation of the house *Prices* is $77,000. How much more expensive than the mean *Price* would you predict for the house in part 1 (that's 1 SD above the mean in *Size*)?

5. Suppose the mean *Price* of all these houses is $168,000. What is the predicted *Price* for the house in part 4 that's 1 SD above the mean in *Size*?

6. The mean *Size* of the houses is 1800 sq ft and the standard deviation is 660 sq ft. Predict the *Price* of a 2460 sq ft house (1 SD above the mean) using the least squares equation in the original units to verify that you get the same prediction as in part 5.

# The Regression Line in Real Units

If you wanted to predict the top speed of a roller coaster given its height, you might prefer to do it without having to think about *z*-scores. In fact, you probably would like an equation that gives a predicted speed directly, based on a known height. In Algebra class you may have once seen lines written in the form $y = mx + b$. Statisticians do exactly the same thing, but with different notation:

$$\hat{y} = b_0 + b_1 x.$$

In this equation, $b_0$ is the **y-intercept**, the value of $\hat{y}$ where the line crosses the *y*-axis, and $b_1$ is the **slope**.[4]

---

[3] James Surowiecki, "The Comeback Conundrum," *The New Yorker*, September 21, 2015.

[4] We changed from $mx + b$ to $b_0 + b_1 x$ for a reason—not just to be difficult. Eventually we'll want to add more *x*'s to the model to make it more realistic and we don't want to use up the entire alphabet. What would we use after *m*? The next letter is *n*, and that one's already taken. *o*? See our point? Sometimes subscripts are the best approach.

First let's find the slope. We know that a change of 1 standard deviation in $x$ corresponds to a change of $r$ standard deviations in (predicted) $y$. That's a "rise over run" of $\frac{rs_y}{s_x}$, so the formula for the slope of the least squares regression line is just $b_1 = \frac{rs_y}{s_x}$! It's really just a matter of converting back from $z$-scores, where each unit is a standard deviation, to the units of the original data.

For the roller coaster data, we have:

$$b_1 = \frac{rs_y}{s_x} = \frac{0.80 \times 8.6 \text{ mph}}{18.7 \text{ ft}} = 0.37 \text{ mph of } Speed \text{ per foot of } Height.$$

$\bar{x} = 59.4$ ft   $\bar{y} = 37.6$ mph
$s_x = 18.7$ ft   $s_y = 8.6$ mph
$r = 0.80$.

Next, how do we find the $y$-intercept, $b_0$? Remember that the line has to go through the mean-mean point $(\bar{x}, \bar{y})$. In other words, the model predicts $\bar{y}$ to be the value that corresponds to $\bar{x}$. We can put the means into the equation and write $\bar{y} = b_0 + b_1 \bar{x}$. Solving for $b_0$, we see that the intercept is just

$$b_0 = \bar{y} - b_1 \bar{x}.$$

For the roller coasters, that comes out to

$$b_0 = 37.6 \text{ mph} - 0.37 \frac{\text{mph of } Speed}{\text{ft of } Height} \times 59.4 \text{ ft} = 15.6 \text{ mph}.$$

Putting this back into the regression equation gives

$$\widehat{Speed} = 15.6 + 0.37 \, Height.$$

### Slope
$b_1 = \dfrac{rs_y}{s_x}$

### Intercept
$b_0 = \bar{y} - b_1 \bar{x}$

(We illustrated these computations using rounded numbers for all five of $\bar{y}$, $s_y$, $\bar{x}$, $s_x$, and $r$. It would be better not to do any intermediate rounding. Happily, computers do all these computations quickly without any intermediate rounding, and you'll almost always be using computers to determine regression equations.)

What does this mean? The slope, 0.37, says that, on average, these roller coasters gain about 0.37 additional miles per hour in top speed for every additional foot tall they are. There are other valid ways to interpret the slope. You could say, for example, that you would *predict* (or *estimate*) an increase of 0.37 miles per hour for each additional foot in roller coaster height. But note two things that must be included in an interpretation of slope that are easy to forget: (1) the slope should be described as an *average*, *estimated*, or *predicted* change in $y$ and (2) the average change in $y$ should be linked to a *change* in $x$, and not to $x$ itself. (It would be incorrect to say that roller coasters gain 0.37 miles per hour of speed for each foot tall they are—that ignores the intercept.)

### Units of $y$ per unit of $x$

Get into the habit of identifying the units by writing down "$y$-units per $x$-unit," with the unit names put in place. You'll find it'll really help you to tell about the line in context.

Recall that changing the units of the variables doesn't change the correlation. For the slope, however, units do matter. The units of the slope are always units of $y$ per unit of $x$.

How about the **intercept** of the roller coaster regression line, 15.6? Algebraically, that's the value that $\hat{y}$ takes when $x$ is zero. Sometimes that interpretation makes sense and can help you understand the model. But there are times when $x = 0$ is meaningless in the context of the data, and the intercept then serves only as a model coefficient without any natural interpretation of its own. That's the case with the linear model for the roller coasters. Our model suggests that a roller coaster with a height of 0 feet would still have a top speed of 15.6 miles per hour. (That would be quite a strange roller coaster!) Clearly, in this case the intercept isn't very meaningful.

So we have a linear model, $\widehat{Speed} = 15.6 + 0.37 \, Height$, which allows us to predict a roller coaster's speed based on its height.

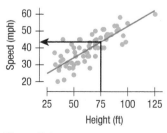

**Figure 7.4**
Roller coasters in their natural units with the regression line

## FOR EXAMPLE

### A Linear Model for Hurricanes

The barometric pressure at the center of a hurricane is often used to measure the strength of the hurricane because it can predict the maximum wind speed of the storm. A scatterplot shows that the relationship is straight, strong, and negative. It has a correlation of −0.898.

Using technology to fit the straight line, we find

$$\widehat{MaxWindSpeed} = 1028.10 - 0.972\, CentralPressure.$$

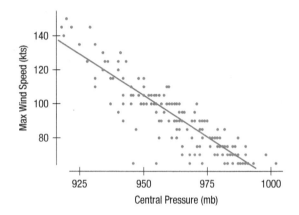

**QUESTION:** Interpret this model. What does the slope mean in this context? Does the intercept have a meaningful interpretation?

**ANSWER:** The negative slope says that lower *CentralPressure* is associated, on average, with higher *MaxWindSpeed*. This is consistent with the way hurricanes work: From physics, we know that low central pressure pulls in moist air, driving the rotation and the resulting destructive winds. The slope's value says that, on average, the maximum wind speed is about 0.972 knots higher for every 1-millibar drop in central pressure. Of course, we can't conclude from the regression that pressure causes wind speed—and, in fact, higher wind speeds also reduce the central pressure. The regression only estimates the slope of the relationship, not its underlying mechanism.

And it's not meaningful to interpret the intercept as the wind speed predicted for a central pressure of zero—that would be a vacuum. Instead, it is merely a starting value for the model.

## JUST CHECKING

Let's look again at the relationship between *Price* (in dollars) and house *Size* (in square feet) in Saratoga, New York. The regression model is

$$\widehat{Price} = 8400 + 88.67\, Size.$$

7. What is the slope and what are its units?
8. What is the interpretation of the slope?
9. Is the model's intercept meaningful? Explain.
10. Your house is 1000 square feet larger than your neighbor's. How much more is its estimated worth (price), based on this linear model?

To use a regression model, we should check the same conditions for regressions as we did for correlation: the **Quantitative Variables Condition**, the **Straight Enough Condition**, and the **Outlier Condition**. In addition, remember that we should use the model to make estimates or predictions only for cases that are similar to the ones in our dataset. Our linear model for predicting roller coaster speeds should not be used for "Extreme" coasters or "Kiddie" roller coasters, for example—we excluded those from our dataset.

## STEP-BY-STEP EXAMPLE

### Calculating a Regression Equation

During the evening rush hour of August 1, 2007, an eight-lane steel truss bridge spanning the Mississippi River in Minneapolis, Minnesota, collapsed without warning, sending cars plummeting into the river, killing 13 and injuring 145. Although similar events had brought attention to our aging infrastructure, this disaster put the spotlight on the problem and raised the awareness of the general public.

How can we tell which bridges are safe?

Most states conduct regular safety checks, giving a bridge a structural deficiency score on various scales. The New York State Department of Transportation uses a scale that runs from 1 to 7, with a score of 5 or less indicating "deficient." Many factors contribute to the deterioration of a bridge, including amount of traffic, material used in the bridge, weather, and bridge design.

New York has more than 17,000 bridges. We have available data on the 193 bridges of Tompkins County.[5]

One natural concern is the age of a bridge. A model that relates age to safety score might help the DOT focus inspectors' efforts where they are most needed.

QUESTION: Is there a relationship between the age of a bridge and its safety rating?

 PLAN State the problem.

VARIABLES Identify the variables and report the W's.

Just as we did for correlation, check the conditions for regression by making a picture. Never fit a regression model without looking at the scatterplot first.

I want to know whether there is a relationship between the age of a bridge in Tompkins County, New York, and its safety rating.

I have data giving the Safety Score and Age at time of inspection for 193 bridges constructed or replaced since 1900.

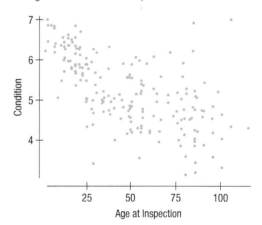

Conditions:

✓ **Quantitative Variables:** Yes, although safety rating has no units. Age is in years.

✓ **Straight Enough:** Scatterplot looks straight.

✓ **No Outliers:** Two points in upper right may deserve attention.

It is OK to use a linear regression to model this relationship.

---

[5]The home county of two of the authors. What a coincidence!

| | |
|---|---|
| **SHOW** — **MECHANICS** Find the equation of the regression line. Summary statistics give the building blocks of the calculation (though we will generally use statistical software to calculate regression slopes and intercepts). <br><br> (We generally report summary statistics to one more digit of accuracy than the data. We do the same for intercept and predicted values, but for slopes we usually report an additional digit. Remember, though, not to round until you finish computing an answer.)[6] | **Age** <br> $\bar{x} = 47.86$ <br> $s_x = 28.70$ <br><br> **Safety Score (points from 1 to 7)** <br> $\bar{y} = 5.24$ <br> $s_y = 0.887$ <br><br> **Correlation** <br> $r = -0.634$ |
| Find the slope, $b_1$. | $b_1 = \dfrac{r s_y}{s_x}$ <br> $= \dfrac{(-0.634)(0.887)}{28.70}$ <br> $= -0.0196$ points per year |
| Find the intercept, $b_0$. | $b_0 = \bar{y} - b_1 \bar{x} = 5.24 - (-0.0196)47.86$ <br> $= 6.18$ |
| Write the equation of the model, using meaningful variable names. | The least squares line is <br> $$\hat{y} = 6.18 - 0.0196x$$ <br> or <br> $$\widehat{Rating} = 6.18 - 0.0196\, Age.$$ |
| **TELL** — **CONCLUSION** Interpret what you have found in the context of the question. Discuss in terms of the variables and their units. | The condition of the bridges in Tompkins County, New York, is generally lower for older bridges. Specifically, we expect bridges that are a year older to have safety ratings that are, on average, 0.02 points lower, on a scale of 1 to 7. The model uses a base of 6.2, which is quite reasonable because a new bridge (0 years of age) should have a safety score near the maximum of 7. <br><br> Because I have data only from one county, I can't tell from these data whether this model would apply to bridges in other counties of New York or in other locations. |

# Residuals Revisited

**Why *e* for "Residual"?**

The flip answer is that *r* is already taken, but the truth is that *e* stands for "error." No, that doesn't mean it's a mistake. Statisticians often refer to variability not explained by a model as error.

The residuals are the part of the data that *hasn't* been modeled. We see that in the definition:

$$Residual = Observed\ values - predicted\ value$$

or, written in symbols,

$$e = y - \hat{y}.$$

When we want to know how well the model fits, we can ask instead what the model missed. No model is perfect, so it's important to know how and where it fails. To see that, we look at the residuals.

---

[6] We warned you that we'll round in the intermediate steps of a calculation to show the steps more clearly, and we've done that here. If you repeat these calculations yourself on a calculator or statistics program, you may get somewhat different results.

## FOR EXAMPLE

### Katrina's Residual

**RECAP:** The linear model relating hurricanes' wind speeds to their central pressures was

$$\widehat{MaxwindSpeed} = 1028.18 - 0.972 \, CentralPressure.$$

Let's use this model to make predictions and see how those predictions do.

**QUESTION:** Hurricane Katrina had a central pressure measured at 920 millibars. What does our regression model predict for her maximum wind speed? How good is that prediction, given that Katrina's actual wind speed was measured at 150 knots?

**ANSWER:** Substituting 920 for the central pressure in the regression model equation gives

$$\widehat{MaxwindSpeed} = 1028.18 - 0.972(920) = 133.94.$$

The regression model predicts a maximum wind speed of 133.94 knots for Hurricane Katrina.

The residual for this prediction is the observed value minus the predicted value:

$$150 - 133.94 = 16.06 \text{ kt}$$

In the case of Hurricane Katrina, the observed wind speed was 16.06 knots greater than the model predicts.

---

Residuals help us to see whether the model makes sense. When a regression model is appropriate, it should capture the underlying relationship. Nothing interesting should be left behind. So after we fit a regression model, we usually plot the residuals against the explanatory variable in the hope of finding . . . nothing.

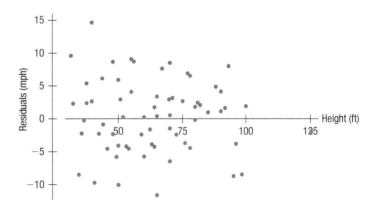

**Figure 7.5**
The residuals for the roller coaster regression look appropriately boring. There are no obvious patterns, which means that whatever information about *Speed* could be "captured" by *Height* has already been captured in the regression model. What's left over is just a "residue" of statistical noise. (The term "residual" is related to "residue"—it's what's "left over" in a response variable after you've explained some of it with a model.)

A scatterplot of the residuals should be the most boring scatterplot you've ever seen. It shouldn't have any interesting features, like a direction or shape. It should show about the same amount of scatter throughout with no bends or outliers. If you see any of these features, find out what the regression model missed.

(Most computer statistics packages plot the residuals against the predicted values $\hat{y}$, but calculators plot them against $x$. When the slope is negative, the two versions are mirror images. When the slope is positive, they're identical except for the axis labels. Because all we care about is the patterns (or, better, lack of patterns) in the residuals, it really doesn't matter which way they are plotted.)

> ### JUST CHECKING
>
> Our linear model for Saratoga homes uses the *Size* (in thousands of square feet) to estimate the *Price* (in thousands of dollars):
>
> $$\widehat{Price} = 8400 + 88.67 \, Size.$$
>
> Suppose you're thinking of buying a home there.
>
> 11. Would you prefer to find a home with a negative or a positive residual? Explain.
> 12. You plan to look for a home of about 3000 square feet. How much should you expect to have to pay?
> 13. You find a nice home that size selling for $300,000. What's the residual?

# The Typical Residual Size (aka Standard Error)

### Why $n - 2$?

Why $n - 2$ rather than $n - 1$? We used $n - 1$ for $s$ when we estimated the mean. Now we're estimating both a slope and an intercept. Looks like a pattern—and it is. We subtract one more for each parameter we estimate.

If the residuals show no interesting patterns, then we're in business and we can look at how big they are. The typical residual size is one way of evaluating how good the model is—it's a rough estimate of how much our model is typically "off" by when we use it to predict or estimate y-values from x-values. But we can't use the average of the residuals as the "typical" size because some of them will be negative.[7]

Back in Chapter 3 we ran into a similar situation. We wanted to express how much a set of quantitative data "typically" deviate from their mean, but instead of averaging the deviations (which include a mix of positive and negative values) we averaged their squares and took the square root. The summary statistic that results is the standard deviation. Here's a reminder of its formula:

$$s = \sqrt{\frac{\sum(y - \bar{y})^2}{n - 1}}.$$

We're going to do pretty much the same thing with our residuals. The only two differences are these: (1) instead of averaging the squared deviations from $\bar{y}$, we'll average the squared deviations from $\hat{y}$; and (2) instead of dividing by $n - 1$ to average the values, we'll divide by $n - 2$.[8] So this is what our formula looks like for the typical residual size, which is more properly called the **standard error**:

$$s_e = \sqrt{\frac{\sum(y - \hat{y})^2}{n - 2}} = \sqrt{\frac{\sum e^2}{n - 2}}.$$

The *standard error* summarizes the typical residual (or error) size just like the *standard deviation* from Chapter 3 summarizes the size of the typical deviation from the mean.

It doesn't hurt to understand how the formula works, but you'll almost always rely on technology to compute the standard error for you. In fact, when you use technology to perform a linear regression, the standard error is one of several summary statistics that is nearly always displayed automatically. Computer output often calls this standard error RMSE, for "root mean squared error"—which exactly describes how it's calculated.

For the roller coasters, the standard error is about 5.2 miles per hour. That means that when this model is used to estimate a coaster's top speed based on its height, the estimates will typically be off by about 5.2 miles per hour. Happily, that's consistent with what we saw in the residual plot (Figure 7.5)—5.2 miles per hour looks like the size of a typical residual.

---

[7] If your model is the least squares regression line, then the residuals will always average to exactly zero.

[8] The reason we divide by $n$ minus *anything* to take an average usually takes years of studying statistics to really understand. (We're not kidding.)

There are two more things that are often worth looking at when it comes to residuals. One is to look at the residual plot and confirm that the residuals don't have a tendency to grow or shrink in size (shrinking is much rarer) as you go from low $x$-values to high ones. If they did, then it wouldn't make sense to talk about "a" typical residual size. For the roller coasters, this is confirmed—the typical residual size seems about the same for both shorter and taller roller coasters.[9]

The other thing that's good to look at is the distribution of residuals. If it's consistent with a Normal distribution (which happens surprisingly often), then it's reasonable to apply the 68–95–99.7 Rule. One implication would be that approximately 95% of the model's *estimated* $y$-values will be within 2 standard errors of their corresponding *actual* $y$-values.

Here's the distribution of residuals for the roller coasters:

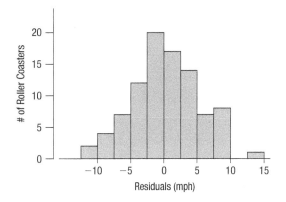

**Figure 7.6**
Both the histogram and the normal probability plot indicate that using a Normal model for the residuals is reasonable.

The histogram of residuals is pretty unimodal and symmetric, so the distribution of the residuals can be reasonably approximated with a Normal model. Because $s_e = 5.2$ miles per hour, we can say that for approximately 95% of these roller coasters, the model predicts a speed based on its height that is within $\pm 10.4$ miles per hour of the coaster's actual speed. If we're willing to extend this model to roller coasters that we believe are similar even though they're not in this dataset, then we can not only estimate a roller coaster's top speed based on its height, but also give a margin of error for our prediction: plus or minus about 10.4 miles per hour. If you think about it, that's pretty powerful!

# Different Samples, Different Regression Lines

The Newfield Covered Bridge is Tompkins County's oldest, completed in 1853 at a cost of $800. It was reconstructed twice: in 1972 and 1998. This photo was taken by one of the authors, who lives nearby and sees the bridge regularly.

The roller coaster data we've been using is a complete population: all the operational roller coasters in the United States classified as "Thrill" rides. Much more often, you'll want to understand things about a population, but you'll have data for only a random sample of that population. Drawing conclusions about a population based on information in a sample is called *inference* and much of the last half of this book is devoted to it. You don't need to know much about inference now, but you should know this: different random samples of data can lead to regression lines that aren't all the same. Let's look at an example.

The Finger Lakes region of central New York boasts 11 major lakes, several smaller ones, and numerous rivers and streams. That means there are a lot of highway bridges— over 700 in the counties bordering the large lakes. Bridges are professionally inspected from time to time and are given safety scores ranging from 1.0 (horrible) to 7.0 (excellent).

Suppose that you wanted to estimate the average rate at which these safety scores decline with age. You'd want current data, but the cost of inspecting every bridge in the population may be prohibitive. So you might want to select a random sample of bridges

---
[9]Perhaps you think that the typical residual size appears to be getting smaller as you go from left to right. If so, be sure that your impression isn't influenced too much by just one or two residuals at the periphery of the scatterplot.

and have them freshly inspected and given a safety score. Assuming the relationship between safety score (y) and age (x) for the bridges in the sample could reasonably be modeled as linear, then the slope of the regression line would be just what you're after: the average rate at which the safety score declines per year.

But what if different samples of bridges lead to different regression lines? We did a simulation to find out whether and how much difference that makes. We started with a dataset containing age and safety score for all the bridges in the region, more than 700 of them. We took a random sample of 20 bridges and determined the regression line predicting safety score based on age. The graph below shows our data points and the regression line, whose equation is $\widehat{Score} = 6.35 - 0.023\, Age$.

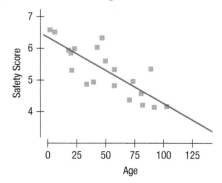

Based on this sample of data, we would estimate that bridges in this region lose an average of 0.023 safety score points per year, or about 1 point in 43 years.

But would a different sample of bridges give the same result? Here's what we got when the computer did the same thing with a different random sample of 20 bridges from the same population:

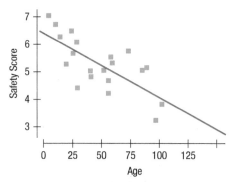

The equation of this regression line is $\widehat{Score} = 6.40 - 0.022\, Age$. So this sample would lead us to estimate that bridges in this region lose an average of 0.022 safety score points per year, or about 1 point in 45 years.

Clearly, the regression line and its slope (and intercept) depend upon the sample. To get an idea of how much the lines vary, we did the same thing with five more random samples of 20 bridges. We won't show all the data points in all the samples because it would clutter up our graph. But here is a graph showing the five different regression lines that resulted, one from each sample of bridges.

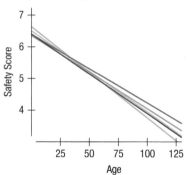

The bad news in this picture is the confirmation that different samples lead to different regression lines with different slopes. So when we do linear regression on a sample of data, we need to keep in mind that the line won't be the same as it would have been for other samples, and probably isn't the same as the line for the whole population either. But there's good news too: the lines in this picture aren't all over the map. They vary a little bit from sample to sample, but *not a whole lot*. In fact, in Chapter 26 of this book, we'll see how to estimate not only the slope and intercept of a regression line, but also how much regression lines are likely to vary from sample to sample. That way, we'll be able to give the slope of the line *and by how much it might reasonably be expected to differ from the population regression line*. But for now, the takeaway message is only that different random samples lead to different regression lines.

## $R^2$—The Variation Accounted For

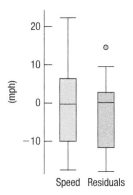

**Figure 7.7**
Compare the variability of *Speed* with that of the residuals from the linear regression. (They've both been centered on zero to make it easier to compare spreads.) The smaller spread in the residuals shows that some part of the original variability in *Speed* can be explained using the regression model on *Height*. In fact, the variance of the residuals is about 0.35 of the variance of the original speeds; about 35 percent of the variability in our coasters' speeds still remains in the residuals.

There are different ways of assessing how well a linear model fits data. The most important is to *make a picture*. Look at a scatterplot of the data with the regression line superimposed. Does it look like the points cluster tightly around the line, or is there a lot of scatter?

Then there are ways of quantifying the fit with a number. The correlation $r$ is one such number. Remember that it must be between $-1$ and $1$. The closer it is to zero, the more the data vary around the regression line, and the closer it is to $-1$ or $1$, the less they vary around the line.

A different way to quantify the fit with a number is to determine what fraction of the variability in the response variable is explained by the regression model. A fraction of 0% would mean that the model explained no variability in $y$ at all and was useless. A fraction of 100% would mean that the linear regression model on $x$ predicted all the $y$-values perfectly, so that all of the variability in $y$ could be explained by the regression model.

Let's see what fraction of the variability in our roller coasters' speeds can be explained by the regression model we constructed based on their heights. The actual speeds can be broken down into a sum of predicted speeds and residuals, which we might think of as "signal" (a real pattern) and "noise" (random variability):

$$Speed = \widehat{Speed} + e$$

An extremely useful property of the least squares regression line is that when it is used to make predictions or estimates, then not only is the equation above true, but so is a similar equation breaking down the *variance* of the roller coasters' speeds into two components: the variance coming from the predictions ("signal") and the variance coming from the residuals ("noise")[10]:

$$Var(Speed) = Var(\widehat{Speed}) + Var(e)$$

For our roller coaster data and the LSRL model predicting *Speed* from *Height*, we computed the following three variances:

$$\begin{array}{ccccc} Var(Speed) & = & Var(\widehat{Speed}) & + & Var(e) \\ 74.22 & = & 47.98 & + & 26.24 \end{array}$$

From this equation we can see that $\frac{47.98}{74.22} = 0.646$, or about 65%, of the variability in the coasters' speeds can be explained by our model based on their heights. And $\frac{26.24}{74.22} = 0.354$, or about 35%, of the variability in their speeds remains unexplained by our model.

In general, when the LSRL model is used to fit bivariate data, the fraction of the variability in the responses that can be explained by the model is denoted $R^2$ (pronounced "R-squared").

$$R^2 = \frac{Var(\hat{y})}{Var(y)}$$

Notice that because variances are necessarily positive, then $Var(y) = Var(\hat{y}) + Var(e)$ implies that $R^2$ must be between 0 and 1. $R^2 = 0$ means that none of the variability in the $y$'s is explained by the model, and $R^2 = 1$ means that all of it is.

TI-*nspire*

**Understanding $R^2$.** Watch the unexplained variability decrease as you drag points closer to the regression line.

---

[10] This equation is ONLY true if the model being used to predict speeds is the least squares regression line. Proving that is, unfortunately, beyond the scope of this text.

There's one more thing you need to know about $R^2$ that's kind of a surprise. It "happens" that $R^2$ is the square of the correlation $r$![11] Of course, this isn't really a coincidence. But connecting the two statistics to one another would require a lot of formulas and computations that would, for most readers, be boring and unhelpful. We want you to take away from this section:

- That $R^2$ is a way of quantifying how well a linear model (and only a linear model) fits data;
- That it does so by expressing the fraction of overall variability in the response variable that is explained by the regression model using the explanatory variable;
- That it can be as low as 0% (the model explains nothing) or as high as 100% (the model explains everything); and
- That it is equal to the square of the correlation.[12]

It doesn't hurt to know the formula, either, but like we said—most of the time you'll be getting $R^2$ with technology, not with the formula.

### Twice as Strong?

Is a correlation of 0.80 twice as strong as a correlation of 0.40? Not if you think in terms of $R^2$. A correlation of 0.80 means an $R^2$ of $0.80^2 = 64\%$. A correlation of 0.40 means an $R^2$ of $0.40^2 = 16\%$—only a quarter as much of the variability accounted for. A correlation of 0.80 gives an $R^2$ *four* times as strong as a correlation of 0.40 and accounts for four times as much of the variability.

### FOR EXAMPLE
#### Interpreting $R^2$

**RECAP:** Our regression model that predicts maximum wind speed in hurricanes based on the storm's central pressure has correlation $r = -0.898$.

**QUESTION:** What is $R^2$ and what does that say about our regression model?

**ANSWER:** $R^2 = (-0.898)^2 = 0.806$. An $R^2$ of 81% indicates that 81% of the variation in maximum wind speed can be accounted for by the hurricane's central pressure. Other factors, such as temperature and whether the storm is over water or land, may explain some of the remaining variation.

### JUST CHECKING

Back to our regression of house *Price* ($) on house *Size* (square feet):

$$\widehat{Price} = 8400 + 88.67 \, Size.$$

The $R^2$ value is reported as 57.8%, and the standard deviation of the residuals is $53,790.

14. What does the $R^2$ value mean about the relationship of price and size?
15. Is the correlation of price and size positive or negative? How do you know?
16. You find that your house is worth $50,000 more than the regression model predicts. You are undoubtedly pleased, but is this actually a surprisingly large residual?
17. If we measure house *Size* in square meters instead, would $R^2$ change? Would the slope of the line change? Explain.

---

[11] Why does one use an uppercase $R$ and the other a lowercase $r$? That is a very good question.
[12] And because $0 \leq r^2 \leq 1$ (100%), it must be true that $-1 \leq r \leq 1$. But you knew that already!

# How Big Should $R^2$ Be?

> **SOME EXTREME TALES**
> One major company developed a method to differentiate between proteins. To do so, they had to distinguish between regressions with $R^2$ of 99.99% and 99.98%. For this application, 99.98% was not high enough.
>
> The president of a financial services company reports that although his regressions give $R^2$ below 2%, they are highly successful because those used by his competition are even lower.

$R^2$ is always between 0% and 100%. But what's a "good" $R^2$ value? The answer depends on the kind of data you are analyzing and on what you want to do with it. Just as with correlation, there is no value for $R^2$ that automatically determines that the regression model is "good." As we've seen, an $R^2$ of 100% is a perfect fit, with no scatter around the line. Then $s_e$ would be zero. All of the variance is accounted for by the model and none is left in the residuals at all. This sounds great, but it's too good to be true for real data.[13] Data from scientific experiments often have $R^2$ in the 80% to 90% range and even higher. Data from observational studies and surveys, though, often show relatively weak associations because there are usually lots of extraneous variables that you can't control like you can in an experiment. An $R^2$ of 50% to 30% or even lower might be taken as evidence of a useful model. $R^2$ is the first part of a regression analysis that many people look at because, along with the scatterplot, it tells whether the regression model is even worth thinking about. The standard deviation of the residuals can give us more contextual information about the usefulness of the model by telling us how much scatter there is around the line.

# More About Regression Assumptions and Conditions

Linear regression models may be the most widely used models in all of Statistics. They have everything we could want in a model: two easily estimated parameters, a meaningful measure of how well the model fits the data, and the ability to predict new values. They even provide a self-check in plots of the residuals to help us avoid silly mistakes.

Like all models, though, linear models are only appropriate if some assumptions are true. We can't confirm assumptions, but we often can check related conditions.

First, be sure that both variables are quantitative. It makes no sense to perform a regression on categorical variables. After all, what could the slope possibly mean? Always check the **Quantitative Variables Condition**.

Because a linear model only makes sense if the underlying relationship is linear, you must consider whether a *Linearity Assumption* is justified. To see, check the associated **Straight Enough Condition**. Just look at the scatterplot of *y* vs. *x*. You don't need a *perfectly* straight plot, but it must be straight enough for the linear model to make sense. If you try to model a curved relationship with a straight line, you'll usually get just what you deserve. If the scatterplot is not straight enough, stop here. You can't always use a linear model for just *any* two variables, even if they are related.

For the standard deviation of the residuals to summarize the scatter of all the residuals, the residuals must share the same spread for each value of *x*. That's an assumption. But if the spread of the scatterplot from the line looks roughly the same everywhere and (often more vividly) if the *residual plot* of residuals *vs.* predicted values also has a consistent vertical spread, then that assumption is reasonable. The most common violation of this equal variance assumption is residuals that spread out more for *larger* values of *x*, so a good nickname for this check is the **Does the Plot Thicken? Condition**.

Outlying points can dramatically change a regression model. They can even change the sign of the slope, which would give a very different impression of the relationship between the variables if you only look at the regression model. So check the **Outlier Condition**. Check both the scatterplot of *y* against *x* and the residual plot to be sure there are no outliers. The residual plot often shows violations more clearly and may reveal other unexpected patterns or interesting quirks in the data. Of course, any outliers are likely to be interesting and informative, so be sure to look into why they are unusual.

---

[13] If you see an $R^2$ of 100%, it's a good idea to figure out what happened. You may have discovered a new law of Physics, but it's much more likely that you accidentally regressed two variables that measure the same thing.

### Make a Picture (Or Two)

You can't check the conditions just by checking boxes. You need to examine both the original scatterplot of y against x before you fit the model, and the plot of residuals afterward. These plots can save you from making embarrassing errors.[14]

To summarize:

Before starting, be sure to check the

- **Quantitative Variable Condition** If either y or x is categorical, you can't make a scatterplot and you can't perform a regression analysis. Stop.

From the scatterplot of y against x, check the

- **Straight Enough Condition** Is the relationship between y and x straight enough to proceed with a linear regression model?
- **Outlier Condition** Are there any outliers that might dramatically influence the fit of the least squares line?
- **Does the Plot Thicken? Condition** Does the spread of the data around the generally straight relationship seem to be consistent for all values of x?

After fitting the regression model, make a plot of residuals and look for

- Any bends that would violate the **Straight Enough Condition**,
- Any outliers that weren't clear before, and
- A tendency of the residuals to grow or shrink in size as you go from left to right.

# A Tale of Two Regressions

Regression equations may not behave exactly the way you'd expect. Our regression model for the roller coasters was $\widehat{Speed} = 15.6 + 0.37\ Height$. That equation allowed us (among other things) to estimate that a 78-foot tall coaster like the Blue Streak would reach a top speed of about 44 mph. Suppose, though, that we wanted to construct a roller coaster reaching a certain speed, and we wanted to estimate how tall it would need to be. It might seem natural to think that by solving our equation for *Height* we'd get a model estimating *Height* from *Speed*. But that doesn't work.

Our original model is $\hat{y} = b_0 + b_1 x$, but the new one needs to evaluate an $\hat{x}$ based on a value of y. There's no y in our original model, only $\hat{y}$, and that makes a big difference. Our model doesn't fit the roller coaster data perfectly, and the least squares criterion focuses on the *vertical* errors (or residuals) the model makes in using x to model y—not the *horizontal* errors related to modeling x.

A quick look at the equations reveals why. Simply solving our equation for x would give a new line whose slope must be reciprocal. To model y in terms of x, our slope is $b_1 = \frac{rs_y}{s_x}$. To model x in terms of y, we'd need to use the slope $b_1 = \frac{rs_x}{s_y}$. Notice that is not the reciprocal of ours.

| Height | Speed |
|---|---|
| $\bar{x}$ = 59.4 ft | $\bar{y}$ = 37.6 mph |
| $s_x$ = 18.7 ft | $s_y$ = 8.6 mph |
| r = 0.80. | |

If we want to estimate *Height* from *Speed*, we need to create a different model. The slope is $b_1 = \frac{(0.80)(18.7)}{8.6} = 1.74$ feet of *Height* per mile per hour of *Speed*. The equation turns out to be $\widehat{Height} = -6.23 + 1.74\ Speed$. Now we'd estimate that a roller coaster whose top speed was 44 mph would need to be 71 feet tall—not the 78 feet that would arise from the first equation.

Moral of the story: *Think*. (Where have you heard *that* before?) Decide which variable you want to use (x) to predict values for the other (y). Then find the model that does that. If, later, you want to make predictions in the other direction, you'll need to start over and create the other model from scratch.

---

[14] Or losing points on an exam!

# CHAPTER 7  Linear Regression

## STEP-BY-STEP EXAMPLE

### Regression

Even if you hit the fast-food joints for lunch, you should have a good breakfast. Nutritionists, concerned about "empty calories" in breakfast cereals, recorded facts about 77 cereals, including their *Calories* per serving and *Sugar* content (in grams).

QUESTION: How can we use sugar content to estimate calories in breakfast cereals?

---

**THINK**

**PLAN** State the problem and determine the role of the variables.

**VARIABLES** Name the variables and report the W's.

Check the conditions for regression by making a picture. Never fit a regression line without looking at the scatterplot first.

I am interested in using cereals' sugar content (x) to estimate calories (y).

✓ **Quantitative Variables Condition**: I have two quantitative variables, *Calories* and *Sugar* content per serving, measured on 77 breakfast cereals. The units of measurement are calories and grams of sugar, respectively.

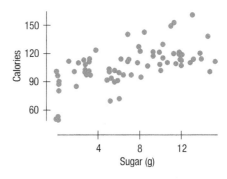

✓ **Outlier Condition**: There are no obvious outliers or groups.

✓ The **Straight Enough Condition** is satisfied; I will fit a regression model to these data.

✓ The **Does the Plot Thicken? Condition** is satisfied. The spread around the line looks about the same throughout.

---

**SHOW**

**MECHANICS** If there are no clear violations of the conditions, fit a straight line model of the form $\hat{y} = b_0 + b_1 x$ to the data. Summary statistics give the building blocks of the calculation.

Calories

$$\bar{y} = 107.0 \text{ calories}$$
$$s_y = 19.5 \text{ calories}$$

Sugar

$$\bar{x} = 7.0 \text{ grams}$$
$$s_x = 4.4 \text{ grams}$$

| | |
|---|---|
| Find the slope. | **Correlation** $$r = 0.564$$ $$b_1 = \frac{rs_y}{s_x} = \frac{0.564(19.5)}{4.4}$$ |
| Find the intercept. | $$= 2.50 \text{ calories per gram of sugar}$$ $$b_0 = \bar{y} - b_1\bar{x} = 107 - 2.50(7) = 89.5 \text{ calories}$$ |
| Write the equation, using meaningful variable names. | So the least squares line is $$\hat{y} = 89.5 + 2.50x \text{ or}$$ $$\widehat{Calories} = 89.5 + 2.50 \text{ Sugar.}$$ |
| State the value of $R^2$. | Squaring the correlation gives $$R^2 = 0.564^2 = 0.318 \text{ or } 31.8\%.$$ |

**TELL** — **CONCLUSION** Describe what the model says in words and numbers. Be sure to use the names of the variables and their units.

The scatterplot shows a positive, linear relationship and no outliers. The slope of the least squares regression line suggests that cereals have about 2.50 Calories more per additional gram of Sugar.

The key to interpreting a regression model is to start with the phrase "$b_1$ $y$-units per $x$-unit," substituting the estimated value of the slope for $b_1$ and the names of the respective units. The intercept is then a starting or base value.

The intercept predicts that sugar-free cereals would average about 89.5 calories.

$R^2$ gives the fraction of the variability of $y$ accounted for by the linear regression model.

The $R^2$ says that 31.8% of the variability in Calories is accounted for by variation in Sugar content.

Find the standard deviation of the residuals, $s_e$, and compare it to the original $s_y$. (We found its value using technology.)

$s_e = 16.2$ calories. That's smaller than the original SD of 19.5, but still fairly large.

**THINK AGAIN** — **CHECK AGAIN** Even though we looked at the scatterplot *before* fitting a regression model, a plot of the residuals is essential to any regression analysis because it is the best check for additional patterns and interesting quirks in the data.

TI-*nspire*

**Residuals plots.** See how the residuals plot changes as you drag points around in a scatterplot.

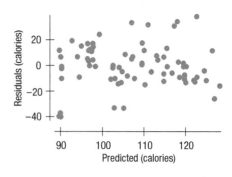

The residuals show a shapeless form and roughly equal scatter for all predicted values. The linear model appears to be appropriate.

# CHAPTER 7 Linear Regression

## TI TIPS

### Regression Lines and Residuals Plots

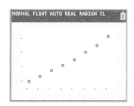

By now you will not be surprised to learn that your calculator can do it all: scatterplot, regression line, and residuals plot. Let's try it using the $CO_2$ data from the last chapter. (TI Tips, p. 159) You should still have that saved in lists named YEAR and CO2. First, re-create the scatterplot.

**FIND THE EQUATION OF THE REGRESSION LINE.** Actually, you already found the line when you used the calculator to get the correlation. But this time we'll be a little fancier so that we can display the line on our scatterplot. We want to tell the calculator to do the regression analysis and save the equation of the model as a graphing variable.

- Under STAT CALC choose LinReg(a + bx).
- Specify that Xlist and Ylist are YEAR and $CO_2$, as before, but . . .
- Now add one more specification to store the regression equation. Press VARS, go to the Y-VARS menu, choose 1:Function, and finally(!) choose Y1.
- To Calculate, hit ENTER.

There's the equation. The calculator tells you that the regression line is $\widehat{CO2} = -3162 + 1.767 \, Year$. Can you explain what the slope and $y$-intercept mean?

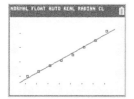

**ADD THE LINE TO THE PLOT.** When you entered the LinReg command, the calculator automatically saved the equation as Y1. Just hit GRAPH to see the line drawn across your scatterplot.

**CHECK THE RESIDUALS.** Remember, you are not finished until you check to see if a linear model is appropriate. That means you need to see if the residuals appear to be randomly distributed. To do that, you need to look at the residuals plot.

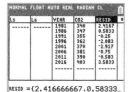

This is made easy by the fact that the calculator has already placed the residuals in a list named RESID. Want to see them? Go to STAT EDIT and look through the lists. (If RESID is not already there, go to the first blank list and import the name RESID from your LIST NAMES menu. The residuals should appear.) Every time you have the calculator compute a regression analysis, it will automatically save this list of residuals for you. (Any previously computed residuals will be overwritten.)

**NOW CREATE THE RESIDUALS PLOT.**

- Set up STAT PLOT Plot2 as a scatterplot with Xlist:YEAR and Ylist:RESID.

- Before you try to see the plot, go to the Y= screen. By moving the cursor around and hitting ENTER in the appropriate places you can turn off the regression line and Plot1, and turn on Plot2.
- ZoomStat will now graph the residuals plot.

Uh-oh! See the curve? The residuals are high at both ends, low in the middle. Looks like a linear model isn't appropriate after all. Notice that the residuals plot makes the curvature much clearer than the original scatterplot did.

***Moral: Always check the residuals plot!***

So a linear model isn't appropriate here. What now? The next two chapters provide techniques for dealing with data like these.

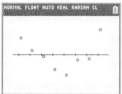

# Reality Check: Is the Regression Reasonable?

> **REGRESSION: ADJECTIVE, NOUN, OR VERB**
>
> You may see the term *regression* used in different ways. There are many ways to fit a line to data, but the term "regression line" or "regression" without any other qualifiers always means least squares. People also speak of *regressing* a *y*-variable on an *x*-variable to mean fitting a linear model.

Statistics don't come out of nowhere. They are based on data. The results of a statistical analysis should reinforce your common sense, not fly in its face. If the results are surprising, then either you've learned something new about the world or your analysis is wrong.

Whenever you perform a regression analysis, think about the coefficients and ask whether they make sense. Is a slope of 2.5 calories per gram of sugar reasonable? That's hard to say right off. We know from the summary statistics that a typical cereal has about 100 calories and 7 grams of sugar. A gram of sugar contributes some calories (actually, 4, but you don't need to know that), so calories should go up with increasing sugar. The direction of the slope seems right.

To see if the *size* of the slope is reasonable, a useful trick is to consider its order of magnitude. We'll start by asking if shrinking the slope by a factor of 10 seems reasonable. Is 0.25 calories per gram of sugar enough? Then the 7 grams of sugar found in the average cereal would contribute less than 2 calories. That seems too small.

Now let's try inflating the slope by a factor of 10. Is 25 calories per gram reasonable? Then the average cereal would have 175 calories from sugar alone. The average cereal has only 100 calories per serving, though, so that slope seems too big.

We have tried inflating the slope by a factor of 10 and deflating it by 10 and found both to be unreasonable. So, like Goldilocks, we're left with the value in the middle that's just right. And an increase of 2.5 calories per gram of sugar is at least *plausible*.

It's easy to take something that comes out of a computer at face value and just go with it. The small effort of asking yourself whether the regression equation makes sense is repaid whenever you catch errors or avoid saying something silly or absurd about the data.

## WHAT CAN GO WRONG?

There are many ways in which data that appear at first to be good candidates for regression analysis may be unsuitable. And there are ways that people use regression that can lead them astray. Here's an overview of the most common problems. We'll discuss these and more in the next chapter.

- **Don't fit a straight line to a nonlinear relationship.** Linear regression is suited only to relationships that are, well, *linear*. Fortunately, we can often improve the linearity easily by using re-expression. We'll come back to that topic in Chapter 9.

- **Don't ignore outliers.** Outliers can have a serious impact on the fitted model. You should identify them and think about why they are extraordinary. If they turn out not to be obvious errors, read the next chapter for advice.

- **Don't infer that *x* causes *y* just because there is a good linear model for their relationship.** When two variables are strongly correlated, it is often tempting to assume a causal relationship between them. Putting a regression line on a scatterplot tempts us even further, but it doesn't make the assumption of causation any more valid. For example, our regression model predicting hurricane wind speeds from the central pressure was reasonably successful, but the relationship is very complex. It is reasonable to say that low central pressure at the eye is responsible for the high winds because it draws moist, warm air into the center of the storm, where it swirls around, generating the winds. But as is often the case, things aren't quite that simple. The winds themselves also contribute to lowering the pressure at the center of the storm as it becomes a hurricane. Understanding causation requires far more work than just finding a correlation or modeling a relationship.

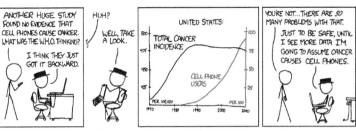

© 2013 Randall Munroe. Reprinted with permission. All rights reserved.

**Think Variation!**

$R^2$ does not mean that height accounts for 66% of the speed of a roller coaster. It is the *variation* in speeds that is accounted for by the linear model.

- **Don't choose a model based on $R^2$ alone.** Although $R^2$ measures the *strength* of a linear association, a high $R^2$ does not demonstrate the *appropriateness* of the regression model. Even a relationship that's actually curved could produce a high $R^2$, as could an outlier even if there's little association in the rest of the data. Or an otherwise strong relationship could have an outlier that makes $R^2$ misleadingly low. Always look at the scatterplot.

- **Don't invert the regression.** The model works in one direction only. If the equation predicts $\hat{y}$ from $x$, it will not correctly predict $\hat{x}$ from $y$. This isn't algebra class[15]; you can't solve the equation for the other variable. If you want to make predictions in the other direction, you'll need to create the model that does that.

## WHAT HAVE WE LEARNED?

We've learned that when the relationship between quantitative variables is fairly straight, a linear model can help summarize that relationship and give us insights about it:
- The regression (best fit) line doesn't pass through all the points, but it is the best compromise in the sense that the sum of squares of the residuals is the smallest possible.

We've learned several things the correlation, $r$, tells us about the regression:
- For each SD of $x$ that we are away from the $x$ mean, we expect to be $r$ SDs of $y$ away from the $y$ mean.
- Because $r$ is always between $-1$ and $+1$, each predicted $y$ is fewer SDs away from its mean than the corresponding $x$ was, a phenomenon called regression to the mean.
- The slope of the line is based on the correlation, adjusted for the units of $x$ and $y$:

$$b_1 = \frac{rs_y}{s_x}.$$

We've learned to interpret the slope in context as predicted change in $y$-units per 1 unit change in $x$.

We've learned that the residuals and $R^2$ reveal how well the model works:
- If a plot of residuals shows a pattern, especially curvature, we should re-examine the data to see why.
- The standard deviation of the residuals, $s_e$, quantifies the amount of scatter around the line.
- The square of the correlation coefficient, $R^2$, gives us the fraction of the variation of the response variable accounted for by the regression model. The remaining $1 - R^2$ of the variation is left in the residuals and not explained by the model.
- $R^2$ is an overall measure of how successful the regression is in <u>linearly</u> relating $y$ to $x$.

---

[15] And that's a Good Thing, isn't it?

Of course, the linear model makes no sense unless the **Linearity Assumption** is satisfied. We check the **Straight Enough Condition** and **Outlier Condition** with a scatterplot, as we did for correlation, and also with a plot of residuals against either the *x* or the predicted values. For the standard deviation of the residuals to make sense as a summary, we have to make the **Equal Variance Assumption**. We check it by looking at both the original scatterplot and the residual plot for the **Does the Plot Thicken? Condition**.

# TERMS

| | |
|---|---|
| **Model** | An equation or formula that simplifies and represents reality. (p. 173) |
| **Linear model** | A linear model is an equation of a line. To interpret a linear model, we need to know the variables (along with their W's) and their units. (p. 173) |
| **Predicted value** | The value of $\hat{y}$ found for a given *x*-value in the data. A predicted value is found by substituting the *x*-value in the regression equation. The predicted values are the values on the fitted line; the points $(x, \hat{y})$ all lie exactly on the fitted line. (p. 173) |
| **Residuals** | Residuals are the differences between data values and the corresponding values predicted by the regression model—or, more generally, values predicted by any model. (p. 173) $$\text{Residual} = \text{Observed value} - \text{Predicted value} = e = y - \hat{y}$$ |
| **Least squares** | The least squares criterion specifies the unique line that minimizes the variance of the residuals or, equivalently, the sum of the squared residuals. (p. 174) |
| **Regression to the mean** | Because the correlation is always less than 1.0 in magnitude, each predicted $\hat{y}$ tends to be fewer standard deviations from its mean than its corresponding *x* was from its mean. This is called regression to the mean. (p. 175) |
| **Regression line (Line of best fit)** | The particular linear equation $$\hat{y} = b_0 + b_1 x$$ that satisfies the least squares criterion is called the least squares regression line (LSRL). Casually, we often just call it the regression line, or the line of best fit. (p. 175) |
| **Slope** | The slope, $b_1$, gives a value in "*y*-units per *x*-unit." Changes of one unit in *x* are associated with changes of $b_1$ units in predicted values of *y*. The slope can be found by $$b_1 = \frac{rs_y}{s_x}.\ (\text{p. 176})$$ |
| **Intercept** | The intercept, $b_0$, gives a starting value in *y*-units. It's the $\hat{y}$-value when *x* is 0. You can find it from $b_0 = \bar{y} - b_1\bar{x}$. (p. 177) |
| **Standard error ($s_e$)** | The standard deviation of the residuals is found by $s_e = \sqrt{\dfrac{\sum e^2}{n-2}}$. When the assumptions and conditions are met, the residuals can be well described by using this standard deviation and the 68–95–99.7 Rule. (p. 182) |
| **$R^2$** | $R^2$ (the square of the correlation between *y* and *x*) gives the fraction of the variability of *y* accounted for by the least squares linear regression model. (p. 185) |
| **Does the Plot Thicken? Condition** | The scatterplot of residuals plot should show consistent (vertical) spread in *y*-values. (p. 187) |

# ON THE COMPUTER
## Regression Analysis

All statistics packages make a table of results for a regression analysis. These tables may differ slightly from one package to another, but all are essentially the same—and all include much more than we need to know for now. Every computer regression table includes a section something like the one below (based on data for other fast food items):

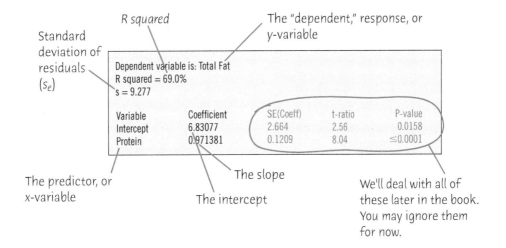

The slope and intercept coefficient are given in a table such as this one. Usually the slope is labeled with the name of the x-variable, and the intercept is labeled "Intercept" or "Constant." So the regression equation shown here is

$$\widehat{Fat} = 6.83077 + 0.97138\ Protein.$$

It is not unusual for statistics packages to give many more digits of the estimated slope and intercept than could possibly be estimated from the data. (The original data were reported to the nearest gram.) Ordinarily, you should round most of the reported numbers to one digit more than the precision of the data, and the slope to two. We will learn about the other numbers in the regression table later in the book. For now, all you need to be able to do is find the coefficients, the $s_e$, and the $R^2$ value.

# EXERCISES

1. **Cereals** For many people, breakfast cereal is an important source of fiber in their diets. Cereals also contain potassium, a mineral shown to be associated with maintaining a healthy blood pressure. An analysis of the amount of fiber (in grams) and the potassium content (in milligrams) in servings of 77 breakfast cereals produced the regression model $\widehat{Potassium} = 38 + 27\ Fiber$. If your cereal provides 9 grams of fiber per serving, how much potassium does the model estimate you will get?

2. **Horsepower** A study that examined the relationship between the fuel economy (mpg) and horsepower for 15 models of cars produced the regression model $\widehat{mpg} = 46.87 - 0.084\ HP$. If the car you are thinking of buying has a 200-horsepower engine, what does this model suggest your gas mileage would be?

3. **More cereal** Exercise 1 describes a regression model that estimates a cereal's potassium content from the amount of fiber it contains. In this context, what does it mean to say that a cereal has a negative residual? What does it mean to say that a cereal has a residual of $-22$?

4. **Horsepower, again** Exercise 2 describes a regression model that uses a car's horsepower to estimate its fuel economy. In this context, what does it mean to say that a certain car has a positive residual? What does it mean to say a certain car has a residual of 3?

5. **Another bowl** In Exercise 1, the regression model $\widehat{Potassium} = 38 + 27\ Fiber$ relates fiber (in grams) and potassium content (in milligrams) in servings of breakfast cereals. Explain what the slope means.

**204** PART II Exploring Relationships Between Variables

**6. More horsepower** In Exercise 2, the regression model $\widehat{mpg} = 46.87 - 0.084\,HP$ relates cars' horsepower to their fuel economy (in mpg). Explain what the slope means.

**7. Cereal again** The correlation between a cereal's fiber and potassium contents is $r = 0.903$. What fraction of the variability in potassium is accounted for by the amount of fiber that servings contain?

**8. Another car** The correlation between a car's horsepower and its fuel economy (in mpg) is $r = -0.869$. What fraction of the variability in fuel economy is accounted for by the horsepower?

**9. Last bowl!** For Exercise 1's regression model predicting potassium content (in milligrams) from the amount of fiber (in grams) in breakfast cereals, $s_e = 30.77$. Explain in this context what that means.

**10. Last tank!** For Exercise 2's regression model predicting fuel economy (in mpg) from the car's horsepower, $s_e = 3.287$. Explain in this context what that means.

**11. Residuals I** Tell what each of the residual plots below indicates about the appropriateness of the linear model that was fit to the data.

**12. Residuals II** Tell what each of the residual plots below indicates about the appropriateness of the linear model that was fit to the data.

**13. Ballpark figures** At a small ballpark, the manager keeps track of ticket sales at each game and also how much revenue is collected for snacks, including food and drinks. At the end of the season, he sees from a scatterplot that these variables (*Food Revenue* and *Number of Tickets Sold*) have approximately a linear relationship.

a) Explain briefly what it means in this context for the relationship between these variables to be linear and not curved.
b) Which of these is most likely to be the slope of the regression line for *Food revenue* (y) on *Number of tickets sold* (x): 0.05, 0.50, 5.00, or 50? Explain your reasoning.
c) Give one plausible source of variability in food sales other than the variable *Number of tickets sold*.

**14. More ballpark figures** The manager of the ballpark described in Exercise 13 also keeps data on the duration of every game.

a) Why might it be reasonable to expect game duration to have a positive association with food and drink revenue?
b) If the manager finds there to be *no* association between game duration and food/drink revenue, what does that indicate about the eating habits of game patrons?

**15. True or false** If false, explain briefly.

a) We choose the linear model that passes through the most data points on the scatterplot.
b) The residuals are the observed y-values minus the corresponding y-values predicted by the linear model.
c) Least squares means that the square of the largest residual is as small as it could possibly be.

**16. True or false II** If false, explain briefly.

a) Some of the residuals from a least squares linear model will be positive and some will be negative.
b) Least squares means that some of the squares of the residuals are minimized.
c) We write $\hat{y}$ to denote the predicted values and $y$ to denote the observed values.

**17. From statistics to equation I** In Chapter 6, we saw that there is a linear relationship between *Fuel Economy* and *Engine Size* for a random sample of 35 cars. For that sample, the correlation between Fuel economy and Engine size is $-0.797$ and the summary statistics for each variable are given in the table below.

| Variable | n | Mean | Std. Dev. | Median | Range | Min | Max | Q1 | Q3 |
|---|---|---|---|---|---|---|---|---|---|
| Engine size (liters) | 35 | 3.017 | 1.1942 | 3 | 4.6 | 1.4 | 6 | 2 | 3.7 |
| Fuel economy (MPG) | 35 | 24.343 | 6.5302 | 23 | 27 | 13 | 40 | 19 | 30 |

a) What is the slope of the regression line to predict *Fuel economy* from *Engine size*? Explain what this means in context.
b) What is the equation of the regression line? Define the variables used in the equation.

**18. From statistics to equation II** In Chapter 6, we learned about the Minnesota Department of Transportation's attempt to weigh moving vehicles. We saw a linear relationship between the weight measured while the vehicle was moving and the weight measured while sitting on a standard scale (static weight). The correlation between the two measures is 0.965 and the summary statistics for each variable are given below.

| Variable | n | Mean | Std. Dev. | Median | Range | Min | Max | Q1 | Q3 |
|---|---|---|---|---|---|---|---|---|---|
| Weight-in-Motion (1000 lb) | 10 | 32.02 | 5.685 | 31.3 | 15.1 | 25.1 | 40.2 | 26 | 36.2 |
| Static Weight (1000 lb) | 10 | 31.28 | 3.757 | 29.95 | 11 | 27 | 38 | 27.9 | 34.5 |

a) What is the slope of the regression line to predict *Static weight* from *Weight-in-motion*? Explain what this means in context.
b) What is the equation of the regression line? Define the variables used in the equation.

**19. Book sales** Recall the data we examined in Chapter 6, Exercise 5, examining the relationship between the amount of sales and the number of salespeople working in a small bookstore.

| Number of Salespeople Working | Sales (in $1000) |
|---|---|
| 2 | 10 |
| 3 | 11 |
| 7 | 13 |
| 9 | 14 |
| 10 | 18 |
| 10 | 20 |
| 12 | 20 |
| 15 | 22 |
| 16 | 22 |
| 20 | 26 |

Here is the regression analysis using *Number of Salespeople Working* to predict *Sales (in $1000)*.

Dependent variable is Sales
R-squared = 93.2%
s = 1.47

| Variable | Coefficient |
|---|---|
| Constant | 8.1006 |
| Salespeople | 0.9134 |

a) Write the regression equation. Define the variables used in your equation.
b) Interpret the slope in context.
c) What does the *y*-intercept mean in this context? Is it meaningful?
d) If 8 Salespeople are working, what is the predicted amount of *Sales*?
e) State and interpret the $R^2$ value in context.

**20. Disk drives 2020 again** Recall the data on disk drives we saw in Chapter 6, Exercise 6. Suppose we want to predict *Price* from *Capacity*.

| Capacity (in TB) | Price (in $) |
|---|---|
| 0.25 | 24.99 |
| 0.5 | 36.79 |
| 1.0 | 47.99 |
| 2.0 | 59.99 |
| 3.0 | 79.99 |
| 4.0 | 97.99 |
| 5.0 | 104.99 |
| 6.0 | 124.99 |
| 8.0 | 218.80 |
| 10.0 | 269.97 |
| 12.0 | 319.99 |
| 16.0 | 519.99 |
| 32.0 | 1399.95 |

Dependent variable is Price
R-squared = 95.7%
s = 80.64

| Variable | Coefficient |
|---|---|
| Constant | −65.851 |
| Capacity | 41.729 |

a) Write the regression equation, defining the variables in your equation.
b) Interpret the slope in context.
c) Interpret the *y*-intercept in context. Is this meaningful?
d) What would your model predict for the price of a 20 TB drive?
e) A 20 TB drive on Amazon.com was listed at $1023.24. According to the model, does this seem like a good buy?
f) The correlation is very high. Does this mean that the model is appropriate? Explain. (The scatterplot is shown below.)

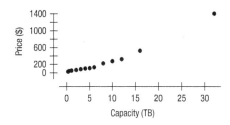

g) Interpret the value of $R^2$ in this context.

**21. Book sales again** Here are the residuals for a regression of *Sales (in $1000)* on *Number of Salespeople working* for Exercise 19.

| Number of Salespeople Working | Residuals |
|---|---|
| 2 | 0.0726 |
| 3 | 0.1592 |
| 7 | −1.4944 |
| 9 | −2.3212 |
| 10 | 0.7654 |
| 10 | 2.7654 |
| 12 | 0.9385 |
| 15 | 0.1983 |
| 16 | −0.7151 |
| 20 | −0.3687 |

a) What are the units of the residuals?
b) Which residual contributes the most to the sum that was minimized according to the least squares criterion to find this regression model?
c) Which residual contributes least to that sum?

**22. Disk drives 2020, residuals** Here are the residuals for a regression of *Price* on *Capacity* for the hard drives of Exercise 20 (based on the hand-computed coefficients).

| Capacity | Residual |
|---|---|
| 0.25 | 80.41 |
| 0.5 | 81.78 |
| 1.0 | 72.11 |
| 2.0 | 42.38 |
| 3.0 | 20.6 |
| 4.0 | −3.08 |
| 5.0 | −37.80 |
| 6.0 | −59.53 |
| 8.0 | −49.18 |
| 10.0 | −81.87 |
| 12.0 | −114.91 |
| 16.0 | −81.82 |
| 32.0 | 130.47 |

a) Which residual contributes the most to the sum that is minimized by the least squares criterion?
b) Seven of the residuals are negative. What does that mean about those drives? Be specific and use the correct units.

23. **Book sales revisited** Suppose the manager of the bookstore in Exercise 19 collects new data on *Sales* and *Number of Salespeople Working* for 10 different days, sees a linear relationship in the scatterplot, and calculates the slope of the LSRL. However, this time the slope is 0.8815, not the same as before. What does this indicate?

24. **Disk drives encore** Your best friend has a part-time job at a local electronics store. She is curious about the cost of memory. On a day when she is bored at work (can you imagine?!), she collects her own data. She finds a slope of $44.3/TB. Why are these values different?

25. **Residual plots** Here are residual plots (residuals plotted against predicted values) for three linear regression models. Indicate which condition appears to be violated (linearity, outlier or equal spread) in each case.

a)
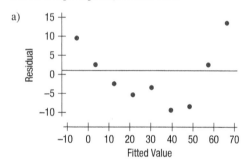

b)
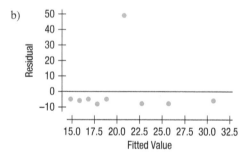

c)
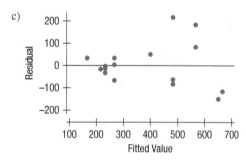

26. **Disk drives 2020, residuals again** Here is a scatterplot of the residuals from the regression of the hard drive prices on their sizes from Exercise 22.

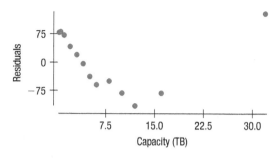

a) Are any assumptions or conditions violated? If so, which ones?
b) What would you recommend about this regression?

27. **Real estate** A random sample of records of sales of homes from February 15 to April 30, 1993, from the files maintained by the Albuquerque Board of Realtors gives the *Price* and *Size* (in square feet) of 117 homes. A regression to predict *Price* (in thousands of dollars) from *Size* has an *R*-squared of 71.4%. The residuals plot indicated that a linear model is appropriate.

a) What are the variables and units in this regression?
b) What units does the slope have?
c) Do you think the slope is positive or negative? Explain.

28. **Baby size** A least squares regression line was calculated to relate the length (cm) of newborn boys to their weight in kg. $R^2 = 64.3\%$.

a) What are the variables and what are the units in this regression?
b) What are the units for the slope?
c) Do you think the slope is positive or negative? Explain.

29. **Real estate again** The regression of *Price* on *Size* of homes in Albuquerque had $R^2 = 71.4\%$, as described in Exercise 27. Write a sentence (in context, of course) summarizing what the $R^2$ says about this regression model.

30. **Baby two** The relationship between *Length* and *Weight* in Exercise 28 has an $R^2$ of 64.3%. Interpret this value in context.

31. **Real estate redux** The regression of *Price* on *Size* of homes in Albuquerque had $R^2 = 71.4\%$, as described in Exercise 27.

a) What is the correlation between *Size* and *Price*? Explain why you chose the sign (+ or −) you did.
b) What would you predict about the *Price* of a home 1 standard deviation above average in *Size*?
c) What would you predict about the *Price* of a home 2 standard deviations below average in *Size*?

32. **Baby three** The regression in Exercise 28 between *Length* and *Weight* has an $R^2$ of 64.3%.

a) Find the correlation coefficient and interpret it in context.
b) What would you predict about the weight of a baby whose initial length was 1 standard deviation below the mean length?
c) What would you predict about the weight of a baby whose initial length was 3 standard deviations above the mean length?

33. **More real estate** Consider the Albuquerque home sales from Exercise 27 again. The regression analysis gives the model $\widehat{Price} = 47.82 + 0.061\ Size$.

a) Explain what the slope of the line says about housing prices and house size.
b) What price would you predict for a 3000-square-foot house in this market?

c) A real estate agent shows a potential buyer a 1200-square-foot home, saying that the asking price is $6000 less than what one would expect to pay for a house of this size. What is the asking price, and what is the $6000 called?

**34. One last baby** A regression equation was found for Exercise 28. It is $\widehat{Weight} = -5.94 + 0.1875\, Length$.

a) Interpret the slope in context.
b) If a baby boy is 52 cm long, what is his predicted weight?
c) The newborn grandson of one of the authors was 48 cm long and weighed 3 kg. According to the regression model, what was his residual? What does that say about him?

**35. Misinterpretations** A Biology student who created a regression model to use a bird's *Height* when perched for predicting its *Wingspan* made these two statements. Assuming the calculations were done correctly, explain what is wrong with each interpretation.

a) My $R^2$ of 93% shows that this linear model is appropriate.
b) A bird 10 inches tall will have a wingspan of 17 inches.

**36. More misinterpretations** A Sociology student investigated the association between a country's *Literacy Rate* and *Life Expectancy*, then drew the conclusions listed below. Explain why each statement is incorrect. (Assume that all the calculations were done properly.)

a) The *Literacy Rate* determines 64% of the *Life Expectancy* for a country.
b) The slope of the line shows that an increase of 5% in *Literacy Rate* will produce a 2-year improvement in *Life Expectancy*.

**37. ESP** People who claim to "have ESP" participate in a screening test in which they have to guess which of several images someone is thinking of. You and a friend both took the test. You scored 2 standard deviations above the mean, and your friend scored 1 standard deviation below the mean. The researchers offer everyone the opportunity to take a retest.

a) Should you choose to take this retest? Explain.
b) Now explain to your friend what their decision should be and why.

**38. SI jinx** Players in any sport who are having great seasons, turning in performances that are much better than anyone might have anticipated, often are pictured on the cover of *Sports Illustrated*. Frequently, their performances then falter somewhat, leading some athletes to believe in a "*Sports Illustrated* jinx." Similarly, it is common for phenomenal rookies to have less stellar second seasons—the so-called "sophomore slump." While fans, athletes, and analysts have proposed many theories about what leads to such declines, a statistician might offer a simpler (statistical) explanation. Explain.

**39. Cigarettes** Is the nicotine content of a cigarette related to the "tars"? A collection of data (in milligrams) on 816 cigarettes produced the scatterplot, regression analysis, and residuals plot shown:

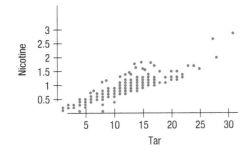

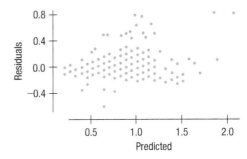

Response variable is: Nicotine
R squared = 81.4%
s = 0.1343

| Variable | Coefficient |
|---|---|
| Intercept | 0.148305 |
| TAR | 0.062163 |

a) Do you think a linear model is appropriate here? Explain.
b) Explain the meaning of $R^2$ in this context.

**40. Attendance 2019, revisited** In Chapter 6, Exercise 43 looked at the relationship between the number of runs scored by American League baseball teams and the average attendance at their home games for the 2019 season. Here are the scatterplot, the residuals plot, and part of the regression analysis for *all* major league teams in 2019 (National League teams in red, American League in blue):

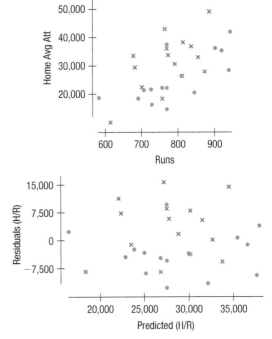

Response variable is: Home Avg Att
R squared = 32.7%
s = 7835

| Variable | Coefficient | SE(Coeff) |
|---|---|---|
| Intercept | −18352.0 | 12723 |
| Runs | 59.66 | 16.16 |

a) Do you think a linear model is appropriate here? Explain.
b) Interpret the meaning of $R^2$ in this context.
c) Do the residuals show any pattern worth remarking on?
d) The point at the top of the plots is the L.A. Dodgers. What can you say about the residual for the Dodgers?

**41. Cigarettes II** Consider again the regression of *Nicotine* content on *Tar* (both in milligrams) for the cigarettes examined in Exercise 39.

a) What is the correlation between *Tar* and *Nicotine*?
b) What would you predict about the average *Nicotine* content of cigarettes that are 2 standard deviations below average in *Tar* content?
c) If a cigarette is 1 standard deviation above average in *Nicotine* content, what do you suspect is true about its *Tar* content?

**42. Attendance 2019, revisited** Consider again the regression of *Home Average Attendance* on *Runs* for the baseball teams examined in Exercise 40.

a) What is the correlation between *Runs* and *Home Average Attendance*?
b) What would you predict about the *Home Average Attendance* for a team that is 2 standard deviations above average in *Runs*?

**43. Cigarettes III** Take another look at the regression analysis of tar and nicotine content of the cigarettes in Exercise 39.

a) Write the equation of the regression line.
b) Predict the *Nicotine* content of cigarettes with 9 milligrams of *Tar*.
c) Interpret the meaning of the slope of the regression line in this context.
d) What does the *y*-intercept mean?
e) If a new brand of cigarette contains 7 milligrams of tar and a nicotine level whose residual is $-0.05$ mg, what is the nicotine content?

**44. Attendance 2019, last inning** Refer again to the regression analysis for home average attendance and games won by baseball teams, seen in Exercise 40.

a) Write the equation of the regression line.
b) Estimate the *Home Average Attendance* for a team with 750 *Runs*.
c) Interpret the meaning of the slope of the regression line in this context.
d) In general, what would a negative residual mean in this context?

**45. Income and housing revisited** In Chapter 6, Exercise 37, we learned that the Office of Federal Housing Enterprise Oversight (OFHEO) collects data on various aspects of housing costs around the United States. Here's a scatterplot (by state) of the *Housing Cost Index* (HCI) versus the *Median Family Income* (MFI) for the 50 states. The correlation is $r = 0.65$. The mean HCI is 338.2, with a standard deviation of 116.55. The mean MFI is \$46,234, with a standard deviation of \$7072.47.

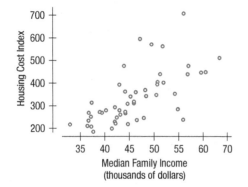

a) Is a regression analysis appropriate? Explain.
b) What is the equation that predicts Housing Cost Index from median family income?
c) For a state with MFI = \$44,993, what would be the predicted HCI?
d) Washington, DC, has an MFI of \$44,993 and an HCI of 548.02. How far off is the prediction in part c from the actual HCI?
e) If we standardized both variables, what would be the regression equation that predicts standardized HCI from standardized MFI?
f) If we standardized both variables, what would be the regression equation that predicts standardized MFI from standardized HCI?

**46. Interest rates and mortgages 2020 again** In Chapter 6, Exercise 38, we saw a plot of mortgages in the United States (in trillions of dollars) vs. the interest rate at various times over the past 25 years. The correlation is $r = -0.867$. The mean mortgage amount is \$8.846T and the mean interest rate is 6.642%. The standard deviations are \$4.794T for mortgage amounts and 2.24% for the interest rates.

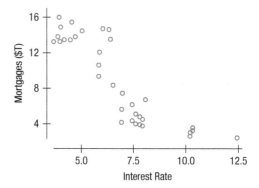

a) Is a linear regression model appropriate for predicting mortgage amount from interest rates? Explain.
b) Regardless of your answer to part a, find the equation that predicts mortgage amount from interest rates.
c) What would you predict the mortgage amount would be if the interest rates climbed to 13%?
d) Do you have any reservations about your prediction in part c? Explain.
e) If we standardized both variables, what would be the regression equation that predicts standardized mortgage amount from standardized interest rates?
f) If we standardized both variables, what would be the regression equation that predicts standardized interest rates from standardized mortgage amount?

**47. Online clothes** An online clothing retailer keeps track of its customers' purchases. For those customers who signed up for the company's credit card, the company also has information on the customer's *Age* and *Income*. A random sample of 500 of these customers shows the following scatterplot of *Total Yearly Purchases* by *Age*:

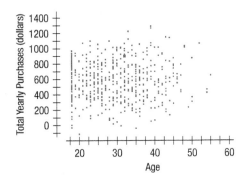

The correlation between *Total Yearly Purchases* and *Age* is $r = 0.037$. Summary statistics for the two variables are:

|  | Mean | SD |
|---|---|---|
| Age | 29.67 yrs | 8.51 yrs |
| Total Yearly Purchase | $572.52 | $253.62 |

a) What is the linear regression equation for predicting *Total Yearly Purchase* from *Age*?
b) Do the assumptions and conditions for regression appear to be met?
c) What is the predicted average *Total Yearly Purchase* for an 18-year-old? For a 50-year-old?
d) What percentage of the variability in *Total Yearly Purchases* is accounted for by this model?
e) Do you think the regression model might be a useful one for the company? Explain.

**T 48. Online clothes II** For the online clothing retailer discussed in the previous problem, the scatterplot of *Total Yearly Purchases* by *Income* shows

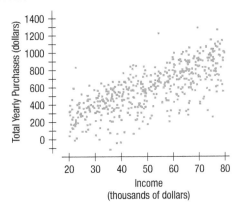

The correlation between *Total Yearly Purchases* and *Income* is 0.722. Summary statistics for the two variables are:

|  | Mean | SD |
|---|---|---|
| Income | $50,343.40 | $16,952.50 |
| Total Yearly Purchase | $572.52 | $253.62 |

a) What is the linear regression equation for predicting *Total Yearly Purchase* from *Income*?
b) Do the assumptions and conditions for regression appear to be met?
c) What is the predicted average *Total Yearly Purchase* for someone with a yearly *Income* of $20,000? For someone with an annual *Income* of $80,000?
d) What percentage of the variability in *Total Yearly Purchases* is accounted for by this model?
e) Do you think the regression model might be a useful one for the company? Comment.

**T 49. SAT scores** The SAT is a test often used as part of an application to college. SAT scores are between 200 and 800, but have no units. In the old SAT, tests were given in both Math and Verbal areas. Doing the SAT-Math problems also involves the ability to read and understand the questions, but can a person's verbal score be used to predict the math score? Verbal and math SAT scores of a high school graduating class are displayed in the scatterplot, with the regression line added.

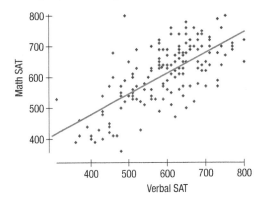

Here is the regression analysis of *Math SAT* vs. *Verbal SAT*.

Dependent variable is Math SAT
R-squared = 46.9%
s = 71.75

| Variable | Coefficient |
|---|---|
| Intercept | 209.5542 |
| Verbal SAT | 0.67507 |

a) Describe the relationship.
b) Are there any students whose scores do not seem to fit the overall pattern?
c) Find the correlation coefficient and interpret this value in context.
d) Write the equation of the regression line, defining any variables used in the equation.
e) Interpret the slope of this line.
f) Predict the math score of a student with a verbal score of 500.
g) Every year some student scored a perfect 1600 on these two parts of the test. Based on this model, what would be that student's Math score residual?

**50. Success in college** Colleges use SAT scores in the admissions process because they believe these scores provide some insight into how a high school student will perform at the college level. Regression analysis was computed on using *SAT* to predict *GPA*.

Dependent variable is GPA
R-squared = 22.1%

| Variable | Coefficient |
|---|---|
| Intercept | −1.262 |
| SAT | 0.00214 |

a) Write the equation of the regression line.
b) Explain what the $y$-intercept of the regression line indicates.
c) Interpret the slope of the regression line.
d) Predict the GPA of a freshman who scored a combined 2100.
e) Based upon these statistics, how effective do you think SAT scores would be in predicting academic success during the first semester of the freshman year at this college? Explain.
f) As a student, would you rather have a positive or a negative residual in this context? Explain.

**51. SAT, take 2** Suppose the AP calculus students complained and insisted that we should use SAT math scores to estimate verbal scores (using the same data from Exercise 49). Here is the regression analysis of *Math SAT* vs. *Verbal SAT*.

Dependent variable is Verbal
s = 72.77

| Variable | Coefficient |
|---|---|
| Intercept | 171.333 |
| Math SAT | 0.6943 |

a) What is the correlation?
b) Write the equation of the line of regression predicting verbal scores from math scores.
c) In general, what would a positive residual mean in this context?
d) A person tells you her math score was 500. Predict her verbal score.
e) Using that predicted verbal score and the equation you created in Exercise 49, predict her math score.
f) Why doesn't the result in part e come out to 500?

**52. Success, part 2** The standard deviation of the residuals in Exercise 50 is 0.275. Interpret this value in context.

**53. Wildfires 2015** The National Interagency Fire Center (www.nifc.gov) reports statistics about wildfires. Here's an analysis of the number of wildfires between 1985 and 2015.

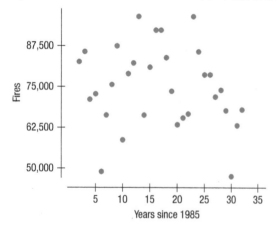

Response variable is Fires
R squared = 2.7%
s = 12397

| Variable | Coefficient |
|---|---|
| Constant | 78791.6 |
| Years since 1985 | −221.575 |

a) Is a linear model appropriate for these data? Explain.
b) Interpret the slope in this context.

c) Can we interpret the intercept? Why or why not?
d) What does the value of $s_e$ say about the size of the residuals? What does it say about the effectiveness of the model?
e) What does $R^2$ mean in this context?

**54. Wildfires 2015—sizes** We saw in Exercise 53 that the number of fires was fairly constant. But has the damage they cause remained constant as well? Here's a regression analysis that examines the trend in *Acres per Fire* (in hundreds of thousands of acres) together with some supporting plots:

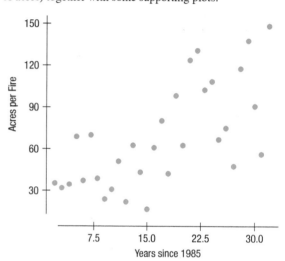

Response variable is Acres/Fire
R squared = 45.5%
s = 27.72

| Variable | Coefficient |
|---|---|
| Intercept | 21.7466 |
| Years since 1985 | 2.73606 |

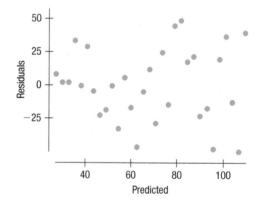

a) Is the regression model appropriate for these data? Explain.
b) Interpret the slope in this context.
c) Interpret the intercept, or explain why it is not meaningful.
d) What interpretation (if any) can you give for the $R^2$ in the regression table?

**55. Used convertibles 2020** Carfax.com lists used cars available for sale in your area. A search for a convertible turns up the data shown in the table on the next page:

a) Make a scatterplot of *Price* vs. *Age*.
b) Describe the association of these variables.
c) Do you think a linear model is appropriate?

d) Computer software says that $R^2 = 24.1\%$. What is the correlation between *Age* and *Price*?
e) Explain the meaning of $R^2$ in this context.
f) Why doesn't this model explain 100% of the variability in *Price*?

| Car | Age (yr) | Price | Mileage |
|---|---|---|---|
| Ford Mustang GT | 1 | $36,900 | 16,834 |
| Chevy Corvette | 12 | 26,500 | 24,446 |
| Mercedes-Benz E- 550 | 4 | 32,598 | 39,732 |
| Ford Mustang Premium | 3 | 26,499 | 16,133 |
| Jaguar F-Type S | 3 | 50,995 | 24,883 |
| Jaguar XJ XJS | 25 | 19,990 | 54,485 |
| BMW Z3 1.9 | 23 | 9,999 | 24,773 |
| Mercedes-Benz CLK | 16 | 18,995 | 40,237 |
| Pontiac G6 GT | 13 | 5,500 | 1,15,321 |
| Ford Mustang GT | 13 | 9,998 | 1,27,829 |
| Volkswagen New Beetle | 12 | 9,989 | 19,703 |
| Chevy Corvette | 12 | 26,500 | 24,446 |
| Volvo C70 T5 | 9 | 16,900 | 75,571 |
| Mini Cooper S | 9 | 6,995 | 1,01,788 |
| Audi A5 Premium | 8 | 16,885 | 50,998 |
| Volvo C70 T5 | 7 | 19,495 | 37,119 |
| Mini Cooper Roadster | 7 | 14,990 | 31,601 |
| Volkswagen Beetle | 6 | 16,995 | 62,012 |
| Audi TT | 4 | 31,995 | 24,434 |
| Jaguar F-Type S | 3 | 43,490 | 41,447 |
| BMW 2 Series 230i | 3 | 32,354 | 19,570 |
| Volkswagen Beetle SE | 3 | 23,903 | 19,940 |
| Chevrolet Camaro Z28 | 18 | 29,995 | 16,501 |
| Chevrolet Corvette LT2 | 13 | 26,999 | 23,523 |
| Smart Fortwo Passion | 11 | 7,995 | 26,255 |
| Chevrolet Cornette LT2 | 6 | 48,900 | 6,508 |

**56. Drug abuse** In the exercises of the last chapter you examined results of a survey conducted in the United States and 10 countries of Western Europe to determine the percentage of teenagers who had used marijuana and other drugs. Below is the scatterplot.

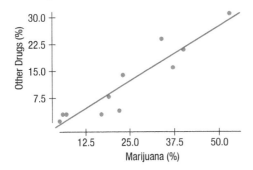

Dependent variable is Other%
R-squared = 87.3%

| Variable | Coefficient |
|---|---|
| Constant | −3.001 |
| Marijuana% | 0.611 |

a) Do you think a linear model is appropriate? Explain.
b) For this regression, $R^2$ is 87.3%. Interpret this statistic in this context.
c) Write the equation you would use to estimate the percentage of teens who use other drugs from the percentage who have used marijuana.
d) Explain in context what the slope of this line means.
e) Do these results confirm that marijuana is a "gateway drug," that is, that marijuana use leads to the use of other drugs?

**57. More used convertibles 2020** Use the advertised prices for used convertibles in Exercise 55 to find a linear model to predict *Price* from *Age*.
a) Find the equation of the regression line.
b) Explain the meaning of the slope of the line.
c) Explain the meaning of the *y*-intercept of the line.
d) What might be a fair price for a 10-year old convertible?
e) You have a chance to buy one of two convertibles. They are about the same age and appear to be in equally good condition. Would you rather buy the one with a positive residual or the one with a negative residual? Explain.
f) The dataset actually (really!) included another car, which we initially set aside. It is a 6-year old Mclaren MP4-12C, offered for sale for a mere $124,998. What does your linear model say about this car and its price?
g) You see a "For Sale" sign on a 10-year-old convertible stating the asking price as $24,500. What is the residual?
h) Would this regression model be useful in establishing a fair price for a 30-year-old convertible? Explain.

**58. Birthrates 2015** The table shows the number of live births per 1000 population in the United States, starting in 1965. (National Center for Health Statistics, www.cdc.gov/nchs/)

| Year | 1965 | 1970 | 1975 | 1980 | 1985 | 1990 | 1995 | 2000 | 2005 | 2010 | 2015 |
|---|---|---|---|---|---|---|---|---|---|---|---|
| Rate | 19.4 | 18.4 | 14.8 | 15.9 | 15.6 | 16.4 | 14.8 | 14.4 | 14.0 | 13.0 | 12.4 |

a) Make a scatterplot and describe the general trend in *Birthrates*. (Enter *Year* as years since 1900: 65, 70, 75, etc.)
b) Find the equation of the regression line.
c) Check to see if the line is an appropriate model. Explain.
d) Interpret the slope of the line.
e) The table gives rates only at intervals. Estimate what the rate was in 1978.
f) In 1978, the birthrate was actually 15.0. How close did your model come?

**59. Burgers** In the last chapter, you examined the association between the amounts of *Fat* and *Calories* in fast-food hamburgers. Here are the data:

| Fat (g) | 19 | 31 | 34 | 35 | 39 | 39 | 43 |
|---|---|---|---|---|---|---|---|
| Calories | 410 | 580 | 590 | 570 | 640 | 680 | 660 |

a) Create a scatterplot of *Calories* vs. *Fat*.
b) Interpret the value of $R^2$ in this context.
c) Write the equation of the line of regression.
d) Use the residuals plot to explain whether your linear model is appropriate.
e) Explain the meaning of the *y*-intercept of the line.
f) Explain the meaning of the slope of the line.
g) A new burger containing 28 grams of fat is introduced. According to this model, its residual for calories is +33. How many calories does the burger have?

**60. Chicken** Chicken sandwiches are often advertised as a healthier alternative to beef because many are lower in fat. Data from tests on 15 different sandwiches randomly selected from the website http://fast-food-nutrition.findthebest.com/d/a/Chicken-Sandwich produced the *Calories* vs. *Fat* scatterplot and the regression analysis below.

Dependent variable is Calories
R-squared = 93.2%
s = 63.73

| Variable | Coefficient | SE |
|---|---|---|
| Constant | 12.7234 | 43.910 |
| Fat | 21.2171 | 1.5944 |

a) Do you think a linear model is appropriate in this situation?
b) Describe the strength of this association.
c) Write the equation of the regression line to estimate calories from the fat content.
d) Explain the meaning of the slope.
e) Explain the meaning of the y-intercept.
f) What does it mean if a certain sandwich has a negative residual?

**61. A second helping of burgers** In Exercise 59 you created a model that can estimate the number of *Calories* in a burger when the *Fat* content is known.

a) Explain why you cannot use that model to estimate the fat content of a burger with 600 calories.
b) Make that estimate using an appropriate model.

**62. A second helping of chicken** In Exercise 60 you created a model to estimate the number of *Calories* in a chicken sandwich when you know the *Fat*.

a) Explain why you cannot use that model to estimate the fat content of a 400-calorie sandwich.
b) Quiznos large mesquite sandwich stands out on the graph with an impressive 53 fat grams and 1190 calories. What effect do you think this value has on the regression equation?

**63. Climate change 2019** The earth's climate is getting warmer. Climate scientists attribute the increase to an increase in atmospheric levels of carbon dioxide ($CO_2$), a greenhouse gas. Here is a scatterplot showing the mean annual temperature anomaly (the difference between the mean global temperature and a base period of 1981 to 2010 in °C) and the $CO_2$ concentration in the atmosphere in parts per million (ppm) at the top of Mauna Loa in Hawaii for the years 1958 to 2019.

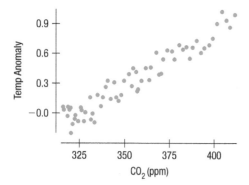

A regression predicting *Temperature* anomaly from $CO_2$ produces the following output table:

Response variable is: Global temp anomaly
R-squared = 91.8%   s = 0.094

| Variable | Coefficient |
|---|---|
| Intercept | −3.4527 |
| $CO_2$(ppm) | 0.0106 |

a) What is the correlation between $CO_2$ and *Temperature*?
b) Explain the meaning of R-squared in this context.
c) Give the regression equation.
d) What is the meaning of the slope in this equation?
e) What is the meaning of the y-intercept of this equation?
f) Here is a scatterplot of the residuals vs. predicted values. Does this plot show evidence of the violation of any assumptions behind the regression? If so, which ones?

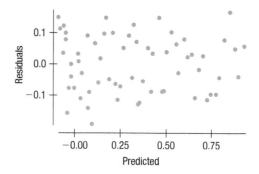

g) $CO_2$ levels will probably reach 450 ppm by 2050. What mean temperature *anomaly* does the regression predict for that concentration of $CO_2$?
h) Does the answer in part g mean that when the $CO_2$ level hits 450 ppm, the temperature anomaly will reach the predicted level? Explain briefly.

**64. $CO_2$ levels.** The level of carbon dioxide in the atmosphere has been measured monthly since 1958 at the top of Mauna Loa in Hawaii. The hope is that measurements in the middle of the Pacific will not be influenced by local traffic or industry. A timeplot of $CO_2$ levels (parts per million) vs. fractional year shows annual cycles and an overall pattern.

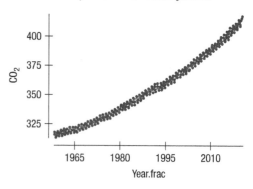

A regression looks like this:
R squared = 97.6%   s = 4.389

| Variable | Coefficient |
|---|---|
| Intercept | −2784.68 |
| Year.frac | 1.57835 |

a) What is the meaning of $R^2$ in this context?
b) Interpret the slope.
c) Do the regression assumptions appear to be satisfied?

Here is a plot of residuals against predicted values:

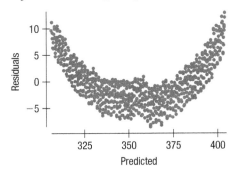

d) What can you conclude from the plot of residuals? Be specific. In what ways does the residual plot indicate deviations from the regression model?

**65. Body fat** It is difficult to determine a person's body fat percentage accurately without immersing them in water. Researchers hoping to find ways to make a good estimate immersed 20 male subjects, then measured their waists and recorded their weights.

| Waist (in.) | Weight (lb) | Body Fat (%) | Waist (in.) | Weight (lb) | Body Fat (%) |
|---|---|---|---|---|---|
| 32 | 175 | 6 | 33 | 188 | 10 |
| 36 | 181 | 21 | 40 | 240 | 20 |
| 38 | 200 | 15 | 36 | 175 | 22 |
| 33 | 159 | 6 | 32 | 168 | 9 |
| 39 | 196 | 22 | 44 | 246 | 38 |
| 40 | 192 | 31 | 33 | 160 | 10 |
| 41 | 205 | 32 | 41 | 215 | 27 |
| 35 | 173 | 21 | 34 | 159 | 12 |
| 38 | 187 | 25 | 34 | 146 | 10 |
| 38 | 188 | 30 | 44 | 219 | 28 |

a) Create a model to predict *%Body Fat* from *Weight*.
b) Do you think a linear model is appropriate? Explain.
c) Interpret the slope of your model.
d) Is your model likely to make reliable estimates? Explain.
e) What is the residual for a person who weighs 190 pounds and has 21% body fat?

**66. Body fat again** Would a model that uses the person's *Waist* size be able to predict the *%Body Fat* more accurately than one that uses *Weight*? Using the data in Exercise 65, create and analyze that model.

**67. Women's heptathlon revisited** We discussed the women's 2016 Olympic heptathlon in Chapter 5. Here are the results from the high jump, 800-meter run, and long jump for the 27 women who successfully completed all three events in the 2016 Olympics:

| Name | High Jump | 800 m | Long Jump |
|---|---|---|---|
| Evelis Aguilar | 1.74 | 134.32 | 6.23 |
| Nadine Broersen | 1.77 | 137.55 | 6.15 |
| Katerina Cachov | 1.77 | 138.95 | 5.91 |
| Vanessa Chefer | 1.68 | 134.2 | 6.1 |
| Ivona Dadic | 1.77 | 135.64 | 6.05 |
| Jessica Ennis-Hill | 1.89 | 129.07 | 6.34 |
| Alysbeth Felix | 1.68 | 135.32 | 6.22 |
| Laura Ikauniece-Admidina | 1.77 | 129.43 | 6.12 |
| Katarina Johnson-Thompson | 1.98 | 130.47 | 6.51 |
| Akela Jones | 1.89 | 161.12 | 6.3 |
| Hanna Kasyanova | 1.77 | 136.58 | 5.88 |
| Eliska Klucinova | 1.8 | 142.81 | 6.08 |
| Xenia Krizsan | 1.77 | 133.46 | 6.08 |
| Heather Miller-Koch | 1.8 | 126.82 | 6.16 |
| Antoinette Nana Djimou Ida | 1.77 | 140.36 | 6.43 |
| Barbara Nwaba | 1.83 | 131.61 | 5.81 |
| Jennifer Oeser | 1.86 | 133.82 | 6.19 |
| Claudia Rath | 1.74 | 127.22 | 6.55 |
| Yorgelis Rodríguez | 1.86 | 134.65 | 6.25 |
| Carolin Schafer | 1.83 | 136.52 | 6.2 |
| Brianne Theisen-Eaton | 1.86 | 129.5 | 6.48 |
| Nafissatou Thiam | 1.98 | 136.54 | 6.58 |
| Anouk Vetter | 1.77 | 137.71 | 6.1 |
| Nadine Visser | 1.68 | 134.47 | 6.35 |
| Kendell Williams | 1.83 | 136.24 | 6.31 |
| Sofia Yfantidou | 1.65 | 150.08 | 5.51 |
| Gyorgyi Zsivoczky-Farkas | 1.86 | 131.76 | 6.31 |

Let's examine the association among these events. Perform a regression to predict long jump performance from 800 m run times.

a) What is the regression equation? What does the slope mean?
b) What percentage of the variability in long jumps can be accounted for by differences in 800-m times?
c) Do good long jumpers tend to be fast runners? (Be careful—low times are good for running events and high distances are good for jumps.)
d) What does the residuals plot reveal about the model?
e) Do you think this is a useful model? Would you use it to predict long jump performance? (Compare the residual standard deviation to the standard deviation of the long jumps.)

**68. Heptathlon revisited again** We saw the data for the women's 2016 Olympic heptathlon in Exercise 67. Are the two jumping events associated? Perform a regression of the long-jump results on the high-jump results.

a) What is the regression equation? What does the slope mean?
b) What percentage of the variability in long jumps can be accounted for by high-jump performances?
c) Do good high jumpers tend to be good long jumpers?
d) What does the residuals plot reveal about the model?
e) Do you think this is a useful model? Would you use it to predict long-jump performance? (Compare the residual standard deviation to the standard deviation of the long jumps.)

**69. Least squares I** Consider the four points (10, 10), (20, 50), (40, 20), and (50, 80). The least squares line is $\hat{y} = 7.0 + 1.1x$. Explain what "least squares" means, using these data as a specific example.

**70. Least squares II** Consider the four points (200, 1950), (400, 1650), (600, 1800), and (800, 1600). The least squares line is $\hat{y} = 1975 - 0.45x$. Explain what "least squares" means, using these data as a specific example.

## JUST CHECKING

### Answers

1. You should expect the price to be 0.76 standard deviations above the mean.
2. You should expect the size to be $2(0.76) = 1.52$ standard deviations below the mean.
3. The home is 1.5 standard deviations above the mean in size no matter how size is measured.
4. $58,520 above the mean price.
5. $226,520.
6. $226,528 (round-off error).
7. 88.67 dollars per square foot.
8. The model suggests that house prices average about $88.67 higher per additional square foot of size.
9. No. No one wants a house with 0 square feet!
10. An increase in home size of 1000 square feet is associated with an increase in price of $88,670, on average.
11. Negative; that indicates it's priced lower than a typical home of its size.
12. About $274,410 on average.
13. $25,590 (positive!).
14. Differences in the size of houses account for about 57.8% of the variation in the house prices.
15. It's positive. The correlation and the slope have the same sign.
16. No, the standard deviation of the residuals is 53.79 thousand dollars. We shouldn't be surprised by any residual smaller than 2 standard deviations, and a residual of $50,000 is less than 1 standard deviation.
17. $R^2$ would not change, but the slope would. Slope depends on the units used but correlation doesn't.

# Regression Wisdom

Regression is used every day throughout the world to predict customer loyalty, numbers of admissions at hospitals, sales of automobiles, and many other things. Because regression is so widely used, it's also widely abused and misinterpreted. This chapter presents examples of regressions in which things are not quite as simple as they may have seemed at first and shows how you can still use regression to discover what the data have to say.

## Getting the "Bends": When the Residuals Aren't Straight

### Straight Enough?

We can't *know* whether the Linearity Assumption is true, but we can see if it's *plausible* by checking the Straight Enough Condition.

No regression analysis is complete without a display of the residuals to check that the linear model is reasonable. Because the residuals are what is "left over" after the model describes the relationship, they often reveal subtleties that were not clear from a plot of the original data. Sometimes these are additional details that help confirm or refine our understanding. Sometimes they reveal violations of the regression conditions that require our attention.

For example, the fundamental assumption in working with a linear model is that the relationship you are modeling is, in fact, linear. That sounds obvious, but you can't take it for granted. It may be hard to detect nonlinearity from the scatterplot you looked at before you fit the regression model. Sometimes you can see important features such as nonlinearity more readily when you plot the residuals. Let us show you an example.

Jessica Meir[1] and Paul Ponganis studied emperor penguins at the Scripps Institution of Oceanography's Center for Marine Biotechnology and Biomedicine at the University of California at San Diego. Meir says:

> *Emperor penguins are the most accomplished divers among birds, making routine dives of 5–12 minutes, with the longest recorded dive over 27 minutes. These birds can also dive to depths of over 500 meters! Since air-breathing animals like penguins must hold their breath while submerged, the duration of any given dive depends on how much oxygen is in the bird's body at the beginning of the dive, how quickly that oxygen gets used, and the lowest level of oxygen the bird can tolerate. The rate of oxygen depletion is primarily determined by the penguin's heart rate. Consequently, studies of heart rates during dives can help us understand how these animals regulate their oxygen consumption in order to make such impressive dives.*[2]

The researchers equip emperor penguins with devices that record their heart rates during dives. Here's a scatterplot of the *Dive Heart Rate* (beats per minute) and the *Duration* (minutes) of dives by these high-tech penguins.

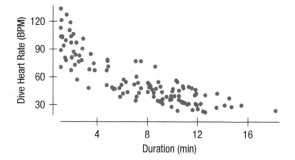

**Figure 8.1**
The scatterplot of *Dive Heart Rate* in beats per minute (bpm) vs. *Duration* (minutes) shows a moderately strong negative association.

The scatterplot shows a moderately strong negative association ($R^2 = 71.5\%$). The linear regression equation

$$\widehat{Dive\ Heart\ Rate} = 96.9 - 5.47\ Duration$$

says that for longer dives, the average *Dive Heart Rate* is lower by about 5.47 beats per dive minute, starting from a value of 96.9 beats per minute.

However, the scatterplot of the *Residuals* against *Duration* is revealing. The Linearity Assumption says we should not see a pattern, but instead there's a bend, starting high on the left, dropping down in the middle of the plot, and rising again at the right. Graphs of residuals often reveal patterns such as this that were easy to miss in the original scatterplot. (That was especially true of the $CO_2$ data in the TI Tips from Chapter 7.)

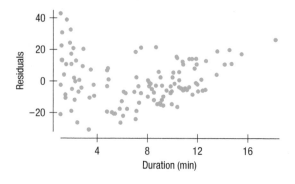

**Figure 8.2**
Plotting the *Residuals* against *Duration* reveals a bend. It was also in the original scatterplot, but here it's easier to see.

---

[1]Since completing this research, Meir has gone on to loftier things: she's now a NASA astronaut!

[2]Excerpt from Research Note on Emperor Penguins, Scripps Institution of Oceanography's Center for Marine Biotechnology and Biomedicine at the University of California at San Diego by Jessica Meir. Published by Meir, Jessica.

Now if you look back at the original scatterplot, you may see that the scatter of points isn't really straight. There's a slight bend to that plot, but that bend is much easier to see in the residuals. Even though it means rechecking the Straight Enough Condition *after* you find the regression model, it's always a good idea to check your residual plot for bends that you might have overlooked in the original scatterplot.

## Extrapolation: Reaching Beyond the Data

Linear models give a predicted value for each case in the data. Put a new *x*-value into the equation, and it gives a predicted value, $\hat{y}$, to go with it. But when the new *x*-value lies far from the data values we used to build the regression, how trustworthy is the prediction?

The simple answer is that the farther the new *x*-value is from $\bar{x}$, the less trust we should place in the predicted value. Once we venture into new *x* territory, such a prediction is called an **extrapolation**. Extrapolations are dubious because they require the very questionable assumption that nothing about the relationship between *x* and *y* changes even at extreme values of *x* and beyond.

Extrapolation is a good way to see just where the limits of our model may be. But it requires caution. When the *x*-variable is *Time*, extrapolation becomes an attempt to peer into the future.[3] People have always wanted to see into the future, and it doesn't take a crystal ball to foresee that they always will. In the past, seers, oracles, and wizards were called on to predict the future. Today, mediums, fortune-tellers, astrologers, and Tarot card readers still find many customers.

> Prediction is difficult, especially about the future.
> —Niels Bohr, Danish physicist

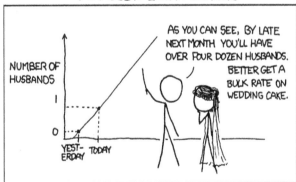

© 2013 Randall Munroe. Reprinted with permission. All rights reserved.

Those with a more scientific outlook may use a linear model as their digital crystal ball. Some physical phenomena do exhibit a kind of "inertia" that allows us to guess that current systematic behavior will continue, but be careful in counting on that kind of regularity in phenomena such as stock prices, sales figures, hurricane tracks, or public opinion.

Extrapolating from current trends is so tempting that even professional forecasters sometimes expect too much from their models—and sometimes the errors are striking. In the mid-1970s, oil prices surged and long lines at gas stations were common. In 1970, oil cost about $22 a barrel (in 2020 dollars)—about what it had cost for 20 years or so. But then, within just a few years, the price surged to over $50. In 1975, a survey of 15 top econometric forecasting models (built by groups that included Nobel prize–winning economists) found predictions for 1985 oil prices that ranged from $300 to over $700 a barrel (in 2020 dollars). How close were these forecasts?

---

[3]Or to peer into the past—at a time before the data were collected.

### When the Data Are Years . . .

. . . we usually don't enter them as four-digit numbers. Here we used 0 for 1970, 10 for 1980, and so on. Or we may simply enter two digits, using 82 for 1982, for instance. Rescaling years like this often makes calculations easier and equations simpler. We recommend you do it, too. But be careful: if 1982 is 82, then 2004 is 104 (not 4), right?

Here's a scatterplot (with regression line) of oil prices from 1970 to 1981 (in 2020 dollars).

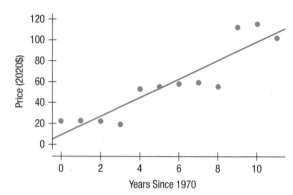

Figure 8.3
The price of a barrel of oil increased by about $8.89 per year from 1970 to 1981.

The regression model

$$\widehat{Price} = 9.06 + 8.89 \, Years \, Since \, 1970$$

says that prices had been going up 8.89 dollars per year, or about $88.87 in 10 years. If you assume that they would *keep going up*, it's not hard to imagine almost any price you want. That would result in a forecast of $337 for the year 2017.

So, how did the forecasters do? Well, in the period from 1982 to 1998 oil prices didn't exactly continue that steady increase. In fact, they went down so much that by 1998, prices (adjusted for inflation) were the lowest they'd been since before World War II.

Not one of the experts' models predicted that.

Of course, these decreases clearly couldn't continue, or oil would have been free by the year 2000. The Energy Information Administration (EIA) has provided the U.S. Congress with both short- and long-term oil price forecasts every year since 1979. In that year they predicted that 1995 prices would be $93 (in today's dollars). As we've just seen, prices were closer to $25. So, 20 years later, what did the EIA learn about forecasting? Let's see how well their 1998 forecasts for 2020 did. Here's a timeplot of the EIA's predictions and the actual prices (in 2020 dollars).

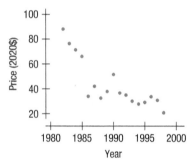

Figure 8.4
Oil prices from 1982 to 1998 *decreased* by about $3.25 per year.

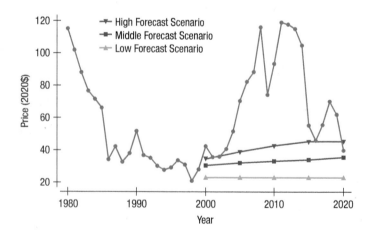

Figure 8.5
EIA forecasts and actual oil prices from 1981 to 2020. Ironically, the EIA 1998 forecast for 2020 is almost correct, but forecasters seem to have missed the run-up in the 20 years in between.

Oops! The EIA forecast the first two years after 1998 reasonably well, then missed the next two decades sometimes by a factor of 5. Ironically, with the drop in prices due to the pandemic in 2020, they may actually have forecast the 2020 price reasonably well!

Where do you think oil prices will go in the next decade? *Your* guess may be as good as anyone's! Figure 8.6 shows the 2019 EIA forecasts for the next 30 years. How confident are you that oil prices will be anywhere near the forecasts?

### Figure 8.6
The 2019 Forecasts for the years 2020 to 2050. The EIA has already missed the downturn in 2020 but expects an essentially linear increase in real prices for the next 30 years.

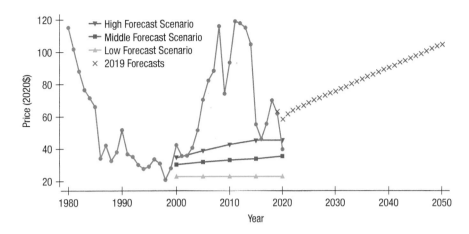

Of course, knowing that extrapolation requires thought and caution doesn't stop people. The temptation to see into the future is hard to resist. So our more realistic advice is this:

*If you extrapolate into the future, at least don't believe blindly that the prediction will come true.*

## FOR EXAMPLE
### Extrapolation: Reaching Beyond the Data

The U.S. Census Bureau (www.census.gov) reports the median age at first marriage for men and women. Here's a regression analysis of median *Age* (at first marriage) for men against *Year* (since 1890) at every census from 1890 to 1940:

R-squared = 92.6%
s = 0.2417

| Variable | Coefficient |
|---|---|
| Intercept | 26.07 |
| Year | −0.04 |

The regression equation is

$$\widehat{Age} = 26.07 - 0.04 \, Year.$$

**QUESTION:** What would this model predict as the age at first marriage for men in the year 2000?

**ANSWER:** When *Year* counts from 0 in 1890, the year 2000 is "110." Substituting 110 for *Year*, we find that the model predicts a first marriage *Age* of $26.07 - 0.04 \times 110 = 21.7$ years old.

**QUESTION:** In the year 2019, the median *Age* at first marriage for men was 29.8 years. What's gone wrong?

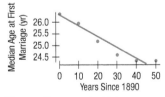

The median age at which men first married fell at the rate of about a year every 25 years from 1890 to 1940.

**ANSWER:** It is never safe to extrapolate beyond the data very far. The regression was calculated for years up to 1940. To see how absurd a prediction from that period can be when extrapolated into the present, look at a scatterplot of the median *Age* at first marriage for men for all the data from 1890 to 2019:

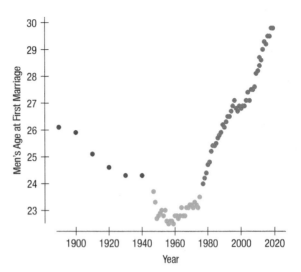

Median *Age* at first marriage (years of age) for men in the United States vs. *Year*. The regression model above was fit only to the first 50 years of the data (shown in purple), which looked nicely linear. But the linear pattern could not have continued, and in fact it changed in direction, steepness, and strength.

Now we can see why the extrapolation failed. Although the trend in *Age* at first marriage was linear and negative for the first part of the century, after World War II, it bottomed out around 1960. Since 1980 or so, it has risen steadily. To characterize age at first marriage, we should probably treat these three time periods separately.

# Outliers, Leverage, and Influence

Ralph Nader

Pat Buchanan

The outcome of the 2000 U.S. presidential election was determined in Florida amid much controversy. The main race was between George W. Bush and Al Gore, but two minor candidates played a significant role. To the political right of the main party candidates was Pat Buchanan, while to the political left was Ralph Nader. Generally, Nader earned more votes than Buchanan throughout the state. We would expect counties with larger vote totals to give more votes to each candidate. The regression model relating the two candidates' vote totals by county is

$$\widehat{Buchanan} = 50.3 + 0.14\, Nader.$$

It says that, in each county, Buchanan received about 0.14 times (or 14% of) the vote Nader received, starting from a base of 50.3 votes.

This seems like a reasonable regression, with an $R^2$ of almost 43%. But we've violated all three Rules of Data Analysis by going straight to the regression table without making a picture.[4] Let's have a look.

The scatterplot on the next page shows the vote for Buchanan in each county of Florida plotted against the vote for Nader. The striking **outlier** is Palm Beach County.

The scatterplot shows a strong, positive, linear association, and one striking point. With Palm Beach removed from the regression, the $R^2$ jumps from 42.8% to 82.1% and the slope of the line changes to 0.1, suggesting that Buchanan received only about 10% of the vote that Nader received. With more than 82% of the variability of the Buchanan vote accounted for, the model when Palm Beach is omitted certainly fits better. Palm Beach County now stands out, not as a Buchanan stronghold, but rather as a clear violation of the model begging for explanation.

[4] Why didn't you stop us?

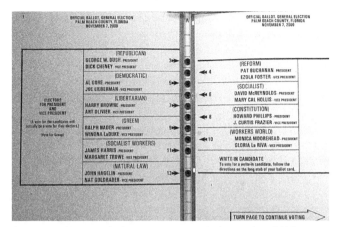

The unusual layout of this ballot may have led more than 2000 voters to inadvertently vote for Buchanan when they intended to vote for Gore; that's more than the number of votes by which Bush defeated Gore in Florida.

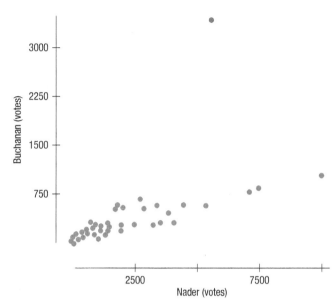

**Figure 8.7**
**Votes received by Buchanan against votes for Nader in all Florida counties in the presidential election of 2000** The red point is Palm Beach County, home of the "butterfly ballot."

The so-called "butterfly ballot," used only in Palm Beach County, was a source of controversy. Many claim that the format of this ballot confused voters so that some who intended to vote for the Democrat, Al Gore, punched the wrong hole next to his name and, as a result, voted for Buchanan.

**Figure 8.8**
The red line shows the effect that one unusual point can have on a regression. Omitting the red point makes the blue line's slope quite different.

**APPLET**
Exploring Regression:
See influential points.

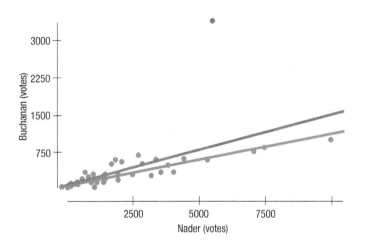

Give me a place to stand and I will move the Earth.

—Archimedes
(287–211 BCE)

One of the great values of models is that, by establishing an idealized behavior, they help us to see when and how data values are unusual. In regression, a point can stand out in two different ways. First, a data value can have a large residual, as Palm Beach County does in this example. Because they seem to be different from the other cases, points whose residuals are large always deserve special attention.

A data point can also be unusual if its *x*-value is far from the mean of the *x*-values. Such a point is said to have high **leverage**. The physical image of a lever is exactly right. We know the line must pass through $(\bar{x}, \bar{y})$, so you can picture that point as the fulcrum of the lever. Just as sitting farther from the hinge on a see-saw gives you more leverage to pull it your way, points with values far from $\bar{x}$ pull more strongly on the regression line.

**TI-nspire**

**Influential points.** Try to make the regression line's slope change dramatically by dragging a point around in the scatterplot.

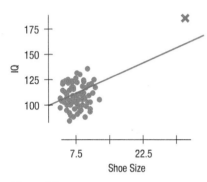

**Figure 8.9**
Bozo's extraordinarily large shoes give his data point high leverage in the regression. Wherever Bozo's IQ falls, the regression line will follow.

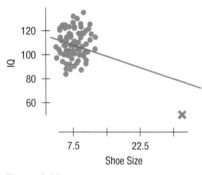

**Figure 8.10**
If Bozo's IQ were low, the regression slope would change from positive to negative. A single influential point can change a regression model drastically.

A point with high leverage has the potential to change the regression line. But it doesn't always use that potential. If the point lines up with the pattern of the other points, then including it doesn't change our estimate of the line. By sitting so far from $\bar{x}$, though, it may strengthen the relationship, inflating the correlation and $R^2$. How can you tell if a high-leverage point actually changes the model? Just fit the linear model twice, both with and without the point in question. We say that a point is **influential** if omitting it from the analysis gives a very different model.[5]

Influence depends on both leverage and residual; a case with high leverage whose *y*-value sits right on the line fit to the rest of the data is not influential. Removing that case won't change the slope, even if it does affect $R^2$. A case with modest leverage but a very large residual (such as Palm Beach County) can be influential. Of course, if a point has enough leverage, it can pull the line right to it. Then it's highly influential, but its residual is small. The only way to be sure is to fit both regressions.

Unusual points in a regression often tell us more about the data and the model than any other points. We face a challenge: The best way to identify unusual points is against the background of a model, but good models are free of the influence of unusual points. Don't give in to the temptation to simply delete points that don't fit the line. You can take points out and discuss what the model looks like with and without them, but arbitrarily deleting points can give a false sense of how well the model fits the data. Your goal should be understanding the data, not making $R^2$ as big as you can.

In 2000, George W. Bush won Florida (and thus the presidency) by only a few hundred votes, so Palm Beach County's residual is big enough to be meaningful. It's the rare unusual point that determines a presidency, but all are worth examining and trying to understand.

A point with so much influence that it pulls the regression line close to it can make its residual deceptively small. Influential points like that can have a shocking effect on the regression. At the left is a plot of *IQ* against *Shoe Size*, again from the fanciful study of intelligence and foot size in comedians we saw in Chapter 6 ("What Can Go Wrong?"). With Bozo there, $R^2 = 24.8\%$. But almost all of the variance accounted for is due to Bozo. Without him, there is little correlation between *Shoe Size* and *IQ*: the $R^2$ value is only 0.7%—a very weak linear relationship (as one might expect!).

What would have happened if Bozo hadn't shown his comic genius on IQ tests? Suppose his measured *IQ* had been only 50. The slope of the line would then drop from 0.96 IQ points/shoe size to −0.69 IQ points/shoe size. No matter where Bozo's *IQ* is, the line tends to follow it because his *Shoe Size*, being so far from the mean *Shoe Size*, makes this a high-leverage point.

Even though this example is far-fetched, similar situations occur all the time in real life. For example, a regression analysis of sales against floor space for hardware stores that looked primarily at small-town businesses could be dominated in a similar way if The Home Depot were included.

### Warning

Influential points can hide in plots of residuals. Points with high leverage pull the line close to them, so they often have small residuals. You'll see influential points more easily in scatterplots of the original data or by finding a regression model with and without the points.

---

[5] Some textbooks use the term *influential point* for any observation that influences the slope, intercept, or $R^2$. We'll reserve the term for points that influence the slope.

## JUST CHECKING

Each of these scatterplots shows an unusual point. For each, tell whether the point is a high-leverage point, whether it would have a large residual, and whether it is influential.

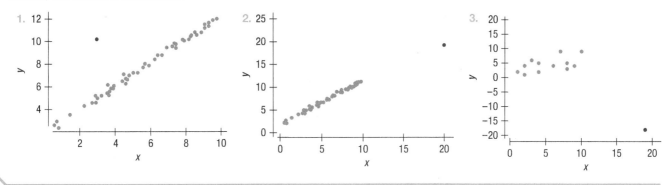

# Lurking Variables and Causation

**Causing Change?**

One common way to interpret a regression slope is to say that "a change of 1 unit in x results in a change of $b_1$ units in y." This way of saying things encourages causal thinking. Beware.

In Chapter 6, we tried to make it clear that no matter how strong the correlation is between two variables, there's no simple way to show that one variable causes the other. Putting a regression line through a cloud of points just increases the temptation to think and to say that the x-variable *causes* the y-variable. Just to make sure, let's repeat the point again: No matter how strong the association, no matter how large the $R^2$ value, no matter how straight the line, there is no way to conclude from a regression analysis alone that one variable *causes* the other. There's always the possibility that some third variable is driving both of the variables you have observed. With observational data, as opposed to data from a designed experiment, there is no way to be sure that a **lurking variable** is not the cause of any apparent association.

Here's an example: The scatterplot shows the *Life Expectancy* (average of men and women, in years) for each of 41 countries of the world, plotted against the square root of the number of *Doctors* per person in the country. (The square root is here to make the relationship satisfy the Straight Enough Condition. In the next chapter we'll talk more about how to "straighten" scatterplots that show curvature.)

**Figure 8.11**
The relationship between *Life Expectancy (years)* and availability of Doctors (measured as $\sqrt{Doctors/person}$) for countries of the world is fairly strong, positive, and linear.

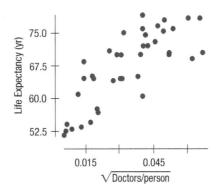

The strong positive association ($R^2 = 62.4\%$) seems to confirm our expectation that more *Doctors* per person improves health care, leading to longer lifetimes and a greater *Life Expectancy*. The strength of the association would *seem* to argue that we should send more doctors to developing countries to increase life expectancy.

That conclusion is about the consequences of a change. Would sending more doctors increase life expectancy? Specifically, do doctors *cause* greater life expectancy? Perhaps, but these are observed data, so there may be another explanation for the association.

In this next scatterplot the *x*-variable is the square root of the number of *Televisions* per person in each country.

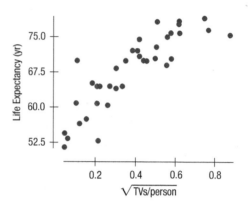

**Figure 8.12**
To increase life expectancy, don't send doctors, send TVs; they're cheaper and more fun. Or maybe that's not the right interpretation of this scatterplot of *Life Expectancy* against availability of TVs (as $\sqrt{TVs/person}$).

The positive association in this scatterplot is even *stronger* than the association in the previous plot ($R^2 = 72.3\%$). We can fit the linear model, and quite possibly use the number of TVs as a way to predict life expectancy. Should we conclude that increasing the number of TVs actually extends lifetimes? If so, we should send TVs instead of doctors to developing countries. Not only is the correlation with life expectancy higher, but TVs are much cheaper than doctors.

What's wrong with this reasoning? Maybe we were a bit hasty earlier when we concluded that doctors *cause* longer lives. Maybe there's a lurking variable here. Countries with higher standards of living have both longer life expectancies *and* more doctors (and more TVs). Could higher living standards cause changes in the other variables? If so, then improving living standards might be expected to prolong lives, increase the number of doctors, and increase the number of TVs.

From this example, you can see how easy it is to fall into the trap of mistakenly inferring causality from a regression analysis. For all we know, doctors (or TVs!) *do* increase life expectancy. But we can't tell that from data like these, no matter how much we'd like to. Resist the temptation to conclude that *x* causes *y* from a regression analysis, no matter how obvious that conclusion seems to you.

## Predict Changes? No!

Not only is it incorrect and dangerous to interpret association as causation, but when using regression there's a more subtle danger. Never interpret a regression slope coefficient as predicting how *y* is likely to change if its *x* value in the data were changed. Here's an example: In Chapter 7, we found a regression model relating calories in breakfast cereals to their sugar content as

$$\widehat{Calories} = 89.5 + 2.50\, Sugar.$$

It might be tempting to interpret this slope as implying that adding 1 gram of sugar is expected to lead to an increase of 2.50 calories. We can't say that. The correct interpretation of the slope is that cereals having a gram more sugar in them tend to have about 2.5 more calories per serving. That is, the regression model describes how the cereals differ, but does not tell us how they might change if circumstances were different. As a matter of fact, if there were no other differences in the cereals, simply adding a gram of sugar would add 3.90 calories—that's the calorie content of a gram of sugar.

To believe that *y* would change in a certain way if we were to change *x* is to believe the relationship is causal. When you interpret a slope, don't go there. Regression models describe the data as they are, not as they might be under other circumstances.

# Working with Summary Values

Scatterplots of statistics summarized over groups tend to show less variability than we would see if we measured the same variable on individuals. This is because the summary statistics themselves vary less than the data on the individuals do—a fact we will make more specific in coming chapters.

In the life expectancy and TVs example, we have no good measure of exposure to doctors or to TV on an individual basis. But if we did, we should expect the scatterplot to show more variability and the corresponding $R^2$ to be smaller. The bottom line is that you should be a bit suspicious of conclusions based on regressions of summary data. They may look better than they really are.

Here is a plot of the percent of people voting for Trump in 2016 against the percent of the population living in a rural area by state. Each data value is the average in that state. The relationship is positive and moderately strong. The $R^2$ value is 43%.

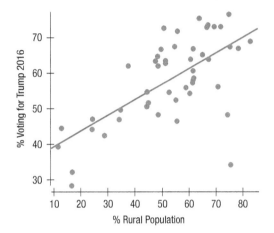

**Figure 8.13**
*% Voting for Trump 2016* vs *% Rural Population* by state in the United States. There's a moderate, positive, linear association with an $R^2$ value of 43%.

Suppose, instead of data on states, we investigated individual counties. The scatterplot would show much more scatter because individuals vary more than averages. The $R^2$ value is now only 24% and the relationship is less linear.

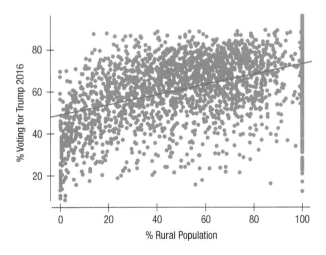

**Figure 8.14**
*Percent voting Trump 2016* against *% Rural Population* by county. Because individuals vary more than averages, the scatterplot shows more variation and the relationship looks weaker. The state averages may have overstated the strength of the relationship. The $R^2$ value is now only 24%.

Scatterplots of summary statistics show less scatter than the baseline data on individuals and can give a false impression of how well a line summarizes the data. There's no simple correction for this phenomenon. Once we're given summary data, there's no simple way to get the original values back.

## FOR EXAMPLE
### Putting it Together

The Environmental Protection Agency (EPA) annually measures numerous characteristics of new cars, including their gas mileage when driven in the city and their gas mileage when driven on highways. The scatterplot below shows the association between these two variables for all 2021 small car models that use regular gasoline, along with some regression output from a computer.

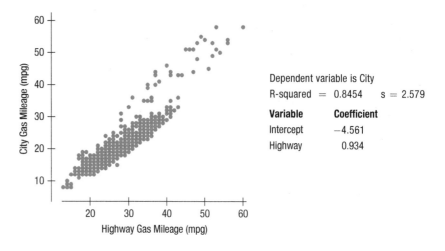

Dependent variable is City
R-squared = 0.8454    s = 2.579

| Variable | Coefficient |
|---|---|
| Intercept | −4.561 |
| Highway | 0.934 |

**QUESTION:** What do you see in the scatterplot?

**ANSWER:** There is a strong positive association between the highway gas mileage and the city gas mileage of these cars. (Not surprisingly, cars that get good gas mileage on highways also tend to get good gas mileage on city streets.) Most of the data points cluster in a roughly linear pattern, but there are about 30 that do not seem to fit the pattern of the others—the city mileage is higher than you'd expect based on the highway gas mileage.

The fact that a number of cars don't fit the pattern of the others suggests that perhaps two different populations are mixed together here. (Remember the roller coasters in Chapter 7?) The points that don't fit the pattern tend to get unusually high gas mileage in the city. If you're knowledgeable about cars, this might suggest to you that they are gas-electric hybrids. (Fully electric vehicles are not included in this dataset.) We checked it out by color-coding the hybrids and the non-hybrids differently. Sure enough, the hybrids are the ones that don't fit the pattern of the others.

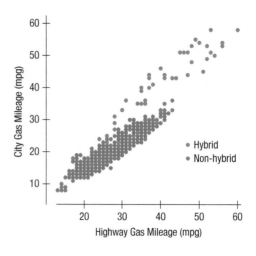

QUESTION: If we look at only the non-hybrid cars, how would we expect the regression to change from before? Describe how the slope, $R^2$, and $s_e$ will change.

ANSWER: Removing the hybrid cars would make the line slant less steeply, so we expect a lower slope. More of the variability in city mileage would be explained by the model, so we would expect $R^2$ to increase. And the tighter clustering around the line should make $s_e$ decrease.

Let's verify these things by performing regression on just the non-hybrid cars. Below are the scatterplot with the regression line, a residual plot, and some regression output from a computer.

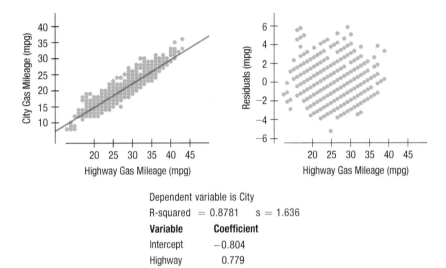

Dependent variable is City
R-squared = 0.8781    s = 1.636

| Variable | Coefficient |
| --- | --- |
| Intercept | −0.804 |
| Highway | 0.779 |

Our predictions about how the slope, $R^2$, and $s_e$ would change were all true. (Predicting and verifying them isn't just an exercise. Good statisticians try to anticipate results and verify them as they go.)

QUESTION: Using the scatterplot of the data, the residual plot, and the (second) regression output, describe the association between *City Mileage* and *Highway Mileage* for these 2021 small non-hybrid cars and what the regression output indicates.

ANSWER: There is a strong, positive linear association between *City Mileage* and *Highway Mileage* for these cars. The residual plot from a linear regression shows no pattern,[6] so a linear model is reasonable. The $R^2$ value of 0.8781 indicates that about 88 percent of the variability in the cars' city gas mileages can be explained by the linear regression on highway mileage. The slope means that for each additional mile per gallon that cars get on the highway, we'd expect them to average about 0.78 more miles per gallon in the city, starting from an initial value of about −0.804 miles per gallon. (The intercept isn't meaningful by itself because a car getting 0 miles per gallon on the highway wouldn't be much of a car.[7]) Finally, the standard deviation of the residuals, $s_e \approx 1.64$, indicates that our model's predictions of city mileage based on highway mileage are typically off by about 1.64 miles per gallon.

---

[6]You remember, of course, that the ideal residual plot would look like just random noise, because a pattern can indicate something about the relationship that isn't captured by the model. Perhaps you're concerned about the way the points in the residual plot form neat diagonal lines. Although this is indeed a clear "pattern," it's not the kind we need to worry about. The pattern we really are looking out for (and hope *not* to see) is a U-shaped curve, or an upside-down U-shaped curve, which would indicate nonlinearity in the data. The weird diagonal rows in our residual plot actually just result from the fact that the original data are rounded to the nearest integer. They're nothing to worry about.

[7]Especially if it got negative 0.804 miles to the gallon in the city.

# WHAT CAN GO WRONG?

This entire chapter has held warnings about things that can go wrong in a regression analysis. So let's just recap. When you make a linear model:

- **Make sure the relationship is straight.** Check the Straight Enough Condition. Always examine the residuals for evidence that the Linearity Assumption has failed. It's often easier to see deviations from a straight line in the residuals plot than in the scatterplot of the original data. Pay special attention to the most extreme residuals because they may have something to add to the story told by the linear model.
- **Be on guard for different groups in your regression.** Check for evidence that the data consist of separate subsets. If you find subsets that behave differently, consider fitting a different linear model to each subset.
- **Beware of extrapolating.** Beware of extrapolation beyond the $x$-values that were used to fit the model. Although it's common to use linear models to extrapolate, the practice is dangerous.

> "I should be 10 feet tall by the time I'm 40!"

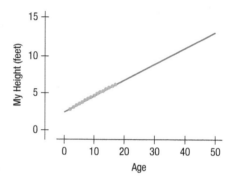

- **Beware especially of extrapolating into the future!** To predict the future with linear models, you must assume that future changes will continue at the same rate you've observed in the past. Predicting the future is particularly tempting and particularly dangerous.
- **Look for unusual points.** Unusual points always deserve attention and may well reveal more about your data than the rest of the points combined. Always look for them and try to understand why they stand apart. A scatterplot of the data is a good way to see high-leverage and influential points. A scatterplot of the residuals against the predicted values is a good tool for finding points with large residuals.
- **Beware of high-leverage points and especially of those that are influential.** Influential points can alter the regression model a great deal. The resulting model may say more about one or two points than about the overall relationship.
- **Consider comparing two regressions.** To see the impact of outliers on a regression, it's often wise to run two regressions, one with and one without the extraordinary points, and then to discuss the differences.
- **Treat unusual points honestly.** If you remove enough carefully selected points, you can always get a regression with a high $R^2$ eventually. But it won't give you much understanding. Some variables are not related in a way that's simple enough for a linear model to fit very well. When that happens, report the failure and stop.
- **Beware of lurking variables.** Think about lurking variables before interpreting a linear model. It's particularly tempting to explain a strong regression by thinking that the $x$-variable *causes* the $y$-variable. A linear model alone can never demonstrate such causation, in part because it cannot eliminate the chance that a lurking variable has caused the variation in both $x$ and $y$.
- **Watch out when dealing with data that are summaries.** Be cautious in working with data values that are themselves summaries, such as means or medians. Such statistics are less variable than the data on which they are based, so they tend to inflate the impression of the strength of a relationship.

- **Don't even *imply* causation.** By now you know (We hope!) that the presence of an association between two variables doesn't mean one causes the other, but it's easy to imply causation when interpreting the slope of the regression model. If an analysis shows that automobile fuel efficiency tends to drop about 0.8 miles per gallon for every extra 100 pounds cars weigh, that does not mean your car will get 0.8 mpg less if you give your sister a ride. Be careful that your interpretation of slope does not predict what will happen to y if x changes.

## WHAT HAVE WE LEARNED?

We've learned to be alert to the many ways in which a dataset may be unsuitable for a regression analysis.

- Watch out for more than one group hiding in your regression analysis. If you find subsets of the data that behave differently, consider fitting a different regression model to each subset.
- The Straight Enough Condition says that the relationship should be reasonably straight to fit a regression line. Somewhat paradoxically, sometimes it's easier to see that the relationship is not straight *after* fitting the regression line by examining the residuals. The same is true of outliers.
- The Outlier Condition actually means two things: Points with large residuals or high leverage (especially both) can influence the regression model significantly. It's a good idea to perform the regression analysis with and without such points to see their impact.

And we've learned that even a good regression model doesn't mean we should believe that the model says more than it really does.

- Extrapolation far from $\bar{x}$ can lead to silly and useless predictions.
- Even an $R^2$ near 100% doesn't indicate that x causes y (or the other way around). Watch out for lurking variables that may affect both x and y.
- Be careful when you interpret regressions based on *summaries* of the datasets. These regressions tend to look stronger than the regression based on all the individual data.

## TERMS

**Extrapolation** — Although linear models provide an easy way to predict values of y for a given value of x, it is unsafe to predict for values of x far from the ones used to find the linear model equation. Such extrapolation may pretend to see into the future, but the predictions should not be trusted. (p. 217)

**Outlier** — Any data point that stands away from the others can be called an outlier. In regression, outliers can be extraordinary in two ways: by having a large residual or by having high leverage. (p. 220)

**Leverage** — Data points whose x-values are far from the mean of x are said to exert leverage on a linear model. High-leverage points pull the line close to them, and so they can have a large effect on the line, sometimes very strongly influencing the slope and intercept. With high enough leverage, their residuals can be deceptively small. (p. 221)

**Influential point** — If omitting a point from the data results in a very different regression model, then that point is called an influential point. (p. 222)

**Lurking variable** — A variable that is not explicitly part of a model but affects the way the variables in the model appear to be related is called a lurking variable. Because we can never be certain that observational data are not hiding a lurking variable that influences both x and y, it is never safe to conclude that a linear model demonstrates a causal relationship, no matter how strong the linear association. (p. 223)

# ON THE COMPUTER

## Regression Diagnosis

Most statistics technology offers simple ways to check whether your data satisfy the conditions for regression. We have already seen that these programs can make a simple scatterplot. They can also help us check the conditions by plotting residuals.

# EXERCISES

**1. Marriage age 2019** Is there evidence that the age at which women get married has changed over the past 100 years? The scatterplot shows the trend in age at first marriage for American women (www.census.gov).

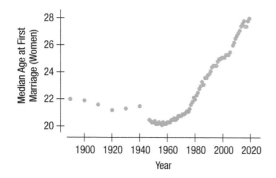

a) Is there a clear pattern? Describe the trend.
b) Is the association strong?
c) Is the correlation high? Explain.
d) Is a linear model appropriate? Explain.

**2. Smoking 2018** The Gallup Poll has tracked cigarette smoking in the United States since World War II. How has the percentage of people who smoke changed since the health dangers of smoking became clear during the last half of the 20th century? The scatterplot shows percentages of smokers among adults, as estimated by surveys, from 1944 to 2018.

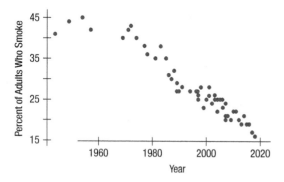

a) Is there a clear pattern? Describe the trend.
b) Is the association strong?
c) Is a linear model appropriate? Explain.

**3. Human Development Index 2018** The United Nations Development Programme (UNDP) uses the Human Development Index (HDI) in an attempt to summarize in one number the progress in health, education, and economics of a country (hdr.undp.org/en/data#). In 2018, the HDI was as high as 0.95 for Norway and as low as 0.38 for Niger. The gross national income per capita (GNI), by contrast, is often used to summarize the *overall* economic strength of a country. Is the HDI related to the GNI? Here is a scatterplot of *HDI* against *GNI*.

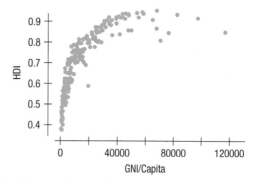

a) Explain why fitting a linear model to these data might be misleading.
b) If you fit a linear model to the data, what do you think a scatterplot of residuals versus predicted *HDI* will look like?

**4. HDI 2018 revisited** As explained in Exercise 3, the Human Development Index (HDI) is a measure that attempts to summarize in one number the progress in health, education, and economics of a country. The percentage of older people (65 and older) in a country is positively associated with its HDI. Can the percentage of older adults be used to predict the HDI? Here is a scatterplot of the two variables.

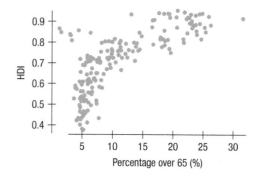

a) Explain why fitting a linear model to these data might be misleading.
b) If you fit a linear model to the data, what do you think a scatterplot of residuals vs. predicted *HDI* will look like?

**5. Good model?** In justifying his choice of a model, a student wrote, "I know this is the correct model because $R^2 = 99.4\%$."
a) Is this reasoning correct? Explain.
b) Does this model allow the student to make accurate predictions? Explain.

**6. Bad model?** A student who has created a linear model is disappointed to find that her $R^2$ value is a very low 13%.
a) Does this mean that a linear model is not appropriate? Explain.
b) Does this model allow the student to make accurate predictions? Explain.

**7. Movie dramas** Here's a scatterplot of the production budgets (in millions of dollars) vs. the running time (in minutes) for major release movies in 2005. Dramas are plotted as red x's and all other genres are plotted as blue dots. (The re-make of *King Kong* is plotted as a black "-". At the time it was the most expensive movie ever made, and not typical of any genre.) A separate least squares regression line has been fitted to each group. For the following questions, just examine the plot:

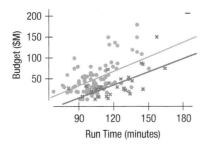

a) What are the units for the slopes of these lines?
b) In what way are dramas and other movies similar with respect to this relationship?
c) In what way are dramas different from other genres of movies with respect to this relationship?

**8. Smoking 2018, all adults** In Exercise 2, we examined the percentage of adults who smoked from 1944 to 2018 according to the Gallup Poll. The Centers for Disease Control and Prevention (CDC) also collects similar data. Here's a scatterplot showing the corresponding percentages for both men and women aged 18–24 along with least squares lines for each:

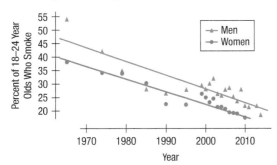

a) In what ways are the trends in smoking behavior similar for men and women?

b) How do the smoking rates for women differ from those for men?
c) Viewed alone, the trend for men may have seemed to violate the Linearity Condition. How about the trend for women? Does the consistency of the two patterns encourage you to think that a linear model for the trend in men might be appropriate? (Note: There is no correct answer to this question; it is raised for you to think about.)

**9. Abalone** Abalones are edible sea snails that include over 100 species. A researcher is working with a model that uses the number of rings in an abalone's shell to predict its age. He finds an observation that he believes has been miscalculated. After deleting this outlier, he redoes the calculation. Does it appear that this outlier was exerting very much influence?

| Before: | | After: | |
|---|---|---|---|
| Dependent variable is Age | | Dependent variable is Age | |
| R-squared = 67.5% | | R-squared = 83.9% | |
| Variable | Coefficient | Variable | Coefficient |
| Intercept | 1.736 | Intercept | 1.56 |
| Rings | 0.45 | Rings | 1.13 |

**10. Abalone again** The researcher in Exercise 9 is content with the second regression. But he has found a number of shells that have large residuals and is considering removing all of them. Is this good practice?

**11. Skinned knees** There is a strong correlation between the temperature and the number of skinned knees on playgrounds. Does this tell us that warm weather causes children to trip?

**12. Cell phones and life expectancy** The correlation between cell phone usage and life expectancy is very high. Should we buy cell phones to help people live longer?

**13. Grading** A team of Calculus teachers is analyzing student scores on a final exam compared to the midterm scores. One teacher proposes that they already have every teacher's class averages and they should just work with those averages. Explain why this is problematic.

**14. Average GPA** An athletic director proudly states that he has used the average GPAs of the university's sports teams and is predicting a high graduation rate for the teams. Why is this method unsafe?

**15. Oakland passengers 2016** The scatterplot below shows the number of passengers at Oakland (CA) airport month by month since 1997 (oaklandairport.com/news/statistics/passenger-history/).

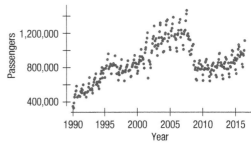

a) Describe the patterns in passengers at Oakland airport that you see in this timeplot.
b) Until 2009, analysts got fairly good predictions using a linear model. Why might that not be the case now?

c) If they considered only the data from 2009 to 2016, might they get reasonable predictions into the future?
d) What impact do you think the years 2020 and 2021 would have on the trend seen in the graph? What do you think will be the trend going forward?

**16. Tracking hurricanes 2018** In Chapter 6, we saw data on the errors (in nautical miles) made by the National Hurricane Center in predicting the path of hurricanes. The scatterplot below shows the trend in the 24-hour tracking errors from 1970 to 2018 (www.nhc.noaa.gov).

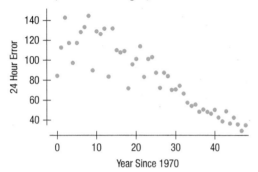

Dependent variable is Error
R-squared = 79.8%   $s_e$ = 15.14

| Variable | Coefficient |
|---|---|
| Intercept | 133.343 |
| Years since 1970 | −2.081 |

a) Interpret the slope and intercept of the model.
b) Interpret $s_e$ in this context.
c) The Center wanted to achieve an average tracking error of 9 nautical miles by 2030. Do you think they'll make it? Defend your response.
d) What if their goal had been an average tracking error of 3 nautical miles?
e) What cautions would you state about your conclusion?

**17. Unusual points** Each of the four scatterplots that follow shows a cluster of points and one "stray" point. (Because the points do not represent real data, we've left off axis labels.) For each, answer these questions:

1) In what way is the point unusual? Does it have high leverage, a large residual, or both?
2) Do you think that point is an influential point?
3) If that point were removed, would the correlation become stronger or weaker? Explain.
4) If that point were removed, would the slope of the regression line increase or decrease? Explain.

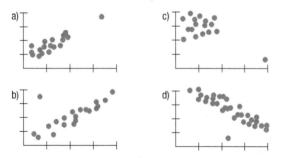

**18. More unusual points** Each of the following scatterplots shows a cluster of points and one "stray" point. (Because the points do not represent real data, we've left off axis labels.) For each, answer these questions:

1) In what way is the point unusual? Does it have high leverage, a large residual, or both?
2) Do you think that point is an influential point?
3) If that point were removed, would the correlation become stronger or weaker? Explain.
4) If that point were removed, would the slope of the regression line increase or decrease? Explain.

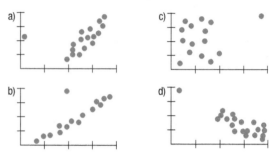

**19. The extra point** The scatterplot shows five blue data points at the left. Not surprisingly, the correlation for these points is $r = 0$. Suppose *one* additional data point is added at one of the five positions suggested below in green. Match each point (a–e) with the correct new correlation from the list given.

1) −0.90   2) −0.40   3) 0.00   4) 0.05   5) 0.75

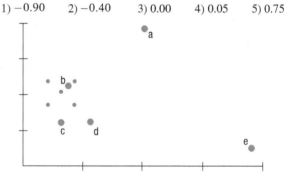

**20. The extra point revisited** The original five points in Exercise 19 produce a regression line with slope 0. Match each of the green points (a–e) with the slope of the line after that one point is added:

1) −0.45   2) −0.30   3) 0.00   4) 0.05   5) 0.85

**21. What's the cause?** Suppose a researcher studying health issues measures blood pressure and the percentage of body fat for several adult males and finds a strong positive association. Describe three different possible cause-and-effect relationships that might be present.

**22. What's the effect?** A researcher studying violent behavior in elementary school children asks the children's parents how much time each child spends playing computer games and has their teachers rate each child on the level of aggressiveness they display while playing with other children. Suppose that the researcher finds a moderately strong positive correlation. Describe three different possible cause-and-effect explanations for this relationship.

**23. Reading** To measure progress in reading ability, students at an elementary school take a reading comprehension test every year. Scores are measured in "grade-level" units; that is, a score of 4.2 means that a student is reading at slightly above the expected level for a fourth grader. The school principal prepares a report to parents that includes a graph showing the mean reading score for each grade. In his comments he points out that the strong positive trend demonstrates the success of the school's reading program.

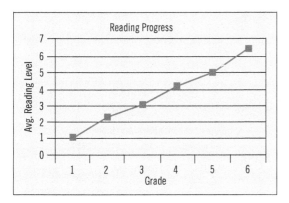

a) Does this graph indicate that students are making satisfactory progress in reading? Explain.
b) What would you estimate the correlation between *Grade* and *Average Reading Level* to be?
c) If, instead of this plot showing average reading levels, the principal had produced a scatterplot of the reading levels of all the individual students, would you expect the correlation to be the same, higher, or lower? Explain.
d) Although the principal did not do a regression analysis, someone as statistically astute as you might do that. (But don't bother.) What value of the slope of that line would you view as demonstrating acceptable progress in reading comprehension? Explain.

24. Grades  A college admissions officer, defending the college's use of SAT scores in the admissions process, produced the following graph. It shows the mean GPAs for last year's freshmen, grouped by SAT scores. How strong is the evidence that *SAT Score* is a good predictor of *GPA*? What concerns you about the graph, the statistical methodology or the conclusions reached?

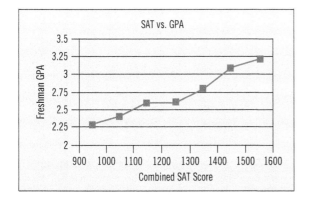

25. Heating  After keeping track of his heating expenses for several winters, a homeowner believes he can estimate the monthly cost ($) from the average daily Fahrenheit temperature with the model $\widehat{Cost} = 133 - 2.13\,Temp$. Here is the residuals plot for his data:

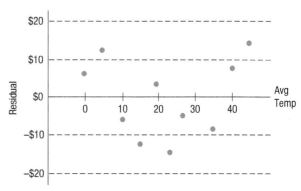

a) Interpret the slope of the line in this context.
b) Interpret the *y*-intercept of the line in this context.
c) During months when the temperature stays around freezing, would you expect cost predictions based on this model to be accurate, too low, or too high? Explain.
d) What heating cost does the model predict for a month that averages 10°?
e) During one of the months on which the model was based, the temperature did average 10°. What were the actual heating costs for that month?
f) Should the homeowner use this model? Explain.
g) Would this model be more successful if the temperature were expressed in degrees Celsius? Explain.

26. Speed  How does the speed at which you drive affect your fuel economy? To find out, researchers drove a compact car for 200 miles at speeds ranging from 35 to 75 miles per hour. From their data, they created the model $\widehat{Fuel\ Efficiency} = 32 - 0.1\,Speed$ and created this residual plot:

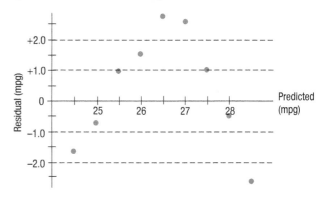

a) Interpret the slope of this line in context.
b) Explain why it's silly to attach any meaning to the *y*-intercept.
c) When this model predicts high *Fuel Efficiency*, what can you say about those predictions?
d) What *Fuel Efficiency* does the model predict when the car is driven at 50 mph?
e) What was the actual *Fuel Efficiency* when the car was driven at 45 mph?
f) Do you think there appears to be a strong association between *Speed* and *Fuel Efficiency*? Explain.
g) Do you think this is the appropriate model for that association? Explain.

27. T Bill rates 2020  Here are a plot and regression output showing the federal rate on 3-month Treasury bills from 1950 to 1980, and a regression model fit to the relationship between the *Rate*

(in %) and *Years Since 1950* (https://fred.stlouisfed.org/series/TB3MS).

a) What is the correlation between *Rate* and *Year*?
b) Interpret the slope and intercept.
c) What does this model predict for the interest rate in the year 2030?
d) Would you expect this prediction to be accurate? Explain.

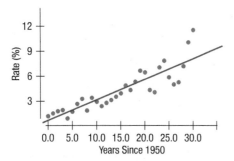

Dependent variable is Rate
R-squared = 77.6%   s = 1.232

| Variable | Coefficient |
| --- | --- |
| Intercept | 0.61149 |
| Year-1950 | 0.24788 |

**28. Marriage age 2019 II** The graph shows the ages of both men and women at first marriage (www.census.gov).

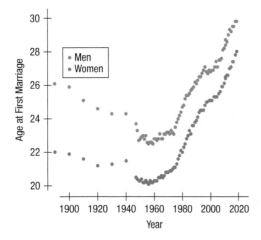

Clearly, the patterns for men and women are similar. But are the two lines getting closer together?

Here are a timeplot showing the *difference* in average age (men's age – women's age) at first marriage, the regression analysis, and the associated residuals plot.

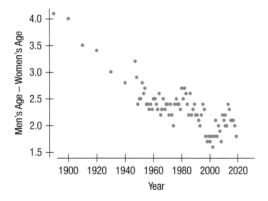

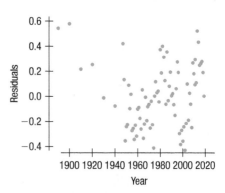

Dependent variable is Age Difference
R-squared = 71.0%   s = 0.2457

| Variable | Coefficient |
| --- | --- |
| Intercept | 29.672 |
| Year | −0.0138 |

a) What is the correlation between *Age Difference* and *Year*?
b) Interpret the slope of this line.
c) Predict the average age difference in 2030.
d) Describe reasons why you might not place much faith in that prediction.

**29. T Bill rates 2020 revisited** In Exercise 27, you investigated the federal rate on 3-month Treasury bills between 1950 and 1980. The scatterplot below shows that the trend changed dramatically after 1980, so we computed a new regression model for the years 1981 to 2019.

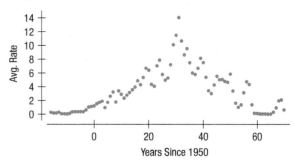

Here's the model for the data from 1981 to 2019 (in years since 1950):

Dependent variable is Rate
R-squared = 75.4%   s = 1.693

| Variable | Coefficient |
| --- | --- |
| Intercept | 16.6154 |
| Year-1950 | −0.2505 |

a) How does this model compare to the one in Exercise 27?
b) What does this model estimate the interest rate will be in 2030? How does this compare to the rate you predicted in Exercise 27?
c) Do you trust this newer predicted value? Explain.
d) Would you use either of these models to predict the T Bill rate in the future? Explain.

**30. Marriage age 2019 III** Has the trend of decreasing difference in age at first marriage seen in Exercise 28 gotten stronger recently? The scatterplot and residual plot for the data from 1980 through 2019, along with a regression for just those years, are below.

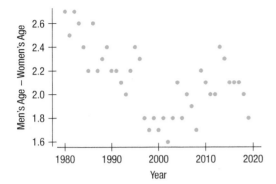

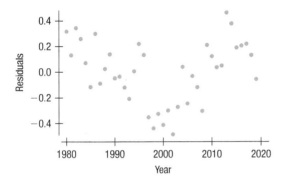

Response variable is Men − Women
R-squared = 29.8%   s = 0.2456

| Variable | Coefficient |
|---|---|
| Intercept | 29.1301 |
| Year | −0.0135 |

a) Is this linear model appropriate for the post-1980 data? Explain.
b) What does the slope say about marriage ages?
c) Explain why it's not reasonable to interpret the y-intercept.

**31. Gestation** For women, pregnancy lasts about 9 months. In other species of animals, the length of time from conception to birth varies. Is there any evidence that the gestation period is related to the animal's life span? The first scatterplot shows *Gestation Period* (in days) vs. *Life Expectancy* (in years) for 18 species of mammals. The red x at the far right represents humans.

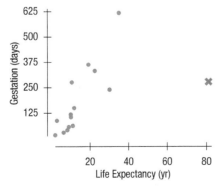

a) For these data, $r = 0.54$, not a very strong relationship. Do you think the association would be stronger or weaker if humans were removed? Explain.

b) Is there reasonable justification for removing humans from the dataset? Explain.
c) Here are the scatterplot and regression analysis for the 17 nonhuman species. Comment on the strength of the association.

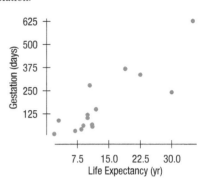

Dependent variable is Gestation
R-Squared = 72.2%

| Variable | Coefficient |
|---|---|
| Constant | −39.5172 |
| LifExp | 15.4980 |

d) Interpret the slope of the line.
e) Some species of monkeys have a life expectancy of about 20 years. Estimate the expected gestation period of one of these monkeys.

**32. Swim the lake 2019** People swam across Lake Ontario from Niagara on the Lake to Toronto (52 km, or about 32.3 mi) 65 times between 1954 and 2019. We might be interested in whether the swimmers are getting any faster or slower. Here are the regression of the crossing *Times* (minutes) against the *Year* since 1954 of the crossing and a plot of the residuals against *Years Since 1954*:

Dependent variable is Time
R-Squared = 0.01%   s = 399.2

| Variable | Coefficient |
|---|---|
| Intercept | 1201.056 |
| Years Since 1954 | 2.053 |

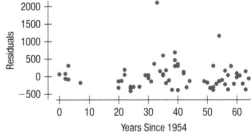

a) What does the $r^2$ mean for this regression?
b) Are the swimmers getting faster or slower? Explain.
c) The outlier seen in the residuals plot is a crossing by Vicki Keith in 1987 in which she swam a round trip, north to south, and then back again. Clearly, this swim doesn't belong with the others. Would removing it change the model a lot? Explain.

**33. Elephants and hippos** We removed humans from the scatterplot in Exercise 31 because our species was an outlier in life expectancy. The resulting scatterplot shows two points that now may be of concern. The point in the upper right corner of this scatterplot is for elephants, and the other point at the far right is for hippos.

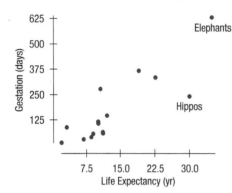

a) By removing one of these points, we could make the association appear to be stronger. Which point? Explain.
b) Would the slope of the line increase or decrease?
c) Should we just keep removing animals to increase the strength of the model? Explain.
d) If we remove elephants from the scatterplot, the slope of the regression line becomes 11.6 days per year. Do you think elephants were an influential point? Explain.

**34. Another swim** In Exercise 32, we saw in the **Swim the Lake 2019** data that Vicki Keith's round-trip swim of Lake Ontario was an obvious outlier among the other one-way times. Here is the new regression after this unusual point is removed:

Dependent variable is Time
R-Squared = 2.3%   s = 299.2

| Variable | Coefficient |
|---|---|
| Intercept | 1146.627 |
| Years Since 1954 | 2.620 |

a) In this new model, the value of $s_e$ is smaller. Explain what that means in this context.
b) Are you more convinced (compared to the previous regression) that Ontario swimmers are getting faster (or slower)?

**35. Marriage age predictions** Again we consider the age at first marriage for women in the U.S. Here is a regression model for the data on women, along with a residuals plot:

Response variable is Women
R-squared = 61.1%   s = 1.474

| Variable | Coefficient |
|---|---|
| Intercept | −112.543 |
| Year | 0.068479 |

a) Based on this model, what would you predict the marriage age will be for women in 2025?
b) How much faith do you place in this prediction? Explain.
c) Would you use this model to make a prediction about your grandchildren, say, 50 years from now? Explain.

Now, let's restrict our model to the years 1975–2015. Here are the regression analysis and residual plot:

Response variable is Women
75 total cases of which 34 are missing
R squared = 98.2%   s = 0.2450

| Variable | Coefficient |
|---|---|
| Intercept | −274.742 |
| Year | 0.149983 |

d) Based on this model, what would you predict the marriage age will be for women in 2025?
e) How much faith do you place in this prediction? Explain.
f) Would you use this model to make a prediction about your grandchildren, say, 50 years from now? Explain.

**36. Hard water** In an investigation of environmental causes of disease, data were collected on the annual mortality rate (deaths per 100,000) for males in 61 large towns in England and Wales. In addition, the water hardness was recorded as the calcium concentration (parts per million, ppm) in the drinking water. The following scatterplot shows the relationship between *Mortality* and *Calcium* concentration for these towns:

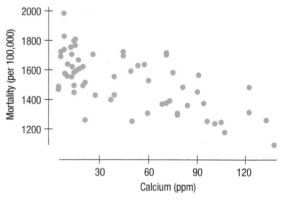

a) Describe the association between these variables in context.
b) Here is the regression analysis for the linear model. Write the equation of the regression line and give the correlation.

Dependent variable is Mortality
R-Squared = 42.9%

| Variable | Coefficient |
|---|---|
| Intercept | 1676.3556 |
| Years Since 1954 | −3.2261 |
| Calcium | 3.2261 |

c) Use the scatterplot and the residuals plot below to determine whether this linear model is appropriate.

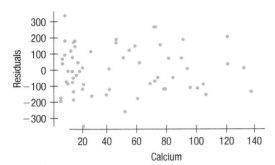

**37. Fertility and life expectancy 2018** The World Bank reports many demographic statistics about countries of the world. The data file holds the *Fertility* rate (births per woman) and the female *Life Expectancy* at birth (in years) for nearly 200 countries of the world.

Response variable is:   Life expectancy
R squared = 74.2%
s = 4.01

| Variable | Coefficient |
|---|---|
| Intercept | 89.421 |
| Fertility | −5.358 |

Here is a scatterplot of the data.

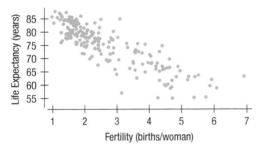

a) Are the conditions for regression satisfied? Discuss. Here is a plot of the residuals.

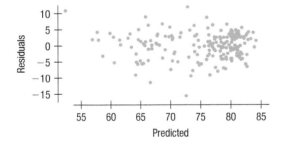

b) Is there an outlier? If so, identify it. Which data value is unusual?
c) Interpret the value of $R^2$.
d) If government leaders want to increase life expectancy, should they encourage women to have fewer children? Explain.

**38. Tour de France 2020** We met the Tour de France dataset in Chapter 1 (in Just Checking). One hundred years ago, the fastest rider finished the course at an average speed of about 25.3 kph (around 15.8 mph). By the 21st century, winning riders were averaging over 40 kph (nearly 25 mph).

a) Make a scatterplot of *Avg Speed* against *Year*. Describe the relationship of *Avg Speed* by *Year*, being careful to point out any unusual features in the plot.
b) Find the regression equation of *Avg Speed* on *Year*.
c) Are the conditions for regression met? Comment.

**39. Inflation 2020** The Consumer Price Index (CPI) tracks the prices of consumer goods in the United States, as shown in the following table. The CPI is reported monthly, but we can look at selected values. The table shows the January CPI at five-year intervals. It indicates, for example, that the average item costing $17.30 in 1925 cost $258.00 in the year 2020.

| Year | JanCPI | Year | JanCPI |
|---|---|---|---|
| 1915 | 10.1 | 1970 | 37.8 |
| 1920 | 19.3 | 1975 | 52.1 |
| 1925 | 17.3 | 1980 | 77.8 |
| 1930 | 17.1 | 1985 | 105.5 |
| 1935 | 13.6 | 1990 | 127.4 |
| 1940 | 13.9 | 1995 | 150.3 |
| 1945 | 17.8 | 2000 | 168.8 |
| 1950 | 23.5 | 2005 | 190.7 |
| 1955 | 26.7 | 2010 | 216.7 |
| 1960 | 29.3 | 2015 | 233.7 |
| 1965 | 31.2 | 2020 | 258.0 |

a) Make a scatterplot showing the trend in consumer prices. Describe what you see.
b) Be an economic forecaster: Project increases in the cost of living in 2030. Justify decisions you make in creating your model.

**40. Second stage 2020** Look once more at the data from the Tour de France. In Exercise 38, we looked at the whole history of the race, but now let's consider just the modern era from 1967 on. Following are a scatterplot, residuals plot, and regression model for average speed on year for races from 1967 to 2020.

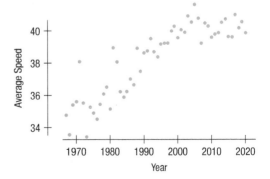

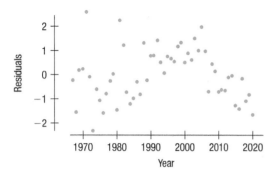

Dependent variable is Ave Speed
R-Squared = 76.5%   s = 1.087

| Variable | Coefficient |
|---|---|
| Intercept | 207.96 |
| Year | 0.1235 |

a) Are the conditions for regression met?
b) Interpret the slope.
c) The years 1999–2005 have been disqualified because of doping. How does this help explain the fact that 9 out of 10 residuals in the final 10 years are negative?
d) Would you extrapolate from this model to predict next year's winning speed?

**41. Tompkins County bridges 2016** An eight-lane steel truss bridge spanning the Mississippi River in Minneapolis, Minnesota, collapsed without warning in 2007, killing 13 and injuring 145. This disaster put a spotlight on the problem of our aging infrastructure. Most states now conduct regular safety checks, giving a bridge a structural deficiency score on various scales. A model that relates age to safety score might help the Department of Transportation focus inspectors' efforts where they are most needed. We have available data on the 195 bridges of Tompkins County in New York. One natural concern is the age of a bridge. Look at the scatterplot and following regression analysis of the association between *Condition* and year *Built* for 193 bridges built since 1900:

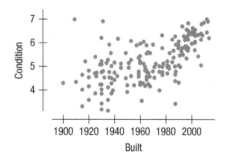

Dependent variable is Condition
R-squared = 40.0%   s = 0.6891

| Variable | Coefficient |
|---|---|
| Intercept | −33.4115 |
| Built | 0.0196 |

a) Interpret the slope and intercept of the model.
b) What does the value of $R^2$ say about the model?
c) Interpret $s_e$ in this context.
d) Tompkins County is the home of the oldest covered bridge in daily use in New York State. It was built in 1853.
   Would you use this model to predict the condition of this bridge? Explain.

This bridge, with a condition of 4.57, and another built in 1865 that has a rating of 5.4 have been added to the scatterplot below, each marked with a red X.

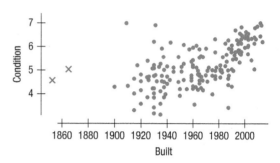

e) What effect would you expect the addition of these two bridges to have on the regression slope? Explain.
f) What effect would you expect the addition of these two bridges to have on $R^2$? Explain.
g) The 1853 bridge was extensively restored in 1972. If we use that date instead as the *Built* date, do you find the condition of the bridge remarkable?

**42.** In Chapter 7, we used the relationship between a roller coaster's height and its speed to introduce the concept of the regression line. Let's recall a few facts.

The correlation between the *Speed* and the *Height* of the roller coasters is 0.80. The mean and standard deviation of *Speed* are 37.6 mph and 8.6 mph, respectively. For *Height* they are 59.4 feet and 18.7 feet. Using these facts, we calculated the equation of the regression line to be $\widehat{Speed} = 15.6 + 0.37\ Height$.

a) In an Algebra class, you might be asked to solve the equation $y = 15.6 + 0.37x$ for $x$. Now you're being asked to do that in a Statistics class.
b) Use the facts we have learned about the regression line to calculate the regression equation to predict a coaster's *Height* from its *Speed*.
c) Why is this different from the equation you got in part a? What does this tell you about using your regression equation to predict $x$ from $y$?

### JUST CHECKING

Answers

1. Not high leverage, not influential, large residual.
2. High leverage, not influential, small residual.
3. High leverage, influential, not large residual.

# 9

# Re-expressing Data: Get It Straight!

The country of Monaco is among the most densely populated regions in the world, with a population density of about 26,337 people per square kilometer. If you don't deal with such data very often, you may not have a very good *feel* for what this means. (And converting it to 68,212 people per square mile may not help much.) You might find it easier to understand by expressing it as how much area each person has, on average. That turns out to be about 409 square feet per person, which is only about 0.7% of the size of a football field. In the United States, by contrast, the average population density is about 35.7 people per square kilometer—just over 5.2 football fields apiece.

Which is a better way to express population density—people per square kilometer or football fields per person? The truth is, neither one is necessarily better than the other. There is no single natural way to express population density. And note that converting from one of these measurements into the other is not merely a conversion of units. "People per square kilometer" would be high in a *densely* populated country, but "football fields per person" would be high in a *sparsely* populated country, so while both measurements are legitimate, they actually go in opposite directions from one another.

So, how does this insight help us understand data? All quantitative data come to us expressed in some way, but sometimes there are better ways to express the information. It's not whether meters are better (or worse) than fathoms or leagues. Those are different units, but they're all units expressing the same thing, length. What we're talking about is a different type of **re-expression**: applying a function, such as a square root, log, or reciprocal to the data. You already use some of them, even though you may not know it. For example, the Richter scale of earthquake strength (logs), the decibel scale for sound intensity (logs), the f/stop scale for camera aperture openings (squares), and the gauges of shotguns (square roots) all include simple functions of this sort.

> BEYOND $b_0 + b_1 x$
> Scan through any science textbook. Most equations have powers, reciprocals, or logs.

Why bother? As with population density, some expressions of data may be easier to think about. But more important, some may be much easier to analyze with statistical methods. We've seen that symmetric distributions are easier to summarize than asymmetric ones and straight scatterplots are easier to model with equations than are curved ones. We often look to re-express our data if doing so makes them more suitable for our methods.

# Straight to the Point

| WHO | 38 cars |
|---|---|
| WHAT | Weight, mpg |
| HOW | mpg measured on a track. Note: not EPA values |
| WHY | Evaluation of performance for ratings |

We know from common sense and from Physics that heavier cars need more fuel, but exactly how does a car's weight affect its fuel efficiency? Here are the scatterplot of *Weight* (in pounds) and *Fuel Efficiency* (in miles per gallon) for 38 cars, and the residual plot.

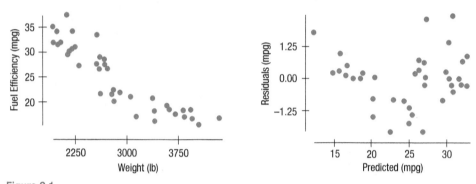

Figure 9.1

**Fuel Efficiency (mpg) vs. Weight (lb) for 38 cars as reported by Consumer Reports** The scatterplot shows a negative direction, roughly linear shape, and strong relationship. However, the residuals from a regression of *Fuel Efficiency* on *Weight* reveal a bent shape when plotted against the predicted values. Looking back at the original scatterplot, you may be able to see the bend.

Hmm.... Even though $R^2$ is 81.6%, the residuals don't show the random scatter we were hoping for. The shape is clearly bent. Looking back at the first scatterplot, you can probably see the slight bending. Think about the regression line through the points. How heavy would a car have to be to have a predicted gas mileage of 0? It looks like the *Fuel Efficiency* would go negative at about 6000 pounds. The Maybach 62 was a luxury car produced by Daimler AG to compete with Rolls-Royce and Bentley. It was the heaviest "production car" in the world at 6184 lb. Although not known for its fuel efficiency (rated at 10 mpg city), it did get more than the *minus* 2.6 mpg predicted from the model. Sadly, after selling only 2100 of them in 10 years, Daimler stopped production in 2013. The 2012 model sold (without extra features) for $427,700. Extrapolation always requires caution, but it can go dangerously wrong when your model is wrong, because wrong models tend to do even worse the farther you get from the middle of the data.

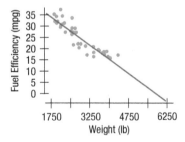

Figure 9.2
Extrapolating the regression line gives an absurd answer for vehicles that weigh as little as 6000 pounds.

The bend in the relationship between *Fuel Efficiency* and *Weight* is the kind of failure to satisfy the conditions for an analysis that we can repair by re-expressing the data. Instead of looking at miles per gallon, we could take the reciprocal and work with gallons per hundred miles.[1]

---

[1] Multiplying by 100 to get gallons per 100 miles simply makes the numbers easier to think about: You might have a good idea of how many gallons your car needs to drive 100 miles, but probably a much poorer sense of how much gas you need to go just 1 mile.

The direction of the association is positive now, because we're measuring gas consumption and heavier cars consume more gas per mile. The relationship is much straighter, as we can see from a scatterplot of the regression residuals.

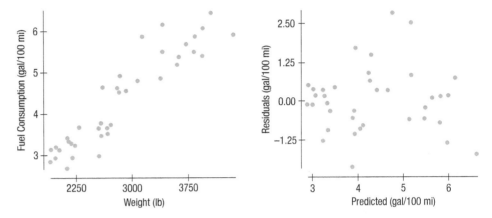

**Figure 9.3**
The reciprocal (1/y) is measured in gallons per mile. Gallons per 100 miles gives more meaningful numbers. The reciprocal is more nearly linear against *Weight* than the original variable, but the re-expression changes the direction of the relationship. The residuals from the regression of *Fuel Consumption* (gal/100 mi) on *Weight* show less of a pattern than before.

This is more the kind of boring residual plot (no direction, no particular shape, no outliers, no bends) that we hope to see, so we have reason to think that the Straight Enough Condition is now satisfied. Now here's the payoff: What does the reciprocal model say about the Maybach? The regression line fit to *Fuel Consumption* vs. *Weight* predicts somewhere near 9.26 for a car weighing 6184 pounds. What does this mean? It means the car is predicted to use 9.26 gallons for every 100 miles, or in other words,

$$\frac{100 \text{ miles}}{9.26 \text{ gallons}} = 10.8 \text{ mpg}.$$

That's a much more reasonable prediction and very close to the reported value of 10 miles per gallon. (Of course, *your* mileage may vary.)

> **A CASE FOR EXPRESSING FUEL EFFICIENCY DIFFERENTLY**
> Some consumer advocates have fought for years to get people to express fuel efficiency in gallons per 100 miles rather than in miles per gallon. Their main reason is that a difference in miles per gallon between two cars is not actually a measure of how much fuel you'll save with the more efficient car. Suppose you need to make a 100-mile trip in either Car A that gets 20 miles per gallon or Car B that gets 25 miles per gallon. The difference of 5 miles per gallon doesn't quantify fuel savings. Car A will use 5 gallons of gas to make the trip and Car B will use only 4 gallons. That difference of one gallon of gas per 100 miles *does* quantify fuel savings: how much gas you save for each 100 miles you drive. So expressing fuel efficiency in gallons per 100 miles makes it much easier to comparison-shop for fuel-efficient vehicles.

# Goals of Re-expression

| WHO | 77 large companies |
|---|---|
| WHAT | Assets, sales, and market sector |
| UNITS | $100,000 |
| HOW | Public records |
| WHEN | 1986 |
| WHY | By *Forbes* magazine in reporting on the *Forbes* 500 for that year |

We re-express data for several reasons. Each of these goals helps make the data more suitable for analysis by our methods.

## Goal 1

***Make the distribution of a variable (as seen in its histogram, for example) more symmetric.***
It's easier to summarize the center of a symmetric distribution, and for nearly symmetric distributions, we can use the mean and standard deviation. If the distribution is unimodal, then the resulting distribution may be closer to the Normal model, allowing us to use the 68–95–99.7 Rule.

Here are a histogram, quite skewed, showing the *Assets* of 77 companies selected from the *Forbes* 500 list (in $100,000) and the more symmetric histogram after taking logs.

Figure 9.4
The distribution of the *Assets* of large companies is skewed to the right. Data on wealth often look like this. Taking logs makes the distribution more nearly symmetric.

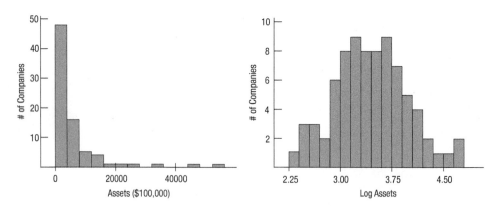

## Goal 2

***Make the spread of several groups (as seen in side-by-side boxplots) more alike, even if their centers differ.*** Groups that share a common spread are easier to compare. We'll see methods later in the book that can be applied only to groups with a common standard deviation. We saw an example of re-expression for comparing groups with boxplots in Chapter 4.

Here are the *Assets* of these companies by *Market Sector*:

Figure 9.5
**Assets of large companies by Market Sector** It's hard to compare centers or spreads, and there seem to be a number of high outliers.

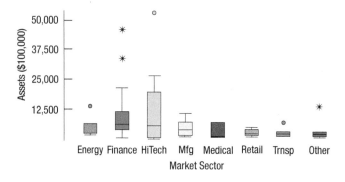

Taking logs makes the individual boxplots more symmetric and gives them spreads that are more nearly equal.

Figure 9.6
After re-expressing by logs, it's much easier to compare across market sectors. The boxplots are more nearly symmetric, most have similar spreads, and the companies that seemed to be outliers before are no longer extraordinary. Two new outliers have appeared in the finance sector. They are the only companies in that sector that are not banks. Perhaps they don't belong there.

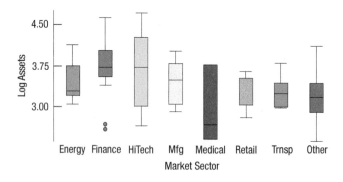

Doing this makes it easier to compare assets across market sectors. It can also reveal problems in the data. Some companies that looked like outliers on the high end turned out to be more typical. But two companies in the finance sector now stand out. Unlike the rest of the companies in that sector, they are not banks. They may have been placed in the wrong sector, but we couldn't see that in the original data.

## Goal 3

***Make the form of a scatterplot more nearly linear.*** Linear scatterplots are easier to model. We saw an example of scatterplot straightening in Chapter 6. The greater value of re-expression to straighten a relationship is that we can fit a linear model once the relationship is straight.

Physical therapists measure a patient's manual dexterity with a simple task. The patient picks up small cylinders from a 4 × 4 frame with one hand, flips them over (still with one hand), and replaces them in the frame. The task is timed for all 16 cylinders. Researchers used this tool to study how dexterity improves with age in children.[2] Figure 9.7 shows the relationship.

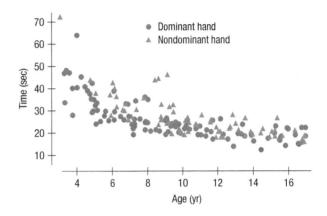

Figure 9.7
The relationship between *Time* and *Age* is curved, and appears to be more variable for younger children.

Often data recorded as time to complete a task can benefit from being re-expressed as a reciprocal. Instead of seconds/task, the re-expressed variable is in units of tasks/second—essentially a measure of speed. When *Time* is re-expressed as 1/*Time*, the result is Figure 9.8.

## Goal 4

***Make the scatter in a scatterplot spread out evenly rather than thickening at one end.***
Having an even scatter is a condition of many methods of statistics. As Figure 9.8 shows, a well-chosen re-expression can even out the spread.

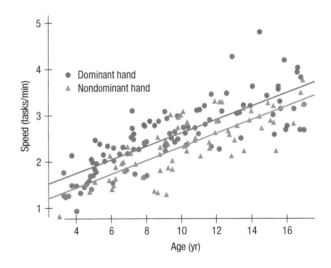

Figure 9.8
Re-expressing *Time* as *Speed* = 60 × 1/*Time* gives a plot that is straight and more nearly equal in variability throughout. In addition, the regression lines for dominant and nondominant hands are now parallel. (Multiplying by 60 converts the units from tasks per second to tasks per minute—and makes the speeds easier to understand.)

## Goal 5

***Make the distribution of the residuals more nearly Normal.*** Although not strictly necessary until we talk about inference for regression, a side benefit of a good re-expression is that when we make the plot of *y* versus *x* more linear and the spread around the line more even, we'll often see the shape of the histogram of the residuals become more symmetric as well.

---

[2]"Hand dexterity in children: Administration and normative values of the Functional Dexterity Test (FDT)," Gloria R. Gogola, MD, Paul F. Velleman, PhD, Shuai Xu, BS, MS, Adrianne M. Morse, BA, Barbara Lacy BS, Dorit Aaron, MA OTR CHT FAOTA, *J Hand Surg Am*. 2013 Dec; 38(12):2426–31. doi: 10.1016/j.jhsa.2013.08.123. Epub 2013 Nov. 1. You can see the task in the photograph at the beginning of Chapter 18.

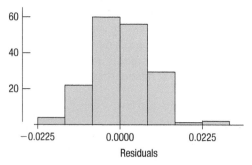

**Figure 9.9**
The residuals for a regression of *Speed* on *Age* are unimodal and nearly symmetric.

## FOR EXAMPLE
### Recognizing when a Re-expression Can Help

In Chapter 8, we saw the awesome ability of emperor penguins to slow their heart rates while diving. Here are three displays relating to the diving heart rates.

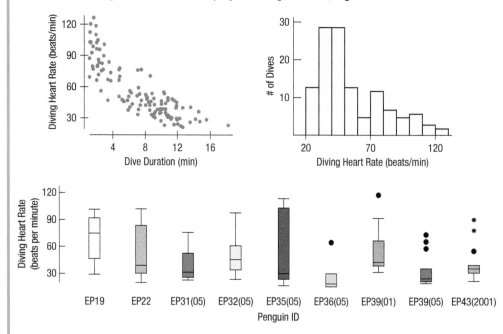

(The boxplots show the diving heart rates for each of the 9 penguins whose dives were tracked. The names are those given by the researchers; EP = emperor penguin.)

QUESTION: What features of each of these displays suggest that a re-expression might be helpful?

ANSWER: The scatterplot shows a curved relationship, concave upward, between the duration of the dives and penguins' heart rates. Re-expressing either variable may help to straighten the pattern.

The histogram of heart rates is skewed to the high end. Re-expression often helps to make skewed distributions more nearly symmetric.

Most of the boxplots show skewness to the high end as well. The medians are low in the boxes, and several show high outliers.

# The Ladder of Powers

How can we pick a re-expression to use? The secret is to choose a re-expression from a simple family of functions that includes powers and the logarithm.[3] We raise each data value to the same power—$\frac{1}{2}$, for example—by taking square roots. Or $-1$, by finding reciprocals. The good news is that the family of re-expressions line up in order, so that the farther you move away from the original data (the "1" position), the greater the effect on any curvature. This fact lets you search systematically for a re-expression that works, stepping a bit farther from "1" or taking a step back toward "1" as you see the results.

Where to start? It turns out that certain kinds of data are more likely to be helped by particular re-expressions. Knowing that gives you a good place to start your search, and from there you can look around a bit for a useful re-expression. We call this collection of re-expressions the **Ladder of Powers**.

| Power | Name | Comment |
|---|---|---|
| 2 | The square of the data values, $y^2$. | Try this for unimodal distributions that are skewed to the left. |
| 1 | The raw data—no change at all. This is "home base." The farther you step from here up or down the ladder, the greater the effect. | Data that can take on both positive and negative values with no bounds are less likely to benefit from re-expression. |
| 1/2 | The square root of the data values, $\sqrt{y}$. | Counts often benefit from a square root re-expression. For counted data, start here. |
| "0" | Although mathematicians define the "0-th" power differently,[4] for us the place is held by the logarithm. You may feel uneasy about logarithms. Don't worry; the computer or calculator does the work.[5] | Measurements that cannot be negative, and especially values that grow by percentage increases such as salaries or populations, often benefit from a log re-expression. When in doubt, start here. If your data have zeros, try adding a small constant to all values before finding the logs. |
| −1/2 | The (negative) reciprocal square root, $(-)1/\sqrt{y}$. | An uncommon re-expression, but sometimes useful. (Changing the sign to take the *negative* of the reciprocal square root preserves the direction of relationships, making things a bit simpler.) |
| −1 | The (negative) reciprocal, $(-)1/y$. | Ratios of two quantities, like miles per hour, often benefit from a reciprocal. (You have about a 50–50 chance that the original ratio was taken in the "wrong" order for simple statistical analysis and would benefit from re-expression.) Often, the reciprocal will have simple units (hours per mile). Change the sign if you want to preserve the direction of relationships. If your data have zeros, try adding a small constant to all values before finding the reciprocal. |

**TI-nspire**

**Re-expression.** See a curved relationship become straighter with each step on the Ladder of Powers.

The Ladder of Powers orders the effects that the re-expressions have on data. If you try, say, taking the square roots of all the values in a variable and it helps, but not enough, then move farther down the ladder to the logarithm or reciprocal root. Those re-expressions will have a similar, but even stronger, effect on your data. If you go too far, you can always back up. But don't forget—when you take a negative power, the *direction* of the relationship will change. That's OK. You can always change the sign of the response variable if you want to keep the same direction. With modern technology, finding a suitable re-expression is no harder than the push of a button.

---

[3] Don't be scared. You may have learned lots of properties of logarithms or done some messy calculations. Relax! You won't need that stuff here.

[4] You may remember that for any nonzero number $y$, $y^0 = 1$. This is not a very exciting transformation for data; every data value would be the same. We use the logarithm in its place.

[5] Your calculator or software package probably gives you a choice between "base 10" logarithms and "natural (base $e$)" logarithms. Don't worry about that. It doesn't matter at all which you use; they have exactly the same effect on the data. If you want to choose, base 10 logarithms can be a bit easier to interpret.

## JUST CHECKING

1. You want to model the relationship between the number of birds counted at a nesting site and the temperature (in degrees Celsius). The scatterplot of *Counts* vs. *Temperature* shows an upwardly curving pattern, with more birds spotted at higher temperatures. What transformation (if any) of the bird counts might you start with?

2. You want to model the relationship between prices for various items in Paris and in Hong Kong. The scatterplot of *Hong Kong Prices* vs. *Parisian Prices* shows a generally straight pattern with a small amount of scatter. What transformation (if any) of the Hong Kong prices might you start with?

3. You want to model the population growth of the United States over the past 200 years. The scatterplot shows a strongly upwardly curved pattern. What transformation (if any) of the population might you start with?

## FOR EXAMPLE

### Trying a Re-expression

**RECAP:** We've seen curvature in the relationship between emperor penguins' diving heart rates and the duration of the dive. Let's start the process of finding a good re-expression. Heart rate is in beats per minute; maybe heart "speed" in minutes per beat would be a better choice. Here are the corresponding displays for this reciprocal re-expression (as we often do, we've changed the sign to preserve the order of the data values).

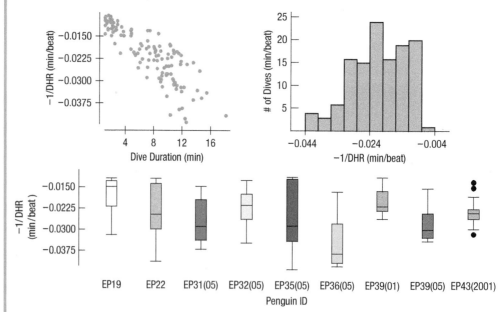

**QUESTION:** Were the re-expressions successful? Compare these graphs to the originals (p. 244).

**ANSWER:** The scatterplot bends less than before, but now may be slightly concave downward. The histogram is now slightly skewed to the low end. Most of the boxplots have no outliers. These boxplots seem better than the ones for the raw heart rates.

Overall, it looks like I may have moved a bit "too far" on the ladder of powers. Halfway between "1" (the original data) and "−1" (the reciprocal) is "0," which represents the logarithm. I'd try that for comparison.

# CHAPTER 9  Re-expressing Data: Get It Straight!   247

## STEP-BY-STEP EXAMPLE
### Re-expressing to Straighten a Scatterplot

Standard (monofilament) fishing line comes in a range of strengths, usually expressed as "test pounds." Five-pound test line, for example, can be expected to withstand a pull of up to five pounds without breaking. The convention in selling fishing line is that the price of a spool doesn't vary with strength. Instead, the length of line on the spool varies. Higher test pound line is thicker, though, and spools of fishing line hold about the same amount of material. Some spools hold line that is thinner and longer, some fatter and shorter. Let's look at the *Length* and *Strength* of spools of monofilament line manufactured by the same company and sold for the same price at one store.

QUESTION: How are the *Length* on the spool and the *Strength* related? And what re-expression will straighten the relationship?

---

**THINK**  PLAN  State the problem.

I want to fit a linear model for the length and strength of monofilament fishing line.

VARIABLES  Identify the variables and report the W's.

I have the *length* and "pound test" *strength* of monofilament fishing line sold by a single vendor at a particular store. Each case is a different strength of line, but all spools of line sell for the same price.

Let *Length* = length (in yards) of fishing line on the spool

*Strength* = the test strength (in pounds).

PLOT  Check that even if there is a curve, the overall pattern does not reach a minimum or maximum and then turn around and go back. An up-and-down curve can't be fixed by re-expression.

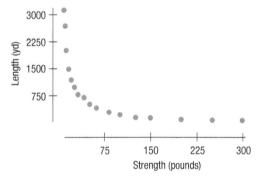

The plot shows a negative direction and an association that has little scatter but is not straight.

---

**SHOW**  MECHANICS  Try a re-expression.

Here's a plot of the square root of *Length* against *Strength*:

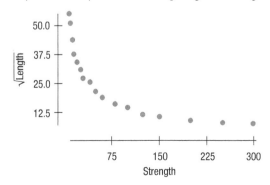

The lesson of the Ladder of Powers is that if we're moving in the right direction but have not had sufficient effect, we should go farther along the ladder. This example shows improvement, but is still not straight.

(Because *Length* is an amount of something and cannot be negative, we probably should have started with logs. This plot is here in part to illustrate how the Ladder of Powers works.)

The plot is less bent, but still not straight.

Stepping from the 1/2 power to the "0" power, we try the logarithm of *Length* against *Strength*.

The scatterplot of the logarithm of *Length* against *Strength* is even less bent:

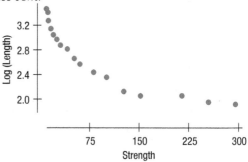

The straightness is improving, so we know we're moving in the right direction. But because the plot of the logarithms is not yet straight, we know we haven't gone far enough. To keep the direction consistent, change the sign and re-express to $-1/Length$.

This is much better, but still not straight, so I'll take another step to the "−1" power, or reciprocal.

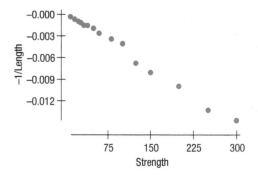

Maybe now I moved too far along the ladder.

We may have to choose between two adjacent re-expressions. For most data analyses, it really doesn't matter which we choose.

A half-step back is the −1/2 power: the reciprocal square root.

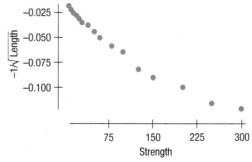

**TELL** **CONCLUSION** Specify your choice of re-expression. If there's some natural interpretation (as for gallons per 100 miles), give that.

Either of these last 2 options may be good enough. I think strength may depend on the cross-sectional area of the line. Area involves square units so I'll choose the −1/2 power.

Now that the re-expressed data satisfy the Straight Enough Condition, we can fit a linear model by least squares. We find that

$$\frac{-1}{\sqrt{Length}} = -0.023 - 0.000373 \, Strength.$$

We can use this model to predict the length of a spool of, say, 35-pound test line:

$$\frac{-1}{\sqrt{Length}} = -0.023 - 0.000373 \times 35 = -0.036.$$

We could leave the result in these units $(-1/\sqrt{yards})$. Sometimes the new units may be as meaningful as the original, but here we want to transform the predicted value back into yards. Fortunately, each of the re-expressions in the Ladder of Powers can be reversed.

To reverse the process, we first take the reciprocal: $\sqrt{\widehat{Length}} = -1/(-0.036) = 27.778$. Then squaring gets us ef back to the original units:

$$\widehat{Length} = 27.778^2 = 771.6 \; yards.$$

This may be the most painful part of the re-expression. Getting back to the original units can sometimes be a little work. Nevertheless, it's worth the effort to always consider re-expression. Re-expressions extend the reach of all of your Statistics tools by helping more data to satisfy the conditions they require. Just think how much more useful this course just became!

## FOR EXAMPLE
### Comparing Re-expressions

**RECAP:** We've concluded that in trying to straighten the relationship between *Diving Heart Rate* and *Dive Duration* for emperor penguins, using the reciprocal re-expression goes a bit "too far" on the Ladder of Powers. Now we try the logarithm. Here are the resulting displays.

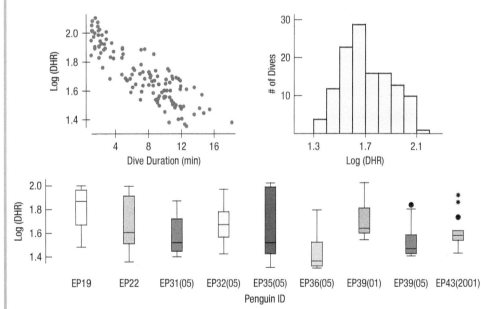

**QUESTION:** Comment on these displays. Now that we've looked at the original data (rung 1 on the Ladder, page 244), the reciprocal (rung −1, page 246), and the logarithm (rung 0), which re-expression of *Diving Heart Rate* would you choose?

**ANSWER:** The scatterplot is now more linear and the histogram is symmetric. The boxplots are still a bit skewed to the high end, but less so than for the original *Diving Heart Rate* values. We don't expect real data to cooperate perfectly, and the logarithm seems like a good compromise re-expression, improving several different aspects of the data.

## TI TIPS

### Re-expressing Data to Achieve Linearity

For a statistics project, a student wanted to explore the relationship between the amount of water in a 5-gallon cylindrical cooler and the time it takes for all the water to drain out from a spigot at the bottom. She repeatedly put different amounts of water in it and timed how long it took for the water flow to become a drip. The table below shows her data.

| Volume (gal) | 0.5 | 1.0 | 1.5 | 2.0 | 2.5 | 3.0 | 3.5 | 4.0 | 4.5 | 5.0 |
|---|---|---|---|---|---|---|---|---|---|---|
| Time (sec) | 20 | 30 | 36 | 43 | 48 | 51 | 57 | 59 | 64 | 67 |

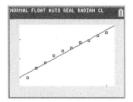

- Enter the data in your calculator in lists called VOL and TIME, and make a scatterplot with *Volume* on the *x*-axis and *Time* on the *y*-axis. Go ahead and add the least squares regression line.

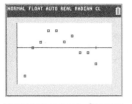

- It's pretty clear from the scatterplot that there's curvature in the data that a line can't capture. Construct a residuals plot and you'll see the curvature even more clearly.

We'd like to transform the data to straighten out the scatterplot. But what re-expression to use?

- Let's try the first re-expression in the Ladder of Powers—squaring the *y*-values. Because we don't yet know whether this will be a good transformation, let's not bother giving this list a special name. You can store the squares of the *Times* as L1 using the command LTIME$^2$ STO L1.

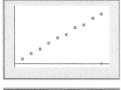

- Check the scatterplot for the re-expressed data by changing your STATPLOT specifications to Xlist:VOL and Ylist:L1. (Don't forget to use 9:ZoomStat to resize the window for the transformed *y*-values.)

The new scatterplot looks quite linear, but we've seen before that the residuals plot can reveal curvature that is easy to miss in the scatterplot.

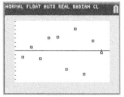

- Every time you do a regression on your calculator, the list called LRESID is replaced with the new residuals. Do a new regression on the transformed data and make a new residuals plot.

Now the residuals look just the way we like them—like random noise. We got lucky. The first transformation in the Ladder of Powers did about as good a job as we could hope for straightening out the scatterplot.

Do you know what the model's equation is? Look at the slope and intercept for the line your calculator determined, and remember that the *y*-values are not *Time*, but *Time* squared:

$$\widehat{Time^2} = -20 + 904\ Volume.$$

If you want to use this model to estimate how long it would take a certain amount of water to drain from the cooler, first use the equation to find $\widehat{Time^2}$, and then (Don't forget!) you need to take the square root.

# Plan B: Attack of the Logarithms

The Ladder of Powers is often successful at finding an effective re-expression. Sometimes, though, the curvature is more stubborn, and we're not satisfied with the residual plots. What then? When none of the data values is zero or negative, logarithms can be a helpful ally in the search for a useful model. Try taking the logs of both the *x*- and *y*-variables. Then re-express the data using some combination of *x* or log(*x*) vs. *y* or log(*y*). You may find that one of these works pretty well.

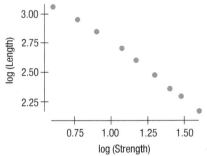

**Figure 9.10**
Plotting log (*Length*) against log (*Strength*) gives a straighter shape.

| Model Name | x-axis | y-axis | Comment |
|---|---|---|---|
| Exponential | x | log(y) | This model is the "0" power in the ladder approach, useful for values that grow by percentage increases. |
| Logarithmic | log(x) | y | A wide range of x-values, or a scatterplot descending rapidly at the left but leveling off toward the right, may benefit from trying this model. |
| Power | log(x) | log(y) | The Goldilocks model: When one of the ladder's powers is too big and the next is too small, this one may be just right. |

When we tried to model the relationship between the length of fishing line and its strength, we were torn between the "−1" power and the "−1/2" power. The first showed slight upward curvature, and the second downward. Maybe there's a better power between those values.

The scatterplot shows what happens when we graph the logarithm of *Length* against the logarithm of *Strength*. Technology reveals that the equation of our log–log model is

$$\widehat{\log(Length)} = 4.49 - 1.08 \log(Strength).$$

A warning, though! Don't expect to be able to straighten every curved scatterplot you find. It may be that there just isn't a very effective re-expression to be had. You'll certainly encounter situations when nothing seems to work the way you wish it would. Don't set your sights too high—you won't find a perfect model. Keep in mind: We seek a *useful* model, not perfection (or even "the best").

## TI TIPS

### Using Logarithmic Re-expressions

If you have any body piercings, you probably know that the *gauge* of your jewelry (a measure of size) is a lower number for larger jewelry and a higher number for smaller jewelry. For example, if you wanted a nose ring, you might initially get a small gauge 16 piercing and then, over time, stretch it to a gauge 14, then a gauge 12, and so on. The table below shows some standard jewelry gauges and gives the approximate corresponding diameters in millimeters (mm). (The gauge itself has no units.)

| Gauge | 0 | 2 | 4 | 6 | 8 | 10 | 12 | 14 | 16 | 18 | 20 |
|---|---|---|---|---|---|---|---|---|---|---|---|
| Diameter (mm) | 8.0 | 6.5 | 5.0 | 4.0 | 3.0 | 2.4 | 2.0 | 1.6 | 1.2 | 1.0 | 0.8 |

Let's try to find a model that will estimate the diameter of the jewelry given the gauge.

◆ First (of course) *make a picture*. Because we're estimating diameter from gauge, *Gauge* should be the *x*-variable and *Diameter* should be the *y*-variable.

You might have some success straightening the data with the Ladder of Powers, but when we tried it, the residuals still showed a pattern even when we found a transformation that seemed to make the data look pretty linear. Here we're going to use logarithms.

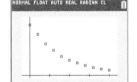

◆ Take the log of the diameter and store it in list L1. Then examine a scatterplot of *Gauge* on the *x*-axis and log(*Diameter*) on the *y*-axis.

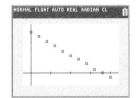

◆ Once again we've gotten lucky with our first re-expression. This transformation makes the data appear very linear. But compute the regression line and check a residuals plot to be sure. (Remember: When you compute a regression line, the residuals are saved automatically as list LRESID.)

The residuals plot shows no clear pattern, just random noise. So our model should be a reasonably good one for expressing the relationship between jewelry gauge and diameter. But what is the model?

- Look at the equation from the calculator, and remember that the *y*-variable is actually log(*Diameter*):

$$\widehat{\log(Diameter)} = 0.900 - 0.050(Gauge).$$

The data table included only even-numbered gauges, but this model could be used to estimate diameters of other gauges. For example, for a 3-gauge piece of jewelry, we would compute:

$$\widehat{\log(Diameter)} = 0.900 - 0.050 \times 3 = 0.75.$$

And then using the definition of a log:

$$\widehat{Diameter} = 10^{0.75} \approx 5.6 \text{ mm}.$$

It's always good to verify that a model's estimate is about what you'd expect. That's easy to do here by verifying that the estimated diameter for a 3-gauge piece of jewelry is between that of a 2-gauge and a 4-gauge—and it is.

Before leaving this example, we want to point out that sometimes a useful re-expression is taking the logs of both *x*- and *y*-variables. We didn't need to try that for our jewelry data because our first re-expression worked well, but if we had tried it, we'd have run into a problem: one of the data values is a jewelry gauge of 0, and you can't take the log of zero.[6] A zero (or any negative values) in the data is a good indicator that taking the log of that variable will not be helpful.

# Why Not Just Use a Curve?

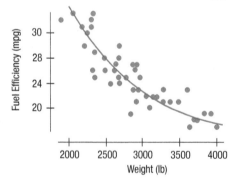

When a clearly curved pattern shows up in the scatterplot, why not just fit a curve to the data? We saw earlier that the association between the *Weight* of a car and its *Fuel Efficiency* was not a straight line. Instead of trying to find a way to straighten the plot, why not find a curve that seems to describe the pattern well?

We can find "curves of best fit" using essentially the same approach that led us to linear models. You won't be surprised, though, to learn that the mathematics and the calculations are considerably more difficult for curved models. Many calculators and computer packages do have the ability to fit curves to data, but this approach has many drawbacks.

Straight lines are easy to understand. We know how to think about the slope and the *y*-intercept, for example. We often want some of the other benefits mentioned earlier, such as making the spread around the model more nearly the same everywhere. In later chapters you will learn more advanced statistical methods for analyzing linear associations.

We give all of that up when we fit a model that is not linear. For many reasons, then, it is usually better to re-express the data to straighten the plot.

> ### AVOID NONLINEAR REGRESSION MODELS
> Besides fitting lines to data, your calculator can fit curves, using functions like `ExpReg`, `PwrReg`, `QuadReg`, and `CubicReg`. These are precisely the ones we urge you to avoid using. While they may appear to give good fits to some curved datasets, interpreting the resulting models is trickier than you might think. And knowing when they're truly appropriate requires some experience and advanced knowledge that is beyond this course. Even professional statisticians usually avoid these regressions if they can straighten their data with re-expressions and use a line.

---

[6] At least no one's succeeded yet.

# WHAT CAN GO WRONG?

**OCCAM'S RAZOR**

If you think that simpler explanations and simpler models are more likely to give a true picture of the way things work, then you should look for opportunities to re-express your data and simplify your analyses.

The general principle that simpler explanations are likely to be the better ones is known as Occam's Razor, after the English philosopher and theologian William of Occam (1284–1347).

- **Don't get seduced by ExpReg and its calculator cousins.** Those so-called "curved" regression options look enticing, but don't go there. This course is about *linear* regression. If you see a curve, re-express the data to achieve linearity and then fit a line. Equations of lines are easier to interpret and will be far easier to work with later on when we do more advanced statistical analyses.

- **Don't expect your model to be perfect.** In Chapter 5 we quoted statistician George Box: "All models are wrong—but some are useful." Be aware that the real world is a messy place and data can be uncooperative. Don't expect to find one elusive re-expression that magically irons out every kink in your scatterplot and produces perfect residuals. You aren't looking for the Right Model, because that mythical creature doesn't exist. Find a useful model and use it wisely.

- **Don't stray too far from the ladder.** It's wise not to stray too far from the powers that we suggest in the Ladder of Powers. Stick to powers between 2 and −2. Even in that interval, you should prefer the simpler powers in the ladder to those in the cracks. A square root is easier to understand than the 0.413 power. That simplicity may compensate for a slightly less straight relationship.

- **Don't choose a model based on $R^2$ alone.** You've tried re-expressing your data to straighten a curved relationship and found a model with a high $R^2$. Beware: That doesn't mean the pattern is straight now. Below is a plot of a relationship with an $R^2$ of 98.3%. The $R^2$ is about as high as we could ask for, but if you look closely, you'll see that there's a consistent bend. Plotting the residuals from the least squares line makes the bend much easier to see.

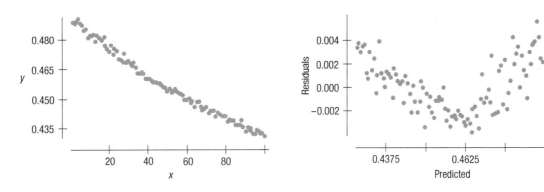

Remember the basic rule of data analysis: *Make a picture*. Before you fit a line, always look at the pattern in the scatterplot. After you fit the line, check for linearity again by plotting the residuals.

- **Beware of multiple modes.** Re-expression can often make a skewed unimodal histogram more nearly symmetric, but it cannot pull separate modes together. A suitable re-expression may, however, make the separation of the modes clearer, simplifying their interpretation and making it easier to separate them to analyze individually.

- **Watch out for scatterplots that turn around.** Re-expression can straighten many bent relationships but not those that go up and then down or down and then up. You should refuse to analyze such data with methods that require a linear form.

**Figure 9.11**
The shape of the scatterplot of *Birthrates* (births per 100,000 women) in the United States shows an oscillation that cannot be straightened by re-expressing the data.

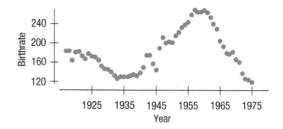

- **Watch out for zero or negative data values.** It's impossible to re-express negative values by any power that is not a whole number on the Ladder of Powers or to re-express values that are zero for negative powers. One possible cure for zeros and small negative values is to add a constant ($\frac{1}{2}$ and $\frac{1}{6}$ are often used) to bring all the data values above zero.

## WHAT HAVE WE LEARNED?

We've learned that when the conditions for linear regression are not met, a simple re-expression of the data may help. There are several reasons to consider a re-expression:
- To make the distribution of a variable more symmetric (as we saw in Chapter 4)
- To make the spread across different groups more similar
- To make the form of a scatterplot straighter
- To make the scatter around the line in a scatterplot more consistent

We've learned that when seeking a useful re-expression, taking logs is often a good, simple starting point. To search further, the Ladder of Powers or the log–log approach can help us find a good re-expression.

We've come to understand that our models won't be perfect, but that re-expression can lead us to a useful model.

### TERMS

**Re-expression**  We re-express data by taking the logarithm, the square root, the reciprocal, or some other mathematical operation of all values of a variable. (p. 239)

**Ladder of Powers**  The Ladder of Powers places in order the effects that many re-expressions have on the data. (p. 245)

## ON THE COMPUTER

### Re-expression

Computers and calculators make it easy to re-express data. Most statistics packages offer a way to re-express and compute with variables. Some packages permit you to specify the power of a re-expression with a slider or other moveable control, possibly while watching the consequences of the re-expression on a plot or analysis. This, of course, is a very effective way to find a good re-expression.

# EXERCISES

1. **Residuals** Suppose you have fit a linear model to some data and now take a look at the residuals. For each of the following possible residual plots, tell whether you would try a re-expression and, if so, why.

2. **More residuals** Suppose you have fit a linear model to some data and now take a look at the residuals. For each of the following possible residual plots, tell whether you would try a re-expression and, if so, why.

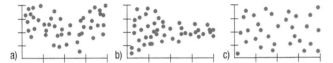

3. **Oakland passengers 2016 revisited** In Chapter 8, Exercise 15, we considered whether a linear model would be appropriate to describe the trend in the number of passengers departing from the Oakland (CA) airport each month since the start of 1997. If we fit a regression model, we obtain this residual plot (plotted against predicted values). We've added lines to show the order of the values in time:

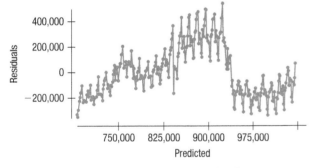

   a) Can you account for the pattern shown here?
   b) Would a re-expression help us deal with this pattern? Explain.

4. **Hopkins winds revisited** In Chapter 4, we examined the wind speeds in the Hopkins Memorial Forest over the course of a year. Here's the scatterplot we saw then:

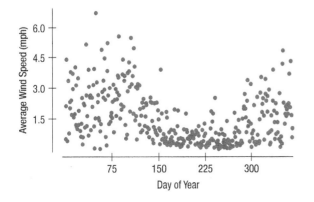

   a) Describe the pattern you see there.
   b) Should we try re-expressing either variable to make this plot straighter? Explain.

5. **Models** For each of the models listed below, predict $y$ when $x = 2$.
   a) $\ln \hat{y} = 1.2 + 0.8x$ 
   b) $\sqrt{\hat{y}} = 1.2 + 0.8x$
   c) $\dfrac{1}{\hat{y}} = 1.2 + 0.8x$
   d) $\hat{y} = 1.2 + 0.8 \ln x$
   e) $\log \hat{y} = 1.2 + 0.8 \log x$

6. **More models** For each of the models listed below, predict $y$ when $x = 2$.
   a) $\hat{y} = 1.2 + 0.8 \log x$
   b) $\log \hat{y} = 1.2 + 0.8x$
   c) $\ln \hat{y} = 1.2 + 0.8 \ln x$
   d) $\hat{y}^2 = 1.2 + 0.8x$
   e) $\dfrac{1}{\sqrt{\hat{y}}} = 1.2 + 0.8x$

7. **Gas mileage** As the example in the chapter indicates, one of the important factors determining a car's *Fuel Efficiency* is its *Weight*. Let's examine this relationship again, for 11 cars.
   a) Describe the association between these variables shown in the scatterplot.

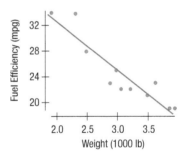

   b) Here is the regression analysis for the linear model. What does the slope of the line say about this relationship?

   Dependent variable is Fuel Efficiency
   R-squared = 85.9%

   | Variable | Coefficient |
   |---|---|
   | Intercept | 47.9636 |
   | Weight | −7.65184 |

   c) Do you think this linear model is appropriate? Use the residual plot to explain your decision.

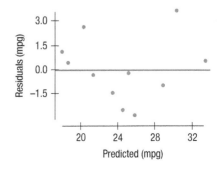

**8. World health** In the struggle to improve the health of people around the world, it is important to know what variables are related to the health of people in a nation. One obvious thing to consider is the amount of money a country has per capita. The plot below shows how *Life Expectancy* was related to *Gross Domestic Product* (*GDP*) per capita for many countries around the world in 2018. (www.gapminder.org)

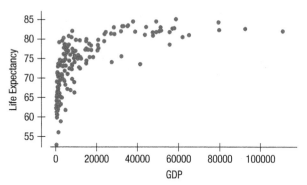

a) Describe the association between the variables shown in the scatterplot.
b) Explain why you should re-express these data before trying to fit a model.
c) What re-expression of GDP would you try as a starting point?

**9. Gas mileage revisited** Let's try the re-expressed variable *Fuel Consumption* (gal/100 mi) to examine the fuel efficiency of the 11 cars in Exercise 7. Here are the revised regression analysis and residual plot:

Dependent variable is Fuel Consumption
R-squared = 89.2%

| Variable | Coefficient |
|---|---|
| Intercept | 0.624932 |
| Weight | 1.17791 |

a) Explain why this model appears to be better than the linear model.
b) Using the regression analysis above, write an equation of this model.
c) Interpret the slope of this line.
d) Based on this model, how many miles per gallon would you expect a 3500-pound car to get?

**10. World health again** In Exercise 8 we looked at data about *GDP* and life expectancy for different countries. For a re-expression, a student tried the square root of *GDP* because they thought the shape looked like the shape of a square root graph. Here is a scatterplot and residuals plot for *Life span* vs. $\sqrt{GDP}$.

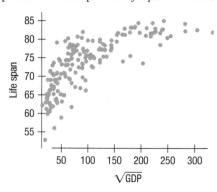

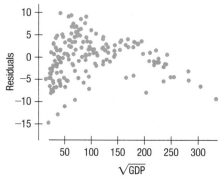

a) Is this a useful re-expression? Refer to features in both the scatterplot and the residuals plot to explain your decision.
b) What re-expression would you suggest this student try next?

Another student tried the natural log of *GDP*, thinking that the graph looked like the shape of a logarithmic graph. Here is a scatterplot and residuals plot for *Life span* vs. ln(*GDP*).

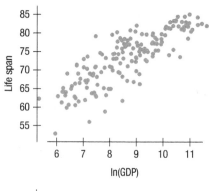

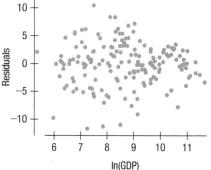

c) Use the scatterplot and residuals plot to comment on the usefulness of this re-expression.
d) Which re-expression was more successful?

T 11. **USGDP 2020** The scatterplot shows the gross domestic product (GDP) of the United States in trillions of 2012 dollars plotted against years since 1950.

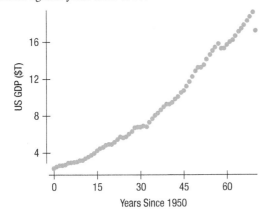

A linear model fit to the relationship looks like this:

Dependent variable is: GDP($T)
R-squared = 97.0%  s = 0.9124

| Variable | Coefficient |
|---|---|
| Intercept | 0.499 |
| Years Since 1950 | 0.248 |

a) Does the $R^2$ value of 97.0% suggest that this is a good model? Explain.
b) Here's a scatterplot of the residuals. Now do you think this is a good model for these data? Explain.

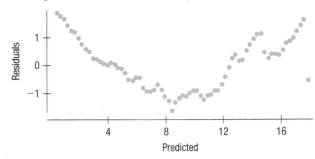

T 12. **T bill rates 2020** The 3-month Treasury bill interest rate discussed in Chapter 8, Exercises 27 and 29, is watched by investors and economists. Here is the scatterplot of the 3-month Treasury bill rate since 1934 that we saw in Exercise 29:

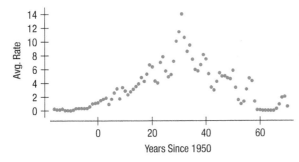

Clearly, the relationship is not linear. Can it be made nearly linear with a re-expression? If so, which one would you suggest? If not, why not?

T 13. **Better GDP model?** Consider again the post-1950 trend in U.S. GDP we examined in Exercise 11. Here are regression output and a residual plot when we use the log of a GDP in the model. Is this a better model for GDP? Explain. Would you want to consider a different re-expression? If so, which one?

Dependent variable is log(GDP)
R-squared = 98.9%  s = 0.0291

| Variable | Coefficient |
|---|---|
| Intercept | 0.4183 |
| Years Since 1950 | 0.0133 |

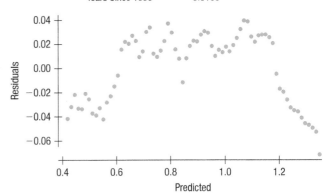

T 14. **T bill rates more recently** A student, in analyzing the scatterplot of Treasury bill rates from Exercise 12, concluded that something changed the trend in rates around 1981. He chose to look at only the more recent data to describe the trend. Here are a scatterplot, residual plot, and regression analysis for the Treasury bill rates since 1982.

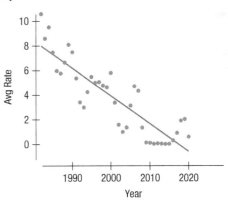

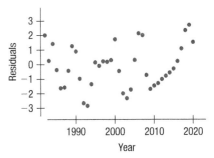

Dependent variable is Rate
R-squared = 75.4%  s = 1.693

| Variable | Coefficient |
|---|---|
| Intercept | 505.16 |
| Year | −0.2505 |

a) Is a linear model appropriate for describing this trend since 1982?
b) Would you expect this trend to continue past 2025?

**15. Brakes** The following table shows stopping distances in feet for a car tested 3 times at each of 5 speeds. We hope to create a model that predicts *Stopping Distance* from the *Speed* of the car. Also given are a scatterplot and residual plot of the data.

| Speed (mph) | Stopping Distances (ft) |
|---|---|
| 20 | 64, 62, 59 |
| 30 | 114, 118, 105 |
| 40 | 153, 171, 165 |
| 50 | 231, 203, 238 |
| 60 | 317, 321, 276 |

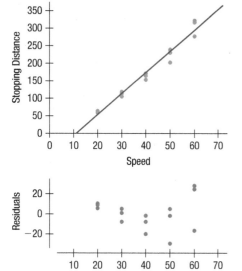

a) Explain why a linear model is not appropriate.

The data are re-expressed plotting $\sqrt{Distance}$ vs. *Speed*. Here are the scatterplot, residual plot, and regression output.

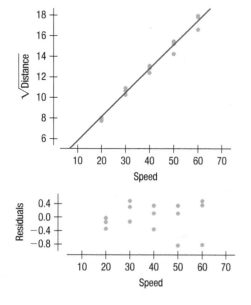

Dependent variable is $\sqrt{Distance}$
R-squared = 98.4%  s = 0.4489

| Variable | Coefficient |
|---|---|
| Intercept | 3.3034 |
| Speed | 0.2355 |

b) Comment on the appropriateness of using a linear model to represent the transformed data.
c) What does the model predict for the stopping distance for a car traveling 45 mph?
d) What does the model predict for the stopping distance for a car traveling 10 mph?
e) Do you have concerns about using this model for either of these predictions? Why?

**16. Pendulum** A student experimenting with a pendulum counted the number of full swings the pendulum made in 20 seconds for various lengths of string. Here are her data, a scatterplot with a least squares regression line, and a residual plot.

| Length (in.) | 6.5 | 9 | 11.5 | 14.5 | 18 | 21 | 24 | 27 | 30 | 37.5 |
|---|---|---|---|---|---|---|---|---|---|---|
| Number of Swings | 22 | 20 | 17 | 16 | 14 | 13 | 13 | 12 | 11 | 10 |

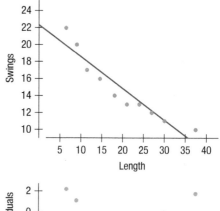

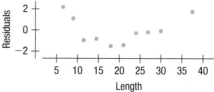

a) Explain why a linear model is not appropriate for using the *Length* of a pendulum to predict the *Number of Swings* in 20 seconds.

The student re-expressed the data plotting the reciprocal of the *Number of Swings* against the *Length*. The scatterplot, residual plot, and regression output are shown.

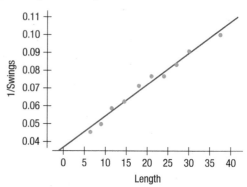

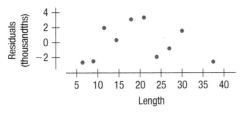

Dependent variable is 1/Swings
R-squared = 98.1%   s = 0.0025

| Variable  | Coefficient |
|-----------|-------------|
| Intercept | 0.0367      |
| Length    | 0.00176     |

b) Comment on the appropriateness of using a linear model to represent the transformed data.
c) What does the model predict for the number of swings for a pendulum with a 20-inch string?
d) What does the model predict for the number of swings for a pendulum with a 60-inch string?
e) How much confidence do you place in each of these predictions? Why?

**17. Coaster speeds** Recall the data about the roller coasters in Chapter 7. Here are boxplots of speed comparing coasters of different scales. Below that, the plot graphs log(*Speed*). Which of the goals of re-expression does this illustrate?

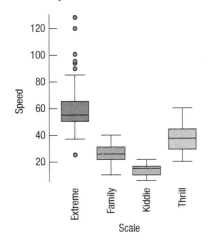

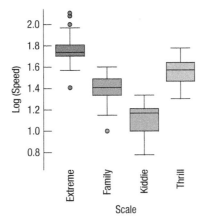

**18. Coasters again** Exercise 17 looked at the distribution of speeds of roller coasters, comparing coasters of different scales. That exercise offered the logarithm as a re-expression of *Speed*. Here are two other alternatives, the square root and the reciprocal. Would you still prefer the log? Explain why.

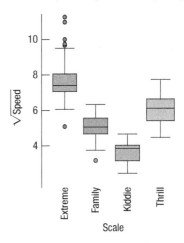

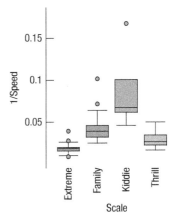

**19. Planets** Here is a table of the 9 sun-orbiting objects formerly known as planets. (All but one of them still are known as planets.) Also shown is a scatterplot of *Length* of the year against the *Distance* from the sun, a regression line, and a residual plot.

| Planet  | Position Number | Distance from Sun (million miles) | Length of Year (Earth years) |
|---------|-----------------|-----------------------------------|------------------------------|
| Mercury | 1               | 36.254                            | 0.24                         |
| Venus   | 2               | 66.931                            | 0.62                         |
| Earth   | 3               | 92.960                            | 1                            |
| Mars    | 4               | 141.299                           | 1.88                         |
| Jupiter | 5               | 483.392                           | 11.86                        |
| Saturn  | 6               | 886.838                           | 29.46                        |
| Uranus  | 7               | 1782.97                           | 84.01                        |
| Neptune | 8               | 2794.37                           | 164.8                        |
| Pluto   | 9               | 3671.92                           | 248                          |

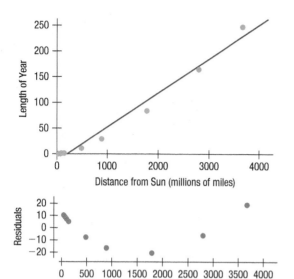

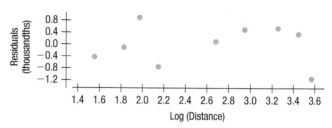

a) Describe the relationship between *Length* of the year and *Distance* from the sun.

Below are two re-expressions; the first is log(*Length*) vs. *Distance* and the second is log(*Length*) vs. log(*Distance*).

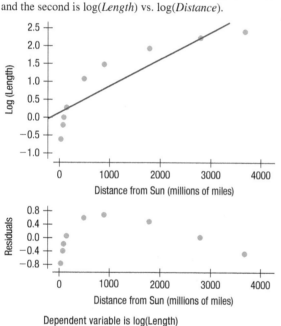

Dependent variable is log(Length)
R-squared = 79.8%   s = 0.540

| Variable | Coefficient |
| --- | --- |
| Intercept | 0.12176 |
| Distance | 0.000745 |

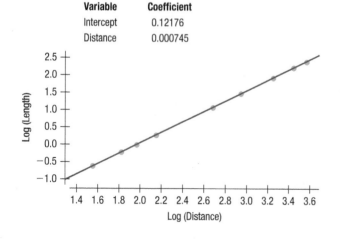

Dependent variable is log(Length)
R-squared = 100.0%   s = 0.0007

| Variable | Coefficient |
| --- | --- |
| Intercept | -2.9546 |
| log(Distance) | 1.50048 |

b) Which expression was more effective in creating a linear model? Justify your choice.

c) Comment on how well this model fits the data.

**20. Is Pluto a planet?** Let's look again at the pattern in the locations of the planets in our solar system seen in the table in Exercise 19. Below is a plot of the *Distance* of each "planet" from the sun vs. its *Position*, along with two re-expressions. Select the most appropriate model and use it to answer the following question: *Would you agree with the International Astronomical Union (IAU) that Pluto is not a planet?* Explain.

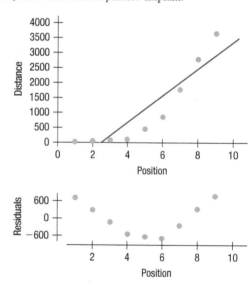

Dependent variable is Distance
R-squared = 82.8%   s = 597.17

| Variable | Coefficient |
| --- | --- |
| Intercept | -1131.9 |
| Position | 447.78 |

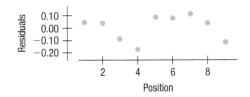

Dependent variable is log(Distance)
R-squared = 98.2%   s = 0.1087

| Variable | Coefficient |
|---|---|
| Intercept | 1.2450 |
| Position | 0.27097 |

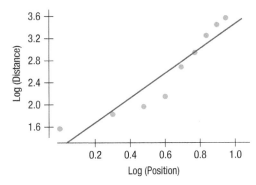

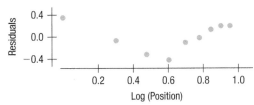

Dependent variable is log(Distance)
R-squared = 88.7%   s = 0.2692

| Variable | Coefficient |
|---|---|
| Intercept | 1.2063 |
| log(Position) | 2.2557 |

**21. Planets and asteroids** The asteroid belt between Mars and Jupiter may be the remnants of a failed planet. If so, then Jupiter is really in position 6, Saturn is in 7, and so on. Using this revised method of numbering the positions, the plot of log(*Distance*) from the sun vs. this new *Adjusted Position* is shown. Which method of identifying *Position* seems to work better?

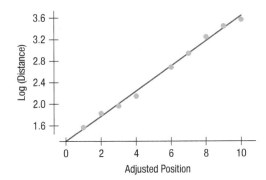

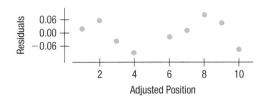

Dependent variable is log(Distance)
R-squared = 99.4%   s = 0.0608

| Variable | Coefficient |
|---|---|
| Intercept | 1.3063 |
| AdjPosition | 0.2328 |

**22. Planets and Eris** In July 2005, astronomers Mike Brown, Chad Trujillo, and David Rabinowitz announced the discovery of a sun-orbiting object, since named Eris,[7] that is 5% larger than Pluto and even farther from the sun. Eris orbits the sun once every 560 earth years at an average distance of about 6300 million miles from the sun. Based on its *Position*, how does Eris's *Distance* from the sun (re-expressed to logs) compare with the prediction made by the model of Exercise 21?

**23. Planets, models, and laws** The model you found in Exercise 19 is a relationship noted in the 17th century by Kepler as his Third Law of Planetary Motion. It was subsequently explained as a consequence of Newton's Law of Gravitation. The models for Exercises 20–22 relate to what is sometimes called the Titius-Bode "law," a pattern noticed in the 18th century but lacking any scientific explanation.

Compare how well the re-expressed data are described by their respective linear models. What aspect of the model of Exercise 19 suggests that we have found a physical law? In the future, we may learn enough about a planetary system around another star to tell whether the Titius-Bode pattern applies there. If you discovered that another planetary system followed the same pattern, how would it change your opinion about whether this is a real natural "law"? What would you think if the next system we find does not follow this pattern?

**24. Weightlifting 2020** Listed below are the world record men's weightlifter's performances as of 2020, followed by some analyses. A weightlifter's achievement (total weight) is the combined weights lifted in the *snatch* and the *clean* and *jerk*.

| Weight Class (kg) | Record Holder | Country | Total Weight (kg) |
|---|---|---|---|
| 56 | Long Qingquan | China | 307 |
| 62 | Kim Un-Guk | North Korea | 327 |
| 69 | Galain Boevski | Bulgaria | 357 |
| 77 | Lu Xiaojun | China | 379 |
| 85 | Kianoush Rostami | Iran | 396 |
| 94 | Ilya Ilyin | Kazakhstan | 418 |
| 105 | Andrei Aramnau | Belarus | 436 |
| 105+ | Lasha Tlakhadze | Georgia | 473 |

---

[7] Eris is the Greek goddess of warfare and strife who caused a quarrel among the other goddesses that led to the Trojan war. In the astronomical world, Eris stirred up trouble when the question of its proper designation led to the raucous meeting of the IAU in Prague where IAU members voted to demote Pluto and Eris to dwarf-planet status (www.gps.caltech.edu/~mbrown/planetlila/#paper).

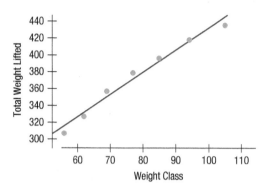

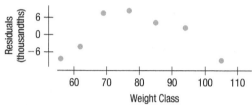

Dependent variable is Total Weight Lifted
R-squared = 97.7%   s = 7.8273

| Variable | Coefficient |
|---|---|
| Intercept | 167.31 |
| Weight Class | 2.644 |

a) What does the residual plot tell you about the need to re-express?

We re-expressed these data two ways, first using the reciprocal and then taking the logs of both variables. Here are the results.

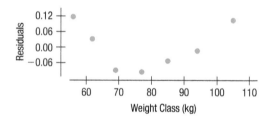

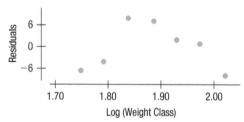

b) What does the residual plot tell you about the success of the re-expressions?

**25. Slower is cheaper?** Researchers studying how a car's *Fuel Efficiency* varies with its *Speed* drove a compact car 200 miles at various speeds on a test track. Their data are shown in the table.

| Speed (mph) | 35 | 40 | 45 | 50 | 55 | 60 | 65 | 70 | 75 |
|---|---|---|---|---|---|---|---|---|---|
| Fuel Eff. (mpg) | 25.9 | 27.7 | 28.5 | 29.5 | 29.2 | 27.4 | 26.4 | 24.2 | 22.8 |

Create a scatterplot for this relationship and describe concerns you have about trying to linearize this model.

**26. Tree growth** A study in Texas examined the growth of grapefruit trees, determining the average trunk *Diameter* (in inches) for trees of varying *Ages*:

| Age (yr) | 2 | 4 | 6 | 8 | 10 | 12 | 14 | 16 | 18 | 20 |
|---|---|---|---|---|---|---|---|---|---|---|
| Diameter (in.) | 2.1 | 3.9 | 5.2 | 6.2 | 6.9 | 7.6 | 8.3 | 9.1 | 10.0 | 11.4 |

a) Fit a linear model to these data. What concerns do you have about the model?
b) If data had been given for individual trees instead of averages, would you expect the fit to be stronger, less strong, or about the same? Explain.

**27. Orange production** The table below shows that as the number of oranges on a tree increases, the fruit tends to get smaller. Create a model for this relationship, and express any concerns you may have.

| Number of Oranges/Tree | Average Weight/Fruit (lb) |
|---|---|
| 50 | 0.60 |
| 100 | 0.58 |
| 150 | 0.56 |
| 200 | 0.55 |
| 250 | 0.53 |
| 300 | 0.52 |
| 350 | 0.50 |
| 400 | 0.49 |
| 450 | 0.48 |
| 500 | 0.46 |
| 600 | 0.44 |
| 700 | 0.42 |
| 800 | 0.40 |
| 900 | 0.38 |

**28. Years to live, 2017** Insurance companies and other organizations use actuarial tables to estimate the remaining life spans of their customers. The data file gives life expectancy and estimated additional years of life for Black men in the United States, according to a 2016 National Vital Statistics Report.

| Age | Years to Live | Age | Years to Live |
|---|---|---|---|
| 0 | 71.9 | 50 | 27.1 |
| 1 | 71.8 | 55 | 23.1 |
| 5 | 67.9 | 60 | 19.6 |
| 10 | 63 | 65 | 16.4 |
| 15 | 58.1 | 70 | 13.4 |
| 20 | 53.4 | 75 | 10.7 |
| 25 | 48.9 | 80 | 8.2 |
| 30 | 44.5 | 85 | 6.1 |
| 35 | 40 | 90 | 4.5 |
| 40 | 35.7 | 95 | 3.3 |
| 45 | 31.3 | 100 | 2.5 |

Here are the results of a re-expression:

Dependent variable is √Years to Live
R-squared = 99.8%   s = 0.10700

| Variable  | Coefficient |
|-----------|-------------|
| Intercept | 8.6925      |
| Age       | −0.07168    |

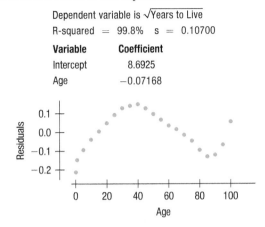

a) Evaluate the success of the re-expression.
b) Predict the life span of an 18-year-old Black man.
c) Are you satisfied that your model could predict the life expectancy of a friend of yours?

**29. Boyle** Scientist Robert Boyle examined the relationship between the volume in which a gas is contained and the pressure in its container. He used a cylindrical container with a moveable top that could be raised or lowered to change the volume. He measured the *Height* in inches by counting equally spaced marks on the cylinder, and measured the *Pressure* in inches of mercury (as in a barometer). Here is Boyle's original table, but you can find the data in the file **Boyle**. Find a suitable re-expression of *Pressure*, fit a regression model, and check whether it is appropriate for modeling the relationship between *Pressure* and *Height*.

**30. Baseball salaries 2019** Ballplayers have been signing ever-larger contracts. The highest salaries (in dollars per season) for each year since 1874 are in the data file **Baseball salaries 2019**.
a) Make a scatterplot of *Salary in 2019* vs. *Year*. Does it look straight?
b) Find the best re-expression of *Salary in 2019* that you can to straighten out the scatterplot.
c) Fit the regression for the re-expression you found in part b, and write a sentence telling what it says.
d) Plot the residuals of the model in part c against *Year*.
e) Comment on the model you found given the residual plot in part d.

## JUST CHECKING

### Answers

1. Counts are often best transformed by using the square root.
2. None. The relationship is already straight.
3. Even though, technically, the population values are counts, you should probably try a stronger re-expression like log(*Population*) because populations grow in proportion to their size.

# Review of Part II

## EXPLORING RELATIONSHIPS BETWEEN VARIABLES

### Quick Review

You have now survived your second major unit of Statistics. Here's a brief summary of the key concepts and skills:

- We treat data two ways: as categorical or as quantitative.
- To explore relationships in categorical data, check out Chapter 2.
- To explore relationships in quantitative data:
  - Make a picture. Use a scatterplot. Put the explanatory variable on the *x*-axis and the response variable on the *y*-axis.
  - Describe the association between two quantitative variables in terms of direction, form, and strength.
  - The amount of scatter determines the strength of the association.
  - If greater values of one variable tend to be associated with greater values of the other, the association is positive. If greater values of one variable tend to be associated with lesser values of the other, then the association is negative.
  - If the form of the association is roughly linear (and *only* if it's roughly linear!), calculate the correlation to measure its strength numerically, and do a regression analysis to model it.
  - Correlations closer to −1 or +1 indicate stronger linear associations. Correlations near 0 indicate weak linear relationships, but other forms of association may still be present.
  - The model used most often for a linear relationship is the least squares regression line, or LSRL, which minimizes the sum of the squared residuals. It is sometimes called simply the regression line or the regression model.
- The regression line predicts values of the response variable from values of the explanatory variable.
- A residual is the difference between the true value of the response variable and the value predicted by the regression model.
- The slope of the line is an average rate of change, best described in "*y*-units" per "*x*-unit."
- $R^2$ gives the fraction of the variation in the response variable that is accounted for by the model.
- The standard deviation of the residuals measures the amount of scatter around the line.
- Outliers and influential points can distort any of our models.
- If you see a curved pattern in the residuals plot, your chosen model is not appropriate; use a different model. You may, for example, straighten the relationship by re-expressing one or both of the variables.
- To straighten bent relationships, re-express the data using logarithms or a power (squares, square roots, reciprocals, etc.).
- Always remember that an association is not necessarily an indication that one of the variables causes the other.

Need more help with some of this? Try rereading some sections of Chapters 6 through 9. Starting right here on this very page are more opportunities to review these concepts and skills.

> One must learn by doing the thing; though you think you know it, you have no certainty until you try.
>
> —Sophocles (495–406 BCE)

## REVIEW EXERCISES

1. **College** Every year, *U.S. News and World Report* publishes a special issue on many U.S. colleges and universities. The scatterplots have *Student/Faculty Ratio* (number of students per faculty member) for the colleges and universities on the *y*-axes plotted against 4 other variables. The correct correlations for these scatterplots appear in this list. Match them.

   −0.98   −0.71   −0.51   0.09   0.23   0.69

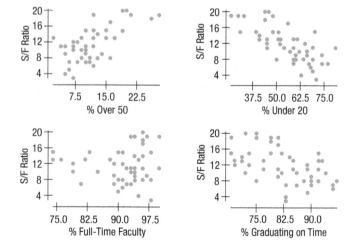

264

2. **Togetherness** Are good grades in high school associated with family togetherness? A random sample of 142 high school students was asked how many meals per week their families ate together. Their responses produced a mean of 3.78 meals per week, with a standard deviation of 2.2. Researchers then matched these responses against the students' grade point averages (GPAs). The scatterplot appeared to be reasonably linear, so they created a line of regression. No apparent pattern emerged in the residuals plot. The equation of the line was $\widehat{GPA} = 2.73 + 0.11\,Meals$.

   a) Interpret the y-intercept in this context.
   b) Interpret the slope in this context.
   c) What was the mean GPA for these students?
   d) If a student in this study had a negative residual, what did that mean?
   e) Upon hearing of this study, a counselor recommended that parents who want to improve the grades their children get should get the family to eat together more often. Do you agree with this interpretation? Explain.

3. **Vineyards** Here are the scatterplot and regression analysis for *Case Prices* of 36 wines from vineyards in the Finger Lakes region of New York State and the *Ages* of the vineyards.

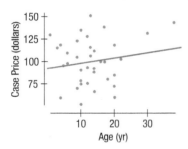

   Dependent variable is Case Price
   R-squared = 2.7%

   | Variable | Coefficient |
   |---|---|
   | Constant | 92.7650 |
   | Age | 0.567284 |

   a) Does it appear that vineyards in business longer get higher prices for their wines? Explain.
   b) What does this analysis tell us about vineyards in the rest of the world?
   c) Write the regression equation.
   d) Explain why that equation is essentially useless.

4. **Vineyards again** Instead of *Age*, perhaps the *Size* of the vineyard (in acres) is associated with the price of the wines. Look at the scatterplot, in which one point is indicated by a red "+" because it is referred to in parts c) and d):

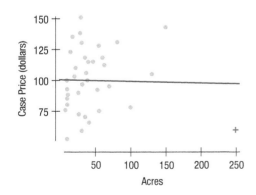

   a) Do you see any evidence of an association?
   b) What concern do you have about this scatterplot?
   c) If the red "+" data point is removed, would the correlation become stronger or weaker? Explain.
   d) If the red "+" data point is removed, would the slope of the line increase or decrease? Explain.

5. **Twins by year 2018** In January 2012, the *New York Times* published a story called "Twin Births in the U.S., Like Never Before," in which they reported a 76% increase in the rate of twin births from 1980 to 2009. Here are the number of twin births each year (per 1000 live births). (www.cdc.gov/nchs/births.htm)

   | Year | Twin Birth (per 1000 live births) | Year | Twin Birth (per 1000 live births) |
   |---|---|---|---|
   | 1980 | 18.92 | 2000 | 29.30 |
   | 1981 | 19.21 | 2001 | 30.12 |
   | 1982 | 19.46 | 2002 | 31.11 |
   | 1983 | 19.86 | 2003 | 31.46 |
   | 1984 | 19.88 | 2004 | 32.15 |
   | 1985 | 20.50 | 2005 | 32.17 |
   | 1986 | 21.30 | 2006 | 32.11 |
   | 1987 | 21.36 | 2007 | 32.19 |
   | 1988 | 21.80 | 2008 | 32.62 |
   | 1989 | 22.41 | 2009 | 33.22 |
   | 1990 | 22.46 | 2010 | 33.15 |
   | 1991 | 23.05 | 2011 | 33.20 |
   | 1992 | 23.35 | 2012 | 33.15 |
   | 1993 | 23.88 | 2013 | 33.7 |
   | 1994 | 24.39 | 2014 | 33.9 |
   | 1995 | 24.86 | 2015 | 33.5 |
   | 1996 | 25.84 | 2016 | 33.4 |
   | 1997 | 26.83 | 2017 | 33.3 |
   | 1998 | 28.08 | 2018 | 32.6 |
   | 1999 | 28.87 | | |

   a) Using the data only up to 2009 (as in the article), find the equation of the regression line for predicting the number of twin births by *Years Since 1980*.
   b) Explain in this context what the slope means.
   c) Predict the number of twin births in the United States for the year 2018. Then compare your prediction to the actual value.
   d) Fit a new regression model to the entire data sequence. Comment on the fit. Now plot the residuals. Are you satisfied with the model? Explain.

6. **Dow Jones 2020** The Dow Jones Industrial Average summarizes the performance of stocks in some of the largest companies on the New York Stock Exchange. A regression of daily Dow prices on fractional year (that is, year plus the fraction of the year for each date) since 2020 for the years 2010 through the middle of 2020 looks like this:

   Response variable is DJIA
   R-squared = 93.2,  s = 1353

   | Variable | Coefficient |
   |---|---|
   | Intercept | −8517.62 |
   | Year fraction | 1733.46 |

a) What is the correlation between *Dow Jones Index* and *Year*?
b) Write the regression equation.
c) Explain in context what the slope of the regression equation means.
d) Here's a scatterplot of the residuals against year. Comment on the model, what it says, and what it failed to say:

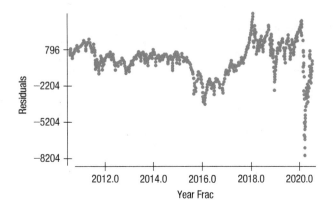

7. **Streams** Biologists studying the effects of acid rain on wildlife collected data from 163 streams in the Adirondack Mountains. They recorded the *pH* (acidity) of the water and the *BCI*, a measure of biological diversity, and they calculated $R^2 = 27\%$. Here's a scatterplot of *BCI* against *pH*:

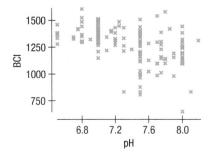

a) What is the correlation between *pH* and *BCI*?
b) Describe the association between these two variables.
c) If a stream has average *pH*, what would you predict about the *BCI*?
d) In a stream where the *pH* is 3 standard deviations above average, what would you predict about the *BCI*?

8. **Manatees 2017** Marine biologists warn that the growing number of powerboats registered in Florida threatens the existence of manatees. The following scatterplot shows data from Florida Fish and Wildlife Conservation Commission (https://myfwc.com/research/manatee/rescue-mortality-response/statistics/mortality/yearly/) and the U.S. Coast Guard Office of Auxiliary and Boating Safety (www.uscgboating.org/library/accident-statistics/Recreational-Boating-Statistics-2017.pdf).

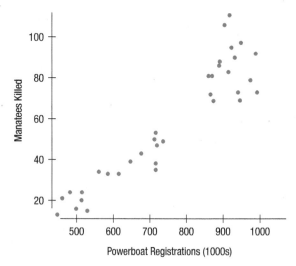

Response variable is Manatees Killed
R-squared = 86.7, s = 10.852

| Variable | Coefficient |
| --- | --- |
| Intercept | −57.129 |
| Registrations | 0.1524 |

a) In this context, why is *Powerboat Registrations* the explanatory variable?
b) Describe the association you see in the scatterplot.
c) Interpret the value of $R^2$.
d) Interpret *s* in this context.
e) Calculate the correlation coefficient *r*. Explain how this supports your description of the association.

9. **Streams II** Exercise 7 examined the correlation between *BCI* and *pH* in streams sampled in the Adirondack Mountains. Here is the corresponding regression model:

Response variable is BCI
R-squared = 27.1%   s = 140.4

| Variable | Coefficient |
| --- | --- |
| Intercept | 2733.37 |
| pH | −197.694 |

a) Write the regression model.
b) What is the interpretation of the coefficient of *pH*?
c) What would you predict the *BCI* would be for a stream with a *pH* of 8.2?

10. **A manatee model 2017** Continue your analysis of the manatee situation from Exercise 8.

a) Create a linear model of the association between *Manatees Killed* and *Powerboat Registrations*.
b) Interpret the slope of your model.
c) Interpret the *y*-intercept of your model.
d) If there were 646,000 powerboat registrations, how many manatees does the model predict would be killed?
e) The data point for 646,000 powerboat registrations has 39 manatees killed. What is the residual for that point and what does it mean?
f) Would it be reasonable to use the fitted regression line to predict the number of manatees that would be killed if there were 1,500,000 powerboat registrations in a year?

**11. Final exam** A statistics instructor created a linear regression equation to predict students' final exam scores from their midterm exam scores. The regression equation was $\widehat{Final} = 10 + 0.9\, Midterm$.

a) If Susan scored a 70 on the midterm, what did the instructor predict for her score on the final?
b) Susan got an 80 on the final. How big is her residual?
c) If the standard deviation of the final was 12 points and the standard deviation of the midterm was 10 points, what is the correlation between the two tests?
d) How many points would someone need to score on the midterm to have a predicted final score of 100?
e) Suppose someone scored 100 on the final. Explain why you can't estimate this student's midterm score from the information given.
f) One of the students in the class scored 100 on the midterm but got overconfident, slacked off, and scored only 15 on the final exam. What is the residual for this student?
g) No other student in the class "achieved" such a dramatic turnaround. If the instructor decides not to include this student's scores when constructing a new regression model, will the $R^2$ value of the regression increase, decrease, or remain the same? Explain.
h) Will the slope of the new line increase or decrease?

**12. Traffic** Highway planners investigated the relationship between traffic *Density* (number of automobiles per mile) and the average *Speed* of the traffic on a moderately large city thoroughfare. The data were collected at the same location at 10 different times over a span of 3 months. They found a mean traffic *Density* of 68.6 cars per mile (cpm) with standard deviation of 27.07 cpm. Overall, the cars' average *Speed* was 26.38 mph, with standard deviation of 9.68 mph. These researchers found the regression line for these data to be
$\widehat{Speed} = 50.55 - 0.352\, Density$.

a) What is the value of the correlation coefficient between *Speed* and *Density*?
b) What percent of the variation in average *Speed* is explained by traffic *Density*?
c) Predict the average *Speed* of traffic on the thoroughfare when the traffic *Density* is 50 cpm.
d) What is the value of the residual for a traffic *Density* of 56 cpm with an observed *Speed* of 32.5 mph?
e) The dataset initially included the point $Density = 125\, cpm, Speed = 55$ mph. This point was considered an outlier and was not included in the analysis. Will the slope increase, decrease, or remain the same if we redo the analysis and include this point?
f) Will the correlation become stronger, weaker, or remain the same if we redo the analysis and include this point (125, 55)?
g) A European member of the research team measured the *Speed* of the cars in kilometers per hour (1 km ≈ 0.62 miles) and the traffic *Density* in cars per kilometer. Find the value of his calculated correlation between speed and density.

**13. Cramming** One Thursday, researchers gave students enrolled in a section of basic Spanish a set of 50 new vocabulary words to memorize. On Friday, the students took a vocabulary test. When they returned to class the following Monday, they were retested—without advance warning. Here are the test scores for the 25 students:

| Fri. | Mon. | Fri. | Mon. | Fri. | Mon. |
|---|---|---|---|---|---|
| 42 | 36 | 48 | 37 | 39 | 41 |
| 44 | 44 | 43 | 41 | 46 | 32 |
| 45 | 46 | 45 | 32 | 37 | 36 |
| 48 | 38 | 47 | 44 | 40 | 31 |
| 44 | 40 | 50 | 47 | 41 | 32 |
| 43 | 38 | 34 | 34 | 48 | 39 |
| 41 | 37 | 38 | 31 | 37 | 31 |
| 35 | 31 | 43 | 40 | 36 | 41 |
| 43 | 32 |  |  |  |  |

a) What is the correlation between *Friday* and *Monday* scores?
b) What does a scatterplot show about the association between the scores?
c) What does it mean for a student to have a positive residual?
d) What would you predict about a student whose *Friday* score was one standard deviation below average?
e) Write the equation of the regression line.
f) Predict the *Monday* score of a student who earned a 40 on Friday.

**14. Cars, horsepower** Can we predict the *Horsepower* of the engine that manufacturers will put in a car by knowing the *Weight* of the car? Here are the regression analysis and residuals plot:

Dependent variable is Horsepower
R-squared = 84.1%

| Variable | Coefficient |
|---|---|
| Intercept | 3.49834 |
| Weight | 34.3144 |

a) Write the equation of the regression line.
b) Do you think the car's *Weight* is measured in pounds or thousands of pounds? Explain.
c) Do you think this linear model is appropriate? Explain.
d) The highest point in the residuals plot, representing a residual of 22.5 horsepower, is for a Chevy weighing 2595 pounds. How much horsepower does this car have?

**15. Colorblind** Although some women are colorblind, this condition is found primarily in men. Why is it wrong to say there's a strong correlation between *Sex* and *Colorblindness*?

**16. Old Faithful again** There is evidence that eruptions of Old Faithful can best be predicted by knowing the duration of the previous eruption.

a) Describe what you see in the scatterplot of *Intervals* between eruptions vs. *Duration* of the previous eruption.

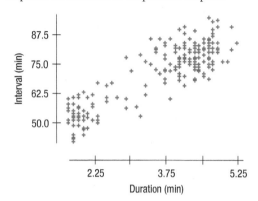

b) Write the equation of the line of best fit. Here's the regression analysis:

Dependent variable is Interval
R-squared = 77.0%   s = 6.16 min

| Variable | Coefficient |
|---|---|
| Intercept | 33.9668 |
| Duration | 10.3582 |

c) Carefully explain what the slope of the line means in this context.
d) Explain what *s* means in this context.
e) How accurate do you expect predictions based on this model to be? Cite statistical evidence.
f) If you just witnessed an eruption that lasted 4 minutes, how long do you predict you'll have to wait to see the next eruption?
g) So you waited, and the next eruption came in 79 minutes. Use this as an example to define a residual.

**17. Crocodile lengths** The ranges inhabited by the Indian gharial crocodile and the Australian saltwater crocodile overlap in Bangladesh. Suppose a very large crocodile skeleton is found there, and we wish to determine the species of the animal. Wildlife scientists have measured the lengths of the heads and the complete bodies of several crocs (in centimeters) of each species, creating the regression analyses below:

**Indian Crocodile**
Dependent variable is IBody
R-squared = 97.2%

| Variable | Coefficient |
|---|---|
| Intercept | −69.3693 |
| IHead | 7.40004 |

**Australian Crocodile**
Dependent variable is ABody
R-squared = 98.1%

| Variable | Coefficient |
|---|---|
| Intercept | −21.3429 |
| AHead | 7.82761 |

a) Do the associations between the sizes of the heads and bodies of the two species appear to be strong? Explain.
b) In what ways are the two relationships similar? Explain.
c) What is different about the two models? What does that mean?
d) The crocodile skeleton found had a head length of 62 cm and a body length of 380 cm. Which species do you think it was? Explain why.

**18. How old is that tree?** One can determine how old a tree is by counting its rings, but that requires either cutting the tree down or extracting a sample from the tree's core. Can we estimate the tree's age simply from its diameter? A forester measured 27 trees of the same species that had been cut down, and counted the rings to determine the ages of the trees.

a) The correlation between *Diameter* and *Age* is 0.89. Does this suggest that a linear model may be appropriate? Explain.

Here are the scatterplot and residual plot of the data for these 27 trees:

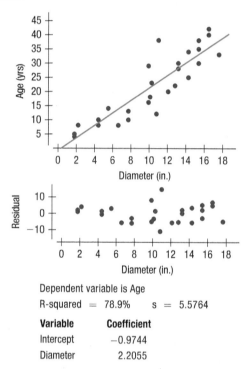

Dependent variable is Age
R-squared = 78.9%   s = 5.5764

| Variable | Coefficient |
|---|---|
| Intercept | −0.9744 |
| Diameter | 2.2055 |

b) Explain why a linear model is probably not appropriate.
c) If you used this model, would it generally overestimate or underestimate the ages of very large trees? Explain.

**19. Improving trees** In the last exercise, you saw that the linear model had some deficiencies. Perhaps the cross-sectional area of a tree would be a better predictor of its age. Because area is measured in square units, we have re-expressed the data by squaring the diameters.

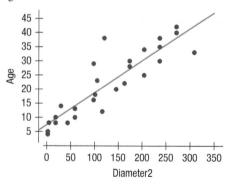

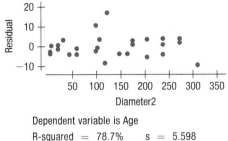

Dependent variable is Age
R-squared = 78.7%    s = 5.598

| Variable | Coefficient |
|---|---|
| Intercept | 7.2396 |
| Diameter2 | 0.1130 |

a) Based on the scatterplot and residual plot, is the squared diameter a better predictor of *Age* than *Diameter*? Explain.
b) A forester wants to estimate the age of a tree and measures the diameter of the trunk to be 16.5 inches. What would you estimate its age to be?
c) How far off would you expect this prediction to be?
d) Explain why the forester should not use this model to estimate the age of a tree that is 26 inches in diameter.

**20. Big screen** An electronics website collects data on the size of new HD flat panel televisions (measuring the diagonal of the screen in inches) to predict the cost (in hundreds of dollars). Which of these is most likely to be the slope of the regression line: 0.03, 0.3, 3, 30? Explain.

**T 21. Happiness 2018** Finland is repeatedly reported to be the happiest country in the world. This is based on the average score of people's response to the following question: *Please imagine a ladder with steps numbered from zero at the bottom to 10 at the top. The top of the ladder represents the best possible life for you and the bottom of the ladder represents the worst possible life for you. On which step of the ladder would you say you personally feel you stand at this time?*

What contributes to people's happiness? The scatterplot below shows the relationship between *Happiness* and *Income per Capita* for the countries whose data was available from 2018.

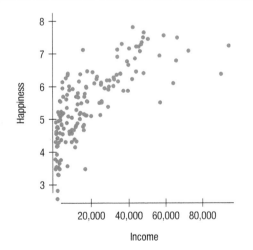

a) Describe the association shown in the graph.
b) Does a linear model seem appropriate for this relationship?

Below are a scatterplot, residuals plot, and regression analysis for *Happiness* vs. *log(Income)*.

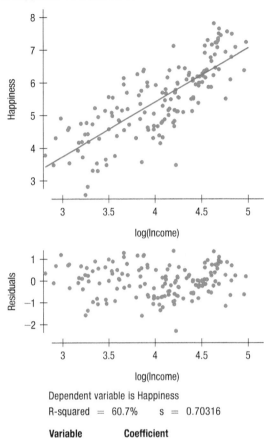

Dependent variable is Happiness
R-squared = 60.7%    s = 0.70316

| Variable | Coefficient |
|---|---|
| Intercept | 1.2714 |
| log(Income) | 1.6673 |

c) Comment on the effectiveness of the transformation.
d) What does *s* mean in this context?
e) What percent of the variation in *Happiness* is explained by the model based on *log(Income)*?
f) What are some other factors that might be associated with *Happiness* in a country?

**T 22. Happier and happier?** The trend in happiness you saw in the previous exercise shows that higher happiness is associated with more income per capita, but will it ever stop?

a) Explain why no model based on the data shown in Exercise 21 could reliably predict the *Happiness* of a country with a per capita income of $150,000.
b) Explain how you would create a model that would predict the amount of *Income per capita* that would result in a perfect 10 on the happiness scale.
c) Comment on the reliability of such a prediction.

**23. Tips** It's commonly believed that people use tips to reward good service. A researcher for the hospitality industry examined tips and ratings of service quality from 2645 dining parties at 21 different restaurants. The correlation between ratings of service and tip percentages was 0.11. (M. Lynn and M. McCall, "Gratitude and Gratuity," *Journal of Socio-Economics* 29: 203–214)

a) Describe the relationship between *Quality of Service* and *Tip Size*.
b) Find and interpret the value of $R^2$ in this context.

**24. U.S. cities** Data from 50 large U.S. cities show the mean *January Temperature* and the *Latitude*. Describe what you see in the scatterplot.

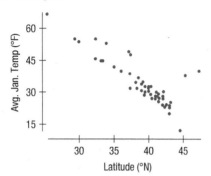

**25. Winter in the city** Summary statistics for the data relating the *Latitude* and average *January Temperature* for 55 large U.S. cities are given below.

| Variable | Mean | StdDev |
|---|---|---|
| Latitude | 39.02 | 5.42 |
| JanTemp | 26.44 | 13.49 |
| Correlation = −0.848 | | |

a) What percentage of the variation in *January Temperature* can be explained by variation in *Latitude*?
b) What is indicated by the fact that the correlation is negative?
c) Write the equation of the line of regression for predicting average *January Temperature* from *Latitude*.
d) Explain what the slope of the line means.
e) Do you think the *y*-intercept is meaningful? Explain.
f) The latitude of Denver is 40°N. Predict the mean January temperature there.
g) What does it mean if the residual for a city is positive?

**26. Depression** The September 1998 issue of the *American Psychologist* published an article by Kraut et al. that reported on an experiment examining "the social and psychological impact of the Internet on 169 people in 73 households during their first 1 to 2 years online." In the experiment, 73 households were offered free Internet access for 1 or 2 years in return for allowing their time and activity online to be tracked. The members of the households who participated in the study were also given a battery of tests at the beginning and again at the end of the study. The conclusion of the study made news headlines: Those who spent more time online tended to be more depressed at the end of the experiment. Although the paper reports a more complex model, the basic result can be summarized in the following regression analysis of *Depression* (at the end of the study, in "depression scale units") vs. *Internet Use* (in mean hours per week):

Dependent variable is Depression
R-squared = 4.6%   s = 0.4563

| Variable | Coefficient |
|---|---|
| Intercept | 0.5655 |
| Internet use | 0.0199 |

The news reports about this study clearly concluded that using the Internet causes depression. Discuss whether such a conclusion can be drawn from this regression. If so, discuss the supporting evidence. If not, say why not.

**T 27. Olympic jumps 2016** How are Olympic performances in various events related? The plot shows winning long-jump and high-jump distances, in meters, for the Summer Olympics from 1912 through 2016:

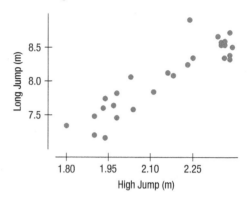

a) Describe the association.
b) Do long-jump performances somehow influence the high-jumpers? How do you account for the relationship you see?
c) The correlation for the given scatterplot is 0.910. If we converted the jump lengths to centimeters by multiplying by 100, would that make the actual correlation higher or lower?
d) What would you predict about the long jump in a year when the high-jumper jumped one standard deviation better than the average high jump?

**T 28. Modeling jumps 2016** Here are the summary statistics for the Olympic jumps displayed in the previous exercise.

| Event | Mean | StdDev |
|---|---|---|
| High Jump | 2.15667 | 0.195271 |
| Long Jump | 8.06222 | 0.507606 |
| Correlation = 0.90992 | | |

a) Write the equation of the line of regression for estimating *High Jump* from *Long Jump*.
b) Interpret the slope of the line.
c) In a year when the long jump is 8.9 m, what high jump would you predict?
d) Why can't you use this line to estimate the long jump for a year when you know the high jump was 2.2 m?
e) Write the equation of the line you need to make that prediction.

**29. French** Consider the association between a student's score on a French vocabulary test and the weight of the student. What direction and strength of correlation would you expect in each of the following situations? Explain.

a) The students are all in third grade.
b) The students are in third through twelfth grades in the same school district.
c) The students are in tenth grade in France.
d) The students are in third through twelfth grades in France.

**30. Twins** Twins are often born at less than 9 months gestation. The graph from the *Journal of the American Medical Association (JAMA)* shows the rate of preterm twin births in the United States over the past 20 years. In this study, *JAMA* categorized the level of prenatal medical care received in each case as inadequate, adequate, or intensive. (Source: *JAMA* 284[2000]: 335–341)

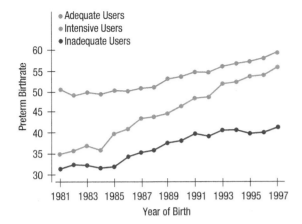

a) Describe similarities in the trend in preterm twin births, depending on the level of prenatal medical care the mother received.
b) Describe any differences you see in this trend, depending on the level of prenatal care.
c) Should expectant parents be advised to cut back on the level of medical care they seek in the hope of avoiding preterm births? Explain.

**31. Lunchtime** Does how long toddlers sit at the lunch table help predict how much they eat? The table and graph show the number of minutes the kids stayed at the table and the number of calories they consumed. Create a model for these data and interpret the slope.

| Calories | Time | Calories | Time |
|---|---|---|---|
| 472 | 21.4 | 450 | 42.4 |
| 498 | 30.8 | 410 | 43.1 |
| 465 | 37.7 | 504 | 29.2 |
| 456 | 33.5 | 437 | 31.3 |
| 423 | 32.8 | 489 | 28.6 |
| 437 | 39.5 | 436 | 32.9 |
| 508 | 22.8 | 480 | 30.6 |
| 431 | 34.1 | 439 | 35.1 |
| 479 | 33.9 | 444 | 33.0 |
| 454 | 43.8 | 408 | 43.7 |

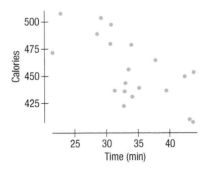

**32. Gasoline** After clean-air regulations dictated the use of unleaded gasoline, the supply of leaded gas in New York state diminished. The following table was given on the August 2001 New York State Math B exam, a statewide achievement test for high school students:

| Year | 1984 | 1988 | 1992 | 1996 | 2000 |
|---|---|---|---|---|---|
| Gallons (1000's) | 150 | 124 | 104 | 76 | 50 |

a) Create a linear model and predict the number of gallons available in 2015. Comment.
b) The exam then asked students to estimate the year when leaded gasoline will first become unavailable, expecting them to use the model from part a to answer the question. Explain why that method is incorrect.
c) Create a model that *would* be appropriate for that task, and make the estimate.
d) The 2015 prediction from part (a), though based on an inappropriate model, is still fairly accurate in this case. Why?

**33. Tobacco and alcohol** Are people who use tobacco products more likely to consume alcohol? Here are data on household spending (in pounds) taken by the British government on 11 regions in Great Britain. Do tobacco and alcohol spending appear to be related? What questions do you have about these data? What conclusions can you draw?

| Region | Alcohol | Tobacco |
|---|---|---|
| North | 6.47 | 4.03 |
| Yorkshire | 6.13 | 3.76 |
| Northeast | 6.19 | 3.77 |
| East Midlands | 4.89 | 3.34 |
| West Midlands | 5.63 | 3.47 |
| East Anglia | 4.52 | 2.92 |
| Southeast | 5.89 | 3.20 |
| Southwest | 4.79 | 2.71 |
| Wales | 5.27 | 3.53 |
| Scotland | 6.08 | 4.51 |
| Northern Ireland | 4.02 | 4.56 |

**34. Williams football 2020** The Sears Cup was established in 1993 to honor institutions that maintain a broad-based athletic program, achieving success in many sports, both men's and women's. In the years following its Division III inception in 1995, the cup was won by Williams College 22 of 24 years. Why did the football team win so much? Was it because they were heavier than their opponents? The table shows the average team weights for selected years from 1973 to 2019.

| Year | Weight (lb) | Year | Weight (lb) |
|---|---|---|---|
| 1973 | 185.5 | 1989 | 202.9 |
| 1975 | 182.4 | 1991 | 206.0 |
| 1977 | 182.1 | 1993 | 198.7 |
| 1979 | 191.1 | 2007 | 218.7 |
| 1981 | 189.4 | 2013 | 219.6 |
| 1983 | 192.0 | 2019 | 217.8 |
| 1987 | 196.9 | | |

a) Fit a straight line to the relationship between *Weight* and *Year*.
b) Does a straight line seem reasonable?
c) Predict the average weight of the team for the year 2025. Does this seem plausible?
d) What about the prediction for the year 2103? Explain.
e) What about the prediction for the year 3003? Explain.

**35. Models** For each model, determine the predicted value of *y* when $x = 10$.
a) $\hat{y} = 2 + 0.8 \ln x$
b) $\log \hat{y} = 5 - 0.23x$
c) $\dfrac{1}{\sqrt{\hat{y}}} = 17.1 - 1.66x$

**T 36. Williams vs. Texas 2020** Here are the average weights of the football team for the University of Texas for various years in the 20th century:

| Year | 1905 | 1919 | 1932 | 1945 | 1955 | 1965 | 2008 | 2013 | 2020 |
|---|---|---|---|---|---|---|---|---|---|
| Weight (lb) | 164 | 163 | 181 | 192 | 195 | 199 | 229 | 228 | 234 |

a) Fit a straight line to the relationship of *Weight* by *Year* for Texas football players.
b) According to these models, in what year will the predicted weight of the Williams College team from Exercise 34 first be more than the weight of the University of Texas team?
c) Do you believe this? Explain.

**37. Vehicle weights** As we saw in Chapter 6, the Minnesota Department of Transportation hoped that they could measure the weights of big trucks without actually stopping the vehicles by using a newly developed "weigh-in-motion" scale. After installation of the scale, a study was conducted to find out whether the scale's readings correspond to the true weights of the trucks being monitored. The scatterplot for the data they collected showed the association to be approximately linear with $R^2 = 93\%$. Their regression equation is $Weight = 10.85 + 0.64\ Scale$, where both the scale reading and the predicted weight of the truck are measured in thousands of pounds.
a) Estimate the weight of a truck if this scale reads 31,200 pounds.
b) If that truck actually weighed 32,120 pounds, what was the residual?
c) If the scale reads 35,590 pounds, and the truck has a residual of −2440 pounds, how much does it actually weigh?
d) If the police plan to use this scale to issue tickets to trucks that appear to be overloaded, will negative or positive residuals be a greater problem? Explain.

**T 38. Companies** How are a company's profits related to its sales? Let's examine data from 71 large U.S. corporations. All amounts are in millions of dollars.
a) Histograms of *Profits* and *Sales* and histograms of the logarithms of *Profits* and *Sales* are as shown. Why are the re-expressed data better for regression?

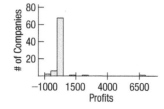

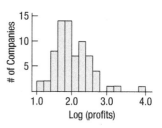

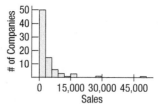

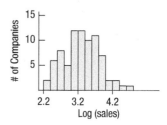

b) Here are the scatterplot and residuals plot for the regression of logarithm of *Profits* vs. log of *Sales*. Do you think this model is appropriate? Explain.

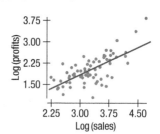

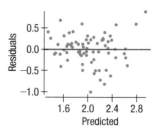

c) Here's the regression analysis. Write the equation.

Dependent variable is Log Profit
R-squared = 48.1%

| Variable | Coefficient |
|---|---|
| Intercept | −0.106259 |
| LogSales | 0.647798 |

d) Use your equation to estimate profits earned by a company with sales of 2.5 billion dollars. (That's 2500 million.)

**39. Down the drain** Most water tanks have a drain plug so that the tank may be emptied when it's to be moved or repaired. How long it takes a certain size of tank to drain depends on the size of the plug, as shown in the table. Create a model.

| Plug Dia (in.) | $\frac{3}{8}$ | $\frac{1}{2}$ | $\frac{3}{4}$ | 1 | $1\frac{1}{4}$ | $1\frac{1}{2}$ | 2 |
|---|---|---|---|---|---|---|---|
| Drain Time (min) | 140 | 80 | 35 | 20 | 13 | 10 | 5 |

**40. Chips** A start-up company has developed an improved electronic chip for use in laboratory equipment. The company needs to project the manufacturing cost, so it develops a spreadsheet model that takes into account the purchase of production equipment, overhead, raw materials, depreciation, maintenance, and other business costs. The spreadsheet estimates the cost of producing 10,000 to 200,000 chips per year, as seen in the table. Develop a regression model to predict *Costs* based on the *Level* of production.

| Chips Produced (1000s) | Cost per Chip ($) | Chips Produced (1000s) | Cost per Chip ($) |
|---|---|---|---|
| 10 | 146.10 | 90 | 47.22 |
| 20 | 105.80 | 100 | 44.31 |
| 30 | 85.75 | 120 | 42.88 |
| 40 | 77.02 | 140 | 39.05 |
| 50 | 66.10 | 160 | 37.47 |
| 60 | 63.92 | 180 | 35.09 |
| 70 | 58.80 | 200 | 34.04 |
| 80 | 50.91 | | |

# PRACTICE EXAM

## I. MULTIPLE CHOICE

(Questions 1–3) Based on data collected over several sessions, a statistically minded trainer of office typists modeled the linear relationship between the number of hours of training a typist receives and the typist's speed (in words per minute) with the equation $\widehat{Speed} = 10.6 + 5.4\ Hour$.

1. Which of these statements best interprets this equation?
   A) Typists increase their speed by 10.6 wpm for every 5.4 hours of training.
   B) Typists increase their speed by 5.4 wpm for every 10.6 hours of training.
   C) A typist who trains for an additional hour will benefit with a speed increase of 5.4 wpm.
   D) On average, typists tend to increase their speed by roughly 5.4 wpm for every hour of training.
   E) For every 5.4 hours of training, typists can increase their speed from 10.6 wpm to faster.

2. Which is the best interpretation of the *y*-intercept for this model?
   A) People who can't type need about 10.6 hours of training.
   B) Before undergoing this training, typists' average speed was about 10.6 words per minute.
   C) The *y*-intercept is meaningless here because no one types at 0 wpm.
   D) The *y*-intercept is meaningless here because none of the typists had 0 hours of training.
   E) In regression models, the slope has meaning, but not the *y*-intercept.

3. After some training, one of the typists was told that the speed he attained had a residual of 4.3 words per minute. How should he interpret this?
   A) He types slower than the model predicted, given the amount of time he spent training.
   B) He types faster than the model predicted, given the amount of time he spent training.
   C) He can't interpret his residual without also knowing the correlation.
   D) He can't interpret his residual without also knowing the size of other people's residuals.
   E) He can't interpret his residual without also knowing the standard deviation of the residuals.

4. The Bureau of Labor Statistics looked at the association between students' GPAs in high school ($gpa\_HS$) and their freshmen GPAs at a University of California school ($gpa\_U$). The resulting least-squares regression equation is $\widehat{gpa\_U} = 0.22 + 0.72 gpa\_HS$. Calculate the residual for a student with a 3.8 in high school who achieved a freshman GPA of 3.5.
   A) −0.844    B) −0.544    C) 2.956
   D) 0.544     E) 0.844

5. In April of 2021, the Centers for Disease Control and Prevention announced that the rates of sexually transmitted diseases (STDs) have reached a record high for the sixth consecutive year, with a nearly 30% increase in reportable STDs between 2015 and 2019. Which of these conclusions is an example of extrapolation in this context?
   A) There was an increasing trend in STD rates at the time of this study.
   B) Time is an explanatory variable in the change STD rates.
   C) By 2025, STD rates will be 70% higher and set new records.
   D) There is a linear relationship between year and STD rates.
   E) None of these is an example of extrapolation.

6. An engineer studying the performance of a certain type of bolt predicts the failure rate (bolts per 1000) from the load (in pounds) using the model $\widehat{\log(Fail)} = 1.04 + 0.0013 Load$. If these bolts are subjected to a load of 600 pounds, what failure rate should we expect?
   A) 0.26    B) 0.60    C) 1.82
   D) 6.17    E) 66.07

7. A researcher analyzing some data created a linear model with $R^2 = 94\%$ and having the residuals plot seen here. What should she conclude?

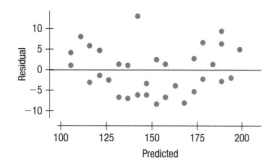

A) The linear model is appropriate, because about the half the residuals are positive and half negative.
B) The linear model is appropriate, because the value of $R^2$ is quite high.
C) The linear model is not appropriate, because the value of $R^2$ is not high enough.
D) The linear model is not appropriate, because the residuals plot shows curvature.
E) The linear model is not appropriate, because the residuals plot identifies an outlier.

8. Researchers at UC San Francisco discovered that high plasma levels of vitamins B, C, D, and E are associated with better cognitive performance. "Each standard deviation higher plasma level for these vitamins predicted a global cognitive score 0.28 standard deviations better," the researchers reported. Which value are the researchers interpreting in this statement?

A) the correlation coefficient between plasma level and cognitive score
B) the $y$-intercept of the regression model predicting cognitive score from plasma level
C) the slope of the regression model predicting cognitive score from plasma level
D) the standard deviation of the regression model's residuals
E) $R^2$ for the regression model

9. This graph shows the relationship the number of days since the NBA season began and the number of injuries, over the course of three different seasons. In 2011–12, the season was shortened by a labor strike.

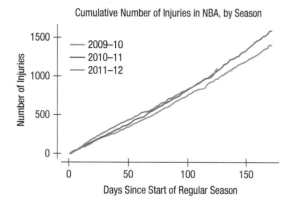

Of statements A–D, which of the following is NOT a correct conclusion that can be drawn from this graph?

A) There is a roughly linear relationship between days since the start of the season and the number of injuries.
B) At first the rate of injuries was higher during the strike-shortened season.
C) Had the strike-shortened season continued there would have been more injuries than in the other two seasons.
D) While the strike-shortened season had more injuries initially, we cannot know for certain if something about the upcoming strike caused the difference or if it was attributable to other variables.
E) All of A–D are correct.

10. Which of statements A–D is true?

A) An influential point always has a large residual.
B) An influential point changes the slope of the regression equation.
C) An influential point decreases the value of $R^2$.
D) An influential point does not affect the $y$-intercept.
E) Statements A–D are all false.

(Questions 11–12) The segmented bar charts below depict the data from the NAAL (National Assessment of Adult Literacy) conducted in 2003.

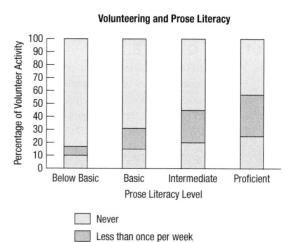

11. Which of the following is greatest?

A) The number of people who volunteer once per week or more and test Below Basic on Prose Literacy.
B) The number of people who volunteer less than once per week and test Basic on Prose Literacy.
C) The number of people who never volunteer and test Proficient on Prose Literacy.
D) The number of people who volunteer less than once per week and test Intermediate in Prose Literacy.
E) It is impossible to determine which is greatest without knowing the actual number of people at each literacy level.

12. Based on the segmented bar graphs, does there appear to be an association between volunteerism and literacy level?

A) Yes, because all three bars have the same number of segments.
B) Yes, because all three bars have the same height.
C) Yes, because the corresponding segments of the three bars have different heights.
D) No, because the corresponding segments of the three bars have different heights.
E) No, because the sums of the 3 proportions in each bar are identical.

13. A TV meteorologist's end-of-year analysis of the local weather showed that among all the years for which records had been kept, the past year's average temperature had a $z$-score of 2.8. What does that mean?

   A) The past year's average temperature was 2.8° higher than the historical mean.
   B) The past year's average temperature was 2.8 standard deviations above the historical mean.
   C) The past year's average temperature was 2.8% higher than the historical mean.
   D) The past year's temperatures had a standard deviation of 2.8°.
   E) The past year had 2.8 times as many days with above average temperatures as is typical for that area.

14. In Statsville there's a city-wide speed limit of 30 mph. If you are caught speeding the fine is $100 plus $10 for every mile per hour you were over the speed limit. For example, if you're ticketed for going 45 mph, your fine is $100 + 10(45 − 30) = $250$. Last month all the drivers who were fined for speeding averaged 42 mph with a standard deviation of 7 mph. What were the mean and standard deviation of the fines?

   A) $120 and $70
   B) $220 and $7
   C) $220 and $70
   D) $220 and $170
   E) $420 and $70

15. Among those Statsville drivers fined for speeding, the fastest 10% were caught exceeding how many miles per hour?

   A) 37.0
   B) 48.3
   C) 51.0
   D) 58.3
   E) It cannot be determined from the information given.

## II. FREE RESPONSE

1. A diligent Statistics student recorded the length of his faithful #2 pencil as he worked away on his homework. He discovered a strong linear relationship between the number of hours that he worked and the length of his pencil. Here is the regression analysis for these data.

   Dependent variable is Length (cm)
   R-squared = 92.3%   R-squared(adj) = 89.5%

   |  | Coefficient | se | t ratio | p value |
   |---|---|---|---|---|
   | Constant | 17.047 | 0.128 | 23.58 | <0.0001 |
   | Time (hr) | −1.914 | 0.047 | 35.28 | <0.0001 |

   a) Write the equation of the least squares regression line.
   b) Interpret $R^2$ in this context.
   c) Interpret the equation in this context.
   d) This student's girlfriend tried out his model on a pencil she had used for 5 hours and found a residual of −0.88 cm. How long was her pencil at that time?
   e) Should she have expected this model to describe the rate for her pencils? Why or why not?

2. Energy drinks come in different-sized packages: pouches, small bottles, large bottles, twin-packs, 6-packs, and so on. How is the price related to the amount of beverage? Data were collected on a variety of packages, resulting in the following regression model.

   Dependent variable is Price
   R-squared = 82.8%   s = 0.26

   | Variable | Coefficient |
   |---|---|
   | Intercept | 1.21424 |
   | Volume | 0.01745 |

   a) Interpret the value of $r$ in context.
   b) Write the equation for the least squares regression line.
   c) Interpret the slope of the least squares regression line in context. Include the proper units in your answer.
   d) Interpret the y-intercept of the least squares regression line in context.
   e) Explain what $s$ means in context.

3. The Pew Research Center conducted two surveys, one in December 2011 and another in November 2012, asking people about their reading habits. Pew reported the percentage of people who read at least one e-book in the past year, given that they had read at least one book. Those percentages, broken down by age group, are shown in the table below.

   | Date of Poll | Age Group | | | | |
   |---|---|---|---|---|---|
   |  | 16–17 | 18–29 | 30–49 | 50–64 | 65+ |
   | December 2011 | 13 | 25 | 25 | 19 | 12 |
   | November 2012 | 28 | 31 | 41 | 23 | 20 |

   a) Create an appropriate graphical display that allows a comparison of responses between the two years and also among the different age groups.
   b) Write a few sentences comparing e-book readership in the two time periods.
   c) Is there an association between age and the growth in e-book readership? Use evidence from the table or your graph to justify your answer.

# 10
# Understanding Randomness

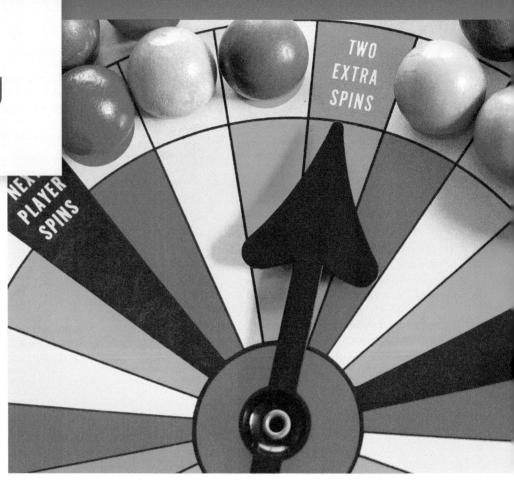

> The most decisive conceptual event of twentieth century physics has been the discovery that the world is not deterministic.... A space was cleared for chance.
>
> —Ian Hocking,
> The Taming of Chance

We all know what it means for something to be random. Or do we?[1] Many children's games rely on chance outcomes. Rolling dice, spinning spinners, and shuffling cards all select at random. Adult games use randomness as well, from card games to lotteries to Bingo. What's the most important aspect of the randomness in these games? It must be fair.

What is it about **random** selection that makes it seem fair? It's really two things. First, nobody can guess the outcome before it happens. Second, when we want things to be fair, usually some underlying set of outcomes will be equally likely (although in many games, some combinations of outcomes are more likely than others).

Randomness is not always what we might think of as "at random." Random outcomes have a lot of structure, especially when viewed in the long run. You can't predict how a fair coin will land on any single toss, but you're pretty confident that if you flipped it thousands of times you'd see about 50% heads. As we will see, randomness is essential in Statistics. Statisticians don't think of randomness as the annoying tendency of things to be unpredictable or haphazard. Statisticians use randomness as a tool. In fact, without deliberately applying randomness, we couldn't do most of Statistics, and this book would stop right about here.[2]

But truly random values are surprisingly hard to get. Just to see how fair humans are at selecting, pick a number at random from the top of the next page. Go ahead. Turn the page, look at the numbers quickly, and pick a number at random.

Ready?

Go.

---

[1] Don't say "random" when you mean "unexpected," as in "I got a random call from Jose last night." *Really*? Your friend Jose was dialing random numbers and just happened to phone *you*? Ri-i-i-i-ight!

[2] Don't get your hopes up.

# 1 2 3 4

## It's Not Easy Being Random

> The generation of random numbers is too important to be left to chance.
>
> —Robert R. Coveyou,
> Oak Ridge National Laboratory

Did you pick 3? If so, you've got company. Almost 75% of all people pick the number 3. About 20% pick either 2 or 4. If you picked 1, well, consider yourself a little different. Only about 5% choose 1. Psychologists have proposed reasons for this phenomenon, but for us, it simply serves as a lesson that we've got to find a better way to choose things at random.

So how should we **generate random numbers**? It's surprisingly difficult to get random values even when they're equally likely. Computers, calculators, and today smartphones have become a way to generate random numbers. Even though they often do much better than humans, they can't generate truly random numbers either. Start a program or app from the same place, and it will always follow exactly the same path, so such generated numbers are not truly random. Technically, "random" numbers generated this way are *pseudorandom* numbers. Fortunately, pseudorandom values are virtually indistinguishable from truly random numbers, and that's usually good enough.

There *are* ways to generate random numbers so that they are both equally likely and truly random. There are published tables of carefully generated random numbers.[3] Or, we can find genuinely random digits on the Internet. Such websites use methods like timing the decay of a radioactive element to generate truly random digits.[4] A string of random digits might look like this:

```
2217726304387410092537086270581997622725849795907032825001108963
3217535822643800292254644943760642389043766557204107354186024508
8906427308645681412198226653885873285801699027843110380420067664
8740522639824530519902027044464984322000946238678577902639002954
8887003319933147508331265192321413908608674496383528968974910533
6944182713168919406022181281304751019321546303870481407676636740
6070204916508913632855351361361043794293428486909462881431793360
7706356513310563210508993624272872250535395513645991015328128202
```

The best ways we know to generate data that give a fair and accurate picture of the world rely on randomness, and the ways in which we draw conclusions from those data depend on randomness, too. If this sounds familiar to you, it should. In previous chapters we've shared with you some simulations we've done that use randomness to consider things like "accidental correlation." In this chapter we're going to let you take the reins and start designing and conducting simulations yourself.

A standard deck of playing cards, like the ones used in bridge and many other card games, consists of 52 cards. There are numbered cards (2 through 10), and face cards (Jack, Queen, King, Ace) whose value depends on the game you are playing. Each card is also marked by one of four suits (clubs, diamonds, hearts, or spades) whose significance is also game-specific.

---

[3]Believe it or not, in the mid-1900s scientists often had entire *books* on their shelves that contained nothing but tables of random digits! You'll find a single table of random digits like that in the back of this book.
[4]For example, www.random.org or www.randomnumbers.info

> **THE ACHILLES' HEEL OF PSEUDORANDOMNESS**
> When people play poker using playing cards, the very visible act of shuffling the deck assures all players that the cards dealt will be unpredictable. But poker machines, found in many U.S. casinos, typically shuffle the virtual deck of cards used in the game using *pseudo*random numbers. These not-truly-random numbers seem unpredictable, but in fact are generated using sophisticated hidden computer code; if you know the code, you can predict all the numbers. In 2014, a group of Russian hackers figured out this hidden code for a collection of poker machines in the United States. Four of them were soon earning about $250,000 a week between them by placing one-cent bets when they knew they would lose and $100 bets when they knew they would win. Interestingly, it was statistics that helped the casinos—and later the cops—detect this behavior! Some of those involved went to jail.[5]

# Let's Simulate!

Suppose a cereal manufacturer puts pictures of famous athletes on cards in boxes of cereal in the hope of boosting sales. The manufacturer announces that 20% of the boxes contain a picture of champion gymnast Simone Biles, 30% a picture of soccer star Megan Rapinoe, and the rest a picture of tennis pro Serena Williams. You want all three pictures. How many boxes of cereal do you expect to have to buy in order to get the complete set?

How can we answer questions like this? Well, one way is to buy hundreds of boxes of cereal to see what might happen. But let's not. Instead, we'll consider using a random model. Why random? When we pick a box of cereal off the shelf, we don't know what picture is inside. We'll assume that the pictures are randomly placed in the boxes and that the boxes are distributed randomly to stores around the country. Why a model? Because we won't actually buy the cereal boxes. We can't afford all those boxes and we don't want to waste food. So we need an imitation of the real process that we can manipulate and control. In short, we're going to simulate reality.

A **simulation** mimics reality by using random numbers to represent the outcomes of real events. Just as pilots use flight simulators to learn about and practice real situations, we can learn a great deal about the real events by carefully modeling the randomness and analyzing the simulation results.

The question we've asked is how many boxes do you expect to buy to get a complete card collection. But we can't answer our question by completing a card collection just once. We want to understand the *typical* number of boxes to open, how that number varies, and, often, the shape of the distribution. So we'll have to do this over and over. We call each time we obtain a simulated answer to our question a **trial**.

For the sports cards, a trial's outcome is the number of boxes. We'll need at least 3 boxes to get one of each card, but with really bad luck, you could empty the shelves of several supermarkets before finally finding the card you lack to get all 3. So, the possible outcomes of a trial are 3, 4, 5, or lots more. But our simulation can't simply pick one of those numbers at random, because they're not equally likely. We'd be surprised if we only needed 3 boxes to get all the cards, but we'd probably be even more surprised to find that it took exactly 7,359 boxes. In fact, the reason we're doing the simulation is that it's hard to guess how many boxes we'd expect to have to open.

## Building a Simulation

We know how to find equally likely random digits—roll a ten-sided die, for example, or look in a random digits table, or use a computer program. How can we get from there to simulating the trial outcomes? We know the relative frequencies of the cards: 20% Simone, 30% Megan, and 50% Serena. So, we can interpret the digits 0 and 1 as finding Simone;

> **IT'S ALL RANDOM!**
> Modern physics has shown that randomness is not just a mathematical game; it is fundamentally the way the universe works.
> *Regardless of improvements in data collection or in computer power, the best we can ever do, according to quantum mechanics . . . is predict the probability that an electron, or a proton, or a neutron, or any other of nature's constituents, will be found here or there. Probability reigns supreme in the microcosmos.*
> —Brian Greene, *The Fabric of the Cosmos: Space, Time, and the Texture of Reality* (p. 91)

---
[5]www.abc.net.au/news/science/2017-06-13/dr-karl-how-russian-cheats-beat-the-pokies/8607598

2, 3, and 4 as finding Megan; and 5 through 9 as finding Serena to simulate opening one box. Opening one box is the basic building block, called a **simulation component**. But the component's outcome isn't the result we want. We need to observe a sequence of components until our card collection is complete. The *trial's* outcome is called the **response variable**; for this simulation that's the *number* of components (boxes) in the sequence.

Let's look at the steps for making a simulation:

**Specify how to model a component outcome using equally likely random digits:**

1. **Identify the component to be repeated.** In this case, our component is the opening of a box of cereal.
2. **Explain how you will model the component's outcome.** The digits from 0 to 9 are equally likely to occur. Because 20% of the boxes contain Simone's picture, we'll use 2 of the 10 digits to represent that outcome. Three of the 10 digits can model the 30% of boxes with Megan's cards, and the remaining 5 digits can represent the 50% of boxes with Serena. One possible assignment of the digits, then, is

    0, 1 Simone   2, 3, 4 Megan   5, 6, 7, 8, 9 Serena.

**Specify how to simulate trials:**

3. **Explain how you will combine the components to model a trial.** We pretend to open boxes (repeat components) until our collection is complete. We do this by looking at each random digit and indicating what picture it represents. We continue until we've found all three.
4. **State clearly what the response variable is.** What are we interested in? We want to find out the number of boxes it might take to get all three pictures.

**Put it all together to run the simulation:**

5. **Run several trials.** For example, consider the third line of random digits shown earlier (p. 277):

    89064273086456814121982266538858732858016990278431103804200670664.

    Let's see what happens.

    The first random digit, 8, means you get Serena's picture. So the first component's outcome is Serena. The second digit, 9, means Serena's picture is also in the next box. Continuing to interpret the random digits, we get Simone's picture (0) in the third, Serena's (6) again in the fourth, and finally Megan (4) on the fifth box. Because we've now found all three pictures, we've finished one trial of our simulation. This trial's outcome is 5 boxes.

    Now we keep going, running more trials by looking at the rest of our line of random digits:

    89064 2730 8645681 41219 822665388587328580 169902 78431 1038 042006 7664.

    It's good to create a chart to keep track of what happens:

| Trial Number | Component Outcomes | Trial Outcomes: $y$ = Number of Boxes |
|---|---|---|
| 1 | 89064 = Serena, Serena, Simone, Serena, Megan | 5 |
| 2 | 2730 = Megan, Serena, Megan, Simone | 4 |
| 3 | 8645681 = Serena, Serena, Megan, ..., Simone | 7 |
| 4 | 41219 = Megan, Simone, Megan, Simone, Serena | 5 |
| 5 | 822665388587328580 = Serena, Megan, ..., Simone | 18 |
| 6 | 169902 = Simone, Serena, Serena, Serena, Simone, Megan | 6 |
| 7 | 78431 = Serena, Serena, Megan, Megan, Simone | 5 |
| 8 | 1038 = Simone, Simone, Megan, Serena | 4 |
| 9 | 042006 = Simone, Megan, Megan, Simone, Simone, Serena | 6 |
| 10 | 7664 ... = Serena, Serena, Serena, Megan ... | ? |

**Analyze the response variable:**

6. **Collect and summarize the results of all the trials.** Summarize and display the simulated values of the response variable as you would a set of data. You'll certainly want to report the shape, center, and spread, and depending on the question asked, you may want to include more.

7. **State your conclusion,** as always, in the context of the question you wanted to answer. Based on this simulation, we estimate that sports fans hoping to complete their card collection should anticipate needing to open around 5, 6, or 7 boxes, but it could take a lot more.

If you fear that these may not be accurate estimates because we ran only nine trials, you are absolutely correct. The more trials the better, and nine is woefully inadequate. How many is enough? We'll explore that question later in this chapter.

### FOR EXAMPLE
### Simulating a Dice Game

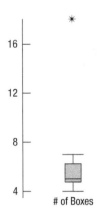

The game of 21 can be played with an ordinary 6-sided die. Competitors each roll the die repeatedly, trying to get the highest total less than or equal to 21. If your total exceeds 21, you lose.

Suppose your opponent has rolled 18 points so far. Your task is to try to beat them by getting more than 18 points without going over 21. How many die rolls will you need to make, and what are your chances of winning? Let's use a simulation to estimate answers to those questions. We'll use a random digits table.

QUESTION: How will you simulate the components?

ANSWER: A component is one roll of the die. I'll simulate each roll by looking at a random digit from a table or an Internet site. The digits 1 through 6 will represent the results on the die; I'll ignore digits 7–9 and 0.

QUESTION: How will you combine components to model a trial? What's the response variable?

ANSWER: I'll add components until my total is greater than 18, counting the number of rolls. If my total is greater than 21, it is a loss; if not, it is a win. There are two response variables. I'll record the number of digits I use for each trial (not counting the ones I ignored), and I'll also record whether I win or lose for each trial.

QUESTION: How would you use these random digits to run trials? Show your method clearly for two trials.

91129 58757 69274 92380 82464 33089

ANSWER: I've marked the discarded digits in color.

Trial #1:  9  **1  1  2**  9  **5**  8  7  **5**  7  **6**
Total:        1  2  4     9           14        20        Outcomes: 6 rolls, won

Trial #2:  9  **2**  7  **4**  9  **2  3**  8  0  8  **2  4  6**
Total:        2        6        8  11           13  17  23  Outcomes: 7 rolls, lost

QUESTION: Suppose you run 30 trials, getting the outcomes tallied here. What is your conclusion?

ANSWER: Based on my simulation, when competing against an opponent who has a score of 18, I expect my turn to usually last 5 or 6 rolls, and I should win about 70% of the time.

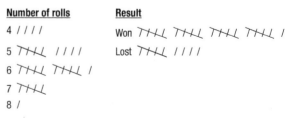

### THE EASY WAY
Some internet sites and (Good News!) your graphing calculator will allow you to simply generate random digits from 1 to 6, just like a die. That makes things much easier, because then you don't need to ignore the bogus "rolls" of 0, 7, 8, and 9 that show up in random digit tables.

# CHAPTER 10 Understanding Randomness

## JUST CHECKING

The baseball World Series consists of up to seven games. The first team to win four games wins the series. The first two are played at one team's home ballpark (Team A), the next three at the other team's park (Team B), and the final two (if needed) are played back at Team A's park. Records over the past century show that there is a home field advantage; in any game the home team has about a 55% chance of winning. Does the current system of alternating ballparks even out the home field advantage? How often will Team A, who begins at home, win the series?

Let's set up the simulation:

1. What is the component to be repeated?
2. How will you model each component from equally likely random digits?
3. How will you model a trial by combining components?
4. What is the response variable?
5. How will you analyze the response variable?

## STEP-BY-STEP EXAMPLE

### Simulation

Fifty-seven students participated in a lottery for a particularly desirable dorm room—a triple with a fireplace and private bath in the tower. Twenty of the participants were members of the same varsity team. When all three winners were members of the team, the other students cried foul.

QUESTION: Could an all-team outcome reasonably be expected to happen if everyone had a fair shot at the room?

 **PLAN** State the problem. Identify the important parts of your simulation.

I'll use a simulation to investigate whether it's unlikely that three varsity athletes would get the great room in the dorm if the lottery were fair.

**COMPONENTS** Identify the components.

A component is the selection of a student.

**OUTCOMES** State how you will model each component using equally likely random digits. You can't just use the digits from 0 to 9 because the outcomes you are simulating are not multiples of 10%.

There are 20 and 37 students in the two groups. This time you must use *pairs* of random digits (and ignore some of them) to represent the 57 students.

I'll look at pairs of random digits.

Let 00–19 represent the 20 varsity applicants.

Let 20–56 represent the other 37 applicants.

Skip 57–99. If I get a number in this range, I'll throw it away and go back for another pair of random digits.

**TRIAL** Explain how you will combine the components to simulate a trial. In each of these trials, you can't choose the same student twice, so you'll need to ignore a random number if it comes up a second or third time. Be sure to mention this in describing your simulation.

Each trial consists of identifying pairs of digits as V (varsity) or N (nonvarsity) until 3 people are chosen, ignoring out-of-range or repeated numbers (I'll indicate these by marking "X")—I can't put the same person in the room twice.

**RESPONSE VARIABLE** Define your response variable.

The response variable is whether or not all three selected students are on the varsity team.

**SHOW** — **MECHANICS** Run several trials. Carefully record the random numbers, indicating
1. the corresponding component outcomes (here, varsity, nonvarsity, or ignored number) and
2. the value of the response variable.

| Trial Number | Component Outcomes | All Varsity? |
|---|---|---|
| 1 | 74 02 94 39 02 77 55<br>X V X N X X N | No |
| 2 | 18 63 33 25<br>V X N N | No |
| 3 | 05 45 88 91 56<br>V N X X N | No |
| 4 | 39 09 07<br>N V V | No |
| 5 | 65 39 45 95 43<br>X N N X N | No |
| 6 | 98 95 11 68 77 12 17<br>X X V X X V V | Yes |
| 7 | 26 19 89 93 77 27<br>N V X X X N | No |
| 8 | 23 52 37<br>N N N | No |
| 9 | 16 50 83 44<br>V N X N | No |
| 10 | 74 17 46 85 09<br>X V N X V | No |

**ANALYZE** Summarize the results across all trials to answer the initial question.

"All varsity" occurred once, or 10% of the time.

**TELL** — **CONCLUSION** Describe what the simulation shows, and interpret your results in the context of the real world.

In my simulation of "fair" room draws, the three people chosen were all varsity team members only 10% of the time. While this result could happen by chance, it is not particularly likely. I'm suspicious, but I'd need many more trials (for a more accurate estimate of the probability) and a smaller frequency of the all-varsity outcome (for more confidence in my conclusion) before I would make an accusation of unfairness.

## TI TIPS

### Generating Random Numbers

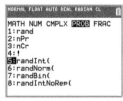

Instead of using coins, dice, cards, or tables of random numbers, you may decide to use your calculator for simulations. There are several random number generators offered in the MATH PROB menu.

randInt( is of particular importance. This command will produce any number of random integers in a specified interval. In the dialog box, the lower and upper bounds determine the interval and n indicates how many random numbers you want to simulate. It's okay to leave n blank if you only want to simulate one number.

Here are some examples showing how to use randInt for simulations:

◆ randInt(0,1) randomly chooses a 0 or a 1. This is an effective simulation of a coin toss. You could let 0 represent tails and 1 represent heads.

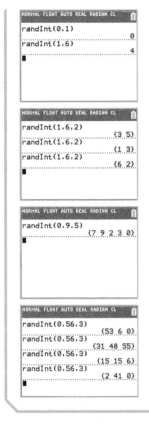

- `randInt(1,6)` produces a random integer from 1 to 6, a good way to simulate rolling a die.

- `randInt(1,6,2)` simulates rolling *two* dice. To do several rolls in a row, just hit `ENTER` repeatedly.

- `randInt(0,9,5)` produces five random integers that might represent the pictures in the cereal boxes. Our run gave us one Simone (0, 1), two Megans (2, 3, 4), and two Serenas (5–9).

- `randInt(0,56,3)` produces three random integers between 0 and 56, a nice way to simulate the dorm room lottery. The window shows 4 trials, but we would skip the third one because one student was chosen twice. In none of the remaining 3 trials did three athletes (0–19) win.

## How Many Trials?

| The Simulation . . . | | | Results So Far . . . | |
|---|---|---|---|---|
| Box Number | Random Digit | Picture Found | Number of Simones | Percent Simone |
| 1 | 8 | Serena | 0 out of 1 | 0% |
| 2 | 3 | Megan | 0 out of 2 | 0% |
| 3 | 3 | Megan | 0 out of 3 | 0% |
| 4 | 0 | Simone | 1 out of 4 | 25% |
| 5 | 6 | Serena | 1 out of 5 | 20% |
| 6 | 1 | Simone | 2 out of 6 | 33% |
| 7 | 9 | Serena | 2 out of 7 | 28% |
| 8 | 2 | Megan | 2 out of 8 | 25% |
| 9 | 9 | Serena | 2 out of 9 | 22% |
| 10 | 1 | Simone | 3 out of 10 | 30% |

When we showed you how to do simulations, first looking for the cereal box pictures and then checking the fairness of dorm room assignments, we ran just 10 trials. Let's see why that's not really enough. How? With a simulation, of course!

As an easy example, we'll just pretend to open cereal boxes looking for pictures of Simone Biles. While she's in 20% of the boxes, that doesn't tell us what we'll actually find as we go box by box. The table shows the results of our first 10 trials.

Remember that the intent of a simulation is to gain insight about situations we don't understand. If we hadn't already known that Simone's picture was in 20% of the boxes, these 10 trials wouldn't tell us that. At best, we might feel comfortable guessing that fewer than half of the boxes contain Simone's picture, but 10 trials just isn't enough to say anything very definitive.

For the homework[6] exercises we suggest you do 20 trials. How much better is that? Let's open more cereal boxes. Look at the graph displaying Simone's percentage after each of the first 10 trials in the table on the left and for 10 more trials.

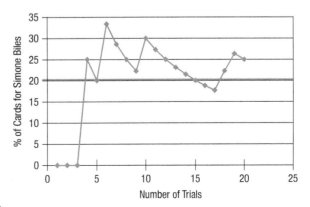

---

[6] Stop making that face. You *knew* there'd be homework.

Now would we conclude that 20% was correct? Probably not, even though the estimated percentages do seem to be settling down a bit. It appears 20 trials is still too few. So let's go big. We used a computer to run 1000 trials. The graph below shows what happened.

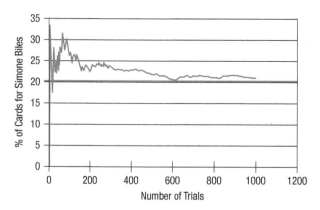

It appears that in this simulation there were quite a few Simone pictures in the first 100 (or so) boxes, but as the number of trials mounted the percentage drifted toward the true value of 20%. With 1,000 trials we might be able to make a pretty good guess about the cereal boxes. Frankly, though, this is a pretty simple situation. In the real world, simulations are used to explore very complex issues like climate change, election outcomes, and even national defense. Those investigations require thousands or even millions of trials![7]

# WHAT CAN GO WRONG?

- **Don't overstate your case.** Let's face it: In some sense, a simulation is *always* wrong. After all, it's not the real thing. We didn't buy any cereal or run a room draw. So beware of confusing what *really* happens with what a simulation suggests *might* happen. Never forget that future results will not match your simulated results exactly.

- **Model outcome chances accurately.** A common mistake in constructing a simulation is to adopt a strategy that may appear to produce the right kind of results, but that does not accurately model the situation. For example, in our room draw, we could have gotten 0, 1, 2, or 3 team members. Why not just see how often these digits occur in random digits from 0 to 9, ignoring the digits 4 and up?

    3 2 1 7 9 0 0 5 9 7 3 7 9 2 5 2 4 1 3 8
    3 2 1 x x 0 0 x x x 3 x x 2 x 2 x 1 3 x

    This "simulation" makes it seem fairly likely that three team members would be chosen. There's a big problem with this approach, though: The digits 0, 1, 2, and 3 occur with equal frequency among random digits, making each outcome appear to happen 25% of the time. In fact, the selections of 0, 1, 2, or all 3 team members are not all equally likely outcomes. In our correct simulation, we estimated that all 3 would be chosen only about 10% of the time. If your simulation overlooks important aspects of the real situation, your model will not be accurate.

- **Run enough trials.** Simulation is cheap and fairly easy to do. Don't try to draw conclusions based on 5 or 10 trials (even though we did for illustration purposes here). We'll get a better handle on how many trials to use in later chapters. For now, err on the side of large numbers of trials.

TI-*nspire*

**Simulations.** Improve your predictions by running thousands of trials.

---

[7] We hope that makes you feel better about doing just 20 trials for the homework. See, we're actually being nice to you!

## WHAT HAVE WE LEARNED?

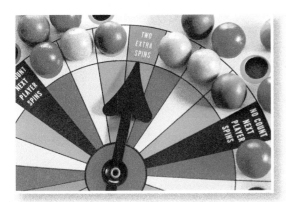

We've learned to harness the power of randomness. We've learned that a simulation model can help us investigate a question for which many outcomes are possible, we can't (or don't want to) collect data, and a mathematical answer is hard to calculate. We've learned how to base our simulation on random values generated by a computer, generated by a randomizing device such as a die or spinner, or found on the Internet. Like all models, simulations can provide us with useful insights about the real world.

### TERMS

**Random**  An outcome is random if we know the possible values it can have, but not which particular value it takes. A random outcome is free of human influence. (p. 276)

**Generating random numbers**  Random numbers are hard to generate. Nevertheless, several Internet sites offer an unlimited supply of equally likely random values. (p. 277)

**Simulation**  A simulation models a real-world situation by using random-digit outcomes to mimic the uncertainty of a response variable of interest. (p. 278)

**Trial**  The sequence of several components representing events that we are pretending will take place. (p. 278)

**Simulation component**  A component uses equally likely random digits to model simple random occurrences whose outcomes may not be equally likely. (p. 279)

**Response variable**  Values of the response variable record the results of each trial with respect to what we were interested in. (p. 279)

## ON THE COMPUTER

### Simulation

Simulations are best done with the help of technology simply because running more trials makes for a better simulation, and computers are fast. There are special computer programs designed for simulation, and most statistics packages and calculators can at least generate random numbers to support a simulation.

All technology-generated random numbers are *pseudorandom*. The random numbers available on the Internet may technically be better, but the differences won't matter for any simulation of modest size. Pseudorandom numbers generate the next random value from the previous one by a specified algorithm. But they have to start somewhere. This starting point is called the "seed." Most programs let you set the seed. There's usually little reason to do this, but if you wish to, go ahead. If you reset the seed to the same value, the programs will generate the same sequence of "random" numbers.

**APPLET**
Generate random numbers

## EXERCISES

1. **Random outcomes** For each of the following scenarios, decide if the outcome is random.
   a) Flip a coin to decide who takes out the trash. Is who takes out the trash random?
   b) A friend asks you to quickly name a professional sports team. Is the sports team named random?
   c) Names are selected out of a hat to decide roommates in a dormitory. Is your roommate for the year random?

2. **More random outcomes** For each of the following scenarios, decide if the outcome is random.
   a) You enter a contest in which the winning ticket is selected from a large drum of entries. Was the winner of the contest random?
   b) When playing a board game, the number of spaces you move is decided by rolling a six-sided die. Is the number of spaces you move random?
   c) Before flipping a coin, your friend asks you to "call it." Is your choice (heads or tails) random?

3. **The lottery** Many states run lotteries, giving away millions of dollars if you match a certain set of winning numbers. How are those numbers determined? Do you think this method guarantees randomness? Explain.

4. **Games** Many kinds of games people play rely on randomness. Cite three different methods commonly used in the attempt to achieve this randomness, and discuss the effectiveness of each.

5. **Pool or spa** Younger people in America enjoy having their own body of water! In fact, 17% of 18–29-year-olds reported having a pool or spa (compared to only 13% of 50–64-year-olds). How would you assign random numbers to conduct a simulation based on this statistic? (Source: statista.com)

6. **Lefties** By some estimates, about 10% of the general population is left-handed. How would you assign random numbers to conduct a simulation based on this statistic?

7. **Geography** An elementary school teacher with 25 students plans to have each of them make a poster about two different states. The teacher first numbers the states (in alphabetical order, from 01-Alabama to 50-Wyoming), then uses a random number table to decide which states each kid gets. Here are the random digits:

   45921 01710 22892 37076

   a) Which two state numbers does the first student get?
   b) Which two state numbers go to the second student?

8. **Get rich** Your state's BigBucks Lottery prize has reached $100,000,000, and you decide to play. You have to pick five numbers between 1 and 60, and you'll win if your numbers match those drawn by the state. You decide to pick your "lucky" numbers using a random number table. Which numbers do you play, based on these random digits?

   43680 98750 13092 76561 58712

9. **Play the lottery** Some people play state-run lotteries by always playing the same favorite "lucky" number. Assuming that the lottery is truly random, is this strategy better, worse, or the same as choosing different numbers for each play? Explain.

10. **Play it again, Sam** In Exercise 8 you imagined playing the lottery by using random digits to decide what numbers to play. Is this a particularly good or bad strategy? Explain.

11. **Bad simulations** Explain why each of the following simulations fails to model the real situation properly:
    a) Use a random integer from 0 through 9 to represent the number of heads when 9 coins are tossed.
    b) A random person is selected from the country of Canada. They are tested to find if they carry the COVID-19 antibody. A random number is selected. Odd numbers represent they have the antibody, even numbers represent they do not have the antibody.
    c) Use random numbers from 1 through 13 to represent the denominations of the cards in a five-card poker hand.

12. **More bad simulations** Explain why each of the following simulations fails to model the real situation:
    a) Use random numbers 2 through 12 to represent the sum of the faces when two dice are rolled.
    b) Use a random integer from 0 through 5 to represent the number of boys in a family of 5 children.
    c) Simulate your classmates' method for traveling to school: 0 = walks, 1 = bus, 2 = car, and 3 = other.

13. **Wrong conclusion** A Statistics student properly simulated the length of checkout lines in a grocery store and then reported, "The average length of the line will be 3.2 people." What's wrong with this conclusion?

14. **Another wrong conclusion** After simulating the spread of a disease, a researcher wrote, "24% of the people contracted the disease." What should the correct conclusion be?

15. **Election** You're pretty sure that your candidate for class president has about 55% of the votes in the entire school. But you're worried that only 100 students will show up to vote. How often will the underdog (the one with 45% support) win? To find out, you set up a simulation.
    a) Describe how you will simulate a component.
    b) Describe how you will simulate a trial.
    c) Describe the response variable.

16. **Two pairs or three of a kind?** When drawing five cards randomly from a deck, which is more likely, two pairs or three of a kind? A pair is exactly two of the same denomination. Three of a kind is exactly 3 of the same denomination. (Don't count three 8's as a pair—that's 3 of a kind. And don't count 4 of the same kind as two pairs—that's 4 of a kind, a very special hand.) How could you simulate 5-card hands? Be careful; once you've picked the 8 of spades, you can't get it again in that hand.
    a) Describe how you will simulate a component.
    b) Describe how you will simulate a trial.
    c) Describe the response variable.

17. **Cereal** In the chapter's example, 20% of the cereal boxes contained a picture of Simone Biles, 30% Megan Rapinoe, and the rest Serena Williams. Suppose you buy five boxes of cereal. Estimate the probability that you end up with a complete set of the pictures. Your simulation should have at least 20 runs.

18. **Cereal again** Suppose you really want the Simone Biles picture. How many boxes of cereal do you need to buy to be pretty sure of getting at least one? Your simulation should use at least 10 trials.

19. **Multiple choice** You take a quiz with 6 multiple choice questions. After you studied, you estimated that you would have about an 80% chance of getting any individual question right. What are your chances of getting them all right? Use at least 20 trials.

20. **Lucky guessing?** A friend of yours who took the multiple choice quiz in Exercise 19 got all 6 questions right, but now claims to have guessed blindly on every question. If each question offered 4 possible answers, do you believe her? Explain, basing your argument on a simulation involving at least 10 trials.

21. **Beat the lottery** Many states run lotteries to raise money. A website advertises that it knows "how to increase YOUR chances of Winning the Lottery." They offer several systems and criticize others as foolish. One system is called *Lucky Numbers*. People who play the *Lucky Numbers* system just pick a "lucky" number to play, but maybe some numbers are luckier than others. Let's use a simulation to see how well this system works.

    To make the situation manageable, simulate a simple lottery in which a single digit from 0 to 9 is selected as the winning number. Pick a single value to bet, such as 1, and keep playing it over and over. You'll want to run at least 100 trials. (If you can program the simulations on a computer, run several hundred. Or generalize the questions to a lottery that chooses two- or three-digit numbers—for which you'll need thousands of trials.)

    a) What proportion of the time do you expect to win?
    b) Would you expect better results if you picked a "luckier" number, such as 7? (Try it if you don't know.) Explain.

22. **Random is as random does** The "beat the lottery" website discussed in Exercise 21 suggests that because lottery numbers are random, it is better to select your bet randomly. For the same simple lottery in Exercise 21 (random values from 0 to 9), generate each bet by choosing a separate random value between 0 and 9. Play many games. What proportion of the time do you win?

23. **It evens out in the end** The "beat the lottery" website of Exercise 21 notes that in the long run we expect each value to turn up about the same number of times. That leads to their recommended strategy. First, watch the lottery for a while, recording the winners. Then bet the value that has turned up the least, because it will need to turn up more often to even things out. If there is more than one "rarest" value, just take the lowest one (because it doesn't matter). Simulating the simplified lottery described in Exercise 21, play many games with this system. What proportion of the time do you win?

24. **Play the winner?** Another strategy for beating the lottery is the reverse of the system described in Exercise 23. Simulate the simplified lottery described in Exercise 21. Each time, bet the number that just turned up. The website suggests that this method should do worse. Does it? Play many games and see.

25. **Driving test** You are about to take the road test for your driver's license. You hear that only 34% of candidates pass the test the first time, but the percentage rises to 72% on subsequent retests. Because most teenagers really want to drive, they keep taking the test until they pass!

    a) Create a plan for a simulation to estimate the average number of tests drivers take in order to get a license.
    b) The histogram shows the results of 100 trials of a simulation. Use these results to estimate the average number of tests drivers take in order to get a license.

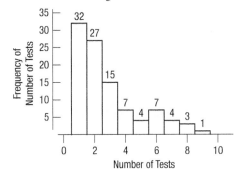

26. **Basketball strategy** Late in a basketball game, the team that is behind often fouls someone in an attempt to get the ball back. Sometimes the rules put the foul shooter in a "one-and-one" situation. This means that if the shooter misses the first free throw, he stops shooting and gets no points. But if he makes the first shot, he gets to shoot again. If he misses the second shot, he has scored one point; if he makes the second shot, he has scored two points. Suppose the opposing player has made 72% of his foul shots this season.

    a) Create a plan for a simulation to estimate the number of points he will score in a one-and-one situation.
    b) Here is a display of the results of 100 trials of a simulation. Use these results to estimate the average number of points he will score in one-and-one situations.

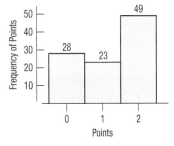

27. **Still learning?** As in Exercise 25, assume that your chance of passing the driver's test is 34% the first time and 72% for subsequent retests. Estimate the percentage of those tested who still do not have a driver's license after two attempts.

28. **Blood donors** A person with type O-positive blood can receive blood only from other type O donors. About 44% of the U.S. population has type O blood. At a blood drive, how many potential donors do you expect to examine in order to get three units of type O blood?

29. **Free groceries** To attract shoppers, a supermarket runs a weekly contest that involves "scratch-off" cards. With each purchase, customers get a card with a black spot obscuring a message. When the spot is scratched away, most of the cards simply

say, "Sorry—please try again." But during the week, 100 customers will get cards that make them eligible for a drawing for free groceries. Ten of the cards say they may be worth $200, 10 others say $100, 20 may be worth $50, and the rest could be worth $20. To register those cards, customers write their names on them and put them in a barrel at the front of the store. At the end of the week the store manager draws cards at random, awarding the lucky customers free groceries in the amount specified on their cards. The drawings continue until the store has given away more than $500 of free groceries. Estimate the average number of winners each week.

30. **Find the ace** A technology store holds a contest to attract shoppers. Once an hour, someone at checkout is chosen at random to play in the contest. Here's how it works: An ace and four other cards are shuffled and placed face down on a table. The customer gets to turn over cards one at a time, looking for the ace. The person wins $100 of store credit if the ace is the first card, $50 if it is the second card, and $20, $10, or $5 if it is the third, fourth, or last card chosen. What is the average dollar amount of store credit given away in the contest? Estimate with a simulation.

31. **The family** Many couples want to have both a boy and a girl. If they decide to continue to have children until they have one child of each sex, what would the average family size be? Assume that boys and girls are equally likely.

32. **Repeat?** You are listening to music on your phone. Your music player is in shuffle mode. After one of your favorite songs is played, you are surprised to hear another song played by the same artist. And then a third! This makes you wonder if shuffle mode really shuffles all that randomly. However, you realize that you are listening to a playlist with only your five favorite artists. So perhaps three in a row with only five different choices (all roughly equal in proportion) isn't that unusual. Run a simulation to untangle this dilemma!

33. **Dice game** You are playing a children's game in which the number of spaces you get to move is determined by the rolling of a die. You must land exactly on the final space in order to win. If you are 10 spaces away, how many turns might it take you to win?

34. **Doubles** In the game of Monopoly you roll two 6-sided dice. If both dice show the same number, that is called doubles. If you roll doubles, you get to roll again. Your friend is lucky and rolls doubles twice in a row. How often do you expect this to happen? (*Hint:* Using technology to generate two numbers between 1 and 6 is the best way to run this simulation.)

35. **The hot hand** A basketball player with a 65% shooting percentage has just made 6 shots in a row. The announcer says this player "is hot tonight! She's in the zone!" Assume the player takes about 20 shots per game. Is it unusual for her to make 6 or more shots in a row during a game?

36. **The World Series** The World Series is a "best of 7" situation. That is, the teams play until one team has won 4 games. That might happen if one team wins 4 in a row. But it might take 7 games until that happens. Suppose that sports analysts consider one team a bit stronger, with a 55% chance to win any individual game. Estimate the likelihood that the underdog (the team with only a 45% chance) wins the series.

37. **Teammates** Four couples at a dinner party play a board game after the meal. They decide to play as teams of two and to select the teams randomly. All eight people write their names on slips of paper. The slips are thoroughly mixed, then drawn two at a time. How likely is it that every person will be teamed with someone other than the person they came to the party with?

38. **Second team** Suppose the couples in Exercise 37 choose the teams by having one member of each couple write their names on the cards and the other people each pick a card at random. How likely is it that every person will be teamed with someone other than the person they came with?

39. **Social distancing** Suppose that your chance of getting COVID-19 from another person is 2%. You decide to see some friends and you spend time (in a not socially distanced setting!) with eight people. Assume that each person, independently, has a 2% chance of sharing COVID-19 with you. Run 30 simulations and report the likelihood that you will contract the disease.

40. **Smartphones** A proud legislator claims that your state's new law banning texting and hand-held phones while driving reduced occurrences of infractions to less than 10% of all drivers. While on a long drive home from your college, you notice a few people seemingly texting. You decide to count everyone using their smartphones illegally who pass you on the expressway for the next 20 minutes. It turns out that 5 out of the 20 drivers were actually using their phones illegally. Does this cast doubt on the legislator's figure of 10%? Use a simulation to estimate the likelihood of seeing at least 5 out of 20 drivers using their phones illegally if the actual usage rate is only 10%. Explain your conclusion clearly.

41. **Ferret nasal spray** During the COVID pandemic of 2020, a group of scientists developed a nasal spray that was intended to block the virus responsible for COVID from attaching to airways and lung cells, thus preventing infection even after exposure to the virus.[8] In order to test the effectiveness of the spray, the following experiment was conducted using ferrets, which can catch viruses through the nose like humans. One ferret was deliberately infected with the COVID virus and placed into a cage. Four other ferrets not infected with the virus were placed into the same cage. Of these four, two had been randomly selected and given the experimental nasal spray while the other two had been given a placebo nasal spray. After 24 hours, the ferrets given the placebo spray had both contracted COVID; the two given the experimental spray had not.

Let's assume that the nasal spray is completely ineffective. Because two of the four ferrets contracted the virus, we'll also assume that the chance of any ferret contracting the virus is 50%.

a) Describe a simulation using a random digit table that can be used to estimate the chances that the two ferrets given the placebo spray would catch the virus and the two given the experimental spray would not.

b) Run 20 trials of your simulation using a random digit generator of your choice and estimate the chances that the observed outcome would happen just by chance if the treatment were ineffective.

---

[8] https://www.nytimes.com/2020/11/05/health/coronavirus-ferrets-vaccine-spray.html

c) Based on your answer to part (b), do you think the experiment provides evidence that the experimental nasal spray is effective? Explain your reasoning.

42. **More ferrets** In Exercise 41, we assumed that each ferret not already infected with COVID had a 50% chance of catching it from the infected ferret. That was a reasonable assumption because 2 of the 4 uninfected ferrets ended up catching COVID, and if the nasal spray was actually ineffective, then every ferret would have about a 50% chance of catching it. However, in your simulation you may have noticed that in many trials, more or fewer than two ferrets caught COVID. There's a different assumption we can make that more closely matches the actual experimental outcome, but it can't be simulated easily with random digits.

   Let's continue to assume that the nasal spray is ineffective. However, this time let's assume that *exactly two* of the four uninfected ferrets are going to get infected, but *which* two is completely random (because we're assuming the nasal spray is ineffective).

   a) Describe how you could use four standard playing cards (see p. 277) to simulate one trial—determining randomly which two of the four ferrets will get infected. Be sure to state which four cards you're going to use and what they represent.
   b) If you have access to playing cards, run 20 trials of your simulation and estimate the chances that if two ferrets were going to get infected at random, that they would by chance be the two that received the placebo. (If you don't have playing cards, you can use index cards and write anything you want on them.)
   c) What does the outcome of your simulation indicate—if anything—about the effectiveness of the nasal spray in the real experiment? Explain your reasoning.

43. **Still more ferrets!** In reality, the researchers used *three* cages for their experiment, each one containing five ferrets, exactly like the experiment described in Exercise 41, and in each one the same thing happened: the two ferrets given the placebo became infected and the two given the experimental treatment did not.

   a) Describe how you could modify your simulation from Exercise 41 and/or Exercise 42 to reflect the two additional cages of ferrets.
   b) Without actually carrying out any more simulations, would you expect the probability of the outcome from the real experiment (with three cages) to be higher, lower, or about the same as the probability estimate you would get from a simulation involving just one cage? Explain your reasoning.[9]

---

[9] https://xkcd.com/2400

## JUST CHECKING

Answers

1. The component is one game.
2. I'll generate pairs of random digits and assign numbers from 00 to 54 to the home team's winning and from 55 to 99 to the visitors' winning.
3. I'll generate components until one team wins 4 games. I'll record which team wins the series.
4. The response is who wins the series.
5. I'll calculate the proportion of wins by the home team.

# 11 Sample Surveys

The year 2020 was challenging for nearly everyone on the planet. Governments around the world were scrambling to figure out how to balance concerns for public safety with keeping their economies running. In the summer of 2020, at the height of the COVID pandemic, polling agencies asked whether respondents thought the United States was headed in the right direction or on the "wrong track." The "right direction/wrong track" question is a common one, used by many polling companies. But the responses varied, as shown in the table below.

All these polls claim that their estimates are close to the true percentages that they would have found if they had asked all U.S. adults. But, their numbers are different. How we understand and account for these differences is a central topic of Statistics and will concern us for most of the rest of this book. To make business decisions, to do science, to choose wise

Table 11.1
Survey responses to the right direction/wrong track question all taken within a month of each other show differences in responses. (RV = Registered voters, LV = Likely voters)

| Polling Data | | | | |
|---|---|---|---|---|
| Poll | Date | Sample | Right Direction | Wrong Track |
| RCP Average | 6/17–7/14 | — | 23.6 | 68.8 |
| Reuters/Ipsos | 7/13–7/14 | 961 RV | 22 | 69 |
| Economist/YouGov | 7/12–7/14 | 1252 RV | 23 | 70 |
| Politico/Morning Consult | 7/10–7/12 | 1992 RV | 28 | 72 |
| NBC News/Wall St. Jrnl | 7/9–7/12 | 900 RV | 19 | 72 |
| Rasmussen Reports | 7/5–7/9 | 2500 LV | 24 | 72 |
| Monmouth | 6/26–6/30 | 733 RV | 18 | 75 |
| USA Today/Suffolk | 6/25–6/29 | 1000 RV | 20 | 67 |
| NY Times/Siena | 6/17–6/22 | 1137 RV | 31 | 58 |
| Harvard-Harris | 6/17–6/18 | 1886 RV | 27 | 64 |

investments, or to understand how voters think they'll vote in the next election, we need to stretch beyond the data we have at hand to the world at large. That step from a small sample to the entire population is made possible by the methods you'll learn in this course.

To make that stretch, we need three ideas. You'll find the first one natural. The second may be more surprising. The third is one of the strange but true facts that often confuse those who don't know Statistics.

# Idea 1: Examine a Part of the Whole

The first idea is to draw a sample. We'd like to know about an entire group of individuals—a **population**—but examining all of them is usually impractical, if not impossible. So we settle for examining a smaller group of individuals—a **sample**—selected from the population.

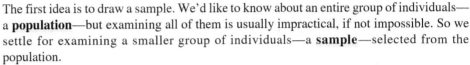

You do this every day. For example, suppose you wonder how the vegetable soup you're cooking for dinner tonight is going to go over with your friends. To decide whether it meets your standards, you need to try only a small amount. You might taste just a spoonful or two. You trust that the taste will *represent* the flavor of the entire pot. You know that a small sample, if selected properly, can represent the entire population.

It's hard to go a day without hearing about the latest opinion poll. These polls are examples of **sample surveys**, designed to ask questions of a small group of people in the hope of learning something about the entire population. How can the pollsters claim that a sample is representative of the entire population? Professional pollsters work quite hard to ensure that the "taste"—the sample that they take—does represent the population.

### The W's and Sampling

The population we are interested in is usually determined by the *Why* of our study. The sample we draw will be the *Who*. *When* and *How* we draw the sample may depend on what is practical.

# Bias

Selecting a sample to represent the population fairly is more difficult than it sounds. Polls or surveys may overlook subgroups that are harder to find (such as people dealing with homelessness) or favor others (such as Internet users who like to respond to online surveys). Sampling methods that, by their nature, tend to over- or underemphasize some characteristics of the population are said to be **biased**. Bias is the bane of sampling—the one thing above all to avoid. Conclusions based on samples drawn with biased methods are inherently flawed. There is usually no way to fix bias after the sample is drawn and no way to salvage useful information from it.

Here's a famous example of a really dismal failure. By the beginning of the 20th century, it was common for newspapers to ask readers to return "straw" ballots on a variety of topics. (Today's Internet surveys are the same idea, gone electronic.)

During the period from 1916 to 1936, the magazine *Literary Digest* regularly surveyed public opinion and forecast election results correctly. During the 1936 presidential campaign between Alf Landon and Franklin Delano Roosevelt, it mailed more than 10 million ballots and got back an astonishing 2.4 million. (Polls were still a relatively novel idea, and many people thought it was important to send back their opinions.) The results were clear: Alf Landon would be the next president by a landslide, 57% to 43%. You remember President Landon? No? In fact, Landon carried only two states. Roosevelt won, 62% to 37%, and, perhaps coincidentally, the *Digest* went bankrupt soon afterward.

What went wrong? The *Digest* used the phone book, as many surveys did for decades.[1] But in 1936, at the height of the Great Depression, telephones were a luxury, so

---

[1] Today, phone numbers are computer-generated to make sure that unlisted numbers are included. But special methods must be used to account for the growing part of the population that will not answer a call from an unknown number.

In 1936 a young pollster named George Gallup used a smaller but representative sample to predict that Roosevelt would get 56% of the vote. (This photo was taken later in his career.)

the *Digest under-*sampled poor voters. The campaign of 1936 focused on the economy, and those who were less well off were more likely to vote for the Democrat. So the *Digest*'s sample was hopelessly biased.

How do modern polls get their samples to *represent* the entire population? You might think that they'd handpick individuals to sample with care and precision. But in fact, they do something quite different: They select individuals to sample *at random*.[2]

## Idea 2: Randomize

Think back to the soup sample. Suppose you add some salt to the pot. If you sample it from the top before stirring, you'll get the misleading idea that the whole pot is salty. If you sample from the bottom, you'll get an equally misleading idea that the whole pot is bland. By stirring, you *randomize* the amount of salt throughout the pot, making each taste more typical of the whole pot.

Not only does randomization protect you against factors that you know are in the data, it can also help protect against factors that you didn't even know were there. Suppose, while you weren't looking, a friend added a handful of peas to the soup. If they're down at the bottom of the pot, and you don't randomize the soup by stirring, your test spoonful won't have any peas. By stirring in the salt, you *also* randomize the peas throughout the pot, making your sample taste more typical of the overall pot *even though you didn't know the peas were there*. So randomizing protects us even in this case.

How do we "stir" people in a survey? We select them at random. **Randomizing** protects us from the influences of *all* the features of our population by making sure that, *on average,* the sample looks like the rest of the population. The importance of deliberately using randomness is one of the great insights of Statistics.

### Why Not Match the Sample to the Population?

Rather than randomizing, we could try to design our sample so that the people we choose are typical in terms of every characteristic we can think of. We might want the income levels of those we sample to match the population. How about age? Political affiliation? Marital status? Having children? Living in the suburbs? We can't possibly think of all the things that might be important. Even if we could, we wouldn't be able to match our sample to the population for all these characteristics.

### FOR EXAMPLE

#### Is a Random Sample Representative?

Here are summary statistics comparing two samples of 8000 drawn at random from a company's database of 3.5 million customers:

| | Mean Age (yr) | Female (%) | Mean # of Children | Mean Income Bracket (1–7) | Mean Wealth Bracket (1–9) | Homeowner? (% Yes) |
|---|---|---|---|---|---|---|
| Sample 1 | 61.4 | 56.2 | 1.54 | 3.91 | 5.29 | 71.36 |
| Sample 2 | 61.2 | 56.4 | 1.51 | 3.88 | 5.33 | 72.30 |

QUESTION: Do you think these samples are representative of the population of the company's customers? Explain.

ANSWER: The two samples look very similar with respect to these seven variables. It appears that randomizing has automatically matched them pretty closely. We can reasonably assume that because the two samples don't differ too much from each other, they don't differ much from the rest of the population either.

---

[2] Because participation in phone polls has plummeted recently, some polling organizations are experimenting with new sampling methods that *do* involve handpicking participants, up to a point. But these methods are not yet well understood or fully accepted by statisticians, who nearly all agree that random sampling is the ideal, and nonrandom samples should be used only when random sampling is essentially impossible.

## Idea 3: It's the Sample Size

**APPLET**
Explore effects of sample size

How large a random sample do we need for the sample to be reasonably representative of the population? Most people think that we need a large percentage, or *fraction,* of the population, but it turns out that what matters is the *number* of individuals in the sample, not what fraction of the population it is. Think about what that means: a random sample of 100 students at a college represents the student body just about as well as a random sample of 100 voters represents the entire electorate of the United States. This is the *third* idea and probably the most surprising one in designing surveys.

How can it be that only the size of the sample, and not the population, matters? Well, let's return one last time to that pot of soup. If you're cooking for a banquet rather than just for a few people, your pot will be bigger, but do you need a bigger spoon to decide how the soup tastes? Of course not. The same-size spoonful is probably enough to make a decision about the entire pot, no matter how large the pot. The *fraction* of the population that you've sampled doesn't matter.[3] It's the **sample size**—the number of individuals in the sample—that's important.

How big a sample do you need? That depends on what you're estimating. To get an idea of what's really in the soup, you'll need a large enough taste to get a *representative* sample from the pot. For a survey that tries to find the proportion of the population falling into a category, you'll usually need several hundred respondents to say anything precise enough to be useful.

### LARGER IS BETTER

A friend who knows that you are taking Statistics asks your advice on their study. What can you possibly say that will be helpful? Just say, "As long as your sample is representative, a larger sample would probably improve your study." Even though a larger sample might not be worth the cost, it will almost always make the results more precise.

### WHAT DO THE POLLSTERS DO?

How do professional polling agencies do their work? The most common polling method today is to contact respondents by telephone. Computers generate random telephone numbers, so pollsters can even call some people with unlisted phone numbers. The person who answers the phone is invited to respond to the survey—if that person qualifies. (For example, only if it's an adult who lives at that address.) If the person answering doesn't qualify, the caller will ask if they may speak to someone who does qualify. (For example, they may ask a child whether they may speak to an adult in the household.)

Do these methods work? The Pew Research Center for the People and the Press, reporting on one survey, says that

> Across five days of interviewing, surveys today are able to make some kind of contact with the vast majority of households (76%). But because of busy schedules, skepticism and outright refusals, interviews were completed in just 38% of households that were reached using standard polling procedures.

If those who respond to the survey give a good snapshot of the larger population, all is well. But if not then the bias resulting from those who don't respond is called **nonresponse bias**, and will be discussed later in this chapter.

### TI-*nspire*

**Populations and Samples.** How well can a sample reveal the population's shape, center, and spread? Explore what happens as you change the sample size.

## Does a Census Make Sense?

Why bother determining the right sample size? Wouldn't it be better to just include everyone and "sample" the entire population? Such a special sample is called a **census**. Although a census would appear to provide the best possible information about the population, there are a number of reasons why it might not.

---

[3] Well, that's not exactly true. If your sample is a big enough chunk of the population—more than about 10%—then, for technical reasons, the fraction and the sample size both matter. But in practice we seldom have the luxury of a sample that big.

First, it can be difficult to complete a census. Some individuals in the population will be hard (and expensive) to locate. Or a census might just be impractical. If you were a taste tester for the Hostess™ Company, you probably wouldn't want to conduct a census by eating *all* the Twinkies on the production line. Not only might this be life-endangering, but the company wouldn't have any left to sell.

Second, populations rarely stand still. In populations of people, babies are born and folks die or leave the country. In opinion surveys, events may cause a shift in opinion during the survey. A census takes longer to complete and the population changes while you work. A sample surveyed in just a few days may give more accurate information.

Third, taking a census can be more complex than sampling. For example, the U.S. Census records too many college students. Many are counted once with their families and are then counted a second time in a report filed by their schools.

> **THE UNDERCOUNT**
> It's particularly difficult to compile a complete census of a population as large, complex, and spread out as the U.S. population. The U.S. Census is known to miss some residents. On occasion, the undercount has been striking. For example, there have been blocks in inner cities in which the number of residents recorded by the Census was smaller than the number of electric meters for which bills were being paid. What makes the problem particularly important is that some groups have a higher probability of being missed than others, such as undocumented immigrants, homeless people, the poor. The Census Bureau proposed the use of random sampling to estimate the number of residents missed by the ordinary census. Unfortunately, the resulting debate has become more political than statistical. In 2020 it even reached the Supreme Court, after the Trump administration cut off census data collection sooner than the statisticians at the Census Bureau recommended.

# Populations and Parameters

**Statistics and Parameters**

Any quantity that we calculate from data could be called a "statistic." But in practice, we usually use a statistic to estimate a population parameter.

**We'll Never Know!**

Remember: Population model parameters are not just unknown—usually they are *unknowable*. We have to settle for sample statistics.

A study found that teens were less likely than adults to "buckle up." The National Center for Chronic Disease Prevention and Health Promotion reports that 21.7% of U.S. teens never or rarely wear seat belts. We're sure they didn't take a census, so what *does* the 21.7% mean? We can't know what percentage of all teenagers wear seat belts. Reality is just too complex. But we can simplify the question by building a model.

Models use mathematics to represent reality. Parameters are the key numbers in those models. A parameter used in a model for a population is sometimes called (redundantly) a **population parameter**.

But let's not forget about the data. We use summaries of the data to estimate the population parameters. As we know, any summary found from the data is a **statistic**. Sometimes you'll see the (also redundant) term **sample statistic**.[4]

We've already met two parameters in Chapter 5: the mean, $\mu$, and the standard deviation, $\sigma$. We'll try to keep denoting population model parameters with Greek letters and the corresponding statistics with Latin letters. Usually, but not always, the letter used for the statistic and the parameter correspond in a natural way. So the standard deviation of the data is $s$, and the corresponding parameter is $\sigma$ (Greek for $s$). In Chapter 6, we used $r$ to denote the sample correlation. The corresponding correlation in a model for the population would be called $\rho$ (rho). In Chapter 7, $b_1$ represented the slope of a linear regression estimated from the data. But when we think about a (linear) *model* for the population, we denote the slope parameter $\beta_1$ (beta).

Get the pattern? Good. But now it breaks down. We denote the mean of a population model with $\mu$ (because $\mu$ is the Greek letter for $m$). It might make sense to denote the sample mean with $m$, but long-standing convention is to put a bar over anything when we average it, so we write $\bar{y}$.

---

[4] Where else besides a sample *could* a statistic come from?

What about proportions? Suppose we want to talk about the proportion of teens who don't wear seat belts. If we use $p$ to denote the proportion from the data, what is the corresponding model parameter? By all rights it should be $\pi$. But statements like $\pi = 0.25$ might be confusing because $\pi$ has been equal to 3.1415926... for so long, and it's worked so *well*. So, once again we violate the rule. We'll use $p$ for the *population* model parameter and $\hat{p}$ for the proportion from the *data* (because, like $\hat{y}$ in regression, it's an estimated value).

Here's a table summarizing the notation:

> **NOTATION ALERT**
> This entire table is a notation alert.

| Name | Statistic | Parameter |
|---|---|---|
| Mean | $\bar{y}$ | $\mu$ (mu, pronounced "myoo," not "moo") |
| Standard deviation | $s$ | $\sigma$ (sigma) |
| Correlation | $r$ | $\rho$ (rho) |
| Regression coefficient | $b$ | $\beta$ (beta, pronounced "baytah"[5]) |
| Proportion | $\hat{p}$ | $p$ (pronounced "pee"[6]) |

We draw samples because we can't work with entire populations, but we want the statistics we compute from a sample to reflect the corresponding parameters accurately. A sample that does this is said to be **representative**. A biased sampling methodology tends to over- or underestimate the parameter of interest.

### JUST CHECKING

1. Various claims are often made for surveys. Why is each of the following claims not correct?
    a) It is always better to take a census than to draw a sample.
    b) Stopping students on their way out of the cafeteria is a good way to sample if we want to know about the quality of the food there.
    c) We drew a sample of 100 from the 3000 students in a school. To get the same level of precision for a town of 30,000 residents, we'll need a sample of 1000.
    d) A poll taken at a Statistics support website garnered 12,357 responses. The majority said they enjoy doing Statistics homework. With a sample size that large, we can be pretty sure that most Statistics students feel this way, too.
    e) The true percentage of all Statistics students who enjoy the homework is called a "population statistic."

## Simple Random Samples

How would you select a representative sample? Most people would say that every individual in the population should have an equal chance of being selected, and certainly that seems fair. But it's not sufficient. There are many ways to give everyone an equal chance that still wouldn't give a representative sample. Consider, for example, a school that has equal numbers of males and females. We could sample like this: Flip a coin. If it comes up heads, select 100 female students at random. If it comes up tails, select 100 males at random. Everyone has an equal chance of selection, but every sample is of only a single sex—hardly representative.

We need to do better. Suppose we insist that every possible *sample* of the size we plan to draw has an equal chance to be selected. This ensures that situations like the one just described are not likely to occur and still guarantees that each person has an equal chance of being selected. What's different is that with this method, each *combination* of people

---

[5] If you're from the United States. If you're British or Canadian, it's "beetah."
[6] Just in case you weren't sure.

has an equal chance of being selected as well. A sample drawn in this way is called a **simple random sample**, usually abbreviated **SRS**. An SRS is the standard against which we measure other sampling methods, and the sampling method on which the theory of working with sampled data is based.

To select a sample at random, we first need to define where the sample will come from. The **sampling frame** is a list of individuals from which the sample is drawn. For example, to draw a random sample of students at a college, we might obtain a list of all registered full-time students and sample from that list. In defining the sampling frame, we must deal with the details of defining the population. Are part-time students included? How about those who are attending school elsewhere and transferring credits back to the college?

There are several ways to draw an SRS from a sampling frame. Two simple ideas are:

- Assign a distinct random number (say, from 0 to 100,000) to each individual. Then sort the random numbers into numerical order, keeping each name with its number. The first $n$ names are then a random sample of that size.
- Assign each individual a single random digit, 0 to 9. Then those with a specific random digit (say, 5) are a 10% SRS.

In practice, simple random samples are often selected automatically by a computer. You provide the sampling frame and the sample size you want, and it produces an SRS.

Samples drawn at random generally differ one from another. Each draw of random numbers selects *different* people for our sample. These differences lead to different values for the variables we measure. We call these sample-to-sample differences **sampling variability**. Sampling variability is unavoidable. This isn't a problem; it's an opportunity. Surprisingly, this sampling variability is *quantifiable*; we can know how large it might be. In future chapters we'll investigate what the variation in a sample can tell us about its population.

### Error Okay, Bias No Way!

Sampling variability is sometimes referred to as *sampling error*, making it sound like it's some kind of mistake. It's not. We understand that samples will vary, so "sampling error" is to be expected. It's *bias* we must strive to avoid. Bias means our sampling method distorts our view of the population, and that will surely lead to mistakes.

### FOR EXAMPLE

#### Using Random Numbers to Get an SRS

There are 80 students enrolled in an Introductory Statistics class; you are to select a sample of 5.

QUESTION: How can you select an SRS of 5 students using these random digits found on the Internet: 05166 29305 77482?

ANSWER: First, I'll number the students from 00 to 79. Taking the random numbers two digits at a time gives me 05, 16, 62, 93, 05, 77, and 48. I'll ignore 93 because the students were numbered only up to 79. And, so as not to pick the same person twice, I'll skip the repeated number 05. My simple random sample consists of students with the numbers 05, 16, 62, 77, and 48.

(Other valid answers are also possible.)

# Stratified Sampling

Simple random sampling is not the only fair way to sample. More complicated designs may save time or money or help avoid sampling problems. All statistical sampling designs have in common the idea that chance, rather than human choice, is used to select the sample.

Sometimes the population is first sliced into homogeneous groups, called *strata*, before the sample is selected.[7] Then simple random sampling is used within each stratum before the results are combined. This common sampling design is called **stratified random sampling**.

---

[7]The word *homogeneous* literally means "same type." So homogeneous groups might be people of similar ages, or apples grown in the same county, or restaurants serving similar foods.

Why would we want to complicate things? Let's look at a fictional, but plausible, example. The chief administrator of State University, President Vazquez, wants to gauge the opinions of the university's varsity athletes regarding the quality of the sports facilities there. A census of all varsity athletes would be difficult, so Vazquez decides to survey an SRS of 10% of its athletes instead. At State University, it happens that the swimming pool is well-maintained while the tennis courts are fairly run-down. Swimmers may by chance end up being overrepresented in an SRS and tennis players underrepresented, which could easily mislead Vazquez into thinking that varsity athletes' opinions are generally higher than they really are. On the other hand, the swimmers might be underrepresented and the tennis players overrepresented, which could mislead Vazquez into thinking the varsity athletes' opinions are generally *lower* than they really are. It's important to understand that the SRS isn't *systematically* favoring good or bad opinions from students, so the SRS sampling method *isn't biased*. But it allows a lot of variability across different possible samples, so that any particular sample could easily be unrepresentative of the population.

But Vazquez could do the following instead: purposefully select 10% of the varsity athletes *from each sport*. That would make sure that if athletes' opinions varied from sport to sport, no sport's athletes would end up constituting an "unfair" share of the sample—either by being under- or overrepresented. Different samples will still lead to different opinions, but no longer because of chance variability in how many athletes from each sport get sampled.

In this example, the strata are the different sports and we say that the stratified random sampling process reduces (or "controls") some of the sampling variability in the responses. Note that stratification on just any old variable wouldn't necessarily do that. For example, using students' birth months as strata wouldn't help because birth months aren't likely to have anything to do with athletes' opinions of the sports facilities.

Compared to simple random sampling, stratified sampling has an additional benefit besides just reducing sampling variability. It also allows President Vazquez to see how students' opinions of the sports facilities varies by sport. (Vazquez might take note that all of the tennis players in the sample express dissatisfaction, and look at upgrading the tennis courts.) But the more important benefit of stratifying is reducing sampling variability. Let's look more closely at how that works using another hypothetical situation and (surprise!) a simulation.

# Stratified Sampling: A Simulation

Every year when the National Council of Teachers of Mathematics (NCTM) holds its convention, about 10,000 math teachers converge on some lucky city.[8] Suppose we want to conduct a poll to find out what fraction of these teachers approve of the programs and policies of the nation's Secretary of Education. From conference registration records we could learn that 60% of the attendees teach in public schools and 40% in private schools. We might suspect that these two groups have different opinions on this issue. In reality, we could not know in advance what that difference might be, but for purposes of our simulation let's assume that the true approval rate is only 30% among the public school teachers and a whopping 90% among the private school crowd. How could this approval gap affect what we might learn from our poll?

We decide to survey 200 randomly selected math teachers. What sampling methodology should we use?

**PLAN A:** We choose a simple random sample of any 200 teachers.

**PLAN B:** We stratify our sample by type of school, randomly choosing 120 public school teachers and 80 private school teachers. This makes the respondents' school type ratio the same as the population's, so neither group can be over- or underrepresented in the sample.

---

[8] Q: What could be more fun than 10,000 math teachers in one place? (A: Only 5000?)

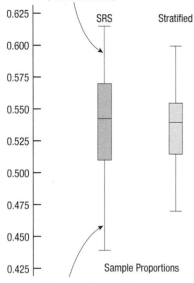

Most of the samples that led to these estimates probably included an overrepresentation of private school teachers

Most of the samples that led to these estimates probably included an overrepresentation of public school teachers

Figure 11.1
Stratified sampling reduces sample-to-sample variability.

With that framework, we ran our simulation. In the first SRS 115 teachers expressed approval of the Secretary of Education, a sample proportion of $115/200 = 0.575$. In the next SRS the proportion was $96/200 = 0.48$. And so on; we simulated 100 simple random samples. Then we tried the stratified sampling plan 100 times, too. In our first stratified sample 38 of 120 public school teachers approved, as did 73 of the 80 from private schools. That made this sample's approval proportion $(38 + 73)/200 = 0.555$. One down, 99 more to go.

What did these simulated samples reveal? The boxplots compare the distributions of the simulated sample proportions. What do you see? For one thing, the medians are about the same. It appears that both SRSs and stratified samples produce estimates centering around 54% approval.[9] It's good to see that both approaches target the same truth in the NCTM population . . . *on average*.

But in the real world we don't have the luxury of choosing 100 samples. We just get one. We'd want to use a sampling method that's more likely to produce a result close to the truth. The boxplots clearly show that when we stratify there's less variation from sample to sample, so we can expect a stratified sample to provide a more accurate estimate.

If there weren't an opinion gap between public and private school teachers, stratifying would be a waste of time; an SRS would work just as well. But when we think different subgroups of the population will tend to have different responses, going to the extra trouble of choosing a stratified sample offers us the important advantage of reducing sampling variability.

Here's another example of how stratification reduces sampling variability. Suppose that during a presidential election period, you want to estimate the proportion of registered U.S. voters who favor the Republican candidate. You could try to take a simple random sample from the population of all registered voters. But you might, just by chance, get an overrepresentation of, say, Texas residents. Because Texans, more than voters in the general population, tend to favor Republicans, this particular sample could easily lead you to overestimate the overall support for the Republican candidate. But you could just as easily have gotten an *under*representation of Texans, which could easily lead you to *under*estimate the overall support for the Republican candidate. If you construct your sample so that Texans are represented in the sample in the same proportion as they are in the population, then both of these things are less likely to happen.[10]

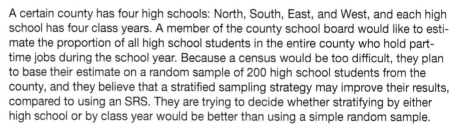

## FOR EXAMPLE
### Stratifying the Sample

A certain county has four high schools: North, South, East, and West, and each high school has four class years. A member of the county school board would like to estimate the proportion of all high school students in the entire county who hold part-time jobs during the school year. Because a census would be too difficult, they plan to base their estimate on a random sample of 200 high school students from the county, and they believe that a stratified sampling strategy may improve their results, compared to using an SRS. They are trying to decide whether stratifying by either high school or by class year would be better than using a simple random sample.

QUESTION: How should the school board member decide whether stratifying on either of these variables would be better than using a simple random sample? And what does "better" mean in this context?

---

[9] Yes, 54% is indeed the true approval rate based on the percentages we used to create this NCTM population. See if you can figure out why. Go ahead, try! We believe in you.

[10] In the examples of stratified sampling that we've shared here, we described sampling proportionally from each stratum. Stratified sampling actually doesn't require that. In fact, it's more common to deliberately sample about the same *number* of people (or objects) from each stratum, which turns out to be a more efficient use of resources. But doing that requires constructing weighted averages and weighted proportions—topics we won't discuss in this text.

ANSWER: Stratified sampling works better the more strongly the strata are associated with the variable of interest. If the school board member suspects that the proportion of students holding a part-time job varies a lot across the four schools, then stratifying by school would be better than using an SRS. If they suspect that the proportion of students holding a part-time job varies a lot across the four class years, then stratifying by class year would be better than using an SRS. In either case, "better" means that the stratified sampling strategy is more likely to produce a sample that's representative of the population of all students than is simple random sampling.

## Cluster and Multistage Sampling

Suppose we wanted to assess the reading level of this textbook based on the length of the sentences. Simple random sampling could be awkward; we'd have to number each sentence, then find, for example, the 576th sentence or the 2482nd sentence, and so on. Doesn't sound like much fun, does it?

It would be much easier to pick a few *pages* at random and count the lengths of the sentences on those pages. That works if we believe that each page is representative of the entire book in terms of reading level. Splitting the population into representative *clusters* can make sampling more practical. Then we could simply select one or a few clusters at random and perform a census within each of them. This sampling design is called **cluster sampling**.

Clusters are generally selected for reasons of efficiency, practicality, or cost. Ideally, if each cluster represents the full population fairly, cluster sampling will be unbiased. But often we just hope that by choosing a sample of clusters, we can obtain a representative sample of the entire population.

### FOR EXAMPLE
#### Cluster Sampling

RECAP: A school board member in a county with four high schools wants to estimate the proportion of high school students in the county who have part-time jobs using a random sample of 200 students. They considered using a SRS and two possible stratified sampling strategies, but all of these methods involve tracking down a lot of students all over the place. Fortunately, the principals at the four high schools will all permit the school board member to survey every student in each of a few homeroom classes—but only a few. Each homeroom has about 20 students at all the schools.

QUESTION: How could the local school board member use this fact to draw a cluster sample? How is this an improvement over the other sampling methods that were considered? Do you have any concerns about this cluster sampling method?

ANSWER: The school board member could randomly sample 10 homeroom classes from among all the homerooms across the county. They could visit each of those homerooms and survey all of its students at the same time. This is an improvement over simple random sampling and stratified random sampling because it's much more convenient to carry out logistically; the school board member needs to visit only 10 homerooms to get all their data. It may produce a sample that's fairly representative of the whole county if the proportions of students holding part-time jobs is fairly constant across all the different homerooms. But if there is a lot of variation across homerooms, then cluster sampling could easily produce unrepresentative samples, leading to an estimate of countywide job participation that is way off.

What's the difference between cluster sampling and stratified sampling? We stratify to ensure that our sample represents different groups in the population, and we sample randomly within each stratum. Strata are internally homogeneous, but differ from one another. By contrast, we select clusters to make sampling more practical or affordable. Clusters can be heterogeneous; we want our randomly selected clusters to provide a representative sample of the population.

### Stratified vs. Cluster Sampling

Boston cream pie consists of layers of yellow cake separated by layers of cream, all topped by chocolate frosting. Suppose you are a professional taster (yes, there really are such people) whose job is to check your company's pies for quality. You'd need to eat small samples of randomly selected pies, tasting all three components: the cake, the creme, and the frosting.

One approach is to cut a thin vertical slice out of the pie. Such a slice will be a lot like the entire pie, so by eating that slice, you'll learn about the whole pie. This vertical slice containing all the different ingredients in the pie would be a *cluster* sample.

Another approach is to sample in *strata:* Select some tastes of the cake at random, some tastes of creme at random, and some bits of frosting at random. You'll end up with a reliable judgment of the pie's quality.

Many populations you might want to learn about are like this Boston cream pie. You can think of the subpopulations of interest as horizontal strata, like the layers of pie. Cluster samples slice vertically across the layers to obtain clusters, each of which is representative of the entire population. Stratified samples represent the population by drawing some from each layer, reducing variability in the results that could arise because of the differences among the layers.

Sometimes we use a variety of sampling methods together. In trying to assess the reading level of this book, we might worry that it starts out easy and then gets harder as the concepts become more difficult. If so, we'd want to avoid samples that selected heavily from early or from late chapters. To guarantee a fair mix of chapters, we could randomly choose one chapter from each of the seven parts of the book and then randomly select a few pages from each of those chapters. If, altogether, that made too many sentences, we might select a few sentences at random from each of the chosen pages. So, what is our sampling strategy? First we stratify by the part of the book and randomly choose a chapter to represent each stratum. Within each selected chapter, we choose pages as clusters. Finally, we consider an SRS of sentences within each cluster. Sampling schemes that combine several methods are called **multistage samples**. Most surveys conducted by professional polling organizations use some combination of stratified and cluster sampling as well as simple random samples. Analyzing data that are collected using multistage sampling requires somewhat advanced statistical tools. It's good to know that multistage sampling is used in practice, but you won't encounter it much more in this text after this chapter.

# Systematic Samples

Some samples select individuals systematically. For example, you might survey every 10th person on an alphabetical list of students. To make it random, you must start the systematic selection at a randomly selected individual and the order of the list must not be associated in any way with the responses sought. Then **systematic sampling** can give a representative sample.

Systematic sampling can be much less expensive than true random sampling. When you use a systematic sample, you should justify the assumption that the systematic method is not associated with any of the measured variables. For example, if you decided to sample students in their dorm rooms by knocking on every other door, a dorm in which male and female rooms were alternated could result in a sample that was all male or all female.

## JUST CHECKING

2. We need to survey a random sample of the 300 passengers on a flight from San Francisco to Tokyo. Name each sampling method described below.
   a) Pick every 10th passenger as people board the plane, starting with a randomly chosen passenger among the first 10.
   b) From the boarding list, randomly choose 5 people flying first class and 25 of the other passengers.
   c) Randomly generate 30 seat numbers and survey the passengers who sit there.
   d) Randomly select a seat position (right window, right center, right aisle, etc.) and survey all the passengers sitting in those seats.

## STEP-BY-STEP EXAMPLE

### Sampling

The assignment says, "Conduct your own sample survey to find out how many hours per week students at your school spend watching TV during the school year." Let's see how we might do this step by step. (Remember, though—actually collecting the data from your sample can be difficult and time consuming.)

QUESTION: How would you design this survey?

**THINK** — **PLAN** State what you want to know.

I want to design a study to find out how many hours of TV students at my school watch.

**POPULATION AND PARAMETER** Identify the W's of the study. The *Why* determines the population and the associated sampling frame. The *What* identifies the parameter of interest and the variables measured. The *Who* is the sample we actually draw. The *How, When,* and *Where* are given by the sampling plan.

Often, thinking about the *Why* will help us see whether the sampling frame and plan are adequate to learn about the population.

The population to be studied is students at our school. I will obtain a list of all students currently enrolled and use it as the sampling frame. The parameter of interest is the number of TV hours watched per week during the school year, which I will attempt to measure by asking students how much TV they watched during the previous week.

**SAMPLING PLAN** Specify the sampling method and the sample size, *n*. Specify how the sample was actually drawn. What is the sampling frame? How was the randomization performed?

A good description should be complete enough to allow someone to replicate the procedure, drawing another sample from the same population in the same manner.

I decided against stratifying by class or sex because I didn't think TV watching would differ much between males and females or across classes. I selected a simple random sample of students from the list. I obtained an alphabetical list of students, assigned each a random digit between 0 and 9, and then selected all students who were assigned a 4. This method identified a sample of 212 students from the population of 2133 students.

**SHOW** — **SAMPLING PRACTICE** Specify *When, Where,* and *How* the sampling was performed. Specify any other details of your survey, such as how respondents were contacted, what incentives were offered to encourage them to respond, how nonrespondents were treated, and so on.

The survey was taken over the period October 15 to October 25. I sent surveys to selected students by e-mail, with the request that they respond by e-mail as well. Students who could not be reached by e-mail were handed the survey in person.

**TELL** **SUMMARY AND CONCLUSION** This report should include a discussion of all the elements. In addition, it's good practice to discuss any special circumstances. Professional polling organizations report the *When* of their samples but will also note, for example, any important news that might have changed respondents' opinions during the sampling process. In this survey, perhaps, a major news story or sporting event might change students' TV viewing behavior.

The question you ask also matters. It's better to be specific ("How many hours did you watch TV last week?") than to ask a general question ("How many hours of TV do you usually watch in a week?").

The report should show a display of the data, provide and interpret the statistics from the sample, and state the conclusions that you reached about the population.

During the period October 15 to October 25, 212 students were randomly selected, using a simple random sample from a list of all students currently enrolled. The survey they received asked the following question: "How many hours did you spend watching television last week?"

Of the 212 students surveyed, 110 responded. It's possible that the nonrespondents differ in the number of TV hours watched from those who responded, but I was unable to follow up on them due to limited time. The 110 respondents reported an average 3.62 hours of TV watching per week. The median was only 2 hours per week. A histogram of the data shows that the distribution is highly right-skewed, indicating that the median might be a more appropriate summary of the typical TV watching of the students.

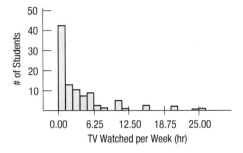

Most of the students (90%) watch between 0 and 10 hours per week, while 30% reported watching less than 1 hour per week. A few watch much more. About 3% reported watching more than 20 hours per week.

# The Valid Survey

It isn't sufficient to just draw a sample and start asking questions. You'll want your survey to be *valid*. A valid survey yields the information you are seeking about the population you are interested in. Before setting out to survey, ask yourself:

- What do I want to know?
- Am I asking the right respondents?
- Am I asking the right questions?
- What would I do with the answers if I had them; would they address the things I want to know?

These questions may sound obvious, but there are a number of pitfalls to avoid:

*Know what you want to know.* Before considering a survey, understand what you hope to learn and about whom you hope to learn it.

*Use the right sampling frame.* A valid survey obtains responses from the appropriate respondents. Be sure you have a suitable *sampling frame*. Have you identified the population of interest and sampled from it appropriately? A company might survey customers who returned warranty registration cards—a readily available sampling frame. But if the company wants to make their product more attractive, the most important population is the customers who rejected their product in favor of one from a competitor.

*Tune your instrument.* You may be tempted to ask questions you don't really need, but longer survey instruments yield fewer responses and thus a greater chance of nonresponse bias.

*Ask specific rather than general questions.* People are not very good at estimating their typical behavior, so it is better to ask "How many hours did you sleep last night?" than "How much do you usually sleep?" Sure, some responses will include some unusual events (My dog was sick; I was up all night), but overall you'll get better data.

*Ask for quantitative results when possible.* "How many magazines did you read last week?" is better than "How much do you read: A lot, A moderate amount, A little, or None at all?" People will interpret these phrases differently.

*Be careful in phrasing questions.* A respondent may not understand the question—or may understand the question differently than the researcher intended it. ("Does anyone in your family ride a motorcycle?" Do you mean just me, my spouse, and my children? Or does "family" include my father, my siblings, and my second cousin once removed? And does a motor scooter count?) Respondents are unlikely (or may not have the opportunity) to ask for clarification. A question like "Do you approve of the recent actions of the Ambassador to Mexico?" is likely not to measure what you want if many respondents don't know who the Ambassador is or what they recently did.

Respondents may even lie or shade their responses if they feel embarrassed by the question ("Did you have too much to drink last night?"), are intimidated or insulted by the question ("Could you understand our new *Instructions for Dummies* manual, or was it too difficult for you?"), or if they want to avoid offending the interviewer ("Would you hire a man with a tattoo?" asked by a tattooed interviewer). Also, be careful to avoid phrases that have double or regional meanings. ("How often do you go to town?" might be interpreted differently by different people and cultures.)

*Even subtle differences in phrasing can make a difference.* In January 2006, the *New York Times* asked half of the 1229 U.S. adults in their sample the following question:

> After 9/11, President Bush authorized government wiretaps on some phone calls in the U.S. without getting court warrants, saying this was necessary to reduce the threat of terrorism. Do you approve or disapprove of this?[12]

They found that 53% of respondents approved. But when they asked the other half of their sample a question with different phrasing,

> After 9/11, George W. Bush authorized government wiretaps on some phone calls in the U.S. without getting court warrants. Do you approve or disapprove of this?[13]

only 46% approved.

*Be careful in phrasing answers.* It's often a good idea to offer choices rather than inviting a free response. Open-ended answers can be difficult to analyze. "How did you like the movie?" may start an interesting debate, but it may be better to give a range of possible responses. Be sure to phrase answers in a neutral way. When asking "Do you support higher school taxes?" positive responses could be worded "Yes," "Yes, it is important for our children," or "Yes, our future depends on it." But those are not equivalent answers.

The best way to protect a survey from unanticipated measurement errors is to perform a pilot survey. A **pilot** is a trial run of the survey you eventually plan to give to a larger group, using a draft of your survey questions administered to a small sample drawn from the same sampling frame you intend to use. By analyzing the results from this smaller survey, you can often discover ways to improve your survey instrument.

---

### A SHORT SURVEY

Given that the *New York Times* reports[11] that statisticians can earn $125,000 at top companies their first year on the job, do you think this course will be valuable to *you*?

### PILOT TESTING A SURVEY QUESTION

A researcher distributed a survey to an organization before some economizing changes were made. She asked how people felt about a proposed cutback in secretarial and administrative support on a seven-point scale from Very Happy to Very Unhappy.

But virtually all respondents were very unhappy about the cutbacks, so the results weren't particularly useful. If she had pretested the question, she might have chosen a scale that ran from unhappy to outraged.

---

[11] www.nytimes.com/2009/08/06/technology/06stats.html
[12] Excerpt from "The New York Times/CBS News Poll, January 20–25, 2006." Published in *The New York Times*, © 2006.
[13] Ibid.

# Lots Can Go Wrong: How to Sample Badly

Bad sample designs yield worthless data. Many of the most convenient forms of sampling can be seriously biased. And there is no way to correct for the bias from a bad sample. So it's wise to pay attention to sample design—and to beware of reports based on poor samples.

## Mistake 1: Sample Volunteers

One of the most common dangerous sampling methods is a voluntary response sample. In a **voluntary response sample**, a large group of individuals is invited to respond, and those who choose to respond are counted. The respondents, rather than the researcher, decide who will be in the sample. This method is used by call-in shows, 900 numbers, Internet polls, and letters written to members of Congress. Voluntary response sampling is almost always biased, so conclusions drawn from such data are almost always wrong.

It's often hard to define the sampling frame of a voluntary response study. Practically, the frames are groups such as Internet users who frequent a particular website or those who happen to be watching a particular TV show at the moment. But those sampling frames don't correspond to the population of interest.

Even within the sampling frame, voluntary response samples are often biased toward those with strong opinions or those who are strongly motivated. People with very negative opinions tend to respond more often than those with equally strong positive opinions. The sample is not representative, even though every individual in the population may have been offered the chance to respond. The resulting **voluntary response bias** invalidates the survey.

> **IF YOU HAD IT TO DO OVER AGAIN, WOULD YOU HAVE CHILDREN?**
> Ann Landers, the advice columnist, asked parents this question. The overwhelming majority—70% of the more than 10,000 people who chose to write in—said no, kids weren't worth it. A more carefully designed survey later showed that about 90% of parents actually are happy with their decision to have children. What accounts for the striking difference in these two results? What parents do you think are most likely to respond to the original question?

### FOR EXAMPLE: Voluntary Response Sample

**RECAP:** A local school board member would like to estimate the proportion of high school students in the county who hold part-time jobs. They are considering posting questionnaires at the exits of the four high schools and inviting students to complete one and drop it in a provided box.

**QUESTION:** What's wrong with this idea?

**ANSWER:** Lots! For one thing, with unobserved questionnaires and drop boxes, there's no way to know that responses are actually coming from high school students, nor that students are submitting only one response. But even if you assume the good intentions of everyone who sees the questionnaire, there's still the problem of voluntary response bias. Perhaps students with part-time jobs tend to be proud of it, and are more likely to respond to a survey about jobs than are those without part-time jobs. Or perhaps those with part-times jobs are *less* likely to respond to any questionnaire because they have less time to spare—they've got to get to work!

Even though we can't know which direction the bias will go in, that doesn't mean it's not there. Results from a survey conducted like this could not be trusted.

## Mistake 2: Sample Conveniently

Another sampling method that doesn't work is convenience sampling. As the name suggests, in **convenience sampling** we simply include the individuals who are convenient for us to sample. Unfortunately, this group may not be representative of the population. Here's an amusing example. Back in 2001, when computer use in the home was not as common as it is today, a survey of 437 potential home buyers in Orange County, California, reached the surprising conclusion that

> *All but 2 percent of the buyers have at least one computer at home, and 62 percent have two or more. Of those with a computer, 99 percent are connected to the Internet. (Source: Jennifer Hieger, "Portrait of Homebuyer Household: 2 Kids and a PC," Orange County Register, 27 July 2001)*

How was the survey conducted? On the Internet!

### Internet Surveys

Internet convenience surveys are worthless. As voluntary response surveys, they have no well-defined sampling frame (all those who use the Internet and visit their site?) and thus report no useful information. Do not believe them.

### FOR EXAMPLE: Convenience Sample

**RECAP:** A local school board member would like to estimate the proportion of high school students in the county who hold part-time jobs during the school year. Knowing that football games between two of the county's high schools draw lots of students from all over the county, they consider simply attending a football game and surveying all the high school attendees that they can.

**QUESTION:** What's wrong with this sampling strategy?

**ANSWER:** This would be a convenience sample, and it's likely to be biased. Perhaps those with part-time jobs are more likely to attend football games because they have more money to spend on admission tickets and snacks at the game. Perhaps those with part-time jobs are *less* likely to attend football games because they're *at* their jobs during football games. It's impossible to be sure of the direction of the bias, but there's no way to know that the sample would be representative of the population and there are plenty of reasons to doubt it.

## Mistake 3: Use a Bad Sampling Frame

An SRS from an incomplete sampling frame introduces bias because the individuals included may differ from the ones not in the frame. People in prison, people dealing with homelessness, and long-term travelers are all likely to be missed in phone polls, for example.

## Mistake 4: Undercoverage

Many survey designs suffer from **undercoverage**, in which some portion of the population is not sampled at all or has a smaller representation in the sample than it has in the population. Undercoverage can arise for a number of reasons, but it's always a potential source of bias.

Telephone surveys are usually conducted when you are likely to be home, such as dinnertime. If you eat out often, you may be less likely to be surveyed, a possible source of undercoverage.

In addition to avoiding making mistakes when sampling, you should also be alert to some potential problems that can arise on their own. We describe two here: nonresponse bias and response bias.

## Nonresponse Bias

A common and serious potential source of bias for most surveys is **nonresponse bias**. No survey succeeds in getting responses from everyone. The problem is that those who don't respond may differ from those who do—and they may differ on precisely the variables we care about. Rather than sending out a large number of surveys for which the response rate will be low, it is often better to design a smaller randomized survey for which you have the resources to ensure a high response rate. One of the problems with nonresponse bias is that it's usually impossible to tell what the nonrespondents might have said.

It turns out that the *Literary Digest* survey was wrong on two counts. First, their list of 10 million people was not representative. There was a selection bias in their sampling frame. There was also a nonresponse bias. We know this because the *Digest* also surveyed a *systematic* sample in Chicago, sending the same question used in the larger survey to every third registered voter. They *still* got a result in favor of Landon, even though Chicago voted overwhelmingly for Roosevelt in the election. This suggests that the Roosevelt supporters were less likely to respond to the *Digest* survey. There's a modern version of this problem: Many people today refuse calls coming from numbers they don't recognize, which would include all polling agencies. People who do that almost certainly differ in many ways from people who answer the call. For one thing, the call-refusers tend to be younger.

## Response Bias

**Response bias**[14] refers to anything in the survey design that influences the responses. Response biases include the tendency of respondents to tailor their responses to try to please the interviewer, the natural unwillingness of respondents to reveal personal facts or admit to illegal or unapproved behavior, and the ways in which the wording of the questions can influence responses.

## How to Think About Biases

- **Look for biases in any survey you encounter.** If you design one of your own, ask someone else to help look for biases that may not be obvious to you. And do this *before* you collect your data. There's no way to recover from a biased sampling method or a survey that asks biased questions.
  Sorry, it just can't be done.
  A bigger sample size for a biased study just gives you a bigger useless study.
  A really big sample gives you a really big useless study. (Think of the 2.4 million *Literary Digest* responses.)
- **Spend your time and resources reducing biases.** No other use of resources is as worthwhile as reducing the biases.
- **Think about the members of the population who could have been excluded from your study.** Be careful not to claim that you have learned anything about them.
- **If you can, pilot-test your survey.** Administer the survey in the exact form that you intend to use it to a small sample drawn from the population you intend to sample. Look for misunderstandings, misinterpretation, confusion, or other possible biases. Then refine your survey instrument.
- **Always report your sampling methods in detail.** Others may be able to detect biases where you did not expect to find them.

> **What's the Sample?**
> The population we want to study is determined by asking *why*. When we design a survey, we use the term *sample* to refer to the individuals selected, from whom we hope to obtain responses. Unfortunately, the real sample is just those we can reach to obtain responses—the *who* of the study. These are slightly different uses of the same term *sample*. The context usually makes clear which we mean, but it's important to realize that the difference between the two could undermine even a well-designed study.

---

[14] Response bias is not the opposite of nonresponse bias. (We don't make these terms up; we just try to explain them.)

## WHAT CAN GO WRONG?

- **Get a sample that looks like the population.** The principal thing that can go wrong in sampling is that the sample can fail to represent the population. Unfortunately, this can happen in many different ways and for many different reasons. We've considered many of them in the chapter.
- **Avoid bias.** Any of the many types of bias we've discussed can render your results meaningless. Pay attention to all of them.

## WHAT HAVE WE LEARNED?

We've learned that a representative sample can offer us important insights about populations. It's the size of the sample—and not its fraction of the larger population—that determines the precision of the statistics it yields.

We've learned several ways to draw samples, all based on the power of randomness to make them representative of the population of interest:

- A simple random sample (SRS) is our standard. Every possible group of n individuals has an equal chance of being our sample. That's what makes it *simple*. One convenient way to select a SRS from a small population is to simply draw well-mixed slips of paper from a hat, but for larger populations we use random numbers.
- Stratified samples can reduce sampling variability by identifying homogeneous subgroups (strata) and then randomly sampling within each. Stratified sampling works better when the variable used to construct the strata is more strongly associated with the response variable.
- Cluster samples randomly select among heterogeneous subgroups (clusters) and conduct a census within each cluster. This approach can make our sampling tasks more manageable.
- Systematic samples can work in some situations and are often the least expensive method of sampling. But we still want to start them randomly.
- Multistage samples combine several random sampling methods.

We've learned that bias can destroy our ability to gain insights from our sample:

- Voluntary response samples are almost always biased and should be avoided and distrusted.
- Convenience samples are likely to be flawed for similar reasons.
- Bad sampling frames can lead to samples that don't represent the population of interest.
- Undercoverage occurs when individuals from a subgroup of the population are selected less often than they should be.
- Nonresponse bias can arise when sampled individuals will not or cannot respond.
- Response bias arises when respondents' answers might be affected by external influences, such as question wording or interviewer behavior.

Finally, we've learned to look for biases in any survey we find and to be sure to report our methods whenever we perform a survey so that others can evaluate the fairness and accuracy of our results.

## TERMS

**Population**
The entire group of individuals or instances about whom we hope to learn. (p. 291)

**Sample**
A (representative) subset of a population, examined in hope of learning about the population. (p. 291)

**Sample survey**
A study that asks questions of a sample drawn from some population in the hope of learning something about the entire population. Polls taken to assess voter preferences are common sample surveys. (p. 291)

**Bias**
Any systematic failure of a sampling method to represent its population is bias. Biased sampling methods tend to over- or underestimate parameters. It is almost impossible to recover from bias, so efforts to avoid it are well spent. (p. 291) Common errors include
- relying on voluntary response,
- undercoverage of the population,
- nonresponse bias, and
- response bias.

**Randomization**
The best defense against bias is randomization, in which each individual is given a fair, random chance of selection. (p. 292)

**Sample size**
The number of individuals in a sample. The sample size determines how well the sample represents the population, not the fraction of the population sampled. (p. 293)

**Census**
A sample that consists of the entire population is called a census. (p. 293)

**Population parameter**
A numerically valued attribute of a model for a population. We rarely expect to know the true value of a population parameter, but we do hope to estimate it from sampled data. For example, the mean income of all employed people in the country is a population parameter. (p. 294)

**Statistic, sample statistic**
Statistics are values calculated for sampled data. Those that correspond to, and thus estimate, a population parameter are of particular interest. For example, the mean income of all employed people in a representative sample can provide a good estimate of the corresponding population parameter. The term "sample statistic" is sometimes used, usually to parallel the corresponding term "population parameter." (p. 294)

**Representative**
A sample is said to be representative if the statistics computed from it accurately reflect the corresponding population parameters. (p. 295)

**Simple random sample (SRS)**
A simple random sample of sample size $n$ is a sample in which each set of $n$ elements in the population has an equal chance of selection. (p. 296)

**Sampling frame**
A list of individuals from whom the sample is drawn is called the sampling frame. Individuals who may be in the population of interest, but who are not in the sampling frame, cannot be included in any sample. (p. 296)

**Sampling variability**
The natural tendency of randomly drawn samples to differ, one from another. Sometimes, unfortunately, called *sampling error*, sampling variability is no error at all, but just the natural result of random sampling. (p. 296)

**Stratified random sample**
A sampling design in which the population is divided into several subpopulations, or **strata.** Random individuals are then drawn from each stratum so that the sample includes individuals from each, often in a representative proportion. If the strata are homogeneous, but are different from each other, stratified sampling can reduce variability in results. (p. 296)

**Cluster sample**
A sampling design in which entire groups, or **clusters,** are chosen at random. Cluster sampling is usually selected as a matter of convenience, practicality, or cost. Clusters are heterogeneous, and a random sample of clusters should be representative of the population. (p. 299)

**Multistage sample**
Sampling schemes that combine several sampling methods are called multistage samples. For example, a national polling service may stratify the country by geographical regions, select a random sample of cities from each region, and then interview a cluster of residents in each city. (p. 300)

| | |
|---|---|
| **Systematic sample** | A sample drawn by selecting individuals systematically from a sampling frame. When there is no relationship between the order of the sampling frame and the variables of interest, a systematic sample can be representative. (p. 300) |
| **Pilot survey** | A small trial run of a survey to check whether questions are clear. A pilot survey can reduce errors due to ambiguous questions. (p. 303) |
| **Voluntary response bias** | Bias introduced to a sample when individuals can choose on their own whether to participate in the sample. Samples based on voluntary response are always invalid and cannot be recovered, no matter how large the sample size. (p. 304) |
| **Convenience sample** | A convenience sample consists of the individuals who are conveniently available. Convenience samples often fail to be representative because every individual in the population is not equally convenient to sample. (p. 305) |
| **Undercoverage** | A sampling scheme that biases the sample in a way that gives a part of the population less representation than it has in the population suffers from undercoverage. (p. 305) |
| **Nonresponse bias** | Bias introduced when a large fraction of those sampled fails to respond. Those who do respond are likely to not represent the entire population. Voluntary response bias is a form of nonresponse bias, but nonresponse may occur for other reasons. For example, those who are at work during the day won't respond to a telephone survey conducted only during working hours. (p. 306) |
| **Response bias** | Anything in a survey design that influences responses falls under the heading of response bias. One typical response bias arises from the wording of questions, which may suggest a favored response. Voters, for example, are more likely to express support of "the President" than support of the particular person holding that office at the moment. (p. 306) |

## ON THE COMPUTER

### Sampling

Computer-generated pseudorandom numbers are usually good enough for drawing random samples. But there is little reason not to use the truly random values available on the Internet.

Here's a convenient way to draw an SRS of a specified size using a computer-based sampling frame. The sampling frame can be a list of names or of identification numbers arrayed, for example, as a column in a spreadsheet, statistics program, or database:

1. Generate random numbers of enough digits so that each exceeds the size of the sampling frame list by several digits. This makes duplication unlikely.

2. Assign the random numbers arbitrarily to individuals in the sampling frame list. For example, put them in an adjacent column.

3. Sort the list of random numbers, carrying along the sampling frame list.

4. Now the first $n$ values in the sorted sampling frame column are an SRS of $n$ values from the entire sampling frame.

## EXERCISES

1. **Roper** Through its *Roper Reports Worldwide*, GfK Roper conducts a global consumer survey to help multinational companies understand different consumer attitudes throughout the world. Within 30 countries, the researchers interview 1000 people aged 13–65. Their samples are designed so that they get 500 men and 500 women in each country. (www.gfkamerica.com)

   a) Are they using a simple random sample? Explain.
   b) What kind of design do you think they are using?

2. **Student center survey** For their class project, a group of Statistics students decide to survey the student body to assess opinions about the proposed new student center. Their sample of 200 contained 50 first-year students, 50 sophomores, 50 juniors, and 50 seniors.

   a) Do you think the group was using an SRS? Why?
   b) What sampling design do you think they used?

**3. Ice Cream** The website www.buzzfeed.com is an entertainment and news website, known for fun and silly content. In 2021, a Buzzfeed poll asked visitors of the site to select between 8 different ice cream choices. Of the 9072 respondents, 17% said that they wanted vanilla ice cream.

   a) What kind of sample was this?
   b) Define the parameter of interest.
   c) How much confidence would you place in using 17% as an estimate of the fraction of people who prefer vanilla ice cream?

**4. Drug tests** Major League Baseball tests players to see whether they are using performance-enhancing drugs. Officials select a team at random, and a drug-testing crew shows up unannounced to test all 40 players on the team. Each testing day can be considered a study of drug use in Major League Baseball.

   a) Who is the population of interest?
   b) What kind of sample is this?
   c) Is that choice appropriate?

**5. Gallup** At its website (www.gallup.com) the Gallup Poll publishes results of a new survey each day. Scroll down to the end, and you'll find a statement that includes words such as these:

   *Results are based on telephone interviews with 1,008 national adults, aged 18 and older, conducted January 3–5, 2013. . . . In addition to sampling error, question wording and practical difficulties in conducting surveys can introduce error or bias into the findings of public opinion polls.*

   a) For this survey, identify the population of interest.
   b) Gallup performs its surveys by phoning numbers generated at random by a computer program. What is the sampling frame?

**6. Gallup World** At its website (www.gallupworldpoll.com) the Gallup World Poll describes its methods. After one report it explained:

   *Results are based on face-to-face interviews with randomly selected national samples of approximately 1,000 adults, aged 15 and older, who live permanently in each of the 21 sub-Saharan African nations surveyed. Those countries include Angola (areas where land mines might be expected were excluded), Benin, Botswana, Burkina Faso, Cameroon, Ethiopia, Ghana, Kenya, Madagascar (areas where interviewers had to walk more than 20 kilometers from a road were excluded), Mali, Mozambique, Niger, Nigeria, Senegal, Sierra Leone, South Africa, Tanzania, Togo, Uganda (the area of activity of the Lord's Resistance Army was excluded from the survey), Zambia, and Zimbabwe. . . . In all countries except Angola, Madagascar, and Uganda, the sample is representative of the entire population.*

   a) Gallup is interested in sub-Saharan Africa. What kind of survey design is it using?
   b) Some of the countries surveyed have large populations. (Nigeria is estimated to have about 130 million people.) Some are quite small. (Togo's population is estimated at 5.4 million.) Nonetheless, Gallup sampled 1000 adults in each country. How does this affect the precision of its estimates for these countries?

**7–10. What did they do?** *For the following reports about statistical studies, identify the following items (if possible). If you can't tell, then say so—this often happens when we read about a survey.*

   a) The population of interest
   b) The population parameter of interest
   c) The sampling frame
   d) The sample
   e) The sampling method, including whether or not randomization was employed
   f) Who (if anyone) was left out of the study
   g) Any potential sources of bias you can detect and any problems you see in generalizing to the population of interest

**7. Medical treatments** Consumers Union, in an attempt to get information about U.S. adults, asked all subscribers whether they had used alternative medical treatments and, if so, whether they had benefited from them. For almost all of the treatments, approximately 20% of those responding reported cures or substantial improvement in their condition. Consumers Union received replies from 12% of its subscribers.

**8. Snack foods** A company packaging snack foods maintains quality control by randomly selecting 10 cases from each day's production and weighing the bags. Then inspectors open one randomly selected bag from each case and inspect the contents. For this exercise, answer the questions both for the bags that are weighed and for the bags that are opened for inspection.

**9. Drinking and driving** In order to determine how adults of legal drinking age in their city feel about whether drinking and driving was a problem, researchers waited outside a bar they had randomly selected from a list of such establishments. They rolled a ten-sided die and it came up 4, so they stopped the fourth person who came out of the bar, then every 10th person after that, and asked whether they thought drinking and driving was a serious problem.

**10. Mayoral race** Hoping to learn what issues may resonate with voters in the coming election, the campaign director for a mayoral candidate randomly selects two blocks from each of the city's election districts. Staff members go there and interview all the residents they can find. The residents were asked to select the three most important issues from a prepared list.

**11. Toxic waste** The Environmental Protection Agency took a map of a region near a former industrial waste dump and placed a grid of 552 squares on it. It randomly selected any 16 of those squares from which to collect soil samples and checked each for evidence of toxic chemicals.

   a) What type of sampling was used?
   b) Is there any sort of bias associated with this sampling procedure?
   c) One researcher suggests that plots closer to the old dump site could contain more contaminants than those farther away. How could the sampling procedure be improved to take this into account?

**12. Social life** A question posted on the gamefaqs.com website on August 1, 2011, asked visitors to the site, "Do you have an active social life outside the Internet?" 22% of the 55,581 respondents said "No" or "Not really, most of my personal contact is online."

   a) Can this survey be used to estimate the proportion of U.S. adults who would say they have an active social life outside the Internet? Why or why not?
   b) Can this survey be used to estimate the proportion of visitors to the site who would say they have an active social life outside the Internet? Why or why not?

**13. Roadblock** State police set up a roadblock to estimate the percentage of cars with up-to-date registration, insurance, and safety inspection stickers. It would be too inconvenient and costly to check every vehicle that passes through a checkpoint, so they decide to stop about 1/20 of the vehicles.

a) Why would a simple random sample be unreasonable for this situation?
b) Identify two possible sampling schemes that could be used. Explain how randomization would be used in each.

**14. Milk samples** Dairy inspectors visit farms unannounced and take samples of the milk to test for contamination. If the milk is found to contain dirt, antibiotics, or other foreign matter, the milk will be destroyed and the farm reinspected until purity is restored.

Would simple random sampling be appropriate for selecting farms for inspection? If so, explain how it would be done. If not, explain why it is not appropriate.

**15. Mistaken poll** A local TV station conducted a "PulsePoll" about the upcoming mayoral election. Evening news viewers were invited to text in their votes, with the results to be announced on the late-night news. Based on the texts, the station predicted that Amabo would win the election with 52% of the vote. They were wrong: Amabo lost, getting only 46% of the vote. Do you think the station's faulty prediction is more likely to be a result of bias or sampling error? Explain.

**16. Another mistaken poll** Prior to the mayoral election discussed in Exercise 15, the newspaper also conducted a poll. The paper surveyed a random sample of registered voters stratified by political party, age, gender, and area of residence. This poll predicted that Amabo would win the election with 52% of the vote. The newspaper was wrong: Amabo lost, getting only 46% of the vote. Do you think the newspaper's faulty prediction is more likely to be a result of bias or sampling error? Explain.

**17. Parent opinion, part 1** In a large city school system with 20 elementary schools, the school board is considering the adoption of a new policy that would require elementary students to pass a test in order to be promoted to the next grade. The PTA wants to find out whether parents agree with this plan. Listed below are some of the ideas proposed for gathering data. For each, indicate what kind of sampling strategy is involved and what (if any) biases might result.

a) Put a big ad in the newspaper asking people to log their opinions on the PTA website.
b) Randomly select one of the elementary schools and contact every parent by phone.
c) Send a survey home with every student, and ask parents to fill it out and return it the next day.
d) Randomly select 20 parents from each elementary school. Send them a survey, and follow up with a phone call if they do not return the survey within a week.

**18. Parent opinion, part 2** Let's revisit the school system described in Exercise 17. Four new sampling strategies have been proposed to help the PTA determine whether parents favor requiring elementary students to pass a test in order to be promoted to the next grade. For each, indicate what kind of sampling strategy is involved and what (if any) biases might result.

a) Run a poll on the local TV news, asking people to dial one of two phone numbers to indicate whether they favor or oppose the plan.
b) Hold a PTA meeting at each of the 20 elementary schools, and tally the opinions expressed by those who attend the meetings.
c) Randomly select one class at each elementary school and contact each of those parents.
d) Go through the district's enrollment records, selecting every 40th parent, starting with a randomly selected parent among the first 40. PTA volunteers will go to those homes to interview the people chosen.

**19. Churches** For your political science class, you'd like to take a survey from a sample of all the Catholic Church members in your city. A list of churches shows 17 Catholic churches within the city limits. Rather than try to obtain a list of all members of all these churches, you decide to pick 3 churches at random. For those churches, you'll ask to get a list of all current members and contact any 100 members selected at random.

a) What kind of design have you used?
b) Describe the steps you would use to carry out this sampling method. Be specific.
c) One possible parameter to measure would be the percentage of members who are active in their church. Describe a potential bias that might arise when attempting measure this parameter.

**20. Playground** Some people have been complaining that the children's playground at a municipal park is too small and is in need of repair. Managers of the park decide to survey city residents to see if they believe the playground should be rebuilt.

a) Define the parameter the managers are attempting to estimate.
b) They hand out questionnaires to parents who bring children to the park. Describe possible biases in this sample and the direction of the bias.
c) Describe a sampling method that will address the problems in part (b).

**21. Roller coasters** An amusement park has opened a new roller coaster. It is so popular that people are waiting for up to 3 hours for a 2-minute ride. Concerned about how patrons (who paid a large amount to enter the park and ride on the rides) feel about this, they survey every 10th person on the line for the roller coaster, starting from a randomly selected individual.

a) What kind of sample is this?
b) What is the sampling frame?
c) Is it likely to be representative?
d) What members of the population of interest are omitted?

**22. Playground, act 2** The survey described in Exercise 20 asked,

*Many people believe this playground is too small and in need of repair. Do you think the playground should be repaired and expanded even if that means raising the entrance fee to the park?*

Describe two ways this question may lead to response bias.

**23. Wording the survey** Two members of the PTA committee in Exercises 17 and 18 have proposed different questions to ask in seeking parents' opinions.

*Question 1: Should elementary school–age children have to pass high-stakes tests in order to remain with their classmates?*

*Question 2: Should schools and students be held accountable for meeting yearly learning goals by testing students before they advance to the next grade?*

a) Do you think responses to these two questions might differ? How? What kind of bias is this?
b) Propose a question with more neutral wording that might better assess parental opinion.

**24. Banning ephedra** An online poll at a website asked:

*A nationwide ban of the diet supplement ephedra went into effect recently. The herbal stimulant has been linked to 155 deaths and many more heart attacks and strokes. Ephedra manufacturer NVE Pharmaceuticals, claiming that the FDA lacked proof that ephedra is dangerous if used as directed, was denied a temporary restraining order on the ban yesterday by a federal judge. Do you think that ephedra should continue to be banned nationwide?*

65% of 17,303 respondents said "yes." Comment on each of the following statements about this poll:

a) With a sample size that large, we can be pretty certain we know the true proportion of Americans who think ephedra should be banned.
b) The wording of the question is clearly very biased.
c) The sampling frame is all Internet users.
d) Results of this voluntary response survey can't be reliably generalized to any population of interest.

**25. Survey questions** Examine each of the following questions for possible bias. If you think the question is biased, indicate how and propose a better question.

a) Should companies that pollute the environment be compelled to pay the costs of cleanup?
b) Given that 18-year-olds are old enough to vote and to serve in the military, is it fair to set the drinking age at 21?

**26. More survey questions** Examine each of the following questions for possible bias. If you think the question is biased, indicate how and propose a better question.

a) Don't you think it is a good idea to prohibit high school students from using distracting cell phones while at school?
b) Given humanity's great tradition of exploration, do you favor continued funding for space flights?

**27. Phone surveys** Anytime we conduct a survey, we must take care to avoid undercoverage. To learn about city residents' views, suppose we plan to select 500 names from a database of phone numbers for city residents call their homes between noon and 4 p.m., and interview whoever answers, anticipating contacts with at least 200 people.

a) Why is it difficult to use a simple random sample here?
b) Describe a more efficient, but still random, sampling strategy.
c) What kinds of households are likely to be included in the eventual sample of opinion? Excluded?
d) Suppose, instead, that we continue calling each number, perhaps in the morning or evening, until an adult is contacted and interviewed. How does this improve the sampling design?
e) Random-digit dialing machines can generate the phone calls for us. How would this improve our design? Is anyone still excluded?

**28. Cell phone survey** There are ways to draw a random sample by calling only cell phone numbers. Discuss the advantages and disadvantages of such a sampling method compared with surveying randomly chosen non–cell telephone numbers. Do you think these advantages and disadvantages have changed over time? How do you expect they'll change in the future?

**29. Arm length** How long is your arm compared with your hand size? Put your right thumb at your left shoulder bone, stretch your hand open wide, and extend your hand down your arm. Put your thumb at the place where your little finger is, and extend down the arm again. Repeat this a third time. Now your little finger will probably have reached the back of your left hand. If the fourth hand width goes past the end of your middle finger, turn your hand sideways and count finger widths to get there.

a) How many hand and finger widths is your arm?
b) Suppose you repeat your measurement 10 times and average your results. What parameter would this average estimate? What is the population?
c) Suppose you now collect arm lengths measured in this way from 9 friends and average these 10 measurements. What is the population now? What parameter would this average estimate?
d) Do you think these 10 arm lengths are likely to be representative of the population of arm lengths in your community? In the country? Why or why not?

**30. Fuel economy** Occasionally, when I fill my car with gas, I figure out how many miles per gallon my car got. I wrote down those results after 6 fill-ups in the past few months. Overall, it appears my car gets 28.8 miles per gallon.

a) What statistic have I calculated?
b) What is the parameter I'm trying to estimate?
c) How might my results be biased?
d) When the Environmental Protection Agency (EPA) checks a car like mine to predict its fuel economy, what parameter is it trying to estimate?

**31. Accounting** Between quarterly audits, a company likes to check on its accounting procedures to address any problems before they become serious. The accounting staff processes payments on about 120 orders each day. The next day, the supervisor rechecks 10 of the transactions to be sure they were processed properly.

a) Propose a sampling strategy for the supervisor.
b) How would you modify that strategy if the company makes both wholesale and retail sales, which require different bookkeeping procedures?

**32. Happy workers?** A manufacturing company employs 14 project managers, 48 supervisors, and 377 laborers. In an effort to keep informed about any possible sources of employee discontent, management wants to conduct job satisfaction interviews with a sample of employees every month.

a) Do you see any potential danger in the company's plan? Explain.
b) Propose a sampling strategy that uses a simple random sample.
c) Why do you think a simple random sample might not provide the representative opinion the company seeks?
d) Propose a better sampling strategy.
e) Listed below are the last names of the project managers. Use random numbers to select two people to be interviewed. Explain your method carefully.

| Barrett | Bowman | Chen |
| DeLara | DeRoos | Grigorov |
| Maceli | Mulvaney | Pagliarulo |
| Rosica | Smithson | Tadros |
| Williams | Yamamoto | |

**33. Quality control** Sammy's Salsa, a small local company, produces 200 cases of salsa a day. Each case contains 12 jars and is imprinted with a code indicating the date and batch number. To help maintain consistency, at the end of each day, Sammy selects jars of salsa, weighs the contents, and tastes the product.

   a) Carefully explain how Sammy would use a cluster sampling strategy, using the cases as clusters. Describe, in detail, how to implement your method.

   b) Sammy is concerned that the quality of his product might decline as the day progresses. Describe a stratified sampling method that will address this concern.

   c) What are the response variables? Are they categorical or quantitative?

**34. A fish story** Concerned about reports of discolored scales on fish caught downstream from a newly sited chemical plant, scientists set up a field station in a shoreline public park.

   a) For one week they asked people who fished there to bring any fish they caught to the field station for a brief inspection. At the end of the week, the scientists said that 18% of the 234 fish that were submitted for inspection displayed the discoloration.

      i) Define the parameter of interest.

      ii) Name this sampling method.

      iii) Describe a bias that concerns you about this method and whether you think the statistic is an under- or overestimate of the parameter.

   b) One scientist suggests that the sampling be conducted for a second week in order to increase the sample size. Will this second week correct the bias discussed in part iii? Explain.

**35. Sampling methods** Consider each of these situations. Do you think the proposed sampling method is appropriate? Explain.

   a) We want to know what percentage of doctors in the county accept Medicaid patients. We call the offices of 50 doctors randomly selected from Yelp, an app that rates businesses.

   b) We want to know what percentage of county businesses anticipate hiring additional employees in the upcoming month. Using the app Yelp, which rates businesses, the first 40 businesses in the app are contacted.

**36. More sampling methods** Consider each of these situations. Do you think the proposed sampling method is appropriate? Explain.

   a) We want to know if there is neighborhood support to turn a vacant lot into a playground. We spend a Saturday afternoon going door-to-door in the neighborhood, asking people to sign a petition.

   b) We want to know if students at our college are satisfied with the selection of food available on campus. We go to the largest cafeteria and interview every 10th person in line, starting with a randomly selected person.

**37. Pep rally** At Central High School, the principal wants to know what percent of the student body like the mandatory school pep rallies that occur on some game days. He wants to select a simple random sample of students and give them a survey. A Statistics student points out that there is probably a large difference in opinion between those who participate in team sports and those who do not, and suggests they do a stratified random sample instead. To help the confused principal understand what she means, she uses a simulation to show how stratified sampling can reduce the variability in the response variable (which in this case is the proportion of students in the sample who like the mandatory pep rallies) compared to a simple random sample. They know that 30% of the students participate in team sports. The student says, "Let's pretend that among students who do participate in sports, 90 percent of them like the pep rallies. Among those who do not participate in team sports, only 20 percent like the pep rallies."

   a) Describe how you could use a random digits table to simulate responses ("Do you like mandatory pep rallies?") from 50 students in a *simple random sample*.

   b) Describe how you could use a random digits table to simulate responses from 50 students in a *stratified random sample*, with the strata being "team sports participants" and "non–team sports participants."

   c) Conduct two trials of the simulation described in part a), and two trials of the simulation described in part b). Each of the two trials in both cases should yield an estimate of the proportion of all students at the school who like the pep rallies.

We performed these two simulations on a computer and did 100 trials of each one. The dotplots below show our results.

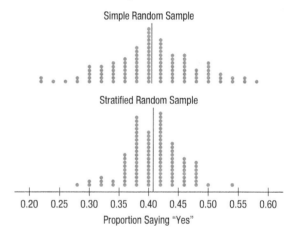

   d) Describe how the two dotplots are similar and how they are different.

   e) In each of the two dotplots, identify the dot that corresponds to the highest estimate of the population proportion who like the pep rallies. Use those two points to help you explain how stratified sampling reduces the variability in the response variable.

**38. Pep rally, again** The principal in Exercise 37 liked the idea of stratified sampling, but suggested that they stratify by grade level instead of by participation in sports, because it would be easier to select a sample of freshmen from the student roster, for example, than to select a sample of sports participants. The student said, "I don't think there's a big difference between grade levels in preference for the pep rallies." The principal asked, "What difference does that make?"

   a) How could the student design a pair of simulations similar to those in Exercise 37 to show the principal why stratifying on grade level would not be helpful if the student's belief is correct that the different grade levels feel about the same about pep rallies. (You might rather use 60 as your sample size because it is divisible by four.)

   b) Conduct two trials of each simulation described in part a). Each of the two trials in both cases should yield an estimate of the proportion of all students at the school who like the pep rallies.

We performed these two simulations on a computer and did 100 trials of each one. The dotplots below show our results.

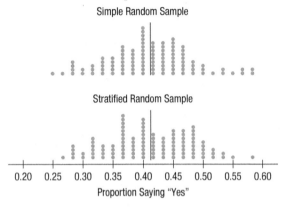

c) How does this pair of plots make the student's point that stratifying on grade level would not be helpful?
d) What do Exercises 37 and 38, taken together, tell you about when a stratified random sample is desirable over a simple random sample?

**39.** Texas A&M Administrators at Texas A&M University were interested in estimating the percentage of students who are the first in their family to go to college. The A&M student body has about 46,000 members.

a) What problems do you see with asking the following question of students? "Are you the first member of your family to seek higher education?"
b) For each scenario, identify the kind of sample used by the university administrators:
   i) Select several dormitories at random and contact everyone living in the selected dorms.
   ii) Using a computer-based list of registered students, contact 200 freshmen, 200 sophomores, 200 juniors, and 200 seniors selected at random from each class.
   iii) Using a computer-based alphabetical list of registered students, select one of the first 25 on the list by random and then contact the student whose name is 50 names later, and then every 50 names beyond that.
c) A professor teaching a large lecture class of 350 students samples her class by rolling a die. Then, starting with the row number on the die (1 to 6), she passes out a survey to every fourth row of the large lecture hall. She says that this is a simple random sample because everyone had an equal opportunity to sit in any seat and because she randomized the choice of rows. What do you think? Be specific.
d) For each of these proposed survey designs, identify the problem and the effect it would have on the estimate of the percentage of students who are the first in their family to go to college.
   i) Publish an advertisement inviting students to visit a website and answer questions.
   ii) Set up a table in the student union and ask students to stop and answer a survey.
e) The president of the university plans a speech to an alumni group. He plans to talk about the proportion of students who responded in the survey that they are the first in their family to attend college, but the first draft of his speech treats that proportion as the actual proportion of current A&M students who are the first in their families to attend college. Explain to the president the difference between the proportion of respondents who are first attenders and the proportion of the entire student body that are first attenders. Use appropriate Statistics terminology.

**40.** Satisfied workers The managers of a large company wished to know the percentage of employees who feel "extremely satisfied" to work there. The company has roughly 24,000 employees. They contacted a random sample of employees and asked them about their job satisfaction, obtaining 437 completed responses.

a) The company's annual report states, "Our survey shows that 87.34% of our employees are 'very happy' working here." Comment on that claim. Use appropriate Statistics terminology.
b) One manager suggested surveying employees by assigning computer-generated random numbers to each employee on a list of all employees and then contacting all those whose assigned random number is divisible by 7. Is this a simple random sample?
c) For each scenario suggested by a different manager, determine the sampling method.
   i) Use the company e-mail directory to contact 150 employees from among those employed for less than 5 years, 150 from among those employed for 5–10 years, and 150 from among those employed for more than 10 years.
   ii) Use the company e-mail directory to contact every 50th employee on the list, starting with a randomly selected person.
   iii) Select several divisions of the company at random. Within each division, draw an SRS of employees to contact.
d) One manager suggested having the head of each corporate division hold a meeting of their employees to ask whether they are happy on their jobs. They will ask people to raise their hands to indicate whether they are happy. What problems do you see with this plan?
e) For each of these designs proposed by a different manager, identify the problem with the method and the effect it would have on the estimate of the percentage of employees who feel "extremely satisfied" to work there.
   i) Leave a stack of surveys out in the employee cafeteria so people can pick them up and return them.
   ii) Stuff a questionnaire in the mailbox of each employee with the request that they fill it out and return it.

## JUST CHECKING

### Answers

1. a) It can be hard to reach all members of a population, and it can take so long that circumstances change, affecting the responses. A well-designed sample is often a better choice.
   b) This sample is probably biased—students who didn't like the food at the cafeteria might choose not to eat there.
   c) No, only the sample size matters, not the fraction of the overall population.
   d) Students who frequent this website might be more enthusiastic about Statistics than the overall population of Statistics students. A large sample cannot compensate for bias.
   e) It's the population "parameter." "Statistics" describe samples.
2. a) Systematic.  b) Stratified.  c) Simple.
   d) Cluster—but only one cluster is sampled, so much faith is being put in the assumption that the selected cluster will be a representative cross-section of all the passengers.

# 12

# Experiments and Observational Studies

Who gets good grades? And, more importantly, why? Is there something schools and parents could do to help weaker students improve their grades? Some people think they have an answer: music! No, not on your phone, but an instrument. In a study conducted at Mission Viejo High School in California, researchers compared the scholastic performance of music students with that of non–music students. Guess what? The music students had a much higher overall grade point average than the non–music students, 3.59 to 2.91. Not only that: A whopping 16% of the music students had all A's compared with only 5% of the non–music students.

As a result of this study and others, many parent groups and educators pressed for expanded music programs in the nation's schools. They argued that the work ethic, discipline, and feeling of accomplishment fostered by learning to play an instrument also enhance a person's ability to succeed in school. They thought that involving more students in music would raise academic performance. What do you think? Does this study provide solid evidence? Or are there other possible explanations for the difference in grades? Is there any way to really prove such a conjecture?

## Observational Studies

This research tried to show an association between music education and grades. But it wasn't a survey. Nor did it assign students to get music education. Instead, it simply observed students "in the wild," recording the choices they made and the outcome. Such studies are called **observational studies**. In observational studies, researchers don't *assign* choices; they simply observe them. In addition, this was a **retrospective study**, because researchers first identified subjects who studied music and then collected data on their past grades.

> **RETROSPECTIVE STUDIES CAN GIVE VALUABLE CLUES**
>
> For rare illnesses, it's not practical to draw a large enough sample to see many ill respondents, so the only option remaining is to develop retrospective data. For example, researchers can interview those who have become ill. The likely causes of both Legionnaires' disease and HIV were initially identified from such retrospective studies of the small populations who were initially infected. But to confirm the causes, researchers needed laboratory-based experiments.

What's wrong with concluding that music education causes good grades? The claim that music study *caused* higher grades depends on there being *no other differences* between the groups that could account for the differences in grades. But there are lots of variables that might cause the groups to perform differently. Students who study music may have better work habits to start with. Music students may have more parental support and that support may have enhanced their academic performance too. Maybe they came from wealthier homes and had other advantages. Or it could be that smarter kids just like to play musical instruments.

Observational studies are used widely in public health and marketing. Those that study rare outcomes, such as specific diseases, are often retrospective. They first identify people with the disease and then look into their history and heritage in search of things that may be related to their condition. But retrospective studies have a restricted view of the world because they are usually limited to a small part of the entire population. And because retrospective records are based on historical data and memories, they can have errors. (Do you recall *exactly* what you ate even yesterday? How about last Wednesday?)

If we identify subjects in advance and collect data as events unfold, we have a **prospective study**. That's a more reliable approach. For example, we might select a cohort of infants to study, follow whether they later take music lessons, and track their academic performance.

Although an observational study may identify important variables related to an outcome of interest, there is no guarantee that we have found the right or the most important related variables. Students who choose to study an instrument might still differ from the others in some important way that we failed to observe. It may be this difference—whether we know what it is or not—rather than music itself that leads to better grades. It's just not possible for observational studies, whether prospective or retrospective, to demonstrate a causal relationship.

### FOR EXAMPLE: Designing an Observational Study

In early 2007, a larger-than-usual number of cats and dogs developed kidney failure; many died. Initially, researchers didn't know why, so they used an observational study to investigate.

**QUESTION:** Suppose you were called on to plan a study seeking the cause of this problem. Would your design be retrospective or prospective? Explain why.

**ANSWER:** I would use a retrospective observational study. Even though the incidence of disease was higher than usual, it was still rare. Surveying all pets would have been impractical. Instead, it makes sense to locate some who were sick and ask about their diets, exposure to toxins, and other possible causes.

# Randomized, Comparative Experiments

> He that leaves nothing to chance will do few things ill, but he will do very few things.
> —Lord Halifax (1633–1695)

Is it *ever* possible to get convincing evidence of a cause-and-effect relationship? Well, yes it is, but we would have to take a different approach. We could take a group of third graders, randomly assign half to take music lessons, and forbid the other half to do so. Then we could compare their grades several years later. This kind of study design is called an **experiment**.

An experiment requires a **random assignment** of subjects to treatments. Only an experiment can justify a claim like "Music lessons cause higher grades." Questions such as "Does taking vitamin C reduce the chance of getting a cold?" and "Does working with computers improve performance in Statistics class?" and "Is this drug a safe and effective treatment for that disease?" require a designed experiment to establish cause and effect.

Experiments study the relationship between two or more variables. An experimenter must identify at least one explanatory variable, called a **factor**, to manipulate and at least one **response variable** to measure. What distinguishes an experiment from other types of investigation is that the experimenter actively and deliberately manipulates the factors to control the details of the possible treatments, and assigns the subjects to those treatments *at random*. The experimenter then observes the response variable and *compares* responses for different groups of subjects who have been treated differently. For example, we might design an experiment to see whether the amount of sleep and exercise you get affects your grades.

The individuals on whom or which we experiment are known by a variety of terms. Humans who are experimented on are commonly called **subjects** or **participants**. Other individuals (rats, days, petri dishes of bacteria) are commonly referred to by the more generic term **experimental units**. (When we recruit subjects for our sleep deprivation experiment by advertising in Statistics class, we'll probably have better luck if we invite them to be participants than if we advertise that we need experimental units.)

The specific values that the experimenter chooses for a factor are called the **levels** of the factor. We might assign our participants to sleep for 4, 6, or 8 hours. Often there are several factors at a variety of levels. (Our subjects will also be assigned to a treadmill for 0 or 30 minutes.) The combination of specific levels from all the factors that an experimental unit receives is known as its **treatment**. (Our experiment has six different treatments—three sleep levels, each at two exercise levels.)

How should we assign our participants to these treatments? Some students prefer 4 hours of sleep, while others need 8. Some exercise regularly; others are couch potatoes. Should we let the students choose the treatments they'd prefer? No. That would not be a good idea. To have any hope of drawing a fair conclusion, we must assign our participants to their treatments *at random*.

It may be obvious to you that we shouldn't let the students choose the treatment they'd prefer, but the need for random assignment is a lesson that was once hard for some to accept. For example, physicians might naturally prefer to assign patients to the therapy that they think best rather than have a random element such as a coin flip determine the treatment. But if anyone knew for sure which treatment was better, we wouldn't be doing an experiment. We've known for more than a century that for the results of an experiment to be valid, we must use deliberate randomization.

Experimental design was advanced in the 19th century by work in psychophysics by Gustav Fechner (1801–1887), the founder of experimental psychology. Fechner designed ingenious experiments that exhibited many of the features of modern designed experiments. Fechner was careful to control for the effects of factors that might affect his results. For example, in his 1860 book *Elemente der Psychophysik* he cautioned readers to group experiment trials together to minimize the possible effects of time of day and fatigue.

## An Experiment

*manipulates* the factor levels to create treatments, *randomly assigns* subjects to these treatment levels, and *compares* the responses of the subject groups across treatment levels.

## THE FDA

No drug can be sold in the United States without first showing, in suitably designed experiments approved by the Food and Drug Administration (FDA), that it's safe and effective. The small print on the booklet that comes with many prescription drugs usually describes the outcomes of those experiments.

## THE WOMEN'S HEALTH INITIATIVE

The Women's Health Initiative is a major 15-year research program funded by the National Institutes of Health to address the most common causes of death, disability, and poor quality of life in older women. It consists of an observational study with more than 93,000 participants as well as several randomized comparative experiments. The goals of this study include

- giving reliable estimates of the extent to which known risk factors predict heart disease, cancers, and fractures;
- identifying "new" risk factors for these and other diseases in women;
- comparing risk factors, presence of disease at the start of the study, and new occurrences of disease during the study across all study components; and
- creating a future resource to identify biological indicators of disease, especially substances and factors found in blood.

That is, the study seeks to identify possible risk factors and assess how serious they might be. It seeks to build up data that might be checked retrospectively as the women in the study continue to be followed. There would be no way to find out these things with an experiment because the task includes identifying new risk factors. If we don't know those risk factors, we could never control them as factors in an experiment.

By contrast, one of the clinical trials (randomized experiments) that received much press attention randomly assigned postmenopausal women to take either hormone replacement therapy or an inactive pill. The results published in 2002 and 2004 concluded that hormone replacement with estrogen carried increased risks of stroke.

> **FOR EXAMPLE**
>
> ### Determining the Treatments and Response Variable
>
> **RECAP:** In 2007, deaths of a large number of pet dogs and cats were ultimately traced to contamination of some brands of pet food. The manufacturer now claims that the food is safe, but before it can be released, it must be tested.
>
> **QUESTION:** In an experiment to test whether the food is now safe for dogs to eat,[1] what would be the treatments and what would be the response variable?
>
> **ANSWER:** The treatments would be ordinary-size portions of two dog foods: the new one from the company (the *test food*) and one that I was certain was safe (perhaps prepared in my kitchen or laboratory). The response would be a veterinarian's assessment of the health of the test animals.

# The Four Principles of Experimental Design

Imagine a universe where there is no variability. All people are identical, all foods are identical, and so on. This would be an extremely boring world to live in, but conducting experiments would be easy. If you wanted to see the effect of a treatment, you could apply it to a single experimental unit while letting a single other experimental unit go untreated. You'd otherwise treat the two experimental units absolutely the same. Any differences you saw in what resulted would have to be due to the treatment, because everything else about the experimental units was identical.

But of course we (thankfully!) don't live in such a boring universe. The variety that makes life interesting also makes experiments harder than what we just described. In the real world, if you used only two experimental units for an experiment, you'd have no way of knowing whether any differences in results were due to the treatments, or perhaps to inherent differences in the experimental units, or maybe to inadvertent differences in the ways the experimental units were handled apart from the treatment. It's because of this natural, unavoidable variability that well-designed experiments adhere to the principles of control, randomization, replication, and (sometimes) blocking. Here's more about these principles.

The deep insight that experiments should use random assignment is quite an old one. It can be attributed to the American philosopher and scientist C. S. Peirce in his experiments with J. Jastrow, published in 1885.

1. **Control.** We control sources of variation other than the factors we are testing by making conditions as similar as possible for all treatment groups. For human subjects, we try to treat them alike. Controlling extraneous sources of variation reduces the variability of the responses, making it easier to detect differences among the treatment groups. (However, there is always a question of degree and practicality.)

   Making generalizations from the experiment to other levels of the controlled factor can be risky. For example, suppose we test two laundry detergents and carefully control the water temperature at 180°F. This would reduce the variation in our results due to water temperature, but what could we say about the detergents' performance in cold water? Not much.

   Although we "control" both experimental factors and other sources of variation, we mean two different things by "control." We control a factor by assigning subjects to different factor levels because we want to see how the response will change at those different levels. We control other sources of variation to *prevent* them from changing and affecting the response variable.

---

[1] It may disturb you (as it does us) to think of deliberately putting dogs at risk in this experiment, but in fact that is what is done. The risk is borne by a small number of dogs so that the far larger population of dogs can be kept safe.

2. **Randomize.** As in sample surveys, randomization allows us to equalize the effects of unknown or uncontrollable sources of variation. It does not eliminate the effects of these sources, but by distributing them equally (on average) across the treatment levels, it makes comparisons among the treatments fair. Assigning experimental units to treatments at random allows us to use the powerful methods of Statistics to draw conclusions from an experiment. Assigning subjects to treatments at random reduces the risk that an imbalance in some uncontrolled—perhaps even unobserved—source of variation will produce an apparent, but misleading, effect. (We'll talk more about a related problem called "confounding" later in this chapter.)

    Experimenters often control factors that are easy or inexpensive to control. They randomize to protect against the effects of other factors, even factors they haven't thought about. How to choose between the two strategies is best summed up in the old adage that says "control what you can, and randomize the rest."

3. **Replicate.** Drawing conclusions about the world is impossible unless we repeat, or replicate, our results. Two kinds of replication show up in comparative experiments. First, we should apply each treatment to a number of subjects. Only with such replication can we estimate the variability of responses. If we have not assessed the variation, the experiment is not complete. The outcome of an experiment on a single subject is an anecdote, not data.

> ### AN ANECDOTAL SNOWBALL
> In February 2015, while senators in Washington, D.C., were discussing policies that might help reduce the effects of climate change, much of the eastern United States experienced an unusual cold spell. One senator who believed that climate change was a hoax brought to the Senate floor a snowball that he'd made outside. He said, "We keep hearing that 2014 has been the warmest year on record. I ask the chair, you know what this is? It's a snowball, and that's just from outside here, so it's very, very cold out, very unseasonable." This anecdotal "evidence" that climate change was not occurring was shortly rejected by another senator. "I want to respond to the presentation . . . suggesting that the continued existence of snow disproves climate change." The mixture of statistical "noise" (unexplained variability, like cold spells) with "signal" (a trend, like gradual ongoing global warming) is why anecdotal evidence should be looked at very skeptically. It could easily be picking up mostly noise.

   A second kind of replication shows up when the entire experiment is repeated on a different population of experimental units. We may believe that what is true of the students in Psych 101 who volunteered for the sleep experiment is true of all humans, but we'll feel more confident if our results for the experiment are *replicated* in another part of the country, with people of different ages, and at different times of the year. Replication of an entire experiment with the controlled sources of variation at different levels is an essential step in science.

4. **Block.** Suppose the participants available for a study of balance include two members of the varsity girls gymnastics team and 10 other students. Randomizing may place both gymnasts in the same treatment group. In the long run, if we could perform the experiment over and over, it would all equalize. But wouldn't it be better to assign one gymnast to each group and five of the other students to each group (at random)? Doing this improves fairness in the *short* run.

    The differences in the students' gymnastics training is likely to affect what we want to measure. Whenever identifiable differences among the experimental units might affect the outcome being studied, it is wise to collect the units into homogeneous groups or **blocks**. Here, the variable "Training" is a blocking variable, and the levels of "Experience" are called blocks.

Grouping similar individuals into blocks can help us account for unwanted variation among the subjects, allowing us to see differences in the treatments that might otherwise be obscured. Blocking is an important addition to the principles of randomization, control, and replication. However, unlike the first three, blocking is not required in an experimental design.

### FOR EXAMPLE
#### Control, Randomize, and Replicate

**RECAP:** We're planning an experiment to see whether the new pet food is safe for dogs to eat. We'll feed some animals the new food and others a food known to be safe, comparing their health after a period of time.

**QUESTION:** In this experiment, how will you implement the principles of control, randomization, and replication?

**ANSWER:** I'd control the portion sizes eaten by the dogs. To reduce possible variability from factors other than the food, I'd standardize other aspects of their environments—housing the dogs in similar pens and ensuring that each got the same amount of water, exercise, play, and sleep time, for example. I might restrict the experiment to a single breed of dog and to adult dogs to further minimize variation.

To equalize traits, pre-existing conditions, and other unknown influences, I would assign dogs to the two feed treatments randomly.

I would replicate by assigning more than one dog to each treatment to allow for variability among individual dogs. If I had the time and funding, I might replicate the entire experiment using, for example, a different breed of dog.

## Diagrams

An experiment is carried out over time with specific actions occurring in a specified order. A diagram of the procedure can help in thinking about experiments.[2]

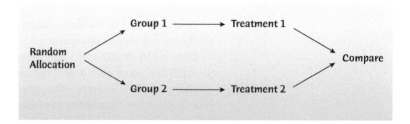

The diagram emphasizes the random allocation of subjects to treatment groups, the separate treatments applied to these groups, and the ultimate comparison of results. It's best to specify the responses that will be compared. (A diagram like this can be useful for outlining an experiment, but if it's all by itself, it's not enough to explain how an experiment is designed. Explanations require complete sentences, not just diagrams.)

---

[2]Diagrams of this sort were introduced by David Moore in his textbooks and are still widely used.

## STEP-BY-STEP EXAMPLE

### Designing an Experiment

An ad for OptiGro plant fertilizer claims that with this product you will grow "juicier, tastier" tomatoes. You'd like to test this claim and also wonder whether you might be able to get by with half the specified dose. How can you set up an experiment to check out the claim?

Of course, you'll have to get some tomatoes, try growing some plants with the product and some without, and see what happens. But you'll need a clearer plan than that. How should you design your experiment?

Let's work through the design, step by step. We'll design the simplest kind of experiment, in which each experimental unit is equally likely to end up in any of the treatment groups. This experiment is a **completely randomized design** in one factor. Because this is a *design* for an experiment, most of the steps are part of the *Think* stage. The statements in the right column are the kinds of things you would need to say in *proposing* an experiment. You'd need to include them in the "methods" section of a report once the experiment is run.

QUESTION: How would you design an experiment to test OptiGro fertilizer?

A completely randomized experiment is the ideal simple design, just as a *simple random sample* is the ideal simple sample—and for many of the same reasons.

---

**THINK**

**PLAN** State what you want to know.

I want to know whether tomato plants grown with OptiGro yield juicier, tastier tomatoes than plants raised in otherwise similar circumstances but without the fertilizer.

**RESPONSE** Specify the response variable.

I'll evaluate the juiciness and taste of the tomatoes by asking a panel of judges to rate them on a scale from 1 to 10 in juiciness and in taste.

**TREATMENTS** Specify the factor levels and the treatments.

The factor is fertilizer, specifically OptiGro fertilizer. I'll grow tomatoes at three different factor levels: some with no fertilizer, some with half the specified amount of OptiGro, and some with the full dose of OptiGro. These are the three treatments.

**EXPERIMENTAL UNITS** Specify the experimental units.

I'll obtain 24 tomato plants of the same variety from a local garden store.

**EXPERIMENTAL DESIGN** Observe the principles of design:

**CONTROL** any sources of variability you know of and can control.

I'll locate the garden plots near each other so that the plants get similar amounts of sun and rain and experience similar temperatures. I will weed the plots equally and otherwise treat the plants alike.

**REPLICATE** results by placing more than one plant in each treatment group.

I'll use 8 plants in each treatment group.

**RANDOMLY ASSIGN** experimental units to treatments, to equalize the effects of unknown or uncontrollable sources of variation.

Describe how the randomization will be accomplished.

To randomly divide the plants into three groups, first I'll label the plants with numbers 00-23. I'll look at pairs of digits across a random number table. The first 8 plants identified (ignoring numbers 24-99 and any repeats) will get no fertilizer, the next 8 a half dose, and the remaining plants the full amount.

**MAKE A PICTURE** A diagram of your design can help you think about it clearly. You still need to write a complete explanation, though.

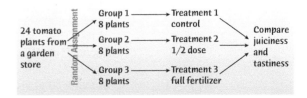

| | |
|---|---|
| Specify any other experiment details. You must give enough details so that another experimenter could exactly replicate your experiment.<br><br>Specify how to measure the response. | I will grow the plants until the tomatoes are mature, as judged by reaching a standard color.<br><br>I'll harvest the tomatoes when ripe and store them for evaluation.<br><br>I'll set up a numerical scale of juiciness and one of tastiness for the taste testers. I will have several people taste slices of tomato and rate them. |
| **SHOW** Once you collect the data, you'll need to display them and compare the results for the three treatment groups. | I will display the results with side-by-side boxplots to compare the three treatment groups.<br><br>I will compare the means of the groups. |
| **TELL** To answer the initial question, we ask whether the differences we observe in the means of the three groups are meaningful. If so, because this is a randomized experiment, we can attribute significant differences to the treatments. (To do this properly, we'll need methods from what is called "statistical inference," the subject of much of the rest of this book.) | If the differences in taste and juiciness among the groups are greater than I would expect by knowing the usual variation among tomatoes, I may be able to conclude that these differences can be attributed to treatment with the fertilizer. |

## Does the Difference Make a Difference?

If the differences among the treatment groups are big enough, we'll attribute the differences to the treatments, but how can we decide whether the differences are big enough?

Would we expect the group means to be identical? Not really. Even if the treatment made no difference at all, there would still be some variation. We assigned the tomato plants to treatments at random. But a different random assignment would have led to different results. Even a repeat of the *same* treatment on a different randomly assigned set of plants would lead to different means. The real question is whether the differences we observed are about as big as we might get just from the randomization alone, or whether they're bigger than that. If we decide that they're bigger, we'll attribute the differences to the treatments. In that case we say the differences are **statistically significant**.

© 2013 Randall Munroe. Reprinted with permission. All rights reserved.

How will we decide if something is different enough to be considered statistically significant? To get some intuition, think about deciding whether a coin is fair. If we flip a fair coin 100 times, we expect, *on average*, to get 50 heads. Suppose we get 54 heads out of 100. That doesn't seem very surprising. It's well within the bounds of ordinary random fluctuations. What if we'd seen 94 heads? That's clearly outside the bounds. We'd be pretty sure that the coin flips were not random. But what about 74 heads? Is that far enough from 50–50 to arouse our suspicions? That's the sort of question we need to ask of our experiment results.

In Statistics terminology, 94 heads would be a statistically significant difference from 50, and 54 heads would not. Whether 74 is *statistically significant* or not would depend on the chance of getting 74 heads in 100 flips of a fair coin and on our tolerance for believing that rare events can happen to us. Stay tuned: You'll learn to calculate that probability soon. For now, the important point is that an outcome is statistically significant if the probability that it happened just by chance is so low that we're convinced there must be another explanation.

Back at the tomato stand, we ask whether the taste differences we see among the treatment groups could have arisen merely from randomization. A good way to get a feeling for that is to look at how much our results vary among plants that get the *same* treatment. Boxplots of our results by treatment group can give us a general idea.

For example, Figure 12.1 shows two pairs of boxplots whose centers differ by exactly the same amount. In the pair on the left, that difference appears to be larger than we'd expect just by chance. Why? Because the variation is quite small *within* treatment groups, so the larger difference *between* the groups is unlikely to be just from the randomization. In the pair on the right, that same difference between the centers looks less impressive. There the variation *within* each group swamps the difference *between* the two medians. We'd say the difference is statistically significant in the pair at the left and not statistically significant in the pair at the right.

In later chapters we'll see statistical tests that quantify this intuition. But we've already been considering statistical significance (without yet calling it that) in some of the simulations we've seen in earlier chapters. For example, in Chapter 2 we simulated passengers on the *Titanic* surviving without any regard to their ticket class. In the simulation, the differences in survival rates among ticket classes were consistently very modest, while the actual historical differences in survival rates were much greater. That indicated that the differences were statistically significant. In Chapter 6 we wondered whether the correlation of $r = 0.221$ between age and height for the Cincinnati Reds was "meaningful." A simulation that randomly paired ages and heights showed that a correlation that large occurred just by chance in about 8% of the trials. While 8% is somewhat unusual, it's not so strikingly unusual that we felt compelled to explain why older baseball players tended to be taller. We were content to believe instead that the correlation was not "meaningful"—that is, not statistically significant.

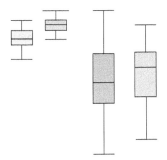

Figure 12.1
The boxplots in both pairs have medians the same distance apart, but when the spreads are large, that observed difference may be due to just random chance, or statistical "noise."

## JUST CHECKING

1. At one time, a method called "gastric freezing" was used to treat people with peptic ulcers. An inflatable bladder was inserted down the esophagus and into the stomach, and then a cold liquid was pumped into the bladder. Now you can find the following notice on the Internet site of a major insurance company:

    *[Our company] does not cover gastric freezing (intragastric hypothermia) for chronic peptic ulcer disease . . . .*

    *Gastric freezing for chronic peptic ulcer disease is a non-surgical treatment which was popular about 20 years ago but now is seldom performed. It has been abandoned due to a high complication rate, only temporary improvement experienced by patients, and a lack of effectiveness when tested by double-blind, controlled clinical trials.*

    What did that "controlled clinical trial" (experiment) probably look like? (Don't worry about "double-blind"; we'll get to that soon.)

    a) What was the factor in this experiment?
    b) What was the response variable?
    c) What were the treatments?
    d) How did researchers decide which subjects received which treatment?
    e) Were the results statistically significant?

# Experiments and Samples

**Not a Random Sample!**

Experiments are rarely performed on random samples from a population. Don't describe the subjects in an experiment as a random sample unless they really are. More likely, the randomization was in assigning subjects to treatments.

Both experiments and sample surveys use randomization, but they do so in different ways and for different purposes. Sample surveys try to estimate population parameters, so the sample needs to be as representative of the population as possible. By contrast, experiments try to assess the effects of treatments. Experimental units are not usually drawn randomly from the population. For example, a medical experiment may deal only with local patients who have the disease being studied. The randomization is in the assignment of their therapy. We want a sample to exhibit the diversity and variability of the population, but for an experiment the more homogeneous the subjects are, the more easily we'll spot differences in the effects of the treatments.

Unless the experimental units are chosen from the population at random, you should be cautious about generalizing experiment results to larger populations until the experiment has been repeated under different circumstances. Results become more persuasive if they remain the same in completely different settings, such as in a different season, in a different country, or for a different species, to name a few.

Nevertheless, experiments can draw stronger conclusions than observational studies. Only by actively manipulating the factors can an experimenter conclude that observed changes are due to the factors themselves. This is the fundamental insight of the scientific method.

# Control Treatments

In the 1970s, a crime-prevention program called "Scared Straight" was begun with the hope of discouraging at-risk teenagers from turning to a life of crime. Young minor offenders were given tours of prisons and talked with inmates, who made explicit how bad life in prison was. Hanna Fry writes about it in the *New Yorker*:[3]

> At first, the program appeared to be a roaring success. Few kids who had been through the program were later involved in crime. . . . The intervention gained public approval and was copied in a number of countries around the world. There was just one tiny problem: no one had stopped to wonder what would have happened to a similar set of kids who hadn't gone through the experience.
>
> When a series of proper [randomized controlled trials] was run, and a direct comparison could be made between kids who went through the Scared Straight program and similar groups of kids who didn't, it became clear that the program was not working as intended. The intervention was, in fact, increasing the chances that kids would become criminals.

Experiments involve treatments, and to truly understand the effect of a treatment, you have to have something to compare it to. Such a baseline measurement is called a **control treatment**, and the experimental units or subjects to whom it is applied are called a **control group**.

This is a use of the word "control" in an entirely different context from its earlier use. Previously, we controlled extraneous sources of variation by keeping them constant. Here, we use a control treatment as another *level* of the factor in order to compare the treatment results to a situation in which "nothing is done." That's what we did in the tomato experiment when we used no fertilizer on the 8 tomatoes in Group 1. While all experiments must control the variables, not every experiment needs a control group.[4]

---

[3]"Experiments on Trial," by Hannah Fry, *The New Yorker*, March 2, 2020.

[4]While every good experiment requires at least two treatment groups, it's not always essential to have a group to which "nothing is done." Suppose a disease that is fatal when untreated has an 80% survival rate when treatment A is applied. It is believed that a new treatment B may raise the survival rate. You can do a controlled, randomized experiment to compare treatments A and B, but including a third group receiving no treatment would be, in this case, unnecessary and unethical.

# Blinding

Humans are notoriously susceptible to errors in judgment.[5] All of us. When we know what treatment was assigned, it's difficult not to let that knowledge influence our assessment of the response, even when we try to be careful.

Suppose you were trying to advise your school on which brand of cola to stock in the school's vending machines. You set up an experiment to see which of the three competing brands students prefer (or whether they can tell the difference at all). But people have brand loyalties. You probably prefer one brand already. So if you knew which brand you were tasting, it might influence your rating. To avoid this problem, it would be better to disguise the brands as much as possible. This strategy is called **blinding** the participants to the treatment.

But it isn't just the subjects who should be blind. Experimenters themselves often subconsciously behave in ways that favor what they believe. Even technicians may treat plants or test animals differently if, for example, they expect them to die. An animal that starts doing a little better than others by showing an increased appetite may get fed a bit more than the experimental protocol specifies.

People are so good at picking up subtle cues about treatments that the best (in fact, the *only*) defense is to keep *anyone* who could affect the outcome or the measurement of the response from knowing which subjects have been assigned to which treatments. So, not only should your cola-tasting subjects be blinded, but also *you,* as the experimenter, shouldn't know which drink is which, either—at least until you're ready to analyze the results.

There are two main classes of individuals who can affect the outcome of the experiment:

- Those who could influence the results (the subjects, treatment administrators, or technicians)
- Those who evaluate the results (judges, treating physicians, etc.)

> ### BLINDING ALL PARTIES
> In the first decade of this century, some Yale University researchers conducted an ingenious experiment on infants 9–12 months old to try to see whether they have an innate sense of right and wrong.[6] While sitting in its parent's lap, each infant was shown a short puppet show in which one puppet was nice and another was mean. (The roles were assigned to the puppets by coin flip.) After watching the same show several times, the infant was then presented with the two puppets and allowed to reach for one. The researchers wanted to see whether the infants would tend to reach for the "nice" puppet. (They did.)
>
> Blinding was critical in this experiment. The experimenter who presented the puppets to the infant had not seen the show and was not aware of which puppet played which role, lest they inadvertently suggest a choice to the infant. The parents holding the infants in their laps were asked to close their eyes when the puppets were presented so that they might not inadvertently influence their child's choice. However, the parents were *not* asked to close their eyes *during* the puppet shows, leaving open at least the *possibility* that the infants preferred the "nice" puppet not because they perceived it to be nice, but because they sensed that their *parents* preferred that puppet.[7]
>
> Of course this does not negate the conclusions of the experiment, which was otherwise exceptionally well planned and executed. But it illustrates how extremely careful one must be to be sure all parties whose awareness of the treatments might influence the response should be blinded to them.

---

[5] For example, here we are in Chapter 12 and you're still reading the footnotes.

[6] https://www.nytimes.com/2010/05/09/magazine/09babies-t.html

[7] https://www.nytimes.com/video/magazine/1247467772000/can-babies-tell-right-from-wrong.html. You'll enjoy this video, unless for some reason you just don't like babies.

When all the individuals in either one of these classes are blinded, an experiment is said to be **single-blind**. When everyone in *both* classes is blinded, we call the experiment **double-blind**. Even if several individuals in one class are blinded—for example, both the patients and the technicians who administer the treatment—the study would still be just single-blind. If only some of the individuals in a class are blind—for example, if subjects are not told of their treatment, but the administering technician is not blind—there is a substantial risk that subjects can discern their treatment from subtle cues in the technician's behavior or that the technician might inadvertently treat subjects differently. Such experiments cannot be considered truly blind.

In our tomato experiment, we certainly don't want the people judging the taste to know which tomatoes got the fertilizer. That makes the experiment single-blind. We might also not want the people caring for the tomatoes to know which ones were being fertilized, in case they might treat them differently in other ways, too. We can accomplish this double-blinding by having some fake fertilizer for them to put on the other plants. Read on.

## FOR EXAMPLE
### Blinding

**RECAP:** In our experiment to see if the new pet food is now safe, we're feeding one group of dogs the new food and another group a food we know to be safe. Our response variable is the health of the animals as assessed by a veterinarian.

**QUESTION:** Should the vet be blinded? Why or why not? How would you do this? (Extra credit: Can this experiment be double-blind? Would that mean that the test animals wouldn't know what they were eating?)

**ANSWER:** Whenever the response variable involves judgment, it is a good idea to blind the evaluator to the treatments. The veterinarian should not be told which dogs ate which foods.

Extra credit: There is a need for double-blinding. In this case, the workers who care for and feed the animals should not be aware of which dogs are receiving which food. We'll need to make the safe food look as much like the test food as possible.

# Placebos

Often, simply applying *any* treatment can induce an improvement. Every parent knows the medicinal value of a kiss to make a toddler's scrape or bump stop hurting. Some of the improvement seen with a treatment—even an effective treatment—can be due simply to the act of treating. To separate these two effects, we can use a control treatment that mimics the treatment itself.

A "fake" treatment that looks just like the treatments being tested is called a **placebo** (pronounced pluh-SEE-bow). Placebos are the best way to blind subjects from knowing whether they are receiving the treatment or not. One common version of a placebo in drug testing is a "sugar pill." Especially when psychological attitude can affect the results, control group subjects treated with a placebo may show an improvement.

The fact is that subjects treated with a placebo sometimes improve. It's not unusual for 20% or more of subjects given a placebo to report reduction in pain, improved movement, or greater alertness, or even to demonstrate improved health or performance. This **placebo effect** highlights both the importance of effective blinding and the importance of comparing treatments with a control. Placebo controls are so effective that you should use them as an essential tool for blinding whenever possible.

In summary, the best experiments are usually

- randomized.
- comparative.
- double-blind.
- placebo-controlled.

> **ACTIVE PLACEBOS**
> The placebo effect is stronger when placebo treatments are administered with authority or by a figure who appears to be an authority. "Doctors" in white coats generate a stronger effect than salespeople in polyester suits. But the placebo effect is not reduced much even when subjects know that the effect exists. People often suspect that they've gotten the placebo if nothing at all happens. So, recently, drug manufacturers have gone so far in making placebos realistic that they cause the same side effects as the drug being tested! Such "active placebos" usually induce a stronger placebo effect. When those side effects include loss of appetite or hair, the practice raises some ethical questions.

© 2015 Randall Munroe. Reprinted with permission. All rights reserved.

## Blocking

Suppose we had wanted to use 24 tomato plants of the same variety for our experiment, but the garden store had only 12 plants left. Then we drove down to the nursery and found just 6 more plants of that variety. Using only 18 plants is okay, but we worry that the tomato plants from the two stores are different somehow, or have received different care.

How can we design the experiment so that the differences between the stores don't mess up our attempts to see differences among fertilizer levels? We can't study the effect of a store the same way as we can the fertilizer because we can't assign it at random. You can't tell a tomato what store to come from.

Because stores may vary in the care they give plants or in the sources of their seeds, the plants from either store are likely to be more similar to each other than they are to the plants from the other store. When groups of experimental units are similar, it's often a good idea together them together into **blocks**. By blocking, we isolate the variability attributable to the differences between the blocks, so that we can see the differences caused by the treatments more clearly. Here, we would define the plants from each store to be a block. The randomization is introduced when we randomly assign treatments *within* each block.

In a completely randomized design, each of the 18 plants would have an equal chance to land in each of the three treatment groups. But we realize that the store may have an effect. To isolate the store effect, we block on store by assigning the plants from each store to treatments at random. So we now have six treatment groups, three for each block. Within each block, we'll randomly assign the same number of plants to each of the three treatments. The experiment is still fair because each treatment is still applied (at random) to the same number of plants and to the same proportion from each store: 4 from store A and 2 from store B. Because the randomization occurs only within the blocks (plants from one store cannot be assigned to treatment groups for the other), we call this a **randomized block design**.[8]

---

[8] In this example, the number of tomato plants from each nursery (12 and 6) happened to be a multiple of three, so for each block we were able to put exactly the same number of tomato plants in each treatment group. When this occurs, the design is called "balanced." A balanced design makes it somewhat easier to analyze the resulting data, but unbalanced designs can be fine too, and they're actually very common.

In effect, we conduct two parallel experiments, one for tomatoes from each store, and then combine the results. The picture tells the story[9]:

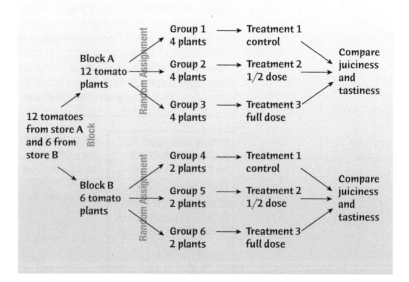

In a retrospective or prospective study, subjects are sometimes paired because they are similar in ways *not* under study. **Matching** subjects in this way can reduce variation in much the same way as blocking. For example, a retrospective study of music education and grades might match each student who studies an instrument with someone of the same gender who is similar in family income but didn't study an instrument. When we compare grades of music students with those of non–music students, the matching would reduce the variation due to income and gender differences.

Blocking in experiments is based on the same idea as stratifying in sampling. Both methods group together subjects that are similar with respect to extraneous variables and randomize within those groups as a way to reduce unwanted variation in the variable(s) of interest. (But be careful to keep the terms straight. Don't say that we "stratify" an experiment or "block" a sample.) We use blocks to reduce variability so we can see the effects of the factors; we're not usually interested in studying the effects of the blocks themselves.

### FOR EXAMPLE

#### Blocking

RECAP: In 2007, pet food contamination put cats at risk, as well as dogs. Our experiment should probably test the safety of the new food on both animals.

QUESTION: Why shouldn't we randomly assign a mix of cats and dogs to the two treatment groups? What would you recommend instead?

ANSWER: Dogs and cats might respond differently to the foods, and that variability could obscure my results. Blocking by species can remove that superfluous variation. I'd randomize cats to the two treatments (test food and safe food) separately from the dogs. I'd measure their responses separately and look at the results afterward.

---

[9] The diagram almost suggests that we're conducting a *completely* separate experiment on each block. But the diagram doesn't tell the whole story. In analyzing the data, the responses measured on tomatoes in one block are actually combined with those of tomatoes in the other block. But the "block effect"—that is, the juiciness/tastiness differential between the two stores—is estimated and accounted for first.

## JUST CHECKING

2. Recall the experiment about gastric freezing, an old method for treating peptic ulcers that you read about in the first Just Checking. Doctors would insert an inflatable bladder down the patient's esophagus and into the stomach and then pump in a cold liquid. A major insurance company now states that it doesn't cover this treatment because "double-blind, controlled clinical trials" failed to demonstrate that gastric freezing was effective.

   a) What does it mean that the experiment was double-blind?
   b) Why would you recommend a placebo control?
   c) Suppose that researchers suspected that the effectiveness of the gastric freezing treatment might depend on whether a patient had recently developed the peptic ulcer or had been suffering from the condition for a long time. How might the researchers have designed the experiment?

# Adding More Factors

There are two kinds of gardeners. Some water frequently, making sure that the plants are never dry. Others let nature take its course and leave the watering to it. The makers of OptiGro want to ensure that their product will work under a wide variety of watering conditions. Maybe we should include the amount of watering as part of our experiment. Can we study a second factor at the same time and still learn as much about fertilizer?

We now have two factors (fertilizer at three levels and irrigation at two levels). We combine them in all possible ways to yield six treatments:

|  | No Fertilizer | Half Fertilizer | Full Fertilizer |
|---|---|---|---|
| **No Added Water** | ① | ② | ③ |
| **Daily Watering** | ④ | ⑤ | ⑥ |

If we allocate the original 24 plants, the experiment now assigns 4 plants to each of these six treatments at random. This experiment is a **completely randomized two-factor experiment** because any plant could end up assigned at random to any of the six treatments (and we have two factors).

It's often important to include several factors in the same experiment in order to see what happens when the factor levels are applied in different *combinations*. For example, we might find out that regular watering allows a full dose of fertilizer to work best, but plants left to the whims of the weather do better on just half the recommended amount. Experiments with more than one factor are both more efficient and provide more information than one-at-a-time experiments. There are many ways to design efficient multifactor experiments. You can take a whole course on the design and analysis of such experiments.

> **THINK LIKE A STATISTICIAN**
>
> With two factors, we can account for more of the variation. That lets us see the underlying patterns more clearly.

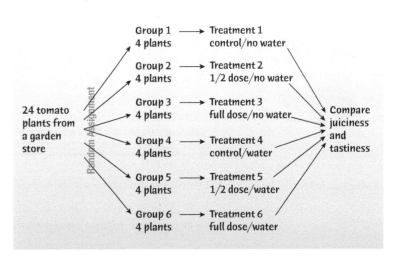

# Confounding

Professor Stephen Ceci of Cornell University performed an experiment to investigate the effect of a teacher's classroom style on student evaluations. He taught a class in developmental psychology during two successive terms to a total of 472 students in two very similar classes. He kept everything about his teaching identical (same text, same syllabus, same office hours, etc.) and modified only his style in class. During the fall term, he maintained a subdued demeanor. During the spring term, he lectured with more enthusiasm, varying his vocal pitch and using more hand gestures. He administered a standard student evaluation form at the end of each term.

The students in the fall term class rated him only an average teacher. Those in the spring term class rated him an excellent teacher, praising his knowledge and accessibility, and even the quality of the textbook. On the question "How much did you learn in the course?" the average response changed from 2.93 to 4.05 on a 5-point scale.[10]

How much of the difference he observed was due to his difference in manner, and how much might have been due to the season of the year? The fall term in Ithaca, New York (home of Cornell University), starts out colorful and pleasantly warm but ends cold and bleak. The spring term starts out bitter and snowy and ends with blooming flowers and singing birds. Might students' overall happiness have been affected by the season and reflected in their evaluations?

Unfortunately, there's no way to tell. Nothing in the data enables us to tease apart these two effects, because all the students who experienced the subdued manner did so during the fall term and all who experienced the expansive manner did so during the spring. When the levels of one factor are associated with the levels of another factor, we say that these two factors are confounded with one another, or simply that they are **confounded**.

In some experiments, such as this one, it's just not possible to avoid some confounding. Professor Ceci could have randomly assigned students to one of two classes during the same term, but then we might question whether mornings or afternoons were better, or whether he really delivered the same class the second time (after practicing on the first class). Or he could have had another professor deliver the second class, but that would have raised more serious issues about differences in the two professors and concern over more serious confounding.

> The word "confound" comes from the roots "con," meaning "together," and "fundere," meaning "pour." (The word "funnel" also shares this root.) So two things are confounded when they are "poured together." Two miscible liquids, once poured together, can't be separated from one another, and the effects of two confounded explanatory variables can't be separated either.

## FOR EXAMPLE

### Confounding

**RECAP:** After many dogs and cats suffered health problems caused by contaminated foods, we're trying to find out whether a newly formulated pet food is safe. Our experiment will feed some animals the new food and others a food known to be safe, and a veterinarian will check the response.

**QUESTION:** Why would it be a bad design to feed the test food to some dogs and the safe food to cats?

**ANSWER:** This would create confounding. We would not be able to tell whether any differences in animals' health were attributable to the food they had eaten or to differences in how the two species responded.

---

[10] But the two classes performed almost identically well on the final exam.

### A Two-Factor Example

Confounding can also arise from a badly designed multifactor experiment. Here's a classic. A credit card bank wanted to test the sensitivity of the market to two factors: the annual fee charged for a card and the annual percentage rate charged. Not wanting to scrimp on sample size, the bank selected 100,000 people at random from a mailing list. It sent out 50,000 offers with a low rate and no fee and 50,000 offers with a higher rate and a $50 annual fee. Guess what happened? That's right—people preferred the low-rate, no-fee card. No surprise. In fact, they signed up for that card at over twice the rate as the other offer. And because of the large sample size, the bank was able to estimate the difference precisely. But the question the bank really wanted to answer was "how much of the change was due to the rate, and how much was due to the fee?" Unfortunately, there's simply no way to separate out the two effects. If the bank had sent out all four possible different treatments—low rate with no fee, low rate with $50 fee, high rate with no fee, and high rate with $50 fee—each to 25,000 people, it could have learned about both factors and could have also seen what happens when the two factors occur in combination.

# Lurking or Confounding?

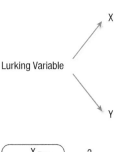

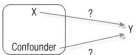

Confounding may remind you of the problem of lurking variables we discussed back in Chapters 6 and 8. Confounding variables and lurking variables are alike in that they interfere with our ability to identify what's causing an observed outcome. But they differ in how the interference comes about.

A lurking variable creates an association between two other variables that tempts us to think that one may cause the other. This can easily happen in an observational study when a lurking variable influences both the explanatory and response variables. For example, countries with more TV sets per capita tend to have longer life expectancies, but we shouldn't conclude it's the TVs "causing" longer life. We suspect instead that a generally higher standard of living may mean that people can afford more TVs and get better health care, too. Our data revealed an association between TVs and life expectancy, but economic conditions were a likely lurking variable. A lurking variable, then, is usually thought of as a variable associated with both $y$ and $x$ that makes it appear that $x$ may be causing $y$.

Confounding can arise in experiments when some other variable associated with a factor has an effect on the response variable. However, in a designed experiment, the experimenter *assigns* treatments (at random) to subjects rather than just observing them. A confounding variable can't be thought of as causing that assignment. Professor Ceci's choice of teaching styles was not caused by the weather, but because he used one style in the fall and the other in spring, he was unable to tell how much of his students' reactions were attributable to his teaching and how much to the weather. A confounding variable, then, is associated in a noncausal way with a factor and affects the response. Because of the confounding, we find that we can't tell whether any effect we see was caused by our factor or by the confounding variable—or even by both working together.

Both confounding and lurking variables are outside influences that make it harder to understand the relationship we are modeling. However, the nature of the causation is different in the two situations. In observational studies, we can only observe associations between variables. Although we can't demonstrate a causal relationship, we often imagine whether $x$ *could* cause $y$. We can be misled by a lurking variable that influences both. In a designed experiment, we often hope to show that the factor causes a response. Here we can be misled by a confounding variable that's associated with the factor and causes or contributes to the differences we observe in the response.

It's worth noting that the role of blinding in an experiment is to combat a possible source of confounding. There's a risk that knowledge about the treatments could lead the subjects or those interacting with them to behave differently or could influence judgments made by the people evaluating the responses. That means we won't know whether the treatments really do produce different results or if we're being fooled by these confounding influences.

### A Legendary Example of Confounding

The following story is often told to Statistics students, so you should hear it too. (But it may be only a legend—we were not able to trace it to a reputable source.) Researchers wanted to test the hypothesis that drinking 8 glasses of water daily would reduce the chance of catching a cold. They recruited a number of volunteer subjects and randomly assigned them either to a treatment group that was to drink 8 glasses of water daily or to a control group that was to drink water like they normally do. The treatment group did get significantly fewer colds than the control group, but was the water responsible? Probably not—drinking a lot of water results in more bathroom trips, and therefore more hand-washing, which was likely the *actual* cause of the reduced colds—and a good example of a confounding factor!

# Statistical Significance, Revisited

Earlier in this chapter we discussed statistical significance. ("Does the Difference Make a Difference?") Before concluding this chapter, let's revisit this important principle and think about what might happen in our tomato experiment (without blocking). We randomly assigned 24 tomato plants into three groups (no fertilizer, half dose, and full dose) and now their fruit is ripe and ready to eat. We've sliced up a nice tomato from each of the 24 plants and had a neighbor who's something of a tomato aficionado taste them. Without knowing which tomatoes came from fertilized plants, she rated each one on a scale from 1 to 10, with 1 representing less than satisfactory taste and 10 meaning absolutely delicious. The data table and boxplots below display her evaluations.

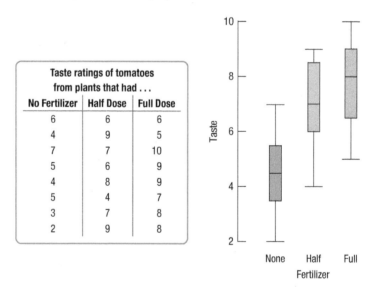

It certainly appears the fertilizer worked. Our tester rated both the tomatoes that got the half dose and those that got the full dose as much tastier than those that went unfertilized. Furthermore, it looks like the full-dose tomatoes were somewhat better than those from plants that got only half the fertilizer. In fact, the mean taste ratings for the three groups were 4.5, 7, and 7.75 out of 10, respectively. But is that ¾-point difference between the two fertilized groups statistically significant? Did the *extra* fertilizer really help, or could we just be seeing natural variability in tomato plants? To find out, we could randomly split those 16 ratings into two other groups of 8 (ignoring the 8 plants that received no fertilizer) and see how big a difference might occur by chance. Yes, it's time for another simulation!

In our first random split, Group A (6, 9, 4, 6, 9, 8, 7, 5) had a mean rating of 6.75, and Group B (7, 6, 9, 8, 10, 9, 7, 8) a mean of 8. That's an even bigger difference than what the extra fertilizer appeared to produce. Hmm.

Our simulation split the evaluations randomly a total of 500 times. The dotplot below shows the differences in means for the random groupings.

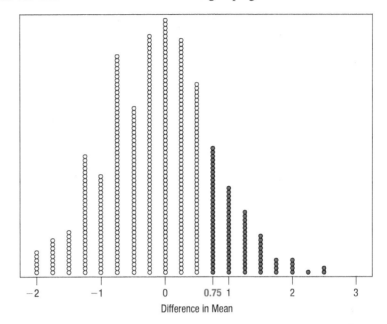

In our actual data the fully fertilized and half-fertilized tomatoes had mean taste ratings that differed by 0.75, in favor of the fully fertilized tomatoes. But that difference doesn't look very impressive now. In our simulation a difference of 0.75 points or more happened by chance a whopping 91 times in 500 trials—nearly 20% of the time. Such a difference is far from unusual; in other words, it's not statistically significant. Our neighbor's ratings don't provide evidence that using a full dose of fertilizer instead of only half a dose will produce tomatoes that taste better.

If right now you're thinking that maybe even the apparently large difference between the tomatoes that got fertilizer and those that had none isn't significant either, good for you! Skepticism is the mark of a good statistician. To ease your mind, we checked that out for you. Instead of a simulation, though, we used one of the statistical tests we told you you'd learn about later in this course. That test revealed that a difference as large as the one we see between the mean taste ratings for tomatoes from fertilized vs. non fertilized plants could arise by chance only 1 time in 2000! Now, *that's* statistically significant.

Based on these ratings, then, the OptiGro company would be on solid ground in asserting that its fertilizer produces tastier tomatoes. But the company might also be tempted to warn customers that using only half the recommended amount wouldn't work as well. When you learn to think like a statistician, such claims made without evidence leave a bad taste in your mouth.[11]

## WHAT CAN GO WRONG?

- **Don't give up just because you can't run an experiment.** Sometimes we can't run an experiment because we can't identify or control the factors. Sometimes it would simply be unethical to run the experiment. (Consider randomly assigning students to take—and be graded in—a Statistics course deliberately taught to be boring and difficult or one that had an unlimited budget to use multimedia, real-world examples, and field trips to make the subject more interesting.) If we can't perform an experiment, often an observational study is a good choice.

---

[11]Sorry.

- **Beware of confounding.** Use randomization whenever possible to ensure that the factors not in your experiment are not confounded with your treatment levels. Be alert to confounding that cannot be avoided, and report it along with your results.

- **Bad things can happen even to good experiments.** Protect yourself by recording additional information. An experiment in which the air conditioning failed for 2 weeks, affecting the results, was saved by recording the temperature (although that was not originally one of the factors) and estimating the effect the higher temperature had on the response.[12]

  It's generally good practice to collect as much information as possible about your experimental units and the circumstances of the experiment. For example, in the tomato experiment, it would be wise to record details of the weather (temperature, rainfall, sunlight) that might affect the plants and any facts available about their growing situation. (Is one side of the field in shade sooner than the other as the day proceeds? Is one area lower and a bit wetter?) Sometimes we can use this extra information during the analysis to reduce biases.

- **Don't spend your entire budget on the first run.** Just as it's a good idea to pretest a survey, it's always wise to try a small pilot experiment before running the full-scale experiment. You may learn, for example, how to choose factor levels more effectively, about effects you forgot to control, and about unanticipated confoundings.

---

[12] R. D. De Veaux and M. Szelewski, "Optimizing Automatic Splitless Injection Parameters for Gas Chromatographic Environmental Analysis." *Journal of Chromatographic Science* 27, no. 9 (1989): 513–518.

# WHAT HAVE WE LEARNED?

We've learned to recognize sample surveys, observational studies, and randomized comparative experiments. We know that these methods collect data in different ways and lead us to different conclusions.

We've learned to identify retrospective and prospective observational studies and understand the advantages and disadvantages of each.

We've learned that only well-designed experiments can allow us to reach cause-and-effect conclusions. We manipulate levels of treatments to see if the factor we have identified produces changes in our response variable.

We've learned the principles of experimental design:

- We want to be sure that variation in the response variable can be attributed to our factor, so we identify and control as many other sources of variability as possible.
- Because there are many possible sources of variability that we cannot identify, we try to equalize those by randomly assigning experimental units to treatments.
- We replicate the experiment on as many subjects as possible.
- We consider blocking to reduce variability from sources we cannot control but we suspect may affect response to treatments.
- We've learned to recognize the factors, their levels, the treatments, and the response variable in a description of a designed experiment.

We've learned the value of having a control group and of using blinding and placebo controls.

We've learned to recognize the problems posed by confounding variables in experiments and lurking variables in observational studies.

Finally, we've learned the differences between experiments and surveys:
- Surveys try to estimate facts (parameter) about a population, so they require a representative random sample from that population.
- Experiments try to estimate the differences in the effects of treatments. They randomize a group of experimental units to treatments, but there is no need for the experimental units to be a representative sample from the population.

## TERMS

**Observational study** — A study based on data in which no manipulation of factors has been employed. (p. 315)

**Retrospective study** — An observational study in which subjects are selected and then their previous conditions or behaviors are determined. Retrospective studies need not be based on random samples and they usually focus on estimating differences between groups or associations between variables. (p. 315)

**Prospective study** — An observational study in which subjects are followed to observe future outcomes. Because no treatments are deliberately applied, a prospective study is not an experiment. Nevertheless, prospective studies typically focus on estimating differences among groups that might appear as the groups are followed during the course of the study. (p. 316)

**Experiment** — An experiment *manipulates* factor levels to create treatments, *randomly assigns* subjects to these treatment levels, and then *compares* the responses of the subject groups across treatment levels. (p. 316)

**Random assignment** — To be valid, an experiment must assign experimental units to treatment groups at random. This is called random assignment. (p. 316)

**Factor** — A variable whose levels are manipulated by the experimenter. Experiments attempt to discover the effects that differences in factor levels may have on the responses of the experimental units. (p. 317)

**Response variable** — A variable whose values are compared across different treatments. In a randomized experiment, large response differences can be attributed to the effect of differences in treatment level. (p. 317)

**Experimental units** — Individuals on whom an experiment is performed. Usually called **subjects** or **participants** when they are human. (p. 317)

**Level** — The specific values that the experimenter chooses for a factor are called the levels of the factor. (p. 317)

**Treatment** — The process, intervention, or other controlled circumstance applied to randomly assigned experimental units. Treatments are the different levels of a single factor or are made up of combinations of levels of two or more factors. (p. 317)

**Principles of experimental design**
- **Control** aspects of the experiment that we know may have an effect on the response, but that are not the factors being studied.
- **Randomize** subjects to treatments to even out effects that we cannot control.
- **Replicate** over as many subjects as possible. Results for a single subject are just anecdotes. If, as often happens, the subjects of the experiment are not a representative sample from the population of interest, replicate the entire study with a different group of subjects, preferably from a different part of the population.
- **Block** to reduce the effects of identifiable attributes of the subjects that may affect their responses but cannot be controlled. (pp. 318, 319)

**Completely randomized design** — In a completely randomized design, all experimental units have an equal chance of receiving any treatment. (p. 321)

| | |
|---|---|
| **Statistically significant** | When an observed difference is too large for us to believe that it is likely to have occurred by chance, we consider the difference to be statistically significant. Subsequent chapters will show specific calculations and give rules, but the principle remains the same. (p. 322) |
| **Control group** | The experimental units assigned to a baseline treatment level, typically either the default treatment, which is well understood, or a null, placebo treatment. Either way, this is called the **control treatment**. Their responses provide a basis for comparison. (p. 324) |
| **Blinding** | Any individual associated with an experiment who is not aware of how subjects have been allocated to treatment groups is said to be blinded. (p. 325) |
| **Single-blind Double-blind** | There are two main classes of individuals who can affect the outcome of an experiment:<br>♦ those who could *influence the results* (the subjects, treatment administrators, or technicians).<br>♦ those who *evaluate the results* (judges, treating physicians, etc.).<br><br>When every individual in *either* of these classes is blinded, an experiment is said to be single-blind. When everyone in *both* classes is blinded, we call the experiment double-blind. (p. 326) |
| **Placebo** | A treatment known to have no effect, administered to one group so that all groups experience the same conditions. Many subjects respond to such a treatment (a response known as a **placebo effect**). Only by comparing with a placebo can we be sure that the observed effect of a treatment is not due simply to the placebo effect. (p. 326) |
| **Placebo effect** | The tendency of many human subjects (often 20% or more of experiment subjects) to show a response even when administered a placebo. (p. 326) |
| **Blocking** | When subgroups of the experimental units differ in ways that may affect their responses to treatments, it is a good idea to gather them together into blocks. By blocking, we isolate the variability attributable to the differences between the blocks so that we can see the differences caused by the treatments more clearly. (p. 327) |
| **Randomized block design** | In a randomized block design, the subjects are randomly assigned to treatments only within blocks. (p. 327) |
| **Matching** | In a retrospective or prospective study, subjects who are similar in ways not under study may be matched and then compared with each other on the variables of interest. Matching, like blocking, reduces unwanted variation. (p. 328) |
| **Confounding** | When the levels of one factor are associated with the levels of another factor in such a way that their effects cannot be separated, we say that these two factors are confounded. (p. 330) |

# ON THE COMPUTER

## Experiments

Most experiments are analyzed with a statistics package. You should almost always display the results of a comparative experiment with side-by-side boxplots. You may also want to display the means and standard deviations of the treatment groups in a table.

The analyses offered by statistics packages for comparative randomized experiments fall under the general heading of Analysis of Variance, usually abbreviated ANOVA. These analyses are beyond the scope of this chapter.

# EXERCISES

1. **Standardized test scores** For his Statistics class experiment, researcher J. Gilbert decided to study how parents' income affects children's performance on standardized tests like the SAT. He proposed to collect information from a random sample of test takers and examine the relationship between parental income and SAT score.
   a) Is this an experiment? If not, what kind of study is it?
   b) If there is relationship between parental income and SAT score, why can't we conclude that differences in score are caused by differences in parental income?

2. **Heart attacks and height** Researchers who examined health records of thousands of males found that men who died of myocardial infarction (heart attack) tended to be shorter than men who did not.
   a) Is this an experiment? If not, what kind of study is it?
   b) Is it correct to conclude that shorter men are at higher risk for heart attack? Explain.

3. **MS and vitamin D** Multiple sclerosis (MS) is an autoimmune disease that strikes more often the farther people live from the equator. Could vitamin D—which most people get from the sun's ultraviolet rays—be a factor? Researchers compared vitamin D levels in blood samples from 150 U.S. military personnel who had developed MS with blood samples of nearly 300 who had not. The samples were taken, on average, five years before the disease was diagnosed. Those with the highest blood vitamin D levels had a 62% lower risk of MS than those with the lowest levels. (The link was only in whites, not in Blacks or Latinos.)
   a) What kind of study was this?
   b) Is that an appropriate choice for investigating this problem? Explain.
   c) Who were the subjects?
   d) What were the variables?

4. **Super Bowl commercials** When spending large amounts to purchase advertising time, companies want to know what audience they'll reach. In January 2011, a poll by *The Hollywood Reporter* asked randomly selected American adults whether they planned to watch the upcoming Super Bowl. Men and women were asked separately whether they were looking forward more to the football game or to watching the commercials. Among the men who planned on watching, 70% were watching for the game. Among women, 60% were looking forward primarily to the game.
   a) Was this a stratified sample or a blocked experiment? Explain.
   b) Was the design of the study appropriate for the advertisers' questions?

5. **Maggot therapy?** People generally don't think "health" when they hear the word "maggot," but one experiment tested the ability of maggots to remove dead tissue from open wounds that would not heal on their own. Sterile maggots were placed in a small pouch which, in turn, was placed on the wound. One hundred men with wounds on their lower limbs were randomly assigned to receive either a traditional surgical treatment or maggot therapy. After eight days, the percentage of dead tissue in the wounds that underwent maggot treatment was 54.5%, compared to 66.5% with the surgical treatment. (The difference decreased with time, and the advantage disappeared by about day 15.) Neither patients nor the doctors evaluating the wounds knew which therapy had been applied. (Patients were blindfolded as bandages were changed.) Surprisingly, the number of patients that reported a crawling sensation in their wound was about the same in both groups! (www.myhealthnewsdaily.com/2030-maggots-clean-wounds-faster-surgeons.html)
   a) What kind of study was this?
   b) Is that an appropriate choice for this investigation?
   c) Who were the subjects?
   d) Identify the treatment and response variables.

6. **Honesty** Coffee stations in offices often just ask users to leave money in a tray to pay for their coffee, but many people cheat. On some days researchers at Newcastle University replaced the picture of flowers on the wall behind the coffee station with a picture of staring eyes. They found that the average contribution increased significantly above the well-established standard when people felt they were being watched, even though the eyes were patently not real. (*New York Times*, 12/10/2006)
   a) Was this a survey, an observational study, or an experiment? How can we tell?
   b) Identify the variables.
   c) What does "increased significantly" mean in a statistical sense?

**7–16.** *What's the design?* Read each brief report of statistical research, and identify
   a) whether it was an observational study or an experiment.
   
   *If it was an observational study, identify (if possible)*
   b) whether it was retrospective or prospective.
   c) the subjects studied and how they were selected.
   d) the parameter of interest.
   e) the nature and scope of the conclusion the study can reach.
   
   *If it was an experiment, identify (if possible)*
   b) the subjects studied.
   c) the factor(s)/explanatory variable(s) in the experiment and the number of levels for each.
   d) the number of treatments.
   e) the response variable measured.
   f) the design (completely randomized, blocked, or matched).
   g) whether it was blind (or double-blind).
   h) the nature and scope of the conclusion the experiment can reach.

7. **Superglue** 130 patients with eligible lacerations randomly assigned to have the wound closed either with Octylcyanoacrylate Tissue Adhesive (essentially superglue) or with traditional sutures. When evaluated at the end of the study, the two treatments worked equally well with regard to scarring, and the adhesive was less painful and worked faster. Physicians who evaluated the scarring did not know which treatment the patients had been given.

**8. Truancy** A group of researchers analyzed three observational studies that followed children's attendance and mental health, among other things. They found that students who missed more school tended to have more incidences of depression. One study followed 20,745 secondary students in a random sample of all secondary schools in the United States. Another tracked 2311 first graders at 18 Baltimore schools who were participating in an intervention program. The third study followed 671 students from first or fifth grade to their senior year in high risk areas of Eugene, Oregon, who had been randomly assigned to an intervention or to no intervention.

**9. Hypertension** In a test of roughly 200 older men and women, those with moderately high blood pressure (averaging 164/89 mm Hg) did worse on tests of memory and reaction time than those with normal blood pressure. (*Hypertension* 36 [2000]: 1079)

**10. Tossing and turning** Is diet or exercise effective in combating insomnia? Some believe that cutting out desserts can help alleviate the problem, while others recommend exercise. Forty volunteers suffering from insomnia agreed to participate in a month-long test. Half were randomly assigned to a special no-desserts diet; the others continued desserts as usual. Half of the people in each of these groups were randomly assigned to an exercise program, while the others did not exercise. Those who ate no desserts and engaged in exercise showed the most improvement.

**11. Alcohol and estrogen** After menopause, some women take supplemental estrogen. There is some concern that if these women also drink alcohol, their estrogen levels will rise too high. Twelve volunteers who were receiving supplemental estrogen were randomly divided into two groups, as were 12 other volunteers not on estrogen. In each case, one group drank an alcoholic beverage, the other a nonalcoholic beverage. An hour later, everyone's estrogen level was checked. Only those on supplemental estrogen who drank alcohol showed a marked increase.

**12. Dioxin** Researchers have linked an increase in the incidence of breast cancer in Italy to dioxin released by an industrial accident in 1976. The study identified 981 women who lived near the site of the accident and were under age 40 at the time. Fifteen of the women had developed breast cancer at an unusually young average age of 45. Medical records showed that they had heightened concentrations of dioxin in their blood and that each tenfold increase in dioxin level was associated with a doubling of the risk of breast cancer. (*Science News*, August 3, 2002)

**13. Boys and girls** In 2002 the journal *Science* reported that a study of women in Finland indicated that having sons shortened the lifespans of mothers by about 34 weeks per son, but that daughters helped to lengthen the mothers' lives. The data came from church records from the period 1640 to 1870.

**14. Herbal remedy** Scientists at a major pharmaceutical firm investigated the effectiveness of an herbal compound to treat the common cold. They exposed each subject to a cold virus, then randomly assigned him or her either the herbal compound or a sugar solution known to have no effect on colds. Several days later they assessed the patient's condition, using a cold severity scale ranging from 0 to 5. They found no evidence of benefits associated with the compound.

**15. Depression** The May 4, 2000 issue of *Science News* reported that, contrary to popular belief, depressed individuals cry no more often in response to sad situations than nondepressed people. Researchers studied 23 men and 48 women with major depression and 9 men and 24 women with no depression. They showed the subjects a sad film about a boy whose father has died, noting whether or not the subjects cried. Women cried more often than men, but there were no significant differences between the depressed and nondepressed groups.

**16. Vitamin C doping** Some people who race greyhounds give the dogs large doses of vitamin C in the belief that the dogs will run faster. Investigators at the University of Florida tried three different diets in random order on each of five racing greyhounds. They were surprised to find that when the dogs ate high amounts of vitamin C they ran more slowly. (*Science News*, July 20, 2002) *Note:* Grayhound racing became illegal in Florida in 2021.

**17. Torn ACL** Having at least one 15-minute warm-up session per week resulted in a drastic reduction in tears in the *anterior cruciate ligament* (ACL). In a study involving about 4500 adolescent girls' soccer players in Sweden, one group was randomly assigned to warm up with a neuromuscular exercise session. This group had 64% fewer ACL tears than the control group.

a) Is this an experiment or an observational study? Explain why.
b) Identify the treatments in this study. What is the response variable?
c) Give one *statistical* advantage of using only Swedish girls who played soccer in this study.
d) Give one *statistical* disadvantage of using only Swedish girls who played soccer in this study.

**18. Losing sleep** In February 2021, a medical study was released that compared the cell phone use and sleep habits of over 4000 Chinese medical students. The study reported that students who used their phones for more than 5 hours a day showed a higher risk of poor sleep. The same result applied for students who used their phones for at least 30 minutes after the lights were turned off. (Source: https://pubmed.ncbi.nlm.nih.gov/33532989/)

a) Is this an experiment? Explain why or why not.
b) Researchers concluded that "Daily cumulative mobile phone use and use with the lights off before sleep are associated with poorer sleep quality." Explain why the word "association" was used.
c) Would you be comfortable applying the results of this study to the students at your school?

**19. Migraines** Some people claim they can get relief from migraine headache pain by drinking a large glass of ice water. Researchers plan to enlist several people who suffer from migraines in a test. Participants will be randomly assigned to a standard pain reliever or a placebo. When a participant experiences a migraine headache, they will take the pill. Half of each group will also drink ice water. Participants will then report the level of pain relief they experience.

a) Identify the factors and levels in this experiment.
b) Identify the treatments and the response variable.
c) Is there any blinding described in the study?
d) No blocking is described in the study. What might be an appropriate variable on which to block? Clearly explain why you think this variable would be appropriate.

**20. Low-cal dog food** Researchers at a dog food company want to compare a new lower calorie food with their standard dog food to see if it's effective in helping inactive dogs maintain a healthy weight. They have found several dog owners willing to participate in the trial. The dogs have been classified as small, medium, or large breeds, and the company will supply some owners of each size of dog with one of the two foods. The owners have agreed not to feed their dogs anything else for a period of 6 months, after which the dogs' weights will be checked.

   a) Identify the treatments, the experimental units, and the response variable.
   b) Describe a method of assigning treatments if this is to be a randomized block design with size of the breed as the blocking variable.
   c) Is blinding important in this experiment? Double-blinding? How could blinding be conducted?

**21. Omega-3** An experiment that showed that high doses of omega-3 fats might be of benefit to people with bipolar disorder involved a control group of subjects who received a placebo.

   a) *Who* are the individuals being studied? *What* is the explanatory variable? *What* is the response variable? Are these variables categorical or quantitative?
   b) Why didn't the experimenters just give everyone the omega-3 fats to see if they improved?

**22. Insomnia** Exercise 10 describes an experiment showing that exercise helped people sleep better. The experiment involved other groups of subjects who didn't exercise.

   a) *Who* are the individuals being studied? *What* is the explanatory variable? *What* is the response variable? Are these variables categorical or quantitative?
   b) Why didn't the experimenters just have everyone exercise and see if their ability to sleep improved?

**23. Omega-3 revisited** Exercise 21 describes an experiment investigating a dietary approach to treating bipolar disorder. Researchers randomly assigned 30 subjects to two treatment groups, one group taking a high dose of omega-3 fats and the other a placebo.

   a) Why was it important to randomize in assigning the subjects to the two groups?
   b) What would be the advantages and disadvantages of using 100 subjects instead of 30?

**24. Insomnia again** Exercises 10 and 22 describe an experiment investigating the effectiveness of exercise in combating insomnia. Researchers randomly assigned half of the 40 volunteers to an exercise program.

   a) Why was it important to randomize in deciding who would exercise?
   b) What would be the advantages and disadvantages of using 100 subjects instead of 40?

**25. Omega-3, finis** Exercises 21 and 23 describe an experiment investigating the effectiveness of omega-3 fats in treating bipolar disorder. Suppose some of the 30 subjects were very active people who walked a lot or got vigorous exercise several times a week, while others tended to be more sedentary, working office jobs and watching a lot of TV. Why might researchers choose to block the subjects by activity level before randomly assigning them to the omega-3 and placebo groups?

**26. Insomnia, at last** Exercises 10, 22, and 24 describe an experiment investigating the effectiveness of exercise in combating insomnia. Suppose some of the 40 subjects had maintained a healthy weight, but others were quite overweight. Why might researchers choose to block the subjects by weight level before randomly assigning some of each group to the exercise program?

**27. Tomatoes** Describe a strategy to randomly split the 24 tomato plants into the three groups for the chapter's completely randomized single factor test of OptiGro fertilizer.

**28. Tomatoes II** The chapter also described a completely randomized two-factor experiment testing OptiGro fertilizer in conjunction with two different routines for watering the plants. Describe a strategy to randomly assign the 24 tomato plants to the six treatments.

**29. COVID Vaccine** In a double-blind study, researchers administered the Moderna mRNA vaccine to 15,210 volunteers. A group of the same size received a placebo injection. A second dose was administered 28 days later. In the vaccinated group, 11 of the participants contracted COVID-19, while 185 participants were infected in the placebo group. (Source: https://pubmed.ncbi.nlm.nih.gov/33378609/)

   a) What was the purpose of a placebo in this experiment?
   b) Their results were reported to be very statistically significant. Explain what that means.
   c) The volunteers in the study included most of the ethnic groups found in the United States. It was also restricted to volunteers 18 years of age or older. Describe how this impacts the conclusion of the experiment.

**30. Swimsuits** A swimsuit manufacturer wants to test the speed of its newly designed suit. The company designs an experiment by having 6 randomly selected Olympic swimmers swim as fast as they can with their old swimsuit first and then swim the same event again with the new, expensive swimsuit. The company will use the difference in times as the response variable. Criticize the experiment and point out some of the problems with generalizing the results.

**31. Hamstrings** Athletes who had suffered hamstring injuries were randomly assigned to one of two exercise programs. Those who engaged in static stretching returned to sports activity in a mean of 15.2 days faster than those assigned to a program of agility and trunk stabilization exercises. (*Journal of Orthopaedic & Sports Physical Therapy* 34 [March 2004]: 3)

   a) Explain why it was important to assign the athletes to the two different treatments randomly.
   b) There was no control group consisting of athletes who did not participate in a special exercise program. Explain the advantage of including such a group.
   c) How might blinding have been used?
   d) One group returned to sports activity in a mean of 37.4 days (SD = 27.6 days) and the other in a mean of 22.2 days (SD = 8.3 days). Do you think this difference is statistically significant? Explain.

**32. Diet and blood pressure** The DASH diet is an approach to healthy eating that's designed to help treat or prevent high blood pressure. An experiment showed that subjects fed the DASH diet were able to lower their blood pressure by an average of 6.7 points compared to a group fed a "control diet." All meals were prepared by dieticians.

a) Why were the subjects randomly assigned to the diets instead of letting people pick what they wanted to eat?
b) Why were the meals prepared by dieticians?
c) Why did the researchers need the control group? If the DASH diet group's blood pressure was lower at the end of the experiment than at the beginning, wouldn't that prove the effectiveness of that diet?
d) What additional information would you want to know in order to decide whether an average reduction in blood pressure of 6.7 points was statistically significant?

**33. Mozart** Will listening to a Mozart piano sonata make you smarter? In a 1995 study published in the journal *Psychological Science*, Rauscher, Shaw, and Ky reported that when students were given a spatial reasoning section of a standard IQ test, those who listened to Mozart for 10 minutes improved their scores more than those who simply sat quietly.

a) These researchers said the differences were statistically significant. Explain what that means in context.
b) Steele, Bass, and Crook tried to replicate the original study. In their study, also published in *Psychological Science* (1999), the subjects were 125 college students who participated in the experiment for course credit. Subjects first took the test. Then they were assigned to one of three groups: listening to a Mozart piano sonata, listening to music by Philip Glass, and sitting for 10 minutes in silence. Three days after the treatments, they were retested. Draw a diagram displaying the design of this experiment.
c) These boxplots show the differences in scores before and after treatment for the three groups. Did the Mozart group show improvement?

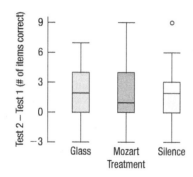

d) Do you think the results prove that listening to Mozart is beneficial? Explain.

**34. Contrast baths** Contrast bath treatments use the immersion of an injured limb alternately in water of two contrasting temperatures. Those who use the method claim that it can reduce swelling. Researchers compared three treatments: (1) contrast baths and exercise, (2) contrast baths alone, and (3) exercise alone. (R. G. Janssen, D. A. Schwartz, and P. F. Velleman, "A Randomized Controlled Study of Contrast Baths on Patients with Carpal Tunnel Syndrome," *Journal of Hand Therapy*, 2009)

They report the following boxplots comparing the change in hand volume after treatment:

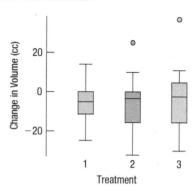

a) The researchers conclude that the differences were not statistically significant. Explain what that means in context.
b) The title says that the study was randomized and controlled. Explain what that probably means for this study.
c) The study did not use a placebo treatment. What was done instead? Do you think that was an appropriate choice? Explain.

**35. Wine** A 2001 Danish study published in the *Archives of Internal Medicine* casts significant doubt on suggestions that adults who drink wine have higher levels of "good" cholesterol and fewer heart attacks. These researchers followed a group of individuals born at a Copenhagen hospital between 1959 and 1961 for 40 years. Their study found that in this group the adults who drank wine were richer and better educated than those who did not.

a) What kind of study was this?
b) It is generally true that people with high levels of education and high socioeconomic status are healthier than others. How does this call into question the supposed health benefits of wine?
c) Can studies such as these prove causation (that wine helps prevent heart attacks, that drinking wine makes one richer, that being rich helps prevent heart attacks, etc.)? Explain.

**36. Swimming** Recently, a group of adults who swim regularly for exercise were evaluated for depression. It turned out that these swimmers were less likely to be depressed than the general population. The researchers said the difference was statistically significant.

a) What does "statistically significant" mean in this context?
b) Is this an experiment or an observational study? Explain.
c) News reports claimed this study proved that swimming can prevent depression. Explain why this conclusion is not justified by the study. Include an example of a possible lurking variable.
d) But perhaps it is true. We wonder if exercise can ward off depression, and whether anaerobic exercise (like weight training) is as effective as aerobic exercise (like swimming). We find 120 volunteers not currently engaged in a regular program of exercise. Design an appropriate experiment.

**37. Ferrets** In Chapter 10, Exercise 41 a researcher was using a nasal spray to stop the spread of COVID-19. He placed five ferrets in each of three cages. One ferret had the COVID virus, two were given the medicated nasal spray, and the remaining two received a placebo nasal spray.

a) How should the researcher randomize the administration of the treatment?
b) None of the ferrets who received the medicated nasal spray contracted the virus, while all the ferrets in the placebo group were infected. Do you think this level of success is statistically significant? Explain.
c) Can the researcher conclude that the nasal spray caused the observed difference? Explain.

**38. Healing** A medical researcher suspects that giving postsurgical patients large doses of vitamin E will speed their recovery times by helping their incisions heal more quickly. Design an experiment to test this conjecture. Be sure to identify the explanatory variable, treatments, response variable, and the role of randomization.

**39. Reading** Some schools teach reading using phonics (the sounds made by letters) and others using whole language (word recognition). Suppose a school district wants to know which method works better.
a) The school district has 30 first- and-second grade teachers. Describe a completely randomized design for an appropriate experiment.
b) Suppose 10 teachers want to teach phonics and the rest of the teachers want to use the whole language method. Is is a problem to allow teachers to self-select a teaching method? Explain.

**40. Gas mileage** Do cars get better gas mileage with premium instead of regular unleaded gasoline? It might be possible to test some engines in a laboratory, but we'd rather use real cars and real drivers in real day-to-day driving, so we get 20 volunteers.
a) Design a completely randomized experiment.
b) What variable might you consider to create a block design? Justify your decision.
c) Is it possible to use a matched pairs design? Explain.

**41. Weekend deaths** A study published in the *New England Journal of Medicine* (August 2001) suggests that it's dangerous to enter a hospital on a weekend. During a 10-year period, researchers tracked over 4 million emergency admissions to hospitals in Ontario, Canada. Their findings revealed that patients admitted on weekends had a much higher risk of death than those who went on weekdays.
a) The researchers said the difference in death rates was "statistically significant." Explain in this context what that means.
b) What kind of study was this? Explain.
c) If you think you're quite ill on a Saturday, should you wait until Monday to seek medical help? Explain.
d) Suggest some possible explanations for this troubling finding.

**42. Shingles** A research doctor has discovered a new ointment that she believes will be more effective than the current medication in the treatment of shingles (a painful skin rash). Eight patients have volunteered to participate in the initial trials of this ointment. You are the statistician hired as a consultant to help design a completely randomized experiment.
a) Describe how you will conduct this experiment.
b) Suppose the eight patients' last names start with the letters A to H. Using the random numbers listed below, show which patients you will assign to each treatment. Explain your randomization procedure clearly.

41098  18329  78458  31685  55259

c) Can you make this experiment double-blind? How?
d) The initial experiment revealed that males and females may respond differently to the ointment. Further testing of the drug's effectiveness is now planned, and many patients have volunteered. What changes in your first design, if any, would you make for this second stage of testing?

**43. Beetles** Hoping to learn how to control crop damage by a certain species of beetle, a researcher plans to test two different pesticides in small plots of corn. A few days after application of the chemicals, he'll check the number of beetle larvae found on each plant. The researcher wants to know whether either pesticide works and whether there is a significant difference in effectiveness between them. Design an appropriate experiment.

**44. SAT Prep** Can special study courses actually help raise SAT scores? One organization says that the 30 students they tutored achieved an average gain of 60 points when they retook the test.
a) Explain why this does not necessarily prove that the special course caused the scores to go up.
b) Propose a design for an experiment that could test the effectiveness of the tutorial course.
c) Suppose you suspect that the tutorial course might be more helpful for students whose initial scores were particularly low. How would this affect your proposed design?

**45. Safety switch** An industrial machine requires an emergency shutoff switch that must be designed so that it can be easily operated with either hand. Design an experiment to find out whether workers will be able to deactivate the machine as quickly with their dominant hands as with their nondominant hands. Be sure to explain the role of randomization in your design.

**46. Washing clothes** A consumer group wants to test the effectiveness of a new "organic" laundry detergent and make recommendations to customers about how to best use the product. They intentionally get grass stains on 30 white T-shirts in order to see how well the detergent will clean them. They want to try the detergent in cold water and in hot water on both the "regular" and "delicates" wash cycles. Design an appropriate experiment, indicating the number of factors, levels, and treatments. Explain the role of randomization in your experiment.

**47. Skydiving, anyone?** A humor piece published in the *British Medical Journal* (Gordon, Smith, and Pell, "Parachute Use to Prevent Death and Major Trauma Related to Gravitational Challenge: Systematic Review of Randomized Control Trials," *BMJ*, 2003:327) notes that we can't tell for sure whether parachutes are safe and effective because there has never been a properly randomized, double-blind, placebo-controlled study of parachute effectiveness in skydiving. (Yes, this is the sort of thing statisticians find funny . . . .) Suppose you were designing such a study:
a) What is the factor in this experiment?
b) What experimental units would you propose?[13]
c) What would serve as a placebo for this study?
d) What would the treatments be?
e) What would the response variable be?
f) What sources of variability would you control?
g) How would you randomize this "experiment"?
h) How would you make the experiment double-blind?

---
[13]Don't include your Statistics instructor!

**48. Steroids** The 1990s and early 2000s could be considered the steroids era in Major League Baseball, as many players have admitted to using the drug to increase performance on the field. A sports writer wanted to compare home run totals from the steroids era to the 1960s.

   a) Is this an observational study or an experiment? Explain.
   b) Could the writer conclude that it was the steroids that caused the increase in home runs? Why or why not? If not, what other variables may have contributed to the increase in home runs?

**49. E-commerce** A business student conjectures that the Internet caused companies to become more profitable, because many transactions previously handled "face-to-face" could now be completed online. The student compares earnings from a sample of companies from the 1980s to a sample from the 2000s.

   a) Is this an observational study or an experiment? Explain.
   b) If indeed profitability increased, can she conclude the Internet was the cause? Why or why not?

**50. Tips** A pizza delivery driver, always trying to increase tips, runs an experiment on his next 40 deliveries. He flips a coin to decide whether or not to call a customer from his mobile phone when he is five minutes away, hoping this slight bump in customer service will lead to a slight bump in tips. After 40 deliveries, he will compare the average tip percentage between the customers he called and those he did not.

   a) What are the experimental units, and how did he randomize treatments?
   b) What are the explanatory variable, the levels, and the response variable?
   c) Name some variables the driver did or should have controlled. Was the experiment randomized and replicated?
   d) Can this experiment use blinding? Double-blinding?
   e) Name some confounding variables that might influence the experiment's results.
   f) The driver wants to know about tipping in general, so he recruits several other drivers to participate in the experiment. Each driver randomly decides whether to phone customers before delivery and records the tip percentage. Is this experiment blocked? Is that a good idea?

## JUST CHECKING

**Answers**

1. a) The factor was type of treatment for peptic ulcer.
   b) The response variable could be a measure of relief from gastric ulcer pain or an evaluation by a physician of the state of the disease.
   c) Treatments would be gastric freezing and some alternative control treatment.
   d) Treatments should be assigned randomly.
   e) No. The website reports "lack of effectiveness," indicating that no large differences in patient healing were noted.

2. a) Neither the patients who received the treatment nor the doctor who evaluated them afterward knew what treatment they had received.
   b) The placebo is needed to accomplish blinding. The best alternative would be using body-temperature liquid rather than the freezing liquid.
   c) The researchers should block the subjects by the length of time they had had the ulcer, then randomly assign subjects in each block to the freezing and placebo groups.

# Review of Part III

## GATHERING DATA

### Quick Review

Before you can make a boxplot, calculate a mean, describe a distribution, or fit a line, you must have meaningful data to work with. Getting good data is essential to any investigation. No amount of clever analysis can make up for badly collected data. Here's a brief summary of the key concepts and skills:

- The way you gather data depends both on what you want to discover and on what is practical.
- To get some insight into what might happen in a real situation, model it with a **simulation** using random numbers, playing cards, dice, coins, or any fair randomizing device.
- To answer questions about a target population, collect information from a sample with a **survey** or poll.
  - Choose the sample randomly. Random sampling designs include simple, stratified, systematic, cluster, and multistage.
  - A simple random sample draws without restriction from the entire target population.
  - When there are subgroups within the population that may respond differently, use a stratified sample.
  - Avoid bias, a systematic distortion of the results. Sample designs that allow undercoverage or response bias and designs such as voluntary response or convenience samples don't faithfully represent the population.
  - Samples will naturally vary one from another. This sample-to-sample variation is called sampling variability (or sampling error). Each sample only approximates the target population.
- **Observational studies** collect information from a sample drawn from a target population.
  - Retrospective studies examine existing data. Prospective studies identify subjects in advance, then follow them to collect data as the data are created, perhaps over many years.
  - Observational studies can spot associations between variables but cannot establish cause and effect. It's impossible to eliminate the possibility of lurking or confounding variables.
- To see how different treatments influence a response variable, design an **experiment**.
  - Assign subjects to treatments randomly. If you don't assign treatments randomly, your experiment is not likely to yield valid results.
  - Control known sources of variation as much as possible. Reduce variation that cannot be controlled by using blocking, if possible.
  - Assign more than one subject or experimental unit to each treatment level. This is one type of replication.
  - If possible, replicate the entire experiment with an entirely different collection of subjects. This is another type of replication.
  - A well-designed experiment can provide evidence that changes in the factors cause changes in the response variable.
- Understand what the different purposes are of random **sampling** in a survey and random **treatment allocation** in an experiment. Random sampling makes it likely that the data represent the population, and conclusions can be extrapolated. Random treatment allocation makes it likely that an observed association between treatment and response is causal.

Now for more opportunities to review these concepts and skills . . .

## REVIEW EXERCISES

**1–18.** *What design? Analyze the design of each research example reported. Is it a sample survey, an observational study, or an experiment? If a sample, what are the population, the parameter of interest, and the sampling procedure? If an observational study, was it retrospective or prospective? If an experiment, describe the factor(s)/explanatory variable(s), treatments, randomization, response variable, and any blocking, matching, or blinding that may be present. In each, what kind of conclusions can be reached?*

1. Researchers identified 242 children in the Cleveland area who had been born prematurely (at about 29 weeks). They examined these children at age 8 and again at age 20, comparing them to another group of 233 children not born prematurely. Their report, published in the *New England Journal of Medicine*, said the "preemies" engaged in significantly less risky behavior than the others. Differences showed up in the use of alcohol and marijuana, conviction of crimes, and teenage pregnancy.

2. The journal *Circulation* reported that among 1900 people who had heart attacks, those who drank an average of 19 cups of tea a week were 44% more likely than nondrinkers to survive at least 3 years after the attack.

3. Researchers at the Purina Pet Institute studied Labrador retrievers for evidence of a relationship between diet and longevity. At 8 weeks of age, 2 puppies of the same sex and weight were randomly assigned to one of two groups—a total of 48 dogs in all. One group was allowed to eat all they wanted, while the other group was fed a diet about 25% lower in calories. The median life span of dogs fed the restricted diet was 22 months longer than that of other dogs. (Source: *Science News* 161, no. 19)

4. The radioactive gas radon, found in some homes, poses a health risk to residents. To assess the level of contamination in their area, a county health department wants to test a few homes. If

the risk seems high, they will publicize the results to emphasize the need for home testing. Officials plan to use the local property tax list to randomly choose 25 homes from various areas of the county.

5. A group of 521,330 people were enrolled in the European Prospective Investigation into Cancer and Nutrition. A study looked at the level of coffee consumption (asked at the time of enrollment in the study) and death rates. The participants in the highest quartile of coffee consumption had lower mortality during the duration of the study than did nonconsumers. (Source: *Annals of Internal Medicine*, July 2017)

6. In the journal *Science*, a research team reported that plants in southern England are flowering earlier in the spring. Records of the first flowering dates for 385 species over a period of 47 years indicate that flowering has advanced an average of 15 days per decade, an indication of climate warming, according to the authors.

7. Fireworks manufacturers face a dilemma. They must be sure that the rockets work properly, but test-firing a rocket essentially destroys it. On the other hand, not testing the product leaves open the danger that they sell a bunch of duds, leading to unhappy customers and loss of future sales. The solution, of course, is to test a few of the rockets produced each day, assuming that if those tested work properly, the others are ready for sale.

8. People who read the last page of a mystery novel first generally like stories better. Researchers recruited 819 college students to read short stories, and for one story, they were given a spoiler paragraph beforehand. On the second and third stories, the spoiler was incorporated as the opening paragraph or not given at all. Overall, participants liked the stories best after first reading spoilers. (Source: *Psychological Science*, August 12, 2011)

9. Does keeping a child's lunch in an insulated bag, even with ice packs, protect the food from warming to temperatures where germs can proliferate? Researchers used an electric temperature gun on 235 lunches at preschools 90 minutes before they were to be eaten. Of the lunches with ice packs, over 90% of them were at unsafe temperatures. The study was of particular interest because preschoolers develop up to four times as many foodborne infections as do adults. (Source: *Science News*, August 9, 2011)

10. Some doctors have expressed concern that men who have vasectomies seemed more likely to develop prostate cancer. Medical researchers used a national cancer registry to identify 923 men who had had prostate cancer and 1224 men of similar ages who had not. Roughly one quarter of the men in each group had undergone a vasectomy, many more than 25 years before the study. The study's authors concluded that there is strong evidence that having the operation presents no long-term risk for developing prostate cancer. (Source: *Science News*, July 20, 2002)

11. Widely used antidepressants may reduce ominous brain plaques associated with Alzheimer's disease. In the study, mice genetically engineered to have large amounts of brain plaque were given a class of antidepressants that boost serotonin in the brain. After a single dose, the plaque levels dropped, and after four months, the mice had about half the brain plaques as mice that didn't take the drug. (Source: *Proceedings of the National Academy of Sciences*, August 22, 2011)

12. An artisan wants to create pottery that has the appearance of age. He prepares several samples of clay with four different glazes and test fires them in a kiln at three different temperature settings.

13. Tests of gene therapy on laboratory rats have raised hopes of stopping the degeneration of tissue that characterizes chronic heart failure. Researchers at the University of California, San Diego, used hamsters with cardiac disease, randomly assigning 30 to receive the gene therapy and leaving the other 28 untreated. Five weeks after treatment the gene therapy group's heart muscles stabilized, while those of the untreated hamsters continued to weaken. (Source: *Science News*, July 27, 2002)

14. People aged 50 to 71 were initially contacted in the mid-1990s to participate in a study about smoking and bladder cancer. Data were collected from more than 280,000 men and 186,000 women from eight states who answered questions about their health, smoking history, alcohol intake, diet, physical activity, and other lifestyle factors. When the study ended in 2006, about half the bladder cancer cases in adults age 50 and older were traceable to smoking. (Source: *Journal of the American Medical Association*, August 17, 2011)

15. An orange-juice processing plant will accept a shipment of fruit only after several hundred oranges selected from various locations within the truck are carefully inspected. If too many show signs of unsuitability for juice (bruised, rotten, unripe, etc.), the whole truckload is rejected.

16. A soft-drink manufacturer must be sure the bottle caps on the soda are fully sealed and will not come off easily. Inspectors pull a few bottles off the production line at regular intervals and test the caps. If they detect any problems, they will stop the bottling process to adjust or repair the machine that caps the bottles.

17. Older Americans with a college education are significantly more likely to be emotionally well-off than are people in this age group with less education. Among those aged 65 and older, 35% scored 90 or above on the Emotional Health Index, but for those with a college degree, the percentage rose to 43% (postgraduate degree, 46%). The results are based on phone interviews conducted between January 2010 and July 2011. (Source: gallup.com, August 19, 2011)

18. Does the use of computer software in Introductory Statistics classes lead to better understanding of the concepts? A professor teaching two sections of Statistics decides to investigate. She teaches both sections using the same lectures and assignments, but gives one class statistics software to help them with their homework. The classes take the same final exam, and graders do not know which students used computers during the semester. The professor is also concerned that students who have had calculus may perform differently from those who have not, so she plans to compare software vs. no-software scores separately for these two groups of students.

19. **Point spread** When taking bets on sporting events, bookmakers often include a "point spread" that awards the weaker team extra points. In theory, this makes the outcome of the bet a toss-up. Suppose a gambler places a $10 bet and picks the winners of five games. If he's right about fewer than three of the games, he loses. If he gets three, four, or all five correct, he's paid $10, $20, and $50, respectively. Estimate the amount such a bettor might expect to lose over many weeks of gambling.

**20. The lottery** Many people spend a lot of money trying to win huge jackpots in state lotteries. Let's play a simplified version using only the numbers from 1 to 20. You bet on three numbers. The state picks five winning numbers. If your three are all among the winners, you are rich!

a) Simulate repeated plays. How long did it take you to win?
b) In real lotteries, there are many more choices (often 54) and you must match all five winning numbers. Explain how these changes affect your chances of hitting the jackpot.

**21. Everyday randomness** Aside from casinos, lotteries, and games, there are other situations you encounter in which something is described as "random" in some way. Give three different examples. Describe how randomness is (or is not) achieved in each.

**22. Cell phone risks** Researchers at the Washington University School of Medicine randomly placed 480 rats into one of three chambers containing radio antennas. One group was exposed to digital cell phone radio waves, the second to analog cell phone waves, and the third group to no radio waves. Two years later, the rats were examined for signs of brain tumors. In June 2002, the scientists said that differences among the three groups were not statistically significant.

a) Is this a study or an experiment? Explain.
b) Explain in this context what "not statistically significant" means.
c) Comment on the fact that this research was funded by Motorola, a manufacturer of cell phones.

**23. Tips** In restaurants, servers rely on tips as a major source of income. Does serving candy after the meal produce larger tips? To find out, two waiters determined randomly whether or not to give candy to 92 dining parties. They recorded the sizes of the tips and reported that guests getting candy tipped an average of 17.8% of the bill, compared with an average tip of only 15.1% from those who got no candy. (Source: "Sweetening the Till: The Use of Candy to Increase Restaurant Tipping," *Journal of Applied Social Psychology* 32, no. 2 [2002]: 300–309)

a) Was this an experiment or an observational study? Explain.
b) Is it reasonable to conclude that the candy caused guests to tip more? Explain.
c) The researchers said the difference was statistically significant. Explain in this context what that means.

**24. Tips, take 2** In another experiment to see if getting candy after a meal would induce customers to leave a bigger tip, a server randomly decided what to do with 80 dining parties. Some parties received no candy, some just one piece, and some two pieces. Others initially got just one piece of candy, and then the server suggested that they take another piece. She recorded the tips received, finding that, in general, the more candy, the higher the tip, but the highest tips (23%) came from the parties who got one piece and then were offered more. (Source: "Sweetening the Till: The Use of Candy to Increase Restaurant Tipping," *Journal of Applied Social Psychology* 32, no. 2 [2002]: 300–309)

a) Diagram this experiment.
b) What are the explanatory variable(s)? How many levels are there?
c) How many treatments are there?
d) What is the response variable?
e) Did this experiment involve blinding? Double-blinding?
f) In what way might the server, perhaps unintentionally, have biased the results?

**25. Alternate-day fasting** A paper published online in 2017 in the *Journal of the American Medical Association (JAMA) Internal Medicine* (jamanetwork.com/journals/jamainternalmedicine/fullarticle/2623528) reported on a study of alternate-day fasting as a weight-loss method. One hundred obese persons were assigned at random to one of three groups: an alternate-day fasting group, a calorie-restricted group, and a control. The alternate-day fasting group alternately consumed 25% of their usual caloric intake during lunch on fasting days and 125% on the alternating days. The calorie-restricted group consumed 75% of baseline energy over three meals each day. The control group ate as usual. The study reports that there was essentially no difference in weight loss between the alternate-day fasting group and the calorie-restricted group, both losing an average of 6.8% of their weight. From this description, identify

a) the participants.
b) the treatments.
c) the response.
d) Was the study blind? If not, should it have been blind?
e) The participants were not a random sample from the population. Is that a problem for this study?

**26. Timing** In August 2011, a sodahead.com voluntary response poll asked site visitors, "Obama Is on Vacation Again: Does He Have the Worst Timing Ever?" 56% of the 629 votes were for "Yes." During the week of the poll, a 5.8 earthquake struck near Washington, D.C., and Hurricane Irene made its way up the East coast. What types of bias may be present in the results of the poll?

**27. Laundry** An experiment to test a new laundry detergent, SparkleKleen, is being conducted by a group of consumer advocates. They would like to compare its performance with that of a laboratory standard detergent they have used in previous experiments. They can stain 16 swatches of cloth with 2 teaspoons of a common staining compound and then use a well-calibrated optical scanner to detect the amount of the stain left after washing. To save time in the experiment, several suggestions have been made. Comment on the possible merits and drawbacks of each one.

a) Because data for the laboratory standard detergent are already available from previous experiments, for this experiment wash all 16 swatches with SparkleKleen, and compare the results with the previous data.
b) Use both detergents with eight separate runs each, but to save time, use only a 10-second wash time with very hot water.
c) To ease bookkeeping, first run all of the standard detergent washes on eight swatches, then run all of the SparkleKleen washes on the other eight swatches.
d) Rather than run the experiment, use data from the company that produced SparkleKleen, and compare them with past data from the standard detergent.

**28. When to stop?** You play a game that involves rolling a die. You can roll as many times as you want, and your score is the total for all the rolls. But ... if you roll a 6 your score is 0 and your turn is over. What might be a good strategy for a game like this?

a) One of your opponents decides to roll 4 times, and then stop (hoping not to get the dreaded 6 before then). Use a simulation to estimate his average score.
b) Another opponent decides to roll until she gets at least 12 points, and then stop. Use a simulation to estimate her average score.
c) Propose another strategy that you would use to play this game. Using your strategy, simulate several turns. Do you think you would beat the two opponents?

**29. Rivets** A company that manufactures rivets believes the shear strength of the rivets it manufactures follows a Normal model with a mean breaking strength of 950 pounds and a standard deviation of 40 pounds.

  a) What percentage of rivets selected at random will break when tested under a 900-pound load?
  b) You're trying to improve the rivets and want to examine some that fail. Use a simulation to estimate how many rivets you might need to test in order to find three that fail at 900 pounds (or below).

**30. Homecoming** A college Statistics class conducted a survey concerning community attitudes about the college's large homecoming celebration. That survey drew its sample in the following manner: Telephone numbers were generated at random by selecting one of the local telephone exchanges (first three digits) at random and then generating a random four-digit number to follow the exchange. If a person answered the phone and the call was to a residence, then that person was taken to be the subject for interview. (Undergraduate students and those under voting age were excluded, as was anyone who could not speak English.) Calls were placed until a sample of 200 eligible respondents had been reached.

  a) Did every telephone number that could occur in that community have an equal chance of being generated?
  b) Did this method of generating telephone numbers result in a simple random sample (SRS) of local residences? Explain.
  c) Did this method generate an SRS of local voters? Explain.
  d) Is this method unbiased in generating samples of households? Explain.

**31. How long is 30 seconds?** Sofie, Ryan, and Alessandra wanted to design an experiment to find out how distraction affects our ability to judge time. The experiment consisted of starting a clock (out of view of the subjects) and then asking the subjects to tell them when they thought 30 seconds had passed. For each of four subjects, they repeated the experiment eight times under different treatment conditions: eyes open or closed, having music on or off, and sitting or moving around.

  a) Identify the factors and the factor levels.
  b) What is the role of the variable *Subject*?
  c) How many runs did they perform?
  d) They suspect that subjects may do slightly better the more times they perform the task. Ryan argues that randomizing the run order but having all four subjects use the same run order is a good idea. Sofie insists that they should randomize the run order for all four subjects. What do you think? Explain briefly.

**32. Cookies** Mary Beth, Nigel, and Molly want to design an experiment to find the recipe for the best chocolate chip cookies. They will try to keep the size of the cookies the same, but use cooking times of 10 and 15 minutes. They will use three different temperatures: 325°F, 375°F, and 425°F, and use either 5 or 10 chips in each cookie. Six of their friends will taste the cookies and rank them in order.

  a) What are the factors and levels?
  b) What does blinding mean in the context of this experiment? Is it double-blind?
  c) How many different cookies will each judge taste?
  d) What kind of variable is the response variable?
  e) What are some of the challenges of carrying out the experiment?

**33. Youthful appearance** *Readers' Digest* (April 2002, p. 152) reported results of several surveys that asked graduate students to examine photographs of men and women and try to guess their ages. Researchers compared these guesses with the number of times the people in the pictures reported having sexual intercourse. It turned out that those who had been more sexually active were judged as looking younger, and that the difference was described as "statistically significant." Psychologist David Weeks, who compiled the research, speculated that lovemaking boosts hormones that "reduce fatty tissue and increase lean muscle, giving a more youthful appearance."

  a) What does "statistically significant" mean in this context?
  b) Explain in statistical terms why you might be skeptical about Dr. Weeks's conclusion. Propose an alternative explanation for these results.

**34. Smoking and Alzheimer's** Medical studies indicate that smokers are less likely to develop Alzheimer's disease than people who never smoked.

  a) Does this prove that smoking may offer some protection against Alzheimer's? Explain.
  b) Offer an alternative explanation for this association.
  c) How would you conduct a study to investigate this?

**35. First step** A group of students at a southern California high school devised a clever plan for their final AP Statistics project. These students were members of the school marching band and as members of this band they were trained to step forward with their left foot first. They were curious if this habit created a difference that extended into everyday habits.

  To test their theory, they recruited 50 band students and 50 students not in band. They had students stand with their toes at the edge of a long tape measure. To disguise their idea, students were told that their stride length was going to be measured. Students walked the length of the tape measure, but the only observation that was recorded was whether the student led with their left or right foot.

  a) Is this an experiment or an observational study? Explain.
  b) The study uses volunteers rather than a random sample of students. Does this make the results invalid? Explain.
  c) The Statistics students were rather pleased with their results. The band students led with their left foot 50 out of 50 times, while only 17 out of 50 non–band students led with their left foot. Does this result sound statistically significant? Explain.

**36. Sex and violence** Does the content of a television program affect viewers' memory of the products advertised in commercials?

  a) Design an experiment to compare the ability of viewers to recall brand names of items featured in commercials during programs with violent content, sexual content, or neutral content.
  b) What are the explanatory variable(s) and the treatments?
  c) What is the response variable in this experiment? Is is quantitative or categorical? What kind of graph could be used to compare the results?

**37. Art and diagnosis** At the University of Texas Health Center, 19 medical students were taught how to look at art—like a kindergartner! The students were taught to use Visual Thinking Strategies (VTSs), a technique originally developed to help kindergartners look at art. Researchers asked students, 'What do you see? What do you see that makes you think that? What more do you?'" Students were also assigned "art patients" with skin conditions, a rash, tattoo removal, etc. The VTS method was used to help students diagnose the conditions. The students were given a pretest and posttest that evaluated their ability to describe both art and the medical images. There were several areas of significant growth, including using more words and an increase in the total number of observations. Also, students used fewer personal narratives and instead improved at using the evidence in front of them to describe their observations. (Source: https://www.sciencedaily.com/releases/2015/03/150331175808.htm)

Let's examine this study from a statistical perspective.

a) What are the W's: *Who, What, When, Where, Why, How?*
b) Is this an observational study or an experiment? Explain.
c) What population do the researchers think the experiment applies to?
d) Can the researchers make a causal conclusion? Explain.
e) The researchers did not use a control group that viewed only medical images and no art. Describe why this might be a helpful control to add to a follow-up study.

**38. Age and party 2016** The Pew Research Center conducts surveys on demographic information several times a year and compiles the results of those surveys. Among the questions asked are age and political affiliation. The table summarizes the numbers of people surveyed that fall into each category.

|       | Republican | Democrat | Other | Total |
|-------|------------|----------|-------|-------|
| 18–29 | 464        | 779      | 55    | 1298  |
| 30–49 | 1035       | 1246     | 234   | 2515  |
| 50–64 | 1178       | 1129     | 127   | 2434  |
| 65+   | 893        | 740      | 233   | 1866  |
| Total | 3570       | 3894     | 649   | 8113  |

a) What sampling strategy do you think the pollsters used? Explain.
b) What percentage of the people surveyed were Democrats?
c) Do you think this is a good estimate of the percentage of voters in the United States who are registered Democrats? Why or why not?
d) In creating this sample design, what question do you think the pollsters were trying to answer?

**39. Bias?** Political analyst Michael Barone has written that "conservatives are more likely than others to refuse to respond to polls, particularly those polls taken by media outlets that conservatives consider biased" (Source: *The Weekly Standard*, March 10, 1997). The Pew Research Foundation tested this assertion by asking the same questions in a national survey run by standard methods and in a more rigorous survey that was a true SRS with careful follow-up to encourage participation. The response rate in the "standard survey" was 42%. The response rate in the "rigorous survey" was 71%.

a) What kind of bias does Barone claim may exist in polls?
b) What is the population for these surveys?
c) On the question of political position, the Pew researchers report the following table:

|              | Standard Survey | Rigorous Survey |
|--------------|-----------------|-----------------|
| Conservative | 37%             | 35%             |
| Moderate     | 40%             | 41%             |
| Liberal      | 19%             | 20%             |

What makes you think these results are incomplete?

d) The Pew researchers report that differences between opinions expressed on the two surveys were not statistically significant. Explain what "not statistically significant" means in this context.

**40. Save the grapes** Vineyard owners have problems with birds that like to eat ripening grapes. Some vineyards use scarecrows to try to keep birds away. Others use netting that covers the plants. Owners would like to know if either method works and, if so, which one is better. One owner has offered to let you use his vineyard this year for an experiment. Propose a design. Carefully indicate how you would set up the experiment, specifying the explanatory variable(s) and response variable.

**41. Bats** It's generally believed that baseball players can hit the ball farther with aluminum bats than with the traditional wooden ones. Is that true? And, if so, how much farther? Players on your local high school baseball team have agreed to help you find out. Design an appropriate experiment.

**42. Acupuncture** Research reported in 2008 brings to light the effectiveness of treating chronic lower back pain with different methods. One-third of nearly 1200 volunteers were administered conventional treatment (drugs, physical therapy, and exercise). The remaining patients got 30-minute acupuncture sessions. Half of these patients were punctured at sites suspected of being useful and half received needles at other spots on their bodies. Comparable shares of each acupuncture group, roughly 45%, reported decreased back pain for at least six months after their sessions ended. This was almost twice as high as those receiving the conventional therapy, leading the researchers to conclude that results were statistically significant.

a) Why did the researchers feel it was necessary to have some of the patients undergo a "fake" acupuncture?
b) Because patients had to consent to participate in this experiment, the subjects were essentially self-selected—a kind of voluntary response group. Explain why that does not invalidate the findings of the experiment.
c) What does "statistically significant" mean in the context of this experiment?

**43. Fuel efficiency** Wayne Collier designed an experiment to measure the fuel efficiency of his family car under different tire pressures. For each run, he set the tire pressure to either 28 or 32 psi and then measured the miles driven on a highway (I-95 between Mills River and Pisgah Forest, North Carolina) until he ran out of fuel using 2 liters of fuel each time. He also used two different types of gasoline (regular and premium). To run the experiment he made some alterations to the normal flow of gasoline to the engine. In Wayne's words, "I inserted a T-junction into the fuel line just before the fuel filter, and a line into the passenger compartment of my car, where it joined with a graduated 2 liter Rubbermaid™ bottle that I mounted in a box where the passenger seat is normally fastened. Then I sealed off the fuel-return line, which under normal operation sends excess fuel from the fuel pump back to the fuel tank."

a) Identify the factors and the levels in the experiment.
b) How many times would he need to drive to make sure all treatments are represented and each treatment combination is replicated?
c) For simplicity, he wants to run all the regular gasoline runs first. Explain why this might not be a good idea. What would you suggest?

**44. NBA draft lottery** Professional basketball teams hold a "draft" each year in which they get to pick the best available college and high school players. In an effort to promote competition, teams with the worst records get to pick first, theoretically allowing them to add better players. To combat the fear that teams with no chance to make the playoffs might try to get better draft picks by intentionally losing late-season games, the NBA's Board of Governors adopted a weighted lottery system in 1990. Under this system, the 11 teams that did not make the playoffs were eligible for the lottery. The NBA prepared 66 cards, each naming one of the teams. The team with the worst win-loss record was named on 11 of the cards, the second-worst team on 10 cards, and so on, with the team having the best record among the nonplayoff clubs getting only one chance at having the first pick. The cards were mixed, then drawn randomly to determine the order in which the teams could draft players. Suppose there are two exceptional players available in this year's draft and your favorite team had the third-worst record. Use a simulation to find out how likely it is that your team gets to pick first or second. Describe your simulation carefully.

**45. Security** There are 20 first-class passengers and 120 coach passengers scheduled on a flight. In addition to the usual security screening, 10% of the passengers will be subjected to a more complete search.

a) Describe a sampling strategy to randomly select those to be searched.
b) Here is the first-class passenger list and a set of random digits. Select two passengers to be searched, carefully demonstrating your process.

65436 71127 04879 41516 20451 02227 94769 23593

| Bergman | Cox | Fontana | Perl |
| Bowman | DeLara | Forester | Rabkin |
| Burkhauser | Delli-Bovi | Frongillo | Roufaiel |
| Castillo | Dugan | Furnas | Swafford |
| Clancy | Febo | LePage | Testut |

c) Explain how you would use a random number table to select the coach passengers to be searched.

**46. Profiling?** Among the 20 first-class passengers on the flight described in Exercise 45, there were four businesspeople from the Middle East. Two of them were the two passengers selected to be searched. They complained of profiling, but the airline claims that the selection was random. What do you think? Support your conclusion with a simulation.

**47. Par 4** In theory, a golfer playing a par-4 hole tees off, hitting the ball in the fairway, then hits an approach shot onto the green. The first putt (usually long) probably won't go in, but the second putt (usually much shorter) should. Sounds simple enough, but how many strokes might it really take? Use a simulation to estimate a fairly good golfer's score based on these assumptions:

- The tee shot hits the fairway 70% of the time.
- A first approach shot lands on the green 80% of the time from the fairway, but only 40% of the time otherwise.
- Subsequent approach shots land on the green 90% of the time.
- The first putt goes in 20% of the time, and subsequent putts go in 90% of the time.

**48. The back nine** Use simulations to estimate more golf scores, similar to the procedure in Exercise 47.

a) On a par 3, the golfer hopes the tee shot lands on the green. Assume that the tee shot behaves like the first approach shot described in Exercise 47.
b) On a par 5, the second shot will reach the green 10% of the time and hit the fairway 60% of the time. If it does not hit the green, the golfer must play an approach shot as described in Exercise 47.
c) For golfers: create a list of assumptions that describe *your* golfing ability, and then simulate your score on a few holes. Explain your simulation clearly.

**49. Internet speed** Carsten, Matt, and Rainer designed an experiment to see how different environments affect the Internet speed around campus. They used their own Mac computer and a PC belonging to the school and tested each in two different libraries, the main and the science library. Other factors included Time of Week (weekday or weekend), Time of Day (before or after 5 p.m.), and how busy the computer was (running other jobs or not). They measured the time it took to download a 50 Mb file from the school course management system.

a) Name the factors and levels.
b) How many runs will they need to run all treatment levels?
c) In what order should they perform all the runs?

# PRACTICE EXAM

## I. MULTIPLE CHOICE

**(Questions 1–2).** Do math skills matter when looking for gainful employment? The latest U.S. Department of Education National Assessment of Adult Literacy was conducted in 2003. A random sample of 18,102 adults (aged 16+) living in U.S. households were tested on Quantitative Literacy and asked whether they thought that their math skills limited their job opportunities. Quantitative Literacy was measured as a respondent's ability to identify and perform computations using data embedded in printed materials such as balancing a checkbook, figuring out a tip, completing an order form, or determining the amount of interest on a loan from an advertisement. Respondents were placed into one of four categories (Below Basic, Basic, Intermediate, or Proficient) based upon their scores on various tasks. The table below summarizes the data that were collected.

**Number of Adults Who Think Their Math Skills Limit Their Ability to Get a Job**

| | Quantitative Proficiency Level | | | | |
|---|---|---|---|---|---|
| Response | Below Basic | Basic | Intermediate | Proficient | Total |
| Not at All | 1654 | 4095 | 4813 | 2175 | 12738 |
| A Little | 662 | 807 | 662 | 196 | 2326 |
| Some | 786 | 807 | 421 | 49 | 2062 |
| A Lot | 1034 | 496 | 120 | 24 | 1675 |
| Total | 4136 | 6204 | 6016 | 2444 | 18801 |

1. What percentage of those with Basic quantitative skills think that their math skills limit their job opportunities A Little?
   A) 4%   B) 12%
   C) 13%   D) 33%
   E) 35%

2. Adults in which proficiency level are most likely to think that their math skills limit them "A Little" in job opportunities?
   A) Below Basic   B) Basic
   C) Intermediate   D) Proficient
   E) Below Basic and Intermediate have the same high likelihood.

3. A school newspaper is going to investigate students' perceptions regarding the campus security guards. They plan to survey 100 students at the school. The editor-in-chief tells his staff that he is worried about undercoverage bias created by chronically truant students. His concern is . . .
   A) unwarranted because some students may refuse to participate and that is inevitable.
   B) justified because chronically truant students will be difficult to survey and will probably have a disproportionately negative view of the security guards.
   C) unwarranted because students who cannot be surveyed are not a part of the population of interest.
   D) unwarranted because students who cannot be surveyed are not a part of the sample.
   E) justified because people need to be forced to answer a survey in order for the results to be valid.

4. Professional NBA basketball scouts are on the lookout for tall players. Ideally, a player makes a great center if he is 7 feet tall or taller. Male height in America is roughly normally distributed with a mean of 69.5" and a standard deviation of 3". If there are 4.6 million males in America who are 18 or 19 years old, about how many prospective centers are available?
   A) 3   B) 5
   C) 22   D) 48
   E) 67

5. A poll of 500 randomly selected likely voters predicted that Candidate A would get 51% of the vote. The opposition insists that only 47% of the voters support Candidate A. You test their claim using a simulation, based on the assumption that 47% of the likely voters will vote for Candidate A. You assign digit pairs 00–46 to represent a voter who would pick Candidate A and 47–99 to represent a voter who would not. You do 200 trials, recording the proportion of voters in each of your simulated samples that would vote for Candidate A. The resulting sample proportions are shown in the plot below.

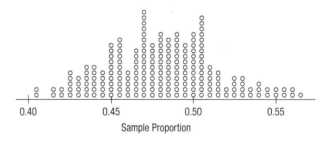

Is the opposition's claim that Candidate A has the support of only 47% of the likely voters plausible?

A) No, the average proportion of voters supporting Candidate A in the samples would be less than 51%, so a sample proportion of 51% would be too unexpected using this model.

B) No, because the majority of sample proportions would be less than 51%, so a sample proportion of 51% would be too unexpected using this model.

C) No, because only 29 of the 200 samples had a proportion of 51% or greater supporting Candidate A, so a sample proportion of 51% would be too unexpected using this model.

D) Yes, because 29 of the 200 samples had a proportion of 51% or greater supporting Candidate A, so a sample proportion of 51% would be plausible using this model.

E) Yes, because a sample proportion of 51% occurred more than once, so it's possible to get a sample proportion of 51% using this model.

6. A small airline runs commuter flights with a plane that holds 10 people. Each ticket-holder has a 10% chance of not showing up, so the airline sells 12 tickets for each flight. Which is an appropriate plan for a simulation that uses a table of random digits to estimate the probability that exactly ten people show up for the flight?

   A) Let digit pairs 01–12 represent the 12 tickets. In the table, select pairs of digits, ignoring repeats and pairs that do not represent a ticket. Continue until you get 10 seats filled. Record the number of pairs needed to get 10 seats filled.
   B) Let digit pairs 01–12 represent the 12 tickets. In the table, select pairs of digits until you find 10 pairs that represent tickets and record the proportion of trials that required ten or fewer.
   C) Let digit pairs 01–12 represent a seat that was filled, and other pairs represent a seat that was not filled. In the table, select 10 pairs of digits, ignoring repeats, and record the number of seats that were filled.
   D) Let digit 0 represent a seat that was filled, and 1–9 a seat that was not filled. In the table, select 10 digits and record the number of seats that were filled.
   E) Let digit 0 represent a ticket-holder that doesn't show up, and 1–9 a ticket-holder who shows up. Select 12 digits and record the number of passengers who show up.

7. A pile of sand on your local beach has no strength and can be knocked down by a toddler with a light kick. Sandstones, however, have a great variety of strength. A coloration difference indicates a difference in the amount of cementation. The more iron oxide cement the darker and the stronger the sandstone. A geologist collected data to study the relationship between porosity and sandstone strength. Based on those data, the least squares regression line is $\hat{y} = 20560 - 1344.4x$, where $x$ is the percent of porosity and $y$ is unconfined compressive sandstone strength measured in psi (pounds per square inch). Which of the following best describes the meaning of the slope of the least squares regression line?

   A) For each increase of 1 psi in strength, the estimated porosity is expected to decrease by 1344.4%.
   B) For each increase of 1% in porosity, the estimated strength is expected to increase by 20560 psi.
   C) For each increase of 1% in porosity, the estimated strength is expected to increase by 1344.4 psi.
   D) For each increase of 1% porosity, the estimated strength is expected to decrease by 1344.4 psi.
   E) For each increase of 1% in porosity, the estimated strength is expected to increase by 19,215.6 psi.

8. Which of these is a main difference between experiments and observational studies?

   A) There is a response variable in an experiment, but not in an observational study.
   B) There is at least one explanatory variable in an experiment, but not in an observational study.
   C) An experiment requires blocking, while an observational study does not.
   D) An experiment can be used to establish a causal relationship, but an observational study cannot.
   E) Observational studies require larger samples than experiments.

9. An analysis of price of crude oil ($/barrel) and gasoline prices at the pump ($/gallon) from 1976 to 2004 found a correlation coefficient of 0.829.

   Which of the following is a true statement?

   A) Because the correlation coefficient is high, there is a linear relationship between crude oil prices and prices at the pump.
   B) Because the correlation coefficient is only moderately high, the relationship between crude oil prices and prices at the pump is probably not linear.
   C) For every one dollar increase in crude oil price per barrel, the gasoline price at the pump is expected to increase by $0.829 per gallon.
   D) 68.7% of the price of a gallon of gasoline can be explained by crude oil prices.
   E) None of the statements A–D is true.

10. Researchers measured gestation time (in days) and brain weight (in grams) for a random sample of 23 unborn infants. They fit two possible regression models to their data.

    Here are the equation and residuals plot for Model I:
    $\widehat{Brain\ Weight} = -78.46 + 0.9176\ Gestation$

    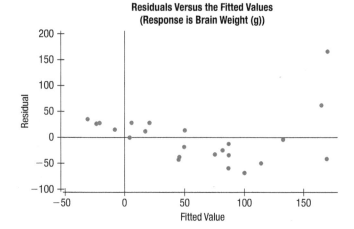

    Here are the equation and residuals plot for Model II:
    $\log(\widehat{Brain\ Weight}) = 0.3457 + 0.007371\ Gestation$

    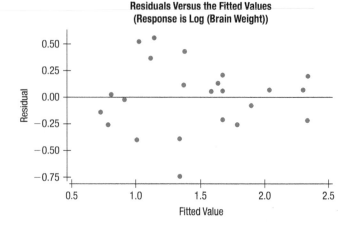

Which of the following conclusions is correct?

A) Model I is appropriate, because the relationship between *Gestation* time and *Brain Weight* is linear except for a couple of points.
B) Model I is appropriate, because the relationship between *Gestation* time and *Brain Weight* appears to be stronger.
C) Model II is appropriate, because the residuals are smaller.
D) Model II is appropriate, because the relationship between *Gestation* time and *Brain Weight* appears to be linear.
E) Model II is appropriate, because the relationship between *Gestation* time and the logarithm of *Brain Weight* appears to be linear.

11. An advertising agency is testing the effectiveness of a new commercial. They are trying out the commercial in three different time lengths (15, 30, and 60 seconds) and two formats (animated vs. live action). Each test audience will rate the product after seeing the commercial shown in one way. Which description of this study is correct?

A) An experiment with 2 factors (length and format), one having 3 levels (15, 30, 60) and the other having 2 levels (animated and live) for 6 treatments.
B) An experiment with 2 factors (length and format) at 5 levels (15, 30, 60, animated, and live) for 5 treatments.
C) An experiment with 1 factor (length) blocked by format with 3 levels (15, 30, 60).
D) A sample survey with two questions and one response.
E) A sample stratified by length and format.

12. Can you tell how old a lion is by looking at its nose? A professor at the University of Wisconsin–Madison conducted a study of data taken from 32 lions and observed the relationship between age (in years) and proportion of blackness in the lion's nose. The equation of the least squares regression line was

$$\hat{y} = 0.8790 + 10.6471x$$

where $\hat{y}$ is the predicted age of the lion, measured in years, and $x$ is the proportion of the lion's nose that is black. A lion whose nose was 11% black was known to be 1.9 years old. What is the residual for the age of this lion?

A) −0.15 years    B) 0.15 years
C) 0.88 years     D) 2.05 years
E) 10.65 years

(*Source*: http://www.stat.wisc.edu/~st571-1/15-regression-4.pdf)

13. A *New York Times* article dated February 19, 2013 reported on a Japanese study involving 35 people with lower back pain. They participated in an aquatic exercise program that included activities such as walking in a pool and (naturally) swimming. Six months after the program began, nearly all of the participants had improved. But the researchers observed that the ones who participated at least twice a week had improved significantly more than those who only went once a week.

Which of the following statements is true?

A) This experiment proves that swimming causes a reduction in lower back pain.
B) This study is a poorly designed experiment because there is no control group and no randomization.
C) Conclusions based on this observational study are suspect because participation in the aquatic program may be confounded with other lifestyle behaviors that may cause the improvement in lower back health.
D) Conclusions based on this observational study are suspect because a sample size of 35 is too small.
E) This study is not an observational study because the aquatic program is a treatment.

14. A high school statistics teacher ran an experiment with his classes in which the students tied rubber bands to a weight. They recorded the number of rubber bands used and the distance, in centimeters that the weight fell when dropped before rebounding. When they created a scatterplot the relationship appeared to be linear, and the correlation was $r = 0.996$. After discussing these results, the teacher instructed the students to convert the distances to inches. What effect will this have on the correlation between the two variables?

A) Because 1 cm = 0.393701 inches, the correlation will become .392126.
B) Because 1 cm = 0.393701 inches, the correlation will become 2.5298.
C) Because only the length measurements have changed, the correlation will decrease substantially.
D) Because changing from centimeters to inches does not affect the value of the correlation, the correlation will remain 0.996.
E) Because inches is a much more common measurement for distance in the United States, the relationship between the data will be stronger and thus the correlation will increase.

15. A local news program decides to conduct a poll to see how their viewers feel about their new programming format. At the end of each program during one week they ask viewers to call in and express their opinions. The station gives one number to dial if you like the new format, and another number to dial if you do not like it. Which of the following is a correct characterization of this sampling approach?

A) This sampling approach will be biased because people might dial the number incorrectly, and therefore not everyone will be correctly represented.
B) This sampling approach will be biased because the people who call in will be those who have stronger feelings about the new format, and therefore the sample will not be representative of the population.
C) This sampling approach will be unbiased because people can call in with either opinion; therefore the sample will be representative of the population.
D) This sampling approach will be unbiased because the station doesn't know who will call in, making it a random sample.
E) It is impossible to tell whether this sampling approach will be biased or not because the station cannot predict who will call in.

16. A city in the midwestern United States is considering a plan to add roundabout intersections in some high traffic areas to reduce the number of accidents. One city council member reached out to his constituents using a phone survey that contained the following question: "Many people object to the city's plan to reduce the unnecessary accidents by turning several intersections into those strange European style roundabouts. Do you also object to it?" He called the homes in his district during his lunch break at his regular job, which is between 12:00 and 12:30 p.m. Which of the following is a likely source of bias in his survey?

A) Calling between 12:00 and 12:30 limits his sample to people who will be at home during those hours, which leaves out everyone who regularly works during that time of day. This might influence the estimate of the proportion of people who object to the plan.

B) The phrase "plan to reduce unnecessary accidents" might influence people to voice support for the plan, making the estimate of the proportion of constituents who object to the plan too low.

C) The phrases "Many people object" and "Do you also object?" may influence people to say they object, inflating the estimate of the proportion of constituents who object.

D) The phrase "strange European style roundabouts" may influence people to say they object, inflating the estimate of the proportion of constituents who object.

E) All of these are possible sources of bias for this survey.

17. A forest ranger researching ecosystem recovery following forest fires collected data on a large growth of young pine trees. The heights of the trees had a mean of 3.2 feet and standard deviation 0.6 feet. A botany journal has asked that the data be expressed in inches. When the data are rescaled, what will be the new mean and standard deviation?

A) 3.2 and 0.6  B) 3.2 and 7.2
C) 38.4 and 0.6  D) 38.4 and 7.2
E) 38.4 and 86.4

18. Statistics teachers often debate the best order in which to teach topics. One group of teachers likes to teach design of studies first. Another group likes to begin with data analysis. To see which order is more effective in preparing students for the AP Exam, an experiment was proposed. A large group of teachers, each of whom teaches two sections of Statistics, volunteered to be a part of the experiment. Each teacher will randomly assign one of their classes to begin with design, and the other to begin with data analysis. Which is the correct description of this design?

A) The experimental units are the classrooms, the blocks are the teachers, and the response variable is the difference in average AP Exam score for each teacher's classes.

B) The experimental units are the teachers, there are no blocks, and the response variable is the average AP Exam score for each teacher.

C) The experimental units are the individual students, the blocks are the classrooms, and the response variable is each individual student's AP Exam score.

D) The experimental units are the individual students, the blocks are the teachers, and the response variable is the average AP Exam score for each classroom.

E) The experimental units are the orders of topics, the blocks are the teachers, and the response variable is the average AP Exam score for all students who used each order of topics.

19. In one study, researchers at McGill University recruited 127 people with high cholesterol and split them into two groups. One took a probiotic supplement twice a day for nine weeks, while the second group took a placebo. The probiotic group saw their total cholesterol drop by 9% and their LDL, or "bad cholesterol," fall almost 12%. In a different study, conducted in Britain, 80 volunteers all were administered probiotics for six weeks, followed by a placebo. The British study did not find a significant difference in cholesterol levels of the study participants when they were taking probiotics versus the placebo.

Which of the following is true?

A) The McGill study is a controlled randomized experiment, while the British study is an observational study.
B) The McGill study is a completely randomized experiment, while the British study is a matched pairs experiment.
C) The results from the McGill study are more valid because it includes blinding while the British study does not.
D) We have reason to believe that the design of one of these experiments was flawed because the results are contradictory.
E) Conclusions from the British study cannot be trusted because there is no control group.

20. The segmented bar graphs below depict data from the NAAL (National Assessment of Adult Literacy) conducted in 2003.

(*Source:* Kutner, M., et al., Literacy in Everyday Life: Results From the 2003 National Assessment of Adult Literacy (NCES 2007–480).U.S. Department of Education. Washington, DC)

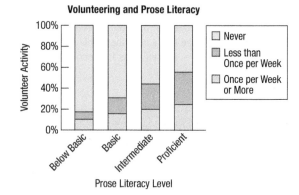

Does there appear to be a relationship between volunteerism and literacy level?

A) Yes, because all three bars have the same number of segments.
B) Yes, because all three bars have the same height.
C) Yes, because the corresponding segments of the three bars have different heights.
D) No, because the corresponding segments of the three bars have different heights.
E) No, because the sums of the 3 proportions in each bar are identical.

## FREE RESPONSE

1. In 2016, the Chicago Cubs had the best record in baseball with 103 wins and 58 losses. The Minnesota Twins had the worst record with 59 wins and 103 losses. To win, a team needs to score runs. Here are the summary statistics for the runs scored for hitters on each team (excluding pitchers and players with fewer than 50 at bats).

   | Team | n | Mean | Variance | StdDev | StdErr | Med | Range | Min | Max | Q1 | Q3 |
   |------|---|------|----------|--------|--------|-----|-------|-----|-----|-----|-----|
   | Cubs | 15 | 42.3 | 768.7 | 27.72 | 7.16 | 30 | 92 | 10 | 102 | 21 | 70 |
   | Twins | 16 | 24.6 | 144.1 | 12.01 | 3.00 | 20 | 43 | 10 | 53 | 16.5 | 30 |

   a) Construct parallel boxplots for the runs scored. (Note: The two best hitters for the Twins were Brian Dozier with 53 runs and Joe Mauer with 46 runs.)
   b) Compare the distributions of runs scored.
   c) Explain why the difference in means for these distributions is so much larger than the difference in the medians.

2. Over the years Olympic racers have been getting faster in most events, and the women's singles 500-meter kayak race is no exception. A scatterplot displaying the data for years since 1948 and time in seconds suggests that a linear model is appropriate. The equation of the least squares regression line is $\widehat{Time} = 144.627 - 0.776\ Years$, and $r^2 = 0.932$.

   a) Interpret the value of $r^2$ in this context.
   b) Compute and interpret the value of $r$ in context.
   c) The Olympics are held every 4 years. What change in the winning time does this model predict from one Olympics to the next?
   d) The residual for the winning time in 1980 was $-1.795$ seconds. Find this gold medal time.

3. A high school administration wants to collect data on the amount of time the students use computers. There are 1200 students in the school, and they have been assigned to 40 different homerooms, 10 homerooms per grade 9–12, with 30 students in each homeroom. The administration wants a sample of 120 students.

   a) Describe a method to select a simple random sample of students.
   b) Describe a method to select students using a stratified sample.
   c) Describe a method to select students using a cluster sample.
   d) Describe any advantages or disadvantages in using the stratified or the cluster sampling method here.

4. In its May 2019 issue, the business magazine *Fast Company* wrote on the use of text messages vs. e-mail to contact clients. While the article states that face-to-face meetings are generally the best way to meet with clients, 69% of respondents said they approve of being reached via a text message and just over half of them said that a text message was the easiest form of communication. We will assume that the survey was a random sample taken in the United States. (Source: https://www.fastcompany.com/90350516/how-to-decide-whether-toemail-text-call-or-talk-in-person)

   a) If you were in a sales position working with clients in the United States, would you be comfortable concluding that 69% is a good estimate of the percentage of your potential clients who approve of receiving a text message? Explain.
   b) If you were in a sales position working with clients internationally, would this change your answer to part (a)? Explain.
   c) It seems reasonable to suspect that age may be associated with a person's willingness to receive text messages. How might the sampling technique be improved by taking this association into account? Explain.
   d) Because of the topic in question, suppose that it was decided to distribute this survey via a text message. Describe how this sampling method would create a bias and how it would affect the estimate of the parameter.

# 13

# From Randomness to Probability

Early humans saw a world filled with random events. To help them make sense of the chaos around them, they sought out seers, consulted oracles, and read tea leaves. As science developed, we learned to recognize some events as predictable. We can now forecast the change of seasons, tell precisely when eclipses will occur, and even make a reasonably good guess at how warm it will be tomorrow. But many other events are still essentially random to us. Will the stock market go up or down today? When will the next car pass this corner?

But we have also learned to understand randomness. The surprising fact is that in the long run, many truly random phenomena settle down in a way that's consistent and predictable. It's this property of random phenomena that makes the next steps we're about to take in Statistics possible. The previous three chapters showed that randomness plays a critical role in gathering data. That fact alone makes it important to understand how random events behave. From here on, randomness will be fundamental to how we think about data.

## Random Phenomena

Every day you drive through the intersection at College and Main. Even though it may seem that the light is never green when you get there, you know this can't really be true. In fact, if you try really hard, you can recall just sailing through the green light once in a while.

What's random here? The light itself is governed by a timer. Its pattern isn't haphazard. In fact, the light may even be red at precisely the same times each day. It's the time you arrive at the light that is *random*. It isn't that your driving is erratic. Even if you try to leave your house at exactly the same time every day, whether the light is red or green as you reach the intersection is a random phenomenon.[1] For us, a **random phenomenon** is a situation in which we know what outcomes can possibly occur, but we don't know which particular outcome will happen. Even though the color of the light is random[2] as you approach it, some *fraction* of the time, the light will be green. How can you figure out what that fraction is?

You might record what happens at the intersection each day and graph the *accumulated percentage* of green lights like this:

**Figure 13.1**
The accumulated percentage of times the light is green settles down as you see more outcomes.

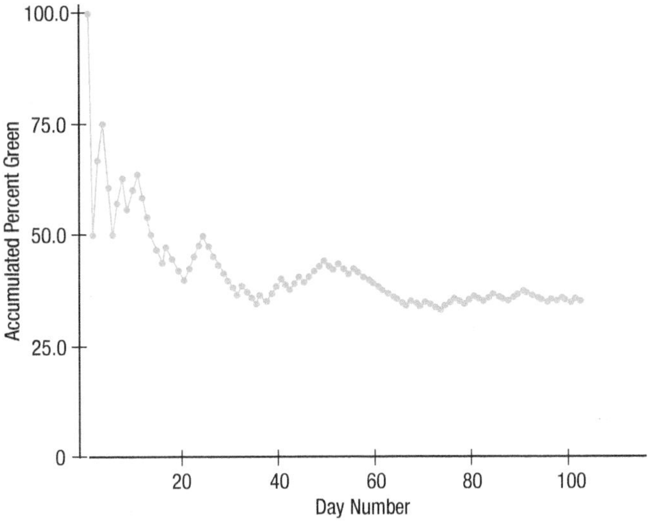

| Day | Light | % Green |
|-----|-------|---------|
| 1 | Green | 100 |
| 2 | Red | 50 |
| 3 | Green | 66.7 |
| 4 | Green | 75 |
| 5 | Red | 60 |
| 6 | Red | 50 |
| ⋮ | ⋮ | ⋮ |

## Trials, Outcomes, and Events

An occurrence of a phenomenon is called a trial. Each trial has an outcome. Outcomes combine to make events.

## The Sample Space

For a random phenomenon, the sample space, **S**, is the set of all possible outcomes of each trial.

The first day you recorded the light, it was green. Then on the next five days, it was red, then green again, then green, red, and red. When you plot the percentage of green lights against days, the graph starts at 100% (because the first time, the light was green, so 1 out of 1, or 100%). Then the next day it was red, so the accumulated percentage drops to 50% (1 out of 2). The third day it was green again (2 out of 3, or 67% green), then green (3 out of 4, or 75%), then red twice in a row (3 out of 5, for 60% green, and then 3 out of 6, for 50%), and so on. As you collect a new data value for each day, each new outcome becomes a smaller and smaller fraction of the accumulated experience, so, in the long run, the graph settles down. As it settles down, you can see that, in fact, the light is green about 35% of the time.

In general, each occasion upon which we observe a random phenomenon is called a **trial**. At each trial, we note the value of the random phenomenon, and call that the trial's **outcome**. (If this language reminds you of Chapter 10, that's *not* unintentional.)

For the traffic light, there are really three possible outcomes: red, yellow, or green. Often we're more interested in a combination of outcomes rather than in the individual ones. When you see the light turn yellow, what do *you* do? If you race through the intersection, then you treat the yellow more like a green light. If you step on the brakes, you treat it more like a red light. Either way, you might want to group the yellow with one or the other. When we combine outcomes like that, the resulting combination is an **event**.[3] We call the collection of *all possible outcomes* the **sample space**. We'll denote the sample space **S**. (Some books are even fancier and use the Greek letter $\Omega$.) For the traffic light, S = {red, green, yellow}. If you flip a coin once, the sample space is very simple:

---

[1] If you somehow managed to leave your house at *precisely* the same time every day and there was *no* variation in the time it took you to get to the light, then there wouldn't be any randomness, but that's not very realistic.

[2] Even though the randomness here comes from the uncertainty in our arrival time, we can think of the light itself as showing a color at random.

[3] Each individual outcome is also an event.

S = { H, T }. If you flip two coins, it's more complicated because now there are four outcomes, so S = { HH, HT, TH, TT }. If *ABC News* takes a sample of 1023 randomly chosen U.S. adults for a poll, the sample space is incomprehensibly enormous because it would list every combination of 1023 adults you could take from the approximately 250 million adults in the United States.

## The Law of Large Numbers

What's the *probability* of a green light at College and Main? Based on the graph, it looks like the relative frequency of green lights settles down to about 35%, so saying that the probability is about 0.35 seems like a reasonable answer. But do random phenomena always behave well enough for this to make sense? Might the relative frequency of an event bounce back and forth between two values forever, never settling on just one number?

Fortunately, a principle called the **Law of Large Numbers** (LLN) gives us the guarantee we need. The LLN says that as we repeat a random process over and over, the proportion of times that an event occurs does settle down to one number.[4] We call this number the **probability** of the event. But the Law of Large Numbers requires two key assumptions. First, the random phenomenon we're studying must not change—the outcomes must have the same chances for each trial. And, the events must be **independent**.[5] Informally, independence means that the outcome of one trial doesn't influence the outcomes of the others. (We'll see a formal definition of independent events in the next chapter.) The LLN says that as the number of independent trials increases, the long-run *relative frequency* of repeated events gets closer and closer to a single value. We call that the *probability* of the event.[6] If the relative frequency of green lights at that intersection settles down to 35% in the long run, we say that the probability of encountering a green light is 0.35, and we write $P(\text{green}) = 0.35$.

> **Probability**
> For any event A,
> $P(A) = \dfrac{\text{\# of times A occurs}}{\text{total \# of trials}}$
> in the long run.

> "For even the most stupid of men . . . is convinced that the more observations have been made, the less danger there is of wandering from one's goal."
> —Jacob Bernoulli, 1713, discoverer of the LLN

Although he could have said it much more gently, Bernoulli is trying to tell us how intuitive the LLN actually is, even though it wasn't formally proved until the 18th century. Most of us would guess that the law is true from our everyday experiences.

## The Nonexistent "Law of Averages"

Even though the LLN seems natural, it is often misunderstood because the idea of the *long run* is hard to grasp. Many people believe, for example, that an outcome of a random event that hasn't occurred in many trials is "due" to occur. Many gamblers bet on numbers that haven't been seen for a while, mistakenly believing that they're likely to come up sooner. A common term for this is the "Law of Averages." After all, we know that in the long run, the relative frequency will settle down to the probability of that outcome, so now we have some "catching up" to do, right?

**APPLET**
Explore what the Law of Large Numbers really says about probability . . .

Wrong. The Law of Large Numbers says nothing about short-run behavior. Relative frequencies even out *only in the long run*. And, according to the LLN, the long run is *really* long (*infinitely* long, in fact).

---

[4] Or as Marty Byrde in Season 3 of *Ozark* said, "I can't tell you which customer is going to cheat us, but thanks to the Law of Large Numbers, I can tell you precisely how many out of the next million will."

[5] There are stronger forms of the Law that don't require independence, but for our purposes, this form is general enough.

[6] There are rebel statisticians (including one of the authors of this book) who argue that the definition of probability should be broadened to include events that aren't random or repeatable, but merely unknown, such as a particular painting being a forgery, or a particular stone with ambiguous marks being a human artifact. This text doesn't adopt that broader definition; doing so requires an extremely different approach to data analysis.

The so-called Law of Averages doesn't exist at all. But you'll hear people talk about it as if it does. Is a good hitter in baseball who has struck out the last six times *due* for a hit his next time up? If you've been doing particularly well in weekly quizzes in Statistics class, are you *due* for a bad grade? No. This isn't the way random phenomena work. There is no Law of Averages for short runs.

> Slump? I ain't in no slump. I just ain't hittin'.
> —Yogi Berra

### There Is No Law of Averages
Don't let yourself think that there's a Law of Averages that promises short-term compensation for recent deviations from expected behavior. A belief in such a "Law" can lead to money lost in gambling and to poor business decisions.

### FOR EXAMPLE
#### Coins and the Supposed Law of Averages

You've just flipped a fair coin and seen six heads in a row.

QUESTION: Does the coin "owe" you some tails? Suppose you spend that coin and your friend gets it in change. When she starts flipping the coin, should she expect a run of tails?

ANSWER: Of course not. Each flip is a new event. The coin can't "remember" what it did in the past, so it can't "owe" any particular outcomes in the future.

Just to see how this works in practice, the authors ran a simulation of 100,000 flips of a fair coin. We collected 100,000 random numbers, letting the numbers 0 to 4 represent heads and the numbers 5 to 9 represent tails. In our 100,000 "flips," there were 2981 streaks of at least 5 heads. The "Law of Averages" suggests that the next flip after a run of 5 heads should be tails more often to even things out. Actually, the next flip was heads more often than tails: 1550 times to 1431 times. That's 51.9% heads. You can perform a similar simulation easily on a computer. Try it!

Of course, sometimes an apparent drift from what we expect means that the probabilities are, in fact, *not* what we thought. If you get 10 heads in a row, maybe the coin has heads on both sides!

**TI-nspire**

**The Law of Large Numbers.** Watch the relative frequency of a random event approach the true probability *in the long run*.

The lesson of the LLN is that sequences of random events don't compensate in the *short* run and don't need to do so to get back to the right long-run probability. If the probability of an outcome doesn't change and the events are independent, the probability of any outcome in another trial is *always* what it was, no matter what has happened in other trials.

> ### BEAT THE CASINO
> Keno is a simple casino game in which numbers from 1 to 80 are chosen. The numbers, as in most lottery games, are supposed to be equally likely. Payoffs are made depending on how many of those numbers you match on your card. A group of graduate students from a Statistics department decided to take a field trip to Reno. They (*very* discreetly) wrote down the outcomes of the games for a couple of days, then drove back to test whether the numbers were, in fact, equally likely. It turned out that some numbers were *more likely* to come up than others. Rather than bet on the "Law of Averages" and put their money on the numbers that were "due," the students put their faith in the LLN—and all their (and their friends') money on the numbers that had come up more frequently before. After they pocketed more than $50,000, they were escorted off the premises and invited never to show their faces in that casino again.

> ### JUST CHECKING
>
> 1. One common proposal for beating the lottery is to note which numbers have come up lately, eliminate those from consideration, and bet on numbers that have not come up for a long time. Proponents of this method argue that in the long run, every number should be selected equally often, so those that haven't come up are due. Explain why this is faulty reasoning.

## Modeling Probability

Probability was first studied extensively by a group of French mathematicians who were interested in games of chance.[7] Rather than *experiment* with the games (and risk losing their money), they developed mathematical models. When the probability comes from a mathematical model and not from observation, it is called **theoretical probability**. To make things simple (as we usually do when we build models), they started by looking at games in which the different outcomes were equally likely. Fortunately, many games of chance are like that. Any of 52 cards is equally likely to be the next one dealt from a well-shuffled deck. Each face of a die is equally likely to land up (or at least it *should be*).

It's easy to find probabilities for events that are made up of several *equally likely* outcomes. We just count all the outcomes that the event contains. The probability of the event is the number of outcomes in the event divided by the total number of possible outcomes. We can write

$$P(A) = \frac{\#\text{ outcomes in A}}{\#\text{ of possible outcomes}}.$$

For example, the probability of drawing a face card (jack, queen, or king) from a deck is

$$P(\text{face card}) = \frac{\#\text{ face cards}}{\#\text{ cards}} = \frac{12}{52} = \frac{3}{13}.$$

> **NOTATION ALERT**
> We often use capital letters—and usually from the beginning of the alphabet—to denote events. We *always* use $P$ to denote probability. So,
>
> $P(A) = 0.35$
>
> means "the probability of the event A is 0.35."
> When being formal, use decimals (or fractions) for the probability values, but sometimes, especially when talking more informally, it's easier to use percentages.

> ### HOW HARD CAN COUNTING BE?
> Finding the probability of any event when the outcomes are equally likely is straightforward, but not necessarily easy. It gets hard when the number of outcomes in the event (and in the sample space) gets big. Think about flipping two coins. The sample space is S = {HH, HT, TH, TT} and each outcome is equally likely. So, what's the probability of getting exactly one head and one tail? Let's call that event **A**. Well, there are two outcomes in the event **A** = {HT, TH} out of the 4 possible equally likely ones in S, so $P(\mathbf{A}) = \frac{2}{4}$, or $\frac{1}{2}$.
> OK, now flip 100 coins. What's the probability of exactly 67 heads? Well, first, how many outcomes are in the sample space? S = {HHHHHHHHHHH ... H, HH ... T, ...} Um .... a lot. In fact, there are 1,267,650,600,228,229,401,496,703,205,376 different outcomes possible when flipping 100 coins. And that's just the denominator of the probability! But don't worry: later in this book we'll see shortcuts for determining probabilities like this one.

Don't get trapped into thinking that random events are always equally likely. The chance of winning a lottery—especially lotteries with very large payoffs—is small. Regardless, people continue to buy tickets. In an attempt to understand why, an interviewer asked

---
[7] OK, gambling.

someone who had just purchased a lottery ticket, "What do you think your chances are of winning the lottery?" The reply was, "Oh, about 50–50." The shocked interviewer asked, "How do you get that?" to which the response was, "Well, the way I figure it, either I win or I don't!"

The moral of this story is that events are *not* always equally likely.

## Subjective Probability

On May 14, 2013, the *New York Times* published a letter from the actress Angelina Jolie, who explained her difficult decision to have a preventive double mastectomy—breast removal and reconstructive surgery in advance of possible breast cancer. Breast cancer was in her family, and she knew she had a faulty gene that is associated with a higher incidence of breast cancer. Jolie explained, "My doctors estimated that I had an 87 percent risk of breast cancer . . . although the risk is different in the case of each woman." She later wrote that following the surgery, "My chances of developing breast cancer have dropped from 87 percent to under 5 percent. I can tell my children that they don't need to fear they will lose me to breast cancer."

What is meant by an "87 percent risk of breast cancer" in her case? The Law of Large Numbers says that when a random trial is repeated many times under similar conditions, then the relative frequency of an event should "settle down" to one number, and that number is called the probability of the event. But the life of a particular individual—Angelina Jolie or anyone else—is not repeatable. How can "87% chance" be associated with a single individual?

In fact, probabilities *can't* be associated with individuals—at least not the kind of probabilities that we're using in this text. Angelina Jolie's doctors were using a different kind of probability called **subjective probability**. The number 87% isn't a fraction of anything. The doctors were using a scale of 0 to 100 to express their personal opinion (surely informed by much experience, data, and mathematical models) that she would eventually develop breast cancer.

Subjective probability might sound somewhat sloppy, but the doctors were not giving Jolie worthless information. Subjective probability was an excellent way for them to communicate expertise to someone without their medical knowledge. You yourself have surely used subjective probability many times without thinking about it. ("I'm 90% sure I'll be able to make it to dinner on time, but I can't promise!") But although subjective probabilities are neither meaningless nor worthless, most statisticians[8] agree that for data analysis it's better to stick to *objective* probabilities, as we're doing in this text, because they are more trusted and there is a lot of historical precedent for using them.

## The First Three Rules for Working with Probability

1. Make a picture.
2. Make a picture.
3. Make a picture.

We're dealing with probabilities now, not data, but the three rules don't change. One common kind of picture to make is called a Venn diagram. We'll use Venn diagrams throughout the rest of this chapter. Even experienced statisticians make Venn diagrams to help them think about probabilities of compound and overlapping events. You should, too.

---

[8]Not all, though. If you're reading the footnotes (and you seem to be) then you might recall the "rebel statisticians" of footnote 6 in this chapter. If you're feeling rebellious yourself, you can research "Bayesian statistics," a branch whose basis is subjective probability.

# Formal Probability

For some people, the phrase "50/50" means something vague like "I don't know" or "whatever." But when we discuss probabilities of outcomes, it takes on the precise meaning of *equally likely*. Speaking vaguely about probabilities will get us into trouble, so whenever we talk about probabilities, we'll need to be precise. And to do that, we'll need to develop some formal rules about how probability works.

**Rule 1.** If the probability is 0, the event *never* occurs, and likewise if it has probability 1, it *always* occurs. Even if you think an event is very unlikely, its probability can't be negative, and even if you're sure it will happen, its probability can't be greater than 1. (Think about relative frequencies.) So we require that

**A probability is a number between 0 and 1.**

**For any event A, $0 \leq P(A) \leq 1$.**

> ### Surprising Probabilities
> We've been careful to discuss probabilities only for situations in which the outcomes were finite, or at least *countably* infinite. (That means you can list the outcomes, although the list might be infinitely long.). But if the outcomes can take on *any* numerical value at all (we say they are *continuous*), things can get surprising. For example, what is the probability that a randomly selected child will be *exactly* 3 feet tall? Well, if we mean 3.00000 . . . feet, the answer is zero. No randomly selected child—even one whose height would be recorded as 3 feet—will be *exactly* 3 feet tall (to an infinite number of decimal places). But, if you've grown taller than 3 feet, there must have been a time in your life when you actually *were* exactly 3 feet tall, even if only for a second. So this is an outcome with probability 0 that not only has happened—it has happened to *you*.

**Rule 2.** If a random phenomenon has only one possible outcome, it's not very interesting (or very random). So we need to distribute the probabilities among all the outcomes a trial can have. How can we do that so that it makes sense? For example, consider what you're doing as you read this book. The possible outcomes might be

**A:** You read to the end of this chapter before stopping.

**B:** You finish this section but stop reading before the end of the chapter.

**C:** You bail out before the end of this section.

When we assign probabilities to these outcomes, the first thing to be sure of is that we distribute all of the available probability. *Something* always occurs, so the probability of the entire sample space is 1.

Making this more formal gives the **Probability Assignment Rule**.

**The set of all possible outcomes of a trial**

**must have probability 1.**

$$P(S) = 1$$

**Rule 3.** Suppose the probability that you get to class on time is 0.8. What's the probability that you don't get to class on time? Yes, it's 0.2. The set of outcomes that are *not* in the event **A** is called the **complement** of **A**, and is denoted $A^C$. This leads to the **Complement Rule**:

**The probability that an event does not occur**

**is 1 minus the probability that it does.**

$$P(A^C) = 1 - P(A)$$

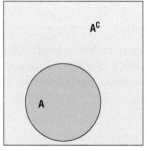

The set **A** and its complement $A^C$. Together, they make up the entire sample space **S**.

### FOR EXAMPLE
#### Applying the Complement Rule

**RECAP:** We opened the chapter by looking at the traffic light at the corner of College and Main, observing that when we arrive at that intersection, the light is green about 35% of the time.

**QUESTION:** If $P(\text{green}) = 0.35$, what's the probability the light isn't green when you get to College and Main?

**ANSWER:** "Not green" is the complement of "green,"
so $P(\text{not green}) = 1 - P(\text{green})$
$= 1 - 0.35 = 0.65$

There's a 65% chance I won't have a green light.

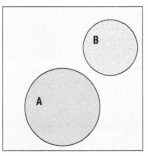

Two disjoint sets, **A** and **B**.

**Rule 4.** Suppose the probability that (**A**) a randomly selected student is a sophomore is 0.20, and the probability that (**B**) they are a junior is 0.30. What is the probability that the student is *either* a sophomore *or* a junior, written $P(\mathbf{A} \cup \mathbf{B})$? If you guessed 0.50, you've deduced the Addition Rule, which says that you can add the probabilities of events that are disjoint. To see whether two events are disjoint, we think about whether both can occur at the same time. **Disjoint (or mutually exclusive)** events have no outcomes in common. The **Addition Rule** states,

**For two disjoint events A and B, the probability that one *or* the other occurs is the sum of the probabilities of the two events.**

$P(\mathbf{A} \cup \mathbf{B}) = P(\mathbf{A}) + P(\mathbf{B})$, **provided that A and B are disjoint**

### FOR EXAMPLE
#### Applying the Addition Rule

**RECAP:** When you get to the light at College and Main, it's either red, green, or yellow. We know that $P(\text{green}) = 0.35$.

**QUESTION:** Suppose we find out that $P(\text{yellow})$ is about 0.04. What's the probability the light is red?

**ANSWER:** To find the probability that the light is green or yellow, I can use the Addition Rule because these are disjoint events: The light can't be both green and yellow at the same time.

$P(\text{green} \cup \text{yellow}) = 0.35 + 0.04 = 0.39$

Red is the only remaining alternative, and the probabilities must add up to 1, so

$P(\text{red}) = P(\text{not}(\text{green} \cup \text{yellow}))$
$= 1 - P(\text{green} \cup \text{yellow})$
$= 1 - 0.39 = 0.61$

The Addition Rule can be extended to any number of disjoint events, and that's helpful for checking probability assignments. Because individual sample space outcomes are always disjoint, we have an easy way to check whether the probabilities we've assigned to the possible outcomes are legitimate. The Probability Assignment Rule tells us that to be a **legitimate assignment of probabilities**, the sum of the probabilities of all possible outcomes must be exactly 1. No more, no less. For example, if we were told that the

> Baseball is 90% mental. The other half is physical.
> —Yogi Berra

probabilities of selecting at random a first-year, sophomore, junior, or senior from all the undergraduates at a college were 0.25, 0.23, 0.22, and 0.20, respectively, we would know that something was wrong. These "probabilities" sum to only 0.90, so this is not a legitimate probability assignment. Either a value is wrong, or we just missed some possible outcomes, like "pre-frosh" or "postgraduate" categories that soak up the remaining 0.10. Similarly, a claim that the probabilities were 0.26, 0.27, 0.29, and 0.30 would be wrong because these "probabilities" sum to more than 1.

But be careful: The Addition Rule doesn't work for events that aren't disjoint. If the probability of owning a cat is 0.50 and the probability of owning a dog is 0.90, the probability of owning either a cat or a dog may be pretty high, but it is *not* 1.40! Why can't you add probabilities like this? Because these events are not disjoint. You *can* own both a cat and a dog. In the next chapter, we'll see how to add probabilities for events like these, but we'll need another rule.

**NOTATION ALERT**
We write $P(A \text{ or } B)$ as $P(A \cup B)$. The symbol $\cup$ means "union," representing the outcomes in event **A** *or* event **B** (or both). The symbol $\cap$ means "intersection," representing outcomes that are in both event **A** *and* event **B**. We write $P(A \text{ and } B)$ as $P(A \cap B)$.

**Rule 5.** Suppose your job requires you to fly from Atlanta to Chicago every Monday morning. The airline's website reports that this flight is on time 85% of the time. What's the chance that it will be on time two weeks in a row? That's the same as asking for the probability that your flight is on time this week *and* it's on time again next week. For independent events, the answer is very simple. Remember that independence means that the outcome of one event doesn't affect the probability of the other. What happens with your flight this week doesn't influence whether it will be on time next week, so it's reasonable to assume that those events are independent. The **Multiplication Rule** says that for independent events, to find the probability that both events occur, we just multiply the probabilities together. Formally,

**For two independent events A and B, the probability that both A and B occur is the product of the probabilities of the two events.**

$$P(A \cap B) = P(A) \times P(B), \text{provided that}$$
**A and B are independent**.

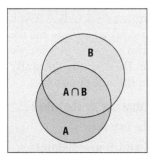

Two sets **A** and **B** that are not disjoint. The event $(A \cap B)$ is their intersection.

This rule can be extended to more than two independent events. What's the chance of your flight being on time for a month—four Mondays in a row? We can multiply the probabilities of it happening each week:

$$0.85 \times 0.85 \times 0.85 \times 0.85 = 0.522$$

or just over 50–50. Of course, to calculate this probability, we have used the assumption that the four events are independent.

Many Statistics methods require an **Independence Assumption**, but *assuming* independence doesn't make it true. Always *Think* about whether that assumption is reasonable before using the Multiplication Rule.

## Checking for Independence

Statistical methods often require you to assume that certain events are independent. Sometimes the reasonableness of such an assumption cannot be validated with data. Then the extent to which you trust the conclusions depends on how reasonable you think the independence assumption is.

There are other times, though, when the reasonableness of the independence assumption can actually be checked using data. Let's look at just one example. When we computed the probability that your Monday morning commuter flight would be on time two or four times in a row, we assumed that arriving on time was independent for any two Mondays. Let's now suppose that over 20 consecutive Mondays, you record which flights are on time and which are late. Here are your data, in consecutive order, with "☺" standing for "on time" and "L" standing for "late":

☺☺☺☺L L L ☺☺☺☺☺☺☺☺☺☺☺☺☺.

First we should note that with 17 out of 20 flights being on time, your data are certainly consistent with the airline's claim that 85% of the flights are on time.[9] So there's nothing to worry about there. But what about the assumption we made that on-time arrivals are independent between any two Mondays? Is that a reasonable assumption that's consistent with the data?

Something jumps out at us about the three late arrivals—they're consecutive. That certainly could occur just by chance even if they're all independent, but an alternative explanation is that late arrivals are *not* independent—that a Monday flight is more likely to be late if the one immediately before or after it is late. Perhaps they all occur during a period of time when many people are traveling and the additional passenger traffic makes delays more likely. Or perhaps weather patterns make delays more likely at certain times of the year. Whatever. We don't have to know the explanation, we only have to know what *a lack of independence* might look like in our data—and it could easily look like what we do see: late arrivals clustering together.

But let's return to our first possible explanation for the clustering: sheer chance. If the late arrivals really are all independent, is it likely that all 3 late arrivals would have occurred consecutively just by chance? Perhaps we can estimate just *how* likely it is with . . . oh, I don't know . . . a *simulation*?

The authors did one trial of such a simulation with a deck of playing cards. We picked 17 black cards to stand for "on time" and 3 red cards to stand for "late." We shuffled the cards and laid them out in order to stand for the 20 consecutive flights. Here's what we saw, with "☺" standing for "black card" (which itself represents "on time") and with "L" standing for "red card" (which represents "late"):

☺ ☺ ☺ ☺ ☺ ☺ ☺ ☺ ☺ L ☺ ☺ L ☺ ☺ ☺ L ☺ ☺ ☺.

That trial did not result in the three late arrivals being consecutive. We shuffled the cards a couple more times and laid them out for two more trials. Here's what we saw:

☺ L ☺ L ☺ ☺ ☺ ☺ ☺ ☺ ☺ L ☺ ☺ ☺ ☺ ☺ ☺ ☺ ☺

☺ ☺ ☺ ☺ ☺ L ☺ ☺ ☺ ☺ ☺ ☺ ☺ ☺ L L ☺ ☺ ☺.

So far, no occurrences yet of three consecutive late arrivals. We did this 997 more times for a total of 1000 trials.[10] Among them there were only 16 trials in which the three late arrivals were consecutive. That means that if the late arrivals were all independent of one another, there's only about a 1.6% chance that we'd see all three late arrivals occur consecutively. Our data are somewhat inconsistent with the independence assumption. That doesn't *prove* that the late arrivals are not independent, but it should lead us to at least be skeptical about the assumption that they are.

With the independence assumption called into question, we can no longer answer the question of how likely it is that two or four consecutive Mondays will have on-time flights. We just don't have enough information.

## FOR EXAMPLE
### Applying the Multiplication Rule (and Others)

**RECAP:** We've determined that the probability that we encounter a green light at the corner of College and Main is 0.35, a yellow light 0.04, and a red light 0.61. Let's think about your morning commute in the week ahead.

**QUESTION:** What's the probability you find the light red both Monday and Tuesday?

---

[9] By the way, your data don't prove that the airline's claimed on-time arrival rate is correct. The data are simply *consistent* with that claim. Know the difference.

[10] Okay, yes, we confess that we used a computer for the others.

ANSWER: Because the color of the light I see on Monday doesn't influence the color I'll see on Tuesday, these are independent events; I can use the Multiplication Rule:

$$P(\text{red Monday} \cap \text{red Tuesday}) = P(\text{Red}) \times P(\text{red})$$
$$= (0.61)(0.61)$$
$$= 0.3721$$

There's about a 37% chance I'll hit red lights both Monday and Tuesday mornings.

QUESTION: What's the probability you don't encounter a red light until Wednesday?

ANSWER: For that to happen, I'd have to see green or yellow on Monday, green or yellow on Tuesday, and then red on Wednesday. I can simplify this by thinking of it as not red on Monday and Tuesday and then red on Wednesday.

$$P(\text{not red}) = 1 - P(\text{red}) = 1 - 0.61 = 0.39, \text{ so}$$
$$P(\text{not red Monday} \cap \text{not red Tuesday} \cap \text{red Wednesday})$$
$$= P(\text{not red}) \times P(\text{not red}) \times P(\text{red})$$
$$= (0.39)(0.39)(0.61)$$
$$= 0.092781$$

There's about a 9% chance that this week I'll see my first red light there on Wednesday morning.

QUESTION: What's the probability that you'll see a red light *at least once* during the week?

ANSWER: Seeing a red light at least once means that I see the light either 1, 2, 3, 4, or 5 times next week. It's easier to think about the complement: never seeing a red light. Seeing a red light at least once means that I didn't make it through the week with no red lights.

$$P(\text{seeing a red light at least once in 5 days})$$
$$= 1 - P(\text{no red lights for 5 days in a row})$$
$$= 1 - P(\text{not red} \cap \text{not red} \cap \text{not red} \cap \text{not red} \cap \text{not red})$$
$$= 1 - (0.39)(0.39)(0.39)(0.39)(0.39)$$
$$= 1 - 0.0090$$
$$= 0.991$$

There's over a 99% chance I'll see at least one red light sometime this week.

### At Least

Note that the phrase "at least" is often a tip-off to think about the complement. Something that happens *at least once* does happen. Happening at least once is the complement of not happening at all, and that's often easier to find.

### JUST CHECKING

2. Opinion polling organizations contact their respondents by telephone. Random telephone numbers are generated, and interviewers try to contact those households. In the 1990s this method could reach about 69% of U.S. households. According to the Pew Research Center for the People and the Press, by 2003 the contact rate had risen to 76%. We can reasonably assume each household's response to be independent of the others. What's the probability that . . .

   a) the interviewer successfully contacts the next household on the list?

   b) the interviewer successfully contacts both of the next two households on the list?

   c) the interviewer's first successful contact is the third household on the list?

   d) the interviewer makes at least one successful contact among the next five households on the list?

# STEP-BY-STEP EXAMPLE

## Probability

The five rules we've seen can be used in a number of different combinations to answer a surprising number of questions. Let's try one to see how we might go about it.

M&M's® Milk Chocolate candies now come in 7 colors, but they've changed over time. In 1995, Americans voted to change tan M&M's (which had replaced violet in 1949) to blue. In 2002, Mars™, the parent company of M&M's, used the Internet to solicit global opinion for a seventh color. To decide which color to add, Mars surveyed kids in nearly every country of the world and asked them to vote among purple, pink, and teal. The global winner was purple!

In the United States, 42% of those who voted said purple, 37% said teal, and only 19% said pink. But in Japan the percentages were 38% pink, 36% teal, and only 16% purple. Let's use Japan's percentages to ask some questions:

1. What's the probability that a Japanese M&M's survey respondent selected at random preferred either pink or teal?
2. If we pick two respondents at random, what's the probability that they both preferred purple?
3. If we pick three respondents at random, what's the probability that *at least one* preferred purple?

---

**THINK** The probability of an event is its long-term relative frequency. It can be determined in several ways: by looking at many replications of an event, by deducing it from equally likely events, or by using some other information. Here, we are told the relative frequencies of the three responses.

The M&M's website reports the proportions of Japanese votes by color. These give the probability of selecting a voter who preferred each of the colors:

$$P(\text{pink}) = 0.38$$
$$P(\text{teal}) = 0.36$$
$$P(\text{purple}) = 0.16$$

Make sure the probabilities are legitimate. Here, they're not. Either there was a mistake, or the other voters must have chosen a color other than the three given. A check of the reports from other countries shows a similar deficit, so probably we're seeing those who had no preference or who wrote in another color.

Each is between 0 and 1, but they don't all add up to 1. The remaining 10% of the voters must have not expressed a preference or may have written in another color. I'll put them together into "no preference" and add $P(\text{no preference}) = 0.10$.

Now, I have a legitimate assignment of probabilities.

---

**QUESTION 1:** What's the probability that a Japanese M&M's survey respondent selected at random preferred either pink or teal?

**THINK** **PLAN** Decide which rules to use and check the conditions they require.

The events "Pink" and "Teal" are individual outcomes (a respondent can't choose both colors), so they are disjoint. I can apply the Addition Rule.

**SHOW** **MECHANICS** Show your work.

$$P(\text{pink} \cup \text{teal}) = P(\text{pink}) + P(\text{teal})$$
$$= 0.38 + 0.36 = 0.74$$

**TELL** **CONCLUSION** Interpret your results in the proper context.

The probability that the respondent said pink or teal is 0.74.

---

**QUESTION 2:** If we pick two respondents at random, what's the probability that they both said purple?

**THINK** **PLAN** The word "both" suggests we want $P(A \text{ and } B)$, which calls for the Multiplication Rule. Think about the assumption.

✓ **Independence Assumption:** It's unlikely that the choice made by one random respondent affected the choice of the other, so the events seem to be independent. I can use the Multiplication Rule.

| | | |
|---|---|---|
| **SHOW** MECHANICS Show your work. For both respondents to pick purple, each one has to pick purple. | | $P(\text{both purple})$ $= P(\text{first purple} \cap \text{second purple})$ $= P(\text{first purple}) \times P(\text{second purple})$ $= 0.16 \times 0.16 = 0.0256$ |
| **TELL** CONCLUSION Interpret your results in the proper context. | | The probability that both respondents pick purple is 0.0256. |

QUESTION 3: If we pick three respondents at random, what's the probability that at least one preferred purple?

| | | |
|---|---|---|
| **THINK** PLAN The phrase "at least ..." often flags a question best answered by looking at the complement, and that's the best approach here. The complement of "At least one preferred purple" is "None of them preferred purple." Think about the assumption. | | $P(\text{at least one purple})$ $= P(\{\text{none purple}\}^c)$ $= 1 - P(\text{none purple})$ $= 1 - P(\text{not purple} \cap \text{not purple} \cap \text{not purple})$ ✓ **Independence Assumption**: These are independent events because they are choices by three random respondents. I can use the Multiplication Rule. |
| **SHOW** MECHANICS First we find $P(\text{not purple})$ with the Complement Rule. Next we calculate $P(\text{none picked purple})$ by using the Multiplication Rule. Then we can use the Complement Rule to get the probability we want. | | $P(\text{not purple}) = 1 - P(\text{purple})$ $= 1 - 0.16 = 0.84$ $P(\text{al least one picked purple})$ $= 1 - P(\text{none purple})$ $= 1 - P(\text{not purple} \cap \text{not purple} \cap \text{not purple})$ $= 1 - (0.84)(0.84)(0.84)$ $= 1 - 0.5927$ $= 0.4073$ |
| **TELL** CONCLUSION Interpret your results in the proper context. | | There's about a 40.7% chance that at least one of the respondents picked purple. |

## WHAT CAN GO WRONG?

- **Beware of probabilities that don't add up to 1.** To be a legitimate probability assignment, the sum of the probabilities for all possible outcomes must total 1. If the sum is less than 1, you may need to add another category ("other") and assign the remaining probability to that outcome. If the sum is more than 1, check that the outcomes are disjoint. If they're not, then you can't assign probabilities by just counting relative frequencies.

- **Don't add probabilities of events if they're not disjoint.** Events must be disjoint to use the Addition Rule. The probability of being under 80 *or* a female is not the probability of being under 80 *plus* the probability of being female. That sum may be more than 1.
- **Don't multiply probabilities of events if they're not independent.** The probability of selecting a student at random who is over 6'10" tall *and* on the basketball team is *not* the probability the student is over 6'10" tall *times* the probability he's on the basketball team. Knowing that the student is over 6'10" changes the probability of his being on the basketball team. You can't multiply these probabilities. The multiplication of probabilities of events that are not independent is one of the most common errors people make in dealing with probabilities.
- **Don't confuse disjoint and independent.** Disjoint events *can't* be independent. If **A** = {you get an A in this class} and **B** = {you get a B in this class}, **A** and **B** are disjoint. Are they independent? If you find out that **A** is true, does that change the probability of **B**? You bet it does! So they can't be independent. We'll return to this issue in the next chapter.

# WHAT HAVE WE LEARNED?

**We've learned that the probability of an event is its long-run frequency of occurrence.**

**We understand that the Law of Large Numbers speaks only of long-run (*very* long) behavior.** We've learned not to fall victim to the short-run false reasoning called the "Law of Averages."

**We've learned some basic probability rules, and how to apply them.**
- A probability is a number between 0 and 1.
- The sum of the probabilities for all outcomes must be 1.
- The **Complement Rule** says that $P(\text{not } \mathbf{A}) = P(\mathbf{A}^C) = 1 - P(\mathbf{A})$.
- The **Addition Rule** says that $P(\mathbf{A} \cup \mathbf{B}) = P(\mathbf{A}) + P(\mathbf{B})$, provided events **A** and **B** are disjoint.
- The **Multiplication Rule** says that $P(\mathbf{A} \cap \mathbf{B}) = P(\mathbf{A}) \times P(\mathbf{B})$, provided events **A** and **B** are independent.

## TERMS

**Random phenomenon**    A phenomenon is random if we know what outcomes could happen, but not which particular values will happen. (p. 355)

**Trial**    A single attempt or realization of a random phenomenon. (p. 355)

**Outcome**    The outcome of a trial is the value measured, observed, or reported for an individual instance of that trial. (p. 355)

**Event**    A collection of outcomes. Usually, we identify events so that we can attach probabilities to them. We denote events with bold capital letters such as **A**, **B**, or **C**. (p. 355)

**Sample space**    The collection of all possible outcome values. The sample space has a probability of 1. (p. 355)

**Law of Large Numbers**    The Law of Large Numbers states that the long-run *relative frequency* of repeated independent events gets closer and closer to the *true* relative frequency as the number of trials increases. (p. 356)

**Probability**    The probability of an event is a number between 0 and 1 that reports the long-run frequency of that event's occurrence. We write $P(\mathbf{A})$ for the probability of the event **A**. (p. 356)

| | |
|---|---|
| **Independence (informally)** | Two events are *independent* if learning that one event occurs does not change the probability that the other event occurs. (p. 356) |
| **Theoretical probability** | When a probability is based on a model (such as equally likely outcomes), it is called a theoretical probability. (p. 358) |
| **Subjective probability** | When a probability represents someone's personal degree of belief, it is called a subjective probability. (p. 359) |
| **Probability Assignment Rule** | The probability of the entire sample space must be 1. $P(\mathbf{S}) = 1$. (p. 360) |
| **Complement Rule** | The probability of an event occurring is 1 minus the probability that it doesn't occur: $$P(\mathbf{A^C}) = 1 - P(\mathbf{A}).$$ (p. 360) |
| **Disjoint (Mutually exclusive)** | Two events are disjoint if they share no outcomes in common. If **A** and **B** are disjoint, then knowing that **A** occurs tells us that **B** cannot occur. Disjoint events are also called "mutually exclusive." (p. 361) |
| **Addition Rule** | If **A** and **B** are disjoint events, then the probability of **A** *or* **B** is $$P(\mathbf{A} \cup \mathbf{B}) = P(\mathbf{A}) + P(\mathbf{B}).$$ (p. 361) |
| **Legitimate probability assignment** | An assignment of probabilities to outcomes is legitimate if<br>♦ each probability is between 0 and 1 (inclusive).<br>♦ the sum of the probabilities is 1. (p. 361) |
| **Multiplication Rule** | If **A** and **B** are independent events, then the probability of **A** *and* **B** is $$P(\mathbf{A} \cap \mathbf{B}) = P(\mathbf{A}) \times P(\mathbf{B}).$$ (p. 362) |
| **Independence Assumption** | We often require events to be independent. (So you should think about whether this assumption is reasonable.) (p. 362) |

# EXERCISES

1. **Sample spaces** For each of the following, list the sample space and tell whether you think the events are equally likely:
   a) Toss 2 coins; record the order of heads and tails.
   b) Roll a four-sided die three times; record the number of 3's.
   c) Flip a coin until you get a head or 3 consecutive tails; record each flip.
   d) Roll two dice; record the larger number.

2. **Sample spaces** For each of the following, list the sample space and tell whether you think the events are equally likely:
   a) Roll two dice; record the sum of the numbers.
   b) Draw three cards from a standard deck without replacing the first before drawing the second; record the number of red cards.
   c) Toss four coins; record the number of tails.
   d) Toss a coin 10 times; record the length of the longest run of heads.

3. **Roulette** A casino claims that its roulette wheel is truly random. What should that claim mean?

4. **Rain** The weather reporter on TV makes predictions such as a 25% chance of rain. What do you think is the meaning of such a phrase?

5. **Winter** Comment on the following quotation:

    *"What I think is our best determination is it will be a colder than normal winter,"* said Pamela Naber Knox, a Wisconsin state climatologist. *"I'm basing that on a couple of different things. First, in looking at the past few winters, there has been a lack of really cold weather. Even though we are not supposed to use the law of averages, we are due."* (Associated Press, fall 1992, quoted by Schaeffer et al.)

6. **Snow** After an unusually dry autumn, a radio announcer is heard to say, "Watch out! We'll pay for these sunny days later on this winter." Explain what he's trying to say, and comment on the validity of his reasoning.

7. **Cold streak** A batter who had failed to get a hit in seven consecutive times at bat then hits a game-winning home run. When talking to reporters afterward, he says he was very confident that last time at bat because he knew he was "due for a hit." Comment on his reasoning.

**8. Crash** Commercial airplanes have an excellent safety record. Nevertheless, there are crashes occasionally, with the loss of many lives. In the weeks following a crash, airlines often report a drop in the number of passengers, probably because people are afraid to risk flying.

a) A travel agent suggests that because the law of averages makes it highly unlikely to have two plane crashes within a few weeks of each other, flying soon after a crash is the safest time. What do you think?

b) If the airline industry proudly announces that it has set a new record for the longest period of safe flights, would you be reluctant to fly? Are the airlines due to have a crash?

**9. Auto insurance** Insurance companies collect annual payments from drivers in exchange for paying for the cost of accidents.

a) Why should you be reluctant to accept a $1500 payment from your neighbor to cover his automobile accidents in the next year?

b) Why can the insurance company make that offer?

**10. Jackpot** On February 11, 2009, the AP news wire released the following story:

(*LAS VEGAS, Nev.* )—A man in town to watch the NCAA basketball tournament hit a $38.7 million jackpot on Friday, the biggest slot machine payout ever. The 25-year-old software engineer from Los Angeles, whose name was not released at his request, won after putting three $1 coins in a machine at the Excalibur hotel-casino, said Rick Sorensen, a spokesman for slot machine maker International Game Technology.

a) How can the Excalibur afford to give away millions of dollars on a $3 bet?

b) Why was the maker willing to make a statement? Wouldn't most businesses want to keep such a huge loss quiet?

**11. Wardrobe** In your dresser are five blue shirts, three red shirts, and two black shirts.

a) What is the probability of randomly selecting a red shirt?

b) What is the probability that a randomly selected shirt is not black?

**12. Playlists** Your list of favorite songs contains 10 rock songs, 7 hip hop songs, and 3 K-pop songs.

a) What is the probability that a randomly played song is a hip hop song?

b) What is the probability that a randomly played song is not K-pop?

**13. Cell phones and surveys 2016** A survey conducted by the National Center for Health Statistics found that about 45.9% of U.S. households used landline phones in 2016. This raises concerns about the accuracy of certain surveys, as they depend on random-digit dialing to households via landlines. We are going to pick five U.S. households at random:

a) What is the probability that all five of them have a landline?

b) What is the probability that at least one of them does not have a landline?

c) What is the probability that at least one of them does have a landline?

**14. Cell phones and surveys II** The survey by the National Center for Health Statistics further found that 71% of adults ages 25–29 had only a cell phone and no landline. We randomly select four 25–29-year-olds:

a) What is the probability that all of these adults have only a cell phone and no landline?

b) What is the probability that none of these adults have only a cell phone and no landline?

c) What is the probability that at least one of these adults has only a cell phone and no landline?

**15. Spinner** The plastic arrow on a spinner for a child's game stops rotating to point at a color that will determine what happens next. Which of the following probability assignments are possible?

| | Probabilities of . . . | | | |
|---|---|---|---|---|
| | Red | Yellow | Green | Blue |
| a) | 0.25 | 0.25 | 0.25 | 0.25 |
| b) | 0.10 | 0.20 | 0.30 | 0.40 |
| c) | 0.20 | 0.30 | 0.40 | 0.50 |
| d) | 0 | 0 | 1.00 | 0 |
| e) | 0.10 | 0.20 | 1.20 | −1.50 |

**16. Scratch off** Many stores run "secret sales": Shoppers receive cards that determine how large a discount they get, but the percentage is revealed by scratching off that black stuff (what *is* that?) only after the purchase has been totaled at the cash register. The store is required to reveal (in the fine print) the distribution of discounts available. Which of these probability assignments are legitimate?

| | Probabilities of . . . | | | |
|---|---|---|---|---|
| | 10% off | 20% off | 30% off | 50% off |
| a) | 0.20 | 0.20 | 0.20 | 0.20 |
| b) | 0.50 | 0.30 | 0.20 | 0.10 |
| c) | 0.80 | 0.10 | 0.05 | 0.05 |
| d) | 0.75 | 0.25 | 0.25 | −0.25 |
| e) | 1.00 | 0 | 0 | 0 |

**17. Electronics** Suppose that 46% of families living in a certain county own a computer and 18% own an HDTV. The Addition Rule might suggest, then, that 64% of families own either a computer or an HDTV. What's wrong with that reasoning?

**18. Homes** Funding for many schools comes from taxes based on assessed values of local properties. People's homes are assessed higher if they have extra features such as garages and swimming pools. Assessment records in a certain school district indicate that 37% of the homes have garages and 3% have swimming pools. The Addition Rule might suggest, then, that 40% of residences have a garage or a pool. What's wrong with that reasoning?

**19. Speeders** Traffic checks on a certain section of highway suggest that 60% of drivers are speeding there. Because $0.6 \times 0.6 = 0.36$, the Multiplication Rule might suggest that there's a 36% chance that two vehicles in a row are both speeding. What's wrong with that reasoning?

**20. Lefties** Although it's hard to be definitive in classifying people as right- or left-handed, some studies suggest that about 14% of people are left-handed. Because $0.14 \times 0.14 = 0.0196$, the Multiplication Rule might suggest that there's about a 2% chance that a brother and a sister are both lefties. What's wrong with that reasoning?

**21. College admissions 2025** For the college class of 2025, admissions to the nation's most selective schools were the most competitive in memory. Harvard accepted about 3.4% of its applicants, Dartmouth 6.2%, and Penn 5.7%. Jorge has applied to all three. Assuming that he's a typical applicant, he figures that his chances of getting into both Harvard and Dartmouth must be about 0.21%.
  a) How has he arrived at this conclusion?
  b) What additional assumption is he making?
  c) Do you agree with his conclusion?

**22. College admissions II** In Exercise 21, we saw that for 2025 college graduates, Harvard accepted about 3.4% of its applicants, Dartmouth 6.2%, and Penn 5.7%. Jorge has applied to all three. He figures that his chances of getting into at least one of the three must be about 15.3%.
  a) How has he arrived at this conclusion?
  b) What assumption is he making?
  c) Do you agree with his conclusion?

**23. Car repairs** A consumer organization estimates that over a one-year period 17% of cars will need to be repaired once, 7% will need repairs twice, and 4% will require three or more repairs. What is the probability that a car chosen at random will need
  a) no repairs?
  b) no more than one repair?
  c) some repairs?

**24. Classroom music** Your teacher has an digital music player filled with music. It has many thousands of songs. You figure that roughly 60% of the songs are cool new music you and your friends like, 25% of the music is annoying childrens's songs chosen by your teacher's young kids, and the rest 80s pop music that only your teacher likes. Some days at the beginning of class, the teacher puts the player on shuffle and you all listen to whatever randomness produces. What is the probability that the first song is
  a) an 80s pop song?
  b) a song liked by either you and your friends or by the teacher's kids?
  c) not a song you and your friends like?

**25. More repairs** Consider again the auto repair rates described in Exercise 23. If you own two cars, what is the probability that
  a) neither will need repair?
  b) both will need repair?
  c) at least one car will need repair?

**26. More music** You listen to two songs, as described in Exercise 24. What is the probability that
  a) neither will be songs that you and your friends like?
  b) both will be annoying children's songs?
  c) at least one will be one of the 80s pop songs?

**27. Repairs, again** You used the Multiplication Rule to calculate repair probabilities for your cars in Exercise 25.
  a) What must be true about your cars in order to make that approach valid?
  b) Do you think this assumption is reasonable? Explain.

**28. Coda** You used the Multiplication Rule to calculate probabilities about the music choices of your teacher's digital music player in Exercise 26.
  a) What must be true about the songs to make that approach valid?
  b) Do you think this assumption is reasonable? Explain.

**29. Independence!** As discussed in this chapter, sometimes the Independence Assumption is reasonable, but other times it is not. Consider the following situations and comment on whether you think it is safe to assume independence.
  a) A pollster asks every 20th person exiting a movie if they liked the movie or not.
  b) A pollster asks people exiting a movie if they liked the movie or not. He interviews the movie-goers in groups as they are exiting.
  c) The amount of rainfall in a given city over the course of 10 days
  d) The amount of rainfall in 20 randomly chosen cities on a given day of the year

**30. Dependent?** Once again (as in Exercise 29) consider whether the Independence Assumption is reasonable in these scenarios.
  a) You want to know if seniors are attending the fall dance or not. You ask 20 groups of students at lunch.
  b) To estimate attendance at the fall dance, you survey 100 randomly chosen students in their English classes.
  c) A manager is tracking the number of customer complaints at his fast food restaurant. He counts the number of complaints every hour for three days.
  d) The manager in part c counts the number of complaints on 10 randomly selected days chosen out of the last three months.

**31. Is it random?** Is it reasonable to think that the following scenarios are independent where random chance is at play? Or do you think the Independence Assumption might be violated? Explain.
  a) You flip a coin 100 times and get 6 heads in a row.
  b) A basketball player shoots 30 times in a game and at one point makes 8 shots in a row.
  c) Your favorite burrito shop serves great food. But then your last four burritos have had more rice than anything else and not enough of your favorite ingredients.

**32. What are the chances?** As in Exercise 31, explain if you think the following observations are just by chance or if you think the Independence Assumption may be in doubt.
  a) After striking out three times in a row, a batter gets five hits. Then he strikes out four more times in a row.
  b) A doctor sees 50 patients a day. Yesterday, 20 of his patients had the flu and she saw 12 of them all in a row.
  c) A poker player is in a long tournament, playing many games. At one point, she is dealt 8 hearts in a row.

**33. Wind energy 2021** A Gallup Poll in March 2021 asked 1010 U.S. adults whether increasing wind as an energy source should receive more emphasis, less emphasis, or about the same emphasis that it currently does. Here are the results:

| Response | Number |
| --- | --- |
| More emphasis | 662 |
| Less emphasis | 164 |
| The same as now | 182 |
| Don't know/refused to answer | 2 |
| Total | 1010 |

If we select a person at random from this sample of 1010 adults,

a) what is the probability that the person responded "More emphasis"?
b) what is the probability that the person responded "The same as now" or did not give an opinion?

**34. Doting parents?** A Pew Research poll in 2019 asked 9834 U.S. adults whether parents of young adults are doing too much, too little, or about the right amount for their young adult children. Here's how they responded:

| Response | Number |
| --- | --- |
| Too much | 5409 |
| Too little | 983 |
| About the right amount | 3344 |
| No response | 98 |
| Total | 9834 |

If we select a respondent at random from this sample of 9834 adults,

a) what is the probability that the selected person responded "Too much"?
b) what is the probability that the person responded the "Too little" or "About the right amount"?

**35. Wind energy** Exercise 33 shows the results of a question about wind energy from a Gallup Poll. Suppose we select three people at random from this sample.

a) What is the probability that all three responded "More emphasis"?
b) What is the probability that none responded "The same as now"?
c) What assumption did you make in computing these probabilities?
d) Explain why you think that assumption is reasonable.

**36. Parents, revisited** Consider again the results of the poll about parenting discussed in Exercise 34. If we select two people at random from this sample,

a) what is the probability that both think that parents do too much for their young adult children?
b) what is the probability that neither thinks parents do too little?

c) what is the probability that the first person said parents do too much and the second one didn't?
d) what assumption did you make in computing these probabilities?
e) Explain why you think that assumption is reasonable.

**37. Polling** As mentioned in the chapter, opinion-polling organizations contact their respondents by sampling random telephone numbers. Although interviewers now can reach about 46% of U.S. households, the percentage of those contacted who agree to cooperate with the survey has fallen from 43% in 1997 to only 6% in 2018 (Pew Research Center). Each household selected, of course, is independent of the others.

a) What is the probability that the next household on the list will be contacted but will refuse to cooperate?
b) What was the probability (in 2018) of failing to contact a household or of contacting the household but not getting them to agree to the interview?
c) Show another way to calculate the probability in part b.

**38. Polling, part II** According to Pew Research, the contact rate (probability of contacting a selected household) was 90% in 1997 and 46% in 2018. However, the cooperation rate (probability of someone at the contacted household agreeing to be interviewed) was 43% in 1997 and dropped to 6% in 2018.

a) What was the probability (in 2018) of obtaining an interview with the next household on the sample list? (To obtain an interview, an interviewer must both contact the household and then get agreement for the interview.)
b) Was it more likely to obtain an interview from a randomly selected household in 1997 or in 2018?

**39. M&M's** Managers at the M&M's plant in Cleveland, Ohio say that yellow candies made up 13.5% of their Milk Chocolate M&M's, red another 13.1%, orange 20.5%, blue 20.7%, and green 19.8%. The rest are brown.

a) If you pick an M&M at random, what is the probability that
  1) it is brown?
  2) it is yellow or orange?
  3) it is not green?
  4) it is striped?
b) If you pick three M&M's in a row, what is the probability that
  1) they are all brown?
  2) you don't get a red the first two tries, but finally get a red on the third selection?
  3) none are yellow?
  4) at least one is green?

**40. Blood** The American Red Cross says that about 45% of the U.S. population has Type O blood, 40% Type A, 11% Type B, and the rest Type AB.

a) Someone volunteers to give blood. What is the probability that this donor
  1) has Type AB blood?
  2) has Type A or Type B?
  3) is not Type O?

b) Among four potential donors, what is the probability that
   1) all are Type O?
   2) no one is Type AB?
   3) they are not all Type A?
   4) at least one person is Type B?

**41. Disjoint or independent?** In Exercise 39 you calculated probabilities of getting various M&M's. Some of your answers depended on the assumption that the outcomes described were *disjoint*; that is, they could not both happen at the same time. Other answers depended on the assumption that the events were *independent*; that is, the occurrence of one of them doesn't affect the probability of the other. Do you understand the difference between disjoint and independent?

a) If you draw one M&M, are the events of getting a red one and getting an orange one disjoint, independent, or neither?
b) If you draw two M&M's one after the other, are the events of getting a red on the first and a red on the second disjoint, independent, or neither?
c) Can disjoint events ever be independent? Explain.

**42. Disjoint or independent?** In Exercise 40 you calculated probabilities involving various blood types. Some of your answers depended on the assumption that the outcomes described were *disjoint*; that is, they could not both happen at the same time. Other answers depended on the assumption that the events were *independent*; that is, the occurrence of one of them doesn't affect the probability of the other. Do you understand the difference between disjoint and independent?

a) If you examine one person, are the events that the person is Type A and that the person is Type B disjoint, independent, or neither?
b) If you examine two people, are the events that the first is Type A and the second Type B disjoint, independent, or neither?
c) Can disjoint events ever be independent? Explain.

**43. Dice** You roll a fair die three times. What is the probability that
a) you roll all 6's?
b) you roll all odd numbers?
c) none of your rolls gets a number divisible by 3?
d) you roll at least one 5?
e) the numbers you roll are not all 5's?

**44. Slot machine** A slot machine has three wheels that spin independently. Each has 10 equally likely symbols: 4 bars, 3 lemons, 2 cherries, and a bell. If you play, what is the probability that
a) you get 3 lemons?
b) you get no fruit symbols?
c) you get 3 bells (the jackpot)?
d) you get no bells?
e) you get at least one bar (an automatic loser)?

**45. Champion bowler** A certain bowler can bowl a strike 70% of the time. What's the probability that she
a) goes three consecutive frames without a strike?
b) doesn't get a strike the first two times, but finally gets a strike on her third try?
c) has at least one strike in the first three frames?
d) bowls a perfect game (12 consecutive strikes)?

**46. The train** To get to work, a commuter must cross train tracks. The time the train arrives varies slightly from day to day, but the commuter estimates he'll get stopped on about 15% of work days. During a certain 5-day work week, what is the probability that he
a) gets stopped on Monday and again on Tuesday?
b) doesn't get stopped the first 3 days, but then is stopped on Thursday?
c) gets stopped every day?
d) gets stopped at least once during the week?

**47. Voters** Suppose that in your city 37% of the voters are registered as Democrats, 29% as Republicans, and 11% as members of other U.S. parties with national reach (Libertarian, Socialist, Green, etc.). Voters not aligned with any official party are termed "Independent." You are conducting a poll by calling registered voters at random. In your first three calls, what is the probability you talk to
a) all Republicans?
b) no Democrats?
c) at least one Independent?

**48. Religion** Census reports for a city indicate that 62% of residents classify themselves as Christian, 12% as Jewish, and 16% as members of other religions (Muslims, Buddhists, etc.). The remaining residents classify themselves as nonreligious. A polling organization seeking information about public opinions wants to be sure to talk with people holding a variety of religious views, and makes random phone calls. Among the first four people they call, what is the probability they reach
a) all Christians?
b) no Jews?
c) at least one person who is nonreligious?

**49. Tires** You bought a new set of four tires from a manufacturer who just announced a recall because 2% of those tires are defective. What is the probability that at least one of yours is defective?

**50. Light bulbs** You purchased a five-pack of new light bulbs that were recalled because 6% of the bulbs did not work. What is the probability that at least one of your bulbs is defective?

**51. 9/11?** On September 11, 2002, the first anniversary of the terrorist attack on the World Trade Center, the New York State Lottery's daily number came up 9–1–1. An interesting coincidence or a cosmic sign?

a) What is the probability that the winning three numbers match the date on any given day?
b) What is the probability that a whole year passes without this happening?
c) What is the probability that the date and winning lottery numbers match at least once during any year?
d) If every one of the 50 states has a three-digit lottery, what is the probability that at least one of them will come up 9–1–1 on September 11?

**52. Red cards** You shuffle a standard deck of 52 playing cards and then start turning them over one at a time. The first one is red. So is the second. And the third. In fact, you are surprised to get 10 red cards in a row. You start thinking, "The next one is due to be black!"

a) Are you correct in thinking that there's a higher probability that the next card will be black than red? Explain.
b) Is this an example of the Law of Large Numbers? Explain.

## JUST CHECKING

Answers

1. The LLN works only in the long run, not in the short run. The random methods for selecting lottery numbers have no memory of previous picks, so there is no change in the probability that a certain number will come up.

2. a) 0.76.
   b) $0.76(0.76) = 0.5776$.
   c) $(1 - 0.76)^2(0.76) = 0.043776$.
   d) $1 - (1 - 0.76)^5 = 0.9992$.

# 14

# Probability Rules!

U.S. bills have either (dead) presidents or other important historical figures on them.
$1 George Washington (Pres #1)
$2 Thomas Jefferson (Pres #3)
$5 Abraham Lincoln (Pres #16)
$10 Alexander Hamilton (First Treasury secretary)
$20 Andrew Jackson (for now– Pres #7)
$50 Ulysses Grant (Pres #18)
$100 Benjamin Franklin (inventor and First Postmaster General).

Pull a bill from your wallet or pocket without looking at it. An outcome of this trial is the kind of bill you select. The sample space is all the bills in circulation: **S** = {$1 bill, $2 bill, $5 bill, $10 bill, $20 bill, $50 bill, $100 bill}.[1] These are *all* the possible outcomes. ($500 and $1000 bills were taken out of circulation in 1969.)

We can combine the outcomes in different ways to make many different events. For example, the event **A** = {$1, $5, $10} represents selecting a $1, $5, or $10 bill. The event **B** = {a bill that does not have a president on it} is the collection of outcomes (Don't look! Can you name them?): {$10 (Hamilton), $100 (Franklin)}. The event **C** = {enough money to pay for a $12 meal with one bill} is the set of outcomes {$20, $50, $100}.

Notice that these outcomes are not equally likely. You'd no doubt be more surprised (and pleased) to pull out a $100 bill than a $1 bill—it's not as likely, though. You probably carry many more $1 than $100 bills, but without information about the probability of each outcome, we can't calculate the probability of an event.

The probability of the event **C** (getting a bill worth more than $12) is *not* 3/7. There are 7 possible outcomes, and 3 of them exceed $12, but they are not *equally likely*.

## The General Addition Rule

Now look at the bill in your hand. There are images of famous buildings in the center of the backs of all but two bills in circulation. The $1 bill has the word ONE in the center, and the $2 bill shows the signing of the Declaration of Independence.

---

[1] Well, technically, the sample space is all the bills in your pocket. You may be quite sure there isn't a $100 bill in there, but *we* don't know that, so humor us that it's at least *possible* that any legal bill could be there.

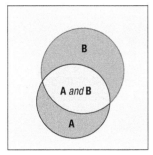

Events **A** and **B** and their intersection.

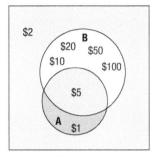

Denominations of bills that are odd (**A**) or that have a building on the reverse side (**B**). The two sets both include the $5 bill, and both exclude the $2 bill.

What's the probability of randomly selecting **A** = {a bill with an odd-numbered value} *or* **B** = {a bill with a building on the reverse}? We know **A** = {$1, $5} and **B** = {$5, $10, $20, $50, $100}. But $P(\mathbf{A}\text{ or }\mathbf{B})$ is not simply the sum $P(\mathbf{A}) + P(\mathbf{B})$, because the events **A** and **B** are not disjoint. The $5 bill is in both sets. So what can we do? We'll need a new probability rule.

As the diagrams show, we can't get the right answer by just adding the two probabilities because the events are not disjoint; they overlap. There's an outcome (the $5 bill) in the *intersection* of **A** and **B**. The Venn diagram represents the sample space. Notice that the $2 bill has neither a building nor an odd denomination, so it sits outside both circles.

The $5 bill plays a crucial role here because it is both odd *and* has a building on the reverse. It's in both **A** and **B**, which places it in the *intersection* of the two circles. The reason we can't simply add the probabilities of **A** and **B** is that we'd count the $5 bill twice.

If we did add the two probabilities, we could compensate by *subtracting* out the probability of that $5 bill. So,

$P$(odd number value *or* building)

$$= P(\text{odd number value}) + P(\text{building}) - P(\text{odd number value }and\text{ building})$$
$$= P(\$1, \$5) + P(\$5, \$10, \$20, \$50, \$100) - P(\$5).$$

This method works in general. We add the probabilities of two events and then subtract out the probability of their intersection. This approach gives us the **General Addition Rule**, which does not require disjoint events:

$$P(\mathbf{A} \cup \mathbf{B}) = P(\mathbf{A}) + P(\mathbf{B}) - P(\mathbf{A} \cap \mathbf{B}).$$

### FOR EXAMPLE
#### Using the General Addition Rule

A survey of college students found that 56% live in a campus residence hall, 62% participate in a campus meal program, and 42% do both.

QUESTION: What's the probability that a randomly selected student either lives or eats on campus?

ANSWER: Let **L** = {student lives on campus} and **M** = {student has a campus meal plan}.

$P$(a student either lives or eats on campus) $= P(\mathbf{L} \cup \mathbf{M})$
$$= P(\mathbf{L}) + P(\mathbf{M}) - P(\mathbf{L} \cap \mathbf{M})$$
$$= 0.56 + 0.62 - 0.42$$
$$= 0.76$$

There's a 76% chance that a randomly selected college student either lives or eats on campus.

### Would You Like Dessert or Coffee?

Natural language can be ambiguous. In this question, is the answer one of the two alternatives, or simply "yes"? Must you decide between them, or may you have both? That kind of ambiguity can confuse our probabilities.

Suppose we had been asked a different question: What is the probability that the bill we draw has *either* an odd value *or* a building but *not both*? Which bills are we talking about now? The set we're interested in would be { $1, $10, $20, $50, $100}. We don't include the $5 bill in the set because it has both characteristics.

Why isn't this the same answer as before? The problem is that when we say the word "or," we usually mean *either* one *or* both. We don't usually mean the *exclusive* version of "or" as in, "Would you like the steak *or* the vegetarian entrée?" Ordinarily when we ask for the probability that **A** or **B** occurs, we mean **A** or **B** or both. And we know *that* probability is $P(\mathbf{A}) + P(\mathbf{B}) - P(\mathbf{A} \text{ and } \mathbf{B})$. The General Addition Rule subtracts the probability of the outcomes in **A** and **B** because we've counted those outcomes *twice*. But they're still there.

If we really mean **A** or **B** but NOT both, we have to get rid of the outcomes in {**A** and **B**}. So $P(\mathbf{A} \text{ or } \mathbf{B} \text{ but } not \text{ both}) = P(\mathbf{A} \cup \mathbf{B}) - P(\mathbf{A} \cap \mathbf{B}) = P(\mathbf{A}) + P(\mathbf{B}) - 2 \times P(\mathbf{A} \cap \mathbf{B})$. Now we've subtracted $P(\mathbf{A} \cap \mathbf{B})$ twice—once because we don't want to double-count these events and a second time because we really didn't want to count them at all.

Confused? *Make a picture*. It's almost always easier to think about such situations by looking at a Venn diagram.

## FOR EXAMPLE
### Using Venn Diagrams

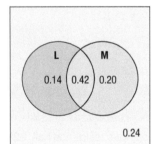

**RECAP:** We return to our survey of college students: 56% live on campus, 62% have a campus meal program, and 42% do both.

**QUESTION:** Based on a Venn diagram, what is the probability that a randomly selected student

a) lives off campus and doesn't have a meal program?
b) lives in a residence hall but doesn't have a meal program?

**ANSWER:** Let **L** = {student lives on campus} and **M** = {student has a campus meal plan}. In the Venn diagram, the intersection of the circles is $P(\mathbf{L} \cap \mathbf{M}) = 0.42$. Because $P(\mathbf{L}) = 0.56$, $P(\mathbf{L} \cap \mathbf{M}^C) = 0.56 - 0.42 = 0.14$. Also, $P(\mathbf{L}^C \cap \mathbf{M}) = 0.62 - 0.42 = 0.20$. Now, $0.14 + 0.42 + 0.20 = 0.76$, leaving $1 - 0.76 = 0.24$ for the region outside both circles.

Now ... $P(\text{off campus and no meal program}) = P(\mathbf{L}^C \cap \mathbf{M}^C) = 0.24$, and

$P(\text{on campus and no meal program}) = P(\mathbf{L} \cap \mathbf{M}^C) = 0.14$

## JUST CHECKING

1. We sampled some pages of this book at random to see whether they held an equation, a graph, or other data display and found the following:

   48% of pages had some kind of data display,
   27% of pages had an equation, and
   7% of pages had both a data display and an equation.

   a) Display these results in a Venn diagram.
   b) What is the probability that a randomly selected sample page had neither a data display nor an equation?
   c) What is the probability that a randomly selected sample page had a data display but no equation?

## STEP-BY-STEP EXAMPLE

### Using the General Addition Rule

You almost certainly qualify as an "online adult"—an adult who uses the Internet. Pew Research tracks the behavior of online adults. Recently they found that 79% were Facebook users, 24% used Twitter, and 22% used both. (www.pewinternet.org/2016/11/11/social-media-update-2016/)

QUESTIONS: For a randomly selected online adult, what is the probability that the person
1. uses either Facebook or Twitter?
2. uses either Facebook or Twitter, but not both?
3. doesn't use Facebook or Twitter?

**THINK**

**PLAN** Define the events we're interested in. There are no conditions to check; the General Addition Rule works for any events!

**PLOT** Make a picture, and use the given probabilities to find the probability for each region.

Let $A = \{\text{respondent uses Facebook}\}$.

Let $B = \{\text{respondent uses Twitter}\}$.

I know that
$$P(A) = 0.79$$
$$P(B) = 0.24$$
$$P(A \cap B) = 0.22$$

So
$$P(A \cap B^C) = 0.79 - 0.22 = 0.57$$
$$P(B \cap A^C) = 0.24 - 0.22 = 0.02$$
$$P(A^C \cap B^C) = 1 - (0.57 + 0.22 + 0.02)$$
$$= 0.19$$

The blue region represents **A** but not **B**. The green intersection region represents **A** and **B**. Note that because $P(A) = 0.79$ and $P(A \cap B) = 0.22$, the probability of **A** but not **B** must be $0.79 - 0.22 = 0.57$.

The yellow region is **B** but not **A**.

The gray region outside both circles represents the outcome neither **A** nor **B**. All the probabilities must total 1, so you can determine the probability of that region by subtraction.

Now, figure out what you want to know. The probabilities can come from the diagram or a formula. Sometimes translating the words to equations is the trickiest step.

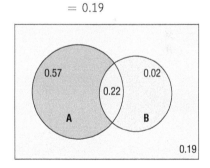

QUESTION 1. What is the probability that the person uses either Facebook or Twitter?

**SHOW**

**MECHANICS** The probability that the respondent uses Facebook or Twitter is $P(A \cup B)$. We can use the General Addition Rule, or we can add the probabilities seen in the diagram.

$$P(A \cup B) = P(A) + P(B) - P(A \cap B)$$
$$= 0.79 + 0.24 - 0.22$$
$$= 0.81$$

OR

$$P(A \cup B) = 0.57 + 0.22 + 0.02 = 0.81$$

**TELL**

**CONCLUSION** Don't forget to interpret your result in context.

81% of online adults use Facebook or Twitter.

**QUESTION 2.** What is the probability that the person uses either Facebook or Twitter, but not both?

| SHOW — MECHANICS We can use the rule, or just add the appropriate probabilities seen in the Venn diagram. | $P(\mathbf{A} \text{ or } \mathbf{B} \text{ but NOT both}) = P(\mathbf{A} \cup \mathbf{B}) - P(\mathbf{A} \cap \mathbf{B})$ $= 0.81 - 0.22 = 0.59$ **OR** $P(\mathbf{A} \text{ or } \mathbf{B} \text{ but NOT both}) = P(\mathbf{A} \cap \mathbf{B}^c) - P(\mathbf{B} \cap \mathbf{A}^c)$ $= 0.57 + 0.02 = 0.59$ |
|---|---|
| TELL — CONCLUSION Interpret your result in context. | 59% of online adults use Facebook or Twitter, but not both. |

**QUESTION 3.** What is the probability that the person doesn't use Facebook or Twitter?

| SHOW — MECHANICS Not using Facebook or Twitter is the complement of using one or the other. Use the Complement Rule or just notice that "not using Facebook or Twitter" is represented by the region outside both circles. | $P(\text{not Facebook or Twitter})$ $= 1 - P(\text{either Facebook or Twitter})$ $= 1 - P(\mathbf{A} \cup \mathbf{B})$ $= 1 - 0.81 = 0.19$ **OR** $P(\mathbf{A}^c \cap \mathbf{B}^c) = 0.19$ |
|---|---|
| TELL — CONCLUSION Interpret your result in context. | Only 19% of online adults use neither Facebook nor Twitter. |

# It Depends . . .

In Chapter 2 we discussed the fates of the passengers and crew on board the *Titanic* when it sank. Here, again, is a contingency table showing counts of those who were saved and those who were lost, grouped by their ticket class or crew status:

|  |  | Class |  |  |  |  |
|---|---|---|---|---|---|---|
|  |  | First | Second | Third | Crew | Total |
| **Survival** | Saved | 203 | 118 | 178 | 212 | 711 |
|  | Lost | 122 | 167 | 528 | 673 | 1490 |
|  | Total | 325 | 285 | 706 | 885 | 2201 |

We also saw in Chapter 2 that a useful graphical tool for comparing conditional distributions across different groups is a mosaic plot. Here's a mosaic plot that makes it easy to compare the distributions of *Survival* status across the four different groups:

Figure 14.1

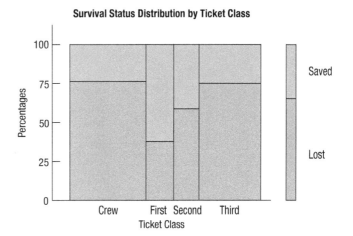

We can see that approximately 25% of the crew were saved, as were about 25% of the third-class passengers, compared to about two-fifths of the second-class passengers and more than half of the first-class passengers. If we want to obtain these percentages more exactly, we could examine a contingency table that shows column percentages along with counts:

|  |  | Class | | | | |
|---|---|---|---|---|---|---|
|  |  | First | Second | Third | Crew | Total |
| Survived | Saved Count | 203 | 118 | 178 | 212 | 711 |
|  | % of Column | 62.5% | 41.4% | 25.2% | 24.0% | 32.3% |
|  | Lost Count | 122 | 167 | 528 | 673 | 1490 |
|  | % of Column | 37.5% | 58.6% | 74.8% | 76.0% | 67.7% |
|  | Total Count | 325 | 285 | 706 | 885 | 2201 |
|  |  | 100% | 100% | 100% | 100% | 100% |

The table is harder to interpret than the mosaic plot because you have to study the numbers carefully. But the table gives precise values for column percentages. Let's look more closely at how they were determined, and see how these distributions of categorical data translate into probabilities.

◆ If you select a person at random from among everyone on the *Titanic*, the probability of selecting a first-class passenger is $P(\text{First}) = \frac{325}{2201} = 0.148$. (This number, a fraction of the total, does not appear in the table above, which only shows column percentages.)

◆ If you select a person at random from among everyone on the *Titanic*, the probability of selecting a first-class passenger *who was saved* is $P(\text{First} \cap \text{Saved}) = \frac{203}{2201} = 0.092$. (This is also a fraction of the total and does not appear in the table above.)

◆ What if we selected someone at random from only among the first-class passengers? Or, equivalently, what if we selected someone at random from everyone on board the *Titanic*, and then *learned* that the person was a first-class passenger? What's the probability that the person was saved? It's $\frac{203}{325} = 0.625$, because the person must be among the 325 first-class passengers, and we know that 203 of them were saved. This is one of the column percentages that's shown in the table above. It's called a *conditional probability* and we write $P(\text{Saved} | \text{First}) = 0.625$. We say that the probability of a person being saved *given that they are a first-class passenger* is 62.5%. We're determining the probability of one event (being saved) *already knowing the outcome of another* (being a first-class passenger).

In general, when we want the probability of an event from a *conditional* distribution, we write $P(\mathbf{B}|\mathbf{A})$ and pronounce it "the probability of **B** *given* **A**." A probability that takes into account a given condition such as this is called a **conditional probability**.

Let's determine a rule for computing conditional probabilities. We just saw that for a randomly selected person on the *Titanic*, $P(\text{Saved}|\text{First}) = \frac{203}{325}$. The denominator is a count of people in first-class and the numerator is a count of people who were in first-class AND were saved. Dividing both numerator and denominator by the total number of people on board won't change the overall proportion, and it converts counts into probabilities:

$$P(\text{Saved}|\text{First}) = \frac{203}{325} = \frac{(203/2201)}{(325/2201)} = \frac{P(\text{First} \cap \text{Saved})}{P(\text{First})}$$

We can generalize this rule to apply to any events **B** and **A**:

$$P(\mathbf{B}|\mathbf{A}) = \frac{P(\mathbf{A} \cap \mathbf{B})}{P(\mathbf{A})}$$

When you're working with conditional probabilities, it's critical to know what event is the *given* condition. For example, if we selected a person at random from those on board the *Titanic*, then $P(\text{First}|\text{Saved})$ would be the chance of them being a first-class passenger given that they were saved, which turns out to be about 28.6%.[2]

So remember, $P(\mathbf{B}|\mathbf{A})$ is not the same thing as $P(\mathbf{A}|\mathbf{B})$! Whenever you're finding a conditional probability, always *Think* about which event is given.

> NOTATION ALERT
> $P(\mathbf{B}|\mathbf{A})$ is the conditional probability of **B** *given* **A**.

---

**OOPS!**

*On August 13, 2017, the New York Times printed a correction to a previously published opinion essay about 55- to 64-year-old women who were looking for work. The essay had erroneously stated that half of the women in that age bracket were among the long-term unemployed. They had meant to write that nearly half of the long-term unemployed in that age bracket were women.*

We're sure *you* see the difference!

---

### FOR EXAMPLE
### Finding a Conditional Probability

**RECAP:** Our survey found that 56% of college students live on campus, 62% have a campus meal program, and 42% do both.

**QUESTION:** While dining in a campus facility open only to students with meal plans, you meet someone interesting. What is the probability that your new acquaintance lives on campus?

**ANSWER:** Let **L** = {student lives on campus} and **M** = {student has a campus meal plan}.

$P(\text{student lives on campus given that the student has a meal plan}) = P(\mathbf{L}|\mathbf{M})$

$$= \frac{P(\mathbf{L} \cap \mathbf{M})}{P(\mathbf{M})}$$
$$= \frac{0.42}{0.62}$$
$$\approx 0.677$$

There's a probability of about 0.677 that a student with a meal plan lives on campus. Notice that this is higher than the probability for all students.

---

[2] You should verify this!

# Independence

## Independence

If we had to pick one idea in this chapter that you should understand and remember, it's the definition and meaning of independence. We'll need this idea in every one of the chapters that follow.

It's time to return to the question of just what it means for events to be independent. We've said informally that what we mean by independence is that the outcome of one event does not influence the probability of the other. With our new notation for conditional probabilities, we can write a formal definition: Events **A** and **B** are **independent** whenever

$$P(\mathbf{B}|\mathbf{A}) = P(\mathbf{B}).$$

In other words, the probability that **B** happens is the same whether **A** happens or not.

In 2012 some researchers in France conducted a survey of nearly 2000 young adults who frequented bars in the western part of the country.[3] Their research was focused on whether there was an association between having body art (tattoos and/or piercings) and greater alcohol consumption. We're going to leave that question aside and just look at whether their research indicates an association between having tattoos and having body piercings.[4]

Table 14.1 shows how many of the young adults they surveyed said they had tattoos, body piercings, or both. Let's determine a couple of probabilities.

- If we select a person at random from the researchers' survey participants, the probability that the person will have a piercing is $P(\text{piercing}) = \frac{303}{1965} = 0.154$.
- If we select a person at random from the researchers' survey participants and learn that the person has a tattoo, then the conditional probability of them also having a piercing is $P(\text{piercing}|\text{tattoo}) = \frac{112}{334} = 0.335$.

While there's only a 15% chance of a survey participant having a piercing, that probability goes up to about 34% if the person has a tattoo. Because these probabilities are different, having a piercing is *not* independent of having a tattoo. (In particular, having a tattoo increases the chance of having a piercing.) When events are not independent, we could say that they're *dependent*, but we usually don't, perhaps because the word sound like it implies causality. Instead, we usually say the events are *associated*. Having one or more tattoos is associated with having one or more piercings.

Incidentally, it might seem obvious, but we'll point it out anyway: If an event **A** is independent of an event **B**, then **B** is also independent of **A**. Independence is symmetric. We just used the fact that $P(\text{piercing}) \neq P(\text{piercing}|\text{tattoo})$ to show that having a tattoo and having a piercing are not independent. But we could have instead used the following two probabilities and compared them:

- $P(\text{tattoo}) = \frac{334}{1965} = 0.170$
- $P(\text{tattoo}|\text{piercing}) = \frac{112}{303} = 0.370$

While about 17% of these young adults have tattoos, about 37% do among those who have piercings. Because these probabilities are different, having a tattoo and having a piercing are *not* independent. Of course we knew that already, and we didn't have to verify it with this additional comparison of probabilities. We just wanted to point out that if you want to see if events **A** and **B** are independent, you can either check for $P(\mathbf{A}) = P(\mathbf{A}|\mathbf{B})$ or you can check for $P(\mathbf{B}) = P(\mathbf{B}|\mathbf{A})$. You don't need to do both comparisons because if either equation is true then the events are independent and the other equation must be true as well. Conversely, if either equation is *false*, then there must be an association between the events, and the other equation must also be false.

|  |  | Piercings | | |
|---|---|---|---|---|
|  |  | Yes | No | Total |
| Tattoos | Yes | 112 | 222 | 334 |
|  | No | 191 | 1440 | 1631 |
|  | Total | 303 | 1662 | 1965 |

**Table 14.1**
Counts of young adults in a French survey who have tattoos, body piercings, or both

---

[3] https://www.news-medical.net/news/20120417/Tattoos-and-piercings-linked-to-heavy-drinking-Study.aspx
[4] In case you're wondering, their research did indicate that young adults with more body art tend to consume more alcohol.

## FOR EXAMPLE
### Checking for Independence

**RECAP:** Our survey told us that 56% of college students live on campus, 62% have a campus meal program, and 42% do both.

**QUESTION:** Are living on campus and having a meal plan independent? Are they disjoint?

**ANSWER:** Let $L$ = {student lives on campus} and $M$ = {student has a campus meal plan}. If these events are independent, then knowing that a student lives on campus doesn't affect the probability that he or she has a meal plan. I'll check to see if $P(M|L) = P(M)$:

$$P(M|L) = \frac{P(L \cap M)}{P(L)}$$
$$= \frac{0.42}{0.56}$$
$$= 0.75, \quad \text{but } P(M) = 0.62.$$

Because $0.75 \neq 0.62$, the events are not independent; students who live on campus are more likely to have meal plans. Living on campus and having a meal plan are not disjoint either; in fact, 42% of college students do both.

## Independent ≠ Disjoint

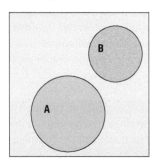

Figure 14.2
Because these events are mutually exclusive, learning that **A** happened tells us that **B** didn't. The probability of **B** has changed from whatever it was to zero. So the disjoint events **A** and **B** are not independent.

Suppose you select a date at random from a 365-day calendar. Consider these two events:

- $A$ = the date you select is in October.
- $B$ = the date you select is in the summer.

Answer this quickly: Are these two events independent?

If you said "yes," then although you're wrong (sorry!), you're in good company. This is a very common mistake. If **A** and **B** were independent, then the following would have to be true: $P(A) = P(A|B)$. But think about what those two probabilities mean. The first is the probability that you select a date in October, which is approximately 1/12. But the second is the probability that you select a date in October, *given that the date you select is in the summer*. That's clearly zero. Because the probability of **A** changes when you know that **B** happened ($P(A) \neq P(A|B)$), **A** and **B** must not be independent.

Still confused? Your eyes are seeing "independent," but you are *thinking* "disjoint." A date being in October and it being in the summer most certainly *are* disjoint events, because they can't both happen at the same time. But as we just showed, they're not independent. Don't get these two terms mixed up. They both have something to do with "separateness," but it's not the same kind of separateness. Disjoint events are "separate" in that they can't occur together. Independent events are "separate" in that knowing the outcome of one event doesn't inform you in any new way about the outcome of the other. *Disjoint* and *independent* don't mean the same thing.

In fact, not only do *disjoint* and *independent* not mean the same thing, disjoint events *can't* be independent! You can see why if you think again about the nonexistent summer day in October. If events **A** and **B** are disjoint, then they can't occur together. So $P(A|B)$ must be zero. (Knowing that **B** occurs immediately tells you that **A** did not; the conditional probability of **A** must be zero.) But of course $P(A)$ isn't itself zero.[5] So it must be true that $P(A) \neq P(A|B)$. **A** and **B** can't be independent.

---

[5] Okay, $P(A)$ could be zero, and then you'd have two events that were both disjoint and independent. But for practical purposes, we can ignore this special case, because statisticians don't concern themselves much with things that never happen.

To be honest, you won't see much more about disjoint events in this text after this chapter. It's a term you need to know, but it pales in importance compared to independence. *Independence*—or actually, *association* (non-independence)—is what this course is largely about: whether and how things are related to one another.

Let's summarize:

◆ Two events could be either independent or disjoint, *but not both*.
◆ And they could be *neither* disjoint nor independent.

One spring day your high school baseball team will be playing for the league championship right after school. You and your friends are planning to go, but when you look outside at lunchtime it's raining. Consider the events **R**: rain at lunchtime and **B**: baseball after school. They're not disjoint: even though it's raining now, the game may still go on as scheduled. And they're not independent either: you thought you were going to the game, but now that you see the rain, you're not so sure anymore. Knowing **R** changes the probability of **B**.

### JUST CHECKING

2. A company's office of Human Resources reports a breakdown of employees by job type and gender as seen in this table.

   a) Are being male and having a supervision job disjoint events?

   b) Is having a supervision job independent of the gender of the employee?

|  |  | Gender | |
|---|---|---|---|
|  |  | Male | Female |
| Job Type | Management | 7 | 6 |
|  | Supervision | 8 | 12 |
|  | Production | 45 | 72 |

## Check for Independence

In earlier chapters we said informally that two events were independent if learning that one occurred didn't change what you thought about the other occurring. Now we can be more formal. Events **A** and **B** are independent if (and only if) the probability of **A** is the same when we are given that **B** has occurred. That is, $P(\mathbf{A}) = P(\mathbf{A}|\mathbf{B})$.

Although sometimes your intuition is enough, now that we have the formal rule, use it whenever you can.

### DEPENDING ON INDEPENDENCE

A note of caution: People often estimate the probability of a compound event by multiplying probabilities together without thinking about whether those probabilities are independent.

For example, experts have assured us that the probability of a major commercial nuclear plant failure is so small that we should not expect such a failure to occur even in a span of hundreds of years. Yet in only a few decades of commercial nuclear power, the world has seen three failures (Chernobyl, Three Mile Island, and Fukushima). How could the estimates have been so wrong?

One simple part of the failure calculation is to test a particular valve and determine that valves such as this one fail only once in, say, 100 years of normal use. For a coolant failure to occur, several valves must fail. So we need the compound probability, $P$(valve 1 fails *and* valve 2 fails *and* . . . ). A simple risk assessment might multiply the small probability of one valve failure together as many times as needed. But if the valves all came from the same manufacturer, a flaw in one might be found in the others. And maybe when the first fails, it puts additional pressure on the next one in line. In either case, the events aren't independent and so we can't simply multiply the probabilities.

Whenever you see probabilities multiplied together, stop and ask whether you think they are really independent before you believe the result.

## Tables, Venn Diagrams, and Probability

In our earlier Facebook–Twitter example, we found that 79% of online adults use Facebook, 24% use Twitter, and 22% use both. That may not look like enough information to answer all the questions you[6] might have about the problem. But often, the right picture

---
[6]or your teacher

can help. Let's try to put what we know into a table. Translating percentages to probabilities, what we know looks like this:

|  | | Use Facebook | | |
|---|---|---|---|---|
|  | | Yes | No | Total |
| Use Twitter | Yes | 0.22 | | 0.24 |
|  | No | | | |
|  | Total | 0.79 | | 1.00 |

Notice that the 0.79 and 0.24 are *marginal* probabilities and so they go into the *margins*. The 0.22 is the probability of using both sites—Facebook and Twitter—so that's a *joint* probability. Those belong in the interior of the table.

Because the cells of the table show disjoint events, the probabilities always add to the marginal totals going across rows or down columns. So, filling in the rest of the table is quick:

|  | | Use Facebook | | |
|---|---|---|---|---|
|  | | Yes | No | Total |
| Use Twitter | Yes | 0.22 | 0.02 | 0.24 |
|  | No | 0.57 | 0.19 | 0.76 |
|  | Total | 0.79 | 0.21 | 1.00 |

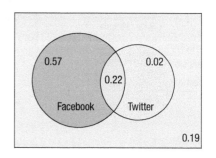

Compare this with the Venn diagram. Notice which entries in the table match up with the sets in this diagram. Whether a Venn diagram or a table is better to use will depend on what you are given and the questions you're being asked. Try both.

## STEP-BY-STEP EXAMPLE

### Are the Events Disjoint? Independent?

Let's take another look at the social networking survey. Researchers report that 79% of online adults use Facebook, 24% use Twitter, and 22% use both.

QUESTIONS:
1. Are using Facebook and using Twitter mutually exclusive?
2. Are Facebooking and Tweeting independent?

| THINK | PLAN Define the events we're interested in. State the given probabilities. | Let **A** = {respondent uses Facebook}. Let **B** = {respondent uses Twitter}. I know that $P(\mathbf{A}) = 0.79$, $P(\mathbf{B}) = 0.24$, and $P(\mathbf{A} \text{ and } \mathbf{B}) = 0.22$ |
|---|---|---|

QUESTION 1. Are using Facebook and using Twitter mutually exclusive?

| SHOW | MECHANICS Disjoint events cannot *both* happen at the same time, so check to see if $P(A \cap B) = 0$. | $P(A \cap B) = 0.22$. Because some adults use both Facebook and Twitter, $P(A \cap B) \neq 0$. The events are not mutually exclusive. |
|---|---|---|
| TELL | CONCLUSION State your conclusion in context. | 22% of online adults use both Facebook and Twitter, so these are not disjoint events. |

QUESTION 2. Are Facebooking and Tweeting independent?

| THINK | PLAN Make a table. | |
|---|---|---|

|  |  | Use Facebook | | |
|---|---|---|---|---|
|  |  | Yes | No | Total |
| Use Twitter | Yes | 0.22 | 0.02 | 0.24 |
|  | No | 0.57 | 0.19 | 0.76 |
|  | Total | 0.79 | 0.21 | 1.00 |

| SHOW | MECHANICS Does using Facebook change the probability of using Twitter? That is, does $P(B\|A) = P(B)$? Because the two probabilities are *not* the same, the events are not independent. | $P(B\|A) = \dfrac{P(A \cap B)}{P(A)} = \dfrac{0.22}{0.79} \approx 0.28$ <br> $P(B) = 0.24$ <br> $P(B\|A) \neq P(B)$ |
|---|---|---|
| TELL | CONCLUSION Interpret your results in context. | Overall, 24% of online adults use Twitter, but 28% of Facebook users use Twitter. Because respondents who use Facebook are more likely to Tweet, the two events are not independent. |

## JUST CHECKING

3. Remember our sample of pages in this book from the earlier Just Checking?

    *48% of pages had a data display,*

    *27% of pages had an equation, and*

    *7% of pages had both a data display and an equation.*

    a) Make a contingency table for the variables *Display* and *Equation*.
    b) What is the probability that a randomly selected sample page with an equation also had a data display?
    c) Are having an equation and having a data display disjoint events?
    d) Are having an equation and having a data display independent events?

# The General Multiplication Rule

> **A LITTLE ALGEBRA**
> We know:
> $P(\mathbf{B}|\mathbf{A}) = \dfrac{P(\mathbf{A} \cap \mathbf{B})}{P(\mathbf{A})}$.
> Multiply both sides of the equation by $P(\mathbf{A})$ to get:
> $P(\mathbf{A} \cap \mathbf{B}) = P(\mathbf{A}) \times P(\mathbf{B}|\mathbf{A})$.

Remember the Multiplication Rule for the probability of **A** *and* **B**? It said

$$P(\mathbf{A} \cap \mathbf{B}) = P(\mathbf{A}) \times P(\mathbf{B}) \text{ when } \mathbf{A} \text{ and } \mathbf{B} \text{ are independent.}$$

Now we can write a more general rule that doesn't require independence. In fact, we've *already* written it down. We just need to rearrange the equation a bit.

The equation in the definition for conditional probability contains the probability of **A** *and* **B**. Rewriting the equation gives

$$P(\mathbf{A} \cap \mathbf{B}) = P(\mathbf{A}) \times P(\mathbf{B}|\mathbf{A}).$$

This is a **General Multiplication Rule** for compound events that does not require the events to be independent. Better than that, it even makes sense. The probability that two events, **A** and **B**, *both* occur is the probability that event **A** occurs multiplied by the probability that event **B** *also* then occurs—that is, by the probability that event **B** occurs *given* that event **A** has occurred.

Notice that this General Multiplication Rule works regardless of whether the events are independent. If they are, then $P(\mathbf{B}|\mathbf{A}) = P(\mathbf{B})$, so $P(\mathbf{A} \cap \mathbf{B}) = P(\mathbf{A}) \times P(\mathbf{B}|\mathbf{A}) = P(\mathbf{A}) \times P(\mathbf{B})$ for independent events. We hope that looks familiar.

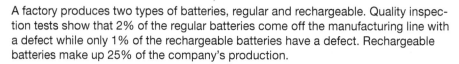

### FOR EXAMPLE Using the General Multiplication Rule

A factory produces two types of batteries, regular and rechargeable. Quality inspection tests show that 2% of the regular batteries come off the manufacturing line with a defect while only 1% of the rechargeable batteries have a defect. Rechargeable batteries make up 25% of the company's production.

QUESTION: What's the probability that if we choose one of the company's batteries at random we get

   a) a defective rechargeable battery?
   b) a regular battery and it's not defective?

ANSWER: Let **R** = rechargeable and **B** = a regular battery. It's given that $P(\mathbf{R}) = 0.25$, so $P(\mathbf{B}) = 0.75$.

Let **D** = defective. It's given that $P(\mathbf{D}|\mathbf{B}) = 0.02$ and $P(\mathbf{D}|\mathbf{R}) = 0.01$.

Now: $P(\mathbf{R} \cap \mathbf{D}) = P(\mathbf{R}) \times P(\mathbf{D}|\mathbf{R}) = (0.25)(0.01) = 0.0025$

If 2% of the regular batteries are defective, then the other 98% aren't; in other words: $P(\mathbf{D}^C|\mathbf{B}) = 0.98$. So:

$$P(\mathbf{B} \cap \mathbf{D}^C) = P(\mathbf{B}) \times P(\mathbf{D}^C|\mathbf{B}) = (0.75)(0.98) = 0.735$$

Only 1/4 of 1% of the company's batteries are rechargeable and defective, while 73.5% are nondefective regular batteries.

# Drawing Without Replacement

Room draw is a process for assigning rooms to students who live on a college campus. When it's time for you and your friend to draw, there are 12 rooms left. Three are in Gold Hall, a very desirable residence hall. You get to draw first, and then your friend will draw. Naturally, you would both like to score rooms in Gold. What are your chances? In particular, what's the chance that you *both* get rooms in Gold?

When you go first, the chance that *you* will draw one of the Gold rooms is 3/12. Suppose you do. Now, with you clutching your prized room assignment, what chance does your friend have? At this point there are only 11 rooms left and just 2 left in Gold, so your friend's chance is now 2/11.

Using our notation, we write
$$P(\text{friend draws Gold}|\text{you draw Gold}) = 2/11.$$

The reason the denominator changes is that we draw these rooms *without replacement*. That is, once one is drawn, it doesn't go back into the pool.

We often sample without replacement. When we draw from a very large population, the change in the denominator is too small to worry about. But when there's a small population to draw from, as in this case, we need to take note of the changing probabilities.

What are the chances that *both* of you will luck out? Well, now we've calculated the two probabilities we need for the General Multiplication Rule, so we can write:

$P(\text{you draw Gold} \cap \text{friend draws Gold})$
$$= P(\text{you draw Gold}) \times P(\text{friend draws Gold}|\text{you draw Gold})$$
$$= 3/12 \times 2/11 = 1/22 = 0.045$$

In this instance, it doesn't matter who went first, or even if the rooms were drawn simultaneously. Even if the room draw was accomplished by shuffling cards containing the names of the residence halls and then dealing them out to 12 applicants (rather than by each student drawing a room in turn), we can still *think* of the calculation as having taken place in two steps:

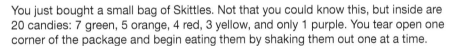

Picturing conditional probabilities this way leads to a more general way of helping us think with pictures—tree diagrams.

### FOR EXAMPLE
### Drawing Without Replacement

You just bought a small bag of Skittles. Not that you could know this, but inside are 20 candies: 7 green, 5 orange, 4 red, 3 yellow, and only 1 purple. You tear open one corner of the package and begin eating them by shaking them out one at a time.

QUESTION: What's the probability that your first 2 Skittles are both orange? That none of your first 3 candies is green?

ANSWER: Getting two orange candies in a row means I draw an orange one first **and** then another one second, with one orange candy already missing from the bag:

$P(2 \text{ orange}) = P(\text{orange first} \cap \text{orange second})$
$$= P(\text{orange first}) \times P(\text{orange second}|\text{orange first})$$
$$= \frac{5}{20} \times \frac{4}{19} = \frac{1}{19}$$

There's a 1 in 19 chance (just over 5%) that I'd shake out 2 orange Skittles in a row. Not getting any green ones means all of the first 3 Skittles were among the colors (13 candies):

$P(3 \text{ non-greens}) = P(\text{green}^c \cap \text{green}^c \cap \text{green}^c)$
$$= \frac{13}{20} \times \frac{12}{19} \times \frac{11}{18} \approx 0.25$$

There's about a 25% chance I won't get any green Skittles among the first 3 I shake out of the bag.

## JUST CHECKING

4. Think some more about that bag of Skittles described in the *For Example* on the previous page (7 green, 5 orange, 4 red, 3 yellow, 1 purple). Write out the fractions you'd multiply together to find the probabilities of these outcomes. (Don't bother multiplying them together—unless you're curious.)

   a) The first two are both red.
   b) You eat three without seeing a yellow one.
   c) The fourth candy out of the bag is the purple one.

# Tree Diagrams

For men, binge drinking is defined as having five or more drinks in a row, and for women as having four or more drinks in a row.[7] (The difference is because men generally weigh more than women.) According to a study by the Harvard School of Public Health, 44% of college students engage in binge drinking, 37% drink moderately, and 19% abstain entirely. Another study, published in the *American Journal of Health Behavior*, finds that among binge drinkers aged 21 to 34, 17% have been involved in an alcohol-related automobile accident, while among non-bingers of the same age, only 9% have been involved in such accidents.

What's the probability that a randomly selected college student will be a binge drinker and has had an alcohol-related car accident?

To start, we see that the probability of selecting a binge drinker is about 44%. To find the probability of selecting someone who is both a binge drinker and a driver with an alcohol-related accident, we would need to pull out the General Multiplication Rule and multiply the probability of one of the events by the conditional probability of the other given the first.

Or we *could* make a picture. Which would you prefer?

We thought so.

Neither tables nor Venn diagrams can handle conditional probabilities. The kind of picture that helps us look at conditional probabilities is called a **tree diagram** because it shows sequences of events as paths that look like branches of a tree. It is a good idea to make a tree diagram almost any time you plan to use the General Multiplication Rule. The number of different paths we can take can get large, so we usually draw the tree starting from the left and growing vine-like across the page, although sometimes you'll see them drawn from the bottom up or top down.

The first branch of our tree (Figure 14.3) separates students according to their drinking habits. We label each branch of the tree with a possible outcome and its corresponding probability.

Notice that we cover all possible outcomes with the branches. The probabilities add up to one. But we're also interested in car accidents. The probability of having an alcohol-related accident *depends* on one's drinking behavior. Because the probabilities are *conditional*, we draw the alternatives separately on each branch of the tree.

On each of the second set of branches, we write the possible outcomes associated with having an alcohol-related car accident (having an accident or not) and the associated probability (Figure 14.4). These probabilities are different because they are *conditional* depending on the student's drinking behavior. (It shouldn't be too surprising that those who binge drink have a higher probability of alcohol-related accidents.) The probabilities add up to one, because given the outcome on the first branch, these outcomes cover all the possibilities. Looking back at the General Multiplication Rule, we can see how the tree depicts the

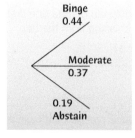

Figure 14.3
We can diagram the three outcomes of drinking and indicate their respective probabilities with a simple tree diagram.

---
[7]According to the U.S. Centers for Disease Control: https://www.cdc.gov/alcohol/fact-sheets/binge-drinking.htm

**Figure 14.4**
Extending the tree diagram, we can show both drinking and accident outcomes. The accident probabilities are conditional on the drinking outcomes, and they change depending on which branch we follow. Because we are concerned only with *alcohol-related* accidents, the conditional probability *P*(accident|abstain) must be 0.

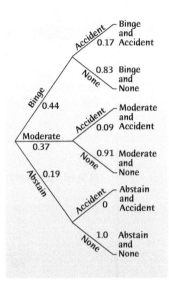

calculation. To find the probability that a randomly selected student will be a binge drinker who has had an alcohol-related car accident, we follow the top branches. The probability of selecting a binger is 0.44. The conditional probability of an accident *given* binge drinking is 0.17. The General Multiplication Rule tells us that to find the *joint* probability of being a binge drinker and having an accident, we multiply these two probabilities together:

$$P(\text{binge and accident}) = P(\text{binge}) \times P(\text{accident}|\text{binge})$$
$$= 0.44 \times 0.17 = 0.075.$$

And we can do the same for each combination of outcomes:

**Figure 14.5**
We can find the probabilities of compound events by multiplying the probabilities along the branch of the tree that leads to the event, just the way the General Multiplication Rule specifies. The probability of abstaining and having an alcohol-related accident is, of course, zero.

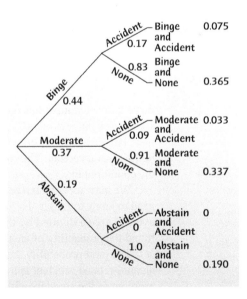

All the outcomes at the far right are disjoint because at each branch of the tree we chose between disjoint alternatives. And they are *all* the possibilities, so the probabilities on the far right must add up to one.

Because the final outcomes are disjoint, we can add up their probabilities to get probabilities for compound events. For example, what's the probability that a selected student has had an alcohol-related car accident? We simply find *all* the outcomes on the far right in which an accident has happened. There are three and we can add their probabilities: $0.075 + 0.033 + 0 = 0.108$—almost an 11% chance.

## FOR EXAMPLE
### Tree Diagrams

**RECAP:** Let's revisit the battery factory. Remember, it produced both regular and rechargeable batteries, with 25% of them rechargeable. History indicates that 2% of the regular batteries and 1% of the rechargeable batteries have some kind of defect.

**QUESTION:** What's the probability that a battery chosen at random from a shipment of this factory's products turns out to be defective?

**ANSWER:** First, I'll create a tree diagram. Batteries are either regular (**B**) or rechargeable (**R**), and each type may be defective (**D**) or **OK**.

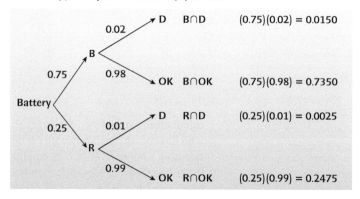

Check to see if the probabilities of all the possible outcomes add up to 1:
$$0.0150 + 0.7350 + 0.0025 + 0.2475 = 1.0000 \text{ (Hooray!)}$$
And now, the probability that a randomly chosen battery is defective is:
$$P(\mathbf{D}) = P(\mathbf{B} \cap \mathbf{D}) + P(\mathbf{R} \cap \mathbf{D}) = 0.0150 + 0.0025 = 0.0175$$
Overall, 1.75% of the batteries produced at this factory are defective.

# Reversing the Conditioning

If we know a student has had an alcohol-related accident, what's the probability that the student is a binge drinker? That's an interesting question, but we can't just read it from the tree. The tree gives us $P(\text{accident}|\text{binge})$, but we want $P(\text{binge}|\text{accident})$—conditioning in the other direction. The two probabilities are definitely *not* the same. We have reversed the conditioning.

We may not have the conditional probability we want, but we do know everything we need to know to find it. To find a conditional probability, we need the probability that both events happen divided by the probability that the given event occurs. We have already found the probability of an alcohol-related accident: $0.075 + 0.033 + 0 = 0.108$.

The joint probability that a student is both a binge drinker and someone who's had an alcohol-related accident is found at the top branch: 0.075. We've restricted the *Who* of the problem to the students with alcohol-related accidents, so we divide the two to find the conditional probability:

$$P(\text{binge}|\text{accident}) = \frac{P(\text{binge and accident})}{P(\text{accident})}$$
$$= \frac{0.075}{0.108} = 0.694.$$

The chance that a student who has an alcohol-related car accident is a binge drinker is more than 69%! As we said, reversing the conditioning is rarely intuitive, but tree diagrams help us keep track of the calculation when there aren't too many alternatives to consider.

# STEP-BY-STEP EXAMPLE

## Reversing the Conditioning

Back in the day when the authors were in college, there were only three requirements for graduation that were the same for all students: You had to be able to tread water for 2 minutes, you had to learn a foreign language, and you had to be free of tuberculosis. For the last requirement, all first-year students had to take a TB screening test that consisted of a nurse jabbing what looked like a corncob holder into your forearm. You were then expected to report back in 48 hours to have it checked. If you were healthy and TB-free, your arm was supposed to look as though you'd never had the test.

Sometime during the 48 hours, one of us had a reaction. When he finally saw the nurse, his arm was about 50% bigger than normal and a very unhealthy red. Did he have TB? The nurse had said that the test was about 99% effective, so it seemed that the chances must be pretty high that he had TB. How high do you think the chances were? Go ahead and guess. Guess low.

We'll call **TB** the event of actually having TB and + the event of testing positive. To start a tree, we need to know P(**TB**), the probability of having TB.[8] We also need to know the conditional probabilities $P(+|\textbf{TB})$ and $P(+|\textbf{TB}^C)$. Diagnostic tests can make two kinds of errors. They can give a positive result for a healthy person (a *false positive*) or a negative result for a sick person (a *false negative*). Being 99% accurate usually means a false-positive rate of 1%. That is, someone who doesn't have the disease has a 1% chance of testing positive anyway. We can write $P(+|\textbf{TB}^C) = 0.01$.

Because a false negative is more serious (because a sick person might not get treatment), tests are usually constructed to have a lower false-negative rate. We don't know exactly, but let's assume a 0.1% false-negative rate. So only 0.1% of sick people test negative. We can write $P(-|\textbf{TB}) = 0.001$.

---

**THINK**

**PLAN** Define the events we're interested in and their probabilities.

Let **TB** = {having TB} and **TB**$^C$ = {no TB}
+ = {testing positive} and
− = {testing negative}

Figure out what you want to know in terms of the events. Use the notation of conditional probability to write the event whose probability you want to find.

I know that $P(+|\textbf{TB}^C) = 0.01$ and $P(-|\textbf{TB}) = 0.001$. I also know that $P(\textbf{TB}) = 0.00005$.

I'm interested in the probability that the author had TB given that he tested positive: $P(\textbf{TB}|+)$.

---

**SHOW**

**PLOT** Draw the tree diagram. When probabilities are very small like these are, be careful to keep all the significant digits.

To finish the tree we need $P(\textbf{TB}^C)$, $P(-|\textbf{TB}^C)$, and $P(+|\textbf{TB})$. We can find each of these from the Complement Rule:

$$P(\textbf{TB}^C) = 1 - P(\textbf{TB}) = 0.99995$$
$$P(-|\textbf{TB}^C) = 1 - P(+|\textbf{TB}^C)$$
$$= 1 - 0.01 = 0.99 \text{ and}$$
$$P(+|\textbf{TB}) = 1 - P(-|\textbf{TB})$$
$$= 1 - 0.001 = 0.999$$

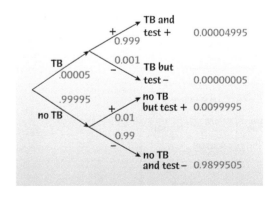

(Check: 0.00004995 + 0.00000005 + 0.0099995 + 0.98995050 = 1)

**MECHANICS** Multiply along the branches to find the probabilities of the four possible outcomes. Check your work by seeing if they total 1.

---

[8] This isn't given, so we looked it up. Although TB is a matter of serious concern to public health officials, it is a fairly uncommon disease, with an incidence of about 5 cases per 100,000 in the United States (see www.cdc.gov/tb/default.htm).

Add up the probabilities corresponding to the condition of interest—in this case, testing positive. We can add because the tree shows disjoint events.

$$P(+) = P(TB \cap +) + P(TB^C \cap +)$$
$$P = 0.00004995 + 0.0099995$$
$$= 0.01004945$$

Divide the probability of both events occurring (here, having TB and a positive test) by the probability of satisfying the condition (testing positive).

$$P(TB|+) = \frac{P(TB \cap +)}{P(+)}$$
$$= \frac{0.00004995}{0.01004945}$$
$$= 0.00497$$

**TELL**    **CONCLUSION** Interpret your result in context.

The chance of actually having TB after you test positive is less than 0.5%.

When we reverse the order of conditioning, we change the *Who* we are concerned with. With events of low probability, the result can be surprising. That's the reason patients who test positive for HIV, for example, are always told to seek medical counseling. They may have only a small chance of actually being infected. That's why global drug or disease testing can have unexpected consequences if people interpret *testing* positive as *being* positive.

## *Bayesian Reasoning

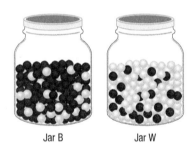

Jar B      Jar W

Please do this "thought experiment" with us. Your teacher shows you two clear glass jars full of marbles. One of the jars (we'll call it Jar B) contains 75 black marbles and 25 white ones. The other (Jar W) contains 75 white marbles and 25 black ones. Your teacher now selects one of the two jars without telling you which one and hides its contents with a cloth. What's the probability that she's selected Jar B?

Without any reason to believe that she would have been more likely to select either particular jar, you'd probably think there's a 50% chance that she's selected Jar B. (We'd agree.)

Let's continue. Still without indicating which jar she has selected, she selects a single marble from it at random and hands it to you. The marble is black. *Now* what is the probability that the jar she has selected is Jar B?

Without doing any calculations at all, you probably intuitively think that the probability must now be greater than 50%. After all, you'd be more likely to draw a black marble from Jar B than you would from Jar W. Your intuition is not wrong. You can use the technique we showed you in the last section to "reverse the conditioning":

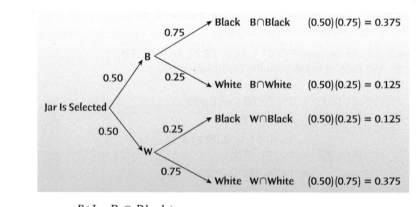

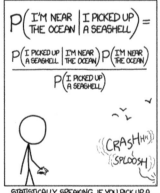

STATISTICALLY SPEAKING, IF YOU PICK UP A SEASHELL AND *DON'T* HOLD IT TO YOUR EAR, YOU CAN PROBABLY HEAR THE OCEAN.

© 2013 Randall Munroe. Reprinted with permission. All rights reserved.

$$P(\text{Jar B}|\text{Black}) = \frac{P(\text{Jar B} \cap \text{Black})}{P(\text{Black})}$$
$$= \frac{P(\text{Jar B} \cap \text{Black})}{P(\text{Jar B} \cap \text{Black}) + P(\text{Jar W} \cap \text{Black})} = \frac{0.375}{0.375 + 0.175} = 0.75$$

So initially there was a 50% chance that Jar B was selected. But after seeing a black marble drawn at random from the jar, there's now a 75% chance that Jar B was selected.

What if your teacher draws a second marble from the jar at random and hands it to you, and it's also black? We won't do any more computations, but you can see that the same principle applies as before: Because a black marble is more likely to be drawn from Jar B than from Jar W, then the chance she selected Jar B must increase. (It actually increases to about 90%.)

Okay, the thought experiment is over. Now we'll confess that we deliberately misled you (sort of) so that we could immerse you in what is called "Bayesian reasoning." Do you remember those "rebel statisticians" we mentioned in the last chapter?[9] So-called "Bayesians" are statisticians who do data analysis using the kind of reasoning that was in this thought experiment. You start with a probability of some hypothesis, such as $P(\text{Jar B}) = 0.5$. Then you collect data; in our thought experiment this was drawing a marble from the jar. Finally, you compute the (updated) probability of the hypothesis *conditional upon the data you observed*. If more data are collected, you keep updating the probability of the hypothesis, always conditional upon what the new data show.

This type of reasoning is called "Bayesian" because a British minister and amateur statistician named Thomas Bayes (1702–1761) described it in a paper published posthumously in 1763. Although some people (maybe including you) find Bayesian reasoning a very natural way to think about things, it is not widely used among statisticians. Why not? Because Bayesian reasoning necessarily uses *subjective probabilities*[10]—not the long-run relative frequencies that we use almost exclusively in this text. Once your teacher had selected a jar, it did not change after that. So when the probability of it being Jar B increased from 50% to 75%, that wasn't because anything about the *jar* had changed. It was because what you *believed* about the jar had changed.

The upshot is that you should understand how to use a probability tree to "reverse the conditioning" when a situation demands it. But although Bayesian reasoning has an increasingly important place in 21st-century Statistics, in this text we'll be showing you more traditional ways of analyzing data.[11]

### FOR EXAMPLE
#### Reversing the Conditioning

RECAP: Remember the battery factory that produced 25% rechargeable batteries and the rest the regular kind? A small number of its batteries had defects: 2% of the regular batteries and only 1% of the rechargeable ones.

QUESTION: A quality control inspector inspects some batteries selected at random from each shipment. If one of those batteries turns out to be defective, what's the probability it's the rechargeable type?

ANSWER: In the tree diagram, **B** = regular and **R** = rechargeable; **D** = defective.

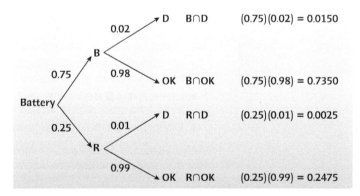

---

[9]Yes, we know *you* do, because you're reading the footnotes. That's where they were mentioned. ☺

[10]We discussed this type of probability in an earlier section about Angelina Jolie and the probability that she would develop breast cancer during her lifetime.

[11]To learn more about how Bayesians think, just sign up for more Statistics courses!

$$P(\mathbf{R}|\mathbf{D}) = \frac{P(\mathbf{R} \cap \mathbf{D})}{P(\mathbf{D})}$$
$$= \frac{0.0025}{0.0150 + 0.0025}$$
$$= 0.143$$

Only 14.3% of all defective batteries are the rechargeable kind.

## WHAT CAN GO WRONG?

- **Don't use a simple probability rule where a general rule is appropriate.** Don't assume independence without reason to believe it. Don't assume that outcomes are disjoint without checking that they are. Remember that the general rules always apply, even when outcomes are in fact independent or disjoint.
- **Don't find probabilities for samples drawn without replacement as if they had been drawn with replacement.** Remember to adjust the denominator of your probabilities. This warning applies only when we draw from small populations or draw a large fraction of a finite population. When the population is very large relative to the sample size, the adjustments make very little difference, and we ignore them.
- **Don't reverse conditioning naively.** As we have seen, the probability of **A** *given* **B** may not, and in general does not, resemble the probability of **B** *given* **A**. The true probability may be counterintuitive.
- **Don't confuse "disjoint" with "independent."** Disjoint events *cannot* happen at the same time. When one happens, you know the other did not, so $P(\mathbf{B}|\mathbf{A}) = 0$. Independent events *must* be able to happen at the same time. When one happens, you know it has no effect on the other, so $P(\mathbf{B}|\mathbf{A}) = P(\mathbf{B})$.

## WHAT HAVE WE LEARNED?

We've learned the general rules of probability and how to apply them:
- The General Addition Rule says that $P(\mathbf{A} \cup \mathbf{B}) = P(\mathbf{A}) + P(\mathbf{B}) - P(\mathbf{A} \cap \mathbf{B})$.
- The General Multiplication Rule says that $P(\mathbf{A} \cap \mathbf{B}) = P(\mathbf{A}) \times P(\mathbf{B}|\mathbf{A})$.

We've learned to work with conditional probabilities and have seen that $P(\mathbf{B}|\mathbf{A}) = \frac{P(\mathbf{A} \cap \mathbf{B})}{P(\mathbf{A})}$.

We've learned to think clearly about independence:
- We can use conditional probability to determine whether events are independent and to work with events that are not independent.
- Events **A** and **B** are independent if the occurrence of one does not effect the probability that the other occurs: $P(\mathbf{B}|\mathbf{A}) = P(\mathbf{B})$.
- Events that are mutually exclusive (disjoint) cannot be independent.

We've learned to organize our thinking about probability with Venn diagrams, tables, and tree diagrams, and to use tree diagrams to solve problems about reverse conditioning.

## TERMS

**General Addition Rule**    For any two events, **A** and **B**, the probability of **A** or **B** is
$$P(\mathbf{A} \cup \mathbf{B}) = P(\mathbf{A}) + P(\mathbf{B}) - \mathbf{P}(\mathbf{A} \cap \mathbf{B}). \text{ (p. 375)}$$

| | Conditional probability | $P(B|A) = \dfrac{P(A \cap B)}{P(A)}$; $P(B|A)$ is read "the probability of **B** given **A**." (p. 379) |
|---|---|---|
| | Independence (used formally) | Events **A** and **B** are independent when $P(B|A) = P(B)$. (p. 381) |
| | General Multiplication Rule | For any two events, **A** and **B**, the probability of **A** and **B** is $$P(A \cap B) = P(A) \times P(B|A).$$ (p. 386) |
| | Tree diagram | A display of conditional events or probabilities that is helpful in thinking through conditioning. (p. 388) |

# EXERCISES

1. **Pet ownership** Suppose that 25% of people have a dog, 29% of people have a cat, and 12% of people own both. What is the probability that a randomly selected person owns a dog or a cat?

2. **Cooking and shopping** Forty-five percent of Americans like to cook and 59% of Americans like to shop, while 23% enjoy both activities. What is the probability that a randomly selected American either enjoys cooking or shopping or both?

3. **Sports fans** In a group of people, preferences for watching football and basketball are shown in the table below. What is the probability that a person selected at random from this group likes to watch football, given that they also like to watch basketball?

|  | Football | No Football |
|---|---|---|
| Basketball | 27 | 13 |
| No Basketball | 38 | 22 |

4. **Sports fans again** From Exercise 3, if a randomly selected person from the group doesn't like to watch basketball, what is the probability that they will be a football fan?

5. **Late to the train** A student always catches their train if class ends on time. However, 30% of classes run late and then there's a 45% chance they'll miss it. What is the probability that they miss the train today?

6. **Field goals** A nervous kicker usually makes 70% of his first field goal attempts. If he makes his first attempt, his success rate rises to 90%. What is the probability that he will make his first two kicks?

7. **Sports fans III** In Exercise 3, 40% of the people in the group like to watch basketball. About 41.5% of the football fans like to watch basketball. Are the events *Likes to watch football* and *Likes to watch basketball* independent?

8. **Births** If the sex of a child is independent of all other births, is the probability of someone giving birth to a girl after having four boys greater than it was on the first birth? Explain.

9. **Snapchat** As of March 2017, roughly 20% of Americans were users of the social media tool Snapchat. The U.S. population is approximately 28% 13- to 34-year-olds. Because Snapchat is popular with younger Americans, 17% of the country are 13- to 34-year-olds who are on Snapchat. Construct a probability table for these data.

10. **Online banking** A national survey indicated that 30% of adults conduct their banking online. It also found that 40% are younger than 50, and that 25% are younger than 50 and conduct their banking online. Make a probability table.

11. **Phones** Recent research suggests that 46% of Americans have a landline, 95% have a cell phone, and 42% of people have both. What is the probability that an American has

    a) a landline or cell phone? (If $L = $ *has a landline* and $C = $ *has a cell pone*, we can write this in symbols as $P(L$ *or* $C)$ or $P(L \cup C)$)
    b) neither a landline nor a cell phone? ($P($ **not** $L$ *and* **not** $C)$ or $P(L^C \cap C^C)$)
    c) a cell phone but no landline? ($P(C$ *and* **not** $L)$)

12. **Devices** Roughly 73% of Americans have a laptop/desktop computer, while only 45% have some sort of tablet. 36% of Americans are lucky enough to have both types of devices in their homes. What is the probability that a randomly selected home has

    a) a tablet, but no laptop/desktop? (If $T = $ *has a tablet* and $L = $ *has a laptop/desktop computer*, we can write this in symbols as $P(T$ *and* **not** $L)$ or $P(T \cap L^C)$)
    b) has either a tablet or a laptop/desktop? $P(T$ *or* $L)$ or $P(T \cup L)$)
    c) has neither computing device? ($P(T^C$ *and* $L^C)$)

13. **Amenities** A check of residential rooms on a large college campus revealed that 38% had refrigerators, 52% had TVs, and 21% had both a TV and a refrigerator. For each question, write the probability symbolically and give the numerical answer:

    What's the probability that a randomly selected residential room has

    a) a TV but no refrigerator?
    b) a TV or a refrigerator, but not both?
    c) neither a TV nor a refrigerator?

14. **Workers** Employment data at a large company reveal that 72% of the workers are married, that 44% are college graduates, and that half of the college grads are married. For each question, write the probability symbolically and give the numerical answer:

    What's the probability that a randomly chosen worker

    a) is neither married nor a college graduate?
    b) is married but not a college graduate?
    c) is married or a college graduate?

**15. Global survey** The marketing research organization GfK Custom Research North America conducts a yearly survey on consumer attitudes worldwide. They collect demographic information on the roughly 1500 respondents from each country that they survey. Here is a table showing the number of people with various levels of education in five countries:

| | \multicolumn{5}{c}{Educational Level by Country} | |
|---|---|---|---|---|---|---|
| | Post-graduate | College | Some high school | Primary or less | No answer | Total |
| China | 7 | 315 | 671 | 506 | 3 | 1502 |
| France | 69 | 388 | 766 | 309 | 7 | 1539 |
| India | 161 | 514 | 622 | 227 | 11 | 1535 |
| U.K. | 58 | 207 | 1240 | 32 | 20 | 1557 |
| U.S. | 84 | 486 | 896 | 87 | 4 | 1557 |
| Total | 379 | 1910 | 4195 | 1161 | 45 | 7690 |

If we select someone at random from this survey,

a) what is the probability that the person is from the United States?
b) what is the probability that the person completed their education before college?
c) find $P(\text{France } or \text{ Post-grad})$
d) what is $P(\text{France} \cap \text{Primary or less})$

**16. Birth order** A survey of students in a large Introductory Statistics class asked about their birth order (1 = oldest or only child) and which college of the university they were enrolled in. Here are the results:

| | | Birth Order | | |
|---|---|---|---|---|
| | | 1 or Only | 2 or More | Total |
| College | Arts & Sciences | 34 | 23 | 57 |
| | Agriculture | 52 | 41 | 93 |
| | Human Ecology | 15 | 28 | 43 |
| | Other | 12 | 18 | 30 |
| | Total | 113 | 110 | 223 |

Suppose we select a student at random from this class. What is the probability that the person is

a) a Human Ecology student?
b) a firstborn student?
c) firstborn *and* a Human Ecology student?
d) firstborn *or* a Human Ecology student?

**17. Cards** You draw a card at random from a standard deck of 52 cards. Find each of the following conditional probabilities:

a) The card is a heart, given that it is red.
b) The card is red, given that it is a heart.
c) $P(\text{Ace}|\text{Red})$.
d) $P(\text{Queen}|\text{Face Card})$.

**18. Pets** In its monthly report, the local animal shelter states that it currently has 24 dogs and 18 cats available for adoption. Eight of the dogs and 6 of the cats are male. Find each of the following conditional probabilities if an animal is selected at random:

a) The pet is male, given that it is a cat.
b) The pet is a cat, given that it is female.
c) $P(\text{Female}|\text{Dog})$.
d) $P(\text{Dog}|\text{Female})$

**19. Men's health** The probabilities that an adult American man has high blood pressure and/or high cholesterol are shown in the table:

| | | Blood Pressure | |
|---|---|---|---|
| | | High | OK |
| Cholesterol | High | 0.11 | 0.21 |
| | OK | 0.16 | 0.52 |

What's the probability that

a) a man has both conditions?
b) a man has high blood pressure?
c) a man with high blood pressure has high cholesterol?
d) a man has high blood pressure if it's known that he has high cholesterol?

**20. Roller coaster varieties** The table shows, for the 674 coasters in our dataset, the type (wood or steel) and scale (Extreme, Family, Kiddie, or Thrill).

| | | Scale | | | |
|---|---|---|---|---|---|
| | | Extreme | Family | Kiddie | Thrill |
| Type | Steel | 212 | 171 | 63 | 125 |
| | Wood | 55 | 18 | 0 | 30 |

a) What's the probability that
 i) a randomly chosen coaster is a family coaster?
 ii) a randomly chosen wood coaster is a thrill coaster?
 iii) a randomly chosen Family coaster is made of steel?
b) A roller coaster aficionado prefers Thrill coasters or Wood coasters. What proportion of these coasters does this include?

**21. Global survey, take 2** Look again at the table in Exercise 15 summarizing some data about education levels of people in five countries.

a) If we select a respondent at random, what's the probability we choose a person from the United States who has done post-graduate study? ($P(\text{U.S. } and \text{ Post-grad})$)
b) Among the respondents who have done post-graduate study, what's the probability the person is from the United States? ($P(\text{U.S.}|\text{Post-grad})$)
c) What's the probability that a respondent from the United States has done post-graduate study? ($P(\text{Post-grad}|\text{U.S.})$)
d) What's the probability that a respondent from China has only a primary-level education? How would you represent this symbolically?
e) What's the probability that a respondent with only a primary-level education is from China? How would you represent this symbolically?

**22. Birth order, take 2** Look again at the data about birth order of Intro Statistics students and their choices of colleges shown in Exercise 16.

a) If we select a student at random, what's the probability the person is an Arts and Sciences student who is a second child (or more)? ($P(\textbf{A\&S} \text{ and } \textbf{2nd})$)
b) Among the Arts and Sciences students, what's the probability a student was a second child (or more)? ($P(\textbf{2nd}|\textbf{A\&S})$)
c) Among second children (or more), what's the probability the student is enrolled in Arts and Sciences? ($P(\textbf{A\&S}|\textbf{2nd})$)
d) What's the probability that a randomly selected first or only child is enrolled in the Agriculture College? How would you represent this symbolically?
e) What is the probability that a randomly selected Agriculture student is a first or only child? How would you represent this symbolically?

**23. Sick kids** Seventy percent of kids who visit a doctor have a fever, and doctors observe that if a kid has a fever, there is a 30% chance they have a sore throat too. What's the probability that a kid who goes to the doctor has a fever and a sore throat?

**24. Sick cars** Twenty percent of cars that are inspected have faulty pollution control systems. The cost of repairing a pollution control system exceeds $100 about 40% of the time. When a driver takes her car in for inspection, what's the probability that she will end up paying more than $100 to repair the pollution control system?

**25. Cards again** You are dealt a hand of three cards from a standard 52-card deck, one at a time. Find the probability of each of the following:
a) You need to draw a heart. You first do so on the third try.
b) Your cards are all red (that is, all diamonds or hearts).
c) You get no spades.
d) You have at least one ace.

**26. Another hand** You pick a set of three cards at random from a standard 52-card deck. Find the probability of each event described below.
a) You get no aces.
b) You get all hearts.
c) You want to draw a red. You do so on the third try.
d) You have at least one diamond.

**27. Batteries** A junk box in your room contains a dozen old batteries, five of which are totally dead. You start picking batteries one at a time and testing them. Find the probability of each outcome.
a) The first two you choose are both good.
b) At least one of the first three works.
c) The first four you pick all work.
d) You have to pick five batteries to find one that works.

**28. Shirts** The soccer team's shirts have arrived in a big box, and people just start grabbing them, looking for the right size. The box contains 4 medium, 10 large, and 6 extra-large shirts. You want a medium for you and one for your sister. Find the probability of each event described.
a) The first two you grab are the wrong sizes.
b) You need a medium. Your first two tries are the wrong size, but on the third try you get your size.
c) The first four shirts you pick are all extra-large.
d) At least one of the first four shirts you check is a medium.

**29. Eligibility** A university requires its biology majors to take a course called BioResearch. The prerequisite for this course is that students must have taken either a Statistics course or a computer course. By the time they are juniors, 52% of the Biology majors have taken Statistics, 23% have had a computer course, and 7% have done both.
a) What percent of the junior Biology majors are ineligible for BioResearch?
b) What's the probability that a junior Biology major who has taken Statistics has also taken a computer course?
c) Are taking these two courses disjoint events? Explain.
d) Are taking these two courses independent events? Explain.

**30. Benefits** Fifty-six percent of all American workers have a workplace retirement plan, 68% have health insurance, and 49% have both benefits. We select a worker at random.
a) What's the probability they have neither employer-sponsored health insurance nor a retirement plan?
b) What's the probability this person has health insurance if they have a retirement plan?
c) Are having health insurance and a retirement plan independent events? Explain.
d) Are having these two benefits mutually exclusive? Explain.

**31. More phones** A survey found that 46% of Americans have a landline, 95% have a cell phone, and 42% of people have both.
a) If a person has a landline, what's the probability that they have a cell phone also?
b) Are having a landline and a cell phone independent events? Explain.
c) Are having a landline and a cell phone mutually exclusive? Explain.

**32. More devices** According to Exercise 12, the probability that an American has a desktop/laptop is 73%, a tablet is 45%, and both is 36%.
a) What is the probability that if a randomly selected American has a tablet, they have a laptop/desktop also?
b) Are having a tablet and having a desktop/laptop disjoint events? Explain.
c) Are having a tablet and having a desktop/laptop independent events? Explain.

**33. Aces** If you draw a card at random from a well-shuffled standard deck, is getting an ace independent of the suit? Explain.

**34. Pets again** The local animal shelter in Exercise 18 reported that it currently has 24 dogs and 18 cats available for adoption; 8 of the dogs and 6 of the cats are male. Are the species and sex of the animals independent? Explain.

**35. Unsafe food** Early in 2010, *Consumer Reports* published the results of an extensive investigation of broiler chickens purchased from food stores in 23 states. Tests for bacteria in the meat showed that 62% of the chickens were contaminated with *campylobacter*, 14% with *salmonella*, and 9% with both.
a) What's the probability that a tested chicken was not contaminated with either kind of bacteria?
b) Are contamination with one kind of bacteria and contamination with the other kind of bacteria disjoint events?
c) Are contamination with one kind of bacteria and contamination with the other kind of bacteria independent events?

**36. Birth order, finis** In Exercises 16 and 22 we looked at the birth orders and college choices of some Intro Statistics students. For these students:

a) Are enrolling in Agriculture and enrolling in Human Ecology disjoint events? Explain.
b) Are enrolling in Agriculture and enrolling in Human Ecology independent events? Explain.
c) Are being firstborn and enrolling in Human Ecology disjoint events? Explain.
d) Are being firstborn and enrolling in Human Ecology independent events? Explain.

**37. Men's health, again** Given the table of probabilities from Exercise 19, are high blood pressure and high cholesterol independent? Explain.

| Cholesterol | Blood Pressure | |
|---|---|---|
| | High | OK |
| High | 0.11 | 0.21 |
| OK | 0.16 | 0.52 |

**38. Independent coasters** Given the table of probabilities from Exercise 20,

a) are the events Extreme and Steel independent? Explain.
b) Identify two events that are mutually exclusive.
c) Are the variables *Type* and *Scale* independent? Explain.

| Type | Scale | | | |
|---|---|---|---|---|
| | Extreme | Family | Kiddie | Thrill |
| Steel | 212 | 171 | 63 | 125 |
| Wood | 55 | 18 | 0 | 30 |

**39. Using social media at work** In June 2016 Pew Research conducted a survey to measure the effects of social media in the workplace environment. (To no one's surprise, it reported that most workers are on social media, even if their employer has policies prohibiting this behavior!) Thirty-seven percent of workers reported they used social media for work-related purposes, but 86% of those workers admitted that social media was distracting. Overall, 58% of workers said that social media was a distraction.

a) If a worker finds social media distracting, what are the chances that they use social media for work-related purposes?
b) Are these events independent?

**40. Snoring** After surveying 995 adults, 81.5% of whom were over 30, the National Sleep Foundation reported that 36.8% of all the adults snored; 32% of the respondents were snorers over the age of 30.

a) What percentage of the respondents were under 30 and did not snore?
b) Is snoring independent of age? Explain.

**41. District policy** A poll of teachers conducted by a school superintendent classified respondents by the level at which they teach, and asked whether they favor or oppose a planned change in policy. Is *Support* for the policy independent of *Level*?

| | Elementary | Middle | High |
|---|---|---|---|
| Favor | 61 | 31 | 22 |
| Oppose | 14 | 19 | 43 |

**42. Cars** A random survey of autos parked in student and staff lots at a large university classified the brands by continent of origin, as seen in the table. Is continent of origin independent of type of driver?

| Origin | Driver | |
|---|---|---|
| | Student | Staff |
| American | 119 | 91 |
| European | 39 | 21 |
| Asian | 46 | 44 |

**43. Making independence** In an imaginary Statistics class, there are 12 juniors and 15 seniors. Of the 27 students in the class, nine of them are interested in an after-school cooking club. The other 18 are not. Insert frequencies into the table if *Year* and *Interest in cooking club* are independent.

| Year | Interest in Cooking Club | | |
|---|---|---|---|
| | Yes | No | Total |
| Junior | | | 12 |
| Senior | | | 15 |
| Total | 9 | 18 | 27 |

**44. Independent sports fans** Here the totals for each row and column have been added to the table from Exercise 3. What numbers in the cells of the table would make *Football* preference and *Basketball* preference independent?

| | Football | No Football | Total |
|---|---|---|---|
| Basketball | | | 40 |
| No Basketball | | | 60 |
| Total | 65 | 35 | 100 |

**45. Fish and prostate cancer mosaic** This mosaic plot, first seen in Chapter 2, Exercise 35, shows the amount of fish consumed by men that did and men that did not develop prostate cancer. Are the variables *Prostate Cancer* and *Fish Consumption* independent? Use the plot to support your answer.

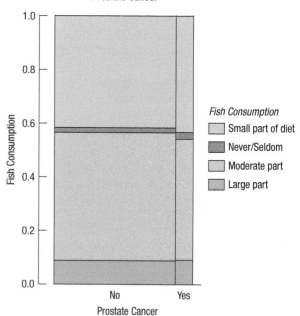

**46. College value mosaic** This mosaic plot, first seen in Chapter 2, Exercise 36, shows how adults and college presidents responded when asked to "rate the job the higher education system is doing in providing value for the money spent by students and their families." Is *Perceived Value* of higher education independent of whether a person is *President* of a college? Use the mosaic plot to explain your answer.

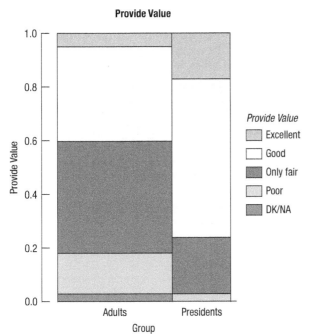

**47. District policy again** Refer to the table from Exercise 41.
   a) Create a mosaic plot to represent the frequencies in the table.
   b) Assuming the same totals in favor of and opposed to the policy, and the same total number of teachers at each level, complete the table that would make *Level* and *Support* for the policy independent.
   c) Create a mosaic plot for the table in part b.

**48. Cars again** Refer to the table from Exercise 42.
   a) Create a mosaic plot to represent the frequencies in the table.
   b) Assuming the same total number of cars from each continent and same total numbers of faculty cars and student cars, complete the table that would make *Origin* and *Driver* independent.
   c) Create a mosaic plot for the table in part b.

**49. Luggage** Elliott is flying from Boston to Denver with a connection in Chicago. The probability their first flight leaves on time is 0.15. If the flight is on time, the probability that their luggage will make the connecting flight in Chicago is 0.95, but if the first flight is delayed, the probability that the luggage will make it is only 0.65.
   a) Are the first flight leaving on time and the luggage making the connection independent events? Explain.
   b) What is the probability that their luggage arrives in Denver with them?

**50. Graduation** A private college report contains these statistics: *70% of incoming first-year students attended public schools. 75% of public school students who enroll as first-year students eventually graduate.* *90% of other first-year students eventually graduate.*
   a) Is there any evidence that a first-year student's chances of graduating may depend upon what kind of high school the student attended? Explain.
   b) What percentage of first-year students eventually graduate?

**51. Late luggage** Remember Elliott (Exercise 49)? Suppose you pick them up at the Denver airport, and their luggage is not there. What is the probability that Elliott's first flight was delayed?

**52. Graduation, part II** What percentage of students who graduate from the college in Exercise 50 attended a public high school?

**53. Absenteeism** A company's records indicate that on any given day about 1% of its day-shift employees and 2% of its night-shift employees will miss work. Sixty percent of the employees work the day shift.
   a) Is absenteeism independent of shift worked? Explain.
   b) What percentage of employees are absent on any given day?

**54. E-readers** Pew Internet reported in January 2014 that 32% of U.S. adults own at least one e-reader, and that 28% of U.S. adults read at least one e-book in the previous year (and thus, presumably, owned an e-reader). Overall, 76% of U.S. adults read at least one book (electronic or otherwise) in the previous year (www.pewinternet.org/2014/01/16/a-snapshot-of-reading-in-america-in-2013/)
   a) Explain how these statistics indicate that owning an e-reader and reading at least one book in the previous year are not independent.
   b) What's the probability that a randomly selected U.S. adult has an e-reader but didn't use it to read an e-book in the previous year?

**55. Absenteeism, part II** At the company described in Exercise 53, what percent of the absent employees are on the night shift?

**56. E-readers II** Given the e-reader data presented in Exercise 54:
   a) If a randomly selected U.S. adult has an e-reader, what is the probability that they haven't read an e-book in the past year?
   b) Is it more or less likely that a randomly selected U.S. adult who does not own an e-reader would have read no books in the past year?

**57. Driving under the influence** Police often set up sobriety checkpoints—roadblocks where drivers are asked a few brief questions to allow the officer to judge whether or not the person may have been drinking. If the officer does not suspect a problem, drivers are released to go on their way. Otherwise, drivers are detained for a Breathalyzer test that will determine whether or not they will be arrested. The police say that based on the brief initial stop, trained officers can make the right decision 80% of the time. Suppose the police operate a sobriety checkpoint after 9 p.m. on a Saturday night, a time when national traffic safety experts suspect that about 12% of drivers have been drinking.
   a) You are stopped at the checkpoint and, of course, have not been drinking. What's the probability that you are detained for further testing?
   b) What's the probability that any given driver will be detained?
   c) What's the probability that a driver who is detained has actually been drinking?
   d) What's the probability that a driver who was released had actually been drinking?

**58. No-shows** An airline offers discounted "advance-purchase" fares to customers who buy tickets more than 30 days before travel and charges "regular" fares for tickets purchased during those last 30 days. The company has noticed that 60% of its customers take advantage of the advance-purchase fares. The "no-show" rate among people who paid regular fares is 30%, but only 5% of customers with advance-purchase tickets are no-shows.

a) What percentage of all ticket holders are no-shows?
b) What's the probability that a customer who didn't show had an advance-purchase ticket?
c) Is being a no-show independent of the type of ticket a passenger holds? Explain.

**59. Dishwashers** Lucia's Diner employs three dishwashers. Shawn washes 40% of the dishes and breaks only 1% of those they handle. Hamsa and Xiong each wash 30% of the dishes, and Hamsa breaks only 1% of hers, but Xiong breaks 3% of the dishes he washes. (He, of course, will need a new job soon . . . .) You are planning to go to Lucia's for supper one night. If you hear a dish break at the sink, what's the probability that Xiong is on the job?

**60. Parts** A company manufacturing electronic components for home entertainment systems buys electrical connectors from three suppliers. The company prefers to use supplier A because only 1% of those connectors prove to be defective, but supplier A can deliver only 70% of the connectors needed. The company must also purchase connectors from two other suppliers, 20% from supplier B, and the rest from supplier C. The rates of defective connectors from B and C are 2% and 4%, respectively. You plan to buy one of these components. If you try to use it and you find that the connector is defective, what's the probability that your component is from supplier A?

**61. HIV testing** In July 2005 the journal *Annals of Internal Medicine* published a report on the reliability of HIV testing. Results of a large study suggested that among people with HIV, 99.7% of tests conducted were (correctly) positive, while for people without HIV 98.5% of the tests were (correctly) negative. A clinic serving an at-risk population offers free HIV testing, believing that 15% of the patients may actually carry HIV. What's the probability that a patient testing negative is truly free of HIV?

**62. Polygraphs** The use of polygraphs to detect lies is a controversial procedure, due to unreliable and often biased results. They are barred from use as evidence in many courts. Nonetheless, many employers use polygraph screening as part of their hiring process in the hope that they can avoid hiring people who might be dishonest. There has been some research, but no agreement, about just how unreliable polygraph tests are. Based on this research, suppose that a polygraph can detect 65% of lies, but incorrectly identifies 15% of true statements as lies.

A certain company believes that 95% of its job applicants are trustworthy. (Assume this is correct.) The company gives everyone a polygraph test, asking, "Have you ever stolen anything from your place of work?" Naturally, all the applicants answer "No," but the polygraph identifies some of those answers as lies, making the person ineligible for a job. What's the probability that a job applicant rejected under suspicion of dishonesty is actually trustworthy?

**63. COVID-19 antibody testing** The first FDA-approved COVID-19 antibody test, Cellex, has a 94% chance of being right (testing positive for antibodies) when used by people with antibodies, and 96% chance of being right (testing negative for antibodies) when used by people without antibodies. Suppose you take a Cellex test, and test positive for antibodies. What's the probability that you actually have antibodies for COVID-19, assuming that 2% of the entire population have antibodies? (https://cellextest.com)

**64. COVID-19 antibody testing, continued.** Consider the COVID-19 antibody test in the previous exercise. If 2% of the entire population have antibodies, and you test negative, what is the probability that you actually do have antibodies?

**65. Theoretical ferrets** The following experiment was described in Chapter 10, Exercise 41:

*During the COVID pandemic of 2020, a group of scientists developed a nasal spray that was intended to block the virus responsible for COVID from attaching to airways and lung cells, thus preventing infection even following exposure to the virus. In order to test the effectiveness of the spray, the following experiment was conducted using ferrets, which can catch viruses through the nose like humans. One ferret was deliberately infected with the COVID virus and placed into a cage. Four other ferrets not infected with the virus were placed into the same cage. Of these four, two had been randomly selected and given the experimental nasal spray while the other two had been given a placebo nasal spray. After 24 hours, the ferrets given the placebo spray had both contracted COVID; the two given the experimental spray had not.*

To determine whether such a result could be reasonably attributed to chance, that exercise asked for the probability in such an experiment that the two ferrets with the placebo will contract COVID, and the two with the nasal spray will not contract COVID if the nasal spray is ineffective.

a) One approach is to assume, because 50% of the ferrets in the real experiment contracted COVID, that each ferret has a 50% chance of catching the disease whether they have the nasal spray or not. Use the Multiplication Rule to determine, based on this assumption, that the two ferrets that receive the placebo will contract COVID and the two receiving the nasal spray will not.

b) Another (probably better) approach is to assume that exactly two ferrets will contract COVID as in the original experiment, but whether or not they are the ones given the experimental nasal spray is left to chance. Use the Multiplication Rule to determine the probability that the two ferrets that receive the placebo will be the two to contract COVID under this assumption.

**66. More theoretical ferrets** In the real experiment described in Exercise 65, the researchers actually had three cages set up the same way, and had the same result in all three cages: both ferrets with the nasal spray avoided getting COVID, and both ferrets with the placebo became infected.

a) Use the Multiplication Rule to determine the probability of getting such a result if we were to repeat the experiment. Do so using the assumption in part a) of Exercise 65, that each ferret has a 50% chance of contracting COVID, whether they receive the nasal spray or not.

b) Now, calculate the probability of getting such a result using the assumption from part b. That is, assume that exactly two ferrets in each cage will contract covid.
c) What do the answers to parts a and b tell you about the importance of the assumptions made in your calculation?
d) What insight do the answers to parts a and b provide about the effectiveness of the vaccine?

**67. Westworld** In Season 1, Episode 3 of the HBO science-fiction series *Westworld*, a robot gunslinger in a futuristic theme park "killed" six of nine other robot gunslingers in his company. The managers of the theme park realized that the six robots he "killed," had "killed" him in previous incarnations. The three robots he spared had not previously harmed him. This led the managers to believe this robot had memories of previous incarnations, which was not supposed to happen!

Determine whether this result could reasonably be attributed to chance. In other words, find the probability that, if the gunslinger were to shoot six robots randomly selected from nine, he would happen to shoot the six that had previously harmed him.

## JUST CHECKING

Answers

1. a)

b) 0.32.   c) 0.41.

2. a) No; 8 people are both male and supervisors.
   b) Yes. $P(\mathbf{F}|\mathbf{S}) = \dfrac{12}{20} = 0.6$ and $P(\mathbf{F}) = \dfrac{90}{150} = 0.6$, so $P(\mathbf{F}|\mathbf{S}) = P(\mathbf{F})$.

3. a)

|  | Equation | | |
|---|---|---|---|
|  | Yes | No | Total |
| Display Yes | 0.07 | 0.41 | 0.48 |
| Display No | 0.20 | 0.32 | 0.52 |
| Total | 0.27 | 0.73 | 1.00 |

b) $P(\mathbf{D}|\mathbf{Eq})P(\mathbf{D} \cap \mathbf{Eq})/P(\mathbf{Eq}) = 0.07/0.27$
$= 0.259$.

c) No, pages can (and 7% do) have both.

d) To be independent, we'd need $P(\mathbf{D}|\mathbf{Eq}) = P(\mathbf{D})$. $P(\mathbf{D}|\mathbf{Eq}) = 0.259$, but $P(\mathbf{D}) = 0.48$. Overall, 48% of pages have data displays, but only about 26% of pages with equations do. They do not appear to be independent.

4. a) $\dfrac{4}{20} \times \dfrac{3}{19}$.   b) $\dfrac{17}{20} \times \dfrac{16}{19} \times \dfrac{15}{18}$.   c) $\dfrac{19}{20} \times \dfrac{18}{19} \times \dfrac{17}{18} \times \dfrac{1}{17}$.

# 15

# Random Variables

Insurance companies make bets. They bet that you're going to live a long life. You bet that you're going to die sooner. Both you and the insurance company want the company to stay in business, so it's important to find a "fair price" for your bet. Of course, the right price for *you* depends on many factors, and nobody can predict exactly how long you'll live. But when the company looks at averages over enough customers, it can make reasonably accurate estimates of the amount it can expect to collect on a policy before it has to pay its benefit.

Here's a simple example. An insurance company offers a "death and disability" policy that pays $10,000 if you die or $5000 if you are permanently disabled. It charges a premium of only $50 a year for this benefit. Is the company likely to make a profit selling such a plan? To answer this question, the company needs to know the *probability* that its clients will die or be disabled in any year. From actuarial information like this, the company can calculate the expected value of this policy.

> ### WHAT IS AN ACTUARY?
> Actuaries are the daring people who put a price on risk, estimating the likelihood and costs of rare events, so they can be insured. That takes financial, statistical, and business skills. It also makes them invaluable to many businesses. Actuaries are rather rare themselves; only about 19,000 work in North America. Perhaps because of this, they are well paid. If you're enjoying this course, you may want to look into a career as an actuary. Contact the Society of Actuaries or the Casualty Actuarial Society (which, despite what you may think, did not pay for this blurb).

# Expected Value: Center

**NOTATION ALERT**
The most common letters for random variables are X, Y, and Z. But be cautious: If you see *any* capital letter, it just might denote a random variable.

We'll want to build a probability model in order to answer the questions about the insurance company's risk. First we need to define a few terms. The amount the company pays out on an individual policy is called a **random variable** because its numeric value is based on the outcome of a random event. We use a capital letter, like X, to denote a random variable. We'll denote a particular value that it can have by the corresponding lowercase letter, in this case x. For the insurance company, x can be $10,000 (if you die that year), $5000 (if you are disabled), or $0 (if neither occurs). Because we can list all the outcomes, we might formally call this random variable a **discrete random variable**. Otherwise, we'd call it a **continuous random variable**. The collection of all the possible values and the probabilities that they occur is called the **probability model** for the random variable.

Suppose, for example, that the death rate in any year is 1 out of every 1000 people, and that another 2 out of 1000 suffer some kind of disability.[1] Then we can display the probability model for this insurance policy in a table like this:

**Table 15.1**
The probability model shows all the possible values of the random variable and their associated probabilities.

| Policyholder Outcome | Payout x | Probability P(x) |
|---|---|---|
| Death | 10,000 | $\frac{1}{1000}$ |
| Disability | 5000 | $\frac{2}{1000}$ |
| Neither | 0 | $\frac{997}{1000}$ |

Given these probabilities, what should the insurance company expect to pay out? It can't know exactly what will happen to any particular insured person, but it can calculate what to expect on average over a large group of insured people. In a probability model, we call that the **expected value** and denote it either of two ways: $E(X)$ or $\mu$. $E(X)$ is just shorthand for the expected value of x. We use $\mu$ when we want to emphasize that it is a parameter of a model. To understand the calculation for the expected value, imagine that the company insures exactly 1000 people. Further imagine that, in perfect accordance with the probabilities, 1 of the policyholders dies, 2 are disabled, and the remaining 997 survive the year unscathed. The company would pay $10,000 to one client and $5000 to each of 2 clients. That's a total of $20,000, or an average of $20,000/1000 = \$20$ per policy. Because it is charging people $50 for the policy, the company expects to make an average profit of $30 per customer.[2] Not bad!

Let's look at this expected value calculation more closely. We imagined that we have exactly 1000 clients. Of those, exactly 1 died and 2 were disabled, corresponding to what the probabilities say. The average payout is:

$$\mu = E(X) = \frac{10,000(1) + 5000(2) + 0(997)}{1000} = \$20 \text{ per policy.}$$

**NOTATION ALERT**
The expected value (or mean) of a random variable is written $E(X)$ or $\mu$.

Instead of writing the expected value as one big fraction, we can rewrite it as separate terms with a common denominator of 1000:

$$E(X) = \$10,000\left(\frac{1}{1000}\right) + \$5000\left(\frac{2}{1000}\right) + \$0\left(\frac{997}{1000}\right)$$
$$= \$20.$$

---

[1] You might reasonably ask, "How can the death rate be only 1 in 1000 people? Wouldn't that mean that people are living 1000 years?" If you were wondering that, good for you. Such a low death rate across an entire population would indeed mean that! In this section we're actually talking about a one-year insurance policy for a large group of people all of the same age. The price of a policy is pegged to a person's age, and the price goes up with age. For young adults, a death rate of 1 in 1000 is reasonable.

[2] Notice that the expected profit of $30 per customer is an *average*. The expected profit of $30 actually doesn't occur for any particular customer at all. The term "expected value" is an unfortunate one, because it sounds like a value that's expected on an individual basis. But it's really a long-run expected average, obeying a principle similar to the Law of Large Numbers.

How convenient! See the probabilities? For each policy, there's a 1/1000 chance that we'll have to pay $10,000 for a death and a 2/1000 chance that we'll have to pay $5000 for a disability. Of course, there's a 997/1000 chance that we won't have to pay anything.

Take a good look at the expression now. It's easy to calculate the expected value of a discrete random variable—just multiply each possible value by the probability that it occurs, and find the sum:

$$\mu = E(X) = \sum x P(x).$$

Be sure that every possible outcome is included in the sum. And verify that you have a valid probability model to start with—the probabilities should each be between 0 and 1 and should sum to 1.

## FOR EXAMPLE
### Love and Expected Values

On Valentine's Day the Quiet Nook restaurant offers a Lucky Lovers Special that could save couples money on their romantic dinners. When the server brings the check, they'll also bring the four aces from a deck of cards. They'll shuffle them and lay them out face down on the table. The couple will then get to turn one card over. If it's a black ace, they'll owe the full amount, but if it's the ace of hearts, the server will give them a $20 Lucky Lovers discount. If they first turn over the ace of diamonds (hey—at least it's red!), they'll then get to turn over one of the remaining cards, earning a $10 discount for finding the ace of hearts the second time.

QUESTION: Construct a probability model for the size of the Lucky Lovers discounts the restaurant will award and use it to determine the expected discount for a couple.

ANSWER: Let $X$ = the Lucky Lovers discount. The probabilities of the three outcomes are:

$$P(X = 20) = P(A\heartsuit) = \frac{1}{4},$$
$$P(X = 10) = P(A\diamondsuit, \text{then } A\heartsuit) = P(A\diamondsuit) \times P(A\heartsuit | A\diamondsuit)$$
$$= \frac{1}{4} \times \frac{1}{3} = \frac{1}{12}, \text{ and}$$
$$P(X = 0) = P(X \neq 20 \text{ or } 10) = 1 - \left(\frac{1}{4} + \frac{1}{12}\right) = \frac{2}{3}.$$

My probability model is:

| Outcome | x | P(X = x) |
|---|---|---|
| A♥ | 20 | $\frac{1}{4}$ |
| A♦, then A♥ | 10 | $\frac{1}{12}$ |
| Black Ace | 0 | $\frac{2}{3}$ |

$$E(X) = 20 \times \frac{1}{4} + 10 \times \frac{1}{12} + 0 \times \frac{2}{3} = \frac{70}{12} \approx 5.83$$

The restaurant can expect to give an average discount of $5.83 per couple.

> ## JUST CHECKING
>
> 1. One of the authors took his minivan in for repair because the air conditioner was cutting out intermittently. The mechanic identified the problem as dirt in a control unit. He said that in about 75% of such cases, drawing down and then recharging the coolant a couple of times cleans up the problem—and costs only $60. If that fails, then the control unit must be replaced at an additional cost of $100 for parts and $40 for labor.
>    a) Define the random variable and construct the probability model.
>    b) What is the expected value of the cost of this repair?
>    c) What does that mean in this context?
>
> Oh—in case you were wondering—the $60 fix worked!

# First Center, Now Spread . . .

Of course, this expected value (or mean) is not what actually happens to any *particular* policyholder. No individual insurance policy actually costs the company $20. We are dealing with random events, so some policyholders receive big payouts, others nothing. Because the insurance company must anticipate this variability, it needs to know the *standard deviation* of the random variable.

For data, we calculated the **standard deviation** by first computing the deviation from the mean and squaring it. We do that with discrete random variables as well. First, we find the deviation of each payout from the mean (expected value):

Table 15.2
Deviations from the mean

| Policyholder Outcome | Payout $x$ | Probability $P(X = x)$ | Deviation $(x - \mu)$ |
|---|---|---|---|
| Death | 10,000 | $\frac{1}{1000}$ | $(10{,}000 - 20) = 9980$ |
| Disability | 5000 | $\frac{2}{1000}$ | $(5000 - 20) = 4980$ |
| Neither | 0 | $\frac{997}{1000}$ | $(0 - 20) = -20$ |

Next, we square each deviation. The **variance** is the expected value of those squared deviations, so we multiply each by the appropriate probability and sum those products. That gives us the variance of $X$. Here's what it looks like:

$$Var(X) = 9980^2 \left(\frac{1}{1000}\right) + 4980^2 \left(\frac{2}{1000}\right) + (-20)^2 \left(\frac{997}{1000}\right) = 149{,}600.$$

Finally, we take the square root to get the standard deviation:

$$SD(X) = \sqrt{149{,}600} \approx \$386.78.$$

The insurance company can expect an average payout of $20 per policy, with a standard deviation of $386.78.

Think about that. The company charges $50 for each policy and expects to pay out $20 per policy. Sounds like an easy way to make $30. In fact, most of the time (probability 997/1000) the company pockets the entire $50. But would you consider selling your neighbor such a policy? The problem is that occasionally the company loses big. With

## Variance and Standard Deviation

$$\sigma^2 = Var(X)$$
$$= \sum(x - \mu)^2 P(X)$$
$$\sigma = SD(X) = \sqrt{Var(X)}$$

probability 1/1000, it will pay out $10,000, and with probability 2/1000, it will pay out $5000. That's probably more risk than you're willing to take on in a single policy with your neighbor. The rather large standard deviation of $386.78 gives an indication that the outcome is no sure thing. That's a pretty big spread (and risk) for an average profit of $30.

Here are the formulas for what we just did. Because these are parameters of our probability model, the variance and standard deviation can also be written as $\sigma^2$ and $\sigma$. You should recognize both kinds of notation.

$$\sigma^2 = Var(X) = \sum(x - \mu)^2 P(x)$$
$$\sigma = SD(X) = \sqrt{Var(X)}$$

### FOR EXAMPLE
#### Finding the Standard Deviation

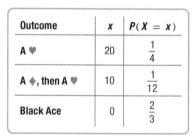

RECAP: Here's the probability model for the Lucky Lovers restaurant discount.

| Outcome | x | $P(X = x)$ |
|---|---|---|
| A ♥ | 20 | $\frac{1}{4}$ |
| A ♦, then A ♥ | 10 | $\frac{1}{12}$ |
| Black Ace | 0 | $\frac{2}{3}$ |

We found that couples can expect an average discount of $\mu = \$5.83$.

QUESTION: What's the standard deviation of the discounts?

ANSWER: First find the variance:

$$Var(X) = \sum(x - \mu)^2 \times P(x)$$
$$= (20 - 5.83)^2 \times \frac{1}{4} + (10 - 5.83)^2 \times \frac{1}{12} + (0 - 5.83)^2 \times \frac{2}{3}$$
$$\approx 74.306.$$

So, $SD(X) = \sqrt{74.306} \approx \$8.62$.

Couples can expect the Lucky Lovers discounts to average $5.83, with a standard deviation of $8.62.

### STEP-BY-STEP EXAMPLE
#### Expected Values and Standard Deviations for Discrete Random Variables

As the head of inventory for Knowway computer company, you were thrilled that you had managed to ship 2 computers to your biggest client the day the order arrived. You are horrified, though, to find that someone had restocked refurbished computers in with the new computers in your storeroom. The shipped computers were selected randomly from the 15 computers in stock, but 4 of those were actually refurbished.

If your client gets 2 new computers, things are fine. If the client gets one refurbished computer, it will be sent back at your expense—$100—and you can replace it. However, if both computers are refurbished, the client will cancel the order this month and you'll lose a total of $1000.

QUESTION: What's the expected value and the standard deviation of the company's loss?

| | | |
|---|---|---|
| **THINK** | **PLAN** State the problem. | I want to find the company's expected loss for shipping refurbished computers and the standard deviation. |
| | **VARIABLE** Define the random variable. | Let $X$ = amount of loss. |
| | **PLOT** Make a picture. This is another job for tree diagrams. If you prefer calculation to drawing, find $P(NN)$ and $P(RR)$, then use the Complement Rule to find $P(NR \text{ or } RN)$. | 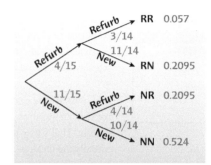 |

| | | |
|---|---|---|
| | **MODEL** List the possible values of the random variable, and determine the probability model. | Outcome \| $x$ \| $P(X = x)$ <br> Two refurbs \| 1000 \| $P(RR) = 0.057$ <br> One refurb \| 100 \| $P(NR \cup RN) = 0.2095 + 0.2095 = 0.419$ <br> Two new \| 0 \| $P(NN) = 0.524$ |

| | | |
|---|---|---|
| **SHOW** | **MECHANICS** Find the expected value. | $E(X) = 0(0.524) + 100(0.419) + 1000(0.057)$ <br> $\phantom{E(X)} = \$98.90$ |
| | Find the variance. | $Var(X) = (0 - 98.90)^2(0.524)$ <br> $\phantom{Var(X) =} + (100 - 98.90)^2(0.419)$ <br> $\phantom{Var(X) =} + (1000 - 98.90)^2(0.057)$ <br> $\phantom{Var(X)} = 51{,}408.79$ |
| | Find the standard deviation. | $SD(X) = \sqrt{51{,}408.79} = \$226.735$ |

| | | |
|---|---|---|
| **TELL** | **CONCLUSION** Interpret your results in context. <br> **REALITY CHECK** Both numbers seem reasonable. The expected value of \$98.90 is between the extremes of \$0 and \$1000, and there's great variability in the outcome values. | The expected cost to the firm of this mistake is \$98.90, with a standard deviation of \$226.74. The large standard deviation reflects the fact that there's a pretty large range of possible losses. |

## TI TIPS

### Finding the Mean and SD of a Random Variable

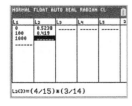

You can easily calculate means and standard deviations for a random variable with your TI. Let's do the Knowway computer example.

- Enter the values of the variable in a list, say, L1: 0, 100, 1000.
- Enter the probability model in another list, say, L2. Notice that you can enter the probabilities as fractions. For example, multiplying along the top branches of the tree gives the probability of a \$1000 loss to be $\frac{4}{15} \times \frac{3}{14}$. When you enter that, the TI will automatically calculate the probability as a decimal!

- Under the STAT CALC menu, choose 1-Var Stats with List:L1, FreqList:L2, then go to Calculate and hit ENTER. (OR on an older calculator just ask for 1-Var Stats L1, L2.)

Now you see the mean and standard deviation (along with some other things). Don't fret that the calculator's mean and standard deviation aren't precisely the same as the ones we found. Such minor differences can arise whenever we round off probabilities to do the work by hand.

When we use the calculator in this way to determine the mean and standard deviation of a random variable, we're actually appropriating a function that was designed to determine the mean and standard deviation of a *dataset*. That's why the mean is denoted $\bar{x}$. You should report the mean as $\mu$, however. *You* know that this is actually the mean of a random variable. The calculator doesn't know that.

# Representations of Random Variables

We've now talked about describing a random variable's "center" and also its "spread." Does that sound familiar? In Chapter 3 we talked about describing distributions of data, and we wanted you to remember to include center, spread . . . and *shape*. A random variable can also be described by the shape of its probability distribution. Let's look at an example.

Suppose we select a household at random from the U.S. and let $X$ (a random variable) be the number of people living in the household. The U.S. Census of 2020 reports the counts for household sizes in the United States as shown in Table 15.3.[3]

Because the Census report does not distinguish between household sizes greater than six people, in this section we will treat households having more than six people as if they all had exactly seven people. Over one and a half million U.S. households fall in that category, but they comprise only about 1.2% of all U.S. households. That makes this approximation reasonable for our purposes: representations of random variables.[4]

Here is a histogram showing the distribution of household sizes in the U.S. in 2020:

Table 15.3

Counts of household sizes in 2020 Census

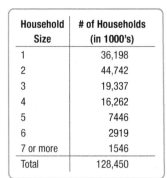

| Household Size | # of Households (in 1000's) |
|---|---|
| 1 | 36,198 |
| 2 | 44,742 |
| 3 | 19,337 |
| 4 | 16,262 |
| 5 | 7446 |
| 6 | 2919 |
| 7 or more | 1546 |
| Total | 128,450 |

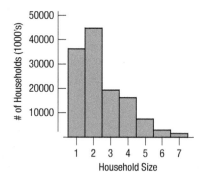

The probability of a randomly selected household having any particular size is simply the number of households of that size divided by the total number of households. So the probability distribution of $X$ is the same as the distribution of the Census data, only rescaled by dividing all the counts of household sizes by the total number of households,

---

[3] https://www.census.gov/data/tables/time-series/demo/families/households.html

[4] For some other purposes, this approximation would be a poor idea. For example, the mean household size would be too low if we computed it using this approximation, and we wouldn't even know how far off it was.

128,450 (thousands). For example, the probability of getting a one-person household is $\Pr(X = 1) = \frac{36,198}{128,450} = 0.282$, or about 28%. So here's the same histogram, re-scaled:

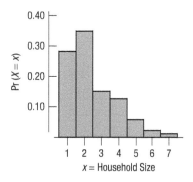

We can look at the histogram of the probability distribution of $X$ and understand properties of the population, which is the household sizes of all U.S. households in 2020. Its center is around 2 or 3, so those are typical household sizes. The spread of the distribution is such that household sizes of 1 and 4 are also common. We can also describe the shape: it's right-skewed. That's not surprising, because households can be large but cannot be smaller than 1 person. We would describe both the distribution of household sizes and the probability distribution of $X$ as right-skewed. The same descriptors of shape that we used for data distributions—like right-skewed, left-skewed, symmetric, bimodal, and so on—also can be applied to probability distributions.

We can also talk about cumulative probability in the same way that we talked about cumulative distributions of data in Chapter 3. Consider again our random variable $X$, the size of a randomly selected U.S. household. The table below expands on Table 15.3 in two ways: we've added a column for the *cumulative* counts—that is, the number of households less than or equal to any particular size—and we've also added columns that re-scale the counts data in the other two columns by dividing by the population size. Those last two columns show the probability distribution of $X$ and the *cumulative probability distribution* of $X$.

| Household Size (x) | # of Households (in 1000's) | Cumulative # of Households (in 1000's) | Pr($X = x$) | Pr($X \leq x$) |
|---|---|---|---|---|
| 1 | 36,198 | 36,198 | 0.282 | 0.282 |
| 2 | 44,742 | 80,940 | 0.348 | 0.630 |
| 3 | 19,337 | 100,277 | 0.150 | 0.780 |
| 4 | 16,262 | 116,539 | 0.127 | 0.907 |
| 5 | 7446 | 123,985 | 0.058 | 0.965 |
| 6 | 2919 | 126,904 | 0.023 | 0.988 |
| 7 or more | 1546 | 128,450 | 0.012 | 1.000 |
| Total | 128,450 | — | 1.000 | — |

The **cumulative probability distribution** of $X$ is the collection of cumulative probabilities associated with $X$. For example, the value in the last column above associated with $x = 4$ is 0.907. That means that if we randomly select a household from the United States, there's about a 90.7% chance we'll get a household with 4 or fewer people. We would write: $P(X \leq 4) = 0.907$. If you subtract that from 1, you get the probability of the complementary event: $P(X \geq 5) = 1 - P(X \leq 4) = 1 - 0.907 = 0.093$. About 9.3% of households in the United States had five or more people in the 2020 census.

At this point you may be wondering, What's really all that different about a probability distribution and a distribution of data? Why bother with random variables and probability distributions at all? We already had means, variances, and standard deviations for data. We already had histograms and cumulative distributions (graphs) for data. This doesn't seem new.

If you're thinking that way, we're glad you're comfortable with these ideas—and that you've made the connection between the probability distribution of a random variable and

the distribution of quantitative data. That's important. The reason we need random variables is that they can represent far more interesting things than just a single measurement (like household size) taken on a single randomly sampled member of a population. Consider the example we began this chapter with: a death and disability insurance policy. The different outcomes aren't measurements, and the probabilities aren't based on random sampling. In the "Lucky Lovers" examples, too, there is no population being sampled from, and the probabilities are based on theory, not data.

Here's an example of a random variable that isn't based on data at all, and that we can understand so well that we'll be able to write a function for its probability distribution. In the game of *Dungeons and Dragons* (D&D), players use a 20-sided die to determine the success or failure of things their characters attempt to do, like hit a monster with a weapon or magic spell. (This is called an "attack roll.") The die faces are numbered 1 to 20 and all are equally likely. If a player rolls a 20 when making an attack roll, it's called a "critical hit." A critical hit is an automatic success and confers extra damage to the monster.

Obviously critical hits aren't extremely common—they only occur one time in 20 on average. No one is upset for not getting a critical hit on their first attack roll of the night. But if your character fights monsters all night long and you make many attack rolls, you may be disappointed if you don't get any critical hits at all. Should you be? Should you be disappointed if it takes 10 attack rolls before you get a critical hit? How many attack rolls will you typically have to make before getting your first critical hit of the night? In an earlier chapter of this book, we'd have suggested answering the question with a simulation.[5] But now we know more about probability, and can determine theoretical probabilities exactly.

Let $X$ be a random variable representing the number of attack rolls that it takes until a critical hit is rolled. If you roll a 20 on your first try, then $X = 1$. If your first critical hit is on your second try, then $X = 2$. And so on. We'd like to determine the probability distribution of $X$ and use it to address the question: What's a typical number of die rolls until you get a critical hit?

Let's begin with $P(X = 1)$. This probability must be $\frac{1}{20}$, or 0.05, because the value $X = 1$ corresponds to the outcome of rolling a 20 on the very first roll.

What about $P(X = 2)$? The outcomes that correspond to this value are those in which something other than a 20 was rolled first, and then a 20 was rolled second. Because successive die rolls are independent, we can multiply probabilities and get $P(X = 2) = (0.95)(0.05) = 0.0475$.

Other probabilities are determined similarly, based on the number of noncritical hits you have to roll before getting your first critical hit. $P(X = 3) = (0.95)^2(0.05)$, for example, and $P(X = 4) = (0.95)^3(0.05)$. There's a pattern here that's not too hard to discern. In order for the random variable $X$ to take on a *particular* numerical value $x$, the outcome of the die rolls must be $(x - 1)$ noncritical hits, followed by a critical hit. All die rolls are independent, so we can multiply probabilities. For this particular random variable:

$$P(X = x) = (0.95)^{x-1}(0.05) \text{ for } x = 1, 2, 3, \ldots$$

When your random variable is based on data, you usually can't determine a function like this because real data are so complex. But in situations like this you can! Here's a histogram showing this probability distribution, just for $x \leq 50$:

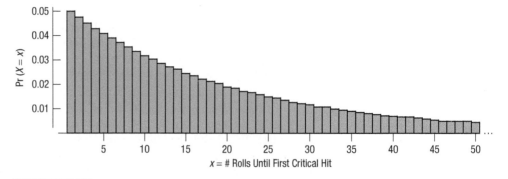

---
[5]Or collecting data while playing D&D!

We can see that the probability distribution of $X$ is right-skewed. Just by "eyeballing" the graph, we see that a typical value appears to be around 15 or so, with other common values being anything from 5 to about 30. What does this mean? It means that if your D&D character fights monsters for a long enough time, you should expect to get your first critical hit probably between the 5th and 30th attack rolls. If it occurs before your fifth roll, you're unusually lucky. If you've rolled 30 times already and not yet gotten a critical hit, then you're unusually unlucky.

The important thing to notice here is that we've determined the probability distribution of a random variable based not on any data, but on theoretical probabilities; and that we were able to write a *function* to describe its probability distribution. We could give it a function symbol of its own if we wanted to, like $p$: $p(x) = \Pr(X = x) = (0.95)^{x-1}(0.05)$.

We could even write a function for its *cumulative distribution*, which is $\Pr(X \leq x)$. Normally this would involve lots of addition—for example, $\Pr(X \leq 3) = \Pr(X = 1) + \Pr(X = 2) + \Pr(X = 3)$. That's troublesome to compute, but in this particular example, there's an available shortcut. First we note that because $X \leq x$ and $X > x$ are complementary events, we can write: $P(X \leq x) = 1 - P(X > x)$. And $X > x$ is synonymous with you *not* getting a critical hit in any of your first $x$ rolls. (Convince yourself of that.) The probability of *not* getting a critical hit in any of your first $x$ rolls is $P(X > x) = (0.95)^x$ thanks to independence and the multiplication rule.

That means that the probability that your first critical hit will occur on or before roll number $x$ must be $1 - (0.95)^x$. This function is called a *cumulative distribution function*, sometimes shortened to just *cdf*. The cumulative distribution function of a random variable $X$ is a function that gives values of $P(X \leq x)$ for given values of $x$.

For our D&D example, the cumulative distribution function looks like this:

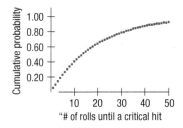

Cumulative distribution functions, as we've seen before with data,[6] can be tricky to read, but they allow for estimating quantiles of a distribution precisely. For example, the graph above shows that at around $X = 13$ or $X = 14$, the cumulative probability hits the 0.50 mark. That means that if you play a long D&D session with lots of attack rolls, there's about a 50% chance that you'll achieve a critical hit on or before your 14th roll.

# More About Means and Variances

Let's think again about the insurance example. Our insurance company expected to pay out an average of $20 per policy, with a standard deviation of about $387. If we take the $50 premium into account, we see the company makes a profit of $50 - 20 = \$30$ per policy. Suppose the company lowers the premium by $5 to $45. It's pretty clear that the expected profit also drops an average of $5 per policy, to $45 - 20 = \$25$.

What about the standard deviation? The differences among payouts haven't changed. We know that adding or subtracting a constant from data shifts the mean but doesn't change the variance or standard deviation. The same is true of random variables:[7]

$$E(X \pm c) = E(X) \pm c \qquad Var(X \pm c) = Var(X).$$

---

[6]In Chapter 3 we looked cumulative distribution graphs for the ages of members of the U.S. Senate and the U.S. House of Representatives

[7]The rules in this section are true for both discrete *and* continuous random variables.

## FOR EXAMPLE
### Adding a Constant

**RECAP:** We've determined that couples dining at the Quiet Nook can expect Lucky Lovers discounts averaging $5.83 with a standard deviation of $8.62. Suppose that for several weeks the restaurant has also been distributing coupons worth $5 off any one meal (one discount per table).

**QUESTION:** If every couple dining there on Valentine's Day brings a coupon, what will be the mean and standard deviation of the total discounts they'll receive?

**ANSWER:** Let $D$ = total discount (Lucky Lovers plus the coupon); then $D = X + 5$.

$$E(D) = E(X + 5) = E(X) + 5 = 5.83 + 5 = \$10.83$$
$$Var(D) = Var(X + 5) = Var(X) = 8.62^2$$
$$SD(D) = \sqrt{Var(X)} = \$8.62$$

Couples with the coupon can expect total discounts averaging $10.83. The standard deviation is still $8.62.

Back to insurance policies . . . What if the company decides to double all the payouts—that is, pay $20,000 for death and $10,000 for disability? This would double the average payout per policy and also increase the variability in payouts. We have seen that multiplying or dividing all data values by a constant changes both the mean and the standard deviation by the same factor. Variance, being the square of the standard deviation, changes by the square of the constant. The same is true of random variables. In general, multiplying each value of a random variable by a constant multiplies the mean by that constant and the variance by the *square* of the constant. Taking the square root of both sides of the variance formula gives us one for standard deviation. (Remember that the square root of $a^2$ is $|a|$, not a.)

$$E(aX) = aE(X) \qquad Var(aX) = a^2 Var(X) \qquad SD(aX) = |a|SD(X).$$

## FOR EXAMPLE
### Double the Love

**RECAP:** On Valentine's Day at the Quiet Nook, couples may get a Lucky Lovers discount averaging $5.83 with a standard deviation of $8.62. When two couples dine together on a single check, the restaurant doubles the discount offer—$40 for the ace of hearts on the first card and $20 on the second.

**QUESTION:** What are the mean, variance, and standard deviation of discounts for such foursomes?

**ANSWER:**
$$E(2X) = 2E(X) = 2(5.83) = \$11.66$$
$$Var(2x) = 2^2 Var(x) = 2^2(8.62)^2 = 297.2176$$
$$SD(2X) = |2|SD(X) = 2(8.62) = \$17.24$$

If the restaurant doubles the discount offer, two couples dining together can expect to save an average of $11.66 with a standard deviation of $17.24.

An insurance company sells policies to more than just one person. How can the company find the total expected value (and standard deviation) of policies taken over all policyholders? Consider a simple case: just two customers, Mr. Ecks and Ms. Wye. With an expected payout of $20 on each policy, we might predict a total of $20 + $20 = $40 to be paid out on the two policies. Nothing surprising there. The expected value of the sum is the sum of the expected values:

$$E(X + Y) = E(X) + E(Y).$$

The variability is another matter. Is the risk of insuring two people the same as the risk of insuring one person for twice as much? We wouldn't expect both clients to die or become disabled in the same year. Because we've spread the risk, the standard deviation should be smaller. Indeed, this is the fundamental principle behind insurance. By spreading the risk among many policies, a company can keep the standard deviation quite small and predict costs more accurately.

But how much smaller is the standard deviation of the sum? It turns out that, if the random variables are independent, there is a simple **Addition Rule for Variances**: *The variance of the sum of two independent random variables is the sum of their individual variances.*

For Mr. Ecks and Ms. Wye, the insurance company can expect their outcomes to be independent, so (using $X$ for Mr. Ecks's payout and $Y$ for Ms. Wye's):

$$Var(X + Y) = Var(X) + Var(Y)$$
$$= 149{,}600 + 149{,}600$$
$$= 299{,}200.$$

If they had insured only Mr. Ecks for twice as much, there would be only one outcome rather than two *independent* outcomes, so the variance would have been

$$Var(2X) = 2^2 Var(X) = 4 \times 149{,}600 = 598{,}400,$$

or twice as big as with two independent policies.

Of course, variances are in squared units. The company would prefer to know standard deviations, which are in dollars. The standard deviation of the payout for two independent policies is $\sqrt{299{,}200} = \$546.99$. But the standard deviation of the payout for a single policy of twice the size is $\sqrt{598{,}400} = \$773.56$, or about 40% more.

If the company has two customers, then it will have an expected annual total payout of $40 with a standard deviation of about $547.

## FOR EXAMPLE
### Adding the Discounts

**RECAP:** The Valentine's Day Lucky Lovers discount for couples averages $5.83 with a standard deviation of $8.62. We've seen that if the restaurant doubles the discount offer for two couples dining together on a single check, they can expect to save $11.66 with a standard deviation of $17.24. Some couples decide instead to get separate checks and pool their two discounts.

**QUESTION:** You and your friend go to this restaurant with another couple and agree to share any benefit from this promotion. Does it matter whether you pay separately or together?

**ANSWER:** Let $X_1$ and $X_2$ represent the two separate discounts, and $T$ the total; then $T = X_1 + X_2$.

$$E(T) = E(X_1 + X_2) = E(X_1) + E(X_2) = 5.83 + 5.83 = \$11.66,$$

so the expected saving is the same either way.

The cards are reshuffled for each couple's turn, so the discounts couples receive are independent. It's okay to add the variances:

$$Var(T) = Var(X_1 + X_2) = Var(X_1) + Var(X_2) = 8.62^2 + 8.62^2 = 148.6088$$
$$SD(T) = \sqrt{148.6088} = \$12.19$$

When two couples get separate checks, there's less variation in their total discount. The standard deviation is $12.19, compared to $17.24 for couples who play for the double discount on a single check. It does, therefore, matter whether they pay separately or together.

We've now seen formulas for the mean, variance, and standard deviation of $X \pm c$, where $X$ is a random variable and $c$ is a constant. We've also seen formulas for the mean, variance, and standard deviation of $aX$, where $X$ is a random variable and $a$ is a constant. Finally, we've seen formulas for the mean, variance, and standard deviation of $X + Y$, where $X$ and $Y$ are independent random variables. It's possible to consolidate all these formulas into three "super-formulas" of which all the ones we've seen are special cases. They look like this:

- If $X$ and $Y$ are independent random variables; and $a$, $b$, and $c$ are constants; and if $Z$ is a random variable defined as $Z = aX + bY + c$, then:
  - $E(Z) = aE(X) + bE(Y) + c$     or equivalently: $\mu_Z = a\mu_X + b\mu_Y + c$
  - $Var(Z) = a^2 Var(X) + b^2 Var(Y)$     or equivalently: $\sigma_Z^2 = a^2 \sigma_X^2 + b^2 \sigma_Y^2$
  - $SD(Z) = \sqrt{a^2 Var(X) + b^2 Var(Y)}$     or equivalently: $\sigma_Z = \sqrt{a^2 \sigma_X^2 + b^2 \sigma_Y^2}$

If you're like most students, then understanding when and how to apply these formulas will be one of the most challenging parts of this course for you. It takes practice, but as with most things, you'll get better with practice.

There's one special case of our "super-formula" that we want to point out because it's going to show up a lot and it's somewhat surprising until you think about it carefully. If $a = 1$, $b = -1$, and $c = 0$, then our random variable $Z$ becomes the difference $X - Y$, and our three formulas become:

- $E(Z) = E(X) - E(Y)$
  - or equivalently: $\mu_Z = \mu_X - \mu_Y$
- $Var(Z) = Var(X) + (-1)^2 Var(Y) = Var(X) + Var(Y)$
  - or equivalently: $\sigma_Z = \sigma_X^2 + \sigma_Y^2$
- $SD(Z) = \sqrt{Var(X) + Var(Y)}$
  - or equivalently: $\sigma_Z = \sqrt{\sigma_X^2 + \sigma_Y^2}$

In other words: Variances of sums *or differences* of independent random variables *add*. The variance of $X - Y$ is not the difference in their variances but the sum. Subtracting random quantities, like adding them, creates more uncertainty, not less.

## Pythagorean Theorem of Statistics[8]

We often use the standard deviation to measure variability, but when we add independent random variables, we use their variances. Think of the Pythagorean Theorem. In a right triangle (only), the *square* of the length of the hypotenuse is the sum of the *squares* of the lengths of the other two sides:

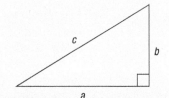

$$c^2 = a^2 + b^2.$$

For independent random variables (only), the *square* of the standard deviation of their sum is the sum of the *squares* of their standard deviations:

$$SD^2(X + Y) = SD^2(X) + SD^2(Y).$$

It's simpler to write this with *variances*:

For independent random variables, $X$ and $Y$, $Var(X + Y) = Var(X) + Var(Y)$.

---

[8] Calling this the "Pythagorean Theorem of Statistics" may seem like just a metaphor inspired by the similarity of two otherwise unrelated math formulas. But in fact there is a profound connection between geometry and random variables that statisticians sometimes exploit to help them better understand statistical phenomena. Understanding this connection requires taking more advanced math and statistics courses.

## FOR EXAMPLE
### Working with Differences

**RECAP:** The Lucky Lovers discount at the Quiet Nook averages $5.83 with a standard deviation of $8.62. Just up the street, the Wise Fool restaurant has a competing Lottery of Love promotion. There a couple can select a specially prepared chocolate from a large bowl and unwrap it to learn the size of their discount. The restaurant's manager says the discounts vary with an average of $10.00 and a standard deviation of $15.00.

**QUESTION:** How much more can you expect to save at the Wise Fool? With what standard deviation?

**ANSWER:** Let $W$ = the discount at the Wise Fool, $X$ = the discount at the Quiet Nook, and $D$ = the difference: $D = W - X$. These are different promotions at separate restaurants, so the outcomes are independent.

$$E(W - X) = E(W) - E(X) = 10.00 - 5.83 = \$4.17$$
$$\begin{aligned}SD(W - X) &= \sqrt{Var(W - X)} \\ &= \sqrt{Var(W) + Var(X)} \\ &= \sqrt{15^2 + 8.62^2} \\ &\approx 17.30\end{aligned}$$

Discounts at the Wise Fool will average $4.17 more than at the Quiet Nook, and the standard deviation of the difference is $17.30.

## For Random Variables, Does $X + X + X = 3X$?

Maybe, but be careful. As we've just seen, insuring one person for $30,000 is not the same risk as insuring three people for $10,000 each. When each instance represents a different outcome for the same random variable, it's easy to fall into the trap of writing all of them with the same symbol. Don't make this common mistake. Make sure you write each instance as a *different* random variable. Just because each random variable describes a similar situation doesn't mean that each random outcome will be the same.

These are *random* variables, not the variables you saw in Algebra. Being random, they take on different values each time they're evaluated. So what you really mean is $X_1 + X_2 + X_3$. Written this way, it's clear that the sum shouldn't necessarily equal 3 times *anything*.

## FOR EXAMPLE
### Summing a Series of Outcomes

**RECAP:** The Quiet Nook's Lucky Lovers promotion offers couples discounts averaging $5.83 with a standard deviation of $8.62. The restaurant owner is planning to serve 40 couples on Valentine's Day.

**QUESTION:** What's the expected total of the discounts the owner will give? With what standard deviation?

**ANSWER:** Let $X_1, X_2, X_3, \ldots, X_{40}$ represent the discounts to the 40 couples, and $T$ the total of all the discounts. Then:

$$T = X_1 + X_2 + X_3 + \cdots + X_{40} \text{ and}$$
$$\begin{aligned}E(T) &= E(X_1 + X_2 + X_3 + \cdots + X_{40}) \\ &= E(X_1) + E(X_2) + E(X_3) + \cdots + E(X_{40}) \\ &= 5.83 + 5.83 + 5.83 + \cdots + 5.83 \\ &= \$233.20\end{aligned}$$

Reshuffling cards between couples makes the discounts independent, so:

$$SD(T) = \sqrt{Var(X_1 + X_2 + X_3 + \cdots + X_{40})}$$
$$= \sqrt{Var(X_1) + Var(X_2) + Var(X_3) + \cdots + Var(X_{40})}$$
$$= \sqrt{8.62^2 + 8.62^2 + 8.62^2 + \cdots + 8.62^2}$$
$$\approx \$54.52$$

The restaurant owner can expect the 40 couples to win discounts totaling $233.20, with a standard deviation of $54.52.

## JUST CHECKING

2. Suppose the time it takes a customer to get and pay for seats at the ticket window of a baseball park is a random variable with a mean of 100 seconds and a standard deviation of 50 seconds. When you get there, you find only two people in line in front of you.
   a) How long do you expect to wait for your turn to get tickets?
   b) What's the standard deviation of your wait time?
   c) What assumption did you make about the two customers in finding the standard deviation?

## STEP-BY-STEP EXAMPLE

### Hitting the Road: Means and Variances

You're planning to spend next year wandering through the mountains of Kyrgyzstan. You plan to sell your used SUV so you can purchase an off-road Honda motor scooter when you get there. Used SUVs of the year and mileage of yours are selling for a mean of $6940 with a standard deviation of $250. Your research shows that scooters in Kyrgyzstan are going for about 65,000 Kyrgyzstan som with a standard deviation of 500 som. One U.S. dollar is worth about 38.5 Kyrgyzstan som (38 som and 50 tylyn).

QUESTION: How much cash can you expect to pocket after you sell your SUV and buy the scooter?

**THINK**    **PLAN** State the problem.

**VARIABLES** Define the random variables.

Write an appropriate equation.

Think about the assumptions.

I want to model how much money I'd have (in som) after selling my SUV and buying the scooter.

Let $A$ = sale price of my SUV (in dollars),
$B$ = price of a scooter (in som), and
$D$ = profit (in som).

$D = 38.5A - B$

✓ **Independence Assumption**: The prices are independent.

| | | |
|---|---|---|
| **SHOW** **MECHANICS** Find the expected value, using the appropriate rules. | | $E(D) = E(38.5A - B)$<br>$\phantom{E(D)} = 38.5E(A) - E(B)$<br>$\phantom{E(D)} = 38.5(6,940) - (65,000)$<br>$E(D) = 202,190$ som |
| Find the variance, using the appropriate rules. Be sure to check the assumptions first! | | Because sale and purchase prices are independent,<br>$Var(D) = Var(38.5A - B)$<br>$\phantom{Var(D)} = Var(38.5A) + Var(B)$<br>$\phantom{Var(D)} = (38.5)^2 Var(A) + Var(B)$<br>$\phantom{Var(D)} = 1482.25(250)^2 + (500)^2$<br>$\phantom{Var(D)} = 92,890,625$ |
| Find the standard deviation. | | $SD(D) = \sqrt{92,890,625} = 9637.98$ som |
| **TELL** **CONCLUSION** Interpret your results in context. (Here that means talking about dollars.) | | I can expect to clear about 202,190 som ($5252) with a standard deviation of 9638 som ($250). |
| REALITY CHECK Given the initial cost estimates, the mean and standard deviation seem reasonable. | | |

# Continuous Random Variables

A company manufactures home theater systems. At the end of the production line, the systems are packaged and prepared for shipping. Stage 1 of this process is called "packing." Workers must collect all the system components (a subwoofer, four speakers, a power cord, some cables, and a remote control), put each in plastic bags, and then place everything inside a protective Styrofoam form. The packed form then moves on to Stage 2, called "boxing." There, workers place the form and a packet of instructions in a cardboard box, close it, then seal and label the box for shipping.

The company says that times required for the packing stage can be described by a Normal model with a mean of 9 minutes and standard deviation of 1.5 minutes. The times for the boxing stage can also be modeled as Normal, with a mean of 6 minutes and standard deviation of 1 minute.

This is a common way to model events. Do our rules for random variables apply here? What's different? We no longer have a list of discrete outcomes, with their associated probabilities. Instead, we have **continuous random variables** that can take on any value. Now any single value won't have a probability. We saw this back in Chapter 5 when we first saw the Normal model (although we didn't talk then about "random variables" or "probability"). We know that the probability that $z = 1.5$ doesn't make sense, but we *can* talk about the probability that $z$ lies *between* 0.5 and 1.5. For a Normal random variable, the probability that it falls within an interval is just the area under the Normal curve over that interval.

Some continuous random variables have Normal models; others may be skewed, uniform, or bimodal. Regardless of shape, every continuous random variable has a mean (which we also call an *expected value*), a standard deviation, and a variance. In this book we won't worry about how to calculate them, but we can still work with models for continuous random variables when we're given these parameters. Happily, nearly everything we've said about how discrete random variables behave is true of continuous random variables as well.

In addition, random variables that have a Normal model have a very useful property that is not true of most other random variables. When two independent continuous random variables have Normal models, so does their sum or difference. This simple fact is a special property of Normal models and is very important. It allows us to apply our knowledge of Normal probabilities to questions about the sum or difference of independent random variables.

## STEP-BY-STEP EXAMPLE

### Packaging Stereos

Consider the company that manufactures and ships home theater systems that we just discussed. Recall that times required to pack the systems can be described by a Normal model with a mean of 9 minutes and standard deviation of 1.5 minutes. The times for the boxing stage can also be modeled as Normal, with a mean of 6 minutes and standard deviation of 1 minute.

QUESTIONS:
1. What is the probability that packing two consecutive systems takes over 20 minutes?
2. What percentage of the theater systems take longer to pack than to box?

QUESTION 1: What is the probability that packing two consecutive systems takes over 20 minutes?

| | |
|---|---|
| **THINK**    **PLAN** State the problem. | I want to estimate the probability that packing two consecutive systems takes over 20 minutes. <br><br> Let $P_1$ = time for packing the first system, <br> $P_2$ = time for packing the second, and <br> $T$ = total time to pack two systems. |
| **VARIABLES** Define your random variables. <br><br> Write an appropriate equation. <br><br> Think about the assumptions. Sums of independent Normal random variables follow a Normal model. Such simplicity isn't true in general. | $T = P_1 + P_2$ <br><br> ✓ **Normal Model Assumption**: We are told that both random variables follow Normal models. <br><br> ✓ **Independence Assumption**: We can reasonably assume that the two packing times are independent. |
| **SHOW**    **MECHANICS** Find the expected value. <br><br><br><br> For sums of independent random variables, variances add. (We don't need the variables to be Normal for this to be true—just independent.) <br><br><br><br><br> Find the standard deviation. | $E(T) = E(P_1 + P_2)$ <br> $\quad\quad = E(P_1) + E(P_2)$ <br> $\quad\quad = 9 + 9 = 18$ minutes <br><br> Because the times are independent, <br> $Var(T) = Var(P_1 + P_2)$ <br> $\quad\quad\quad = Var(P_1) + Var(P_2)$ <br> $\quad\quad\quad = 1.5^2 + 1.5^2$ <br> $Var(R) = 4.50$ <br> $SD(T) = \sqrt{4.50} \approx 2.12$ minutes |

| | |
|---|---|
| Now we use the fact that these independent random variables both follow Normal models to say that their sum is also Normal.<br><br>Sketch a picture of the Normal model for the total time, shading the region representing over 20 minutes.<br><br><br><br>Find the z-score for 20 minutes.<br><br>Use technology (or Table Z in the appendices) to find the probability. | I'll model T with N(18, 2.12).<br><br>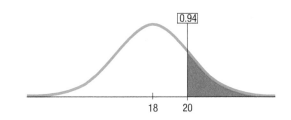<br><br>$z = \dfrac{20 - 18}{2.12} = 0.94$<br><br>$P(T > 20) = P(z > 0.94) = 0.1736$ |
| **TELL**   **CONCLUSION** Interpret your result in context. | There's a little more than a 17% chance that it will take a total of over 20 minutes to pack two consecutive home theater systems. |

QUESTION 2: What percent of the home theater systems take longer to pack than to box?

| | |
|---|---|
| **THINK**   **PLAN** State the question.<br><br>**VARIABLES** Define your random variables.<br><br><br><br><br>Write an appropriate equation.<br><br>What are we trying to find? Notice that we can tell which of two quantities is greater by subtracting and asking whether the difference is positive or negative.<br><br>Don't forget to think about the assumptions. | I want to estimate the percentage of the systems that take longer to pack than to box.<br><br>Let   $P$ = time for packing a system,<br>        $B$ = time for boxing a system, and<br>        $D$ = difference in times to pack and box a system.<br><br>$D = P - B$<br><br>The probability that it takes longer to pack than to box a system is the probability that the difference $P - B$ is greater than zero.<br><br>✓ **Normal Model Assumption**: We are told that both random variables follow Normal models.<br><br>✓ **Independence Assumption**: We can assume that the times it takes to pack and to box a system are independent. |
| **SHOW**   **MECHANICS** Find the expected value.<br><br><br><br>For the difference of independent random variables, variances add.<br><br><br><br><br><br>Find the standard deviation. | $E(D) = E(P - B)$<br>          $= E(P) - E(B)$<br>          $= 9 - 6 = 3$ minutes<br><br>Because the times are independent,<br>$Var(D) = Var(P - B)$<br>            $= Var(P) + Var(B)$<br>            $= 1.5^2 + 1^2$<br>$Var(D) = 3.25$<br>$SD(D) = \sqrt{3.25} \approx 1.80$ minutes |

| | |
|---|---|
| State what model you will use.<br><br>Sketch a picture of the Normal model for the difference in times, and shade the region representing a difference greater than zero. | I'll model D with N (3, 1.80).<br>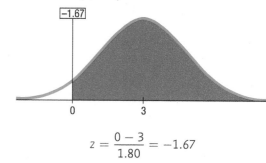 |
| Find the z-score for 0 minutes; then use Table Z or technology to find the probability. | $z = \dfrac{0-3}{1.80} = -1.67$<br><br>$P(D > 0) = P(z > -1.67) = 0.9525$ |
| **TELL** CONCLUSION Interpret your result in context. | About 95% of all the home theater systems will require more time for packing than for boxing. |

## WHAT CAN GO WRONG?

- **Probability models are still just models.** Models can be useful, but they are not reality. Think about the assumptions behind your models. Are your dice really perfectly fair? (They are probably pretty close.) But when you hear that the probability of a nuclear accident is 1/10,000,000 per year, is that likely to be a precise value? Question probabilities as you would data.

- **If the model is wrong, so is everything else.** Before you try to find the mean or standard deviation of a random variable, check to make sure the probability model is reasonable. As a start, the probabilities in your model should add up to 1. If not, you may have calculated a probability incorrectly or left out a value of the random variable. For instance, in the insurance example, the description mentions only death and disability. Good health is by far the most likely outcome, not to mention the best for both you and the insurance company (who gets to keep your money). Don't overlook that.

- **Don't assume everything's Normal.** Just because a random variable is continuous or you happen to know a mean and standard deviation doesn't mean that a Normal model will be useful. You must *Think* about whether the Normality Assumption is justified. Using a Normal model when it really does not apply will lead to wrong answers and misleading conclusions.

    To find the expected value of the sum or difference of random variables, we simply add or subtract means. Center is easy; spread is trickier. Watch out for some common traps.

- **Watch out for variables that aren't independent.** You can add expected values of *any* two random variables, but you can only add variances of independent random variables. Suppose a survey includes questions about the number of hours of sleep people get each night and also the number of hours they are awake each day. From their answers, we find the mean and standard deviation of hours asleep and hours awake. The expected total must be 24 hours; after all, people are either asleep or awake.[9] The means still add just fine. Because all the totals are exactly 24 hours, however, the standard deviation of the total will be 0. We can't add variances here because the number of hours you're awake depends on the number of hours you're asleep. Be sure to check for independence before adding variances.

---

[9] Although some students do manage to attain a state of consciousness somewhere between sleeping and wakefulness during Statistics class.

- Don't forget: Variances of independent random variables add. Standard deviations don't.
- Don't forget: Variances of independent random variables add, even when you're looking at the difference between them.
- Don't write independent instances of a random variable with notation that looks like they are the same variables. Make sure you write each instance as a different random variable. Just because each random variable describes a similar situation doesn't mean that each random outcome will be the same. These are *random* variables, not the variables you saw in Algebra. Write $X_1 + X_2 + X_3$ rather than $X + X + X$.

## WHAT HAVE WE LEARNED?

We've learned to work with random variables. We can use the probability model for a discrete random variable to find its expected value and its standard deviation.

We've learned that the mean of the sum or difference of two random variables, discrete or continuous, is just the sum or difference of their means. And we've learned the Pythagorean Theorem of Statistics: *For independent random variables,* the variance of their sum or difference is always the *sum* of their variances.

Finally, we've learned that Normal models are once again special. Sums or differences of normally distributed independent random variables also follow Normal models.

## TERMS

**Random variable**  
A random variable assumes any of several different numeric values as a result of some random event. Random variables are denoted by a capital letter such as $X$. (p. 403)

**Discrete random variable**  
A random variable that can take one of a finite number[10] of distinct outcomes is called a discrete random variable. (p. 403)

**Continuous random variable**  
A random variable that can take any numeric value within an interval of values is called a continuous random variable. The interval may be infinite or bounded at either or both ends. (p. 403)

**Probability model**  
The probability model is a function that associates a probability $P$ with each value of a discrete random variable $X$, denoted $P(X = x)$, or with any interval of values of a continuous random variable. (p. 403)

**Cumulative probability distribution**  
For a random variable $X$, the collection of cumulative probabilities $P(X \leq x)$ is its cumulative probability distribution. If that distribution is summarized in a functional form then it is called the cumulative distribution function, or *cdf*, associated with $X$. (p. 409)

**Expected value**  
The expected value of a random variable is its theoretical long-run average value, the center of its model. Denoted $\mu$ or $E(X)$, it is found (if the random variable is discrete) by summing the products of variable values and probabilities:
$$\mu = E(X) = \sum xP(x). \text{ (p. 403)}$$

**Variance**  
The variance of a random variable is the expected value of the squared deviation from the mean. For discrete random variables, it can be calculated as:
$$\sigma^2 = Var(X) = \sum (x - \mu)^2 P(x). \text{ (p. 405)}$$

---

[10] Actually, there could be an infinite number of outcomes, as long as they're *countable*. Essentially that means we can imagine listing them all in order, like the counting numbers 1, 2, 3, 4, 5, . . . . For example, in D&D, the number of attack rolls until you get a critical hit is a discrete random variable with an infinite number of possible outcomes.

**Standard deviation**  The standard deviation of a random variable describes the spread in the model, and is the square root of the variance:

$$\sigma = SD(X) = \sqrt{Var(X)} \text{ (p. 405)}$$

**Changing a random variable by a constant**

$$E(X \pm c) = E(X) \pm c \qquad Var(X \pm c) = Var(X) \text{ (p. 411)}$$
$$E(aX) = aE(X) \qquad Var(aX) = a^2 Var(X) \text{ (p. 412)}$$

**Addition Rule for Expected Values of Random Variables**

$$E(X \pm Y) = E(X) \pm E(Y) \text{ (p. 413)}$$

**Addition Rule for Variances**  If $X$ and $Y$ are *independent*: $Var(X \pm Y) = Var(X) + Var(Y)$,

and $SD(X \pm Y) = \sqrt{Var(X) + Var(Y)}$. (p. 413)

**The Pythagorean Theorem of Statistics** (p. 396)

## ON THE COMPUTER

### Random Variables

Statistics packages deal with data, not with random variables. Nevertheless, the calculations needed to find means and standard deviations of random variables are little more than weighted means. Most packages can manage that, but then they are just being overblown calculators. For technological assistance with these calculations, we recommend you pull out your calculator.

## EXERCISES

**1. Expected value**  Find the expected value of each random variable:

a)
| x | 10 | 20 | 30 |
|---|----|----|----|
| P(X = x) | 0.3 | 0.5 | 0.2 |

b)
| x | 2 | 4 | 6 | 8 |
|---|---|---|---|---|
| P(X = x) | 0.3 | 0.4 | 0.2 | 0.1 |

**2. Expected value**  Find the expected value of each random variable:

a)
| x | 0 | 1 | 2 |
|---|---|---|---|
| P(X = x) | 0.2 | 0.4 | 0.4 |

b)
| x | 100 | 200 | 300 | 400 |
|---|-----|-----|-----|-----|
| P(X = x) | 0.1 | 0.2 | 0.5 | 0.2 |

**3. Swing club**  Leslie the park manager wants to determine if there are enough swings to go around in one of the local parks. She goes to the park and develops the following probability model for the usage level of the six swings.

| Number of Swings in Use | 0 | 1 | 2 | 3 | 4 | 5 | 6 |
|---|---|---|---|---|---|---|---|
| Probability | 0.12 | 0.19 | 0.26 | 0.28 | 0.10 | 0.03 | 0.02 |

a) Draw a histogram of the probability distribution and describe its shape.
b) How many swings are in use, on average?
c) What is the median number of swings in use?
d) How do the mean and median compare? Is this what you would expect based on the shape of the distribution?

**4. Caffeinated**  A coffee shop tracks sales and has observed the distribution in the following table.

| # of Sales | 145 | 150 | 155 | 160 | 170 |
|---|---|---|---|---|---|
| Probability | 0.15 | 0.22 | 0.37 | 0.19 | 0.07 |

a) Draw a histogram of the probability distribution and describe its shape.
b) What is the average daily sales that it can expect?
c) What is the median amount of daily sales?
d) How do the mean and median compare? Is this what you would expect based on the shape of the distribution?

**5. Swing some more**  What is the standard deviation of Exercise 3?

**6. Caffeinated again**  What is the standard deviation for Exercise 4?

**7. Make a bet?**  Your crafty teacher makes every student in the class a simple offer. Your teacher has an octahedron-of-chance (an eight-sided die). Every student rolls once. If you roll an 8, you get five extra credit points on the next test. If you roll anything else you roll again. If, on the second roll, you roll a number 3 or higher, you get a two-point bonus on the test, but if not, you lose one point! (This is not legal in most states.)

a) Create a probability model for the number of points a student in the class will receive/lose.
b) Find the expected number of points the teacher will award to students in this class on average.
c) Would you play? Why or why not?

8. **You bet!** You roll a die. If it comes up a 6, you win $100. If not, you get to roll again. If you get a 6 the second time, you win $50. If not, you lose.

a) Create a probability model for the amount you win.
b) Find the expected amount you'll win.
c) What would you be willing to pay to play this game?

9. **Right or wrong?** Mr Chen, a math teacher, has an app that randomly chooses a student in the class to "volunteer" to answer a question. He will stop choosing students once a student gets the right answer, but if he is given four wrong answers in a row, he will stop and help the class himself. He estimates that 85% of the students in the class know the correct answer (smart class!).

a) Create a probability model for the number of students Mr Chen will call on until he receive the correct answer (or gives up in frustration!).
b) Find the expected number of students called on.
c) Find the expected number of wrong answers that will be offered.

10. **Carnival** A carnival game offers a $100 cash prize for anyone who can break a balloon by throwing a dart at it. It costs $5 to play, and you're willing to spend up to $20 trying to win. You estimate that you have about a 10% chance of hitting the balloon on any throw.

a) Create a probability model for this carnival game.
b) Find the expected number of darts you'll throw.
c) Find your expected winnings.

11. **Software** A small software company bids on two contracts. It anticipates a profit of $60,000 if it gets the larger contract and a profit of $20,000 on the smaller contract. The company estimates there's a 30% chance it will get the larger contract and a 60% chance it will get the smaller contract. Assuming the contracts will be awarded independently, what's the expected profit?

12. **Racehorse** A man buys a racehorse for $20,000 and enters it in two races. He plans to sell the horse afterward, hoping to make a profit. If the horse wins both races, its value will jump to $100,000. If it wins one of the races, it will be worth $50,000. If it loses both races, it will be worth only $10,000. The man believes there's a 20% chance that the horse will win the first race and a 30% chance it will win the second one. Assuming that the two races are independent events, find the man's expected profit.

13. **Variation 1** Find the standard deviations of the random variables in Exercise 1.

14. **Variation 2** Find the standard deviations of the random variables in Exercise 2.

15. **Students bet** Find the standard deviation of the number of points in Exercise 7.

16. **The die** Find the standard deviation of the amount you might win rolling a die in Exercise 8.

17. **Get It right?** Find the standard deviation of the number of students called on in Exercise 9.

18. **Darts** Find the standard deviation of your winnings throwing darts in Exercise 10.

19. **Repairs** The probability model below describes the number of repair calls that an appliance repair shop may receive during an hour.

| Repair Calls | 0 | 1 | 2 | 3 |
|---|---|---|---|---|
| Probability | 0.1 | 0.3 | 0.4 | 0.2 |

a) How many calls should the shop expect per hour?
b) What is the standard deviation?

20. **Red lights** Felana must pass through five traffic lights on her way to work and will have to stop at each one that is red. She estimates the probability model for the number of red lights she hit, as shown below.

| X = # of Red | 0 | 1 | 2 | 3 | 4 | 5 |
|---|---|---|---|---|---|---|
| P(X = x) | 0.05 | 0.25 | 0.35 | 0.15 | 0.15 | 0.05 |

a) How many red lights should Felana expect to hit each day?
b) What's the standard deviation?

21. **Defects** A consumer organization inspecting new cars found that many had appearance defects (dents, scratches, paint chips, etc.). While none had more than three of these defects, 7% had three, 11% two, and 21% one defect. Find the expected number of appearance defects in a new car and the standard deviation.

22. **Insurance** An insurance policy costs $100 and will pay policyholders $10,000 if they suffer a major injury (resulting in hospitalization) or $3000 if they suffer a minor injury (resulting in lost time from work). The company estimates that each year 1 in every 2000 policyholders may have a major injury, and 1 in 500 a minor injury only.

a) Create a probability model for the profit on a policy.
b) What's the company's expected profit on this policy?
c) What's the standard deviation?

23. **Cancelled flights** Mary is deciding whether to book the cheaper flight home from college after her final exams, but she's unsure when her last exam will be. She thinks there is only a 20% chance that the exam will be scheduled after the last day she can get a seat on the cheaper flight. If it is and she has to cancel the flight, she will lose $150. If she can take the cheaper flight, she will save $100.

a) If she books the cheaper flight, what can she expect to gain, on average?
b) What is the standard deviation?

24. **Day trading** An option to buy a stock is priced at $200. If the stock closes above 30 on May 15, the option will be worth $1000. If it closes below 20, the option will be worth nothing, and if it closes between 20 and 30 (inclusively), the option will be worth $200. A trader thinks there is a 50% chance that the stock will close in the 20–30 range, a 20% chance that it will close above 30, and a 30% chance that it will fall below 20 on May 15.

a) Should she buy the stock option?
b) How much does she expect to gain?
c) What is the standard deviation of her gain?

**25. Contest** You play two games against the same opponent. The probability you win the first game is 0.4. If you win the first game, the probability you also win the second is 0.2. If you lose the first game, the probability that you win the second is 0.3.
  a) Are the two games independent? Explain.
  b) What's the probability you lose both games?
  c) What's the probability you win both games?
  d) Let random variable $X$ be the number of games you win. Find the probability model for $X$.
  e) What are the expected value and standard deviation?

**26. Contracts** Your company bids for two contracts. You believe the probability you get contract 1 is 0.8. If you get contract 1, the probability you also get contract 2 will be 0.2, and if you do not get 1, the probability you get 2 will be 0.3.
  a) Are the two contracts independent? Explain.
  b) Find the probability you get both contracts.
  c) Find the probability you get no contract.
  d) Let $X$ be the number of contracts you get. Find the probability model for $X$.
  e) Find the expected value and standard deviation.

**27. Batteries** In a group of 10 batteries, 3 are dead. You choose 2 batteries at random.
  a) Create a probability model for the number of good batteries you get.
  b) What's the expected number of good ones you get?
  c) What's the standard deviation?

**28. Kittens** In a litter of seven kittens, three are female. You pick two kittens at random.
  a) Create a probability model for the number of male kittens you get.
  b) What's the expected number of males?
  c) What's the standard deviation?

**29. Random variables** Given independent random variables with means and standard deviations as shown, find the mean and standard deviation of:
  a) $3X$
  b) $Y + 6$
  c) $X + Y$
  d) $X - Y$
  e) $X_1 + X_2$

|   | Mean | SD |
|---|---|---|
| X | 10 | 2 |
| Y | 20 | 5 |

**30. Random variables** Given independent random variables with means and standard deviations as shown, find the mean and standard deviation of:
  a) $X - 20$
  b) $0.5Y$
  c) $X + Y$
  d) $X - Y$
  e) $Y_1 + Y_2$

|   | Mean | SD |
|---|---|---|
| X | 80 | 12 |
| Y | 12 | 3 |

**31. Random variables** Given independent random variables with means and standard deviations as shown, find the mean and standard deviation of:
  a) $0.8Y$
  b) $2X - 100$
  c) $X + 2Y$
  d) $3X - Y$
  e) $Y_1 + Y_2$

|   | Mean | SD |
|---|---|---|
| X | 120 | 12 |
| Y | 300 | 16 |

**32. Random variables** Given independent random variables with means and standard deviations as shown, find the mean and standard deviation of:
  a) $2Y + 20$
  b) $3X$
  c) $0.25X + Y$
  d) $X - 5Y$
  e) $X_1 + X_2 + X_3$

|   | Mean | SD |
|---|---|---|
| X | 80 | 12 |
| Y | 12 | 3 |

**33. Dice**
  a) Create a probability distribution for rolling a fair six-sided die.
  b) Calculate the expected value and standard deviation of this random variable.
  c) Using the properties of random variables, calculate the expected value and standard deviation of rolling a die and adding 5 to the result.
  d) Create a probability distribution for rolling a die and adding 5 to the result. Calculate the expected value and standard deviation.
  e) Compare to your calculations from part b.

**34. Twice dice**
  a) Create a probability distribution for rolling a fair six-sided die. (Or, if you did Exercise 33, refer to the one you created there.)
  b) Calculate the expected value and standard deviation of this random variable. (Unless you already did it in Exercise 33)
  c) Using the properties of random variables, calculate the expected value and standard deviation of rolling a die and doubling the result.
  d) Create a probability distribution for rolling a die and doubling the result. Calculate the expected value and standard deviation.
  e) Compare to your calculations from part b.

**35. Dice twice**
  a) Create a probability distribution for rolling a fair six-sided die. (Or, if you did Exercise 33 or 34, refer to the one you created there.)
  b) Calculate the expected value and standard deviation of this random variable. (Unless you already did it in Exercise 33)
  c) Using the properties of random variables, calculate the expected value and standard deviation of rolling a die twice and adding the results.
  d) Create a probability distribution for rolling a die twice and adding the results. Calculate the expected value and standard deviation.
  e) Compare to your calculations from part b.

**36. Dice twice again**
  a) Create a probability distribution for rolling a fair six-sided die. (Or, if you did Exercise 33, 34, or 35, refer to the one you created there.)
  b) Calculate the expected value and standard deviation of this random variable. (Unless you already did it.)
  c) Using the properties of random variables, calculate the expected value and standard deviation of rolling a die twice and subtracting the second roll from the first.
  d) Create a probability distribution for rolling a die twice and subtracting the second roll from the first. Calculate the expected value and standard deviation.
  e) Compare to your calculations from part b.

**37. Salary** An employer pays a mean salary for a 5-day workweek of $1250 with a standard deviation of $129. On the weekends, their salary expenses have a mean of $450 with a standard deviation of $57. What is the mean and standard deviation of their total weekly salaries?

**38. Golf scores** A golfer keeps track of their score for playing nine holes of golf (half a normal golf round). Their mean score is 85 with a standard deviation of 11. Assuming that the second 9 has the same mean and standard deviation, what is the mean and standard deviation of their total score if they play a full 18 holes?

**39. Eggs** A grocery supplier believes that in a dozen eggs, the mean number of broken ones is 0.6 with a standard deviation of 0.5 eggs. You buy 3 dozen eggs without checking them.
   a) How many broken eggs do you expect to get?
   b) What's the standard deviation?
   c) What assumptions did you have to make about the eggs in order to answer this question?

**40. Garden** A company selling vegetable seeds in packets of 20 estimates that the mean number of seeds that will actually grow is 18, with a standard deviation of 1.2 seeds. You buy 5 different seed packets.
   a) How many good seeds do you expect to get?
   b) What's the standard deviation?
   c) What assumptions did you make about the seeds? Do you think that assumption is warranted? Explain.

**41. Eggs again** In Exercise 39 you bought 3 dozen eggs.
   a) How many good eggs do you expect?
   b) What's the standard deviation of the number of good eggs?
   c) Why does it make sense for the standard deviation not to change?

**42. Garden grows** In Exercise 40 you bought 5 seed packets.
   a) How many bad seeds do you expect?
   b) What is the standard deviation of the bad seed count?
   c) Why does it make sense that the standard deviation is not different?

**43. SAT or ACT revisited** Remember back in Chapter 5 when we used the equation $SAT = 40 \times ACT + 150$ to convert an ACT score into a SAT score. Let's use this transformation again, now with random variables.
   a) Suppose your school has a mean ACT score of 29. What would its equivalent mean SAT score be?
   b) If your school has a standard deviation of 5 ACT points, what is the standard of its equivalent SAT score?

**44. Colder?** We used the formula $°F = 9/5°C + 32$ to convert Celsius to Fahrenheit in Chapter 5. Let's put it to use in a random variable setting.
   a) Suppose your town has a mean January temperature of 11°C. What is the mean temperature in °F?
   b) Fortunately your local meteorologist has recently taken a statistics course and is keen to show off their newfound knowledge. They report that January has a standard deviation of 6°C. What is the standard deviation in °F?

**45. Reading** Pew Research Center tracks, among other things, reading habits of American adults. In an article from 2019, they said, "Overall, Americans read an average (mean) of 12 books per year, while the typical (median) American has read four books in the past 12 months."
   a) Explain how the "average" number of books read, and the number of books read by the "typical" American can be so different.
   b) What does this imply about the shape of the distribution of the number of books read by U.S. adults?
   c) If you were to randomly select an American adult, what can you say about the probability that you would select a person who has read more than four books in the past year?
   d) What can you say about the probability that the randomly selected person will have read more than 12 books in the past year?

**46. Household size** This probability distribution, which was given in this chapter, shows the probability of getting a particular household size when selecting a single U.S. household at random.

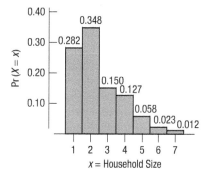

   a) Describe the shape of the probability distribution. Based on this, how would you expect the mean and median to compare?
   b) If $X$ is the random variable *Household Size*, what is $Med(X)$, the median value of $X$?
   c) What is $E(X)$ based on the graph?
   d) Recall that the category labeled "7" in the graph actually includes households with more than seven people. If the histogram had more complete information, what effect would that have on the shape of the graph, on $E(X)$ and on $Med(X)$?

**47. Repair calls** Suppose that the appliance shop in Exercise 19 plans an 8-hour day.
   a) Find the mean and standard deviation of the number of repair calls they should expect in a day.
   b) What assumption did you make about the repair calls?
   c) Use the mean and standard deviation to describe what a typical 8-hour day will be like.
   d) At the end of a day, a worker comments, "Boy, I'm tired. Today was sure unusually busy!" How many repair calls would justify such an observation?

**48. Stop!** Suppose the commuter in Exercise 20 has a 5-day workweek.
   a) Find the mean and standard deviation of the number of red lights the commuter should expect to hit in her week.
   b) What assumption did you make about the days?
   c) Use the mean and standard deviation to describe a typical week.
   d) Upon arriving home on Friday, the commuter remarks, "Wow! My commute was quick all week." How many red lights would it take to warrant a feeling of good luck?

**49. Tickets** A delivery company's trucks occasionally get parking tickets, and based on past experience, the company plans that the trucks will average 1.3 tickets a month, with a standard deviation of 0.7 tickets.

a) If they have 18 trucks, what are the mean and standard deviation of the total number of parking tickets the company will have to pay this month?
b) What assumption did you make in answering?

**50. Donations** Organizers of a televised fundraiser know from past experience that most people donate small amounts ($10–$25), some donate larger amounts ($50–$100), and a few people make very generous donations of $250, $500, or more. Historically, pledges average about $32 with a standard deviation of $54.

a) If 120 people call in pledges, what are the mean and standard deviation of the total amount raised?
b) What assumption did you make in answering this question?

**51. Fire!** An insurance company estimates that it should make an annual profit of $150 on each homeowner's policy written, with a standard deviation of $6000.

a) Why is the standard deviation so large?
b) If it writes only two of these policies, what are the mean and standard deviation of the annual profit?
c) If it writes 10,000 of these policies, what are the mean and standard deviation of the annual profit?
d) Is the company likely to be profitable? Explain.
e) What assumptions underlie your analysis? Can you think of circumstances under which those assumptions might be violated? Explain.

**52. Casino** A casino knows that people play the slot machines in hopes of hitting the jackpot but that most of them lose their dollar. Suppose a certain machine pays out an average of $0.92, with a standard deviation of $120.

a) Why is the standard deviation so large?
b) If you play 5 times, what are the mean and standard deviation of the casino's profit?
c) If gamblers play this machine 1000 times in a day, what are the mean and standard deviation of the casino's profit?
d) Is the casino likely to be profitable? Explain.

**53. U.S. cars** In the United States, about 9% of households are "car-free." The relative frequencies of households with different numbers of vehicles are shown below.

| Number of Vehicles | Relative Frequency of U.S. Households |
|---|---|
| 0 | 0.09 |
| 1 | 0.33 |
| 2 | 0.37 |
| 3 or more | 0.21 |

a) If you were to randomly select a U.S. household, what are the expected value and standard deviation of the random variable *Number of Vehicles*? (Treat "3 or more" as exactly 3 for these calculations.)
b) Make a histogram of this probability distribution and describe it. Remember to describe the shape, center, and spread in context.

c) If the category "3 or more" were expanded to show the percent of households with 3, 4, 5, and 6 *or more* vehicles, what effect would you expect this have on your description of the shape of the probability distribution?
d) If the category "3 or more" were expanded to show the percentage of households with 3, 4, 5, and 6 or more vehicles, what effect would this have on the expected value and standard deviation of the random variable?

**54. Belgian cars** The relative frequency distribution of the number of vehicles per household in Belgium is shown in the graph below. (Note that if you randomly select a Belgian household, this is also a probability distribution for the number of vehicles.)

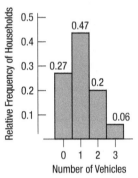

a) If you were to randomly select a Belgian household, calculate the expected value and standard deviation of the probability distribution of the random variable *Number of Vehicles*. (Treat "3 or more" as exactly 3 for these calculations.)
b) Write a couple of sentences comparing this probability distribution to that for the United States, as in the previous exercise.
c) There is more information available about the distribution of vehicles per household in Belgium.

| Number of Vehicles | 3 | 4 | 5 | More than 5 |
|---|---|---|---|---|
| Relative Frequency | 0.044 | 0.013 | 0.002 | 0.001 |

Use this additional information to make a more complete graph of the probability distribution. How does this affect the shape of the graph?
d) Recalculate the expected value and standard deviation using the new information above. For "More than 5" use 6 as the value. Compare to your answers from part a.
e) What does this tell you about the instruction to treat "3 or more" as exactly 3 in part a? Would similar additional information have more or less influence on these values for the United States?

**55. Cereal** The amount of cereal that can be poured into a small bowl varies with a mean of 1.5 ounces and a standard deviation of 0.3 ounces. A large bowl holds a mean of 2.5 ounces with a standard deviation of 0.4 ounces. You open a new box of cereal and pour one large and one small bowl.

a) How much more cereal do you expect to be in the large bowl?
b) What's the standard deviation of this difference?
c) If the difference follows a Normal model, what's the probability the small bowl contains more cereal than the large one?
d) What are the mean and standard deviation of the total amount of cereal in the two bowls?

e) If the total follows a Normal model, what's the probability you poured out more than 4.5 ounces of cereal in the two bowls together?
f) The amount of cereal the manufacturer puts in the boxes is a random variable with a mean of 16.3 ounces and a standard deviation of 0.2 ounces. Find the expected amount of cereal left in the box and the standard deviation.

**56. Pets** The American Veterinary Association claims that the annual cost of medical care for dogs averages $100, with a standard deviation of $30, and for cats averages $120, with a standard deviation of $35.
  a) What's the expected difference in the cost of medical care for dogs and cats?
  b) What's the standard deviation of that difference?
  c) If the costs can be described by Normal models, what's the probability that medical expenses are higher for someone's dog than for their cat?
  d) What concerns do you have?

**57. More cereal** In Exercise 55 we poured a large and a small bowl of cereal from a box. Suppose the amount of cereal that the manufacturer puts in the boxes is a random variable with mean 16.2 ounces and standard deviation 0.1 ounces.
  a) Find the expected amount of cereal left in the box.
  b) What's the standard deviation?
  c) If the weight of the remaining cereal can be described by a Normal model, what's the probability that the box still contains more than 13 ounces?

**58. More pets** You're thinking about getting two dogs and a cat. Assume that annual veterinary expenses are independent and have a Normal model with the means and standard deviations described in Exercise 56.
  a) Define appropriate variables and express the total annual veterinary costs you may have.
  b) Describe the model for this total cost. Be sure to specify its name, expected value, and standard deviation.
  c) What's the probability that your total expenses will exceed $400?

**59. Backpack backpack** Two students wanted to collect data to estimate how much longer it would take someone to run 50 yards while wearing a heavy backpack than it would without.
  One student recruited 8 volunteers, randomly divided them into two groups of four, and timed all the students running 50 yards. (The subjects in one of the treatment groups were wearing a heavy backpack.)
  Another student recruited only 4 volunteers and asked each of them to run 50 yards twice—once with a backpack and once without. Because the student had been paying attention in class, he knew it was best to randomly assign the order of the treatments—some ran with the backpack first, some second.
  Both students planned to compute the difference between the average time with and the average time without, to estimate how much time the backpack added to the run.
  a) Describe which plan you prefer and explain why.
  b) One of the two students has a plan that will result in an estimate of the backpack effect that has less variability in it than the other student's plan. Which student is it? Explain, without doing computations, why that student's estimate will have less variability in it.

**60. Rinse repeat** At a shampoo bottling factory, workers perform a number of repetitive tasks in bottling the product as it comes off the assembly line. The time it takes each worker to complete the task varies and sometimes the process is slower than other times. The company is wondering if fatigue is a problem. To test this theory, the company is going to insert 60-second breaks in-between groups of products. It hopes that a short break will actually improve overall speed.
  One supervisor proposes that two different groups of workers be chosen. One group will be randomly assigned to have the 60-second breaks inserted; the other will work as they typically have.
  Another supervisor proposes that each worker will have a period of time working with the 60-second break and a time without. A coin will be flipped to determine if the worker has the 60-second break before or after lunch.
  a) Describe which of these experimental designs is more effective. Justify your answer.
  b) One of these designs has less variability than the other. Can you explain, without computation, why the variability in one design is less than the variability in the other?

**61. Medley** In the 4 × 100 medley relay event, four swimmers swim 100 yards, each using a different stroke. A college team preparing for the conference championship looks at the times their swimmers have posted and creates a model based on the following assumptions:
  - The swimmers' performances are independent.
  - Each swimmer's times follow a Normal model.
  - The means and standard deviations of the times (in seconds) are as shown:

| Swimmer | Mean | SD |
| --- | --- | --- |
| 1 (backstroke) | 50.72 | 0.24 |
| 2 (breaststroke) | 55.51 | 0.22 |
| 3 (butterfly) | 49.43 | 0.25 |
| 4 (freestyle) | 44.91 | 0.21 |

  a) What are the mean and standard deviation for the relay team's total time in this event?
  b) The team's best time so far this season was 3:19.48. (That's 199.48 seconds.) Do you think the team is likely to swim faster than this at the conference championship? Explain.

**62. Bikes** Bicycles arrive at a bike shop in boxes. Before they can be sold, they must be unpacked, assembled, and tuned (lubricated, adjusted, etc.). Based on past experience, the shop manager makes the following assumptions about how long this may take:
  - The times for each setup phase are independent.
  - The times for each phase follow a Normal model.
  - The means and standard deviations of the times (in minutes) are as shown:

| Phase | Mean | SD |
|---|---|---|
| Unpacking | 3.5 | 0.7 |
| Assembly | 21.8 | 2.4 |
| Tuning | 12.3 | 2.7 |

a) What are the mean and standard deviation for the total bicycle setup time?
b) A customer decides to buy a bike like one of the display models but wants a different color. The shop has one, still in the box. The manager says they can have it ready in half an hour. Do you think the bike will be set up and ready to go as promised? Explain.

**63.** Farmers' market A farmer has 100 lb of apples and 50 lb of potatoes for sale. The market price of apples (per pound) for any given day is a random variable with a mean of 0.5 dollars and a standard deviation of 0.2 dollars. Similarly, for a pound of potatoes, the mean price is 0.3 dollars and the standard deviation is 0.1 dollars. It also costs him 2 dollars to bring all the apples and potatoes to a market. This market is busy with eager shoppers, so we can assume that he'll be able to sell all of each type of produce at that day's price.
a) Define your random variables, and use them to express the farmer's net income.
b) Find the mean.
c) Find the standard deviation of the net income.
d) Do you need to make any assumptions in calculating the mean? How about the standard deviation?
e) Are these assumptions reasonable? Explain.

**64.** Bike sale The bicycle shop in Exercise 52 will be offering 2 specially priced children's models at a sidewalk sale. The basic model will sell for $120 and the deluxe model for $150. Past experience indicates that sales of the basic model will have a mean of 5.4 bikes with a standard deviation of 1.2, and sales of the deluxe model will have a mean of 3.2 bikes with a standard deviation of 0.8 bikes. The cost of setting up for the sidewalk sale is $200.
a) Define random variables and use them to express the bicycle shop's net income.
b) What's the mean of the net income?
c) What's the standard deviation of the net income?
d) Do you need to make any assumptions in calculating the mean? How about the standard deviation?
e) Are these assumptions reasonable? Explain.

**65.** Coffee and doughnuts At a certain coffee shop, all the customers buy a cup of coffee; some also buy a doughnut. The shop owner believes that the number of cups they sell each day is normally distributed with a mean of 320 cups and a standard deviation of 20 cups. They also believe that the number of doughnuts they sell each day is independent of the coffee sales and is normally distributed with a mean of 150 doughnuts and a standard deviation of 12.
a) The shop is open every day but Sunday. Assuming day-to-day sales are independent, what's the probability they'll sell over 2000 cups of coffee in a week?
b) If they make a profit of 50 cents on each cup of coffee and 40 cents on each doughnut, can they reasonably expect to have a day's profit of over $300? Explain.
c) What's the probability that on any given day they'll sell a doughnut to more than half of their coffee customers?

**66.** Weightlifting The Atlas BodyBuilding Company (ABC) sells "starter sets" of barbells that consist of one bar, two 20-pound weights, and four 5-pound weights. The bars weigh an average of 10 pounds with a standard deviation of 0.25 pounds. The weights average the specified amounts, but the standard deviations are 0.2 pounds for the 20-pounders and 0.1 pounds for the 5-pounders. We can assume that all the weights are normally distributed and independent.
a) ABC ships these starter sets to customers in two boxes: The bar goes in one box and the six weights go in another. What's the probability that the total weight in that second box exceeds 60.5 pounds?
b) It costs ABC $0.40 per pound to ship the box containing the weights. Because it's an odd-shaped package, though, shipping the bar costs $0.50 a pound plus a $6.00 surcharge. Find the mean and standard deviation of the company's total cost for shipping a starter set.
c) Suppose a customer puts a 20-pound weight at one end of the bar and the four 5-pound weights at the other end. Although they expect the two ends to weigh the same, this might not be exactly the case. What's the probability the difference in weight between the two ends is more than a quarter of a pound?

## JUST CHECKING

**Answers**

1. a)

| Outcome | X = Cost | Probability |
|---|---|---|
| Recharging works | $60 | 0.75 |
| Replace control unit | $200 | 0.25 |

b) $60(0.75) + 200(0.25) = \$95$.
c) Car owners with this problem will spend an average of $95 to get it fixed.

2. $100 + 100 = 200$ seconds.
a) $\sqrt{50^2 + 50^2} = 70.7$ seconds.
b) We assumed the times for the two customers are independent.

# 16

# Probability Models

Suppose a cereal manufacturer puts pictures of famous athletes on cards in boxes of cereal, in the hope of increasing sales. The manufacturer announces that 20% of the boxes contain a picture of Simone Biles, 30% a picture of Megan Rapinoe, and the rest a picture of Serena Williams.

Sound familiar? In Chapter 10 we simulated to find the number of boxes we'd need to open to get one of each card. That's a fairly complex question and one well suited for simulation. But many important questions can be answered more directly by using probability models.

## Searching for Simone: Bernoulli Trials

Suppose you're a huge Simone Biles fan. You don't care about completing the whole sports card collection, but you've just *got* to have Simone's picture. How many boxes do you expect you'll have to open before you find her? This isn't the same question that we asked before, but this situation is simple enough for a probability model.

We'll keep the assumption that pictures are distributed at random and we'll trust the manufacturer's claim that 20% of the cards are Simone. So, when you open the box, the probability that you succeed in finding Simone is 0.20. Now we'll call the act of opening *each* box a trial, and note that:

- There are only two possible outcomes (called *success* and *failure*) on each trial. Either you get Simone's picture (success), or you don't (failure).
- In advance, the probability of success, denoted $p$, is the same on every trial. Here $p = 0.20$ for each box.
- As we proceed, the trials are independent. Finding Simone in the first box does not change what might happen when you reach for the next box.

429

Daniel Bernoulli (1700–1782) was the nephew of Jacob, whom you saw in Chapter 14. He was the first to work out the mathematics for what we now call Bernoulli trials.

Situations like this occur often and are called **Bernoulli trials**. Common examples of Bernoulli trials include tossing a coin, looking for defective products rolling off an assembly line, or blindly guessing on a true/false question you don't know the answer to.

Back to finding Simone. We want to know how many boxes we'll need to open to find her card. Let's call this random variable $Y = \#\ of\ Boxes$, and build a probability model for it. What's the probability you find her picture in the first box of cereal? It's 20%, of course. We could write $P(Y = 1) = 0.20$.

How about the probability that you don't find Simone until the second box? Well, that means you fail on the first trial and then succeed on the second. With the probability of success 20%, the probability of failure, denoted $q$, is $1 - 0.2 = 80$. Because the trials are independent, the probability of getting your first success on the second trial is $P(Y = 2) = (0.8)(0.2) = 0.16$.

Of course, you could have a run of bad luck. Maybe you won't find Simone until the fifth box of cereal. What are the chances of that? You'd have to fail 4 straight times and then succeed, so $P(Y = 5) = (0.8)^4(0.2) = 0.08192$.

How many boxes might you expect to have to open? We could reason that because Simone's picture is in 20% of the boxes, or 1 in 5, we expect to find her picture, on average, in the fifth box; that is, $E(Y) = \dfrac{1}{0.2} = 5$ boxes. That's correct, but not easy to prove.[1]

## The 10% "Rule"

One of the important requirements for Bernoulli trials is that the trials be independent. Sometimes that's a reasonable assumption—when tossing a coin or rolling a die, for example. But that becomes a problem when (often!) we're looking at situations involving samples chosen without replacement. We said that whether we find a Simone Biles card in one box has no effect on the probabilities in other boxes. This is *almost* true. Technically, if exactly 20% of the boxes have Simone cards, then when you find one, you've reduced the number of remaining Simone cards. With a few million boxes of cereal, though, the difference is hardly worth mentioning. But if you knew there were 2 Simone Biles cards hiding in the 10 boxes of cereal you see on the market shelf, then finding one in the first box you try would clearly change your chances of finding his picture in the next box.

If we had an infinite number of boxes, there wouldn't be a problem. It's selecting from a finite population that causes the probabilities to change, making the trials not independent. Obviously, taking 2 out of 10 boxes changes the probability. Taking even a few hundred out of millions, though, makes very little difference. Fortunately, it turns out that if we look at less than 10% of the population, we can pretend that the trials are independent and still calculate probabilities that are quite accurate. (Close enough for statisticians!) That's our 10% rule of thumb:

> **The 10% Condition:** Bernoulli trials must be independent. When sampling without replacement the independence condition is violated. But it's still okay to proceed as though the trials were independent as long as we randomly sample less than 10% of the population.[2]

### JUST CHECKING

1. Think about each of these situations. Are these random variables based on Bernoulli trials? If you don't think so, explain why not.

    a) The waitstaff at a small restaurant consists of 5 men and 8 women. They write their names on slips of paper and the boss chooses 4 people at random to work overtime on a holiday weekend. We count the number of women who are chosen.

---

[1] See the Math Box, coming soon to a textbook near you.

[2] The 10% Condition almost makes it sound as though having more data is a bad thing. It's not! Having more data—assuming they're responsibly collected—is always a good thing in Statistics. But data that amount to more than 10% of the population require more advanced methods than you'll see in this book.

b) In the United States about 1 in every 90 pregnant women gives birth to twins. Among a group of pregnant women who work in the same office, we count the number of them who gave birth to twins.
c) We count the number of times a woman who has been pregnant 3 times gave birth to twins.
d) We pick 40 M&M's at random from a large bag, counting how many of each color we get.
e) A small town's merchant's association says that 26% of all businesses there are owned by women. You call 15 businesses randomly chosen from the 77 listed on the website for merchant association, counting the number owned by women.

# The Geometric Model: Waiting for Success

**TI-nspire**
**Geometric probabilities.** See what happens to a geometric model as you change the probability of success.

*NOTATION ALERT*
Now we have two more reserved letters. Whenever we deal with Bernoulli trials, $p$ represents the probability of success, and $q$ the probability of failure. (Of course, $q = 1 - p$.)

We want to model how long it will take to achieve the first success in a series of Bernoulli trials. The model that tells us this probability is called the **Geometric probability model**. Geometric models are completely specified by one parameter, $p$, the probability of success, and are denoted Geom($p$). Achieving the first success on trial number $x$ requires first experiencing $x - 1$ failures, so the probabilities are easily expressed by a formula.

### Geometric Probability Model for Bernoulli Trials: Geom($p$)

$p$ = probability of success (and $q = 1-p$ = probability of failure)
$X$ = number of trials until the first success occurs

$$P(X = x) = q^{x-1}p$$

Expected value: $E(X) = \mu = \dfrac{1}{p}$  Standard deviation: $\sigma = \sqrt{\dfrac{q}{p^2}} = \dfrac{\sqrt{1-p}}{p}$

Does this probability model look familiar? In Chapter 15 we saw that if someone playing *Dungeons and Dragons* makes many "attack rolls," then the probability of achieving their first "critical hit" on roll number $x$ was $P(X = x) = (0.95)^{x-1}(0.05)$. That was in fact an instance of the Geometric model. The probability of a critical hit on any particular roll is 0.05 because it's one face on a 20-sided die. In the more general model above we've replaced 0.05 with the unspecified parameter $p$, but otherwise it's the same probability formula. In Chapter 15 we didn't bring up expected value or standard deviation. Now that we have these formulas we can say that the expected number of attack rolls required until achieving a critical hit is $\dfrac{1}{p} = \dfrac{1}{0.05} = 20$. If an avid D&D player always counts the number of attack rolls needed to get their first critical hit, then that count would average 20 over many sessions of playing. Unsurprising, really, since it's a 20-sided die, yes?

### FOR EXAMPLE
#### Spam and the Geometric Model

Postini is a global company specializing in communications security. The company monitors over 1 billion Internet messages per day and recently reported that 91% of e-mails are spam!

Let's assume that your e-mail is typical—91% spam. We'll also assume you aren't using a spam filter, so every message gets dumped in your inbox. And, because spam comes from many different sources, we'll consider your messages to be independent.

QUESTION: Overnight your inbox collects e-mail. When you first check your e-mail in the morning, about how many spam e-mails should you expect to have to wade through and discard before you find a real message? What's the probability that the fourth message in your inbox is the first one that isn't spam?

**ANSWER:** When I check my e-mails one by-one:
- There are two possible outcomes each time: a real message (success) or spam (failure).
- Because 91% of all e-mails are spam, the probability of success is
$$p = 1 - 0.91 = 0.09$$
- My messages arrive in random order from many different sources and are far fewer than 10% of all e-mail messages. I can treat them as independent.

Let $X$ = the number of e-mails I'll check until I find a real message. I can use the model Geom(0.09).

$$E(X) = \frac{1}{p} = \frac{1}{0.09} = 11.1$$

$$P(X = 4) = (0.91)^3(0.09) = 0.0678$$

On average, I expect to have to check just over 11 e-mails before I find a real message. There's slightly less than a 7% chance that my first real message will be the fourth one I check.

Note that this probability calculation isn't new. It's simply Chapter 13's Multiplication Rule used to find $P(\text{spam} \cap \text{spam} \cap \text{spam} \cap \text{real})$.

## MATH BOX

We want to find the mean (expected value) of random variable $X$, using a Geometric model with probability of success $p$.

First, write the probabilities:

| $x$ | 1 | 2 | 3 | 4 | ... |
|---|---|---|---|---|---|
| $P(X = x)$ | $p$ | $qp$ | $q^2p$ | $q^3p$ | ... |

The expected value is: $E(X) = 1p + 2qp + 3q^2p + 4q^3p + \cdots$

Let $p = 1 - q$: $\quad = (1-q) + 2q(1-q) + 3q^2(1-q) + 4q^3(1-q) + \cdots$

Simplify: $\quad = 1 - q + 2q - 2q^2 + 3q^2 - 3q^3 + 4q^3 - 4q^4 + \cdots$

$\quad = 1 + q + q^2 + q^3 + \cdots$

That's an infinite geometric series, with first term 1 and common ratio $q$:

$\quad = \dfrac{1}{1-q}$

So, finally... $E(X) = \dfrac{1}{p}$.

## STEP-BY-STEP EXAMPLE

### Working with a Geometric Model

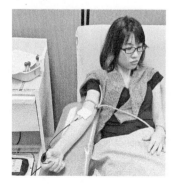

People with O-negative blood are called "universal donors" because O-negative blood can be given to anyone else, regardless of the recipient's blood type. Only about 6% of people have O-negative blood.

**QUESTIONS:**

1. If donors line up at random for a blood drive, how many do you expect to examine before you find someone who has O-negative blood?

2. What's the probability that the first O-negative donor found is one of the first four people in line?

| | |
|---|---|
| **THINK**    **PLAN** State the questions. | I want to estimate how many people I'll need to check to find an O-negative donor, and the probability that 1 of the first 4 people is O-negative. |
| Check to see that these are Bernoulli trials. | ✓ There are two outcomes: $$\text{Success} = \text{O-negative}$$ $$\text{Failure} = \text{other blood types}$$ ✓ The probability of success for each person is $p = 0.06$, because they lined up randomly. ✓ **10% Condition**: Trials aren't independent because the population is finite, but the donors lined up are fewer than 10% of all possible donors. |
| **VARIABLE** Define the random variable. **MODEL** Specify the model. | Let $X$ = number of donors until one is O-negative. I can model $X$ with Geom(0.06). |
| **SHOW**    **MECHANICS** Find the mean. Calculate the probability of success on one of the first four trials. That's the probability that $X = 1, 2, 3,$ or $4$.[3] | $E(X) = \dfrac{1}{0.06} \approx 16.7$ $P(X \le 4) = P(X=1) + P(X=2) + \\ \phantom{P(X \le 4) =} P(X=3) + P(X=4)$ $\phantom{P(X \le 4) } = (0.06) + (0.94)(0.06) + \\ \phantom{P(X \le 4) = } (0.94)^2(0.06) + (0.94)^3(0.06)$ $\phantom{P(X \le 4) } \approx 0.2193$ |
| **TELL**    **CONCLUSION** Interpret your results in context. | Blood drives such as this one expect to examine an average of 16.7 people to find a universal donor. About 22% of the time there will be one within the first 4 people in line. |

## TI TIPS

### Finding Geometric Probabilities

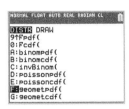

Your TI knows the Geometric model. Just as you saw back in Chapter 5 with the Normal model, commands to calculate probability distributions are found in the 2nd DISTR menu. Have a look. After many others (Yes, there's still more to learn!) you'll see two Geometric probability functions at the bottom of the list.

- geometpdf(.

  The "pdf" stands for "probability density function."[4] This command allows you to find the probability of any *individual* outcome. You need only specify *p*, which defines the Geometric model, and *x*, which indicates the number of trials until you get a success. The format is geometpdf(p,x).

---

[3] Alternatively, you could compute the probability that *none* of the first four people in line are universal donors, and subtract that from 1. Because the trials are independent, you can multiply $P(failure) = 0.94$ four times, once for each of the first four people in line: $P(X \le 4) = 1 - (0.94)^4 = 0.2193$.

[4] It is more properly called a probability *mass* function, or pmf, for *discrete* random variables like this one. But some technology platforms use the shortcut notation "pdf" for both continuous and discrete random variables.

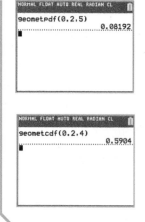

For example, suppose we want to know the probability that we find our first Simone Biles picture in the fifth box of cereal. Because Simone is in 20% of the boxes, we enter `geometpdf(` with `p:0.2`, `X value:5`, then go to Paste and hit ENTER (twice). The calculator says there's about an 8% chance.

◆ `geometcdf(`.

This is the "cumulative distribution function," meaning that it finds the accumulated sum of the probabilities of several consecutive possible outcomes. In general, the command `geometcdf(p,x)` calculates the probability of finding the first success *on or before* the *x*th trial.

Let's find the probability of getting a Simone Biles picture by the time we open the fourth box of cereal—in other words, the probability our first success comes on the first box, or the second, or the third, or the fourth. Again we specify $p = 0.2$, and now use $x = 4$. The command `geometcdf` with `p:0.2`, `x value:4` calculates all the probabilities and sums them.[5] There's about a 59% chance that our quest for a Simone photo will succeed by the time we open the fourth box.

# The Binomial Model: Counting Successes

We can use the Bernoulli trials to answer other common questions. Suppose you buy 5 boxes of cereal. What's the probability you get *exactly* 2 pictures of Simone Biles? Before, we asked how long it would take until our first success. Now we want to find the probability of getting 2 successes among the 5 trials. We are still talking about Bernoulli trials, but we're asking a different question.

This time we're interested in the *number of successes* in the 5 trials, so we'll call it $X$ = number of successes. We want to find $P(X = 2)$. This is an example of a **Binomial probability**. It takes two parameters to define this **Binomial model**: the number of trials, $n$, and the probability of success, $p$. We denote this model $Binom(n, p)$. Here, $n = 5$ trials, and $p = 0.2$, the probability of finding a Simone Biles card in any trial.

Exactly 2 successes in 5 trials means 2 successes and 3 failures. It seems logical that the probability should be $(0.2)^2(0.8)^3$. Too bad: It's not quite that easy. That calculation would give you the probability of finding Simone in the first 2 boxes and not in the next 3— *in that order*. But you could find Simone in, say, the third and fifth boxes and still have 2 successes. The probability of those outcomes in that particular order is $(0.8)(0.8)(0.2)(0.8)(0.2)$. That's also $(0.2)^2(0.8)^3$. In fact, the probability of two successes and three failures will always be the same, no matter what order the successes and failures occur in. Anytime we get 2 successes in 5 trials, regardless of the order, the probability will be $(0.2)^2(0.8)^3$. We just need to count all the possible orders in which the outcomes can occur.

Fortunately, these possible orders are *disjoint*. (For example, if your two successes came on the first two trials, they couldn't come on the last two.) So we could use the Addition Rule and add up the probabilities, but because they're are all the same, we really only need to know how many orders are possible. For small *n*'s, we can just make a tree diagram and count the branches. For larger numbers this isn't practical: fortunately, there's a formula for that.

Each different order in which we can have $k$ successes in $n$ trials is called a "combination." The total number of ways that can happen is written $\binom{n}{k}$ or $_nC_k$ and pronounced "*n* choose *k*."

---

[5] Well, that's *equivalent* to what the command does. The command actually invokes a single preprogrammed formula for the cumulative probability, one that doesn't require summing individual probabilities. We're not giving that formula here (just use the built-in calculator function if you need it) but if you're curious, look back at the D&D example in Chapter 15. A formula was given there for a cumulative distribution function, and you should be able to see how to generalize it for any Geometric model.

$$\binom{n}{k} = \frac{n!}{k!(n-k)!} \text{ where } n!(\text{pronounced "}n\text{ factorial"}) = n \times (n-1) \times \cdots \times 1$$

For 2 successes in 5 trials,

$$\binom{5}{2} = \frac{5!}{2!(5-2)!} = \frac{5 \times 4 \times 3 \times 2 \times 1}{2 \times 1 \times 3 \times 2 \times 1} = \frac{5 \times 4}{2 \times 1} = 10$$

So there are 10 ways to get exactly 2 Simone pictures in 5 boxes, and the probability of each is $(0.2)^2(0.8)^3$. Now we can find what we wanted:

$$P(\#\text{success} = 2) = 10(0.2)^2(0.8)^3 = 0.2048$$

In general, the probability of exactly $k$ successes in $n$ trials is $\binom{n}{k}p^k q^{n-k}$.

> **NOTATION ALERT**
> Now punctuation! Throughout mathematics $n!$, pronounced "$n$ factorial," is the product of all the integers from 1 to $n$. For example, $4! = 4 \times 3 \times 2 \times 1 = 24$.

It's not hard to find the expected value for a binomial random variable. If we have 5 boxes, and Simone's picture is in 20% of them, then we would expect to have $5(0.2) = 1$ success. If we had 100 trials with probability of success 0.2, how many successes would you expect? Can you think of any reason not to say 20? It seems so simple that most people wouldn't even stop to think about it. You just multiply the probability of success by $n$. In other words, $E(X) = np$. Not fully convinced? We prove it in the next Math Box.

The standard deviation is less obvious; you can't just rely on your intuition. Fortunately, the formula for the standard deviation also boils down to something simple: $SD(X) = \sqrt{npq}$. (If you're curious about where that comes from, it's in the Math Box too!) In 100 boxes of cereal, we expect to find 20 Simone Biles cards, with a standard deviation of $\sqrt{100 \times 0.8 \times 0.2} = 4$ pictures.

Time to summarize. A Binomial probability model describes the number of successes in a specified number of trials. It takes two parameters to specify this model: the number of trials $n$ and the probability of success $p$.

> **TI-nspire**
> **Binomial probabilities.** Do-it-yourself Binomial models! Watch the probabilities change as you control $n$ and $p$.
>
>
> **APPLET**
> Explore the Binomial model

### Binomial Probability Model for Bernoulli Trials: Binom($n, p$)

$n$ = number of independent trials
$p$ = probability of success (and $q = 1 - p$ = probability of failure)
$X$ = number of successes in $n$ trials

$$P(X = x) = \binom{n}{x}p^x q^{n-x}, \text{ where } \binom{n}{x} = \frac{n!}{x!(n-x)!}$$

Mean: $\mu = np$
Standard Deviation: $\sigma = \sqrt{npq}$

### MATH BOX

To derive the formulas for the mean and standard deviation of a Binomial model, we start with the most basic situation.

Consider a single Bernoulli trial with probability of success $p$. Let's find the mean and variance of the number of successes.

Here's the probability model for the number of successes:

| $x$ | 0 | 1 |
|---|---|---|
| $P(X = x)$ | $q$ | $p$ |

Find the expected value:
$E(X) = 0q + 1p$
$E(X) = p$

And now the variance:
$Var(X) = (0 - p)^2 q + (1 - p)^2 p$
$= p^2 q + q^2 p$
$= pq(p + q)$
$= pq(1)$
$Var(X) = pq$

What happens when there is more than one trial, though? A Binomial model simply counts the number of successes in a series of $n$ independent Bernoulli trials. That makes it easy to find the mean and standard deviation of a binomial random variable, $Y$.

$$\text{Let } Y = X_1 + X_2 + X_3 + \cdots + X_n$$
$$E(Y) = E(X_1 + X_2 + X_3 + \cdots + X_n)$$
$$= E(X_1) + E(X_2) + E(X_3) + \cdots + E(X_n)$$
$$= p + p + p + \cdots + p \text{ (There are } n \text{ terms.)}$$

So, as we thought, the mean is $E(Y) = np$. And because the trials are independent, the Pythagorean Theorem of Statistics tells us that the variances add:

$$Var(Y) = Var(X_1 + X_2 + X_3 + \cdots + X_n)$$
$$= Var(X_1) + Var(X_2) + Var(X_3) + \cdots + Var(X_n)$$
$$= pq + pq + pq + \cdots + pq \text{ (Again, } n \text{ terms.)}$$
$$Var(Y) = npq$$

Voilà! The standard deviation is $SD(Y) = \sqrt{npq}$.

## FOR EXAMPLE
### Spam and the Binomial Model

**RECAP:** The communications monitoring company Postini has reported that 91% of e-mail messages are spam. Suppose your inbox contains 25 messages.

**QUESTION:** What are the mean and standard deviation of the number of real messages you should expect to find in your inbox? What's the probability that you'll find only 1 or 2 real messages?

**ANSWER:** I assume that messages arrive independently and at random, with the probability of success (a real message) $p = 1 - 0.91 = 0.09$. Let $X$ = the number of real messages among 25. I can use the model $Binom(25, 0.09)$.

$$E(X) = np = 25(0.09) = 2.25$$
$$SD(X) = \sqrt{npq} = \sqrt{25(0.09)(0.91)} = 1.43$$
$$P(X = 1 \text{ or } 2) = P(X = 1) + P(X = 2)$$
$$= \binom{25}{1}(0.09)^1(0.91)^{24} + \binom{25}{2}(0.09)^2(0.91)^{23}$$
$$= 0.2340 + 0.2777$$
$$= 0.5117$$

Among 25 e-mail messages, I expect to find an average of 2.25 that aren't spam, with a standard deviation of 1.43 messages. There's just over a 50% chance that 1 or 2 of my 25 e-mails will be real messages.

## STEP-BY-STEP EXAMPLE
### Working with a Binomial Model

Suppose 20 donors come to a blood drive. Recall that 6% of people are "universal donors."

**QUESTION:**
1. What are the mean and standard deviation of the number of universal donors among them?
2. What is the probability that there are 2 or 3 universal donors?

| | | |
|---|---|---|
| **THINK** PLAN State the question. | | I want to know the mean and standard deviation of the number of universal donors among 20 people and the probability that there are 2 or 3 of them. |
| Check to see that these are Bernoulli trials. | | ✓ There are two outcomes: |
| | | $\qquad$ Success = O-negative |
| | | $\qquad$ Failure = other blood types |
| | | ✓ $p = 0.06$, because people have lined up at random. |
| | | ✓ **10% Condition**: Trials are not independent, because the population is finite, but fewer than 10% of all possible donors are lined up. |
| VARIABLE Define the random variable. | | Let $X$ = number of O-negative donors among $n = 20$ people. |
| MODEL Specify the model. | | I can model $X$ with $Binom(20, 0.06)$. |
| **SHOW** MECHANICS Find the expected value and standard deviation. | | $E(X) = np = 20(0.06) = 1.2$ |
| | | $SD(X) = \sqrt{npq} = \sqrt{20(0.06)(0.94)} \approx 1.06$ |
| Calculate the probability. | | $P(X = 2 \text{ or } 3) = P(X = 2) + P(X = 3)$ |
| | | $= \binom{20}{2}(0.06)^2(0.94)^{18}$ |
| | | $\quad + \binom{20}{3}(0.06)^3(0.94)^{17}$ |
| | | $\approx 0.2246 + 0.0860$ |
| | | $= 0.3106$ |
| **TELL** CONCLUSION Interpret your results in context. | | In groups of 20 randomly selected blood donors, I expect to find an average of 1.2 universal donors, with a standard deviation of 1.06. About 31% of the time, I'd find 2 or 3 universal donors among the 20 people. |

## TI TIPS

### Finding Binomial Probabilities

Remember how the calculator handles Geometric probabilities? Well, the commands for finding Binomial probabilities are essentially the same. Again you'll find them in the 2nd DISTR menu.

◆ `binompdf(`

This function allows you to find the probability of an *individual* outcome. You need to define the Binomial model by specifying $n$ and $p$, and then indicate the desired number of successes, $x$. The format is `binompdf(n,p,X)`.

For example, recall that Simone Biles's picture is in 20% of the cereal boxes. Suppose that we want to know the probability of finding Simone exactly twice among 5 boxes of cereal. We enter `binompdf(` with `trials:5, p:0.2, x value:2`, then go to `Paste` and hit `ENTER` (twice). There's about a 20% chance of getting two pictures of Simone in five boxes of cereal.

- `binomcdf(`

  Need to add several consecutive Binomial probabilities? To find the total probability of getting $x$ or fewer successes among the $n$ trials, use the cumulative Binomial density function `binomcdf(n,p,X)`.

  For example, suppose we have 10 boxes of cereal and wonder about the probability of finding up to 4 pictures of Simone. That's the probability of 0, 1, 2, 3 or 4 successes, so we specify the command binomcdf( with `trials:10, p:0.2, x value:4`. Pretty likely!

  Of course "up to 4" allows for the possibility that we end up with none. What's the probability we get at least 4 pictures of Simone in 10 boxes? Well, "at least 4" means "not 3 or fewer." That's the complement of 0, 1, 2, or 3 successes. Have your TI evaluate `1-binomcdf(10,.2,3)`. There's about a 12% chance we'll find at least 4 pictures of Simone in 10 boxes of cereal.

### JUST CHECKING

2. The Pew Research Center reports that it is able to contact only 76% of randomly selected households drawn for telephone surveys. Suppose a pollster has a list of 12 calls to make.

   a) Why can these phone calls be considered Bernoulli trials?
   b) Find the probability that the fourth call is the first one that makes contact.
   c) Find the expected number of successful calls out of the 12.
   d) Find the standard deviation of the number of successful calls.
   e) Find the probability that exactly 9 of the 12 calls are successful.
   f) Find the probability that at least 9 of the calls are successful.

# The Normal Model to the Rescue!

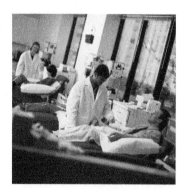

Suppose the Tennessee Red Cross anticipates the need for at least 1850 units of O-negative blood this year. It estimates that it will collect blood from 32,000 donors. How great is the risk that the Tennessee Red Cross will fall short of meeting its need? We've just learned how to calculate such probabilities. We can use the Binomial model with $n = 32,000$ and $p = 0.06$. The probability of getting *exactly* 1850 units of O-negative blood from 32,000 donors is $\binom{32000}{1850} \times 0.06^{1850} \times 0.94^{30150}$. Very few calculators can calculate that first factor (it has more than 100,000 digits). And that's just the beginning. The problem said *at least* 1850, so we have to do it again for 1851, for 1852, and all the way up to 32,000. No thanks.

There's one thing we can often do when faced with a problem that's computationally difficult: Use a simulation. Simulating 32,000 blood donors would be next to impossible using physical objects like playing cards, and using a random digits table would take a horrifically long time. But computers are great at this sort of thing. We wrote a very short program (one line long) that simulates 32,000 blood donors, each of which has a 6% chance of being a universal donor, and that tells you how many universal donors there were among them. We got 1936 universal donors, slightly more than the 6% you'd expect in the *very* long run.

Even though *each* of the 32,000 donors was a single trial in the Binomial probability model, the simulation we did involved just *one trial* in the larger context of simulating a whole year's worth of blood donations in Tennessee. With one trial we can't tell what's really a typical overall outcome, so we did a lot more. In fact, we repeated a simulation of 32,000 donors again and again[6] until we had done it 10,000 times in all, each time counting how many universal blood donors there were among all of Tennessee's 32,000 blood donors. Among the 10,000 trials we simulated, we found that 9530 of them resulted in 1850 or more universal donors. That's about 95%. So the probability appears to be quite high—approximately 95%—that if Tennessee gets 32,000 blood donations in a year, it will meet its need for at least 1850 units of O-negative blood, the universal donor type.

We've used our simulation results to address the question of interest, but before we leave it behind, let's take a look at the distribution of all 10,000 outcomes from our simulation. The histogram shows the results of all 10,000 simulated instances of a year's worth of Tennessee blood donors.

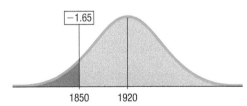

Whoa! This histogram looks very much like a Normal distribution! It's certainly unimodal, symmetric, and bell-curve-shaped. Although our simulation on a computer made it easy to count exactly how many outcomes met the 1850-universal-donor goal, this histogram suggests that we could have used the Normal model for the outcomes instead. Let's do that now and see whether we get a similar answer.

The center of this distribution appears to be around 1925 or so. But we needn't guess—this histogram is showing us outcomes of a Binomial probability model, and we have a formula for the mean: $\mu = np = (32,000)(0.06) = 1920$. In fact, we have a formula for the standard deviation too: $\sigma = \sqrt{npq} = \sqrt{(32,000)(0.06)(0.94)} = 42.5$. That means that even though the actual distribution of the count of universal donors is $Binom(32000, 0.06)$, it looks like a very good approximation would be $N(1920, 42.5)$.

In fact, we're really glad to see the Normal model pop up here, because to determine the probability of seeing at least 1850 universal donors, we need to find only a single area—the area to the right of 1850 under our bell curve. Using the Binomial probability model directly would have involved adding up all those different outcomes, which we've already acknowledged would be a real pain.

We haven't done this in a little while, so let's refresh our memories. The $z$-score associated with the count of interest, 1850, is $z = \dfrac{X - \mu}{\sigma} = \dfrac{1850 - 1920}{42.5} = -1.65$. Sketch a picture:

Now we use technology (or the standard Normal table) to determine the area to the right of a $z$-score of $-1.65$:

$$P(X \geq 1850) = P(z \geq -1.65) = 0.95.$$

Happily, that's just what we found when we looked more directly at the output from our computer simulation.

Can we always use a Normal model to make estimates of Binomial probabilities? No. Consider the Simone Biles situation—pictures in 20% of the cereal boxes. If we buy 5 boxes, the actual Binomial probabilities that we get 0, 1, 2, 3, 4, or 5 pictures of Simone are 33%, 41%, 20%, 5%, 1%, and 0.03%, respectively. The first histogram at left shows that this probability model is skewed. That makes it clear that we should not try to estimate these probabilities by using a Normal model.

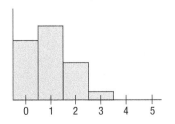

---

[6]This wasn't hard—we just changed a single number in our original single-line computer code! It took less than a second to run the code. (We used a computer language called R that's especially useful for doing simulations.)

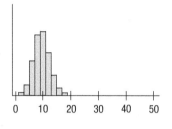

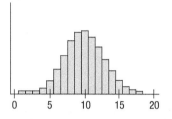

**TI-*nspire***

**How close to Normal?** How well does a Normal curve fit a Binomial model? Check out the Success/Failure Condition for yourself.

Now suppose we open 50 boxes of this cereal and count the number of Simone pictures we find. The second histogram shows this probability model. It is centered at $np = 50(0.2) = 10$ pictures, as expected, and it appears to be fairly symmetric around that center. Let's have a closer look.

The third histogram again shows *Binom*(50, 0.2), this time magnified somewhat and centered at the expected value of 10 pictures of Simone. It looks close to Normal, for sure. With this larger sample size, it appears that a Normal model might be a useful approximation.

A Normal model, then, is a close enough approximation only for a large enough number of trials. And what we mean by "large enough" depends on the probability of success. We'd need a larger sample if the probability of success were very low (or very high). It turns out that a Normal model works pretty well if we expect to see at least 10 successes and 10 failures. We call this the **Success/Failure Condition**.

**The Success/Failure Condition:** A Binomial model can be approximated by a Normal model if we expect at least 10 successes and 10 failures:

$$np \geq 10 \text{ and } nq \geq 10.$$

### MATH BOX

Let's see where the magic number 10 comes from. A Normal model extends infinitely in both directions. But a Binomial model must have between 0 and $n$ successes, so if we use a Normal to approximate a Binomial, we have to cut off its tails. That's not very important if the center of the Normal model is so far from 0 and $n$ that the lost tails have only a negligible area. More than three standard deviations should do it, because a Normal model has little probability past that.

So the mean needs to be at least 3 standard deviations from 0 and at least 3 standard deviations from $n$. Let's just look at the 0 end.

| We require: | $\mu - 3\sigma > 0$ |
| Or in other words: | $\mu > 3\sigma$ |
| For a Binomial, that's: | $np > 3\sqrt{npq}$ |
| Squaring yields: | $n^2 p^2 > 9npq$ |
| Now simplify: | $np > 9q$ |
| Because $q \leq 1$, we can require: | $np > 9$ |

For simplicity, we usually require that $np$ (and $nq$ for the other tail) be at least 10 to use the Normal approximation, the Success/Failure Condition.

### FOR EXAMPLE

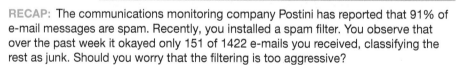

Spam and the Normal Approximation to the Binomial

**RECAP:** The communications monitoring company Postini has reported that 91% of e-mail messages are spam. Recently, you installed a spam filter. You observe that over the past week it okayed only 151 of 1422 e-mails you received, classifying the rest as junk. Should you worry that the filtering is too aggressive?

**QUESTION:** What's the probability that no more than 151 of 1422 e-mails is a real message?

**ANSWER:** I assume that messages arrive randomly and independently, with a probability of success (a real message) $p = 0.09$. The model *Binom*(1422, 0.09) applies, but will be hard to work with. Checking conditions for the Normal approximation, I see that:

✓ These messages represent less than 10% of all e-mail traffic.

✓ I expect $np = (1422)(0.09) = 127.98$ real messages and $nq = (1422)(0.91) = 1294.02$ spam messages, both far greater than 10.

It's okay to approximate this Binomial probability by using a Normal model.

$$\mu = np = 1422(0.09) = 127.98$$
$$\sigma = \sqrt{npq} = \sqrt{1422(0.09)(0.91)} \approx 10.79$$
$$P(x \leq 151) = P\left(z \leq \frac{151 - 127.98}{10.79}\right)$$
$$= P(z \leq 2.13)$$
$$= 0.9834$$

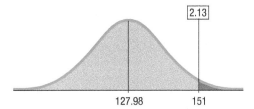

Among my 1422 e-mails, there's over a 98% chance that no more than 151 of them were real messages, so the filter may be working properly.

## A Word About Continuous Random Variables

Besides the problem of trimming the tails, there's another problem with approximating a Binomial model with a Normal model. The Binomial is discrete, giving probabilities for specific counts, but the Normal models a *continuous* random variable that can take on *any value*. For continuous random variables, we can no longer list all the possible outcomes and their probabilities, as we could for discrete random variables.[7]

As we saw in the previous chapter, models for continuous random variables give probabilities for *intervals* of values. So, when we use the Normal model, we no longer calculate the probability that the random variable equals a *particular* value, but only that it lies *between* two values. We won't calculate the probability of getting exactly 1850 units of blood, but we have no problem approximating the probability of getting 1850 *or more*, which was, after all, what we really wanted.[8]

### JUST CHECKING

3. Let's think about the Pew Research pollsters one more time. They tell us they are successful in contacting 76% of the households randomly selected for telephone surveys. When surveying public opinion, they hope to poll at least 1000 adults. Suppose they have compiled a list of 1300 phone numbers to call. What's the probability that they'll reach enough people?

   a) Despite the fact that pollsters sample people without replacement, can we think of these calls as independent trials?
   b) Find the mean and standard deviation of $X =$ the number of adults Pew may successfully contact.
   c) We want to find $P(X \geq 1000)$. Can we use a Normal model to approximate this Binomial probability?
   d) Find the approximate probability Pew is able to contact at least 1000 voters on its list.

---

[7] In fact, some people use an adjustment called the "continuity correction" to help with this problem. It's related to the suggestion we make in the next footnote and is discussed in more advanced textbooks.

[8] If we really had been interested in a single value, we might have approximated it by finding the probability of getting between 1849.5 and 1850.5 units of blood.

# Should I Be Surprised? A First Look at Statistical Significance

You watch a friend toss a coin 100 times and get 67 heads. That's more heads than you'd expect, but is it enough more that you should think she might be cheating somehow? You probably wouldn't consider 52 or 53 heads instead of a "perfect" 50 to be unusual, but if she tossed heads 90 times out of 100 you'd be really suspicious. How about 67? After all, random outcomes do vary, sometimes ending up higher or lower than expected. Is 67 heads too strange to be explained away as just random chance?

For quite a while now we've been thinking about *statistical significance*. The results of an experiment or a sample are said to be **statistically significant** if it's not reasonable to believe they occurred just by chance.

Let's think about your friend's 67 heads in 100 tosses. Coin tosses are Bernoulli trials; here there are $n = 100$ trials with probability of success $p = 0.5$. We can model the random variable $X$ = number of heads with $Binom(100, 0.5)$. For our model, the mean is $np = 100(0.5) = 50$. OK, on average we expect 50 heads (duh!), but we know it won't be *exactly* 50 every time. The standard deviation is $\sqrt{npq} = \sqrt{100(0.5)(0.5)} = 5$ heads, and that's our clue about how much variation is reasonable.

Add one more key insight and we're ready to go: because we expect more than 10 successes (50) and more than 10 failures ($nq$ is also 50), a Normal model is useful here. Her 67 heads is 17 more than we expected. Because the SD = 5, we know her results are over 3 standard deviations above the mean. (To be exact, $z = \dfrac{67 - 50}{5} = 3.4$) Remember the 68–95–99.7 Rule? More than 99.7% of the time, the result should be within 3 standard deviations of the mean, but hers isn't. If her coin-tossing method is fair, this would be an exceedingly rare outcome. Such an unusual result is statistically significant—friend or not, we should be very suspicious.

This is a real breakthrough! (Drumroll, please!) For the first time we've been able to decide whether what we've observed is just a chance occurrence or is strong evidence that something unusual is afoot without using a simulation. We'll explore this kind of reasoning in greater detail in the chapters ahead. For now it's enough to recognize that when a Normal model is useful,[9] outcomes more than 2 standard deviations from the expected value should be considered surprising.

## STEP-BY-STEP EXAMPLE

### Looking for Statistical Significance

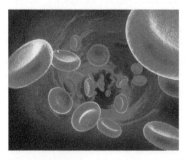

Before a blood drive, a local Red Cross agency puts out a plea for universal donors, hoping that it will get more than the usual 6% among the donors who show up. That day they collected 202 units of blood, and among them 17 units were Type O-negative.

**QUESTION:** Does this suggest that making a public plea is an effective way to get more O-negative donors to come to blood drives?

| | | |
|---|---|---|
| **THINK** | **PLAN** State the question. | I expect 6% of all blood donors to be O-negative. I want to decide whether getting 17 O-negative donors among 202 people is statistically significant evidence that the Red Cross's public plea may have worked. |
| | **VARIABLE** Define the random variable. | $X$ = number of O-negative donors |

---

[9]*Always* check the conditions to be sure!

# CHAPTER 16  Probability Models

| | | |
|---|---|---|
| **CHECK THE CONDITIONS** We've already confirmed that these are Bernoulli trials (p. 433), but it's critical to be sure that a Normal model applies. | ✓ **10% Condition:** $202 < 10\%$ of all possible donors. ✓ **Success/Failure Condition:** Among 202 donors with $P = 0.06$ I expect: $$np = (202)(0.06) = 12.12 \text{ successes,}$$ $$\text{and}$$ $$nq = (202)(0.94) = 189.88 \text{ failures.}$$ Both are at least 10. | |
| **MODEL** Name your model. | OK to use a Normal model. | |
| **SHOW** — **MECHANICS** Find the mean and standard deviation. Find the z-score for the observed result. | $n = 202 \quad p = 0.06$ $E(X) = np = 202(0.6) = 12.12$ $SD(X) = \sqrt{npq} = \sqrt{202(0.06)(0.94)} = 3.375$ $z = \dfrac{17 - 12.12}{3.375} = 1.45$ | |
| Use the 68–95–99.7 Rule to think about whether that z-score seems unusual. We shouldn't be surprised unless the outcome is more than 2 standard deviations above or below the mean. | 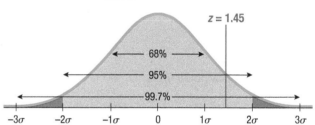 This doesn't look unusual; it's within 2 standard deviations of the mean. | |
| **TELL** — **CONCLUSION** Explain (in context, of course) whether or not you consider the outcome to be statistically significant. | Although it was a good turnout, getting 17 Type O-negative donors among 202 people is only about 1.5 standard deviations more than expected. This could have been just random chance, so it's not strong evidence that the Red Cross's public plea raised the number of universal donors who came to the blood drive.[10] | |

## WHAT CAN GO WRONG?

- **Be sure you have Bernoulli trials.** Be sure to check the requirements first: two possible outcomes per trial ("success" and "failure"), a constant probability of success, and independence. Remember to check the 10% Condition when sampling without replacement.

- **Don't confuse Geometric and Binomial models.** Both involve Bernoulli trials, but the issues are different. If you are repeating trials until your first success, that's a Geometric probability. You don't know in advance how many trials you'll need—theoretically, it could take forever. If you are counting the number of successes in a specified number of trials, that's a Binomial probability.

- **Don't use the Normal approximation with small $n$.** To use a Normal approximation in place of a Binomial model, there must be at least 10 expected successes and 10 expected failures.

---

[10] Even if the number of Type O-negative donors *had* been surprisingly high (statistically significant), we still couldn't be sure it was high *because* of the public plea. Remember this principle from Chapter 12: association does not imply causation. But it would at least give the Red Cross a good reason to use the strategy again, especially when lives are at stake.

# WHAT HAVE WE LEARNED?

We've learned that Bernoulli trials show up in lots of places. Depending on the random variable of interest, we can use one of three models to estimate probabilities for Bernoulli trials:

- a Geometric model when we're interested in the number of Bernoulli trials until the next success,
- a Binomial model when we're interested in the number of successes in a certain number of Bernoulli trials, and
- a Normal model to approximate a Binomial model when we expect at least 10 successes and 10 failures.

We've learned (yet again) the importance of checking assumptions and conditions before proceeding.

And we've learned to use a Normal model to help us think about statistical significance. We consider observations more than 2 standard deviations from what's expected to be unusual.

## TERMS

**Bernoulli trials**  A collection of real or simulated data are called Bernoulli trials if: (1) each trial (observation) has exactly two possible outcomes, often identified as "success" and "failure"; (2) the probability of success is the same for every trial; and (3) the trials are all independent. (p. 430)

**10% Condition**  When sampling without replacement, trials are not independent. It's still okay to think of them as independent as long as the random sample is smaller than 10% of the population. (p. 430)

**Geometric probability model**  A Geometric model is appropriate for a random variable that counts the number of Bernoulli trials until the first success.

$X$ = the number of independent trials until the first success

$P(x) = q^{x-1}p$, where $p$ = the probability of success and $q = 1 - p$

$\mu = E(X) = \dfrac{1}{p}$ and $\sigma = SD(X) = \sqrt{\dfrac{q}{p^2}} = \dfrac{\sqrt{1-p}}{p}$   (p. 431)

**Binomial probability model**  A Binomial model is appropriate for a random variable that counts the number of successes in a fixed number of Bernoulli trials.

$X$ = the number of successes in $n$ trials

$P(x) = \binom{n}{x} p^x q^{n-x}$, where $p$ = the probability of success and $q = 1 - p$

$\mu = E(X) = np$ and $\sigma = SD(X) = \sqrt{npq}$ (p. 434)

**Success/Failure Condition**  For a Normal model to be a good approximation of a Binomial model, we must expect at least 10 successes and 10 failures. That is, $np \geq 10$ and $nq \geq 10$. (p. 440)

**Statistically significant**  The results of a study are considered statistically significant if there's a very low probability they could have occurred by chance. (p. 442)

# ON THE COMPUTER

## The Binomial Model

Most statistics packages offer functions that compute Binomial probabilities. Some technology solutions automatically use the Normal approximation for the Binomial when the exact calculations become unmanageable.

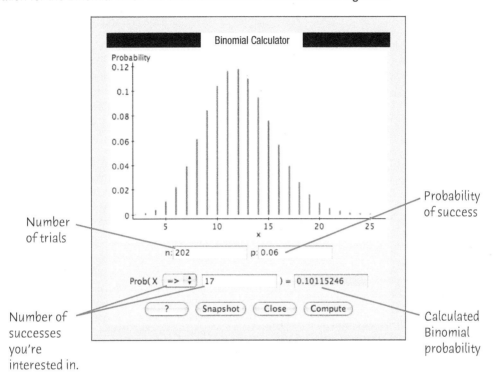

# EXERCISES

1. **Bernoulli** Do these situations involve Bernoulli trials? Explain.
   a) We roll 50 dice to find the distribution of the number of spots on the faces.
   b) How likely is it that in a group of 120 the majority may have Type A blood, given that Type A is found in 43% of the population?
   c) We deal 7 cards from a deck and get all hearts. How likely is that?
   d) We wish to predict the outcome of a vote on the school budget, and poll 500 of the 3000 likely voters to see how many favor the proposed budget.
   e) A company realizes that about 10% of its packages are not being sealed properly. In a case of 24, is it likely that more than 3 are unsealed?

2. **Bernoulli 2** Do these situations involve Bernoulli trials? Explain.
   a) You are rolling 5 dice and need to get at least two 6's to win the game.
   b) We record the distribution of eye colors found in a group of 500 people.
   c) A manufacturer recalls a doll because about 3% have buttons that are not properly attached. Customers return 37 of these dolls to the local toy store. Is the manufacturer likely to find any dangerous buttons?
   d) A city council of 11 Republicans and 8 Democrats picks a committee of 4 at random. What's the probability they choose all Democrats?
   e) A 2002 Rutgers University study found that 74% of high school students have cheated on a test at least once. Your local high school principal conducts a survey in homerooms and gets responses that admit to cheating from 322 of the 481 students.

3. **Simulating the model** Think about the Simone Biles picture search again. You are opening boxes of cereal one at a time looking for her picture, which is in 20% of the boxes. You want to know how many boxes you might have to open to find Biles.

a) Describe how you would simulate the search for Biles using random numbers.
b) Run at least 30 trials.
c) Based on your simulation, estimate the probabilities that you might find your first picture of Simone Biles in the first box, the second, etc.
d) Calculate the actual probability model.
e) Compare the distribution of outcomes in your simulation to the probability model.

4. **Simulation II** You are one space short of winning a child's board game and must roll a 1 on a die to claim victory. You want to know how many rolls it might take.

a) Describe how you would simulate rolling the die until you get a 1.
b) Run at least 30 trials.
c) Based on your simulation, estimate the probabilities that you might win on the first roll, the second, the third, etc.
d) Calculate the actual probability model.
e) Compare the distribution of outcomes in your simulation to the probability model.

5. **Simone again** Let's take one last look at the Simone Biles picture search. You know her picture is in 20% of the cereal boxes. You buy five boxes to see how many pictures of Biles you might get.

a) Describe how you would simulate the number of pictures of Simone Biles you might find in five boxes of cereal.
b) Run at least 30 trials.
c) Based on your simulation, estimate the probabilities that you get no pictures of Biles, 1 picture, 2 pictures, etc.
d) Find the actual probability model.
e) Compare the distribution of outcomes in your simulation to the probability model.

6. **Seat belts** Suppose 75% of all drivers always wear their seat belts. Let's investigate how many of the drivers might be belted among five cars waiting at a traffic light.

a) Describe how you would simulate the number of seat belt–wearing drivers among the five cars.
b) Run at least 30 trials.
c) Based on your simulation, estimate the probabilities there are no belted drivers, exactly one, two, etc.
d) Find the actual probability model.
e) Compare the distribution of outcomes in your simulation to the probability model.

7. **On time** A Department of Transportation report about air travel found that, nationwide, 76% of all flights are on time. Suppose you are at the airport and your flight is one of 50 scheduled to take off in the next two hours. Can you consider these departures to be Bernoulli trials? Explain.

8. **Lost luggage** A Department of Transportation report about air travel found that airlines misplace about 5 bags per 1000 passengers. Suppose you are traveling with a group of people who have checked 22 pieces of luggage on your flight.
Can you consider the fate of these bags to be Bernoulli trials? Explain.

9. **Hoops** A basketball player has made 80% of their foul shots during the season. Assuming the shots are independent, find the probability that in tonight's game

a) their first miss occurs on their fifth attempt.
b) the first shot they make occurs on their fifth attempt.
c) they make their first basket on one of their first 3 shots.

10. **Chips** Suppose a computer chip manufacturer rejects 2% of the chips produced because they fail presale testing.

a) What's the probability that the fifth chip you test is the first bad one you find?
b) What's the probability you find a bad one within the first 10 you examine?

11. **More hoops** For the basketball player in Exercise 9,

a) What's the expected number of shots until they miss?
b) If the player shoots 10 foul shots in the fourth quarter, how many shots do you expect them to make?
c) What is the standard deviation of the number of made shots out of 10?

12. **Chips ahoy** For the computer chips described in Exercise 10,

a) How many do you expect to test before finding a bad one?
b) In a random sample of 400 chips, what is the mean number of chips that are expected to fail?
c) What is the standard deviation of that same sample?

13. **Customer center operator** Raaj works at the customer service call center of a major credit card bank. Cardholders call for a variety of reasons, but regardless of their reason for calling, if they hold a platinum card, Raaj is instructed to offer them a double-miles promotion. About 10% of all cardholders hold platinum cards, and about 50% of those will take the double-miles promotion. On average, how many calls will Raaj have to take before finding the first cardholder to take the double-miles promotion?

14. **Cold calls** Justine works for an organization committed to raising money for Alzheimer's research. From past experience, the organization knows that about 20% of all potential donors will agree to give something if contacted by phone. They also know that of all people donating, about 5% will give $100 or more. On average, how many potential donors will she have to contact until she gets her first $100 donor?

15. **Blood** Only 4% of people have Type AB blood.

a) On average, how many donors must be checked to find someone with Type AB blood?
b) What's the probability that there is a Type AB donor among the first 5 people checked?
c) What's the probability that the first Type AB donor will be found among the first 6 people?
d) What's the probability that we won't find a Type AB donor before the 10th person?

16. **Colorblindness** About 8% of males are colorblind. A researcher needs some colorblind subjects for an experiment and begins checking potential subjects.

a) On average, how many men should the researcher expect to check to find one who is colorblind?
b) What's the probability that they won't find anyone colorblind among the first 4 men they checks?
c) What's the probability that the first colorblind man found will be the sixth person checked?
d) What's the probability that they find someone who is colorblind before checking the 10th man?

**17. Smartphones** According to a 2019 Common Sense Media study, 84% of teenagers (age 13–18) have a smartphone. (https://www.commonsensemedia.org/research/the-common-sense-census-media-use-by-tweens-and-teens-2019) If we select teenagers at random, find the probability of each outcome described below.

a) You select teenagers one at a time until you get one with a smartphone. The first three people chosen do not own a smartphone but the fourth person does.
b) You select 8 teenagers and there is at least one smartphone owner among them.
c) You select teenagers one at a time. The first smartphone owner is either the second or third person chosen.
d) You select 8 teenagers and there are exactly six smartphone owners among them.
e) You select 8 teenagers and there are at least six smartphone owners among them.
f) You select 8 teenagers and there are no more than six smartphone owners among them.

**18. Arrows** An Olympic archer is able to hit the bull's-eye 80% of the time. Assume each shot is independent of the others. If the archer shoots arrows, what's the probability of each of the following results?

a) They start shooting and their first bull's-eye occurs on the third shot.
b) They start shooting and their first bull's-eye occurs on the fourth or fifth shot.
c) They shoot six arrows and miss the bull's-eye at least once.
d) They shoot six arrows and get exactly four bull's-eyes.
e) They shoot six arrows and get at least four bull's-eyes.
f) They shoot six arrows and get at most four bull's-eyes.

**19. Smartphones redux** Consider randomly selecting a group of 8 teens from Exercise 17.

a) How many smartphone owners do you expect in the group?
b) What is the standard deviation?
c) If you keep picking people until we find a smartphone, how many people do you expect it will take until you find one?

**20. More arrows** Consider our archer from Exercise 18.

a) How many bull's-eyes do you expect them to get if they shoots six arrows?
b) With what standard deviation?
c) If they keep shooting arrows until they hit the bull's-eye, what is the expected number of shots it will take?

**21. Still more smartphones** Suppose we choose 20 teens from Exercise 17.

a) Find the mean and standard deviation of the number of non–smartphone owners in the group.
b) What's the probability that
  i) at least one person of the 20 does not own a smartphone?
  ii) there are no more than 15 smartphones owners in the group?
  iii) there are exactly 10 smartphone owners and 10 non–smartphone owners?
  iv) the majority don't have a smartphone?

**22. Still more arrows** Suppose our archer from Exercise 18 shoots 10 arrows.

a) Find the mean and standard deviation of the number of bull's-eyes they may get.
b) What's the probability that
  i) they never miss?
  ii) there are no more than 8 bull's-eyes?
  iii) there are exactly 8 bull's-eyes?
  iv) they hit the bull's-eye more often than they miss?

**23. Vision** It is generally believed that nearsightedness affects about 12% of all children. A school district tests the vision of 169 incoming kindergarten children. How many would you expect to be nearsighted? With what standard deviation?

**24. International students** At a certain college, 6% of all students come from outside the United States. Incoming students there are assigned at random to first-year student dorms, where students live in residential clusters of 40 first-years sharing a common lounge area. How many international students would you expect to find in a typical cluster? With what standard deviation?

**25. Tennis, anyone?** A certain tennis player makes a successful serve 70% of the time. Assume that each serve is independent of the others. If they serve 6 times, what's the probability

a) all 6 serves are successful?
b) exactly 4 serves are successful?
c) at least 4 serves are successful?
d) no more than 4 serves are successful?

**26. Frogs** A wildlife biologist examines frogs for a genetic trait they suspect may be linked to sensitivity to industrial toxins in the environment. Previous research had established that this trait is usually found in 1 of every 8 frogs. The biologist collects and examines a dozen frogs. If the frequency of the trait has not changed, what's the probability they find the trait in

a) none of the 12 frogs?
b) at least 2 frogs?
c) 3 or 4 frogs?
d) no more than 4 frogs?

**27. Second serve** Consider the tennis player in Exercise 25 who successfully serves 70% of the time.

a) What are the four conditions that need to be met to justify your answers for Exercise 25?
b) Do you think those conditions are valid? Explain.

**28. Easy being green** The biologist in Exercise 26 studied frogs with a 1 in 8 chance of having a certain trait.

a) What are the conditions that must be met to justify your answers for Exercise 26?
b) Do you think those conditions are satisfied? Explain.

**29. And more tennis** Suppose the tennis player in Exercise 25 serves 80 times in a match.

a) What are the mean and standard deviation of the number of successful serves expected?
b) Verify that you can use a Normal model to approximate the distribution of the number of successful serves.

c) Use the 68–95–99.7 Rule to describe this distribution.
d) What's the probability they make at least 65 successful serves?

30. **More arrows** The archer in Exercise 18 will be shooting 200 arrows in a large competition.
    a) What are the mean and standard deviation of the number of bull's-eyes they might get?
    b) Is a Normal model appropriate here? Explain.
    c) Use the 68–95–99.7 Rule to describe the distribution of the number of bull's-eyes they may get.
    d) Would you be surprised if they made only 140 bull's-eyes? Explain.

31. **Apples** An orchard owner knows that about 6% of the apples harvested will have to be used for cider because they will have bruises or blemishes. A tree is expected to produce about 300 apples.
    a) Describe an appropriate model for the number of cider apples that may come from that tree. Justify your model.
    b) Find the probability there will be no more than a dozen cider apples.
    c) Is it likely there will be more than 50 cider apples? Explain.

32. **Frogs, part III** Based on concerns raised by their preliminary research, the biologist in Exercise 26 decides to collect and examine 150 frogs.
    a) Assuming the frequency of the trait is still 1 in 8, determine the mean and standard deviation of the number of frogs with the trait they should expect to find in their sample.
    b) Verify that they can use a Normal model to approximate the distribution of the number of frogs with the trait.
    c) They found the trait in 22 of their frogs. Do you think this proves that the trait has become more common? Explain.

33. **Lefties** A lecture hall has 200 seats with folding arm tablets, 30 of which are designed for left-handers. The typical size of classes that meet there is 188, and we can assume that about 13% of students are left-handed. What's the probability that a right-handed student in one of these classes is forced to use a lefty arm tablet?

34. **No-shows** An airline, believing that 5% of passengers fail to show up for flights, overbooks (sells more tickets than there are seats). Suppose a plane will hold 265 passengers, and the airline sells 275 tickets. What's the probability the airline will not have enough seats, so someone gets bumped?

35. **Annoying phone calls** A newly hired telemarketer is told they will probably make a sale on about 12% of their phone calls. The first week they called 200 people, but only made 10 sales. Should they suspect they were misled about the true success rate? Explain.

36. **The euro** Shortly after the introduction of the euro coin in Belgium, newspapers around the world published articles claiming the coin is biased. The stories were based on reports that someone had spun the coin 250 times and gotten 140 heads—that's 56% heads. Do you think this is evidence that spinning a euro is unfair? In other words, do you think this provides evidence that the probability that a euro coin will land on heads is different from 0.5? Explain.

37. **Seat belts II** Police estimate that 80% of drivers now wear their seat belts. They set up a safety roadblock, stopping cars to check for seat belt use.
    a) How many cars do they expect to stop before finding a driver whose seat belt is not buckled?
    b) What's the probability that the 6th car stopped is the first to contain an unbelted driver?
    c) What's the probability that the first 10 drivers are all wearing their seat belts?
    d) If they stop 30 cars each hour, find the mean and standard deviation of the number of drivers wearing seat belts each hour.
    e) If they stop 120 cars during this safety check, what's the probability they find at least 20 drivers not wearing their seat belts?

38. **Rickets** Vitamin D is essential for strong, healthy bones. Our bodies produce vitamin D naturally when sunlight falls upon the skin, or it can be taken as a dietary supplement. Although the bone disease rickets was largely eliminated in England during the 1950s, some people there are concerned that this generation of children is at increased risk because they are more likely to watch TV or play computer games than spend time outdoors. Recent research indicated that about 20% of British children are deficient in vitamin D. Suppose doctors test a group of elementary school children. Assume we can treat each test as a Bernoulli trial.
    a) If they test children one at a time, what's the probability that the first vitamin D–deficient child is the 8th one tested?
    b) What's the probability that the first 10 children tested are all okay?
    c) How many kids do they expect to test before finding one who has this vitamin deficiency?
    d) They will test 50 students at the third-grade level. Find the mean and standard deviation of the number who may be deficient in vitamin D.
    e) If they test 320 children at this school, what's the probability that no more than 50 of them have the vitamin deficiency?
    f) You were told to assume that we can treat each test as a Bernoulli trial. Is that a reasonable assumption?

39. **ESP** Scientists wish to test the mind-reading ability of a person who claims to "have ESP." They use five cards with different and distinctive symbols (square, circle, triangle, line, squiggle). Someone picks a card at random and thinks about the symbol. The "mind reader" must correctly identify which symbol was on the card. If the test consists of 100 trials, how many would this person need to get right in order to convince you that they might actually have ESP? Explain.

40. **True-false** A true-false test consists of 50 questions. How many does a student have to get right to convince you that they are not merely guessing? Explain.

41. **Hot hand** A basketball player who ordinarily makes about 55% of their free throw shots has made 4 in a row. Is this evidence that they have a "hot hand" tonight? That is, is this streak so unusual that it means the probability that they make a shot must have changed? Explain.

**42. New bow** Our archer in Exercise 18 purchases a new bow, hoping that it will improve their success rate to more than 80% bull's-eyes. They are delighted when they first test their new bow and hit 6 consecutive bull's-eyes. Do you think this is compelling evidence that the new bow is better? In other words, is a streak like this unusual for them? Explain.

**43. Hotter hand** Our basketball player in Exercise 41 has new sneakers, which they think improve their game. Over their past 40 shots, they've made 32—much better than the 55% they usually shoot. Do you think their chances of making a shot really increased? In other words, is making at least 32 of 40 shots really unusual for them? (Do you think it's the new sneakers?)

**44. New bow, again** The archer in Exercise 42 continues shooting arrows, ending up with 45 bull's-eyes in 50 shots. Now are you convinced that the new bow is better? Explain.

## JUST CHECKING

### Answers

1. a) No; the probability of choosing a woman changes with each name drawn.
   b) Yes.
   c) No; women who have had twins are more likely to have them again.
   d) No; there are more than two possible outcomes (colors).
   e) No; the sample is more than 10% of the population.

2. a) There are 2 outcomes (contact or not); $p = 0.76$; fewer than 10% of the population are being contacted randomly.
   b) $(0.24)^3(0.76) = 0.011$.
   c) $\mu = np = 12(0.76) = 9.12$.
   d) $\sigma = \sqrt{npq} = \sqrt{12(0.76)(0.24)} \approx 1.48$.
   e) $\binom{12}{9}(0.76)^9(0.24)^3 \approx 0.26$.
   f) $\binom{12}{9}(0.76)^9(0.24)^3 + \binom{12}{10}(0.76)^{10}(0.24)^2 + \binom{12}{11}(0.76)^{11}(0.24)^1 + (0.76)^{12} \approx 0.68$.

3. a) $1300 < 10\%$ of all households.
   b) $\mu = 1300(0.76) = 988$; $\sigma = \sqrt{1300(0.76)(0.24)} = 15.4$.
   c) Yes; $np = 1300(0.76) = 988$ and $nq = 1300(0.24) = 312$ are both at least 10.
   d) $P(z > 0.78) = 0.22$.

# Review of Part IV

## RANDOMNESS AND PROBABILITY

### Quick Review

Here's a brief summary of the key concepts and skills in probability and probability modeling:

- The Law of Large Numbers says that the more times we try something, the closer the results (cumulatively) will come to theoretical perfection.
  - Don't mistakenly misinterpret the Law of Large Numbers as the "Law of Averages." There's no such thing.
- Basic rules of probability can handle most situations:
  - To find the probability that an event OR another event happens, add their probabilities and subtract the probability that both happen.
  - To find the probability that an event AND another *independent* event both happen, multiply probabilities.
  - Conditional probabilities tell you how likely one event is to happen, knowing that another event happens.
  - Mutually exclusive events (also called "disjoint") cannot both happen at the same time.
  - Two events are independent if the occurrence of one doesn't change the probability that the other happens.

- A probability model for a random variable describes the theoretical distribution of outcomes.
  - The mean of a random variable is its "expected value," which is a technical term and not necessarily a value you'd literally expect to observe. In fact, it may be a value that cannot *possibly* occur.
  - For sums or differences of *independent* random variables, variances add.
  - To estimate probabilities involving quantitative variables, you may be able to use a Normal model—but only if the distribution of the variable is unimodal and symmetric.
  - To estimate the probability you'll get your first success on a certain trial, use a Geometric model.
  - To estimate the probability you'll get a certain number of successes in a specified number of independent trials, use a Binomial model.

Ready? Here are some opportunities to check your understanding of these ideas.

## REVIEW EXERCISES

**1. Quality control** A consumer organization estimates that 29% of new cars have a cosmetic defect, such as a scratch or a dent, when they are delivered to car dealers. This same organization believes that 7% have a functional defect—something that does not work properly—and that 2% of new cars have both kinds of problems.

a) If you buy a new car, what's the probability that it has some kind of defect?
b) What's the probability it has a cosmetic defect but no functional defect?
c) If you notice a dent on a new car, what's the probability it has a functional defect?
d) Are the two kinds of defects disjoint events? Explain.
e) Do you think the two kinds of defects are independent events? Explain.

**2. Workers** A company's human resources officer reports a breakdown of employees by job type and sex as shown in the table.

|  | | Sex | |
|---|---|---|---|
|  | | Men | Women |
| **Job Type** | Management | 7 | 6 |
|  | Supervision | 8 | 12 |
|  | Production | 45 | 72 |

a) What's the probability that a worker selected at random is
  i) a woman?
  ii) a woman or a production worker?
  iii) a woman, if the person works in production?
  iv) a production worker, if the person is a woman?
b) Do these data suggest that job type is independent of being a man or a woman? Explain.

**3. Airfares** Each year a company must send 3 officials to a meeting in China and 5 officials to a meeting in France. Airline ticket prices vary from time to time, but the company purchases all tickets for a country at the same price. Past experience has shown that tickets to China have a mean price of $1000, with a standard deviation of $150, while the mean airfare to France is $500, with a standard deviation of $100.

a) Define random variables and use them to express the total amount the company will have to spend to send these delegations to the two meetings.
b) Find the mean and standard deviation of this total cost.
c) Find the mean and standard deviation of the difference in price of a ticket to China and a ticket to France.
d) Do you need to make any assumptions in calculating these means? How about the standard deviations?

450

**4. Autism** Psychiatrists estimate that about 1 in 100 adults has been diagnosed as being on the autism disorder spectrum. What's the probability that, in a city of 20,000, there are more than 300 people with this condition? Be sure to verify that a Normal model can be used here.

**5. A game** To play a game, you must pay $5 for each play. There is a 10% chance you will win $5, a 40% chance you will win $7, and a 50% chance you will win only $3.
   a) What are the mean and standard deviation of your net winnings?
   b) You play twice. Assuming the plays are independent events, what are the mean and standard deviation of your total winnings?

**6. Emergency switch** Safety engineers must determine whether industrial workers can operate a machine's emergency shutoff device. Among a group of test subjects, 66% were successful with their left hands, 82% with their right hands, and 51% with either hand.
   a) What percentage of these workers could not operate the switch with either hand?
   b) Are successes with right and left hands independent events? Explain.
   c) Are successes with right and left hands mutually exclusive? Explain.

**7. Snapchat** According to a UBS Evidence Lab survey, 34% of teens aged 13–17 use Snapchat at least monthly. That age group makes up 23% of all Snapchat users.
   a) If we randomly select 10 teens aged 13–17, what is the probability that at least one of them is not on Snapchat?
   b) If we randomly select 10 Snapchat users, what is the probability that at least one of them is a teen aged 13–17?

**8. Twins** In the United States, the probability of having twins (usually about 1 in 90 births) rises to about 1 in 10 for women who have been taking the fertility drug Clomid. Among a random sample of 10 pregnant women, what's the probability that
   a) at least one will have twins if none were taking a fertility drug?
   b) at least one will have twins if all were taking Clomid?
   c) at least one will have twins if half were taking Clomid?

**9. Deductible** A car owner may buy insurance that will pay the full price of repairing the car after an at-fault accident, or save $12 a year by getting a policy with a $500 deductible. Their insurance company says that about 0.5% of drivers in their area have an at-fault auto accident during any given year. Based on this information, should they buy the policy with the deductible or not? How does the value of their car influence this decision?

**10. More Snapchat** Using the percentages from Exercise 7, suppose there is a group of 5 teens. What's the probability that
   a) all will be on Snapchat?
   b) exactly 1 will be on Snapchat?
   c) at least 3 will be on Snapchat?

**11. At fault** The car insurance company in Exercise 9 believes that about 0.5% of drivers have an at-fault accident during a given year. Suppose the company insures 1355 drivers in that city.
   a) What are the mean and standard deviation of the number who may have at-fault accidents?
   b) Can you describe the distribution of these accidents with a Normal model? Explain.

**12. Snap me?** One hundred fifty-eight teens are standing in line for a big movie premiere night. (See Exercise 7.)
   a) If we can assume this is essentially a random sample of teens, what are the mean and standard deviation of the number of Snapchat users we might expect to find among this group of teens?
   b) Can we use a Normal model in this situation?
   c) What would be the probability that no more than 42 of the teens are Snapchat users?
   d) Is it actually reasonable to assume this is essentially a random sample of teens?

**13. Child's play** In a board game you determine the number of spaces you may move by spinning a spinner and rolling a die. The spinner has three regions: Half of the spinner is marked "5," and the other half is equally divided between "10" and "20." The six faces of the die show 0, 0, 1, 2, 3, and 4 spots. When it's your turn, you spin and roll, adding the numbers together to determine how far you may move.
   a) Create a probability model for the outcome on the spinner.
   b) Find the mean and standard deviation of the spinner results.
   c) Create a probability model for the outcome on the die.
   d) Find the mean and standard deviation of the die results.
   e) Find the mean and standard deviation of the number of spaces you get to move.

**14. Language** Neurological research has shown that in about 80% of people, language abilities reside in the brain's left side. Another 10% display right-brain language centers, and the remaining 10% have two-sided language control. (The latter two groups are mainly left-handers; *Science News*, 161 no. 24 [2002].)
   a) Assume that a first-year composition class contains 25 randomly selected people. What's the probability that no more than 15 of them have left-brain language control?
   b) In a randomly chosen group of 5 of these students, what's the probability that no one has two-sided language control?
   c) In the entire freshman class of 1200 students, how many would you expect to find of each type?
   d) What are the mean and standard deviation of the number of these freshmen who might be right-brained in language abilities?
   e) If an assumption of Normality is justified, use the 68–95–99.7 Rule to describe how many students in the freshman class might have right-brain language control.

**15. Play again** If you land in a "penalty zone" on the game board described in Exercise 13, your move will be determined by subtracting the roll of the die from the result on the spinner. Now what are the mean and standard deviation of the number of spots you may move?

**16. Beanstalks** In some cities tall people who want to meet and socialize with other tall people can join Beanstalk Clubs. To qualify, a man must be over 6'2" tall, and a woman over 5'10".

According to the National Health Survey, heights of adults may have a Normal model with mean heights of 69.1″ for men and 64.0″ for women. The respective standard deviations are 2.8″ and 2.5″.

a) You're probably not surprised to learn that men are generally taller than women, but what does the greater standard deviation for men's heights indicate?
b) Are men or women more likely to qualify for Beanstalk membership?
c) Beanstalk members believe that height is an important factor when people select their spouses. To investigate, we select at random a married man and, independently, a married woman. Define two random variables, and use them to express how many inches taller the man is than the woman.
d) What's the mean of this difference?
e) What's the standard deviation of this difference?
f) What's the probability that the man is taller than the woman (that the difference in heights is greater than 0)?
g) Suppose a survey of heterosexual married couples reveals that 92% of the husbands were taller than their wives. Based on your answer to part f, do you believe that people's choice of spouses is independent of height? Explain.

**17. Stocks** Since the stock market began in 1872, stock prices have risen in about 73% of the years. Assuming that market performance is independent from year to year, what's the probability that

a) the market will rise for 3 consecutive years?
b) the market will rise 3 years out of the next 5?
c) the market will fall during at least 1 of the next 5 years?
d) the market will rise during a majority of years over the next decade?

**18. Multiple choice** A multiple choice test has 50 questions, with 4 answer choices each. You must get at least 30 correct to pass the test, and the questions are very difficult.

a) Are you likely to be able to pass by guessing on every question? Explain.
b) Suppose, after studying for a while, you believe you have raised your chances of getting each question right to 70%. How likely are you to pass now?
c) Assuming you are operating at the 70% level and the instructor arranges questions randomly, what's the probability that the third question is the first one you get right?

**19. Stock strategy** Many investment advisors argue that after stocks have declined in value for 2 consecutive years, people should invest heavily because the market rarely declines 3 years in a row.

a) Since the stock market began in 1872, there have been two consecutive losing years 10 times. In 8 of those cases, the market rose during the following year. Does this confirm the advice?
b) Overall, stocks have risen in value during 104 of the 145 years since the market began in 1872. How is this fact relevant in assessing the statistical reasoning of the advisors?

**20. Insurance** A 65-year-old woman takes out a $100,000 term life insurance policy. The company charges an annual premium of $520. Estimate the company's expected profit on such policies if mortality tables indicate that only 2.6% of women age 65 die within a year.

**21. Teen smoking** The Centers for Disease Control say that in 2016 about 8% of students said they currently smoke tobacco (down from a high of 38% in 1997). Suppose you randomly select high school students to survey them on their attitudes toward scenes of smoking in the movies. What's the probability that

a) none of the first 4 students you interview is a smoker?
b) the first student interviewed who smokes occurs on the sixth interview?
c) there are no more than 2 smokers among 10 people you choose?

**22. Passing stats** Molly's college offers two sections of Statistics 101. From what she has heard about the two professors listed, Molly estimates that her chances of passing the course are 0.80 if she gets Professor Scedastic and 0.60 if she gets Professor Kurtosis. The registrar uses a lottery to randomly assign the 120 enrolled students based on the number of available seats in each class. There are 70 seats in Professor Scedastic's class and 50 in Professor Kurtosis's class.

a) What's the probability that Molly will pass Statistics?
b) At the end of the semester, we will randomly select one student from among those who failed. If Molly's chances of passing are typical, what's the probability the selected student will have had Professor Kurtosis?

**23. Teen smoking II** Suppose that, as reported by the Centers for Disease Control, about 8% of high school students smoke tobacco. You randomly select 150 high school students to survey them on their attitudes toward scenes of smoking in the movies.

a) What's the expected number of smokers?
b) What's the standard deviation of the number of smokers?
c) The number of smokers among 150 randomly selected students will vary from group to group. Explain why that number can be described with a Normal model.
d) Using the 68–95–99.7 Rule, create and interpret a model for the number of smokers among your group of 150 students.

**24. Language again** Neurological research has shown that in about 80% of people language abilities reside in the brain's left side. Another 10% display right-brain language centers, and the remaining 10% have two-sided language control. (The latter two groups are mainly left-handers.) (*Science News*, 161, no. 24 [2002])

a) We select 60 people at random. Is it reasonable to use a Normal model to describe the possible distribution of the proportion of the group that has left-brain language control? Explain.
b) What's the probability that our group has at least 75% left-brainers?
c) If the group had consisted of 100 people, would that probability be higher, lower, or about the same? Explain why, without actually calculating the probability.
d) How large a group would almost certainly guarantee at least 75% left-brainers? Explain.

**25. Random variables** Given independent random variables with means and standard deviations as shown, find the mean and standard deviation of each of these variables:

a) $X + 50$
b) $10Y$
c) $X + 0.5Y$
d) $X - Y$
e) $X_1 + X_2$

|   | Mean | SD |
|---|------|----|
| X | 50   | 8  |
| Y | 100  | 6  |

**26. Merger** Explain why the facts you know about variances of independent random variables might encourage two small insurance companies to merge. (*Hint:* Think about the expected amount and potential variability in payouts for the separate and the merged companies.)

**27. Youth survey** According to a 2019 Common Sense Media survey, 84% of teens (ages 13–18) and 41% of tweens (ages 8–12) own a smartphone. The survey also found that 56% of tweens say they stream videos on services like YouTube, compared with 69% of teens.

a) For teens, the cited percentages are 84% owning a smartphone and 69% streaming videos every day. That total is 153%, so there is obviously a mistake in the report. No? Explain.
b) Based on these results, do you think owning a smartphone and streaming videos daily are mutually exclusive among teens? Explain.
c) Do you think whether a child streams videos every day is independent of being a teen or a tween? Explain.
d) Suppose that in fact 84% of the teens in your area do own a smartphone. You want to interview a few who do not, so you start contacting teens at random. What is the probability that it takes you 5 interviews until you find the first teen who does not own a smartphone?

**28. Meals** A college student on a seven-day meal plan reports that the amount of money they spend daily on food varies with a mean of $13.50 and a standard deviation of $7.

a) What are the mean and standard deviation of the amount they might spend in two consecutive days?
b) What assumption did you make in order to find that standard deviation? Are there any reasons you might question that assumption?
c) Estimate their average weekly food costs, and the standard deviation.
d) Do you think it likely they might spend less than $50 in a week? Explain, including any assumptions you make in your analysis.

**29. Travel to Kyrgyzstan** Your pocket copy of *Kyrgyzstan on 4237 ± 360 Som a Day* claims that you can expect to spend about 4237 som each day with a standard deviation of 360 som. How well can you estimate your expenses for the trip?

a) Your budget allows you to spend 90,000 som. To the nearest day, how long can you afford to stay in Kyrgyzstan, on average?
b) What's the standard deviation of your expenses for a trip of that duration?
c) You doubt that your total expenses will exceed your expectations by more than two standard deviations. How much extra money should you bring? On average, how much of a "cushion" will you have per day?

**30. Picking melons** Two stores sell watermelons. At the first store the melons weigh an average of 22 pounds, with a standard deviation of 2.5 pounds. At the second store the melons are smaller, with a mean of 18 pounds and a standard deviation of 2 pounds. You select a melon at random at each store.

a) What's the mean difference in weights of the melons?
b) What's the standard deviation of the difference in weights?
c) If a Normal model can be used to describe the difference in weights, what's the probability that the melon you got at the first store is heavier?

**31. Home, sweet home 2020** In 2020, 65.8% of U.S. households owned the home they lived in. (Source: https://www.statista.com/statistics/184902/homeownership-rate-inthe-us-since-2003/) A mayoral candidate conducts a survey of 820 randomly selected homes in your city and finds only 498 owned by the current residents. The candidate then attacks the incumbent mayor, saying that there is an unusually low level of homeownership in the city. Do you agree? Explain.

**32. Buying melons** The first store in Exercise 30 sells watermelons for 32 cents a pound. The second store is having a sale on watermelons—only 25 cents a pound. Find the mean and standard deviation of the difference in the price you may pay for melons randomly selected at each store.

**33. Who's the boss?** The website census.gov revealed that 19.9% of all firms that employed people in the United States were owned by women in 2018. You call some randomly selected firms doing business locally, assuming that the national percentage is true in your area. Based on your assumption,

a) what's the probability that the first 3 you call are all owned by women?
b) what's the probability that none of your first 4 calls finds a firm that is owned by a woman?
c) suppose none of your first 5 calls found a firm owned by a woman. What's the probability that your next call does?

**34. Jerseys** A Statistics professor comes home to find that all four of his children got white team shirts from soccer camp this year. He concludes that this year, unlike other years, the camp must not be using a variety of colors. But then he finds out that in each child's age group there are 4 teams, only 1 of which wears white shirts. Each child just happened to get on the white team at random.

a) Why was he so surprised? If each age group uses the same 4 colors, what's the probability that all four kids would get the same-color shirt?
b) What's the probability that all 4 would get white shirts?
c) We lied. Actually, in the oldest child's group there are 6 teams instead of the 4 teams in each of the other three groups. How does this change the probability you calculated in part b?

**35. When to stop?** In Exercise 28 of the Review Exercises for Part III, we posed this question:

*You play a game that involves rolling a die. You can roll as many times as you want, and your score is the total for all the rolls. But . . . if you roll a 6, your score is 0 and your turn is over. What might be a good strategy for a game like this?*

You attempted to devise a good strategy by simulating several plays to see what might happen. Let's try calculating a strategy.

a) On what roll would you expect to get a 6 for the first time?
b) So, roll *one time less* than that. Assuming all those rolls were not 6's, what's your expected score?
c) What's the probability that you can roll that many times without getting a 6?

**36. Plan B** Here's another attempt at developing a good strategy for the dice game in Exercise 35. Instead of stopping after a certain number of rolls, you could decide to stop when your score reaches a certain number of points.

a) How many points would you expect a roll to *add* to your score?
b) In terms of your current score, how many points would you expect a roll to *subtract* from your score?
c) Based on your answers in parts a and b, at what score will another roll "break even"?
d) Describe the strategy this result suggests.

**37. Technology on campus 2015** Every 5 years, the Conference Board of the Mathematical Sciences surveys college math departments. In 2015, the board reported that 67% of all math departments used graphing calculators in introductory statistics courses and 48% use statistics software. Suppose that 25% used both calculators and statistics software.

a) What percent used neither calculators nor statistics software?
b) What percent used calculators but not statistics software?
c) What percent of the calculator sections used statistics software?
d) Based on this survey, do calculator and statistics software use appear to be independent events? Explain.

**38. Dogs** A census by the county dog control officer found that 18% of homes kept one dog as a pet, 4% had two dogs, and 1% had three or more. If a salesperson visits two homes selected at random, what's the probability they encounter . . .

a) no dogs?
b) some dogs?
c) dogs in each home?
d) more than one dog in each home?

**39. Socks** In your sock drawer you have 4 blue socks, 5 grey socks, and 3 black ones. Half asleep one morning, you grab 2 socks at random and put them on. Find the probability you end up wearing

a) 2 blue socks.
b) no gray socks.
c) at least 1 black sock.
d) a green sock.
e) matching socks.

**40. U.S. cars** As stated in Chapter 15, Exercise 53, the relative frequencies of numbers of vehicles in U.S. households are shown below. Doing that exercise, you would have found $E(X) = 1.7$, $\sigma(X) = 0.9$, using exactly 3 vehicles for "3 or more."

| Number of Vehicles | Relative Frequency of U.S. Households |
|---|---|
| 0 | 0.09 |
| 1 | 0.33 |
| 2 | 0.37 |
| 3 or more | 0.21 |

a) Use properties of random variables to determine the expected value and standard deviation of the total number of vehicles in *two* randomly selected households.
b) What assumptions must you make in part a, and are they justified?
c) Create a probability distribution for the total number of vehicles in two randomly selected households.
d) Calculate the expected value and standard deviation directly from this probability distribution, and compare with your answers from part a.
e) Sketch a histogram of this distribution and describe its shape, center, and spread in context.

**41. Gym time** A local gym charges a $150 one-time fee to join. After joining, there is a $40 per month charge to continue your membership.

a) The gym has tracked its members over time and estimates that the customers have a mean membership length of 21 months. What is the mean total of the money collected?
b) The standard deviation of the length of stay is 5 months. What is the standard deviation of the total money collected?

**42. Coins** A coin is to be tossed 36 times.

a) What are the mean and standard deviation of the number of heads?
b) Suppose the resulting number of heads is unusual, two standard deviations above the mean. How many "extra" heads were observed?
c) If the coin were tossed 100 times, would you still consider the same number of extra heads unusual? Explain.
d) In the 100 tosses, how many extra heads would you need to observe in order to say the results were unusual?
e) Explain how these results refute the "Law of Averages" but confirm the Law of Large Numbers.

**43. The Drake equation** In 1961 astronomer Frank Drake developed an equation to try to estimate the number of extraterrestrial civilizations in our galaxy that might be able to communicate with us via radio transmissions. Now largely accepted by the scientific community, the Drake equation has helped spur efforts by radio astronomers to search for extraterrestrial intelligence. Here is the equation:

$$N_C = N \cdot f_p \cdot n_e \cdot f_l \cdot f_i \cdot f_c \cdot f_L.$$

OK, it looks a little messy, but here's what it means:

| Factor | What It Represents | Possible Value |
|---|---|---|
| $N$ | Number of stars in the Milky Way Galaxy | 200–400 billion |
| $f_p$ | Probability that a star has planets | 20%–50% |
| $n_e$ | Number of planets in a solar system capable of sustaining earth-type life | 1? 2? |
| $f_l$ | Probability that life develops on a planet with a suitable environment | 1%–100% |
| $f_i$ | Probability that life evolves intelligence | 50%? |
| $f_c$ | Probability that intelligent life develops radio communication | 10%–20% |
| $f_L$ | Fraction of the planet's life for which the civilization survives | $\frac{1}{1,000,000}$? |
| $N_c$ | Number of extraterrestrial civilizations in our galaxy with which we could communicate | ? |

So, how many ETs are out there? That depends; values chosen for the many factors in the equation depend on ever-evolving scientific knowledge and one's personal guesses. But now, some questions.

a) What quantity is calculated by the first product, $N \cdot f_p$?
b) What quantity is calculated by the product, $N \cdot f_p \cdot n_e \cdot f_l$?
c) What probability is calculated by the product, $f_l \cdot f_i$?
d) Which of the factors in the formula are conditional probabilities? Restate each in a way that makes the condition clear.

*Note:* If you're interested in this topic, do a quick internet search and you'll find sites where you can play with the Drake equation yourself.

**44. Recalls** In a car rental company's fleet, 70% of the cars are American brands, 20% are Japanese, and the rest are German. The company notes that manufacturers' recalls seem to affect 2% of the American cars, but only 1% of the others.

a) What's the probability that a randomly chosen car is recalled?
b) What's the probability that a recalled car is American?

**45. Pregnant?** Suppose that 70% of the women who suspect they may be pregnant and purchase an in-home pregnancy test are actually pregnant. Further suppose that the test is 98% accurate. What's the probability that a woman whose test indicates that she is pregnant actually is?

**46. Door prize** You are among 100 people attending a charity fundraiser at which a large-screen TV will be given away as a door prize. To determine who wins, 99 white balls and 1 red ball have been placed in a box and thoroughly mixed. The guests will line up and, one at a time, pick a ball from the box. Whoever gets the red ball wins the TV, but if the ball is white, it is returned to the box. If none of the 100 guests gets the red ball, the TV will be auctioned off for additional benefit of the charity.

a) What's the probability that the first person in line wins the TV?
b) You are the third person in line. What's the probability that you win the TV?
c) What's the probability that the charity gets to auction the TV because no one wins?
d) Suppose you get to pick your spot in line. Where would you want to be in order to maximize your chances of winning?
e) After hearing some protest about the plan, the organizers decide to award the prize by not returning the white balls to the box, thus ensuring that 1 of the 100 people will draw the red ball and win the TV. Now what position in line would you choose in order to maximize your chances?

# PRACTICE EXAM

## I. MULTIPLE CHOICE

**1.** Researchers who were interested in the types of movies preferred by children of different age groups asked students in the sixth grade and in the eighth grade if they would prefer to see an animated feature like *The Lion King* or an action feature like *The Avengers*. The results are summarized in the table:

|  | Animated | Action |
|---|---|---|
| Sixth Grade | 45 | 35 |
| Eighth Grade | 40 | 60 |

Which proportions represent the conditional distribution of grade for children who preferred an action feature?

A) 0.194 and 0.333
B) 0.368 and 0.632
C) 0.40 and 0.60
D) 0.444 and 0.556
E) 0.529 and 0.471

**2.** Iron is an essential nutrient. Iron deficiency has been linked with symptoms such as anemia, rapid heartbeat, increased risk of infections, and lightheadedness. At the other end of the spectrum is iron overload, described in an August 2012 *New York Times* article. Excess iron is deposited in the liver, heart, and pancreas and can cause cirrhosis, liver cancer, cardiac arrhythmias, and diabetes. According to a Framingham Heart Study researcher, "About one person in 250 inherits a genetic disorder called hemochromatosis that increases iron absorption and results in a gradual, organ-damaging buildup of stored iron." Suppose we have a random sample of 1000 adults, and want to find the probability that at least 5 of them have this disorder. Which of these statements is true?

I. We would expect 4 of these people to have hemochromatosis.
II. We can calculate this probability using a Binomial model.
III. We can approximate this probability using a Normal model.

A) None
B) I only
C) I and II only
D) I and III only
E) I, II, and III

3. A bookstore asked a sample of adults how many e-books they had downloaded to their e-book readers during the last three months. Some of the data are shown below:

| Values Below Q1 | Q1 | Median | Q3 | Values Above Q3 |
|---|---|---|---|---|
| 8, 11, 14 | 16 | 18 | 22 | 25, 28, 30, 33 |

Which values will a boxplot identify as outliers?

A) None  B) 8 only  C) 33 only
D) 8 and 33 only  E) 8, 30, and 33

4. A university reports that 80% of its students enroll there as first-year students, while the rest transfer in from other 2- or 4-year colleges. The eventual graduation rate is 85% among the transfers, but only 70% among those who arrived as first-year students. What's the probability that a former student who never graduated was a transfer student?

A) 0.03  B) 0.11  C) 0.15
D) 0.27  E) 0.77

5. A company that supplies LP gas for heating keeps data on the low temperature for each day of each month. It summarizes these data by finding the mean, median, standard deviation, and interquartile range. The company assumes that people use the LP gas for heat when the temperature is below 70°, so it creates a second set of data by subtracting 70 from each daily low temperature. Which of the four summary statistics will change for this second dataset?

A) Mean and median only
B) Mean and standard deviation only
C) Median and IQR only
D) Standard deviation and IQR only
E) Mean, median, standard deviation, and IQR

6. The bar graph below summarizes information from a New York City (NYC) Survey of Elementary School Children (NYC DOHMH and DOE 2003). In a random sample of 400 Latinx NYC schoolchildren, what are the expected number who are obese and the standard deviation?

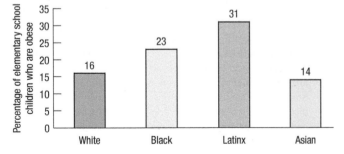

A) Mean 124; standard deviation 7.7
B) Mean 148; standard deviation 7.7
C) Mean 124; standard deviation 9.25
D) Mean 148; standard deviation 9.66
E) Mean 124; standard deviation 11.1355

7. Luis is a star high jumper for his high school track team. His Statistics teacher tells him that his jumps last year had an average $z$-score of $+1.8$, while this year his jumps have an average $z$-score of $+2.1$. Which of these statements can Luis know for certain?

A) He is jumping higher this year than last year.
B) The competing jumpers are not jumping as high this year as they did last year.
C) He is winning the high jump competitions more often this year.
D) His jumps average 0.3 feet higher this year.
E) His jumps this year are higher relative to the other jumpers.

8. An electronics retailer is developing a model for insurance policies on new cell phone purchases. It estimates that 60% of customers never make a claim, 25% of customers require a small repair costing an average of $50, and 15% of customers request a full refund costing $200. What is the long-term average cost the retailer should expect to pay to its customers per claim?

A) $42.50  B) $50  C) $83.33
D) $125  E) None of these

9. A study conducted by students in an AP Psychology class at South Kent School (SKS) in Connecticut discovered a correlation of $-0.38$ between students' hours of sleep ($x$) and GPAs ($y$). The scatterplot below displays the relationship. What conclusion can be reasonably drawn from these data?

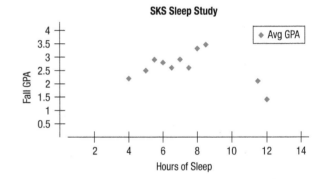

A) Students who get more sleep tend to earn higher GPAs.
B) If a student wishes to earn higher grades, they should get more sleep.
C) Based on the pattern in the scatterplot and the correlation, these data indicate that as students sleep longer, they tend to earn lower GPAs.
D) The scatterplot and the correlation coefficient contradict one another.
E) The scatterplot shows two influential points that affect the value of the correlation.

10. In 2012 *American Idol* completed its 11th season with Jennifer Lopez, Steven Tyler, and Randy Jackson as judges. Although the judges expressed their opinions about the contestants' performances, viewers voted to keep their favorites in the competition. During each episode, host Ryan Seacrest explained how viewers could vote via text, phone call, or online. Nothing prevented a viewer from submitting multiple votes. Each week the contestant with the lowest number of votes was removed from the competition. Why are the results not a true depiction of all Americans' musical preferences?

A) Undercoverage bias
B) Voluntary response bias
C) The judges' comments influenced voters' opinions.
D) Because some voters may have voted multiple times, one vote would not have corresponded to one person.
E) All of the above

11. In 2009 the Organization for Economic Cooperation and Development (OECD) conducted a study of 34 member countries called the Programme for International Student Assessment (PISA). The OECD looked at reading scores among 15-year-olds in each country as well as the students' socioeconomic backgrounds. PISA reported that higher socioeconomic status was associated with higher reading scores, with their socioeconomic index explaining 14% of the variability in reading scores. (Source: Pasi Sahlberg, "A Model Lesson: Finland Shows Us What Equal Opportunity Looks Like," *American Educator* [Spring 2012]: 25.)

    Which of the following can be correctly concluded from this information?

    A) For every additional point in the mean reading score in an OECD country, the PISA socioeconomic index is expected to increase by 14% on average.
    B) For every additional 1% increase in the PISA socioeconomic index, the mean reading score is expected to increase by 14% on average.
    C) If the U.S. government wants to increase the mean reading score for American 15-year-olds, it would be wise to institute programs to improve students' socioeconomic status.
    D) The correlation between the PISA socioeconomic index and reading scores for these OECD countries is 0.374.
    E) 14% of the variation in reading scores is caused by student socioeconomic status.

12. In the 2017 National Survey of the Use of Booster Seats (NSUBS), it was reported that 68.5% of 4- to 7-year-old children were placed in appropriate car seats in the United States. 28% were restrained by seat belts (which is not considered adequate for children) and 10.6% were completely unrestrained. What is the probability that researchers found their first unrestrained child aged 4 to 7 by the fifth car they checked?

    A) 0.00001   B) 0.0677   C) 0.4289
    D) 0.5711   E) 0.99999

13. Two student organizations conducted surveys to ascertain student opinion on a proposed plan to make the campus more environmentally responsible.
    - A political action group asked respondents whether they agreed with "the proposal to stop spewing pollution into the air."
    - A fiscal watchdog group asked whether respondents "really want tuition dollars to be spent on such non-educational renovations."

    One of these polls found 36% agreement with the proposal; the other found 74% agreement. Which of the following statements is true?

    A) The political action poll found 36% agreement, and the fiscal watchdog poll found 56% agreement.
    B) The political action poll includes a leading question, while the fiscal watchdog poll uses neutral wording.
    C) The fiscal watchdog poll includes a leading question, while the political action poll uses neutral wording.
    D) Both polls use neutral wording, so the different results reflect sampling error.
    E) Both polls contained leading questions, which may explain the very different results.

14. The weight of large bags of one brand of potato chips is a random variable with a mean of 18 ounces and a standard deviation of 0.7 ounces. The amount of chips a professional caterer pours into a certain size serving bowl can be described as a random variable with a mean of 4 ounces and a standard deviation of 0.3 ounces. The mean amount left in the bag after the caterer has filled two bowls is 10 ounces. Assuming these two variables are independent, what is the standard deviation of this remaining weight?

    A) 0.10 oz   B) 0.36 oz   C) 0.56 oz
    D) 0.82 oz   E) 0.92 oz

15. A quality control procedure at a manufacturing facility involves selecting 5 items at random from a large batch, and then accepting the entire batch if at least 3 of selected items pass inspection. If in reality 80% of all items produced would individually pass inspection, what is the probability that the batch will be accepted?

    A) 0.2048   B) 0.7373   C) 0.8000
    D) 0.9421   E) 0.9488

16. Which of these is the best description of a block?

    A) A random sample of a population who serve as subjects in an experiment
    B) Any of the different groups randomly selected in a stratified sample
    C) Any subgroup of the subjects in an experiment
    D) A subgroup of experimental subjects randomly assigned the same treatment
    E) A subgroup of experimental subjects that are the same with regard to some source of variation

*Use the following information for Exercises 17 and 18.*

A local bookseller carefully collects data on the customers that enter their store. They use the term "unit" to describe either a customer that comes in alone or a group of customers that come in together. Based on past experience, they estimate that 8% of the units who enter their store will make some type of purchase. It appears that the units are independent.

17. What is the probability that three units in a row will make a purchase?

    A) 0.000512   B) 0.068   C) 0.08
    D) 0.203   E) 0.24

18. In one hour, 10 units enter their store. What is the probability that at least one of them makes a purchase?

    A) 0.038   B) 0.378   C) 0.434
    D) 0.566   E) 0.812

19. A pet store sells fancy handcrafted nametags to go on dogs' and cats' collars. The store's profit is $6 for each dog nametag and $5 for each cat nametag sold. Each week the store sells an average of 12 dog tags with a standard deviation of 4 tags, and an average of 15 cat tags with a standard deviation of 3. The store's expected weekly profit on these products is $147. Assuming sales are independent, what's the standard deviation of this weekly profit?

    A) $11.87   B) $14.80   C) $28.30
    D) $39   E) $55

20. Once a month a local theater group stages a live theater production. The group is able to sell enough tickets so that the theater is almost full each month. However, the number of adult tickets and children's tickets that are sold vary depending on the play being performed. For these productions, which statement best describes the correlation between the number of adult tickets and the number of children's tickets sold?

   A) The correlation will be exactly 1.
   B) The correlation will be negative.
   C) The correlation will be 0.
   D) The correlation will be positive and less than 1.
   E) The correlation cannot be described based on the information given.

21. The Substance Abuse and Mental Health Services Administration reports that in 2011, 20,783 people were treated for medical emergencies related to energy drinks, largely attributable to the very high doses of caffeine in drinks of this type. The table below shows the age distribution for these patients. What is the probability that a person who visited the emergency department for an energy drink–related emergency was under 40 given that the person was at least 18 years old?

   A) 0.676
   B) 0.729
   C) 0.806
   D) 0.904
   E) 0.928

   | Age Range | Number of Patients |
   |---|---|
   | 12–17 | 1499 |
   | 18–25 | 7322 |
   | 26–39 | 6729 |
   | 40 or older | 5233 |

22. Suppose you wish to compare the ages at inauguration of Democratic and Republican presidents. Which is the most appropriate type of technique for gathering the needed data?

   A) Census
   B) Sample survey
   C) Experiment
   D) Prospective observational study
   E) None of these methods is appropriate.

23. A set of paired data has a least squares regression line with equation $\hat{y} = 0.50x + 2.0$ and a correlation coefficient of $r = 0.80$. Suppose we convert the data for each variable to z-scores and then compute the new regression line. What will the equation be?

   A) $\hat{z}_y = 0.50 z_x$  B) $\hat{z}_y = 0.64 z_x$  C) $\hat{z}_y = 0.80 z_x$
   D) $\hat{z}_y = 0.50 z_x + 20$  E) $\hat{z}_y = 0.80 z_x + 20$

24. A forest ranger has data on the heights of a large growth of young pine trees. The mean height is 3.2 feet and the standard deviation 0.6 feet. A histogram shows that the distribution of heights is approximately normal. Approximately what fraction of the trees should we expect to be between 4.0 and 4.4 feet tall?

   A) 2%  B) 7%  C) 9%  D) 91%  E) 98%

25. A researcher examined a sample of rainbow trout taken from the Spokane River in Washington State, recording their lengths (mm) and weights (grams). For example, one trout was 360 mm long and weighed 469 g. Because a scatterplot using length to predict weight showed an exponential relationship, the researcher took the log of weight and successfully linearized the relationship. Use their regression model $\log(\widehat{Weight}) = 1.491 + 0.00331 \, Length$ to predict the weight of a 400 mm rainbow trout.

   A) 2.815 g  B) 16.7 g  C) 509 g
   D) 598 g  E) 653 g

## II. FREE RESPONSE

1. In the National Football League every team must submit an injury list prior to each game. The list contains the names of the players who are "out," meaning they will not play in the game. There are also three other categories: "Doubtful," "Questionable," and "Probable." The guidelines state that "Doubtful" should mean about a 25% chance that the player will play, and "Questionable" means about a 50% chance that the player will play. The table below shows what happened with the three categories of players for a particular week.

   |  | Doubtful | Questionable | Probable |
   |---|---|---|---|
   | Played | 4 | 16 | 52 |
   | Did Not Play | 16 | 20 | 12 |

   a) What percent of these players who did play had been listed as probable?
   b) What percent of these players were listed as probable and did play?
   c) Create a graph that compares the relative frequencies of players who got into the game for the three status categories.
   d) Based on this information, is whether or not a player gets into the game independent of their pre-game status? Explain.

2. In the United States, homes are measured by the total number of square feet of area of all floors of the house. Data from www.census.gov show how the median size of a home in the Northeast changed from 1973 through 2010. Here are the regression results, and on the next page you'll find a plot of the residuals against predicted values.

   Dependent variable is NEsqft
   No Selector
   R squared = 94.8%   R squared(adjusted) = 94.6%
   S = 72.07 with 38 − 2 = 36 degrees of freedom

   | Source | Sum of Squares | df | Mean Square | F-ratio |
   |---|---|---|---|---|
   | Regression | 3403782 | 1 | 3403782 | 655 |
   | Residual | 186979 | 36 | 5193.87 | |

   | Variable | Coefficient | s.e. of Coeff | t-ratio | prob |
   |---|---|---|---|---|
   | Constant | −52410.9 | 2123 | −24.7 | ≤ 0.0001 |
   | Year | 27.2927 | 1.066 | 25.6 | ≤ 0.0001 |

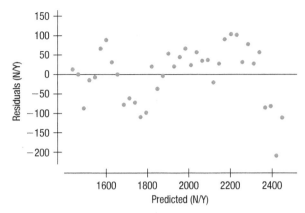

a) Write the equation of the regression line.
b) Do you think this linear model is appropriate? Explain.
c) What does this residual plot tell you about estimates based on this model?
d) According to this model, how has median home size changed per decade?
e) Estimate the *actual* median home size in 2009.

3. Fifty patients suffering from knee pain have volunteered to participate in an experiment to examine the effects of pain relief medication. Half will be assigned to take 400 mg of the pain reliever every six hours, and the other half to take 200 mg every 3 hours. The patients will report their pain levels on a 1–10 scale at the beginning of the experiment and again after 8 days on the medication.

   a) Define the experimental units and the response variable.
   b) What are the treatments?
   c) Describe an appropriate method to assign the volunteers to treatments.
   d) An alternate design could include a control group whose members take a placebo. Describe an advantage of adding a control group.

4. In 2012 the Centers for Disease Control and Prevention reported that the rate of autism in children has risen to 1 in 50. A study conducted in Norway suggests that the risk of autism may be significantly lower if women take folic acid supplements during pregnancy. Suppose a large medical center enlists 900 expectant mothers in a trial; the women agree to take the folic acid for the duration of their pregnancies. The researchers will track the children after birth to see how many develop autism.

   a) Explain why the number of autism cases can be treated as a binomial random variable.
   b) If folic acid supplements actually have no effect, what is the mean and standard deviation of the number of autism cases that these researchers might find?
   c) How low would the number of autism cases among these 900 children have to be in order to convince you that the folic acid supplements may be effective? Explain your reasoning.

5. An online retailer sells its $25 gift cards at supermarkets. The retailer knows that 20% of people who purchase these cards spend the full value, 70% leave a balance of $5 that is never spent, and the rest never use the cards at all. Because people pay cash for the cards, the amounts that are not spent are pure profit for the retailer.

   a) If you buy three of these cards for three friends who have birthdays coming up, are the conditions met to treat these as Bernoulli trials if the variable is whether the cards are completely used or not?
   b) Assuming we *can* treat these as Bernoulli trials, what is the probability that none of the three cards will be completely used?
   c) What is the online retailer's expected profit per card?
   d) What is the standard deviation of the retailer's profit per card?
   e) A service club makes a proposal to the retailer. The club wants to buy 100 of the cards for only $2000, so it can sell them as a fundraiser. Should the retailer worry that it might suffer a loss if it sells the club these cards at this discount? Explain.

# 17 Sampling Distribution Models

The National Center for Health Statistics published data on all 3,945,192 U.S. live births in 1998 (www.cdc.gov/nchs/data_access/vitalstatsonline.htm), recording information on the babies (such as birthweight and Apgar score) and on the mother. They may have missed a few, but we'll treat this as the population of 1998 babies. Babies born before the 37th week of gestation are considered pre-term and, according to the Mayo Clinic (mayoclinic.org), are at risk for both short- and long-term health problems. In 1998, 11.6% of births in the United States were premature (preemies). Figure 17.1 shows the distribution of gestation times.

## The Sampling Distribution of a Proportion

**Probability or Proportion?**

If we say that 11.6% of American babies were born prematurely, that 11.6% is a proportion. Based on this observed proportion, we might estimate that the probability that a randomly selected American baby is born prematurely is 11.6%.

From any one random sample, the best guess of the true proportion is the observed sample proportion. But since each random sample is different, our challenge—and the challenge faced by anyone who has only a sample—is not only to guess the true proportion in the entire population, but also to guess *how good that guess is*. Quantifying the uncertainty in our guesses is the core of statistical thinking.

Imagine throwing a dart at the bull's eye of a dartboard after putting a plain sheet of paper in front of the target to record where the dart hits. (The bull's eye is the center of the circular dart board. Check out the picture at the start of this chapter.) Now, take the paper off, show it to a friend, and ask them to guess the location of the bull's eye. Their best guess has to be where the dart hit.

But that's not likely to be exactly right. Instead, let them draw a circle that they think will contain the bull's eye. How large a circle should they draw? They have no idea how close your throw was to the bull's eye, so they'll find it hard to judge the right size for their circle.

Figure 17.1
Gestation times (in weeks) of 3,904,759 babies born in the United States in 1998. The distribution is slightly skewed to the left. About 11.6% of the babies are preemies (with gestation times less than 37 weeks), highlighted in purple.

| WHO | U.S. babies |
| --- | --- |
| WHAT | Gestation times |
| WHEN | 1998 |
| WHERE | United States |
| WHY | Public health |

**IMAGINE**
We see only the sample that we actually drew, but by simulating or modeling, we can *imagine* what we might have seen had we drawn other possible random samples.

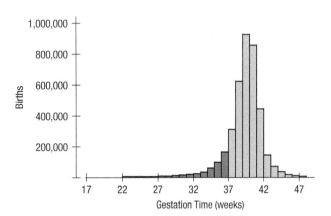

What if you'd thrown several darts? They'd probably guess that the bull's eye is in the middle of the cluster of holes, and by looking at your consistency (or lack of), they now have a much better idea of how big a circle to draw.

It is unusual to have an entire population to work with. Usually, we draw a sample at random. But, as we have seen, samples vary from one to the next. To get a handle on how consistently we can estimate the proportion of preemies from a sample of births, it may seem that we'll need many "darts"—samples drawn at random from our population. Suppose we could examine *all* possible samples of size 100 from the population of 4,000,000. Then we could look at the distribution of those proportions and see how much they vary. Of course, we can't do that, but we can *imagine* doing it.

That distribution of the proportions from all possible samples is called the **sampling distribution** of the proportion.[1] Unfortunately, the number of all possible samples of size 100 drawn from a population of 3,945,192 is about $4 \times 10^{501}$, a number so big we can't even imagine it.[2] But we've seen in earlier chapters that a simulation can often provide a good approximation of a distribution, sometimes even with just a few thousand trials. *Here comes another simulation!*

We used a computer to simulate a random sample of 100 newborn babies from a large database in which 11.6% of them were born prematurely. In the first sample, 13 of the 100 babies sampled were born prematurely: 13%. Slightly higher than you'd expect. In our second sample, eight. Slightly lower. We simulated a few thousand such samples and collected together all the sample proportions. The graph below shows the distribution of all the sample proportions we collected.

Figure 17.2

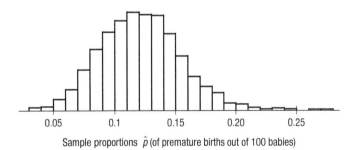

---

[1] A word of caution. Until now, we've been plotting the *distribution of the sample*, a display of the actual data that were collected in that one sample. But now we're plotting a *sampling distribution*; a display of summary statistics ($\hat{p}$'s in this case) for many different samples. "Sample distribution" and "sampling distribution" sound a lot alike, but they refer to very different things. (Sorry about that—we didn't make up the terms.)
And the distinction is critical. Whenever you read or write something about one of these, choose your terms carefully.

[2] And neither can you, probably. The number of atoms in the universe is "only" about $4 \times 10^{81}$.

**NOTATION ALERT**

The letter $p$ is our choice for the *parameter* of the model for proportions. It violates our "Greek letters for parameters" rule, but if we stuck to that, our natural choice would be $\pi$, which could be confusing.

So, we'll use $p$ for the model parameter (the probability of a success) and $\hat{p}$ for the observed proportion in a sample. We'll also use $q$ for the probability of a failure ($q = 1 - p$) and $\hat{q}$ for its observed value.

In Chapter 7, we introduced $\hat{y}$ as the predicted value for $y$. The "hat" here plays a similar role. It indicates that $\hat{p}$—the observed proportion in our data—is our *estimate* of the parameter $p$.

But be careful. We've already used capital $P$ for a general probability. And we'll soon see another use of $P$ later in the chapter! There are a lot of $p$'s in this course; you'll need to think clearly about the context to keep them straight.

Let's recall three things that together can describe a distribution: shape, center, and spread. It's pretty clear that the shape here is approximately normal. The center appears to be at about the 0.116 that we'd anticipate. (The population proportion was $p = 0.116$.) As for the spread? It's a little tough to tell eyeballing the graph, but because we *have* all of the simulated sample proportions, we were able to compute their standard deviation to be approximately 0.032.

If the conditions are right, we can use a Normal model for sample proportions, *and* we can know just what the mean and standard deviation of the Normal model should be. Read on.

## Which Normal?

To use a Normal model, we need to specify two parameters: its mean and standard deviation. The center of the sampling distribution of a proportion is naturally at $p$, the true proportion, so that's what we'll use for the mean of the Normal model.

What about the standard deviation? We saw from the simulation of premature births that proportions in samples of size 100 with a population proportion of 11.6% had a standard deviation of 3.2%. Sample proportions are very special. There's a mathematical fact that lets us find the standard deviation of their sampling distribution without a simulation.[3] It gives an exact answer if we know the true proportion. If we write $p$ for the true proportion, $q$ as shorthand for $1 - p$, and $n$ for the sample size, then the standard deviation of $\hat{p}$ is

$$\sigma(\hat{p}) = SD(\hat{p}) = \sqrt{\frac{pq}{n}}.$$

Be careful! What is this standard deviation? It's not the standard deviation of the data. It's the standard deviation of the proportions of all possible samples of $n$ values from the population. Because we have the population, we know that the true proportion is $p = 11.6\%$ preemies. The formula then says that the standard deviation of proportions from samples of size 100 is[4]

$$\sqrt{\frac{pq}{n}} = \sqrt{\frac{(0.116)(0.884)}{100}} = 0.032.$$

Figure 13.3 puts together the facts that the sampling distribution is Normal, centered at $p$, and has standard deviation $\sqrt{\frac{pq}{n}}$.

**Figure 13.3**
The sampling distribution model of a proportion is Normal with mean $p$ and standard deviation $\sqrt{\frac{pq}{n}}$.

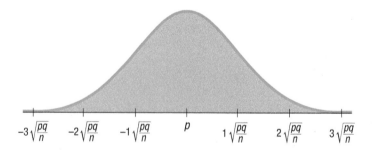

Here's the payoff. Because we have a Normal model, we can use the 68–95–99.7 Rule or look up exact probabilities using a table or technology. For example, 95% of Normally distributed values are within roughly two standard deviations of the mean. So about

---

[3]You actually have the tools to prove this formula. A sample proportion of successes is a random variable that is the same as the sample *count* of successes (binomial) divided by $n$. When you divide a random variable by a constant (like $n$), its standard deviation gets divided by the same constant. What's the standard deviation of a binomial random variable?

[4]The standard deviation is 0.032 or 3.2%. Remember that the standard deviation always has the same units as the data. Here our units are %, or "percentage points." The standard deviation isn't 3.2% of anything, it is just 3.2 percentage points. If that's confusing, try writing the units out as "percentage points" instead of using the symbol %. Many polling agencies now do that too.

95% of samples will have sample proportion within two standard deviations of the true population proportion. Knowing that, we should not be surprised if various polls that may appear to ask the same question report a variety of results, but we *would* be surprised if there was too much variety, because we now know how much variation to expect from sample to sample. Such sample-to-sample variation is sometimes called **sampling error**. It's not really an *error* at all, but just *variability* you'd expect to see from one sample to another. A better term would be **sampling variability**.[5]

## Can We Always Use a Normal Model?

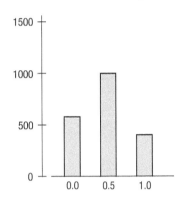

Figure 17.4
Proportions from samples of size 2 can take on only three possible values. A Normal model does not work well.

So. Under the right conditions, the sampling distribution of sample proportions can be modeled well by a Normal model. But under what conditions? Will it always work? Well, no. For example, if we drew samples of size 2, the only possible proportion values would be 0, 0.5, and 1. There's no way a histogram consisting only of those values could look like a Normal model.

The Normal model becomes a better and better approximation of the sampling distribution as the size of the samples gets bigger. Samples of size 1 or 2 won't work at all, but the distributions of sample proportions of larger samples are modeled well by the Normal model as long as $p$ isn't too close to 0 or 1.

Populations with a true proportion, $p$, close to 0 or 1 can be a problem. Suppose a basketball coach surveys students to see how many male high school seniors are over 6'6". What will the proportions of samples of size 1000 look like? If the true proportion of students that tall is 0.001, then the coach is likely to get only a few seniors over 6'6" in any random sample of 1000. Most samples will have proportions of $\frac{0}{1000}, \frac{1}{1000}, \frac{2}{1000}, \frac{3}{1000}, \frac{4}{1000}$ with only a very few samples having a higher proportion. A simulation of 2000 surveys of size 1000 with $p = 0.001$ shows a sampling distribution for $\hat{p}$ that's skewed to the right because $p$ is so close to 0. (Had $p$ been very close to 1, it would have been skewed to the left.) So, even though $n$ is large, $p$ is too small, and so the Normal model still won't work well.

### Assumptions and Conditions

When does the Normal model with mean $p$ and standard deviation $\sqrt{\frac{pq}{n}}$ work well as a model for the sampling distribution of a sample proportion? Before proceeding we must check the following assumptions and conditions:

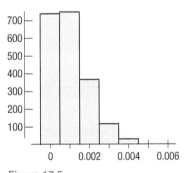

Figure 17.5
The distribution of sample proportions for 2000 samples of size 1000 with $p = 0.001$. Because the true proportion is so small, the sampling distribution is skewed to the right and the Normal model won't work well.

**The Independence Assumption:** The individuals in the sample must be independent of each other. You can't know if an assumption is true, but you *can* check two conditions to see whether the data were collected in a way that makes this assumption reasonable.

**Randomization Condition:** If your data come from an experiment, subjects should have been randomly assigned to treatments. If you have a survey, your sample should be a simple random sample of the population. If some other sampling design was used, be sure the sampling method was not biased and that the data are representative of the population. If they aren't, then your inferences may not be valid.

**10% Condition:** If your sample represents a large fraction of the population, then the Normal model may not be reasonable. Once you've sampled more than about 10% of the population, the remaining individuals are no longer really independent of each other.[6]

---

[5] Of course, polls differ for many reasons in addition to sampling variability. These reasons can include minor differences in question wording and in the ways that the random samples are selected. But polling agencies report their "margin of error" based on sampling variability. We'll see how they find that later in the chapter.

[6] There are special formulas that you can use to adjust for sampling a large part of a population but, as the saying goes, they are "beyond the scope of this course."

## Successes and Failures

The terms "success" and "failure" for the outcomes that have probability $p$ and $q$ are common in Statistics. But they are completely arbitrary labels. When we say that a disease occurs with probability $p$, we certainly don't mean that getting sick is a "success" in the ordinary sense of the word.

We've seen that a Normal model becomes more useful for describing the sampling distribution of $\hat{p}$ as the sample size increases. That's the basis for another assumption we must think about.

**The Sample Size Assumption:** The sample size, $n$, must be large enough. How large? That depends on the value of $p$; the closer $p$ is to 0 or 1, the larger $n$ must be. We check a familiar condition:

**Success/Failure Condition:** The sample size must be large enough that we expect to see at least 10 successes and at least 10 failures. (We check that $np \geq 10$ and $nq \geq 10$.)

The 10% Condition and the Success/Failure Condition may seem to conflict with each other. The **Success/Failure Condition** wants sufficient data. How much depends on $p$. If $p$ is near 0.5, we need a sample of only 20 or so. If $p$ is only 0.01, however, we'd need 1000. But the **10% Condition** says that a sample should be no larger than 10% of the population. If you're thinking, "Wouldn't a larger sample be better?" you're right of course. It's just that if the sample were more than 10% of the population, we'd need to use different methods to analyze the data. Fortunately, this is almost *never* a problem in practice. Often, as in polls that sample from all U.S. adults or industrial samples from a day's production, the populations are much larger than 10 times the sample size.

# A Sampling Distribution Model for a Proportion

We've arrived at a key moment in this course. (Drumroll, please!) We have changed our point of view in a very important way. No longer is a proportion something we just compute for a set of data. We now see a sample proportion to be a random variable that has a probability distribution, and we have a model for that distribution. We call that the **sampling distribution model for a sample proportion**, and we'll make good use of it.

### The Sampling Distribution Model for a Sample Proportion

Provided that the sampled values are independent and the sample size is large enough, the sampling distribution of $\hat{p}$ can be reasonably modeled by a Normal model with mean $E(\hat{p}) = p$ and standard deviation $SD(\hat{p}) = \sqrt{\dfrac{pq}{n}}$.

Without sampling distribution models, the rest of Statistics just wouldn't exist.[7] Sampling models are what make Statistics work. They inform us about the amount of variation we should expect when we sample. A sampling distribution model tells us how surprising a sample statistic is and enables us to make informed decisions about how precise our estimate of the true value of a parameter might be. That's exactly what we'll be doing for the rest of this book.

Sampling distribution models enable us to say something about the population when all we have are data from a sample. This is the huge leap of Statistics. By imagining what *might* happen if we were to draw many, many samples from the same population, we can learn a lot about how close the statistics computed from our one particular sample may be to the corresponding population parameters they estimate. That's the path to the *margin of error* you hear about in polls and surveys. We'll see how to determine that in the next chapter.

---

[7] Actually, without sampling distribution models, you could still learn much of what you wanted using simulations, if you were a good computer programmer.

## FOR EXAMPLE

### Using the Sampling Distribution Model for Proportions

The Centers for Disease Control and Prevention report that 22% of 18-year-old women in the United States have a body mass index (BMI)[8] of 25 or more—a value considered by the National Heart Lung and Blood Institute to be associated with increased health risks.

As part of a routine health check at a large college, the physical education department usually requires students to come in to be measured and weighed. This year, the department decided to try out a self-report system. It asked 200 randomly selected female students to report their heights and weights (from which their BMIs could be calculated). Only 31 of these students had BMIs greater than 25.

QUESTION: Is this proportion of high-BMI students unusually small?

ANSWER: First, check the conditions:

✓ **Randomization Condition:** The department drew a random sample, so the respondents should be independent and randomly selected from the population.

✓ **10% Condition:** 200 respondents is less than 10% of all the female students at a "large college."

✓ **Success/Failure Condition:** The department expected $np = 200(0.22) = 44$ "successes" and $nq = 200(0.78) = 156$ "failures," both at least 10.

It's okay to use a Normal model to describe the sampling distribution of the proportion of respondents with BMIs above 25.

The phys ed department observed $\hat{p} = \dfrac{31}{200} = 0.155$.

The department expected $E(\hat{p}) = p = 0.22$, with $SD(\hat{p}) = \sqrt{\dfrac{pq}{n}}$

$= \sqrt{\dfrac{(0.22)(0.78)}{200}} = 0.029$, so $z = \dfrac{\hat{p} - p}{SD(\hat{p})} = \dfrac{0.155 - 0.22}{0.029} = -2.24$.

By the 68–95–99.7 Rule, I know that values more than 2 standard deviations below the mean of a Normal model show up less than 2.5% of the time. Perhaps women at this college differ from the general population, or self-reporting may not provide accurate heights and weights.

## JUST CHECKING

1. You want to poll a random sample of 100 students at a large university to see if they are in favor of the proposed location for the new student center. Of course, you'll get just one number, your sample proportion, $\hat{p}$. But if you imagined all the possible samples of 100 students you could draw and imagined the histogram of all the sample proportions from these samples, what shape would it have?

2. Where would the center of that histogram be?

3. If you think that about half the students are in favor of the plan, what would the standard deviation of the sample proportions be?

---

[8] BMI = weight in kg/(height in m)$^2$.

# STEP-BY-STEP EXAMPLE

## Working with Sampling Distribution Models for Proportions

Suppose that about 13% of the population is left-handed.[9] A 200-seat school auditorium has been built with 15 "lefty seats," seats that have the built-in desk on the left rather than the right arm of the chair. (For the right-handed readers among you, have you ever tried to take notes in a chair with the desk on the left side?)

QUESTION: In a class of 90 students, what's the probability that there will not be enough seats for the left-handed students?

**THINK**

**PLAN** State what we want to know.

I want to find the probability that in a group of 90 students, more than 15 will be left-handed. Because 15 out of 90 is 16.7%, I need the probability of finding more than 16.7% left-handed students out of a sample of 90 if the proportion of lefties in the population is 13%.

**MODEL** Think about the assumptions and check the conditions.

You might be able to think of cases where the **Independence Assumption** is not plausible—for example, if the students are all related, or if they were selected for being left- or right-handed. Sampling randomly is the key to independence.

✓ **Independence Assumption**: It is reasonable to assume that the enrollment of one left-handed student does not affect the likelihood that other lefties will enroll in the class.

✓ **Randomization Condition**: The 90 students in the class can be thought of as a random sample of all students.

✓ **10% Condition**: 90 is surely less than 10% of the population of all students. (Even if the school itself is small, I'm thinking of the population of all *possible* students who could have gone to the school.)

✓ **Success/Failure Condition**:

$$np = 90(0.13) = 11.7 \geq 10$$
$$nq = 90(0.87) = 78.3 \geq 10$$

State the parameters and the sampling distribution model.

The population proportion is $p = 0.13$. The conditions are satisfied, so I'll model the sampling distribution of $\hat{p}$ with a Normal model with mean 0.13 and a standard deviation of

$$SD(\hat{p}) = \sqrt{\frac{pq}{n}} = \sqrt{\frac{(0.13)(0.87)}{90}} \approx 0.035.$$

My model for $\hat{p}$ is $N(0.13, 0.035)$.

**SHOW**

**PLOT** Make a picture. Sketch the model and shade the area we're interested in, in this case the area to the right of 16.7%.

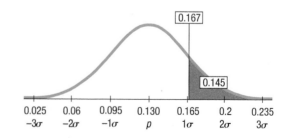

---

[9] Actually, it's quite difficult to get an accurate estimate of the proportion of lefties in the population. Estimates range from 8% to 15%.

**MECHANICS** Use the standard deviation as a ruler to find the z-score of the cutoff proportion. We see that 16.7% lefties would be just over one standard deviation above the mean.

Find the resulting probability from a table of Normal probabilities, a computer program, or a calculator.

$$z = \frac{\hat{p} - p}{SD(\hat{p})} = \frac{0.167 - 0.13}{0.035} = 1.06$$

$$P(\hat{p} > 0.167) = P(z > 1.06) = 0.1446$$

**TELL** **CONCLUSION** Interpret the probability in the context of the question.

There is about a 14.5% chance that there will not be enough seats for the left-handed students in the class.

# The Sampling Distributions of Other Statistics

We're lucky that the Normal model can provide a good approximation for the sampling distribution of a sample proportion—we use proportions a lot and the Normal model is well understood. But it might be useful to know the sampling distribution for *any* statistic that we can calculate, not just the sample proportion. Is the Normal model a good model for all statistics? Would you expect that a Normal model would be a good approximation for the sampling distribution of the minimum or the maximum or the variance of a sample? What about the median? The IQR?

### Simulating the Sampling Distributions of Other Statistics

A study of body measurements of 250 men found the median weight to be 176.125 lb and the variance to be 730.9 lb². (That's pounds squared, not a footnote. Remember: this is a variance, whose units are peculiar.) Treating these 250 men as a population, we can draw repeated random samples of 10 and compute the median, the variance, and the minimum for each sample.

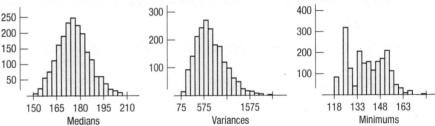

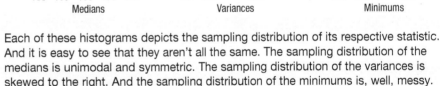

Each of these histograms depicts the sampling distribution of its respective statistic. And it is easy to see that they aren't all the same. The sampling distribution of the medians is unimodal and symmetric. The sampling distribution of the variances is skewed to the right. And the sampling distribution of the minimums is, well, messy.

We can simulate to get a look at the sampling distribution of *any* statistic we like: the maximum, the IQR, the 37th percentile, anything. Both the proportion and the mean (as we'll see in the next section) have sampling distributions that can be well approximated by a Normal model. That's good news because these are the two summary statistics that interest us most.

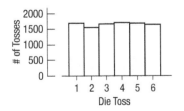

Figure 17.6

# Simulating the Sampling Distribution of a Mean

Here's a simple simulation. Let's start with one fair six-sided die. If we toss this die 10,000 times, what should the histogram of the numbers on the face of the die look like? Figure 17.6 shows the results of a simulated 10,000 tosses.

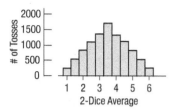

Figure 17.7

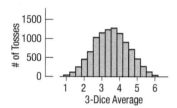

Figure 17.8

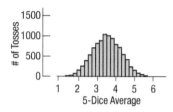

Figure 17.9

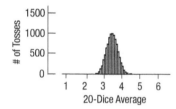

Figure 17.10

Now let's toss a *pair* of dice and record the average of the two. If we repeat this (or at least simulate repeating it) 10,000 times, recording the average of each pair, what will the histogram of the 10,000 averages look like? Before you look, think a minute. Is getting an average of 1 on *two* dice as likely as getting an average of 3 or 3.5? Not at all, right? We're much more likely to get an average near 3.5 than we are to get one near 1 or 6. After all, the *only* way to get an average of 1 is to get two 1's. To get a total of 7 (for an average of 3.5), though, there are many more possibilities. This distribution even has a name: the *triangular* distribution, as seen in Figure 17.7.

What if we average 3 dice? We simulated 10,000 tosses of 3 dice and took their average. Figure 17.8 shows the result.

What's happening? First notice that it's getting harder to have averages near the ends. Getting an average of 1 or 6 with 3 dice requires all three to come up 1 or 6, respectively. That's less likely than for 2 dice to come up both 1 or both 6. The distribution is being pushed toward the middle.

Let's continue this simulation to see what happens with larger samples. Figure 17.9 shows a histogram of the averages for 10,000 tosses of 5 dice.

The pattern is becoming clearer. Two things continue to happen. The first fact we knew already from the Law of Large Numbers. It says that as the sample size (number of dice) gets larger, each sample average is more likely to be closer to the population mean. So, we see the distribution continuing to tighten around 3.5. But the shape is the surprising part. It's approaching the Normal model.

Let's skip ahead and try 20 dice. The histogram of averages for 10,000 throws of 20 dice is Figure 17.10.

Now we see the Normal shape clearly (and notice how much smaller the spread is). But can we count on this happening for situations other than dice throws? It turns out that Normal models work well amazingly often.

## The Fundamental Theorem of Statistics

> The theory of probabilities is at bottom nothing but common sense reduced to calculus.
> —Pierre-Simon Laplace, in *Théorie analytique des probabilités*, 1812

What we saw with dice is true for means of repeated samples for almost every situation. When we looked at the sampling distribution of a proportion, we had to check only a few conditions. For means, the result is even more remarkable. *There are almost no conditions at all.*

Let's say that again: The sampling distribution of *any* sample mean becomes more nearly Normal as the sample size grows. All we need is for the observations to be independent and collected with randomization. We don't even care about the shape of the population distribution![10] This surprising fact was proven by a French mathematician named Pierre Laplace in a fairly general form in 1810. At the time, Laplace's theorem caused quite a stir (at least in mathematics circles) because it is so unintuitive. Laplace's result is called the **Central Limit Theorem**[11] (CLT).

Why should the Normal model show up again for the sampling distribution of means as well as proportions? We're not going to try to persuade you that it is obvious,

---

[10] OK, one technical condition. The data must come from a population with a finite variance. You probably can't imagine a population with an infinite variance, but statisticians can construct such things, so we have to discuss them in footnotes like this. It really makes no difference in how you think about the important stuff, so you can just forget we mentioned it.

[11] The word "central" in the name of the theorem means "fundamental." It doesn't refer to the center of a distribution.

### THE GREATNESS OF LAPLACE

Laplace was one of the greatest scientists and mathematicians of his time. In addition to his contributions to probability and statistics, he published many new results in mathematics, physics, and astronomy (where his nebular theory was one of the first to describe the formation of the solar system in much the way it is understood today). He also played a leading role in establishing the metric system of measurement.

**TI-*nspire***

**The Central Limit Theorem.** See the sampling distribution of sample means take shape as you choose sample after sample.

**APPLET**
Explore the meaning of the amazing Central Limit Theorem

clear, simple, or straightforward. In fact, the CLT is surprising and a bit weird. Not only does the distribution of means of many random samples get closer and closer to a Normal model as the sample size grows, *this is true regardless of the shape of the population distribution!* Even if we sample from a skewed or bimodal population, the Central Limit Theorem tells us that means of repeated random samples will more closely follow a Normal model as the sample size grows. For practical purposes, this means that if the sample size is "large enough," then the sampling distribution of $\bar{y}$ can be reasonably approximated using a Normal model. (We'll talk shortly about "large enough.") So although the Central Limit Theorem is technically about something that happens only in the very long run, its usefulness comes from the fact that it begins to work its magic even in the short run.

And here's a bonus. Consider a population that consists of only 0's and 1's. The CLT says that even means of samples from this population will follow a Normal sampling distribution model. But wait. Suppose we have a categorical variable and we assign a 1 to each individual in the category and a 0 to each individual not in the category. And then we find the mean of these 0's and 1's. That's the same as counting the number of individuals who are in the category and dividing by $n$. That mean will be . . . the *sample proportion*, $\hat{p}$, of individuals who are in the category (a "success"). So maybe it wasn't so surprising after all that proportions, like means, have Normal sampling distribution models; they are actually just a special case of Laplace's remarkable theorem. Of course, for such an extremely bimodal population, we'll need a reasonably large sample size—and that's where the special conditions for proportions come in.

### The Central Limit Theorem (CLT)

The mean of a random sample is a random variable whose sampling distribution can be approximated by a Normal model. The larger the sample, the better the approximation will be.

## Assumptions and Conditions

The CLT requires essentially the same assumptions as we saw for modelling proportions:

**Independence Assumption:** The sampled values must be independent of each other.
**Sample Size Assumption:** The sample size must be sufficiently large.

We can't check these directly, but we can think about whether the **Independence Assumption** is plausible. We can also check some related conditions:

**Randomization Condition:** The data values must be sampled randomly, or the concept of a sampling distribution makes no sense.

**10% Condition:** When the sample is drawn without replacement (as is usually the case), the sample size, $n$, should be no more than 10% of the population.

**Large Enough Sample Condition:** Although the CLT tells us that a Normal model is useful in thinking about the behavior of sample means when the sample size is large enough, it doesn't tell us how large a sample we need. The truth is, it depends; there's no one-size-fits-all rule. If the population is unimodal and symmetric, even a fairly small sample is okay. If the population is strongly skewed, like the compensation for CEOs we looked at in Chapter 4, it can take a pretty large sample to allow use of a Normal model to describe the distribution of sample means. For now you'll just need to think about your sample size in the context of what you know about the population, and then tell whether you believe the Large Enough Sample Condition has been met.

## But Which Normal?

The CLT says that the sampling distribution of any mean or proportion is approximately Normal. But which Normal model? Any Normal model is specified by its mean and standard deviation. For proportions, the sampling distribution is centered at the population proportion. For means, it's centered at the population mean. What else would we expect?

What about the standard deviations, though? We noticed in our dice simulation that the histograms got narrower as we averaged more and more dice together. This shouldn't be surprising. Means vary less than the individual observations. Think about it for a minute. Which would be more surprising, having *one* person in your Statistics class who is over 6′6″ tall or having the *mean* of all students taking the course be over 6′6″? The first event is fairly rare. You may have seen somebody this tall in one of your classes sometime. But finding a whole class whose mean height is over 6′6″ tall just won't happen. Why? Because *sample means have smaller standard deviations than individuals.*

How much smaller? Well, we have good news and bad news. The good news is that the standard deviation of $\bar{y}$ falls as the sample size grows. The bad news is that it doesn't drop as fast as we might like. Like proportions, it only goes down by the *square root* of the sample size. The Math Box will show you that the Normal model for the sampling distribution of the mean has a standard deviation equal to

$$SD(\bar{y}) = \frac{\sigma}{\sqrt{n}},$$

where $\sigma$ is the standard deviation of the population. To emphasize that this is a standard deviation *parameter* of the sampling distribution model for the sample mean, $\bar{y}$, we write $SD(\bar{y})$ or $\sigma(\bar{y})$.

### The Sampling Distribution Model for a Sample Mean (CLT)

When random samples of size *n* are drawn from any population with mean $\mu$ and standard deviation $\sigma$, the sample means, $\bar{y}$, have a sampling distribution with the same mean $\mu$ but whose *standard deviation* is $\frac{\sigma}{\sqrt{n}}$ (and we write $\sigma(\bar{y}) = SD(\bar{y}) = \frac{\sigma}{\sqrt{n}}$).
No matter what population the random sample comes from, the *shape* of the sampling distribution is approximately Normal as long as the sample size is large enough. The larger the sample used, the more closely the Normal model approximates the sampling distribution for the mean.

### MATH BOX

We know that $\bar{y}$ is a sum divided by *n*:

$$\bar{y} = \frac{y_1 + y_2 + y_3 + \cdots + y_n}{n}.$$

As we saw in Chapter 15, when a random variable is divided by a constant its variance is divided by the *square* of the constant:

$$Var(\bar{y}) = \frac{Var(y_1 + y_2 + y_3 + \cdots + y_n)}{n^2}.$$

To get our sample, we draw the *y*'s randomly, ensuring they are independent. For independent random variables, variances add:

$$Var(\bar{y}) = \frac{Var(y_1) + Var(y_2) + Var(y_3) + \cdots + Var(y_n)}{n^2}.$$

All *n* of the *y*'s were drawn from our population, so they all have the same variance, $\sigma^2$:

$$Var(\bar{y}) = \frac{\sigma^2 + \sigma^2 + \sigma^2 + \cdots + \sigma^2}{n^2} = \frac{n\sigma^2}{n^2} = \frac{\sigma^2}{n}.$$

The standard deviation of $\bar{y}$ is the square root of this variance:

$$SD(\bar{y}) = \sqrt{\frac{\sigma^2}{n}} = \frac{\sigma}{\sqrt{n}}.$$

We now have two closely related sampling distribution models that we can use when the appropriate assumptions and conditions are met. Which one we use depends on which kind of data we have:

◆ When we have categorical data, we calculate a sample proportion, $\hat{p}$; the sampling distribution of this random variable has a Normal model with a mean at the true proportion ("Greek letter") $p$ and a standard deviation of $SD(\hat{p}) = \sqrt{\frac{pq}{n}} = \frac{\sqrt{pq}}{\sqrt{n}}$.

We'll use this model in Chapters 18 through 21.

◆ When we have quantitative data, we calculate a sample mean, $\bar{y}$; the sampling distribution of this random variable has a Normal model with a mean at the true mean, $\mu$, and a standard deviation of $SD(\bar{y}) = \frac{\sigma}{\sqrt{n}}$. We'll use this model in Chapters 22, 23, and 24.

The means of these models are easy to remember, so all you need to be careful about is the standard deviations. Remember that these are standard deviations of the *statistics* $\hat{p}$ and $\bar{y}$. They both have a square root of $n$ in the denominator. That tells us that the larger the sample, the less either statistic will vary.

> **NOTATION ALERT**
> To avoid confusion over what standard deviations we're talking about, when working with sampling models we always write:
>
> • $SD(\hat{p})$ for categorical data;
> • $SD(\bar{y})$ for quantitative data.

### FOR EXAMPLE
### Using the CLT for Means

**RECAP:** A college physical education department asked a random sample of 200 female students to self-report their heights and weights, but the percentage of students with body mass indexes over 25 seemed suspiciously low. One possible explanation may be that the respondents "shaded" their weights down a bit. The CDC reports that the mean weight of 18-year-old women is 143.74 lb, with a standard deviation of 51.54 lb, but these 200 randomly selected women reported a mean weight of only 140 lb.

**QUESTION:** Based on the Central Limit Theorem and the 68–95–99.7 Rule, does the mean weight in this sample seem exceptionally low, or might this just be random sample-to-sample variation?

**ANSWER:** The conditions check out okay:

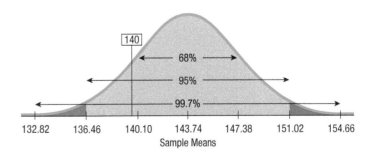

✓ **Randomization Condition:** The women were a random sample and their weights can be assumed to be independent.

✓ **10% Condition:** They sampled fewer than 10% of all women at the college.

✓ **Large Enough Sample Condition:** The distribution of college women's weights is likely to be unimodal and reasonably symmetric, so the CLT would apply even to means of small samples; 200 values is plenty.

The sampling model for sample means is approximately Normal with $E(\bar{y}) = 143.7$ and $SD(\bar{y}) = \dfrac{\sigma}{\sqrt{n}} = \dfrac{51.54}{\sqrt{200}} = 3.64$.

Under this model, the observed sample mean of 140 pounds has a z-score of
$$z = \dfrac{\bar{y} - E(\bar{y})}{SD(\bar{y})} = \dfrac{140 - 143.7}{3.64} = -1.02.$$ The 68–95–99.7 Rule says that approximately 16% of the values in a Normal distribution have z-scores that are at least this far below average.

The 68–95–99.7 Rule suggests that although the reported mean weight of 140 pounds is somewhat lower than expected, it does not appear to be unusual. Such variability is not all that extraordinary for samples of this size.

## STEP-BY-STEP EXAMPLE

### Working with the Sampling Distribution Model for a Mean

The Centers for Disease Control and Prevention reports that the mean weight of adult men in the United States is 190 lb with a standard deviation of 59 lb.[12]

**QUESTION:** An elevator in our building has a weight limit of 10 persons or 2500 lb. What's the probability that if 10 men get on the elevator, they will overload its weight limit?

| | |
|---|---|
| **THINK** **PLAN** State what we want to know. | Asking the probability that the total weight of a sample of 10 men exceeds 2500 pounds is equivalent to asking the probability that their mean weight is greater than 250 pounds. |
| **MODEL** Think about the assumptions and check the conditions. | ✓ **Independence Assumption:** It's reasonable to think that the weights of 10 randomly sampled men will be independent of each other. (But there could be exceptions—for example, if they were all from the same family or if the elevator were in a building with a diet clinic!) |
| | ✓ **Randomization Condition:** I'll assume that the 10 men getting on the elevator are a random sample from the population. |
| | ✓ **10% Condition:** 10 men is surely less than 10% of the population of possible elevator riders. |

---

[12] Cynthia L. Ogden, Cheryl D. Fryar, Margaret D. Carroll, and Katherine M. Flegal, *Mean Body Weight, Height, and Body Mass Index, United States 1960–2002, Advance Data from Vital and Health Statistics Number 347*, October 27, 2004. www.cdc.gov/nchs

Note that if the sample were larger we'd be less concerned about the shape of the distribution of all weights.

State the parameters and the sampling model.

✓ **Large Enough Sample Condition**: I suspect the distribution of weights in the population is roughly unimodal and symmetric, so my sample of 10 men seems large enough.

The mean for all weights is $\mu = 190$ and the standard deviation is $\sigma = 59$ pounds. Because the conditions are satisfied, the CLT says that the sampling distribution of $\bar{y}$ has a Normal model with mean 190 and standard deviation

$$SD(\bar{y}) = \frac{\sigma}{\sqrt{n}} = \frac{59}{\sqrt{10}} \approx 18.66.$$

**SHOW** **PLOT** Make a picture. Sketch the model and shade the area we're interested in. Here the mean weight of 250 pounds appears to be far out on the right tail of the curve.

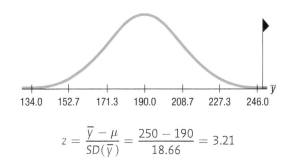

**MECHANICS** Use the standard deviation as a ruler to find the z-score of the cutoff mean weight. We see that an average of 250 pounds is more than 3 standard deviations above the mean.

Find the resulting probability from a table of Normal probabilities such as Table Z, a computer program, or a calculator.

$$z = \frac{\bar{y} - \mu}{SD(\bar{y})} = \frac{250 - 190}{18.66} = 3.21$$

$$P(\bar{y} > 250) = P(z > 3.21) = 0.0007$$

**TELL** **CONCLUSION** Interpret your result in the proper context, being careful to relate it to the original question.

The chance that the weights of a random collection of 10 men will average over 250 pounds is only 0.0007. So, if they are a random sample, it is quite unlikely that 10 people will exceed the total weight limit for the elevator.

# The CLT when the Population Is Very Skewed

The great power of the Central Limit Theorem is that it applies to proportions or means of samples drawn from *any* population. The farther the population distribution is from Normal, the larger the sample we'll need. To see this in action, let's revisit the compensations of the *Fortune* 500 CEOs we saw back in Chapter 4. Here's the distribution; the boxplot below helps illustrate just how highly skewed these data are.

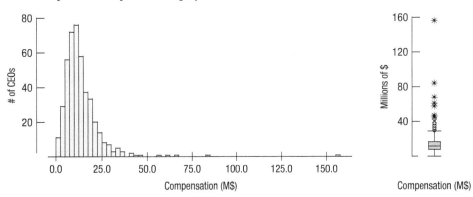

If we sample from this distribution, there's a good chance that we'll get an exceptionally large value. So some samples will have sample means much larger than others. Here is the simulated sampling distribution of the means from 1000 samples of size 10:

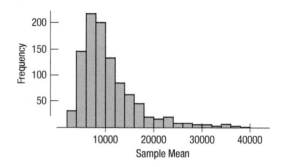

This distribution is not *as* skewed as the population's distribution, but still strongly right skewed.

What happens if we take a larger sample? Here is the simulated sampling distribution of means from samples of size 50:

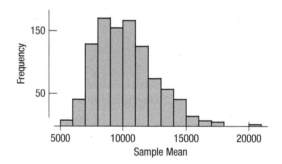

This distribution is less skewed than the corresponding distribution from smaller samples and its mean is again near 10,000.

Will this continue as we increase the sample size? Let's try samples of size 100:

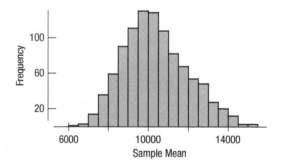

Now the simulated sampling distribution of sample means is even more symmetric, but still skewed enough that we would not want to apply a Normal model.

As we take larger samples, the distribution of means becomes more and more symmetric.

By the time we get to samples of size 200, the distribution is quite symmetrical and, of course, has a mean quite close to 10,000. Now a Normal model will work!

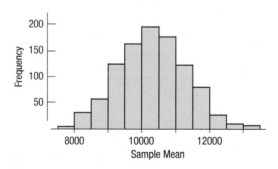

We hope you're amazed. Even though the compensations are extremely skewed, these simulations demonstrate that with large enough samples a Normal model can help us understand the behavior of sample means. The Central Limit Theorem deserves its reputation as one of the most stunning results in all of mathematics.

## About Variation

Sample means vary less than individual data values. That makes sense. If the same test is given to many sections of a large course and the class average is, say, 80%, some students may score 95% because individual scores vary a lot. But we'd be shocked (and pleased!) if the *average* score of the students in any section was 95%. Averages are much less variable. Not only do group averages vary less than individual values, but common sense suggests that averages should be more consistent for larger groups. We've already seen this hunch confirmed in earlier formulas: Both $SD(\hat{p}) = \sqrt{\frac{pq}{n}}$ and $SD(\bar{y}) = \frac{\sigma}{\sqrt{n}}$ have $n$ in the denominator, showing that the variability of sample means decreases as the sample size increases. There's a catch, though. The standard deviation of the sampling distribution declines only with the square root of the sample size and not, for example, with $1/n$.

> The $n$'s justify the means.
> —Apocryphal statistical saying

The mean of a random sample of 4 has half $\left(\frac{1}{\sqrt{4}} = \frac{1}{2}\right)$ the standard deviation of an individual data value. To cut the standard deviation in half again, we'd need a sample of 16, and a sample of 64 to halve it once more.

If only we had a much larger sample, we could get the standard deviation of the sampling distribution *really* under control so that the sample mean could tell us still more about the unknown population mean, but larger samples cost more and take longer to survey. And while we're gathering all those extra data, the population itself may change, or a news story may alter opinions. There are practical limits to most sample sizes. As we shall see, that nasty square root limits how much we can make a sample tell about the population. This is an example of something that's known as the Law of Diminishing Returns.

---

### A BILLION DOLLAR MISUNDERSTANDING?

In the late 1990s the Bill and Melinda Gates Foundation began funding an effort to encourage the breakup of large schools into smaller schools. Why? It had been noticed that smaller schools were more common among the best-performing schools than one would expect. In time, the Annenberg Foundation, the Carnegie Corporation, the Center for Collaborative Education, the Center for School Change, Harvard's Change Leadership Group, the Open Society Institute, Pew Charitable Trusts, and the U.S. Department of Education's Smaller Learning Communities Program all supported the effort. Well over a billion dollars was spent to make schools smaller.

But was it all based on a misunderstanding of sampling distributions? Statisticians Howard Wainer and Harris Zwerling[13] looked at the mean test scores of schools in Pennsylvania. They found that indeed 12% of the top-scoring 50 schools were from the smallest 3% of Pennsylvania schools—substantially more than the 3% we'd naively expect. But then they looked at the *bottom* 50. There they found that 18% were small schools! The explanation? Mean test scores are, well, means. We are looking at a rough real-world simulation in which each school is a trial. Even if all Pennsylvania schools were equally good at educating students, we'd expect their mean scores to vary. How much? The CLT tells us that means of test scores vary according to $\frac{\sigma}{\sqrt{n}}$. Smaller schools have (by definition) smaller $n$'s, so the sampling distributions of their mean scores naturally have larger standard deviations. It's natural, then, that small schools have both higher and lower mean scores.

---

[13]H. Wainer and H. Zwerling, "Legal and Empirical Evidence That Smaller Schools Do Not Improve Student Achievement," *The Phi Delta Kappan* 2006 87:300–303. Discussed in Howard Wainer, "The Most Dangerous Equation," *American Scientist*, May–June 2007, pp. 249–256; also at www.Americanscientist.org

> On October 26, 2005, *The Seattle Times* reported:
>
> *[T]he Gates Foundation announced last week it is moving away from its emphasis on converting large high schools into smaller ones and instead giving grants to specially selected school districts with a track record of academic improvement and effective leadership. Education leaders at the Foundation said they concluded that improving classroom instruction and mobilizing the resources of an entire district were more important first steps to improving high schools than breaking down the size.*

# The Real World vs. the Model World

Be careful. We have been slipping smoothly between the real world, in which we draw random samples of data, and a magical mathematical model world, in which we describe how the sample means and proportions we observe in the real world behave as random variables in all the random samples that we might have drawn. Now we have *two* distributions to deal with. The first is the real-world distribution of the sample, which we might display with a histogram (for quantitative data) or with a bar chart or table (for categorical data). The second is the math world *sampling distribution model* of the statistic, a Normal model based on the Central Limit Theorem. Don't confuse the two.

For example, don't mistakenly think the CLT says that the *data* are Normally distributed as long as the sample is large enough. In fact, as samples get larger, we expect the distribution of the data to look more and more like the population from which they are drawn—skewed, bimodal, whatever—but not necessarily Normal. You can collect a sample of CEO salaries for the next 1000 years,[14] but the histogram will never look Normal. It will be skewed to the right. The Central Limit Theorem doesn't talk about the distribution of the data from the sample. It talks about the sample *means* and sample *proportions* of many different random samples drawn from the same population. Of course, the CLT does require that the sample be big enough when the population shape is not unimodal and symmetric, but the fact that, even then, a Normal model is useful is still a very surprising and powerful result.

## JUST CHECKING

4. Human gestation times have a mean of about 266 days, with a standard deviation of about 10 days. If we record the times of a sample of 100 gestations, do we know that a histogram of the times will be well modeled by a Normal model?

5. Suppose we look at the *average* gestation times for a sample of 100 women. If we imagined all the possible random samples of 100 women we could take and looked at the histogram of all the sample means, what shape would it have?

6. Where would the center of that histogram be?

7. What would be the standard deviation of that histogram?

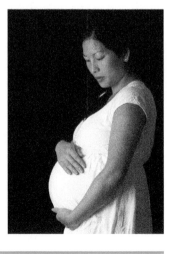

---

[14]Don't forget to adjust for inflation.

# Sampling Distribution Models

Let's summarize what we've learned about sampling distributions. At the heart is the idea that *the statistic itself is a random variable*. We can't know in advance what our statistic will be because it will come from a random sample. It will be something that happens for our particular random sample, but a different random sample would give a different result. This sample-to-sample variability is what generates the sampling distribution of the statistic. The sampling distribution of the statistic shows us the distribution of possible values that the statistic could have had.

We could simulate that distribution by pretending to take lots of samples. Fortunately, for the mean and the proportion, the CLT tells us that under the right conditions we can model their sampling distribution directly with a Normal model.

The two basic truths about sampling distributions are:

1. Sampling distributions arise because samples vary. Each random sample will contain different cases and, so, a different value of the statistic.
2. Although we can always simulate a sampling distribution, the Central Limit Theorem saves us the trouble for means and proportions.

Here's a picture showing the process going into the sampling distribution model:

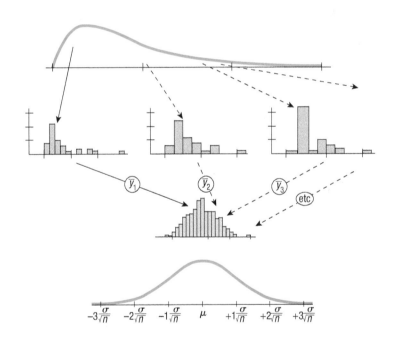

Figure 17.11
We start with a population model, which can have any shape. It can even be bimodal or skewed (as this one is). We label the mean of this model $\mu$ and its standard deviation, $\sigma$.

We draw one real sample (solid line) of size *n* and show its histogram and summary statistics. We imagine (or simulate) drawing many other samples the same size (dotted lines), which have their own histograms and summary statistics.

We (imagine) gathering all the means into a histogram.

The CLT tells us we can model the shape of this histogram with a Normal model. The mean of this Normal is $\mu$, and the standard deviation is $SD(\bar{y}) = \frac{\sigma}{\sqrt{n}}$.

## WHAT CAN GO WRONG?

- **Don't confuse the sampling distribution of a statistic with the distribution of the sample.** It's important to be clear on which of these you're thinking about. The distribution of the data in a sample and the sampling distribution model of a sample statistic are two completely different things. For example, let's have a look using the CEO data once more. On the left at the top of the next page is the distribution of the compensations of a sample of 200 randomly selected CEOs. To its right is the simulated sampling distribution of sample means that we saw earlier in this chapter.

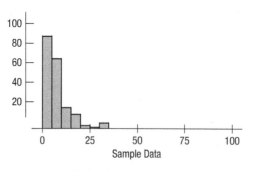
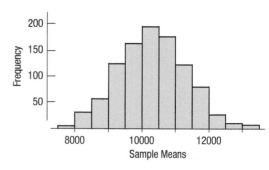

Obviously, they're very different. Be careful not to confuse these two important ideas:

- The larger the sample, the more the sample should look like the population—symmetric, skewed, bimodal, whatever.
- The larger the sample, the more the sampling distribution of sample proportions or means will look like a Normal model.

- **Beware of observations that are not independent.** The CLT depends crucially on the assumption of independence. If our elevator riders are related, are all from the same school (for example, an elementary school), or in some other way aren't a random sample, then the statements we try to make about the mean are going to be wrong. Unfortunately, this isn't something you can check in your data. You have to think about how the data were gathered. Good sampling practice and well-designed randomized experiments ensure independence.

- **Watch out for small samples from skewed populations.** The CLT assures us that the sampling distribution model is Normal if $n$ is large enough. If the population is nearly Normal, even small samples (like our 10 elevator riders) work. If the population is very skewed, then $n$ will have to be large before the Normal model will work well. If we sampled 15 or even 20 CEOs and used $\bar{y}$ to make a statement about the mean of all CEOs' compensation, we'd likely get into trouble because the underlying data distribution is so skewed. Unfortunately, there's no good rule of thumb.[15] It just depends on how skewed the data distribution is. Always plot the data to check.

## WHAT HAVE WE LEARNED?

We've learned to model the variation in statistics from sample to sample with a sampling distribution.
- The Central Limit Theorem tells us that the sampling distributions of both the sample proportion and the sample mean are approximately Normal for large enough samples.

We've learned that, usually, the mean of a sampling distribution is the value of the parameter estimated.
- For the sampling distribution of $\hat{p}$, the mean is $p$.
- For the sampling distribution of $\bar{y}$, the mean is $\mu$.

We've learned about the standard deviation of a sampling distribution.
- The standard deviation of a sampling model is the most important information about it.
- The standard deviation of the sampling distribution of a proportion is $\sqrt{\frac{pq}{n}}$, where $q = 1 - p$.
- The standard deviation of the sampling distribution of a mean is $\frac{\sigma}{\sqrt{n}}$, where $\sigma$ is the population standard deviation.

---

[15] For proportions, of course, there is a rule: the **Success/Failure Condition.** That works for proportions because the standard deviation of a proportion is linked to its mean.

We've learned about the Central Limit Theorem, the most important theorem in Statistics.
- ◆ The sampling distribution of a sample mean tends toward Normal, *no matter what the underlying distribution of the data is*.
- ◆ The CLT says that this happens in the limit, as the sample size grows. The Normal model applies sooner when sampling from a unimodal, symmetric population and more gradually when the population is very non-Normal.

## TERMS

**Sampling variability (sampling error)**  The variability we expect to see from one random sample to another. It is sometimes called sampling error, but sampling variability is the better term. (p. 463)

**Sampling distribution**  Different random samples give different values for a statistic. The distribution of the statistics over all possible samples is called the sampling distribution. The sampling distribution model shows the behavior of the statistic over all the possible samples for the same size *n*. (p. 461)

**Sampling distribution model**  Because we can never see all possible samples, we often use a model as a practical way of describing the theoretical sampling distribution. (p. 461)

**Sampling distribution model for a sample proportion**  If assumptions of independence and random sampling are met, and we expect at least 10 successes and 10 failures, then the sampling distribution of a sample proportion is modeled by a Normal model with a mean equal to the true proportion value, $p$, and a standard deviation equal to $\sqrt{\dfrac{pq}{n}}$. (p. 464)

**Central Limit Theorem**  The Central Limit Theorem (CLT) states that the sampling distribution model of the sample mean (and proportion) from a random sample is approximately Normal for large *n*, *regardless of the distribution of the population, as long as the observations are independent*. (p. 468)

**Sampling distribution model for a sample mean**  If assumptions of independence and random sampling are met, and the sample size is large enough, the sampling distribution of the sample mean is modeled by a Normal model with a mean equal to the population mean, $\mu$, and a standard deviation equal to $\dfrac{\sigma}{\sqrt{n}}$. (p. 470)

## EXERCISES

**1. Send money**  When it sends out its fundraising letter, a philanthropic organization typically gets a return from about 5% of the people on its mailing list. To see what the response rate might be for future appeals, it did a simulation using samples of size 20, 50, 100, and 200. For each sample size, it simulated 1000 mailings with success rate $p = 0.05$ and constructed the histogram of the 1000 sample proportions. Explain how these four histograms demonstrate what the Central Limit Theorem says about the sampling distribution model for sample proportions. Be sure to talk about shape, center, and spread.

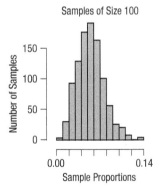

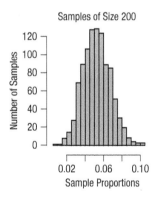

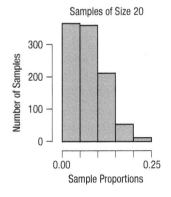

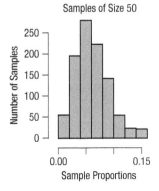

**2. Character recognition**  An automatic character recognition device can successfully read about 85% of handwritten credit card applications. To estimate what might happen when this device reads a stack of applications, the company did a simulation using samples of size 20, 50, 75, and 100. For each sample size, it simulated 1000 samples with success rate $p = 0.85$ and constructed the histogram of the 1000 sample proportions, shown on the next page. Explain how these four histograms demonstrate what the Central Limit Theorem says about the sampling distribution model for sample proportions. Be sure to talk about shape, center, and spread.

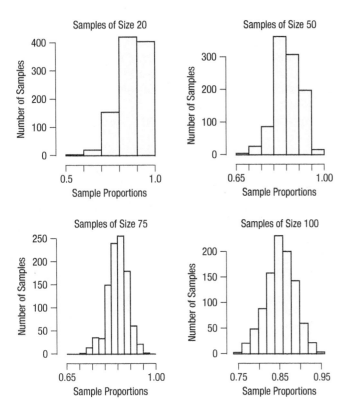

3. **AP exam** A small class of five statistics students received the following scores on their AP exam: 5, 4, 4, 3, 1.

   a) Calculate the mean and standard deviation of these five scores.
   b) List all possible sets of 2 scores that could be chosen from this class. (*Hint*: There are 10 possible sets.)
   c) Calculate the mean of each of these sets of 2 scores and make a dotplot of the sampling distribution of the sample mean.
   d) Calculate the mean and standard deviation of this sampling distribution. How do they compare to those of the individual scores? Is the sample mean an unbiased estimator of the population mean?

4. **AP exam II** For the small class described in Exercise 3 (mean = 3.4, SD = 1.517), you will be selecting samples of size 3 (sampling without replacement).

   a) List all possible samples of size 3 that could be chosen from this class.
   b) Construct the sampling distribution of the sample mean for samples of size 3.
   c) How do the mean and standard deviation of this sampling distribution compare to the mean and standard deviation of the population? To the mean and standard deviation of the sampling distribution of the sample mean from Exercise 3, part d?

5. **Marriage** According to a Pew Research survey, about 27% of American adults are pessimistic about the future of marriage and the family. This is based on a random sample, so we'll assume that this percentage is correct for all American adults.

   a) Using a Binomial model, what is the probability that, in a sample of 20 American adults, 25% or fewer of the people in the sample are pessimistic about the future of marriage and family?
   b) Now use a Normal model to compute that probability. How does this compare to your answer from part a?
   c) Using a Binomial model, what is the probability that, in a sample of 700 American adults, 25% or fewer of the people of the people in the sample are pessimistic about the future of marriage and family?
   d) Now use a Normal model to compute that probability. How does this compare to your answer from part c?
   e) What do these answers tell you about the importance of check that $np$ and $nq$ are both at least 10?

6. **Wow. Just wow** According to a 2013 poll from Public Policy Polling, 4% of American voters believe that shape-shifting reptilian people control our world by taking on human form and gaining power. Yes, you read that correctly! (This was a poll about conspiracy theories.) Assume that's the actual proportion of Americans who hold that belief.

   a) Use a Binomial model to calculate the probability that, in a random sample of 100 people, at least 6% of those in the sample believe the thing about reptilian people controlling our world.
   b) Use a Normal model to calculate the same probability. How does this compare with the answer in part a?
   c) That same poll found that 51% of American voters believe there was a larger conspiracy responsible for the assassination of President Kennedy. Use a Binomial model to calculate the probability that, in a random sample of 100 people, at least 57% of those in the sample believe in the JFK conspiracy theory.
   d) Use a Normal model to calculate the same probability. How does this compare with the answer in part c?
   e) What do these answers tell you about the importance of checking that $np$ and $nq$ are both at least 10?

7. **Send money, again** The philanthropic organization in Exercise 1 expects about a 5% success rate when it sends fundraising letters to the people on its mailing list. In Exercise 1 you looked at the histograms showing distributions of sample proportions from 1000 simulated mailings for samples of size 20, 50, 100, and 200. The sample statistics from each simulation were as follows:

   | n | Mean | SD |
   | --- | --- | --- |
   | 20 | 0.0497 | 0.0479 |
   | 50 | 0.0516 | 0.0309 |
   | 100 | 0.0497 | 0.0215 |
   | 200 | 0.0501 | 0.0152 |

   a) According to the Central Limit Theorem, what should the theoretical mean and standard deviations be for these sample sizes?
   b) How close are those theoretical values to what was observed in these simulations?
   c) Looking at the histograms in Exercise 1, at what sample size would you be comfortable using the Normal model as an approximation for the sampling distribution?
   d) What does the Success/Failure Condition say about the choice you made in part c?

8. **Character recognition, again** The automatic character recognition device discussed in Exercise 2 successfully reads about 85% of handwritten credit card applications. In Exercise 2 you looked at the histograms showing distributions of sample proportions

from 1000 simulated samples of size 20, 50, 75, and 100. The sample statistics from each simulation were as follows:

| n | Mean | SD |
|---|------|-----|
| 20 | 0.8481 | 0.0803 |
| 50 | 0.8507 | 0.0509 |
| 75 | 0.8481 | 0.0406 |
| 100 | 0.8488 | 0.0354 |

a) According to the Central Limit Theorem, what should the theoretical mean and standard deviations be for these sample sizes?
b) How close are those theoretical values to what was observed in these simulations?
c) Looking at the histograms in Exercise 2, at what sample size would you be comfortable using the Normal model as an approximation for the sampling distribution?
d) What does the Success/Failure Condition say about the choice you made in part c?

9. Sample maximum The distribution of scores on a Statistics test for a particular class is skewed to the left. The professor wants to predict the maximum score and so wants to understand the distribution of the sample maximum. They simulate the distribution of the maximum of the test for 30 different tests (with $n = 5$). The histogram below shows a simulated sampling distribution of the sample maximum from these tests.

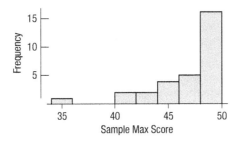

a) Would a Normal model be a useful model for this sampling distribution? Explain.
b) The mean of this distribution is 46.3 and the SD is 3.5. Would you expect about 95% of the samples to have their maximums within 7 of 46.3? Why or why not?

10. Soup A machine is supposed to fill cans with 16 oz of soup. Of course, there will be some variation in the amount actually dispensed, and measurement errors are often approximately normally distributed. The manager would like to understand the variability of the variances (*Reminder*: Variance is the square of the standard deviation.) of the samples, so they collect information from the last 250 randomly selected batches of size 10 and plot a histogram of the variances:

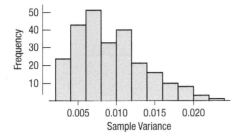

a) Would a Normal model be a useful model for this sampling distribution? Explain.
b) The mean of this distribution is 0.009 and the SD is 0.004. Would you expect about 95% of the samples to have their variances within 0.008 of 0.009? Why or why not?

11. Coin tosses In a large class of Introductory Statistics students, the professor has each person toss a coin 16 times and calculate the proportion of their tosses that were heads. The students then report their results, and the professor plots a histogram of these several proportions.

a) What shape would you expect this histogram to be? Why?
b) Where do you expect the histogram to be centered?
c) How much variability would you expect among these proportions?
d) Explain why a Normal model should not be used here.

12. M&M's The candy company claims that 16% of the Milk Chocolate M&M's it produces are green. Suppose that the candies are thoroughly mixed and then packaged in small bags containing about 50 M&M's. A class of elementary school students learning about percents opens several bags, counts the various colors of the candies, and calculates the proportion that are green.

a) If we plot a histogram showing the proportions of green candies in the various bags, what shape would you expect it to have?
b) Can that histogram be approximated by a Normal model? Explain.
c) Where should the center of the histogram be?
d) What should the standard deviation of the sampling distribution be?

13. More coins Suppose the class in Exercise 11 repeats the coin-tossing experiment.

a) The students toss the coins 25 times each. Use the 68–95–99.7 Rule to describe the sampling distribution model.
b) Confirm that you can use a Normal model here.
c) They increase the number of tosses to 64 each. Draw and label the appropriate sampling distribution model. Check the appropriate conditions to justify your model.
d) Explain how the sampling distribution model changes as the number of tosses increases. (Note: A complete response will discuss center, shape, and spread.)

14. Bigger bag Suppose the class in Exercise 12 buys bigger bags of candy, with 200 M&M's each. Again the students calculate the proportion of green candies they find.

a) Explain why it's appropriate to use a Normal model to describe the distribution of the proportion of green M&M's they might expect.
b) Use the 68–95–99.7 Rule to describe how this proportion might vary from bag to bag.
c) How would this model change if the bags contained even more candies? Discuss center, shape, and spread in your answer.

15. Just (un)lucky? One of the students in the Introductory Statistics class in Exercise 13 claims to have tossed her coin 200 times and found only 42% heads. What do you think of this claim? Explain.

**16. Too many green ones?** In a really large bag of M&M's, the students in Exercise 14 found 500 candies, and 18% of them were green. Is this an unusually large proportion of green M&M's? Explain.

**17. Speeding** State police believe that 70% of the drivers traveling on a major interstate highway exceed the speed limit. They plan to set up a radar trap and check the speeds of 80 cars.
 a) Using the 68–95–99.7 Rule, draw and label the distribution of the proportion of these cars the police will observe speeding.
 b) Do you think the appropriate conditions necessary for your analysis are met? Explain.

**18. Smoking** Public health statistics for 2015 indicate that 15.1% of American adults smoke cigarettes. Using the 68–95–99.7 Rule, describe the sampling distribution model for the proportion of smokers among a randomly selected group of 120 adults. Be sure to discuss your assumptions and conditions.

**19. Vision** It is generally believed that nearsightedness affects about 12% of all children. A school district has registered 170 incoming kindergarten children.
 a) Can you apply the Central Limit Theorem to describe the sampling distribution model for the sample proportion of children who are nearsighted? Check the conditions and discuss any assumptions you need to make.
 b) Sketch and clearly label the sampling model, based on the 68–95–99.7 Rule.
 c) How many of the incoming students might the school expect to be nearsighted? Explain.

**20. Student debt** In August 2017, Student Loan Hero reported that former college students were defaulting in record numbers; 11.1% of student loans were delinquent (more than 90 days late or in default). Suppose a journalist randomly selects a sample of 2000 student loans.
 a) Can you apply the Central Limit Theorem to describe the sampling distribution model for the sample proportion of delinquencies? Check the conditions and discuss any assumptions you need to make.
 b) Sketch and clearly label the sampling model, based on the 68–95–99.7 Rule.
 c) How many of these former students might the auditor expect will default on their loans? Explain.

**21. Loans** Based on past experience, a bank believes that 7% of the people who receive loans will not make payments on time. A bank auditor randomly selects 200 loans.
 a) What are the mean and standard deviation of the proportion of clients in this group who may not make timely payments?
 b) What assumptions underlie your model? Are the conditions met? Explain.
 c) What's the probability that over 10% of these clients will not make timely payments?

**22. Teens with phones** Pew Research reported that, in 2018, 84% of all teens used their cell phone to connect with other people. Assume this estimate is correct. (Note: Some teens reported they use their cell phones to *avoid* other people!)
 a) We randomly pick 100 teens. Let $\hat{p}$ represent the proportion of teens in this sample who own a smartphone. What's the appropriate model for the distribution of $\hat{p}$? Specify the name of the distribution, the mean, and the standard deviation. Be sure to verify that the conditions are met.
 b) What's the approximate probability that more than 90% of this sample use their cell phone to connect with other people?

**23. Back to school?** Best known for its testing program, ACT, Inc., also compiles data on a variety of issues in education. The company reported that the national college fist-year-to-second-year retention rate held steady at 74% over the previous four years. Consider random samples of 400 first-year students who took the ACT. Use the 68–95–99.7 Rule to describe the sampling distribution model for the percentage of those students we expect to return to that school for their second year. Do you think the appropriate conditions are met?

**24. Binge drinking** A national study found that 44% of college students engage in binge drinking (5 drinks at a sitting for men, 4 for women). Use the 68–95–99.7 Rule to describe the sampling distribution model for the proportion of students in a randomly selected group of 200 college students who engage in binge drinking. Do you think the appropriate conditions are met?

**25. Back to school, again** Based on the 74% national retention rate described in Exercise 23, does a college where 522 of the 603 first-year students returned the next year as second-year students have a right to brag that it has an unusually high retention rate? Explain.

**26. Binge sample** After hearing of Exercise 24's national result that 44% of students engage in binge drinking (5 drinks at a sitting for men, 4 for women), a professor surveyed a random sample of 244 students and found that 96 of them admitted to binge drinking in the past week. Should he be surprised at this result? Explain.

**27. Polling** Just before a referendum on a school budget, a local newspaper polls 400 randomly selected voters in an attempt to predict whether the budget will pass. Suppose that the budget actually has the support of 52% of the voters. What's the probability the newspaper's sample will lead them to predict defeat? Be sure to verify that the assumptions and conditions necessary for your analysis are met.

**28. Seeds** Information on a packet of seeds claims that the germination rate is 92%.
 a) What's the probability that more than 95% of the 160 seeds in the packet will germinate? Be sure to discuss your assumptions and check the conditions that support your model.
 b) Given that all the seeds came from the same packet, how confident are you about your answer?

**29. Biological instinct** A study that showed that heterosexual women, during ovulation, were significantly better at correctly identifying the sexual orientation of a man from a photograph of his face than women who were not ovulating. Near ovulation, on average women correctly identified the orientations of about 65% of the 100 men shown to them. If this is the probability of correctly identifying the orientation of a man in any given photograph, what is the probability a woman would correctly classify 80 or more of the men (as two women in the study did)?

**30. Genetic defect** It's believed that 4% of children have a gene that may be linked to juvenile diabetes. Researchers hoping to track 20 of these children for several years test 732 newborns for the presence of this gene. What's the probability that they find enough subjects for their study?

**31. "No Children" section** Some restaurant owners, at the request of some of their less tolerant customers, have stopped allowing children into their restaurant. This, naturally, outrages other customers. One restaurateur hopes to please both sets of customers by having a "no children" section. She estimates that in her 120-seat restaurant, about 30% of her seats, on average, are taken by families with children. How many seats should be in the "children allowed" area in order to be very sure of having enough seating there? Comment on the assumptions and conditions that support your model, and explain what "very sure" means to you.

**32. Meals** A restaurateur anticipates serving about 180 people on a Friday evening and believes that about 20% of the patrons will order the chef's steak special. How many of those meals should he plan on serving in order to be pretty sure of having enough steaks on hand to meet customer demand? Justify your answer, including an explanation of what "pretty sure" means to you.

**33. Sampling** A sample is chosen randomly from a population that can be described by a Normal model.

a) What's the sampling distribution model for the sample mean? Describe shape, center, and spread.
b) If we choose a larger sample, what's the effect on this sampling distribution model?

**34. Sampling, part II** A sample is chosen randomly from a population that was strongly skewed to the left.

a) Describe the sampling distribution model for the sample mean if the sample size is small.
b) If we make the sample larger, what happens to the sampling distribution model's shape, center, and spread?
c) As we make the sample larger, what happens to the expected distribution of the data in the sample?

**35. Waist size** A study measured the waist size of 250 men, finding a mean of 36.33 inches and a standard deviation of 4.02 inches. Here is a histogram of these measurements

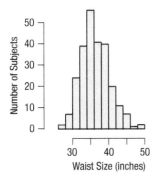

a) Describe the histogram of *Waist Size*.
b) To explore how the mean might vary from sample to sample, they simulated by drawing many samples of size 2, 5, 10, and 20, with replacement, from the 250 measurements. Look at the histograms of the sample means for each simulation. Explain how these histograms demonstrate what the Central Limit Theorem says about the sampling distribution model for sample means.

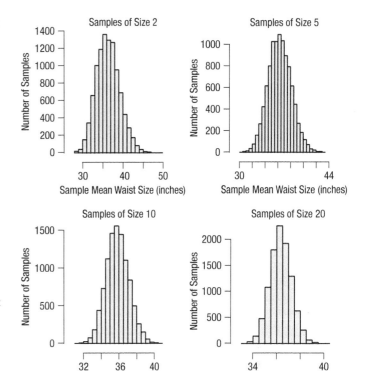

**36. CEO compensation** In Chapter 5 we saw the distribution of the total compensation of the chief executive officers (CEOs) of the top 480 companies. The average compensation (in millions of dollars) is $13.941 and the standard deviation is $9.825. Here is a histogram of their annual compensations:

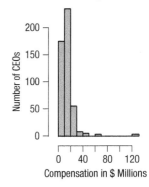

a) Describe the histogram of *Total Compensation*.
A research organization simulated sample means by drawing samples of 30, 50, 100, and 200, with replacement, from the 480 CEOs. The histograms show the distributions of means for many samples of each size.

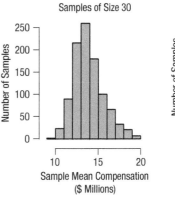
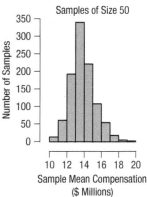

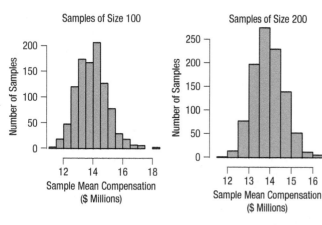

b) Explain how these histograms demonstrate what the Central Limit Theorem says about the sampling distribution model for sample means. Be sure to talk about shape, center, and spread.

c) Comment on an oft-cited "rule of thumb" that "With a sample size of at least 30, the sampling distribution of the mean is approximately Normal."

**37. Waist size revisited** Researchers measured the waist sizes of 250 men in a study on body fat. The true mean and standard deviation of the waist sizes for the 250 men are 36.33 in and 4.019 inches, respectively. In Exercise 35 you looked at the histograms of simulations that drew samples of sizes 2, 5, 10, and 20 (with replacement). The summary statistics for these simulations were as follows:

| n | Mean | SD |
|---|------|-----|
| 2 | 36.314 | 2.855 |
| 5 | 36.314 | 1.805 |
| 10 | 36.341 | 1.276 |
| 20 | 36.339 | 0.895 |

a) According to the Central Limit Theorem, what should the theoretical mean and standard deviation be for each of these sample sizes?
b) How close are the theoretical values to what was observed in the simulation?
c) Looking at the histograms in Exercise 35, at what sample size would you be comfortable using the Normal model as an approximation for the sampling distribution?
d) What about the shape of the distribution of *Waist Size* explains your choice of sample size in part c?

**38. CEOs revisited** In Exercise 36 you looked at the annual compensation for 480 CEOs, for which the true mean and standard deviation were (in millions of dollars) $13.941 and $9.825, respectively. A simulation drew samples of sizes 30, 50, 100, and 200 (with replacement) from the total annual compensations of these 480 CEOs. The summary statistics for these simulations were as follows:

| n | Mean | SD |
|---|------|-----|
| 30 | 13.902 | 1.772 |
| 50 | 13.856 | 1.351 |
| 100 | 13.926 | 1.028 |
| 200 | 13.923 | 0.701 |

a) According to the Central Limit Theorem, what should the theoretical mean and standard deviation be for each of these sample sizes?
b) How close are the theoretical values to what was observed from the simulation?
c) Looking at the histograms in Exercise 36, at what sample size would you be comfortable using the Normal model as an approximation for the sampling distribution?
d) What about the shape of the distribution of *Total Compensation* explains your answer in part c?

**39. GPAs** A college's data about the incoming first-year students indicates that the mean of their high school GPAs was 3.4, with a standard deviation of 0.35; the distribution was roughly mound-shaped and only slightly skewed. The students are randomly assigned to first-year student writing seminars in groups of 40. What might the mean GPA of one of these seminar groups be? Describe the appropriate sampling distribution model—shape, center, and spread—with attention to assumptions and conditions. Make a sketch using the 68–95–99.7 Rule.

**40. Home values** Assessment records indicate that the value of homes in a small city is skewed right, with a mean of $140,000 and standard deviation of $60,000. To check the accuracy of the assessment data, officials plan to conduct a detailed appraisal of 100 homes selected at random. Using the 68–95–99.7 Rule, draw and label an appropriate sampling model for the mean value of the homes selected.

**T 41. Lucky spot?** A reporter working on a story about the New York lottery contacted one of the authors of this book, wanting help analyzing data to see if some ticket sales outlets were more likely to produce winners. His data for each of the 966 New York lottery outlets are graphed below; the scatterplot shows the ratio *Total Paid/Total Sales* vs. *Total Sales* for the state's "instant winner" games for all of 2007.

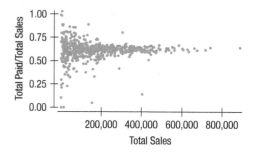

The reporter thinks that by identifying the outlets with the highest fraction of bets paid out, players might be able to increase their chances of winning. (Typically—but not always—instant winners are paid immediately [instantly] at the store at which they are purchased. However, the fact that tickets may be scratched off and then cashed in at any outlet may account for some outlets paying out more than they take in. The few with very low payouts may be on interstate highways where players may purchase cards but then leave.)

a) Explain why the plot has this funnel shape.
b) Explain why the reporter's idea wouldn't have worked anyway.

**42. Safe cities** Allstate Insurance Company identified the 10 safest and 10 least-safe U.S. cities from among the 200 largest cities in the United States, based on the mean number of years drivers went between automobile accidents. The cities on both lists were all smaller than the 10 largest cities. Using facts about the sampling distribution model of the mean, explain why this is not surprising.

**43. Pregnancy** Assume that the duration of human pregnancies can be described by a Normal model with mean 266 days and standard deviation 16 days.

  a) What percentage of pregnancies should last between 270 and 280 days?
  b) At least how many days should the longest 25% of all pregnancies last?
  c) Suppose a certain obstetrician is currently providing prenatal care to 60 pregnant women. Let $\bar{y}$ represent the mean length of their pregnancies. According to the Central Limit Theorem, what's the distribution of this sample mean, $\bar{y}$? Specify the model, mean, and standard deviation.
  d) What's the probability that the mean duration of these patients' pregnancies will be less than 260 days?

**44. Rainfall** Statistics from Cornell's Northeast Regional Climate Center indicate that Ithaca, New York, gets an average of 35.4″ of rain each year, with a standard deviation of 4.2″. Assume that a Normal model applies.

  a) During what percentage of years does Ithaca get more than 40″ of rain?
  b) Less than how much rain falls in the driest 20% of all years?
  c) A Cornell University student is in Ithaca for 4 years. Let $\bar{y}$ represent the mean amount of rain for those 4 years. Describe the sampling distribution model of this sample mean, $\bar{y}$.
  d) What's the probability that those 4 years average less than 30″ of rain?

**45. Pregnant again** The duration of human pregnancies may not actually follow the Normal model described in Exercise 43.

  a) Explain why it may be somewhat skewed to the left.
  b) If the correct model is in fact skewed, does that change your answers to parts a, b, and c of Exercise 43? Explain why or why not for each.

**46. At work** Some business analysts estimate that the length of time people work at a job has a mean of 6.2 years and a standard deviation of 4.5 years.

  a) Explain why you suspect this distribution may be skewed to the right.
  b) Explain why you could estimate the probability that 100 people selected at random had worked for their employers an average of 10 years or more, but you could not estimate the probability that an individual had done so.

**47. Dice and dollars** You roll a die, winning nothing if the number of spots is odd, $1 for a 2 or a 4, and $10 for a 6.

  a) Find the expected value and standard deviation of your prospective winnings.
  b) You play twice. Find the mean and standard deviation of your total winnings.
  c) You play 40 times. What's the probability that you win at least $100?

**48. New game** You pay $10 and roll a die. If you get a 6, you win $50. If not, you get to roll again. If you get a 6 this time, you get your $10 back.

  a) Create a probability model for this game.
  b) Find the expected value and standard deviation of your prospective winnings.
  c) You play this game five times. Find the expected value and standard deviation of your average winnings.
  d) 100 people play this game. What's the probability the person running the game makes a profit?

**49. AP Stats 2019** The College Board reported the score distribution shown in the table for all students who took the 2019 AP Statistics exam.

| Score | Percent of Students |
|---|---|
| 5 | 14.7 |
| 4 | 18.4 |
| 3 | 26.6 |
| 2 | 19.3 |
| 1 | 21.0 |

  a) Find the mean and standard deviation of the scores.
  b) If we select a random sample of 40 AP Statistics students, would you expect their scores to follow a Normal model? Explain.
  c) Consider the mean scores of random samples of 40 AP Statistics students. Describe the sampling model for these means (shape, center, and spread).

**50. Museum membership** A museum offers several levels of membership, as shown in the table.

| Member Category | Amount of Donation ($) | Percent of Members |
|---|---|---|
| Individual | 50 | 41 |
| Family | 100 | 37 |
| Sponsor | 250 | 14 |
| Patron | 500 | 7 |
| Benefactor | 1000 | 1 |

  a) Find the mean and standard deviation of the donations.
  b) During its annual membership drive, the museum hopes to sign up 50 new members each day. Would you expect the distribution of the donations for a day to follow a Normal model? Explain.
  c) Consider the mean donation of the 50 new members each day. Describe the sampling model for these means (shape, center, and spread).

**51. AP Stats 2019, again** An AP Statistics teacher had 63 students preparing to take the AP exam discussed in Exercise 49. Though they were obviously not a random sample, he considered his students to be "typical" of all the national students. What's the probability that his students will achieve an average score of at least 3?

**52. Joining the museum** One of the museum's phone volunteers sets a personal goal of getting an average donation of at least $100 from the new members she enrolls during the membership drive described in Exercise 50. If she gets 80 new members and they can be considered a random sample of all the museum's members, what is the probability that she can achieve her goal?

**53. Pollution** Carbon monoxide (CO) emissions for a certain kind of car vary with mean 2.9 g/mi and standard deviation 0.4 g/mi. A company has 80 of these cars in its fleet. Let $\bar{y}$ represent the mean CO level for the company's fleet.
   a) What's the approximate model for the distribution of $\bar{y}$? Explain.
   b) Estimate the probability that $\bar{y}$ is between 3.0 and 3.1 g/mi.
   c) There is only a 5% chance that the fleet's mean CO level is greater than what value?

**54. Potato chips** The weight of potato chips in a medium-size bag is stated to be 10 ounces. The amount that the packaging machine puts in these bags is believed to have a Normal model with mean 10.2 ounces and standard deviation 0.12 ounces.
   a) What fraction of all bags sold are underweight?
   b) Some of the chips are sold in "bargain packs" of 3 bags. What's the probability that none of the 3 is underweight?
   c) What's the probability that the mean weight of the 3 bags is below the stated amount?
   d) What's the probability that the mean weight of a 24-bag case of potato chips is below 10 ounces?

**55. Tips** A server believes the distribution of their tips has a model that is slightly skewed to the right, with a mean of $9.60 and a standard deviation of $5.40.
   a) Explain why you cannot determine the probability that a given party will tip them at least $20.
   b) Can you estimate the probability that the next 4 parties will tip an average of at least $15? Explain.
   c) Is it likely that his 10 parties today will tip an average of at least $15? Explain.

**56. Groceries** A grocery store's receipts show that Sunday customer purchases have a skewed distribution with a mean of $32 and a standard deviation of $20.
   a) Explain why you cannot determine the probability that the next Sunday customer will spend at least $40.
   b) Can you estimate the probability that the next 10 Sunday customers will spend an average of at least $40? Explain.
   c) Is it likely that the next 50 Sunday customers will spend an average of at least $40? Explain.

**57. More tips** The server in Exercise 55 usually waits on about 40 parties over a weekend of work.
   a) Estimate the probability that they will earn at least $500 in tips.
   b) How much do they earn on the best 10% of such weekends?

**58. More groceries** Suppose the store in Exercise 56 had 312 customers this Sunday.
   a) Estimate the probability that the store's revenues were at least $10,000.
   b) If, on a typical Sunday, the store serves 312 customers, how much does the store take in on the worst 10% of such days?

**59. IQs** Suppose that IQs of East State University's students can be described by a Normal model with mean 130 and standard deviation 8 points. Also suppose that IQs of students from West State University can be described by a Normal model with mean 120 and standard deviation 10.
   a) We select a student at random from East State. Find the probability that this student's IQ is at least 125 points.
   b) We select a student at random from each school. Find the probability that the East State student's IQ is at least 5 points higher than the West State student's IQ.
   c) We select 3 West State students at random. Find the probability that this group's average IQ is at least 125 points.
   d) We also select 3 East State students at random. What's the probability that their average IQ is at least 5 points higher than the average for the 3 West Staters?

**60. Milk** Although most of us buy milk by the quart or gallon, farmers measure daily production in pounds. Ayrshire cows average 47 pounds of milk a day, with a standard deviation of 6 pounds. For Jersey cows, the mean daily production is 43 pounds, with a standard deviation of 5 pounds. Assume that Normal models describe milk production for these breeds.
   a) We select an Ayrshire at random. What's the probability that she averages more than 50 pounds of milk a day?
   b) What's the probability that a randomly selected Ayrshire gives more milk than a randomly selected Jersey?
   c) A farmer has 20 Jerseys. What's the probability that the average production for this small herd exceeds 45 pounds of milk a day?
   d) A neighboring farmer has 10 Ayrshires. What's the probability that his herd average is at least 5 pounds higher than the average for part c's Jersey herd?

**61. Boba Tea** A retail manager tracks the sales at their Boba Tea House (Boba tea, or Bubble Tea, is a tea-based drink originating in Taiwan that usually has chewy tapioca balls in the bottom). Over the course of time, the manager has estimated that 58% of customers order the large size, while the rest of the customers order the small size.
   a) If 25 customers make an order, what are the chances that exactly 14 of them order a large size Boba?
   b) Suppose the manager collects more of these data, collecting 25 customers each day for many months. Each day they calculate the percentage that order a large size Boba. Assuming that 58% of customers continue to order the large size, describe the sampling distribution of the sample proportion.

c) Given your answer to part (b), how high or low would the percentage of large size orders need to be to convince the manager that their customer's habits are changing?

**62.** More Boba Tea  Our manager is, of course, also interested in daily sales revenue. Over time they estimate that daily sales have a mean of $2300, a standard deviation of $589, and are approximately normally distributed over the course of a year.
   a) If the manager randomly chooses one day of the year, what are the chances that this day will have over $3000 in sales?
   b) The manager collects daily sales in 90-day increments (quarters) and finds the quarterly average. Describe the sampling distribution of the sample mean for the quarterly average.
   c) If the manager wanted to see an indication of the quarterly growth, what average would they want to see that would indicate a significant increase in average daily sales?

## JUST CHECKING

Answers
1. A Normal model (approximately).
2. At the actual proportion of all students who are in favor.
3. $SD(\hat{p}) = \sqrt{\dfrac{(0.5)(0.5)}{100}} = 0.05$.
4. No, this is a histogram of individuals. It may or may not be approximately Normal, but we can't tell from the information provided.
5. A Normal model (approximately).
6. 266 days.
7. $\dfrac{10}{\sqrt{100}} = 1.0$ day.

# 18

## Confidence Intervals for Proportions

| WHO | Sea fans |
|---|---|
| WHAT | Percent infected |
| WHEN | June 2000 |
| WHERE | Las Redes Reef, Akumal, Mexico, 40 feet deep |
| WHY | Research |

Coral reef communities are home to one quarter of all marine plants and animals worldwide. These reefs support large fisheries by providing breeding grounds and safe havens for young fish of many species. Coral reefs are seawalls that protect shorelines against tides, storm surges, and hurricanes, and are sand "factories" that produce the limestone and sand of which beaches are made. Beyond the beach, these reefs are major tourist attractions for snorkelers and divers, driving a tourist industry worth tens of billions of dollars.

But marine scientists say that as much as 27% of the world's reef systems have been destroyed in recent times. At current rates of loss, 70% of the reefs could be gone in 40 years. Pollution, global warming, outright destruction of reefs, and increasing acidification of the oceans are all likely factors in this loss.

Dr. Drew Harvell's lab studies corals and the diseases that affect them. They sampled sea fans[1] at 19 randomly selected reefs along the Yucatan peninsula and diagnosed whether the animals were affected by the disease *aspergillosis*.[2] In specimens collected at a depth of 40 feet at the Las Redes Reef in Akumal, Mexico, these scientists found that 54 of 104 sea fans sampled were infected with that disease.

Of course, we care about much more than these particular 104 sea fans. We care about the health of coral reef communities throughout the Caribbean. What can this study tell us about the prevalence of the disease among sea fans?

We have a sample proportion, which we write as $\hat{p}$, of 54/104, or 51.9%. Our first guess might be that this observed proportion is close to the population proportion, $p$. But

---

[1] That's a sea fan in the picture. Although they look like trees, they are actually colonies of genetically identical animals.

[2] K. M. Mullen, C. D. Harvell, A. P. Alker, D. Dube, E. Jordán-Dahlgren, J. R. Ward, and L. E. Petes, "Host Range and Resistance to Aspergillosis in Three Sea Fan Species from the Yucatan," *Marine Biology* (2006), Springer-Verlag.

we also know that because of natural sampling variability, if the researchers had drawn a second sample of 104 sea fans at roughly the same time, the proportion infected from that sample probably wouldn't have been exactly 51.9%.

What *can* we say about the population proportion, $p$? To start to answer this question, think about how different the sample proportion might have been if we'd taken another random sample from the same population. But wait. Remember—we aren't actually going to take more samples. We just want to *imagine* how the sample proportions might vary from sample to sample. In other words, we want to know about the *sampling distribution* of the sample proportion of infected sea fans.

# A Confidence Interval

Let's look at our model for the sampling distribution. What do we know about it? We know it's approximately Normal (under certain assumptions, which we must be careful to check for reasonableness) and that its mean is the proportion of all infected sea fans on the Las Redes Reef. Is the infected proportion of *all* sea fans 51.9%? No, that's just $\hat{p}$, our estimate. We don't know the proportion, $p$, of all the infected sea fans; that's what we're trying to find out. We do know, though, that the sampling distribution model of $\hat{p}$ is centered at $p$, and we know that the standard deviation of the sampling distribution is $\sqrt{\frac{pq}{n}}$.

Now we have a problem: Because we don't know $p$, we can't find the true standard deviation of the sampling distribution model. We do know the observed proportion, $\hat{p}$, so, of course we just use what we know, and we estimate. That may not seem like a big deal, but it gets a special name. Whenever we estimate the standard deviation of a sampling distribution, we call it a **standard error**.[3] For a sample proportion, $\hat{p}$, the standard error is

$$SE(\hat{p}) = \sqrt{\frac{\hat{p}\hat{q}}{n}}.$$

For the sea fans, then

$$SE(\hat{p}) = \sqrt{\frac{\hat{p}\hat{q}}{n}} = \sqrt{\frac{(0.519)(0.481)}{104}} = 0.049 = 4.9\%.$$

Now, so long as we can verify that the Normal model is reasonable (which we'll do later), we know that the sampling model for $\hat{p}$ should look like this:

> **NOTATION ALERT**
> Remember that $\hat{p}$ is our sample-based estimate of the true proportion $p$. Recall also that $q$ is just shorthand for $1 - p$, and $\hat{q} = 1 - \hat{p}$.
> When we use $\hat{p}$ to estimate the standard deviation of the sampling distribution model, we call that the standard error and write $SE(\hat{p}) = \sqrt{\frac{\hat{p}\hat{q}}{n}}$.

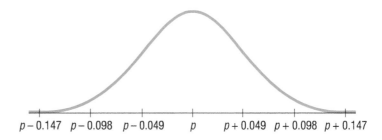

Figure 18.1
The sampling distribution model for $\hat{p}$ is Normal with a mean of $p$ and a standard deviation we estimate to be 0.049.

Great. What does that tell us? Well, because it's Normal, it says that about 68% of all samples of 104 sea fans will have $\hat{p}$'s within 1 $SE$, 0.049, of $p$. And about 95% of all these samples will be within $p \pm 2$ $SE$s. But where is *our* sample proportion in this picture? And what value does $p$ have? We still don't know!

---

[3]This isn't such a great name because it isn't standard and nobody made an error. But it's much shorter and more convenient than saying, "the estimated standard deviation of the sampling distribution of the sample statistic."

We do know that for 95% of random samples, $\hat{p}$ will be no more than 2 SEs away from $p$.[4] So let's look at this from $\hat{p}$'s point of view. If I'm $\hat{p}$, there's a 95% chance that $p$ is no more than 2 SEs away from me. If I reach out 2 SEs, or $2 \times 0.049$, away from me on both sides, I'm 95% sure that $p$ will be within my grasp. Now I've got it! Probably. Of course, even if my interval does catch $p$, I still don't know its true value. The best I can do is to produce an interval, and even then I can't be positive it contains $p$.

**Figure 18.2**
Reaching out 2 SEs on either side of $\hat{p}$ makes us 95% confident that we'll trap the true proportion, $p$.

So what can we really say about $p$? Here's a list of things we'd *like* to be able to say and the reasons we *can't* say most of them:

1. **"51.9% of *all* sea fans on the Las Redes Reef are infected."** It would be nice to be able to make absolute statements about population values with certainty, but we just don't have enough information to do that. There's no way to be sure that the population proportion is the same as the sample proportion; in fact, it almost certainly isn't. Observations vary. Another sample would almost certainly yield a different sample proportion.

2. **"It is *probably* true that 51.9% of all sea fans on the Las Redes Reef are infected."** No. In fact, we can be pretty sure that whatever the true proportion is, it's not exactly 51.900000 . . . %. So the statement is not true.

3. **"We don't know exactly what proportion of sea fans on the Las Redes Reef is infected, but we *know* that it's within the interval $51.9\% \pm 2 \times 4.9\%$. That is, it's between 42.1% and 61.7%."** This is getting closer, but we still can't be certain. We can't know *for sure* that the true proportion is in this interval—or in any particular interval.

4. **"There's a 95% chance that the proportion of infected sea fans on the Las Redes Reef is between 42.1% and 61.7%."** This is the most respectable so far of these statements we'd like to make—it appropriately gives an interval estimate rather than a single number, and it acknowledges that there's a little bit of uncertainty even still. But this sentence, too, has a major problem: It makes a probability statement ("95% chance") about our *particular interval*. This is extremely tempting to do, but is incorrect. The interval from 42.1% to 61.7% either does or does not include the population proportion $p$. Hit or miss. 100% chance or 0% chance, nothing in between. The 95% says that 95% of *all* possible random samples would have led us to an interval that includes $p$. But it doesn't actually say anything at all about any *particular* random sample or resulting interval.

The problem we have is that *before* we collect any data, we can say that there is a 95% chance that our data *will* lead us to an interval containing the true population proportion $p$. But then after the data are collected, the notion of probability evaporates, because there's nothing random about our particular sample or its resulting particular interval. But that's okay, because probability was never exactly what we wanted in the first place. We wanted an estimate of $p$ that we could trust, or have confidence in. And we *can* have confidence in

> Far better an approximate answer to the right question, . . . than an exact answer to the wrong question.
> —John W. Tukey

---
[4]Or, at least, no more than 2 standard *deviations* away from $p$. But as we'll see in the last section of this chapter, for practical purposes you can pretty much treat these as if they're the same thing.

our particular interval: it resulted from a procedure that begins with random sampling and ends in an interval that, for 95% of the possible samples, really does contain $p$. When the procedure as a whole "works" for 95% of all samples, we can have high confidence in any particular interval we obtain—in fact, we can go so far as to say we have "95% confidence" in our interval. Those in the know will get what this really means, and those *not* in the know will at least not be misled.[5]

So what *should* we say?

5. **"We are 95% confident that between 42.1% and 61.7% of Las Redes sea fans are infected."** Statements like this describe **confidence intervals**. Unlike some unfortunately named items in the statistical toolbox (I'm looking at you, expected value), confidence intervals are well-named. We can't associate 95% probability with a particular interval, but we can have a lot of confidence in one.

Each confidence interval discussed in this book has a name. You'll see many different kinds of confidence intervals in the following chapters. Some will be about more than *one* population or treatment, some will be about parameters other than *proportions,* and some will use models other than the Normal. The interval calculated and interpreted here is often called a **one-proportion $z$-interval**.

### JUST CHECKING

A Pew Research study regarding cell phones asked questions about cell phone experience. One growing concern is unsolicited advertising in the form of text messages. Pew asked cell phone owners, "Have you ever received unsolicited text messages on your cell phone from advertisers?" and 17% reported that they had. Pew estimates a 95% confidence interval to be $0.17 \pm 0.04$, or between 13% and 21%.

Are the following statements about people who have cell phones correct? Explain.

1. In Pew's sample, somewhere between 13% and 21% of respondents reported that they had received unsolicited advertising text messages.
2. We can be 95% confident that 17% of U.S. cell phone owners have received unsolicited advertising text messages.
3. We are 95% confident that between 13% and 21% of all U.S. cell phone owners have received unsolicited advertising text messages.
4. We know that between 13% and 21% of all U.S. cell phone owners have received unsolicited advertising text messages.
5. 95% of all U.S. cell phone owners have received unsolicited advertising text messages.

## What Does "95% Confidence" Really Mean?

What do we mean when we say we have 95% confidence that our interval contains the true proportion? Formally, what we mean is that "95% of samples of this size will produce confidence intervals that capture the true proportion." This is correct, but a little long winded, so we sometimes say, "We are 95% confident that the true proportion lies in our interval." Our uncertainty is about whether the particular sample we have at hand is one of the successful ones or one of the 5% that fail to produce an interval that captures the true value.

---

[5]Imagine flipping 5 coins at once. Before you flip them, you can say there's a high probability that at least one will show heads. After you flip them (but before looking at them), you can only say that you're very *confident* that at least one is showing heads. You'd still have this confidence even if you tossed the coins on the ground, walked away without ever looking at them, and *never knew for sure* whether at least one head had been showing. That's the position we're in.

Back in Chapter 17 we saw that proportions vary from sample to sample. If other researchers select their own samples of sea fans, they'll also find some infected by the disease, but each person's sample proportion will almost certainly differ from ours. When they each try to estimate the true rate of infection in the entire population, they'll center *their* confidence intervals at the proportions they observed in their own samples. Each of us will end up with a different interval.

Our interval guessed the true proportion of infected sea fans to be between about 42% and 62%. Another researcher whose sample contained more infected fans than ours did might guess between 46% and 66%. Still another who happened to collect fewer infected fans might estimate the true proportion to be between 23% and 43%. And so on. Every possible sample would produce yet another confidence interval. Although wide intervals like these can't pin down the actual rate of infection very precisely, we expect that most of them should be winners, capturing the true value. Nonetheless, some will be duds, missing the population proportion entirely.

Below you'll see confidence intervals produced by simulating 20 different random samples. The red dots are the proportions of infected fans in each sample, and the blue segments show the confidence intervals found for each. The green line represents the true rate of infection in the population, so you can see that most of the intervals caught it—but a few missed. (And notice again that it is the *intervals* that vary from sample to sample; the green line doesn't move.)

**APPLET**
Explore how often confidence intervals hit the target.

TI-*nspire*

**Confidence intervals.** Generate confidence intervals from many samples to see how often they capture the true proportion.

**Figure 18.3**
The horizontal green line shows the true percentage of all sea fans that are infected. Most of the 20 simulated samples produced confidence intervals that captured the true value, but a few missed.

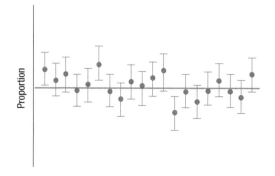

Of course, there's a huge number of possible samples that *could* be drawn, each with its own sample proportion. These are just some of them. Each sample proportion can be used to make a confidence interval. That's a large pile of possible confidence intervals, and ours (the one based on real data) is just one of those in the pile. Did *our* confidence interval "work"? We can never be sure, because we'll never know the true proportion of all the sea fans that are infected. However, the Central Limit Theorem assures us that 95% of the intervals in the pile are winners,[6] covering the true value, and only 5% are duds. *That's* why we're 95% confident that our interval is a winner!

## So, What *Can* I Say?

Technically, we should say, "I am 95% confident that the interval from 42.1% to 61.7% captures the true proportion of sea fans on the Las Redes Reef that are infected." That formal phrasing emphasizes that *our confidence (and our uncertainty) is about the interval, not the true proportion.* But you may choose a more casual phrasing like "I am 95% confident that between 42.1% and 61.7% of Las Redes sea fans are infected." Because you've made it clear that the uncertainty is yours and you didn't suggest that the randomness is in the true proportion, this is OK. Keep in mind that it's the interval that's random and is the focus of both our confidence and doubt.

---

[6] For this reason, the confidence level of an interval (like 95%) is sometimes called the interval's *capture rate*. It is understood that this means the procedure's relative frequency of success, over all possible samples, in "capturing" the population parameter.

## FOR EXAMPLE
### Polls and Confidence Intervals

In January and February 2016, Pew Research polled 4654 U.S. adults, asking them, "How often do you read any newspapers in print?" For the first time, a majority of adults responded "Hardly ever" or "Never."

QUESTION: It is standard among pollsters to use a 95% confidence level unless otherwise stated. Given that, what do these researchers mean by their confidence interval in this context?

ANSWER: If this polling were done repeatedly, 95% of all random samples would yield confidence intervals that contain the true proportion of all U.S. adults who hardly ever or never read a newspaper in print.

# Margin of Error: Degree of Certainty vs. Precision

We've just claimed that with a certain confidence we've captured the true proportion of all infected sea fans. Our confidence interval had the form

$$\hat{p} \pm 2\, SE(\hat{p})$$

The extent of the interval on either side of $\hat{p}$ is called the **margin of error** (*ME*).

We'll want to use the same approach for many other situations besides estimating proportions. In fact, almost any population parameter—a proportion, a mean, or a regression slope, for example—can be estimated with some margin of error. The margin of error is a way to describe our uncertainty in estimating the population value. We'll see how to find a margin of error for each of these values and for others.

For all of those statistics, regardless of how we calculate the margin of error, we'll be able to construct a confidence interval that looks like this:

$$Estimate \pm ME.$$

The margin of error for our 95% confidence interval was 2 *SE*. What if we wanted to be more confident? To be more confident, we'll need to capture *p* more often, and to do that we'll need to make the interval wider. For example, if we want to be 99.7% confident, the margin of error will have to be 3 *SE*.

Figure 18.4
Reaching out 3 *SE* on either side of $\hat{p}$ makes us 99.7% confident we'll trap the true proportion *p*. Compare with Figure 18.2.

The more confident we want to be, the larger the margin of error must be. We can be 100% confident that the proportion of infected sea fans is between 0% and 100%, but that isn't very useful, is it? On the other hand, we could give a confidence interval from 51.8% to 52.0%, but we can't be very confident about a precise statement like this. Every confidence interval is a balance between degree of certainty and precision.

The tension between degree of certainty of precision is always there. Fortunately, in most cases we can be both sufficiently certain and sufficiently precise to make useful statements. There is no simple answer to the conflict. You must choose a confidence level yourself. The data can't do it for you. The choice of confidence level is somewhat arbitrary. The most commonly chosen confidence levels are 90%, 95%, and 99%, but any percentage can be used.

### FOR EXAMPLE
#### Finding the Margin of Error (Take 1)

**RECAP:** The Pew Research poll asking "How often do you read any newspapers in print?" reported a margin of error of 1.5%. It is a convention among pollsters to use a 95% confidence level and to report the "worst case" margin of error, based on $p = 0.5$.

**QUESTION:** How did the researchers calculate their margin of error?

**ANSWER:** Assuming $p = 0.5$, for random samples of $n = 4654$,

$$SD(\hat{p}) = \sqrt{\frac{pq}{n}} = \sqrt{\frac{(0.5)(0.5)}{4654}} = 0.0073.$$

For a 95% confidence level, $ME = 2(0.0073) = 0.0146$, so their margin of error is just a bit under 1.5%.

# Critical Values

**NOTATION ALERT**
We'll put an asterisk on a letter to indicate a critical value, so $z^*$ is always a critical value from a Normal model.

In our sea fans example we used 2 *SE* to give us a 95% confidence interval. To change the confidence level, we'd need to change the *number* of *SE*s so that the size of the margin of error corresponds to the new level. This number of *SE*s is called the **critical value**. Here it's based on the Normal model, so we denote it $z^*$. For any confidence level, we can find the corresponding critical value from a computer, a calculator, or a Normal probability table, such as Table Z.

For a 95% confidence interval, you'll find the precise critical value is $z^* = 1.96$. That is, 95% of a Normal model is found within $\pm 1.96$ standard deviations of the mean. We've been using $z^* = 2$ from the 68–95–99.7 Rule because it's easy to remember.

**Figure 18.5**
For a 90% confidence interval, the critical value is 1.645, because, for a Normal model, 90% of the values are within 1.645 standard deviations from the mean.

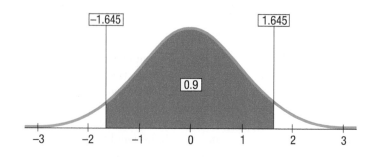

## FOR EXAMPLE
### Finding the Margin of Error (Take 2)

**RECAP:** In the poll on reading newspapers in print, Pew Research found that 52% of adults reported "Hardly ever" or "Never." They reported a 95% confidence interval with a margin of error of 1.5%.

**QUESTION:** Using the critical value of z and the standard error based on the observed proportion, what would be the margin of error for a 90% confidence interval? What's good and bad about this change?

With $n = 4654$ and $\hat{p} = 0.52$, $SE(\hat{p}) = \sqrt{\dfrac{\hat{p}\hat{q}}{n}} = \sqrt{\dfrac{(0.52)(0.48)}{4654}} = 0.0073$.

For a 90% confidence level, $z^* = 1.645$, so $ME = 1.645(0.0073) = 0.0120$.

**ANSWER:** Now the margin of error is only about 1.2%, producing a narrower interval. What's good about the change is that we now have a smaller interval, but what's bad is that we are less certain that the interval actually contains the true proportion of adults who never or hardly ever read a newspaper in print.

## JUST CHECKING

Think some more about the 95% confidence interval originally created for the proportion of U.S. adults who never or hardly ever read a newspaper in print.

6. If the researchers wanted to be 98% confident, would their confidence interval need to be wider or narrower?

7. The study's margin of error was about 1.5%. If the researchers wanted to reduce it to 1%, would their level of confidence be higher or lower?

8. If the researchers had polled more people, would the interval's margin of error have been larger or smaller?

# Assumptions and Conditions

We've just made some pretty sweeping statements about sea fans. Those statements were possible because we used a Normal model for the sampling distribution. But is that model appropriate?

We've also said that the same basic ideas will work for other statistics. One difference for those statistics is that some will have sampling distribution models that are different than the Normal. But the background theory (which we won't bother with) is so similar for all of those models that they share the same basic assumptions and conditions about independence and sample size. Then for each statistic, we usually tack on a special-case assumption or two. We'll deal with those as we get to them in later chapters.

We saw the assumptions and conditions for using the Normal to model the sampling distribution for a proportion in the last chapter. Because they are so crucial to making sure our confidence interval is useful, we'll repeat them here.

## Independence Assumption

**Independence Assumption:** The data values must be independent. To think about whether this assumption is plausible, we often look for reasons to suspect that it fails. We wonder whether there is any reason to believe that the data values somehow affect each other. (For example, might the disease in sea fans be contagious?) Whether you decide that the **Independence Assumption** is plausible depends on your knowledge of the situation. It's not one you can check by looking at the data.

However, now that we have data, there are two conditions that we can check:

**Randomization Condition:** Were the data sampled at random or generated from a properly randomized experiment? Proper randomization can help ensure independence.

**10% Condition:** If you sample more than 10% of a population, the formula for the standard error won't be quite right. There is a special formula (found in advanced books) that corrects for this, but it isn't a common problem unless your population is small.

## Sample Size Assumption

The model we use for inference for proportions is based on the Central Limit Theorem. We need to know whether the sample is large enough to make the sampling model for the sample proportions approximately Normal. It turns out that we need more data as the proportion gets closer and closer to either extreme (0 or 1). That's why we check the following:

**Success/Failure Condition:** We must expect at least 10 "successes" and at least 10 "failures." Recall that by tradition we arbitrarily label one alternative (usually the outcome being counted) as a "success" even if it's something bad (like getting a disease). The other alternative is, of course, then a "failure."

Without knowing the actual value of the population proportion $p$, we can't know how many successes and failures to "expect" (in the "expected value" sense). We check instead how many successes and failures we *do* see in the data. Both should be at least 10.

Let's now verify that the Normal model is a reasonable approximation for the sampling distribution of $\hat{p}$ for our sea fan data. (You *should* check the conditions *before* diving in and doing all the analysis we did earlier. Because we wanted to focus on the "big picture," we put off those checks. Now here they are.)

- We need to check the Randomization Condition. The sea fans were sampled from 19 randomly selected reefs, so it seems reasonable to treat the fans as a random sample from all the sea fans in that area.
- We need to check the 10% Condition. There were 104 sea fans in the sample and it seems clear that this is less than 10% of all the sea fans in the area, so that condition is satisfied.
- And we need to check the Success/Failure Condition. We saw 54 sea fans infected with the disease and therefore $100 - 54 = 50$ healthy sea fans. Both are greater than 10.

With all of these conditions satisfied, the Independence Assumption seems reasonable as does the Sample Size Assumption. Using the Normal model turns out to be justified, and our earlier statements about sea fans are validated.

### One-Proportion z-Interval

When the conditions are met, we are ready to find a level C confidence interval for the population proportion, $p$. The confidence interval is $\hat{p} \pm z^* \times SE(\hat{p})$, where the standard deviation of the proportion is estimated by $SE(\hat{p}) = \sqrt{\dfrac{\hat{p}\hat{q}}{n}}$ and the critical value, $z^*$, specifies the number of SEs needed for C% of random samples to yield confidence intervals that capture the true parameter value.

## STEP-BY-STEP EXAMPLE

### A Confidence Interval for a Proportion

In October 2015, the Gallup Poll[7] asked 1015 randomly sampled adults the question "Generally speaking, do you believe the death penalty is applied fairly or unfairly in this country today?" Of these, 53% answered "Fairly," 41% said "Unfairly," and 6% said they didn't know.

[7] www.gallup.com/poll/1606/death-penalty.aspx

| WHO | Adults in the United States |
|---|---|
| WHAT | Response to a question about the death penalty |
| WHEN | October 2015 |
| WHERE | United States |
| HOW | 1015 adults were randomly sampled and asked by the Gallup Poll |
| WHY | Public opinion research |

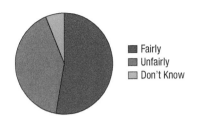

QUESTION: From this survey, what can we conclude about the opinions of *all* adults?

To answer this question, we'll build a confidence interval for the proportion of all U.S. adults who believe the death penalty is applied fairly. There are four steps to building a confidence interval for proportions: Plan, Model, Mechanics, and Conclusion.

---

**THINK**

**PLAN** State the problem and the W's. Identify the *parameter* you wish to estimate. Identify the *population* about which you wish to make statements.

Choose and state a confidence level.

**MODEL** Think about the assumptions and check the conditions.

I want to find an interval that is likely, with 95% confidence, to contain the true proportion, $p$, of U.S. adults who think the death penalty is applied fairly. I have a random sample of 1015 U.S. adults.

 **Randomization Condition**: Gallup drew a random sample from all U.S. adults. I can be confident that the respondents are independent.

✓ **10% Condition**: The sample is certainly less than 10% of the population.

✓ **Success/Failure Condition**:
$$n\hat{p} = 1015(53\%) = 538 \geq 10 \text{ and}$$
$$n\hat{q} = 1015(47\%) = 477 \geq 10,$$

so the sample appears to be large enough to use the Normal model.

State the sampling distribution model for the statistic.
Choose your method.

The conditions are satisfied, so I can use a Normal model to find a one-proportion z-interval.

---

**SHOW**

**MECHANICS** Construct the confidence interval.

First find the standard error. (*Remember*: It's called the "standard error" because we don't know $p$ and have to use $\hat{p}$ instead.)

$$n = 1015, \hat{p} = 0.53, \text{ so}$$
$$SE(\hat{p}) = \sqrt{\frac{\hat{p}\hat{q}}{n}} = \sqrt{\frac{(0.53)(0.47)}{1015}} = 0.0157.$$

Because the sampling model is Normal, for a 95% confidence interval, the critical value $z^* = 1.96$.

Next find the margin of error. We could informally use 2 for our critical value, but 1.96 (found from a table or technology) is more accurate.

The margin of error is
$$ME = z^* \times SE(\hat{p}) = 1.96(0.0157) = 0.031.$$

Write the confidence interval (CI).

So the 95% confidence interval is
$$0.53 \pm 0.031 \text{ or } (0.499, 0.561).$$

**REALITY CHECK** The CI is centered at the sample proportion and about as wide as we might expect for a sample of about 1000.

---

**TELL**

**CONCLUSION** Interpret the confidence interval in the proper context. We're 95% confident that our interval captured the true proportion.

I am 95% confident that between 49.9% and 56.1% of all U.S. adults think that the death penalty is applied fairly.

## TI TIPS

### Finding Confidence Intervals

It will come as no surprise that your TI can calculate a confidence interval for a population proportion. Remember the sea fans? Of 104 sea fans, 54 were diseased. To find the resulting confidence interval, we first take a look at a whole new menu.

- Under STAT go to the TESTS menu. Quite a list! Commands are found here for the inference procedures you will learn through the coming chapters.
- We're using a Normal model to find a confidence interval for a proportion based on one sample. Scroll down the list and select 1-PropZInt.
- Enter the number of successes observed and the sample size.
- Specify a confidence level and then Calculate.

And there it is! Note that the TI calculates the sample proportion for you, but the important result is the interval itself, 42% to 62%. The calculator did the easy part—just Show. Tell is harder. It's your job to interpret that interval correctly.

Beware: You may run into a problem. When you enter the value of x, you need a *count*, not a percentage. Suppose the marine scientists had reported that 52% of the 104 sea fans were infected. You can enter x:.52*104, and the calculator will evaluate that as 54.08. Wrong. Unless you fix that result, you'll get an error message. Think about it—the number of infected sea fans must have been a whole number, evidently 54. When the scientists reported the results, they rounded off the actual percentage (54 ÷ 104 = 51.923%) to 52%. Simply change the value of x to 54 and you should be able to Calculate the correct interval.

### A Confidence Interval for a Count

If you know the size of your population, then you can convert a confidence interval for a population proportion into a confidence interval for a population count. Just multiply by the population size. For example, in June 2021 the Pew Research Center conducted a poll of U.S. adults and found that 10 percent of us believe that "UFOs are a major threat to U.S. national security."[8] Let's suppose that the margin of error for this estimate is ±2 percentage points.[9] That would mean we can be 95% confident that between 8% and 12% of U.S. adults believe UFOs to be a major national security threat. Multiply that by the number of U.S. adults (209 million) and you get the interval from about 16.7 million to 25.1 million. The poll implies that with a high degree of confidence, between 16 and 25 million adults in the United States consider UFOs to be among our major national security threats.

## Choosing Your Sample Size

The question of how large a sample to take is an important step in planning any study. We weren't ready to make that calculation when we first looked at study design in Chapter 11, but now we can—and we always should.

---

[8] https://www.pewresearch.org/fact-tank/2021/06/30/most-americans-believe-in-intelligent-life-beyond-earth-few-see-ufos-as-a-major-national-security-threat/

[9] The Pew Organization's report doesn't include margins of error, but a margin of error of 2 percentage points is consistent with the report.

## What Do I Use Instead of $\hat{p}$?

Often we have an estimate of the population proportion based on experience or perhaps a previous study. If so, use that value as $\hat{p}$ in calculating what size sample you need. If not, the cautious approach is to use $p = 0.5$ in the sample size calculation; that will determine the largest sample necessary regardless of the true proportion.

Suppose a candidate is planning a poll and wants to estimate voter support within 3% with 95% confidence. How large a sample do they need?

Let's look at the margin of error:

$$ME = z^* \sqrt{\frac{\hat{p}\hat{q}}{n}}$$

$$0.03 = 1.96 \sqrt{\frac{\hat{p}\hat{q}}{n}}.$$

We want to find $n$, the sample size. To find $n$ we need a value for $\hat{p}$. We don't know $\hat{p}$ because we don't have a sample yet, but we can probably guess a value. The worst case—the value that makes $\hat{p}\hat{q}$ (and therefore $n$) largest—is 0.50, so if we use that value for $\hat{p}$, we'll certainly be safe. Our candidate probably expects $\hat{p}$ to be near 50% anyway.

Our equation, then, is

$$0.03 = 1.96 \sqrt{\frac{(0.5)(0.5)}{n}}.$$

To solve for $n$, we first multiply both sides of the equation by $\sqrt{n}$ and then divide by 0.03:

$$0.03\sqrt{n} = 1.96\sqrt{(0.5)(0.5)}$$

$$\sqrt{n} = \frac{1.96\sqrt{(0.5)(0.5)}}{0.03} \approx 32.67.$$

Notice that evaluating this expression tells us the *square root* of the sample size. We need to square that result to find $n$:

$$n \approx (32.67)^2 \approx 1067.1.$$

To be safe, we round up and conclude that we need at least 1068 respondents to keep the margin of error as small as 3% with a confidence level of 95%.

### FOR EXAMPLE: Choosing a Sample Size

In a March 2019 survey by Gallup of 1019 U.S. adults, 65% believe that global warming is due more to human activity than natural changes in the environment. The poll reports a margin of error of 4%.

Suppose an environmental group planning a follow-up survey of voters' opinions on global warming wants to determine a 95% confidence interval with a margin of error of no more than ±2%.

**QUESTION:** How large a sample does the group need? (You could take $p = 0.5$, but we have data that indicate $p$ is likely to be near 0.65, so we can use that.)

**ANSWER:**

$$ME = z^* \sqrt{\frac{\hat{p}\hat{q}}{n}}$$

$$0.02 = 1.96 \sqrt{\frac{(0.65)(0.35)}{n}}$$

$$\sqrt{n} = 1.96 \frac{\sqrt{(0.65)(0.35)}}{0.02} \approx 46.74$$

$$n = 46.74^2 = 2184.91$$

The environmental group's survey will need at least 2185 respondents.

Unfortunately, bigger samples cost more money and more effort. Because the standard error declines only with the *square root* of the sample size, to cut the standard error (and thus the margin of error) in half, we must *quadruple* the sample size.

Generally a margin of error of 5% or less is acceptable, but different circumstances call for different standards. For a pilot study, a margin of error of 10% may be fine, so a sample of 100 will do quite well. In a close election, a polling organization might want to get the margin of error down to 2%. Drawing a large sample to get a smaller margin of error, however, can run into trouble. It takes time to survey 2400 people, and a survey that extends over a week or more may be trying to hit a target that moves during the time of the survey. An important event can change public opinion in the middle of the survey process.

Keep in mind that the sample size for a survey is the number of respondents, not the number of people to whom questionnaires were sent or whose phone numbers were dialed. And also keep in mind that a low response rate turns any study essentially into a voluntary response study, which is of little value for inferring population values. It's almost always better to spend resources on increasing the response rate than on surveying a larger group. A full or nearly full response by a modest-size sample can yield useful results.

Surveys are not the only place where proportions pop up. Banks sample huge mailing lists to estimate what proportion of people will accept a credit card offer. Even pilot studies may mail offers to over 50,000 customers. Most don't respond; that doesn't make the sample smaller—they simply said, "No thanks." Those who do respond want the card. To the bank, the response rate[10] is $\hat{p}$. With a typical success rate around 0.5%, the bank needs a very small margin of error—often as low as 0.1%—to make a sound business decision. That calls for a large sample, and the bank must take care in estimating the size needed. For our election poll calculation we used $p = 0.5$, both because it's safe and because we honestly believed $p$ to be near 0.5. If the bank used 0.5, they'd get an absurd answer. Instead, they base their calculation on a proportion closer to the one they expect to find.

### FOR EXAMPLE
#### Sample Size Revisited

A credit card company is about to send out a mailing to test the market for a new credit card. From that sample, the company's managers want to estimate the true proportion of people who will sign up for the card nationwide. A pilot study suggests that about 0.5% of the people receiving the offer will accept it.

QUESTION: To be within a tenth of a percentage point (0.001) of the true rate with 95% confidence, how big does the test mailing have to be?

ANSWER: Using the estimate $\hat{p} = 0.5\%$:

$$ME = 0.001 = z^* \sqrt{\frac{\hat{p}\hat{q}}{n}} = 1.96 \sqrt{\frac{(0.005)(0.995)}{n}}$$

$$(0.001)^2 = 1.96^2 \frac{(0.005)(0.995)}{n} \Rightarrow n = \frac{1.96^2 (0.005)(0.995)}{(0.001)^2}$$
$$= 19{,}111.96 \text{ or } 19{,}112.$$

That's a lot, but it's actually a reasonable size for a trial mailing such as this. Note, however, that if they had assumed the "worst case" value of 0.50 for $p$, they would have found

$$ME = 0.001 = z^* \sqrt{\frac{pq}{n}} = 1.96 \sqrt{\frac{(0.5)(0.5)}{n}}$$

$$(0.001)^2 = 1.96^2 \frac{(0.5)(0.5)}{n} \Rightarrow n = \frac{1.96^2 (0.5)(0.5)}{(0.001)^2} = 960{,}400.$$

Quite a different (and unreasonable) result.

## The Proof Is in the Pudding

Earlier in this chapter we did a little switcheroo. We knew that the standard *deviation* of $\hat{p}$ was $SD(\hat{p}) = \sqrt{\frac{pq}{n}}$, but because we didn't know the population parameter $p$ (nor did we know $q$), we estimated them with $\hat{p}$ and $\hat{q}$ and called the resulting quantity the standard

---
[10] In marketing studies every mailing yields a response—"yes" or "no"—and "response rate" means the proportion of customers who accept an offer. That's not the way we use the term for survey response.

*error* of $\hat{p}$: $SE(\hat{p}) = \sqrt{\dfrac{\hat{p}\hat{q}}{n}}$. But how do we know that substituting the *SE* for the *SD* will work? Also, besides the use of $SE(\hat{p})$ in place of $SD(\hat{p})$, there's also our use of the Normal model to approximate the sampling distribution of $\hat{p}$. Are we being too sloppy? How can we be sure our confidence interval procedure really works?

Well you know what they say—the proof is in the pudding.[11] A confidence interval procedure "works" if its claimed parameter "capture rate" (that's the confidence level) really is the fraction of resulting intervals that contain the parameter of interest, when it's applied over many random samples. This sure sounds like a simulation to us!

We decided to get really fancy with this simulation, so we need to explain carefully what we did before we show you the result. We were interested in seeing how well our "one-proportion *z*-interval" works under different circumstances—in particular, for different population proportions *p* and different sample sizes *n*. We also would have liked to try it with different confidence levels, but that would have gotten needlessly complicated, so we just used a 95% confidence level throughout our simulation.

We looked at possible values of the population proportion *p* ranging from 0.05 up to 0.95, in increments of 0.05. The Success/Failure Condition requires us to see at least 10 successes and at least 10 failures, so we didn't consider any sample sizes less than 20. We looked at sample sizes of 20, 50, 100, 200, 500, 1000, 2000, 5000, and 10,000. We programmed our computer to go through each combination of a *p* and an *n*, and for each one simulated a million samples of size *n* from a population whose proportion of successes is *p*. Then for each of those million samples the program computed the 95% confidence interval given by $\hat{p} \pm 1.96\sqrt{\dfrac{\hat{p}\hat{q}}{n}}$. Finally, we determined what fraction of the million confidence intervals did indeed capture *p*.

Got all that? (If not, seeing the results explained may help.) Ready for the results? To the nearest tenth of a percent, here are the relative frequencies with which our confidence intervals captured the appropriate population parameter *p*. We've highlighted all the values that are less than 94%. (A lower-than-desired capture rate is bad, but a capture rate *greater* than 95% would actually be a good thing!)

|  | Sample Size *n* | | | | | | | | |
|---|---|---|---|---|---|---|---|---|---|
| Population Proportion *p* | 20 | 50 | 100 | 200 | 500 | 1000 | 2000 | 5000 | 10000 |
| 0.05 | 0.639 | 0.920 | 0.877 | 0.926 | 0.931 | 0.942 | 0.946 | 0.948 | 0.951 |
| 0.10 | 0.876 | 0.880 | 0.933 | 0.927 | 0.942 | 0.953 | 0.946 | 0.952 | 0.951 |
| 0.15 | 0.818 | 0.941 | 0.933 | 0.945 | 0.947 | 0.947 | 0.947 | 0.950 | 0.950 |
| 0.20 | 0.921 | 0.938 | 0.933 | 0.941 | 0.949 | 0.947 | 0.949 | 0.950 | 0.950 |
| 0.25 | 0.895 | 0.940 | 0.946 | 0.937 | 0.942 | 0.946 | 0.947 | 0.950 | 0.949 |
| 0.30 | 0.948 | 0.935 | 0.951 | 0.944 | 0.949 | 0.950 | 0.950 | 0.950 | 0.950 |
| 0.35 | 0.936 | 0.945 | 0.940 | 0.946 | 0.951 | 0.949 | 0.951 | 0.950 | 0.950 |
| 0.40 | 0.928 | 0.940 | 0.948 | 0.949 | 0.950 | 0.947 | 0.950 | 0.950 | 0.950 |
| 0.45 | 0.923 | 0.934 | 0.944 | 0.946 | 0.947 | 0.947 | 0.949 | 0.950 | 0.950 |
| 0.50 | 0.959 | 0.935 | 0.943 | 0.944 | 0.946 | 0.946 | 0.948 | 0.951 | 0.949 |
| 0.55 | 0.923 | 0.934 | 0.944 | 0.945 | 0.947 | 0.947 | 0.949 | 0.950 | 0.950 |
| 0.60 | 0.928 | 0.941 | 0.948 | 0.949 | 0.951 | 0.947 | 0.950 | 0.950 | 0.950 |
| 0.65 | 0.936 | 0.945 | 0.941 | 0.946 | 0.951 | 0.949 | 0.951 | 0.950 | 0.950 |
| 0.70 | 0.947 | 0.935 | 0.950 | 0.943 | 0.949 | 0.950 | 0.949 | 0.950 | 0.951 |
| 0.75 | 0.895 | 0.940 | 0.946 | 0.937 | 0.942 | 0.947 | 0.947 | 0.950 | 0.949 |
| 0.80 | 0.921 | 0.937 | 0.933 | 0.941 | 0.949 | 0.947 | 0.949 | 0.950 | 0.950 |
| 0.85 | 0.819 | 0.941 | 0.933 | 0.944 | 0.946 | 0.947 | 0.946 | 0.950 | 0.950 |
| 0.90 | 0.876 | 0.878 | 0.932 | 0.927 | 0.943 | 0.953 | 0.946 | 0.952 | 0.950 |
| 0.95 | 0.639 | 0.919 | 0.877 | 0.926 | 0.932 | 0.942 | 0.947 | 0.948 | 0.951 |

---

[11] Maybe they never say that around you. This expression means, "You can tell if something works by trying it and seeing if it works!"

Here are two main things to take away from this simulation result regarding the one-proportion z-interval procedure:

- For samples of size $n = 500$ or greater, the procedure is quite successful at capturing $p$ at very close to the 95% rate that is the intended capture rate (confidence level). This is especially true for samples of size $n \geq 1000$.
- The capture rate tends to be closest to the intended 95% when both $np$ and $nq$ are large. That's why the procedure does so poorly when the sample size is only $n = 20$. Most of the intervals simulated for the first column of the table resulted from simulated data that didn't satisfy the Success/Failure Condition (at least 10 successes and at least 10 failures). The procedure also doesn't work as well for larger sample sizes if either $p$ or $q$ is very close to zero. But these are also instances where the Success/Failure Condition was probably not met in many of the simulated samples.

The upshot is that if the one-proportion z-interval is applied only when its required conditions are met, then the probability that it will successfully capture the population proportion $p$ is very close to the intended capture rate (confidence level) of 95%.[12]

Usually you'll be using technology to compute confidence intervals from data or summary statistics. But calculators and computers generally don't warn you if you're about to apply a confidence interval procedure when you shouldn't. Nothing will appear to go wrong. But if the Success/Failure Condition is not met, then you should not have 95% confidence in the resulting interval. In fact, you shouldn't be using the one-proportion z-interval procedure at all.

## WHAT CAN GO WRONG?

Confidence intervals are powerful tools. Not only do they tell what we know about the parameter value, but—more important—they also tell what we *don't* know. In order to use confidence intervals effectively, you must be clear about what you say about them.

### Don't Misstate What the Interval Means

- **Don't suggest that the parameter varies.** A statement like "There is a 95% chance that the true proportion is between 42.7% and 51.3%" sounds as though you think the population proportion wanders around and sometimes happens to fall between 42.7% and 51.3%. When you interpret a confidence interval, make it clear that *you* know that the population parameter is fixed and that it is the interval that varies from sample to sample.
- **Don't claim that other samples will agree with yours.** Keep in mind that the confidence interval makes a statement about the true population proportion. An interpretation such as "In 95% of samples of U.S. adults, the proportion who think marijuana should be decriminalized will be between 42.7% and 51.3%" is just wrong. The interval isn't about sample proportions but about the population proportion.
- **Don't be certain about the parameter.** Saying "Between 42.1% and 61.7% of sea fans are infected" asserts that the population proportion cannot be outside that interval. Of course, we can't be absolutely certain of that. (Just pretty sure.)
- **Don't forget: It's about the parameter.** Don't say, "I'm 95% confident that $\hat{p}$ is between 42.1% and 61.7%." Of course you are—in fact, we calculated that $\hat{p} = 51.9\%$ of the fans in our sample were infected. So we already *know* the sample proportion. The confidence interval is about the (unknown) population parameter, $p$.

---

[12]We've demonstrated this for a 95% confidence level, but the same is true for other confidence levels.

- **Don't claim to know too much.** Don't say, "I'm 95% confident that between 42.1% and 61.7% of all the sea fans in the world are infected." You didn't sample from all 500 species of sea fans found in coral reefs around the world. Just those of this type on the Las Redes Reef.
- **Do take responsibility.** Confidence intervals are about *uncertainty*. *You* are the one who is uncertain, not the parameter. You have to accept the responsibility and consequences of the fact that not all the intervals you compute will capture the true value. In fact, about 5% of the 95% confidence intervals you find will fail to capture the true value of the parameter. You *can* say, "I am 95% confident that between 42.1% and 61.7% of the sea fans on the Las Redes Reef are infected."[13]
- **Do treat the whole interval equally.** Although a confidence interval is a set of plausible values for the parameter, don't think that the values in the middle of a confidence interval are somehow "more plausible" than the values near the edges. Your interval provides no information about where in your current interval (if at all) the parameter value is most likely to be hiding.

## Beware of a Margin of Error Too Large to Be Useful

We know we can't be exact, but how precise do we need to be? A confidence interval that says that the percentage of infected sea fans is between 10% and 90% wouldn't be of much use. Most likely, you have some sense of how large a margin of error you can tolerate. What can you do?

One way to make the margin of error smaller is to reduce your level of confidence. But that may not be a useful solution. It's a rare study that reports confidence levels lower than 80%. Levels of 95% or 99% are more common.

The time to think about whether your margin of error is small enough to be useful is when you design your study. Don't wait until you compute your confidence interval. To get a narrower interval without giving up confidence, you need to have less variability in your sample proportion. How can you do that? Choose a larger sample.

## Look for Violations of Assumptions

Confidence intervals and margins of error are often reported along with poll results and other analyses. But it's easy to misuse them and wise to be aware of other ways things can go wrong.

- **Watch out for biased sampling.** Don't forget about the potential sources of bias in surveys that we discussed in Chapter 11. Just because we have more statistical machinery now doesn't mean we can forget what we've already learned. A questionnaire that finds that 85% of people enjoy filling out surveys still suffers from nonresponse bias even though now we're able to put confidence intervals around this (biased) estimate.
- **Think about independence.** The assumption that the values in our sample are mutually independent is one that we usually cannot check. It always pays to think about it, though. For example, the disease affecting the sea fans might be contagious, so that fans growing near a diseased fan are more likely themselves to be diseased. Such contagion would violate the Independence Assumption and could severely affect our sample proportion. It could be that the proportion of infected sea fans on the entire reef is actually quite small, and the researchers just happened to find an infected area. To avoid this, the researchers should be careful to sample sites far enough apart to make contagion unlikely.

---

[13] When we are being very careful we say, "95% of samples of this size will produce confidence intervals that capture the true proportion of infected sea fans on the Las Redes Reef."

# WHAT HAVE WE LEARNED?

We've learned to construct a confidence interval for a proportion, $p$, as the statistic, $\hat{p}$, plus and minus a margin of error.
- The margin of error consists of a critical value based on the sampling model times a standard error based on the sample.
- The critical value is found from the Normal model.
- The standard error of a sample proportion is calculated as $\sqrt{\dfrac{\hat{p}\hat{q}}{n}}$.

We've learned to interpret a confidence interval correctly.
- You can claim to have the specified level of confidence that the interval you have computed actually covers the true value.

We've come to understand the relationship of the sample size, $n$, to both the certainty (confidence level) and precision (margin of error).
- For the same sample size and true population proportion, more certainty means less precision (wider interval) and more precision (narrower interval) implies less certainty.

We've learned to check the assumptions and conditions for finding and interpreting confidence intervals:
- Independence Assumption or Randomization Condition,
- 10% Condition, and
- Success/Failure Condition.

We've learned to find the sample size required, given a proportion, a confidence level, and a desired margin of error.

## TERMS

**Standard error**   When we estimate the standard deviation of a sampling distribution using statistics found from the data, the estimate is called a standard error:

$$SE(\hat{p}) = \sqrt{\dfrac{\hat{p}\hat{q}}{n}}.$$ (p. 489)

**Confidence interval**   A level C confidence interval for a model parameter is an interval of values usually of the form

*Estimate* ± *Margin of Error*

found from data in such a way that C% of all random samples will yield intervals that capture the true parameter value. (p. 491)

**One-proportion z-interval**   A confidence interval for the true value of a proportion. The confidence interval is

$$\hat{p} \pm z^{*}SE(\hat{p}),$$

where $z^*$ is a critical value from the Standard Normal model corresponding to the specified confidence level. (p. 491)

**Margin of error**   In a confidence interval, the extent of the interval on either side of the observed statistic value is called the margin of error. A margin of error is typically the product of a critical value from the sampling distribution and a standard error from the data. A small margin of error corresponds to a confidence interval that pins down the parameter precisely. A large margin of error corresponds to a confidence interval that gives relatively little information about the estimated parameter. For a proportion,

$$ME = z^{*}\sqrt{\dfrac{\hat{p}\hat{q}}{n}}.$$ (p. 493)

**Critical value**   The number of standard errors to move away from the sample statistic to specify an interval that corresponds to the specified level of confidence. The critical value, denoted $z^*$, is usually found from a table or with technology. (p. 494)

## ON THE COMPUTER

### Confidence Intervals for Proportions

Confidence intervals for proportions are so easy and natural that many statistics packages don't offer special commands for them. Most statistics programs want the "raw data" for computations. For proportions, the raw data are the "success" and "failure" status for each case. Usually, these are given as 1 or 0, but they might be category names like "yes" and "no." Other software and graphing calculators allow you to create confidence intervals from summaries of the data—all you need to enter are the number of successes and the sample size.

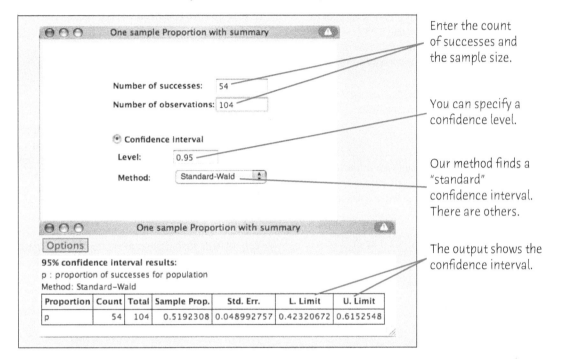

## EXERCISES

1. **Margin of error** A TV newscaster reports the results of a poll of voters, and then says, "The margin of error is plus or minus 4%." Explain carefully what that means.

2. **Margin of error** Medical researchers estimate the percentage of children exposed to lead-based paint, adding that they believe their estimate has a margin of error of about 3%. Explain what the margin of error means.

3. **Conditions** For each situation described below, identify the population and the sample, explain what $p$ and $\hat{p}$ represent, and tell whether the methods of this chapter can be used to create a confidence interval.

   a) Police set up an auto checkpoint at which drivers are stopped and their cars inspected for safety problems. They find that 14 of the 134 cars stopped have at least one safety violation. They want to estimate the percentage of all cars that may be unsafe.

   b) A TV talk show asks viewers to register their opinions on prayer in schools by logging on to a website. Of the 602 people who voted, 488 favored prayer in schools. We want to estimate the level of support among the general public.

   c) A school is considering requiring students to wear uniforms. The PTA surveys parent opinion by sending a questionnaire home with all 1245 students; 380 surveys are returned, with 228 families in favor of the change.

   d) A college admits 1632 first-year students one year, and four years later 1388 of them graduate on time. The college wants to estimate the percentage of all their first-year enrollees who graduate on time.

4. **More conditions** Consider each situation described. Identify the population and the sample, explain what $p$ and $\hat{p}$ represent, and tell whether the methods of this chapter can be used to create a confidence interval.

   a) A consumer group hoping to assess customer experiences with auto dealers surveys 167 people who recently bought new cars; 3% of them expressed dissatisfaction with the salesperson.

b) What percent of college students have cell phones? 2883 students were asked as they entered a football stadium, and 243 said they had phones with them.
c) 240 potato plants in a field in Maine are randomly checked, and only 7 show signs of blight. How severe is the blight problem for the U.S. potato industry?
d) 12 of the 309 employees of a small company suffered an injury on the job last year. What can the company expect in future years?

**5. Conclusions** A catalog sales company promises to deliver orders placed on the Internet within 3 days. Follow-up calls to a few randomly selected customers show that a 95% confidence interval for the proportion of all orders that arrive on time is 88% ± 6%. What does this mean? Are these conclusions correct? Explain.

a) Between 82% and 94% of all orders arrive on time.
b) 95% of all random samples of customers will show that 88% of orders arrive on time.
c) 95% of all random samples of customers will show that 82% to 94% of orders arrive on time.
d) We are 95% sure that between 82% and 94% of the orders placed by the sampled customers arrived on time.
e) On 95% of the days, between 82% and 94% of the orders will arrive on time.

**6. More conclusions** In January 2002, two students made worldwide headlines by spinning a Belgian euro 250 times and getting 140 heads—that's 56%. That makes the 90% confidence interval (51%, 61%). What does this mean? Are these conclusions correct? Explain.

a) Between 51% and 61% of all euros are unfair.
b) We are 90% sure that in this experiment this euro landed heads on between 51% and 61% of the spins.
c) We are 90% sure that spun euros will land heads between 51% and 61% of the time.
d) If you spin a euro many times, you can be 90% sure of getting between 51% and 61% heads.
e) 90% of all spun euros will land heads between 51% and 61% of the time.

**7. Confidence intervals** Several factors are involved in the creation of a confidence interval. Among them are the sample size, the level of confidence, and the margin of error. Which statements are true?

a) For a given sample size, higher confidence means a smaller margin of error.
b) For a specified confidence level, larger samples provide smaller margins of error.
c) For a fixed margin of error, larger samples provide greater confidence.
d) For a given confidence level, halving the margin of error requires a sample twice as large.

**8. Confidence intervals, again** Several factors are involved in the creation of a confidence interval. Among them are the sample size, the level of confidence, and the margin of error. Which statements are true?

a) For a given sample size, reducing the margin of error will mean lower confidence.
b) For a certain confidence level, you can get a smaller margin of error by selecting a bigger sample.
c) For a fixed margin of error, smaller samples will mean lower confidence.
d) For a given confidence level, a sample 9 times as large will make a margin of error one third as big.

**9. Living online** Pew Research, in 2018, polled a random sample of 1058 U.S. teens (ages 13–17) about Internet use. 45% reported that they were online "almost constantly"—a fact of great interest to advertisers. (https://www.pewresearch.org/internet/2018/05/31/teens-social-media-technology-2018/)

a) Explain the meaning of $\hat{p} = 0.45$ in the context of this situation.
b) Calculate the standard error of $\hat{p}$.
c) Explain what this standard error means in the context of this situation.

**10. How's life?** Gallup regularly conducts a poll using a "Cantril scale," which asks respondents to imagine a ladder with 10 rungs. Rung 0 represents the worst possible life, and rung 10 represents the best possible life. Respondents are asked what rung they would say they are on. Responses are classified as "Thriving" (standing on rung 7 or higher, and expecting to be on rung 8 or higher five years from now), "Suffering" (standing on rung 4 or lower and expecting to be on rung 4 or lower five years from now), or "Struggling" (not thriving or suffering). In May of 2020, Gallup found that the index had reached the lowest level since the Great Recession of 2008, at 48.8% thriving, with a sample size of 20,006. (https://news.gallup.com/poll/308276/life-ratings-plummet-year-low.aspx)

a) Explain the meaning of $\hat{p} = 0.488$ in the context of this situation.
b) Calculate the standard error of $\hat{p}$.
c) Explain what this standard error means in the context of this situation.

**11. Still living online** The 95% confidence interval for the proportion of teens in Exercise 9 who reported that they went online several times daily is from 42% to 48%.

a) Interpret the interval in this context.
b) Explain the meaning of "95% confident" in this context.

**12. Family variety** In a 2019 Pew Research study on trends in marriage and family, 30% of randomly surveyed adults say that the growing variety of family arrangements is a "good thing." The 95% confidence interval is from 29.1% to 30.9% ($n = 9942$).

a) Interpret the interval in this context.
b) Explain the meaning of "95% confident" in this context.

**13. Wrong direction** A Morning Consult/Politico poll of 1997 registered voters in July 2020 asked a standard polling question of whether the United States was headed in the right direction or was on the wrong track. 75% said that things are on the wrong track vs. 25% who said right direction.

a) Calculate the margin of error for the proportion of all U.S. adults who think things are on the wrong track for 90% confidence.
b) Explain in a simple sentence what your margin of error means.

**14. California Exodus?** A poll conducted by the UC Berkeley institute of Governmental Studies in 2019 found that 52% of 4527 respondents said they considered moving out of the state.

a) Compute a 90% confidence interval for the proportion of all Californians who considered moving out of state.
b) Interpret your interval in context.
c) Since high cost of housing has often been cited as a reason why residents move out, suppose the state government will consider investing in new affordable housing initiatives if it can be reasonably sure that more than 50% of Californians would consider moving. What should the state government conclude, based on the data?

**15. Wrong direction again** Consider the poll of Exercise 13.

a) Are the assumptions and conditions met?
b) Would the margin of error be larger or smaller for 95% confidence? Explain.

**16. More Exodus** in Exercise 14 we saw that 52% of surveyed Californians considered moving out of the state.

a) Are the conditions for constructing a confidence interval met?
b) Would the margin of error be larger or smaller for 95% confidence?

**17. Cars** What fraction of cars sold in the United States is made in Japan? The computer output below summarizes the results of a random sample of 50 autos sold in the United States. Explain carefully what it tells you.

z-Interval for proportion
With 90.00% confidence,
0.29938661 < p(Japan) < 0.46984416

**18. Parole** A study of 902 decisions made by the Nebraska Board of Parole produced the following computer output. Assuming these cases can be treated as a random sample of all cases that may come before the Board, what can you conclude?

z-Interval for proportion
With 95.00% confidence,
0.56100658 < p(parole) < 0.62524619

**19. Mislabeled seafood** In 2013 the environmental group Oceana (usa.oceana.org) analyzed 1215 samples of seafood purchased across the United States and genetically compared the pieces to standard gene fragments that can identify the species. Laboratory results indicated that 33% of the seafood was mislabeled according to U.S. Food and Drug Administration guidelines.

a) Construct a 95% confidence interval for the proportion of all seafood sold in the United States that is mislabeled or misidentified. Make sure to check the conditions for inference.
b) Explain what your confidence interval says about seafood sold in the United States.
c) A 2009 report by the Government Accountability Office says that the Food and Drug Administration has spent very little time recently looking for seafood fraud. Suppose an official said, "That's only 1215 packages out of the billions of pieces of seafood sold in a year. With the small number tested, I don't know that one would want to change one's buying habits." (An official was quoted similarly in a different but similar context.) Is this argument valid? Explain.

**20. Mislabeled seafood, second course** A Consumer Reports study similar to the study described in Exercise 19 found that 12 of the 22 "red snapper" packages tested were a different kind of fish.

a) Are the conditions for creating a confidence interval satisfied? Explain.
b) Construct a 95% confidence interval.
c) Explain what your confidence interval says about "red snapper" sold in these three states.

**21. Baseball fans** In a poll taken in December 2012, Gallup asked 1006 U.S. adults whether they were baseball fans; 48% said they were. Almost five years earlier, in February 2008, only 35% of a similar-size sample had reported being baseball fans.

a) Find the margin of error for the 2012 poll if we want 90% confidence in our estimate of the percent of national adults who are baseball fans.
b) Explain what that margin of error means.
c) If we wanted to be 99% confident, would the margin of error be larger or smaller? Explain.
d) Find that margin of error.
e) In general, if all other aspects of the situation remain the same, will smaller margins of error produce greater or less confidence in the interval?

**22. Contributions, please** The Paralyzed Veterans of America is a philanthropic organization that relies on contributions. It sends free mailing labels and greeting cards to potential donors on its list and asks for a voluntary contribution. To test a new campaign, it recently sent letters to a random sample of 100,000 potential donors and received 4781 donations.

a) Give a 95% confidence interval for the true proportion of its entire mailing list who may donate. Make sure to check the conditions for inference.
b) A staff member thinks that the true rate is 5%. Given the confidence interval you found, do you find that percentage plausible?

**23. Take the offer** First USA, a major credit card company, is planning a new offer for its current cardholders. The offer will give double airline miles on purchases for the next 6 months if the cardholder goes online and registers for the offer. To test the effectiveness of the campaign, First USA recently sent out offers to a random sample of 50,000 cardholders. Of those, 1184 registered.

a) Give a 95% confidence interval for the true proportion of those cardholders who will register for the offer. Also, please check the conditions for inference.
b) If the acceptance rate is only 2% or less, the campaign won't be worth the expense. Given the confidence interval you found, what would you say?

**24. Still living online** The Pew Research poll described in Exercise 9 found that 45% of a sample of 1058 teens are online almost constantly. (Treat this as a simple random sample.)

a) Find the margin of error for this poll if we want 95% confidence in our estimate of the percent of American teens who go online several times a day.

b) Explain what that margin of error means.
c) If we only need to be 90% confident, will the margin of error be larger or smaller? Explain.
d) Find that margin of error.
e) In general, if all other aspects of the situation remain the same, would smaller samples produce smaller or larger margins of error?

**25. Teenage drivers** An American insurance company checks police records on 582 accidents and notes that teenagers were at the wheel in 91 of them. The data were collected randomly and from across the entire United States.

a) Create a 95% confidence interval for the percentage of all auto accidents that involve teenage drivers. Make sure to check the conditions for inference.
b) Explain what your interval means.
c) Explain what "95% confidence" means.
d) A politician urging tighter restrictions on drivers' licenses issued to teens says, "In one of every five auto accidents, a teenager is behind the wheel." Does your confidence interval refute this statement? Explain.

**26. Junk mail** Direct mail advertisers send solicitations (a.k.a. "junk mail") to thousands of potential customers in the hope that some will buy the company's product. The acceptance rate is usually quite low. Suppose a company wants to test the response to a new flyer, and sends it to 1000 people randomly selected from its mailing list of over 200,000 people. It gets orders from 123 of the recipients.

a) Create a 90% confidence interval for the percentage of people the company contacts who may buy something. Also, please check the conditions for inference.
b) Explain what this interval means.
c) Explain what "90% confidence" means.
d) The company must decide whether to now do a mass mailing. The mailing won't be cost-effective unless it produces at least a 5% return. What does your confidence interval suggest? Explain.

**27. Safe food** Some food retailers propose subjecting food to a low level of radiation in order to improve safety, but sale of such "irradiated" food is opposed by many people. Suppose a grocer wants to find out what their customers think. They have cashiers distribute surveys at checkout and ask customers to fill them out and drop them in a box near the front door. They get responses from 122 customers, of whom 78 oppose the radiation treatments. What can the grocer conclude about the opinions of all their customers? Consider carefully the conditions for inference.

**28. Local news** The mayor of a small city has suggested that the state locate a new prison there, arguing that the construction project and resulting jobs will be good for the local economy. A total of 183 residents show up for a public hearing on the proposal, and a show of hands finds only 31 in favor of the prison project. What can the city council conclude about public support for the mayor's initiative? Consider carefully the conditions for inference.

**29. Death penalty, again** In the chapter, you read about one survey on the death penalty. In a different survey, the Gallup Poll actually split the sample at random, asking 510 respondents the question, "Generally speaking, do you believe the death penalty is applied fairly or unfairly in this country today?" The other 510 were asked "Generally speaking, do you believe the death penalty is applied unfairly or fairly in this country today?" Seems like the same question, but sometimes the order of the choices matters. Asked the first question, 58% said the death penalty was fairly applied; only 54% said so with the second wording.

a) What kind of bias may be present here?
b) If we combine the two groups of respondents, considering the overall group to be one larger random sample of 1020 respondents, what is a 95% confidence interval for the proportion of the general public that thinks the death penalty is being fairly applied?
c) How does the margin of error based on this pooled sample compare with the margins of error from the separate groups? Why?

**30. Gambling** A city ballot includes a local initiative that would legalize gambling. The issue is hotly contested, and two groups decide to conduct polls to predict the outcome. The local newspaper finds that 53% of 1200 randomly selected voters plan to vote "yes," while a college Statistics class finds 54% of 450 randomly selected voters in support. Both groups will create 95% confidence intervals.

a) Without finding the confidence intervals, explain which one will have the larger margin of error.
b) Find both confidence intervals.
c) Which group concludes that the outcome is too close to call? Why?

**31. Rickets** Vitamin D, whether ingested as a dietary supplement or produced naturally when sunlight falls on the skin, is essential for strong, healthy bones. The bone disease rickets was largely eliminated in England during the 1950s, but now there is concern that a generation of children more likely to watch TV or play computer games than spend time outdoors is at increased risk. A recent study of 2700 children randomly selected from all parts of England found 20% of them deficient in vitamin D.

a) Find a 98% confidence interval.
b) Explain carefully what your interval means.
c) Explain what "98% confidence" means.

**32. Speech transcriptions** In a study comparing speech-to-text transcription services, researchers estimated the accuracy rate of YouTube's auto-transcription service to be 72% (28% of words were incorrectly transcribed), with a 95% confidence interval of (67.8%, 75.5%).

a) Interpret this interval in context.
b) "Explain what "95% confidence" means.
c) Do these data refute the claim that the transcription service is 75% accurate? Explain.

**33. Higher ed post-COVID** In June 2020, a survey of 97 college presidents by *Inside Higher Ed* and Hanover Research showed that 68% were concerned about faculty readiness to conduct online learning or hybrid learning.

a) Assuming this was a representative sample of college presidents, compute a 95% confidence interval for the proportion of all college presidents who may be concerned about faculty readiness to conduct online learning or hybrid learning. Make sure to check the conditions for inference.

b) Interpret your interval in context.
c) Do the survey results support the idea that more than 60% of all college presidents were concerned about faculty readiness to conduct online learning or hybrid learning?

34. **Back to campus** ACT, Inc., reported that 74% of 1644 randomly selected college first-year students returned to college the next year. The study was stratified by type of college—public or private. The retention rates were 71.9% among 505 students enrolled in public colleges and 74.9% among 1139 students enrolled in private colleges.
   a) Will the 95% confidence interval for the true national retention rate in private colleges be wider or narrower than the 95% confidence interval for the retention rate in public colleges? Explain.
   b) Do you expect the margin of error for the overall retention rate to be larger or smaller? Explain.

35. **Deer ticks** Wildlife biologists inspect 153 deer taken by hunters and find 32 of them carrying ticks that test positive for Lyme disease.
   a) Create a 90% confidence interval for the percentage of deer that may carry such ticks.
   b) If the scientists want to cut the margin of error in half, how many deer must they inspect?
   c) What concerns do you have about this sample?

36. **Back to campus again** Suppose ACT, Inc., wants to update its information from Exercise 34 on the percentage of first-year students that return for a second year of college.
   a) It wants to cut the stated margin of error in half. How many college freshmen must be surveyed?
   b) Do you have any concerns about this sample? Explain.

37. **Graduation** It's believed that as many as 25% of adults over 50 never graduated from high school. We wish to see if this percentage is the same among the 25-year-old to 30-year-old age group.
   a) How many of this younger age group must we survey in order to estimate the proportion of non-grads to within 6% with 90% confidence?
   b) Suppose we want to cut the margin of error to 4%. What's the necessary sample size?
   c) What sample size would produce a margin of error of 3%?

38. **Hiring** In preparing a report on the economy, we need to estimate the percentage of businesses that plan to hire additional employees in the next 60 days.
   a) How many randomly selected employers must we contact in order to create an estimate in which we are 98% confident with a margin of error of 5%?
   b) Suppose we want to reduce the margin of error to 3%. What sample size will suffice?
   c) Why might it not be worth the effort to try to get an interval with a margin of error of only 1%?

39. **Graduation, again** As in Exercise 37, we hope to estimate the percentage of adults aged 25 to 30 who never graduated from high school. What sample size would allow us to increase our confidence level to 95% while reducing the margin of error to only 2%?

40. **Better hiring info** Editors of the business report in Exercise 38 are willing to accept a margin of error of 4% but want 99% confidence. How many randomly selected employers will they need to contact?

41. **Pilot study** A state's environmental agency worries that many cars may be violating clean air emissions standards. The agency hopes to check a sample of vehicles in order to estimate that percentage with a margin of error of 3% and 90% confidence. To gauge the size of the problem, the agency first randomly picks 60 cars and finds 9 with faulty emissions systems. How many should be sampled for a full investigation?

42. **Another pilot study** During routine screening, a doctor notices that 22% of her adult patients show higher than normal levels of glucose in their blood—a possible warning signal for diabetes. Hearing this, some medical researchers decide to conduct a large-scale study, hoping to estimate the proportion to within 4% with 98% confidence. How many randomly selected adults must they test?

43. **Approval rating** A newspaper reports that the governor's approval rating stands at 65%. The article adds that the poll is based on a random sample of 972 adults and has a margin of error of 2.5%. What level of confidence did the pollsters use?

44. **Amendment** A TV news reporter says that a proposed constitutional amendment is likely to win approval in the upcoming election because a poll of 1505 likely voters indicated that 52% would vote in favor. The reporter goes on to say that the margin of error for this poll was 3%.
   a) Explain why the poll is actually inconclusive.
   b) What confidence level did the pollsters use?

45. **Trading cards** Joe wanted to estimate the proportion of his Magic: The Gathering card collection that were "land cards," but he didn't want to sort through his thousands of cards to count exactly. Instead, he decided to use what he'd learned in AP Statistics and he randomly sampled 15 cards and saw that 5 of them were land cards, then used his calculator to compute a 95% confidence interval for $p$, the proportion of all of his cards that are land cards.
   a) Why should Joe not have done that?
   b) Joe did it anyway. What interval did he come up with?
   c) Is there anything about this interval that should make it obvious to Joe that he did something wrong?
   d) If Joe had found that 2 of his 15 cards were land cards, what would have been his interval? Would anything about that interval have tipped him off that he'd done something wrong?
   e) Clearly, Joe has missed a few things from his AP Statistics class. What is the consequence of his error on his interval estimate?

46. **More cards** Joe, from Exercise 45, explained his estimation process to his friend Maya, who pays more attention in Statistics class. Maya tried to explain the problem with Joe's process, but in the end had to do a simulation to demonstrate it. She said, "Let's say 40% of your collection are land cards. You randomly selected 15 cards, and created a confidence interval for $p$. Let's see what happens if we do that 100 times when $p = 0.4$."

The plot below shows these 100 intervals. Those with a blue dot in the center captured the value of $p$, and those with a red dot did not. Using this plot as evidence, summarize Maya's criticism of Joe's process.

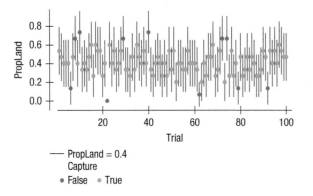

## JUST CHECKING

### Answers

1. While true, we know that in the sample 17% said "yes"; there's no need for a margin of error.
2. No. We are 95% confident that the percentage falls in some interval, not exactly on a particular value.
3. Yes. That's what the confidence interval means.
4. No. We don't know for sure that's true; we are only 95% confident.
5. No. That's our level of confidence, not the proportion of people receiving unsolicited text messages. The sample suggests the proportion is much lower.
6. Wider.
7. Lower.
8. Smaller.

# 19

# Testing Hypotheses About Proportions

Ingots are huge pieces of metal, sometimes weighing more than 20,000 pounds, made in a giant mold. They must be cast in one large piece for use in fabricating large structural parts for cars and planes. As the liquid metal cools, cracks can develop on the surface of the ingot, which can propagate into the zone required for the part, compromising its integrity. Airplane manufacturers insist that metal for their planes be defect-free, so the ingot must be made over if any cracking is detected.

Even though the metal from the cracked ingot is recycled, the cost runs into the tens of thousands of dollars to recast an ingot, not to mention the energy waste. About two-thirds of all aluminum ingots produced in the United States use a process called the "direct chill" method designed to reduce recasting. Metal manufacturers would like to avoid cracking if at all possible. But the casting process is complicated and not everything is completely under control. It's estimated that about 5% of aluminum ingots need to be recast because of cracking. That rate depends on the size of the ingot cast. In one plant that specializes in very large (over 30,000 lb) ingots designed for the airplane industry, about 20% of the ingots have had some kind of surface crack. In an attempt to reduce the cracking proportion, the plant engineers and chemists recently tried out some changes in the casting process. Since then, 400 ingots have been cast and only 68 (17%) of them have cracked. Has the new method worked? Has the cracking rate really decreased, or was 17% just due to luck? Is the 17% cracking rate merely a result of natural sampling variability, or is this enough evidence to justify a change to the new method?

People want to make informed decisions like this all the time. Does the new website design increase our click-through rate? Has the "click-it or ticket" campaign increased compliance with seat belt laws? Did the Super Bowl ad we bought actually increase sales? To answer such questions so that we can make informed decisions, we test *hypotheses*.

# Hypotheses

> **HYPOTHESIS**
> n.; pl. {Hypotheses}. A supposition; a proposition or principle which is supposed or taken for granted, in order to draw a conclusion or inference for proof of the point in question; something not proved, but assumed for the purpose of argument.
> —Webster's Unabridged Dictionary, 1913

If the changes made lowered the cracking rate from 20%, management will need to decide whether the costs of the new method warrant the changes. Managers are naturally cautious, and because humans are natural skeptics, they assume the new method makes no difference—but they hope the data can convince them otherwise. The starting hypothesis to be tested is called the **null hypothesis**—null because it assumes that nothing has changed. We denote it $H_0$. It specifies a parameter—here the proportion of cracked ingots—and a value—that the cracking rate is 20%. We usually write this in the form $H_0$: *parameter = hypothesized value*. So, for the ingots we would write $H_0: p = 0.20$.

The **alternative hypothesis**, which we denote $H_A$, is not a single value, but contains all the other values of the parameter that we'd consider plausible if we reject the null hypothesis. We can write $H_A: p \neq 0.20$.

What would convince you that the cracking rate had actually changed? If the rate dropped from 20% to 19.8%, would that convince you? After all, observed proportions do vary, so even if the changes had no effect, you'd expect some difference. What if it dropped to 5%? Would random fluctuation account for a change that big? That's the crucial question in a hypothesis test. As usual in statistics, when thinking about the size of a change, we *naturally* think of using its standard deviation to measure that change. We ask how many standard deviations the observed value is from the hypothesized value, and we know how to find the standard deviation of a proportion:

$$SD(\hat{p}) = \sqrt{\frac{pq}{n}} = \sqrt{\frac{(0.20)(0.80)}{400}} = 0.02.$$

> **NOTATION ALERT**
> Capital H is the standard letter for hypotheses. $H_0$ always labels the null hypothesis, and $H_A$ labels the alternative hypothesis.

### Why Is This a Standard Deviation and Not a Standard Error?

Remember that we reserve the term "standard error" for the *estimate* of the standard deviation of the sampling distribution of some statistic. But we're not estimating here—we have a value of $p$ from our null hypothesis model. To remind us that the parameter value comes from the null hypothesis, it is sometimes written as $p_0$ and the standard deviation as $SD(\hat{p}) = \sqrt{p_0 q_0 / n}$. That's different than when we found a confidence interval for $p$. In that case we couldn't assume that we knew its value, so we estimated the standard deviation from the sample value $\hat{p}$. (That was one of the reasons why, at the end of the last chapter, we used a simulation to verify that the one-proportion z-interval procedure "worked.")

If the changes have no effect, then the true cracking rate is still 0.20, and for samples of 400, the standard deviation of $\hat{p}$ is 0.02. We know from the sampling distribution of $\hat{p}$ that in 95% of samples of this size, the engineers will see a cracking rate within 0.04 of 0.20 just by chance. In other words, they expect to see between 64 and 96 cracked ingots (see Figure 19.1). The engineers saw 68 cracked ingots out of 400 (a rate of 0.17). Given an assumed rate of 0.20, is that a surprising number? Would you say that the cracking rate has changed?

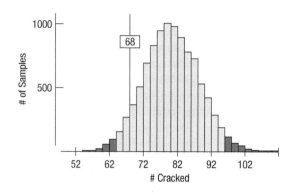

**Figure 19.1**
A simulation of 10,000 samples of 400 ingots with a cracking rate of 20% shows how we should expect the number of cracked ingots to vary.

Now that we have the Central Limit Theorem we don't need to rely on simulation. The CLT tells us that we can use the Normal model to find the probability instead. The engineers observed a cracking rate of 0.17—a difference of 0.03 from the standard (null hypothesis) value of 0.20. Using the fact that the standard deviation is 0.02, we can find the area in the two tails of the Normal model that lie more than 0.03 away from 0.20 (see Figure 19.2). That tells us how rare a rate like our observed one is.

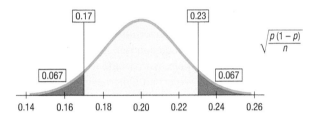

**Figure 19.2**
How likely is it to have a sample proportion less than 17% or more than 23% when the true proportion is 20%? Each red area is 6.7% of the total area, and they sum to 0.13. That's not *very* unlikely.

You know how to find that probability using either technology or Table Z at the back of the book. (See Chapter 5.) As you can see in Figure 19.2, the probability comes to about 0.13. In other words, a sample of 400 ingots with a cracking rate this far from 0.20 would happen about 13% of the time just by chance. That doesn't seem very unusual, so the observed proportion of 0.17, even though it's lower, doesn't provide convincing evidence that the new method changed the cracking rate. Or in other words, the difference between the observed cracking rate of 17% and the old cracking rate of 20% is not statistically significant.

### FOR EXAMPLE
### Framing Hypotheses

Summit Projects is a full-service interactive agency, based in Hood River, Oregon, that offers companies a variety of website services. One of Summit's clients, SmartWool®, produces and sells wool apparel, including the famous SmartWool socks. After Summit redesigned SmartWool's apparel website, analysts at SmartWool wondered whether traffic had changed since the new website went live. In particular, they wanted to know if the proportion of visits resulting in a sale had increased, decreased, or stayed pretty much the same since the new site went online. They also wondered if the average sale amount had changed.

QUESTION: If the old site's proportion was 15%, what are appropriate null and alternative hypotheses for the proportion?

ANSWER: Let $p$ = proportion of visits that result in a sale. Then

$$H_0: p = 0.15 \text{ vs. } H_A: p \neq 0.15.$$

## A Jury Trial as a Hypothesis Test

Managers would be really interested to learn that the engineers' changes changed the cracking rate. But to test it, they assumed that the rate had *not* changed. Does this reasoning seem backward? That could be because we usually prefer to think about getting things right rather than getting them wrong. But, we've seen this reasoning before in a different context. This is the logic of jury trials.

Let's suppose a defendant has been accused of robbery. In British common law and those systems derived from it (including U.S. law), the null hypothesis is that the defendant is innocent. Instructions to juries are quite explicit about this.

How is the null hypothesis tested? The prosecution first collects evidence. ("If the defendant were innocent, wouldn't it be remarkable that the police found him at the scene of the crime with a bag full of money in his hand, a mask on his face, and a getaway car parked outside?") For us, the data are the evidence.

> **BEYOND A REASONABLE DOUBT**
> We ask whether the data were unlikely beyond a reasonable doubt. We've just calculated that probability. The probability that the observed statistic value (or an even more extreme value) could occur if the null model were true—in this case, 0.13—is the P-value.

The next step is to judge the evidence. Evaluating the evidence is the responsibility of the jury in a trial, but it falls on your shoulders in hypothesis testing. The jury considers the evidence in light of the *presumption* of innocence and judges whether the evidence against the defendant would be plausible *if the defendant were in fact innocent*.

Like the jury, you ask, "Could these data plausibly have happened by chance if the null hypothesis were true?" If they are very unlikely to have occurred, then the evidence raises a reasonable doubt about the null hypothesis.

Ultimately, you must make a decision. The standard of "beyond a reasonable doubt" is wonderfully ambiguous because it leaves the jury to decide the degree to which the evidence contradicts the hypothesis of innocence. Juries don't explicitly use probability to help them decide whether to reject that hypothesis. But when you ask the same question of your null hypothesis, you have the advantage of being able to quantify exactly how surprising the evidence would be if the null hypothesis were true.

How unlikely is unlikely? Some people set rigid standards, like 1 time out of 20 or 1 time out of 100. But if *you* have to make the decision, you must judge for yourself in each situation whether the probability of observing your data is small enough to constitute "reasonable doubt."

## P-Values: Are We Surprised?

> **NOTATION ALERT**
> We have many P's to keep straight. We use an uppercase P for probabilities, as in $P(\mathbf{A})$, and for the special probability we care about in hypothesis testing, the P-value.
> We use lowercase $p$ to denote our model's underlying proportion parameter and $\hat{p}$ to denote our observed proportion statistic.

To test a hypothesis we must answer the question "Would our data be surprising if the null hypothesis were true?" So we need *probability*—specifically, the probability of seeing data like these (or something even less likely) *given* that the null hypothesis actually is true. In the ingots example, this came to 0.13. This probability is the value on which we base our decision, so statisticians give it a special name: the **P-value**. Usually, you'll use a sampling distribution model or a simulation to find P-values. Either way, the computer will do the heavy lifting.

When a P-value is very low, there are only two possible explanations.[1] Either the null hypothesis is correct and we've just seen something remarkable, or the null hypothesis is wrong (and the reason for a low P-value is that the model was wrong). Now we have a choice. Should we decide that a rare event has happened to us, or should we trust that the data were not unusual and that our null model was wrong? We don't believe in rare events,[2] so a low enough P-value leads us to reject the null hypothesis. Even then, we have not proven the null hypothesis false, but we have evidence that suggests so. There is no hard-and-fast rule about how low the P-value has to be, a decision threshold that can depend on the consequences of our decision.

When the P-value is high, we haven't seen anything unlikely or surprising at all. The data are consistent with the model from the null hypothesis, and we have no reason to reject it. Does that mean we've proved it? No. Many other models could be consistent with the data we've seen, so *we haven't proven anything*. The most we can say is that the null model doesn't appear to be false. Formally, we "fail to reject" the null hypothesis. That's a pretty weak conclusion, but it's all we can do with a high P-value.

> *"If the People fail to satisfy their burden of proof, you must find the defendant not guilty."*
> —NY state jury instruction

### FOR EXAMPLE
#### Using P-Values to Make Decisions

**QUESTION:** The SmartWool analysts in our earlier example collected a representative sample of visits after the new website went online and found that the P-value for the test of whether the proportion of visits resulting in a sale had changed was 0.0015 and the P-value for the test of whether the mean sale amount had changed was 0.3740. What conclusions should they have drawn?

---
[1] Actually there are three. All hypothesis tests are carried out under various assumptions, and if the assumptions are unreasonable then a true null hypothesis might easily lead to a small (and inaccurate) P-value.
[2] Or at least we think that they don't happen to us.

**ANSWER:** The P-value of 0.0015 is very low. That means that there is good evidence that the proportion of website visits resulting in a sale changed. However, the P-value of 0.3740 is quite high. That indicates that there is very little evidence that the mean sale amount changed.

## What to Do with an "Innocent" Defendant

Back to the jury trial. The jury assumes the defendant is innocent, but when the evidence is not strong enough to reject that hypothesis, they say, "Not guilty." They do not claim that the null hypothesis is true and say that the defendant is innocent. All they say is that they have not seen sufficient evidence to convict. The defendant may, in fact, be innocent, but the jury has no way to be sure.

In the same way, when the P-value is large, the most we can do is to "fail to reject" our null hypothesis. We never declare the null hypothesis to be true (or "accept" the null), because we simply do not know whether it's true or not. (But, unlike a jury trial, there is no "double jeopardy" in science. More data may become available in the future.)

> The null hypothesis is never proved or established, but is possibly disproved, in the course of experimentation. Every experiment may be said to exist only in order to give the facts a chance of disproving the null hypothesis.
> —Sir Ronald Fisher, *The Design of Experiments*

### Don't "Accept" the Null Hypothesis

Think about the null hypothesis $H_0$: All swans are white. Does collecting a sample of 100 white swans prove the null hypothesis? The data are *consistent* with this hypothesis and seem to lend support to it, but they don't *prove* it. In fact, all we can do is disprove the null hypothesis—for example, by finding just one non-white swan.

### JUST CHECKING

1. A research team wants to know if aspirin helps to thin blood. The null hypothesis says that it doesn't. They test 12 patients, observe the proportion with thinner blood, and get a P-value of 0.32. They proclaim that aspirin doesn't work. What would you say?

2. An allergy drug has been tested and found to give relief to 75% of the patients in a large clinical trial. Now the scientists want to see if the new, improved version works even better. What would the null hypothesis be?

3. The new drug is tested and the P-value is 0.0001. What would you conclude about the new drug?

# Alternative Alternatives

Tests on the ingot data can be viewed in two different ways. We know the old cracking rate is 20%, so the null hypothesis is

$$H_0: p = 0.20.$$

But we have a choice of alternative hypotheses. A metallurgist working for the company might be interested in *any* change in the cracking rate due to the new process. Even if the rate got worse, they might learn something useful from it. In that case, they'd be interested in possible changes on both sides of the null hypothesis. So they would write their alternative hypothesis as

$$H_A: p \neq 0.20.$$

An alternative hypothesis such as this is known as a **two-sided alternative** because we are equally interested in deviations on either side of the null hypothesis value. For two-sided alternatives, the P-value is the probability of deviating in *either* direction from the null hypothesis value.

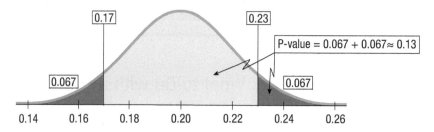

But, unlike the metallurgist, the management team wants to know only if the cracking rate has *decreased* to below 20%. Knowing how to *increase* the cracking rate probably doesn't interest them. To make that explicit, they could write their alternative hypothesis as

$$H_A: p < 0.20.$$

An alternative hypothesis that focuses on deviations from the null hypothesis value in only one direction is called a **one-sided alternative**.[3]

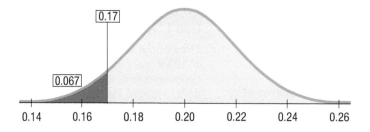

For a hypothesis test with a one-sided alternative, the P-value is the probability of deviating *only in the direction of the alternative* away from the null hypothesis value. For the same data, the one-sided P-value is half the two-sided P-value. So a one-sided test will reject the null hypothesis more easily, assuming that the one-sidedness was well-chosen—that a parameter deviation in the other direction is out of the question. Imposing this "out-of-the-question" belief on one of the two directions is a subjective choice. It may be a very reasonable choice. But it must be made (if it is made) before any data are collected, and for reasons that are defensible on their own.

In published research, opting for a one-sided test opens a researcher up to the criticism that they did so to cut their P-value in half, making their evidence against the null appear stronger. To avoid even the appearance of such impropriety, many researchers stick exclusively to two-sided alternative hypotheses.

# The Reasoning of Hypothesis Testing

Hypothesis tests follow a carefully structured path. To avoid getting lost as we navigate down it, we divide that path into four distinct sections.

### 1. Hypotheses

First, we state the hypotheses. The null hypothesis is usually the skeptical claim that nothing's different. Are we considering a (New! Improved!) possibly better method? The null hypothesis says, "Oh yeah? Convince me!" To convert a skeptic, we must pile up enough evidence against the null hypothesis that we can reasonably reject it.

---

[3] These are also called **two-** and **one-tailed alternatives**, because the probabilities we care about are found in the tails of the sampling distribution.

### How to Say It

You might think that the 0 in $H_0$ should be pronounced as "zero" or "O," but it's actually pronounced "naught" as in "all is for naught."

In statistical hypothesis testing, hypotheses are almost always about model parameters. To assess how unusual our data may be, we need a null model. The null hypothesis specifies a particular parameter value to use in our model. In the usual shorthand, we write $H_0$: *parameter = hypothesized value*. The alternative hypothesis, $H_A$, contains the values of the parameter we consider plausible when we reject the null.

> **FOR EXAMPLE**
> ### Writing Hypotheses
>
> A large city's Department of Motor Vehicles claimed that 80% of candidates pass driving tests, but a newspaper reporter's survey of 90 randomly selected local teens who had taken the test found only 61 who passed.
>
> QUESTION: Does this finding suggest that the passing rate for teenagers is lower than the DMV reported? Write appropriate hypotheses.
>
> ANSWER: I'll assume that the passing rate for teenagers is the same as the DMV's overall rate of 80%, unless there's strong evidence that it's lower.
>
> $$H_0: p = 0.80$$
> $$H_A: p < 0.80$$

## 2. Model

To plan a statistical hypothesis test, specify the *model* you will use to test the null hypothesis and the parameter of interest. Of course, all models require assumptions, so you will need to state them and check any corresponding conditions.

Your Model step should end with a statement such as

*Because the conditions are satisfied, I can model the sampling distribution of the sample proportion with a Normal model.*

Watch out, though. Your Model step could end with

*Because the conditions are not satisfied, I can't proceed with the test.*

If that's the case, stop and reconsider.

### When the Conditions Fail . . .

You might proceed with caution, explicitly stating your concerns. Or you may need to do the analysis with and without an outlier, or on different subgroups, or after re-expressing the response variable. Or you may not be able to proceed at all.

After you check the assumptions and conditions and specify the model, then name the test you will use. We'll see many tests in the chapters that follow. Some will be about more than one population or treatment, some will involve parameters other than proportions, and some will use models other than the Normal (and so will not use $z$-scores). The test about proportions is commonly called a **one-proportion $z$-test**.[4]

### One-Proportion $z$-Test

The conditions for the one-proportion $z$-test are almost the same as for the one-proportion $z$-interval. The only difference is that for the one-proportion $z$-interval, the Success/Failure Condition requires us to observe at least 10 actual successes and at least 10 actual failures in the data, while for the one-proportion $z$-test, it requires that the expected number of successes and failures—$np_0$ and $nq_0$, respectively—are both at least 10.

---

[4] It's also called a "one-sample test for a proportion."

We test the hypothesis $H_0: p = p_0$ using the statistic $z = \dfrac{(\hat{p} - p_0)}{SD(\hat{p})}$. We use the hypothesized proportion to find the standard deviation, $SD(\hat{p}) = \sqrt{\dfrac{p_0 q_0}{n}}$. When the conditions are met and the null hypothesis is true, this statistic follows the standard Normal model, so we can use that model to obtain a P-value.

### FOR EXAMPLE
#### Checking the Conditions

**RECAP:** A large city's DMV claimed that 80% of candidates pass driving tests. A reporter has results from a survey of 90 randomly selected local teens who had taken the test.

**QUESTION:** Are the conditions for inference satisfied?

✓ **Randomization Condition:** The 90 teens surveyed were a random sample of local teenage driving candidates.

✓ **10% Condition:** 90 is fewer than 10% of the teenagers who take driving tests in a large city.

✓ **Success/Failure Condition:** We expect $np_0 = 90(0.80) = 72$ successes and $nq_0 = 90(0.20) = 18$ failures. Both are at least 10.

**ANSWER:** The conditions are satisfied, so it's okay to use a Normal model and perform a one-proportion z-test.

## 3. Mechanics

> **Conditional Probability**
>
> Did you notice that a P-value is a conditional probability? It's the probability that the observed results could have happened *if (or given that) the null hypothesis were true.*

Under "Mechanics," we place the actual calculation of our test statistic from the data. Different tests we encounter will have different formulas and different test statistics. Usually, the mechanics are handled by a statistics program or calculator, but it's good to have the formulas recorded for reference and to know what's being computed. The ultimate goal of the calculation is to find out how surprising data like ours would be if the null hypothesis were true. We measure this by the P-value—the probability that the observed statistic value (or an even more extreme value) occurs if the null model is correct. If the P-value is small enough, we'll reject the null hypothesis. To justify our decision, we show our work, give the value of the test statistic (for now, that's the z-score), and report the P-value.

### FOR EXAMPLE
#### Finding a P-Value

**RECAP:** A large city's DMV claimed that 80% of candidates pass driving tests, but a survey of 90 randomly selected local teens who had taken the test found only 61 who passed.

**QUESTION:** What's the P-value for the one-proportion z-test?

**ANSWER:** I have $n = 90$, $x = 61$, and a hypothesized $p = 0.80$.

$$\hat{p} = \frac{61}{90} \approx 0.678$$

$$SD(\hat{p}) = \sqrt{\frac{p_0 q_0}{n}} = \sqrt{\frac{(0.8)(0.2)}{90}} \approx 0.042$$

$$z = \frac{\hat{p} - p_0}{SD(\hat{p})} = \frac{0.678 - 0.800}{0.042} \approx -2.90$$

$$\text{P-value} = P(z < -2.90) = 0.002$$

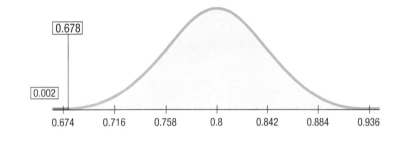

## 4. Conclusion in Context

The conclusion in a hypothesis test always links the P-value to a statement about the null hypothesis. The conclusion must state why the P-value leads us to either reject or fail to reject the null hypothesis. And, as always, the conclusion should be stated in context.

Your conclusion about the null hypothesis should never be the end of a testing procedure. Often, there are actions to take or policies to change. In our ingot example, management must decide whether to continue the changes proposed by the engineers. The decision always includes the practical consideration of whether the new method is worth the cost. Suppose management decides to reject the null hypothesis of 20% cracking in favor of the alternative that the percentage has been reduced. They must still evaluate how much the cracking rate has been reduced and how much it cost to accomplish the reduction. The *size of the effect* is always a concern when we test hypotheses. A good way to look at the **effect size** is to examine a confidence interval.

> . . . They make things admirably plain,
> But one hard question will remain:
> If one hypothesis you lose,
> Another in its place you choose . . .
>
> —James Russell Lowell,
> *Credidimus Jovem Regnare*

### FOR EXAMPLE
### Stating the Conclusion

**RECAP:** A large city's DMV claimed that 80% of candidates pass driving tests. Data from a reporter's survey of randomly selected local teens who had taken the test produced a P-value of 0.002.

**QUESTION:** What can the reporter conclude? And how might the reporter explain what the P-value means for the newspaper story?

**ANSWER:** Because the P-value of 0.002 is very small, I reject the null hypothesis. These survey data provide strong evidence that the passing rate for teenagers taking the driving test is lower than 80%.

If the passing rate for teenage driving candidates were actually 80%, we'd expect to see success rates this low in only about 1 in 500 (0.2%) samples of this size. This seems quite unlikely, casting doubt that the DMV's stated success rate applies to teens.

### How Much Does It Cost?

Formal tests of a null hypothesis base the decision of whether to reject the null hypothesis solely on the size of the P-value. But in real life, we want to evaluate the costs of our decisions as well. How much would you be willing to pay for a faster computer? Shouldn't your decision depend on how much faster? And on how much more it costs? Costs are not just monetary either. Would you use the same standard of proof for testing the safety of an airplane as for the speed of your new computer?

## STEP-BY-STEP EXAMPLE

### Testing a Hypothesis

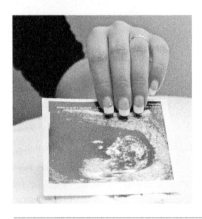

Advances in medical care such as prenatal ultrasound examination now make it possible to determine a child's sex early in a pregnancy. There is a fear that in some cultures some parents may use this technology to select the sex of their children. A study from Punjab, India (E. E. Booth, M. Verma, and R. S. Beri, "Fetal Sex Determination in Infants in Punjab, India: Correlations and Implications," *BMJ* 309 [12 November 1994]: 1259–1261), reports that, in 1993, in one hospital, 56.9% of the 550 live births that year were boys. It's a medical fact that male babies are slightly more common than female babies. The study's authors report a baseline for this region of 51.7% male live births.

QUESTION: Is there evidence that the proportion of male births is different for this hospital?

---

**THINK**

**PLAN** State what we want to know.
Define the variables and discuss the W's.

**HYPOTHESES** The null hypothesis makes the claim of no difference from the baseline.

Before seeing the data, we were interested in any change in male births, so the alternative hypothesis is two-sided.

**MODEL** Think about the assumptions and check the appropriate conditions.

For testing proportions, the conditions are the same ones we had for making confidence intervals, except that we check the **Success/Failure Condition** with the *hypothesized* proportions rather than with the *observed* proportions.

Specify the sampling distribution model.
Name the test you plan to use.

I want to know whether the proportion of male births in this hospital is different from the established baseline of 51.7%. The data are the recorded sexes of the 550 live births from a hospital in Punjab, India, in 1993, collected for a study on fetal sex determination. The parameter of interest, $p$, is the proportion of male births:

$$H_0: p = 0.517$$
$$H_A: p \neq 0.517.$$

✓ **Independence Assumption:** There is no reason to think that the sex of one baby can affect the sex of other babies, so births can reasonably be assumed to be independent with regard to the sex of the child.

**Randomization Condition:** The 550 live births are not a random sample, so I must be cautious about any general conclusions. I hope that this is a representative year, and I think that the births at this hospital may be typical of this area of India (assuming the null hypothesis is true).

**10% Condition:** The births are not a random sample from a larger population, so the 10% Condition does not apply.

✓ **Success/Failure Condition:** Both $np_0 = 550(0.517) = 284.35$ and $nq_0 = 550(0.483) = 265.65$ are greater than 10; I expect the births of at least 10 boys and at least 10 girls, so the sample is large enough.

The conditions are satisfied, so I can use a Normal model and perform a **one-proportion z-test**.

**SHOW** — **MECHANICS** The null model gives us the mean, and (because we are working with proportions) the mean gives us the standard deviation.

The null model is a Normal distribution with a mean of 0.517 and a standard deviation of

$$SD(\hat{p}) = \sqrt{\frac{p_0 q_0}{n}} = \sqrt{\frac{(0.517)(1-0.517)}{550}}$$
$$= 0.0213.$$

The observed proportion, $\hat{p}$, is 0.569, so

$$z = \frac{\hat{p} - p_0}{SD(\hat{p})} = \frac{0.569 - 0.517}{0.0213} = 2.44.$$

The sample proportion lies 2.44 standard deviations above the mean.

We find the z-score for the observed proportion to find out how many standard deviations it is from the hypothesized proportion.

Make a picture. Sketch a Normal model centered at $p_0 = 0.517$. Shade the region to the right of the observed proportion, and because this is a two-tailed test, also shade the corresponding region in the other tail.

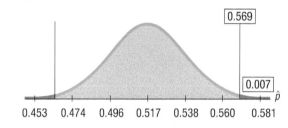

From the z-score, we can find the P-value, which tells us the probability of observing a value that extreme (or more). Because this is a two-tail test, the P-value is the probability of observing an outcome more than 2.44 standard deviations from the mean of a Normal model *in either direction*. We must therefore *double* the probability we find in the upper tail.

$P = 2P(z > 2.44) = 2(0.0073) = 0.0146$

**TELL** — **CONCLUSION** State your conclusion in context.

This P-value is roughly 1 time in 70. That's clearly significant, but don't jump to other conclusions. We can't be sure how this deviation came about. For instance, we don't know whether this hospital is typical, or whether the time period studied was selected at random. And we certainly can't conclude that ultrasound played any role.

The P-value of 0.0146 says that if the true proportion of male babies was 51.7%, then an observed proportion at least as different as 56.9% male babies would occur at random only about 15 times in 1000. With a P-value this small, I reject $H_0$. This is strong evidence that the proportion of boys is not equal to the baseline for the region. It appears that the proportion of boys may be higher.

## TI TIPS

### Testing a Hypothesis

By now probably nothing surprises you about your calculator. Of course it can help you with the mechanics of a hypothesis test. But that's not much. It cannot write the correct hypotheses, check the appropriate conditions, interpret the results, or state a conclusion. You still have to do the tough stuff!

Let's do the mechanics of the Step-By-Step Example about the post-ultrasound male birthrate. Based on historical evidence, we hypothesized that 51.7% of babies would be males, but one year at one hospital the rate was 56.9% among 550 births. Is that unusual?

- Go to the STAT TESTS menu. Scroll down and select 1-PropZTest.
- Specify the hypothesized proportion $p_0$.
- Enter x, the observed number of males. Because you don't know the actual count, enter 550*.569 there and then round off the resulting 312.95 to a whole number.
- Specify the sample size.
- Because this is a two-tailed test, indicate that you want to see if the observed proportion is significantly different ($\neq$) from what was hypothesized.
- Calculate the result.

Okay, the rest is up to you. The calculator reports a z-score of 2.445 and a P-value of 0.0145. Such a small P-value indicates that this higher rate of male births is unlikely to be just sampling error. Be careful how you state your conclusion. Remember: you need linkage, a decision, and context.

# P-Values and Decisions: What to Tell About a Hypothesis Test

Hypothesis tests are particularly useful when we must make a decision. (Is the defendant guilty or not? Should we choose print advertising or television?) The absolute nature of the hypothesis test decision, however, makes some people (including the authors) uneasy. Whenever possible, it's a good idea to report a confidence interval for the parameter of interest as well.

How small should the P-value be to reject the null hypothesis? A jury needs enough evidence to show the defendant guilty "beyond a reasonable doubt." How does that translate to P-values? The answer is that there is no good, universal answer. How small the P-value has to be to reject the null hypothesis is highly context-dependent. When we're screening for a disease and want to be sure we treat all those who are sick, we may be willing to reject the null hypothesis of no disease with a P-value as large as 0.10. That would mean that 10% of the healthy people would be considered as sick and subjected to further testing. We might rather treat (or recommend further testing for) the occasional healthy person than fail to treat someone who was really sick. But a long-standing hypothesis, believed by many to be true, needs stronger evidence (and a correspondingly small P-value) to reject it.

See if you require the same P-value to reject each of the following null hypotheses:

- A renowned musicologist claims that she can distinguish between the works of Mozart and Haydn simply by hearing a randomly selected 20 seconds of music from any work by either composer. What's the null hypothesis? If she's just guessing, she'll get 50% of the pieces correct, on average. So our null hypothesis is that $p$ equals 50%. If she's for real, she'll get more than 50% correct. Now, we present her with 10 pieces of Mozart or Haydn chosen at random. She gets 9 out of 10 correct. It turns out that the P-value associated with that result is 0.011. (In other words, if you tried to just guess, you'd get at least 9 out of 10 correct only about 1% of the time.) What would *you* conclude? Most people would probably reject the null hypothesis and be convinced that she has some ability to do as she claims. Why? Because the P-value is small and we don't have any particular reason to doubt the alternative.

- On the other hand, imagine a student who bets that he can make a flipped coin land the way he wants just by thinking hard. To test him, we flip a fair coin 10 times. Suppose he gets 9 out of 10 right. This also has a P-value of 0.011. Are you willing now to

### DON'T WE WANT TO REJECT THE NULL?

Often the folks who collect the data or perform the experiment hope to reject the null. (They hope the new drug is better than the placebo, or the new ad campaign is better than the old one.) But when we practice Statistics, we can't allow that hope to affect our decision. The essential attitude for a hypothesis tester is skepticism. Until we become convinced otherwise, we cling to the null's assertion that there's nothing unusual, no effect, no difference, etc. As in a jury trial, the burden of proof rests with the alternative hypothesis—innocent until proven guilty. When you test a hypothesis, you must act as judge and jury, but you are not the prosecutor.

> An extraordinary claim requires extraordinary proof. [5]
> 
> —Marcello Truzzi

reject this null hypothesis? Are you convinced that he's not just lucky? What amount of evidence *would* convince you? We require more evidence if rejecting the null hypothesis would contradict long-standing beliefs or other scientific results. Of course, with sufficient evidence we would revise our opinions (and scientific theories). That's how science makes progress.

Another factor in choosing a P-value is the importance of the issue being tested. Consider the following two tests:

- A researcher claims that the proportion of college students who hold part-time jobs now is higher than the proportion known to hold such jobs a decade ago. You might be willing to believe the claim (and reject the null hypothesis of no change) with a P-value of 0.05.
- An engineer claims that even though there were several problems with the rivets holding the wing on an airplane in their fleet, they've retested the proportion of faulty rivets and now the P-value is small enough to reject the null hypothesis that the proportion is the same and thus declare the plane safe. What P-value would be small enough to get you to fly on that plane? Probably something lower than 0.05, yes? This time the reason for requiring a very low P-value (i.e., requiring very strong evidence) isn't because the alternative hypothesis is inherently implausible; it's because the consequences of an erroneous decision could be catastrophic.

Your conclusion about any null hypothesis should always be linked to the P-value of the test and, ideally, accompanied by a confidence interval to report the size of the effect. Don't just declare the null hypothesis rejected or not rejected. Report the P-value to show the strength of the evidence against the hypothesis and a confidence interval to show the effect size. This will let each reader decide whether or not to reject the null hypothesis and whether or not to consider the result important.

When you reject a null hypothesis you conclude that the parameter value lies in the alternative. But the set of alternative parameter values is absurdly large—usually every possible value except the one stated in the null. A confidence interval is based on the observed data and provides a much more useful set of plausible values. In fact, it's likely that you didn't believe the null value anyway. (Is the coin *exactly* fair? Is $P(\text{head}) = 0.5000000\ldots$ and not $0.50000001$?) So what are hypothesis tests good for? Well, sometimes we need to make a decision. Setting an arbitrary threshold for the P-value provides a "bright line" decision rule. And the P-value provides useful information about how inconsistent the data are with the null hypothesis.

P-values have become controversial because some people base decisions solely on their P-values without regard to assumptions, conditions, effect size, or cost. The American Statistical Association recently published a statement about P-values.[6] It recommends six principles underlying the proper interpretation of P-values:

1. P-values can indicate how incompatible the data are with a specified statistical model.
2. P-values do not measure the probability that the studied hypothesis is true, or the probability that the data were produced by random chance alone.
3. Scientific conclusions and business or policy decisions should not be based only on whether a P-value passes a specific threshold.
4. Proper inference requires full reporting and transparency.
5. A P-value, or statistical significance, does not measure the size of an effect or the importance of a result.
6. By itself, a P-value does not provide a good measure of evidence regarding a model or hypothesis.

---

[5] This saying is often quoted by scientists without attributing it to Truzzi. But he appears to have published it first (in "On the Extraordinary: An Attempt at Clarification," *Zetetic Scholar* (1978): vol. 1, no. 1, p. 11).
[6] For more details, see the paper at www.amstat.org/asa/files/pdfs/P-ValueStatement.pdf

## JUST CHECKING

4. A bank is testing a new method for getting delinquent customers to pay their past-due credit card bills. The standard way was to send a letter asking the customer to pay. That worked 30% of the time. They want to test a new method that involves sending a text to customers encouraging them to contact the bank and set up a payment plan. They are concerned that this may offend some of their customers, so they would prefer not to send texts unless doing so would lead to a substantial increase in customers paying their past-due bills. What is the parameter of interest? What are the null and alternative hypotheses?

5. The bank sets up an experiment to test the effectiveness of the text. They send it to several thousands of randomly selected delinquent customers and keep track of how many actually do contact the bank to arrange payments. The bank's statistician calculates a P-value of 0.003. What does this P-value suggest about the text?

6. The statistician tells the bank's management that the results are clear and that they should switch to the text method. Do you agree? What else might you want to know?

## STEP-BY-STEP EXAMPLE

### Tests and Intervals

Anyone who plays or watches sports has heard of the "home field advantage." Tournaments in many sports are designed to try to neutralize the advantage of the home team or player. Most people believe that teams tend to win more often when they play at home. But do they?

If there were no home field advantage, the home teams would win about half of all games played. To test this, we'll use the games in the Major League Baseball 2017 season. That year, there were 2430 regular-season games. It turns out that the home team won 1312 of the 2430 games, or 53.99% of the time.

QUESTION: Could this deviation from 50% be explained just from natural sampling variability, or is it evidence to suggest that there really is a home field advantage, at least in professional baseball?

**THINK**   **PLAN** State what we want to know.

Define the variables and discuss the W's.

I want to know whether the home team in professional baseball is more likely to win. The data are all 2430 games from the 2017 Major League Baseball season. The variable is whether or not the home team won. The parameter of interest is the proportion of home team wins. If there's no advantage, I'd expect that proportion to be 0.50.

**HYPOTHESES** The null hypothesis makes the claim of no difference from the baseline. Here, that means no home field advantage.

$H_0: p = 0.50$

$H_A: p \neq 0.50$

**MODEL** Think about the assumptions and check the appropriate conditions. This is not a random sample. If we wanted to talk only about this season there would be no inference. So, we view the 2430 games here not as a random sample, but as a representative collection of games. Our inference is about all years of Major League Baseball.

✓ **Independence Assumption:** Generally, the outcome of one game has no effect on the outcome of another game. But this may not be strictly true. For example, if a key player is injured, the probability that the team will win in the next couple of games may decrease slightly, but independence is still roughly true. The data come from one entire season, but it may be reasonable to assume that other seasons would be similar with regard to whether or not a home field advantage exists.

| | |
|---|---|
| | I'm not just interested in 2017, and those games, while not randomly selected, should be a reasonable representative sample of all Major League Baseball games in the recent past and near future. |
| | ✓ **10% Condition**: We are interested in home field advantage for Major League Baseball for all seasons. While not a random sample, these 2430 games are fewer than 10% of all games played over the years. |
| | ✓ **Success/Failure Condition**: Both $np_0 = 2430(0.50) = 1215$ and $nq_0 = 2430(0.50) = 1215$ are at least 10. |
| Specify the sampling distribution model.<br>Name the test you plan to use. | Because the conditions are satisfied, I'll use a Normal model for the sampling distribution of the proportion and do a **one-proportion z-test**. |
| **SHOW**   **MECHANICS** The null model gives us the mean, and (because we are working with proportions) the mean gives us the standard deviation. | The null model is a Normal distribution with a mean of 0.50 and a standard deviation of $$SD(\hat{p}) = \sqrt{\frac{p_0 q_0}{n}} = \sqrt{\frac{(0.5)(1-0.5)}{2430}}$$ $$= 0.010143$$ The observed proportion, $\hat{p}$, is 0.53992. |
| Next, we find the z-score for the observed proportion, to find out how many standard deviations it is from the hypothesized proportion. | So the z-value is $$z = \frac{0.53992 - 0.5}{0.010143} = 3.935$$ |
| From the z-score, we can find the P-value, which tells us the probability of observing a value that extreme (or more). | The sample proportion lies 3.935 standard deviations above the mean. |
| The probability of observing a value 3.935 or more standard deviations away from the mean of a Normal model can be found to be less than 0.0001. | The corresponding P-value is less than 0.0001. |
| **TELL**   **CONCLUSION** State your conclusion about the parameter—in context, of course! | The P-value of 0.0001 says that if the true proportion of home team wins were 0.50, then an observed value of 0.53992 (or **more extreme**) would occur less than one time in 1000. With a P-value so small, I reject $H_0$. I have reasonable evidence that the true proportion of home team wins is **not** 50%. |

QUESTION: OK, but how big a difference are we talking about? Just knowing that there is an effect is only part of the answer. Let's find a confidence interval for the home field advantage.

| | |
|---|---|
|    **MODEL** Think about the assumptions and check the conditions. | ✓ **Success/Failure Condition**: There were 1312 home team wins and 1118 losses, both at least 10. |
| The conditions are identical to those for the hypothesis test, with one difference: Now we are not given a hypothesized proportion, $p_0$, so we must instead work with the observed results. | |
| Specify the sampling distribution model.<br>Tell what method you plan to use. | The conditions are satisfied, so I can model the sampling distribution of the proportion with a Normal model and find a **one-proportion z-interval**. |

> **SHOW** **MECHANICS** We can't find the sampling model standard deviation from the null model proportion. (In fact, we've just rejected it.) Instead, we find the standard error of $\hat{p}$ from the *observed* proportions. Other than that substitution, the calculation looks the same as for the hypothesis test.
>
> With this large a sample size, the difference between the standard error and the standard deviation is negligible, but in smaller samples, it could matter.

$$SE(\hat{p}) = \sqrt{\frac{\hat{p}\hat{q}}{n}} = \sqrt{\frac{(0.53992)(1 - 0.53992)}{2431}}$$
$$= 0.01011$$

The sampling model is Normal, so for a 95% confidence interval, the critical value $z^* = 1.96$.

The margin of error is

$$ME = z^* \times SE(\hat{p}) = 1.96 \times 0.01011 = 0.0198.$$

So the 95% confidence interval is

$$0.53992 \pm 0.0198 \text{ or } (0.5201, 0.5597).$$

> **TELL** **CONCLUSION** Confidence intervals help us think about the size of the effect. Here we can see that the home field advantage may affect enough games to make a real difference. (In that 2017 season, the Milwaukee Brewers were just one win short of making the playoffs.)

I am 95% confident that, in professional baseball, home teams win between 52.01% and 55.97% of the games.

In a season of 162 games, the low end of this interval, 52.01% of the 81 home games, would mean just over one and a half extra home victories, on average. The upper end, 55.79%, would mean almost 5 extra wins.

# Hypothesis Testing by Simulation

Some types of hypothesis test are so commonly used that they are identified by a name and are associated with a well-understood sampling distribution model. The one-proportion $z$-test is such an example; it uses the Normal model to approximate the sampling distribution of $\hat{p}$. And in the coming chapters of this text we'll look at several others among the most common hypothesis tests in the business.

But sometimes a real situation seems to invite a hypothesis test and there's no obvious sampling distribution model to use. The data may not even *be* a sample. In such a situation, it is sometimes possible to look at what the distribution of a test statistic would look like under the null hypothesis by using a simulation. The example we're about to share with you is one that has been discussed in Statistics and sports literature for many years. It is about the "hot hand" in basketball.

Suppose you're watching the Lakers play basketball and LeBron James has made all five of his most recent field goal shots. It's nearly halftime and one more field goal would put the Lakers ahead by one point. The crowd is aching for LeBron to get the ball—he's a great player, after all, but he also seems right now to have a *hot hand*—he hasn't missed any of his last five shots.

Is there really such a thing as a hot hand? More specifically in this situation, if LeBron James makes five shots in a row at some point during a game, is that evidence that he's "in the zone"? Let's think about this in terms of hypothesis testing. As is often the case, we hope to indirectly provide evidence for something (the "hot hand") by showing that the data (five consecutive successful field goal attempts) are unlikely to occur under a null model (no "hot hand"). So we could write the following:

$H_0$: There is no hot hand. Each attempted field goal shot by LeBron James has the same chance of succeeding throughout the game, regardless of how he did on recent attempts.

$H_A$: LeBron James has a hot hand. He is more likely to make shots now because he has made several recent ones and is "in the zone."

We have our hypotheses. Next we need a test statistic—some summary statistic that we expect to look different if there's no hot hand than if there is. The basketball story invites us to think about runs of consecutive successful field goal attempts. Is five in a row an unusually long sequence? If not, what would be unusual? Let's let our test statistic be the length of the *longest* run of successful field goal attempts by LeBron James in a game. It's a random variable; let's call it $L$ for "longest" (or "LeBron"). If $H_A$ is true, then we would expect $L$ to typically take higher values than if $H_0$ is true.

We have no reason to believe that the probability distribution of $L$ under $H_0$ should be Normal, or that it should follow any other particular probability distribution model. But we can simulate it! Under $H_0$, all of James's field goal attempts are independent of one another, so we can just simulate a game's worth of Bernoulli trials and count how many successes are in the longest run. We'll need to approximate the chance of success on each individual shot. Because we're trying to mimic LeBron's shooting field goal shots as realistically as we can, we'll use a recent field goal percentage: 51.3% in the 2020–2021 NBA season. And we'll need to approximate the number of field goal attempts he usually takes in a game. We'll assume it is 18, which is typical in recent years.

To summarize:

- We'll simulate 18 attempts with $p = 0.513$.
- We'll count how many successes are in the longest run and call it $L$.
- We'll repeat this many times to approximate the probability distribution of $L$ over many games.
- We'll estimate a P-value for our actual data by computing the relative frequency with which $L \geq 5$ in our simulation.
- We'll draw an appropriate conclusion in context based on our estimated P-value.

Although you could do this with a calculator or even a random digits table, it would be tedious and slow. We used a computer. Here's the distribution we got after simulating 10,000 basketball games:

As the graph makes clear, 5 in a row is not at all unusual; in fact, it's extremely common! We counted 3091 trials in our simulation in which $L \geq 5$, giving us an estimated P-value of $P(L \geq 5) = 0.309$. That is not low enough to indicate anything inconsistent between $H_0$ (no hot hand) and our data (five consecutive field goals). Looking at the graph, we can see that LeBron making even 6 or 7 field goals in a row would be unsurprising ... *even if there's no such thing as a hot hand.*

Of course, this doesn't prove that the hot hand doesn't exist—absence of evidence isn't the same as evidence of absence. But in fact, statisticians have examined data from many players over many games and many seasons—and they never saw anything that is particularly inconsistent with the null hypothesis of "no hot hand."[7]

The basketball audience's *perception* of the "hot hand" is likely part of a phenomenon we've mentioned a couple of times before in this text: humans have a tendency to discern patterns even when looking at nothing but randomness. Among the most fundamental reasons that hypothesis testing gained such widespread use in science in the 20th century was that it allowed us to step away from such human biases and quantify just how unusual something *really* is, instead of how unusual it *seems*.

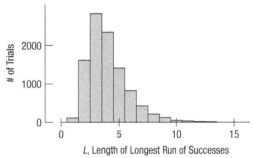

© 2013 Randall Munroe. Reprinted with permission. All rights reserved.

[7]For example, see Thomas Gilovich, Robert Vallone, and Amos Tversky, *The Hot Hand in Basketball: On the Misperception of Random Sequences* (Academic Press, 1985), p. 295. psych.cornell.edu

> ### PROBABILITY DISTRIBUTIONS THAT AREN'T SAMPLING DISTRIBUTIONS?
> It's true that most common hypothesis tests involve a null hypothesis that is a statement about a population parameter, that the test statistic is based on a random sample, and that the test statistic has a *sampling distribution* under the null hypothesis. But the null hypothesis can be a statement about how the data came about, even if the data do not represent a random sample. (Random *sampling* isn't the only way randomness can show up in the world.) For example, a null hypothesis could be that Charlie was just randomly guessing when he answered the five multiple choice questions on his last test, or that airlines' flight delays are completely unrelated to one another if they occur on different days. These are questions that can legitimately be addressed with hypothesis tests, even though the data we would use wouldn't be a random sample of anything.

## WHAT CAN GO WRONG?

Hypothesis tests are so widely used—and so widely misused—that we've devoted all of the next chapter to discussing the pitfalls involved, but there are a few issues that we can talk about already.

- **Don't base your null hypothesis on what you see in the data.** You are not allowed to look at the data first and then adjust your null hypothesis so that it will be rejected. When your sample value turns out to be $\hat{p} = 51.8\%$, with a standard deviation of 1%, don't form a null hypothesis like $H_0: p = 49.8\%$, knowing that you can reject it. You should always *Think* about the situation you are investigating and make your null hypothesis describe the "nothing interesting" or "nothing has changed" scenario. No peeking at the data!

- **Don't base your alternative hypothesis on the data, either.** Again, you need to *Think* about the situation. Are you interested only in knowing whether something has *increased*? Then write a one-sided (upper-tailed) alternative. Or would you be equally interested in a change in either direction? Then you want a two-sided alternative. You should decide whether to do a one- or two-sided test based on what results would be of interest to you, not what you see in the data.

- **Don't make your null hypothesis what you want to show to be true.** Remember, the null hypothesis is the status quo, the nothing-is-strange-here position a skeptic would take. You wonder whether the data cast doubt on that. You can reject the null hypothesis, but you can never "accept" or "prove" the null.

- **Don't forget to check the conditions.** The reasoning of inference depends on randomization. No amount of care in calculating a test result can recover from biased sampling. The probabilities we compute depend on the Independence Assumption. And the sample must be large enough to justify the use of a Normal model.

- **Don't accept the null hypothesis.** You may not have found enough evidence to reject it, but you surely have *not* proven it's true!

- **Don't say you proved the alternative hypothesis.** Strong evidence is not certainty.

- **If you fail to reject the null hypothesis, don't think that a bigger sample would be more likely to lead to rejection.** If the results you looked at were "almost" significant, it's enticing to think that because you would have rejected the null had these same observations come from a larger sample, then a larger sample would surely lead to rejection. Don't be misled. Remember, each sample is different, and a larger sample won't necessarily duplicate your current observations. Indeed, the Central Limit Theorem tells us that statistics will vary *less* in larger samples. We should therefore expect such results to be less extreme. Maybe they'd be statistically significant but maybe (perhaps even probably) not. Even if you fail to reject the null hypothesis, it's a good idea to examine a confidence interval. If none of the plausible parameter values in the interval would matter to you (for example, because none would be *practically* significant), then even a larger study with a correspondingly smaller standard error is unlikely to be worthwhile.

# WHAT HAVE WE LEARNED?

We've learned to use what we see in a random sample to test a hypothesis about the world. Hypothesis tests go hand in hand with confidence intervals.
- A hypothesis test makes a decision about the plausibility of a parameter value.
- A confidence interval estimates a set of plausible values for the parameter.

We've learned that testing a hypothesis requires four important steps:
- writing **hypotheses**;
- determining what **test** to use by checking appropriate assumptions and conditions;
- completing the **mechanics** of the test by finding a z-score and a P-value; and
- linking the P-value to a decision and stating our **conclusion** in the proper context.

We've learned to formulate appropriate hypotheses.
- The null hypothesis specifies the parameter of a model we'll test using our data. It has the form $H_0: p = p_0$.
- The alternative hypothesis states what we'll have evidence for if we reject the null. It can be one-sided or two-sided, depending on what we want to investigate.

We've learned to confirm that we can use a Normal model and to name the test we'll perform.
- We check the Independence Assumption, the Randomization Condition, the 10% Condition, and the Success/Failure Condition.
- If all of these check out, we use a Normal model to perform a one-proportion z-test.

We've learned to complete the mechanics of the test.
- Based on our assumption that the null hypothesis is true, we find the standard deviation of the sampling model for the sample proportion: $SD(\hat{p}) = \sqrt{\frac{p_0 q_0}{n}}$.
- We calculate the test statistic $z = \frac{\hat{p} - p_0}{SD(\hat{p})}$ and use a Normal model to find the P-value. The P-value is the probability of observing an outcome at least as extreme as ours if the null hypothesis is actually true.

We've learned to state an appropriate conclusion.
- If the P-value is large, then it's plausible that the results we've observed may be just sampling error. We'll fail to reject the null hypothesis and conclude there's not enough evidence to suggest that the null hypothesis is false.
- If the P-value is very small, then it's highly unlikely we'd observe results like ours if the null hypothesis were true. We'll reject the null hypothesis and conclude there's strong evidence to suggest that the null hypothesis is false.

## TERMS

**Null hypothesis**  The claim being assessed in a hypothesis test is called the null hypothesis. Usually, the null hypothesis is a statement of "no change from the traditional value," "no effect," "no difference," or "no relationship." For a claim to be a testable null hypothesis, it must specify a value for some population parameter that can form the basis for assuming a sampling distribution for a test statistic. (p. 512)

**Alternative hypothesis**  The alternative hypothesis proposes what we should conclude if we find the null hypothesis to be unlikely. (p. 512)

**Two-sided alternative (Two-tailed alternative)**  An alternative hypothesis is two-sided ($H_A: p \neq p_0$) when we are interested in deviations in *either* direction away from the hypothesized parameter value. (p. 515)

**One-sided alternative (One-tailed alternative)**  An alternative hypothesis is one-sided (e.g., $H_A: p > p_0$ or $H_A: p < p_0$) when we are interested in deviations in *only one* direction away from the hypothesized parameter value. (p. 516)

**P-value**  The conditional probability of observing a value for a test statistic at least as far from the hypothesized value as the statistic value actually observed if the null hypothesis is true. A small P-value indicates either that the observation is improbable or that the probability calculation was based on incorrect assumptions. The assumed truth of the null hypothesis is the assumption under suspicion. (p. 514)

**One-proportion z-test**  A test of the null hypothesis that the proportion of a single sample equals a specified value ($H_0: p = p_0$) by referring the statistic $z = \dfrac{\hat{p} - p_0}{SD(\hat{p})}$ to a Standard Normal model. (p. 517)

**Effect size**  The difference between the null hypothesis value and the actual value of population parameter. (p. 519)

## ON THE COMPUTER
### Hypothesis Tests for a Proportion

You can conduct a hypothesis test for a proportion using a graphing calculator or a statistics software package on a computer. Often all you need to do is enter information about the hypotheses, the observed number of successes, and the sample size. Some programs want the original data, in which success and failure may be coded as 1 and 0 or "yes" and "no." The technology will report the z-score and the P-value.

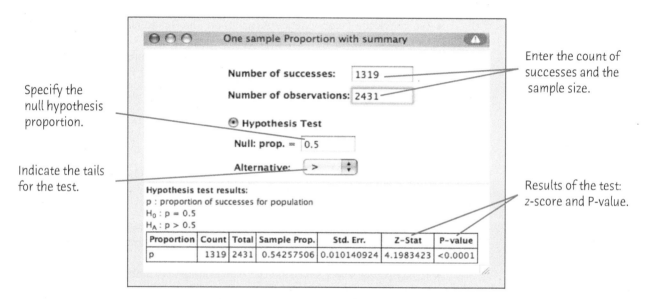

## EXERCISES

1. **Hypotheses** Write the null and alternative hypotheses you would use to test each of the following situations:
   a) A governor is concerned about their "negatives"—the percentage of state residents who express disapproval of their job performance. Their political committee pays for a series of TV ads, hoping that they can keep the negatives below 30%. They will use follow-up polling to assess the ads' effectiveness.
   b) Is a coin fair?
   c) Only about 20% of people who try to quit smoking succeed. Sellers of a motivational tape claim that listening to the recorded messages can help people quit.

2. **More hypotheses** Write the null and alternative hypotheses you would use to test each situation.
   a) In the 1950s only about 40% of high school graduates went on to college. Has the percentage changed?
   b) 20% of cars of a certain model have needed costly transmission work after being driven between 50,000 and 100,000 miles. The manufacturer hopes that a redesign of a transmission component has solved this problem.
   c) We field-test a new-flavor soft drink, planning to market it only if we are sure that over 60% of the people like the flavor.

**3. Negatives** After the political ad campaign described in Exercise 1a, pollsters check the governor's negatives. They test the hypothesis that the ads produced no change against the alternative that the negatives are now below 30% and find a P-value of 0.22. Which conclusion is appropriate? Explain.

a) There's a 22% chance that the ads worked.
b) There's a 78% chance that the ads worked.
c) There's a 22% chance that their poll is correct.
d) There's a 22% chance that natural sampling variation could produce poll results like these if there's really no change in public opinion.

**4. Dice** The seller of a loaded die claims that it will favor the outcome 6. We don't believe that claim, and roll the die 200 times to test an appropriate hypothesis. Our P-value turns out to be 0.03. Which conclusion is appropriate? Explain.

a) There's a 3% chance that the die is fair.
b) There's a 97% chance that the die is fair.
c) There's a 3% chance that a loaded die could randomly produce the results we observed, so it's reasonable to conclude that the die is fair.
d) There's a 3% chance that a fair die could randomly produce the results we observed, so it's reasonable to conclude that the die is loaded.

**5. Relief** A company's old antacid formula provided relief for 70% of the people who used it. The company tests a new formula to see if it is better and gets a P-value of 0.27. Is it reasonable to conclude that the new formula and the old one are equally effective? Explain.

**6. Cars** A survey investigating whether the proportion of today's high school seniors who own their own cars is higher than it was a decade ago finds a P-value of 0.017. Is it reasonable to conclude that more high-schoolers have cars? Explain.

**7. Coin tossing** A friend of yours claims that when he tosses a coin he can control the outcome. You are skeptical and want him to prove it. You tel him to toss it so that it will land on heads, and he does. Then you tell him to toss it so that it will land on tails, and he does.

a) Do two losses in a row convince you that he really can control the toss? Explain.
b) You tell him a third time how to make the coin land and again he succeeds. What's the probability of succeeding three tosses in a row if the outcome is really random?
c) Would three successes in a row convince you that your friend can control the coin? Explain.
d) How many times in a row would he have to succeed in order for you to be pretty sure that this friend really can control the toss? Justify your answer by calculating a probability and explaining what it means.

**8. Candy** Someone hands you a box of a dozen chocolate-covered candies, telling you that half are vanilla creams and the other half peanut butter. You pick candies at random and discover the first three you eat are all vanilla.

a) If there really were 6 vanilla and 6 peanut butter candies in the box, what is the probability that you would have picked three vanillas in a row?

b) Do you think there really might have been 6 of each? Explain.
c) Would you continue to believe that half are vanilla if the fourth one you try is also vanilla? Explain.

**9. Better than aspirin?** A very large study showed that aspirin reduced the rate of first heart attacks by 44%. A pharmaceutical company thinks it has a drug that will be more effective than aspirin, and plans to do a randomized clinical trial to test the new drug.

a) What is the null hypothesis the company will use?
b) What is the company's alternative hypothesis?

The company conducted the study and found that the group using the new drug had somewhat fewer heart attacks than those in the aspirin group.

c) The P-value from the hypothesis test was 0.28. What do you conclude?
d) What would you have concluded if the P-value had been 0.004?

**10. Psychic** A friend of yours claims to be psychic. You are skeptical. To test this you take a stack of 100 playing cards and have your friend try to identify the suit (hearts, diamonds, clubs, or spades), without looking, of course!

a) State the null hypothesis for your experiment.
b) State the alternative hypothesis.

You did the experiment and your friend correctly identified more than 25% of the cards.

c) A hypothesis test gave a P-value of 0.014. What do you conclude?
d) What would you conclude if the P-value had been 0.245?

**11. Smartphones** Many people have trouble setting up all the features of their smartphones, so a company has developed what it hopes will be easier instructions. The goal is to have at least 96% of customers succeed. The company tests the new system on 200 people, of whom 188 were successful. Is this strong evidence that the new system fails to meet the company's goal? A student's test of this hypothesis is shown. How many mistakes can you find?

$H_0: \hat{p} = 0.96$

$H_A: \hat{p} \neq 0.96$

SRS, $0.96(200) > 10$

$\frac{188}{200} = 0.94;\ SD(\hat{p}) = \sqrt{\frac{(0.94)(0.06)}{200}} = 0.017$

$z = \frac{0.96 - 0.94}{0.017} = 1.18$

$P = P(z > 1.18) = 0.12$

There is strong evidence the new instructions don't work.

**12. Obesity 2018** In 2018, the Centers for Disease Control and Prevention reported that 42.4% of adults in the United States are obese. A county health service planning a new awareness campaign polls a random sample of 750 adults living there. In this sample, 228 people were found to be obese based on their answers to a health questionnaire.

Do these responses provide strong evidence that the 42.4% figure is not accurate for this region? Correct the mistakes you find in a student's attempt to test an appropriate hypothesis.

$H_0: \hat{p} = 0.424$

$H_A: \hat{p} < 0.424$

SRS, $750 \geq 10$

$\frac{228}{750} = 0.304; \quad SD(\hat{p}) = \sqrt{\frac{(0.304)(0.696)}{750}} = 0.017$

$z = \frac{0.304 - 0.424}{0.017} = -7.06$

P-value $= P(z > -7.06) = 0.9998$

There is more than a 99.98% chance that the stated percentage is correct for this region.

**13. Dowsing** In a rural area, only about 30% of the wells that are drilled find adequate water at a depth of 100 feet or less. A local resident claims to be able to find water by "dowsing"—using a forked stick to indicate where the well should be drilled. You randomly select 80 of their over 1000 former customers and find that 27 have wells less than 100 feet deep. What do you conclude about their claim?

a) Write appropriate hypotheses.
b) Check the necessary assumptions.
c) Perform the mechanics of the test. What is the P-value?
d) Explain carefully what the P-value means in context.
e) What's your conclusion?

**14. Abnormalities** In the 1980s it was generally believed that congenital abnormalities affected about 5% of the nation's children. Some people believe that the increase in the number of chemicals in the environment has led to an increase in the incidence of abnormalities. A recent study examined 384 children and found that 46 of them showed signs of an abnormality. Is this strong evidence that the risk has increased?

a) Write appropriate hypotheses.
b) Check the necessary assumptions.
c) Perform the mechanics of the test. What is the P-value?
d) Explain carefully what the P-value means in context.
e) What's your conclusion?
f) Do environmental chemicals cause congenital abnormalities?

**15. Absentees** The National Center for Education Statistics monitors many aspects of elementary and secondary education. Its 1996 numbers are often used as a baseline to assess changes. In 1996 34% of students had not been absent from school even once during the previous month. In the 2000 survey, responses from 8302 randomly chosen students showed that this figure had slipped to 33%. Do these figures give evidence of a change in student attendance?

a) Write appropriate hypotheses.
b) Check the assumptions and conditions.
c) Perform the test and find the P-value.
d) State your conclusion.
e) Do you think this difference is meaningful? Explain.

**16. Educated mothers** The National Center for Education Statistics monitors many aspects of elementary and secondary education. Its 1996 numbers are often used as a baseline to assess changes. In 1996, 31% of students reported that their mothers had graduated from college. In 2000, responses from 8368 randomly chosen students found that this figure had grown to 32%. Is this evidence of a change in education level among mothers?

a) Write appropriate hypotheses.
b) Check the assumptions and conditions.
c) Perform the test and find the P-value.
d) State your conclusion.
e) Do you think this difference is meaningful? Explain.

**17. Contributions, please, part II** In Chapter 18, Exercise 22, you learned that the Paralyzed Veterans of America is a philanthropic organization that relies on contributions. It sends free mailing labels to millions of households and greeting cards to potential donors on its list and asks for a voluntary contribution. To test a new campaign, the organization recently sent letters to a random sample of 100,000 potential donors and received 4781 donations. It has had a contribution rate of 5% in past campaigns, but a staff member worries that the rate will be lower if it runs this campaign as currently designed.

a) What are the hypotheses?
b) Are the assumptions and conditions for inference met?
c) Do you think the rate would drop? Explain.

**18. Take the offer, part II** In Chapter 18, Exercise 23, you learned that First USA, a major credit card company, is planning a new offer for its current cardholders. First USA will give double airline miles on purchases for the next 6 months if the cardholder goes online and registers for this offer. To test the effectiveness of this campaign, the company recently sent out offers to a random sample of 50,000 cardholders. Of those, 1184 registered. A staff member suspects that the success rate for the full campaign (sent to 2 million cardholders) will be comparable to the standard 2% rate that it is used to seeing in similar campaigns. What do you predict?

a) What are the hypotheses?
b) Are the assumptions and conditions for inference met?
c) Do you think the rate would change if it uses this fundraising campaign? Explain.

**19. Law school** According to the Law School Admission Council, in the fall of 2007, 66% of law school applicants were accepted to some law school.[8] The training program LSATisfaction claims that 163 of the 240 students trained in 2006 were admitted to law school. You can safely consider these trainees to be representative of the population of law school applicants. Has LSATisfaction demonstrated a real improvement over the national average?

a) What are the hypotheses?
b) Check the conditions and find the P-value.
c) Would you recommend this program based on what you see here? Explain.

---

[8]As reported by the Cornell Office of Career Services in Cornell's Class of 2007 Postgraduate Report.

20. **Med school 2011** According to the Association of American Medical Colleges, only 46% of medical school applicants were admitted to a medical school in the fall of 2011.[9] Upon hearing this, the trustees of Striving College expressed concern that only 77 of the 180 students in their class of 2011 who applied to medical school were admitted. The college president assured the trustees that this was just the kind of year-to-year fluctuation in fortunes that is to be expected and that, in fact, the school's success rate was consistent with the national average. Who is right?
    a) What are the hypotheses?
    b) Check the conditions and find the P-value.
    c) Are the trustees right to be concerned, or is the president correct? Explain.

21. **Pollution** A company with a fleet of 150 cars found that the emissions systems of 7 out of the 22 it randomly selected and tested failed to meet pollution control guidelines. Is this strong evidence that more than 20% of the fleet might be out of compliance? Explain (in detail!) why you would have concerns in running a hypothesis test for these data.

22. **Scratch and dent** An appliance manufacturer stockpiles washers and dryers in a very large warehouse for shipment to retail stores. Sometimes in handling them the appliances get damaged. Even though the damage may be minor, the company must sell those machines at drastically reduced prices. The company goal is to keep the level of damaged machines below 2%. One day an inspector randomly checks 60 washers and finds that 5 of them have scratches or dents. Is this strong evidence that the warehouse is failing to meet the company goal? Explain, very specifically, why running a test would be inappropriate.

23. **Twins** A national vital statistics report indicated that about 3% of all births produced twins. Is the rate of twin births the same among very young mothers? Data from a large city hospital found that only 7 sets of twins were born to 469 teenage girls. Test an appropriate hypothesis and state your conclusion. Be sure the appropriate assumptions and conditions are satisfied before you proceed.

24. **Football 2016** During the first 15 weeks of the 2016 season, the home team won 137 of the 238 regular-season National Football League games. Is this strong evidence of a home field advantage in professional football? Test an appropriate hypothesis and state your conclusion. Be sure the appropriate assumptions and conditions are satisfied before you proceed.

25. **WebZine** A magazine is considering the launch of an online edition. The magazine plans to go ahead only if it's convinced that more than 25% of current readers would subscribe. The magazine contacted a simple random sample of 500 current subscribers, and 137 of those surveyed expressed interest. What should the company do? Test an appropriate hypothesis and state your conclusion. Be sure the appropriate assumptions and conditions are satisfied before you proceed.

26. **Seeds** A garden center wants to store leftover packets of vegetable seeds for sale the following spring, but the center is concerned that the seeds may not germinate at the same rate a year later. The manager selects seeds at random from last year's green bean seeds and plants them as a test (he had thousands to choose from). Although the packet claims a germination rate of 92%, only 171 of 200 test seeds sprout. Is this evidence that the seeds have lost viability during a year in storage? Test an appropriate hypothesis and state your conclusion. Be sure the appropriate assumptions and conditions are satisfied before you proceed.

27. **Pick me!** Companies pay large amounts of money to make sure their websites are highly ranked on Google. *Advanced Web Ranking* reported that if your website is the very first search result on Google, desktop users will click on that result 34% of the time (this is called a click-through rate). Suppose that a certain company asks its Web manager to report the company's click-through rate. He conducts a random sample of 567 users and reports a click-through rate of 45%. Does this sample give the company reason to believe its results are working better than usual? Or could this just be a lucky result?

28. **Don't bother me!** In a July 2017 report, Pew Research stated that 41% of Americans reported being harassed while online. An AP Statistics student decided to conduct her own survey at her high school of 3330 students. She randomly selected 233 students to be surveyed and 103 of them stated that they had been harassed online. Does her survey suggest that the problem is worse at her school than it is for a typical American? (www.pewinternet.org/2017/07/11/online-harassment-2017/)

29. **Dropouts 2020** The National Center for Education Statistics reported that the high school dropout rate for the year 2020 was 5.3%. One school district whose dropout rate has always been very close to the national average reports that 130 of their 1782 high school students dropped out last year. Is this evidence that their dropout rate may be unusually high? Explain.

30. **Acid rain** A study of the effects of acid rain on trees in the Hopkins Memorial Forest shows that 25 of 100 trees randomly sampled exhibited some sort of damage from acid rain. This rate seemed to be higher than the 15% quoted in a recent *Environmetrics* article on the average proportion of damaged trees in the Northeast. Does the sample suggest that trees in the Hopkins Memorial Forest are more susceptible than trees from the rest of the region? Comment, and write up your own conclusions based on an appropriate confidence interval as well as a hypothesis test. Include any assumptions you made about the data.

31. **Lost luggage** An airline's public relations department says that the airline rarely loses passengers' luggage. It further claims that on those occasions when luggage is lost, 90% is recovered and delivered to its owner within 24 hours. A consumer group randomly surveys a large number of air travelers and finds that only 103 of 122 people who lost luggage on that airline were reunited with the missing items by the next day. Does this cast doubt on the airline's claim? Explain.

---
[9] www.aamc.org/data/facts/applicantmatriculant/

**32. TV ads** A start-up company is about to market a new computer printer. It decides to gamble by running commercials during the Super Bowl. The company hopes that name recognition will be worth the high cost of the ads. The goal of the company is that over 40% of the public recognize its brand name and associate it with computer equipment. The day after the game, a pollster contacts 420 randomly chosen adults and finds that 181 of them know that this company manufactures printers. Would you recommend that the company continue to advertise during Super Bowls? Explain.

**33. John Wayne** Like a lot of other Americans, John Wayne died of cancer. But is there more to this story? In 1955 Wayne was in Utah shooting the film *The Conqueror*. Across the state line, in Nevada, the United States military was testing atomic bombs. Radioactive fallout from those tests drifted across the filming location. A total of 46 of the 220 people working on the film eventually died of cancer. Cancer experts estimate that one would expect only about 30 cancer deaths in a group this size.

a) Is the death rate among the movie crew unusually high?
b) Does this prove that exposure to radiation increases the risk of cancer?

**34. AP Stats 2019** The College Board reported that 59.7% of all students who took the 2019 AP Statistics exam earned scores of 3 or higher. One teacher wondered if the performance of her school was better. She believed that year's students to be typical of those who will take AP Stats at that school and was pleased when 34 of her 54 students achieved scores of 3 or better.

a) How many standard derivations above the national rate did her students score? Does that seem like a lot? Explain.
b) Can she claim that her school is better? Explain.

**35. Trapped!** A pest control company is running tests on a new trap that attracts cockroaches. The goal for the trap is that at least 70% of cockroaches that pass within 20 cm of the trap will enter. A trial is run and a hypothesis test is calculated to evaluate if the goal has been met. The researchers find a test statistic of $z = 2.12$.

a) What is the correct P-value for this test?
b) Given this test statistic and P-value, what is the proper conclusion?

**36. Pick me** An advertising company runs an ad for a new chewing gum. The ads are played on a small video screen that is placed in the dairy section of supermarkets. Their marketing plan has an aim that at least 20% of gum purchases will be for their product, when the product is featured at a supermarket store check-out counter. Their test statistic is $z = 1.14$.

a) What is the P-value for this hypothesis test?
b) What is the correct conclusion for this hypothesis test?

**37. Stop!** A small-town cop decides to measure what percentage of cars actually come to a full and complete stop at the only stop sign in the center of town. She takes a random sample from different times of the day. Then she puts her Statistics education to good use and creates a 95% confidence interval to estimate the true percentage of cars that will come to a full and complete stop. Her interval is (29.3%, 38.7%).

The town mayor has set a goal of at least 45% of drivers to come to a full and complete stop. The mayor ran a hypothesis test using the random sample described above. Use the confidence interval to explain the conclusion that the mayor will reach.

**38. Putt-putt** A golfer is working on his putting game. His goal is that at least 85% of putts that he attempts within 15 feet successfully drop in the cup. After many hours of practice, his coach collects a random sample of his attempts. A 95% confidence interval is made and the coach reports that he makes between 82.3% and 89.8% of his putts within 15 feet.

If the golfer ran a hypothesis test using this data, what is his conclusion? Use the confidence interval to answer this question.

## JUST CHECKING

**Answers**

1. You can't conclude that the null hypothesis is true. You can conclude only that the experiment was unable to reject the null hypothesis. They were unable, on the basis of 12 patients, to show that aspirin was effective.

2. The null hypothesis is $H_0: p = 0.75$.

3. With a P-value of 0.0001, this is very strong evidence against the null hypothesis. We can reject $H_0$ and conclude that the improved version of the drug gives relief to a higher proportion of patients.

4. The parameter of interest is the proportion, $p$, of all delinquent customers who will pay their bills. $H_0: p = 0.30$ and $H_A: p > 0.30$.

5. The very low P-value leads us to reject the null hypothesis. There is strong evidence that the text is more effective in getting people to start paying their debts than just sending a letter had been.

6. All we know is that there is strong evidence to suggest that $p > 0.30$. We don't know how much higher than 30% the new proportion is. We'd like to see a confidence interval to tell whether the increase in loan repayments is great enough to offset any possible drawbacks of sending texts to customers.

# 20

# More About Tests and Intervals

| WHO | Florida motorcycle riders aged 20 and younger involved in motorcycle accidents |
|---|---|
| WHAT | % wearing helmets |
| WHEN | 2001–2003 |
| WHERE | Florida |
| WHY | Assessment of injury rates commissioned by the National Highway Traffic Safety Administration (NHTSA) |

All motorcycle riders were once required to wear helmets in Florida. In 2000, the law was changed to allow riders 21 and older to ride helmetless. Even though the law did not apply to those under 21, a report by the Preusser Group[1] suggests that helmet use may have declined in this group as well.

It isn't very practical to survey young motorcycle riders. (How could you construct a sampling frame? If you contacted licensed riders, would they admit to riding illegally without a helmet?) To avoid these problems, the researchers adopted a different strategy. They looked only at police reports of motorcycle accidents, which record whether the rider wore a helmet and give the rider's age.

Before the change in the helmet law, 60% of youths involved in a motorcycle accident had been wearing their helmets. During the three years following the law change, 396 of 781 young riders who were involved in accidents were wearing helmets. That's only 50.7%. Is this evidence of a decline in helmet wearing, or just the natural fluctuation of such statistics?

## Zero In on the Null

How do we choose the null hypothesis? The appropriate null arises directly from the context of the problem. It is dictated, not by the data, but by the situation. One good way to identify both the null and alternative hypotheses is to think about why the study is being done and what we hope to learn from the test. Typical null hypotheses might be that the

[1] one.nhtsa.gov/people/injury/pedbimot/motorcycle/FlaMCReport/pages/Index.htm

proportion of patients recovering after receiving a new drug is the same as we would expect of patients receiving a placebo, or that the mean strength attained by athletes training with new equipment is the same as with the old equipment. The alternative hypotheses would be that the new drug cures a higher proportion of patients or that the new equipment results in a greater mean strength.[2]

To write a null hypothesis, identify a parameter and choose a null value that relates to the question at hand. Even though the null usually means "no difference" or "no change," you can't automatically interpret "null" to mean *zero*. A claim that nobody wears a motorcycle helmet would be absurd. The null hypothesis for the Florida study is that the true rate of helmet use remained the same at $p = 0.60$ among young riders after the law changed. The alternative is that the proportion has decreased. Both the value for the parameter in the null hypothesis and the nature of the alternative arise from the context of the problem.

There is a temptation to state your *claim* as the null hypothesis. As we have seen, however, you cannot prove a null hypothesis true any more than a trial proves a defendant innocent. So it makes more sense to use what you want to show as the *alternative*. This way, if you reject the null because the data are inconsistent with it, you are left with what you want to show.

## FOR EXAMPLE
### Writing Hypotheses

The diabetes drug Avandia® was approved to treat Type 2 diabetes in 1999. But an article in the *New England Journal of Medicine* (*NEJM*)[3] raised concerns that the drug might carry an increased risk of heart attack. This study combined results from a number of other separate studies to obtain an overall sample of 4485 diabetes patients taking Avandia. People with Type 2 diabetes are known to have about a 20.2% chance of suffering a heart attack within a seven-year period. According to the article's author, Dr. Steven E. Nissen,[4] the risk found in the *NEJM* study was equivalent to a 28.9% chance of heart attack over seven years. The FDA is the government agency responsible for relabeling Avandia to warn of the risk if it is judged to be unsafe. Although the statistical methods they use are more sophisticated, we can get an idea of their reasoning with the tools we have learned.

QUESTION: What hypotheses about the seven-year heart attack risk would you test? Explain.

ANSWER:
$$H_0: p = 0.202$$
$$H_A: p \neq 0.202$$

The parameter of interest is the proportion of diabetes patients suffering a heart attack in seven years. The FDA is interested in any deviation from the expected rate of heart attacks.[5]

---

[2]In the last chapter, we pointed out that scientists hoping to publish their research often stick to two-tailed tests to avoid even the appearance of impropriety. In this text we will usually use a one-sided test when the context naturally invites it. But proper statistical practice *always* requires that the null and alternative hypotheses be formulated before data are collected.

[3]Steven E. Nissen, M.D., and Kathy Wolski, M.P.H., "Effect of Rosiglitazone on the Risk of Myocardial Infarction and Death from Cardiovascular Causes," *NEJM* 2007; 356.

[4]Interview reported in the *New York Times* (May 26, 2007).

[5]If you think the one-sided alternative hypothesis $p > 0.202$ is appropriate here, we don't disagree with you in principle. But testing the safety of drugs is one of those situations when it is better to be conservative and use the two-sided alternative, to avoid any appearance of trying to stack the deck in your favor.

# How to Think About P-Values

### Which Conditional?

Suppose that as a political science major you are offered the chance to be a White House intern. There would be a very high probability that next summer you'd be in Washington, D.C. That is, $P(\text{Washington}|\text{Intern})$ would be high. But if we find a student in Washington, D.C., is it likely that they're a White House intern? Almost surely not; $P(\text{Intern}|\text{Washington})$ is low. You can't switch around conditional probabilities. The P-value is $P(\text{data}|H_0)$. We might wish we could report $P(H_0|\text{data})$, but these two quantities are NOT the same.

A P-value is a conditional probability. It tells us the probability of getting results at least as unusual as the observed statistic, *given* that the null hypothesis is true. We can write P-value $= P(\text{observed statistic value [or even more extreme]}|H_0)$.

Writing the P-value this way helps to make clear that the P-value is *not* the probability that the null hypothesis is true. It is a probability about the data. Let's say that again:

*The P-value is **not** the probability that the null hypothesis is true.*

The P-value is not even the conditional probability that the null hypothesis is true given the data. We would write that probability as $P(H_0|\text{observed statistic value})$. This is a conditional probability but in reverse. It would be nice to know this probability, but we can't. The null hypothesis isn't a random event about which probability statements even make sense. It's a conjecture that must in fact be either true or false, we just don't know which.

We can find the P-value, $P(\text{observed statistic value}|H_0)$, because $H_0$ gives the parameter values that we need to calculate the required probability. But we can't find $P(H_0|\text{observed statistic value})$. As tempting as it may be to say that a P-value of 0.03 means there's a 3% chance that the null hypothesis is true, that just isn't right. All we can say is that, given the null hypothesis, there's a 3% chance of observing the statistic value that we have actually observed (or one even more inconsistent with $H_0$).

## What to Do with a Small P-Value

We know that a small P-value means that a result like the one we just observed is unlikely to occur if the null hypothesis is true. So we have evidence against the null hypothesis. An even smaller P-value implies stronger evidence against the null hypothesis, but it doesn't mean that the null hypothesis is "less true" (see "How Guilty Is the Suspect" on page 539).

How small the P-value has to be for you to reject the null hypothesis depends on a lot of things, not all of which can be precisely quantified. Your belief in the null hypothesis will influence your decision. Your trust in the data, in the experimental method if the data come from a planned experiment, in the survey protocol if the data come from a designed survey, all influence your decision. The P-value should serve as a measure of the strength of the evidence against the null hypothesis, but should never serve as a hard-and-fast rule for decisions. You have to take that responsibility on yourself.

As a review, let's look at the helmet law example from the chapter opener. Did helmet wearing among young riders decrease after the law allowed older riders to ride without helmets? What is the evidence?

## STEP-BY-STEP EXAMPLE

### Another One-Proportion z-Test

**QUESTION:** Has helmet use in Florida declined among riders under the age of 21 subsequent to the change in the helmet laws?

| | | |
|---|---|---|
| **THINK** | **PLAN** State the problem and discuss the variables and the W's.<br><br>**HYPOTHESES** The null hypothesis is established by the rate set before the change in the law. The study was concerned with safety, so they'll want to know of any decline in helmet use, making this a lower-tailed test. | I want to know whether the rate of helmet wearing among Florida's motorcycle riders under the age of 21 decreased after the law changed to allow older riders to go without helmets. The proportion before the law was passed was 60% so I'll use that as my null hypothesis value. The alternative is one-sided because I'm interested only in seeing if the rate decreased. I have data from accident records showing 396 of 781 young riders were wearing helmets.<br><br>$H_0: p = 0.60$<br>$H_A: p < 0.60$ |
| **SHOW** | **MODEL** Check the conditions.<br><br><br><br><br><br><br><br><br><br><br>Specify the sampling distribution model and name the test. | ✓ **Independence Assumption**: The data are for riders involved in accidents during a three-year period. Individuals are independent of one another.<br><br>✗ **Randomization Condition**: No randomization was applied, but we are considering these riders involved in accidents to be a representative sample of all riders. We should take care in generalizing our conclusions.<br><br>✓ **10% Condition**: These 781 riders are a small sample of a larger population of all young motorcycle riders.<br><br>✓ **Success/Failure Condition**: We'd expect<br>$np = 781(0.6) = 468.6$ helmeted riders and<br>$nq = 781(0.4) = 312.4$ non-helmeted. Both are at least 10.<br><br>The conditions are satisfied, so I can use a Normal model and perform a one-proportion z-test. |
| **SHOW** | **MECHANICS** Find the standard deviation of the sampling model using the hypothesized proportion.<br><br><br>Find the z-score for the observed proportion.<br><br>Make a picture. Sketch a Normal model centered at the hypothesized helmet rate of 60%. This is a lower-tail test, so shade the region to the left of the observed rate.<br><br><br><br><br>Given this z-score, the P-value is obviously very low. | There were 396 helmet wearers among the 781 accident victims.<br><br>$$\hat{p} = \frac{396}{781} = 0.507$$<br><br>$$SD(\hat{p}) = \sqrt{\frac{p_0 q_0}{n}} = \sqrt{\frac{(0.60)(0.40)}{781}} = 0.0175$$<br><br>$$z = \frac{\hat{p} - p_0}{SD(\hat{p})} = \frac{0.507 - 0.60}{0.0175} = -5.31$$<br><br>[Normal curve centered at 0.60, with 0.507 marked to the left]<br><br>The observed helmet rate is 5.31 standard deviations below the former rate. The corresponding P-value is less than 0.001. |
| **TELL** | **CONCLUSION** Link the P-value to your decision about the null hypothesis, and then state your conclusion in context. | The very small P-value says that if the overall rate of helmet wearing among riders under 21 were still 60%, the probability of observing a rate no higher than 50.7% in a sample like this is less than 1 chance in 1000, so I reject the null hypothesis. There is strong evidence that there has been a decline in helmet use among riders under 21. |

The P-value in the helmet example is quite small—less than 0.001. That's strong evidence to suggest that the rate has decreased since the law was changed. But it doesn't say that it was "a lot lower." To answer that question, you'd need to construct a confidence interval:

$$\hat{p} \pm z^*\sqrt{\frac{\hat{p}\hat{q}}{n}} = 0.507 \pm 1.96(0.0175) = (0.473, 0.541)$$

(using 95% confidence).

The confidence interval provides the additional information. Now we see that even in the best case, the rate is below 55%. Whether a change that large makes an important difference in safety is a judgment that depends on the situation, but not on the P-value. In fact, Florida made the decision to require a motorcycle "endorsement" for all riders. For riders under 21, that requires a motorcycle safety course. The percentage of unendorsed riders involved in crashes dropped considerably afterward.[6]

### How Guilty Is the Suspect?

The smaller the P-value is, the more confident we can be in rejecting the null hypothesis. However, it does not make the null hypothesis any more false. Think again about a jury trial (p. 513). Our null hypothesis is that the defendant is innocent. But the bank's security camera showed the robber was male and about the same size as the defendant. We're starting to question his innocence a little. Then witnesses add that the robber wore a blue jacket just like the one the police found in a garbage can behind the defendant's house. Well, if he's innocent, then that doesn't seem very likely, does it? If he's really innocent, the probability that all of these could have happened is getting pretty low. As the evidence rolls in, our P-value may become small enough to be called "beyond a reasonable doubt" and lead to a conviction. Each new piece of evidence strains our belief in the null a bit more. The more compelling the evidence—the more unlikely it would be were he innocent—the more convinced we become that he's guilty.

But even though it may make *us* more confident in declaring him guilty, additional evidence does not make *him* any guiltier. Either he robbed the bank or he didn't. Additional evidence just makes us more confident that we did the right thing when we convicted him. The lower the P-value, the more comfortable we feel about our decision to reject the null hypothesis, but the null hypothesis doesn't get any more false.

> The wise man proportions his belief to the evidence.
>
> —David Hume, "Enquiry Concerning Human Understanding," 1748

### FOR EXAMPLE
#### Thinking About the P-Value

RECAP: A *New England Journal of Medicine* paper reported that the seven-year risk of heart attack in diabetes patients taking the drug Avandia was increased from the baseline of 20.2% to an estimated risk of 28.9% and said the P-value was 0.03.

QUESTION: How should the P-value be interpreted in this context?

ANSWER: The *P-value* = $P(\hat{p} \geq 28.9\% | p = 20.2\%)$. That is, it's the probability of seeing such a high heart attack rate among the people studied if, in fact, taking Avandia really didn't increase the risk at all.

---
[6] www.ridesmartflorida.com

## What to Do with a High P-Value

Therapeutic touch (TT), taught in many schools of nursing, is a therapy in which the practitioner moves their hands near, but does not touch, a patient in an attempt to manipulate a "human energy field." Therapeutic touch practitioners believe that by adjusting this field they can promote healing. However, no instrument has ever detected a human energy field, and no experiment has ever shown that TT practitioners can detect such a field.

In 1998, the *Journal of the American Medical Association* published a paper reporting work by a then nine-year-old girl.[7] She had performed a simple experiment in which she challenged 15 TT practitioners to detect whether her unseen hand was hovering over their left or right hand (selected by the flip of a coin).

The practitioners "warmed up" with a period during which they could see the experimenter's hand, and each said that they could detect the girl's human energy field. Then a screen was placed so that the practitioners could not see the girl's hand, and they attempted 10 trials each. Overall, of 150 trials, the TT practitioners were successful only 70 times—a success proportion of 46.7%.

The null hypothesis here is that the TT practitioners were just guessing. If that were the case, because the hand was chosen using a coin flip, the practitioners would guess correctly 50% of the time. So the null hypothesis is that $p = 0.5$ and the alternative that they could actually detect a human energy field is (one-sided) $p > 0.5$.

What would constitute evidence that they weren't guessing? Certainly, a very high proportion of correct guesses out of 150 would convince most people. Exactly how high the proportion of correct guesses has to be for you to reject the null hypothesis depends on how small a P-value you need to be convinced (which, in turn, depends on how often you're willing to make mistakes—a topic we'll discuss later in the chapter).

But let's look again at the TT practitioners' proportion. Does it provide any evidence that they weren't guessing? The proportion of correct guesses is 46.7%—that's *less* than the hypothesized value, not greater! When we find $SD(\hat{p}) = 0.041$ (or 4.1%) we can see that 46.7% is almost 1 SD *below* the hypothesized proportion:

$$SD(\hat{p}) = \sqrt{\frac{p_0 q_0}{n}} = \sqrt{\frac{(0.5)(0.5)}{150}} \approx 0.041$$

The observed proportion, $\hat{p}$, is 0.467.

$$z = \frac{\hat{p} - p_0}{SD(\hat{p})} = \frac{0.467 - 0.5}{0.041} = -0.805$$

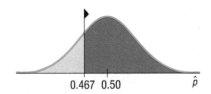

The observed success rate is 0.805 standard deviations below the hypothesized mean.

$$\text{P-value} = p(z > -0.805) = 0.790$$

If the practitioners had been highly successful, we would have seen a low P-value. In that case, we would then have concluded that they could actually detect a human energy field.

But that's not what happened. What we observed was a $\hat{p} = 0.467$ success rate. The P-value for this proportion is greater than 0.5 because the observed value is on the "wrong" side of the null hypothesis value. To convince us, the practitioners should be doing better than guessing, not worse!

---

[7] L. Rosa, E. Rosa, L. Sarner, and S. Barrett, "A Close Look at Therapeutic Touch," *JAMA* 279(13) [1 April 1998]: 1005–1010.

Obviously, we won't be rejecting the null hypothesis; for us to reject it, the P-value would have to be quite small.

Big P-values just mean that what we've observed isn't surprising. That is, the results are consistent with our assumption that the null hypothesis models the world, so we have no reason to reject it. A big P-value doesn't prove that the null hypothesis is true, but it certainly offers no evidence that it's *not* true. When we see a large P-value, all we can say is that we "don't reject the null hypothesis."

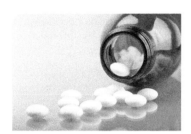

### FOR EXAMPLE
#### More About P-Values

**RECAP:** The question of whether the diabetes drug Avandia increased the risk of heart attack was raised by a study in the *New England Journal of Medicine*. This study estimated the seven-year risk of heart attack to be 28.9% and reported a P-value of 0.03 for a test of whether this risk was higher than the baseline seven-year risk of 20.2%. An earlier study (the ADOPT study) had estimated the seven-year risk to be 26.9% and reported a P-value of 0.27.

**QUESTION:** The researchers in the ADOPT study did not express any alarm about the increased heart-attack risk they observed. Should they have?

**ANSWER:** No. A P-value of 0.27 means that a heart attack rate at least as high as the one they observed could be expected in 27% of similar experiments even if, in fact, there were no increased risk from taking Avandia. That's not remarkable enough to reject the null hypothesis. In other words, the ADOPT study wasn't convincing.

# Alpha Levels

> **NOTATION ALERT**
> The first Greek letter, $\alpha$, is used in Statistics for the threshold value of a hypothesis test. You'll hear it referred to as the alpha level. Common values are 0.10, 0.05, 0.01, and 0.001.

Up to now, we have avoided the question of how small the P-value needs to be in order to reject the null. One strategy is to set an arbitrary threshold before collecting the data. Then, if the P-value falls below that "bright line," reject the null hypothesis and call the result **statistically significant**. The threshold is called an **alpha level** and labeled with the Greek letter $\alpha$. Common $\alpha$ levels are 0.10, 0.05, 0.01, and 0.001. There's nothing special or memorable about the letter $\alpha$, but this is the standard terminology and notation, so we're stuck with it.

It can be hard to justify a particular choice of $\alpha$, though, so 0.05 is often used. But you have to select the alpha level *before* you look at the data. Otherwise you can be accused of cheating by tuning your alpha level to suit the data.

> **WHERE DID THE VALUE 0.05 COME FROM?**
> In 1931, in a famous book called *The Design of Experiments*, Sir Ronald Fisher discussed the amount of evidence needed to reject a null hypothesis. He said that it was *situation dependent*, but remarked, somewhat casually, that for many scientific applications, 1 out of 20 *might be* a reasonable value. Since then, some people—indeed some entire disciplines—have treated the number 0.05 as sacrosanct.

Sir Ronald Fisher (1890–1962) was one of the founders of modern Statistics.

The alpha level is also called the **significance level**. When we reject the null hypothesis, we say that the test is "significant at that level." For example, we might say that we reject the null hypothesis "at the 5% level of significance."

> **IT COULD HAPPEN TO YOU!**
> Of course, if the null hypothesis *is* true, no matter what alpha level you choose, you still have a probability of rejecting the null hypothesis by mistake. This is the rare event we want to protect ourselves against. When we do reject the null hypothesis, no one ever thinks that *this* is one of those rare times. As statistician Stu Hunter notes, "*The statistician says 'rare events do happen—but not to me!'*"[8]

The automatic nature of the reject/fail-to-reject decision when we use an alpha level may make you uncomfortable. If your P-value falls just slightly above your alpha level, you're not allowed to reject the null. Yet a P-value just barely below the alpha level leads to rejection. If this bothers you, you're in good company. Many statisticians think it better to report the P-value than to base a decision on an arbitrary alpha level.

> ### It's in the Stars
> Some disciplines carry the idea further and code P-values by their size. In this scheme, a P-value between 0.05 and 0.01 gets highlighted by *. A P-value between 0.01 and 0.001 gets **, and a P-value less than 0.001 gets ***. This can be a convenient summary of the weight of evidence against the null hypothesis if it's not taken too literally. But we warn you against taking the distinctions too seriously and against making a black-and-white decision near the boundaries. The boundaries are a matter of tradition, not science; there is nothing special about 0.05. A P-value of 0.051 should be looked at very seriously and not casually thrown away just because it's larger than 0.05, and one that's 0.009 is not very different from one that's 0.01.

# Practical vs. Statistical Significance

You've probably heard or read discussions in which a statistical result is called "significant." Sounds important and scientific, doesn't it? But really, all that means is that the test statistic had a P-value lower than the specified alpha level. And if you weren't told that alpha level, it's hard to know exactly what "significant" means. A test with a small P-value may be surprising (if you believe the null value), but a small P-value says nothing about the size of the effect—and that's what determines whether the result actually makes a difference. Don't be lulled into thinking that statistical significance carries with it any sense of practical importance or impact.

When we reject a null hypothesis, what we really care about is the actual change or difference in the data. The difference between the value you see in your data and the null value is called the **effect size**. For large samples, even a small, unimportant effect size can be statistically significant. On the other hand, if the sample is not large enough, even a large financially or scientifically important effect may not be statistically significant.

It's good practice to report a confidence interval for the parameter along with the P-value to indicate the interval of plausible values for the parameter. The confidence interval is centered on the observed effect and puts bounds on how big or small the effect size may actually be.

> **STATISTICALLY SIGNIFICANT? YES, BUT IS IT IMPORTANT?**
> A large insurance company mined its data and found a statistically significant ($P = 0.04$) difference between the mean value of policies sold in 2013 and 2014. The difference in the mean values was $9.83. Even though it was statistically significant, management did not see this as an important difference when a typical policy sold for more than $1000. On the other hand, even a clinically important improvement of 10% in cure rate with a new treatment is not likely to be statistically significant in a study of fewer than 225 patients. A small clinical trial would probably not be conclusive.

---
[8] Personal communication with co-author Dick DeVeaux; also heard by co-author Paul Velleman.

# Confidence Intervals and Hypothesis Tests

A hypothesis test examines whether some particular "candidate" parameter value is consistent with the data. But a confidence interval gives the *entire set* of candidate parameter values that are consistent with the data. Then why do a hypothesis test at all? Wouldn't it be simpler to just construct the confidence interval and see whether the particular candidate parameter value of interest is in the confidence interval or not?

The short answer is (you may be surprised) . . . *Yes*. In fact, it's not only simpler, but because we'd have wanted to report the confidence interval anyway had the hypothesis test rejected the null, cutting out the hypothesis test altogether saves extra work.

But the longer answer (sorry) is that it's a little bit more complicated than that, and there are good reasons for carrying out the hypothesis test anyway. Here are some of them:

- A two-sided hypothesis test done at a 5% significance level should reject any value of the parameter that isn't included in a 95% confidence interval. But a *one*-sided hypothesis test does not have such a confidence interval counterpart.[9]
- Knowing that a particular parameter value lies outside a 95% confidence interval tells you that a (two-sided) hypothesis test would reject that parameter value at a significance level of $\alpha = 0.05$. But it doesn't give you a P-value, so it doesn't let you know just how strong the evidence is against that particular parameter value, which many people care about.
- Recall that when computing a confidence interval for a population proportion, the standard error $SE(\hat{p}) = \sqrt{\dfrac{\hat{p}\hat{q}}{n}}$ is used to estimate the variability in $\hat{p}$. But when a hypothesis test is done for a population proportion, the variability in $\hat{p}$ is assumed to be $SD(\hat{p}) = \sqrt{\dfrac{p_0 q_0}{n}}$. The difference between these expressions usually doesn't matter because they're so close—but sometimes it does matter. And later we'll see other situations in which the computations differ somewhat between the confidence interval and the hypothesis test because of different assumptions.
- Finally, there are actually plenty of hypothesis tests in which $H_0$ isn't a statement about a population parameter, so there's no confidence interval counterpart at all. We saw an example in the last chapter when we used a simulation to see whether there was any evidence of a "hot hand" in basketball. We'll see others later in this text, some of which are used quite often. So being able to conduct and understand hypothesis tests is important if you want to draw conclusions based on data.

The upshot of this is that while both confidence intervals and hypothesis tests are designed to address the consistency between parameter values and data, and there is a (limited) parallel between them, we encourage you to use a hypothesis test when the situation focuses on the plausibility of a *single* parameter value, and a confidence interval when it doesn't. You may end up reporting a confidence interval anyway, to communicate the effect size, but you'll have begun your report in the right place, focusing on the parameter value of greatest interest.

## JUST CHECKING

1. A researcher reports that teens are spending more time on average online now than they were 10 years ago and reports that the P-value for the difference is 0.001. Is the result "statistically significant"? Is it practically significant?

---

[9]Okay, that's not quite true. There are such things as "one-sided confidence intervals." They're not very widely used, and we're not going to discuss them further.

2. Suppose that zoologists were concerned that one of their chimpanzees was spending too much time sleeping and were worried about its health. After observing the chimp at randomly sampled times, they conducted a hypothesis test in which the null was that the chimp was sleeping as much as normal chimps do and the alternative was that it was sleeping more than that. Their test yielded a P-value of 0.19. Should they be more concerned or heave a sigh of relief? Or neither?

3. If the null hypothesis is, in fact, true, how often will a test give a statistically significant result just by chance if the researchers use an $\alpha$-level of 0.05? *(Think carefully!)*

# *A 95% Confidence Interval for Small Samples

When the **Success/Failure Condition** fails, all is not lost. A simple adjustment to the calculation lets us make a 95% confidence interval anyway.

All we do is add four *phony* observations—two to the successes, two to the failures. So instead of the proportion $\hat{p} = \frac{y}{n}$, we use the adjusted proportion $\tilde{p} = \frac{y+2}{n+4}$ and, for convenience, we write $\tilde{n} = n + 4$. We modify the interval by using these adjusted values:

$$\tilde{p} \pm z^* \sqrt{\frac{\tilde{p}(1-\tilde{p})}{\tilde{n}}}$$

Called a "plus-four" interval, this adjusted form gives better performance overall[10] and works much better for proportions near 0 or 1. It has the additional advantage that we no longer need to check the Success/Failure Condition that $n\hat{p}$ and $n\hat{q}$ are greater than 10. Because of these properties, the use of plus-four intervals is becoming increasingly common in modern statistics.

### *FOR EXAMPLE

#### An Agresti-Coull "Plus-Four" Interval

Surgeons examined their results to compare two methods for a surgical procedure used to alleviate pain on the outside of the wrist. A new method was compared with the traditional "freehand" method for the procedure. Of 45 operations using the "freehand" method, three were unsuccessful, for a failure rate of 6.7%. With only 3 failures, the data don't satisfy the **Success/Failure Condition**, so we can't use a standard confidence interval.

QUESTION: What's the confidence interval using the "plus-four" method?

ANSWER: There were 42 successes and 3 failures. Adding 2 "pseudo-successes" and 2 "pseudo-failures," we find

$$\tilde{p} = \frac{3+2}{45+4} = 0.102$$

---

[10] By "better performance," we mean that 95% confidence intervals have more nearly a 95% chance of covering the true population proportion. In Chapter 18 (p. 501) our simulations showed that our original, simpler confidence interval is in fact less likely than 95% to cover the true population proportion when the sample size is small or the proportion very close to 0 or 1. The original idea for fixing this can be attributed to E. B. Wilson. The "plus-4" approach discussed here was proposed by Agresti and Coull (A. Agresti and B. A. Coull, "Approximate Is Better Than 'Exact' for Interval Estimation of Binomial Proportions," *The American Statistician*, 52[1998]: 119–129).

> A 95% confidence interval is then
>
> $$0.102 \pm 1.96 \sqrt{\frac{0.102(1 - 0.102)}{49}} = 0.102 \pm 0.085 \text{ or } (0.017, 0.187).$$
>
> Notice that although the observed failure rate of 0.067 is contained in the interval, it is not at the center of the interval—something we haven't seen with any of the other confidence intervals we've considered.

# Making Errors

Nobody's perfect. Even with lots of evidence, we can still make the wrong decision. In fact, when we perform a hypothesis test, we can make mistakes in *two* ways:

**I.** The null hypothesis is true, but we reject it.
**II.** The null hypothesis is false, but we fail to reject it.

These two types of errors are known as **Type I** and **Type II errors**. One way to keep the names straight is to remember that we start by assuming the null hypothesis is true, so a Type I error is the first kind of error we could make.

In medical disease testing, the null hypothesis is usually the assumption that a person is healthy. The alternative is that they have the disease we're testing for. So a Type I error is a *false positive:* A healthy person is diagnosed with the disease. A Type II error, in which an infected person is diagnosed as disease-free, is a *false negative*. These terms are so much clearer than the nondescriptive "Type I error" and "Type II error" that they are often used in contexts unrelated to medicine or disease testing. A Type I error is also sometimes called a *false alarm*.

Which type of error is more serious depends on the situation. In a jury trial, a Type I error occurs if the jury convicts an innocent person. A Type II error occurs if the jury lets a guilty person go free. Which seems more serious? In medical diagnosis, a false negative could mean that a sick patient goes untreated. A false positive might mean that a healthy person must undergo further tests. Which of these errors seems more serious? It depends on the situation, the cost, and your point of view.

Here's an illustration of the possible outcomes of a hypothesis test:

> **FALSE POSITIVES**
>
> Some false-positive results mean no more than an unnecessary chest X-ray. But for a drug test or a disease like AIDS, a false-positive result that is not kept confidential could have serious consequences.

|  | | The Truth | |
|---|---|---|---|
| | | $H_0$ True | $H_0$ False |
| **My Decision** | Reject $H_0$ | Type I Error (false positive) | OK (and interesting!) |
| | Fail to reject $H_0$ | OK (and boring.) | Type II Error (false negative) |

In our diagram we deliberately tried to make each outcome look different from the others, because all four are dramatically different. If they were personalities, here's what they'd be like.

- Upper left corner: Chicken Little. In an old fable, an acorn hits Chicken Little on the head and she runs around shouting that the sky is falling. This is a classic false alarm—a "discovery" of something that is not true. Chicken Little personifies the Type I error.
- Upper right corner: Sherlock Holmes. Sir Arthur Conan Doyle's character examined clues carefully and always uncovered the truth. This is where we most want to be with a hypothesis test: using data to discern actual phenomena.

- Lower left corner: Bilbo Baggins. "Sorry! I don't want any adventures, thank you. Not today."[11] Hobbiton is very unexciting and its residents like it that way. Mild-mannered Bilbo Baggins (before his adventures) personifies this dullest of outcomes: nothing discovered because there's nothing to be discovered.
- Lower right corner: Sad Sack. This inept old comic book character always had bad luck and never could seem to succeed at anything. A Type II error is a bit like that: An interesting discovery is out there . . . but you fail to detect it.

Recall that we use the Greek letter $\alpha$ as a threshold for deciding whether a P-value is small enough to warrant rejecting a null hypothesis, and it's called the significance level of the test. Interestingly, the threshold we choose and call $\alpha$ is also the probability that we'll make a Type I error, should the null hypothesis actually be true. It's not obvious why this should be the case, so let's do a thought experiment. Suppose you're doing a hypothesis test at a significance level of $\alpha = 0.05$ and $H_0$ is true. What are the chances that you'll get a P-value less than 0.05 and will reject this true null? Well, the P-value will be less than 0.05 precisely whenever the test statistic you're using (such as $\hat{p}$) falls somewhere in the outer 0.05 of the sampling distribution that's based on $H_0$ being true. And the probability of that happening is . . . well, it's 5% if $H_0$ is true, right? (There's a 5% chance of *any* random variable falling in the outer 5% of its distribution.) So if $H_0$ is true, then there's a 5% chance you'll reject it and (unknowingly) make a Type I error.

You might need to read that paragraph again to convince yourself.

If $\alpha$ is the (conditional)[12] probability of making a Type I error, do we have a symbol for the (conditional)[13] probability of making a Type II error? Yes. It is always denoted by the Greek letter beta: $\beta$. Note that we choose $\alpha$ ourselves, but we can't dictate $\beta$. Its value depends on several things, of which some we can control and some we can't. Of course we want $\beta$ to be small, but achieving a small $\beta$ isn't as simple as declaring it to be small.

In the next section we'll talk about what things influence the value of $\beta$ and what you can do to make it smaller.

**APPLET**
Explore the risk of rejecting a hypothesis that's actually true.

> **CHOOSING $\alpha$; LIVING WITH $\beta$**
>
> The significance level $\alpha$ is a threshold that you *choose*. The probability of a Type II error, $\beta$, is not. You have some *indirect* control over $\beta$, but you don't get to set it to a value of your choice. To a certain extent you just have to live with it.

> **NOTATION ALERT**
>
> In Statistics, $\alpha$ is almost always saved for the alpha level. But $\beta$ is also used for the parameters of a linear model.

### FOR EXAMPLE
#### Thinking About Errors

**RECAP:** A published study found the risk of heart attack to be increased in patients taking the diabetes drug Avandia. The issue of the *New England Journal of Medicine* (*NEJM*) in which that study appeared also included an editorial that said, in part, "A few events either way might have changed the findings for myocardial infarction[14] or for death from cardiovascular causes. In this setting, the possibility that the findings were due to chance cannot be excluded."

**QUESTION:** What kind of error would the researchers have made if, in fact, their findings were due to chance? What could be the consequences of this error?

**ANSWER:** The null hypothesis said the risk didn't change, but the researchers rejected that model and claimed evidence of a higher risk. If these findings were just due to chance, they rejected a true null hypothesis—a Type I error.

If, in fact, Avandia carried no extra risk, then patients might be deprived of its benefits for no good reason.

---

[11] J. R. R. Tolkien, *The Hobbit.* 75th anniversary edition, Houghton Mifflin Harcourt (September 18, 2012), p. 7.
[12] If the null hypothesis is actually true.
[13] If the null hypothesis is actually false.
[14] Doctorese for "heart attack."

> **WHAT IF WE TEST MANY HYPOTHESES?**
>
> With a significance level of $\alpha = 0.05$, there's 1 chance in 20 that we'll commit a Type I error and conclude that we've found something significant when there's nothing more going on than sampling error. The more hypotheses we test, the greater the risk becomes. If you see what's funny about this cartoon, then you understand the concept.

© 2013 Randall Munroe. Reprinted with permission. All rights reserved.

# Power

> Because power is the probability of rejecting $H_0$, and $\beta$ is the probability of *failing* to reject $H_0$, then Power $= 1 - \beta$.

The **power** of a hypothesis test is the probability of rejecting a null hypothesis that actually is false.[15] Because we want to reject a null hypothesis when it's false, then we want power to be *high*. Is there anything we can do to increase the power of a test? What affects the power?

It turns out that four things affect the power of a hypothesis test. And they are . . .

## . . . the Significance Level $\alpha$, . . .

Manufacturing companies test their own products all the time for quality control purposes. If something goes wrong with some part of the manufacturing process, then it needs to be addressed as soon as possible. But a few errors are bound to slip through even if the whole process is humming along smoothly. So manufacturers are always checking to see whether error rates exceed a certain threshold. They are, in effect, continually performing hypothesis tests on the proper functioning of their manufacturing process, in which the hypotheses are:

$H_0$: The manufacturing process is working properly with the usual error rate.
$H_A$: The manufacturing process is making too many errors.

If the observed error rate is too high to be attributable to random chance, then $H_0$ will be rejected, and a quality-control manager will begin trying to track down the problem.

There is some threshold error rate ($p^*$) that, when observed, will trigger the alarm. Should the company choose to set that to be a high rate or a low rate? Let's think about why the company might like each of these. If it sets a high threshold, that will be good at preventing false alarms. It won't often send the quality-control manager out on a wild goose chase. And so for that reason the company might like a high threshold.

But on the other hand, if the company sets a high threshold it will be harder to detect when the process *really is* malfunctioning. The company could end up manufacturing a lot of defective products. So for that reason the company might like a *low* threshold for triggering the alarm. It's a tradeoff that—all other things being equal—can't be avoided. If you raise the threshold for a false alarm, you're necessarily also raising the chance that you'll fail to detect a real problem.

Exactly the same thing is true with hypothesis testing. If you want a very low chance of a false alarm—a rejection of a true null—then you want to choose a very low significance level $\alpha$. However, this raises the bar for the amount of evidence needed to reject $H_0$, and therefore makes it also harder to reject $H_0$ even when it's false. In other words, it lowers the power of the hypothesis test.

Similarly, choosing a higher significance level $\alpha$ relaxes the amount of evidence required for rejecting $H_0$, and so it increases the power of the test. But the tradeoff, of course, is that with a greater $\alpha$ comes a greater chance of a "false alarm," rejecting a true null. If $\alpha$ is *way too high*, then you won't be able to convince anyone (even yourself) of anything. Why should anyone believe that you've found evidence of something, when there's a very good chance that you're just seeing random noise?

In practice, $\alpha$ is chosen for reasons we discussed earlier in the chapter that have nothing to do with power. It's important to know how it will affect power, but you don't choose $\alpha$ with power in mind.

> **NOTATION ALERT**
> We've attached symbols to many of the *p*'s. Let's keep them straight. *p* is a true proportion parameter. $p_0$ is a hypothesized value of *p*. $\hat{p}$ is an observed proportion. $p^*$ is a critical value of a proportion corresponding to a specified $\alpha$.

## . . . the Effect Size, . . .

You might think that winning a game of Rock, Paper, Scissors is pretty much as likely as losing—isn't it completely random? Participants in any of numerous Rock, Paper, Scissors tournaments around the world would disagree.[16] Avid players say that it's a

---
[15]Technically, power is also defined when the null hypothesis is true—it's still the probability of rejecting the null hypothesis. But we're almost exclusively interested in situations when $H_0$ is false, so it's okay to think of power as the chance of correctly discerning that the null hypothesis is false.
[16]wrpsa.com

psychological game, and that you win by reading your opponent, or by tricking them into misreading you.

We're going to set aside the question of whether winning Rock, Paper, Scissors either is or isn't completely random.[17] Instead, we want to consider how easy (or hard) it would be to detect that someone's long-run winning rate is something other than fifty percent.

Let's imagine that someone—we'll call them Champ—claims to be able to win in a game of Rock, Paper, Scissors more often than not. We want to put that claim to the test. If $p$ is the probability that Champ will beat an opponent, then our hypotheses are:

- $H_0$: $p = 0.50$ (Champ's chance of winning is 50%.)
- $H_A$: $p \neq 0.50$ (Champ's chance of winning is not 50%.)[18]

Suppose we pitted Champ against a series of 25 Rock, Paper, Scissors opponents. We'll let $\hat{p}$ be the proportion of opponents whom Champ defeats. Since this is an imaginary study, we feel okay generating some imaginary data. That's right—we're about to do another simulation! The question we're addressing isn't whether Rock, Paper, Scissors is a game of skill. For that we'd need *real data*, of course! No, we're trying to find out *how easy it would be to detect* that $p$ is anything other than 50% using a study with 25 opponents. For that kind of question, a simulation is the *perfect* tool.

As you surely suspect, the ease or difficulty of detecting any deviation from 50% depends on just *how* different from 50% $p$ really is. For example, if Champ's probability of winning each game is actually $p = 0.90$, then our imaginary study would almost surely provide clear evidence—Champ would almost surely win so many of their 25 games that we'd reject our null hypothesis of $p = 0.50$. The power of the test would be high. On the other hand, if Champ's probability of winning is $p = 0.51$, then even though the null is false, our power to detect this small effect would be low. Champ probably wouldn't win an astonishingly high number of games.

We wrote a computer program to see what would happen if Champ's probability of winning were $p = 0.80$.

- We generated 25 imaginary games (using $p = 0.80$) and checked to see if the proportion of wins by Champ was high enough (or low enough!) to lead us to reject the hypothesis that $p = 0.50$. That was one trial.
- Next we ran 1000 such trials. In about 90% of those, the simulated data led us to reject the null hypothesis (which was the correct decision, because $0.80 \neq 0.50$). That's our simulation's estimate of the power of this test to detect an effect when the true value of $p$ is 0.80.
- And then we looked at what would happen for other values of $p$. We ran the same simulation using $p = 0$, $p = 0.05$, $p = 0.10$, and so on, up to $p = 1.00$. And for each one we estimated the power of detecting that $p \neq 0.50$.

Here's a graph showing the estimated power for each of our values of $p$:

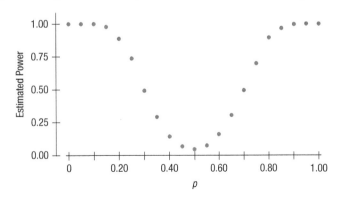

### FISHER AND $\alpha = 0.05$

Why did Sir Ronald Fisher suggest 0.05 as a criterion for testing hypotheses? It turns out that he had in mind small initial studies. Small studies have relatively little power. Fisher was concerned that they might make too many Type II errors—failing to discover an important effect—if too strict a criterion were used. Once a test failed to reject a null hypothesis, it was unlikely that researchers would return to that hypothesis to try again.

On the other hand, the increased risk of Type I errors arising from a generous criterion didn't concern him as much for exploratory studies because these are ordinarily followed by a replication or a larger study. The probability of a Type I error is $\alpha$—in this case, 0.05. The probability that two independent studies would both make Type I errors is $0.05 \times 0.05 = 0.0025$, so Fisher was confident that Type I errors in initial studies were not a major concern.

The widespread use of the relatively generous 0.05 criterion even in large studies is most likely not what Fisher had in mind.

### TI-*nspire*

**Errors and power.** Explore the relationships among Type I and Type II errors, sample size, effect size, and the power of a test.

---

[17]For what it's worth, we're persuaded that it's a game of skill.

[18]Because we're testing Champ's claim of better-than-even odds of winning, we might reasonably have chosen a one-sided alternative hypothesis here, namely: $H_A$: $p > 0.50$. But that would mean conceding from the outset that Champ's chance of winning *cannot* be less than 0.50, which we're not willing to concede. Perhaps Rock, Paper, Scissors *is* a game of skill—but Champ is a poor player, and often gets psyched out by their opponent.

Let's be clear that although its shape resembles an upside-down bell curve, that's an irrelevant coincidence. The graph shows how easy it would be for us to detect that Rock, Paper, Scissors is actually a game of skill, for each value of $p$ going from 0.00 to 1.00 in increments of 0.05. A graph like this shows how power varies depending on how far from the null the truth lies. And although there's more to this *power graph* than just the Big Idea, we're only going to point out to you the Big Idea:

Big effect sizes are easier to detect than small effect sizes.

That seems pretty obvious, doesn't it? We can see from the power graph that if Champ's probability of winning was, say, 70%, then we could expect to have about a 50% chance of rejecting $H_0$. And if Champ's probability of winning was 80%, then the probability would be about 90% that a series of 25 games would confirm that Champ's chance of winning was not fifty-fifty. But if Champ's probability of winning were 60%, then even though $H_0: p = 0.50$ would be false, our experiment would only have a slim chance of picking up on that—about a 16% chance, based on the simulation.

So the difference between $p = 0.50$ and $p = 0.60$ would be pretty unlikely to reveal itself in an experiment with $n = 25$ opponents. That's probably not very important in this context (unless you're Champ), but in many *real* studies, such a difference can be very consequential. For example, suppose the best known treatment for a life-threatening illness has a 50% chance of saving the person's life, and that a proposed new treatment may save as many as 60% of the lives of those with the illness. The difference of 10% might represent thousands of lives saved.

If it were important to show that $p \neq 0.50$, you'd want a study that had a high chance of rejecting the null of $p = 0.5$—even if the real value of $p$ was 0.60. But we just saw in our simulation that our imaginary experiment would be unlikely to detect that difference. What can we do?

We can conduct an experiment with more than 25 trials.

### . . . the Sample Size, . . .

Let's continue thinking about our imaginary experiment in which "Champ" plays a series of Rock, Paper, Scissors games against different opponents. And let's suppose—a little arbitrarily—that Champ's actual probability of winning each game is $p = 0.56$. (Tough to detect, but potentially big enough to matter in many real contexts.)

Let's repeat the earlier simulation, but this time instead of holding the sample size constant at $n = 25$ and seeing how the parameter $p$ affects the power, let's hold the parameter value constant at $p = 0.56$ and see how the sample size $n$ affects the power. Drum roll, please . . .

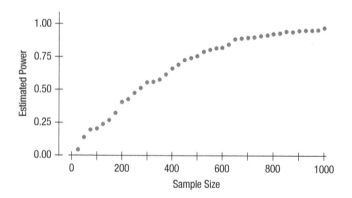

> **WE LIKE LARGER SAMPLES!**
>
> Not only can we achieve greater power, but we also have the opportunity to decrease the chances of both Type I and Type II errors. What's not to like?

This time here are two Big Ideas we want you to take away from the graph. The first is the most obvious: larger sample sizes lead to greater power. The second is a little subtler: there's a law of diminishing returns at work—you can increase your power by increasing the sample size, but eventually the additional data won't increase power enough to be worth the effort.

Data can be costly. Sometimes even a single data point can cost thousands of dollars or more to obtain. Given that subtle effects can still be important, and that large samples can sometimes be costly, it would be good if there was some other way to increase power besides collecting more data. Fortunately, there is: reducing unexplained variability.

## . . . and the Amount of Unexplained Variability.

Suppose you are a U.S. Senator trying to gauge the support among your constituents for an upcoming bill. You personally are not fond of the bill, but you're willing to vote for it if you can be convinced that more than half of your constituents support it. You consider commissioning a poll of your constituents and testing $H_0$: $p = 0.5$ against $H_A$: $p > 0.5$. A simple random sample should work if you can get a high enough response rate. But what about a stratified random sample? Is there any reason you might prefer that sampling method?

Maybe you know that the bill in question is generally more favored by people living in more rural parts of your state, and less favored by people in more urban parts of your state. If you were to take a simple random sample and ask them about the bill, you might by chance get a disproportionately large number of rural voters, making overall support for the bill likely to appear greater than it really is. Or you might just as easily, also by chance, get a disproportionately large number of *urban* voters, making overall support for the bill appear *less* than it really is. By using a *stratified* random sample, with the strata being "rural" and "urban," we can deliberately sample people from both regions proportional to their numbers in the population. We'll then still have sampling variability due to chance variation based on *which* set of rural constituents and *which* set of urban constituents we happen to sample. But we won't have the *additional* sampling variability that can come from having "too many" rural constituents or "too many" urban constituents in the sample. This is, you'll recall, the whole point of stratified sampling: to reduce variability.

With less sampling variability in $\hat{p}$, even a small deviation from the null value of $p_0 = 0.5$ may appear unusual—perhaps unusual enough to reject $H_0$ and lead you to vote for the bill. The same exact value of $\hat{p}$ might not have appeared all that unusual had there been more variability in its distribution, and you might not have rejected $H_0$. Stratifying, then, can potentially increase the power of a significance test by reducing the sampling variability in the test statistic.

Anything that reduces variability in a test statistic increases the power of an associated test. For example, in an experiment, carefully controlling extraneous factors increases the power by making it less likely that the "signal" of a treatment effect will be drowned out by the "noise" of extraneous factors. A blocked experimental design can increase power by attributing some of what would otherwise have been "noise" to the effect of the block. Doing a study on only cocker spaniels would likely lead to greater statistical power than using the same number of dogs of many breeds. (But your conclusions would apply only to cocker spaniels.)

In summary, power is greater when any of the following is true:

- The significance level $\alpha$ is greater. (But that also leads to a greater chance of a Type I error, a false alarm.)
- The effect size is greater. (But you have no control over the effect size.)
- The sample size is greater. (But more data usually cost more money and resources.)
- Variability from extraneous sources is reduced. (This is done by changing the way that data are collected.)

**APPLET**
Explore the relationship among alpha, errors, power, effect size, and sample size.

Researchers are concerned with estimating what the power of a hypothesis test will be before they ever collect any data. They want to be sure their research will have a reasonably good chance of yielding a meaningful, data-supported conclusion. But in this text we won't discuss how to compute power. It's sufficient to know what it is and to know how it is affected by the various choices you make when you collect and analyze data.

## FOR EXAMPLE
### Sample Size, Errors, and Power

**RECAP:** The meta-analysis of the risks of heart attacks in patients taking the diabetes drug Avandia combined results from 47 smaller studies. As GlaxoSmithKline (GSK), the drug's manufacturer, pointed out in their rebuttal, "Data from the ADOPT clinical trial did show a small increase in reports of myocardial infarction among the *Avandia*-treated group . . . however, the number of events is too small to reach a reliable conclusion about the role any of the medicines may have played in this finding."

**QUESTION:** Why would this smaller study have been less likely to detect the difference in risk? What are the appropriate statistical concepts for comparing the smaller studies? What are the possible drawbacks of a very large sample?

**ANSWER:** Smaller studies are subject to greater sampling variability; that is, the sampling distributions they estimate have a larger standard deviation for the sample proportion. That gives small studies less power: They'd be less able to discern whether an apparently higher risk was merely the result of chance variation or evidence of real danger. The FDA doesn't want to restrict the use of a drug that's safe and effective (Type I error), nor do they want patients to continue taking a medication that puts them at risk (Type II error). Larger sample sizes can reduce the risk of both kinds of error. Greater power (the probability of rejecting a false null hypothesis) means a better chance of spotting a genuinely higher risk of heart attacks. Of course, larger studies are more expensive. Their larger power also makes them able to detect differences that may not be clinically or financially significant. This tension among the power, cost, and sample size lies at the heart of most clinical trial designs.

On September 23, 2010, "The U.S. Food and Drug Administration announced that it will significantly restrict the use of the diabetes drug Avandia (rosiglitazone) to patients with Type 2 diabetes who cannot control their diabetes on other medications. These new restrictions are in response to data that suggest an elevated risk of cardiovascular events, such as heart attack and stroke, in patients treated with Avandia." (https://www.recallreport.org/dangerous-drugs-products/avandia/)

## JUST CHECKING

Suppose a large fast-food chain develops a new look to its menu, hoping that it will encourage more customers to order combo meals. Before switching to the menu in all of its thousands of restaurants, it first decides to try it out in a random sample of restaurants, and compare combo meal sales among them with combo meal sales among a randomly selected control group of restaurants. Define $p_T$ ($T$ is for "treatment") to be the proportion, across all of the restaurants, of customers who would order a combo meal when presented with the new menu; and define $p_C$ ($C$ is for "control") to be the proportion, across all of the restaurants, of customers who would order a combo meal when presented with the old menu. The hypotheses the fast-food chain wants to consider are:

$$H_0: p_T = p_C$$
$$H_A: p_T > p_C$$

4. Describe an advantage to the fast-food chain of using a significance level of $\alpha = 0.05$ compared to a significance level of $\alpha = 0.01$. Describe a disadvantage.

5. Describe an advantage to the fast-food chain of including 200 of its restaurants in the study, compared to including only 50 restaurants. Describe a disadvantage.

6. Describe an advantage to the fast-food chain of including in its study only restaurants located in large cities. Describe a disadvantage.

7. Suppose the effect of the new menu is to entice one additional customer out of every two hundred, on average, to buy a combo meal when they otherwise wouldn't have. What would be the most likely outcome of the fast-food chain's study?

## WHAT CAN GO WRONG?

- **Don't interpret the P-value as the probability that $H_0$ is true.** The P-value is about the data, not the hypothesis. It's the probability of observing data this unusual, *given* that $H_0$ is true, not the other way around.
- **Don't believe too strongly in arbitrary alpha levels.** There's not really much difference between a P-value of 0.051 and a P-value of 0.049, but sometimes it's regarded as the difference between night (having to refrain from rejecting $H_0$) and day (being able to shout to the world that your results are "statistically significant"). It may just be better to report the P-value and a confidence interval and let the world decide along with you.
- **Don't confuse practical and statistical significance.** A large sample size can make it easy to discern even a trivial change from the null hypothesis value. On the other hand, an important difference can be missed if your test lacks sufficient power.
- **Don't forget that in spite of all your care, you might make a wrong decision.** We can never reduce the probability of a Type I error ($\alpha$) or of a Type II error ($\beta$) to zero (but increasing the sample size helps).
- **Don't test multiple hypotheses.** Even though the risk of Type I error may be small for a single hypothesis test, the more tests you run the more likely it becomes that that statistical significance will arise purely by random chance.

## WHAT HAVE WE LEARNED?

We've learned that there's a lot more to hypothesis testing than a simple yes/no decision.
- We've learned that the P-value can indicate evidence against the null hypothesis when it's small, but it does not tell us the probability that the null hypothesis is true.
- We've learned that the alpha level of the test establishes the level of proof we'll require. That determines the critical value of z that will lead us to reject the null hypothesis.
- We've also learned more about the connection between hypothesis tests and confidence intervals; they're really two ways of looking at the same question. The hypothesis test gives us the answer to a decision about a parameter; the confidence interval tells us the plausible values of that parameter.

We've learned about the two kinds of errors we might make, and we've seen why in the end we're never sure we've made the right decision.
- If the null hypothesis is really true and we reject it, that's a Type I error; the alpha level of the test is the probability that this could happen.
- If the null hypothesis is really false but we fail to reject it, that's a Type II error.
- The power of the test is the probability that we reject the null hypothesis when it's false. The larger the size of the effect we're testing for, the greater the power of the test to detect it.
- We've seen that tests with a greater likelihood of Type I error have more power and less chance of a Type II error. We can increase power while reducing the chances of both kinds of error by increasing the sample size.

## TERMS

**Statistically significant**  When the P-value falls below the alpha level, we say that the test is "statistically significant" at that alpha level. (p. 541)

**Alpha level**  The threshold P-value that determines when we reject a null hypothesis. If we observe a statistic whose P-value based on the null hypothesis is less than $\alpha$, we reject that null hypothesis. (p. 541)

**Significance level**  The alpha level is also called the significance level, most often in a phrase such as a conclusion that a particular test is "significant at the 5% significance level". (p. 541)

**Effect size**  The difference between the null hypothesis value and true value of a model parameter is called the effect size. (p. 542)

**Type I error**  The error of rejecting a null hypothesis when in fact it is true (also called a "false positive"). The probability of a Type I error is $\alpha$. (p. 545)

**Type II error**  The error of failing to reject a null hypothesis when in fact it is false (also called a "false negative"). The probability of a Type II error is commonly denoted $\beta$ and depends on the effect size. (p. 545)

**Power**  The probability that a hypothesis test will correctly reject a false null hypothesis is the power of the test. The larger the effect size, the greater the power of the test to detect it. For any specific effect size, the power is $1 - \beta$. (p. 548)

## ON THE COMPUTER

### Hypothesis Tests

Reports about hypothesis tests generated by technologies don't follow a standard form. Most will name the test and provide the test statistic value, its standard deviation, and the P-value. But these elements may not be labeled clearly. For example, the expression "*Prob > |z|*" is a fancy (and not very clear) way of saying two-tailed P-value. In some packages, you can specify that the test be one-sided. Others might report three P-values, covering the ground for both one-sided tests and the two-sided test.

Sometimes a confidence interval and hypothesis test are automatically given together. The CI ought to be for the corresponding confidence level: $1 - \alpha$ for 2-tailed tests, $1 - 2\alpha$ for 1-tailed tests.

Often, the standard deviation of the statistic is called the "standard error," and usually that's appropriate because we've had to estimate its value from the data. That's not the case for proportions, however: We get the standard deviation for a proportion from the null hypothesis value. Nevertheless, you may see the standard deviation called a "standard error" even for tests with proportions.

It's common for statistics packages and calculators to report more digits of "precision" than could possibly have been found from the data. You can safely ignore them. Round values such as the standard deviation to one digit more than the number of digits reported in your data.

The example of results shown below is not from any program or calculator we know of, but it displays some of the things you might see in typical computer output.

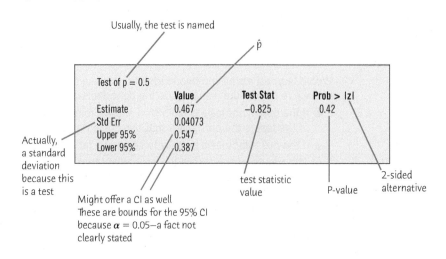

# EXERCISES

1. **Parameters and hypotheses** For each of the following situations, define the parameter and write the null and alternative hypotheses in terms of parameter values. Example: We want to know if the proportion of up days in the stock market is 50%. Answer: Let $p$ = the proportion of up days. $H_0$: $p = 0.5$ vs. $H_A$: $p \neq 0.5$.

   a) A casino wants to know if a slot machine really delivers the 1 in 100 win rate that it claims.
   b) A pharmaceutical company wonders if its new drug has a cure rate different from the 30% reported by subjects who had been given a placebo.
   c) A bank wants to know if the percentage of customers using its website has changed from the 40% that used it before its system crashed last week.

2. **Hypotheses and parameters** As in Exercise 1, for each of the following situations, define the parameter and write the null and alternative hypotheses in terms of parameter values.

   a) Observed seat belt use in 2016 was 92% in states with primary enforcement laws (where you can be pulled over for not wearing a seat belt). A state with secondary enforcement laws (where you can be ticketed for seat belt violations only if pulled over for something else) wants to see if its compliance is lower.
   b) Last year, a survey found that 45% of a company's employees were willing to pay for on-site day care. The company wants to know if that has changed.
   c) Regular card customers have a default rate of 6.7%. A credit card bank wants to know if that rate is different for its Gold card customers.

3. **True or false** Which of the following are true? If false, explain briefly.

   a) A P-value of 0.01 means that the null hypothesis is false.
   b) A P-value of 0.01 means that the null hypothesis has a 0.01 chance of being true.
   c) A P-value of 0.01 is evidence against the null hypothesis.
   d) A P-value of 0.01 means we should definitely reject the null hypothesis.

4. **False or true** Which of the following are true? If false, explain briefly.

   a) If the null hypothesis is true, you'll get a high P-value.
   b) If the null hypothesis is true, a P-value of 0.01 will occur about 1% of the time.
   c) A P-value of 0.90 means that the null hypothesis has a good chance of being true.
   d) A P-value of 0.90 is strong evidence that the null hypothesis is true.

5. **P-values** Which of the following are true? If false, explain briefly.

   a) A very high P-value is strong evidence that the null hypothesis is false.
   b) A very low P-value proves that the null hypothesis is false.
   c) A high P-value shows that the null hypothesis is true.
   d) A P-value below 0.05 is always considered sufficient evidence to reject a null hypothesis.

6. **More P-values** Which of the following are true? If false, explain briefly.

   a) A very low P-value provides evidence against the null hypothesis.
   b) A high P-value is strong evidence in favor of the null hypothesis.
   c) A P-value above 0.10 shows that the null hypothesis is true.
   d) If the null hypothesis is true, you can't get a P-value below 0.01.

7. **Hypotheses** For each of the following, write out the null and alternative hypotheses, being sure to state whether the alternative is one-sided or two-sided.

   a) A company knows that last year 40% of its reports in accounting were on time. Using a random sample this year, it wants to see if that proportion has changed.
   b) Last year, 42% of employees enrolled in at least one wellness class at a company's site. Using a survey, it wants to see whether a greater percentage is planning to take a wellness class this year.
   c) A political candidate wants to know from recent polls if she's going to garner a majority of votes in next week's election.

8. **More hypotheses** For each of the following, write out the alternative hypothesis, being sure to indicate whether it is one-sided or two-sided.

   a) *Consumer Reports* discovered that 20% of a certain computer model had warranty problems over the first three months. From a random sample, the manufacturer wants to know if a new model has improved that rate.
   b) The last time a philanthropic agency requested donations, 4.75% of people responded. From a recent pilot mailing, it wonders if that rate has increased.
   c) A student wants to know if other students on her campus prefer Coke or Pepsi.

9. **Alpha true and false** Which of the following statements are true? If false, explain briefly.

   a) Using an alpha level of 0.05, a P-value of 0.04 results in rejecting the null hypothesis.
   b) The alpha level depends on the sample size.
   c) With an alpha level of 0.01, a P-value of 0.10 results in rejecting the null hypothesis.
   d) Using an alpha level of 0.05, a P-value of 0.06 means the null hypothesis is true.

10. **Alpha false and true** Which of the following statements are true? If false, explain briefly.

    a) It is better to use an alpha level of 0.05 than an alpha level of 0.01.
    b) If we use an alpha level of 0.01, then a P-value of 0.001 is statistically significant.

c) If we use an alpha level of 0.01, then we reject the null hypothesis if the P-value is 0.001.
d) If the P-value is 0.01, we reject the null hypothesis for any alpha level greater than 0.01.

11. **Errors** For each of the following situations, state whether a Type I, a Type II, or neither error has been made. Explain briefly.

   a) A bank wants to see if the enrollment on its website is above 30% based on a small sample of customers. It tests $H_0: p = 0.3$ vs. $H_A: p > 0.3$ and rejects the null hypothesis. Later the bank finds out that actually 28% of all customers enrolled.
   b) A student tests 100 students to determine whether other students on her campus prefer Coke or Pepsi and finds no evidence that preference for Coke is not 0.5. Later, a marketing company tests all students on campus and finds no difference.
   c) A pharmaceutical company tests whether a drug lifts the headache relief rate from the 25% achieved by the placebo. The test fails to reject the null hypothesis because the P-value is 0.465. Further testing shows that the drug actually relieves headaches in 38% of people.

12. **More errors** For each of the following situations, state whether a Type I, a Type II, or neither error has been made.

   a) A test of $H_0: p = 0.8$ vs. $H_A: p < 0.8$ fails to reject the null hypothesis. Later it is discovered that $p = 0.9$.
   b) A test of $H_0: p = 0.5$ vs. $H_A: p \neq 0.5$ rejects the null hypothesis. Later it is discovered that $p = 0.65$.
   c) A test of $H_0: p = 0.7$ vs. $H_A: p < 0.7$ fails to reject the null hypothesis. Later it is discovered that $p = 0.6$.

13. **One sided or two?** In each of the following situations, is the alternative hypothesis one-sided or two-sided? What are the hypotheses?

   a) A business student conducts a taste test to see whether students prefer Diet Coke or Diet Pepsi.
   b) PepsiCo recently reformulated Diet Pepsi in an attempt to appeal to teenagers. The company runs a taste test to see if the new formula appeals to more teenagers than the standard formula.
   c) A budget override in a small town requires a two-thirds majority to pass. A local newspaper conducts a poll to see if there's evidence it will pass.
   d) One financial theory states that the stock market will go up or down with equal probability. A student collects data over several years to test the theory.

14. **Which alternative?** In each of the following situations, is the alternative hypothesis one-sided or two-sided? What are the hypotheses?

   a) A college dining service conducts a survey to see if students prefer plastic or metal cutlery.
   b) In recent years, 10% of college juniors have applied for study abroad. The dean's office conducts a survey to see if that's changed this year.
   c) A pharmaceutical company conducts a clinical trial to see if more patients who take a new drug experience headache relief than the 22% who claimed relief after taking the placebo.
   d) At a small computer peripherals company, only 60% of the hard drives produced passed all their performance tests the first time. Management recently invested a lot of resources into the production system and now conducts a test to see if it helped.

15. **P-value** A medical researcher tested a new treatment for poison ivy against the traditional ointment. He concluded that the new treatment is more effective. Explain what the P-value of 0.047 means in this context.

16. **Another P-value** Have harsher penalties and ad campaigns increased seat belt use among drivers and passengers? Observations of randomly selected cars failed to find evidence of a significant change compared with three years ago. Explain what the study's P-value of 0.17 means in this context.

17. **Alpha** A researcher developing scanners to search for hidden weapons at airports has concluded that a new device is significantly better than the current scanner. They made this decision based on a test using $\alpha = 0.05$. Would they have made the same decision at $\alpha = 0.10$? How about $\alpha = 0.01$? Explain.

18. **Alpha again** Environmentalists concerned about the impact of high-frequency radio transmissions on birds found that there was no evidence of a higher mortality rate among hatchlings in nests near cell towers. They based this conclusion on a test using $\alpha = 0.05$. Would they have made the same decision at $\alpha = 0.10$? How about $\alpha = 0.01$? Explain.

19. **Significant?** Public health officials believe that 90% of children have been vaccinated against measles. A random survey of medical records at many schools across the country found that, among more than 13,000 children, only 89.4% had been vaccinated. A statistician would reject the 90% hypothesis with a P-value of $P = 0.011$.

   a) Explain what the P-value means in this context.
   b) The result is statistically significant, but is it important? Comment.

20. **Significant again?** A new reading program may reduce the number of elementary school students who read below grade level. The company that developed this program supplied materials and teacher training for a large-scale randomized experiment involving nearly 8500 children in several different school districts. Statistical analysis of the results showed that the percentage of students who did not meet the grade-level goal was reduced from 15.9% to 15.1%. The hypothesis that the new reading program produced no improvement was rejected with a P-value of 0.023.

   a) Explain what the P-value means in this context.
   b) Even though this reading method has been shown to be significantly better, why might you not recommend that your local school adopt it?

21. **Video Gamers 2020** The Entertainment Software Association releases an annual report on the status of the video game industry. Using a random sample of 4000 gamers, the 2020 report states 38% of gamers are between the ages of 18 and 34. (Source: https://www.theesa.com/wp-content/uploads/2021/03/Final-Edited-2020-ESA_Essential_facts.pdf)

   a) Estimate the percentage of gamers who are between the ages of 18 and 34. Use a 98% confidence interval. Check the conditions first.

b) A video game store owner believed that as many as 35% of gamers are 18 to 34, and targets his advertising accordingly. He wishes to conduct a hypothesis test to see if the fraction is in fact higher than 35%. What does your confidence interval indicate? Explain.

c) What is the level of significance of this test? Explain.

22. Game with me? The survey in Exercise 21 also reported that 65% of gamers play games with others when they are gaming. The random sample was 4000 gamers.

a) Estimate the true proportion of gamers that play with others. Use a 99% confidence interval. Don't forget to check the conditions.

b) Does your confidence interval provide evidence that two-thirds of gamers play with others? Explain.

c) What is the significance level of this test? Explain.

23. Approval 2017 In September 2017, Donald Trump's approval rating stood at 40% in a CNN poll of 1500 randomly surveyed U.S. adults.

a) Make a 95% confidence interval for his approval rating by all U.S. adults.

b) Based on the confidence interval, test the null hypothesis that this is no worse than Obama's approval rating of 52.7% at the same point in his presidency.

24. Hard times In June 2010, a poll of 800 randomly selected working men found that 9% had taken on a second job to help pay the bills. (www.careerbuilder.com)

a) Estimate the true percentage of men that are taking on second jobs by constructing a 95% confidence interval.

b) A pundit on a TV news show claimed that only 6% of working men had a second job. Use your confidence interval to test whether their claim is plausible given the poll data.

25. Dogs Canine hip dysplasia is a degenerative disease that causes pain in many dogs. Sometimes advanced warning signs appear in puppies as young as 6 months. A veterinarian checked 42 puppies whose owners brought them to a vaccination clinic, and she found 5 with early hip dysplasia. She considers this group to be a random sample of all puppies.

a) Explain why we cannot use this information to construct a confidence interval for the rate of occurrence of early hip dysplasia among all 6-month-old puppies even if we can treat this as a random sample.

b) Give an example of one way in which treating this as a random sample of puppies might be problematic.

26. Fans A survey of 81 randomly selected people standing in line to enter a football game found that 73 of them were home team fans. Explain why we cannot use this information to construct a confidence interval for the proportion of all people at the game who are fans of the home team.

27. Loans Before lending someone money, banks must decide whether they believe the applicant will repay the loan. One strategy used is a point system. Loan officers assess information about the applicant, totaling points they award for the person's income level, credit history, current debt burden, and so on. The higher the point total, the more convinced the bank is that it's safe to make the loan. Any applicant with a lower point total than a certain cutoff score is denied a loan.

We can think of this decision as a hypothesis test. Because the bank makes its profit from the interest collected on repaid loans, its null hypothesis is that the applicant will repay the loan and therefore should get the money. Only if the person's score falls below the minimum cutoff will the bank reject the null and deny the loan. This system is reasonably reliable, but, of course, sometimes there are mistakes.

a) When a person defaults on a loan, which type of error did the bank make?

b) Which kind of error is it when the bank misses an opportunity to make a loan to someone who would have repaid it?

c) Suppose the bank decides to lower the cutoff score from 250 points to 200. Is that analogous to choosing a higher or lower value of $\alpha$ for a hypothesis test? Explain.

d) What impact does this change in the cutoff value have on the chance of each type of error?

28. Spam Spam filters try to sort your e-mails, deciding which are real messages and which are unwanted. One method used is a point system. The filter reads each incoming e-mail and assigns points to the sender, the subject, key words in the message, and so on. The higher the point total, the more likely it is that the message is unwanted. The filter has a cutoff value for the point total; any message rated lower than that cutoff passes through to your inbox, and the rest, suspected to be spam, are diverted to the junk mailbox.

We can think of the filter's decision as a hypothesis test. The null hypothesis is that the e-mail is a real message and should go to your inbox. A higher point total provides evidence that the message may be spam; when there's sufficient evidence, the filter rejects the null, classifying the message as junk. This usually works pretty well, but, of course, sometimes the filter makes a mistake.

a) When the filter allows spam to slip through into your inbox, which kind of error is that?

b) Which kind of error is it when a real message gets classified as junk?

c) Some filters allow the user (that's you) to adjust the cutoff. Suppose your filter has a default cutoff of 50 points, but you reset it to 60. Is that analogous to choosing a higher or lower value of $\alpha$ for a hypothesis test? Explain.

d) What impact does this change in the cutoff value have on the chance of each type of error?

29. Second loan Exercise 27 describes the loan score method a bank uses to decide which applicants it will lend money. Only if the total points awarded for various aspects of an applicant's financial condition fail to add up to a minimum cutoff score set by the bank will the loan be denied.

a) In this context, what is meant by the power of the test?

b) What could the bank do to increase the power?

c) What's the disadvantage of doing that?

30. Spam, spam, spam, spam! Consider again the points-based spam filter described in Exercise 28. When the points assigned to various components of an e-mail exceed the cutoff value you've set, the filter rejects its null hypothesis (that the message is real) and diverts that e-mail to a junk mailbox.

a) In this context, what is meant by the power of the test?

b) What could you do to increase the filter's power?

c) What's the disadvantage of doing that?

**31. Homeowners 2019** In 2019, the U.S. Census Bureau reported that 65.1% of American families owned their homes—up slightly from the 20-year low of 63.8%. Census data reveal that the ownership rate in one small city is much lower. The city council is debating a plan to offer tax breaks to first-time home buyers in order to encourage people to become homeowners. It decides to adopt the plan on a 2-year trial basis and use the data it collects to make a decision about continuing the tax breaks. Because this plan costs the city tax revenues, it will continue to use it only if there is strong evidence that the rate of homeownership is increasing.

a) In words, what will the city council's hypotheses be?
b) What would a Type I error be?
c) What would a Type II error be?
d) For each type of error, tell who would be harmed.
e) What would the power of the test represent in this context?

**32. Alzheimer's** Testing for Alzheimer's disease can be a long and expensive process, consisting of lengthy tests and medical diagnosis. Recently, a group of researchers devised a 7-minute test to serve as a quick screen for the disease for use in the general population of senior citizens. A patient who tested positive would then go through the more expensive battery of tests and medical diagnosis. The authors reported a false positive rate of 4% and a false negative rate of 8%.

a) Put this in the context of a hypothesis test. What are the null and alternative hypotheses?
b) What would a Type I error mean?
c) What would a Type II error mean?
d) Which is worse here, a Type I or Type II error? Explain.
e) What is the power of this test?

**33. Testing cars** A clean air standard requires that vehicle exhaust emissions not exceed specified limits for various pollutants. Many states require that cars be tested annually to be sure they meet these standards. Suppose state regulators double-check a random sample of cars that a suspect repair shop has certified as okay. They will revoke the shop's license if they find significant evidence that the shop is certifying vehicles that do not meet standards.

a) In this context, what is a Type I error?
b) In this context, what is a Type II error?
c) Which type of error would the shop's owner consider more serious?
d) Which type of error might environmentalists consider more serious?

**34. Quality control** Production managers on an assembly line must monitor the output to be sure that the level of defective products remains small. They periodically inspect a random sample of the items produced. If they find a significant increase in the proportion of items that must be rejected, they will halt the assembly process until the problem can be identified and repaired.

a) In this context, what is a Type I error?
b) In this context, what is a Type II error?
c) Which type of error would the factory owner consider more serious?
d) Which type of error might customers consider more serious?

**35. Cars again** As in Exercise 33, state regulators are checking up on repair shops to see if they are certifying vehicles that do not meet pollution standards.

a) In this context, what is meant by the power of the test the regulators are conducting?
b) Will the power be greater if they test 20 or 40 cars? Why?
c) Will the power be greater if they use a 5% or a 10% level of significance? Why?
d) Will the power be greater if the repair shop's inspectors are only a little out of compliance or a lot? Why?

**36. Production** Consider again the task of the quality control inspectors in Exercise 34.

a) In this context, what is meant by the power of the test the inspectors conduct?
b) They are currently testing 5 items each hour. Someone has proposed that they test 10 instead. What are the advantages and disadvantages of such a change?
c) Their test currently uses a 5% level of significance. What are the advantages and disadvantages of changing to an alpha level of 1%?
d) Suppose that, as a day passes, one of the machines on the assembly line produces more and more items that are defective. How will this affect the power of the test?

**37. Equal opportunity?** A high school is forming a committee of juniors and seniors that will provide feedback to the school's administration. The seniors at the school are frustrated because only 40% of the committee is made up of seniors, when the number of seniors and and juniors at the school is about the same. They ask the administration to reconsider this underrepresentation. As a stats student, you run a test to see if the low percentage of seniors could just be attributed to random variation.

a) Is this a one-tailed or a two-tailed test? Why?
b) In this context, what would a Type I error be?
c) In this context, what would a Type II error be?
d) In this context, what is meant by the power of the test?
e) If the hypothesis is tested at the 5% level of significance instead of 1%, how will this affect the power of the test?
f) The students' complaint is based on a committee size of 40 students. Is the power of the test higher than, lower than, or the same as it would be if it were based on 80 students?

**38. Stop signs** Highway safety engineers test new road signs, hoping that increased reflectivity will make them more visible to drivers. Volunteers drive through a test course with several of the new- and old-style signs and rate which kind shows up the best.

a) Is this a one-tailed or a two-tailed test? Why?
b) In this context, what would a Type I error be?
c) In this context, what would a Type II error be?
d) In this context, what is meant by the power of the test?
e) If the hypothesis is tested at the 1% level of significance instead of 5%, how will this affect the power of the test?
f) The engineers hoped to base their decision on the reactions of 50 drivers, but time and budget constraints may force them to cut back to 20. How would this affect the power of the test? Explain.

**39. Dropouts** A Statistics professor has observed that for several years about 13% of the students who initially enroll in his Introductory Statistics course withdraw before the end of the semester. A salesperson suggests that he try a statistics software package that gets students more involved with computers, predicting that it will cut the dropout rate. The software is expensive, and the salesperson offers to let the professor use it for a semester to see if the dropout rate goes down significantly. The professor will have to pay for the software only if he chooses to continue using it.

a) Is this a one-tailed or two-tailed test? Explain.
b) Write the null and alternative hypotheses.
c) In this context, explain what would happen if the professor makes a Type I error.
d) In this context, explain what would happen if the professor makes a Type II error.
e) What is meant by the power of this test?

40. Ads A company is willing to renew its advertising contract with a local radio station only if the station can prove that more than 20% of the residents of the city have heard the ad and recognize the company's product. The radio station conducts a random phone survey of 400 people.

a) What are the hypotheses?
b) The station plans to conduct this test using a 10% level of significance, but the company wants the significance level lowered to 5%. Why?
c) What is meant by the power of this test?
d) For which level of significance will the power of this test be higher? Why?
e) The station finally agrees to use $\alpha = 0.05$, but the company proposes that the station call 600 people instead of the 400 initially proposed. Will that make the risk of Type II error higher or lower? Explain.

41. Dropouts, part II Initially, 203 students signed up for the Stats course in Exercise 39. They used the software suggested by the salesperson, and only 11 dropped out of the course.

a) Should the professor spend the money for this software? Support your recommendation with an appropriate test.
b) Explain what your P-value means in this context.

42. Testing the ads The company in Exercise 40 contacts 600 people selected at random, and only 133 remember the ad.

a) Should the company renew the contract? Support your recommendation with an appropriate test.
b) Explain what your P-value means in this context.

43. Two coins In a drawer are two coins. They look the same, but one coin produces heads 90% of the time when spun while the other one produces heads only 30% of the time. You select one of the coins. You are allowed to spin it *once* and then must decide whether the coin is the 90%- or the 30%-head coin. Your null hypothesis is that your coin produces 90% heads.

a) What is the alternative hypothesis?
b) Given that the outcome of your spin is tails, what would you decide? What if it were heads?
c) How large is $\alpha$ in this case?
d) How large is the power of this test? (*Hint:* How many possibilities are in the alternative hypothesis?)
e) How could you lower the probability of a Type I error and increase the power of the test at the same time?

44. Faulty or not? You are in charge of shipping computers to customers. You learn that a faulty disk drive was put into some of the machines. There's a simple test you can perform, but it's not perfect. All but 4% of the time, a good disk drive passes the test, but unfortunately, 35% of the bad disk drives pass the test, too. You have to decide on the basis of one test whether the disk drive is good or bad. Make this a hypothesis test.

a) What are the null and alternative hypotheses?
b) Given that a computer fails the test, what would you decide? What if it passes the test?
c) How large is $\alpha$ for this test?
d) What is the power of this test? (*Hint:* How many possibilities are in the alternative hypothesis?)

45. Hoops A basketball player with a poor foul-shot record practices intensively during the off-season. He tells the coach that he has raised his proficiency from 60% to 80%. Dubious, the coach asks him to take 10 shots, and is surprised when the player hits 9 out of 10. Did the player prove that he has improved?

a) Suppose the player really is no better than before—still a 60% shooter. What's the probability he can hit at least 9 of 10 shots anyway? (*Hint:* Use a Binomial model.)
b) If that is what happened, now the coach thinks the player has improved when he has not. Which type of error is that?
c) If the player really can hit 80% now, and it takes at least 9 out of 10 successful shots to convince the coach, what's the power of the test?
d) List two ways the coach and player could increase the power to detect any improvement.

46. Pottery An artist experimenting with clay to create pottery with a special texture has been experiencing difficulty with these special pieces. About 40% break in the kiln during firing. Hoping to solve this problem, she buys some more expensive clay from another supplier. She plans to make and fire 10 pieces and will decide to use the new clay if at most one of them breaks.

a) Suppose the new, expensive clay really is no better than her usual clay. What's the probability that this test convinces her to use it anyway? (*Hint:* Use a Binomial model.)
b) If she decides to switch to the new clay and it is no better, what kind of error did she commit?
c) If the new clay really can reduce breakage to only 20%, what's the probability that her test will not detect the improvement?
d) How can she improve the power of her test? Offer at least two suggestions.

47. Skin care A pharmaceutical company is developing a topical cream to address a particular type of skin rash. This condition goes away on its own within a month in about 40% of cases. The company has set up a preliminary study (not a randomized experiment yet—it just wants to see if the cure rate with the new cream is better than 40%).

a) State the null and alternative hypotheses.
b) The company will test the drug on 50 subjects. What are the mean and standard deviation of the proportion of patients who would be cured within one month?
c) You would expect that in 95% of trials, between ____% and ____% of patients would be cured within one month.

The model below shows what would be expected if the study were performed repeatedly and the null hypothesis were true. The vertical lines appear at 1.96 standard deviations above and below $p_0 = 0.4$.

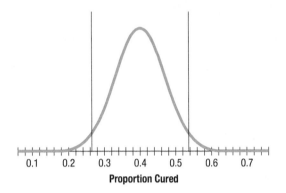

As it turns out, the new medication does not affect the cure rates at all! We have conducted a simulation to see what might happen if the researchers were to repeat their study with many sets of 50 subjects. The results of 200 runs of the simulation are shown below.

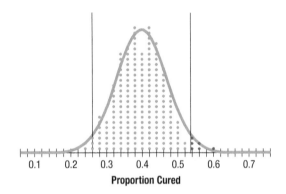

d) If the actual experiment resulted in the proportion cured represented by any green dot, would the researchers reject the null hypothesis? Would this be the right decision? If not, what type of error would they be making? According to the simulation results, what is the approximate probability that they will reach this conclusion?
e) If the actual experiment resulted in the proportion cured represented by any red dot, would the researchers reject the null hypothesis? Would this be the right decision? If not, what type of error would they be making? According to the simulation results, what is the approximate probability that they will reach this conclusion?

**48.** More skin care In Exercise 47 we looked at a medication that did not affect the cure rate for a skin rash. The company now has a new formula that it is pretty sure does work. The developer of the formula claims that with this new medication, the skin rash clears up in 60% of people within a month. The company wants to conduct another preliminary study to see whether the new formula actually is better than doing nothing. It plans to use 50 subjects again, so it performed another simulation.

The null hypothesis is still represented by the Normal model. The researchers performed a simulation, this time with 200 trials based on the developer's claim that the rash will clear up in 60% of cases if they use the new medication.

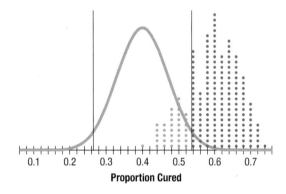

a) Suppose the developer is correct. If the actual experiment resulted in the proportion cured represented by any green dot, would the researchers reject the null hypothesis? Would this be the right decision? If not, what type of error would they be making? According to the simulation results, what is the approximate probability that they will reach this conclusion?
b) If the actual experiment resulted in the proportion cured represented by any red dot, would the researchers reject the null hypothesis? Would this be the right decision? If not, what type of error would they be making? According to the simulation results, what is the approximate probability that they will reach this conclusion?
c) Which part, a or b, is asking for the power of the test? Which is asking for $\beta$?
d) The plot below shows both the model for $H_0$ and 200 simulated results for $p = 0.6$, but with $n = 100$. How does this affect the probabilities asked for in parts a and b?

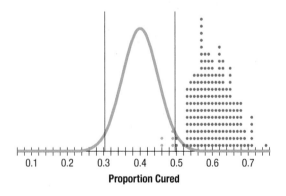

## JUST CHECKING

#### Answers

1. The result is statistically significant because the P-value is very small. There's no way of knowing whether the difference is practically significant, given just the P-value. We need to know the effect size.

2. They should neither be more concerned nor heave a sigh of relief. A P-value of 0.19 is low enough to suggest that it's a bit unusual to see a chimp sleeping this much. But it's also not *so* low to suggest that the chimp's sleep is an extremely unusual, or alarming, outlier.

3. At an alpha level of 0.05, a true null hypothesis is rejected with probability 0.05, because that's the chance of getting a test statistic that's among the most unusual 5% of possible outcomes.

4. An advantage of using a higher significance level would be that the test would have greater power: if the new menu is effective, the fast food chain will be more likely to find significant evidence of it in their study. A disadvantage is that increasing $\alpha$ would make it more likely for the experiment to apparently provide evidence that the new menu is effective, even if it really isn't (a Type I error).

5. Using more restaurants in the study would increase the power of the hypothesis test, making it more likely that if the new menu is effective, the study will find evidence of it. But using more restaurants also requires a greater expenditure of money and effort to conduct the study. (A very large fast food chain would probably have little difficulty conducting such a study with 200 of its restaurants.)

6. It's possible that the proportion of customers who buy combo meals tends to differ in large cities compared to smaller towns and rural areas. Conducting the study only in large cities would reduce sampling variability, thereby making the test more powerful—that is, more likely to find evidence of a menu effect if there is one. But focusing only on large cities would mean that any conclusions about the menu might not extrapolate to stores in less urban areas.

7. An increase of one customer out of every two hundred is an increase of only half of a percentage point, which seems pretty small. With the effect being real but very subtle, it seems likely that the study will conclude with a Type II error, a failure to discern a menu effect even though there is one.

# 21 Comparing Two Proportions

| WHO | 6971 male drivers |
|---|---|
| WHAT | Seat belt use |
| WHY | Highway safety |
| WHEN | 2007 |
| WHERE | Massachusetts |

Do men take more risks than women? Psychologists have documented that in many situations, men choose riskier behavior than women do. An observational study on seat belt use in Massachusetts found that, not surprisingly, male drivers wear seat belts less often than women do.

But the study also found an unexpected pattern. Men's belt-wearing jumped more than 16 percentage points when they were sitting next to a female passenger. Seat belt use was recorded at 161 locations in Massachusetts, using random-sampling methods developed by the National Highway Traffic Safety Administration (NHTSA).[1] Female drivers wore belts more than 70% of the time, regardless of the gender of their passengers. Of 4208 male drivers with female passengers, 2777 (66.0%) were belted. But among 2763 male drivers with male passengers only, 1363 (49.3%) wore seat belts. This was a random sample, but it suggests there may be a shift in men's risk-taking behavior when women are present.

Comparisons between two percentages are much more common than questions about isolated percentages. And they are more interesting. We often want to know how two groups differ, whether a treatment is better than a placebo control, or whether this year's results are better than last year's.

## Another Ruler

We know the difference between the proportions of men wearing seat belts seen in the *sample*. It's 16.7%. But what's the difference for *all* men? We know that our estimate probably isn't exactly right. To be able to say more, we need a new ruler—the standard

---

[1] Because cars were observed in passing, gender classifications were apparently made based on dress, hair style, and other cues, and thus do not reflect personal identities See, for example, bjs.gov/content/pub/pdf/amtireg.pdf

deviation of the sampling distribution model for the difference in the sample proportions. Now we have two proportions, and each will vary from sample to sample. We are interested in the difference between them. So what is the correct standard deviation?

The answer comes to us from Chapter 15. Remember the Pythagorean Theorem of Statistics?

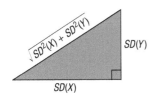

*The variance of the sum **or difference** of two independent random variables is the sum of their variances.*

This is such an important (and powerful) idea in Statistics that it's worth pausing a moment to review the reasoning. Here's some intuition about why variation increases even when we subtract two random quantities.

Grab a full box of cereal. The box claims to contain 16 ounces of cereal. We know that's not exact: There's some small variation from box to box. Now pour a bowl of cereal. Of course, your 2-ounce serving will not be exactly 2 ounces. There'll be some variation there, too. How much cereal would you guess was left in the box? Do you think your guess will be as close as your guess for the full box? *After* you pour your bowl, the amount of cereal in the box is still a random quantity (with a smaller mean than before), but it is even *more variable* because of the additional variation in the amount you poured.

### Variation Grows

Combining independent random quantities always *increases* the overall variation, so even for *differences* of independent random variables, **variances add**.

According to our rule, the variance of the amount of cereal left in the box would now be the *sum* of the two *variances*.

We want a standard deviation, not a variance, but that's just a square root away. We can write symbolically what we've just said:

$$Var(X - Y) = Var(X) + Var(Y), \text{ so}$$
$$SD(X - Y) = \sqrt{SD^2(X) + SD^2(Y)} = \sqrt{Var(X) + Var(Y)}.$$

Be careful, though—this simple formula applies only when $X$ and $Y$ are independent. Just as the Pythagorean Theorem works only for right triangles, our formula works only for independent random variables. Always check for independence before using it.

# The Standard Deviation of the Difference Between Two Proportions

Fortunately, proportions observed in independent random samples *are* independent, so we can put those two proportions in for $X$ and $Y$ and add their variances. We just need to use careful notation to keep things straight.

When we have two samples, each can have a different size and proportion value, so we keep them straight with subscripts. Often we choose subscripts that remind us of the groups. For our example, we might use "$_M$" and "$_F$", but generically we'll just use "$_1$" and "$_2$". We will represent the two sample proportions as $\hat{p}_1$ and $\hat{p}_2$, and the two sample sizes as $n_1$ and $n_2$.

The standard deviations of the sample proportions are $SD(\hat{p}_1) = \sqrt{\dfrac{p_1 q_1}{n_1}}$ and $SD(\hat{p}_2) = \sqrt{\dfrac{p_2 q_2}{n_2}}$, so the variance of the difference in the proportions is

### Remember

For *independent* random variables, variances add.

$$Var(\hat{p}_1 - \hat{p}_2) = \left(\sqrt{\dfrac{p_1 q_1}{n_1}}\right)^2 + \left(\sqrt{\dfrac{p_2 q_2}{n_2}}\right)^2 = \dfrac{p_1 q_1}{n_1} + \dfrac{p_2 q_2}{n_2}.$$

The standard deviation is the square root of that variance:

$$SD(\hat{p}_1 - \hat{p}_2) = \sqrt{\dfrac{p_1 q_1}{n_1} + \dfrac{p_2 q_2}{n_2}}.$$

We usually don't know the true values of $p_1$ and $p_2$. When we have the sample proportions in hand from the data, we use them to estimate the variances. So the standard error is

$$SE(\hat{p}_1 - \hat{p}_2) = \sqrt{\frac{\hat{p}_1\hat{q}_1}{n_1} + \frac{\hat{p}_2\hat{q}_2}{n_2}}.$$

### FOR EXAMPLE
#### Finding the Standard Error of a Difference in Proportions

The COVID-19 outbreak resulted in mass closures of businesses such as restaurants all across the United States. A Pew Research survey conducted in April 2020 asked whether respondents had ordered groceries online as a result of the coronavirus outbreak. 23% of 2698 college-educated adult respondents answered affirmatively. Compare that to 19% of 696 adult respondents without a college education who responded affirmatively.[2]

QUESTION: What's the standard error of the difference in the sample proportions?

ANSWER: The individual respondents were selected at random, so it is reasonable to assume that their responses were independent and that the two groups are independent as well. So we can add the variances:

$$SE(\hat{p}_{NC}) = \sqrt{\frac{0.19 \times (1 - 0.19)}{696}} = 0.01487$$

$$SE(\hat{p}_C) = \sqrt{\frac{0.23 \times (1 - 0.23)}{2698}} = 0.008102$$

$$SE(\hat{p}_{NC} - \hat{p}_C) = \sqrt{0.008102^2 + 0.01487^2} = 0.01693$$

## Assumptions and Conditions

As always, we need to check assumptions and conditions. The first one is new, and very important.

### Independence Assumptions

Because we are comparing two groups, we need a new Independence Assumption. In fact, this is the most important of these assumptions. If it is violated, these methods just won't work.

**Independent Groups Assumption**: The two groups we're comparing must be independent *of each other*. Usually, the independence of the groups from each other is evident from the way the data were collected.

Why is the Independent Groups Assumption so important? If we compare husbands with their wives, or a group of subjects before and after some treatment, we can't just add the variances. Married people may have attitudes similar to their spouses'. Subjects' performance after a treatment might very well be related to their performance before the treatment. That means the proportions are not independent and the Pythagorean-style variance formula does not hold. We'll see a way to compare a common kind of nonindependent sample in a later chapter.

---

[2] pewresearch.org/fact-tank/2020/04/30/from-virtual-parties-to-ordering-food-how-americans-areusingthe-internet-during-covid-19/

You'll recognize the rest of the assumptions and conditions.

**Independence Assumption:** Within each group, the data should be based on results for independent individuals. We can't check that for certain, but we *can* check the following two conditions:

**Randomization Condition:** The data in each group should be drawn independently and at random from a homogeneous population or generated by a randomized comparative experiment.

**The 10% Condition:** If the data are sampled without replacement, the sample should not exceed 10% of the population.

## Sample Size Condition

*Each* of the groups must be big enough. As with individual proportions, we need larger groups to estimate proportions that are near 0% or 100%. We usually check the Success/Failure Condition for each group.

**Success/Failure Condition:** Both groups are big enough that at least 10 successes and at least 10 failures have been observed in each.

### FOR EXAMPLE
### Checking Assumptions and Conditions

RECAP: Among randomly sampled adults, 23% of 2698 college grads had ordered groceries online during the COVID-19 outbreak compared to 19% of 696 repondents without a college degree.

QUESTION: Can we use these results to make inferences about the online grocery-ordering behavior of all American adults during the COVID-19 outbreak?

ANSWER:

✓ **Randomization Condition:** Both samples (those with and without college degrees) were chosen randomly.

✓ **Independent Groups Assumption:** Because the samples were selected at random, it's reasonable to believe the online shopping behaviors of the two cohorts are independent.

✓ **The 10% Condition:** There are millions of adults among both those who do and do not hold a college degree. The few thousands of people in the samples both represent far less than 10% of their respective populations.

✓ **Success/Failure Condition:** $696(0.19) = 132$ college grads ordered groceries online and $696(0.81) = 564$ did not. $2698(0.23) = 621$ of those without college degrees ordered groceries online and $2698(0.77) = 2077$ did not. All counts are greater than 10.

Because all assumptions and conditions are satisfied, it's okay to proceed with inference for the difference in proportions.

(Note: The observed counts of successes and failures were rounded off to whole numbers. We're using the reported percentages to approximate the actual counts, which were not reported.)

# A Confidence Interval

We're almost there. We just need one more fact about proportions. We already know that for large enough samples, each of our sample proportions has an approximately Normal sampling distribution. The same is true of their difference.

### Why Normal?

In Chapter 15 you learned that sums and differences of Independent Normal random variables also follow a Normal model. That's the reason we use a Normal model for the difference of two independent sample proportions.

### The Sampling Distribution Model for a Difference Between Two Independent Sample Proportions

Provided that the sampled values are independent, the samples are independent, and the sample sizes are large enough, the sampling distribution of $\hat{p}_1 - \hat{p}_2$ is modeled by a Normal model with mean $\mu = P_1 - P_2$ and standard deviation

$$SD(\hat{p}_1 - \hat{p}_2) = \sqrt{\frac{p_1 q_1}{n_1} + \frac{p_2 q_2}{n_2}}.$$

The sampling distribution model and the standard deviation give us all we need to find a margin of error for the difference in proportions—or at least they would if we knew the true proportions, $p_1$ and $p_2$. However, we don't know the true values, so we'll work with the observed proportions, $\hat{p}_1$ and $\hat{p}_2$, and use $SE(\hat{p}_1 - \hat{p}_2)$ to estimate the standard deviation. The rest is just like a one-proportion z-interval.

### A Two-Proportion z-Interval

When the conditions are met, we are ready to find the confidence interval for the difference of two proportions, $p_1 - p_2$. The confidence interval is

$$(\hat{p}_1 - \hat{q}_1) \pm z^* \times SE(\hat{p}_1 - \hat{p}_2)$$

where we find the standard error of the difference,

$$SE(\hat{p}_1 - \hat{p}_2) = \sqrt{\frac{\hat{p}_1 \hat{q}_1}{n_1} + \frac{\hat{p}_2 \hat{q}_2}{n_2}},$$

from the observed proportions.

The critical value $z^*$ depends on the particular confidence level, C, that we specify.

### FOR EXAMPLE

#### Finding a Two-Proportion z-Interval

**RECAP:** Among randomly sampled adults, 19% of 696 of respondents without a college degree had ordered groceries online during the COVID-19 outbreak compared to 23% of 2698 college grads. We calculated the standard error of the difference in sample proportions to be 0.01693.

**QUESTION:** What does a confidence interval say about the difference in the proportions that ordered groceries online?

**ANSWER:** A 95% confidence interval for $p_C - p_{NC}$ is
$(\hat{p}_C - \hat{p}_{NC}) \pm z^* SE(\hat{p}_C - \hat{p}_{NC})$:

$$(0.23 - 0.19) \pm 1.96(0.01693) = (0.0068, 0.0732)$$

We can be 95% confident that among American adults, the proportion of college grads who ordered groceries online during the COVID-19 outbreak was between 0.68 and 7.32 percentage points higher than the proportion of those without college degrees who did.

## STEP-BY-STEP EXAMPLE

### A Two-Proportion z-Interval

Now we are ready to be more precise about the passenger-based gap in male drivers' seat belt use. We'll estimate the difference with a confidence interval using a method called the two-proportion z-interval and follow the four confidence interval steps.

**QUESTION:** How much difference is there in the proportion of male drivers who wear seat belts when sitting next to a male passenger and the proportion who wear seat belts when sitting next to a female passenger?

---

**THINK**    **PLAN** State what you want to know. Discuss the variables and the W's.

Identify the parameter you wish to estimate. (It usually doesn't matter in which direction we subtract, so, for convenience, we usually choose the direction with a positive difference.)

Choose and state a confidence level.

**MODEL** Think about the assumptions and check the conditions.

I want to know the difference in the population proportion, $p_M$, of male drivers who wear seat belts when sitting next to a man and $p_F$, the proportion who wear seat belts when sitting next to a woman. The data are from a random sample of drivers in Massachusetts in 2007, observed according to procedures developed by the NHTSA. The parameter of interest is the difference $p_F - p_M$.

I will find a 95% confidence interval for this difference.

✓ **Independent Groups Assumption:** There's no reason to believe that seat belt use among drivers with male passengers and those with female passengers are not independent.

✓ **Independence Assumption:** Driver behavior was independent from car to car.

✓ **Randomization Condition:** The NHTSA methods are more complex than an SRS, but they result in a suitable random sample.

✓ **10% Condition:** The samples include far fewer than 10% of all male drivers accompanied by male or by female passengers.

The Success/Failure Condition must hold for each group.

✓ **Success/Failure Condition:** Among male drivers with female passengers, 2777 wore seat belts and 1431 did not; of those driving with male passengers, 1363 wore seat belts and 1400 did not. Each group contained far more than 10 successes and 10 failures.

State the sampling distribution model for the statistic.

Choose your method.

Under these conditions, the sampling distribution of the difference between the sample proportions is approximately Normal, so I'll find a **two-proportion z-interval**.

---

**SHOW**    **MECHANICS** Construct the confidence interval.

I know
$$n_F = 4208, n_M = 2763.$$

The observed sample proportions are

$$\hat{p}_F = \frac{2777}{4208} = 0.660, \hat{p}_M = \frac{1363}{2763} = 0.493$$

As often happens, the key step in finding the confidence interval is estimating the standard deviation of the sampling distribution model of the statistic. Here the statistic is the difference between the sample proportion of men who wear seat belts when they have a female passenger and the sample proportion who do so with a male passenger.

I'll estimate the SD of the difference with

$$SE(\hat{p}_F - \hat{p}_M) = \sqrt{\frac{\hat{p}_F \hat{q}_F}{n_F} + \frac{\hat{p}_M \hat{p}_M}{n_M}}$$

$$= \sqrt{\frac{(0.660)(0.340)}{4208} + \frac{(0.493)(0.507)}{2763}}$$

$$= 0.012$$

The sampling distribution is Normal, so the critical value for a 95% confidence interval, $Z^*$, is 1.96. The margin of error is the critical value times the SE.

$$ME = z^* \times SE(\hat{p}_F - \hat{p}_M)$$
$$= 1.96(0.012) = 0.024$$

The observed difference in proportions is

$$\hat{p}_F - \hat{p}_M = 0.660 - 0.493 = 0.167,$$

so the 95% confidence interval is

The confidence interval is the statistic ±ME.

$$0.167 \pm 0.024$$
or 14.3% to 19.1%

**TELL** **CONCLUSION** Interpret your confidence interval in the proper context. (Remember: We're 95% confident that our interval captured the true difference.)

I am 95% confident that the proportion of male drivers who wear seat belts when driving next to a female passenger is between 14.3 and 19.1 percentage points higher than the proportion who wear seat belts when driving next to a male passenger.

This is an interesting result—but be careful not to try to say too much! In Massachusetts, overall seat belt use is lower than the national average, so we can't be certain that these results generalize to other states. And these were two different groups of men, so we can't say that any *individual* man is more likely to buckle up when he has a female passenger than when he has a male passenger. You can probably think of several alternative explanations; we'll suggest just a couple. Perhaps age is a lurking variable: Maybe younger men are less likely to wear seat belts and are also more likely to be driving with only male passengers. Or maybe men who don't wear seat belts are seen as reckless so female friends decline to ride with them driving!

## TI TIPS

### Finding the Confidence Interval

You can use a routine in the STAT TESTS menu to create confidence intervals for the difference of two proportions. Remember, the calculator can do only the mechanics—checking conditions and writing conclusions are still up to you.

A Gallup Poll asked whether the attribute "intelligent" described men in general. The poll revealed that 28% of 506 men thought it did, but only 14% of 520 women agreed. We want to estimate the size of the gender gap by creating a 95% confidence interval.

◆ Go to the STAT TESTS menu. Scroll down the list and select 2-PropZInt.

◆ Enter the observed number of males: .28*506. Remember that the actual number of males must be a whole number, so be sure to round off.

◆ Enter the sample size: 506 males.

◆ Repeat those entries for women: .14*520 agreed, and the sample size was 520.

◆ Specify the desired confidence level.

◆ Calculate the result.

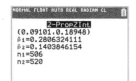

And now explain what you see: We are 95% confident that the proportion of men who think the attribute "intelligent" describes males in general is between 9 and 19 percentage points higher than the proportion of women who think so.

## JUST CHECKING

A public broadcasting station plans to launch a special appeal for additional contributions from current members. Unsure of the most effective way to contact people, it runs an experiment. It randomly selects two groups of current members. It sends the same request for donations to everyone, but the request goes to one group by e-mail and to the other group by regular mail. The station was successful in getting contributions from 26% of the members it e-mailed but only from 15% of those who received the request by regular mail. A 90% confidence interval estimated the difference in donation rates to be 11% ± 7%.

1. Interpret the confidence interval in this context.

2. Based on this confidence interval, what conclusion would we reach if we tested the hypothesis that there's no difference in the response rates to the two methods of fundraising? Explain.

# Screen Time Before Bed: A Cultural Difference?

| | |
|---|---|
| WHO | Randomly sampled adults in different countries |
| WHAT | Proportion who use their laptops or computers in the hour before sleep |
| WHEN | 2013 |
| WHERE | United States, Japan, Canada, Germany, Mexico, United Kingdom |
| WHY | To study sleep behaviors |

The 2013 International Bedroom Poll[3] surveyed about 250 randomly sampled adult residents in each of six countries asking about their sleep habits. Sleep researchers generally report that

> people who regularly use their computers or laptops in the hour before trying to go to sleep are *less* likely to report getting a good night's sleep . . . , more likely to be categorized as "sleepy" . . . , and more likely to drive drowsy . . . than their counterparts.[4]

The poll found that 128, or 51%, of 251 U.S. respondents reported using a computer, laptop, or electronic tablet in the hour before trying to go to sleep almost every night. By contrast, 162, or 65%, of 250 Japanese respondents made the same admission. Is this difference of 14 percentage points real, or is it likely to be due only to sampling variability?

The question calls for a hypothesis test. Now the parameter of interest is the true *difference* between the (reported) pre-sleep habits of the two groups.

What's the appropriate null hypothesis? That's easy here. We hypothesize that there is no difference in the proportions. This is such a natural null hypothesis that we rarely consider any other. But instead of writing $H_0: p_1 = p_2$, we treat the difference as the parameter that we are testing:

$$H_0: p_1 - p_2 = 0.$$

We'll reject the null hypothesis if we see a difference in our sample proportions that's so large it's unlikely to be sampling error. How can we decide if $\hat{p}_1 - \hat{p}_2$ is unusually large? The same way we always do: we'll use a standard error to find a z-score.

---
[3]sleepfoundation.org/sites/default/files/RPT495a.pdf
[4]Excerpt from "The 2011 Sleep in America Poll" from National Sleep Foundation. Copyright © 2011, published by National Sleep Foundation.

# Everyone into the Pool

### Rounding

When we have only proportions and not the counts, we have to reconstruct the number of successes by multiplying the sample sizes by the proportions:

$Success_1 = n_1\hat{p}_1$ and $Success_2 = n_2\hat{p}_2$.

If these calculations don't come out to whole numbers, round them first. There must have been a whole number of successes, after all. (This is the *only* time you should round values in the middle of a calculation.)

Our hypothesis is about a new parameter: the *difference* in proportions. We need a standard error for that statistic. Wait—don't we know that SE already? Yes and no. We know that the standard error of the difference in sample proportions is

$$SE(\hat{p}_1 - \hat{p}_2) = \sqrt{\frac{\hat{p}_1\hat{q}_1}{n_1} + \frac{\hat{p}_2\hat{q}_2}{n_2}},$$

and we could just plug in the given numbers, but that presents a logical dilemma. Our null hypothesis says we're assuming the two proportions $p_1$ and $p_2$ are equal. Why would we then substitute the unequal values $\hat{p}_1$ and $\hat{p}_2$ as estimates? That's like saying, "I think Tony and Maria are the same age. I'd guess he's 15 and she's 18." To avoid such a contradiction, we need to come up with a single value for $\hat{p}$ in the SE formula.

Let's see how to do this for the laptop-use-before-bed example. If the null hypothesis is true, then the two groups aren't really different, so we can think of our two samples as really just parts of one bigger sample of the combined population. Overall, we saw $128 + 162 = 290$ people who use their laptops before bed out of a total of $251 + 250 = 501$ adults who responded to this question. The overall proportion of people using their laptops before bed was $290/501 = 0.5788$.

Combining the counts like this to get an estimated overall proportion is called **pooling**. Whenever we have data from different sources or different groups but we hypothesize that they really came from the same underlying population, we pool them to get better estimates.

Using the counts for each group, we can find the pooled proportion as

$$\hat{p}_{pooled} = \frac{Success_1 + Success_2}{n_1 + n_2},$$

where $Success_1$ is the number of successes in group 1 and $Success_2$ is the number of successes in group 2. That's our best estimate of the population proportion of success.

We then put this pooled value into the formula, substituting it for *both* sample proportions in the standard error formula:

$$SE_{pooled}(\hat{p}_1 - \hat{p}_2) = \sqrt{\frac{\hat{p}_{pooled}\hat{q}_{pooled}}{n_1} + \frac{\hat{p}_{pooled}\hat{q}_{pooled}}{n_2}}$$

$$\sqrt{\frac{0.5788 \times (1 - 0.5788)}{251} + \frac{0.5788 \times (1 - 0.5788)}{250}}.$$

This comes out to 0.044.

### Improving the Success/Failure Condition

The vaccine Gardasil® was introduced to prevent the strains of human papillomavirus (HPV) that are responsible for almost all cases of cervical cancer. In randomized placebo-controlled clinical trials,[5] only 1 case of HPV was diagnosed among 7897 women who received the vaccine, compared with 91 cases diagnosed among 7899 who received a placebo. The one observed HPV case ("success") doesn't meet the at-least-10-successes criterion. Surely, though, we should not refuse to test the effectiveness of the vaccine just because it failed so rarely; that would be absurd.

---

[5] *Quadrivalent Human Papillomavirus Vaccine: Recommendations of the Advisory Committee on Immunization Practices (ACIP)*, National Center for HIV/AIDS, Viral Hepatitis, STD and TB Prevention (May 2007).

For that reason, in a two-proportion z-test, the proper Success/Failure test uses the *expected* frequencies, which we can find from the pooled proportion. In this case,

$$\hat{p}_{pooled} = \frac{91 + 1}{7899 + 7897} = 0.0058$$

$$n_1\hat{p}_{pooled} = 7899(0.0058) = 46$$

$$n_2\hat{p}_{pooled} = 7897(0.0058) = 46,$$

so we can proceed with the hypothesis test.

Often it is easier just to check the observed numbers of successes and failures. If they are both greater than 10, you don't need to look further. But keep in mind that the correct test uses the expected frequencies rather than the observed ones. If you're communicating a two-proportion z-test in writing, you should compute these and show that they're both at least 10.

# The Two-Proportion z-Test

At last, we're ready to test a hypothesis about the difference of two proportions. We use the pooled estimate of the population proportion to find the standard error. That provides the yardstick we need to first calculate a z-score and then determine the P-value that allows us to decide whether the difference we see in the sample proportions is statistically significant.

### Two-Proportion z-Test

The conditions for the two-proportion z-test are the same as for the two-proportion z-interval. We are testing the hypothesis

$$H_0: p_1 - p_2 = 0.$$

Because we hypothesize that the proportions are equal, we pool the groups to find

$$\hat{p}_{pooled} = \frac{Success_1 + Success_2}{n_1 + n_2}$$

and use that pooled value to estimate the standard error:

$$SE_{pooled}(\hat{p}_1 - \hat{p}_2) = \sqrt{\frac{\hat{p}_{pooled}\hat{q}_{pooled}}{n_1} + \frac{\hat{p}_{pooled}\hat{q}_{pooled}}{n_2}}.$$

Now we find the test statistic,

$$z = \frac{(\hat{p}_1 - \hat{p}_2) - 0}{SE_{pooled}(\hat{p}_1 - \hat{p}_2)}.$$

When the conditions are met and the null hypothesis is true, this statistic follows the standard Normal model, so we can use that model to obtain a P-value.

## STEP-BY-STEP EXAMPLE

### A Two-Proportion z-Test

QUESTION: Is there a difference between the proportions of U.S. and Japanese adults who use laptops in the hour before bed?

**THINK**  **PLAN** State what you want to know. Discuss the variables and the W's.

I want to know whether the proportion of U.S. adults who use laptops before bed is different from the proportion of Japanese adults who use laptops before bed. The data are from a 2013 international survey of adults. 501 adults in the two countries responded to a question about whether they use a laptop in the hour before they go to bed.

**HYPOTHESES** We're interested only in whether there's a difference in proportions. A direction is not specified or implied so a two-sided hypothesis is appropriate.

$H_0$: There is no difference in laptop use rates between the two countries:
$$p_{U.S.} - p_{Japan} = 0.$$
$H_A$: The rates are different: $p_{U.S.} - p_{Japan} \neq 0$.

**MODEL** Think about the assumptions and check the conditions.

✓ **Independent Groups Assumption**: The two groups are independent of each other because the U.S. and Japanese respondents were selected at random separately from one another.

✓ **Independence Assumption**: The international bedroom poll selected respondents at random, so they should be independent.

✓ **Randomization Condition**: The respondents were selected at random.

✓ **10% Condition**: The number of adults surveyed in each country is certainly far less than 10% of that country's population.

✓ **Success/Failure Condition**: In the United States, 128 people used laptops before bed and 123 didn't. In Japan, 162 people used laptops before bed and 88 didn't. The observed numbers of both successes and failures are much more than 10 for both groups. We'll confirm that the expected counts are all also greater than 10 when we get to the "Show" section.

State the null model.
Choose your method.

Because the conditions are satisfied, I'll use a Normal model and perform a **two-proportion z-test**.

**SHOW** **MECHANICS** The hypothesis is that the proportions are equal, so pool the sample data.

Use the pooled SE to estimate $SD(\hat{p}_{U.S.} - \hat{p}_{Japan})$.

$n_{U.S.} = 251$, $Y_{U.S.} = 128$, $\hat{p}_{U.S.} = 0.510$

$n_{Japan} = 250$, $Y_{Japan} = 162$, $\hat{p}_{Japan} = 0.648$

$$\hat{p}_{pooled} = \frac{Y_{U.S.} + Y_{Japan}}{n_{U.S.} + n_{Japan}} = \frac{128 + 162}{251 + 250} = 0.5788$$

The expected counts of successes and failures among the two groups, using pooling, are:

$n_{U.S.}\hat{p}_{pooled} = (251)(0.5788) \approx \mathbf{145}$,

$n_{U.S.}(1 - \hat{p}_{pooled}) = (251)(1 - 0.5788) \approx \mathbf{106}$

$n_{Japan}\hat{p}_{pooled} = (250)(0.5788) \approx \mathbf{145}$,

$n_{Japan}(1 - \hat{p}_{pooled}) = (250)(1 - 0.5788) \approx \mathbf{105}$

All four expected counts are greater than 10, verifying the Success/Failure Condition. So we continue:

$SE_{pooled}(\hat{p}_{U.S.} - \hat{p}_{Japan})$

$$= \sqrt{\frac{\hat{p}_{pooled}\hat{q}_{pooled}}{n_{U.S.}} + \frac{\hat{p}_{pooled}\hat{q}_{pooled}}{n_{Japan}}}$$

$$= \sqrt{\frac{(0.5788)(0.4212)}{251} + \frac{(0.5788)(0.4212)}{250}}$$

$$\approx 0.044118$$

The observed difference in sample proportions is

$$\hat{p}_{U.S.} - \hat{p}_{Japan} = 0.510 - 0.648 = -0.138.$$

Make a picture. Sketch a Normal model centered at the hypothesized difference of 0. Shade the region to the left of the observed difference, and because this is a two-tailed test, also shade the corresponding region in the other tail.

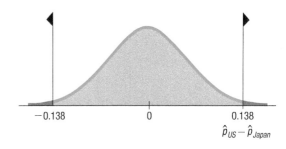

Find the z-score for the observed difference in proportions, −0.138.

Find the P-value using technology (or Table Z). Because this is a two-tailed test, we must *double* the probability we find in the upper tail.

$$z = \frac{(\hat{p}_{U.S.} - \hat{p}_{Japan}) - 0}{SE_{pooled}(\hat{p}_{U.S.} - \hat{p}_{Japan})} = \frac{(-0.138) - 0}{0.044118} = -3.13$$

$$P = 2P(z \leq -3.13) = 0.0017$$

**TELL**    **CONCLUSION** Link the P-value to your decision about the null hypothesis, and state your conclusion in context.

The P-value of 0.0017 says that if there really were no difference between the proportions of U.S. and Japanese adults who use laptops before bed, then a difference as big as the one we observed would be highly unlikely—less than two-tenths of a percent chance. That is so small that I reject the null hypothesis of no difference and conclude that there is evidence of a difference in the proportions of U.S. and Japanese adults who use laptops before bed. It appears that Japanese adults are more likely to use them before bed than are U.S. adults.

## TI TIPS

### Testing the hypothesis

Yes, of course, there's a STAT TESTS routine to test a hypothesis about the difference of two proportions. Let's do the mechanics for the test about laptops. Of 251 U.S. adults, 128 used laptops; among 250 Japanese adults, 162 did.

- In the STAT TESTS menu select 2-PropZTest.

- Enter the observed numbers of laptop users and the sample sizes for both groups.
- Because this is a two-tailed test, indicate that you want to see if the proportions are unequal. When you choose this option, the calculator will automatically include both tails as it determines the P-value.
- Calculate the result. Don't worry; for this procedure the calculator will pool the proportions automatically.

Now it is up to you to interpret the result and state a conclusion. We see a z-score of −3.13 and the P-value is 0.0018. Such a small P-value indicates that the observed difference is unlikely to be due to sampling error. What does that mean about laptop use among U.S. and Japanese adults? Here's a great opportunity to follow up with a confidence interval so you can Tell even more! (You'll find that if you use your calculator to perform the mechanics of confidence interval construction, it will remember the numbers you just used for the hypothesis test. How *convenient!*)

> **JUST CHECKING**
>
> 3. A February 2014 public opinion poll asked 1000 randomly selected adults whether the United States should decrease the amount of immigration allowed; 36% of those responding said, "Yes." In June 1995, a random sample of 1000 had found that 65% of adults thought immigration should be curtailed. To see if that percentage has decreased, why can't we just use a one-proportion $z$-test of $H_0$: $p = 0.65$ and see what the P-value for $\hat{p} = 0.36$ is?
>
> 4. For opinion polls like this, which has more variability: the percentage of respondents answering "yes" in either year or the difference in the percentages between the two years?

# *A Permutation Test for Two Proportions

One of the assumptions for the two-proportion $z$-interval and the two-proportion $z$-test is that the groups being compared are independent of one another. If the groups are samples taken at random from two different populations, then this assumption is certainly reasonable. But if the setting is an experiment, and the groups result from random *allocation* of subjects into treatment groups, then the assumption is, unfortunately, not reasonable. (The inclusion of someone in one group necessarily excludes them from the other, so the groups aren't independent.) Happily, the procedures we described in this chapter are still reasonable to use; they are good approximations of more appropriate procedures that are based on random permutations (rearrangements). Here we give an example and encourage you to compare its conclusion to the one you would obtain using the two-proportion $z$-test described in the last section.

It's estimated that worldwide 50,000 pregnant women die each year of eclampsia, a condition involving high blood pressure and seizures. In 2002 the medical journal *Lancet* reported on an experiment that involved nearly 10,000 at-risk women at 175 hospitals in 33 countries. The good news: Researchers found that treating women with magnesium sulfate significantly reduced the occurrence of eclampsia, an important advance in women's health.

However, there was one cause for concern. Unfortunately, 27.5% women (11 out of 40) who developed eclampsia despite receiving the magnesium sulfate treatment died. In the placebo group, only 20.8% of the women (20 out of 96) who developed eclampsia died. Does this indicate that even though this treatment dramatically reduces the occurrence of eclampsia, women who develop the condition anyway may face a greater risk of death?

You now know how to do a two-proportion $z$-test to see if the observed difference in mortality rates is statistically significant, so we won't do that.[6] Instead, we'll attack the question through simulation.

We'll do what's called a permutation test. The actual experiment saw a total of 31 deaths among 136 women. We wonder if the higher mortality rate in the magnesium sulfate group could have arisen by chance, or was a result of the treatment. To find out we create a list of 136 subjects denoting 31 as having died and 105 as having survived. Then we randomly divide them into a group of 40 and a group of 96, and look at the differences in mortality rates. That gives us a peek at what sort of differences could arise just because of the random allocation of subjects to the treatments—treatments that actually have the same effect.

---

[6] Try it, though. It's good practice, and you'll be able to compare that result to the clever alternative we're showing you here.

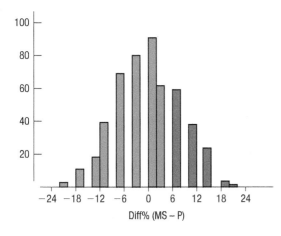

In our first simulation trial, 9 of the deaths randomly landed in the magnesium sulfate group and the other 22 in the placebo group. The simulated mortality rates came out $\frac{9}{40} = 22.5\%$ and $\frac{22}{96} = 22.9\%$, almost identical. Maybe the actual experiment's observed difference of $27.5 - 20.8 = 6.7\%$ is significant?

As you know, though, one trial does not a simulation make. In our next trial the random assignment resulted in a simulated death rate of $\frac{12}{40} = 30.0\%$ in the magnesium sulfate group and only $\frac{19}{96} = 19.8\%$ in the placebo group. This difference of $30.0 - 19.8 = 10.2\%$ is even larger than the 6.7% actually observed, and it arose purely by chance. Could something like that happen often? To find out, we simulated 500 trials.

On the left is a histogram of the differences in sample proportions that arose by chance. In over 25% of our trials (127 out of 500) the simulated difference in mortality rates was at least as large as that seen in the actual experiment. Because such an apparently large difference isn't unusual, it's not evidence of a heightened risk for women receiving preventative treatment for eclampsia.

Permutation tests like this are pretty cool. Although they're not a required topic in this course,[7] the now widespread use of computer simulations makes them an increasingly common analytical tool.

## WHAT CAN GO WRONG?

- **Don't use two-sample proportion methods when the samples aren't independent.** These methods give wrong answers when this assumption of independence is violated. Good random sampling is usually the best insurance of independent groups. Make sure there is no relationship between the two groups. For example, you can't compare the proportion of respondents who own SUVs with the proportion of those same respondents who think the tax on gas should be eliminated. The responses are not independent because you've asked the same people. To use these methods to estimate or test the difference, you'd need to survey two different groups of people.

  Alternatively, if you have a random sample, you can split your respondents according to their answers to one question and treat the two resulting groups as independent samples. So, you could test whether the proportion of SUV owners who favored eliminating the gas tax was the same as the corresponding proportion among non-SUV owners.

- **Don't apply inference methods where there was no randomization.** If the data do not come from representative random samples or from a properly randomized experiment, then the inference about the differences in proportions will be wrong.

- **Don't interpret a significant difference in proportions causally.** It turns out that people with higher incomes are more likely to snore. Does that mean money affects sleep patterns? Probably not. We have seen that older people are more likely to snore, and they are also likely to earn more. In a prospective or retrospective study, there is always the danger that other lurking variables not accounted for are the real reason for an observed difference. Be careful not to jump to conclusions about causality.

---

[7] Yes, you can exhale. It won't be on the test.

# WHAT HAVE WE LEARNED?

We've learned how to extend our understanding of statistical inference to create confidence intervals and test hypotheses about the difference in two proportions.

- We've learned that inference for the difference in two proportions is based on Normal models. In addition to the usual assumptions and conditions, we've learned to check the assumption that the groups are independent so that we can use the Pythagorean Theorem of Statistics to find the standard error for the difference in the two proportions.
- We've learned that when the null hypothesis assumes the proportions are equal we must pool the sample data to estimate the true proportion. We don't pool for confidence intervals because there's no such assumption.

Perhaps most important, we've learned that the concepts, reasoning, and interpretations of statistical inference remain the same; only the mechanics change.

## TERMS

**Independent Groups Assumption**  
The two groups we're comparing are independent of each other. This is usually determined by the way the data were collected. (p. 564)

**Variances of independent random variables add**  
The variance of a sum or difference of independent random variables is the sum of the variances of those variables. (p. 563)

**Sampling distribution of the difference between two independent sample proportions**  
The sampling distribution of $\hat{p}_1 - \hat{p}_2$ is, under appropriate assumptions, modeled by a Normal model with mean $\mu = p_1 - p_2$ and standard deviation
$$SD(\hat{p}_1 - \hat{p}_2) = \sqrt{\frac{p_1 q_1}{n_1} + \frac{p_2 q_2}{n_2}}.$$ (p. 566)

**Two-proportion z-interval**  
A two-proportion z-interval gives a confidence interval for the true difference in proportions, $p_1 - p_2$, in two independent groups.

The confidence interval is $(\hat{p}_1 - \hat{p}_2) \pm z^* \times SE(\hat{p}_1 - \hat{p}_2)$, where $z^*$ is a critical value from the standard Normal model corresponding to the specified confidence level. (p. 567)

**Pooling**  
When we believe a proportion is the same in two different groups, we can get a better estimate of this common proportion by combining the data from our two samples.
$$\hat{p}_{pooled} = \frac{x_1 + x_2}{n_1 + n_2}$$
The resulting standard error is based on more data and hence more reliable (if the null hypothesis is true). (p. 570)

**Two-proportion z-test**  
Test the null hypothesis $H_0: p_1 - p_2 = 0$ by referring the statistic
$$z = \frac{\hat{p}_1 + \hat{p}_2}{SE_{pooled}(\hat{p}_1 - \hat{p}_2)}$$
to a standard Normal model. (p. 571)

# ON THE COMPUTER
## Inferences for the Difference Between Two Proportions

It is so common to test against the null hypothesis of no difference between the two true proportions that most statistics programs simply assume this null hypothesis. And most will automatically use the pooled standard deviation. If you wish to test a different null (say, that the true difference is 0.3), you may have to search for a way to do it. Here's some typical computer software output for confidence intervals or hypothesis tests for the difference in proportions from two independent groups.

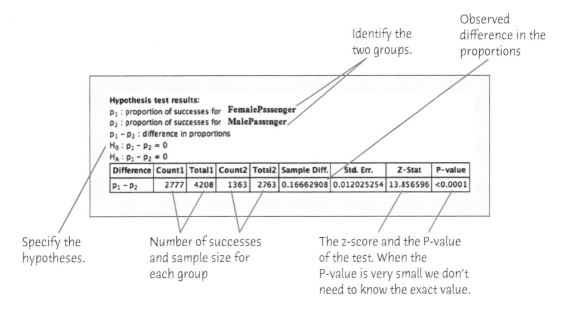

Note that often statistics packages don't offer special commands for inference for differences between proportions. As with inference for single proportions, many statistics programs want the "success" and "failure" status for each case. Computer packages don't usually deal with summary statistics. Calculators typically do a better job.

# EXERCISES

1. **Canada** Suppose an advocacy organization surveys 960 Canadians and 192 of them reported being born in another country (www.unitednorthamerica.org/simdiff.htm). Similarly, 170 out of 1250 Americans reported being foreign-born. Find the standard error of the difference in sample proportions.

2. **Nonprofits** Do people who work for nonprofit organizations differ from those who work at for-profit companies when it comes to personal job satisfaction? Separate random samples were collected by a polling agency to investigate the difference. Data collected from 422 employees at nonprofit organizations revealed that 377 of them were "highly satisfied." From the for-profit companies, 431 out 518 employees reported the same level of satisfaction. Find the standard error of the difference in sample proportions.

3. **Canada, deux** The information in Exercise 1 was used to create a 95% confidence interval for the difference between the proportions of Canadians and Americans who were born in foreign countries. Interpret this interval with a sentence in context. A 95% confidence interval for

$$p_{Canadians} - p_{Americans} \text{ is } (3.24\%, 9.56\%).$$

4. **Nonprofits, part 2** The researchers from Exercise 2 created a 95% confidence interval for the difference in proportions who are "highly satisfied" among people who work at nonprofits versus people who work at for-profit companies.

Interpret the interval with a sentence in context. A 95% confidence interval for

$$p_{Nonprofits} - p_{For\text{-}profits} = (1.77\%, 10.50\%).$$

5. **Canada, trois** For the interval given in Exercise 3, explain what "95% confidence" means.

6. **Nonprofits, part 3** For the interval given in Exercise 4, explain what "95% confidence" means.

**7. Canada, encore** If the information in Exercise 1 is to be used to make inferences about the proportion all Canadians and all U.S. citizens born in other countries, what conditions must be met before proceeding? Are they met? Explain.

**8. Nonprofits, again** If the information in Exercise 2 is to be used to make inferences about all people who work at non-profits and for-profit companies, what conditions must be met before proceeding? List them and explain if they are met.

**9. Canada, test** The researchers from Exercise 1 want to test if the proportions of foreign born are the same in the United States and Canada.
   a) What is the difference in the proportions of foreign-born residents in both countries?
   b) What is the value of the $z$-statistic?
   c) What do you conclude at $\alpha = 0.05$?

**10. Nonprofits test** Complete the analysis begun in Exercise 2.
   a) What is the difference in the proportions of the two types of companies?
   b) What is the value of the $z$-statistic?
   c) What do you conclude at $\alpha = 0.05$?

**11. Social media** In a study in 2020, the Pew Research Center on Journalism & Media surveyed 9220 U.S. adults and found that 18%, or about 1660 respondents, said that their most common way to get political and election news was social media. They were compared to the 16%—about 1475—respondents who get their news from cable TV using a "knowledge index" comprised of nine questions about the news. Researchers found that 57% of the Social Media group exhibited low political knowledge compared with only 35% of the Cable news consumers. What does it mean to say that this difference is statistically significant?

**12. Conspiracy theory** In the same study as cited in Exercise 11, Pew found that 16% of respondents who get their news primarily from network TV had heard about the (false) conspiracy theory that powerful people intentionally planned the coronavirus outbreak. By contrast, 26% of those who get their news primarily from social media had heard this conspiracy theory. What does it mean that the difference is significant?

**13. Name recognition** A political candidate runs a weeklong series of TV ads designed to attract public attention to his campaign. Polls taken before and after the ad campaign show some increase in the proportion of voters who now recognize this candidate's name, with a P-value of 0.033. Is it reasonable to believe the ads may be effective?

**14. Origins** In a 1993 Gallup poll, 47% of the respondents agreed with the statement "God created human beings pretty much in their present form at one time within the last 10,000 years or so." When Gallup asked the same question in 2008, only 44% of those respondents agreed. Is it reasonable to conclude that there was a change in public opinion given that the P-value is 0.17? Explain.

**15. Revealing information** 886 randomly sampled teens were asked which of several personal items of information they thought it okay to share with someone they had just met. 44% said it was okay to share their e-mail addresses, but only 29% said they would give out their cell phone numbers. A researcher claims that a two-proportion $z$-test could tell whether there was a real difference among all teens. Explain why that test would not be appropriate for these data.

**16. Regulating access** When a random sample of 935 parents were asked about rules in their homes, 77% said they had rules about the kinds of TV shows their children could watch. Among the 790 of those parents whose teenage children had Internet access, 85% had rules about the kinds of Internet sites their teens could visit. That looks like a difference, but can we tell? Explain why a two-sample $z$-test would not be appropriate here.

**17. Gender gap** A presidential candidate fears they have a problem with women voters. His campaign staff plans to run a poll to assess the situation. They'll randomly sample 300 men and 300 women, asking if they have a favorable impression of the candidate. Obviously, the staff can't know this, but suppose the candidate has a positive image with 59% of males but with only 53% of females.
   a) What sampling design is his staff planning to use?
   b) What difference would you expect the poll to show?
   c) Of course, sampling error means the poll won't reflect the difference perfectly. What's the standard deviation for the difference in the proportions?
   d) Sketch a sampling model for the size difference in proportions of men and women with favorable impressions of this candidate that might appear in a poll like this.
   e) Could the campaign be misled by the poll, concluding that there really is no gender gap? Explain.

**18. Buy it again?** A consumer magazine plans to poll car owners to see if they are happy enough with their vehicles that they would purchase the same model again. It will randomly select 450 owners of American-made cars and 450 owners of Japanese models. Obviously, the actual opinions of the entire population couldn't be known, but suppose 76% of owners of American cars and 78% of owners of Japanese cars would purchase another.
   a) What sampling design is the magazine planning to use?
   b) What difference would you expect its poll to show?
   c) Of course, sampling error means the poll won't reflect the difference perfectly. What's the standard deviation for the difference in the proportions?
   d) Sketch a sampling model for the difference in proportions that might appear in a poll like this.
   e) Could the magazine be misled by the poll, concluding that owners of American cars are much happier with their vehicles than owners of Japanese cars? Explain.

**19. Arthritis** The Centers for Disease Control and Prevention reported a survey of randomly selected Americans age 65 and older, which found that 411 of 1012 men and 535 of 1062 women suffered from some form of arthritis.
   a) Are the assumptions and conditions necessary for inference satisfied? Explain.
   b) Create a 95% confidence interval for the difference in the proportions of senior men and women who have this disease.
   c) Interpret your interval in this context.
   d) Does this confidence interval suggest that arthritis is more likely to afflict women than men? Explain.

**20. Graduation** The U.S. Department of Commerce reported the results of a large-scale survey on high school graduation. Researchers contacted more than 25,000 randomly chosen Americans aged 24 years to see if they had finished high school; 84.9% of the 12,460 males and 88.1% of the 12,678 females indicated that they had high school diplomas.

  a) Are the assumptions and conditions necessary for inference satisfied? Explain.
  b) Create a 95% confidence interval for the difference in graduation rates between males and females.
  c) Interpret your confidence interval.
  d) Does this provide strong evidence that girls are more likely than boys to complete high school? Explain.

**21. Pets** Researchers at the National Cancer Institute released the results of a study that investigated the effect of weed-killing herbicides on house pets. They examined 827 dogs from randomly selected homes where an herbicide was used on a regular basis, diagnosing malignant lymphoma in 473 of them. Of the 130 dogs from homes where no herbicides were used, only 19 were found to have lymphoma.

  a) What's the standard error of the difference in the two proportions?
  b) Construct a 95% confidence interval for this difference. (Don't forget to check conditions!)
  c) State an appropriate conclusion.

**22. Carpal tunnel** The painful wrist condition called carpal tunnel syndrome can be treated with surgery or less invasive wrist splints. *Time* magazine reported on a study of 176 randomly selected patients. Among the half that had surgery, 80% showed improvement after three months, but only 54% of those who used the wrist splints improved.

  a) What's the standard error of the difference in the two proportions?
  b) Construct a 95% confidence interval for this difference. (Don't forget to check conditions!)
  c) State an appropriate conclusion.

**23. Prostate cancer** There has been debate among doctors over whether surgery can prolong life among men suffering from prostate cancer, a type of cancer that typically develops and spreads very slowly. Recently, *The New England Journal of Medicine* published results of some Scandinavian research. Men diagnosed with prostate cancer were randomly assigned to either undergo surgery or not. Among the 347 men who had surgery, 16 eventually died of prostate cancer, compared with 31 of the 348 men who did not have surgery.

  a) Was this an experiment or an observational study? Explain.
  b) Create a 95% confidence interval for the difference in rates of death for the two groups of men.
  c) Based on your confidence interval, is there evidence that surgery may be effective in preventing death from prostate cancer? Explain.

**24. COVID and blood type** In Chapter 3 we saw evidence that a person's blood type (A, B, AB, O) may predict their susceptibility to the Corona virus that causes COVID-19. Researchers found that the proportion of blood group O in patients with COVID-19 was 25.80% (458 patients) among 1775 patients as compared to a background of 33.84% (1250) of 3694 people without COVID-19.

  a) Create a 95% confidence interval for the difference between the percentage of patients of blood group O with COVID-19 and the percentage of noninfected people with blood group O.
  b) Does this suggest a real (i.e., medically important) difference?

**25. Ear infections** A new vaccine was recently tested to see if it could prevent the painful and recurrent ear infections that many infants suffer from. *The Lancet*, a medical journal, reported a study in which babies about a year old were randomly divided into two groups. One group received vaccinations; the other did not. During the following year, only 333 of 2455 vaccinated children had ear infections, compared to 499 of 2452 unvaccinated children in the control group.

  a) Are the conditions for inference satisfied?
  b) Find a 95% confidence interval for the difference in rates of ear infection.
  c) Use your confidence interval to explain whether you think the vaccine is effective.

**26. Anorexia** The *Journal of the American Medical Association* reported on an experiment intended to see if the drug Prozac® could be used as a treatment for the eating disorder anorexia nervosa. The subjects, women being treated for anorexia, were randomly divided into two groups. Of the 49 who received Prozac, 35 were deemed healthy a year later, compared to 32 of the 44 who got the placebo.

  a) Are the conditions for inference satisfied?
  b) Find a 95% confidence interval for the difference in outcomes.
  c) Use your confidence interval to explain whether you think Prozac is effective.

**27. Another ear infection** In Exercise 25 you used a confidence interval to examine the effectiveness of a vaccine against ear infections in babies. Suppose that instead you had conducted a hypothesis test. (Answer these questions *without* actually doing the test.)

  a) What hypotheses would you test?
  b) State a conclusion based on your confidence interval.
  c) What alpha level did your test use?
  d) If that conclusion is wrong, which type of error did you make?
  e) What would be the consequences of such an error?

**28. Anorexia again** In Exercise 26 you used a confidence interval to examine the effectiveness of Prozac in treating anorexia nervosa. Suppose that instead you had conducted a hypothesis test. (Answer these questions *without* actually doing the test.)

  a) What hypotheses would you test?
  b) State a conclusion based on your confidence interval.
  c) What alpha level did your test use?
  d) If that conclusion is wrong, which type of error did you make?
  e) What would be the consequences of such an error?

**29. Teen smoking, part I** A Vermont study published by the American Academy of Pediatrics examined parental influence on teenagers' decisions to smoke. A group of students who had never smoked were questioned about their parents' attitudes toward smoking. These students were questioned again two years later to see if they had started smoking. The researchers found that, among the 284 students who indicated that their parents disapproved of kids smoking, 54 had become established smokers. Among the 51 students who initially said their parents were lenient about smoking, 14 became smokers. Do these data provide strong evidence that parental attitude influences teenagers' decisions about smoking?

a) What kind of design did the researchers use?
b) Write appropriate hypotheses.
c) Are the assumptions and conditions necessary for inference satisfied?
d) Test the hypothesis and state your conclusion.
e) Explain in this context what your P-value means.
f) If that conclusion is actually wrong, which type of error did you commit?

**30. Depression** A study published in the *Archives of General Psychiatry* examined the impact of depression on a patient's ability to survive cardiac disease. Researchers identified 450 people with cardiac disease, evaluated them for depression, and followed the group for 4 years. Of the 361 patients with no depression, 67 died. Of the 89 patients with minor or major depression, 26 died. Among people who suffer from cardiac disease, are depressed patients more likely to die than nondepressed ones?

a) What kind of design was used to collect these data?
b) Write appropriate hypotheses.
c) Are the assumptions and conditions necessary for inference satisfied?
d) Test the hypothesis and state your conclusion.
e) Explain in this context what your P-value means.
f) If your conclusion is actually incorrect, which type of error did you commit?

**31. Teen smoking, part II** Consider again the Vermont study discussed in Exercise 29.

a) Create a 95% confidence interval for the difference in the proportion of children who may smoke and have lenient parents and those who may smoke and have disapproving parents.
b) Interpret your interval in this context.
c) Carefully explain what "95% confidence" means.

**32. Depression revisited** Consider again the study of the association between depression and cardiac disease survivability in Exercise 30.

a) Create a 95% confidence interval for the difference in survival rates.
b) Interpret your interval in this context.
c) Carefully explain what "95% confidence" means.

**33. Pregnancy** A San Diego reproductive clinic reported 42 live births to 157 women under the age of 38, but only 7 live births for 89 clients aged 38 and older. Is this strong evidence of a difference in the effectiveness of the clinic's methods for older women?

a) Was this an experiment? Explain.
b) Test an appropriate hypothesis and state your conclusion in context.
c) If you concluded there was a difference, estimate that difference with a confidence interval and interpret your interval in context.

**34. Birth weight** In 2003 the *Journal of the American Medical Association* reported a study examining the possible impact of air pollution caused by the 9/11 attack on New York's World Trade Center on the weight of babies. Researchers found that 8% of 182 babies born to mothers who were exposed to heavy doses of soot and ash on September 11 were classified as having low birth weight. Only 4% of 2300 babies born in another New York City hospital whose mothers had not been near the site of the disaster were similarly classified. Does this indicate a possibility that air pollution might be linked to a significantly higher proportion of low-weight babies?

a) Are the assumptions needed for a two-proportion z-test reasonable?
b) Test an appropriate hypothesis and state your conclusion in context.
c) If you concluded there is a difference, estimate that difference with a 90% confidence interval and interpret that interval in context.

**35. Political scandal!** One month before the election, a poll of 630 randomly selected voters showed 54% planning to vote for a certain candidate. A week later, it became known that he had tweeted inappropriate pictures of himself, and a new poll showed only 51% of 1010 voters supporting him. Do these results indicate a decrease in voter support for his candidacy?

a) Test an appropriate hypothesis and state your conclusion.
b) If your conclusion turns out to be wrong, did you make a Type I or Type II error?

**36. Shopping** A survey of 430 randomly chosen adults found that 21% of the 222 men and 18% of the 208 women had purchased books online.

a) Is there evidence that men are more likely than women to make online purchases of books? Test an appropriate hypothesis and state your conclusion in context.
b) If your conclusion in fact proves to be wrong, did you make a Type I or Type II error?
c) Estimate this difference with a confidence interval.
d) Interpret your interval in context.

**37. Mammograms** It's widely believed that regular mammogram screening may detect breast cancer early, resulting in fewer deaths from that disease. One study that investigated this issue over a period of 18 years was published during the 1970s. Among 30,565 women who had never had mammograms, 196 died of breast cancer, while only 153 of 30,131 who had undergone screening died of breast cancer.

a) Do these results suggest that mammograms may be an effective screening tool to reduce breast cancer deaths?
b) If your conclusion is incorrect, what type of error have you committed?

**38. Mammograms redux** A 9-year study in Sweden compared 21,088 women who had mammograms with 21,195 who did not. Of the women who underwent screening, 63 died of breast cancer, compared to 66 deaths among the control group. (*The New York Times*, December 9, 2001)

a) Do these results support the effectiveness of regular mammograms in preventing deaths from breast cancer?
b) If your conclusion is incorrect, what kind of error have you committed?

**39. Pain** Researchers comparing the effectiveness of two pain medications randomly selected a group of patients who had been complaining of a certain kind of joint pain. They randomly divided these people into two groups, then administered the pain killers. Of the 112 people in the group who received medication A, 84 said this pain reliever was effective. Of the 108 people in the other group, 66 reported that pain reliever B was effective.

a) Write a 95% confidence interval for the percent of people who may get relief from this kind of joint pain by using medication A. Interpret your interval.
b) Write a 95% confidence interval for the percent of people who may get relief by using medication B. Interpret your interval.
c) Do the intervals for A and B overlap? What do you think this means about the comparative effectiveness of these medications?
d) Find a 95% confidence interval for the difference in the proportions of people who may find these medications effective. Interpret your interval.
e) Does this interval contain zero? What does that mean?
f) Why do the results in parts c and e seem contradictory? If we want to compare the effectiveness of these two pain relievers, which is the correct approach? Why?

**40.** Gender gap Candidates for political office realize that different levels of support among men and women may be a crucial factor in determining the outcome of an election. One candidate finds that 52% of 473 men polled say they will vote for him, but only 45% of the 522 women in the poll express support.

a) Write a 95% confidence interval for the percent of male voters who may vote for this candidate. Interpret your interval.
b) Write and interpret a 95% confidence interval for the percent of female voters who may vote for him.
c) Do the intervals for males and females overlap? What do you think this means about the gender gap?
d) Find a 95% confidence interval for the difference in the proportions of males and females who will vote for this candidate. Interpret your interval.
e) Does this interval contain zero? What does that mean?
f) Why do the results in parts c and e seem contradictory? If we want to see if there is a gender gap among voters with respect to this candidate, which is the correct approach? Why?

**41.** Food preference GfK Roper Consulting gathers information on consumer preferences around the world to help companies monitor attitudes about health, food, and healthcare products. It asked people in many different cultures how they felt about the following statement:

*I have a strong preference for regional or traditional products and dishes from where I come from.*

In a random sample of 800 respondents, 417 of 646 people who live in urban environments agreed (either completely or somewhat) with that statement, compared to 78 out of 154 people who live in rural areas.

Based on this sample, is there evidence that the percentage of people agreeing with the statement about regional preferences differs between all urban and rural dwellers?

**42.** Fast food The global survey we learned about in Exercise 41 also asked respondents how they felt about the statement "I try to avoid eating fast foods." The random sample of 800 included 411 people 35 years old or younger, and of those, 197 agreed (completely or somewhat) with the statement. Of the 389 people over 35 years old, 246 people agreed with the statement. Is there evidence that the percentage of people avoiding fast food is different in the two age groups?

**43.** Online activity checks Are more parents checking up on their teens' online activities? A Pew survey in 2004 found that 33% of 868 randomly sampled teens said that their parents checked to see what websites they visited. In 2006 the same question posed to 811 teens found 41% reporting such checks. Do these results provide evidence that more parents are checking?

**44.** Computer gaming Who plays online or electronic games? A survey in 2006 found that 69% of 223 boys aged 12–14 said they "played computer or console games like Xbox or PlayStation . . . or games online." Of 248 boys aged 15–17, only 62% played these games. Is this evidence of a real age-based difference?

**45.** Convention bounce Political pundits talk about the "bounce" that a presidential candidate gets after his party's convention. In the past 40 years, it has averaged about 6 percentage points. Just before the 2004 Democratic convention, Rasmussen Reports polled 1500 likely voters at random and found that 47% favored John Kerry. Just afterward, they took another random sample of 1500 likely voters and found that 49% favored Kerry. That's a two percentage point increase, but the pollsters claimed that there was no bounce. Explain.

**46.** Students vs. teachers A school newspaper took a survey of the campus attitudes about homework (of course, they used a random sampling method!). The paper reported that 31 of 161 the teachers and 20 of 358 students responded "Yes" to the question "Do you like the food in the cafeteria?" How big is the difference in proportions in the two populations?

a) Construct and interpret a 95% confidence interval.
b) At a different high school the same survey is completed, but that paper contacted 1302 students and staff and claims a margin of error of ±2.9%. Why is the margin of error different for your confidence interval?

**47.** Sensitive men An article from *Time* magazine, reporting on a survey of men's attitudes, noted that "Young men are more comfortable than older men talking about their problems." The survey reported that 80 of 129 surveyed 18- to 24-year-old men and 98 of 184 25- to 34-year-old men said they were comfortable. What do you think? Is *Time*'s interpretation justified by these numbers?

**48.** Carbs Recently, the Gallup Poll asked 1005 randomly selected U.S. adults if they actively try to avoid carbohydrates in their diet. That number increased to 27% from 20% in a similar 2002 poll. Is this a statistically significant increase? Explain.

**49.** Manners A curious parent keeps track of the manners of the trick-or-treaters who come to door on Halloween. They suspect that the children whose parents are with them at the door will be more polite than those who are not with their parents. Sure enough, while 12 out of 25 children with their parents say, "Thank you!" only 8 out of 25 children without their parents do the same. This small sample size causes problems in performing a hypothesis test. But we can perform a simulation instead.

We have 20 "successes" in total and 30 "failures." We are curious if 12 out of 25 is statistically significantly more than 8 out of 25. Or could a difference of this size happen by chance? So we randomly sort these piles of 20 thank-yous and 30 no-manners into two piles of 25: The first pile will represent the children with their parents; the second pile will represent those without their parents. We find the difference between the two piles (with–without).

We run this simulation 100 times to see how big the differences are, just by chance. Here is a graph of the simulation:

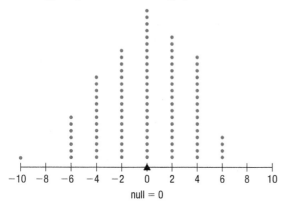

a) How likely is it to get a difference of four or more, just by chance?
b) Given the probability you calculated in part a, what do you conclude?

**50. Watching for Woodpeckers** Ana, a birdwatcher, has two bird feeders in her backyard—one is designed to attract many types of birds and the other is designed to attract woodpeckers specifically. Ana likes to attract woodpeckers, but she's not sure that the feeder designed for woodpeckers is really necessary, because whenever a woodpecker shows up, it often eats from both feeders. So Ana conducts an experiment. For a month, she always keeps seed in the bird feeder designed for many types of birds, but she flips a coin to decide whether to put out the woodpecker feeder or not. Ana observes that on 12 out of the 15 days when the woodpecker feeder is out, a woodpecker shows up. On only 5 out of the 15 days when there is no woodpecker feeder does a woodpecker show up.

As in Exercise 49, we run a simulation because the sample size is so small. We randomly mix the 17 "successes" (days when a woodpecker appears) and the 13 "failures" (no woodpecker), and then allocate them to two piles representing the days with or without the woodpecker feeder. We compute the difference (feeder − no feeder). Here are the results of 100 runs of the simulation:

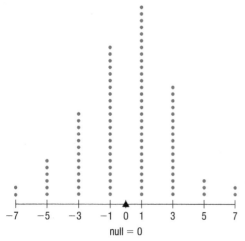

a) Using this simulation, estimate the probability of getting a difference of seven or more, simply by chance.
b) Given the probability you calculated in part a, what do you conclude?

## JUST CHECKING

### Answers

1. We're 90% confident that if members are contacted by e-mail, the donation rate will be between 4 and 18 percentage points higher than if they received regular mail.
2. Because a difference of 0 is not in the confidence interval, we'd reject the null hypothesis. There is evidence that a higher proportion of members will donate if contacted by e-mail.
3. The proportion from the sample in 1995 has variability, too. If we do a one-proportion $z$-test, we won't take that variability into account and our P-value will be incorrect.
4. The difference in the proportions between the two years has more variability than either individual proportion. The variance of the difference is the sum of the two variances.

# Review of Part V

## FROM THE DATA AT HAND TO THE WORLD AT LARGE

### Quick Review

What do samples really tell us about the populations from which they are drawn? Are the results of an experiment meaningful, or are they just sampling error? Statistical inference based on our understanding of sampling models can help answer these questions. Here's a brief summary of the key concepts and skills:

◆ Sampling models describe how sample statistics vary due to variation among samples. A particularly useful sampling model for sample means or proportions owes itself to the rather amazing Central Limit Theorem.
  - When samples are sufficiently large, proportions found in different samples vary according to an approximately Normal model.
  - When samples are sufficiently large, the means of different samples vary according to an approximately Normal model.
  - The variability of sample statistics decreases as sample size increases.
  - An inference procedure is only as valid as the underlying assumptions are reasonable. Always check the conditions before proceeding.

◆ A confidence interval uses a sample statistic (such as a proportion) to estimate a range of plausible values for the parameter of a population model.
  - All confidence intervals involve an estimate of the parameter, a margin of error, and a level of confidence.
  - For confidence intervals based on a given sample, the greater the margin of error, the higher the confidence.
  - At a given level of confidence, the larger the sample, the smaller the margin of error.

◆ A hypothesis test proposes a model for the population, then examines the observed statistics to see if that model is plausible.
  - A null hypothesis suggests a parameter value for the population model. Usually, we assume there is nothing interesting, unusual, or different about the sample results.
  - The alternative hypothesis states what we will believe if the sample results turn out to be inconsistent with our null model.
  - We compare the difference between the statistic and the hypothesized value with the standard deviation of the statistic. It's the sampling distribution of this ratio that gives us a P-value.
  - The P-value of the test is the conditional probability that the null model could produce results at least as extreme as those observed in the sample or the experiment just as a result of sampling error.
  - A low P-value indicates inconsistency between the data and the null model. If it is sufficiently low, we reject the null model.
  - A high P-value indicates that the sample results are not inconsistent with the null model, so we cannot reject it. However, this does *not* prove the null model is true, nor even provide evidence that it is true.
  - Sometimes we will mistakenly reject the null hypothesis even though it's actually true—that's called a Type I error. If we fail to reject a false null hypothesis, that's called a Type II error.
  - The power of a test is the chance that it will reject the null hypothesis. Higher power means a greater chance of detecting that the null is false.

◆ Power is higher when . . .
  - the significance level is higher (but this raises the chance of a false alarm, or a Type I error, should there really be no effect), or
  - the effect size is larger (but you don't get to control this), or
  - the sample size is larger (but that costs more time, money, or resources), or
  - there's less statistical noise (which may be achieved through the design of the study).

And now for some opportunities to review these concepts and skills . . .

## REVIEW EXERCISES

**1. Crohn's disease** Omega-3 fatty acids have been tested as a means to prevent relapse of Crohn's disease. Two large, randomized, placebo-controlled studies have shown no such benefit from omega-3 fatty acids. Suppose you are asked to design an experiment to further study this claim. Imagine that you have collected data on Crohn's relapses in subjects who have used these omega-3 fatty acids and similar subjects who have not used them and that you can measure incidences of relapse for these subjects. State the null and alternative hypotheses you would use in your study.

**2. Colorblind** Medical literature says that about 8% of males are colorblind. A university's introductory psychology course is taught in a large lecture hall. Among the students, there are 325 males. Each semester when the professor discusses visual perception, he shows the class a test for colorblindness. The percentage of males who are colorblind varies from semester to semester.

a) Is the sampling distribution model for the sample proportion likely to be Normal? Explain.

b) What are the mean and standard deviation of this sampling distribution model?

c) Sketch the sampling model, using the 68–95–99.7 Rule.
d) Write a few sentences explaining what the model says about this professor's class.

3. **Birth days** During a 2-month period in 2002, 72 babies were born at the Tompkins Community Hospital in upstate New York. The table shows how many babies were born on each day of the week.

a) If births are uniformly distributed across all days of the week, how many would you expect on each day?
b) Only 7 births occurred on a Monday. Does this indicate that women might be less likely to give birth on a Monday? Explain.
c) Do the 17 births on Tuesdays seem to be unusually many? Explain.
d) Can you think of any reasons why births may not occur completely at random?

| Day | Births |
|-----|--------|
| Sun. | 9 |
| Mon. | 7 |
| Tues. | 17 |
| Wed. | 8 |
| Thurs. | 12 |
| Fri. | 9 |
| Sat. | 10 |

4. **Polling 2016** The 2016 U.S. presidential election was unusual in several ways. First, the candidate who won the most electoral votes, Donald Trump, did not win the most popular votes. Second, several minor-party candidates received enough votes to possibly affect the outcome. The official results showed that Hillary Clinton received 48.04% of the popular vote, Donald Trump received 45.95%, Gary Johnson got 3.28%, and Jill Stein won 1.06%. After the election, there was much discussion about the polls, which had indicated that Clinton would win. Suppose you had taken a simple random sample of 800 voters in an exit poll and asked them for whom they had voted.

a) Would you always get 384 votes for Clinton and 368 votes for Trump?
b) In 95% of such polls, your sample proportion of voters for Trump should be between what two values?
c) What might be a problem in finding a 95% confidence interval for the true proportions of Stein voters from this sample?
d) Would you expect the sample proportion of Johnson votes to vary more than, less than, or about the same as the sample proportion of Trump votes? Why?

5. **Leaky gas tanks** According to the website RightingInjustice.com, about 77% of all underground storage tanks, or USTs, have "confirmed releases" or leaks. Researchers in California want to know if the percentage is lower in their state. To do this, they randomly sample 47 service stations in California and determine whether there is any evidence of leakage. In their sample, 33 of the stations exhibited leakage. Is there evidence that the California percentage is really lower?

a) What are the null and alternative hypotheses?
b) Check the assumptions necessary for inference.
c) Test the null hypothesis.
d) What do you conclude (in plain English)?
e) If California's percentage is actually lower, have you made an error? What kind?
f) What two things could you do to decrease the probability of making this kind of error?
g) What are the advantages and disadvantages of taking those two courses of action?

6. **Surgery and germs** Joseph Lister (for whom Listerine is named!) was a British physician who was interested in the role of bacteria in human infections. He suspected that germs were involved in transmitting infection, so he tried using carbolic acid as an operating room disinfectant. In 75 amputations, he used carbolic acid 40 times. Of the 40 amputations using carbolic acid, 34 of the patients lived. Of the 35 amputations without carbolic acid, 19 patients lived. The question of interest is whether carbolic acid is effective in increasing the chances of surviving an amputation.

a) What kind of a study is this?
b) What do you conclude? Support your conclusion by testing an appropriate hypothesis.
c) What reservations do you have about the design of the study?

7. **Scrabble** Using a computer to play many simulated games of Scrabble, researcher Charles Robinove found that the letter "A" occurred in 54% of the hands. This study had a margin of error of $\pm 10\%$. (*Chance*, 15, no. 1 [2002])

a) Explain what the margin of error means in this context.
b) Why might the margin of error be so large?
c) Probability theory predicts that the letter "A" should appear in 63% of the hands. Does this make you concerned that the simulation might be faulty? Explain.

8. **Dice** When one die is rolled, the number of spots showing has a mean of 3.5 and a standard deviation of 1.7. Suppose you roll 10 dice. What's the approximate probability that your total is between 30 and 40 (that is, the average for the 10 dice is between 3 and 4)? Specify the model you use and the assumptions and conditions that justify your approach.

9. **Getting news** In 2016, the Pew Research Foundation sampled 4654 U.S. adults and asked about their choice of news sources. 38% of those surveyed said they often get news from online sources.

a) Pew reports a margin of error of $\pm 1.4$ percentage points and a 95% confidence level. Explain what the margin of error means.
b) Pew's survey included 4339 respondents contacted via the Web and 315 contacted by mail. If the percentage of people who often get news online is the same in both groups and Pew estimated those percentages separately, which group would have the larger margin of error? Explain.
c) Pew reports that 80% of the 2141 respondents who prefer to watch their news, rather than read it, often get their news from television. Find a 95% confidence interval for this parameter.
d) How does the margin of error for your confidence interval compare with the values in parts a? Explain.

**10. Afghanistan 2021** A research firm conducts a randomized survey asking U.S. adults about the withdrawal of American troops from Afghanistan in August of 2021. The researchers asked if the troop pullout was the right decision or the wrong decision. The firm reports a 95% confidence interval for the percentage of U.S. adults that said the withdrawal was the right decision.

a) The reported interval was (50.9%, 57.1%). Find the point estimate and the margin of error.
b) Would a higher confidence level result in a smaller or larger margin of error?
c) Can the firm run a 2-proportion $z$-test to compare the percentage of adults that said it was the right decision to the percentage who said it was the wrong decision? Explain why or why not.

**11. Bimodal** We are sampling randomly from a distribution known to be bimodal.

a) As our sample size increases, what's the expected shape of the sample's distribution?
b) What's the expected value of our sample's mean? Does the size of the sample matter?
c) How is the variability of sample means related to the standard deviation of the population? Does the size of the sample matter?
d) How is the shape of the sampling distribution model affected by the sample size?

**12. Vitamin D 2012** In 2012, the *American Journal of Clinical Nutrition* reported that 31% of Australian adults over age 25 have a vitamin D deficiency. The data came from the AusDiab study of 11,218 Australians.

a) Do these data meet the assumptions necessary for inference? What would you like to know that you don't?
b) Create a 95% confidence interval.
c) Interpret the interval in this context.
d) Explain in this context what "95% confidence" means.

**13. Archery** A champion archer can generally hit the bull's-eye 80% of the time. Suppose she shoots 200 arrows during competition. Let $\hat{p}$ represent the percentage of bull's-eyes she gets (the sample proportion).

a) What are the mean and standard deviation of the sampling distribution model for $\hat{p}$?
b) Is a Normal model appropriate here? Explain.
c) Sketch the sampling model, using the 68–95–99.7 Rule.
d) What's the probability that she gets at least 85% bull's-eyes?

**14. Black Lives Matter** In 2016, Pew research conducted a survey of 3769 adults asking whether people "Strongly Support, Support, Oppose, or Strongly Oppose" the Black Lives Matter movement. Of those asked, 43% said they support the movement at some level. We know that if we could ask the entire population of American adults, we would not find that exactly 43% of Americans would say they support the movement. Construct a 95% confidence interval for the true percentage of American adults who support the Black Lives Matter movement.

**15. Twins** There is some indication in medical literature that doctors may have become more aggressive in inducing labor or doing preterm cesarean sections when a woman is carrying twins. Records at a large hospital show that, of the 43 sets of twins born in 1990, 20 were delivered before the 37th week of pregnancy. In 2000, 26 of 48 sets of twins were born preterm. Does this indicate an increase in the incidence of early births of twins? Test an appropriate hypothesis and state your conclusion.

**16. Eclampsia** It's estimated that 50,000 pregnant women worldwide die each year of eclampsia, a condition involving elevated blood pressure and seizures. A research team from 175 hospitals in 33 countries investigated the effectiveness of magnesium sulfate in preventing the occurrence of eclampsia in at-risk patients. Results are summarized below. (*Lancet*, June 1, 2002)

| | | Reported Side Effects | Developed Eclampsia | Deaths | Total Subjects |
|---|---|---|---|---|---|
| Treatment | Magnesium sulfate | 1201 | 40 | 11 | 4999 |
| | Placebo | 228 | 96 | 20 | 4993 |

a) Write a 95% confidence interval for the increase in the proportion of women who may develop side effects from this treatment. Interpret your interval.
b) Is there evidence that the treatment may be effective in preventing the development of eclampsia? Test an appropriate hypothesis and state your conclusion.

**17. Eclampsia again** Refer again to the research summarized in Exercise 16. Is there any evidence that when eclampsia does occur, the magnesium sulfate treatment may help prevent the woman's death?

a) Write an appropriate hypothesis.
b) Check the assumptions and conditions.
c) Find the P-value of the test.
d) What do you conclude about the magnesium sulfate treatment?
e) If your conclusion is wrong, which type of error have you made?
f) Name two things you could do to increase the power of this test.
g) What are the advantages and disadvantages of those two options?

**18. Eggs** The ISA Babcock Company supplies poultry farmers with hens, advertising that a mature B300 Layer produces eggs with a mean weight of 60.7 grams. Suppose that egg weights follow a Normal model with standard deviation 3.1 grams.

a) What fraction of the eggs produced by these hens weigh more than 62 grams?
b) What's the probability that a dozen randomly selected eggs average more than 62 grams?
c) Using the 68–95–99.7 Rule, sketch a model of the total weights of a dozen eggs.

**19. Polling Details** In January of 2022, a Quinnipiac University poll estimated that 81 percent of adults in the United States believed that the coronavirus then circulating around the country would be "the new normal," and never eliminated. Quinnipiac reports that their polls are conducted using random digit dialing, and that if a call elicits no response, the same number is called again at least three more times. They also report that their polls routinely include more than 1,000 respondents. Let's suppose that the poll about the coronavirus included 1,100 respondents and (critically) that nearly every phone number called eventually elicited a response.

a) What margin of error would you associate with the poll estimate of 81 percent if you wanted to be 95 percent confident in the accuracy of your resulting confidence interval?

b) Quinnipiac University also reported that the view of the coronavirus as "the new normal" was held by 82 percent of those polled who were 34 years old or younger. Without doing any further calculations, would the margin of error associated with this group be greater than, less than, or equal to that of the entire group? (Assume the pollsters continue to use a 95 percent confidence level.)

c) In parts a) and b), we assumed that nearly every phone number called eventually elicited a response. In fact, this is highly unlikely. Many people never answer calls from numbers they do not recognize. Quinnipiac University also reported this statement about their polls: "Gender, age, education, race, and region are the demographics that are weighted to reflect Census information." Explain what it means to "weight" demographics, and why that's important in light of low response rates.

20. Enough eggs? One of the important issues for poultry farmers is the production rate—the percentage of days on which a given hen actually lays an egg. Ideally, that would be 100% (an egg every day), but realistically, hens tend to lay eggs on about 3 of every 4 days. ISA Babcock wants to advertise the production rate for the B300 Layer (see Exercise 18) as a 95% confidence interval with a margin of error of $\pm 2\%$. How many hens must it collect data on?

21. Teen deaths Traffic accidents are the leading cause of death among people aged 15 to 20. In 2014, 36% of drivers 15 to 20 years old who were involved in fatal crashes were speeding at the time of the crash. An insurance company surveyed 825 randomly selected drivers age 15 to 20, and 99 of the teens said they speed frequently or occasionally. Is this strong evidence that the rate of speeding among drivers age 15 to 20 is lower than the rate of speeding among those in that age group involved in fatal crashes?

a) Test an appropriate hypothesis and state your conclusion.
b) Explain what your P-value means in this context.

22. Multivitamin A medical experiment is conducted to compare the effects of a multivitamin on general health. A placebo is used as a control in a double-blind, completely randomized experiment. At the end of the experiment, a 2-sample test is run to measure if a higher percentage of the multivitamin users report better health outcomes.

a) Is this a one- or two-sided test? Explain.
b) The researchers found a test statistic of $z = \pm 0.32$. Calculate the P-value that corresponds with this test.
c) What is the conclusion for this test?

23. Largemouth bass Organizers of a fishing tournament believe that the lake holds a sizable population of largemouth bass. They assume that the weights of these fish have a model that is skewed to the right with a mean of 3.5 pounds and a standard deviation of 2.2 pounds.

a) Explain why a skewed model makes sense here.
b) Explain why you cannot determine the probability that a largemouth bass randomly selected ("caught") from the lake weighs over 3 pounds.

c) Each participant in the tournament catches 5 fish each day. Can you determine the probability that someone's catch averages over 3 pounds? Explain.

d) The 12 competing participants each caught the limit of 5 fish. What's the probability that the total catch of 60 fish averaged more than 3 pounds?

24. Cheating A Rutgers University study found that many high school students cheat on tests. The researchers surveyed a random sample of 4500 high school students nationwide; 74% of them said they had cheated at least once.

a) Create a 90% confidence interval for the level of cheating among high school students. Don't forget to check the appropriate conditions.
b) Interpret your interval.
c) Explain what "90% confidence" means.
d) Would a 95% confidence interval be wider or narrower? Explain without actually calculating the interval.

25. Language Neurological research has shown that in about 80% of people language abilities reside in the brain's left side. Another 10% display right-brain language centers, and the remaining 10% have two-sided language control. (The latter two groups are mainly left-handers.) (*Science News*, 161, no. 24 [2002])

a) We select 60 people at random. Is it reasonable to use a Normal model to describe the possible distribution of the proportion of the group that has left-brain language control? Explain.
b) What's the probability that our group has at least 75% left-brainers?
c) If the group had consisted of 100 people, would that probability be higher, lower, or about the same? Explain why, without actually calculating the probability.
d) How large a group would almost certainly guarantee at least 75% left-brainers? Explain.

26. Cigarettes 2009 In 1999, researchers at the Centers for Disease Control and Prevention estimated that about 34.8% of high school students smoked cigarettes. They established a national health goal of reducing that figure to 16% by the year 2010. To that end, they would be on track if they achieved a reduction to 17.7% by 2009. In 2009, they released a research study in which 19.5% of a random sample of 5080 high school students said they were current smokers. Is this evidence that progress toward the goal was off track?

a) Write appropriate hypotheses.
b) Verify that the appropriate assumptions are satisfied.
c) Find the P-value of this test.
d) Explain what the P-value means in this context.
e) State an appropriate conclusion.
f) Of course, your conclusion may be incorrect. If so, which kind of error did you commit?

27. Crohn's disease In 2002 the medical journal *The Lancet* reported that 335 of 573 patients suffering from Crohn's disease responded positively to injections of the arthritis-fighting drug infliximab.

a) Create a 95% confidence interval for the effectiveness of this drug.
b) Interpret your interval in context.
c) Explain carefully what "95% confidence" means in this context.

**28. Teen e-smoking 2016** Researchers at the Centers for Disease Control and Prevention say that about 11.3% of high school students have smoked an electronic cigarette in the past 30 days (up 1.5 percentage points from 2011). In a random sample of 522 high school students, would it be likely that more than 15% of them have tried an electronic cigarette in the past 30 days?

**29. Alcohol abuse** Growing concern about binge drinking among college students has prompted one large state university to conduct a survey to assess the size of the problem on its campus. The university plans to randomly select students and ask how many have been drunk during the past week. If the school hopes to estimate the true proportion among all its students with 90% confidence and a margin of error of ±4%, how many students must be surveyed?

**30. Errors** An auto parts company advertises that its special oil additive will make the engine "run smoother, cleaner, longer, with fewer repairs." An independent laboratory decides to test part of this claim. It arranges to use a taxicab company's fleet of cars. The cars are randomly divided into two groups. The company's mechanics will use the additive in one group of cars but not in the other. At the end of a year the laboratory will compare the percentage of cars in each group that required engine repairs.

a) What kind of a study is this?
b) Will the company do a one-tailed or a two-tailed test?
c) Explain in this context what a Type I error would be.
d) Explain in this context what a Type II error would be.
e) Which type of error would the additive manufacturer consider more serious?
f) If the cabs with the additive do indeed run significantly better, can the company conclude it is an effect of the additive? Can the company generalize this result and recommend the additive for all cars? Explain.

**31. Preemies** Among 242 Cleveland-area children born prematurely at low birth weights between 1977 and 1979, only 74% graduated from high school. Among a comparison group of 233 children of normal birth weight, 83% were high school graduates. ("Outcomes in Young Adulthood for Very-Low-Birth-Weight Infants," *New England Journal of Medicine*, 346, no. 3 [2002])

a) Create a 95% confidence interval for the difference in graduation rates between children of normal and children of very low birth weights. Be sure to check the appropriate assumptions and conditions.
b) Does this provide evidence that premature birth may be a risk factor for not finishing high school? Use your confidence interval to test an appropriate hypothesis.
c) Suppose your conclusion is incorrect. Which type of error did you make?

**32. Safety** Observers in Texas watched children at play in eight communities. Of the 814 children seen biking, roller skating, or skateboarding, only 14% wore a helmet.

a) Create and interpret a 95% confidence interval.
b) What concerns do you have about this study that might make your confidence interval unreliable?
c) Suppose we want to do this study again, picking various communities and locations at random, and hope to end up with a 98% confidence interval having a margin of error of ±4%. How many children must we observe?

**33. Fried PCs** A computer company recently experienced a disastrous fire that ruined some of its inventory. Unfortunately, during the panic of the fire, some of the damaged computers were sent to another warehouse, where they were mixed with undamaged computers. The engineer responsible for quality control would like to check out each computer in order to decide whether it's undamaged or damaged. Each computer undergoes a series of 100 tests. The number of tests it fails will be used to make the decision. If it fails more than a certain number, it will be classified as damaged and then scrapped. From past history, the distribution of the number of tests failed is known for both undamaged and damaged computers. The relative frequencies of each outcome are listed in the table below:

| Number of tests failed | 0 | 1 | 2 | 3 | 4 | 5 | >5 |
|---|---|---|---|---|---|---|---|
| Undamaged (%) | 80 | 13 | 2 | 4 | 1 | 0 | 0 |
| Damaged (%) | 0 | 10 | 70 | 5 | 4 | 1 | 10 |

The table indicates, for example, that 80% of the undamaged computers have no failures, while 70% of the damaged computers have 2 failures.

a) To the engineers, this is a hypothesis-testing situation. State the null and alternative hypotheses.
b) Someone suggests classifying a computer as damaged if it fails any of the tests. Discuss the advantages and disadvantages of this test plan.
c) What number of tests would a computer have to fail in order to be classified as damaged if the engineers want to have the probability of a Type I error equal to 5%?
d) What's the power of the test plan in part c?
e) A colleague points out that by increasing $\alpha$ just 2%, the power can be increased substantially. Explain.

**34. Power** We are replicating an experiment. How will each of the following changes affect the power of our test? Indicate whether it will increase, decrease, or remain the same, assuming that all other aspects of the situation remain unchanged.

a) We increase the number of subjects from 40 to 100.
b) We require a higher standard of proof, changing from $\alpha = 0.05$ to $\alpha = 0.01$.

**35. Approval 2016** President Obama was very popular at the end of his eight years in office. A CNN/ORC poll of 1000 U.S. adults conducted in the week before the end of his term found that 63% of Americans said they held a favorable view of the President (elections.huffingtonpost.com/pollster/polls/cnn-27029). Commentators noted that, although popular, Obama was not as popular as Bill Clinton had been at the end of his presidency. Clinton had set a mark of 66% approval. What do you think? Was Obama less popular than indicated in the mark set by Clinton?

**36. Grade inflation 2012** In 1996, 20% of the students at a major university had an overall grade point average of 3.5 or higher (on a scale of 4.0). In 2012, a random sample of 1100 student records found that 25% had a GPA of 3.5 or higher. Does this provide evidence that the proportion of all the university's students whose GPA was 3.5 or higher was greater in 2012 than in 1996?

**37. Name recognition** An advertising agency won't sign an athlete to do product endorsements unless it is sure the person is known to more than 25% of its target audience. The agency always conducts a poll of 500 people to investigate the athlete's name recognition before offering a contract. Then it tests $H_0: p = 0.25$ against $H_A: p > 0.25$ at a 5% level of significance.

a) Why does the company use upper tail tests in this situation?
b) Explain what Type I and Type II errors would represent in this context, and describe the risk that each error poses to the company.
c) The company is thinking of changing its test to use a 10% level of significance. How would this change the company's exposure to each type of risk?

**38. Name recognition, part II** The advertising company described in Exercise 37 is thinking about signing a WNBA star to an endorsement deal. In its poll, 27% of the respondents could identify her.

a) Fans who never took Statistics can't understand why the company did not offer this WNBA player an endorsement contract even though the 27% recognition rate in the poll is above the 25% threshold. Explain it to them.
b) Suppose that further polling reveals that this WNBA star really is known to about 30% of the target audience. Did the company initially commit a Type I or Type II error in not signing her?
c) Would the power of the company's test have been higher or lower if the player were more famous? Explain.

**39. NIMBY** In March 2007, the Gallup Poll split a sample of 1003 randomly selected U.S. adults into two groups at random. Half ($n = 502$) of the respondents were asked,

> *Overall, do you strongly favor, somewhat favor, somewhat oppose, or strongly oppose the use of nuclear energy as one of the ways to provide electricity for the U.S.?*

It found that 53% were either "somewhat" or "strongly" in favor. The other half ($n = 501$) were asked,

> *Overall, would you strongly favor, somewhat favor, somewhat oppose, or strongly oppose the construction of a nuclear energy plant in your area as one of the ways to provide electricity for the U.S.?*

Only 40% were somewhat or strongly in favor. This difference is an example of the *NIMBY* (Not In My BackYard) phenomenon and is a serious concern to policy makers and planners. How large is the difference between the proportion of American adults who think nuclear energy is a good idea and the proportion who would be willing to have a nuclear plant in their area? Construct and interpret an appropriate confidence interval.

**40. Women** The U.S. Census Bureau reports that 26% of all U.S. businesses are owned by women. A Colorado consulting firm surveys a random sample of 410 businesses in the Denver area and finds that 115 of them have women owners. Should the firm conclude that its area is unusual? Test an appropriate hypothesis and state your conclusion.

**41. Skin cancer** In February 2012, MedPage Today reported that researchers used vemurafenib to treat metastatic melanoma (skin cancer). Out of 152 patients, 53% had a partial or complete response to vemurafenib.

a) Write a 95% confidence interval for the proportion helped by the treatment, and interpret it in this context.
b) If researchers subsequently hope to produce an estimate (with 95% confidence) of treatment effectiveness for metastatic melanoma that has a margin of error of only 6%, how many patients should they study?

**42. Streams** Researchers in the Adirondack Mountains collect data on a random sample of streams each year. One of the variables recorded is the substrate of the stream—the type of soil and rock over which they flow. The researchers want to estimate the proportion of streams that have a substrate of shale to within a margin of error of 7% (with 95% confidence). How many streams must they sample?

# PRACTICE EXAM

## I. MULTIPLE CHOICE

1. A chemist knows that when he uses his 50 mL pipet there will be some measurement error. The distribution of the errors is very close to a normal distribution with a mean of 0.05 mL and a standard deviation of 0.01 mL. If the chemist measures out 150 mL by using this pipet three times and combining the quantities, the mean error will be 0.15 mL. What's the standard deviation of that error?

A) 0.010  B) 0.017  C) 0.030
D) 0.087  E) 0.090

2. A conclusion that differences in the explanatory variable actually causes differences in the response variable would require a design that uses

A) a random sample.
B) random samples taken from at least two different populations.
C) a random sample of individuals who are then randomly assigned to treatment groups.
D) a control group.
E) random assignment of individuals to treatment groups.

3. A survey of 1025 teens found that 20% of students aged 14 to 18 plan to borrow no money to pay for college. What's the margin of error for a 90% confidence interval for the proportion of all students aged 14 to 18 who plan to borrow no money to pay for college?

A) $\pm 1.25\%$
B) $\pm 1.60\%$
C) $\pm 2.06\%$
D) $\pm 2.45\%$
E) $\pm 21.1\%$

4. Over 50% of family households in the United States have no children under 18 living at home and fewer than 10% have 3 or more, so the distribution of the number of children living in family households is skewed to the right. The mean is 0.96 children, with a standard deviation of 1.26 children. What's the probability that in a random sample of 250 family households the mean number of children living in those homes will be greater than 1?

A) 0.056   B) 0.308   C) 0.487
D) It cannot be determined, because the sample is not large enough.
E) It cannot be determined, because the population distribution is not normal.

5. A company's human resources director randomly selected 100 employees to complete a confidential survey about the effectiveness of a new incentive program. Only 70 of those selected returned the survey. Should the director be concerned about making a conclusion about the incentive program based on the results of the survey?

A) No, the employees were selected randomly.
B) No, 70 is a large sample.
C) No, it was not an experiment.
D) Yes, if the decision to not return the survey is related to one's opinion about the program.
E) Yes, if 70 is more than 10% of the company's employees.

6. Researchers conducted an experiment to compare two treatments for high blood pressure. Their null hypothesis was that the proportion of people whose blood pressure would improve is the same for both treatments. The two-sided test resulted in a P-value of 0.07. Which of the following statements is true?

A) There is a 7% chance that the drugs are equally effective.
B) There is a 7% chance the drugs are not equally effective.
C) The null hypothesis should be rejected at the 0.05 level.
D) 0 would be contained in the 95% confidence interval for the difference in proportions.
E) 0 would be contained in the 90% confidence interval for the difference in proportions.

7. In describing the findings of a 2010 Public Religion Research Institute poll, researchers reported a 3% margin of error at a 95% confidence level. If instead they had used a 99% confidence level, their margin of error would have been:

A) still 3%, because the same sample is used.
B) more than 3%, because they reported greater confidence.
C) less than 3%, because they reported greater confidence.
D) more than 3%, because this sample is too small.
E) less than 3%, because this sample is too small.

8. Which of the following is *not* a correct description of the use of a placebo in experiments?

a) A placebo is a part of every well-designed experiment
b) A placebo is a form of blinding
c) Experiments with plants do not need placebos
d) A placebo is one level of the factor
e) All of these are correct

9. Sometimes a drug that cures a disease turns out to have a nasty side effect. For example, some antidepressant drugs may cause suicidal thoughts in younger patients. A researcher conducts a study of such a drug to look for evidence of such side effects.

He tests the hypothesis that there's no side effect at the 0.05 significance level and finds a P-value of 0.03. Which of the following statements is true?

A) He could have committed a Type I error by concluding there's no such side effect if, in fact, there really is.
B) He could have committed a Type II error by concluding there's no such side effect if, in fact, there really is.
C) He could have committed a Type I error by concluding there is such a side effect if, in fact, there really isn't.
D) He could have committed a Type II error by concluding there is such a side effect if, in fact, there really isn't.
E) With a P-value as low as 0.03 he could not have committed either type of error.

10. Based on data from a study of the association between miles driven and gallons of gasoline used, let $r_1$ represent the correlation coefficient for (miles, gallons) data points, and let $r_2$ represent the correlation coefficient for (gallons, miles) data points. Which of the following must be true?

A) $r_1 + r_2 = 0$   B) $r_1 - r_2 = 0$   C) $r_1 + r_2 = 1$
D) $r_1 \times r_2 = 1$   E) $r_1 \times r_2 = -1$

11. The staff of a school newspaper plans to investigate students' opinions regarding the school's security guards. They decide to survey 100 students at the school. The editor-in-chief recommends that they survey 25 randomly selected students from each class (first-year students, sophomores, juniors, seniors). This is an example of a

A) blocked random sample.
B) multistage random sample.
C) simple random sample.
D) stratified random sample.
E) random cluster sample.

12. The city of Redlands, California is debating a new slow-growth initiative. Proponents of the plan want to restrict the construction of office buildings and apartments more than three stories high. A local newspaper conducts a poll and finds that 53% of those surveyed approve of these restrictions. Which of the following hypotheses would we use to determine if a majority of citizens approve of these restrictions?

A) $H_0: p = 0.50$   $H_A: p \neq 0.50$
B) $H_0: p = 0.50$   $H_A: p < 0.50$
C) $H_0: p = 0.50$   $H_A: p > 0.50$
D) $H_0: p = 0.53$   $H_A: p > 0.53$
E) $H_0: \hat{p} = 0.53$   $H_A: \hat{p} \neq 0.53$

13. A group of friends are planning on taking a road trip. One member of the group has collected a variety of data on the cost of gas for road trips. She even goes so far as to calculate a regression equation that predicts the cost of gas for the number of miles driven. Here is her regression equation.

$$\widehat{cost} = 1.73 + 0.165(miles)$$

The group took a trip that covered 550 miles. They ended up spending $13 more on gas than the regression equation predicted. What was their actual gas cost (rounded to the nearest dollar)?

A) $68   B) $79   C) $92
D) $105   E) none of these

14. Use the same context as in question 13. Which of the following is the correct interpretation of the slope in context?
    A) The start of the trip costs a predicted $1.73.
    B) The start of the trip costs a predicted $0.165.
    C) For every one more mile driven, the cost is predicted to increase by $1.73.
    D) For every 1.73 miles driven, the cost is predicted to increase by $0.165.
    E) For every one more mile driven, the cost is predicted to increase by $0.165.

15. Airport security personnel screen carry-on luggage for items that are not allowed on planes. Most bags pass through without a problem, but occasionally one is pulled aside for further inspection. Those inspections actually find some item that must removed in 2 of every 5 bags that are pulled aside. If 7 of the people on a certain flight had luggage that was subjected to further inspection, what's the probability that at least 2 of them had items removed?
    A) 0.159    B) 0.261    C) 0.420
    D) 0.580    E) 0.841

16. Bank officers are considering targeting young parents with an offer to start a college savings plan. They have decided to pursue this new marketing program only if at least 30% of such parents might be interested in this kind of plan. They surveyed 250 randomly selected parents, of whom 38% indicated some interest. Which formula below correctly computes the $z$-score they should use to see if this survey result provides evidence that the true proportion is over 30%?

    A) $z = \dfrac{0.38 - 0.30}{\sqrt{\dfrac{(0.30)(0.70)}{250}}}$

    B) $z = \dfrac{0.30 - 0.38}{\sqrt{\dfrac{(0.30)(0.70)}{250}}}$

    C) $z = \dfrac{0.38 - 0.30}{\sqrt{\dfrac{(0.38)(0.62)}{250}}}$

    D) $z = \dfrac{0.38 - 0.30}{\sqrt{250(0.30)(0.70)}}$

    E) $z = \dfrac{0.30 - 0.38}{\sqrt{250(0.38)(0.62)}}$

17. In 2012 the *New York Times* surveyed adults about whether they had gone back to school for additional training and, if so, whether they felt this training helped them get new jobs or promotions. In the article that presented the results, the newspaper included the following information: "The nationwide telephone poll was conducted May 31 to June 3 with 976 adults, of whom 229 said they went to school in the last five years. Margin of sampling error for all adults is plus or minus 3 percentage points; for adults who went back to school, plus or minus 6 percentage points. Of those who went back to school, 84% reported that the training was a good investment of time and money." What is the most logical reason for the larger margin of error that was indicated for results about adults who went back to school?

    A) The researchers used a smaller confidence level.
    B) The researchers used a larger confidence level.
    C) The sample size was smaller.
    D) The sample size was larger.
    E) Adults who returned to school were more variable in terms of employment.

    (Source: M. Connelly, M. Stefan, and A. Kayda, "Is It Worth It?" *Education Life; New York Times*, July 22, 2012, p. 31.)

18. In an experiment to investigate how the dosage impacts a medication's effect, researchers randomly assign 60 volunteers to receive 10 mg, 20 mg, or 30 mg of the medication. What is one of the limitations of this experiment?
    A) The conclusions will be questionable because the design has no blocking.
    B) The conclusions will be questionable because no group gets a placebo.
    C) The conclusions cannot be generalized to a population because the subjects are volunteers.
    D) The conclusions will be questionable because there are only 20 subjects in each group.
    E) The conclusions will be questionable because each subject is not given all of the doses.

19. A marketing researcher for a phone company conducted a survey of 500 people and then constructed a confidence interval for the proportion of customers who are likely to switch providers when their contract expires. After seeing this interval, the company CEO insisted that the researcher repeat the survey and provide an interval with a margin of error half as large. What sample size should the researcher use for the new survey?
    A) 250    B) 354    C) 708    D) 1000    E) 2000

20. An entomologist observes that the relationship between the anntenna length of a certain species of grasshopper and its age appears to be linear, and calculates an $R^2$ of 78%. This value tells us that:
    A) Errors in predicting the age of a grasshopper by using its anntenna length will not exceed 22%.
    B) 78% of a grasshopper's anntenna length can be explained by its age.
    C) The regression model makes accurate predictions for 78% of all grasshoppers.
    D) The variability in antenna length depends on the age 78% of the time.
    E) There is a moderately strong relationship between anntenna length and age.

21. The Bureau of Justice conducted a study of 272,111 former inmates released from prisons in 15 states in 1994. A classification of the former prisoners by "most serious offense for which released" yielded the following: 22.5% violent; 33.5% property; 32.6% drug; 11.4% other. Of those prisoners whose most serious crime was violent, 61.7% were rearrested within 3 years of release. For the other categories, the recidivism rate (defined as rearrested within 3 years of release) were: property 73.8%, drug 66.7%, and other 62.6%. If a released prisoner from these 15 states is rearrested within 3 years of release, what is the probability that this prisoner's most serious offense was violent?
    A) 0.139    B) 0.206    C) 0.223    D) 0.617    E) 0.690

    (Source: D. Levine and P. Mangan, "Recidivism of Prisoners Released in 1994"; NCJ193427; Bureau of Justice; File name rpr94bxl.csv http://bjs.ojp.usdoj.gov/index.cfm?ty=pbdetail&iid=1134)

22. In which of the following situations is a one-sided alternative hypothesis appropriate?

   A) A business student conducts a blind taste test to see whether students prefer Diet Coke or Diet Pepsi.
   B) A budget override in a small town requires a two-thirds majority to pass. A local newspaper conducts a poll to see if there's evidence it will pass.
   C) In recent years, 10% of college juniors have applied for study abroad. The dean's office conducts a preliminary survey to see if that has changed this year.
   D) A taxi company checks gasoline usage records to see if a recent change in the brand of tires installed on the cars has any impact on their fleet's fuel economy.
   E) The Centers for Disease Control and Prevention wants to see if a controversial new advertising campaign has had any effect on teen smoking rates.

23. A study conducted by students in an AP Psychology class at South Kent School in Connecticut discovered a correlation of $-0.38$ between hours of sleep ($x$) and GPA ($y$). If we change the variable on the horizontal axis to hours awake ($24 - x$), but make no change to the GPA data, which of the following would be true about the new scatterplot?

   A) It slopes down, and the correlation is $-0.38$.
   B) It slopes down, and the correlation is $+0.38$.
   C) It slopes up, and the correlation is $-0.38$.
   D) It slopes up, and the correlation is $+0.38$.
   E) None of the above choices is correct.

   (Source: http://www.cardinalnewsnetwork.org/south-kent-community/sleep-study-sks-style/)

24. A study of over 3000 network and cable programs found that nearly 60% featured violence. Suppose you want to simulate a collection of five randomly selected programs. Which of the following assignments of the digits 0 through 9 would be appropriate for modeling whether individual programs feature violence for each trial of 5 programs?

   A) Assign the digits 0, 1, 2, 3, 4, and 5 as featuring violence and 6, 7, 8, and 9 as not featuring violence.
   B) Assign the digits 0, 1, 2, 3, 4, and 5 as featuring violence and 6, 7, 8, and 9 as not featuring violence, and ignore repeats.
   C) Assign 6 as featuring violence and the digits 0, 1, 2, 3, 4, 5, 7, 8, and 9 as not featuring violence.
   D) Assign the digits 1, 2, 3, 4, 5, and 6 as featuring violence and 7, 8, and 9 as not featuring violence, ignoring 0.
   E) Assign 0 as featuring violence, the digits 1, 2, 3, 4, and 5 as not featuring violence, and ignore 6, 7, 8, and 9.

25. For her final project, Stacy plans on surveying a random sample of 50 students on whether they plan to leave town for Spring Break. Based on past years, she guesses that about 85% of the students will go somewhere. Is it appropriate for her to use a Normal model for the sampling distribution of the sample proportion?

   A) Yes, because $np = 42.5$, which is greater than 10.
   B) Yes, because a Normal model always applies to sample proportions.
   C) No, because a Normal model is not appropriate for a binomial situation.
   D) No, because $nq = 7.5$, which is less than 10.
   E) We don't know, because we don't know the shape of the population distribution.

26. An employer is ready to give her employees a raise and is considering two plans. Plan A is to give each person an $8.00 per day increase. Plan B is to give each person a 10% increase. Data on the current pay of these employees shows that the median pay is $80 per day with an interquartile range of $10. Which of the following datasets will have the same median?

   A) The current pay and Plan A only
   B) The current pay and Plan B only
   C) Plan A and Plan B only
   D) The current pay, Plan A, and Plan B
   E) None; the current pay, Plan A, and Plan B will all have different medians.

27. Using the same context as Exercise 26, which of the following datasets will have the same interquartile range?

   A) The current pay and Plan A only
   B) The current pay and Plan B only
   C) Plan A and Plan B only
   D) The current pay, Plan A, and Plan B
   E) None; the current pay, Plan A, and Plan B will all have different IQRs.

28. A researcher is reporting characteristics of the subjects she used in a recent study. Two of the variables are hair color and age. Which of these are appropriate choices to summarize these data?

   A) Bar charts for both hair color and age.
   B) Histograms for both hair color and age.
   C) A bar chart for hair color and a histogram for age.
   D) A histogram for hair color and a bar chart for age.
   E) Either bar charts or histograms are good choices for both hair color and age.

29. Researchers collected data on spending habits of first-year students versus senior students at five different high schools. They asked students if they spent less than $10 on fast food in week, $10–$20, or $20 or more. Students were told to count only the money they spent out of their own personal funds, not money spent for them by their parents or other adults. Here is a summary of the data:

   |  | <$10 | 10–$20 | >$20 | Total |
   |---|---|---|---|---|
   | First-year students | 23 | 12 | 8 | 43 |
   | Seniors | 25 | 19 | 28 | 72 |
   | Total | 48 | 31 | 36 | 115 |

   If one of these students is selected at random, what is the probability that person is a senior or spent less than $10?

   A) 0.217   B) 0.347   C) 0.521
   D) 0.783   E) 0.826

30. A teacher gives a test and the distribution of scores turns out to be bimodal. One-third of the class earned scores between 60% and 75%, while the rest of the class scored between 88% and 98%. Which of the following are the most plausible estimates of the mean and the median?

   A) The median is about 80 and the mean is about 85.
   B) The median and the mean are both about 85.

C) The median is about 85 and the mean is about 90.
D) The median is about 90 and the mean is about 85.
E) The median and the mean are both about 90.

## II. FREE RESPONSE

1. The table below summarizes combined city/highway fuel economy for new cars classified as "large" models.

| N | MEAN | STDEV | SEMEAN | TRMEAN | MIN | Q1 | MEDIAN | Q3 | MAX |
|---|---|---|---|---|---|---|---|---|---|
| 65 | 19.6 | 3.36 | 0.417 | 19.5 | 14 | 17 | 19 | 21 | 28 |

   a) What do the summary statistics suggest about the overall shape of the distribution? Explain.
   b) Are there any outliers in the data? Explain how you determined this.
   c) In advertising for its own vehicle, a competitor described one large car that gets 17 mpg as exceptionally gas-thirsty. Based on the summary statistics above, comment on that characterization.

2. Based on data collected at a store where tomatoes are sold by the pound, the association between the number of tomatoes purchased and the total cost (in dollars) appeared to be linear. The output below shows the regression analysis.

   Dependent variable is Cost
   R-squared = 92.4%

   | Variable | Coefficient |
   |---|---|
   | Intercept | 0.03 |
   | Tomatoes | 0.47 |

   a) Predict the cost of 12 tomatoes.
   b) Interpret the value of $R^2$ in this context.
   c) Interpret the meaning of the slope in this context.
   d) Is there any meaningful interpretation of the $y$-intercept in this context?

3. The management team at a company sent out the following e-mail to all 75 employees:

   > Several complaints have come in that morale in this company is low, and that people are not happy with the working conditions. Naturally, as managers we are concerned about this. Please reply to this e-mail with an answer to the following question: Are you happy with the working conditions here?

   Based on answers received from 24 employees who replied to the e-mail, the managers concluded there is no morale problem at their company.

   a) Explain how bias could have been introduced by the wording of the survey, and how that might mislead the company managers.
   b) Explain how bias could be introduced by conducting the survey by e-mail, and how that might mislead the company managers.
   c) Explain how bias could have been introduced by the fact that only 24 of the 75 employees responded, and how that might mislead the company managers.

4. The website VGChartz tracks the weekly sales of the various videogame hardware consoles. The table below shows the weekly sales for the week of August 14, 2021. The table shows number of consoles sold in each of three different regions as well as the five most popular consoles. The values are in thousands of units sold.

| | North America | Europe | Japan | Total |
|---|---|---|---|---|
| Nintendo Switch | 107.9 | 77.8 | 89.8 | 275.5 |
| PlayStation5 | 63.3 | 85.8 | 9.1 | 158.2 |
| Xbox Series S | 60.6 | 42.8 | 1.3 | 104.7 |
| PlayStation 4 | 11 | 13.6 | 1.2 | 25.8 |
| Xbox One | 8.5 | 2.4 | NA | 10.9 |
| Total | 251.3 | 222.4 | 101.4 | 575.1 |

(Source: https://www.vgchartz.com/tools/hw_date.php)

   a) Identify the variables collected in this table.
   b) What percentage of sales were for the Nintendo Switch?
   c) What percentage of North American sales were for the Nintendo Switch?
   d) What percentage of sales in Japan were for the Nintendo Switch?
   e) Using your answers to parts (a), (b), and (c), explain if region and console sales are independent.

5. In 1986 a catastrophic nuclear accident in the Ukranian city of Chernobyl exposed the surrounding area to large amounts of radiation. Since then, scientists have been investigating the after effects on plants, animals, and humans. A 2007 study examined 841 barn swallows there, noting the incidence of deformed toes and beaks, abnormal tail feathers, deformed eyes, tumors, and albinism (a condition that involves discoloration of the birds' plumage). Previous studies involving thousands of barn swallows in Spain, Italy, and Denmark had seen albinism in 4.35% of this species, yet 112 of the Chernobyl swallows displayed the condition.

   a) Is there evidence that the albinism rate is higher among Chernobyl barn swallows?
   b) What does this study tell us about the effect of nuclear contamination on barn swallows?

6. Members of the We-Luv-Plants club are interested in comparing the tendency of boys vs. girls to choose a vegetarian diet. They randomly sampled 75 boys and 75 girls at their school. The data showed that 23 of the girls were vegetarians whereas only 12 of the boys were.

   a) Construct and interpret a 95% confidence interval for the difference in the proportions between girls and boys who are vegetarians.
   b) Does this confidence interval provide evidence that girls are more likely than boys to choose a vegetarian diet? Explain.
   c) To what population(s) do your conclusions in parts a and b apply?

# 22

# Inferences About Means

Psychologists Jim Maas and Rebecca Robbins, in their book *Sleep for Success!*, say that

> In general, high school and college students are the most pathologically sleep-deprived segment of the population. Their alertness during the day is on par with that of untreated narcoleptics and those with untreated sleep apnea. Not surprisingly, teens are also 71 percent more likely to drive drowsy and/or fall asleep at the wheel compared to other age groups. (Males under the age of twenty-six are particularly at risk.)

They report that adults require between 7 and 9 hours of sleep each night and claim that college students require 9.25 hours of sleep to be fully alert. They note that "There is a 19 percent memory deficit in sleep-deprived individuals" (p. 35).

A student surveyed other students at a small college in the northeastern United States and asked, among other things, how much they had slept the previous night. Here's a histogram and the data for 25 of the students selected at random from the survey.

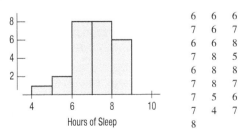

We're interested both in estimating the mean amount slept by all the college's students and in testing whether it is less than the minimum recommended amount of 7 hours. These data were collected in a suitably randomized survey so we can treat them as representative of students at that college.

(We're assuming that the night before the survey was conducted wasn't an unusual one. If it had been, say, the night of a popular college event that ran late in the evening, then we wouldn't expect students' reported hours of sleep the next day to be typical.)

These data differ from data on proportions in one important way. Proportions are summaries of individual responses, which had two possible values such as "yes" and "no," "male" and "female," or "1" and "0." Quantitative data, though, report a quantitative value for each individual. When you have quantitative data, you should remember the three rules of data analysis and plot the data, as we have done.

# Getting Started: The Central Limit Theorem (Again)

You've learned how to create confidence intervals and test hypotheses about proportions. We always center confidence intervals at our best guess for the unknown parameter. Then we add and subtract a margin of error. For proportions, we write $\hat{p} \pm ME$.

We found the margin of error as the product of the standard error, $SE(\hat{p})$, and a critical value, $z^*$, from the Normal model. So we had $\hat{p} \pm z^* SE(\hat{p})$.

We knew we could use $z$ because the Central Limit Theorem told us (back in Chapter 17) that the sampling distribution model for proportions is Normal.

Now we want to do exactly the same thing for means, and fortunately, the Central Limit Theorem (still in Chapter 17) told us that a Normal model also works as the sampling distribution for means.

### The Central Limit Theorem

When random samples are drawn from any population with mean $\mu$ and standard deviation $\sigma$, the sample means, $\bar{y}$, have a sampling distribution with the same *mean* $\mu$ but whose *standard deviation* is $\frac{\sigma}{\sqrt{n}}$ (and we write $\sigma(\bar{y}) = SD(\bar{y}) = \frac{\sigma}{\sqrt{n}}$).

No matter what population the random sample comes from, the *shape* of the sampling distribution is approximately Normal as long as the sample size is large enough. The larger the sample used, the more closely the Normal approximates the sampling distribution for the mean.

## FOR EXAMPLE
### Using the Central Limit Theorem (If We Know $\sigma$)

Based on weighing thousands of animals, the American Angus Association reports that mature Angus cows have a mean weight of 1309 pounds with a standard deviation of 157 pounds. This result was based on a very large sample of animals from many herds over a period of 15 years, so let's assume that these summaries are the population parameters and that the distribution of the weights was unimodal and reasonably symmetric.

QUESTION: What does the Central Limit Theorem (CLT) predict about the mean weight seen in random samples of 100 mature Angus cows?

ANSWER: It's given that weights of all mature Angus cows have $\mu = 1309$ and $\sigma = 157$ pounds. Because $n = 100$ animals is a fairly large sample, I can apply the Central Limit Theorem. I expect the resulting sample means $\bar{y}$ will average 1309 pounds and have a standard deviation of

$$SD(\bar{y}) = \frac{\sigma}{\sqrt{n}} = \frac{157}{\sqrt{100}} = 15.7 \text{ pounds.}$$

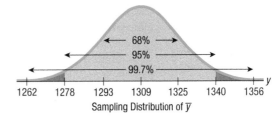

The CLT also says that the distribution of sample means follows a Normal model, so the 68–95–99.7 Rule applies. I'd expect that

✓ in 68% of random samples of 100 mature Angus cows, the mean weight will be between $1309 - 15.7 = 1293.3$ and $1309 + 15.7 = 1324.7$ pounds;

✓ in 95% of such samples, $1277.6 \leq \bar{y} \leq 1340.4$ pounds;

✓ in 99.7% of such samples, $1261.9 \leq \bar{y} \leq 1356.1$ pounds.

The CLT says that all we need to model the sampling distribution of $\bar{y}$ is a sufficiently large random sample of quantitative data.

And the true population standard deviation, $\sigma$.

Uh oh. That's a big problem. How are we supposed to know $\sigma$? With proportions, we had a link between the proportion value and the standard deviation of the sample proportion: $SD(\hat{p}) = \sqrt{\frac{pq}{n}}$. And there was an obvious way to estimate the standard deviation from the data: $SD(\hat{p}) = \sqrt{\frac{\hat{p}\hat{q}}{n}}$. But for means, $SD(\bar{y}) = \frac{\sigma}{\sqrt{n}}$, so knowing $\bar{y}$ doesn't tell us anything about $SD(\bar{y})$. We know $n$, the sample size, but the population standard deviation, $\sigma$, could be *anything*. So what should we do? We do what any sensible person would do: We estimate the population parameter $\sigma$ with $s$, the sample standard deviation based on the data. The resulting standard error is $SE(\bar{y}) = \frac{s}{\sqrt{n}}$.

A century ago, people used this standard error with the Normal model, assuming it would work. And for large sample sizes it *did* work pretty well. But they began to notice problems with smaller samples. A sample standard deviation, $s$, like any other statistic, varies from sample to sample. And this extra variation in the standard error was messing up the P-values and margins of error.

William S. Gosset is the man who investigated this fact. He realized that not only do we need to allow for the extra variation with larger margins of error and P-values, but we even need a new sampling distribution model. In fact, we need a whole *family* of models, depending on the sample size, $n$. These models are unimodal, symmetric, bell-shaped models, but the smaller our sample, the more we must stretch out the tails. Gosset's work transformed Statistics, but most people who use his discovery don't even know his name.

### Standard Error

Because we estimate the standard deviation of the sampling distribution model from the data, it's a *standard error*. So we use the $SE(\bar{y})$ notation. Remember, though, that it's just the estimated standard deviation of the sampling distribution model for means.

## The Story of Gosset's *t*

Gosset had a job that made him the envy of many. He was the chief Experimental Brewer for the Guinness Brewery in Dublin, Ireland. The brewery was a pioneer in scientific brewing and Gosset's job was to meet the demands of the brewery's many discerning customers by developing the best stout (a thick, dark beer) possible.

Gosset's experiments often required as much as a day to make the necessary chemical measurements or a full year to grow a new crop of hops. For these reasons and others, his samples sizes were small—often as small as 3 or 4.

When he calculated means of these small samples, Gosset wanted to compare them to a target mean to judge the quality of the batch. To do so, he followed common statistical practice of the day, which was to calculate *z*-scores and compare them to the Normal model. But Gosset noticed that with samples of this size, his tests weren't quite right. He knew this because when the batches that he rejected were sent back to the laboratory for more extensive testing, too often they turned out to be OK. In fact, about 3 times more often than he expected. Gosset knew something was wrong, and it bugged him.

Guinness granted Gosset time off to earn a graduate degree in the emerging field of Statistics, and naturally he chose this problem to work on. He figured out that when he used the standard error, $\frac{s}{\sqrt{n}}$, as an estimate of the standard deviation of the mean, $\frac{\sigma}{\sqrt{n}}$, the shape of the sampling model changed. He even figured out what the new model should be.

W. S. Gosset (1876–1937) is thought by many Statistics to be the founder of modern statistics because of his research showing how to perform inference for a mean when the standard deviation of the population was estimated from data. To find the sampling distribution of $\frac{\bar{y} - \mu}{s/\sqrt{n}}$, Gosset simulated it *by hand*.

### A Century-Old Simulation

Gosset did his research near the beginning of the 20th century, so he didn't have computers at his disposal. Even so, he made his most important and lasting discovery using simulations. He planned a simulation that would require more data than were available on beer chemistry, so he chose instead a population for which lots of data were available at the time: the heights of male prisoners. He used 3000 cards and wrote on each one the height of a prisoner. Then he shuffled the cards, drew a sample of four, and computed the mean and standard deviation of the sample, from which he could then compute the critical statistic of interest to him (and us): $\frac{\bar{y} - \mu}{s/\sqrt{n}}$. He did this for hundreds of repeated random samples and then studied the shape of the distribution of his results.

We decided to follow in Gosset's footsteps here, except we used a computer instead of a stack of cards. We programmed the computer to simulate four men's heights, each one drawn from a population that is approximately $N(69.1, 2.9)$—that is, the heights are approximately normally distributed with a mean of $\mu = 69.1$ inches and a standard deviation of $\sigma = 2.9$ inches. Our four simulated prisoners turned out to be 64.1 inches tall, 68.3 inches tall, 72.9 inches tall, and 66.7 inches tall. The mean of these is $\bar{y} = 68.0$ inches and the standard deviation is about $s = 3.697$ inches. Remember that our estimate of the standard deviation of $\bar{y}$ is $s/\sqrt{n}$, which in our case is $3.697/\sqrt{4} \approx 1.848$. That means our first sample's mean of 68.0 inches is about $\frac{\bar{y} - \mu}{s/\sqrt{n}} = \frac{68.0 - 69.1}{3.697/\sqrt{4}} = -0.595$ standard errors from the actual population mean of 69.1 inches.

Gosset was studying this quantity because he suspected that the standard Normal model wasn't working very well for it. Our first trial's outcome is perfectly consistent with the standard Normal distribution. But of course, one trial does not a simulation make. We could have programmed the computer to do another hundred thousand trials very easily, but to honor Gosset's work, we did the same number of trials he did with his stack of cards: 750. The histogram below shows the distribution of our 750 different samples' standardized means.

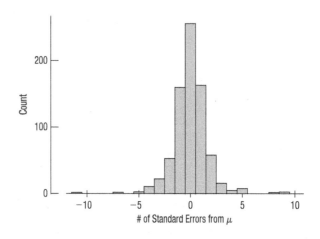

Now we see why Gosset kept rejecting batches of beer that were actually okay. He (and everyone else) had been treating values of magnitude greater than 2 as if they were cause for alarm. In fact, a better threshold for "alarm" might have been around 3, 4, or even 5.

So it's less unusual for $\bar{y}$ to be 4 or 5 standard *errors* $(s/\sqrt{n})$ from the population mean than to be 4 or 5 standard *deviations* $(\sigma/\sqrt{n})$ from the population mean. That's because the occasional random sample will have values unusually close to one another, which reduces the sample standard deviation $s$ in the *denominator* of the standardized mean. A smaller denominator means a value of greater magnitude.

## Student's *t* Distribution Family

The Guinness Company may have been ahead of its time in using statistical methods to manage quality, but it also had a policy that forbade its employees to publish research that might help competing breweries. Gosset pleaded that his results were of no specific value to brewers and was eventually allowed to publish under the pseudonym "Student." This important result is still widely known as **Student's *t***. (We called this section The Story of *Gosset's t* to restore his due credit.)

Gosset's sampling distribution model is always bell-shaped, but the details change with different sample sizes. When the sample size is very large, the model is nearly Normal, but when it's small the tails of the distribution are much fatter than in a Normal distribution. That means that values far from the mean are more common, especially for small samples (see Figure 22.1). So the Student's *t*-models form a whole *family* of related distributions that depend on a parameter known as **degrees of freedom**. The degrees of freedom of a distribution represent the number of independent quantities that are left *after* we've estimated the model's other parameters but before we've estimated $\sigma$ with $s$. Here it's simply the number of data values, $n$, minus the number of estimated parameters. For means, that's just $n - 1$. (The one parameter we estimated was $\mu$.) We often denote degrees of freedom as *df* and the model as $t_{df}$, with the degrees of freedom as a subscript.[1]

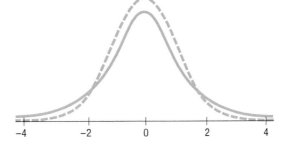

**Figure 22.1**
The *t*-model (solid curve) on 2 degrees of freedom has fatter tails than the Normal model (dashed curve). So the 68–95–99.7 Rule doesn't work for *t*-models with only a few degrees of freedom. It may not look like a big difference, but a *t* with 2 df is more than 4 times as likely to have a value greater than 2 than is a standard Normal distribution.

### Don't Divide by *n*

Some calculators offer an alternative button for standard deviation that divides by *n* instead of by $n - 1$. Why don't you stick a wad of gum over the "*n*" button so you won't be tempted to use it? Use $n - 1$.

### Degrees of Freedom We've Seen Before

Recall that the formula for standard deviation divides by $n - 1$ rather than by $n$. The reason is closely tied to the concept of degrees of freedom.

If only we knew the true population mean, $\mu$, we would use it in our formula for the sample standard deviation:

$$s = \sqrt{\frac{\sum (y - \mu)^2}{n}}.$$

---

[1] If "degrees of freedom" seems tough to understand, that's because it's *extremely* tough to understand, and we've not provided here anything like a full explanation, which would involve multidimensional spaces and orthogonal subspaces and such. (We said it was tough.) In truth, even among researchers who use *t*-distributions all the time, most just memorize how many degrees of freedom there are for each of the several statistical tools they use most often. You're fine doing the same.

But we don't know $\mu$, so we naturally use $\bar{y}$ in its place. And that causes a small problem. For any sample, the data values will generally be closer to their *own* sample mean, $\bar{y}$, than to the population mean, $\mu$. So, when we calculate $s$ using $\sum(y - \bar{y})^2$ instead of $\sum(y - \mu)^2$, our standard deviation estimate is too small. The amazing mathematical fact is that we can fix it by dividing by $n - 1$ instead of by $n$. This difference is much more important when $n$ is small than when the sample size is large. The $t$-distribution inherits this same number and we call $n - 1$ the degrees of freedom.

# A Confidence Interval for Means

To make confidence intervals or test hypotheses for means, we need to use Gosset's model. Which one? Well, for means, it turns out the right value for degrees of freedom is $df = n - 1$.

> **NOTATION ALERT**
> Ever since Gosset, $t$ has been reserved in Statistics for his distribution.

### A Practical Sampling Distribution Model for Means

When certain assumptions and conditions are met, the standardized sample mean,

$$t = \frac{\bar{y} - \mu}{SE(\bar{y})},$$

follows a Student's $t$-model with $n - 1$ degrees of freedom. We estimate the standard deviation with

$$SE(\bar{y}) = \frac{s}{\sqrt{n}}.$$

When Gosset corrected the model to take account of the extra uncertainty, the margin of error got bigger, as you might guess. When you use Gosset's model instead of the Normal model, your confidence intervals will be just a bit wider and your P-values just a bit larger. That's the correction you need. By using the $t$-model, you've compensated for the extra variability in precisely the right way.[2]

> **NOTATION ALERT**
> When we found critical values from a Normal model, we called them $z^*$. When we use a Student's $t$-model, we'll denote the critical values $t^*$.

### One-Sample $t$-Interval for the Mean

When the assumptions and conditions[3] are met, we are ready to find the confidence interval for the population mean, $\mu$. The confidence interval is

$$\bar{y} \pm t^*_{n-1} \times SE(\bar{y}),$$

where the standard error of the mean is $SE(\bar{y}) = \frac{s}{\sqrt{n}}.$

The critical value $t^*_{n-1}$ depends on the particular confidence level, $C$, that you specify and on the number of degrees of freedom, $n - 1$, which we get from the sample size.

**APPLET**
See how effective $t$-intervals are at capturing the true mean.

---

[2] Gosset, as the first to recognize the consequence of using $s$ rather than $\sigma$, was also the first to give the sample standard deviation, $s$, a different letter than the population standard deviation, $\sigma$.

[3] They're coming in the next section. You'll probably find they look pretty familiar.

### FOR EXAMPLE
#### A One-Sample *t*-Interval for the Mean

In 2004, a team of researchers published a study of contaminants in farmed salmon.[4] Fish from many sources were analyzed for 14 organic contaminants. The study expressed concerns about the level of contaminants found. One of those was the insecticide mirex, which has been shown to be carcinogenic and is suspected to be toxic to the liver, kidneys, and endocrine system. One farm in particular produced salmon with very high levels of mirex. After those outliers are removed, summaries for the mirex concentrations (in parts per million) in the rest of the farmed salmon are:

$$n = 150 \quad \bar{y} = 0.0913 \text{ ppm} \quad s = 0.0495 \text{ ppm}.$$

QUESTION: What does a 95% confidence interval say about mirex?

ANSWER:

$$df = 150 - 1 = 149$$

$$SE(\bar{y}) = \frac{s}{\sqrt{n}} = \frac{0.0495}{\sqrt{150}} = 0.0040 \quad \begin{array}{l} t^*_{149} \approx 1.977 \text{ (from Table T, using 140 df)} \\ \left(\text{actually, } t^*_{149} \approx 1.976 \text{ from technology}\right) \end{array}$$

So the confidence interval for $\mu$ is $\bar{y} \pm t^*_{149} \times SE(\bar{y}) = 0.0913 \pm 1.977 \, (0.0040)$
$$= 0.0913 \pm 0.0079$$
$$= (0.0834, 0.0992)$$

I'm 95% confident that the mean level of mirex concentration in farm-raised salmon is between 0.0834 and 0.0992 parts per million.

---

**TI-*nspire***

**The *t*-models.** See how *t*-models change as you change the degrees of freedom.

Student's *t*-models are unimodal, symmetric, and bell-shaped, just like Normal models. But *t*-models with only a few degrees of freedom have longer tails and a larger standard deviation than Normal models. (That's what makes the margin of error bigger.) As the degrees of freedom increase, the *t*-models look more and more like the standard Normal. In fact, the *t*-model with infinite degrees of freedom is exactly Normal.[5] This is great news if you happen to have an infinite number of data values, but that's not likely. Fortunately, after about fifty degrees of freedom, the difference is pretty negligible. Of course, in the rare situation that we *know* $\sigma$, it would be foolish not to use that information. And if we don't have to estimate $\sigma$, we can use a Normal model.

**z or t**

If you know $\sigma$, use z. (That's rare!) Whenever you use *s* to estimate $\sigma$, use *t*.

> **WHEN $\sigma$ IS KNOWN**
>
> Administrators of a hospital were concerned about the prenatal care given to mothers in their part of the city. To study this, they examined the gestation times of babies born there. They drew a sample of 25 babies born in their hospital in the previous 6 months. Human gestation times for healthy pregnancies are thought to be well-modeled by a Normal distribution with a mean of 280 days and a standard deviation of 10 days. The hospital administrators wanted to test the mean gestation time of their sample of babies against the known standard. For this test, they should use the established value for the standard deviation, 10 days, rather than estimating the standard deviation from their sample. Because they use the model parameter value for $\sigma$, they should base their test on the Normal model rather than Student's *t*.

---

[4] Ronald A. Hites, Jeffery A. Foran, David O. Carpenter, M. Coreen Hamilton, Barbara A. Knuth, and Steven J. Schwager, "Global Assessment of Organic Contaminants in Farmed Salmon," *Science* (January 9, 2004): vol. 303, no. 5655, pp. 226–229.

[5] Formally, in the limit as *n* goes to infinity.

## TI TIPS

### Finding *t*-Model Probabilities and Critical Values

**FINDING PROBABILITIES** You already know how to use your TI to find probabilities for Normal models using *z*-scores and `normalcdf`. What about *t*-models? Yes, the calculator can work with them, too.

You know from your experience with confidence intervals that $z = 1.645$ cuts off the upper 5% in a Normal model. Use the TI to check that. From the DISTR menu, enter `normalcdf(1.645,999,0,1)`. Only 0.04998? Close enough for statisticians!

We might wonder about the probability of observing a *t*-value greater than 1.645, but there's not just one such probability. There's only one Normal model, but there are many *t*-models, depending on the number of degrees of freedom. We need to be more specific.

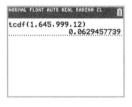

Let's find the probability of observing a *t*-value greater than 1.645 when there are 12 degrees of freedom. That we can do. Look in the DISTR menu again. See it? Yes, `tcdf`. That function works essentially like `normalcdf`. Use `tcdf(` with `lower:1.645`, `upper:999`, `df:12`, then choose Paste and hit ENTER twice (or on an older calculator, enter the command `tcdf(1.645,999,12)`).

The upper tail probability for $t_{12}$ is 0.063, higher than the Normal model's 0.05. That should make sense to you—remember, *t*-models are a bit fatter in the tails, so more of the distribution lies beyond the 1.645 cutoff. (That means we'll have to go a little wider to make a 90% confidence interval.)

Check out what happens when there are more degrees of freedom, say, 25. The command `tcdf(1.645,999,25)` yields a probability of 0.056. That's closer to 0.05, for a good reason: *t*-models look more and more like the Normal model as the number of degrees of freedom increases.

**FINDING CRITICAL VALUES** Your calculator can also determine the critical value of *t* that cuts off a specified percentage of the distribution, using `invT`. It works just like `invNorm`, but for *t* we also have to specify the number of degrees of freedom (of course).

Suppose we have 6 degrees of freedom and want to create a 98% confidence interval. A confidence level of 98% leaves 1% in each tail of our model, so we need to find the value of *t* corresponding to the 99th percentile. If a Normal model were appropriate, we'd use $z = 2.33$. (Try it: `invNorm(.99)`). Now think. How should the critical value for *t* compare?

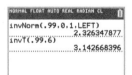

If you thought, "It'll be larger, because *t*-models are more spread out," you're right. Check with your TI. Use `invT(` with `area:0.99`, `df:6`, then choose Paste and hit ENTER twice (or on an older calculator, enter the command `invT(.99,6)`). Were you surprised, though, that the critical value of *t* is so much larger?

So think once more. How would the critical value of *t* differ if there were 60 degrees of freedom instead of only 6? When you think you know, check it out on your TI.

**UNDERSTANDING *t*** Use your calculator to play around with `tcdf` and `invT` a bit. Try to develop a clear understanding of how *t*-models compare to the more familiar Normal model. That will help you as you learn to use *t*-models to make inferences about means.

# Assumptions and Conditions

Sir Ronald Fisher (1890–1962) was one of the founders of modern Statistics.

Gosset initially found the *t*-model by simulation and then made calculations to derive its form. He made some assumptions so his math would work out. Years later, when Sir Ronald A. Fisher showed mathematically that Gosset was right, he confirmed that Gosset's assumptions were necessary for the math to work. These are the assumptions we need to use the Student's *t*-models.

## Independence Assumption

The data values should be mutually independent. There's really no way to check independence of the data by looking at the sample, but you should think about whether the assumption is reasonable.

**Randomization Condition:** This condition is satisfied if the data arise from a random sample or suitably randomized experiment. Randomly sampled data—and especially data from a Simple Random Sample—are almost surely independent. If the data don't satisfy the Randomization Condition then you should think about whether the values are likely to be independent for the variables you are concerned with and whether the sample you have is likely to be representative of the population you wish to learn about.

When we sample without replacement (essentially always), technically the selections are not independent. We should confirm that we have not sampled so much of the population that it matters. We check the ...

**10% Condition:** The sample is less than 10% of the population.

In practice, the 10% Condition is almost always satisfied when we're looking at means. For the rare situation when a sample comprises more than 10% of the population, there are modifications to our formulas that correct for that.[6]

## Normal Population Assumption

Student's *t*-models won't work for small samples that are badly skewed. How skewed is too skewed? Well, formally, we assume that the data are from a population that follows a Normal model. Practically speaking, though, we can't be sure this is true.

And it's almost certainly *not* true. Models are idealized; real data are, well, real—*never* exactly Normal. The good news, however, is that even for small samples, it's sufficient to check the ...

**Nearly Normal Condition:** The data come from a distribution that appears to be unimodal and symmetric, without strong skewness or outliers.

Check this condition by making a histogram or Normal probability plot. Normality is less important for larger sample sizes. Just our luck: It matters most when it's hardest to check.[7]

For small samples ($n < 30$ is a good rule of thumb), the data should be unimodal and reasonably symmetric. The larger the sample size is, the looser we may be in judging "reasonably symmetric." Make a histogram.

When the sample size is larger than 30, the *t* methods are safe to use unless the data are extremely skewed. Be sure to make a histogram. If you find outliers in the data, it's always a good idea to perform the analysis twice—once with and once without the outliers, even for large samples. Outliers may well hold additional information about the data, but you may decide to give them individual attention and then summarize the rest of the data. If you find multiple modes, you may well have different groups that should be analyzed and understood separately.

---

> ### We Don't *Want* to Stop
>
> We check conditions hoping that we can make a meaningful analysis of our data. The conditions serve as *disqualifiers*—we keep going unless there's a serious problem. If we find minor issues, we note them and express caution about our results.
>
> - If the sample is not an SRS but we believe it's representative of some populations, we limit our conclusions accordingly.
> - If there are outliers, rather than stop, we perform the analysis both with and without them.
> - If the sample looks bimodal, we try to analyze subgroups separately.
>
> Only when there's major trouble—like a strongly skewed small sample or an obviously nonrepresentative sample—are we unable to proceed at all.

---

[6]Let's not get into that here. If you're just dying to know, take a second Statistics course.

[7]There are formal tests of Normality, but they don't really help. When we have a small sample—just when we really care about checking Normality—these tests are not very effective. So it doesn't make much sense to use them in deciding whether to perform a *t*-test. We don't recommend that you use them.

## FOR EXAMPLE
### Checking Assumptions and Conditions for Student's $t$

**RECAP:** Researchers purchased whole farmed salmon from 51 farms in eight regions in six countries. The histogram shows the concentrations of the insecticide mirex in 150 farmed salmon.

**QUESTION:** Are the assumptions and conditions for inference satisfied?

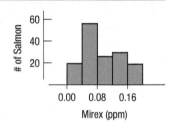

**ANSWER:**

✓ **Independence Assumption:** The fish were not truly a random sample because no simple population existed to sample from. But they were raised in many different places, and samples were purchased independently from several sources, so they were likely to be independent and to represent the population of farmed salmon worldwide.

✓ **Nearly Normal Condition:** The histogram of the data is unimodal. Although it may be somewhat skewed to the right, this is not a concern with a sample size of 150, because that's much greater than 30.

It's okay to use these data for inference about farm-raised salmon.

## JUST CHECKING

Every 10 years, the United States takes a census. The census tries to count every resident. There have been two forms, known as the "short form," answered by most people, and the "long form," slogged through by about one in six or seven households chosen at random. (Since the 2010 Census, the long form has been replaced by the American Community Survey.) According to the Census Bureau (www.census.gov), "[E]ach estimate based on the long form responses has an associated confidence interval."

1. Why does the Census Bureau need a confidence interval for long-form information but not for the questions that appear on both the long and short forms?
2. Why must the Census Bureau base these confidence intervals on $t$-models?

The Census Bureau goes on to say, "These confidence intervals are wider . . . for geographic areas with smaller populations and for characteristics that occur less frequently in the area being examined (such as the proportion of people in poverty in a middle-income neighborhood)."

3. Why is this so? For example, why should a confidence interval for the mean amount families spend monthly on housing be wider for a sparsely populated area of farms in the Midwest than for a densely populated area of an urban center? How does the formula show this will happen?

To deal with this problem, the Census Bureau reports long-form data only for "geographic areas from which about two hundred or more long forms were completed—which are large enough to produce good quality estimates. If smaller weighting areas had been used, the confidence intervals around the estimates would have been significantly wider, rendering many estimates less useful."

4. Suppose the Census Bureau decided to report on areas from which only 50 long forms were completed. What effect would that have on a 95% confidence interval for, say, the mean cost of housing? Specifically, which values used in the formula for the margin of error would change? Which would change a lot and which would change only slightly?
5. Approximately how much wider would that confidence interval based on 50 forms be than the one based on 200 forms?

# STEP-BY-STEP EXAMPLE

## A One-Sample *t*-Interval for the Mean

Let's build a 90% confidence interval for the mean amount of sleep that students get each night at the college in the northeastern United States that surveyed its students.

QUESTION: What can we say about the mean amount of sleep that college students get?

---

**THINK**  **PLAN** State what we want to know. Identify the parameter of interest.

Identify the variables and review the W's.

Make a picture. Check the distribution shape and look for skewness, multiple modes, and outliers.

**REALITY CHECK** The histogram centers around 7 hours, and the data lie between 4 and 9 hours. We'd expect a confidence interval to place the population mean within an hour or so of 7.

**MODEL** Think about the assumptions and check the conditions.

Because this was a randomized survey, we check the Randomization Condition.

State the sampling distribution model for the statistic.

Choose your method.

I want to find a 90% confidence interval for the mean, $\mu$, of hours slept by the college's students. I have data on the number of hours that 25 students slept.

Here's a histogram of the 25 observed amounts that students slept.

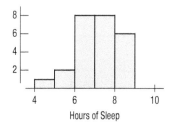

✓ **Randomization Condition**: These are data from a randomized survey, so respondents are likely to be independent.

✓ **10% Condition**: Unless the college has fewer than 250 students (which would be *very* small) then our sample represent fewer than 10% of the student body there.

✓ **Nearly Normal Condition**: Because $n = 25 < 30$ we need the data to be unimodal and reasonably symmetric. The histogram of the *Hours of Sleep* is unimodal and slightly skewed, but not enough to be a concern.

The conditions are satisfied, so I will use a Student's t-model with

$$n - 1 = 24 \text{ degrees of freedom}$$

and find a **one-sample t-interval for the mean**.

---

**SHOW**  **MECHANICS** Construct the confidence interval.

Be sure to include the units along with the statistics.

Calculating from the data (see page 593):

$$n = 25 \text{ students}$$
$$\bar{y} = 6.64 \text{ hours}$$
$$s = 1.075 \text{ hours}$$

| | |
|---|---|
| Find the standard error of the sample mean. | The standard error of $\bar{y}$ is $$SE(\bar{y}) = \frac{s}{\sqrt{n}} = \frac{1.075}{\sqrt{25}} = 0.215 \text{ hours.}$$ |
| The critical value we need to make a 90% interval comes from a Student's $t$ table such as Table T at the back of the book, a computer program, or a calculator. We have $25 - 1 = 24$ degrees of freedom. The selected confidence level says that we want 90% of the probability to be caught in the middle, so we exclude 5% in *each* tail, for a total of 10%. The degrees of freedom and 5% tail probability are all we need to know to find the critical value.<br><br>**REALITY CHECK** The result looks plausible and in line with what we thought. | The 90% critical value is $t^*_{24} = 1.711$, so the margin of error is $$ME = t^*_{24} \times SE(\bar{y})$$ $$= 1.711(0.215)$$ $$= 0.368 \text{ hours.}$$ The 90% confidence interval for the mean number of sleep hours is $6.64 \pm 0.368$ hours $= (6.272, 7.008)$ hours. |
| **TELL**    **CONCLUSION** Interpret the confidence interval in the proper context.<br><br>When we construct confidence intervals in this way, we expect 90% of them to cover the true mean and 10% to miss the true value. That's what "90% confident" means. | I am 90% confident that the interval from 6.272 to 7.008 hours contains the mean number of hours nightly that this college's students sleep. |

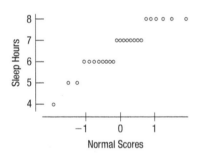

Figure 22.2
A Normal probability plot of sleep hours is reasonably straight. The data are integers. That is why there are flat regions of the plot. Nevertheless, the plot looks reasonably straight.

## Make a Picture, Make a Picture, Make a Picture

The only reasonable way to check the Nearly Normal Condition is with graphs of the data. In the Step-by-Step Example we made a histogram of the data to verify that its distribution is unimodal and reasonably symmetric and that it has no outliers. You could also make a Normal probability plot to see that it's reasonably straight. You'll be able to spot deviations from the Normal model more easily with a Normal probability plot, but it's easier to understand the particular nature of the deviations from a histogram.

# Using Table T to Find *t*-Values

Usually we find critical values and margins of error for Student's $t$-based intervals and tests with technology. But you can also use tables such as Table T at the back of this book. The tables run down the page for as many degrees of freedom as can fit. As the degrees of freedom increase, the $t$-model gets closer and closer to the Normal, so the tables give a final row with the critical values from the Normal model and label it "$\infty$ df."

### Higher Degrees of Freedom and the *t*-Table

As degrees of freedom increase, the shape of Student's $t$-models changes more gradually. Table T at the back of the book includes degrees of freedom between 100 and 1000 selected so that you can pin down the P-value for just about any df. If your df's aren't listed, take the cautious approach by using the next lower value, or better still, use technology.

On the next page is the part of the Student's $t$ table that gives the critical value we needed for the confidence interval in the Step-by-Step Example. To find a critical value, locate the row of the table corresponding to the degrees of freedom and the column corresponding to the probability you want. Our 90% confidence interval leaves 5% of the values on either side, so look for 0.05 at the top of the column or 90% at the bottom. The value in the table at that intersection is the critical value we need: 1.711.

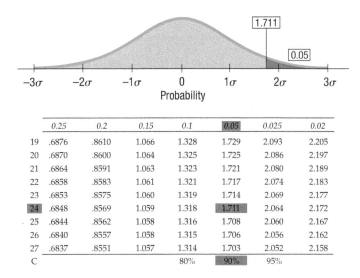

| | 0.25 | 0.2 | 0.15 | 0.1 | 0.05 | 0.025 | 0.02 |
|---|---|---|---|---|---|---|---|
| 19 | .6876 | .8610 | 1.066 | 1.328 | 1.729 | 2.093 | 2.205 |
| 20 | .6870 | .8600 | 1.064 | 1.325 | 1.725 | 2.086 | 2.197 |
| 21 | .6864 | .8591 | 1.063 | 1.323 | 1.721 | 2.080 | 2.189 |
| 22 | .6858 | .8583 | 1.061 | 1.321 | 1.717 | 2.074 | 2.183 |
| 23 | .6853 | .8575 | 1.060 | 1.319 | 1.714 | 2.069 | 2.177 |
| 24 | .6848 | .8569 | 1.059 | 1.318 | 1.711 | 2.064 | 2.172 |
| 25 | .6844 | .8562 | 1.058 | 1.316 | 1.708 | 2.060 | 2.167 |
| 26 | .6840 | .8557 | 1.058 | 1.315 | 1.706 | 2.056 | 2.162 |
| 27 | .6837 | .8551 | 1.057 | 1.314 | 1.703 | 2.052 | 2.158 |
| C | | | | 80% | 90% | 95% | |

For confidence intervals, the values in the table are usually enough to cover most cases of interest. If you can't find a row for the df you need, just use the next smaller df in the table.[8] Of course, you can also create the confidence interval with computer software or a calculator. If you do, you may not know the actual $t^*$. That's OK.

## TI TIPS

### Finding a Confidence Interval for a Mean

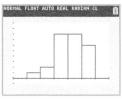

Yes, your calculator can create a confidence interval for a mean. And it's so easy we'll do two!

**FIND A CONFIDENCE INTERVAL GIVEN A SET OF DATA**

◆ Enter the data for the number of hours the 25 students slept in `L1`. Go ahead; we'll wait.

        67676  77786  66888  54678  58767

◆ Set up a `STATPLOT` to create a histogram of the data using `Xscl = 1` so you can check the Nearly Normal Condition. Looks okay—unimodal and not too skewed.

◆ Under `STAT TESTS` choose `TInterval`.

◆ Choose `Inpt:Data`, then specify that your data is `List:L1`.
◆ For these data the frequency is `1`. (If your data have a frequency distribution stored in another list, you would specify that.)
◆ Choose the confidence level you want.
◆ `Calculate` the interval.

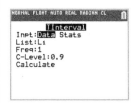

There's the 90% confidence interval. That was easy—but remember, the calculator only does the *Show*. Now you have to *Tell* what it means.

**NO DATA? FIND A CONFIDENCE INTERVAL GIVEN THE SAMPLE'S MEAN AND STANDARD DEVIATION** Sometimes instead of the original data you just have the summary statistics. For instance, suppose a random sample of 53 lengths of fishing line had a mean strength of 83 pounds and standard deviation of 4 pounds. Let's make a 95% confidence interval for the mean strength of this kind of fishing line.

---
[8] You can also find tables on the Internet. Search for terms like "statistical tables *z t*."

Without the data you can't check the Nearly Normal Condition. But 53 is a moderately large sample, so assuming there were no outliers, it's okay to proceed. You need to say that.

- Go back to STAT TESTS and choose TInterval again. This time indicate that you wish to enter the summary statistics. To do that, select Stats, then hit ENTER.
- Specify the sample mean, standard deviation, and sample size.
- Choose a confidence level and Calculate the interval.

We can be 95% confident that this kind of line has a mean strength between 81.9 and 84.1 pounds.

# Be Careful when Interpreting Confidence Intervals

### So What *Should* We Say?

Because 90% of random samples yield an interval that captures the true mean, we *should* say, "I am 90% confident that the interval from 6.272 and 7.008 hours per night contains the mean amount that students sleep." It's also okay to say something less formal: "I am 90% confident that the average amount that students sleep is between 6.272 and 7.008 hours per night." Remember: *Our uncertainty is about the interval, not the true mean.* The interval varies randomly. The mean sleep time is neither variable nor random—just unknown.

Confidence intervals for means offer new, tempting, wrong interpretations. Here are some things you *shouldn't* say:

- ***Don't say***, "90% of all the college's students sleep between 6.272 and 7.008 hours per night." The confidence interval is about the *mean* amount of sleep, not about the sleep of *individual* students.
- ***Don't say***, "We are 90% confident that *a randomly selected student* will sleep between 6.272 and 7.008 hours per night." This false interpretation is also about individual students rather than about the *mean*. We are 90% confident that the *mean* amount of sleep is between 6.272 and 7.008 hours per night.
- ***Don't say***, "The mean amount students sleep is 6.64 hours 90% of the time." That's about means, but still wrong. It implies that the true mean varies, when in fact it is the confidence interval that would have been different had we gotten a different sample.
- Finally, ***don't say***, "90% *of all samples* will have mean sleep between 6.272 and 7.008 hours per night." There's no reason *this* interval's boundaries should become a special standard for other random samples. It's the population parameters that would dictate any such boundaries, but we can't know the values of those parameters without a census.
- ***Do say***, "90% of intervals that could be found in this way would cover the true value." Or make it more personal and contextual and say, "I am 90% confident that the true mean amount that students sleep is between 6.272 and 7.008 hours per night."

# A Hypothesis Test for the Mean

We are told that adults need between 7 and 9 hours of sleep. Do students at this college get what they need? Can we say that the mean amount its students sleep is at least 7 hours? A question like this calls for a hypothesis test called the **one-sample *t*-test for the mean**.

You already know enough to construct this test. The test statistic looks just like the others we've seen. It compares the difference between the observed statistic and a hypothesized value to the standard error of the observed statistic. We've seen that, for means, the appropriate probability model to use for P-values is Student's $t$ with $n - 1$ degrees of freedom.

## One-Sample *t*-Test for the Mean

The assumptions and conditions for the one-sample *t*-test for the mean are the same as for the one-sample *t*-interval. We test the hypothesis $H_0: \mu = \mu_0$ using the statistic

$$t_{n-1} = \frac{\bar{y} - \mu_0}{SE(\bar{y})}.$$

The standard error of $\bar{y}$ is $SE(\bar{y}) = \dfrac{s}{\sqrt{n}}$.

When the conditions are met and the null hypothesis is true, this statistic follows a Student's *t*-model with $n - 1$ degrees of freedom. We use that model to obtain a P-value.

### FOR EXAMPLE
#### A One-Sample *t*-Test for the Mean

**RECAP:** Researchers tested 150 farm-raised salmon for organic contaminants. They found the mean concentration of the carcinogenic insecticide mirex to be 0.0913 parts per million, with standard deviation 0.0495 ppm. As a safety recommendation to recreational fishers, the Environmental Protection Agency's (EPA) recommended "screening value" for mirex is 0.08 ppm.

**QUESTION:** Are farmed salmon contaminated beyond the level permitted by the EPA?

**ANSWER:** (We've already checked the conditions; see page 599.)

$$H_0: \mu = 0.08$$
$$H_A: \mu > 0.08$$

These data satisfy the conditions for inference; I'll do a one-sample *t*-test for the mean:

$$n = 150,\ df = 149$$
$$\bar{y} = 0.0913,\ s = 0.0495$$
$$SE(\bar{y}) = \frac{0.0495}{\sqrt{150}} = 0.0040$$
$$t_{149} = \frac{0.0913 - 0.08}{0.0040} = 2.825$$
$$P(t_{149} > 2.825)$$
$$= 0.0027 \text{ (from technology)}.$$

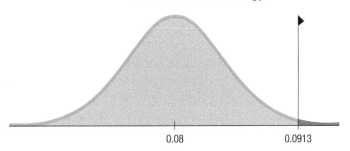

With a P-value that low, I reject the null hypothesis and conclude that, in farm-raised salmon, the mean mirex contamination level does exceed the EPA screening value.

# STEP-BY-STEP EXAMPLE

## A One-Sample *t*-Test for the Mean

Let's apply the one-sample *t*-test to the student sleep survey. It is clear at a glance that students at this college don't get as much sleep as Maas and Robbins recommend. Do they even make it to the minimum for adults? The Sleep Foundation (www.sleepfoundation.org) says that adults should get at least 7 hours of sleep each night.

QUESTION: Is the mean amount that this college's students sleep at least as much as the 7-hour minimum recommended for adults?

---

**THINK** **PLAN** State what we want to test. Make clear what the population and parameter are.

Identify the variables and review the W's.

**HYPOTHESES** The null hypothesis is that the mean sleep time is equal to the minimum recommended. Because we're interested in whether students get enough sleep, the alternative is one-sided.

Make a picture. Check the distribution for skewness, multiple modes, and outliers.

**REALITY CHECK** The histogram of the observed hours of sleep is clustered around a value less than 7. But is this enough evidence to suggest that the mean for all college students is less than 7?

**MODEL** Think about the assumptions and check the conditions.

I want to know whether the mean amount that students at this college sleep meets or exceeds the recommended minimum of 7 hours per night. I have a sample of 25 student reports of their sleep amounts.

$H_0$: Mean sleep, $\mu = 7$ hours
$H_A$: Mean sleep, $\mu < 7$ hours.

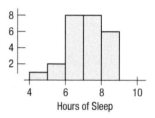

✓ **Randomization Condition**: The students were sampled in a randomized survey, so the amounts they sleep are likely to be mutually independent.

✓ **10% Condition**: Unless the college has as few as 250 students (which would be very small) then the sample represents less than 10% of the student body.

✓ **Nearly Normal Condition**: The histogram of the speeds is unimodal and reasonably symmetric.

State the sampling distribution model. (Be sure to include the degrees of freedom.)

Choose your method.

The conditions are satisfied, so I'll use a Student's *t*-model with $(n - 1) = 24$ degrees of freedom to do a **one-sample *t*-test for the mean**.

---

**SHOW** **MECHANICS** Be sure to include the units when you write down what you know from the data.

We use the null model to find the P-value. Make a picture of the *t*-model centered at $\mu = 7$. Because this is a lower-tail test, shade the region to the left of the observed mean speed.

The *t*-statistic calculation is just a standardized value, like *z*. We subtract the hypothesized mean and divide by the standard error.

From the data,

$$n = 25 \text{ students}$$
$$\bar{y} = 6.64 \text{ hours}$$
$$s = 1.075 \text{ hours}$$
$$SE(\bar{y}) = \frac{s}{\sqrt{n}} = \frac{1.075}{\sqrt{25}} = 0.215 \text{ hours.}$$
$$t = \frac{\bar{y} - \mu_0}{SE(\bar{y})} = \frac{6.64 - 7.0}{0.215} = -1.67$$

**REALITY CHECK** The *t*-statistic is negative because the observed mean is below the hypothesized value. That makes sense because we suspect students get too little sleep, not too much.

The observed mean is 1.67 standard errors below the hypothesized value.

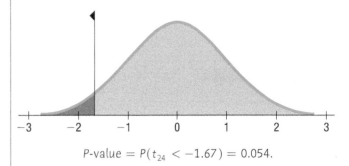

P-value = $P(t_{24} < -1.67) = 0.054$.

The P-value is the probability of observing a sample mean as small as 6.64 (or smaller) *if* the true mean were 7, as the null hypothesis states. We can find this P-value from a table, calculator, or computer program.

---

**TELL**    **CONCLUSION** Link the P-value to your decision about $H_0$, and state your conclusion in context.

The confidence interval will often offer additional insights.

The P-value of 0.054 says that if the mean student nightly sleep amount were 7 hours, samples of 25 students can be expected to have an observed mean of 6.64 hours or less by random chance about 54 times in 1000. If we use 0.05 as the cutoff value, then this P-value is not quite small enough to provide reasonable evidence to reject the hypothesis that the true mean is at least 7 hours. The 90% confidence interval (6.272, 7.008) shows the plausible values for the mean number of hours that college students get, and it does include 7. Although we are unable to reject the hypothesis that the mean is 7 hours, the P-value is very close to 0.05 and the confidence interval shows that there are plausible values well below 7. If those values would result in negative health impacts to students, it might be worth collecting more data to reduce the margin of error.

---

**APPLET**
Investigate the roles of sample size, effect size, errors, and power in *t*-tests.

### Significant and/or Important?

Remember that *statistically significant* does not necessarily mean *important*. A small pilot test for a medical treatment could suggest that the treatment is very promising, with a potentially large effect size—yet still not provide statistically significant evidence of its efficacy because of the small sample size and/or the amount of unexplained variability (statistical noise) in the data. On the other hand, a very large study of a medical treatment involving thousands of subjects may be able to demonstrate the treatment's efficacy with a high degree of statistical significance even if the treatment's actual superiority to existing treatments is very small.

---

## TI TIPS

### Testing a Hypothesis About a Mean

**TESTING A HYPOTHESIS GIVEN A SET OF DATA** Still have the student sleep data in L1? Good. Let's use the TI to see if the mean is significantly lower than 7 hours (you've already checked the histogram to verify the Nearly Normal Condition, of course).

◆ Go to the STAT TESTS menu, and choose T-Test.

- Tell it you want to use the stored `Data`.
- Enter the mean of the null model, and indicate where the data are.
- Because this is a lower tail test, choose the $< \mu_0$ option.
- `Calculate`.

There's everything you need to know: the summary statistics, the calculated value of $t$, and the P-value of 0.054. ($t$ and P differ slightly from the values in our worked example because when we did it by hand we rounded off the mean and standard deviation. No harm done.)

As always, the *Tell* is up to you.

**TESTING A HYPOTHESIS GIVEN THE SAMPLE'S MEAN AND STANDARD DEVIATION** Don't have the actual data? Just summary statistics? No problem, assuming you can verify the necessary conditions. In the last TI Tips we created a confidence interval for the strength of fishing line. We had test results for a random sample of 53 lengths of line showing a mean strength of 83 pounds and a standard deviation of 4 pounds. Is there evidence that this kind of fishing line exceeds the "80-lb test" as labeled on the package?

We bet you know what to do even without our help. Try it before you read on.

- Go back to `T-Test`.
- You're entering `Stats` this time.
- Specify the hypothesized mean and the sample statistics.
- Choose the alternative being tested (upper tail here).
- `Calculate`.

The results of the calculator's mechanics show a large $t$ and a really small P-value (0.0000007). We have very strong evidence that the mean breaking strength of this kind of fishing line is over the 80 pounds claimed by the manufacturer.

# Intervals and Tests

The 90% confidence interval for the mean number of sleep hours was 6.64 hours $\pm$ 0.368, or (6.272 to 7.008 hours). If someone hypothesized that the mean sleep amount was really 8 hours, how would you feel about it? How about 7 hours?

Because the confidence interval includes 7 hours, it certainly looks like 7 hours might be a plausible value for the true mean sleep time. A hypothesized mean of 7 hours lies *within the confidence interval*. It's only one of the plausible values for the mean.

Confidence intervals and significance tests are built from the same calculations. In fact, they are complementary ways of looking at the same question. Here's the connection: The confidence interval contains all the null hypothesis values we can't reject with these data.

How is the confidence level related to the P-value? To be precise, a level C confidence interval contains *all* of the plausible null hypothesis values that would *not* be rejected if you use a (two-sided) P-value of $(1 - C)$ as the cutoff for deciding to reject $H_0$.

Confidence intervals are naturally two-sided, so they correspond to two-sided P-values. When, as in our example, the hypothesis is one-sided, the interval contains values that would not be rejected using a cutoff P-value of $(1 - C)/2$.

> **FAIL TO REJECT**
> Our 90% confidence interval was 6.272 to 7.008 hours. If any of these values had been the null hypothesis for the mean, then the corresponding hypothesis test using a P-value cutoff of 0.05 (because $\frac{1 - 0.90}{2} = 0.05$) would not have been able to reject the null. So, we would not reject any hypothesized value between 6.272 and 7.008 hours.

Confidence intervals and hypothesis tests look at the same problem from two different perspectives. A hypothesis test starts with a *proposed parameter value* and asks if the *data* are consistent with that value. If the observed statistic is too far from the proposed parameter value, that makes it less plausible that the proposed value is the truth. So we reject the null hypothesis. By contrast, a confidence interval starts with the *data* and finds an interval of plausible values for where the parameter may lie.

### JUST CHECKING

In discussing estimates based on the long-form samples, the Census Bureau notes, "The disadvantage . . . is that . . . estimates of characteristics that are also reported on the short form will not match the [long-form estimates]." The short-form estimates are values from a complete census, so they are the "true" values—something we don't usually have when we do inference.

6. Suppose we use long-form data to make 95% confidence intervals for the mean age of residents for each of 100 of the Census-defined areas. How many of these 100 intervals should we expect will fail to include the true mean age (as determined from the complete short-form Census data)?

7. Based only on the long-form sample, we might test the null hypothesis about the mean household income in a region. Would the power of the test increase or decrease if we used an area with more long forms?

# Choosing the Sample Size

How large a sample do we need? The simple answer is "more." But more data costs money, effort, and time, so how much is enough? Suppose your computer just took half an hour to download a movie you wanted to watch. You're not happy. You hear about a program that claims to download movies in less than 15 minutes. You're interested enough to spend $29.95 for it, but only if it really delivers. So you get the free evaluation copy and test it by downloading that movie 5 different times. Of course, the mean download time is not exactly 15 minutes as claimed. Observations vary. If the margin of error were 8 minutes, you might find it hard to decide whether the software is worth the money. Doubling the sample size could take up to 4 hours of testing and would reduce your margin of error to just less than 6 minutes. You'll need to decide whether that's worth the effort.

Armed with the *ME* that you decide is necessary to draw a conclusion and confidence level, you can find the sample size you need. Approximately.

For a mean, $ME = t^*_{n-1} \times SE(\bar{y})$ and $SE(\bar{y}) = \frac{s}{\sqrt{n}}$, so to find the sample size, solve this equation for $n$:

$$ME = t^*_{n-1} \frac{s}{\sqrt{n}}.$$

But the equation needs $s$ and before we collect the data, we don't know $s$. A guess is often good enough, but if you have no idea what the standard deviation might be, or if the sample size really matters (for example, because each additional individual is very expensive to sample or experiment on), a small *pilot study* can provide some information about the standard deviation.

That's not all. Without knowing $n$, you don't know the degrees of freedom so we can't find the critical value, $t^*_{n-1}$. One common approach is to use the corresponding $z^*$ value from the Normal model. If you've chosen a 95% confidence level, then just use 2, following the 68–95–99.7 Rule. If your estimated sample size is, say, 60 or more, it's probably okay—$z^*$ was a good guess. If it's smaller than that, you may want to add a step, using $z^*$ at first, finding $n$, and then replacing $z^*$ with the corresponding $t^*_{n-1}$ and calculating the sample size once more.

Sample size calculations are *never* exact. But it's always a good idea to know whether your sample size is large enough to give you a good chance of being able to tell you what you want to know before you collect your data.

## FOR EXAMPLE
### Finding Sample Size

A company claims its program will allow your computer to download movies quickly. We'll test the free evaluation copy by downloading a movie several times, hoping to estimate the mean download time with a margin of error of only 8 minutes. We think the standard deviation of download times is about 10 minutes.

**QUESTION:** How many trial downloads must we run if we want 95% confidence in our estimate with a margin of error of only 8 minutes?

**ANSWER:** Using $z^* = 2$ (from the 68–95–99.7 Rule), solve

$$8 = 2\frac{10}{\sqrt{n}}$$
$$\sqrt{n} = \frac{20}{8} = 2.5$$
$$n = 2.5^2 = 6.25$$

That's a small sample size, so I'll use $6 - 1 = 5$ degrees of freedom[9] to substitute an appropriate $t^*$ value. At 95%, $t_5^* = 2.571$. Solving the equation one more time:

$$8 = 2.571\frac{10}{\sqrt{n}}$$
$$\sqrt{n} = \frac{2.571 \times 10}{8} \approx 3.214$$
$$n = (3.214)^2 = 10.33$$

To make sure the *ME* is no larger, we'll round *up*, which gives $n = 11$ runs. So, to get an *ME* of 8 minutes, we'll find the downloading times for 11 movies.

## WHAT CAN GO WRONG?

The most fundamental issue you face is knowing when to use Student's *t* methods.

- **Don't confuse proportions and means.** When you treat your data as categorical, counting successes and summarizing with a sample proportion, make inferences using the Normal model methods you learned about in Chapters 18 and 19. When you treat your data as quantitative, summarizing with a sample mean, make your inferences using Student's *t* methods.

Student's *t* methods work only when the Normality Assumption is true. Naturally, many of the ways things can go wrong turn out to be different ways that the Normality Assumption can fail. It's always a good idea to look for the most common kinds of failure. It turns out that you can even fix some of them.

- **Beware of multimodality.** The Nearly Normal Condition clearly fails if a histogram of the data has two or more modes. When you see this, look for the possibility that your data come from two groups. If so, your best bet is to try to separate the data into different groups. (Use the variables to help distinguish the modes, if possible. For example, if the modes seem to be composed mostly of men in one and women in the other, split the data according to sex.) Then you could analyze each group separately.

---

[9] Ordinarily, we'd round the sample size *up*. But at this stage of the calculation, rounding *down* is the safer choice. Can you see why?

- **Beware of skewed data.** Make a Normal probability plot and a histogram of the data. If the data are very skewed, you might try re-expressing the variable. Re-expressing may yield a distribution that is unimodal and symmetric, more appropriate for Student's $t$ inference methods for means. Re-expression cannot help if the sample distribution is not unimodal. Some people may object to re-expressing the data, but unless your sample is very large, you just can't use the methods of this chapter on skewed data.

- **Set outliers aside.** Student's $t$ methods are built on the mean and standard deviation, so we should beware of outliers when using them. When you make a histogram to check the Nearly Normal Condition, be sure to check for outliers as well. If you find some, consider doing the analysis twice, both with the outliers excluded and with them included in the data, to get a sense of how much they affect the results.

> **Don't Ignore Outliers**
>
> As tempting as it is to get rid of annoying values, you can't just throw away outliers and not discuss them. It isn't appropriate to lop off the highest or lowest values just to improve your results.

The suggestion that you can perform an analysis with outliers removed may be controversial in some disciplines. Setting aside outliers is seen by some as "cheating." But an analysis of data with outliers left in place is *always* wrong. The outliers violate the Nearly Normal Condition and also the implicit assumption of a homogeneous population, so they invalidate inference procedures. An analysis of the nonoutlying points, along with a separate discussion of the outliers, is often much more informative and can reveal important aspects of the data.

How can you tell whether there are outliers in your data? The "outlier nomination rule" of boxplots can offer some guidance, but it's just a rule of thumb and not an absolute definition. The best practical definition is that a value is an outlier if removing it substantially changes your conclusions about the data. You won't want a single value to determine your understanding of the world unless you are very, very sure that it is correct and similar in nature to the other cases in your data. Of course, when the outliers affect your conclusion, this can lead to the uncomfortable state of not really knowing what to conclude. Such situations call for you to use your knowledge of the real world and your understanding of the data you are working with.[10]

Of course, Normality issues aren't the only risks you face when doing inferences about means. Remember to *Think* about the usual suspects.

- **Watch out for bias.** Measurements of all kinds can be biased. If your observations differ from the true mean in a systematic way, your confidence interval may not capture the true mean. And there is no sample size that will save you. A bathroom scale that's 5 pounds off will be 5 pounds off even if you weigh yourself 100 times and take the average. We've seen several sources of bias in surveys, and measurements can be biased, too. Be sure to think about possible sources of bias in your measurements.

- **Make sure cases are independent.** Student's $t$ methods also require the sampled values to be mutually independent. We check for random sampling. You should also think hard about whether there are likely violations of independence in the data collection method. If there are, be very cautious about using these methods.

- **Make sure that data are from an appropriately randomized sample.** Ideally, all data that we analyze are drawn from a simple random sample or generated by a randomized experiment. When they're not, be careful about making inferences from them. You may still compute a confidence interval correctly, or get the mechanics of the P-value right, but this might not save you from making a serious mistake in inference.

- **Interpret your confidence interval correctly.** Many statements that sound tempting are, in fact, misinterpretations of a confidence interval for a mean. You might want to have another look at some of the common mistakes, explained on page 606. Keep in mind that a confidence interval is about the mean of the population, not about the means of samples, individuals in samples, or individuals in the population.

---

[10]An important reason for *you* to know Statistics rather than let someone else analyze your data.

# WHAT HAVE WE LEARNED?

We first learned to create confidence intervals for and test hypotheses about proportions. Now we have extended these ideas to means and seen that the concepts remain the same; only the model and mechanics have changed.

- We've learned to check the usual assumptions and conditions. The Normality Assumption is of special importance when using a t-model. We've learned to check the Nearly Normal Condition to see if it's plausible that our sample could have come from a population that's normally distributed. A histogram of the sample data may appear only roughly unimodal and symmetric, but we should not proceed with inference if the sample is strongly skewed or there are outliers present. This concern is very critical in small samples, but less important with sample sizes of 30 to 40 or more.
- We've learned that to use the Central Limit Theorem for the mean in practical applications we must estimate the standard deviation with a *standard error*:

$$SE(\bar{y}) = \frac{s}{\sqrt{n}}.$$

- We've learned that a Student's t-model with $n - 1$ degrees of freedom accounts for the extra uncertainty that arises from using the SE.
- We've learned to construct confidence intervals for the true mean $\mu$, in the form

$$\bar{y} \pm ME \text{ where } ME = t^*_{df} SE(\bar{y}).$$

- We've learned to find the sample size needed to produce a desired margin of error.
- We've learned to test hypotheses about the mean, using the test statistic $t_{df} = \dfrac{\bar{y} - \mu}{\frac{s}{\sqrt{n}}}$.
- We've learned to write clear conclusions, interpreting a confidence interval or the results of a hypothesis test in context.

Above all, we've learned that the reasoning of inference remains the same regardless of whether we are investigating means or proportions.

## TERMS

**Student's t**  A family of distributions indexed by its degrees of freedom. The t-models are unimodal, symmetric, and bell shaped, but have fatter tails and a narrower center than the Normal model. As the degrees of freedom increase, t-distributions approach the Normal. (p. 597)

**Degrees of freedom for a Standardized Mean**  When a mean is standardized using its estimated standard error (which is almost always the case), the degrees of freedom are equal to $n - 1$, where $n$ is the sample size. (p. 597)

**Nearly Normal Condition**  Using a t-model requires the Normality Assumption. We look at a histogram of the sample data to see if it's plausible that our sample could have come from a population that's normally distributed. If the sample size is small ($n < 30$) then the data should be unimodal and roughly symmetric. For samples larger than that, the t-procedures work well unless there are extreme deviations from normality. (p. 601)

**One-sample t-interval for the mean**  A one-sample t-interval for the population mean is

$$\bar{y} \pm t_{n-1} \times SE(\bar{y}), \text{ where } SE(\bar{y}) = \frac{s}{\sqrt{n}}.$$

The critical value $t^*_{n-1}$ depends on the particular confidence level, C, that you specify and on the number of degrees of freedom, $n - 1$. (p. 603)

**One-sample t-test for the mean**  The one-sample t-test for the mean tests the hypothesis $H_0: \mu = \mu_0$ using the statistic

$$t_{n-1} = \frac{\bar{y} - \mu_0}{SE(\bar{y})}. \text{ (p. 606)}$$

The standard error of $\bar{y}$ is

$$SE(\bar{y}) = \frac{s}{\sqrt{n}}.$$

# ON THE COMPUTER
## Inference for Means

Statistics packages offer convenient ways to make histograms of the data. Even better for assessing near-Normality is a Normal probability plot. When you work on a computer, there is simply no excuse for skipping the step of plotting the data to check that it is nearly Normal.

Any standard statistics package can compute a hypothesis test. Here's what the package output might look like in general (although no package we know gives the results in exactly this form):[11]

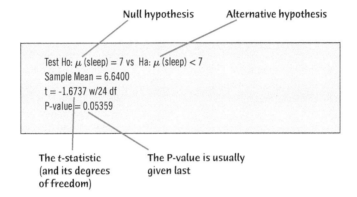

The package computes the sample mean and sample standard deviation of the variable and finds the P-value from the t-distribution based on the appropriate number of degrees of freedom. All modern statistics packages report P-values. The package may also provide additional information such as the sample mean, sample standard deviation, t-statistic value, and degrees of freedom. These are useful for interpreting the resulting P-value and telling the difference between a meaningful result and one that is merely statistically significant. Statistics packages that report the estimated standard deviation of the sampling distribution usually label it "standard error" or "SE."

Inference results are also sometimes reported in a table. You may have to read carefully to find the values you need. Often, test results and the corresponding confidence interval bounds are given together. And often you must read carefully to find the alternative hypotheses. Here's an example of that kind of output:

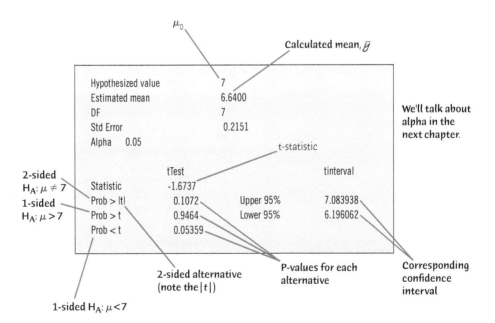

---

[11] Many statistics packages keep as many as 16 digits for all intermediate calculations. If we had kept as many, our results in the Step-by-Step Example would have been closer to these.

# EXERCISES

1. **Salmon** A specialty food company sells whole King Salmon to various customers. The mean weight of these salmon is 35 pounds with a standard deviation of 2 pounds. The company ships them to restaurants in boxes of 4 salmon, to grocery stores in cartons of 16 salmon, and to discount outlet stores in pallets of 100 salmon. To forecast costs, the shipping department needs to estimate the standard deviation of the mean weight of the salmon in each type of shipment

   a) Find the standard deviations of the mean weight of the salmon in each type of shipment.
   b) The distribution of the salmon weights turns out to be skewed to the high end. Would the distribution of shipping weights be better characterized by a Normal model for the boxes or pallets? Explain.

2. **LSAT** The LSAT (a test taken for law school admission) has a mean score of 151 with a standard deviation of 9 and a unimodal, symmetric distribution of scores. A test preparation organization teaches small classes of 9 students at a time. A larger organization teaches classes of 25 students at a time. Both organizations publish the mean scores of all their classes.

   a) What would you expect the distribution of mean class scores to be for each organization?
   b) If either organization has a graduating class with a mean score of 160, they'll take out a full-page ad in the local school paper to advertise. Which organization is more likely to have that success? Explain.
   c) Both organizations advertise that if any class has an average score below 145, they'll pay for everyone to retake the LSAT. Which organization is at greater risk to have to pay?

3. **t-models, part I** Using the $t$ tables, software, or a calculator, estimate

   a) the critical value of $t$ for a 90% confidence interval with df = 17.
   b) the critical value of $t$ for a 98% confidence interval with df = 88.
   c) the P-value for $t \geq 2.09$ with 4 degrees of freedom.
   d) the P-value for $|t| > 1.78$ with 22 degrees of freedom.

4. **t-models, part II** Using the $t$ tables, software, or a calculator, estimate

   a) the critical value of $t$ for a 95% confidence interval with df = 7.
   b) the critical value of $t$ for a 99% confidence interval with df = 102.
   c) the P-value for $t \leq 2.19$ with 41 degrees of freedom.
   d) the P-value for $|t| > 2.33$ with 12 degrees of freedom.

5. **t-models, part III** Describe how the shape, center, and spread of $t$-models change as the number of degrees of freedom increases.

6. **t-models, part IV (last one!)** Describe how the critical value of $t$ for a 95% confidence interval changes as the number of degrees of freedom increases.

7. **Cattle** Livestock are given a special feed supplement to see if it will promote weight gain. Researchers report that the 77 cows studied gained an average of 56 pounds, and that a 95% confidence interval for the mean weight gain this supplement produces has a margin of error of $\pm 11$ pounds. Some students wrote the following conclusions. Did anyone interpret the interval correctly? Explain any misinterpretations.

   a) 95% of the cows studied gained between 45 and 67 pounds.
   b) We're 95% sure that a cow fed this supplement will gain between 45 and 67 pounds.
   c) We're 95% sure that the average weight gain among the cows in this study was between 45 and 67 pounds.
   d) The average weight gain of cows fed this supplement will be between 45 and 67 pounds 95% of the time.
   e) If this supplement is tested on another sample of cows, there is a 95% chance that their average weight gain will be between 45 and 67 pounds.

8. **Teachers** Software analysis of the salaries of a random sample of 288 Nevada teachers produced the confidence interval shown below. Which conclusion is correct? What's wrong with the others?

   $t$-Interval for $\mu$: with 90.00% Confidence, $43454 < \mu(\text{TchPay}) < 45398$

   a) If we took many random samples of 288 Nevada teachers, about 9 out of 10 of them would produce this confidence interval.
   b) If we took many random samples of Nevada teachers, about 9 out of 10 of them would produce a confidence interval that contained the mean salary of all Nevada teachers.
   c) About 9 out of 10 Nevada teachers earn between $43,454 and $45,398.
   d) About 9 out of 10 of the teachers surveyed earn between $43,454 and $45,398.
   e) We are 90% confident that the average teacher salary in the United States is between $43,454 and $45,398.

9. **Meal plan** After surveying students at Dartmouth College, a campus organization calculated that a 95% confidence interval for the mean cost of food for one term (of three in the Dartmouth trimester calendar) is ($1372, $1562). Now the organization is trying to write its report and is considering the following interpretations. Comment on each.

   a) 95% of all students pay between $1372 and $1562 for food.
   b) 95% of the sampled students paid between $1372 and $1562.
   c) We're 95% sure that students in this sample averaged between $1372 and $1562 for food.
   d) 95% of all samples of students will have average food costs between $1372 and $1562.
   e) We're 95% sure that the average amount all students pay is between $1372 and $1562.

**10. Snow** Based on meteorological data for the past century, a local TV weather forecaster estimates that the region's average winter snowfall is 23″, with a margin of error of ±2 inches. Assuming he used a 95% confidence interval, how should viewers interpret this news? Comment on each of these statements:

a) During 95 of the last 100 winters, the region got between 21″ and 25″ of snow.
b) There's a 95% chance the region will get between 21″ and 25″ of snow this winter.
c) There will be between 21″ and 25″ of snow on the ground for 95% of the winter days.
d) Residents can be 95% sure that the area's average snowfall is between 21″ and 25″.
e) Residents can be 95% confident that the average snowfall during the last century was between 21″ and 25″ per winter.

**11. Pulse rates** A medical researcher measured the pulse rates (beats per minute) of a sample of randomly selected adults and found the following Student's *t*-based confidence interval:

With 95.00% Confidence,
$70.887604 < \mu(\text{Pulse}) < 74.497011$

a) Explain carefully what the software output means.
b) What's the margin of error for this interval?
c) If the researcher had calculated a 99% confidence interval, would the margin of error be larger or smaller? Explain.

**12. Crawling** Data collected by child development scientists produced this confidence interval for the average age (in weeks) at which babies begin to crawl:

*t*-Interval for $\mu$
(95.00% Confidence): $29.202 < \mu(\text{age}) < 31.844$

a) Explain carefully what the software output means.
b) What is the margin of error for this interval?
c) If the researcher had calculated a 90% confidence interval, would the margin of error be larger or smaller? Explain.

**13. Home sales** Housing prices in the United States in 2021 had the largest jump in 30 years. Recently, in one large community, realtors randomly sampled 36 bids from potential buyers to estimate the average gain in home value. The sample showed the average gain was $9560 with a standard deviation of $1500.

a) What assumptions and conditions must be checked before finding a confidence interval? How would you check them?
b) Find a 95% confidence interval for the mean gain in value per home.
c) Interpret this interval and explain what 95% confidence means in this context.

**14. Home sales again** In the previous exercise, you found a 95% confidence interval to estimate the average gain in home value.

a) Suppose the standard deviation of the gains had been $3000 instead of $1500. What would the larger standard deviation do to the width of the confidence interval (assuming the same level of confidence)?
b) Your classmate suggests that the margin of error in the interval could be reduced if the confidence level were changed to 90% instead of 95%. Do you agree with this statement? Why or why not?
c) Instead of changing the level of confidence, would it be more statistically appropriate to draw a bigger sample?

**15. CEO compensation** A sample of 20 CEOs from the *Forbes* 500 shows total annual compensations ranging from a minimum of $0.1 to $62.24 million. The average for these 20 CEOs is $7.946 million. Here's a histogram:

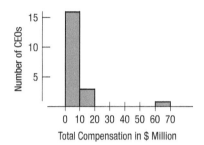

Based on these data, a computer program found that a 95% confidence interval for the mean annual compensation of all *Forbes* 500 CEOs is (1.69, 14.20) $ million. Why should you be hesitant to trust this confidence interval?

**16. Credit card charges** A credit card company takes a random sample of 100 cardholders to see how much they charged on their card last month. Here's a histogram:

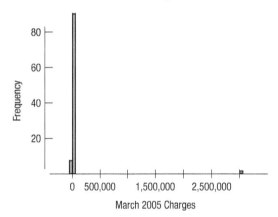

A computer program found that the resulting 95% confidence interval for the mean amount spent in March 2005 is (−$28366.84, $90691.49). Explain why the analysts didn't find the confidence interval useful, and explain what went wrong.

**17. Normal temperature** The researcher described in Exercise 11 also measured the body temperatures of that randomly selected group of adults. Here are summaries of the data he collected. We wish to estimate the average (or "normal") temperature among the adult population.

| Summary | Temperature |
|---|---|
| Count | 52 |
| Mean | 98.285 |
| Median | 98.200 |
| MidRange | 98.600 |
| StdDev | 0.6824 |
| Range | 2.800 |
| IntQRange | 1.050 |

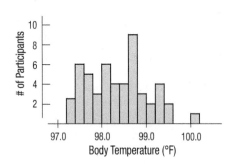

a) Check the conditions for creating a *t*-interval.
b) Find a 98% confidence interval for mean body temperature.
c) Explain the meaning of that interval.
d) Explain what "98% confidence" means in this context.
e) 98.6°F is commonly assumed to be "normal." Do these data suggest otherwise? Explain.

**18. Parking** Hoping to lure more shoppers downtown, a city builds a new public parking garage in the central business district. The city plans to pay for the structure through parking fees. During a two-month period (44 weekdays), daily fees collected averaged $126, with a standard deviation of $15.

a) What assumptions must you make in order to use these statistics for inference?
b) Write a 90% confidence interval for the mean daily income this parking garage will generate.
c) Interpret this confidence interval in context.
d) Explain what "90% confidence" means in this context.
e) The consultant who advised the city on this project predicted that parking revenues would average $130 per day. Based on your confidence interval, do you think the consultant was correct? Why?

**19. Home sales revisited** For the confidence interval you found in Exercise 13, interpret this interval and explain what 95% confidence means in this context.

**20. Police salaries** A survey finds that a 95% confidence interval for the mean salary of a police patrol officer in Fresno, California, in 2016 is $52,516 to $53,509. A student is surprised that so few police officers make more than $53,509. Explain what is wrong with the student's interpretation.

**21. Shoe sizes** A *last* is a form, traditionally made of wood, in the shape of the human foot. Lasts of various sizes are used by shoemakers to make shoes. In the United States, shoe sizes are defined differently for men and women:
U.S. men's shoe size = (last size in inches × 3) − 24.
U.S. women's shoe size = (last size in inches × 3) − 22.5.
But in Europe, they are both: Euro size = last size in cm × 3/2.
A student looked at the European shoe sizes of a sample of men who attended a local university (converted from their reported U.S. shoe sizes). A 95% confidence interval for the mean shoe size shows:

95 percent confidence interval:
44.3071   44.9900

The student knows that European shoes are sized only in whole and half sizes, so is surprised that most men wear size 44.5 shoes. What is wrong with the student's reasoning?

**22. Bird counts** A biology class conducts a bird count every week during the semester. Using the number of species counted each week, a student finds the following confidence interval for the mean number of species counted:

65 percent confidence interval:
16.34   18.69

Knowing that species have to be whole numbers, the student reports that 95% of the bird counts saw 16, 17, or 18 species. Comment on the student's report.

**23. Normal temperatures, part II** Consider again the statistics about human body temperature in Exercise 17.

a) Would a 90% confidence interval be wider or narrower than the 98% confidence interval you calculated before? Explain. (Don't compute the new interval.)
b) What are the advantages and disadvantages of the 98% confidence interval?
c) If we conduct further research, this time using a sample of 500 adults, how would you expect the 98% confidence interval to change? Explain.
d) How large a sample might allow you to estimate the mean body temperature to within 0.1 degrees with 98% confidence?

**24. Parking II** Suppose that, for budget planning purposes, the city in Exercise 18 needs a better estimate of the mean daily income from parking fees.

a) Someone suggests that the city use its data to create a 95% confidence interval instead of the 90% interval first created. How would this interval be better for the city? (You need not actually create the new interval.)
b) How would the 95% interval be worse for the planners?
c) How could they achieve an interval estimate that would better serve their planning needs?
d) How many days' worth of data should they collect to have 95% confidence of estimating the true mean to within $3?

**25. Wisconsin schools** A Wisconsin statistics student, for a class project, decided to find the average reported value of spending per pupil for school districts in Wisconsin. From the Greatschools.org website, they copied and pasted a list of the 456 school districts listed on the website, randomly chose 10 distinct integers between 1 and 456 with a random number generator, and looked up the per pupil spending of those 10 schools. Here is a plot of those 10 values.

a) What concern do you have that the conditions for constructing a confidence interval may not be met?
b) To address their own concerns, the student randomly selected twenty additional districts. Seeing the new plot, shown below, does it fix any concerns you had in part a?

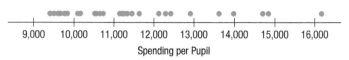

**26. Confindence in schools.** The student in exercise 25 constructed a 95% confidence interval and got ($11151.58, $12793.66). They said, "We are 95% confident that the mean amount spent per pupil in Wisconsin schools is between $11,151.58 and $12,793.66." Explain why this isn't really the parameter they estimated.

**27. Speed of light** In 1882 Michelson measured the speed of light (usually denoted $c$ as in Einstein's famous equation $E = mc^2$). His values are in km/sec and have 299,000 subtracted from them. He reported the results of 23 trials with a mean of 756.22 and a standard deviation of 107.12.

a) Find a 95% confidence interval for the true speed of light from these statistics.
b) State in words what this interval means. Keep in mind that the speed of light is a physical constant that, as far as we know, has a value that is true throughout the universe.
c) What assumptions must you make in order to use your method?

**28. Better light** After his first attempt to determine the speed of light (described in Exercise 27), Michelson conducted an "improved" experiment. In 1897 he reported results of 100 trials with a mean of 852.4 and a standard deviation of 79.0.

a) What is the standard error of the mean for these data?
b) Without computing it, how would you expect a 95% confidence interval for the second experiment to differ from the confidence interval for the first? Note at least three specific reasons why they might differ, and indicate the ways in which these differences would change the interval.
c) According to Stigler (who reports these values), the true speed of light is 299,710.5 km/sec, corresponding to a value of 710.5 for Michelson's 1897 measurements. What does this indicate about Michelson's two experiments? Explain, using your confidence interval.

**29. Flights on time 2019** What are the chances your flight will leave on time? The Bureau of Transportation Statistics of the U.S. Department of Transportation publishes information about airline performance. Here are a histogram and summary statistics for the percentage of flights departing on time each month from January 1994 through September 2019. (www.transtats.bts.gov/HomeDrillChart.asp)

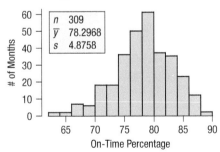

There is no evidence of a trend over time.

a) Check the assumptions and conditions for inference.
b) Find a 90% confidence interval for the percentage of all flights that depart on time.
c) Interpret this interval for a traveler planning to fly.

**30. Flights on time 2019 revisited** Will your flight get you to your destination on time? The Bureau of Transportation Statistics reported the percentage of flights that were delayed each month from 1994 through September of 2019. Have a look at the histogram, along with some summary statistics:

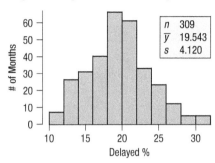

We can consider these data to be a representative sample of all months. There is no evidence of a time trend ($r = -0.07$).

a) Check the assumptions and conditions for inference about the mean.
b) Find a 99% confidence interval for the true percentage of flights that arrive late.
c) Interpret this interval for a traveler planning to fly.

**31. Home prices** In 2021, the average home in the region of the country studied in Exercise 13 gained $9010. Was the community studied in Exercise 13 unusual? Use a $t$-test to decide if the average gain observed in that city was significantly different from the regional average.

**32. Home prices II** Suppose the standard deviation of home price gains had been $3000, as in Exercise 14? What would your conclusion be then?

**33. For Example, 2nd look** This chapter's For Examples looked at mirex contamination in farmed salmon. We first found a 95% confidence interval for the mean concentration to be 0.0834 to 0.0992 parts per million. Later we rejected the null hypothesis that the mean did not exceed the EPA's recommended safe level of 0.08 ppm based on a P-value of 0.0027. Explain how these two results are consistent. Your explanation should discuss the confidence level, the P-value, and the decision.

**34. Hot dogs** A nutrition lab tested 40 hot dogs to see if their mean sodium content was less than the 325 mg upper limit set by regulations for "reduced sodium" franks. The lab failed to reject the hypothesis that the hot dogs did not meet this requirement, with a P-value of 0.142. A 90% confidence interval estimated the mean sodium content for this kind of hot dog at 317.2 to 326.8 mg. Explain how these two results are consistent. Your explanation should discuss the confidence level, the P-value, and the decision.

**35. Pizza** A researcher tests whether the mean cholesterol level among those who eat frozen pizza exceeds the value considered to indicate a health risk. She gets a P-value of 0.07. Explain in this context what the "7%" represents.

**36. Golf balls** The United States Golf Association (USGA) sets performance standards for golf balls. For example, the initial velocity of the ball may not exceed 250 feet per second when measured by an apparatus approved by the USGA. Suppose a manufacturer introduces a new kind of ball and provides a sample for testing. Based on the mean speed in the test, the USGA comes up with a P-value of 0.34. Explain in this context what the "34%" represents.

**37. TV safety** The manufacturer of a metal stand for home TV sets must be sure that its product will not fail under the weight of the TV. Because some larger sets weigh nearly 300 pounds, the company's safety inspectors have set a standard of ensuring that the stands can support an average of over 500 pounds. Their inspectors regularly subject a random sample of the stands to increasing weight until they fail. They test the hypothesis $H_0: \mu = 500$ against $H_A: \mu > 500$, using the level of significance $\alpha = 0.01$. If the sample of stands fails to pass this safety test, the inspectors will not certify the product for sale to the general public.

a) Is this an upper-tail or lower-tail test? In the context of the problem, why do you think this is important?
b) Explain what will happen if the inspectors commit a Type I error.
c) Explain what will happen if the inspectors commit a Type II error.

**38. Catheters** During an angiogram, heart problems can be examined via a small tube (a catheter) threaded into the heart from a vein in the patient's leg. It's important that the company that manufactures the catheter maintain a diameter of 2.00 mm. (The standard deviation is quite small.) Each day, quality control personnel make several measurements to test $H_0: \mu = 2.00$ against $H_A: \mu \neq 2.00$ at a significance level of $\alpha = 0.05$. If they discover a problem, they will stop the manufacturing process until it is corrected.

a) Is this a one-sided or two-sided test? In the context of the problem, why do you think this is important?
b) Explain in this context what happens if the quality control people commit a Type I error.
c) Explain in this context what happens if the quality control people commit a Type II error.

**39. TV safety revisited** The manufacturer of the metal TV stands in Exercise 37 is thinking of revising its safety test.

a) If the company's lawyers are worried about being sued for selling an unsafe product, should they increase or decrease the value of $\alpha$? Explain.
b) In this context, what is meant by the power of the test?
c) If the company wants to increase the power of the test, what options does it have? Explain the advantages and disadvantages of each option.

**40. Catheters again** The catheter company in Exercise 38 is reviewing its testing procedure.

a) Suppose the significance level is changed to $\alpha = 0.01$. Will the probability of a Type II error increase, decrease, or remain the same?
b) What is meant by the power of the test the company conducts?
c) Suppose the manufacturing process is slipping out of proper adjustment. As the actual mean diameter of the catheters produced gets farther and farther above the desired 2.00 mm, will the power of the quality control test increase, decrease, or remain the same?
d) What could they do to improve the power of the test?

**41. Marriage** In 1960, census results indicated that the age at which American men first married had a mean of 23.3 years. It is widely suspected that young people today are waiting longer to get married. We want to find out if the mean age of first marriage has increased during the past 40 years.

a) Write appropriate hypotheses.
b) We plan to test our hypothesis by selecting a random sample of 40 men who married for the first time last year. Do you think the necessary assumptions for inference are satisfied? Explain.
c) Describe the approximate sampling distribution model for the mean age in such samples.
d) The men in our sample married at an average age of 24.2 years, with a standard deviation of 5.3 years. What's the P-value for this result?
e) Explain (in context) what this P-value means.
f) What's your conclusion?

**42. Saving gas** Congress regulates corporate fuel economy and sets an annual gas mileage for cars. A company with a large fleet of cars hoped to meet a 2018 goal of 30.2 mpg or better for their fleet of cars. To see if the goal is being met, they check the gasoline usage for 50 company trips chosen at random, finding a mean of 32.12 mpg and a standard deviation of 4.83 mpg. Is this strong evidence that they have attained their fuel economy goal?

a) Write appropriate hypotheses.
b) Are the necessary assumptions to make inferences satisfied?
c) Describe the sampling distribution model of mean fuel economy for samples like this.
d) Find the P-value.
e) Explain what the P-value means in this context.
f) State an appropriate conclusion.

**T 43. Ruffles** Students investigating the packaging of potato chips purchased 6 randomly selected bags of Lay's Ruffles™ marked with a net weight of 28.3 grams. They carefully weighed the contents of each bag, recording the following weights (in grams): 29.3, 28.2, 29.1, 28.7, 28.9, 28.5.

a) Do these data satisfy the assumptions for inference? Explain.
b) Find the mean and standard deviation of the weights.
c) Create a 95% confidence interval for the mean weight of such bags of chips.
d) Explain in context what your interval means.

**T 44. Doritos** Some students checked 6 bags of Doritos™ (randomly chosen) marked with a net weight of 28.3 grams. They carefully weighed the contents of each bag, recording the following weights (in grams): 29.2, 28.5, 28.7, 28.9, 29.1, 29.5.

a) Do these data satisfy the assumptions for inference? Explain.
b) Find the mean and standard deviation of the weights.
c) Create a 95% confidence interval for the mean weight of such bags of chips.
d) Explain in context what your interval means.

**T 45. Popcorn** Yvon Hopps ran an experiment to test optimum power and time settings for microwave popcorn. His goal was to find a combination of power and time that would deliver high-quality popcorn with less than 10% of the kernels left unpopped, on average. After experimenting with several bags, he determined that power 9 at 4 minutes was the best combination.

a) He concluded that this popping method achieved the 10% goal. If it really does not work that well, what kind of error did Hopps make?

b) To be sure that the method was successful, he popped 8 more bags of popcorn (selected at random) at this setting. All were of high quality, with the following percentages of uncooked popcorn: 7, 13.2, 10, 6, 7.8, 2.8, 2.2, 5.2. Does this provide evidence that he met his goal of an average of no more than 10% uncooked kernels? Explain.

46. **Ski wax** Bjork Larsen was trying to decide whether to use a new racing wax for cross-country skis. He decided that the wax would be worth the price if he could average less than 55 seconds on a course he knew well, so he planned to test the wax by racing on the course 8 times.
    a) Suppose that he eventually decides not to buy the wax, but it really would lower his average time to below 55 seconds. What kind of error would he have made?
    b) His 8 race times were 56.3, 65.9, 50.5, 52.4, 46.5, 57.8, 52.2, and 43.2 seconds. Should he buy the wax? Explain.

47. **Chips Ahoy** In 1998, as an advertising campaign, the Nabisco Company announced a "1000 Chips Challenge," claiming that every 18-ounce bag of their Chips Ahoy™ cookies contained at least 1000 chocolate chips. Dedicated Statistics students at the Air Force Academy (no kidding) purchased some randomly selected bags of cookies, and counted the chocolate chips. Some of their data are given below. (*Chance*, 12, no. 1[1999])

    | 1219 | 1214 | 1087 | 1200 | 1419 | 1121 | 1325 | 1345 |
    |------|------|------|------|------|------|------|------|
    | 1244 | 1258 | 1356 | 1132 | 1191 | 1270 | 1295 | 1135 |

    a) Check the assumptions and conditions for inference. Comment on any concerns you have.
    b) Create a 95% confidence interval for the average number of chips in bags of Chips Ahoy cookies.
    c) What does this evidence say about Nabisco's claim? Use your confidence interval to test an appropriate hypothesis and state your conclusion.

48. **Yogurt** *Consumer Reports* tested 11 brands of vanilla yogurt and found these numbers of calories per serving:

    130  160  150  120  120  110  170  160  110  130  90

    a) Check the assumptions and conditions for inference.
    b) Create a 95% confidence interval for the average calorie content of vanilla yogurt.
    c) A diet guide claims that you will get an average of 120 calories from a serving of vanilla yogurt. What does this evidence indicate? Use your confidence interval to test an appropriate hypothesis and state your conclusion.

49. **Jelly** A consumer advocate wants to collect a sample of jelly jars and measure the actual weight of the product in the container. Enough data must be collected to construct a confidence interval with a margin of error of no more than 2 grams with 99% confidence. The standard deviation of these jars is usually 4 grams. What do you recommend for the sample size?

50. **A good book** An English professor is attempting to estimate the mean number of novels that the student body reads during their time in college. To gather data about this, an exit survey will be conducted with seniors. The estimate is to have a margin of error of 3 books with 95% confidence. From previous studies, a lot of variability is expected, so the professor will assume a standard deviation of 10 books. How many students should take the survey?

51. **Maze** Psychology experiments sometimes involve testing the ability of rats to navigate mazes. The mazes are classified according to difficulty, as measured by the mean length of time it takes rats to find the food at the end. One researcher needs a maze that will take rats an average of about one minute to solve. One maze was tested on several rats, resulting in the data shown.

    | Time (sec) | |
    |------|------|
    | 38.4 | 57.6 |
    | 46.2 | 55.5 |
    | 62.5 | 49.5 |
    | 38.0 | 40.9 |
    | 62.8 | 44.3 |
    | 33.9 | 93.8 |
    | 50.4 | 47.9 |
    | 35.0 | 69.2 |
    | 52.8 | 46.2 |
    | 60.1 | 56.3 |
    | 55.1 | |

    a) Plot the data. Do you think the conditions for inference are satisfied? Explain.
    b) Test the hypothesis that the mean completion time for this maze is 60 seconds. What is your conclusion?
    c) Eliminate the outlier, and test the hypothesis again. What is your conclusion?
    d) Do you think this maze meets the "one-minute average" requirement? Explain.

52. **Braking** A tire manufacturer is considering a newly designed tread pattern for its all-weather tires. Tests have indicated that these tires will provide better gas mileage and longer tread life. The last remaining test is for braking effectiveness. The company hopes the tire will allow a car traveling at 60 mph to come to a complete stop within an average of 125 feet after the brakes are applied. They will adopt the new tread pattern unless there is strong evidence that the tires do not meet this objective. The distances (in feet) for 10 stops on a test track were 129, 128, 130, 132, 135, 123, 102, 125, 128, and 130. Should the company adopt the new tread pattern? Test an appropriate hypothesis and state your conclusion. Explain how you dealt with the outlier and why you made the recommendation you did.

53. **Arrows** A team of anthropologists headed by researcher Nicole Waguespack studied the difference between stone-tipped and wooden-tipped arrows. Stone arrow tips are tougher, but also take longer to make. Many cultures used both types of arrow tips, including the Apache in North America and the Tiwi in Australia. The researchers set up a compound bow with 60 lb of force. They shot arrows of both types into a hide-covered ballistics gel. (Nicole Waguespack et al., *Antiquity*, 2009)
    a) Here are the data for seven shots at the target with a wooden tip. They measured the penetration depth in mm. Find and interpret a 95% confidence interval for the penetration depth.

       216  211  192  208  203  210  203

    b) Here are the penetration depths (mm) for seven shots with a stone tip. Find and interpret a 95% confidence interval for the penetration depth.

       240  208  213  225  232  214  240

54. **Accuracy** The researchers in the previous problem also measured the accuracy of the two types of tips. The bow was aimed at a target and the distance was measured from the center.
    a) Here are the data from the six wooden-tipped shots. Find and interpret a 95% confidence interval for the measure of accuracy (measured in cm).

       9.3  16.7  7.1  14  1  1.2

b) Here are the data from the six stone-tipped shots. Find and interpret a 95% interval for the measure of accuracy (measured in cm).

4.9  21.1  7  1.8  5.4  8.6

**55. Sue me!** Business professor Richard Posthuma examined the number of lawsuits filed in all 50 states from 1996 to 2003. He collected data on lawsuits filed in federal court regarding employment issues. Some states had a few hundred lawsuits, while other states had thousands. Here are the summary statistics. (*Business Horizons*, 2012, 55)

a) Find and interpret a 90% confidence interval for the mean number of employment-related lawsuits that states might expect.
b) What are the shortcomings of this interval?

| n | 50 |
|---|---|
| MEAN | 5734.18 |
| SD | 6387.56 |
| MED | 3626 |
| MIN | 178 |
| Q1 | 1303 |
| Q3 | 7897 |
| MAX | 2631 |

**56. Sued again** Dr. Posthuma (see Exercise 55) also tabulated the total amount of the lawsuits, in 1000's of dollars. Here are the statistics.

a) Find and interpret a 90% confidence interval for the expected average cost of lawsuits for states.
b) What are risks associated with using this confidence interval?

| n | 50 |
|---|---|
| MEAN | 67.1674 |
| SD | 110.237 |
| MED | 35.105 |
| MIN | 1.13 |
| Q1 | 16.14 |
| Q3 | 65.35 |
| MAX | 624.86 |

**57. More Ruffles** Recall from Exercise 43 that students investigated the packaging of potato chips. They purchased 6 bags of Lay's Ruffles marked with a net weight of 28.3 grams. They carefully weighed the contents of each bag, recording the following weights (in grams): 29.3, 28.2, 29.1, 28.7, 28.9, 28.5.

a) Do these data satisfy the assumptions for inference? Explain.
b) Find the mean and standard deviation of the weights.
c) Test the hypothesis that the net weight is as claimed.

**58. More Doritos** We saw in Exercise 44 that some students checked 6 bags of Doritos marked with a net weight of 28.3 grams. They carefully weighed the contents of each bag, recording the following weights (in grams): 29.2, 28.5, 28.7, 28.9, 29.1, 29.5.

a) Do these data satisfy the assumptions for inference? Explain.
b) Find the mean and standard deviation of the weights.
c) Test the hypothesis that the net weight is as claimed.

**59. Fuel economy 2019 revisited** In Chapter 6, Exercise 39, we examined the average fuel economy of a random sample of 35 2019 model vehicles. This sample had a mean gas mileage of 24.3 mpg and a standard deviation of 6.53 mpg.

a) Find and interpret a 95% confidence interval for the gas mileage of 2019 vehicles.
b) Do you think that this confidence interval captures the mean gas mileage for all 2019 vehicles?

**60. Computer lab fees** The technology committee has stated that the average time spent by students per lab visit has increased, and the increase supports the need for increased lab fees. To substantiate this claim, the committee randomly samples 12 student lab visits and notes the amount of time spent using the computer. The times in minutes are as follows:

| Time | |
|---|---|
| 52 | 74 |
| 57 | 53 |
| 54 | 136 |
| 76 | 73 |
| 62 | 8 |
| 52 | 62 |

a) Plot the data. Are any of the observations outliers? Explain.
b) The previous mean amount of time spent using the lab computer was 55 minutes. Find a 95% confidence interval for the true mean. What do you conclude about the claim? If there are outliers, find intervals with and without the outliers present.

**61. Driving distance 2021** How far do professional golfers drive a ball? (For non-golfers, the drive is the shot hit from a tee at the start of a hole and is typically the longest shot.) Below is a histogram of the average driving distances of the 217 leading professional golfers by end of August 2021 along with summary statistics (www.pgatour.com).

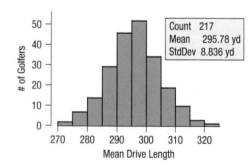

a) Find a 95% confidence interval for the mean drive distance.
b) Interpreting this interval raises some problems. Discuss.
c) The data are the mean driving distance for each golfer. Is that a concern in interpreting the interval? (*Hint:* Review the What Can Go Wrong warnings of Chapter 8. Chapter 8?! Yes, Chapter 8.)

**62. Wind power** Should you generate electricity with your own personal wind turbine? That depends on whether you have enough wind on your site. To produce enough energy, your site should have an annual average wind speed above 8 miles per hour, according to the Wind Energy Association. One candidate site was monitored for a year, with wind speeds recorded every 6 hours. A total of 1114 readings of wind speed averaged 8.019 mph with a standard deviation of 3.813 mph. You've been asked to make a statistical report to help the landowner decide whether to place a wind turbine at this site.

a) Discuss the assumptions and conditions for using Student's $t$ inference methods with these data. Here are some plots that may help you decide whether the methods can be used:

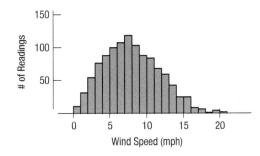

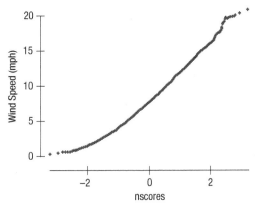

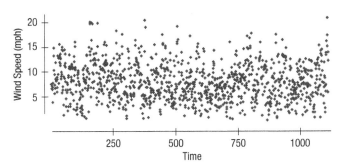

b) What would you tell the landowner about whether this site is suitable for a small wind turbine? Explain.

## JUST CHECKING

### Answers

1. Questions on the short form are answered by everyone in the population. This is a census, so means or proportions *are* the true population values. The long forms are given just to a sample of the population. When we estimate parameters from a sample, we use a confidence interval to take sample-to-sample variability into account.

2. They don't know the population standard deviation, so they must use the sample SD as an estimate. The additional uncertainty is taken into account by $t$-models.

3. The margin of error for a confidence interval for a mean depends, in part, on the standard error,

$$SE(\bar{y}) = \frac{s}{\sqrt{n}}.$$

Because $n$ is in the denominator, smaller sample sizes lead to larger SEs and correspondingly wider intervals. Long forms returned by one in every six or seven households in a less populous area will be a smaller sample.

4. The critical values for $t$ with fewer degrees of freedom would be slightly larger. The $\sqrt{n}$ part of the standard error changes a lot, making the SE much larger. Both would increase the margin of error.

5. The smaller sample is one fourth as large, so the confidence interval would be roughly twice as wide.

6. We expect 95% of such intervals to cover the true value, so 5 of the 100 intervals might be expected to miss.

7. The power would increase if we have a larger sample size.

# 23

# Comparing Means

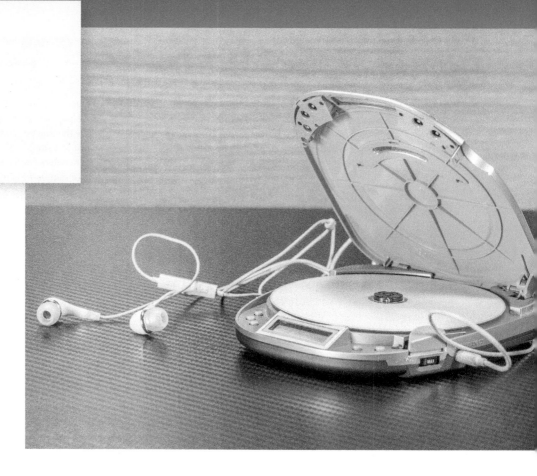

| WHO | AA alkaline batteries |
| --- | --- |
| WHAT | Length of battery life while playing a CD continuously |
| UNITS | Minutes |
| WHY | Class project |
| WHEN | 1998 |

Should you buy generic rather than brand-name batteries? A Statistics student designed a study to test battery life. He wanted to know whether there was any real difference between brand-name batteries and a generic brand. To estimate the difference in mean lifetimes, he kept a battery-powered CD player continuously playing the same CD,[1] with the volume control fixed at 5, and measured the time until no more music was heard through the headphones. (He ran an initial trial to find out approximately how long that would take so that he didn't have to spend the first 3 hours of each run listening to the same CD.) For his trials he used six sets of AA alkaline batteries from two major battery manufacturers: a well-known brand name and a generic brand. He measured the time in minutes until the sound stopped. To account for changes in the CD player's performance over time, he randomized the run order by choosing sets of batteries at random. The table shows his data (times in minutes).

| Brand Name | Generic |
| --- | --- |
| 194.0 | 190.7 |
| 205.5 | 203.5 |
| 199.2 | 203.5 |
| 172.4 | 206.5 |
| 184.0 | 222.5 |
| 169.5 | 209.4 |

Studies that compare two groups are common throughout both science and industry. We might want to compare the effects of a new drug with the traditional therapy, the fuel efficiency of two car engine designs, or the sales of new products in two different test cities. In fact, battery manufacturers themselves do research like this on their products and competitors' products.

---

[1] Did you know that today there is a niche interest in collecting *vinyl records*? Will CDs someday be collectible?

## Plot the Data

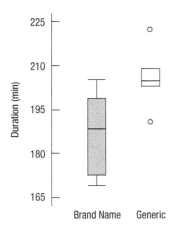

**Figure 23.1**
Boxplots comparing the brand-name and generic batteries suggest a difference in duration.

The natural display for comparing two groups is boxplots of the data for the two groups, placed side by side. Although we can't make a confidence interval or test a hypothesis from the boxplots themselves, they're a great way to visualize quantitative data coming from different groups. Have a look at the boxplots of the battery test data at the left.

It sure looks like the generic batteries lasted longer. And we can see that they were also more consistent. But is the difference large enough to change our battery-buying behavior? Can we be confident that the difference is more than just random fluctuation? That's why we need statistical inference.

The boxplot for the generic data identifies two possible outliers. That's interesting, but with only six measurements in each group, the outlier nomination rule is not very reliable. Both of the extreme values are plausible results, and the range of the generic values is smaller than the range of the brand-name values, even with the outliers. So we're probably better off just leaving these values in the data.

## Comparing Two Means

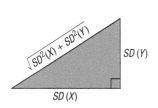

The Pythagorean Theorem of Statistics

Comparing two means is not very different from comparing two proportions. In fact, it's not different in concept from any of the methods we've seen. Now the population model parameter of interest is the difference between the *mean* battery lifetimes of the two brands, $\mu_1 - \mu_2$.

The rest is the much same as before. The statistic of interest is the difference in the two observed means, $\bar{y}_1 - \bar{y}_2$. We'll start with this statistic to build our confidence interval, but we'll need to know its standard deviation and its sampling model. Then we can create confidence intervals and find P-values for hypothesis tests.

We know that, for independent random variables, the variance of their *difference* is the *sum* of their individual variances, $Var(Y - X) = Var(Y) + Var(X)$. To find the standard deviation of the difference between the two independent sample means, we add their variances and then take a square root:

$$SD(\bar{y}_1 - \bar{y}_2) = \sqrt{Var(\bar{y}_1) + Var(\bar{y}_2)}$$
$$= \sqrt{\left(\frac{\sigma_1}{\sqrt{n_1}}\right)^2 + \left(\frac{\sigma_2}{\sqrt{n_2}}\right)^2}$$
$$= \sqrt{\frac{\sigma_1^2}{n_1} + \frac{\sigma_2^2}{n_2}}.$$

Of course, we still don't know the true standard deviations of the two populations, $\sigma_1$ and $\sigma_2$, so as usual, we'll use the estimates, $s_1$ and $s_2$. Using the estimates gives us the *standard error*:

$$SE(\bar{y}_1 - \bar{y}_2) = \sqrt{\frac{s_1^2}{n_1} + \frac{s_2^2}{n_2}}.$$

We'll use the standard error to see how big the difference really is. Because we are working with means and estimating the standard error of their difference using the data, we shouldn't be surprised that the sampling model is a Student's $t$.

## FOR EXAMPLE
### Finding the Standard Error of the Difference in Independent Sample Means

|   | Ordinary Bowl | Refilling Bowl |
|---|---|---|
| $n$ | 27 | 27 |
| $\bar{y}$ | 8.5 oz | 14.7 oz |
| $s$ | 6.1 oz | 8.4 oz |

Can you tell how much you are eating from how full you are? Or do you need visual cues? Researchers[2] constructed a table with two ordinary 18 oz soup bowls and two identical-looking bowls that had been modified to slowly, imperceptibly refill as they were emptied. They assigned experiment participants to the bowls randomly and served them tomato soup. Those eating from the ordinary bowls had their bowls refilled by ladle whenever they were one-quarter full. If people judge their portions by internal cues, they should eat about the same amount. How big a difference was there in the amount of soup consumed? The table summarizes their results.

QUESTION: How much variability do we expect in the difference between the two means? Find the standard error.

ANSWER: Participants were randomly assigned to bowls, so the two groups should be independent. It's okay to add variances.

$$SE(\bar{y}_{refill} - \bar{y}_{ordinary}) = \sqrt{\frac{s_r^2}{n_r} + \frac{s_o^2}{n_o}} = \sqrt{\frac{8.4^2}{27} + \frac{6.1^2}{27}} = 2.0 \text{ oz}$$

---

The confidence interval we build is called a **two-sample $t$-interval** (for the difference in means). The corresponding hypothesis test is called a **two-sample $t$-test**. The interval looks just like all the others we've seen—the statistic plus or minus an estimated margin of error:

$$(\bar{y}_1 - \bar{y}_2) \pm ME$$
$$\text{where } ME = t^* \times SE(\bar{y}_1 - \bar{y}_2).$$

This formula is almost the same as the one for the confidence interval for the difference of two proportions we saw in Chapter 21. It's just that here we use a Student's $t$-model instead of a Normal model to find the critical $t^*$-value corresponding to our chosen confidence level.

What are we missing? Only the degrees of freedom for the Student's $t$-model. Unfortunately, *that* formula is strange.

The deep, dark secret is that the sampling model isn't *really* Student's $t$, but only something close. The trick is that by using a special, adjusted degrees-of-freedom value, we can make it so close to a Student's $t$-model that nobody can tell the difference. The adjustment formula is straightforward but doesn't help our understanding much, so we leave it to the computer or calculator. (If you are curious and really want to see the formula, look in the footnote.[3])

### z or t?
If you know $\sigma$, use z. (That's rare!) Whenever you use s to estimate $\sigma$, use t.

---

[2] Brian Wansink, James E. Painter, and Jill North, "Bottomless Bowls: Why Visual Cues of Portion Size May Influence Intake," *Obesity Research*, Vol. 13, No. 1, January 2005.

[3] $$df = \frac{\left(\frac{s_1^2}{n_1} + \frac{s_2^2}{n_2}\right)^2}{\frac{1}{n_1 - 1}\left(\frac{s_1^2}{n_1}\right)^2 + \frac{1}{n_2 - 1}\left(\frac{s_2^2}{n_2}\right)^2}$$

Are you sorry you looked? This formula usually doesn't even give a whole number. If you are using a table, you'll need a whole number, so round down to be safe. If you are using technology, it's even easier. Computers and calculators deal with degrees of freedom automatically.

## An Easier Rule?

The formula for the degrees of freedom of the sampling distribution of the difference in sample means is long, but the number of degrees of freedom is always at *least* the smaller of the two *n*'s, minus 1. Using that easier value is a poor choice because it can give fewer than *half* the degrees of freedom you're entitled to from the correct formula.

## A Sampling Distribution for the Difference Between Two Sample Means

When the conditions are met, the sampling distribution of the standardized sample difference between the means of two independent groups,

$$t = \frac{(\bar{y}_1 - \bar{y}_2) - (\mu_1 - \mu_2)}{SE(\bar{y}_1 - \bar{y}_2)},$$

can be modeled by a Student's *t*-model with a number of degrees of freedom found with a special formula. We estimate the standard error with

$$SE(\bar{y}_1 - \bar{y}_2) = \sqrt{\frac{s_1^2}{n_1} + \frac{s_2^2}{n_2}}.$$

# Assumptions and Conditions

Now we've got everything we need. Before we can make a two-sample *t*-interval or perform a two-sample *t*-test, though, we have to check the assumptions and conditions.[4]

## Independent Groups Assumption

**Independent Groups Assumption:** To use the two-sample *t*-methods, the two groups we are comparing must be independent of each other. In fact, this test is sometimes called the two *independent samples t*-test. No statistical test can verify this assumption. You have to think about how the data were collected. The assumption would be violated, for example, if one group consisted of married people and the other group their spouses. Whatever we measure on couples might naturally be related. Similarly, if we compared subjects' performances before some treatment with their performances afterward, we'd expect a relationship of each "before" measurement with its corresponding "after" measurement. In cases such as these, where the observational units in the two groups are related or matched, *the two-sample methods of this chapter can't be applied*. When this happens, we need a different procedure that we'll see in the next chapter.

## Independence Assumption

**Independence Assumption:** The data in each group must be drawn independently and at random, or generated by a randomized comparative experiment. Without randomization of some sort, there are no sampling distribution models and no inference. We can check two conditions:

   **Randomization Condition:** Were the data collected with suitable randomization? For surveys, are they a representative random sample? For experiments, was the experiment randomized?

   **10% Condition:** Both of the two samples should represent less than 10% of their respective populations. This doesn't apply to randomized experiments because then there's no sampling.

## Normal Population Assumption

As we did before with Student's *t*-models, we should check the assumption that the underlying populations are *each* Normally distributed. We check the . . .

---

[4]No surprise there, eh?

**Nearly Normal Condition:** This condition is like the Nearly Normal Condition for a single mean, only it applies to *both* samples. If either sample size is less than 30, then the data from that sample should be unimodal and reasonably symmetric. For samples 30 or larger, the shape of the distribution won't indicate trouble unless there's severe deviation from normality.

## FOR EXAMPLE
### Checking Assumptions and Conditions

**RECAP:** Researchers reported that they had randomly assigned pairs of subjects so that at a table with four soup bowls two were eating soup from ordinary bowls and two from bowls that secretly refilled. They asked whether subjects eating from refilling bowls would eat more than those who had accurate visual cues of the amount they consumed.

**QUESTION:** Does this experiment satisfy the assumptions and conditions to draw inferences about the role of visual cues in determining how much people eat?

**ANSWER:** Assuming that the experiment was performed as described:

✓ **Randomization Condition:** Subjects were assigned to bowls at random

✓ **Independent Groups Assumption:** Due to the random assignment, the two treatment groups (refilling bowls and ordinary bowls) should be independent. However, there may be some relationship among those eating soup at the same time if something affected everyone's meal. Nothing of that sort is reported in the study.

✓ **Nearly Normal Condition:** If histograms of the amounts consumed by subjects in each group were nearly Normal, we'd feel confident using Student's *t*-methods. Unfortunately, the researchers provided neither appropriate graphs nor the original data. We might proceed with caution—with sample sizes so close to 30, it isn't as important to satisfy the Nearly Normal Condition, but we haven't been able to check for outliers. The published paper uses methods that assume Normality.

## Two-Sample *t*-Interval for the Difference Between Means

When the conditions are met, we are ready to find the confidence interval for the difference between means of two independent groups, $\mu_1 - \mu_2$. The confidence interval is

$$(\bar{y}_1 - \bar{y}_2) \pm t^*_{df} \times SE(\bar{y}_1 - \bar{y}_2),$$

where the standard error of the difference of the means

$$SE(\bar{y}_1 - \bar{y}_2) = \sqrt{\frac{s_1^2}{n_1} + \frac{s_2^2}{n_2}}.$$

The critical value $t^*_{df}$ depends on the particular confidence level, *C*, that you specify and on the number of degrees of freedom, which we get from a special formula.

## FOR EXAMPLE
### Finding a Confidence Interval for the Difference in Sample Means

|   | Ordinary Bowl | Refilling Bowl |
|---|---|---|
| *n* | 27 | 27 |
| $\bar{y}$ | 8.5 oz | 14.7 oz |
| *s* | 6.1 oz | 8.4 oz |

**RECAP:** Researchers studying the role of internal and visual cues in determining how much people eat conducted an experiment in which some people ate soup from bowls that secretly refilled. The results are summarized in the table.

We've already checked the assumptions and conditions, and have found the standard error for the difference in means to be $SE(\bar{y}_{\text{refill}} - \bar{y}_{\text{ordinary}}) = 2.0$ oz.

QUESTION: What does a 95% confidence interval say about the difference in mean amounts eaten?

ANSWER: The observed difference in means is $\bar{y}_{\text{refill}} - \bar{y}_{\text{ordinary}} = (14.7 - 8.5) = 6.2$ oz.

From technology: df $= 47.46 \quad t^*_{47.46} = 2.011$

$$ME = t^* \times SE(\bar{y}_{\text{refill}} - \bar{y}_{\text{ordinary}}) = 2.011(2.0) = 4.02 \text{ oz}$$

The 95% confidence interval for $\mu_{\text{refill}} - \mu_{\text{ordinary}}$ is $6.2 \pm 4.02$, or $(2.18, 10.22)$ oz. I am 95% confident that people eating from a subtly refilling bowl will eat an average of between 2.18 and 10.22 more ounces of soup than those eating from an ordinary bowl. Visual cues do appear to make a difference when people are deciding when they're full.

## STEP-BY-STEP EXAMPLE

### A Two-Sample *t*-Interval

Judging from the boxplot, the generic batteries seem to have lasted about 20 minutes longer than the brand-name batteries. Before we change our buying habits, what should we expect to happen with the next batteries we buy?

QUESTION: How much longer might the generic batteries last?

**THINK**

**PLAN** State what we want to know.

Identify the *parameter* you wish to estimate. Here our parameter is the difference in the means, not the individual group means.

Identify the *population(s)* about which you wish to make statements. We hope to make decisions about purchasing batteries, so we're interested in all the AA batteries of these two brands.

Identify the variables and review the W's.

**REALITY CHECK** From the boxplots, it appears our confidence interval should be centered near a difference of 20 minutes.

I have measurements of the lifetimes (in minutes) of 6 sets of generic and 6 sets of brand-name AA batteries from a randomized experiment. I want to find an interval that is likely, with 95% confidence, to contain the true difference $\mu_G - \mu_B$ between the mean lifetime of the generic AA batteries and the mean lifetime of the brand-name batteries.

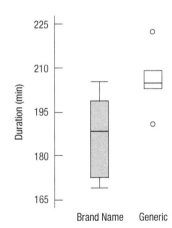

**MODEL** Think about the appropriate assumptions and check the conditions to be sure that a Student's *t*-model for the sampling distribution is appropriate.

✓ **Independent Groups Assumption**: Batteries manufactured by two different companies and purchased in separate packages should be independent.

✓ **Independence Assumption**: The batteries were packaged together, so they may not be independent. For example, a storage problem might affect all the batteries in the same pack. Repeating the study for several different packs of batteries would make the conclusions stronger.

For very small samples like these, we often don't worry about the 10% Condition.

✓ **Randomization Condition:** The batteries were selected at random from those available for sale. Not exactly an SRS, but a reasonably representative random sample.

✓ **10% Condition:** Both samples are certainly less than than 10% of the thousands of batteries produced of each type.

✓ **Nearly Normal Condition:** The samples are small, but the histograms look unimodal and reasonably symmetric:

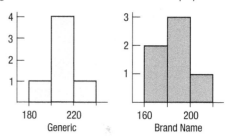

Make a picture. Boxplots are the display of choice for comparing groups, but now we want to check the *shape* of distribution of each group. Histograms or Normal probability plots do a better job there.

State the sampling distribution model for the statistic. The degrees of freedom come from that messy approximation formula.

Specify your method.

Under these conditions, it's okay to use a Student's $t$-model.

I'll use a **two-sample $t$-interval**.

---

**SHOW** **MECHANICS** Construct the confidence interval.

Be sure to include the units along with the statistics. Use meaningful subscripts to identify the groups.

Use the sample standard deviations to find the standard error of the sampling distribution.

I know $n_G = 6$ $\quad n_B = 6$

$\bar{y}_G = 206.0$ min $\quad \bar{y}_B = 187.4$ min

$s_G = 10.3$ min $\quad s_B = 14.6$ min

The groups are independent, so

$$SE(\bar{y}_G - \bar{y}_B) = \sqrt{SE^2(\bar{y}_G) + SE^2(\bar{y}_B)}$$

$$= \sqrt{\frac{s_G^2}{n_G} + \frac{s_B^2}{n_B}}$$

$$= \sqrt{\frac{10.3^2}{6} + \frac{14.6^2}{6}}$$

$$= \sqrt{\frac{106.09}{6} + \frac{213.16}{6}}$$

$$= \sqrt{53.208}$$

$$= 7.29 \text{ min.}$$

The computer or calculator automatically uses the approximation formula for df. This gives a fractional degree of freedom (here df = 8.98).

Technology will use the fractional df to find the critical value for the confidence interval. Here it's $t^* = 2.263$, but your computer or calculator probably won't tell you that. When showing your work, it's okay to just leave $t^*$ in the formulas and not try to find the actual value yourself.

df (from technology[5]) = 8.98

The corresponding critical value for a 95% confidence level is $t^* = 2.263$.

So the margin of error is

$$ME = t^* \times SE(\bar{y}_G - \bar{y}_B)$$

$$= 2.263\,(7.29)$$

$$= 16.50 \text{ min.}$$

---

[5] If you try to find the degrees of freedom with that messy approximation formula (We dare you! It's in the footnote on page 626) using the values above, you'll get 8.99. The minor discrepancy is because we rounded the standard deviations to the nearest 10th.

The 95% confidence interval is

$$(206.0 - 187.4) \pm 16.5 \text{ min}$$
$$\text{or } 18.6 \pm 16.5 \text{ min}$$
$$= (2.1, 35.1) \text{ min}.$$

**TELL** **CONCLUSION** Interpret the confidence interval in the proper context.

Less formally, you could say, "I'm 95% confident that generic batteries last an average of 2.1 to 35.1 minutes longer than brand-name batteries."

I am 95% confident that generic batteries would last, on average, between 2.1 and 35.1 minutes longer than brand-name batteries for this task. If generic batteries are cheaper, there seems little reason not to use them. If it is more trouble or costs more to buy them, then I'd consider whether the additional performance is worth it.

## Another One Just Like the Other Ones?

Yes. That's been our point all along. Once again we see a statistic plus or minus the margin of error. And the ME is just a critical value times the standard error. Just look out for that crazy degrees of freedom formula.

### TI TIPS

#### Creating the Confidence Interval

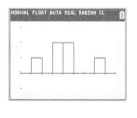

If you have been successful using your TI to make confidence intervals for proportions and 1-sample means, then you can probably already use the 2-sample function just fine. But please humor us while we do one.

**FIND A CONFIDENCE INTERVAL FOR THE DIFFERENCE IN MEANS, GIVEN DATA FROM TWO INDEPENDENT SAMPLES**

Let's do the batteries. Always think about whether the samples are independent. If not, stop right here. These procedures are appropriate only for independent groups.

♦ Enter the data into two lists.

| NameBrand in L1: | 194.0 | 205.5 | 199.2 | 172.4 | 184.0 | 169.5 |
| Generic in L2:   | 190.7 | 203.5 | 203.5 | 206.5 | 222.5 | 209.4 |

♦ Make histograms of the data to check the Nearly Normal Condition. We framed our window for L1 to cover the interval from 150 to 220 minutes, with bins of width 10 minutes.

♦ It's your turn to try this. Check L2. Go on, do it.

♦ Under STAT TESTS choose 2-SampTInt.

♦ Specify that you are using the Data in L1 and L2, specify 1 for both frequencies, and choose the confidence level you want.

♦ Pooled? We'll discuss this issue later in the chapter, but the easy advice is: Just Say No.

♦ To Calculate the interval, you may need to scroll down one more line.

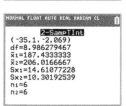

Now you have the 95% confidence interval. See df? The calculator did that messy degrees of freedom calculation for you. You have to love that!

Notice that the interval bounds are negative. That's because the TI is doing $\mu_1 - \mu_2$, and the generic batteries (L2) lasted longer. No harm done—you just need to be careful to interpret that result correctly when you *Tell* what the confidence interval means.

**No Data? Find a Confidence Interval Using the Sample Statistics** In many situations we don't have the original data, but must work with the summary statistics from the two groups. As we saw in the last chapter, you can still have your TI create the confidence interval with `2-SampTInt` by choosing the `Inpt:Stats` option. Enter both means, standard deviations, and sample sizes, then `Calculate`. We show you the details in the next TI Tips.

### JUST CHECKING

Carpal tunnel syndrome (CTS) causes pain and tingling in the hand, sometimes bad enough to keep sufferers awake at night and restrict their daily activities. Researchers studied the effectiveness of two alternative surgical treatments for CTS (Mackenzie, Hainer, and Wheatley, *Annals of Plastic Surgery*, 2000). Patients were randomly assigned to have endoscopic or open-incision surgery. Four weeks later the endoscopic surgery patients demonstrated a mean pinch strength of 9.1 kg compared to 7.6 kg for the open-incision patients.

1. Why is the randomization of the patients into the two treatments important?
2. A 95% confidence interval for the difference in mean strength is about (0.04 kg, 2.96 kg). Explain what this interval means.
3. Why might we want to examine such a confidence interval in deciding between these two surgical procedures?
4. Why might you want to see the data before trusting the confidence interval?

## A Test for the Difference Between Two Means

If you bought a used camera in good condition from a friend, would you pay the same as you would if you bought the same item from a stranger? A researcher at Cornell University[6] wanted to know how friendship might affect simple sales such as this. She randomly divided subjects into two groups and gave each group descriptions of items they might want to buy. One group was told to imagine buying from a friend whom they expected to see again. The other group was told to imagine buying from a stranger.

Here are the prices they offered for a used camera in good condition:

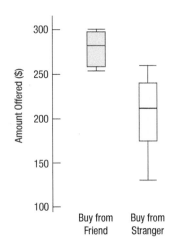

| WHO | University students |
| --- | --- |
| WHAT | Prices offered for a used camera |
| UNITS | $ |
| WHY | Study of the effects of friendship on transactions |
| WHEN | 1990s |
| WHERE | Cornell University |

| Price Offered for a Used Camera ($) ||
| --- | --- |
| Buying from a Friend | Buying from a Stranger |
| 275 | 260 |
| 300 | 250 |
| 260 | 175 |
| 300 | 130 |
| 255 | 200 |
| 275 | 225 |
| 290 | 240 |
| 300 | |

---

[6] J. J. Halpern, "The Transaction Index: A Method for Standardizing Comparisons of Transaction Characteristics Across Different Contexts," *Group Decision and Negotiation*, 6: 557–572.

The researcher who designed this study had a specific concern. Previous theories had doubted that friendship had a measurable effect on pricing. She hoped to find an effect of friendship. This calls for a hypothesis test—in this case a **two-sample *t*-test for the difference between means**.[7]

You already know enough to construct this test. The test statistic looks just like the others we've seen. It finds the difference between the observed group means and compares this with a hypothesized value for that difference. We'll call that hypothesized difference $\Delta_0$ ("delta naught"). It's so common for that hypothesized difference to be zero that we often just assume $\Delta_0 = 0$. We then compare the difference in the means with the standard error of that difference. We already know that for a difference between independent means, we can find P-values from a Student's *t*-model on that same special number of degrees of freedom.

> NOTATION ALERT
> $\Delta_0$—delta naught—isn't so standard that you can assume everyone will understand it. We use it because it's the Greek letter (good for a parameter) "D" for "difference." You should say "delta naught" rather than "delta zero"—that's standard for parameters associated with null hypotheses.

### Two-Sample *t*-Test for the Difference Between Means

The conditions for the two-sample *t*-test for the difference between the means of two independent groups are the same as for the two-sample *t*-interval. We test the hypothesis

$$H_0: \mu_1 - \mu_2 = \Delta_0$$

where the hypothesized difference is almost always 0, using the statistic

$$t = \frac{(\bar{y}_1 - \bar{y}_2) - \Delta_0}{SE(\bar{y}_1 - \bar{y}_2)}.$$

The standard error of $\bar{y}_1 - \bar{y}_2$ is

$$SE(\bar{y}_1 - \bar{y}_2) = \sqrt{\frac{s_1^2}{n_1} + \frac{s_2^2}{n_2}}.$$

When the conditions are met and the null hypothesis is true, this statistic can be closely modeled by a Student's *t*-model with a number of degrees of freedom given by a special formula. We use that model to obtain a P-value.

## STEP-BY-STEP EXAMPLE

### A Two-Sample *t*-Test for the Difference Between Two Means

The usual null hypothesis is that there's no difference in means. That's just the right null hypothesis for the camera purchase prices.

QUESTION: Is there a difference in the price people would offer a friend rather than a stranger?

| THINK | PLAN State what we want to know. | I want to know whether people are likely to offer a different amount for a used camera when buying from a friend than when buying from a stranger. I wonder whether the difference between mean amounts is zero. I have bid prices from 8 subjects buying from a friend and 7 buying from a stranger, found in a randomized experiment. |
|---|---|---|

Identify the *parameter* you wish to estimate. Here our parameter is the difference in the means, not the individual group means.

Identify the variables and check the W's.

---

[7] Because it is performed so often, this test is usually just called a "two-sample *t*-test."

**HYPOTHESES** State the null and alternative hypotheses. The research claim is that friendship changes what people are willing to pay. The natural null hypothesis is that friendship makes no difference.

We didn't start with any knowledge of whether friendship might increase or decrease the price, so we choose a two-sided alternative.

**MODEL** Think about the assumptions and check the conditions.

Note that because this is a randomized experiment, we haven't sampled at all, so the 10% Condition does not apply.

Make a picture. Boxplots are the display of choice for comparing groups, as seen on page 633. We also want to check the shapes of the distribution. Histograms or Normal probability plots do a better job for that.

State the sampling distribution model.
Specify your method.

$H_0$: The difference in mean price offered to friends and the mean price offered to strangers is zero:
$$\mu_F - \mu_S = 0.$$

$H_A$: The difference in mean prices is not zero:
$$\mu_F - \mu_S \neq 0.$$

✓ **Independent Groups Assumption**: Randomizing the experiment gives independent groups.

✓ **Independence Assumption**: This is an experiment, so there is no need for the subjects to be randomly selected from any particular population. All we need to check is whether they were assigned randomly to treatment groups.

✓ **Randomization Condition**: The experiment was randomized. Subjects were assigned to treatment groups at random.

✓ **Nearly Normal Condition**: Histograms of the two sets of prices are roughly unimodal and symmetric:

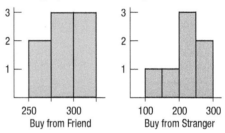

The assumptions are reasonable and the conditions are okay, so I'll use a Student's $t$-model to perform a **two-sample $t$-test**.

---

**SHOW MECHANICS** List the summary statistics. Be sure to use proper notation.

From the data:

$$n_F = 8 \qquad n_S = 7$$
$$\bar{y}_F = \$281.88 \qquad \bar{y}_S = \$211.43$$
$$s_F = \$18.31 \qquad s_S = \$46.43$$

Use the null model to find the P-value. First determine the standard error of the difference between sample means.

For independent groups,

$$SE(\bar{y}_F - \bar{y}_S) = \sqrt{SE^2(\bar{y}_F) + SE^2(\bar{y}_S)}$$
$$= \sqrt{\frac{s_F^2}{n_F} + \frac{s_S^2}{n_S}}$$
$$= \sqrt{\frac{18.31^2}{8} + \frac{46.43^2}{7}}$$
$$= 18.70$$

The observed difference is

$$(\bar{y}_F - \bar{y}_S) = 281.88 - 211.43 = \$70.45$$

Find the $t$-value.

$$t = \frac{(\bar{y}_F - \bar{y}_S) - (0)}{SE(\bar{y}_F - \bar{y}_S)} = \frac{70.45}{18.70} = 3.77$$

Make a picture. Sketch the *t*-model centered at the hypothesized difference of zero. Because this is a two-tailed test, shade the region to the right of the *t*-value for the observed difference and the corresponding region in the other tail.

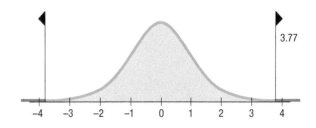

A statistics program or graphing calculator finds the P-value using the fractional degrees of freedom from the approximation formula.

$df = 7.62$ (from technology)
P-value $= 2P(t_{7.62} > 3.77) = 0.006$

 **CONCLUSION** Link the P-value to your decision about the null hypothesis, and state the conclusion in context.

Be cautious about generalizing to items whose prices are much different from those in this study.

If there were no difference in the mean prices, a difference this large would occur only 6 times in 1000. That's too rare to believe, so I reject the null hypothesis. There's evidence that people would not offer friends and strangers the same average amount for a used camera. It appears they'd offer a friend more (and possibly for other, similar items, too).

## TI TIPS

### Testing a Hypothesis About a Difference in Means

Now let's use the TI to do a hypothesis test for the difference of two means—independent, of course! (Have we said that enough times yet?)

**TEST A HYPOTHESIS WHEN YOU KNOW THE SAMPLE STATISTICS** We'll demonstrate by using the statistics from the camera-pricing example. A sample of 8 people suggested they'd sell the camera to a friend for an average price of $281.88 with standard deviation $18.31. An independent sample of 7 other people would charge a stranger an average of $211.43 with standard deviation $46.43. Does this represent a significant difference in prices?

- From the `STAT TESTS` menu select `2-SampTTest`.
- Specify `Inpt:Stats`, and enter the appropriate sample statistics.

- You have to scroll down to complete the specifications. This is a two-tailed test, so choose alternative $\neq \mu 2$.
- `Pooled?` Just say `No`. (We did promise to explain that and we will, coming up next.)
- Ready . . . set . . . `Calculate`!

The TI reports a calculated value of $t = 3.77$ and a P-value of 0.006. It's hard to tell who your real friends are.

**BY NOW WE PROBABLY DON'T HAVE TO TELL YOU HOW TO DO A `2-SampTTEST` STARTING WITH DATA IN LISTS**
So we won't.

## JUST CHECKING

Recall the experiment comparing patients 4 weeks after surgery for carpal tunnel syndrome. The patients who had endoscopic surgery demonstrated a mean pinch strength of 9.1 kg compared to 7.6 kg for the open-incision patients.

5. What hypotheses would you test?
6. The P-value of the test was less than 0.05. State a brief conclusion.
7. The study reports work on 36 "hands," but there were only 26 patients. In fact, 7 of the endoscopic surgery patients had both hands operated on, as did 3 of the open-incision group. Does this alter your thinking about any of the assumptions? Explain.

## FOR EXAMPLE
### A Two-Sample $t$-test

Many office "coffee stations" collect voluntary payments for the food consumed. Researchers at the University of Newcastle upon Tyne performed an experiment to see whether the image of eyes watching would change employee behavior.[8] They alternated pictures (seen here) of eyes looking at the viewer with pictures of flowers each week on the cupboard behind the "honesty box." They measured the consumption of milk to approximate the amount of food consumed and recorded the contributions (in £) each week per liter of milk. The table summarizes their results.

QUESTION: Do these results provide evidence that there really is a difference in honesty even when it's only photographs of eyes that are "watching"?

ANSWER: $H_0: \mu_{eyes} - \mu_{flowers} = 0$
$H_A: \mu_{eyes} - \mu_{flowers} \neq 0$

|  | Eyes | Flowers |
|---|---|---|
| $n$ (# weeks) | 5 | 5 |
| $\bar{y}$ | 0.417 £/l | 0.151 £/l |
| s | 0.1811 £/l | 0.067 £/l |

✓ **Independent Groups Assumption:** The same workers were recorded each week, but week-to-week independence is plausible.

✓ **Independence Assumption:** The amount paid by one person should be independent of the amount paid by others.

✓ **Randomization Condition:** This study was observational. Treatments alternated a week at a time and were applied to the same group of office workers.

✓ **Nearly Normal Condition:** I don't have the data to check, but it seems unlikely there would be outliers in either group. I could be more certain if I could see histograms for both groups.

It's okay to do a two-sample $t$-test for the difference in means:

$$SE(\bar{y}_{eyes} - \bar{y}_{flowers}) = \sqrt{\frac{s^2_{eyes}}{n_{eyes}} + \frac{s^2_{flowers}}{n_{flowers}}} = \sqrt{\frac{0.1811^2}{5} + \frac{0.067^2}{5}} = 0.0864$$

$$df = 5.07$$

$$t_5 = \frac{(\bar{y}_{eyes} - \bar{y}_{flowers}) - 0}{SE(\bar{y}_{eyes} - \bar{y}_{flowers})} = \frac{0.417 - 0.151}{0.0864} = 3.08$$

$$P(|t_5| > 3.08) = 0.027$$

---

[8] Melissa Bateson, Daniel Nettle, and Gilbert Roberts, "Cues of Being Watched Enhance Cooperation in a Real-World Setting," *Biol. Lett.* doi:10.1098/rsbl.2006.0509

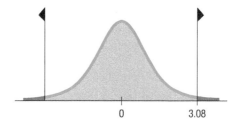

Assuming the data were free of outliers, the very low P-value leads me to reject the null hypothesis. This study provides evidence that people will leave higher average voluntary payments for food if pictures of eyes are "watching."

(**Note:** In Table T we can see that at 5 df, $t = 3.08$ lies between the two-tail critical values for $P = 0.02$ and $P = 0.05$, so we could report $P < 0.05$.)

## Back into the Pool?

Remember that when we know a proportion, we know its standard deviation. When we tested the null hypothesis that two proportions were equal, that connection meant we could assume their variances were equal as well. This led us to pool our data to estimate a standard error for the hypothesis test.

For means, there is also a pooled $t$-test. Like the two-proportions $z$-test, this test assumes that the variances in the two groups are equal. But be careful: Knowing the mean of some data doesn't tell you anything about their variance. And knowing that two means are equal doesn't say anything about whether their variances are equal. If we were willing to *assume* that their variances are equal, we could pool the data from two groups to estimate the common variance. We'd estimate this pooled variance from the data, so we'd still use a Student's $t$-model. This test is called a **pooled $t$-test (for the difference between means)**.

Pooled $t$-tests have a couple of advantages. They often have a few more degrees of freedom than the corresponding two-sample test and a much simpler degrees of freedom formula. But these advantages come at a price: You have to pool the variances and think about another assumption. The assumption of equal variances is a strong one, is often not true, and is difficult to check. For these reasons, we recommend that you use a two-sample $t$-test instead. It's never wrong *not* to pool.

## *The Pooled *t*-Test (In Case You're Curious)

Termites cause billions of dollars of damage each year, to homes and other buildings, but some tropical trees seem to be able to resist termite attack. A researcher extracted a compound from the sap of one such tree and tested it by feeding it at two different concentrations to randomly assigned groups of 25 termites.[9] After 5 days, 8 groups fed the lower dose had an average of 20.875 termites alive, with a standard deviation of 2.23. But 6 groups fed the higher dose had an average of only 6.667 termites alive, with a standard deviation of 3.14. Is this a large enough difference to declare the sap compound effective in killing termites? In order to use the pooled $t$-test, we must make the **Equal Variance Assumption** that the variances of the two populations from which the samples have been drawn are equal. That is, $\sigma_1^2 = \sigma_2^2$. (Of course, we could think about the standard deviations

---

[9] Adam Messer, Kevin McCormick, Sunjaya, H. H. Hagedorm, Ferny Tumbel, and J. Meinwald, "Defensive Role of Tropical Tree Resins: Antitermitic Sesquiterpenes from Southeast Asian Dipterocarpaceae," *Journal of Chemical Ecology*, 16:122, pp. 3333–3352.

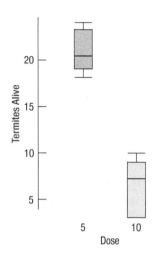

$df = 8 + 6 - 2 = 12$

$t = \dfrac{20.875 - 6.667}{1.43} = 9.935$

being equal instead.) The corresponding **Similar Spreads Condition** really just consists of looking at the boxplots to check that the spreads are not wildly different. We were going to make boxplots anyway, so there's really nothing new here.

Once we decide to pool, we estimate the common variance by combining numbers we already have:

$$s^2_{pooled} = \dfrac{(n_1 - 1)s_1^2 + (n_2 - 1)s_2^2}{(n_1 - 1) + (n_2 - 1)} = \dfrac{(8-1)2.23^2 + (6-1)3.14^2}{(8-1) + (6-1)} = 7.01$$

(If the two sample sizes are equal, this is just the average of the two variances.)

Now we just substitute this pooled variance in place of each of the variances in the standard error formula.

$$SE_{pooled}(\bar{y}_1 - \bar{y}_2) = \sqrt{\dfrac{7.01}{8} + \dfrac{7.01}{6}} = 1.43 \quad SE_{pooled}(\bar{y}_1 - \bar{y}_2) = \sqrt{\dfrac{s^2_{pooled}}{n_1} + \dfrac{s^2_{pooled}}{n_2}}$$

The formula for degrees of freedom for the Student's $t$-model is simpler, too. It was so complicated for the two-sample $t$ that we stuck it in a footnote.[10] Now it's just $df = n_1 + n_2 - 2$.

Substitute the pooled-$t$ estimate of the standard error and its degrees of freedom into the steps of the confidence interval or hypothesis test, and you'll be using the pooled-$t$ method. For the termites, $\bar{y}_1 - \bar{y}_2 = 14.208$, giving a $t$-value = 9.935 with 12 df and a P-value $\leq 0.0001$.

Of course, if you decide to use a pooled-$t$ method, you must defend your assumption that the variances of the two groups are equal.

---

**\*POOLED $t$-TEST AND CONFIDENCE INTERVAL FOR MEANS**

The conditions for the pooled $t$-test for the difference between the means of two independent groups (commonly called a "pooled $t$-test") are the same as for the two-sample $t$-test with the additional assumption that the variances of the two groups are the same. We test the hypothesis

$$H_0: \mu_1 - \mu_2 = \Delta_0$$

where the hypothesized difference, $\Delta_0$, is almost always 0, using the statistic

$$t = \dfrac{(\bar{y}_1 - \bar{y}_2) - \Delta_0}{SE_{pooled}(\bar{y}_1 - \bar{y}_2)}.$$

The standard error of $\bar{y}_1 - \bar{y}_2$ is

$$SE_{pooled}(\bar{y}_1 - \bar{y}_2) = \sqrt{\dfrac{s^2_{pooled}}{n_1} + \dfrac{s^2_{pooled}}{n_2}}$$

where the pooled variance is

$$s^2_{pooled} = \dfrac{(n_1 - 1)s_1^2 + (n_2 - 1)s_2^2}{(n_1 - 1) + (n_2 - 1)}.$$

When the conditions are met, we can model this statistic's sampling distribution with a Student's $t$-model with $(n_1 - 1) + (n_2 - 1)$ degrees of freedom. We use that model to obtain a P-value for a test or a margin of error for a confidence interval. The corresponding confidence interval is

$$(\bar{y}_1 - \bar{y}_2) \pm t^*_{df} \times SE_{pooled}(\bar{y}_1 - \bar{y}_2).$$

---

[10]But not this one. See page 626.

## Is the Pool All Wet?

We're testing whether the means are equal, so we admit that we don't *know* whether they are equal. Doesn't it seem a bit much to just *assume* that the variances are equal? Well, yes—but there are some special cases to consider. So when *should* you use pooled-*t*-methods rather than two-sample *t*-methods?

Never.

What, never?

Well, hardly ever.

You see, when the variances of the two groups are in fact equal, the two methods give pretty much the same result. (For the termites, the two-sample *t*-statistic is barely different— 9.436 with 8 df—and the P-value is still < 0.001.) Pooled methods have a small advantage (slightly narrower confidence intervals, slightly more powerful tests) mostly because they usually have a few more degrees of freedom, but the advantage is slight.

When the variances are *not* equal, the pooled methods are just not valid and can give poor results. You have to use the two-sample methods instead.

Pooling may make sense in a randomized comparative experiment. We start by assigning our experimental units to treatments at random. We know that at the start of the experiment each treatment group comes from the same population,[11] so each treatment group starts out with the same population variance. When we test whether the true means are equal, we may be willing to go a bit further and say that the treatments made no difference *at all*. Then it's not much of a stretch to assume that the variances have remained equal. It's still an assumption, and there are conditions that need to be checked (make the boxplots, make the boxplots, make the boxplots), but at least it's a plausible assumption.

### When Should I Use the Pooled *t*-Test?

Because the advantages of pooling are small, and you are allowed to pool only rarely (when the Equal Variances Assumption is met), *don't*.

It's never wrong *not* to pool.

## *A Permutation Test for Two Means

At the end of Chapter 21 we included an optional section describing a so-called permutation test. Permutation tests are technically the right way to study data from experiments, because random allocation of treatments means group membership isn't *really* independent for the subjects. (Being assigned to one group necessarily means you're not in the other group.) The Independent Groups Assumption is violated. Fortunately, the methods we use for observational studies and surveys that require the Independent Groups Assumption are good approximations of the corresponding permutation tests.

Let us describe how we conducted a permutation test for two means using our battery data. Here's a good way to think about how it works. Imagine that you and an opponent are playing a card game. There are 12 cards; each has one of the battery lifetimes written on it. Your opponent shuffles the cards and deals 6 to you and 6 to themself. The prize will be determined by how much higher the winner's 6-card average is than the loser's. Of course, if they were to deal you the 6 lowest cards while they got the 6 highest, you'd be pretty suspicious that the game wasn't all that fair. But a clever cheater wouldn't be that brazen. Oh, they'd be sure he won, and by quite a bit, of course; just not by enough to attract attention.

So it's game on! He shuffles, he deals, you both calculate your averages, and . . . you lose. You lose by 18.6 points. Seems like a lot. Do you think they cheated? How likely is it that a difference this large would happen just by chance?

| Brand Name | Generic |
|---|---|
| 194.0 | 190.7 |
| 205.5 | 203.5 |
| 199.2 | 203.5 |
| 172.4 | 206.5 |
| 184.0 | 222.5 |
| 169.5 | 209.4 |

---

[11]That is, the population of experimental subjects. Remember that to be valid, experiments do not need a representative sample drawn from a population because we are not trying to estimate a population model parameter.

Our simulation plays that card game. It plays fair. It randomly splits the cards into two hands, computes the average for each, and looks at how big the difference is. Then it does it again. And again. 200 trials in all. Here's the histogram of the differences.

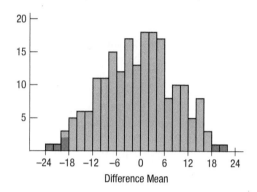

We see that if the two hands are randomly (fairly) created, usually the difference in means would be pretty small. In fact, only 6 times in 200 "games" would it be at least as large as 18.6 in either person's favor. That makes your 18.6-point loss seem pretty unusual.

Back to the batteries. The generic batteries were dealt the winning hand. This permutation test shows that if both types of batteries really perform equally well on average, a difference in means this large would happen by chance only 6 times in 200. Think: What does the proportion $6/200 = 0.03$ represent?

We hope you recognize we've tested the hypothesis that the mean life spans of the two types of batteries are the same, and that our simulation estimated a P-value of 0.03. Now grab your calculator and run a 2-sample $t$-test for these data. Check that P-value. Go ahead; it'll take about one minute. We think you'll be impressed with the permutation test.

## WHAT CAN GO WRONG?

- **Watch out for paired data.** The Independent Groups Assumption deserves special attention. If the groups are not independent, you can't use these two-sample methods. This is probably the main thing that can go wrong when using these two-sample methods. The methods of this chapter can be used *only* if the observations in the two groups are *independent*. Matched-pairs designs in which the observations are deliberately related arise often and are important. The next chapter deals with them.

- **Look at the plots.** The usual cautions about checking for outliers and non-Normal distributions apply, of course. A good defense is to make and examine boxplots. You may be surprised how often this simple step saves you from the wrong or even absurd conclusions that can be generated by a single undetected outlier. You don't want to conclude that two groups have very different means just because one observation is atypical.

### Do What We Say, Not What We Do . . .

Precision machines used in industry often have a bewildering number of parameters that have to be set, so experiments are performed in an attempt to try to find the best settings. Such was the case for a hole-punching machine used by a well-known computer manufacturer to make printed circuit boards. The data were analyzed by one of the authors, but because he was in a hurry, he didn't look at the boxplots first and just performed $t$-tests on the experimental factors. When he found extremely small P-values even for factors that made no sense, he plotted the data. Sure enough, there was one observation 1,000,000 times bigger than the others. It turns out that it had been recorded in microns (millionths of an inch), while all the rest were in inches.

- **Be cautious if you apply inference methods where there was no randomization.** If the data do not come from representative random samples or from a properly randomized experiment, then the inference about the differences between the groups may be wrong.
- **Don't interpret a significant difference in proportions or means causally.** Studies find that people with higher incomes are more likely to snore. Would surgery to increase your snoring be a wise investment? Probably not. It turns out that older people are more likely to snore, and they are also likely to earn more. In a prospective or retrospective study, there is always the danger that other lurking variables not accounted for are the real reason for an observed difference. Be careful not to jump to conclusions about causality.

## WHAT HAVE WE LEARNED?

Are the means of two groups the same? If not, how different are they? We've learned to use statistical inference to compare the means of two independent groups.
- We've seen that confidence intervals and hypothesis tests about the difference between two means, like those for an individual mean, use $t$-models.
- Once again we've seen the importance of checking assumptions that tell us whether our method will work.
- We've seen that, as when comparing proportions, finding the standard error for the difference in sample means depends on believing that our data come from independent groups. Unlike proportions, however, pooling is usually not the best choice here.
- And we've seen once again that we can add variances of independent random variables to find the standard deviation of the difference in two independent means.
- Finally, we've learned that the reasoning of statistical inference remains the same; only the mechanics change.

## TERMS

**Two-sample $t$-methods**  Two-sample $t$-methods allow us to draw conclusions about the difference between the means of two independent groups. The two-sample methods make relatively few assumptions about the underlying populations, so they are usually the method of choice for comparing two sample means. However, the Student's $t$-models are only approximations for their true sampling distribution. To make that approximation work well, the two-sample $t$-methods have a special rule for estimating degrees of freedom. (p. 626)

**Two-sample $t$-interval for the difference between means**  A confidence interval for the difference between the means of two independent groups found as

$$(\bar{y}_1 - \bar{y}_2) \pm t^*_{df} \times SE(\bar{y}_1 - \bar{y}_2)$$

where

$$SE(\bar{y}_1 - \bar{y}_2) = \sqrt{\frac{s_1^2}{n_1} + \frac{s_2^2}{n_2}} \text{ (p. 626)}$$

and the number of degrees of freedom is given by a special formula (see footnote 4 on page 627).

**Two-sample $t$-test for the difference between means**  A hypothesis test for the difference between the means of two independent groups. It tests the null hypothesis

$$H_0: \mu_1 - \mu_2 = \Delta_0,$$

where the hypothesized difference, $\Delta_0$, is almost always 0, using the statistic

$$t_{df} = \frac{(\bar{y}_1 - \bar{y}_2) - \Delta_0}{SE(\bar{y}_1 - \bar{y}_2)},$$

with the number of degrees of freedom given by the special formula. (p. 633)

**\*Pooled-$t$-methods**   Pooled-$t$-methods provide inferences about the difference between the means of two independent populations under the assumption that both populations have the same standard deviation. When the assumption is justified, pooled-$t$-methods generally produce slightly narrower confidence intervals and more powerful significance tests than two-sample $t$-methods. When the assumption is not justified, they generally produce worse results—sometimes substantially worse.

We recommend that you use unpooled two-sample $t$-methods instead. (p. 637)

## ON THE COMPUTER

### Inference for the Difference of Means

Here's some typical computer software output for confidence intervals or hypothesis tests for the difference in means based on two independent groups.

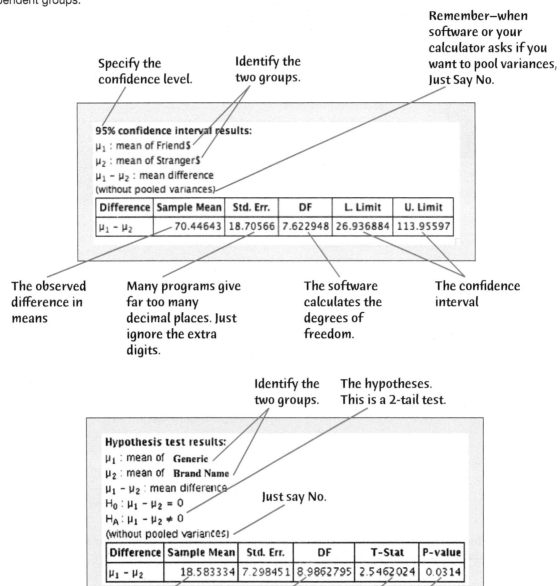

# EXERCISES

1. **Dogs and calories** *Consumer Reports* examined the calorie content of two kinds of hot dogs: meat (usually a mixture of pork, turkey, and chicken) and all beef. The researchers purchased samples of several different brands. The meat hot dogs averaged 111.7 calories, compared to 135.4 for the beef hot dogs. A test of the null hypothesis that there's no difference in mean calorie content yields a P-value of 0.124. Would a 95% confidence interval for $\mu_{Meat} - \mu_{Beef}$ include 0? Explain.

2. **Dogs and sodium** The *Consumer Reports* article described in Exercise 1 also listed the sodium content (in mg) for the various hot dogs tested. A test of the null hypothesis that beef hot dogs and meat hot dogs don't differ in the mean amounts of sodium yields a P-value of 0.11. Would a 95% confidence interval for $\mu_{Meat} - \mu_{Beef}$ include 0? Explain.

3. **Dogs and fat** The *Consumer Reports* article described in Exercise 1 also listed the fat content (in grams) for samples of beef and meat hot dogs. The resulting 90% confidence interval for $\mu_{Meat} - \mu_{Beef}$ is $(-6.5, -1.4)$.
   a) The endpoints of this confidence interval are negative numbers. What does that indicate?
   b) What does the fact that the confidence interval does not contain 0 indicate?
   c) If we use this confidence interval to test the hypothesis that $\mu_{Meat} - \mu_{Beef} = 0$, what's the corresponding alpha level?

4. **Washers** *Consumer Reports* examined top-loading and front-loading washing machines, testing samples of several different brands of each type. One of the variables the article reported was "cycle time," the number of minutes it took each machine to wash a load of clothes. Among the machines rated good to excellent, the 98% confidence interval for the difference in mean cycle time $(\mu_{Top} - \mu_{Front})$ is $(-40, -2)$.
   a) The endpoints of this confidence interval are negative numbers. What does that indicate?
   b) What does the fact that the confidence interval does not contain 0 indicate?
   c) If we use this confidence interval to test the hypothesis that $\mu_{Top} - \mu_{Front} = 0$, what's the corresponding alpha level?

5. **Dogs and fat, second helping** In Exercise 3, we saw a 90% confidence interval of $(-6.5, -1.4)$ grams for $\mu_{Meat} - \mu_{Beef}$, the difference in mean fat content for meat vs. all-beef hot dogs. Explain why you think each of the following statements is true or false:
   a) If I eat a meat hot dog instead of a beef dog, there's a 90% chance I'll consume less fat.
   b) 90% of meat hot dogs have between 1.4 and 6.5 grams less fat than a beef hot dog.
   c) I'm 90% confident that meat hot dogs average 1.4–6.5 grams less fat than the beef hot dogs.
   d) If I were to get more samples of both kinds of hot dogs, 90% of the time the meat hot dogs would average 1.4–6.5 grams less fat than the beef hot dogs.

   e) If I tested many samples, I'd expect about 90% of the resulting confidence intervals to include the true difference in mean fat content between the two kinds of hot dogs.

6. **Second load of wash** In Exercise 4, we saw a 98% confidence interval of $(-40, -22)$ minutes for $\mu_{Top} - \mu_{Front}$, the difference in time it takes top-loading and front-loading washers to do a load of clothes. Explain why you think each of the following statements is true or false:
   a) 98% of top loaders are 22 to 40 minutes faster than front loaders.
   b) If I choose the laundromat's top loader, there's a 98% chance that my clothes will be done faster than if I had chosen the front loader.
   c) If I tried more samples of both kinds of washing machines, in about 98% of these samples I'd expect the top loaders to be an average of 22 to 40 minutes faster.
   d) If I tried more samples, I'd expect about 98% of the resulting confidence intervals to include the true difference in mean cycle time for the two types of washing machines.
   e) I'm 98% confident that top loaders wash clothes an average of 22 to 40 minutes faster than front loaders.

7. **Learning math** The Core Plus Mathematics Project (CPMP) is an innovative approach to teaching Mathematics that engages students in group investigations and mathematical modeling. After field tests in 36 high schools over a three-year period, researchers compared the performances of CPMP students with those taught using a traditional curriculum. In one test, students had to solve applied Algebra problems using calculators. Scores for 320 CPMP students were compared to those of a control group of 273 students in a traditional Math program. Computer software was used to create a confidence interval for the difference in mean scores. (*Journal for Research in Mathematics Education*, 31, no. 3[2000])

   Conf level: 95%    Variable: Mu(CPMP) – Mu(Ctrl)
   Interval: (5.573, 11.427)

   a) What's the margin of error for this confidence interval?
   b) If we had created a 98% CI, would the margin of error be larger or smaller?
   c) Explain what the calculated interval means in context.
   d) Does this result suggest that students who learn Mathematics with CPMP will have significantly higher mean scores in Algebra than those in traditional programs? Explain.

8. **Stereograms** Stereograms appear to be composed entirely of random dots. However, they contain separate images that a viewer can "fuse" into a three-dimensional (3D) image by staring at the dots while defocusing the eyes. An experiment was performed to determine whether knowledge of the form of the embedded image affected the time required for subjects to fuse the images. One group of subjects (group NV) received no information or just verbal information about the shape of the embedded object. A second group (group VV) received both verbal information and visual information (specifically, a drawing of the object).

The experimenters measured how many seconds it took for the subject to report that they saw the 3D image.

2-Sample t-Interval for $\mu_1 - \mu_2$
Conf level = 90, df = 70
$\mu(NV) - \mu(VV)$ interval: (0.55, 5.47)

a) Interpret your interval in context.
b) Does it appear that viewing a picture of the image helps people "see" the 3D image in a stereogram?
c) What's the margin of error for this interval?
d) Explain what the 90% confidence level means.
e) Would you expect a 99% confidence level to be wider or narrower? Explain.
f) Might that change your conclusion in part b? Explain.

9. **CPMP, again** During the study described in Exercise 7, students in both CPMP and traditional classes took another Algebra test that did not allow them to use calculators. The table below shows the results. Are the mean scores of the two groups significantly different?

| Math Program | n | Mean | SD |
|---|---|---|---|
| CPMP | 312 | 29.0 | 18.8 |
| Traditional | 265 | 38.4 | 16.2 |

*Performance on Algebraic Symbolic Manipulation Without Use of Calculators*

a) Write an appropriate hypothesis.
b) Do you think the assumptions for inference are satisfied? Explain.
c) Here is computer output for this hypothesis test. Explain what the P-value means in this context.

2-Sample t-Test of $\mu_1 - \mu_2 \neq 0$
t-Statistic = −6.451 w/574.8761 df
P < 0.0001

d) State a conclusion about the CPMP program.

10. **CPMP and word problems** The study of the new CPMP Mathematics methodology described in Exercise 7 also tested students' abilities to solve word problems. This table shows how the CPMP and traditional groups performed. What do you conclude?

| Math Program | n | Mean | SD |
|---|---|---|---|
| CPMP | 320 | 57.4 | 32.1 |
| Traditional | 273 | 53.9 | 28.5 |

11. **Cost of shopping** Do consumers spend more on a trip to Walmart or Target? Suppose researchers interested in this question collected data from well-designed systematic samples of 85 Walmart customers and 80 Target customers by asking them for their purchase amounts as they left the stores. The data collected are summarized in the table below.

| | Walmart | Target |
|---|---|---|
| n | 85 | 80 |
| $\bar{y}$ | $45 | $53 |
| s | $21 | $19 |

To perform inference on these two samples, what conditions must be met? Are they? Explain.

12. **Athlete ages** A sports reporter suggests that professional baseball players must, on average, be older than professional football players, because football is a contact sport and players are more susceptible to concussions and serious injuries (www.sports.yahoo.com). One player was selected at random from each team in both professional baseball (MLB) and professional football (NFL). The data are summarized below.

| | MLB | NFL |
|---|---|---|
| n | 30 | 32 |
| $\bar{y}$ | 27.5 | 26.16 |
| s | 3.94 | 2.78 |

To perform inference on these two samples, what conditions must be met? Are they? Explain.

13. **Cost of shopping, again** Using the summary statistics provided in Exercise 11, researchers calculated a 95% confidence interval for the mean difference between Walmart and Target purchase amounts. The interval was (−$14.15, −$1.85). Explain in context what this interval means.

14. **Athlete ages, again** Using the summary statistics provided in Exercise 12, the sports reporter calculated the following 95% confidence interval for the mean difference between major league baseball players and professional football players. The 95% interval for $\mu_{MLB} - \mu_{NFL}$ was (−0.41, 3.09). Summarize in context what the interval means.

15. **Commuting** A man who moves to a new city sees that there are two routes he could take to work. A neighbor who has lived there a long time tells him Route A will average 5 minutes faster than Route B. The man decides to experiment. Each day he flips a coin to determine which way to go, driving each route 20 days. He finds that Route A takes an average of 40 minutes, with standard deviation 3 minutes, and Route B takes an average of 43 minutes, with standard deviation 2 minutes. Histograms of travel times for the routes are roughly symmetric and show no outliers.

a) Find a 95% confidence interval for the difference in average commuting time for the two routes.
b) Should the man believe the old-timer's claim that he can save an average of 5 minutes a day by always driving Route A? Explain.

16. **Memory** Does ginkgo biloba enhance memory? In an experiment to find out, subjects were assigned randomly to take ginkgo biloba supplements or a placebo. Their memory was tested to see whether it improved. The numbers reported are the number of items recalled before and after taking a supplement. Here are boxplots comparing the two groups. At the top of the next page is some computer output for the data.

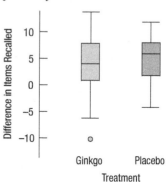

**Summary statistics:**

| Column | n | Mean | Std. dev. | Median | Min | Max | Q1 | Q3 |
|---|---|---|---|---|---|---|---|---|
| Gingko | 104 | 4.2307692 | 5.0323821 | 4 | −10 | 14 | 1 | 8 |
| Placebo | 99 | 5.2222222 | 4.1170515 | 6 | −4 | 12 | 2 | 8 |

a) What do the boxplots suggest about differences between sujects who took ginkgo those who had the placebo?
b) Is it appropriate to analyze these data using the methods of inference discussed in this chapter? Explain.
c) Create a 90% confidence interval for the difference in mean number of items recalled.
d) Does the confidence interval confirm your answer to part a? Explain.

**17. Cereal** The data below show the sugar content (as a percentage of weight) of several randomly selected national brands of children's and adults' cereals. Create and interpret a 95% confidence interval for the difference in mean sugar content. Be sure to check the necessary assumptions and conditions.

**Children's cereals:** 40.3, 55, 45.7, 43.3, 50.3, 45.9, 53.5, 43, 44.2, 44, 47.4, 44, 33.6, 55.1, 48.8, 50.4, 37.8, 60.3, 46.6

**Adults' cereals:** 20, 30.2, 2.2, 7.5, 4.4, 22.2, 16.6, 14.5, 21.4, 3.3, 6.6, 7.8, 10.6, 16.2, 14.5, 4.1, 15.8, 4.1, 2.4, 3.5, 8.5, 10, 1, 4.4, 1.3, 8.1, 4.7, 18.4

**T 18. Penguins** From 2007 to 2009, marine biologist Dr. Kristen Gorman collected data on three species of peguins living on three islands in the Palmer Archipelago, Antarctica. Among other things, Dr. Gorman collected body measurements on hundreds of panguins. Below are the culmen (bill) lengths of a random sample of Adélie and Gentoo penguins from the 344 penguins in her study.

a) Are these data appropriate for inference? Explain.
b) Create and interpret a 95% confidence interval for the difference in mean culmen length between these two species.
c) Do these data provide evidence that the mean length of penguins' bills are different for the two species? Explain.

| Bill Length in mm | | | |
|---|---|---|---|
| Adélie | | Gentoo | |
| 32.1 | 39.6 | 41.7 | 48.2 |
| 33.1 | 39.6 | 42.0 | 48.2 |
| 33.5 | 39.6 | 43.2 | 48.5 |
| 35.0 | 39.7 | 43.8 | 48.7 |
| 35.9 | 39.8 | 44.0 | 48.7 |
| 36.0 | 40.2 | 44.4 | 49.0 |
| 36.0 | 40.2 | 44.5 | 49.1 |
| 36.2 | 40.3 | 44.5 | 49.2 |
| 36.7 | 40.6 | 44.9 | 49.5 |
| 36.7 | 40.9 | 45.0 | 49.5 |
| 37.0 | 41.0 | 45.1 | 49.6 |
| 37.2 | 41.4 | 45.1 | 49.9 |
| 37.5 | 41.4 | 45.1 | 50.0 |
| 37.6 | 41.5 | 45.2 | 50.0 |
| 37.8 | 41.8 | 45.3 | 50.4 |
| 37.8 | 42.1 | 45.5 | 50.5 |
| 37.8 | 42.2 | 45.8 | 50.5 |
| 38.1 | 42.2 | 46.1 | 50.8 |
| 38.1 | 42.3 | 46.2 | 51.1 |
| 38.8 | 43.2 | 46.2 | 51.1 |
| 39.0 | 43.2 | 46.5 | 52.2 |
| 39.0 | 45.6 | 46.5 | 52.5 |
| 39.5 | 46.0 | 46.7 | 53.4 |
| 39.6 | | 46.8 | 54.3 |
| | | 47.5 | 59.6 |
| | | 47.7 | |

**T 19. Reading** An educator believes that new reading activities for elementary school children will improve reading comprehension scores. She randomly assigns third graders to an eight-week program in which some will use these activities and others will experience traditional teaching methods. At the end of the experiment, both groups take a reading comprehension exam. Their scores are shown in the back-to-back stem-and-leaf display. Do these results suggest that the new activities are better? Test an appropriate hypothesis and state your conclusion.

```
New Activities | Control
             1 | 0 7
           4 2 | 0 6 8
           3 3 | 3 7 7
       9 6 3 3 3 4 | 1 2 2 2 2 2 3 8
     9 8 7 6 4 3 2 5 | 3 5 5
           7 2 1 6 | 0 2
               1 7 |
                 8 | 5
```

**T 20. Streams** Researchers collected random samples of water from streams in the Adirondack Mountains to look for any differences in the effects of acid rain. They measured the pH (acidity) of the water and classified the streams with respect to the kind of substrate (type of rock over which they flow). A lower pH means the water is more acidic. Here is a plot of the pH of the streams by substrate (limestone, mixed, or shale):

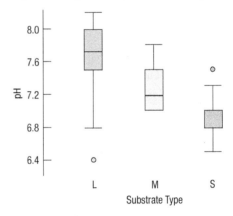

Here are selected parts of a software analysis comparing the pH of streams with limestone(1) and shale(2) substrates:

2-Sample $t$-Test of $\mu_1 - \mu_2$
Difference Between Means = 0.735
$t$-Statistic = 16.30 w/133 df
$p \leq 0.0001$

a) State the null and alternative hypotheses for this test.
b) From the information you have, do the assumptions and conditions appear to be met?
c) What conclusion would you draw?

**21. Home runs 2019** American League baseball teams play their games with the designated hitter rule, meaning that pitchers do not bat. The league believes that replacing the pitcher, traditionally a weak hitter, with another player in the batting order produces more runs and generates more interest among fans. On the next page are the average numbers of home runs hit per game in

American League and National League stadiums for the 2019 season.

| American | | National | |
|---|---|---|---|
| 1.31 | 1.90 | 1.36 | 1.49 |
| 1.51 | 1.89 | 1.54 | 1.33 |
| 1.13 | 1.59 | 1.58 | 1.01 |
| 1.38 | 1.49 | 1.40 | 1.35 |
| 0.93 | 1.34 | 1.38 | 1.03 |
| 1.78 | 1.38 | 1.72 | 1.30 |
| 1.00 | 1.52 | 0.90 | 1.43 |
| 1.36 | | 1.54 | |

a) Create an appropriate display of these data. What do you see?
b) With a 95% confidence interval, estimate the mean number of home runs hit in American League games.
c) Coors Field, in Denver, stands a mile above sea level, an altitude far greater than that of any other major league ballpark. Some believe that the thinner air makes it harder for pitchers to throw curve balls and easier for batters to hit the ball a long way. Do you think the 1.38 home runs hit per game at Coors is unusual? (Denver is a National League team.) Explain.
d) Explain why you should not use two separate confidence intervals to decide whether the two leagues differ in average number of runs scored.
e) Using a 95% confidence interval, estimate the difference between the mean number of home runs hit in American and National League games.
f) Interpret your interval.
g) Does this interval suggest that the two leagues may differ in average number of home runs hit per game?

22. **Tallest country** The Netherlands is the tallest country in the world. To compare them to the United States, two random samples were selected. A random sample of sixteen 18-year-old Dutch men and a random sample of twenty 18-year-old American men were selected. Summaries and plots are shown below.

| | Netherlands | USA |
|---|---|---|
| Number of subjects | 16 | 20 |
| Height: | | |
| Mean | 181.9 | 172.0 |
| SD | 6.61 | 8.08 |

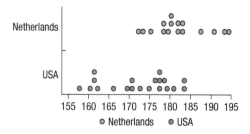

a) The distributions of these samples overlap. Does this suggest that the confidence intervals for the population means will overlap?

b) Find 95% confidence intervals for the average height of 18-year-old American men, and one for 18-year-old Dutch men. Do they overlap?
c) Find a 95% confidence interval for the difference in the mean height of 18-year-old Dutch men and 18-year-old American men.
d) Which of these methods is the correct one for estimating the mean difference in height? Provide a correct interpretation for that method.

23. **Double header** Look again at the data in Exercise 21.
a) Explain why you should not use two separate confidence intervals to decide whether the two leagues differ in average number of runs scored home runs per game.
b) Using a 95% confidence interval, estimate the difference between the mean number of home runs hit in American and National League games.
c) Interpret your interval.
d) Does this interval suggest that the two leagues may differ in average number of home runs hit per game?

24. **Hard water** In an investigation of environmental causes of disease, data were collected on the annual mortality rate (deaths per 100,000) for males in 61 large towns in England and Wales. In addition, the water hardness was recorded as the calcium concentration (parts per million, ppm) in the drinking water. The dataset also notes, for each town, whether it was south or north of Derby. Is there a significant difference in mortality rates in the two regions? Here are the summary statistics:

| Summary of: | | Mortality | | |
|---|---|---|---|---|
| For categories in: | | Derby | | |
| Group | Count | Mean | Median | StdDev |
| North | 34 | 1631.59 | 1631 | 138.470 |
| South | 27 | 1388.85 | 1369 | 151.114 |

a) Test appropriate hypotheses and state your conclusion.
b) The boxplots of the two distributions show an outlier among the data north of Derby. What effect might that have had on your test?

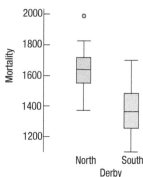

25. **Job satisfaction** A company institutes an exercise break for its workers to see if this will improve job satisfaction, as measured by a questionnaire that assesses workers' satisfaction. Scores for 10 randomly selected workers before and after implementation of the exercise program are shown. The company wants to assess the effectiveness of the exercise program. Explain why you can't use the methods discussed in this chapter to do

that. (Don't worry, we'll give you another chance to do this the right way.)

| Worker Number | Job Satisfaction Index | |
|---|---|---|
| | Before | After |
| 1 | 34 | 33 |
| 2 | 28 | 36 |
| 3 | 29 | 50 |
| 4 | 45 | 41 |
| 5 | 26 | 37 |
| 6 | 27 | 41 |
| 7 | 24 | 39 |
| 8 | 15 | 21 |
| 9 | 15 | 20 |
| 10 | 27 | 37 |

26. **Summer school** Having done poorly on their math final exams in June, six students repeat the course in summer school, then take another exam in August. If we consider these students representative of all students who might attend this summer school in other years, do these results provide evidence that the program is worthwhile?

| June | 54 | 49 | 68 | 66 | 62 | 62 |
|---|---|---|---|---|---|---|
| Aug. | 50 | 65 | 74 | 64 | 68 | 72 |

27. **Sex and violence** The *Journal of Applied Psychology* reported on a study that examined whether the content of TV shows influenced the ability of viewers to recall brand names of items featured in the commercials. The researchers randomly assigned volunteers to watch one of three programs, each containing the same nine commercials. One of the programs had violent content, another sexual content, and the third neutral content. After the shows ended, the subjects were asked to recall the brands of products that were advertised. Here are summaries of the results:

| | Program Type | | |
|---|---|---|---|
| | Violent | Sexual | Neutral |
| No. of Subjects | 108 | 108 | 108 |
| Brands Recalled | | | |
| Mean | 2.08 | 1.71 | 3.17 |
| SD | 1.87 | 1.76 | 1.77 |

a) Do these results indicate that viewer memory for ads may differ depending on program content? A test of the hypothesis that there is no difference in ad memory between programs with sexual content and those with violent content has a P-value of 0.136. State your conclusion.

b) Is there evidence that viewer memory for ads may differ between programs with sexual content and those with neutral content? Test an appropriate hypothesis and state your conclusion.

28. **Ad campaign** You are a consultant to the marketing department of a business preparing to launch an ad campaign for a new product. The company can afford to run ads during one TV show, and has decided not to sponsor a show with sexual content. You read the study described in Exercise 27, then use a computer to create a confidence interval for the difference in mean number of brand names remembered between the groups watching violent shows and those watching neutral shows.

```
TWO-SAMPLE T
95% CI FOR MUviol − MUneut: (−1.578, − 0.602)
```

a) At the meeting of the marketing staff, you have to explain what this output means. What will you say?
b) What advice would you give the company about the upcoming ad campaign?

29. **Sex and violence II** In the study described in Exercise 27, the researchers also contacted the subjects again, 24 hours later, and asked them to recall the brands advertised. Results are summarized in the table below.

| | Program Type | | |
|---|---|---|---|
| | Violent | Sexual | Neutral |
| No. of Subjects | 101 | 106 | 103 |
| Brands Recalled | | | |
| Mean | 3.02 | 2.72 | 4.65 |
| SD | 1.61 | 1.85 | 1.62 |

a) Is there a significant difference in viewers' abilities to remember brands advertised in shows with violent vs. neutral content?
b) Find a 95% confidence interval for the difference in mean number of brand names remembered between the groups watching shows with sexual content and those watching neutral shows. Interpret your interval in this context.

30. **Ad recall** In Exercises 27 and 29, we see the number of advertised brand names people recalled immediately after watching TV shows and 24 hours later. Strangely enough, it appears that they remembered more about the ads the next day. Should we conclude this is true in general about people's memory of TV ads?

a) Suppose one analyst conducts a two-sample hypothesis test to see if memory of brands advertised during violent TV shows is higher 24 hours later. If his P-value is 0.00013, what might he conclude?
b) Explain why his procedure was inappropriate. Which of the assumptions for inference was violated?
c) How might the design of this experiment have tainted the results?
d) Suggest a design that could compare immediate brand-name recall with recall one day later.

31. **View of the water** How much extra is having a waterfront property worth? A student took a random sample of 170 recently sold properties in Upstate New York to examine the question. Here are her summaries and boxplots of the two groups of prices:

| Non-Waterfront Properties | | Waterfront Properties | |
|---|---|---|---|
| $n$ | 100 | $n$ | 70 |
| $\bar{y}$ | $219,896.60 | $\bar{y}$ | $319,906.40 |
| $s$ | $94,627.15 | $s$ | $153,303.80 |

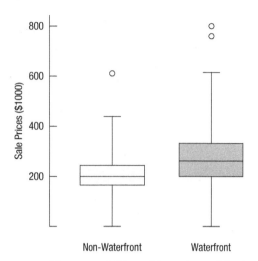

Construct and interpret a 95% confidence interval for the mean additional amount that waterfront property is worth. (From technology, df = 105.48.)

**32. New construction** The sample of house sales we looked at in Exercise 31 also listed whether homes were new construction or not. Find and interpret a 95% confidence interval for how much more an agent can expect to sell a new home for. (From technology, df = 197.8.) Here are the summaries and boxplots of the *Sale Prices*:

|  | Old Construction |  | New Construction |
|---|---|---|---|
| $n$ | 100 | $n$ | 100 |
| $\bar{y}$ | $201,707.50 | $\bar{y}$ | $267,878.10 |
| $s$ | $96,116.88 | $s$ | $93,302.18 |

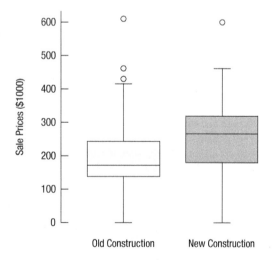

**33. More arrows** In Chapter 22, Exercise 53, we looked at the penetration of stone-tipped versus wooden-tipped arrows. Do these data suggest that stone-tipped arrows penetrate farther than wooden-tipped?

| Wooden (mm) | 216 | 211 | 192 | 208 | 203 | 210 | 203 |
|---|---|---|---|---|---|---|---|
| Stone (mm) | 240 | 208 | 213 | 225 | 232 | 214 | 240 |

Perform a test complete with hypotheses and checks of the conditions.

**34. Final shot** In Exercise 54 in Chapter 22, we considered the accuracy of stone-tipped versus wooden-tipped arrows. Do these data give statistically significant evidence of a difference in accuracy between the two types?

| Wooden | 9.3 | 16.7 | 7.1 | 14 | 1 | 1.2 |
|---|---|---|---|---|---|---|
| Stone | 4.9 | 21.1 | 7 | 1.8 | 5.4 | 8.6 |

**35. Fast or not?** In 2009, *Antiquity* published isotope analyses of the human skeletal tissue excavated from 1957 to 1967 at Whithorn Cathedral Priory, Scotland. These analyses sought to use new isotope methods to test common assumptions about the lifestyle and diet of the bishops and clerics compared to lay individuals buried at the same site.

Specifically, Dr. Muldner (and others) tested an isotope that would indicate whether or not individuals ate seafood regularly. It is believed that the bishops and priests had access to seafood for fast days and other holy days, while the lay individuals did not. A higher measurement of % collagen indicates more seafood in a person's diet. Here are the data:

| Percent Collagen ||
|---|---|
| Priests/Bishops | Lay Individuals |
| 9.6 | 6.1 |
| 7.3 | 8.1 |
| 8.1 | 5.6 |
| 7.5 | 5.3 |
| 7.8 | 5.2 |
| 9.8 | 3.5 |
| 9.8 | 5.8 |

Is there statistically significant evidence of a higher collagen level among priests and bishops?

**36. I can relate** Professor Jody Gittell analyzed workers in the health care system. Specifically, she was interested to see if relational coordination (communicating and relating for the purpose of task integration) makes a worker more resilient in response to external threats that require a coordinated response across multiple roles in the organization. She rated whether physicians and nurses were selected for teamwork qualities, on a scale of 0 to 2. Do these summaries show evidence that teamwork qualities are valued differently in doctors than in nurses? Assume the conditions for inference have been met. (*Journal of Applied Behavioral Science*, March 2008)

| Position | Mean | SD | $n$ |
|---|---|---|---|
| Physicians | 0.44 | 0.88 | 9 |
| Nurses | 1.44 | 0.73 | 9 |

**37. Hungry?** Researchers investigated how the size of a bowl affects how much ice cream people tend to scoop when serving themselves.[12] At an "ice cream social," people were randomly

---
[12] Brian Wansink, Koert van Ittersum, and James E. Painter, "Ice Cream Illusions: Bowls, Spoons, and Self-Served Portion Sizes," *American Journal of Preventive Medicine* 2006.

given either a 17 oz or a 34 oz bowl (both large enough that they would not be filled to capacity). They were then invited to scoop as much ice cream as they liked. Did the bowl size change the selected portion size? Here are the summaries:

| | Small Bowl | | Large Bowl |
|---|---|---|---|
| $n$ | 26 | $n$ | 22 |
| $\bar{y}$ | 5.07 oz | $\bar{y}$ | 6.58 oz |
| $s$ | 1.84 oz | $s$ | 2.91 oz |

Test an appropriate hypothesis and state your conclusions. Assume any assumptions and conditions that you cannot test are sufficiently satisfied to proceed.

**38. Thirsty?** Researchers randomly assigned participants either a tall, thin "highball" glass or a short, wide "tumbler," each of which held 355 ml. Participants were asked to pour a shot (1.5 oz = 44.3 ml) into their glass. Did the shape of the glass make a difference in how much liquid they poured?[13] Here are the summaries:

| | Highball | | Tumbler |
|---|---|---|---|
| $n$ | 99 | $n$ | 99 |
| $\bar{y}$ | 42.2 ml | $\bar{y}$ | 60.9 ml |
| $s$ | 16.2 ml | $s$ | 17.9 ml |

Test an appropriate hypothesis and state your conclusions. Assume any assumptions and conditions that you cannot test are sufficiently satisfied to proceed.

**39. Lower scores?** Newspaper headlines recently announced a decline in science scores among high school seniors. In 2000, a total of 15,109 seniors tested by The National Assessment of Educational Progress (NAEP) scored a mean of 147 points. Four years earlier, 7537 seniors had averaged 150 points. The standard error of the difference in the mean scores for the two groups was 1.22.

a) Have the science scores declined significantly? Cite appropriate statistical evidence to support your conclusion.
b) The sample size in 2000 was almost double that in 1996. Does this make the results more convincing or less? Explain.

**40. The Internet** The NAEP report described in Exercise 39 compared science scores for students who had home Internet access to the scores of those who did not, as shown in the graph. They report that the differences are statistically significant.

a) Explain what "statistically significant" means in this context.
b) If their conclusion is incorrect, which type of error did the researchers commit?
c) Does this prove that using the Internet at home can improve a student's performance in science?

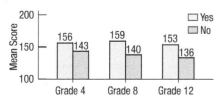

[13]Brian Wansink and Koert van Ittersum, "Shape of Glass and Amount of Alcohol Poured: Comparative Study of Effect of Practice and Concentration," *BMJ* 2005; 331;1512–1514.

**41. Running heats London** In Olympic running events, preliminary heats are determined by random draw, so we should expect the abilities of runners in the various heats to be about the same, on average. Look at the table below showing times (in seconds) for the 400-m women's run in the 2012 Olympics in London for preliminary heats 2 and 5. Is there any evidence that the mean time to finish is different for randomized heats? Explain. Be sure to include a discussion of assumptions and conditions for your analysis. (Note: One runner in heat 2 did not finish and one runner in heat 5 did not start.)

| Country | Name | Time | Heat |
|---|---|---|---|
| BOT | Amantle Montsho | 50.4 | 2 |
| JAM | Christine Day | 51.05 | 2 |
| GBR | Shana Cox | 52.01 | 2 |
| GUY | Aliann Pompey | 52.1 | 2 |
| ANT | Afia Charles | 54.25 | 2 |
| NCA | Ingrid Narvaez | 59.55 | 2 |
| BAH | Shaunae Miller | DNF | 2 |
| RUS | Antonina Krivoshapka | 50.75 | 5 |
| UKR | Alina Lohvynenko | 52.08 | 5 |
| GBR | Lee McConnell | 52.23 | 5 |
| SWE | Moa Hjelmer | 52.86 | 5 |
| MAW | Ambwene Simukonda | 54.2 | 5 |
| FIJ | Danielle Alakija | 56.77 | 5 |
| GRN | Kanika Beckles | DNS | 5 |

**42. Swimming heats London** In Exercise 41, we looked at the times in two different heats for the 400-m women's run from the 2012 Olympics. Unlike track events, swimming heats are not determined at random. Instead, swimmers are seeded so that better swimmers are placed in later heats. Here are the times (in seconds) for the women's 400-m freestyle from heats 2 and 5. Do these results suggest that the mean times of seeded heats are not equal? Explain. Include a discussion of assumptions and conditions for your analysis.

| Country | Name | Time | Heat |
|---|---|---|---|
| BUL | Nina Rangelova | 251.7 | 2 |
| CHI | Kristel Köbrich | 252.0 | 2 |
| JPN | Aya Takano | 252.3 | 2 |
| LIE | Julia Hassler | 253.0 | 2 |
| MEX | Susana Escobar | 254.8 | 2 |
| THA | Nathanan Junkrajang | 256.5 | 2 |
| SIN | Lynette Lim | 258.6 | 2 |
| KOR | Kim Ga-Eul | 283.5 | 2 |
| FRA | Camille Muffat | 243.3 | 5 |
| USA | Allison Schmitt | 243.3 | 5 |
| NZL | Lauren Boyle | 243.6 | 5 |
| DEN | Lotte Friis | 244.2 | 5 |
| ESP | Melanie Costa | 246.8 | 5 |
| AUS | Bronte Barratt | 248.0 | 5 |
| HUN | Boglarka Kapas | 250.0 | 5 |
| RUS | Elena Sokolova | 252.2 | 5 |

**43. Tees** Does it matter what kind of tee a golfer places the ball on? The company that manufactures "Stinger" tees claims that the thinner shaft and smaller head will lessen drag, reducing spin and allowing the ball to travel farther. In August 2003, Golf Laboratories, Inc., compared the distance traveled by golf balls hit off regular wooden tees to those hit off Stinger tees. All the balls were struck by the same golf club using a robotic device set to swing the club head at approximately 95 miles per hour. Summary statistics from the test are shown in the table. Assume that 6 balls were hit off each tee and that the data were suitable for inference.

|  |  | Total Distance (yards) | Ball Velocity (mph) | Club Velocity (mph) |
|---|---|---|---|---|
| Regular Tee | Avg. | 227.17 | 127.00 | 96.17 |
|  | SD | 2.14 | 0.89 | 0.41 |
| Stinger Tee | Avg. | 241.00 | 128.83 | 96.17 |
|  | SD | 2.76 | 0.41 | 0.52 |

Is there evidence that balls hit off the Stinger tees would have a higher initial velocity?

**44. Golf again** Given the test results on golf tees described in Exercise 43, is there evidence that balls hit off Stinger tees would travel farther? Again, assume that 6 balls were hit off each tee and that the data were suitable for inference.

**45. Arrow Tips** In Chapter 22, Exercise 53, researcher Nicole Waguespack studied the difference between wood-tipped and stoned-tipped arrows. Many cultures used both types of arrow tips, including the Apache in North America and the Tiwi in Australia. The researchers set up a compound bow with 60 lb of force. They shot arrows of both types into a hide-covered ballistics gel and measured the penetration depths in mm. (Nicole Waguespack et al., Antiquity, 2009)

Here are the penetration depths for seven shots from each type of arrow.

Wood-tipped: 216 211 192 208 203 210 203
Stone-tipped: 240 208 213 225 232 214 240

Construct and interpret a 95% confidence interval for the mean difference in penetration depths for the two arrow types. Be sure to check the conditions for inference.

**46. Last Look at Arrow Tips** The researchers in Exercise 45 also compared the accuracy of the two types of arrowheads. The bow was aimed at a target and the distance was measured from the center in cm. The distances from the center for the six shots using each of the two types of arrowheads are given below.

Wood-tipped: 9.3 16.7 7.1 14.0 1.0 1.2
Stone-tipped: 4.9 21.1 7.0 1.8 5.4 8.6

Construct and interpret a 95% confidence interval for the mean difference in accuracies for the two arrow types.

**47. Music and memory** Is it a good idea to listen to music when studying for a big test? In a study conducted by some Statistics students, 62 people were randomly assigned to listen to rap music, music by Mozart, or no music while attempting to memorize objects pictured on a page. They were then asked to list all the objects they could remember. Here are summary statistics:

|  | Rap | Mozart | No Music |
|---|---|---|---|
| Count | 29 | 20 | 13 |
| Mean | 10.72 | 10.00 | 12.77 |
| SD | 3.99 | 3.19 | 4.73 |

a) Does it appear that it is better to study while listening to Mozart than to rap music? Test an appropriate hypothesis and state your conclusion.
b) Create a 90% confidence interval for the mean difference in memory score between students who study to Mozart and those who listen to no music at all. Interpret your interval.

**48. Rap** Using the results of the experiment described in Exercise 47, does it matter whether one listens to rap music while studying, or is it better to study without music at all?

a) Test an appropriate hypothesis and state your conclusion.
b) If you concluded there is a difference, estimate the size of that difference with a confidence interval and explain what your interval means.

**49. Cuckoos** Cuckoos lay their eggs in the nests of other (host) birds. The eggs are then adopted and hatched by the host birds. But the potential host birds lay eggs of different sizes. Does the cuckoo change the size of her eggs for different foster species? The numbers in the table are lengths (in mm) of cuckoo eggs found in nests of three different species of other birds. The data are drawn from the work of O. M. Latter in 1902 and were used in a fundamental textbook on statistical quality control by L. H. C. Tippett (1902–1985), one of the pioneers in that field.

| Cuckoo Egg Length (mm) | | |
|---|---|---|
| Foster Parent Species | | |
| Sparrow | Robin | Wagtail |
| 20.85 | 21.05 | 21.05 |
| 21.65 | 21.85 | 21.85 |
| 22.05 | 22.05 | 21.85 |
| 22.85 | 22.05 | 21.85 |
| 23.05 | 22.05 | 22.05 |
| 23.05 | 22.25 | 22.45 |
| 23.05 | 22.45 | 22.65 |
| 23.05 | 22.45 | 23.05 |
| 23.45 | 22.65 | 23.05 |
| 23.85 | 23.05 | 23.25 |
| 23.85 | 23.05 | 23.45 |
| 23.85 | 23.05 | 24.05 |
| 24.05 | 23.05 | 24.05 |
| 25.05 | 23.05 | 24.05 |
|  | 23.25 | 24.85 |
|  | 23.85 |  |

Investigate the question of whether the mean length of cuckoo eggs is the same for different species, and state your conclusion.

**50. Penguins revisited** In Exercise 18, data was provided on culmen (bill) lengths for two species of penguins. A random sample of culmen lengths of a third species, Chinstrap penguins, is provided here. Determine whether there is evidence of a difference in the mean length of the culmen in Chinstrap penguins and in Gentoo penguins.

| Culmen Length (mm) | |
|---|---|
| 42.5 | 50.3 |
| 45.4 | 50.6 |
| 45.7 | 50.7 |
| 46.4 | 51.4 |
| 47.0 | 51.5 |
| 48.5 | 52.0 |
| 49.0 | 52.0 |
| 49.2 | 52.0 |
| 49.8 | 52.8 |
| 50.0 | 53.5 |
| 50.2 | |

**T 51. Attendance 2019 revisited** We have seen data on ballpark attendance in Chapters 6, and 7. Now we find that National League teams drew in, on average, nearly 60,000 more fans per season than American League teams. That translates to over $1,000,000 a year. To see whether that difference is statistically significant:

a) Make a boxplot of the *Home Attendance by League.*
b) Look at this histogram of 1000 differences in means of Home Attendance by League obtained by shuffling the League label among the 30 teams. What does it say about whether 60,000 is a statistically significant difference?

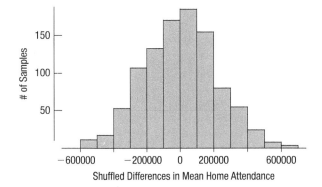

**52. Hard water revisited** In Exercise 24, we saw that the towns in the region south of Derby seemed to have fewer deaths than the towns in the north. To see whether that difference is statistically significant, look at this histogram of 1000 differences in mean *Mortality* by *Region*. What does it say about whether a difference of 242.7 is statistically significant?

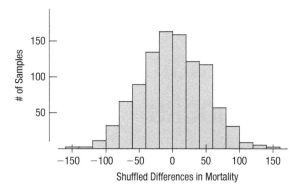

**T 53. Running heats London randomized** In Exercise 41 we looked at the times in two different heats for the 400-m women's run from the 2012 Olympics. You were asked to see if the data provide evidence that the mean time to finish is different for randomized heats. Answer the same question using a randomization test to test the difference. Do your conclusions change?

**T 54. Swimming heats London randomized** In Exercise 42 we looked at the times in two different heats for the women's 400-m freestyle swimming event. You were asked to see if the data provide evidence that the mean time to finish is different for seeded heats. Answer the same question using a randomization test to test the difference. Do your conclusions change?

## JUST CHECKING

### Answers

1. Randomization should balance unknown sources of variability in the two groups of patients and helps us believe the two groups are independent.

2. We can be 95% confident that after 4 weeks endoscopic surgery patients will have a mean pinch strength between 0.04 kg and 2.96 kg higher than open-incision patients.

3. The lower bound of this interval is close to 0, so the difference may not be great enough that patients could actually notice the difference. We may want to consider other issues such as cost or risk in making a recommendation about the two surgical procedures.

4. Without data, we can't check the Nearly Normal Condition.

5. $H_0$: Mean pinch strength is the same after both surgeries. ($\mu_E - \mu_O = 0$).

   $H_A$: Mean pinch strength is different after the two surgeries. ($\mu_E - \mu_O \neq 0$).

6. With a P-value this low, we reject the null hypothesis. There is evidence that mean pinch strength differs after 4 weeks in patients who undergo endoscopic surgery vs. patients who have open-incision surgery. Results suggest that the endoscopic surgery patients may be stronger, on average.

7. If some patients contributed two hands to the study, then the groups may not be internally independent. It is reasonable to assume that two hands from the same patient might respond in similar ways to similar treatments.

# 24

# Paired Samples and Blocks

| WHO | Children in an experiment |
|---|---|
| WHAT | Speed of inverting cylinders |
| UNITS | Cylinders/second |
| WHEN | 2013 |
| WHERE | Israel |
| WHY | To study development of dexterity |

Only about 1% of people are truly ambidextrous. The vast majority of us have a dominant hand. (If you're not sure which is your dominant hand, think about which one you use to hold a soup spoon.) But how unbalanced is your dexterity? One common test, the Functional Dexterity Test (FDT), measures the time it takes to place a series of cylinders into a set of holes. In one experiment researchers in Israel measured the time it took 93 children aged 5 to 18 to complete the task using each hand. (We have converted their results, by taking the reciprocal of the time to obtain the children's *Speed* in cylinders per second.)

Here are the data for the first seven of 93 children tested:

|  | Speed (Cylinders/sec) | | |
|---|---|---|---|
| Age (mo) | Dominant Hand | Nondominant Hand | Gender |
| 117 | 0.353 | 0.216 | male |
| 101 | 0.257 | 0.343 | male |
| 135 | 0.537 | 0.497 | male |
| 119 | 0.444 | 0.496 | male |
| 124 | 0.483 | 0.388 | female |
| 127 | 0.524 | 0.422 | female |
| 101 | 0.455 | 0.381 | male |
| ⋮ | ⋮ | ⋮ | ⋮ |

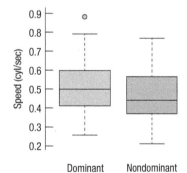

Figure 24.1
Using boxplots to compare dominant and nondominant hands shows little because it ignores the fact that the measurements are in pairs.

How much faster is the dominant hand, on average? We have two samples, so we can do a *t*-test to compare the two means. Or can we? The *t*-test assumes that the groups are independent. We have both hands measured for each child. Some children are naturally more dextrous than others, and 18-year-olds are generally much faster than 5-year-olds. In fact, the correlation between the measurements on the two hands is 0.49. These measurements are clearly *not* independent.

652

## Paired Data

Observations such as these are called **paired data**. We have the speed for each hand for every child. We want to compare the mean speeds for dominant and nondominant hands across all the children, so what we're interested in is the *difference* in speeds for each child.

Paired data arise in a number of ways. Perhaps the most common way is to compare subjects with themselves before and after a treatment. When pairs arise from an experiment, the pairing is a type of *blocking*. When they arise from an observational study, it is a form of *matching*.

### FOR EXAMPLE
#### Identifying Paired Data

Do flexible schedules reduce the demand for resources? The Lake County, Illinois, Health Department experimented with a flexible four-day workweek. For a year, the department recorded the mileage driven by 11 field workers on an ordinary five-day workweek. Then it changed to a flexible four-day workweek and recorded mileage for another year.[1] The data are shown.

| Name | 5-Day Mileage | 4-Day Mileage |
|---|---|---|
| Jeff | 2798 | 2914 |
| Betty | 7724 | 6112 |
| Roger | 7505 | 6177 |
| Tom | 838 | 1102 |
| Aimee | 4592 | 3281 |
| Greg | 8107 | 4997 |
| Larry G. | 1228 | 1695 |
| Tad | 8718 | 6606 |
| Larry M. | 1097 | 1063 |
| Leslie | 8089 | 6392 |
| Lee | 3807 | 3362 |

QUESTION: Why are these data paired?

ANSWER: The mileage data are paired because each driver's mileage is measured before and after the change in schedule. I'd expect drivers who drove more than others before the schedule change to continue to drive more afterward, so the two sets of mileages can't be considered independent.

> **Paired or Independent?**
>
> It matters. A lot. *You cannot use paired t-methods when the groups are independent, nor 2-sample t-methods when the data are paired.* And the data can't tell you what to do; it's all about the study design. Think about how the data were collected. Is there some connection between the two variables that links each value of one to a value of the other? Would it be okay to rearrange the order of the values in just one of the datasets without rearranging the other dataset, too?

Pairing isn't a problem; it's an opportunity. If you know the data are paired, you can take advantage of that fact—in fact, you *must* take advantage of it. The data in the previous "For Example" are paired because each row corresponds to one person measured at two different times. Remember: The two-sample *t-test* requires the two samples to be independent. Paired data aren't. There is no test to determine whether your data are paired. You must determine that from understanding how they were collected and what they mean (check the W's).

---
[1] Charles S. Catlin, "Four-Day Work Week Improves Environment," *Journal of Environmental Health*, Denver, 59:7.

Once we recognize that the hand dexterity data are matched pairs, it makes sense to consider the difference in speeds. So we look at the *pairwise* differences:

| Child | Speed (cylinders/sec) Dominant | Nondominant | Difference |
|---|---|---|---|
| 1 | 0.353 | 0.216 | 0.137 |
| 2 | 0.257 | 0.343 | −0.086 |
| 3 | 0.537 | 0.497 | 0.039 |
| 4 | 0.444 | 0.496 | −0.052 |
| 5 | 0.483 | 0.388 | 0.095 |
| 6 | 0.524 | 0.422 | 0.102 |
| 7 | 0.455 | 0.381 | 0.074 |
| ⋮ | ⋮ | ⋮ | ⋮ |

Because it is the *differences* we care about, we'll treat them as if *they* were the data, ignoring the original two columns. Now that we have only one column of values to consider, we can use a one-sample *t*-test. Mechanically, a **paired *t*-test** is just a one-sample *t*-test for the means of these pairwise differences. The sample size is the number of pairs. (Here there are 93.)

The mechanics of the paired *t*-test are not new. They're the same as a one-sample *t*-test using the differences as the data. You've already seen the *Show*!

# Assumptions and Conditions

## Paired Data Condition

**Paired Data Condition:** The data must be paired. You can't just decide to pair data when in fact the samples are independent. When you have two groups with the same number of observations, it may be tempting to match them up. Don't, unless you are prepared to justify your claim that there is a reason to pair them.

On the other hand, be sure to recognize paired data when you have them. Remember, two-sample *t*-methods aren't valid without independent groups, and paired groups aren't independent. Although this is a strictly required assumption, it is one that can be easy to check if you understand how the data were collected.

## Independence Assumption

**Independence Assumption:** If the data are paired, the *groups* are not independent. For these methods, it's the *differences* that must be independent of each other. There's no reason to believe that the difference in dexterity speed of one child could affect the difference in the speed for another child.

**Randomization Condition:** Randomness can arise in many ways. The pairs may be a random sample. In an experiment, the order of the two treatments may be randomly assigned, or the treatments may be randomly assigned to one member of each pair. In a before-and-after study, we may believe that the observed differences are a representative sample from a population of interest. In the hand dexterity example, the children tested were a representative (and possibly random) sample of children, and which hand speed (dominant or nondominant) was measured first was randomized for each child.

**10% Condition:** The 93 children in the study were certainly less than 10% of similarly aged children in Israel.

> **10% OF WHAT?**
> A fringe benefit of checking the 10% Condition is that it forces us to think about what population we're hoping to make inferences about.

## Normal Population Assumption

We need to assume that the population of *differences* follows a Normal model. We don't need to check the individual groups.

**Nearly Normal Condition:** This condition can be checked with a histogram or Normal probability plot of the *differences*—but not of the individual groups. As with the one-sample *t*-methods, this assumption matters less the more pairs we have to consider. You may be pleasantly surprised when you check this condition. Even if your original measurements are skewed or bimodal, the *differences* may be nearly Normal. After all, the individual who was way out in the tail on an initial measurement is likely to still be out there on the second one, giving a perfectly ordinary difference.

### FOR EXAMPLE
### Checking Assumptions and Conditions

RECAP: Field workers for a health department compared driving mileage on a five-day work schedule with mileage on a new four-day schedule. To see if the new schedule changed the amount of driving they did, we'll look at paired differences in mileages before and after.

| Name | 5-Day Mileage | 4-Day Mileage | Difference |
|---|---|---|---|
| Jeff | 2798 | 2914 | −116 |
| Betty | 7724 | 6112 | 1612 |
| Roger | 7505 | 6177 | 1328 |
| Tom | 838 | 1102 | −264 |
| Aimee | 4592 | 3281 | 1311 |
| Greg | 8107 | 4997 | 3110 |
| Larry G. | 1228 | 1695 | −467 |
| Tad | 8718 | 6606 | 2112 |
| Larry M. | 1097 | 1063 | 34 |
| Leslie | 8089 | 6392 | 1697 |
| Lee | 3807 | 3362 | 445 |

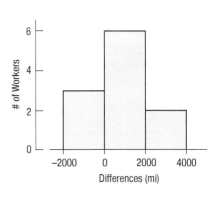

QUESTION: Is it okay to use these data to test whether the new schedule changed the amount of driving?

ANSWER:

✓ **Paired Data Condition:** The data are paired because each value is the mileage driven by the same person before and after a change in work schedule.

✓ **Independence Assumption:** The driving behavior of any individual worker is independent of the others, so the differences are mutually independent.

✓ **Randomization Condition:** The mileages are the sums of many individual trips, each of which experienced random events that arose while driving. Repeating the experiment in two new years would give randomly different values.

✓ **10% Condition:** These 11 field workers probably represent fewer than 10% of all workers who do similar field work.

✓ **Nearly Normal Condition:** The histogram of the mileage differences is unimodal and symmetric.

Because the assumptions and conditions are satisfied, it's okay to use paired *t*-methods for these data.

The steps in testing a hypothesis for paired differences are very much like the steps for a one-sample *t*-test for a mean.

> ### The Paired *t*-Test
> When the conditions are met, we are ready to test whether the mean of paired differences is significantly different from zero. We test the hypothesis
> $$H_0: \mu_d = \Delta_0,$$
> where the *d*'s are the pairwise differences and $\Delta_0$ is almost always 0.
> We use the statistic
> $$t_{n-1} = \frac{\bar{d} - \Delta_0}{SE(\bar{d})},$$
> where $\bar{d}$ is the mean of the pairwise differences, *n* is the number of *pairs*, and
> $$SE(\bar{d}) = \frac{s_d}{\sqrt{n}}.$$
>
> $SE(\bar{d})$ is the ordinary standard error for the mean, applied to the differences.
> When the conditions are met and the null hypothesis is true, we can model the sampling distribution of this statistic with a Student's *t*-model with $n - 1$ degrees of freedom, and use that model to obtain a P-value.

## STEP-BY-STEP EXAMPLE

### A Paired *t*-Test

Speed-skating races are run in pairs. Two skaters start at the same time, one on the inner lane and one on the outer lane. Halfway through the race, they cross over, switching lanes so that each will skate the same distance in each lane. Even though this seems fair, it's not obviously so. Perhaps it's harder to skate the tighter curves (inner lane) after one is tired, for example. At the 2018 Olympics at Pyeongchang, South Korea, there were 16 pairs of speed-skaters in the men's 1500 meter race.[2] Their times (and the differences for each pair—we're going to need them!) are shown in the table below.

| Inner Lane Skater | Time (sec) | Outer Lane Skater | Time (sec) | Difference (Inner − Outer) |
|---|---|---|---|---|
| Tai William | 110.63 | Alexis Contin | 107.33 | 3.30 |
| Denis Kuzin | 109.14 | Shota Nakamura | 107.38 | 1.76 |
| Patrick Roest (Silver medalist) | 104.86 | Reyon Kay | 107.81 | −2.95 |
| Mathias Vosté | 107.34 | Joo Hyong-jun | 106.65 | 0.69 |
| Konrád Nagy | 109.01 | Xiakaini Aerchenghazi | 110.16 | −1.15 |
| Fyodor Mezentsev | 108.23 | Sergey Trofimov | 106.69 | 1.54 |
| Ben Donnelly | 109.68 | Shane Williamson | 106.21 | 3.47 |
| Peter Michael | 106.39 | Livio Wenger | 107.76 | −1.37 |
| Shani Davis | 106.74 | Bart Swings | 105.49 | 1.25 |
| Brian Hansen | 106.44 | Zbigniew Bródka | 106.31 | 0.13 |
| Jan Szymański | 106.48 | Andrea Giovannini | 107.82 | −1.34 |
| Konrad Niedźwiedzki | 107.07 | Denny Morrison | 106.36 | 0.71 |
| Takuro Oda | 105.44 | Kjeld Nuis (Gold medalist) | 104.01 | 1.43 |
| Kim Min-seok (Bronze medalist) | 104.93 | Haralds Silovs | 105.25 | −0.32 |
| Sindre Henriksen | 105.64 | Vincent De Haître | 107.32 | −1.68 |
| Sverre Lunde Pedersen | 106.12 | Joey Mantia | 105.86 | 0.26 |

---

[2] We're only including pairs in which both skaters completed the race.

QUESTION: Is there a difference in mean speeds between skaters who skate the inner lane first and those who skate the outer lane first?

## THINK

**PLAN** State what we want to know.

Identify the *parameter* we wish to estimate. Here our parameter is the mean difference in race times.

Identify the variables and check the W's.

**HYPOTHESES** State the null and alternative hypotheses.

Although we can speculate why starting in the inner or outer lane may disadvantage a skater, there's no clear reason to hold either particular lane suspect, so we test a two-sided alternative.

**MODEL** Think about the assumptions and check the conditions.

State why you think the data are paired. Simply having the same number of individuals in each group and displaying them in side-by-side columns doesn't make them paired.

Think about what we hope to learn and where the randomization comes from. Here, the randomization comes from the racer pairings and lane assignments.

Make a picture—just one. Don't plot separate distributions of the two groups—that entirely misses the pairing. For paired data, it's the distribution of the *differences* that we care about. Treat those paired differences as you would a single variable, and check the Nearly Normal Condition with a histogram (or a Normal probability plot).

I want to know whether starting on the inner lane versus starting on the outer lane made a difference in skaters' times at the 2018 Olympics. I have data for 16 pairs of racers at the men's 1500-m race.

$H_0$: Neither lane offered an advantage:
$$\mu_d = 0.$$
$H_A$: The mean difference is different from zero:
$$\mu_d \neq 0.$$

✓ **Paired Data Condition:** The data are paired because racers compete in pairs.

✓ **Independence Assumption:** Each race is independent of the others, so the differences are mutually independent.

✓ **Randomization Condition:** Skaters are assigned to lanes at random.

✓ **Nearly Normal Condition:** The histogram of the differences is unimodal and roughly symmetric:

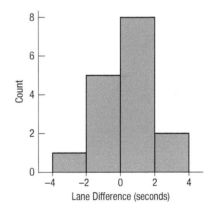

Specify the sampling distribution model.

Choose the method.

(Because this was essentially an experiment—a "natural experiment"—the 10% Condition doesn't apply.)

The conditions are met, so I'll use a Student's *t*-model with $(n - 1) = 15$ degrees of freedom, and perform a **paired t-test**.

## SHOW

**MECHANICS**

*n* is the number of *pairs*—in this case, the number of races.

$\bar{d}$ is the mean difference (*Inner* − *Outer*).

$s_d$ is the standard deviation of the differences.

The data give

$$n = 16 \text{ pairs}$$
$$\bar{d} = 0.358 \text{ seconds}$$
$$s_d = 1.78 \text{ seconds.}$$

Find the standard error and the *t*-score of the observed mean difference. There is nothing new in the mechanics of the paired *t*-methods. These are the mechanics of the *t*-test for a mean applied to the differences.

I estimate the standard deviation of $\bar{d}$ with its standard error:

$$SE(\bar{d}) = \frac{s_d}{\sqrt{n}} = \frac{1.78}{\sqrt{16}} = 0.445$$

So $t_{15} = \dfrac{\bar{d} - 0}{SE(\bar{d})} = \dfrac{0.358}{0.445} = 0.804.$

Make a picture. Sketch a *t*-model centered at the hypothesized mean of 0. Because this is a two-tailed test, shade both the region to the right of the calculated *t*-value and the corresponding region in the lower tail.

Find the P-value, using technology.

REALITY CHECK  The mean difference is 0.359 seconds. That may not seem like much, but the times of the silver and bronze medalists differed by less than this. The standard error is about this big, so a *t*-value less than 1.0 isn't surprising—nor is a large P-value.

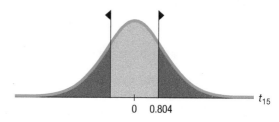

P-value = $2P(t_{15} > 0.804) = 0.43$

**TELL**  CONCLUSION Link the P-value to your decision about $H_0$, and state your conclusion in context.

The P-value is large. Events that can happen more than a third of the time by chance alone are not remarkable. So, even though there is an observed difference between the lanes, I can't conclude that it's due to anything more than just random chance. There's insufficient evidence to declare any lack of fairness due to which lane a skater starts on.

### FOR EXAMPLE
#### Doing a Paired *t*-Test

**RECAP:** We want to test whether a change from a five-day workweek to a four-day workweek could change the amount driven by field workers of a health department. We've already confirmed that the assumptions and conditions for a paired *t*-test are met.

**QUESTION:** Is there evidence that a four-day workweek would change how many miles workers drive?

**ANSWER:** $H_0$: The change in the health department workers' schedules didn't change the mean mileage driven; the mean difference is zero:

$$\mu_d = 0.$$

$H_A$: The mean difference is different from zero:

$$\mu_d \neq 0.$$

The conditions are met, so I'll use a Student's *t*-model with $(n - 1) = 10$ degrees of freedom and perform a paired *t*-test.

The data give

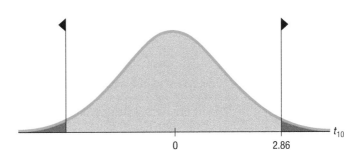

$n = 11$ pairs
$\bar{d} = 982$ miles
$s_d = 1139.6$ miles.

$$SE(\bar{d}) = \frac{s_d}{\sqrt{n}} = \frac{1139.6}{\sqrt{11}} = 343.6$$

So $t_{10} = \dfrac{\bar{d} - 0}{SE(\bar{d})} = \dfrac{982.0}{343.6} = 2.86$

P-value $= 2P(t_{10} > 2.86) = 0.017$

The P-value is small, so I reject the null hypothesis and conclude that the change in workweek did lead to a change in average driving mileage. It appears that changing the work schedule may reduce the mileage driven by workers.

**Note:** We should propose a course of action, but it's hard to tell from the hypothesis test whether the reduction matters. Is the difference in mileage important in the sense of reducing air pollution or costs, or is it merely statistically significant? To help make that decision, we should look at a confidence interval. If the difference in mileage proves to be large in a practical sense, then we might recommend a change in schedule for the rest of the department.

## TI TIPS

### Testing a Hypothesis With Paired Data

Because the inference procedures for matched data are essentially just the one-sample $t$ procedures, you already know what to do . . . once you have the list of paired differences, that is. That list is not hard to create.

**TEST A HYPOTHESIS ABOUT THE MEAN OF PAIRED DIFFERENCES**

Think: Are the samples independent or paired. Independent? Go back to the last chapter! Paired? Read on.

♦ Enter the driving data from page 655 into two lists, say *5-Day Mileage* in L1, *4-Day Mileage* in L2.

♦ Create a list of the differences. We want to take each value in L1, subtract the corresponding value in L2, and store the paired difference in L3. The command is L1−L2 → L3. (The arrow is the STO button.) Now take a look at L3. See—it worked!

♦ Make a histogram of the differences, L3, to check the Nearly Normal Condition. Notice that we do not look at the histograms of the *5-day mileage* or the *4-day mileage*. Those are not the data that we care about now that we are using a paired procedure. Note also that the calculator's first histogram is not close to Normal. More work to do . . .

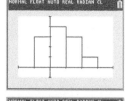

- As you have seen before, small samples often produce ragged histograms, and these may look very different after a change in bar width. We set our WINDOW to Xmin = −3000, Xmax = 6000, and Xscl = 1500. The new histogram looks okay.

- Under STAT TESTS simply use T-Test, as you've done before for hypothesis tests about a mean.
- Specify that the hypothesized difference is 0, you're using the Data in L3, and it's a two-tailed test.
- Calculate.

The small P-value shows strong evidence that on average the change in the workweek reduces the number of miles workers drive.

# Confidence Intervals for Matched Pairs

In many countries, the average age of women is higher than that of men. After all, women tend to live longer. But if we look at *married opposite-sex couples*, husbands tend to be slightly older than wives. How much older, on average, are husbands? We have data from a random sample of 200 British couples, the first 7 of which are shown below. Only 170 couples provided ages for both husband and wife, so we can work only with that many pairs. Let's form a confidence interval for the mean difference of husband's and wife's ages based on these 170 couples. Here are the first 7 pairs:

| WHO | 170 randomly sampled couples |
|---|---|
| WHAT | Ages |
| UNITS | Years |
| WHEN | Recently |
| WHERE | Britain |

| Wife's Age | Husband's Age | Difference (Husband − Wife) |
|---|---|---|
| 43 | 49 | 6 |
| 28 | 25 | −3 |
| 30 | 40 | 10 |
| 57 | 52 | −5 |
| 52 | 58 | 6 |
| 27 | 32 | 5 |
| 52 | 43 | −9 |
| ⋮ | ⋮ | ⋮ |

Clearly, these data are paired. The survey selected *couples* at random, not individuals. We're interested in the mean age difference within couples. How would we construct a confidence interval for the true mean difference in ages?

> ### Paired *t*-Interval
> When the conditions are met, we are ready to find the confidence interval for the mean of the paired differences. The confidence interval is
> $$\bar{d} \pm t^*_{n-1} \times SE(\bar{d}),$$
> where the standard error of the mean difference is $SE(\bar{d}) = \dfrac{s_d}{\sqrt{n}}.$
>
> The critical value $t^*$ from the Student's *t*-model depends on the particular confidence level, *C*, that you specify and on the degrees of freedom, $n - 1$, which is based on the number of pairs, *n*.

Making confidence intervals for matched pairs follows exactly the steps for a one-sample *t*-interval.

## STEP-BY-STEP EXAMPLE

### A Paired *t*-Interval

Using the Functional Dexterity Test data discussed at the beginning of the chapter, we can try to assess how much more dexterous a typical person is with the dominant hand.

QUESTION: How big a difference is there, on average, between the dexterity of dominant and nondominant hands?

**THINK**

**PLAN** State what we want to know.

Identify the variables and check the W's.

Identify the parameter you wish to estimate. For a paired analysis, the parameter of interest is the mean of the differences. The population of interest is the population of differences.

**MODEL** Think about the assumptions and check the conditions.

Make a picture. We focus on the differences, so a histogram or Normal probability plot is best here.

**REALITY CHECK** The histogram shows that dominant hands are generally faster at the FDT task by about 0.05 cylinders/second.

State the sampling distribution model.

Choose your method.

I want to estimate the mean difference in dexterity between dominant and nondominant hands.

✓ **Paired Data Condition**: The data are paired because they are measured on hands for the same individuals.

✓ **Independence Assumption**: The data are from an experiment. Dexterity of one individual should be independent of dexterity for any other.

✓ **Nearly Normal Condition**: The histogram of the dominant − nondominant differences is unimodal and symmetric:

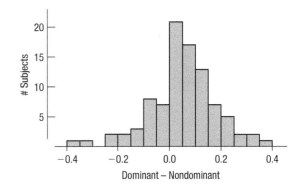

The conditions are met, so I can use a Student's *t*-model with $(n - 1) = 92$ degrees of freedom and find a **paired *t*-interval**.

| | |
|---|---|
| **SHOW** **MECHANICS** $n$ is the number of *pairs*—here, the number of children tested. | $n = 93$ participants<br>$\bar{d} = 0.054$ cylinders/sec<br>$s_d = 0.1318$ cylinders/sec |
| $\bar{d}$ is the mean difference.<br>$s_d$ is the standard deviation of the differences.<br>Be sure to include the units along with the statistics.<br>The critical value we need to make a 95% interval comes from Table T, a computer program such as the interactive table at astools.datadesk.com, or a calculator.<br>REALITY CHECK This result makes sense. Our everyday experience is that we are generally more dexterous with our dominant hand. | I estimate the standard error of $\bar{d}$ as<br>$$SE(\bar{d}) = \frac{s_d}{\sqrt{n}} = \frac{0.1318}{\sqrt{93}} = 0.0137$$<br>The $df$ for the $t$-model is $n - 1 = 92$.<br>The 95% critical value for $t_{92}$ (from technology) is 1.986.<br>The margin of error is<br>$ME = t^*_{92} \times SE(\bar{d}) = 1.986(0.0137) = 0.0272$. So the 95% confidence interval is $0.054 \pm 0.0272$ cylinders/sec, or an interval of (0.027, 0.081) cylinders/sec. |
| **TELL** **CONCLUSION** Interpret the confidence interval in context. | I am 95% confident that the mean difference in dexterity measured on the FDT between dominant and nondominant hands is between 0.027 and 0.081 cylinders/sec. |

## TI TIPS

### Creating a Confidence Interval

Now let's get the TI to create a confidence interval for the mean of paired differences.

We'll demonstrate by using the statistics about the ages of the British married couples. (If we had all the data, we could enter that, of course. All 170 couples? Um, no thanks.) The husbands in the sample were an average of 2.2 years older than their wives, with a standard deviation of 4.1 years. We've already seen that the data are paired and that a histogram of the differences satisfies the Nearly Normal Condition. (With a sample this large, we could proceed with inference even if we didn't have the actual data and were unable to make the histogram.)

- Once again, we treat the paired differences just like data from one sample. A confidence interval for the mean difference, then, like that for a mean, uses the STAT TESTS one-sample procedure TInterval.
- Specify Inpt:Stats, and enter the statistics for the paired differences.
- Calculate.

Done. Finding the interval was the easy part. Now it's time for you to *Tell* what it means. Don't forget to talk about a mean and married couples in Britain.

# Effect Size

When we examined the speed-skating times, we failed to reject the null hypothesis, so we couldn't be certain whether there really was a difference between the lanes. Maybe there wasn't any difference, or maybe whatever difference there might have been was just too small for us to have detected it with our somewhat small, somewhat noisy dataset.[3] Should skating enthusiasts be concerned?

---

[3] Remember the things that affect power (Chapter 20)? Three of them come into play here. The effect size is the advantage, in seconds, arising from starting in one particular lane. If it's subtle (small), it will be hard to detect. It will also be hard to detect if the sample size is small or if the data are noisy (p. 551).

We can't tell from the hypothesis test, but using the same summary statistics, we can find that the corresponding 95% confidence interval for the mean difference is $(-0.59 < \mu_d < 1.31)$ seconds.

A confidence interval is a good way to get a sense for the size of the effect we're trying to understand. That gives us a plausible range of values for the true mean difference in lane times. If differences of 1.31 seconds were too small to matter in 1500-m Olympic speed skating, we'd be pretty sure there was no need for concern.

But in fact, *every* gap between a skater's time and the next-faster one was less than 1.31 seconds. Most were less than 0.2 seconds.

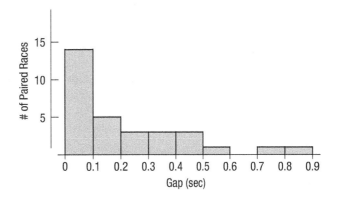

So even though we were unable to discern a real difference, the confidence interval shows that the effects we're considering may be big enough to be important. We may want to continue this investigation by checking out other races on this ice and being alert for possible differences at other venues.

### FOR EXAMPLE
### Looking at Effect Size with a Paired *t* Confidence Interval

**RECAP:** We know that, on average, the switch from a five-day workweek to a four-day workweek reduced the amount driven by field workers in that Illinois health department. However, finding that there is a significant difference doesn't necessarily mean that difference is meaningful or worthwhile. To assess the size of the effect, we need a confidence interval. We already know the assumptions and conditions are met.

**QUESTION:** By how much, on average, might a change in workweek schedule reduce the amount driven by workers?

**ANSWER:**
$$\bar{d} = 982 \text{ mi} \quad SE(\bar{d}) = 343.6 \quad t^*_{10} = 2.228 \text{ (for 95\%)}$$
$$ME = t^*_{10} \times SE(\bar{d}) = 2.228(343.6) = 765.54$$

So the 95% confidence interval for $\mu_d$ is $982 \pm 765.54$ or $(216.46, 1747.54)$ fewer miles.

With 95% confidence, I estimate that by switching to a four-day workweek employees would drive an average of between 216 and 1748 fewer miles per year. This could save money spent on gas and reduce air pollution and carbon emissions.

## Blocking

Because the sample of British husbands and wives includes both older and younger couples, there's a lot of variation in the ages of the men and in the ages of the women. In fact, that variation is so great that boxplots of the two groups' ages look about the same. But that would be the wrong plot. It's the age *difference* we care about. Pairing isolates the extra variation and allows us to focus on the individual differences. In Chapter 12, we saw how to design an experiment with blocking to isolate the variability between identifiable groups of subjects. Blocking makes it easier to see the variability among treatment groups that is attributable to their responses to the treatment. A paired design is an example of blocking.

When we pair, we have roughly half the degrees of freedom of a two-sample test. You may see discussions that suggest that in "choosing" a paired analysis we "give up" these degrees of freedom. This isn't really true, though. If the data are paired, then there never were additional degrees of freedom, and we have no "choice." The fact of the pairing determines how many degrees of freedom are available.

Matching pairs generally removes so much extra variation that it more than compensates for having only half the degrees of freedom, so it is usually a good choice when you design a study. Of course, creating pairs based on something that's not related to the response variable (say, by matching on the first letter of the last name of subjects) would cost degrees of freedom without the benefit of reducing the variance. When you design a study or experiment, you should consider using a paired design if you can pair subjects or experimental units based on something that is likely to have an effect on the response variable.

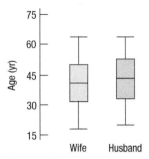

Figure 24.2
This display is worthless. It does no good to compare all the wives as a group with all the husbands. We care about the paired differences.

## JUST CHECKING

Think about each of the 5 situations described here.

- Would you use a two-sample $t$ or paired $t$-method (or neither)? Why? or Why not?
- Would you perform a hypothesis test or find a confidence interval?

1. Random samples of 50 men and 50 women are asked to imagine buying a birthday present for their best friend. We want to estimate the difference in how much they are willing to spend.
2. Parents of twins were surveyed and asked how often in the past month strangers had asked whether the twins were identical.
3. Are parents equally strict with boys and girls? Among those families in a random sample that included one brother and one sister, researchers asked them both to rate how strict their parents were.
4. Forty-eight overweight subjects are randomly assigned to either aerobic or stretching exercise programs. They are weighed at the beginning and at the end of the experiment to see how much weight they lost.
   a) We want to estimate the mean amount of weight lost by those doing aerobic exercise.
   b) We want to know which program is more effective at reducing weight.
5. Opposite-sex couples at a dance club were separated and each person was asked to rate the band. Do men or women like this band more?

## WHAT CAN GO WRONG?

- **Don't use a two-sample $t$-test when you have paired data.** See the What Can Go Wrong? discussion in Chapter 23.
- **Don't use a paired $t$-method when the samples aren't paired.** Just because two groups have the same number of observations doesn't mean they can be paired, even if they are shown side by side in a table. We might have 25 men and 25 women in our study, but they could be completely independent of one another. If they were siblings or spouses, we might consider them paired. Remember that you cannot *choose* which method to use based on your preferences. If the data are from two independent samples, use two-sample $t$-methods. If the data are from an experiment in which observations were paired, you must use a paired method. If the data are from an observational study, you must be able to defend your decision to use matched pairs or independent groups.

- **Don't forget to look for outliers.** The outliers we care about now are in the differences. A subject who is extraordinary both before and after a treatment may still have a perfectly typical difference. But one outlying difference can completely distort your conclusions. Be sure to plot the differences (even if you also plot the data).
- **Don't look for the difference between the means of paired groups with side-by-side boxplots or histograms.** The point of the paired analysis is to remove extra variation. Separate displays of each group still contain that variation. Comparing them is likely to be misleading. Always graph the differences.

## WHAT HAVE WE LEARNED?

When we looked at study designs in Chapters 11 and 12 we saw that pairing can be a very effective strategy. Because pairing helps control variability between individual subjects, paired methods are usually more powerful than methods that compare independent groups. Now we've learned that analyzing data from matched pairs requires different inference procedures.

- We've learned to think about the design of the study that collected the data to recognize when data are paired or matched, and when they are not. Paired data cannot be analyzed using independent $t$-procedures.
- We've learned (again) the importance of checking the appropriate assumptions and conditions before proceeding with inference.
- We've learned that paired $t$-methods look at pairwise differences, analyzing them using the mechanics of one-sample $t$-methods we saw in Chapter 22.
- We've learned to construct a confidence interval for the mean difference based on paired data.
- We've learned to test a hypothesis about the mean difference based on paired data.

And once again we've learned that the reasoning of inference and the proper interpretation of confidence intervals and P-values remain the same.

## TERMS

**Paired data**  Data are paired when the observations are collected in pairs or the observations in one group are naturally related to observations in the other. The simplest form of pairing is to measure each subject twice—often before and after a treatment is applied. More sophisticated forms of pairing in experiments are a form of blocking and arise in other contexts. Pairing in observational and survey data is a form of matching. (p. 653)

**Paired $t$-test**  A hypothesis test for the mean of the pairwise differences of two groups. It tests the null hypothesis
$$H_0: \mu_d = \Delta_0,$$
where the hypothesized difference is almost always 0, using the statistic
$$t = \frac{\bar{d} - \Delta_0}{SE(\bar{d})}$$
with $n - 1$ degrees of freedom, where $SE(\bar{d}) = \frac{s_d}{\sqrt{n}}$, and $n$ is the number of pairs. (p. 654)

**Paired $t$ confidence interval**  A confidence interval for the mean of the pairwise differences between paired groups found as
$$\bar{d} \pm t^*_{n-1} \times SE(\bar{d}), \text{ where } SE(\bar{d}) = \frac{s_d}{\sqrt{n}} \text{ and } n \text{ is is the number of pairs.} \quad \text{(p. 641)}$$

## ON THE COMPUTER

### Paired t Inference

Most statistics programs can compute paired $t$ analyses. The computer, of course, cannot verify that the variables are naturally paired. Most programs will automatically omit any pair that is missing a value for either variable (as we did with the British couples). You must look carefully to see whether that has happened.

As we've seen with other inference results, some packages pack a lot of information into a simple table, but you must locate what you want for yourself. Here's a generic example with comments:

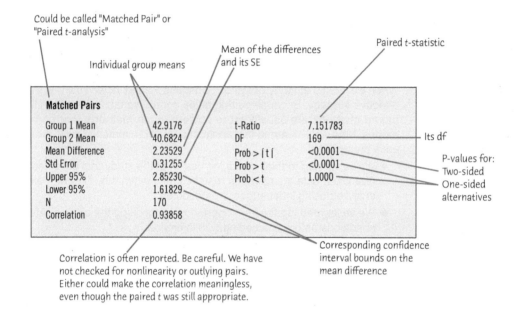

Other packages try to be more descriptive. It may be easier to find the results, but you may get less information from the output table.

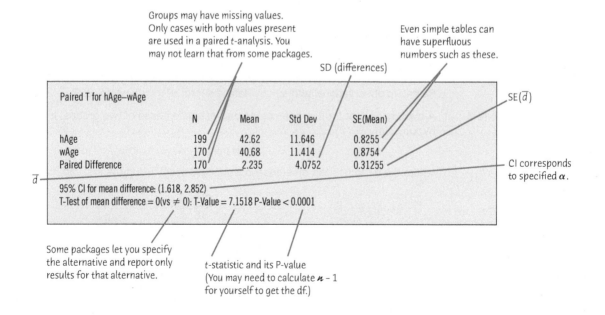

# EXERCISES

**1. Which method?** Which of the following scenarios should be analyzed as paired data?

a) Students take a MCAT prep course. Their before and after scores are compared.
b) The flipper lengths of 20 Adélie penguins and 20 Chinstrap penguins in Dr. Gorman's study from Chapter 23 Exercise 18 were measured. We compare their lengths.
c) A group of first-year college students are asked about the quality of the university cafeteria. A year later, the same students are asked about the cafeteria again. Do students' opinions change during their time at school?

**2. Which method II?** Which of the following scenarios should be analyzed as paired data?

a) Breeding pairs of ospreys (a species of large birds of prey) are studied to compare the time spent by males and females on duties related to raising their young. Which sex spends more time raising their chicks?
b) 50 insomnia patients are given a placebo and 50 are given a mild sedative. Which subjects sleep more hours?
c) A group of first-year college students and a group of second-year students are asked about the quality of the university cafeteria. Do students' opinions change during their time at school?

**3. More eggs?** Can a food additive increase egg production? Agricultural researchers want to design an experiment to find out. They have 100 hens available. They have two kinds of feed: the regular feed and the new feed with the additive. They plan to run their experiment for a month, recording the number of eggs each hen produces.

a) Design an experiment that will require a two-sample $t$ procedure to analyze the results.
b) Design an experiment that will require a matched-pairs $t$ procedure to analyze the results.
c) Which experiment would you consider the stronger design? Why?

**4. Music** Some students do homework with music playing in their headphones. (Anyone come to mind?) Some researchers want to see if people can work as effectively with as without distraction. The researchers will time some volunteers to see how long it takes them to complete some relatively easy crossword puzzles. During some of the trials, the room will be quiet; during other trials in the same room, subjects will wear headphones and listen to a Pandora channel.

a) Design an experiment that will require a two-sample $t$ procedure to analyze the results.
b) Design an experiment that will require a matched-pairs $t$ procedure to analyze the results.
c) Which experiment would you consider the stronger design? Why?

**5. Sex sells?** Ads for many products use sexual images to try to attract attention to the product. But do these ads bring people's attention to the item that was being advertised? We want to design an experiment to see if the presence of sexual images in an advertisement affects people's ability to remember the product.

a) Describe an experimental design requiring a matched-pairs $t$ procedure to analyze the results.
b) Describe an experimental design requiring an independent sample procedure to analyze the results.

**6. First-year 15?** Many people believe that students gain weight during their first year of college. Suppose we plan to conduct a study to see if this is true.

a) Describe a study design that would require a matched-pairs $t$ procedure to analyze the results.
b) Describe a study design that would require a two-sample $t$ procedure to analyze the results.

**7. Women** Values for the labor force participation rate of women (LFPR) are published by the U.S. Bureau of Labor Statistics. We are interested in whether there was a difference between women's participation in 1968 and 1972, a time of rapid change for women. We check LFPR values for 19 randomly selected cities for 1968 and 1972. Shown below is software output for two possible tests:

> Paired $t$-Test of $\mu(1-2)$
> Test $H_0: \mu(1972-1968) = 0$ vs $H_a: \mu(1972-1968) \neq 0$
> Mean of Paired Differences = 0.0337
> $t$-Statistic = 2.458 w/18 df
> $p = 0.0244$
>
> 2-Sample $t$-Test of $\mu_1 - \mu_2$
> $H_0: \mu_1 - \mu_2 = 0$ vs Ha: $\mu_1 - \mu_2 \neq 0$
> Test Ho: $\mu(1972) - \mu(1968) = 0$ vs
> $H_a: \mu(1972) - \mu(1968) \neq 0$
> Difference Between Means = 0.0337
> $t$-Statistic = 1.496 w/35 df
> $p = 0.1434$

a) Which of these tests is appropriate for these data? Explain.
b) Using the test you selected, state your conclusion.

**8. Cloud seeding** Simpson, Alsen, and Eden (*Technometrics* 1975) report the results of trials in which clouds were seeded and the amount of rainfall recorded. The authors report on 26 seeded and 26 unseeded clouds in order of the amount of rainfall, largest amount first. Here are two possible tests to study the question of whether cloud seeding works. Which test is appropriate for these data? Explain your choice. Using the test you select, state your conclusion.

> Paired $t$-Test of $\mu(1-2)$
> Mean of Paired Differences = $-277.39615$
> $t$-Statistic = $-3.641$ w/25 df
> $p = 0.0012$
>
> 2-Sample $t$-Test of $\mu_1 - \mu_2$
> Difference Between Means = $-277.4$
> $t$-Statistic = $-1.998$ w/33 df
> $p = 0.0538$

**9. Friday the 13th, traffic** In 1993 the *British Medical Journal* published an article titled, "Is Friday the 13th Bad for Your Health?" Researchers in Britain examined how Friday the 13th affects human behavior. One question was whether people tend to stay at home more on Friday the 13th. The data below are the number of cars passing Junctions 9 and 10 on the M25 motorway for consecutive Fridays (the 6th and 13th) for five different periods.

| Year | Month | 6th | 13th |
|---|---|---|---|
| 1990 | July | 134,012 | 132,908 |
| 1991 | September | 133,732 | 131,843 |
| 1991 | December | 121,139 | 118,723 |
| 1992 | March | 124,631 | 120,249 |
| 1992 | November | 117,584 | 117,263 |

Here are summaries of two possible analyses:

Paired *t*-Test of mu(1 − 2) = 0 vs. mu(1 − 2) > 0
Mean of Paired Differences: 2022.4
*t*-Statistic = 2.9377 w/4 df
P = 0.0212

2-Sample *t*-Test of mu1 = mu2 vs. mu1 > mu2
Difference Between Means: 2022.4
*t*-Statistic = 0.4273 w/7.998 df
P = 0.3402

a) Which of the tests is appropriate for these data? Explain.
b) Using the test you selected, state your conclusion.
c) Are the assumptions and conditions for inference met?

**10. Friday the 13th, accidents** The researchers in Exercise 9 also examined the number of people admitted to emergency rooms for vehicular accidents on 12 Friday evenings (6 each on the 6th and 13th).

| Year | Month | 6th | 13th |
|---|---|---|---|
| 1989 | October | 9 | 13 |
| 1990 | July | 6 | 12 |
| 1991 | September | 11 | 14 |
| 1991 | December | 11 | 10 |
| 1992 | March | 3 | 4 |
| 1992 | November | 5 | 12 |

Based on these data, is there evidence that more people are admitted, on average, on Friday the 13th? Here are two possible analyses:

Paired *t*-Test of mu(2 − 1) = 0 vs. mu(2 − 1) > 0
Mean of Paired Differences = 3.333
*t*-Statistic = 2.7116 w/5 df
P = 0.0211

2-Sample *t*-Test of mu2 = mu1 vs. mu2 > mu1
Difference Between Means = 3.333
*t*-Statistic = 1.6644 w/9.940 df
P = 0.0636

a) Which of these tests is appropriate for these data? Explain.
b) Using the test you selected, state your conclusion.
c) Are the assumptions and conditions for inference met?

**11. Online insurance I** After seeing countless commercials claiming one can get cheaper car insurance from an online company, a local insurance agent was concerned that he might lose some customers. To investigate, he randomly selected profiles (type of car, coverage, driving record, etc.) for 10 of his clients and checked online price quotes for their policies. The comparisons are shown in the table below. His statistical software produced the following summaries (where PriceDiff = Local − Online):

| Variable | Count | Mean | StdDev |
|---|---|---|---|
| Local | 10 | 799.200 | 229.281 |
| Online | 10 | 753.300 | 256.267 |
| PriceDiff | 10 | 45.9000 | 175.663 |

| Local | Online | PriceDiff |
|---|---|---|
| 568 | 391 | 177 |
| 872 | 602 | 270 |
| 451 | 488 | −37 |
| 1229 | 903 | 326 |
| 605 | 677 | −72 |
| 1021 | 1270 | −249 |
| 783 | 703 | 80 |
| 844 | 789 | 55 |
| 907 | 1008 | −101 |
| 712 | 702 | 10 |

At first, the insurance agent wondered whether there was some kind of mistake in this output. He thought the Pythagorean Theorem of Statistics should work for finding the standard deviation of the price differences—in other words, that $SD(Local - Online) = \sqrt{SD^2(Local) + SD^2(Online)}$. But when he checked, he found that $\sqrt{(229.281)^2 + (256.267)^2}$ = 343.864, not 175.663 as given by the software. Tell him where his mistake is.

**12. Wind speed, part I** To select the site for an electricity-generating wind turbine, wind speeds were recorded at several potential sites every 6 hours for a year. Two sites not far from each other looked good. Each had a mean wind speed high enough to qualify, but we should choose the site with a higher average daily wind speed. Because the sites are near each other and the wind speeds were recorded at the same times, we should view the speeds as paired. Here are the summaries of the speeds (in miles per hour):

| Variable | Count | Mean | StdDev |
|---|---|---|---|
| site2 | 1114 | 7.452 | 3.586 |
| site4 | 1114 | 7.248 | 3.421 |
| site2–site4 | 1114 | 0.204 | 2.551 |

Is there a mistake in this output? Why doesn't the Pythagorean Theorem of Statistics work here? In other words, shouldn't $SD(site2 - site4) = \sqrt{SD^2(site2) + SD^2(site4)}$? But $\sqrt{(3.586)^2 + (3.421)^2} = 4.956$, not 2.551 as given by the software. Explain why this happened.

**13. Online insurance II** In Exercise 11, we saw summary statistics for 10 drivers' car insurance premiums quoted by a local agent and an online company. Here are displays for each company's quotes and for the difference (*Local − Online*):

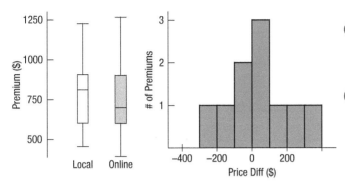

a) Which of the summaries would help you decide whether the online company offers cheaper insurance? Why?
b) The standard deviation of *PriceDiff* is quite a bit smaller than the standard deviation of prices quoted by either the local or online companies. Discuss why.
c) Using the information you have, discuss the assumptions and conditions for inference with these data.

**14. Wind speed, part II** In Exercise 12, we saw summary statistics for wind speeds at two sites near each other, both being considered as locations for an electricity-generating wind turbine. The data, recorded every 6 hours for a year, showed each of the sites had a mean wind speed high enough to qualify, but how can we tell which site is best? Here are some displays:

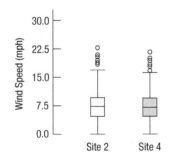

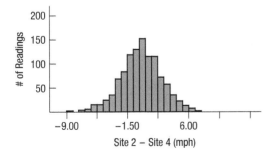

a) The boxplots show outliers for each site, yet the histogram shows none. Discuss why.
b) Which of the summaries would you use to select between these sites? Why?
c) Using the information you have, discuss the assumptions and conditions for paired $t$ inference for these data. (*Hint:* Think hard about the Independence Assumption in particular.)

**15. Online insurance III** Exercises 11 and 13 give summaries and displays for car insurance premiums quoted by a local agent and an online company. Test an appropriate hypothesis to see if there is evidence that drivers might save money by switching to the online company.

**16. Wind speed, part III** Exercises 12 and 14 give summaries and displays for two potential sites for a wind turbine. Test an appropriate hypothesis to see if there is evidence that either of these sites has a higher average wind speed.

**17. Cars and trucks** We have data on the city and highway fuel efficiency of 633 cars and trucks.
a) Would it be appropriate to use paired $t$-methods to compare the cars and the trucks?
b) Would it be appropriate to use paired $t$-methods to compare the city and highway fuel efficiencies of these vehicles?
c) A histogram of the differences in fuel efficiency (highway − city) is shown below. Are the conditions for inference satisfied?

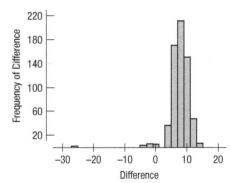

**18. Weighing trucks** One kind of scale for weighing trucks can measure their weight as they drive across a plate. Is this method consistent with the traditional method of static weighing? Are the conditions for matched-pairs inference satisfied? Weights are in 1000's of pounds.

| Weight-in-Motion | Static Weight | Diff (Motion − Static) |
|---|---|---|
| 26.0 | 27.9 | −1.9 |
| 29.9 | 29.1 | 0.8 |
| 39.5 | 38.0 | 1.5 |
| 25.1 | 27.0 | −1.9 |
| 31.6 | 30.3 | 1.3 |
| 36.2 | 34.5 | 1.7 |
| 25.1 | 27.8 | −2.7 |
| 31.0 | 29.6 | 1.4 |
| 35.6 | 33.1 | 2.5 |
| 40.2 | 35.5 | 4.7 |

**19. Cars and trucks again** In Exercise 17, after deleting an outlying value of −27, the mean difference in fuel efficiencies for the 632 vehicles was 7.37 mpg with a standard deviation of 2.52 mpg. Find a 95% confidence interval for this difference and interpret it in context.

**20. Weighing trucks II** Find a 98% confidence interval of the weight differences in Exercise 18. Interpret this interval in context.

21. **Blocking cars and trucks** Thinking about the data on fuel efficiency in Exercise 17, why is the blocking accomplished by a matched-pairs-analysis particularly important for a sample that has both cars and trucks?

22. **Weighing trucks III** Consider the weights from Exercise 18. The side-by-side boxplots below show little difference between the two groups. Should this be sufficient to draw a conclusion about the accuracy of the weight-in-motion scale?

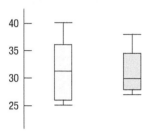

23. **Cataracts** Researchers in Korea tested the effect of patients listening to Korean traditional music on the reported pain level of the surgery. 52 patients who were having surgery on both eyes were randomly assigned to listen to music during the first surgery and not the second, or to the reverse order. After each surgery, patients recorded the pain level using a Visual Analog Scale ranging from 0 (no pain) to 10 (unbearable pain).
    a) Are conditions for inference met?
    b) The study reported that the difference was statistically significant with a *P*-value of 0.013, with lower pain reports for surgeries with music. Assuming a 2-sided test, what was the value of the *t*-statistic?
    c) Write a conclusion for this significance test.

24. **Cats** To test a theory about why cats like to rub their faces and heads against catnip and silver vine plants, researchers applied nepetalactol, a chemical found in the leaves of such plants, to the heads of six cats, and six with a placebo solvent. Treatments were randomly assigned. They then randomly paired the cats, one from each group, anethsetized them, and placed their heads into test cages with mosquitos. They then counted the number of mosquitos that landed on the head of each cat. The data are below.

| Cage | Nepetalactol Treatment (500 $\mu$g) | Mosquito Landings | Control | Mosquito Landings |
|---|---|---|---|---|
| 1 | Cat A | 4 | Cat N | 23 |
| 2 | Cat G | 8 | Cat S | 14 |
| 3 | Cat J | 11 | Cat U | 24 |
| 4 | Cat L | 13 | Cat D | 13 |
| 5 | Cat C | 6 | Cat T | 18 |
| 6 | Cat I | 16 | Cat E | 18 |

Find and interpret a 95% confidence interval to estimate the mean difference in the number of mosquitos landing on the cats who had been treated with nepetalactol and those who had received the placebo solvent. (Source: https://www.science.org/doi/10.1126/sciadv.abd9135)

25. **Push-ups** Every year the students at Gossett High School take a physical fitness test during their gym classes. One component of the test asks them to do as many push-ups as they can. Results for one mixed-grade class (9th and 10th grade) are shown below, separately for 9th and 10th grades. Assuming that students at Gossett are assigned to gym classes at random, create a 90% confidence interval for how many more push-ups grade 10 students can do than grade 9 students, on average, at that high school.

| G10 | 17 | 27 | 31 | 17 | 25 | 32 | 28 | 23 | 25 | 16 | 11 | 34 |
|---|---|---|---|---|---|---|---|---|---|---|---|---|
| G9 | 24 | 7 | 14 | 16 | 2 | 15 | 19 | 25 | 10 | 27 | 31 | 8 |

26. **Cats II** The researchers in the Exercise 24 wanted to test their theory about catnip and silver vine plants providing cats with mosquito resistance in a more natural setting. They had six cats who had rubbed their heads against silver vine leaves and six who had not. The cats were again randomly paired and placed in testing cages. The results of this experiment are given below.

| Assay | Silver Vine Response | Mosquito Landings | Control | Mosquito Landings |
|---|---|---|---|---|
| 1 | Cat A | 6 | Cat N | 15 |
| 2 | Cat G | 12 | Cat S | 15 |
| 3 | Cat J | 5 | Cat U | 18 |
| 4 | Cat L | 9 | Cat D | 15 |
| 5 | Cat C | 3 | Cat T | 15 |
| 6 | Cat I | 8 | Cat E | 8 |

Use a hypothesis test to determine whether rubbing against the silver vine plant reduced the number of mosquitos landing on the heads of these cats.

27. **Job satisfaction** (When you first read about this exercise break plan in Chapter 23, you did not have an inference method that would work. Try again now.) A company institutes an exercise break for its workers to see if it will improve job satisfaction, as measured by a questionnaire that assesses workers' satisfaction. Scores for 10 randomly selected workers before and after the implementation of the exercise program are shown in the table.
    a) Identify the procedure you would use to assess the effectiveness of the exercise program, and check to see if the conditions allow the use of that procedure.
    b) Test an appropriate hypothesis and state your conclusion.
    c) If your conclusion turns out to be incorrect, what kind of error occurred?

| Worker Number | Job Satisfaction Index | |
|---|---|---|
| | Before | After |
| 1 | 34 | 33 |
| 2 | 28 | 36 |
| 3 | 29 | 50 |
| 4 | 45 | 41 |
| 5 | 26 | 37 |
| 6 | 27 | 41 |
| 7 | 24 | 39 |
| 8 | 15 | 21 |
| 9 | 15 | 20 |
| 10 | 27 | 37 |

**28. Summer school** (When you first read about the summer school issue in Chapter 23, you did not have an inference method that would work. Try again now.) Having done poorly on their Math final exams in June, six students repeat the course in summer school and take another exam in August.

| June   | 54 | 49 | 68 | 66 | 62 | 62 |
|--------|----|----|----|----|----|----|
| August | 50 | 65 | 74 | 64 | 68 | 72 |

a) If we consider these students to be representative of all students who might attend this summer school in other years, do these results provide evidence that the program is worthwhile?
b) This conclusion, of course, may be incorrect. If so, which type of error occurred?

**29. Yogurt** Is there a significant difference in calories between servings of strawberry and vanilla yogurt? Based on the data shown in the table, test an appropriate hypothesis and state your conclusion. Don't forget to check assumptions and conditions!

|       |                      | Calories per Serving |         |
|-------|----------------------|----------------------|---------|
|       |                      | Strawberry           | Vanilla |
| Brand | America's Choice     | 210                  | 200     |
|       | Breyer's Lowfat      | 220                  | 220     |
|       | Columbo              | 220                  | 180     |
|       | Dannon Light 'n Fit  | 120                  | 120     |
|       | Dannon Lowfat        | 210                  | 230     |
|       | Dannon la Crème      | 140                  | 140     |
|       | Great Value          | 180                  | 80      |
|       | La Yogurt            | 170                  | 160     |
|       | Mountain High        | 200                  | 170     |
|       | Stonyfield Farm      | 100                  | 120     |
|       | Yoplait Custard      | 190                  | 190     |
|       | Yoplait Light        | 100                  | 100     |

**30. Gasoline** Many drivers of cars that can run on regular gas actually buy premium in the belief that they will get better gas mileage. To test that belief, we use 10 cars from a company fleet in which all the cars run on regular gas. Each car is filled first with either regular or premium gasoline, decided by a coin toss, and the mileage for that tankful is recorded. Then the mileage is recorded again for the same cars for a tankful of the other kind of gasoline. We don't let the drivers know about this experiment.
Here are the results (miles per gallon):

| Car #   | 1  | 2  | 3  | 4  | 5  | 6  | 7  | 8  | 9  | 10 |
|---------|----|----|----|----|----|----|----|----|----|----|
| Regular | 16 | 20 | 21 | 22 | 23 | 22 | 27 | 25 | 27 | 28 |
| Premium | 19 | 22 | 24 | 24 | 25 | 25 | 26 | 26 | 28 | 32 |

a) Is there evidence that cars get significantly better fuel economy with premium gasoline?
b) How big might that difference be? Check a 90% confidence interval.
c) Even if the difference is significant, why might the company choose to stick with regular gasoline?

d) Suppose you had done a "bad thing." (We're sure you didn't.) Suppose you had mistakenly treated these data as two independent samples instead of matched pairs. What would the significance test have found? Carefully explain why the results are so different.

**31. Braking test** A tire manufacturer tested the braking performance of one of its tire models on a test track. The company tried the tires on 10 different cars, recording the stopping distance for each car on both wet and dry pavement. Results are shown in the table.

|       | Stopping Distance (ft) |              |
|-------|------------------------|--------------|
| Car # | Dry Pavement           | Wet Pavement |
| 1     | 150                    | 201          |
| 2     | 147                    | 220          |
| 3     | 136                    | 192          |
| 4     | 134                    | 146          |
| 5     | 130                    | 182          |
| 6     | 134                    | 173          |
| 7     | 134                    | 202          |
| 8     | 128                    | 180          |
| 9     | 136                    | 192          |
| 10    | 158                    | 206          |

a) Write a 95% confidence interval for the mean dry pavement stopping distance. Be sure to check the appropriate assumptions and conditions, and explain what your interval means.
b) Write a 95% confidence interval for the mean increase in stopping distance on wet pavement. Be sure to check the appropriate assumptions and conditions, and explain what your interval means.

**32. Braking test II** For another test of the tires in Exercise 31, a car made repeated stops from 60 miles per hour. The test was run on both dry and wet pavement, with results as shown in the table. (Note that actual *braking distance*, which takes into account the driver's reaction time, is much longer, typically nearly 300 feet at 60 mph!)

a) Write a 95% confidence interval for the mean dry pavement stopping distance. Be sure to check the appropriate assumptions and conditions, and explain what your interval means.
b) Write a 95% confidence interval for the mean increase in stopping distance on wet pavement. Be sure to check the appropriate assumptions and conditions, and explain what your interval means.

| Stopping Distance (ft) |              |
|------------------------|--------------|
| Dry Pavement           | Wet Pavement |
| 145                    | 211          |
| 152                    | 191          |
| 141                    | 220          |
| 143                    | 207          |
| 131                    | 198          |
| 148                    | 208          |
| 126                    | 206          |
| 140                    | 177          |
| 135                    | 183          |
| 133                    | 223          |

**33. Tuition 2020** How much more do public colleges and universities charge out-of-state students for tuition per year? A random sample of 19 public colleges and universities listed at www.collegeboard.com yielded the following data for incoming first-year students in the fall of 2020.

| Institution | In-State | Out-of-State |
|---|---|---|
| University of Akron | 11,463 | 15,500 |
| Ball State University | 9,896 | 26,408 |
| Bloomsburg University of Pennsylvania | 10,958 | 22,782 |
| University of Califomia-Irvine | 13,700 | 42,692 |
| Central State University | 6,346 | 8,346 |
| Clarion University of Pennsylvania | 11,175 | 16,054 |
| Oakota State University | 9,276 | 12,249 |
| Fairmont State University | 7,514 | 16,324 |
| Lock Haven University | 10,878 | 20,702 |
| New College of Florida | 6,916 | 29,944 |
| Oakland University | 12,606 | 241,230 |
| University of Pittsburgh | 19,080 | 32,052 |
| Savannah State University | 5,743 | 16,204 |
| Louisiana State University | 11,950 | 28,627 |
| West Liberty University | 7,680 | 15,620 |
| Central Texas College | 2,700 | 7,050 |
| Woncester State University | 10,161 | 16,241 |
| Wright State University | 9,254 | 18,398 |
| Wayne State University | 13,097 | 27,991 |
| University of Wyoming | 5,400 | 17,490 |

a) Check the conditions for inference on the mean difference in tuition.
b) Create a 90% confidence interval for the mean difference in cost by taking account of what you found in part a, and interpret your interval in context.
c) A national magazine claims that public institutions charge state residents an average of $7000 less than out-of-staters for tuition each year. What does your confidence interval indicate about this assertion?

**34. Sex sells, part II** In Exercise 5 you considered the question of whether sexual images in ads affected people's abilities to remember the item being advertised. To investigate, a group of Statistics students cut ads out of magazines. They were careful to find two ads for each of 10 similar items, one with a sexual image and one without. They arranged the ads in random order and had 39 subjects look at them for one minute. Then they asked the subjects to list as many of the products as they could remember. Their data are shown in the table. Is there evidence that the sexual images mattered?

| Subject Number | Ads Remembered Sexual Image | Ads Remembered No Sex | Subject Number | Ads Remembered Sexual Image | Ads Remembered No Sex |
|---|---|---|---|---|---|
| 1 | 2 | 2 | 21 | 2 | 3 |
| 2 | 6 | 7 | 22 | 4 | 2 |
| 3 | 3 | 1 | 23 | 3 | 3 |
| 4 | 6 | 5 | 24 | 5 | 3 |
| 5 | 1 | 0 | 25 | 4 | 5 |
| 6 | 3 | 3 | 26 | 2 | 4 |
| 7 | 3 | 5 | 27 | 2 | 2 |
| 8 | 7 | 4 | 28 | 2 | 4 |
| 9 | 3 | 7 | 29 | 7 | 6 |
| 10 | 5 | 4 | 30 | 6 | 7 |
| 11 | 1 | 3 | 31 | 4 | 3 |
| 12 | 3 | 2 | 32 | 4 | 5 |
| 13 | 6 | 3 | 33 | 3 | 0 |
| 14 | 7 | 4 | 34 | 4 | 3 |
| 15 | 3 | 2 | 35 | 2 | 3 |
| 16 | 7 | 4 | 36 | 3 | 3 |
| 17 | 4 | 4 | 37 | 5 | 5 |
| 18 | 1 | 3 | 38 | 3 | 4 |
| 19 | 5 | 5 | 39 | 4 | 3 |
| 20 | 2 | 2 | | | |

**35. Strikes** Advertisements for an instructional video claim that the techniques will improve the ability of Little League pitchers to throw strikes and that, after undergoing the training, players will be able to throw strikes on at least 60% of their pitches. To test this claim, we have 20 Little Leaguers throw 50 pitches each, and we record the number of strikes. After the players participate in the training program, we repeat the test. The table shows the number of strikes each player threw before and after the training.

a) Is there evidence that after training players can throw strikes more than 60% of the time?
b) Is there evidence that the training is effective in improving a player's ability to throw strikes?

| Number of Strikes (out of 50) | | Number of Strikes (out of 50) | |
|---|---|---|---|
| Before | After | Before | After |
| 28 | 35 | 33 | 33 |
| 29 | 36 | 33 | 35 |
| 30 | 32 | 34 | 32 |
| 32 | 28 | 34 | 30 |
| 32 | 30 | 34 | 33 |
| 32 | 31 | 35 | 34 |
| 32 | 32 | 36 | 37 |
| 32 | 34 | 36 | 33 |
| 32 | 35 | 37 | 35 |
| 33 | 36 | 37 | 32 |

**36. First-year 15, revisited** In Exercise 6 you thought about how to design a study to see if it's true that students tend to gain weight during their first year in college. Well, Cornell Professor of Nutrition David Levitsky did just that. He recruited students from two large sections of an Introductory Health course. Although they were volunteers, they appeared to match the rest of the first-year class in terms of demographic variables such as sex and ethnicity. The students were weighed during the first week of the semester, then again 12 weeks later. Based on Professor Levitsky's data, estimate the mean weight gain in the first-semester of students' first-year of college and comment on the "first-year 15." (Weights are in pounds.)

| Subject Number | Initial Weight | Terminal Weight | Subject Number | Initial Weight | Terminal Weight |
|---|---|---|---|---|---|
| 1 | 171 | 168 | 35 | 148 | 150 |
| 2 | 110 | 111 | 36 | 164 | 165 |
| 3 | 134 | 136 | 37 | 137 | 138 |
| 4 | 115 | 119 | 38 | 198 | 201 |
| 5 | 150 | 155 | 39 | 122 | 124 |
| 6 | 104 | 106 | 40 | 146 | 146 |
| 7 | 142 | 148 | 41 | 150 | 151 |
| 8 | 120 | 124 | 42 | 187 | 192 |
| 9 | 144 | 148 | 43 | 94 | 96 |
| 10 | 156 | 154 | 44 | 105 | 105 |
| 11 | 114 | 114 | 45 | 127 | 130 |
| 12 | 121 | 123 | 46 | 142 | 144 |
| 13 | 122 | 126 | 47 | 140 | 143 |
| 14 | 120 | 115 | 48 | 107 | 107 |
| 15 | 115 | 118 | 49 | 104 | 105 |
| 16 | 110 | 113 | 50 | 111 | 112 |
| 17 | 142 | 146 | 51 | 160 | 162 |
| 18 | 127 | 127 | 52 | 134 | 134 |
| 19 | 102 | 105 | 53 | 151 | 151 |
| 20 | 125 | 125 | 54 | 127 | 130 |
| 21 | 157 | 158 | 55 | 106 | 108 |
| 22 | 119 | 126 | 56 | 185 | 188 |
| 23 | 113 | 114 | 57 | 125 | 128 |
| 24 | 120 | 128 | 58 | 125 | 126 |
| 25 | 135 | 139 | 59 | 155 | 158 |
| 26 | 148 | 150 | 60 | 118 | 120 |
| 27 | 110 | 112 | 61 | 149 | 150 |
| 28 | 160 | 163 | 62 | 149 | 149 |
| 29 | 220 | 224 | 63 | 122 | 121 |
| 30 | 132 | 133 | 64 | 155 | 158 |
| 31 | 145 | 147 | 65 | 160 | 161 |
| 32 | 141 | 141 | 66 | 115 | 119 |
| 33 | 158 | 160 | 67 | 167 | 170 |
| 34 | 135 | 134 | 68 | 131 | 131 |

**37. White House staff salaries** Is the mean White House staff salary higher in 2020 than it was in 2016? Consider the following sample of 14 job positions, and the annual salary for each.

| Position Title | 2016 | 2020 |
|---|---|---|
| Asst to the President and Cabinet Secretary | $176,461 | $183,000 |
| Asst to the President and Chief of Staff | 176,461 | 183,000 |
| Asst to the Presidentand Deputy National Security Advisor | 174,714 | 183,000 |
| Asst to the Presidentand Director of Presidential Personnel | 172,200 | 183,000 |
| Asst to the President and National Security Advisor | 176,461 | 183,000 |
| Asst to the President and Press Secretary | 176,461 | 183,000 |
| Chief Calligrapher | 100,110 | 109,200 |
| Deputy Asst to the President and Deputy Director of Communications | 142,044 | 158,000 |
| Deputy Asstto the President and Director of Intergovernmental Affairs | 157,299 | 168,000 |
| Deputy Director and Senior Advisor for Records Management | 127,008 | 149,600 |
| Director of Records Management | 145,162 | 170,800 |
| Director of Stenography | 117,846 | 98,900 |
| Executive Clerk | 145,162 | 170,800 |
| Special Asst to the President and Deputy Director of Advance | 103,489 | 145,000 |

a) Is there evidence that the average pay grew higher in 2020 vs. 2016?
b) Create a 95% confidence interval for the average pay increase from 2016 to 2020.
c) The Consumer Price Index (CPI) reports that the purchasing power of $1 in January 2016 required $1.09 in January 2020. Adjust the 2016 salaries by multiplying them by 1.09 and then repeat parts a) and b). Does that change your conclusions in a) and b)?
d) How would your approach to the questions change if we were comparing pay for a set of 14 random White House jobs in 2016 to another set of 14 random White House jobs in 2020?

**38. Tattoos, no sweat!** Sweating is an important function of the skin. With the rise in popularity of tattoos, a group of researchers investigated the effects of tattoos on the skin's ability to sweat. (Luetkemeier et al, *Skin Tattoos Alter Sweat Rate and Na+ Concentration, Medicine & Science in Sports and Exercise*, July 1 2017.) Because the rate of sweating varies greatly between individuals, as well as locations on the body, they compared the rate of sweating by tattooed skin of ten healthy men with the rate of sweating at the identical location on the other side of the same person's body where no tattoo existed. (Sweat rates are measured in $mg/cm^2/min$.)

The data are shown below, along with some summary statistics and a dotplot of the differences.

| Sweat Rate Tattoo | Sweat Rate Non-Tattoo | Difference (Non-Tattoo − Tattoo) |
|---|---|---|
| 0.13 | 0.20 | 0.07 |
| 0.10 | 0.20 | 0.10 |
| 0.09 | 0.13 | 0.04 |
| 0.15 | 0.26 | 0.11 |
| 0.16 | 0.42 | 0.26 |
| 0.13 | 0.41 | 0.28 |
| 0.18 | 0.28 | 0.10 |
| 0.18 | 0.33 | 0.15 |
| 0.10 | 0.23 | 0.13 |
| 0.59 | 0.99 | 0.40 |

| Variable | Count | Mean | StdDev |
|---|---|---|---|
| Sweat Rate (Non-Tattoo) | 10 | 0.345 | 0.245 |
| Sweat Rate (Tattoo) | 10 | 0.181 | 0.147 |
| Difference (Non-Tattoo − Tattoo) | 10 | 0.164 | 0.113 |

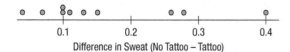
Difference in Sweat (No Tattoo − Tattoo)

a) Explain why the data in this study must be analyzed using a matched-pairs approach.
b) Explain why it may *not* be reasonable to construct a confidence interval of the mean difference using a *t*-procedure.

Below are a dotplot and summary statistics of the natural log of the differences.

| Variable | Count | Mean | StdDev |
|---|---|---|---|
| *ln*(Non-tattoo − Tattoo) | 10 | −2.016 | 0.690 |

*ln* (Non − Tattoo − Tattoo)

c) Explain why it is reasonable to construct a confidence interval of the natural log of the mean difference using a *t*-procedure.
d) Construct a 95% confidence interval for the natural log of the mean difference between the sweat rates of non-tattooed and tattooed skin. Then convert the interval into an interval of differences by applying the exponential function to the interval's endpoints. Interpret the meaning of this second interval in this context.

## JUST CHECKING

Answers

1. These are independent groups sampled at random, so use a two-sample *t* confidence interval to estimate the size of the difference.
2. There is only one sample. Use a one-sample *t*-interval.
3. A brother and sister from the same family represent a matched pair. The question calls for a paired *t*-test.
4. a) A before-and-after study calls for paired *t*-methods. To estimate the loss, find a confidence interval for the before − after differences.
   b) The two treatment groups were assigned randomly, so they are independent. Use a two-sample *t*-test to assess whether the mean weight losses differ.
5. Sometimes it just isn't clear. Most likely, couples would discuss the band or even decide to go to the club because they both like a particular band. If we think that's likely, then these data are paired. But maybe not. If we asked them their opinions of, say, the decor or furnishings at the club, the fact that they were couples might not affect the independence of their answers.

# Review of Part VI

## LEARNING ABOUT THE WORLD

### Quick Review

We continue to explore how to answer questions about the statistics we get from samples and experiments. In this part, those questions have been about means—means of one sample, two independent samples, or matched pairs. Here's a brief summary of the key concepts and skills:

- A confidence interval uses a sample statistic to estimate a range of possible values for a parameter of interest.
- A hypothesis test proposes a model, then examines the plausibility of that model by seeing how surprising our observed data would be if the model were true.
- Statistical inference procedures for proportions are based on the Central Limit Theorem. We can make inferences about a single proportion or the difference of two proportions using Normal models.
- Statistical inference procedures for means are also based on the Central Limit Theorem, but we don't usually know the population standard deviation. Student's $t$-models take into account the additional uncertainty of independently estimating the standard deviation.
  - We can make inferences about one mean, the difference of two independent means, or the mean of paired differences using $t$-models.
  - No inference procedure is valid unless the underlying assumptions are true. Always check the conditions before proceeding.
- Because $t$-models assume that samples are drawn from Normal populations, data in the sample should appear to be nearly Normal. Skewness and outliers are particularly problematic, especially for small samples.
- When there are two variables, you must think carefully about how the data were collected. You may use two-sample $t$-procedures only if the groups are independent.
- Unless there is some obvious reason to suspect that two independent populations have the same standard deviation, you should not pool the variances. It is never wrong to use unpooled $t$-procedures.
- If the two groups are somehow paired, the data are *not* from independent groups. You must use matched-pairs $t$-procedures.

Now for some opportunities to review these concepts. Be careful. You have a lot of thinking to do. These review exercises mix questions about proportions and means. You have to determine which of our inference procedures is appropriate in each situation. Then you have to check the proper assumptions and conditions. Keeping track of those can be difficult, so first we summarize the many procedures with their corresponding assumptions and conditions on the next page. Look them over carefully . . . then on to the Exercises!

| Quick Guide to Inference | | | | | | | |
|---|---|---|---|---|---|---|---|
| **Think** | | | **Show** | | | | **Tell?** |
| Inference about? | One group or two? | Procedure | Model | Parameter | Estimate | SE | Chapter |
| Proportions | One sample | 1-Proportion z-Interval | $z$ | $p$ | $\hat{p}$ | $\sqrt{\dfrac{\hat{p}\hat{q}}{n}}$ | 18 |
| | | 1-Proportion z-Test | | | | $\sqrt{\dfrac{p_0 q_0}{n}}$ | 19, 20 |
| | Two independent groups | 2-Proportion z-Interval | $z$ | $p_1 - p_2$ | $\hat{p}_1 - \hat{p}_2$ | $\sqrt{\dfrac{\hat{p}_1\hat{q}_1}{n_1} + \dfrac{\hat{p}_2\hat{q}_2}{n_2}}$ | 21 |
| | | 2-Proportion z-Test | | | | $\sqrt{\dfrac{\hat{p}\hat{q}}{n_1} + \dfrac{\hat{p}\hat{q}}{n_2}}$, $\hat{p} = \dfrac{y_1 + y_2}{n_1 + n_2}$ | 21 |
| Means | One sample | $t$-Interval $t$-Test | $t$ df $= n - 1$ | $\mu$ | $\bar{y}$ | $\dfrac{s}{\sqrt{n}}$ | 22 |
| | Two independent groups | 2-Sample $t$-Test 2-Sample $t$-Interval | $t$ df from technology | $\mu_1 - \mu_2$ | $\bar{y}_1 - \bar{y}_2$ | $\sqrt{\dfrac{s_1^2}{n_1} + \dfrac{s_2^2}{n_2}}$ | 23 |
| | Matched pairs | Paired $t$-Test Paired $t$-Interval | $t$ df $= n - 1$ | $\mu_d$ | $\bar{d}$ | $\dfrac{s_d}{\sqrt{n}}$ | 24 |

| Assumptions for Inference | And the Conditions That Support or Override Them |
|---|---|
| **Proportions (z)** | |
| • One sample | |
|   1. Individuals are independent. | 1. SRS and $n < 10\%$ of the population. |
|   2. Sample is sufficiently large. | 2. Successes and failures each $\geq 10$. |
| • Two groups | |
|   1. Groups are independent. | 1. (Think about how the data were collected.) |
|   2. Data in each group are independent. | 2. Both are SRSs and $n < 10\%$ of populations OR random allocation. |
|   3. Both groups are sufficiently large. | 3. The expected number of successs and failures each $\geq 10$ for both groups. |
| **Means (t)** | |
| • One sample (df = $n - 1$) | |
|   1. Individuals are independent. | 1. SRS and $n < 10\%$ of the population. |
|   2. Population has a Normal model. | 2. Histogram is unimodal and reasonably symmetric.* |
| • Matched pairs (df = $n - 1$) | |
|   1. Data are matched. | 1. (Think about the design.) |
|   2. Individuals are independent. | 2. SRS and $n < 10\%$ OR random allocation. |
|   3. Population of differences is Normal. | 3. Histogram of differences is unimodal and reasonably symmetric.* |
| • Two independent groups (df from technology) | |
|   1. Groups are independent. | 1. (Think about the design.) |
|   2. Data in each group are independent. | 2. SRSs and $n < 10\%$ OR random allocation. |
|   3. Both populations are Normal. | 3. Both histograms are unimodal and reasonably symmetric.* |
| | (*Less critical as $n$ increases) |

# REVIEW EXERCISES

**1. Crawling** A study found that babies born at different times of the year may develop the ability to crawl at different ages! The author of the study suggested that these differences may be related to the temperature at the time the infant is 6 months old. (Benson and Janette, *Infant Behavior and Development* [1993])

a) The study found that 32 babies born in January crawled at an average age of 29.84 weeks, with a standard deviation of 7.08 weeks. Among 21 July babies, crawling ages averaged 33.64 weeks, with a standard deviation of 6.91 weeks. Is this difference significant?

b) For 26 babies born in April the mean and standard deviation were 31.84 and 6.21 weeks, while for 44 October babies the mean and standard deviation of crawling ages were 33.35 and 7.29 weeks. Is this difference significant?

c) Are these results consistent with the researcher's conjecture?

**2. Mazes and smells** Can pleasant smells improve learning? Researchers timed 21 subjects as they tried to complete paper-and-pencil mazes. Each subject attempted a maze both with and without the presence of a floral aroma. Subjects were randomized with respect to whether they did the scented trial first or second. Is there any evidence that the floral scent improved the subjects' ability to complete the mazes? (A. R. Hirsch and L. H. Johnston, "Odors and Learning." Chicago: Smell and Taste Treatment and Research Foundation)

| Time to Complete the Maze (sec) | | Time to Complete the Maze (sec) | |
|---|---|---|---|
| Unscented | Scented | Unscented | Scented |
| 25.7 | 30.2 | 61.5 | 48.4 |
| 41.9 | 56.7 | 44.6 | 32.0 |
| 51.9 | 42.4 | 35.3 | 48.1 |
| 32.2 | 34.4 | 37.2 | 33.7 |
| 64.7 | 44.8 | 39.4 | 42.6 |
| 31.4 | 42.9 | 77.4 | 54.9 |
| 40.1 | 42.7 | 52.8 | 64.5 |
| 43.2 | 24.8 | 63.6 | 43.1 |
| 33.9 | 25.1 | 56.6 | 52.8 |
| 40.4 | 59.2 | 58.9 | 44.3 |
| 58.0 | 42.2 | | |

**3. Women** The U.S. Census Bureau reports that 26% of all U.S. businesses are owned by women. A Colorado consulting firm surveys a random sample of 410 businesses in the Denver area and finds that 115 of them have women owners. Should the firm conclude that its area is unusual? Test an appropriate hypothesis and state your conclusion.

4. **Drugs** In a full-page ad that ran in many U.S. newspapers in August 2002, a Canadian discount pharmacy listed costs of drugs that could be ordered from a website in Canada. The table compares prices (in US$) for commonly prescribed drugs.

|  | Cost per 100 Pills | | |
|---|---|---|---|
| **Drug Name** | **United States** | **Canada** | **Percent Savings** |
| Cardizem | 131 | 83 | 37 |
| Celebrex | 136 | 72 | 47 |
| Cipro | 374 | 219 | 41 |
| Pravachol | 370 | 166 | 55 |
| Premarin | 61 | 17 | 72 |
| Prevacid | 252 | 214 | 15 |
| Prozac | 263 | 112 | 57 |
| Tamoxifen | 349 | 50 | 86 |
| Vioxx | 243 | 134 | 45 |
| Zantac | 166 | 42 | 75 |
| Zocor | 365 | 200 | 45 |
| Zoloft | 216 | 105 | 51 |

a) Give a 95% confidence interval for the average savings in dollars.
b) Give a 95% confidence interval for the average savings in percent.
c) Which analysis is more appropriate? Why?
d) In small print the newspaper ad says, "Complete list of all 1500 drugs available on request." How does this comment affect your conclusions above?

5. **Pottery** Archaeologists can use the chemical composition of clay found in pottery artifacts to determine whether different sites were populated by the same ancient people. They collected five samples of Romano–British pottery from each of two sites in Great Britain and measured the percentage of aluminum oxide in each. Based on these data, do you think the same people used these two kiln sites? Base your conclusion on a 95% confidence interval for the difference in aluminum oxide content of pottery made at the sites. (A. Tubb, A. J. Parker, and G. Nickless, "The Analysis of Romano–British Pottery by Atomic Absorption Spectrophotometry." *Archaeometry*, 22[1980]:153–171)

| Ashley Rails | 19.1 | 14.8 | 16.7 | 18.3 | 17.7 |
| New Forest | 20.8 | 18.0 | 18.0 | 15.8 | 18.3 |

6. **Streams** Researchers in the Adirondack Mountains collect data on a random sample of streams each year. One of the variables recorded is the substrate of the streams—the type of soil and rock over which they flow. The researchers found that 69 of the 172 sampled streams had a substrate of shale. Construct a 95% confidence interval for the proportion of Adirondack streams with a shale substrate. Clearly interpret your interval in context.

7. **Gehrig** Ever since Lou Gehrig developed amyotrophic lateral sclerosis (ALS), this deadly condition has been commonly known as Lou Gehrig's disease. Some believe that ALS is more likely to strike athletes or the very fit. Columbia University neurologist Lewis P. Rowland recorded personal histories of 431 patients he examined between 1992 and 2002. He diagnosed 280 as having ALS; 38% of them had been varsity athletes. The other 151 had other neurological disorders, and only 26% of them had been varsity athletes. (*Science News*, September 28 [2002])

a) Is there evidence that ALS is more common among athletes?
b) What kind of study is this? How does that affect the inference you made in part a?

8. **Teen drinking** A study of the health behavior of school-aged children asked a sample of 15-year-olds in several European countries (and the U.S. and Israel) if they had been drunk at least twice. The results are shown in the table, by gender. Give a 95% confidence interval for the difference in the rates for males and females. Be sure to check the assumptions that support your chosen procedure, and explain what your interval means. (*Health and Health Behavior Among Young People*. Copenhagen: World Health Organization, 2000)

|  | Percent of 15-Year-Olds Drunk at Least Twice | |
|---|---|---|
| **Country** | **Female** | **Male** |
| Denmark | 63 | 71 |
| Wales | 63 | 72 |
| Greenland | 59 | 58 |
| England | 62 | 51 |
| Finland | 58 | 52 |
| Scotland | 56 | 53 |
| No. Ireland | 44 | 53 |
| Slovakia | 31 | 49 |
| Austria | 36 | 49 |
| Canada | 42 | 42 |
| Sweden | 40 | 40 |
| Norway | 41 | 37 |
| Ireland | 29 | 42 |
| Germany | 31 | 36 |
| Latvia | 23 | 47 |
| Estonia | 23 | 44 |
| Hungary | 22 | 43 |
| Poland | 21 | 39 |
| USA | 29 | 34 |
| Czech Rep. | 22 | 36 |
| Belgium | 22 | 36 |
| Russia | 25 | 32 |
| Lithuania | 20 | 32 |
| France | 20 | 29 |
| Greece | 21 | 24 |
| Switzerland | 16 | 25 |
| Israel | 10 | 18 |

**9. Babies** The National Perinatal Statistics Unit of the Sydney Children's Hospital reports that the mean birth weight of all babies born in Australia in 1999 was 3361 grams—about 7.41 pounds. A Missouri hospital reports that the average weight of 112 babies born there last year was 7.68 pounds, with a standard deviation of 1.31 pounds. If we believe the Missouri babies fairly represent American newborns, is there any evidence that U.S. babies and Australian babies do not weigh the same amount at birth?

**10. Petitions** To get a voter initiative on a state ballot, petitions that contain at least 250,000 valid voter signatures must be filed with the Elections Commission. The board then has 60 days to certify the petitions. A group wanting to create a statewide system of universal health insurance has just filed petitions with a total of 304,266 signatures. As a first step in the process, the Board selects an SRS of 2000 signatures and checks them against local voter lists. Only 1772 of them turn out to be valid.

a) What percent of the sample signatures were valid?
b) What percent of the petition signatures submitted must be valid in order to have the initiative certified by the Elections Commission?
c) What will happen if the Elections Commission commits a Type I error?
d) What will happen if the Elections Commission commits a Type II error?
e) Does the sample provide evidence in support of certification? Explain.
f) What could the Elections Commission do to increase the power of the test?

**11. Feeding fish** In the midwestern United States, a large aquaculture industry raises largemouth bass. Researchers wanted to know whether the fish would grow better if fed a natural diet of fathead minnows or an artificial diet of food pellets. They stocked six ponds with bass fingerlings weighing about 8 grams. For one year, the fish in three of the ponds were fed minnows, and the others were fed the commercially prepared pellets. The fish were then harvested, weighed, and measured. The bass fed a natural food source had a higher average length (19.6 cm) and weight (95.9 g) than those fed the commercial fish food (17.3 cm and 72.0 g, respectively). The researchers reported P-values for differences in both measurements to be less than 0.001.

a) Explain to someone who has not studied Statistics what the P-values mean here.
b) What advice should the researchers give the people who raise largemouth bass?
c) If that advice turns out to be incorrect, what type of error occurred?

**T 12. Risk** A study of auto safety determined the number of driver deaths per million vehicle sales, classified by type of vehicle. The data below are for 6 midsize models and 6 SUVs. Wondering if there is evidence that drivers of SUVs are safer, we hope to create a 95% confidence interval for the difference in driver death rates for the two types of vehicles. Are these data appropriate for this inference? Explain. (Ross and Wenzel, *An Analysis of Traffic Deaths by Vehicle Type and Model*, March 2002)

| Midsize | 47 | 54 | 64 | 76 | 88 | 97 |
|---------|----|----|----|----|----|----|
| SUV     | 55 | 60 | 62 | 76 | 91 | 109 |

**13. Age** In a study of how depression may affect one's ability to survive a heart attack, researchers reported the ages of the two groups they examined. The mean age of 2397 patients without cardiac disease was 69.8 years (SD = 8.7 years), while for the 450 patients with cardiac disease, the mean and standard deviation of the ages were 74.0 and 7.9, respectively.

a) Create a 95% confidence interval for the difference in mean ages of the two groups.
b) How might an age difference confound these research findings about the relationship between depression and ability to survive a heart attack?

**14. Smoking** In the depression and heart attack research described in Exercise 13, 32% of the diseased group were smokers, compared with only 23.7% of those free of heart disease.

a) Create a 95% confidence interval for the difference in the proportions of smokers in the two groups.
b) Is this evidence that the two groups in the study were different? Explain.
c) Could this be a problem in analyzing the results of the study? Explain.

**15. Computer use** Ofcom, the regulator of communication services in the UK, found in a national poll that 38% of 8-11 year-old online gamers use online chat features to talk to others within the games they are playing, and 58% of 12- to 15-year-old gamers do so. 500 gamers in each age group were surveyed.

a) What kind of sampling design was used?
b) Give a 95% confidence interval for the difference in use of in-game chat features by age group.
c) Does your confidence interval suggest that among all UK gamers a higher percentage of 12- to 15-year-old gamers than 8- to 11-year-old gamers use in-game chat features?

**16. Recruiting** In September 2002, CNN reported on a method of grad student recruiting by the Haas School of Business at U.C.-Berkeley. The school notifies applicants by formal letter that they have been admitted, and also e-mails the accepted students a link to a website that greets them with personalized balloons, cheering, and applause. The director of admissions says this extra effort at recruiting has really worked well. The school accepts 500 applicants each year, and the percentage that actually choose to enroll at Berkeley increased from 52% the year before the web greeting to 54% this year.

a) Create a 95% confidence interval for the change in enrollment rates.
b) Based on your confidence interval, are you convinced that this new form of recruiting has been effective? Explain.

**17. Bimodal** We are sampling randomly from a distribution known to be bimodal.

a) As our sample size increases, what's the expected shape of the sample's distribution?
b) What's the expected value of our sample's mean? Does the size of the sample matter?
c) How is the variability of sample means related to the standard deviation of the population? Does the size of the sample matter?
d) How is the shape of the sampling distribution model affected by the sample size?

**18. Eggs** The ISA Babcock Company supplies poultry farmers with hens, advertising that a mature B300 Layer produces eggs with a mean weight of 60.7 grams. Suppose that egg weights follow a Normal model with standard deviation 3.1 grams.

   a) What fraction of the eggs produced by these hens weigh more than 62 grams?
   b) What's the probability that a dozen randomly selected eggs average more than 62 grams?
   c) Using the 68–95–99.7 Rule, sketch a model of the total weights of a dozen eggs.

**T 19. Hearing** Fitting someone for a hearing aid requires assessing the patient's hearing ability. In one method of assessment, the patient listens to a tape of 50 English words. The tape is played at low volume, and the patient is asked to repeat the words. The patient's hearing ability score is the number of words perceived correctly. Four tapes of equivalent difficulty are available so that each ear can be tested with more than one hearing aid. These lists were created to be equally difficult to perceive in silence, but hearing aids must work in the presence of background noise. Researchers had 24 subjects with normal hearing compare two of the tapes when a background noise was present, with the order of the tapes randomized. Is it reasonable to assume that the two lists are still equivalent for purposes of the hearing test when there is background noise? Base your decision on a confidence interval for the mean difference in the number of words people might misunderstand. (Faith Loven, *A Study of the Interlist Equivalency of the CID W-22 Word List Presented in Quiet and in Noise*. University of Iowa [1981])

| Subject | List A | List B |
|---|---|---|
| 1 | 24 | 26 |
| 2 | 32 | 24 |
| 3 | 20 | 22 |
| 4 | 14 | 18 |
| 5 | 32 | 24 |
| 6 | 22 | 30 |
| 7 | 20 | 22 |
| 8 | 26 | 28 |
| 9 | 26 | 30 |
| 10 | 38 | 16 |
| 11 | 30 | 18 |
| 12 | 16 | 34 |
| 13 | 36 | 32 |
| 14 | 32 | 34 |
| 15 | 38 | 32 |
| 16 | 14 | 18 |
| 17 | 26 | 20 |
| 18 | 14 | 20 |
| 19 | 38 | 40 |
| 20 | 20 | 26 |
| 21 | 14 | 14 |
| 22 | 18 | 14 |
| 23 | 22 | 30 |
| 24 | 34 | 42 |

**20. Cesareans** Where a person lives, and even which hospital they attend, can affect the likelihood of certain medical procedures being performed. For example, a survey of several randomly selected hospitals found that 16.6% of 223 recent births in Vermont involved cesarean deliveries, compared to 18.8% of 186 births in New Hampshire. Is this evidence that the rate of cesarean births in the two states is different?

**21. Pooping dogs** In the interest of determining whether animals are affected by the earth's magnetic field, researchers recorded the direction dogs faced while pooping. Yes, they really researched this! They used a null hypothesis that dogs are equally like to face any direction versus an alternative that they have a preference for north-south alignment (along the magnetic field lines). With a statistical test for points on a circle that is not covered in this course, they found that, under stable magnetic field conditions, the P-value was $6.2 \times 10^{-8}$.

   a) Interpret this P-value in the context of this study.
   b) Write a conclusion for this test.

**T 22. Meals** A college student is on a "meal program." His budget allows him to spend an average of $10 per day for the semester. He keeps track of his daily food expenses for 2 weeks; the data are given in the table. Is there strong evidence that he will overspend his food allowance? Explain.

| Date | Cost ($) | Date | Cost ($) |
|---|---|---|---|
| 7/29 | 15.20 | 8/5 | 8.55 |
| 7/30 | 23.20 | 8/6 | 20.05 |
| 7/31 | 3.20 | 8/7 | 14.95 |
| 8/1 | 9.80 | 8/8 | 23.45 |
| 8/2 | 19.53 | 8/9 | 6.75 |
| 8/3 | 6.25 | 8/10 | 0 |
| 8/4 | 0 | 8/11 | 9.01 |

**23. Health care 2017** In the years leading up to 2017, a great deal of debate took place in the United States around who is responsible for providing health care to its citizens. In 2017, in a random sample of 2504 American adults, 1502 felt the federal government is responsible for ensuring health care coverage for all Americans. We know that if we could ask the entire population of American adults, we would not find that exactly 60% think the federal government is responsible for ensuring health care coverage for all Americans. Construct a 95% confidence interval for the true percentage of American adults who did think so then.

**24. Power** We are reproducing an experiment. How will each of the following changes affect the power of our test? Indicate whether it will increase, decrease, or remain the same, assuming that all other aspects of the situation remain unchanged.

   a) We increase the number of subjects from 40 to 100.
   b) We require a higher standard of proof, changing from $\alpha = 0.05$ to $\alpha = 0.01$.

25. **Herbal cancer** A report in the *New England Journal of Medicine* notes growing evidence that the herb *Aristolochia fangchi* can cause urinary tract cancer in those who take it. Suppose you are asked to design an experiment to study this claim. Imagine that you have data on urinary tract cancers in subjects who have used this herb and similar subjects who have not used it and that you can measure incidences of cancer and precancerous lesions in these subjects. State the null and alternative hypotheses you would use in your study.

26. **Free throws 2017** At the middle of the 2016–2017 NBA season, James Hardin led the league by making 468 of 544 free throws, for a success rate of 86%. But Russell Westbrook was close behind with 425 of 517 (82.2%).
    
    a) Find a 95% confidence interval for the difference in their free throw percentages.
    
    b) Based on your confidence interval, is it certain that Hardin is better than Westbrook at making free throws?

27. **Rain and fire** At the University of California, Riverside, Dr. Richard Minnich collected data on the rainfall in the areas east of Los Angeles. He noted that a decrease in rainfall was responsible for an increase in wildfires over these years. Here is the rainfall data from three regions in the area. (*SMCMA Quarterly* 42(3), Richard A. Minnich)

    | Annual Precipitation (cm) | | | |
    |---|---|---|---|
    | Year | Victorville | 29 Palms | Mitchell's Cavern |
    | 1976–77 | 166 | 140 | 65 |
    | 1977–78 | 257 | 228 | 194 |
    | 1978–79 | 180 | 137 | 113 |
    | 1979–80 | 191 | 221 | 270 |
    | 1980–81 | 58 | 111 | 92 |
    | 1981–82 | 120 | 52 | 90 |
    | 1982–83 | 257 | 220 | 178 |
    | 1983–84 | 74 | 215 | 119 |
    | 1984–85 | 135 | 179 | 187 |

    a) Create and interpret a 95% confidence interval for the mean rainfall difference between Victorville and 29 Palms.
    
    b) Create and interpret a 95% confidence interval for the mean rainfall difference between Victorville and Mitchell's Cavern.
    
    c) Create and interpret a 95% confidence interval for the mean rainfall difference between 29 Palms and Mitchell's Cavern.
    
    d) Does it appear, based on these intervals, that these regions receive different average levels of rainfall?

28. **Teach for America** Several programs attempt to address the shortage of qualified teachers by placing uncertified instructors in schools with acute needs—often in under-resourced areas. A 1999–2000 study compared students taught by certified teachers to others taught by uncertified teachers in the same schools. Reading scores of the students of certified teachers averaged 35.62 points with standard deviation 9.31. The scores of students instructed by uncertified teachers had mean 32.48 points with standard deviation 9.43 points on the same test. There were 44 students in each group. The appropriate *t*-procedure has 86 degrees of freedom. Is there evidence of lower scores with uncertified teachers? Discuss. (*The Effectiveness of "Teach for America" and Other Under-certified Teachers on Student Academic Achievement: A Case of Harmful Public Policy*. Education Policy Analysis Archives [2002])

29. **Legionnaires' disease** In 1974, the Bellevue-Stratford Hotel in Philadelphia was the scene of an outbreak of what later became known as legionnaires' disease. The cause of the disease was finally discovered to be bacteria that thrived in the air-conditioning units of the hotel. Owners of the Rip Van Winkle Motel, hearing about the Bellevue-Stratford, replaced their air-conditioning system. The following data are the bacteria counts in the air of eight rooms, before and after a new air-conditioning system was installed (measured in colonies per cubic foot of air). Has the new system succeeded in lowering the bacterial count? Base your analysis on a confidence interval. Be sure to list all your assumptions, methods, and conclusions.

    | Room Number | Before | After |
    |---|---|---|
    | 121 | 11.8 | 10.1 |
    | 163 | 8.2 | 7.2 |
    | 125 | 7.1 | 3.8 |
    | 264 | 14 | 12 |
    | 233 | 10.8 | 8.3 |
    | 218 | 10.1 | 10.5 |
    | 324 | 14.6 | 12.1 |
    | 325 | 14 | 13.7 |

30. **Teach for America, part II** The study described in Exercise 28 also looked at scores in mathematics and language. Here are software outputs for the appropriate tests. Explain what they show.

    **Mathematics**
    T-TEST OF Mu(1) − Mu(2) = 0
    Mu(Cert) − Mu(NoCert) = 4.53  t(86) = 2.95  p = 0.002

    **Language**
    T-TEST OF Mu(1) − Mu(2) = 0
    Mu(Cert) − Mu(NoCert) = 2.13  t(84) = 1.71  p = 0.045

31. **Electronic cigarettes 2013** In a 2013 randomized controlled study in New Zealand, patients who wanted to quit smoking were randomly assigned to smoke e-cigarettes that deliver nicotine, to use a nicotine patch, or to use a placebo e-cigarette that does not deliver nicotine. The e-cigarette group had a lower rate of quitting than expected by the researchers: 21 of 289 quit smoking.

    a) Write a 95% confidence interval and interpret it in context.
    
    b) If researchers want to determine the level of abstinence (quitting smoking) for e-cigarettes to within 1.5 percentage points, how many subjects should get that treatment?

32. **Online testing** The Educational Testing Service is now administering several of its standardized tests online—the CLEP and GMAT exams, for example. Because taking a test on a computer is different from taking a test with pencil and paper, one wonders if the scores will be the same. To investigate this question, researchers created two versions of an SAT-type test and got 20 volunteers to participate in an experiment. Each volunteer took both versions of the test, one with pencil and paper and the other online. Subjects were randomized with respect to the order in which they sat for the tests (online/paper) and which form they took (Test A, Test B) in which environment. The scores (out of a possible 20) are summarized in the table.

| Subject | Paper Test A | Online Test B | Subject | Paper Test B | Online Test A |
|---|---|---|---|---|---|
| 1 | 14 | 13 | 11 | 8 | 13 |
| 2 | 10 | 13 | 12 | 11 | 13 |
| 3 | 16 | 8 | 13 | 15 | 17 |
| 4 | 15 | 14 | 14 | 11 | 13 |
| 5 | 17 | 16 | 15 | 13 | 14 |
| 6 | 14 | 11 | 16 | 9 | 9 |
| 7 | 9 | 12 | 17 | 15 | 9 |
| 8 | 12 | 12 | 18 | 14 | 15 |
| 9 | 16 | 16 | 19 | 16 | 12 |
| 10 | 7 | 14 | 20 | 8 | 10 |

a) Were the two forms (A/B) of the test equivalent in terms of difficulty? Test an appropriate hypothesis and state your conclusion.
b) Is there evidence that the testing environment (paper/online) matters? Test an appropriate hypothesis and state your conclusion.

**33. Bread** Clarksburg Bakery is trying to predict how many loaves of bread to bake. In the last 100 days, the bakery has sold between 95 and 140 loaves per day. Here are a histogram and the summary statistics for the number of loaves sold for the last 100 days.

| Summary of Sales | |
|---|---|
| Mean | 103 |
| Median | 100 |
| SD | 9.000 |
| Min | 95 |
| Max | 140 |
| $Q_1$ | 97 |
| $Q_3$ | 105.5 |

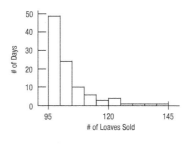

a) Can you use these data to estimate the number of loaves sold on the busiest 10% of all days? Explain.
b) Explain why you can use these data to construct a 95% confidence interval for the mean number of loaves sold per day.
c) Calculate a 95% confidence interval and carefully interpret what that confidence interval means.
d) If the bakery would have been satisfied with a confidence interval whose margin of error was twice as wide, how many days' data could they have used?
e) When the bakery opened, the owners estimated that they would sell an average of 100 loaves per day. Does your confidence interval provide strong evidence that this estimate was incorrect? Explain.

**34. Penguin flippers** Can measurements of the flipper length of penguins be of value when you need to determine the species of a certain penguin? Here are the summary statistics from measurements of the flipper length of two species of penguins studied by Dr. Kristen Gorman (as described in Chapter 23 Exercise 18).

| | Species | |
|---|---|---|
| | Chinstrap | Gentoo |
| Count | 68 | 123 |
| Mean | 195.8 | 217.2 |
| Median | 196 | 216 |
| SD | 7.132 | 6.485 |
| Min | 178 | 203 |
| Max | 212 | 231 |
| Lower Quartile | 191 | 212 |
| Upper Quartile | 201 | 221 |

a) Make parallel boxplots of flipper lengths for the two species.
b) Describe the differences seen in the boxplots.
c) Write a 95% confidence interval for the difference in mean flipper length.
d) Explain what your interval means.
e) Based on your confidence interval, is there evidence of a difference in mean flipper length? Explain.

**35. Insulin and diet** A study published in the *Journal of the American Medical Association* examined people to see if they showed any signs of IRS (insulin resistance syndrome) involving major risk factors for Type 2 diabetes and heart disease. Among 102 subjects who consumed dairy products more than 35 times per week, 24 were identified with IRS. In comparison, IRS was identified in 85 of 190 individuals with the lowest dairy consumption, fewer than 10 times per week.

a) Is this strong evidence that IRS risk is different in people who frequently consume dairy products than in those who do not?
b) Does this indicate that dairy consumption influences the development of IRS? Explain.

**36. Speeding** A newspaper report in August 2002 raised the issue of racial bias in the issuance of speeding tickets. The following facts were noted:

- 16% of drivers registered in New Jersey are Black.
- Of the 324 speeding tickets issued in one month on a 65-mph section of the New Jersey Turnpike, 25% went to Black drivers.
  a) Is the percentage of speeding tickets issued to Black drivers unusually high compared to registrations?
  b) Does this suggest that racial profiling may be present?
  c) What other statistics would you like to know about this situation?

**37. Rainmakers?** In an experiment to determine whether seeding clouds with silver iodide increases rainfall, researchers randomly assigned clouds to be seeded or not. The table summarizes the resulting rainfall (in acre-feet). Create a 95% confidence interval for the average amount of additional rain created by seeding clouds. Explain what your interval means.

| | Unseeded Clouds | Seeded Clouds |
|---|---|---|
| Count | 26 | 26 |
| Mean | 164.588 | 441.985 |
| Median | 44.200 | 221.600 |
| SD | 278.426 | 650.787 |
| IntQRange | 138.600 | 337.600 |
| 25 %ile | 24.400 | 92.400 |
| 75 %ile | 163 | 430 |

**38. Fritos** As a project for an Introductory Statistics course, students checked 6 bags of Fritos marked with a net weight of 35.4 grams. They carefully weighed the contents of each bag, recording the following weights (in grams): 35.5, 35.3, 35.1, 36.4, 35.4, 35.5. Is there evidence that the mean weight of bags of Fritos is less than advertised?

a) Write appropriate hypotheses.
b) Check the assumptions for inference.
c) Test your hypothesis using all 6 weights.
d) Retest your hypothesis with the one unusually high weight removed.
e) What would you conclude about the stated weight?

**39. Color or text?** In an experiment, 32 volunteer subjects are briefly shown seven cards, each displaying the name of a color printed in a different color (example: red, blue, and so on). The subject is asked at random to perform one of two tasks: memorize the order of the words or memorize the order of the colors. Researchers record the number of cards remembered correctly. Then the cards are shuffled and the subject is asked to perform the other task. The table displays the results for each subject. Is there any evidence that either the color or the written word dominates perception?

| Subject | Color | Word | Subject | Color | Word |
|---|---|---|---|---|---|
| 1 | 4 | 7 | 17 | 4 | 3 |
| 2 | 1 | 4 | 18 | 7 | 4 |
| 3 | 5 | 6 | 19 | 4 | 3 |
| 4 | 1 | 6 | 20 | 0 | 6 |
| 5 | 6 | 4 | 21 | 3 | 3 |
| 6 | 4 | 5 | 22 | 3 | 5 |
| 7 | 7 | 3 | 23 | 7 | 3 |
| 8 | 2 | 5 | 24 | 3 | 7 |
| 9 | 7 | 5 | 25 | 5 | 6 |
| 10 | 4 | 3 | 26 | 3 | 4 |
| 11 | 2 | 0 | 27 | 3 | 5 |
| 12 | 5 | 4 | 28 | 1 | 4 |
| 13 | 6 | 7 | 29 | 2 | 3 |
| 14 | 3 | 6 | 30 | 5 | 3 |
| 15 | 4 | 6 | 31 | 3 | 4 |
| 16 | 4 | 7 | 32 | 6 | 7 |

a) What role does randomization play in this experiment?
b) Test appropriate hypotheses and state your conclusion.

**40. And it means?** Every statement about a confidence interval contains two parts: the level of confidence and the interval. Suppose that an insurance agent estimating the mean loss claimed by clients after home burglaries created the 95% confidence interval ($1644, $2391).

a) What's the margin of error for this estimate?
b) Carefully explain what the interval means.
c) Carefully explain what the confidence level means.

**41. Batteries** We work for the "Watchdog for the Consumer" consumer advocacy group. We've been asked to look at a battery company that claims its batteries last an average of 100 hours under normal use. There have been several complaints that the batteries don't last that long, so we decide to test them. To do this, we select 16 batteries and run them until they die. They lasted a mean of 97 hours, with a standard deviation of 12 hours.

a) One of the editors of our newsletter (who does not know statistics) says that 97 hours is a lot less than the advertised 100 hours, so we should reject the company's claim. Explain to him the problem with doing that.
b) What are the null and alternative hypotheses?
c) What assumptions must we make in order to proceed with inference?
d) At a 5% level of significance, what do you conclude?
e) Suppose that, in fact, the average life of the company's batteries is only 98 hours. Has an error been made in part d? If so, what kind?

**42. Hamsters** How large are hamster litters? Among 47 golden hamster litters recorded, there were an average of 7.72 baby hamsters, with a standard deviation of 2.5.

a) Create and interpret a 90% confidence interval.
b) Would a 98% confidence interval have a larger or smaller margin of error? Explain.
c) How many litters must be used to estimate the average litter size to within 1 baby hamster with 95% confidence?

**43. Cramming** Students in two basic Spanish classes were required to learn 50 new vocabulary words. One group of 45 students received the list on Monday and studied the words all week. Statistics summarizing this group's scores on Friday's quiz are given. The other group of 25 students did not get the vocabulary list until Thursday. They also took the quiz on Friday, after "cramming" Thursday night. Then, when they returned to class the following Monday, they were retested—without advance warning. Both sets of test scores for these students are shown.

**Group 1**
Fri.
Number of students = 45
Mean = 43.2 (of 50)
StDev = 3.4
Students passing (score ≥ 40) = 33%

**Group 2**

| Fri. | Mon. | Fri. | Mon. | Fri. | Mon. | Fri. | Mon. |
|---|---|---|---|---|---|---|---|
| 42 | 36 | 50 | 47 | 35 | 31 | 40 | 31 |
| 44 | 44 | 34 | 34 | 43 | 32 | 41 | 32 |
| 45 | 46 | 38 | 31 | 48 | 37 | 48 | 39 |
| 48 | 38 | 43 | 40 | 43 | 41 | 37 | 31 |
| 44 | 40 | 39 | 41 | 45 | 32 | 36 | 41 |
| 43 | 38 | 46 | 32 | 47 | 44 | | |
| 41 | 37 | 37 | 36 | | | | |

a) Did the week-long study group have a mean score significantly higher than that of the overnight crammers?
b) Was there a significant difference in the percentages of students who passed the quiz on Friday?
c) Is there any evidence that when students cram for a test, their "learning" does not last for 3 days?

**44. Surgery and germs** Joseph Lister (for whom Listerine is named!) was a British physician who was interested in the role of bacteria in human infections. He suspected that germs were involved in transmitting infection, so he tried using carbolic acid as an operating room disinfectant. In 75 amputations, he used carbolic acid 40 times. Of the 40 amputations using carbolic acid, 34 of the patients lived. Of the 35 amputations without carbolic acid, 19 patients lived. The question of interest is whether carbolic acid is effective in increasing the chances of surviving an amputation.

   a) What kind of a study was this?
   b) What do you conclude? Support your conclusion by testing an appropriate hypothesis.
   c) What reservations do you have about the design of the study?

**45. Juvenile offenders** According to a 2011 article in the *Journal of Consulting and Clinical Psychology*, Charles Borduin pioneered a treatment called Multisystemic Therapy (MST) as a way to prevent serious mental health problems in adolescents. The therapy involves a total support network including family and community, rather than the more common individual therapy (e.g., visits to a therapist). After a 22-year-long study, one notable fact was that while 15.5% of juveniles who received individual therapy were arrested for a violent felony, only 4.3% of the juveniles treated with MST had been arrested for a violent felony.

   a) Suppose the results are based on sample sizes of 232 juveniles in each group. Create a 99% confidence interval for the reduction in violent felony rate when comparing MST to the traditional individual therapy.
   b) Using your interval, is there evidence of a true reduction for the whole population? Which population is the study investigating?

# PRACTICE EXAM

## I. MULTIPLE CHOICE

1. In December 2001 the research firm GfK surveyed baby boomers (Americans born between 1946 and 1964) with $100,000 or more investable assets. GfK contacted 1006 randomly selected baby boomers by cell phone or landline. The vast majority, 71%, reported that they have provided support to their adult children in the form of helping them pay for college tuition or loans. GfK reported a margin of error of plus or minus 3%. What confidence level were the pollsters using?

   *Source: New York Times, May 5, 2012.*

   A) 90%   B) 95%   C) 96%   D) 98%   E) 99%

2. For the survey described in Exercise 1, if GfK had wanted a smaller margin of error of only 1%, with the same confidence level, how many randomly selected baby boomers would the pollsters need to survey?

   A) 96   B) 336   C) 1865   D) 3010   E) 9054

3. Collecting data from all persons in the population of interest is called
   A) a block.    B) a census.    C) a cluster.
   D) a sample.   E) a stratum.

4. The least squares regression line for predicting the price (in cents) of boxes of cereal from their net weight (in ounces) has a slope of 20. The predicted price for 10-ounce boxes is $4.50. What is the predicted price for 12-ounce boxes?

   A) $4.70   B) $4.74   C) $4.90   D) $5.40   E) $6.90

5. Based on data a coffee shop owner has collected, she believes that 12% of her customers will buy a cookie to go with their coffee and that these purchases are independent. One day as she's getting ready to close, 6 customers enter the shop and she has only 2 cookies left. What is the probability that no more than 2 of these last 6 customers will want a cookie?

   A) 0.026   B) 0.130   C) 0.156   D) 0.610   E) 0.974

6. A store had a sale on a popular soft drink, with a limit of 5 packs per customer. The table shows the probability model for the random variable $X$ = number of packs a customer purchases.

| X | 1 | 2 | 3 | 4 | 5 |
|---|---|---|---|---|---|
| P(X) | 0.20 | 0.16 | 0.10 | 0.24 | 0.30 |

What is the expected number of packs a customer purchases?

A) 2.50   B) 3.00   C) 3.28   D) 3.54   E) 4.00

7. Josie transformed bivariate data by taking the square root of the $y$ values and found the least-squares regression line $\sqrt{\hat{y}} = 7.6 - 0.4x$ to be a useful model. Predict $y$ for $x = 9$.

A) 2   B) 3.5   C) 4   D) 11.5   E) 16

8. Researchers surveyed samples of grade 9 students and grade 12 students about their spending habits. Among several questions, they asked the students how much they spent on fast food during the past week. Here is a summary of the responses:

|  | <$10 | $10–20 | >$20 | Total |
|---|---|---|---|---|
| Grade 9 | 23 | 12 | 8 | 43 |
| Grade 12 | 25 | 19 | 28 | 72 |
| Totals | 48 | 31 | 36 | 115 |

Which best describes the two variables in this study, money spent on fast food and high school class?

A) Disjoint and independent
B) Disjoint but not independent
C) Independent but not disjoint
D) Neither independent nor disjoint
E) Independence cannot be determined because the sample sizes are unequal.

9. Fred is constructing a 95% confidence interval to estimate the average length (in minutes) of movies he watches. His random sample of 15 movies averaged 114 minutes long with a standard deviation of 11 minutes. What critical value and standard error of the mean should he use?

A) $t^* = 2.131, SE = 2.84$
B) $t^* = 2.131, SE = 2.94$
C) $t^* = 2.131, SE = 11$
D) $t^* = 2.145, SE = 2.84$
E) $t^* = 2.145, SE = 2.94$

10. In Exercise 9, Fred constructed a 95% confidence interval to estimate the average length (in minutes) of the movies he watches, with a random sample of 15 movies. He plans to continue collecting data until he has a random sample of 40 movies, and then create a new confidence interval. Which of these statements is most accurate?

A) Both the standard error of the mean and the critical value will probably decrease for the larger sample.
B) The critical value will increase for the larger sample, but the standard error will probably decrease.
C) Because 40 is a large sample size, he will be able to use the Normal distribution.
D) With this larger sample he can be more sure that his 95% confidence interval captures the true mean length of all movies he watches.
E) He should not perform inference at all, because he will probably continue watching movies for many years to come.

11. The weight of a jar of mild salsa has a standard deviation of 5 g and the weight of a jar of hot salsa has a standard deviation of 6 g. A combination pack has 2 jars of mild salsa and 3 jars of hot salsa. What is the variance of the total weight of salsa in the combination pack?

A) 28   B) 61   C) 74   D) 158   E) 424

12. When working with bivariate data, which of these are useful when deciding whether it's appropriate to use a linear model?

I) The scatterplot
II) The residuals plot
III) The correlation coefficient

A) I only
B) II only
C) III only
D) I and II only
E) I, II, and III

13. Consider the following scenarios.

I) A group of musicians count how many times they can snap their fingers in 10 seconds with their dominant hands and with their nondominant hands. They will test to see if musicians can snap faster with dominant hands than nondominant hands.
II) A group of musicians is trying to measure the effectiveness of practice time. Half of them practice a difficult piece of music for one hour each day while the other half practice two hours each day. After a week, each musician plays the piece as a judge counts the number of errors. They will test to see if more practice results in fewer errors.

Which are the proper choices for the hypothesis tests?

A) Both scenarios require a 2-sample $t$-test.
B) Both scenarios require a matched-pairs $t$-test.
C) Scenario I calls for a 2-sample $t$-test, and Scenario II calls for a matched-pairs $t$-test.
D) Scenario I calls for a matched-pairs $t$-test, and Scenario II calls for a 2-sample $t$-test.
E) The proper tests cannot be determined until we see the actual data that are collected.

14. A team of researchers studied the relationship between parenting styles on internet gaming disorder (IGd) among middle-school children who play video games in Iran. They compared the proportion of middle-school children of authoriatrian parents who have IGD to the proportion of middle-school children with permissive parents who have IGD. Which of the following is NOT a condition that the team should check before creating a confidence interval for the difference in population proportions?

A) The samples are each approximately Normal.
B) There at least 10 successes and 10 failures in each sample.
C) The authoritarian sample and permissive sample are independent.
D) The people in each sample were selected at random.
E) No more than 10% of each population was sampled.

**15.** The least squares regression line for a set of (*Age, Skill Set*) data is $\hat{y} = 5.0x + 0.7$. The data point for age 6 has residual $-1.4$. What is the skill score for age 6?

A) $-4.6$  B) 4.6  C) 29.3  D) 30.7  E) 32.1

**16.** Proponents of Neuro-Linguistic Programming (NLP) claim that certain eye movements are reliable indicators of lying. According to this notion, a person looking up to their right suggests a lie whereas looking up to their left is indicative of truth telling. In 2012, British researchers tested the claim that you can spot a lie by watching a person's eyes. They recruited 50 male volunteers and randomly assigned some of them to be trained in using the NLP technique. Then all participants watched 32 video clips that each showed two interviews, one including a true statement and the other a lie. After each clip, the participants identified the interviewee they thought was lying. The 21 NLP-trained participants averaged 16.33 correct decisions with a standard deviation of 3.53, compared to the 29 participants in the control group who averaged 16.59 correct with standard deviation 3.84.

What type of test should be conducted to answer the question, "Do these data provide evidence that you can spot a lie by watching a person's eyes?"

A) One proportion *z*-test
B) One sample *t*-test
C) Matched-pairs *t*-test
D) *z*-test for the difference of two proportions
E) Two sample *t*-test for the difference of two means

*Source:* R. Wiseman, C. Watt, L. ten Brinke, S. Porter, S.-L. Couper et al., "The Eyes Don't Have It: Lie Detection and Neuro-Linguistic Programming," *PLoS ONE* (2012), 7(7): e40259.

**17.** Two Statistics classes took a practice exam. This computer output shows the summary statistics:

| Group   | Count | Mean | Median | StdDev |
|---------|-------|------|--------|--------|
| Class 1 | 32    | 80.4 | 78.5   | 6.1    |
| Class 2 | 24    | 76.3 | 74.2   | 7.0    |

What is the overall mean for all of the students on this exam?

A) 76.35  B) 76.657  C) 77.28
D) 78.35  E) 78.643

**18.** One day in gym class students took a physical fitness test by doing push-ups and sit-ups. The standard deviation of the number of sit-ups they were able to do was 7 and the standard deviation of the number of push-ups was 2. A Statistics student used these data to create a least squares regression line to predict the number of sit-ups a student was able to do based on the number of push-ups the student did. Which of the following could NOT be the slope of that line?

A) $-2$  B) $-0.5$  C) 1
D) 3  E) 4

**19.** Using the data collected in the British study of NLP training as described in Exercise 16, the formula for the critical value of the test statistic is:

A) $z = \dfrac{0.5103 - 0.5}{\sqrt{\dfrac{(0.5)(0.5)}{21}}}$

B) $z = \dfrac{0.5103 - 0.5184}{\sqrt{(0.515)(0.485)\left(\dfrac{1}{21} + \dfrac{1}{29}\right)}}$

C) $t = \dfrac{16.33 - 16}{\dfrac{3.53}{\sqrt{32}}}$

D) $z = \dfrac{16.33 - 16.59}{\dfrac{3.84}{\sqrt{21}}}$

E) $t = \dfrac{16.33 - 16.59}{\sqrt{\dfrac{3.53^2}{21} + \dfrac{3.84^2}{29}}}$

**20.** A torn meniscus is a common type of knee injury. In the case of minor tears, there's some question about whether initial surgery followed by physical therapy (PT) results in a better outcome than just physical therapy alone. To find out, experimenters will randomly assign some subjects with this type of injury to have surgery followed by PT and others to just do PT, and then compare the recovery experiences of the two groups. Which of statements A–D is false?

A) If the researchers mistakenly conclude that the surgery was beneficial, they will commit a Type I error.
B) The power of this test is its ability to detect that there really is a benefit of surgery.
C) The more participants used in this experiment, the higher the power of the test will be.
D) Demanding stronger evidence by using $\alpha = 0.01$ instead of $\alpha = 0.05$ would give the researchers greater power.
E) None; statements A–D are all true.

**21.** A researcher re-expressed a dataset of scientific measurements. In the display below, the upper box and whisker plot summarizes original data and the lower box and whisker plot summarizes the re-expressed data. Which re-expression might the researcher have used?

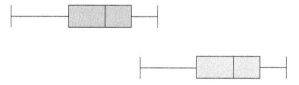

A) Subtract 5 from each data value
B) Add 5 to each data value
C) Multiply each data value by 5
D) Divide each data value by 5
E) Convert the data values to *z*-scores

**22.** Which of these describes an advantage of using blocking in an experiment?

A) It increases the sample size.
B) It ensures that treatments are assigned randomly.
C) It reduces the variability due to differences between blocks.
D) It allows a conclusion of cause and effect.
E) It allows the results to be generalized to a larger group.

**23.** A medical study compares the time it takes for a wound to heal with two different medications. At the end of the study, the conclusion is stated as a 95% confidence interval. The average difference between the two treatments (in days) was reported to be $(-2.45, 3.86)$. From this interval we can conclude that

A) One of the medications is statistically significantly better than the other, because zero is contained in the interval.
B) The experiment must have made a Type I error, because zero is in the interval.
C) Because zero is in the interval, there is no statistically significant difference between the two medications.
D) Because more of the interval is above zero than below it, there must be a slight statistical difference between the treatments.
E) The researcher made a mistake because the difference should not result in a negative number of days.

24. Cam and Denise wish to check the data from their lab experiment to see if the distribution is approximately Normal. Which of the following would be most useful for assessing normality?
A) Boxplot  B) Stem-and-leaf plot  C) Bar graph
D) Scatterplot  E) Residuals plot

25. The senior Class of 2022 at Rancho High strongly believes that it has better grades than its predecessors. The class president, a Statistics student, calculates the mean GPA for the 725 seniors to be 2.97 with a standard deviation of 0.67. The Class of 2021 had a mean of 2.83 with a standard deviation of 0.81. Which of the following is correct?

A) The means should be compared with a one-sample $t$-test, using 2021's mean as the null hypothesis.
B) The means should be compared using a two-sample $t$-test.
C) The means should be compared using a confidence interval for the difference of means.
D) The means should be compared by a matched-pairs $t$-test on the mean difference in GPAs.
E) Inference is not necessary because the data are for both populations.

26. An ecologist who analyzes water samples tests the null hypothesis that any contaminants in the water are below dangerous concentrations. Because the ecologist uses $\alpha = 0.05$, a set of samples from a small lake that produced a P-value of 0.07 led to the conclusion that the evidence did not point to unsafe water conditions. Which is true?

A) There's a 7% chance the lake's water really is safe.
B) There's a 93% chance the lake's water really is safe.
C) There's a 7% chance their sample would have shown as much contamination as it did even if the lake's water really is safe.
D) If the lake's water really is unsafe, there's a 5% chance the ecologist wouldn't notice.
E) If the ecologist had taken more samples they probably would have rejected the null and concluded that the water was unsafe.

27. The distribution of a large set of temperatures is approximately Normal with a mean of 60° and a standard deviation of 5. Estimate the interquartile range for these data.
A) 6.7°  B) 7.5°  C) 10.0°  D) 13.4°  E) 15.0°

28. A clothing store has two locations in a city. They randomly selected sales receipts from 20 cusomers from the first branch and 20 customers from the second branch. They wish to test the belief that on average purchases at the first store are larger than at the second. Which alternate hypothesis is correct?

A) $H_A: \mu_d > 0$; the paired difference in store 1 minus store 2 spending is greater than zero.
B) $H_A: \mu_1 - \mu_2 > 0$; the average purchase at store 1 minus the average purchase at store 2 is greater than zero.
C) $H_A: \mu \neq \mu_m$; the average purchase at store 1 is different from the average purchase at store 2.
D) $H_A: \bar{x}_f > \bar{x}_m$; the average purchase for the 20 customers from store 1 is greater than the average for the 20 customers from store 2.
E) $H_A: \bar{x}_f - \bar{x}_m \neq 0$; the average purchase for the 20 customers from store 1 is different than the average for the 20 customers from store 2.

29. A random sample of 50 grade 9 students and another random sample of 50 grade 12 students were asked on the first day of the school year how they were feeling about the upcoming year. They selected from five options the one word that best described how they felt at the time. The results are summarized in the mosaic plot below:

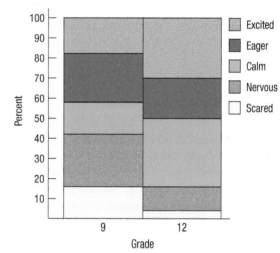

Based on what you see here, which of these statements is (are) true?
 I. There is an association between *Grade* and *Emotion*.
 II. Grade 12 students and grade 9 students are approximately equally likely to say they are calm.
 III. Grade 9 students are approximately equally likely to say they are scared or that they are calm.

A) I only
B) II only
C) III only
D) I and II only
E) I and III only

30. A random sample of students at a college shows that 54 of 200 students had part-time jobs. Which of the following is the correct formula for a 90% confidence interval for the proportion of all students at this college with part-time jobs?

A) $0.27 \pm 1.28\sqrt{\dfrac{(0.27)(0.73)}{200}}$

B) $0.27 \pm 1.28\sqrt{\dfrac{(0.5)(0.5)}{200}}$

C) $0.27 \pm 1.645\sqrt{\dfrac{(0.27)(0.73)}{200}}$

D) $0.27 \pm 1.645\sqrt{\dfrac{(0.5)(0.5)}{200}}$

E) $0.27 \pm 1.96\sqrt{\dfrac{(0.27)(0.73)}{200}}$

31. A two-factor experiment will investigate the best way to make cookies by trying 3 different oven temperatures and 4 different baking times. How many different treatments are there?

    A) 2    B) 7    C) 9    D) 12    E) 20

32. Let Q1 represent the first quartile, Q2 represent the second quartile (or median), and Q3 represent the third quartile. Which of the following computes an important value when considering which data values may be outliers?

    A) Q2 + (Q3 − Q1)
    B) Q2 + 1.5(Q3 − Q1)
    C) Q3 + (Q2 − Q1)
    D) Q3 + 1.5(Q2 − Q1)
    E) Q3 + 1.5(Q3 − Q1)

33. Every scale has some measurement error, and such errors are roughly normally distributed. A certain deli scale is correctly calibrated, but the standard deviation of the errors is 0.15 ounces. What is the probability that a measurement on this scale is within 0.30 ounces of the correct weight?

    A) 0.118   B) 0.236   C) 0.477   D) 0.954   E) 0.977

34. Joanne needs to test boxes of pasta to see if they contain the correct amount of product. Each of the 500 boxes in a recent batch has a unique serial number from 1001 to 1500 stamped on it. She will pick a random sample of 50 boxes by generating a random integer between 1001 and 1010 to select the first box, and then selecting every tenth number in sequence. Her method is called

    A) cluster sampling.          B) convenience sampling.
    C) multistage sampling.       D) stratified sampling.
    E) systematic sampling.

35. On a college admissions test where the scores were approximately normally distributed, Amy's score of 31 was at the 98th percentile. If the mean of the test scores was 20, the standard deviation was approximately

    A) 4.7.   B) 5.4.   C) 5.6.   D) 6.7.   E) 9.7.

## II. FREE RESPONSE

1. Summary statistics for data relating the latitude (°North) and average January low temperature (°F) for 55 large U.S. cities are given below. A scatterplot suggests a linear model is appropriate.

   |            | Latitude | Avg Jan Low |
   |------------|----------|-------------|
   | Mean       | 39.02    | 26.44       |
   | StDev      | 5.42     | 13.49       |
   | Correlation |         | −0.848      |

   a) Write the equation of the least squares regression line that predicts a U.S. city's average January low temperature based on its latitude.
   b) Interpret the slope in context.
   c) Do you think the y-intercept is meaningful? Explain.
   d) For a certain city the residual is −4.2°. Explain what that means.

2. Customers using the drive up window at fast food restaurants are sometimes greeted by a message encouraging them to purchase an item that's currently "on special." Marketing researchers at one restaurant chain want to test the effectiveness of such a message. They have selected 10 of their restaurants in various locations for an experiment, and are considering two different designs.

   *Design I*: 5 of the restaurants will use the message and the other 5 will not.
   *Design II*: Each restaurant will alternate between playing or not playing the message as customers arrive at the drive-thru order station.

   a) Describe a method of assigning the restaurants to the groups for Design I using this list of random digits: 08530 08629 32279 29478 50228
   b) Which do you think is the better design? Explain why.
   c) For the design you chose in part b, describe the data you would collect and the method of analysis you would use to determine whether there is evidence that the message improves sales of the special item.

3. One of the ways contestants on a television game show can win cash is by playing "Bucks-in-the-Box." In this game the host offers the contestant 2 boxes that appear to be identical. The contestant reaches into one of the boxes and draws out an envelope, winning the money inside it. While the contestant can't tell which box is which, the contents are quite different. One box has 5 envelopes; 4 contain $100 and the other contains $1000. The other box has only 4 envelopes, with $100 in one of them and $1000 in each of the other 3.

   a) What is the probability that a Bucks-in-the-Box contestant will win $1000?
   b) If a contestant wins $1000, what's the probability it was the only $1000 envelope in that box?
   c) What are Bucks-in-the-Box contestants' expected winnings?

4. A bank is considering a marketing campaign that would urge parents of preschool children to start college savings program, targeting families with annual incomes above $40,000. Bank officials will proceed with the campaign only if it appears that at least 10% of such families might be interested. When a pilot test presented the campaign materials to a random sample of 200 parents, 26 expressed some interest in the college savings plan.

   a) At the 5% level of significance, do these sample results provide evidence that the bank should pursue the marketing campaign?
   b) Describe two ways that the bank could increase the power of the test, and explain a disadvantage of each.

5. Bicycle frames can be made from carbon, steel, or other materials. Carbon frames are much more expensive than steel, but they are also much lighter. Could this lighter frame have a significant impact on speed? In 2010 a doctor in England, who commutes daily to work 27 miles roundtrip on a bicycle, ran his own randomized experiment to investigate.

   Each day for several months Dr. Groves flipped a coin to determine whether he would ride his steel frame bicycle or his carbon frame bicycle to work and back. He rode a total of

30 journeys on the steel frame bike and 26 journeys on his carbon frame bicycle, recording his total commuting time for each day. The summary statistics are shown below:

|  | Steel | Carbon |
|---|---|---|
| # of Commutes | 30 | 26 |
| Mean (min) | 107.80 | 109.35 |
| SD (min) | 4.90 | 6.25 |

*Source:* Jeremy Groves, "Bicycle Weight and Commuting Time: A Randomized Trial," *Significance: Statistics Making Sense* (June 2011), vol. 8, no. 2: 95–97.

a) Why would you prefer to see the actual data before proceeding with inference?
b) Construct and interpret a 95% confidence interval for the difference in average commuting times.
c) Explain in this context what "95% confidence" means.
d) Based on your confidence interval, do the doctor's data provide evidence that there's a difference in average commuting time for the two bikes? Explain.

6. Hank likes to simulate baseball seasons by playing the manager of his favorite team. In real life the team won 97 of 162 games, but in Hank's simulation his team won 104 games. Hank claims his success is due to his excellent ability as manager, but his friend Joe says he was just lucky. In attempts to settle the argument, Joe finds a $z$-score and Hank runs a computer simulation.

a) What's Joe's best estimate of the probability this team can win a game?
b) What did Joe calculate to be the $z$-score for 104-win season?
c) What conclusion did Joe reach about Hank's managerial prowess?
d) Hank ran 50 simulated seasons letting the computer's randomness act as the team's manager. The number and frequency of games won are shown in the histogram below.

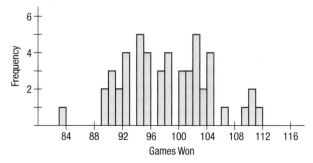

Does the histogram provide Hank with evidence that his 104-win season was due to more than luck? Explain.

# 25

# Comparing Counts

| WHO | Executives of *Fortune* 400 companies |
|---|---|
| WHAT | Zodiac birth sign |
| WHY | Maybe the researcher was a Gemini and naturally curious? |

Does your zodiac sign predict how successful you will be later in life? *Fortune* magazine collected the zodiac signs of 256 heads of the largest 400 companies. The table shows the number of births for each sign.

| Births | Sign | Births | Sign |
|---|---|---|---|
| 23 | Aries | 18 | Libra |
| 20 | Taurus | 21 | Scorpio |
| 18 | Gemini | 19 | Sagittarius |
| 23 | Cancer | 22 | Capricorn |
| 20 | Leo | 24 | Aquarius |
| 19 | Virgo | 29 | Pisces |

Birth counts by sign for 256 *Fortune* 400 executives

We can see some variation in the number of births per sign, and there *are* more Pisces, but is that enough to claim that successful people are more likely to be born under some signs than others?

## Goodness-of-Fit Tests

If these 256 births were distributed uniformly across the year, we would expect about 1/12 of them to occur under each sign of the zodiac. That suggests 256/12, or about 21.3 births per sign. How closely do the observed numbers of births fit this simple "null" model? A hypothesis test to address this question is called a test of **"goodness-of-fit."** The name suggests a certain badness-of-grammar, but it is quite standard. After all, we are asking whether the model that births are uniformly distributed over the signs fits the data good, . . . er, well.

Goodness-of-fit involves testing a hypothesis. We have specified a model for the distribution and want to know whether it fits. There is no single parameter to estimate, so a confidence interval wouldn't make any sense. A one-proportion $z$-test won't work because we have 12 hypothesized proportions, one for each sign. We need a test that considers all of them together and gives an overall idea of whether the observed distribution differs from the hypothesized one.

### FOR EXAMPLE
#### Finding Expected Counts

Birth month may not be related to success as a CEO, but what about on the ball field? It has been proposed by some researchers that children who are the older ones in their class at school perform better in sports initially because they're bigger, and that these children then get more coaching and encouragement, making skill gaps increase over the years. Could that make a difference in who makes it to the professional level in sports?

| Month | Ballplayer Count | National Birth % | Month | Ballplayer Count | National Birth % |
|---|---|---|---|---|---|
| 1 | 137 | 8% | 7 | 102 | 9% |
| 2 | 121 | 7% | 8 | 165 | 9% |
| 3 | 116 | 8% | 9 | 134 | 9% |
| 4 | 121 | 8% | 10 | 115 | 9% |
| 5 | 126 | 8% | 11 | 105 | 8% |
| 6 | 114 | 8% | 12 | 122 | 9% |
|   |     |    | Total | 1478 | 100% |

Baseball is a remarkable sport, in part because so much data are available, including the birth date of every player who ever played in a major league game. Because the effect we're suspecting may be due to relatively recent policies (and to keep the sample size moderate), we'll consider the birth months of 1478 major league players born since 1975. We can also look up the national demographic statistics to find what percentage of people were born in each month. Let's test whether the observed distribution of ballplayers' birth months shows just random fluctuations or whether it represents a real deviation from the national pattern.

QUESTION: How can we find the expected counts?

ANSWER: There are 1478 players in this set of data. I found the national percentage of births in each month. Based on the national birth percentages, I'd expect 8% of players to have been born in January, and $1478(0.08) = 118.24$. I won't round off, because expected "counts" needn't be integers. Multiplying 1478 by each of the birth percentages gives the expected counts shown in the table to the right.

| Month | Expected | Month | Expected |
|---|---|---|---|
| 1 | 118.24 | 7 | 133.02 |
| 2 | 103.46 | 8 | 133.02 |
| 3 | 118.24 | 9 | 133.02 |
| 4 | 118.24 | 10 | 133.02 |
| 5 | 118.24 | 11 | 118.24 |
| 6 | 118.24 | 12 | 133.02 |

### JUST CHECKING

Some people can roll their tongues, like the child in the picture. (Can you?) Some people's earlobes are attached to their necks, while others dangle freely. You wouldn't think these two traits have anything to do with each other, but they're actually controlled by the same gene!

Genetic theory predicts that people will have neither, one, or both of these traits in the ratio 1:3:3:9, as described in the table.

| Tongue | Earlobes | Predicted Fraction |
|---|---|---|
| Non-curling | Attached | 1/16 |
| Non-curling | Free | 3/16 |
| Curling | Attached | 3/16 |
| Curling | Free | 9/16 |

1. The 124 students in a college Biology class plan to collect data on themselves about tongue rolling and earlobes. How many people should they expect to find in each of the four groups?

## Assumptions and Conditions

These data are organized in tables as we saw in Chapter 2, and the assumptions and conditions reflect that. Rather than having an observation for each individual, we typically work with summary counts in categories. In our example, we don't see the birth signs of each of the 256 executives, only the totals for each sign.

**Counted Data Condition** The values in each **cell** must be *counts* for the categories of a categorical variable. This might seem a simplistic, even silly condition. But we can't apply these methods to proportions, percentages, or measurements just because they happen to be organized in a table.

**Independence Assumption** The responses counted in the cells should be independent. The easiest case is when the individuals who are counted in the cells are sampled independently from some population. That's what we'd like to have if we want to draw conclusions about that population. Randomness can arise in other ways, though. For example, these Fortune 400 executives are not a random sample of company executives, but it's reasonable to think that their birth dates should be randomly distributed throughout the year.

If we want to generalize to a large population, we should check two more conditions:

- **Randomization Condition:** The individuals who have been counted should be a random sample from the population of interest.
- **10% Condition:** Our sample is less than 10% of the population.

**Sample Size Assumption** We must have enough data for the methods to work, so we usually check the

- **Expected Cell Frequency Condition:** We should expect to see more than 5 individuals in each cell.

This is quite similar to the condition that $np$ and $nq$ be greater than 10 when we tested proportions. In our astrology example, assuming equal births in each zodiac sign leads us to expect 21.3 births per sign, so the condition is easily met here.

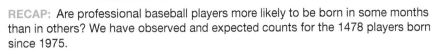

### FOR EXAMPLE
### Checking Assumptions and Conditions

RECAP: Are professional baseball players more likely to be born in some months than in others? We have observed and expected counts for the 1478 players born since 1975.

QUESTION: Are the assumptions and conditions met for performing a goodness-of-fit test?

ANSWER:

✓ **Counted Data Condition:** I have month-by-month counts of ballplayer births.

✓ **Independence Assumption:** These births were independent.

> ✓ **Randomization Condition:** Although they are not a random sample, we can take these players to be representative of players past and future, at least with respect to birthdays.
>
> ✓ **10% Condition:** These 1478 players are less than 10% of the population of 16,804 players who have ever played (or will play) major league baseball.
>
> ✓ **Expected Cell Frequency Condition:** The expected counts extend from 103.46 to 133.02, all much greater than 5.
>
> It's okay to use these data for a goodness-of-fit test.

## JUST CHECKING

A Biology class of 124 students collected data on themselves to check the genetic theory about the frequency of tongue-rolling and free-hanging earlobes.

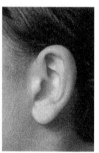
Free

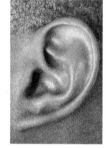
Attached

Their results are summarized in the table.

| Tongue | Earlobes | Observed Count | Expected Count |
|---|---|---|---|
| Non-curling | Attached | 12 | 7.75 |
| Non-curling | Free | 22 | 23.25 |
| Curling | Attached | 31 | 23.25 |
| Curling | Free | 59 | 69.75 |

2. Is it okay to proceed with inference? Check the assumptions and conditions.

> **NOTATION ALERT**
>
> We compare the counts *observed* in each cell with the counts we *expect* to find. The usual notation uses $O$'s and $E$'s or abbreviations such as those we've used here. The method for finding the expected counts depends on the model.

> **NOTATION ALERT**
>
> In Statistics the Greek letter $\chi$ (chi) is used to represent both a test statistic and the associated sampling distribution. This is another violation of our "rule" that Greek letters represent population parameters. Here we are using a Greek letter to name a family of distribution models and a statistic.

## Calculations

Are the discrepancies between what we observed and what we expected just natural sampling variability, or are they so large that they indicate something important? It's natural to look at the *differences* between these observed and expected counts, denoted $(Obs - Exp)$. Just adding up these differences won't work because some are positive; others negative. We've been in this predicament before, and we handle it the same way now: We square them. That gives us positive values and focuses attention on any cells with large differences from what we expected. Because the differences between observed and expected counts generally get larger the more data we have, we also need to get an idea of the *relative* sizes of the differences. To do that, we divide each squared difference by the expected count for that cell.

The test statistic, called the **chi-square** (or chi-squared) **statistic**, is found by adding up the sum of the squares of the deviations between the observed and expected counts divided by the expected counts:

$$\chi^2 = \sum_{all\ cells} \frac{(Obs - Exp)^2}{Exp}.$$

The chi-square statistic is denoted $\chi^2$, where $\chi$ is the Greek letter chi (pronounced "ky" as in "sky"). It refers to a family of sampling distribution models we have not seen before called (remarkably enough) the **chi-square models**.

This family of models, like the Student's $t$-models, differ only in the number of degrees of freedom. The number of degrees of freedom for a goodness-of-fit test is $n - 1$. Here, however, $n$ is *not* the sample size, but instead is the number of categories. For the zodiac example, we have 12 signs, so our $\chi^2$ statistic has 11 degrees of freedom.

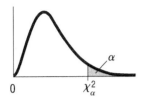

TI-*nspire*

**The $\chi^2$ Models.** See what a $\chi^2$ model looks like, and watch it change as you change the degrees of freedom.

If the observed counts perfectly matched the expected, the $\chi^2$ value would be 0. The greater the differences, positive or negative, the larger $\chi^2$ becomes. If the calculated value is large enough, we'll reject the null hypothesis. What's "large enough" depends on the degrees of freedom. There are tables you can use to estimate a P-value based on the test statistic and the degrees of freedom. But they usually give only ballpark estimates and are, frankly, pretty archaic. We recommend using technology to find a P-value.

But what does rejecting the null tell us? Because squaring the differences makes all the deviations positive whether the observed counts were higher or lower than expected, there's no direction to the rejection of the null. Because we only worry about unexpectedly large $\chi^2$ values, this behaves like a one-sided test, but it's really *many*-sided. With so many proportions, there are many ways the null model can be wrong. All we know is that it doesn't fit.

### FOR EXAMPLE
#### Doing a Goodness-of-Fit Test

**RECAP:** The birth months data for major league baseball players are appropriate for performing a $\chi^2$ test.

**QUESTIONS:** What are the hypotheses, and what does the test show?

**ANSWER:** $H_0$: The distribution of birth months for major league ballplayers is the same as that for the general population.

$H_A$: The distribution of birth months for major league ballplayers differs from that of the rest of the population.

$$df = 12 - 1 = 11$$
$$\chi^2 = \sum \frac{(Obs - Exp)^2}{Exp}$$
$$= \frac{(137 - 118.24)^2}{118.24} + \frac{(121 - 103.46)^2}{103.46} + \cdots$$
$$= 26.48 \text{ (by technology)}$$

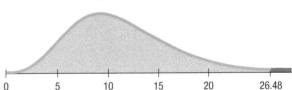

P-value $= P(\chi^2_{11} \geq 26.48) = 0.0055$ (by technology)

Because of the small P-value, I reject $H_0$; there's evidence that birth months of major league ballplayers have a different distribution from the rest of us.

### STEP-BY-STEP EXAMPLE
#### A Chi-Square Test for Goodness-of-Fit

We have counts of 256 executives in 12 zodiac sign categories. The natural null hypothesis is that birth dates of executives are divided equally among all the zodiac signs.

**QUESTION:** Are CEOs more likely to be born under some zodiac signs than others?

# 694 PART VII Inference when Variables Are Related

**THINK** **PLAN** State what you want to know.

Identify the variables and check the W's.

**HYPOTHESES** State the null and alternative hypotheses. For $\chi^2$ tests, it's usually easier to do that in words than in symbols.

**MODEL** Make a picture. The null hypothesis is that the frequencies are equal, so a bar chart (with a line at the hypothesized "equal" value) is a good display.

I want to know whether births of successful people are uniformly distributed across the signs of the zodiac. I have counts of 256 Fortune 400 executives, categorized by their birth sign.

$H_0$: Births are uniformly distributed over zodiac signs.[1]

$H_A$: Births are not uniformly distributed over zodiac signs.

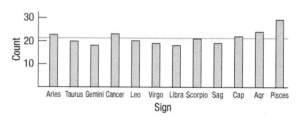

The bar chart shows some variation from sign to sign, and Pisces is the most frequent. But it is hard to tell whether the variation is more than I'd expect from random chance.

Think about the assumptions and check the conditions.

✓ **Counted Data Condition:** I have counts of the number of executives in 12 categories.

✓ **Independence Assumption:** The birth dates of executives should be independent of each other.

✓ **Randomization Condition:** These executives are not a random sample of any population. However, the chi-square goodness-of-fit test is still appropriate because our null hypothesis model is essentially saying that each CEO's sign is "selected" at random from among the 12 signs.

✓ **Expected Cell Frequency Condition:** The null hypothesis expects that 1/12 of the 256 births, or 21.333, should occur in each sign. These expected values are all greater than 5, so the condition is satisfied.

Specify the sampling distribution model.

Name the test you will use.

The conditions are satisfied, so I'll use a $\chi^2$ model with $12 - 1 = 11$ degrees of freedom and do a **chi-square goodness-of-fit test**.

**SHOW** **MECHANICS** Each cell contributes an $\frac{(Obs - Exp)^2}{Exp}$ value to the chi-square sum. We add up these components for each zodiac sign. If you do it by hand, it can be helpful to arrange the calculation in a table. We show that after this Step-by-Step Example.

The $\chi^2$ models are skewed to the high end, and change shape depending on the degrees of freedom. The P-value considers only the right tail. Large $\chi^2$ statistic values correspond to small P-values, which lead us to reject the null hypothesis.

The P-value is the area in the upper tail of the $\chi^2$ model above the computed $\chi^2$ value.

The expected value for each zodiac sign is 21.333.

$$\chi^2 = \sum \frac{(Obs - Exp)^2}{Exp} = \frac{(23 - 21.333)^2}{21.333}$$
$$+ \frac{(20 - 21.333)^2}{21.333} + \cdots$$
$$= 5.094 \text{ for all 12 signs.}$$

P-value $= P(\chi^2_{11} > 5.094) = 0.926$

---

[1] It may seem that we have broken our rule of thumb that null hypotheses should specify parameter values. If you want to get formal about it, the null hypothesis is that

$$p_{Aries} = p_{Taurus} = \cdots = p_{Pisces}.$$

That is, we hypothesize that the true proportions of births of CEOs under each sign are equal. The role of the null hypothesis is to specify the model so that we can compute the test statistic. That's what this one does.

> **TELL** **CONCLUSION** Link the P-value to your decision. Remember to state your conclusion in terms of what the data mean, rather than just making a statement about the distribution of counts.

The P-value of 0.926 says that if the zodiac signs of executives were in fact distributed uniformly, an observed chi-square value of 5.09 or higher would occur about 93% of the time. This certainly isn't unusual, so I fail to reject the null hypothesis, and conclude that these data show virtually no evidence that executives are more likely to have certain zodiac signs.

## The Chi-Square Calculation

In practice, you almost never compute the chi-square statistic yourself; technology gives it to you, along with the P-value. But it can still be helpful to understand how the statistic is computed, so let's make the chi-square procedure very clear. Here are the steps:

1. **Find the expected values.** These come from the null hypothesis model. Every model gives a hypothesized proportion for each cell. The expected value is the product of the total number of observations times this proportion.

    For our example, the null model hypothesizes *equal* proportions. With 12 signs, 1/12 of the 256 executives should be in each category. The expected number for each sign is 21.333.
2. **Compute the residuals.** Once you have expected values for each cell, find the residuals, $Observed - Expected$.
3. **Square the residuals.**
4. **Compute the components.** Now find the **component**, $\dfrac{(Observed - Expected)^2}{Expected}$, for each cell.
5. **Find the sum of the components.** That's the chi-square statistic.
6. **Find the degrees of freedom.** It's equal to the number of cells minus one. For the zodiac signs, that's $12 - 1 = 11$ degrees of freedom.
7. **Test the hypothesis.** Large chi-square values mean lots of deviation from the hypothesized model, so they give small P-values. Look up the critical value from a table of chi-square values, or use technology to find the P-value directly.

The steps of the chi-square calculations are often laid out in tables using one row for each category, with columns for observed counts, expected counts, residuals, squared residuals, and the contributions to the chi-square total like this:

| Sign | Observed | Expected | Residual = (Obs − Exp) | (Obs − Exp)² | Component = $\dfrac{(Obs - Exp)^2}{Exp}$ |
|---|---|---|---|---|---|
| Aries | 23 | 21.333 | 1.667 | 2.778889 | 0.130262 |
| Taurus | 20 | 21.333 | −1.333 | 1.776889 | 0.083293 |
| Gemini | 18 | 21.333 | −3.333 | 11.108889 | 0.520737 |
| Cancer | 23 | 21.333 | 1.667 | 2.778889 | 0.130262 |
| Leo | 20 | 21.333 | −1.333 | 1.776889 | 0.083293 |
| Virgo | 19 | 21.333 | −2.333 | 5.442889 | 0.255139 |
| Libra | 18 | 21.333 | −3.333 | 11.108889 | 0.520737 |
| Scorpio | 21 | 21.333 | −0.333 | 0.110889 | 0.005198 |
| Sagittarius | 19 | 21.333 | −2.333 | 5.442889 | 0.255139 |
| Capricorn | 22 | 21.333 | 0.667 | 0.444889 | 0.020854 |
| Aquarius | 24 | 21.333 | 2.667 | 7.112889 | 0.333422 |
| Pisces | 29 | 21.333 | 7.667 | 58.782889 | 2.755491 |
| | | | | | $\Sigma = 5.094$ |

### *HOW BIG IS BIG?

When we calculated $\chi^2$ for the zodiac sign example, we got 5.094. That value would have been big for $z$ or $t$, leading us to reject the null hypothesis. Not here, though. Were you surprised that $\chi^2 = 5.094$ had a huge P-value of 0.926? What *is* big for a $\chi^2$ statistic, anyway?

Think about how $\chi^2$ is calculated. In every cell, any deviation from the expected count contributes to the sum. Large deviations generally contribute more, but if there are a lot of cells, even small deviations can add up to make the $\chi^2$ value large. So the more cells there are, the higher the value of $\chi^2$ has to get before it becomes noteworthy. For $\chi^2$, then, the decision about how big is big depends on the number of degrees of freedom.

Unlike the Normal and $t$ families, $\chi^2$ models are skewed—although they are more symmetric for higher degrees of freedom. Here, for example, are the $\chi^2$ curves for 5 and 9 degrees of freedom.

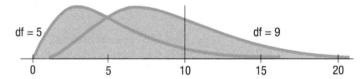

Notice that the value $\chi^2 = 10$ might seem somewhat extreme when there are 5 degrees of freedom, but appears to be rather ordinary for 9 degrees of freedom. Here are two simple facts to help you think about $\chi^2$ models:

- The mode is at $\chi^2 = df - 2$. (Look back at the curves; their peaks are at 3 and 7, see?)
- The expected value (mean) of a $\chi^2$ model is its number of degrees of freedom. That's a bit to the right of the mode—as we would expect for a skewed distribution.

Our test for zodiac birthdays had 11 df, so the relevant $\chi^2$ curve peaks at 9 and has a mean of 11. Knowing that, we might have easily guessed that the calculated $\chi^2$ value of 5.094 wasn't going to be significant.

## TI TIPS

### Testing Goodness-of-Fit

As always, the TI makes doing the mechanics of a goodness-of-fit test pretty easy, but it does take a little work to set it up. Let's use the zodiac data to run through the steps for a $\chi^2$ GOF-Test.

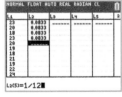

- Enter the counts of executives born under each star sign in L1.

    Those counts were: 23 20 18 23 20 19 18 21 19 22 24 29

- Enter the expected percentages (or fractions, here 1/12) in L2. In this example they are all the same value, but that's not always the case. (Remember the distribution for tongue-rolling and free-hanging ear lobe genes?)

- Convert the expected percentages to expected counts by multiplying each of them by the total number of observations. We use the calculator's summation command in the LIST MATH menu to find the total count for the data summarized in L1 and then multiply that sum by the percentages stored in L2 to produce the expected counts. The command is sum(L1)*L2 → L2. (We don't ever need the percentages again, so we can replace them by storing the expected counts in L2 instead.)

- Choose $\chi^2$ GOF-Test from the STATS TESTS menu.
- Specify the lists where you stored the observed and expected counts, and enter the number of degrees of freedom, here 11.
- Ready, set, Calculate...

- ...and there are the calculated value of $\chi^2$ and your P-value.
- Notice, too, there's a list of values called CNTRB. You can scroll across them, or use LIST NAMES to display them as a data list (as seen below). Those are the cell-by-cell components of the $\chi^2$ calculation. We aren't very interested in them this time, because our data failed to provide evidence that the zodiac sign mattered. However, in a situation where we rejected the null hypothesis, we'd want to look at the components to see where the biggest effects occurred. You'll read more about doing that later in this chapter.

**BY HAND?** If there are only a few cells, you may find that it's just as easy to write out the formula and then simply use the calculator to help you with the arithmetic. After you have found $\chi^2 = 5.09375$ you can use your TI to find the P-value, the probability of observing a $\chi^2$ value at least as high as the one you calculated from your data. As you probably expect, that process is akin to normalcdf and tcdf. You'll find what you need in the DISTR menu at $\chi^2$ cdf. Just specify the left and right boundaries and the number of degrees of freedom.

- Enter $\chi^2$ cdf(5.09375,999,11), as shown. (Why 999? Unlike $t$ and $z$, chi-square values can get pretty big, especially when there are many cells. You may need to go a long way to the right to get to where the curve's tail becomes essentially meaningless. You can see what we mean by looking at Table C, showing chi-square values.)

And there's the P-value, a whopping 0.93! There's nothing at all unusual about these data. (So much for the zodiac's predictive power.)

## JUST CHECKING

Here's the table summarizing the frequency of two traits in that Biology class. Students are checking the genetic theory that the ratio of people with none, one, or both traits is 1:3:3:9.

3. Write the null and alternative hypotheses.
4. How many degrees of freedom are there?
5. Calculate the component of $\chi^2$ for the bottom cell.
6. For these data $\chi^2 = 6.64$. What's the P-value?
7. What should the students conclude?

| Tongue | Earlobes | Observed Count | Expected Count |
|---|---|---|---|
| Non-curling | Attached | 12 | 7.75 |
| Non-curling | Free | 22 | 23.25 |
| Curling | Attached | 31 | 23.25 |
| Curling | Free | 59 | 69.75 |

# Homogeneity: Comparing Observed Distributions

Many universities survey graduating classes to determine the plans of the graduates. We might wonder whether the plans of students are the same at different colleges within a large university. Below is a **two-way table** for graduates from several colleges at one such university. Each cell of the table shows how many students from a particular college made a certain choice.

**Table 25.1**
**Postgraduation activities of the students from several colleges of a large university** (Note: ILR is the School of Industrial and Labor Relations.)

|  |  | College | | | | |
|--|--|--|--|--|--|--|
|  |  | Agriculture | Arts & Sciences | Engineering | ILR | Total |
| Choice | Employed | 209 | 198 | 177 | 101 | 685 |
|  | Grad School | 104 | 171 | 158 | 33 | 466 |
|  | Other | 135 | 115 | 39 | 16 | 305 |
|  | Total | 448 | 484 | 374 | 150 | 1456 |

Because the colleges' sizes are so different, we can see differences better by examining the proportions within each college rather than the counts:

**Table 25.2**
**Activities of graduates as a percentage of respondents from each college**

|  | Agriculture | Arts & Sciences | Engineering | ILR | Combined |
|--|--|--|--|--|--|
| Employed | 46.7% | 40.9% | 47.3% | 67.3% | 47.0% |
| Grad School | 23.2% | 35.3% | 42.2% | 22.0% | 32.0% |
| Other | 30.1% | 23.8% | 10.4% | 10.7% | 20.9% |
| Total | 100.0% | 100.0% | 100.0% | 100.0% | 100.0% |

| WHO | Graduates from four colleges at an upstate New York university |
|--|--|
| WHAT | Postgraduation activities |
| WHEN | 2011 |
| WHY | Survey for general information |

We already know how to test whether *two* proportions are the same. For example, we could use a two-proportion $z$-test to see whether the proportion of students choosing graduate school is the same for Agriculture students as for Engineering students. But now we have more than two groups. We want to test whether the students' choices are the same across all four colleges. The $z$-test for two proportions generalizes to a **chi-square test of homogeneity**.

Chi-square again? It turns out that the mechanics of this test are *identical* to the chi-square test for goodness-of-fit that we just saw. (How similar can you get?) Why a different name, then? The tests are really quite different. The goodness-of-fit test compared counts with a theoretical model. But here we're asking whether the distribution of choices is the same among different groups, so we find the expected counts for each category directly from the data. As a result, we count the degrees of freedom slightly differently as well.

The term "homogeneity" means that things are the same. Here, we ask whether the postgraduation choices made by students are the *same* for these four colleges. The homogeneity test comes with a built-in null hypothesis: We hypothesize that the distribution does not change from group to group. The test looks for differences too large to reasonably arise from random sample-to-sample variation. It can reveal a large deviation in a single category or small, but persistent, differences over all the categories—or anything in between.

## Assumptions and Conditions

The assumptions and conditions are the same as for the chi-square test for goodness-of-fit. The **Counted Data Condition** says that these data must be counts. You can't do a test of homogeneity on proportions, so you have to work with the counts of graduates given in the first table. Also, you can't do a chi-square test on measurements. For example, if we had recorded GPAs for these same groups, we wouldn't be able to determine whether the mean GPAs were different using this test.[2]

---
[2] To do that, you'd use a method called Analysis of Variance (Chapter 27 online).

Ideally, when we compare the proportions across several groups, we would like the cases within each group to be selected randomly. We need to know whether the **Independence Assumption** both within and across groups is reasonable. As usual, check the **Randomization Condition** and the **10% Condition** to make the assumption plausible.

We still must be sure we have enough data for this method to work. The **Expected Cell Frequency Condition** says that the expected count in each cell must be greater than 5. We'll confirm that as we do the calculations.

### Homogeneity Calculations

The null hypothesis says that the distribution of the proportions of graduates choosing each alternative is the same for all four colleges, so we can estimate those overall proportions by pooling our data from the four colleges together. Within each college, the expected proportion for each choice is just the overall proportion of all students making that choice. The expected counts are those proportions applied to the number of students graduating from each college.

For example, overall, 685, or about 47.0%, of the 1456 students who responded to the survey were employed. If the distributions are homogeneous (as the null hypothesis asserts), then 47% of the 448 Agriculture school graduates (or about 210.76 students) should be employed. Similarly, 47% of the 374 Engineering grads (or about 175.95) should be employed.

Working in this way, we (or, more likely, the computer) can fill in expected values for each cell. Because these are theoretical values, they don't have to be integers. The expected values look like this:

**Table 25.3**
Expected values for the 2011 graduates

| EXPECTED | Agriculture | Arts & Sciences | Engineering | ILR | Total |
|---|---|---|---|---|---|
| Employed | 210.769 | 227.706 | 175.955 | 70.570 | **685** |
| Grad School | 143.385 | 154.907 | 119.701 | 48.008 | **466** |
| Other | 93.846 | 101.387 | 78.345 | 31.422 | **305** |
| Total | **448** | **484** | **374** | **150** | **1456** |

Now check the **Expected Cell Frequency Condition**. Indeed, there are more than 5 individuals expected in each cell.

Following the pattern of the goodness-of-fit test, we compute the component for each cell of the table. For the highlighted cell, employed students graduating from the Agriculture school, that's

$$\frac{(Obs - Exp)^2}{Exp} = \frac{(209 - 210.769)^2}{210.769} = 0.0148$$

Summing these components across all cells gives

$$\chi^2 = \sum_{all\ cells} \frac{(Obs - Exp)^2}{Exp} = 93.66$$

How about the degrees of freedom? We don't really need to calculate all the expected values in the table. We know there is a total of 685 employed students, so once we find the expected values for three of the colleges, we can determine the expected number for the fourth by just subtracting. Similarly, we know how many students graduated from each college, so after filling in two rows, we can find the expected values for the remaining row by subtracting. To fill out the table, we need to know the counts in only $R - 1$ rows and $C - 1$ columns. So the table has $(R - 1)(C - 1)$ degrees of freedom.

In our example, we need to calculate only 2 choices in each column and counts for 3 of the 4 colleges, for a total of $2 \times 3 = 6$ degrees of freedom. We'll need the degrees of freedom to find a P-value for the chi-square statistic.

> **NOTATION ALERT**
> For a contingency table, $R$ represents the number of rows and $C$ the number of columns.

# STEP-BY-STEP EXAMPLE

## A Chi-Square Test for Homogeneity

We have reports from four colleges on the postgraduation activities of their 2011 graduating classes.

**QUESTION:** Are the distributions of students' choices of postgraduation activities the same across all the colleges?

**THINK**

**PLAN** State what you want to know.

Identify the variables and check the W's.

I want to test whether postgraduation choices are the same for students from each of four colleges. I have a table of counts classifying each college's Class of 2011 respondents according to their activities.

**HYPOTHESES** State the null and alternative hypotheses.

$H_0$: Students' postgraduation activities are distributed in the same way for all four colleges.
$H_A$: Students' plans do not have the same distribution.

**MODEL** Make a picture: A side-by-side bar chart shows the four distributions of postgraduation activities. Plot column percents to remove the effect of class size differences. A split bar chart would also be an appropriate choice.

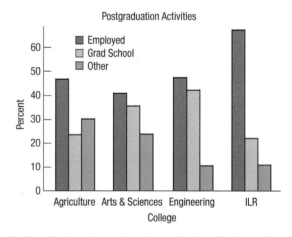

A side-by-side bar chart shows how the distributions of choices differ across the four colleges.

Think about the assumptions and check the conditions. Because we don't want to make inferences about other colleges or other classes, there is no need to check for a random sample.

✓ **Counted Data Condition:** I have counts of the number of students in categories.

✓ **Independence Assumption:** Even though this isn't a random sample, student plans should be largely independent of each other. The occasional friends who decide to join Teach for America together or couples who make grad school decisions together are too rare to affect this analysis.

✓ **Expected Cell Frequency Condition:** The expected values (shown below) are all greater than 5.

State the sampling distribution model and name the test you will use.

The conditions seem to be met, so I can use a $\chi^2$ model with $(3-1) \times (4-1) = 6$ degrees of freedom and do a **chi-square test of homogeneity**.

**SHOW** **MECHANICS** Show the expected counts for each cell of the data table. You could make separate tables for the observed and expected counts, or put both counts in each cell as shown here. While observed counts must be whole numbers, expected counts rarely are—don't be tempted to round those off.

|  | Ag | A&S | Eng | ILR |
|---|---|---|---|---|
| **Employed** | 209 | 198 | 177 | 101 |
|  | 210.769 | 227.706 | 175.955 | 70.570 |
| **Grad School** | 104 | 171 | 158 | 33 |
|  | 143.385 | 154.907 | 119.701 | 48.008 |
| **Other** | 135 | 115 | 39 | 16 |
|  | 93.846 | 101.387 | 78.345 | 31.422 |

Calculate $\chi^2$.

$$\chi^2 = \sum_{\text{all cells}} \frac{(Obs - Exp)^2}{Exp}$$
$$= \frac{(209 - 210.769)^2}{210.769} + \cdots$$
$$= 93.66$$

The shape of a $\chi^2$ model depends on the degrees of freedom. A $\chi^2$ model with 6 df is skewed to the high end.

The P-value considers only the right tail. Here, the calculated value of the $\chi^2$ statistic is off the scale, so the P-value is quite small.

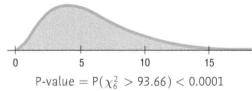

P-value = $P(\chi_6^2 > 93.66) < 0.0001$

**TELL** **CONCLUSION** Link the P-value to your decision, and then state your conclusion in the context of the data. You should specifically talk about whether the distributions for the groups appear to be different.

The P-value is very small, so I reject the null hypothesis and conclude that there's evidence that the postgraduation activities of students from these four colleges don't have the same distribution.

If you find that simply rejecting the hypothesis of homogeneity is a bit unsatisfying, you're in good company. OK, so the postgraduation plans are different. What we'd really like to know is what the differences are, where they're the greatest, and where they're smallest. For that, it's helpful to *make a picture*. A mosaic plot can show how the distributions differ across groups. For example, the mosaic plot below shows how the future plans differ for students from the four different colleges in our example.

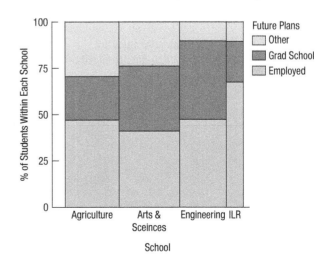

The plot shows that ILR students are the most likely to be planning for employment upon graduation, and the engineers are the most likely to be planning for graduate school. (There are other differences as well—these are just a couple that stand out.)

## JUST CHECKING

Tiny black potato flea beetles can damage potato plants in a vegetable garden. These pests chew holes in the leaves, causing the plants to wither or die. They can be killed with an insecticide, but a canola oil spray has been suggested as a nonchemical "natural" method of controlling the beetles. To conduct an experiment to test the effectiveness of the natural spray, we gather 500 beetles and place them in three Plexiglas® containers. Two hundred beetles go in the first container, where we spray them with the canola oil mixture. Another 200 beetles go in the second container; we spray them with the insecticide. The remaining 100 beetles in the last container serve as a control group; we simply spray them with water. Then we wait 6 hours and count the number of surviving beetles in each container.

8. Why do we need the control group?
9. What would our null hypothesis be?
10. After the experiment is over, we could summarize the results in a table as shown. How many degrees of freedom does our $\chi^2$ test have?

|          | Natural Spray | Insecticide | Water | Total |
|----------|---------------|-------------|-------|-------|
| Survived |               |             |       |       |
| Died     |               |             |       |       |
| Total    | 200           | 200         | 100   | 500   |

11. Suppose that, all together, 125 beetles survived. (That's the first-row total.) What's the expected count in the first cell—survivors among those sprayed with the natural spray?
12. If it turns out that only 40 of the beetles in the first container survived, what's the calculated component of $\chi^2$ for that cell?
13. If the total calculated value of $\chi^2$ for this table turns out to be around 10, would you expect the P-value of our test to be large or small? Explain.

# Chi-Square Test of Independence

A study from the University of Texas Southwestern Medical Center examined 626 people being treated for non–blood-related diseases to see whether the risk of hepatitis C was related to whether people had tattoos and to where they got their tattoos. Hepatitis C causes about 10,000 deaths each year in the United States, but often goes undetected for years after infection.

The data from this study can be summarized in a two-way table, as follows:

Table 25.4
Counts of patients classified by their hepatitis C test status according to whether they had a tattoo from a tattoo parlor or from another source, or had no tattoo

|                  | Hepatitis C | No Hepatitis C | Total |
|------------------|-------------|----------------|-------|
| Tattoo, Parlor   | 17          | 35             | 52    |
| Tattoo, Elsewhere| 8           | 53             | 61    |
| None             | 22          | 491            | 513   |
| Total            | 47          | 579            | 626   |

| WHO   | Patients being treated for non–blood-related disorders |
|-------|--------------------------------------------------------|
| WHAT  | Tattoo status and hepatitis C status                   |
| WHEN  | 1991, 1992                                             |
| WHERE | Texas                                                  |

These data differ from the kinds of data we've considered before in this chapter because they categorize subjects from a single group on two categorical variables rather than on only one. The categorical variables here are *Hepatitis C Status* ("Hepatitis C" or "No Hepatitis C") and *Tattoo Status* ("Parlor," "Elsewhere," "None"). We've seen counts classified by two categorical variables displayed like this in Chapter 2, so we know such tables are called contingency tables. **Contingency tables** categorize counts on two (or more) variables so that we can see whether the distribution of counts on one variable is contingent—or depends—on the other.

The natural question to ask of these data is whether the chance of having hepatitis C is *independent* of tattoo status. Recall that for events **A** and **B** to be independent, $P(\mathbf{A})$ must equal $P(\mathbf{A}|\mathbf{B})$. Here, this means the probability that a randomly selected patient has hepatitis C should be the same regardless of the patient's tattoo status. If *Hepatitis C Status* is independent of tattoos, we'd expect the proportion of people testing positive for hepatitis C to be the same for the three levels of *Tattoo Status*. Of course, for real data we won't expect them to be exactly the same, so we look to see how close they are. This sounds a lot like the test of homogeneity. In fact, the mechanics of the calculation are identical. The difference is that now we have two categorical variables measured on a single population. For the homogeneity test, we had a single categorical variable measured independently on two or more populations. But now we ask a different question: "Are the variables independent?" rather than "Are the groups homogeneous?" These are subtle differences, but they are important when we state hypotheses and draw conclusions. When we ask whether two variables measured on the same population are independent we're performing a **chi-square test of independence**.

### Look at the Design

Homogeneity? Or independence? Look at how the data were collected. When our data are categories of a single variable gathered from more than one group, we wonder whether the groups are the same; the question is homogeneity. When our data cross-categorize two variables gathered from a single group, we wonder whether there's an association; the question is independence.

## FOR EXAMPLE

### Which $\chi^2$ Test?

The American Statistical Association (ASA) is a professional organization of statisticians. To promote statistics education in schools, it oversees the "Census at School," a voluntary and anonymous survey that thousands of U.S. students participate in each year. The ASA then makes the data publicly available.[3]

Among the questions asked of students is this one: "Which of these five superpowers would you most like to have—Flying, Telepathy, Freezing time, Invisibility, or Super strength?"[4] We downloaded a random sample of responses from 500 of the students who took the survey in recent years. Of those, 411 included a response to the superpower question. The table below shows how many students chose each superpower, and it also shows whether the students were in grades 4–6 (which we're calling elementary school), grades 7–9 (middle school) or grades 10–12 (high school).

|  |  | Superpower | | | | | |
|---|---|---|---|---|---|---|---|
|  |  | Flying | Telepathy | Freezing Time | Invisibility | Super Strength | Total |
| **Age group** | Elementary school | 21 | 18 | 22 | 23 | 12 | 96 |
|  | Middle school | 15 | 25 | 22 | 15 | 9 | 86 |
|  | High school | 63 | 59 | 62 | 36 | 9 | 229 |
|  | Total | 99 | 102 | 106 | 74 | 30 | 411 |

QUESTION: Which test would be appropriate for seeing whether, among all the participants in the Census at School survey, there is any relationship between a student's age group and their preferred superpower? What are the hypotheses?

---

[3]https://ww2.amstat.org/censusatschool/

[4]One of the authors had questions. If you freeze time, would you continue to age as normal? If you were super strong, would you *look* super strong? If you were invisible, would your clothes become invisible too? These things make a difference!

> ANSWER: These data represent one random sample of students, categorized on both of two different variables—age group and superpower. I'll do a chi-square test of independence.
>
> $H_0$: The superpower someone would most like to have is independent of their age group.
>
> $H_A$: The superpower someone would most like to have is somehow associated with their age group.

## Assumptions and Conditions

Of course, we still need counts and enough data so that the expected values are more than 5 in each cell.

If we're interested in the independence of variables, we usually want to generalize from the data to some population. In that case, we'll need to check that the data are a representative random sample from, and fewer than 10% of, that population.

### STEP-BY-STEP EXAMPLE

#### A Chi-Square Test for Independence

We have counts of 626 individuals categorized according to their tattoo status and their hepatitis C status.

QUESTION: Are tattoo status and hepatitis C status independent?

| | |
|---|---|
| **THINK** **PLAN** State what you want to know. Identify the variables and check the W's. **HYPOTHESES** State the null and alternative hypotheses. We perform a test of independence when we suspect the variables may not be independent. We are on the familiar ground of making a claim (in this case, that knowing *Tattoo Status* will change probabilities for *Hepatitis C Status*) and testing the null hypothesis that it is *not* true. **MODEL** Make a picture. Because there are only two categories—Hepatitis C and No Hepatitis C—a simple bar chart of the distribution of tattoo sources for hepatitis C patients shows all the information. | I want to test whether the categorical variables *Tattoo Status* and *Hepatitis C Status* are statistically independent. I have a contingency table of 626 Texas patients under treatment for a non-blood-related disease. $H_0$: *Tattoo Status* and *Hepatitis C Status* are independent.[5] $H_A$: *Tattoo Status* and *Hepatitis C Status* are not independent. 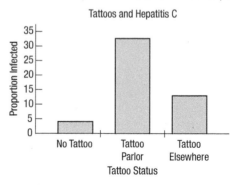 |

---

[5] Once again, parameters are hard to express. The hypothesis of independence itself tells us how to find expected values for each cell of the contingency table. That's all we need.

| | |
|---|---|
| Think about the assumptions and check the conditions. | The bar chart suggests strong differences in hepatitis C risk based on tattoo status.<br><br>✓ **Counted Data Condition**: I have counts of individuals categorized on two variables.<br><br>✓ **Independence Assumption**: The people in this study are likely to be independent of each other.<br><br>✓ **Randomization Condition**: These data are from a retrospective study of patients being treated for something unrelated to hepatitis C. Although they are not an SRS, they were selected to avoid biases.<br><br>✓ **10% Condition**: These 626 patients are far fewer than 10% of all those with tattoos or hepatitis C.<br><br>✗ **Expected Cell Frequency Condition**: The expected values do not meet the condition that all are greater than 5. |
| This table shows both the observed and expected counts for each cell. The expected counts are calculated exactly as they were for a test of homogeneity; in the first cell, for example, we expect $\frac{52}{626}$ (that's 8.3%) of 47.<br><br>*Warning:* Be wary of proceeding when there are small expected counts. If we see expected counts that fall *far* short of 5, or if *many* cells violate the condition, we should not use $\chi^2$. But if the conditions are *almost* met, then you can "proceed cautiously." That means that the decision to reject the null is probably appropriate if your P-value is unambiguously small. | |

|  | Hepatitis C | No Hepatitis C | Total |
|---|---|---|---|
| Tattoo, Parlor | 17 | 35 | 52 |
|  | 3.904 | 48.096 |  |
| Tattoo, Elsewhere | 8 | 53 | 61 |
|  | 4.580 | 56.420 |  |
| None | 22 | 491 | 513 |
|  | 38.516 | 474.484 |  |
| Total | 47 | 579 | 626 |

| | |
|---|---|
| Specify the model.<br><br>Name the test you will use. | Although the Expected Cell Frequency Condition is not satisfied, the values are close to 5. I'll go ahead, but I'll be cautious about my conclusions, especially if the P-value isn't very small. I'll use a $\chi^2$ model with $(3-1) \times (2-1) = 2$ df and do a **chi-square test of independence**. |

---

**SHOW**    **MECHANICS** Calculate $\chi^2$.

$$\chi^2 = \sum_{all\ cells} \frac{(Obs - Exp)^2}{Exp}$$

$$= \frac{(17 - 3.904)^2}{3.904} + \cdots = 57.91$$

The shape of a chi-square model depends on its degrees of freedom. With 2 df, the model looks quite different, as you can see here. We still care only about the right tail.

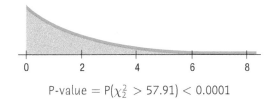

P-value = $P(\chi^2_2 > 57.91) < 0.0001$

---

**TELL**    **CONCLUSION** Link the P-value to your decision. State your conclusion about the independence of the two variables (in context, of course).

(We should be wary of this conclusion because of the small expected counts. Because the P-value is so extremely small, however, the decision to reject the null in this case is probably fine.)

The P-value is very small, so I reject the null hypothesis and conclude that *Hepatitis C Status* is not independent of *Tattoo Status*.

## TI TIPS

### Testing Homogeneity or Independence

Yes, the TI will do chi-square tests of homogeneity and independence. Let's use the tattoo data. Here goes.

**TEST A HYPOTHESIS OF HOMOGENEITY OR INDEPENDENCE**

Stage 1: You need to enter the data as a matrix. A "matrix" is just a formal mathematical term for a table of numbers.

- Push the MATRIX button, and choose to EDIT matrix [A].
- First specify the dimensions of the table, rows × columns.
- Enter the appropriate counts, one cell at a time. The calculator automatically asks for them row by row.

Stage 2: Do the test.

- In the STAT TESTS menu choose $\chi^2$-Test.
- The TI now confirms that you have placed the observed frequencies in [A]. It also tells you that when it finds the expected frequencies it will store those in [B] for you. Now Calculate the mechanics of the test.

The TI reports a calculated value of $\chi^2 = 57.91$ and an exceptionally small P-value.

Stage 3: Check the expected counts.

- Go back to MATRIX EDIT and choose [B].

Notice that two of the cells fail to meet the condition that expected counts be greater than 5. We noted this in our Step-by-Step Example.

## FOR EXAMPLE

### Chi-Square Mechanics

**RECAP:** A random sample of 500 responses from the Census at School included 411 that identified the superpower the student would most like to have. We want to know if their choices are associated with their grade level.

**QUESTIONS:** What are the degrees of freedom for this test? Are all of the expected cell counts greater than 5? What is the P-value? What is the conclusion?

|  |  | Superpower | | | | | |
|---|---|---|---|---|---|---|---|
|  |  | Flying | Telepathy | Freezing Time | Invisibility | Super Strength | Total |
| Age group | Elementary school | 21 | 18 | 22 | 23 | 12 | 96 |
|  | Middle school | 15 | 25 | 22 | 15 | 9 | 86 |
|  | High school | 63 | 59 | 62 | 36 | 9 | 229 |
|  | Total | 99 | 102 | 106 | 74 | 30 | 411 |

ANSWER: This is a 3 × 5 contingency table, so $df = (3 - 1)(5 - 1) = 10$.

We used technology to generate a matrix of expected counts. They're shown in parentheses below the observed counts:

|  |  | Superpower |  |  |  |  |
|---|---|---|---|---|---|---|
|  |  | Flying | Telepathy | Freezing Time | Invisibility | Super Strength | Total |
| Age group | Elementary school | 21 (23.1) | 18 (23.8) | 22 (24.8) | 23 (17.3) | 12 (7.0) | 96 |
|  | Middle school | 15 (20.7) | 25 (21.3) | 22 (22.2) | 15 (15.5) | 9 (6.3) | 86 |
|  | High school | 63 (55.2) | 59 (56.8) | 62 (59.1) | 36 (41.2) | 9 (16.7) | 229 |
|  | Total | 99 | 102 | 106 | 74 | 30 | 411 |

All of the expected cell counts are greater than 5, so the chi-square test of independence is appropriate.

Also using technology, we determined the P-value of the test to be $P \approx 0.038$. Because the P-value is small (in particular, less than 0.05), we conclude that there is some association between Census at School participant's age group and their superpower of choice. The mosaic plot below suggests that the high school students are less likely to choose Super Strength compared to the younger students, and that elementary school students are less likely to choose Telepathy compared to the older students. The differences are not huge, but the chi-square test of independence indicates that they're not likely to be just the result of a fluke sample.

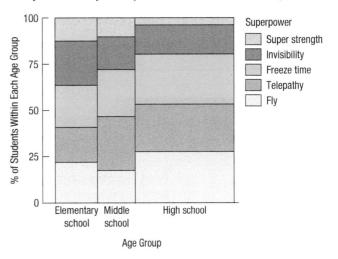

# Chi-Square and Causation

Chi-square tests are common. Tests for independence are especially widespread. Unfortunately, many people interpret a small P-value as proof of causation. You know better, right? Just as correlation between quantitative variables does not demonstrate causation, a failure of independence between two categorical variables does not show a cause-and-effect relationship between them, nor should we say that one variable *depends* on the other.

The chi-square test for independence treats the two variables symmetrically. There is no way to differentiate the direction of any possible causation from one variable to the other. In our example, it is unlikely that having hepatitis C causes one to crave a tattoo, but other situations are not so clear.

In this case, it's easy to imagine that lurking variables are responsible for the observed lack of independence. Perhaps the lifestyles of some people include both tattoos and behaviors that put them at increased risk of hepatitis C, such as body piercings or even drug use. Even a small subpopulation of people with such a lifestyle among those with tattoos might be enough to create the observed result. After all, we observed only 25 patients with both tattoos and hepatitis C.

## JUST CHECKING

Which of the three chi-square tests—goodness-of-fit, homogeneity, or independence—would you use in each of the following situations?

14. A restaurant manager wonders whether customers who dine on Friday nights have the same preferences among the four "chef's special" entrées as those who dine on Saturday nights. One weekend they have the wait staff record which entrées were ordered each night. Assuming these customers to be typical of all weekend diners, they'll compare the distributions of meals chosen Friday and Saturday.

15. Company policy calls for parking spaces to be assigned to everyone at random, but you suspect that may not be so. There are three lots of equal size: lot A, next to the building; lot B, a bit farther away; and lot C, on the other side of the highway. You gather data about employees at middle management level and above to see how many were assigned parking in each lot.

16. Is a student's social life affected by where the student lives? A campus survey asked a random sample of students whether they lived in a dormitory, in off-campus housing, or at home, and whether they had been out on a date 0, 1–2, 3–4, or 5 or more times in the past two weeks.

## WHAT CAN GO WRONG?

- **Don't use chi-square methods unless you have counts.** All three of the chi-square tests apply only to counts of categorical data. Other kinds of data can be arrayed in two-way tables. Just because numbers are in a two-way table doesn't make them suitable for chi-square analysis. Data reported as proportions or percentages can be suitable for chi-square procedures, *but only after they are converted to counts*. If you try to do the calculations without first finding the counts, your results will be wrong.

- **Beware large samples.** Beware *large* samples?! That's not the advice you're used to hearing. The chi-square tests, however, are unusual. Be wary of chi-square tests performed on very large samples. No hypothesized distribution fits perfectly, no two groups are exactly homogeneous, and two variables are rarely perfectly independent. The degrees of freedom for chi-square tests don't grow with the sample size. With a sufficiently large sample size, a chi-square test can always reject the null hypothesis. But we have no measure of how far the data are from the null model. There are no confidence intervals to help us judge the effect size.

- **Don't say that one variable "depends" on the other just because they're not independent.** Dependence suggests a pattern and implies causation, but variables can fail to be independent in many different ways. When variables fail the test for independence, you might just say they are "associated."

# WHAT HAVE WE LEARNED?

We've learned how to test hypotheses about categorical variables. We use one of three related methods. All look at counts of data in categories and rely on chi-square models, a new family indexed by degrees of freedom.

We've learned to look at the study design to see which test is appropriate.
- Goodness-of-fit tests compare the observed distribution of a single categorical variable to an expected distribution based on a theory or a model.
- Tests of homogeneity compare the observed distributions in several groups for a single categorical variable.
- Tests of independence examine observed counts from a single group for evidence of an association between two categorical variables.

We've learned to write appropriate hypotheses for each test, and especially to understand how these differ for tests of homogeneity and independence.

We've learned again to check assumptions and conditions before proceeding with inference.
- **Counted Data Condition:** We have observed counts for the categories.
- **Independence Assumption:** Randomization makes independence more plausible.
- **Expected Cell Frequency Condition:** Expect greater than 5 observations in each cell.
- **10% Condition:** The sample is less than 10% of the population.

We've learned that mechanically the three tests are almost identical. We've learned to find the expected counts, determine the number of degrees of freedom, calculate the value of $\chi^2$, and determine the *P*-value.

We've learned to interpret the results of the inference test.
- These tests are conceptually many-sided, because there are many ways that observed counts can deviate significantly from what we hypothesized.
- Failing to reject the null hypothesis does not confirm that the data do fit the model.
- If we reject the null hypothesis, we can examine a mosaic plot to better understand the association between the two variables.

## TERMS

**Chi-square test of goodness-of-fit**  A test of whether the distribution of counts in one categorical variable matches the distribution predicted by a model is called a test of goodness-of-fit. In a chi-square goodness-of-fit test, the expected counts come from the predicting model. The test finds a P-value from a chi-square model with $n - 1$ degrees of freedom, where $n$ is the number of categories in the categorical variable. (p. 689)

**Cell**  A cell is one element of a table corresponding to a specific row and a specific column. Table cells can hold counts, percentages, or measurements on other variables. Or they can hold several values. (p. 691)

**Chi-square statistic**  The chi-square statistic can be used to test whether the observed counts in a frequency distribution or contingency table match the counts we would expect according to some model. It is calculated as

$$\chi^2 = \sum_{\text{all cells}} \frac{(Obs - Exp)^2}{Exp}.$$

Chi-square statistics differ in how expected counts are found, depending on the question asked. (p. 692)

**Chi-square model**  Chi-square models are skewed to the right. They are parameterized by their degrees of freedom and become less skewed with increasing degrees of freedom. (p. 692)

| | |
|---|---|
| **Chi-square component** | The components of a chi-square calculation are $$\frac{(Obs - Exp)^2}{Exp},$$ found for each cell of the table. (pp. 695, 702) |
| **Two-way table** | Each *cell* of a two-way table shows counts of individuals. One way classifies a sample according to a categorical variable. The other way can classify different groups of individuals according to the same variable or classify the same individuals according to a different categorical variable. (p. 698) |
| **Chi-square test of homogeneity** | A test comparing the distribution of counts for *two or more groups* on the same categorical variable. A chi-square test of homogeneity finds expected counts based on the overall frequencies, adjusted for the totals in each group under the (null hypothesis) assumption that the distributions are the same for each group. We find a P-value from a chi-square distribution with $(\#Rows - 1) \times (\#Cols - 1)$ degrees of freedom, where #Rows gives the number of categories and #Cols gives the number of independent groups. (p. 698) |
| **Contingency table** | A two-way table that classifies individuals according to two categorical variables. (p. 702) |
| **Chi-square test of independence** | A test of whether two categorical variables are independent examines the distribution of counts for *one group of individuals* classified according to both variables. A chi-square test of *independence* finds expected counts by assuming that knowing the marginal totals tells us the cell frequencies, assuming that there is no association between the variables. This turns out to be the same calculation as a test of homogeneity. We find a P-value from a chi-square distribution with $(\#Rows - 1) \times (\#Cols - 1)$ degrees of freedom, where #Rows gives the number of categories in one variable and #Cols gives the number of categories in the other. (p. 703) |

# ON THE COMPUTER

## Chi-Square

Most statistics packages perform chi-square tests on contingency tables. It's up to you to properly interpret them as tests of homogeneity or independence. Some packages want to create the contingency table from the actual data, while others allow you to enter the summary counts. Goodness-of-fit tests may be missing.

Some software packages display the standardized residuals, others the components.

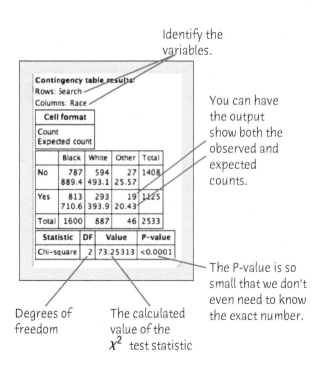

# EXERCISES

**1. Human births** If there is no seasonal effect on human births, we would expect equal numbers of children to be born in each season (winter, spring, summer, and fall). A student takes a census of their Statistics class and finds that of the 120 students in the class, 25 were born in winter, 35 in spring, 32 in summer, and 28 in fall. They wonder wonders if the excess in the spring is an indication that births are not uniform throughout the year.

a) What is the expected number of births in each season if there is no "seasonal effect" on births?
b) Compute the $\chi^2$ statistic.
c) How many degrees of freedom does the $\chi^2$ statistic have?

**2. Bank cards** At a major credit card bank, the percentages of people who historically apply for the Silver, Gold, and Platinum cards are 60%, 30%, and 10%, respectively. In a recent sample of customers responding to a promotion, of 200 customers, 110 applied for Silver, 55 for Gold, and 35 for Platinum. Is there evidence to suggest that the percentages for this promotion may be different from the historical proportions?

a) What is the expected number of customers applying for each type of card in this sample if the historical proportions are still true?
b) Compute the $\chi^2$ statistic.
c) How many degrees of freedom does the $\chi^2$ statistic have?

**3. Human births, again** For the births in Exercise 1,

a) If there is no seasonal effect, about how big, on average, would you expect the $\chi^2$ statistic to be (what is the mean of the $\chi^2$ distribution)?
b) Does the statistic you computed in Exercise 1 seem large in comparison to this mean? Explain briefly.
c) What does that say about the null hypothesis?

**4. Bank cards, again** For the customers in Exercise 2,

a) If the customers apply for the three cards according to the historical proportions, about how big, on average, would you expect the $\chi^2$ statistic to be (what is the mean of the $\chi^2$ distribution)?
b) Does the statistic you computed in Exercise 2 seem large in comparison to this mean? Explain briefly.
c) What does that say about the null hypothesis?

**5. Customer ages** An analyst at a local bank wonders if the age distribution of customers coming for service at his branch in town is the same as at the branch located near the mall. He selects 100 transactions at random from each branch and researches the age information for the associated customer. Here are the data:

|  | Age | | | |
| --- | --- | --- | --- | --- |
|  | Less Than 30 | 30–55 | 56 or Older | Total |
| In-Town Branch | 20 | 40 | 40 | 100 |
| Mall Branch | 30 | 50 | 20 | 100 |
| Total | 50 | 90 | 60 | 200 |

a) What is the null hypothesis?
b) What type of test is this?
c) What are the expected numbers for each cell if the null hypothesis is true?
d) Find the $\chi^2$ statistic.
e) How many degrees of freedom does it have?
f) Find the P-value.
g) What do you conclude?

**6. Bank cards, once more** A market researcher working for the bank in Exercise 2 wants to know if the distribution of applications by card is the same for the past three mailings. She takes a random sample of 200 from each mailing and counts the number applying for Silver, Gold, and Platinum. The data follow:

|  | Type of Card | | | |
| --- | --- | --- | --- | --- |
|  | Silver | Gold | Platinum | Total |
| Mailing 1 | 120 | 50 | 30 | 200 |
| Mailing 2 | 115 | 50 | 35 | 200 |
| Mailing 3 | 105 | 55 | 40 | 200 |
| Total | 340 | 155 | 105 | 600 |

a) What is the null hypothesis?
b) What type of test is this?
c) What are the expected numbers for each cell if the null hypothesis is true?
d) Find the $\chi^2$ statistic.
e) How many degrees of freedom does it have?
f) Find the P-value.
g) What do you conclude?

**7. Human births, last time** For the data in Exercise 1, births in each of the four seasons were not perfectly uniform. Yet the P-value was large and the null hypothesis was not rejected. How would you explain this to a friend who has not taken Statistics?

**8. Bank cards, last time** For the data in Exercise 2, the P-value was small and evidence was found for the alternative hypothesis. Examine the differences between the observed and expected counts and specifically explain where there is a significant difference.

**9. Iliad injuries 800 B.C.E.** Homer's *Iliad* is an epic poem, compiled around 800 B.C.E., that describes several weeks of the last year of the 10-year siege of Troy (Ilion) by the Achaeans. The story centers on the rage of the great warrior Achilles. But it includes many details of injuries and outcomes, and is thus the oldest record of Greek medicine. Here is a table of 146 recorded injuries for which both injury site and outcome are provided in the *Iliad*. Are some kinds of injuries more lethal than others?

|  | | Lethal? | | |
|---|---|---|---|---|
|  | | Fatal | Not Fatal | Total |
| **Injury Site** | Body | 61 | 6 | 67 |
|  | Head/Neck | 44 | 1 | 45 |
|  | Limb | 13 | 21 | 34 |
|  | Total | 118 | 28 | 146 |

a) Under the null hypothesis, what are the expected values?
b) Compute the $\chi^2$ statistic.
c) How many degrees of freedom does it have?
d) Find the P-value.
e) What do you conclude?

**T 10. Iliad weapons** The *Iliad* also reports the cause of many injuries. Here is a table summarizing those reports for the 152 injuries for which the *Iliad* provides that information. Is there an association?

|  | | Injury Site | | | |
|---|---|---|---|---|---|
|  | | Body | Head/Neck | Limb | Total |
| **Weapon** | Arrow | 5 | 2 | 5 | 12 |
|  | Ground/Rock | 1 | 5 | 5 | 11 |
|  | Sword | 61 | 38 | 23 | 122 |
|  | Total | 67 | 45 | 33 | 145 |

a) Under the null hypothesis, what are the expected values?
b) Compute the $\chi^2$ statistic.
c) How many degrees of freedom does it have?
d) Find the P-value.
e) What do you conclude?

**11. Which test?** For each of the following situations, state whether you'd use a chi-square goodness-of-fit test, a chi-square test of homogeneity, a chi-square test of independence, or some other statistical test:

a) A brokerage firm wants to see whether the type of account a customer has (Silver, Gold, or Platinum) affects the type of trades that customer makes (in person, by phone, or on the Internet). It collects a random sample of trades made for its customers over the past year and performs a test.
b) That brokerage firm also wants to know if the type of account affects the size of the account (in dollars). It performs a test to see if the mean size of the account is the same for the three account types.
c) The academic research office at a large community college wants to see whether the distribution of courses chosen (Humanities, Social Science, or Science) is different for its residential and nonresidential students. It assembles last semester's data and performs a test.

**12. Which test again?** For each of the following situations, state whether you'd use a chi-square goodness-of-fit test, a chi-square test of homogeneity, a chi-square test of independence, or some other statistical test:

a) Is the quality of a car affected by what day it was built? A car manufacturer examines a random sample of the warranty claims filed over the past two years and uses their records to determine the day of the week when the car was built. They want to test whether defects are randomly distributed across days of the week.
b) A medical researcher wants to know if blood cholesterol level is related to heart disease. She examines a database of 10,000 patients, testing whether the cholesterol level (in milligrams) is related to whether or not a person has heart disease.
c) A student wants to find out whether political leaning (liberal, moderate, or conservative) is related to choice of major. He surveys 500 randomly chosen students and performs a test.

**T 13. Dice** After getting trounced by your little brother in a children's game, you suspect the die he gave you to roll may be unfair. To check, you roll it 60 times, recording the number of times each face appears. Do these results cast doubt on the die's fairness?

| Face | Count |
|---|---|
| 1 | 11 |
| 2 | 7 |
| 3 | 9 |
| 4 | 15 |
| 5 | 12 |
| 6 | 6 |

a) If the die is fair, how many times would you expect each face to show?
b) To see if these results are unusual, will you test goodness-of-fit, homogeneity, or independence?
c) State your hypotheses.
d) Check the conditions.
e) How many degrees of freedom are there?
f) Find $\chi^2$ and the P-value.
g) State your conclusion.

**14. M&M's** As noted in an earlier chapter, the Masterfoods Company says that until very recently yellow candies made up 20% of its milk chocolate M&M's, red another 20%, and orange, blue, and green 10% each. The rest are brown. On his way home from work the day he was writing these exercises, one of the authors bought a bag of plain M&M's. He got 29 yellow ones, 23 red, 12 orange, 14 blue, 8 green, and 20 brown. Is this sample consistent with the company's stated proportions? Test an appropriate hypothesis and state your conclusion.

a) If the M&M's are packaged in the stated proportions, how many of each color should the author have expected to get in his bag?
b) To see if his bag was unusual, should he test goodness-of-fit, homogeneity, or independence?
c) State the hypotheses.
d) Check the conditions.
e) How many degrees of freedom are there?
f) Find $\chi^2$ and the P-value.
g) State a conclusion.

**15. Pizza Perfection** A group of AP Stats students decides on a project to study pizza toppings. Their favorite combination is pepperoni, sausage, and cashews. In their view, the perfect pizza should have the same number of each topping. However, they are suspicious about this uniformity. They order several pizzas and collect data on the three toppings. Their count reveals 45 pepperoni pieces, 38 sausage pieces, and 53 cashews.

a) What is the expected number of each topping if the distribution were uniform?
b) Compute the $\chi^2$ statistic.
c) How many degrees of freedom does the $\chi^2$ statistic have?

**16.** Xmas lights Another group of the Exercise 15 students set their project sights on the balance of color for outdoor Christmas lights. After some discussion, the group decided that the "perfect" display will have 35% red lights and 35% green, with blue and yellow only 15% each. The students counted lights on the principal's house and found that he has 67 red lights, 73 green, 29 yellow, and 43 blue.

a) What is the expected light distribution at the principal's home if his colors were consistent the group's desires?
b) Compute the $\chi^2$ statistic.
c) How many degrees of freedom does the $\chi^2$ statistic have?

**17.** Nuts A company says its premium mixture of nuts contains 10% Brazil nuts, 20% cashews, 20% almonds, and 10% hazelnuts, and the rest are peanuts. You buy a large can and separate the various kinds of nuts. Upon weighing them, you find there are 112 grams of Brazil nuts, 183 grams of cashews, 207 grams of almonds, 71 grams of hazelnuts, and 446 grams of peanuts. You wonder whether your mix is significantly different from what the company advertises.

a) Explain why the chi-square goodness-of-fit test is not an appropriate way to find out.
b) What might you do instead of weighing the nuts in order to use a $\chi^2$ test?

**18.** Mileage A salesperson who is on the road visiting clients thinks that, on average, they drive the same distance each day of the week. They keep track of their mileage for several weeks and discovers that they average 122 miles on Mondays, 203 miles on Tuesdays, 176 miles on Wednesdays, 181 miles on Thursdays, and 108 miles on Fridays. They wonder if this evidence contradicts their belief in a uniform distribution of miles across the days of the week. Explain why it is not appropriate to test their hypothesis using the chi-square goodness-of-fit test.

**19.** NYPD and race Census data for New York City indicate that 29.2% of the under-18 population is white, 28.2% black, 31.5% Latino, 9.1% Asian, and 2% other ethnicities. The New York Civil Liberties Union points out that, of 26,181 police officers, 64.8% are white, 14.5% black, 19.1% Latino, and 1.4% Asian. Do the police officers reflect the ethnic composition of the city's youth? Test an appropriate hypothesis and state your conclusion.

**20.** Age complaints? A local mayor looks up the age breakdown of her city. She is curious to see if the age distribution is similar to the number of e-mails and letters she receives with ideas (and grumblings . . . ). The four voting age groups in her town are as follows: 18–35 (22%), 36–50 (34%), 51–65 (32%), and 65+ (12%). The number of e-mails and letters she received over a 6-month period are listed at the top of the next column. Run the appropriate test to examine if the distribution of letters and e-mails is the same as the age distribution in the town.

| Age Group | # of Letters/e-mails |
|---|---|
| 18–35 | 21 |
| 36–50 | 41 |
| 51–65 | 49 |
| 65+ | 68 |

**21.** Fruit flies Offspring of certain fruit flies may have yellow or ebony bodies and normal wings or short wings. Genetic theory predicts that these traits will appear in the ratio 9:3:3:1 (9 yellow, normal; 3 yellow, short; 3 ebony, normal; 1 ebony, short). A researcher checks 100 such flies and finds the distribution of the traits to be 59, 20, 11, and 10, respectively.

a) Are the results this researcher observed consistent with the theoretical distribution predicted by the genetic model?
b) If the researcher had examined 200 flies and counted exactly twice as many in each category—118, 40, 22, 20—what conclusion would he have reached?
c) Why is there a discrepancy between the two conclusions?

**22.** Pi Many people know the mathematical constant $\pi$ is approximately 3.14. But that's not exact. To be more precise, here are 20 decimal places: 3.14159265358979323846. Still not exact, though. In fact, the actual value is irrational, a decimal that goes on forever without any repeating pattern. But notice that there are no 0's and only one 7 in the 20 decimal places above. Does that pattern persist, or do all the digits show up with equal frequency? The table shows the number of times each digit appears in the first million digits. Test the hypothesis that the digits 0 through 9 are uniformly distributed in the decimal representation of $\pi$.

| The First Million Digits of $\pi$ | |
|---|---|
| Digit | Count |
| 0 | 99,959 |
| 1 | 99,758 |
| 2 | 100,026 |
| 3 | 100,229 |
| 4 | 100,230 |
| 5 | 100,359 |
| 6 | 99,548 |
| 7 | 99,800 |
| 8 | 99,985 |
| 9 | 100,106 |

**23.** Hurricane frequencies The National Hurricane Center provides data that list the numbers of large (category 3, 4, or 5) hurricanes that have struck the United States, by decade since 1851 (www.nhc.noaa.gov/dcmi.shtml). The data are summarized below.

| Decade | Count | Decade | Count |
|---|---|---|---|
| 1851–1860 | 6 | 1941–1950 | 10 |
| 1861–1870 | 1 | 1951–1960 | 9 |
| 1871–1880 | 7 | 1961–1970 | 6 |
| 1881–1890 | 5 | 1971–1980 | 4 |
| 1891–1900 | 8 | 1981–1990 | 4 |
| 1901–1910 | 4 | 1991–2000 | 5 |
| 1911–1920 | 7 | 2001–2010 | 7 |
| 1921–1930 | 5 | 2011–2020 | 5 |
| 1931–1940 | 8 | | |

Recently, there's been some concern that perhaps the number of large hurricanes has been increasing. The natural null hypothesis would be that the frequency of such hurricanes has remained constant.

a) With 101 large hurricanes observed over the 17 periods, what are the expected value(s) for each cell?
b) What kind of chi-square test would be appropriate?

c) State the null and alternative hypotheses.
d) How many degrees of freedom are there?
e) The value of $\chi^2$ is 12.95. What's the P-value?
f) State your conclusion.

**24. Lottery numbers** The fairness of the South African lottery was recently challenged by one of the country's political parties. The lottery publishes historical statistics at its website (http://www.nationallottery.co.za/lotto/statistics.aspx). Here is a table of the number of times each of the 49 numbers has been drawn in the main lottery and as the "bonus ball" number as of June 2007:

| Number | Count | Bonus | Number | Count | Bonus |
|---|---|---|---|---|---|
| 1 | 81 | 14 | 26 | 78 | 12 |
| 2 | 91 | 16 | 27 | 83 | 16 |
| 3 | 78 | 14 | 28 | 76 | 7 |
| 4 | 77 | 12 | 29 | 76 | 12 |
| 5 | 67 | 16 | 30 | 99 | 16 |
| 6 | 87 | 12 | 31 | 78 | 10 |
| 7 | 88 | 15 | 32 | 73 | 15 |
| 8 | 90 | 16 | 33 | 81 | 14 |
| 9 | 80 | 9 | 34 | 81 | 13 |
| 10 | 77 | 19 | 35 | 77 | 15 |
| 11 | 84 | 12 | 36 | 73 | 8 |
| 12 | 68 | 14 | 37 | 64 | 17 |
| 13 | 79 | 9 | 38 | 70 | 11 |
| 14 | 90 | 12 | 39 | 67 | 14 |
| 15 | 82 | 9 | 40 | 75 | 13 |
| 16 | 103 | 15 | 41 | 84 | 11 |
| 17 | 78 | 14 | 42 | 79 | 8 |
| 18 | 85 | 14 | 43 | 74 | 14 |
| 19 | 67 | 18 | 44 | 87 | 14 |
| 20 | 90 | 13 | 45 | 82 | 19 |
| 21 | 77 | 13 | 46 | 91 | 10 |
| 22 | 78 | 17 | 47 | 86 | 16 |
| 23 | 90 | 14 | 48 | 88 | 21 |
| 24 | 80 | 8 | 49 | 76 | 13 |
| 25 | 65 | 11 | | | |

We wonder if all the numbers are equally likely to be the "bonus ball."

a) What kind of test should we perform?
b) There are 655 bonus ball observations. What are the appropriate expected values for the test?
c) State the null and alternative hypotheses.
d) How many degrees of freedom are there?
e) The value of $\chi^2$ is 34.5. What's the P-value?
f) State your conclusion.

**25. Amazon split testing** There are many different merchants on Amazon.com trying to compete for sales. The website junglescout.com offers to help merchants by using a method they call A/B split testing. In this method a merchant would list their product using one version of their sales page for a period of time, Version A, and then a different version afterwards, Version B. Junglescout describes this process as a randomized experiment and claims that this split test will help merchants increase their profits in the long run. Suppose that one merchant ran a split test and gathered these results: (Source: https://www.junglescout.com/blog/amazon-split-testing/)

| | | Version? | |
|---|---|---|---|
| | | A | B |
| Converted Sale? | Yes | 54 | 67 |
| | No | 349 | 789 |

a) What kind of test would be appropriate?
b) State the null and alternative hypotheses.

**26. Does your doctor know?** A survey[6] of articles from the *New England Journal of Medicine (NEJM)* classified them according to the principal statistics methods used. The articles recorded were all noneditorial articles appearing during the indicated years. Let's just look at whether these articles used statistics at all.

| | Publication Year | | | Total |
|---|---|---|---|---|
| | 1978–79 | 1989 | 2004–05 | |
| No stats | 90 | 14 | 40 | 144 |
| Stats | 242 | 101 | 271 | 614 |
| Total | 332 | 115 | 311 | 758 |

Has there been a change in the use of statistics?

a) What kind of test would be appropriate?
b) State the null and alternative hypotheses.

**27. Amazon splits, part 2** In Exercise 25, the table shows results of Version A vs. Version B of two different Amazon webpages. Both versions are selling the same product and the number of converted sales (out of the total number of views) is being compared. We're planning to do a chi-square test.

a) How many degrees of freedom are there?
b) The smallest expected count will be in the Version A/Yes converted cell. What is it?
c) Check the assumptions and conditions for inference.

**28. Does your doctor know? (part 2)** The table in Exercise 26 shows whether *NEJM* medical articles during various time periods included statistics or not. We're planning to do a chi-square test.

a) How many degrees of freedom are there?
b) The smallest expected count will be in the 1989/No Stats cell. What is it?
c) Check the assumptions and conditions for inference.

**29. Amazon, part 3** In Exercises 25 and 27, we've begun to examine the possible impact of different website versions on successful sales conversions.

a) Calculate the component of chi-square for the Version A/Yes cell.
b) For this test, $\chi^2 = 9.79$. What's the P-value?
c) State your conclusion.

---
[6]Suzanne S. Switzer and Nicholas J. Horton, "What Your Doctor Should Know about Statistics (but Perhaps Doesn't)" *Chance*, 20:1, 2007.

**30. Does your doctor know? (part 3)** In Exercises 26 and 28, we've begun to examine whether the use of statistics in *NEJM* medical articles has changed over time.
   a) Calculate the component of chi-square for the 1989/No Stats cell.
   b) For this test, $\chi^2 = 25.28$. What's the P-value?
   c) State your conclusion.

**31. Amazon, part 4** In Exercises 25, 27, and 29, we've looked at an experiment testing two versions of an Amazon sales page. You might have noticed that about twice as many potential customers viewed Version B than Version A. Does this raise a concern in your mind about the conditions for inference?

**32. Does your doctor know? (part 4)** In Exercises 26, 28, and 30, we considered data on articles in the *NEJM*. The original study listed 23 different statistics methods. (The list read: *t*-tests, contingency tables, linear regression, . . . .) Why would it not be appropriate to use a chi-square test on the 23 × 3 table with a row for each method?

**33. Roll 'em** Fans of various role-playing games will use dice of all sorts. Players use regular 6-sided dice, but will also use dice that have 4, 8, 10, 12, and 20 sides. These dice are sold in sets and are made in all variety of colors and materials. One player is curious if his dice are made uniformly so that every option is truly equally likely. To begin testing, he takes his 12-sided-dodecahedron-of-chance and rolls it 240 times. The data are below. Run the appropriate test to evaluate the claim of uniformity.

| Number | Frequency |
|---|---|
| 1 | 19 |
| 2 | 16 |
| 3 | 23 |
| 4 | 25 |
| 5 | 18 |
| 6 | 15 |
| 7 | 19 |
| 8 | 24 |
| 9 | 20 |
| 10 | 24 |
| 11 | 22 |
| 12 | 15 |

**34. Roll some more** Our curious game player also tests his 20-sided-icosohedron-of-chance, also rolling a total of 240 times. Here are the data.

| Number | Frequency | Number | Frequency |
|---|---|---|---|
| 1 | 10 | 11 | 12 |
| 2 | 14 | 12 | 14 |
| 3 | 15 | 13 | 15 |
| 4 | 11 | 14 | 13 |
| 5 | 12 | 15 | 9 |
| 6 | 13 | 16 | 12 |
| 7 | 11 | 17 | 12 |
| 8 | 8 | 18 | 11 |
| 9 | 11 | 19 | 13 |
| 10 | 10 | 20 | 14 |

   a) Run the appropriate test to find if this die is made uniformly.
   b) If you compare the 12-sided die in Exercises 33 to the 20-sided die in this exercise, you notice something interesting. Both dice deviate from the expected counts by a total of 36 (that is, if you subtract the observed counts from the expected counts and make all those values positive, they all add up to 36. For both samples!). But the two dice have different P-values. Can you explain why this would be the case?
   c) A fundamental idea of this course is that random variation happens. The data in this exercise might seem odd to you. As might the P-value. One might even argue that this data is "suspiciously uniform." If these data were presented to you, would you be suspicious?

**35. Internet use poll** A Pew Research poll in April 2009 from a random sample of U.S. adults asked the questions "Did you use the Internet yesterday?" and "Are you White, Black, or Hispanic/Other?" Is the response to the question about the Internet independent of race?

|  |  | Did You Use the Internet Yesterday? | |
|---|---|---|---|
|  |  | Yes | No |
| Ethnicity | White | 2546 | 856 |
|  | Black | 314 | 146 |
|  | Hispanic/Other | 431 | 174 |

   a) Under the null hypothesis, what are the expected values?
   b) Compute the $\chi^2$ statistic.
   c) How many degrees of freedom does it have?
   d) Find the P-value.
   e) What do you conclude?

**36. Internet use poll, II** The same poll as in Exercise 35 also asked the questions "Did you use the Internet yesterday?" and "What is your educational level?" Is the response to the question about the Internet independent of educational level?

|  |  | Did You Use the Internet Yesterday? | |
|---|---|---|---|
|  |  | Yes | No |
| Education | Less Than High School | 209 | 131 |
|  | High School | 932 | 550 |
|  | Some College | 958 | 346 |
|  | College Grad | 1447 | 247 |

   a) Under the null hypothesis, what are the expected values?
   b) Compute the $\chi^2$ statistic.
   c) How many degrees of freedom does it have?
   d) Find the P-value.
   e) What do you conclude?

**37. Titanic** Here is a table we first saw in Chapter 2 showing who survived the sinking of the *Titanic* based on whether they were crew members, or passengers booked in first-, second-, or third-class staterooms:

|  | Crew | First | Second | Third | Total |
|---|---|---|---|---|---|
| Saved | 212 | 202 | 118 | 178 | **710** |
| Lost | 673 | 123 | 167 | 528 | **1491** |
| Total | **885** | **325** | **285** | **706** | **2201** |

a) If we draw an individual at random, what's the probability that we will draw a member of the crew?
b) What's the probability of randomly selecting a third-class passenger who survived?
c) What's the probability of a randomly selected passenger surviving, given that the passenger was a first-class passenger?
d) If someone's chances of surviving were the same regardless of their status on the ship, how many members of the crew would you expect to have lived?
e) State the null and alternative hypotheses.
f) Give the degrees of freedom for the test.
g) The chi-square value for the table is 187.8, and the corresponding P-value is barely greater than 0. State your conclusions about the hypotheses.

**38. NYPD and sex discrimination** The table below shows the rank attained by male and female officers in the New York City Police Department (NYPD). Do these data summaries indicate that male and female are equitably represented at all levels of the department?

|  | | Sex | | |
|---|---|---|---|---|
|  | | Male | Female | Total |
| **Rank** | Officer | 21,900 | 4,281 | 26,181 |
|  | Detective | 4,058 | 806 | 4,864 |
|  | Sergeant | 3,898 | 415 | 4,313 |
|  | Lieutenant | 1,333 | 89 | 1,422 |
|  | Captain | 359 | 12 | 371 |
|  | Higher Ranks | 218 | 10 | 228 |
|  | Total | 31,766 | 5,613 | 37,379 |

a) What's the probability that a person selected at random from the NYPD is a female?
b) What's the probability that a person selected at random from the NYPD is a detective?
c) Assuming no bias in promotions, how many female detectives would you expect the NYPD to have?
d) To see if there is evidence of differences in ranks attained by males and females, will you test goodness-of-fit, homogeneity, or independence?
e) State the hypotheses.
f) Test the conditions.
g) How many degrees of freedom are there?
h) The chi-square value for the table is 290.1 and the P-value is less than 0.0001. State your conclusion about the hypotheses.

**39. Titanic again** Examine and comment on the chi-square results from Exercise 37. Where did the differences occur that caused the large chi-square value? Be specific.

**40. NYPD again** Examine and comment chi-square results from Exercise 38. Explain what differences caused the P-value to be so small. Be specific.

**41. Cranberry juice** It's common folk wisdom that drinking cranberry juice can help prevent urinary tract infections in women. In 2001 the *British Medical Journal* reported the results of a Finnish study in which 150 women were randomly allocated into three groups of 50 each and were monitored for these infections over 6 months. One group drank cranberry juice daily, another group drank a lactobacillus drink, and the third drank neither of those beverages, serving as a control group. In the control group, 18 women developed at least one infection, compared to 20 of those who consumed the lactobacillus drink and only 8 of those who drank cranberry juice. Does this study provide supporting evidence for the value of cranberry juice in warding off urinary tract infections?

a) Is this a survey, a retrospective study, a prospective study, or an experiment? Explain.
b) Will you test goodness-of-fit, homogeneity, or independence?
c) State the hypotheses.
d) Test the conditions.
e) How many degrees of freedom are there?
f) Find $\chi^2$ and the P-value.
g) State your conclusion.

**42. Cars** A random survey of autos parked in the student lot and the staff lot at a large university classified the brands by country of origin, as seen in the table. Are there differences in the national origins of cars driven by students and staff?

|  | | Driver | |
|---|---|---|---|
|  | | Student | Staff |
| **Origin** | American | 107 | 105 |
|  | European | 33 | 12 |
|  | Asian | 55 | 47 |

a) Is this a test of independence or homogeneity?
b) Write appropriate hypotheses.
c) Check the necessary assumptions and conditions.
d) Find the P-value of your test.
e) State your conclusion and analysis.

**43. Montana** A poll conducted by the University of Montana classified respondents by whether they were male or female and political party, as shown in the table. We wonder if there is evidence of an association between being male or female and party affiliation.

|  | Democrat | Republican | Independent |
|---|---|---|---|
| Male | 36 | 45 | 24 |
| Female | 48 | 33 | 16 |

a) Is this a test of homogeneity or independence?
b) Write an appropriate hypothesis.
c) Are the conditions for inference satisfied?
d) Find the P-value for your test.
e) State a complete conclusion.

**44. Fish diet** Medical researchers followed 6272 Swedish men for 30 years to see if there was any association between the amount of fish in their diet and prostate cancer. ("Fatty Fish Consumption and Risk of Prostate Cancer," *Lancet*, June 2001)

| Fish Consumption | Total Subjects | Prostate Cancers |
|---|---|---|
| Never/Seldom | 124 | 14 |
| Small Part of Diet | 2621 | 201 |
| Moderate Part | 2978 | 209 |
| Large Part | 549 | 42 |

a) Is this a survey, a retrospective study, a prospective study, or an experiment? Explain.
b) Is this a test of homogeneity or independence?

c) Do you see evidence of an association between the amount of fish in a man's diet and his risk of developing prostate cancer?
d) Does this study prove that eating fish does not prevent prostate cancer? Explain.

45. Montana revisited The poll described in Exercise 43 also investigated the respondents' party affiliations based on what area of the state they lived in. Test an appropriate hypothesis about this table and state your conclusions.

|  | Democrat | Republican | Independent |
|---|---|---|---|
| West | 39 | 17 | 12 |
| Northeast | 15 | 30 | 12 |
| Southeast | 30 | 31 | 16 |

46. Working parents In April 2009, Gallup published results from data collected from a large sample of adults in the 27 European Union member states. One of the questions asked was, "Which is the most practicable and realistic option for child care, taking into account the need to earn a living?" The counts below are representative of the entire collection of responses.

|  | Male | Female |
|---|---|---|
| Both Parents Work Full Time | 161 | 140 |
| One Works Full Time, Other Part Time | 259 | 308 |
| One Works Full Time, Other Stays Home for Kids | 189 | 161 |
| Both Parents Work Part Time | 49 | 63 |
| No Opinion | 42 | 28 |

Source: www.gallup.com/poll/117358/Work-Life-Balance-Tilts-Against-Women-Single-Parents.aspx

a) Is this a survey, a retrospective study, a prospective study, or an experiment?
b) Will you test goodness-of-fit, homogeneity, or independence?
c) Based on these results, do you think men and women have differing opinions when it comes to raising children?

47. Maryland lottery In the Maryland Pick-3 Lottery, three random digits are drawn each day. A fair game depends on every value (0 to 9) being equally likely to show up in all three positions. If not, someone who detects a pattern could take advantage of that. The table shows how many times each of the digits was drawn during a recent 32-week period, and some of them—4 and 7, for instance—seem to come up a lot. Could this just be a result of randomness, or is there evidence the digits aren't equally likely to occur?

| Digit | Count |
|---|---|
| 0 | 62 |
| 1 | 55 |
| 2 | 66 |
| 3 | 64 |
| 4 | 75 |
| 5 | 57 |
| 6 | 71 |
| 7 | 74 |
| 8 | 69 |
| 9 | 61 |

48. Stock market Some investors believe that stock prices show weekly patterns, claiming for example that Fridays are more likely to be "up" days. From the trading sessions since October 1, 1928 we selected a random sample of 1000 days on which the Dow Jones Industrial Average (DJIA) showed a gain in stock prices. The table shows how many of these fell on each day of the week. Sure enough, more of them are Fridays—and Tuesday looks like a bad day to own stocks. Can this be explained as just randomness, or is there evidence here to help an investor?

| Day of the Week | Number of "Up" Days |
|---|---|
| Mon | 192 |
| Tues | 189 |
| Wed | 202 |
| Thu | 199 |
| Fri | 218 |

49. Grades Two different professors teach an Introductory Statistics course. The table shows the distribution of final grades they reported. We wonder whether one of these professors is an "easier" grader.

|  | Prof. Alpha | Prof. Beta |
|---|---|---|
| A | 3 | 9 |
| B | 11 | 12 |
| C | 14 | 8 |
| D | 9 | 2 |
| F | 3 | 1 |

a) Will you test goodness-of-fit, homogeneity, or independence?
b) Write appropriate null hypotheses.
c) Find the expected counts for each cell, and explain why the chi-square procedures are not appropriate.

50. Full moon Some people believe that a full moon elicits unusual behavior in people. The table shows the number of arrests made in a small town during weeks of six full moons and six other randomly selected weeks in the same year. We wonder if there is evidence of a difference in the types of illegal activity that take place.

|  | Full Moon | Not Full |
|---|---|---|
| Violent (murder, assault, rape, etc.) | 2 | 3 |
| Property (burglary, vandalism, etc.) | 17 | 21 |
| Drugs/Alcohol | 27 | 19 |
| Domestic Abuse | 11 | 14 |
| Other Offenses | 9 | 6 |

a) Will you test goodness-of-fit, homogeneity, or independence?
b) Write appropriate null hypotheses.
c) Find the expected counts for each cell, and explain why the chi-square procedures are not appropriate.

51. Grades again In some situations where the expected cell counts are too small, as in the case of the grades given by Professors Alpha and Beta in Exercise 49, we can complete an analysis anyway. We can often proceed after combining cells in some way that makes sense and also produces a table in which the conditions are satisfied. Here we create a new table displaying the same data, but calling D's and F's "Below C":

|  | Prof. Alpha | Prof. Beta |
|---|---|---|
| A | 3 | 9 |
| B | 11 | 12 |
| C | 14 | 8 |
| Below C | 12 | 3 |

a) Find the expected counts for each cell in this new table, and explain why a chi-square procedure is now appropriate.
b) With this change in the table, what has happened to the number of degrees of freedom?
c) Test your hypothesis about the two professors, and state an appropriate conclusion.

**52. Full moon, next phase** In Exercise 50 you found that the expected cell counts failed to satisfy the conditions for inference.

a) Find a sensible way to combine some cells that will make the expected counts acceptable.
b) Test a hypothesis about the full moon and state your conclusion.

**53. Racial steering** A subtle form of racial discrimination in housing is "racial steering." Racial steering occurs when real estate agents show prospective buyers only homes in neighborhoods already dominated by that family's race. This violates the Fair Housing Act of 1968. According to an article in *Chance* magazine (vol. 14, no. 2 [2001]), tenants at a large apartment complex recently filed a lawsuit alleging racial steering. The complex is divided into two parts: Section A and Section B. The plaintiffs claimed that white potential renters were steered to Section A, while African-Americans were steered to Section B. The table describes the data that were presented in court to show the locations of recently rented apartments. Do you think there is evidence of racial steering?

|  | New Renters | | |
|---|---|---|---|
|  | White | Black | Total |
| Section A | 87 | 8 | 95 |
| Section B | 83 | 34 | 117 |
| Total | 170 | 42 | 212 |

**54. Titanic, redux** Newspaper headlines at the time, and traditional wisdom in the succeeding decades, have held that women and children escaped the *Titanic* in greater proportions than men. Here's a summary of the relevant data. Do you think that survival was independent of whether the person was male or female? Explain.

|  |  | Sex | | |
|---|---|---|---|---|
|  |  | Female | Male | Total |
| Survival | Saved | 343 | 367 | 710 |
|  | Lost | 127 | 1364 | 1491 |
|  | Total | 470 | 1731 | 2201 |

**55. Steering revisited** You could have checked the data in Exercise 53 for evidence of racial steering using two-proportion $z$ procedures.

a) Find the $z$-value for this approach, and show that when you square your $z$-value, you get the value of $\chi^2$ you calculated in Exercise 53.
b) Show that the resulting P-values are the same.

**56. Survival on the Titanic, one more time** In Exercise 54 you could have checked for a difference in the chances of survival for men and women using two-proportion $z$ procedures.

a) Find the $z$-value for this approach.
b) Show that the square of your calculated value of $z$ is the value of $\chi^2$ you calculated in Exercise 54.
c) Show that the resulting P-values are the same.

**57. Pregnancies** Most pregnancies are full term, but some are preterm (less than 37 weeks). Of those that are preterm, the Centers for Disease Control and Prevention classifies them as early (less than 34 weeks) and late (34 to 36 weeks). A December 2010 National Vital Statistics Report examined those outcomes in the United States broken down by age of the mother. The table shows counts consistent with that report. Is there evidence that the outcomes are not independent of age group?

|  | Early Preterm | Late Preterm |
|---|---|---|
| Under 20 | 129 | 270 |
| 20 to 29 | 243 | 612 |
| 30 to 39 | 165 | 424 |
| 40 or Over | 18 | 39 |

*Source*: www.cdc.gov/nchs/data/nvsr/nvsr59/nvsr59_01.pdf

**58. Education by age** Use the survey results in the table to investigate differences in education level attained among different age groups in the United States.

|  |  | Age Group | | | | |
|---|---|---|---|---|---|---|
|  |  | 25–34 | 35–44 | 45–54 | 55–64 | $\geq 65$ |
| Education | Not HS Graduate | 27 | 50 | 52 | 71 | 101 |
|  | HS | 82 | 19 | 88 | 83 | 59 |
|  | 1–3 Years College | 43 | 56 | 26 | 20 | 20 |
|  | $\geq$ 4 Years College | 48 | 75 | 34 | 26 | 20 |

**59. Superpowers redux** In Chapter 2, Exercise 34, we wondered if boys' and girls' favorite superpower differed. We had a hard time telling if the differences we saw were just random variation or if the genders really picked different powers. Now we can finally answer this question! Here are 100 randomly selected students from Census at Schools. Is there evidence that boys and girls differ in their preferences? (Source: ww2.amstat.org/CensusAtSchool/)

|  | Female | Male | Total |
|---|---|---|---|
| Fly | 15 | 16 | 31 |
| Freeze Time | 6 | 15 | 21 |
| Invisibility | 14 | 9 | 23 |
| Super Strength | 1 | 2 | 3 |
| Telepathy | 18 | 4 | 22 |
| Total | 54 | 45 | 100 |

## JUST CHECKING

Answers

1. 7.75, 23.25, 23.25, 69.75.
2. Yes. These are counts; student traits are independent of each other; although not a random sample, these Biology students should be representative of all students; 124 is fewer than 10% of all students; all expected counts are greater than 5.
3. $H_0$: The traits occur in the predicted proportions 1:3:3:9.
   $H_A$: The proportions of some of the traits are not as predicted.
4. 3 df.
5. 1.66.
6. 0.084.
7. At the 5% level, there's not enough evidence to suggest the genetic theory is incorrect.
8. We need to know how well beetles can survive 6 hours in a Plexiglas® box so that we have a baseline to compare the treatments.
9. There's no difference in survival rate in the three groups.
10. $(2-1)(3-1) = 2$ df.
11. 50.
12. 2.
13. The mean value for a $\chi^2$ with 2 df is 2, so 10 seems pretty large. The P-value is probably small.
14. This is a test of homogeneity. The clue is that the question asks whether the distributions are alike.
15. This is a test of goodness-of-fit. We want to test the model of equal assignment to all lots against what actually happened.
16. This is a test of independence. We have responses on two variables for the same individuals.

# 26

# Inferences for Regression

It is no secret today that the earth is warming due to greenhouse gas accumulations in our atmosphere. A primary culprit (though not the only one) is carbon dioxide, a by-product of burning fossil fuels. The following graph shows, for the years from 1958 to 2019, the association between the carbon dioxide concentration in the earth's atmosphere and the deviation of the average global temperature from a historic baseline.

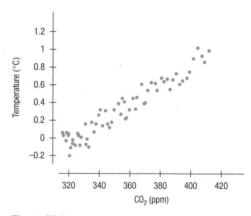

Figure 26.1
Global temperatures and $CO_2$ levels have a strong, positive, linear association.

| WHO | 62 years |
|---|---|
| WHAT | $CO_2$ concentrations and global temperatures |
| UNITS | ppm, and the difference between the mean global temperature and a historic baseline. |
| WHEN | 1958–2019 |
| WHERE | Everywhere on earth |
| WHY | To study climate change |

The graph clearly shows that the association between these two variables is positive, strong, and linear. Over these years, whenever $CO_2$ concentrations have been higher, the average global temperature has tended to be reliably higher as well, by a predictable amount.

Before we explore this dataset further, we want to point out—we hope this has occurred to you already—that these data are observational, not the results of an experiment. If all we had were these observations, it would be reasonable to question whether $CO_2$ accumulations actually *caused* rising temperatures, or were just an accidental association. Both have been rising over the years, and although $CO_2$ concentrations are demonstrably due to burning fossil fuels, the rising temperatures could plausibly have been caused by something else.

But these observations are not all we have. Much research, including controlled experiments, confirms that carbon dioxide is a "greenhouse gas."[1] It acts as a kind of blanket around the earth that allows the sun's warmth to enter while preventing its escape back into space. And $CO_2$ stays around a long time.

Back in Chapter 7, we modeled relationships like this by fitting a least squares line. The plot of Temperatures and $CO_2$ is clearly straight, so we can fit that line. The equation of the least squares line for these data is

$$\widehat{Temp} = -3.45 + 0.0106(CO_2)$$

The slope says that on average, global temperatures are higher by about 0.01 degrees Celsius for each additional 1 part per million (ppm) of $CO_2$ in the atmosphere. A hundredth of a degree doesn't sound like much, but the atmospheric $CO_2$ level is already well over 100 ppm greater than it was before the industrial era began and, with it, the burning of fossil fuels. And it's rising faster than ever.

Just how hot will it get? Our model can't answer that question, because it depends on how well we bring under control our fossil fuel consumption. But our model *can* help us predict how hot it will get *if* we continue adding $CO_2$ to the atmosphere. For that we'll want to make confidence intervals for and test hypotheses about the slope of a regression line.

# The Model and the Data

In this text we've often said that parameters like $\mu$ and $p$ are population characteristics—usually unknowable, but nevertheless fixed, constant numbers. The truth is that population characteristics like a mean $\mu$ or a proportion $p$ are rarely constant, because populations are often changing even as we collect data, and sometimes populations are not even well-defined. More accurately, parameters are *model* characteristics, and one very useful *model* is that of a fixed population—often one with a normal distribution.

But other models are useful too. The data in this chapter are not a random sample from a population, but it is reasonable to model the global temperature as being the sum of two components: an underlying mean that may depend on $CO_2$ levels, plus a random "error" that depends on everything else. That underlying mean doesn't really exist, but the model is still useful. In fact, $R^2$ for our Temperature and $CO_2$ data is about 0.92. So real or not, our model explains about 92% of the variations in global temperatures over the last six decades.

Let's be more precise with what mean by "our model." The linear regression model says that an observed response variable $y$ is the sum of an underlying mean plus some error, and that the underlying mean may depend linearly on $x$. We write:

$$y = (\beta_0 + \beta_1 x) + \varepsilon$$

The error term $\varepsilon$ is what gives a scatterplot its scatter: the observed data never all lie exactly on a line. If we take the error term out, the equation no longer describes an *observed* $y$-value, but instead, the *mean* $y$-value that you would expect for any particular $x$:

$$\mu_y = \beta_0 + \beta_1 x$$

> **NOTATION ALERT**
> This time we used one Greek letter for two things. Lowercase Greek $\beta$ (beta) is the natural choice to correspond to the $b$'s in the regression equation. We used $\beta$ before for the probability of a Type II error, but there's little chance of confusion here.

---
[1] For example, "On the causal structure between $CO_2$ and global temperature," A. Stips et al, *Nature Scientific Reports*, 22 February 2016; also "Sensitivity of a Global Climate Model to an Increase of $CO_2$ Concentration in the Atmosphere," S. Manabe and R Stouffer, *Journal of Geophysical Research*, October 20, 1980.

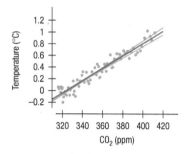

**Figure 26.2**
The LSRL for the data is shown in red. The other lines have slightly different values for their intercept and slope; all four of them are reasonably consistent with the data.

The two model parameters $\beta_0$ and $\beta_1$ are certainly not observable, but we can estimate them from data—and we have before, back in Chapter 7. When we find a regression line, $\hat{y} = b_0 + b_1 x$, the intercept and slope terms are estimates of the model's intercept and slope, $\beta_0$ and $\beta_1$.

What we're still missing is "wiggle room" in our estimates of the parameters. The least squares regression line, or LSRL, gives us *one* estimate of the intercept and *one* estimate of the slope. Although they are in some sense the "best" estimates of $\beta_0$ and $\beta_1$, they're not the only ones for which the data would be reasonably consistent with the model.

Think back to the random samples of bridges that we simulated in Chapter 7. (See pp. 191–192.) Different random samples from the same population yielded somewhat different (but not wildly different) regression lines. Our situation here is similar. The graph on the left shows the least squares regression line along with four other lines that are also reasonably consistent with our data; had any of them been the "real" model, our data could still have easily resulted.

# Assumptions and Conditions

Back in Chapter 7 when we fit lines to data, we needed to check only the Straight Enough Condition. Now, when we want to make inferences about the coefficients of the line, we'll have to make more assumptions. Fortunately, we can check conditions to help us judge whether these assumptions are reasonable for our data. And as we did before, we'll need to wait to make some checks until *after* we find the regression equation.

Also, we need to be careful about the order in which we check conditions. If our initial assumptions are not true, it makes no sense to check the later ones. So now we number the assumptions to keep them in order.

**Check the Scatterplot**
The form must be linear or we can't use linear regression at all.

## 1. Linearity Assumption

If the true relationship is far from linear and we use a straight line to fit the data, our entire analysis will be useless, so we always check this first.

The **Straight Enough Condition** is satisfied if a scatterplot looks generally straight. It's usually not a good idea to draw a line through the scatterplot when checking. That can fool your eyes into seeing the plot as more straight than it is. Sometimes it's easier to see violations of the Straight Enough Condition by looking at a scatterplot of the residuals against $x$ or against the predicted values, $\hat{y}$. That plot will have a horizontal direction and should have no pattern if the condition is satisfied.

If the scatterplot is straight enough, we can go on to some assumptions about the errors. If not, stop here, or consider re-expressing the data (see Chapter 9) to make the scatterplot more nearly linear. For the Temperature and $CO_2$ data, the scatterplot is beautifully linear.

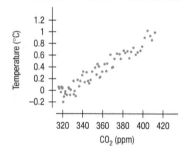

**Check the Residuals Plot (1)**
The residuals should appear to be randomly scattered.

## 2. Independence Assumption

**Independence Assumption**: The errors in the underlying regression model (the $\varepsilon$'s) must be independent of each other. As usual, there's no way to be sure that the Independence Assumption is true, but there are some things we can think about.

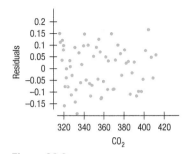

Figure 26.3
The residuals show only random scatter when plotted against $CO_2$.

### Check the Residuals Plot (2)
The vertical spread of the residuals should be roughly the same everywhere.

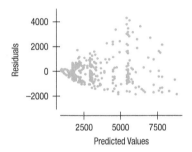

Figure 26.4
A scatterplot of residuals against predicted values can reveal plot thickening. In this plot of the residuals from a regression model of diamond prices on carat weight, we see that larger diamonds have more price variation than smaller diamonds. When the Equal Spread Assumption is violated, we can't summarize how the residuals vary with a single number.

### Check a Histogram of the Residuals
The distribution of the residuals should be unimodal and symmetric.

If we hope to apply our regression model to a larger population we can check the **Randomization Condition** that the individuals are a random sample from that population. We may settle for a sample that we believe to be representative.

We can also check the **Random Residuals Condition** by looking at the residuals plot for evidence of patterns, trends, or clumping, any of which would suggest a failure of independence. In the special case when the $x$-variable is related to time, a common violation of the Independence Assumption is for the errors to be related. (The error our model makes today may be similar to the one it made yesterday.) This violation can be checked by plotting the residuals against the $x$-variable and looking for patterns.

For our Temperature and $CO_2$ data, there *is* a danger of residuals associated with similar $CO_2$ levels being themselves similar to one another, because $CO_2$ levels are related to time. ($CO_2$ is rising pretty consistently.) Fortunately, the residual plot doesn't show any patterns.

## 3. Equal Variance Assumption

The variability of $y$ should be about the same for all values of $x$. In Chapter 7, we looked at the standard deviation of the residuals ($s_e$) to measure the size of the scatter. Now we'll need this standard deviation to build confidence intervals and test hypotheses. The standard deviation of the residuals is the building block for the standard errors of all the regression parameters. But it makes sense only if the scatter of the residuals is the same everywhere. In effect, the standard deviation of the residuals "pools" information across all of the individual distributions at each $x$-value, and pooled estimates are appropriate only when they combine information for groups with the same variance.

As a practical approach, what we can check is the **Does the Plot Thicken? Condition**. The scatterplot offers a visual check: Make sure the spread around the line is nearly constant. Often it is better to look at the residuals plot: Be alert for a "fan" shape or other tendency for the variation to grow or shrink in one part of the scatterplot (as in the plot on the left). That's not a problem for the Temperature and $CO_2$ data, where the residuals plot (Figure 26.3) shows that the spread of Temperatures is remarkably constant across different $CO_2$ levels from 320 ppm to 420 ppm.

If the plot is straight enough, the data are independent, and the plot doesn't thicken, you can now move on to the final assumption.

## 4. Normal Errors Assumption

We assume the errors around the idealized regression line at each value of $x$ follow a Normal model. We need this assumption so that we can use a Student's $t$-model for inference.

As we have at other times when we've used Student's $t$, we'll settle for the residuals satisfying the **Nearly Normal Condition** and the **Outlier Condition**. Look at a histogram or Normal probability plot of the residuals.[2]

The histogram of residuals in the Temperature regression does look reasonably Normal (See Figure 26.5). As we have noted before, the Normality Assumption becomes less important as the sample size grows, because the model is about means and the Central Limit Theorem takes over.

---

[2]*This* is why we have to check the conditions in order. We have to check that the residuals are independent and that the variation is the same for all $x$'s so that we can lump all the residuals together for a single check of the Nearly Normal Condition.

If all four assumptions were true, the idealized regression model would look like the one below. The red line shows the relationship between $\mu_y$ and $x$.

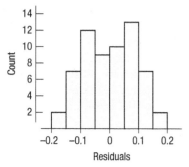

Figure 26.5
A histogram of the residuals is one way to check whether they are nearly Normal. Alternatively, we can look at a Normal probability plot. These residuals from the Temperature and $CO_2$ regression look nearly Normal.

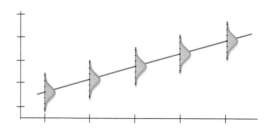

Figure 26.6
The regression model has a distribution of $y$-values for each $x$-value. These distributions follow a Normal model with means lined up along the line and with the same standard deviations.

At each value of $x$, there is a distribution of $y$-values that follows a Normal model, and each of these Normal models is centered on the line and has the same standard deviation. Of course, we don't expect the assumptions to be exactly true, and we know that all models are wrong, but the linear model is often close enough to be very useful.

### FOR EXAMPLE
### Checking Assumptions and Conditions

Look at the moon with binoculars or a telescope, and you'll see craters formed by thousands of impacts. The earth, being larger, has been hit even more often. Meteor Crater in Arizona was the first recognized impact crater and was identified as such only in the 1920s. With the help of satellite images, more and more craters have been identified; now more than 180 are known. These, of course, are only a small sample of all the impacts the earth has experienced: Only 29% of earth's surface is land, and many craters have been covered or eroded away. Astronomers have recognized a roughly 35-million-year cycle in the frequency of cratering, although the cause of this cycle is not fully understood. Here's a scatterplot of the known impact craters from the most recent 35 million years.[3] We've taken logs of both age (in millions of years ago) and diameter (km) to make the relationship more linear. (See Chapter 9.)

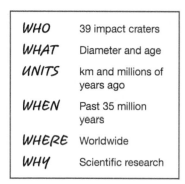

| WHO | 39 impact craters |
| --- | --- |
| WHAT | Diameter and age |
| UNITS | km and millions of years ago |
| WHEN | Past 35 million years |
| WHERE | Worldwide |
| WHY | Scientific research |

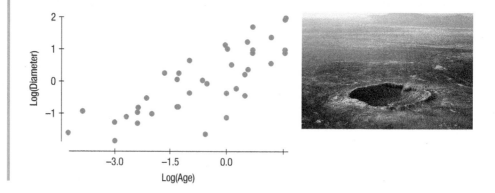

---

[3] Data, pictures, and much more information can be found at the earth Impact Database at www.passc.net/EarthImpactDatabase/index.html

QUESTION: Are the assumptions and conditions satisfied for fitting a linear regression model to these data?

ANSWER:

✓ **Linearity Assumption:** The scatterplot satisfies the Straight Enough Condition.

✓ **Independence Assumption:** Sizes of impact craters are likely to be generally independent.

✗ **Randomization Condition:** These are the only known craters, and may differ from others that may have disappeared or not yet been found. I'll be careful not to generalize my conclusions too broadly.

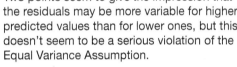

✓ **Random Residuals Condition:** The residuals appear to be randomly scattered.

✓ **Does the Plot Thicken? Condition:** After fitting a linear model, I find the residuals shown.

Two points seem to give the impression that the residuals may be more variable for higher predicted values than for lower ones, but this doesn't seem to be a serious violation of the Equal Variance Assumption.

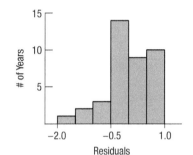

✓ **Nearly Normal Condition:** A histogram suggests a bit of skewness in the distribution of residuals.

There are no violations severe enough to stop my regression analysis, but I'll be cautious about my conclusions.

## Which Come First: The Conditions or the Residuals?

> Truth will emerge more readily from error than from confusion.
> 
> —Francis Bacon (1561–1626)

In regression, there's a little catch. The best way to check many of the conditions is with the residuals, but we get the residuals only *after* we compute the regression analysis. Before we do the analysis, however, we should check at least one of the conditions.

So we work in this order:

1. Make a scatterplot of the data to check the Straight Enough Condition. (If the relationship is curved, try re-expressing the data. Or stop.)
2. If the data are straight enough, fit a regression line and find the predicted values, $\hat{y}$, and the residuals, $e$.
3. Make a scatterplot of the residuals against $x$ or against the predicted values. This plot should have no pattern. Check in particular for any bend (which would suggest that the data weren't all that straight after all), for any thickening (or thinning), and, of course, for any outliers. (If there are outliers, and you can correct them or justify removing them, do so and go back to step 1, or consider performing two regressions—one with and one without the outliers.)
4. If the data are measured over time, plot the residuals against time to check for evidence of patterns that might suggest they are not independent.
5. If the scatterplots look OK, then make a histogram (or Normal probability plot) of the residuals to check the Nearly Normal Condition.
6. If all the conditions seem to be reasonably satisfied, go ahead with inference.

**726** PART VII Inference when Variables Are Related

# STEP-BY-STEP EXAMPLE

## Regression Inference

If our data can jump through all these hoops, we're ready to do regression inference. Let's see how much more we can learn about Temperatures and $CO_2$ levels from a regression model.

**QUESTIONS:** What is the relationship between Temperatures and $CO_2$ levels? What model best predicts Temperature from $CO_2$ levels, and how well does it do the job?

**THINK**

**PLAN** Specify the question of interest.

Name the variables and report the W's.

Identify the parameters you want to estimate.

**MODEL** Think about the assumptions and check the conditions.

Make pictures. For regression inference, you'll need a scatterplot, a residuals plot, and either a histogram or a Normal probability plot of the residuals.

I have 62 years' worth of atmospheric $CO_2$ concentrations and average global temperatures. I want to understand the relationship between them.

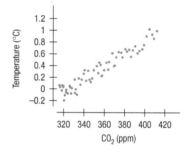

✓ **Straight Enough Condition:** There's no obvious bend in the original scatterplot of the data or in the residuals plot.

✓ **Independence Assumption:** $CO_2$ levels rise over time, raising the possibility of consecutive Temperatures having errors that are associated with one another. Fortunately, the residuals plot does not show any patterns consistent with this.

✗ **Randomization condition:** The data are not a random sample, but it seems reasonable to model Temperatures as being the sum of an underlying mean and unrelated random "noise."

✓ **Random Residuals Condition:** The residuals appear to be randomly scattered:

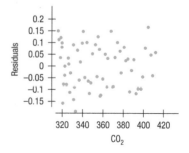

✓ **Does the Plot Thicken? Condition:** The residuals plot shows no evidence of a change in the spread about the line for different $CO_2$ levels.

You could also check the Nearly Normal Condition by making a Normal probability plot; it's reasonably straight.

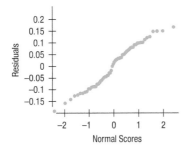

✓ **Nearly Normal Condition, Outlier Condition:** A histogram of the residuals is symmetric and mound-shaped. Although it appears bimodal, this minor deviation from normality isn't a problem when there are $n = 62$ data values.

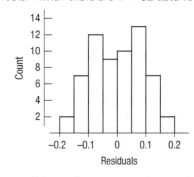

Name your method.

Under these conditions a linear **regression model** is appropriate.

 **MECHANICS** Let's just "push the button" and see what the regression looks like.

The by-hand formula for the regression equation can be found in Chapter 7, and the standard error formulas will be shown a bit later, but regressions are almost always computed with a computer program or calculator.

Write the regression equation.

Dependent variable is Temperature
R-squared = 91.8%
$s = 0.0914$ with $62 - 2 = 60$ degrees of freedom

| Variable | Coeff | SE(Coeff) | t-ratio | P-value |
|---|---|---|---|---|
| Intercept | −3.4527 | 0.1469 | −23.51 | <0.0001 |
| $CO_2$ | 0.0106 | 0.0004 | 25.84 | <0.0001 |

The estimated regression equation is
$$\widehat{Temp} = -3.4527 + 0.0106(CO_2)$$

 **CONCLUSION** Interpret your results in context.

The $R^2$ for regression is 91.8%. This linear model based on $CO_2$ levels seems to explain about 92% of the variability in average global temperatures over the last six decades. The standard deviation of $s = 0.0914$ means that Temperature estimates made by this model will typically deviate from the actual Temperature by about 0.09 degrees Celsius.

The slope of the line says that a rise in $CO_2$ in the atmosphere of 1 ppm is associated with a rise in average global temperatures of about 0.01 degrees Celsius.

The standard error of the slope is much less than the slope itself, so it looks like the estimate is reasonably precise.

**MORE INTERPRETATION** We haven't worked it out in detail yet, but the output gives us numbers labeled as *t*-statistics and corresponding P-values, and we have a general idea of what those mean. Now it's time to learn more about regression inference.

## Intuition About Regression Inference

Wait a minute! We've just pulled a fast one. We've pushed the "regression button" on our computer or calculator but haven't discussed where the standard errors for the slope or intercept come from. We know that if the random "noise" component of each year's Temperature (that's $\varepsilon$ in our model) had been a *different* amount of random noise then the slope and intercept of our regression line would have been different. Even with the model $\mu_y = \beta_0 + \beta_1 x$ holding true, the random nature of the noise would generate different estimates for $b_0$ and $b_1$. We've seen this variability in parameter estimates before; when our model is sampling from a population, we call it sampling variability. In our Temperature and $CO_2$ context, it doesn't represent *variability* so much as it represents *uncertainty*.

We have to be careful here. We're still assuming that the model's parameters are constants and that the randomness comes from the data. But as we saw in Figure 26.2, there are different underlying lines that would have all been reasonably consistent with the data. The standard error of the parameters quantifies their "wiggle room"—that is, how much they might differ from their estimated values and still yield a line that's consistent with the observed data.

Let's look at three things that affect the standard error ("wiggle room") of the parameters.

- **Spread around the line.** Here are two situations in which we might do regression. Which situation would be more likely to yield a consistent slope? That is, if we were to sample over and over from the two underlying populations that these samples come from[4] and compute all the slopes, which group of slopes would vary less?

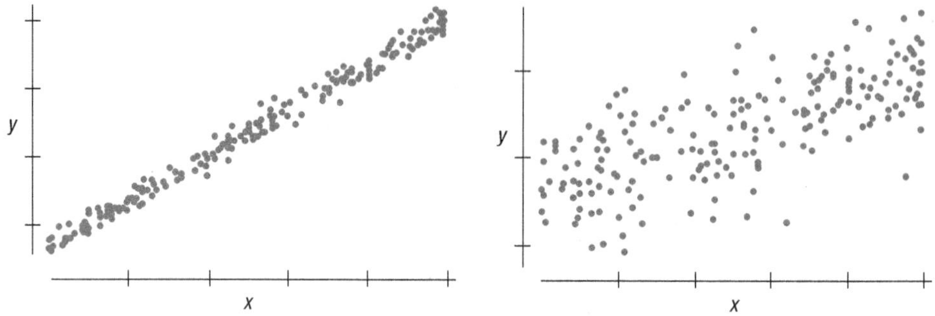

**Figure 26.7**
Which of these scatterplots shows a situation that would give the more consistent regression slope estimate if we were to sample repeatedly from its underlying population?

Clearly, data like those in the left plot will give more consistent slope estimates.

Less scatter around the line means the slope will be more consistent from sample to sample. The spread around the line is measured with the **residual standard deviation**, $s_e$. You can always find $s_e$ in the regression output, often just labeled $s$.[5] It estimates the typical distance actual $y$-values lie from values predicted by the model. Also, $s_e$ estimates the standard deviation of the $y$-values at any specific value of $x$. You're not likely to calculate the residual standard deviation by hand. When we first saw this formula in Chapter 7, we said that it looks a lot like the standard deviation of $y$, only subtracting the predicted values rather than the mean and dividing by $n-2$ instead of $n-1$:

$$s_e = \sqrt{\frac{\sum(y-\hat{y})^2}{n-2}}.$$

### $n-2$?

For the standard deviation (in Chapter 3), we divided by $n-1$ because we didn't know the true mean and had to estimate it. Now it's later in the course and there's even more we don't know. Here we don't know *two* things: the slope and the intercept. If we knew them both, we'd divide by $n$ and have $n$ degrees of freedom. When we *estimate* both, however, we adjust by subtracting 2, so we divide by $n-2$ and (as we will see soon) have 2 fewer degrees of freedom.

The less scatter around the line, the smaller the residual standard deviation and the stronger the relationship between $x$ and $y$.

Some people prefer to assess the strength of a regression model by looking at $s_e$ rather than $R^2$. After all, $s_e$ has the same units as $y$, and because it's the standard deviation of the errors around the line, it tells you how close the data are to our model. By contrast, $R^2$ is the proportion of the variation of $y$ accounted for by $x$. We say, why not look at both?

- **Spread of the $x$'s.** Here are two more situations. Which of these would yield more consistent slope estimates?

**Figure 26.8**
Which of these scatterplots shows a situation that would give the more consistent regression slope estimates if we were to sample repeatedly from the underlying population?

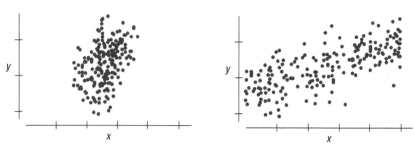

---
[4]Or simulate random data over and over using the two underlying models.
[5]It may be labeled RMSE, for "root mean squared error."

A plot like the one on the right has a broader range of $x$-values. We'd expect the slopes in samples from populations like that to vary less from sample to sample. If $s_x$, the standard deviation of $x$ is large, then the regression line slope will be more stable.
- **Sample size.** What about these two?

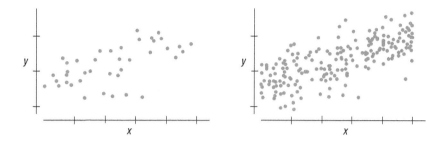

**Figure 26.9**
Which of these scatterplots shows a situation that would give the more consistent regression slope estimates if we were to sample repeatedly from the underlying population?

It shouldn't be a surprise that having a larger sample size, $n$, gives more consistent estimates from sample to sample.

## Standard Error for the Slope

Three aspects of the scatterplot, then, affect the **standard error of the slope**:

- Spread around the line: $s_e$
- Spread of $x$ values: $s_x$
- Sample size: $n$

These are in fact the *only* things that affect the standard error of the slope. Although you'll probably never have to calculate it by hand, the formula for the standard error is

$$SE(b_1) = \frac{s_e}{\sqrt{n-1}\, s_x}.$$

The error standard deviation, $s_e$, is in the *numerator* because spread around the line *increases* the slope's variability. The denominator has both a sample size term, $\sqrt{n-1}$, and $s_x$, because increasing either of these *decreases* the slope's variability.

We know the $b_1$'s vary from sample to sample. As you'd expect, their sampling distribution model is centered at $\beta_1$, the slope of the idealized regression line. Now we can estimate its standard deviation with $SE(b_1)$. What about its shape? Here the Central Limit Theorem and "Wild Bill" Gosset come to the rescue again. When we standardize the slopes by subtracting the model mean and dividing by their standard error, we get a Student's $t$-model, this time with $n-2$ degrees of freedom.

> NOTATION ALERT
> Don't confuse the standard deviation of the residuals, $s_e$, with the standard error of the slope, $SE(b_1)$. The first, $s_e$, measures the scatter around the line, telling us how reliably we can estimate $y$-values. The second, $SE(b_1)$, indicates how much slopes vary from sample to sample, telling us how reliably we can estimate the slope $\beta_1$ that's in the model.

### A Sampling Distribution for Regression Slopes

When the conditions are met, the standardized estimated regression slope,

$$t = \frac{b_1 - \beta_1}{SE(b_1)},$$

follows a Student's $t$-model with $n-2$ degrees of freedom. We estimate the standard deviation with

$$SE(b_1) = \frac{s_e}{\sqrt{n-1}\, s_x},$$

where $s_e = \sqrt{\dfrac{\sum(y-\hat{y})^2}{n-2}}$.

### What About the Intercept?

The same reasoning applies for the intercept. We could write

$$\frac{b_0 - \beta_0}{SE(b_0)} \sim t_{n-2}$$

and use it to construct confidence intervals and test hypotheses, but often the value of the intercept isn't something we care about—or have a natural null hypothesis for. The intercept usually isn't interesting. Most hypothesis tests and confidence intervals for regression are about the slope.

# Regression Inference

> **TI-*nspire***
>
> **Regression Inference.** How big must a slope be in order to be considered statistically significant? See for yourself by exploring the natural sample-to-sample variability in slopes.

Now that we have the standard error of the slope and its sampling distribution, we can test a hypothesis about it and make confidence intervals. The usual null hypothesis in a ***t*-test for the regression slope** is that the true slope is equal to 0. Why? Well, a slope of zero would say that $y$ doesn't tend to change linearly when $x$ changes—in other words, that there is no linear association between the two variables. If the slope were zero, there wouldn't be much left of our regression equation.

So a null hypothesis of a zero slope questions the entire claim of a linear relationship between the two variables—and often that's just what we want to know. In fact, every software package or calculator that does regression simply assumes that you want to test the null hypothesis that the slope is really zero.

To test $H_0: \beta_1 = 0$, we find

$$t_{n-2} = \frac{b_1 - 0}{SE(b_1)}.$$

> **WHAT IF THE SLOPE WERE 0?**
>
> If $b_1 = 0$, our prediction is $\hat{y} = b_0 + 0x$. The equation collapses to just $\hat{y} = b_0$. Now $x$ is nowhere in sight, so $y$ doesn't depend on $x$ at all.
>
> And $b_0$ would turn out to be $\bar{y}$. Why? We know that $b_0 = \bar{y} - b_1\bar{x}$, but when $b_1 = 0$, that becomes simply $b_0 = \bar{y}$. It turns out, then, that when the slope is 0, the equation is just $\hat{y} = \bar{y}$; at every value of $x$, we always predict the mean value for $y$.

This is just like every *t*-test we've seen: a difference between the statistic and its hypothesized value, divided by its standard error.

For our Temperature data, the computer found the slope (0.0106), its standard error (0.0004), and the ratio of the two: $\frac{0.0106 - 0}{0.0004} = 25.84$. Nearly 26 standard errors from the hypothesized value certainly seems big.[6] The P-value ($<0.0001$) confirms that a *t*-ratio this large would be very unlikely to occur if the true slope were zero.

Maybe the standard null hypothesis isn't all that interesting here. Did you have any doubts that Temperature is related to $CO_2$? A more informative use of these same values might be to make a confidence interval for the slope instead.

We can build a confidence interval in the usual way, as an estimate plus or minus a margin of error. As always, the margin of error is just the product of the standard error and a critical value. Here the critical value comes from the *t*-distribution with $n - 2$ degrees of freedom, so a 95% **confidence interval for $\beta$** is

$$b_1 \pm t^*_{n-2} \times SE(b_1).$$

For the Temperature data, $t^*_{60} = 2.000$ (coincidentally), so that comes out to $0.0106 \pm 2.000 \times 0.0004$, or an interval from about $0.0098°C$ to $0.0114°C$ of temperature rise for each additional ppm of $CO_2$.

---

[6] If you compute this yourself using the numbers here you'll get 26.5, not 25.84. That's because we've rounded the slope and its standard error to four decimal places. The computer output, seen on p. 727, computes the ratio based on the unrounded values.

## FOR EXAMPLE

### Interpreting a Regression Model

**RECAP:** On a log scale, there seems to be a linear relationship between the diameter and the age of recent terrestrial impact craters. We have regression output from statistics software:

Dependent variable is LogDiam
R-squared = 63.6%
s = 0.6362 with 39 − 2 = 37 degrees of freedom

| Variable | Coefficient | SE(coeff) | t-ratio | P-value |
|---|---|---|---|---|
| Intercept | 0.358262 | 0.1106 | 3.24 | 0.0025 |
| LogAge | 0.526674 | 0.0655 | 8.05 | ≤0.0001 |

**QUESTION:** What's the regression model, and what can it tell us?

**ANSWER:** For terrestrial impact craters younger than 35 million years, the logarithm of *Diameter* grows linearly with the logarithm of *Age*:

$$\widehat{logDiam} = 0.358 + 0.527 \; logAge.$$

The P-value for each coefficient's *t*-statistic is very small, so I'm quite confident that neither coefficient is zero. Based on my model, I conclude that, on average, the older a crater is, the larger it tends to be. This model accounts for 63.6% of the variation in log(*Diameter*).

Although it is possible that impacts (and their craters) are getting smaller, it is more likely that I'm seeing the effects of age on craters. Small craters are probably more likely to erode or become buried or otherwise be difficult to find as they age. Larger craters may survive the huge expanses of geologic time more successfully.

## JUST CHECKING

Researchers in Food Science studied how big people's mouths tend to be. They measured mouth volume by pouring water into the mouths of subjects who lay on their backs. Unless this is your idea of a good time, it would be helpful to have a model to estimate mouth volume more simply. Fortunately, mouth volume is related to height. (Mouth volume is measured in cubic centimeters and height in meters.)

The data were checked and deemed suitable for regression. Take a look at the computer output to answer these questions:

1. What does the *t*-ratio of 3.27 for the slope tell about this relationship? How does the P-value help your understanding?
2. Would you say that measuring a person's height could reliably be used as a substitute for the wetter method of determining how big a person's mouth is? What numbers in the output helped you reach that conclusion?
3. What does the value of $s_e$ add to this discussion?
4. Interpret the value of the standard error of the slope in this context.

| Summary of | Mouth Volume |
|---|---|
| Mean | 60.2704 |
| StdDev | 16.8777 |

Dependent variable is Mouth Volume
R-squared = 15.3%
s = 15.66 with 61 − 2 = 59 degrees of freedom

| Variable | Coefficient | SE(coeff) | t-ratio | P-value |
|---|---|---|---|---|
| Intercept | −44.7113 | 32.16 | −1.39 | 0.1697 |
| Height | 61.3787 | 18.77 | 3.27 | 0.0018 |

# Another Example: Breaking Up Is Hard to Predict

Every spring, Nenana, Alaska, hosts a contest in which participants try to guess the exact minute that a wooden tripod placed on the frozen Tanana River will fall through the breaking ice. The contest started in 1917 as a diversion for railroad engineers, with a jackpot of $800 for the closest guess. It has grown into an event in which hundreds of thousands of entrants enter their guesses on the Internet[7] and vie for as much as $300,000.

Because so much money and interest depends on the time of breakup, it has been recorded to the nearest minute with great accuracy ever since 1917. And because a standard measure of breakup has been used throughout this time, the data are consistent. An article in *Science*[8] used the data to investigate global warming—whether greenhouse gasses and other human actions have been making the planet warmer. (Others might just want to make a good prediction of next year's breakup time for betting purposes.)

Of course, we can't use regression to tell the *causes* of any change. But we can estimate the *rate* of change (if any) and use it to make better predictions.

Here are some of the data:

| WHO | Years |
| --- | --- |
| WHAT | Year, day, and hour of ice breakup |
| UNITS | $x$ is in years since 1900. $y$ is in days after midnight December 31. |
| WHEN | 1917–present |
| WHERE | Nenana, Alaska |
| WHY | Wagering, but proposed to look at global warming |

| Year (since 1900) | Breakup Date (days after Jan. 1) | Year (since 1900) | Breakup Date (days after Jan. 1) |
| --- | --- | --- | --- |
| 17 | 119.4792 | 27 | 127.7938 |
| 18 | 130.3979 | 28 | 129.3910 |
| 19 | 122.6063 | 29 | 121.4271 |
| 20 | 131.4479 | 30 | 127.8125 |
| 21 | 130.2792 | 31 | 119.5882 |
| 22 | 131.5556 | 32 | 134.5639 |
| 23 | 128.0833 | 33 | 120.5403 |
| 24 | 131.6319 | 34 | 131.8361 |
| 25 | 126.7722 | 35 | 125.8431 |
| 26 | 115.6688 | 36 | 118.5597 |

## STEP-BY-STEP EXAMPLE
### A Regression Slope *t*-Test

The slope of the regression line gives the change in Nenana ice breakup date per year.

QUESTIONS: Is there sufficient evidence to claim that ice breakup times are changing? If so, how rapid is the change?

---

**THINK**   **PLAN** State what you want to know.

Identify the *parameter* you wish to estimate. Here our parameter is the slope.

Identify the variables and review the W's.

**HYPOTHESES** Write your null and alternative hypotheses.

I wonder whether the date of ice breakup in Nenana has changed over time. The slope of that change might indicate climate change. I have the date of ice breakup annually for 95 years starting in 1917, recorded as the number of days and fractions of a day until the ice breakup.

$H_0$: There is no change in the date of ice breakup: $\beta_1 = 0$.

$H_A$: Yes, there is: $\beta_1 \neq 0$.

---

[7] www.nenanaakiceclassic.com
[8] "Climate Change in Nontraditional Data Sets," *Science* 294 [26 October 2001]: 811.

**MODEL** Think about the assumptions and check the conditions.

Make pictures. Because the scatterplot seems straight enough, we can find and plot the residuals.

Usually, we check for suggestions that the Independence Assumption fails by plotting the residuals against the predicted values. Patterns and clusters in that plot raise our suspicions. But when the data are measured over time, it is always a good idea to plot residuals against time to look for trends and oscillations.

✓ **Straight Enough Condition**: I have quantitative data with no obvious bend in the scatterplot.

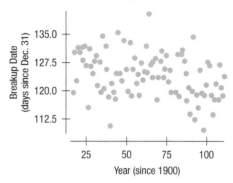

✓ **Independence Assumption**: These data are a time series, which raises my suspicions that they may not be independent. To check, here's a plot of the residuals against time.

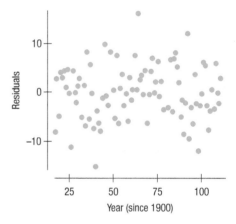

✓ **Random Residuals Condition**: I see a hint that the data oscillate up and down, which suggests some failure of independence, but not so strongly that I can't proceed with the analysis.

✗ **Randomization Condition**: These data are not a random sample, but it seems reasonable to model the deviations from a line as random noise. The residual plot shows no patterns that would make that assumption questionable.

✓ **Does the Plot Thicken? Condition**: The residuals plot shows no obvious changes in the spread.

✓ **Nearly Normal Condition, Outlier Condition**: A histogram of the residuals is unimodal and symmetric.

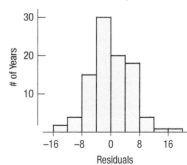

| | | |
|---|---|---|
| | State the sampling distribution model. | Under these conditions, the sampling distribution of the regression slope can be modeled by a Student's t-model with $(n - 2) = 93$ degrees of freedom. |
| | Choose your method. | I'll do a **regression slope t-test**. |

**SHOW** **MECHANICS** The regression equation can be found from the formulas in Chapter 7, but regressions are almost always found from a computer program or calculator.

The P-values given in the regression output table are from the Student's $t$-distribution on $(n - 2) = 93$ degrees of freedom. They are appropriate for two-sided alternatives.

Here's the computer output for this regression:

Dependent variable is Breakup Date
R-squared = 11.1%
s = 5.600 with 95 − 2 = 93 degrees of freedom

| Variable | Coefficient | SE(Coeff) | t-ratio | P-value |
|---|---|---|---|---|
| Intercept | 128.732 | 1.459 | 88.2 | ≤0.0001 |
| Year Since 1900 | −0.071436 | 0.0210 | −3.41 | 0.0010 |

The estimated regression equation is
$$\widehat{Date} = 128.732 - 0.071\ Year\ Since\ 1900.$$

**TELL** **CONCLUSION** Link the P-value to your decision and state your conclusion in the proper context.

The low P-value of 0.0010 means that the association we see in the data is unlikely to have occurred by chance even though the $R^2$ is not particularly strong. I reject the null hypothesis, and conclude that there is strong evidence that, on average, the ice breakup is occurring earlier each year. But the hint of an oscillation pattern in the residuals raises concerns.

**SHOW MORE** **CREATE A CONFIDENCE INTERVAL FOR THE TRUE SLOPE**

A 95% confidence interval for $\beta_1$ is
$$b_1 \pm t^*_{93} \times SE(b_1)$$
$$-0.071 \pm (1.986)(0.0210)$$
or $(-0.11, -0.03)$ days per year.

**TELL MORE** **INTERPRET THE INTERVAL** Simply rejecting the standard null hypothesis doesn't guarantee that the size of the effect is large enough to be important. Whether we want to know the breakup time to the nearest minute or are interested in global warming, a change measured in hours each year is big enough to be interesting.

I am 95% confident that the ice has been breaking up, on average, between 0.03 days (about 40 minutes) and 0.11 days (just over 2.5 hours) earlier each year since 1900.

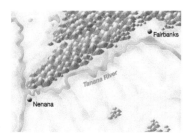

### BUT IS IT GLOBAL WARMING?

So the ice is breaking up earlier. Temperatures are higher. Must be global warming, right?

   Maybe.

   An article challenging the original analysis of the Nenana data proposed a possible confounding variable. It noted that the city of Fairbanks is upstream from Nenana and suggested that the growth of Fairbanks could have warmed the river. So maybe it's not global warming.

   Or maybe global warming is a lurking variable, leading more people to move to a now balmier Fairbanks and also leading to generally earlier ice breakup in Nenana.

Or maybe there's some other variable or combination of variables at work. We can't set up an experiment, so we may never really know.

Only one thing is for sure. When you try to explain an association in observational data by claiming cause and effect, you may to be on thin ice.[9]

## TI TIPS

### Doing Regression Inference

The TI will easily do almost everything you need for inference for regression: scatterplots, residuals plots, histograms of residuals, and $t$-tests and confidence intervals for the slope of the regression line. OK, it won't tell you $SE(b)$, but it will give you enough information to easily figure it out for yourself. Not bad.

As an example we'll use data from *Chance* magazine (Vol. 12, No. 4, 1999) for 11 of the top performances in women's marathons during the 1990s. Let's examine the influence of temperature on the times for elite runners.

| °F  | 44    | 46    | 47    | 50    | 51    | 52    | 54    | 55    | 57    | 60    | 65    |
|-----|-------|-------|-------|-------|-------|-------|-------|-------|-------|-------|-------|
| Min | 142.7 | 142.1 | 143.4 | 143.6 | 144.0 | 143.4 | 142.4 | 143.1 | 143.7 | 143.4 | 143.4 |

**TEST A HYPOTHESIS ABOUT THE ASSOCIATION**

- Enter the temperatures (nearest degree Fahrenheit) in `L1` and the runners' times (nearest tenth of a minute) in `L2`.
- Check the scatterplot. It's not obviously nonlinear, so go ahead.
- Under `STAT TESTS` choose `LinRegTTest`.
- Specify the two data lists (with `Freq:1`).
- Choose the two-tailed option. (We are interested in whether higher temperatures enhance or interfere with a runner's performance.)
- Tell it to store the regression equation in `Y1` (`VARS`, `Y-VARS`, `Function`... remember?), then `Calculate`.

The TI creates so much information you have to scroll down to look at it all! See:
- The calculated value of `t` and the P-value.
- The coefficients of the regression equation, `a` and `b`.
- The value of `s`, our sample estimate of the common standard deviation of errors around the true line.
- The values of `r`$^2$ and `r`.

Wait, where's $SE(b)$? It's not there. No problem—if you need it, you can figure it out. Remember that the $t$-value is $b$ divided by $SE(b)$. So $SE(b)$ must be $b$ divided by $t$. Here $SE(b) = 0.0325 \div 1.1358 = 0.0286$.

**CREATE A CONFIDENCE INTERVAL FOR THE SLOPE**

- Back to `STAT TESTS`; this time you want `LinRegTInt`.
- The specifications for the data lists and the regression equation remain what you entered for the hypothesis test.
- Choose a confidence level, say 95%, and `Calculate`.

(A question for you: How is this confidence interval consistent with the P-value for the hypothesis test?)

---

[9]How *do* scientists sort out such messy situations? Even though they can't conduct an experiment, they *can* look for replications elsewhere. A number of studies of ice on other bodies of water have also shown earlier ice breakup times in recent years. That suggests they need an explanation that's more comprehensive than just Fairbanks and Nenana.

**CHECK THE CONDITIONS** Beware!!! Before you try to interpret any of this, you must check the conditions to see if inference for regression is allowed.

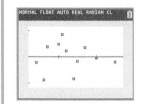

- We already looked at the scatterplot; it was reasonably linear.
- To create the residuals plot, set up another scatterplot with RESID (from LIST NAMES) as your Ylist. OK, it looks fairly random.
- The residuals plot may show a slight hint of diminishing scatter, but with so few data values it's not very clear.
- The histogram of the residuals is unimodal and roughly symmetric.

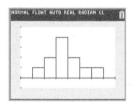

**WHAT DOES IT ALL MEAN?** Because the conditions check out okay, we can try to summarize what we have learned. With a P-value over 28%, it's quite possible that any perceived relationship could be just sampling error. The confidence interval suggests the slope could be positive or negative, so it's plausible that as temperatures increase, women marathoners may run faster—or slower. Based on these 11 races there appears to be little evidence of a linear association between temperature and women's performances in the marathon.

# Different Samples, Different Slopes

You've had your blood pressure taken. The measurement gave you two readings, such as 118/72.[10] The bigger number (systolic) is the pressure on your artery walls when your heart beats, pushing blood through your body. The smaller number (diastolic) is the pressure when your heart has relaxed between beats. Blood pressures vary from person to person, and even yours varies somewhat throughout each day, so the two numbers change. It's reasonable to ask how the systolic and diastolic pressures are related.

To investigate, we'd need to get readings from several different people, make a scatterplot, and examine the relationship. And then we'd face our perpetual dilemma: "OK, I see what's happening in *these* sample data. But what does that tell me about the true relationship in the whole population?"

Let's explore that a bit. We start with a dataset containing 1406 blood pressure readings. Rather than analyze *them*, let's pretend they are our entire population. We'll play the Statistics game: draw a sample from this population and use it to guess at the truth. So, here's what we see in our first random sample of 25.

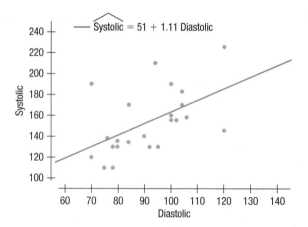

The scatterplot shows an association that's straight enough for regression. The slope of the line suggests we might expect systolic pressure to rise an average of about 1.11 points for each 1-point increase in diastolic pressure.

---

[10] The units are "millimeters of mercury," indicating how high the fluid would rise in a glass tube in the now old-fashioned sphygmomanometers (Yeah, that's the word!) used to measure blood pressure.

That seems pretty straightforward. But keep in mind, that's just where *this* sample leads us. Had we randomly selected a different group of 25 people, we might reach a different conclusion, right? Let's see. Here's our second random sample.

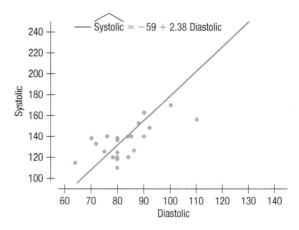

Hmmm. The relationship still looks linear, but this sample suggests the slope is 2.38 systolic points per diastolic point. That seems quite different. So what can a sample tell us about the true slope?

In the real world we generally get only one sample, but the beauty of asking "What if?" is that we get to play around some more. We simulated this process 200 times. Here's a histogram showing the slopes of the 200 regression lines generated by those samples.

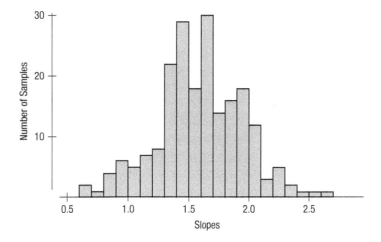

Now, stop and think. What important concept is on display here? Don't read farther until you know what this histogram represents.

Got it? This is a simulated sampling distribution of sample slopes.

We've encountered a lot of sampling distributions in this course. They form the basis for much statistical inference. The sampling distribution of sample proportions approaches a Normal model. The Central Limit Theorem (yes, *that* theorem again!) tells us the sampling distribution of sample means is also Normal (but not knowing the population standard deviation, we must use a *t*-model to do inference). And now . . . what? Let's think about the sampling distribution of sample slopes in the usual way: shape, center, and spread.

- ◆ **Shape:** The histogram above looks roughly unimodal and symmetric. If you're thinking "Normal," good for you. Once again, though, we don't know the population standard deviation, so (as we told you earlier in this chapter), we'll need a *t*-model. This one has 23 degrees of freedom, right?
- ◆ **Center:** Would you say a bit over 1.5? Good guess. The mean of these 200 sample slopes is 1.59. And now, because we're playing "What If?", we get to cheat. We get to actually peek at the population. The true slope for all 1406 BPs is (wait for it . . . ) 1.563.

Random samples are producing slopes that target the right number. That's good news: sample slope ($b_1$) is an unbiased estimator of population slope ($\beta_1$).

♦ **Spread:** For small samples like this, slopes can apparently vary quite a bit. The standard deviation of these 200 sample slopes is 0.355, providing a good estimate of how all sample slopes would vary. Any individual sample allows us to make a similar estimate. You know this as $SE(b_1)$, the standard error of sample slopes.

When we're looking at only one sample, its slope ($b_1$) is our best guess for the true slope ($\beta_1$). And then we attach a margin of error. How big should that be? Big enough to capture the true slope starting with 95% of all samples. Look carefully at the histogram and see if you agree that the middle 95% of our 200 samples have slopes varying from about 0.9 to 2.3. That suggests a typical margin of error should be about $\frac{1}{2}(2.3 - 0.9) = 0.7$.

How well would our first two samples have worked? For Sample #1, the confidence interval would have been $1.11 \pm 0.7 = (0.41, 1.81)$, capturing the true slope of 1.563. Hooray! A hit!

For Sample #2, $2.38 \pm 0.7 = (1.68, 3.08)$. With this sample we struck out.

But in Statistics, overall we bat 0.950! Hall of Fame, baby!

## WHAT CAN GO WRONG?

In this chapter we've added inference to the regression explorations that we did in Chapters 7 and 8. Everything covered in those chapters that could go wrong with regression can still go wrong. It's probably a good time to review Chapter 8. Take your time; we'll wait.

With inference, we've put numbers on our estimates and predictions, but these numbers are only as good as the model. Here are the main things to watch out for:

♦ **Don't fit a linear regression to data that aren't straight.** This is the most fundamental assumption. If the relationship between $x$ and $y$ isn't approximately linear, there's no sense in fitting a straight line to it.

♦ **Watch out for the plot thickening.** The common part of confidence and prediction intervals is the estimate of the error standard deviation, the spread around the line. If it changes with $x$, the estimate won't make sense. Imagine making a prediction interval for these data.

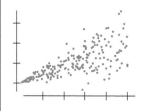

When $x$ is small, we can predict $y$ precisely, but as $x$ gets larger, it's much harder to pin $y$ down. Unfortunately, if the spread changes, the single value of $s_e$ won't pick that up. We'll be too pessimistic about our precision for low $x$-values and too optimistic for high $x$-values. A re-expression of $y$ is often a good fix for changing spread.

♦ **Make sure the errors are Normal.** The size of $s_e$ can help us judge how accurate our predictions may be, but only if the errors come from a Normal model. Check the histogram and Normal probability plot of the residuals to see if this assumption looks reasonable.

♦ **Watch out for extrapolation.** Our predictions are reasonable only if our model is true, and there's no reason to assume that the same pattern extends indefinitely. That's why it's always dangerous to predict for $x$-values that lie far from the center of the data.

♦ **Watch out for influential points and outliers.** We always have to be on the lookout for a few points that have undue influence on our estimated model—and regression is certainly no exception.

♦ **Watch out for one-tailed tests.** Because tests of hypotheses about regression coefficients are usually two-tailed, software packages report two-tailed P-values. If you are using software to conduct a one-tailed test about slope, you'll need to divide the reported P-value in half.

# WHAT HAVE WE LEARNED?

In Chapters 6, 7, and 8 we learned to examine the relationship between two quantitative variables by looking at a scatterplot. We've learned (if it's linear) to quantify the strength and direction with a correlation and model it with least squares regression. Now we have learned to apply inference to these regression models.

We've learned that we can interpret the standard deviation of the residuals, $s_e$, two ways:
- $s_e$ estimates the standard deviation of the y's at any value of x;
- $s_e$ estimates the typical error between predicted values and actual values.

We've learned that the standard error of the slope, $SE(b_1)$, estimates the sample-to-sample variability in slopes of regression lines.

We've learned the assumptions for inference (and how to check them, in order):
- the Linearity Assumption (Straight Enough Condition);
- the Independence Assumption (Random Residuals Condition);
- the Equal Variances Assumption (Does the Plot Thicken? Condition);
- the Normality Assumption (Nearly Normal and Outlier Conditions).

We've learned that when these assumptions are met, the sampling distribution for the slope of a regression line can be described by a $t$-model with $n - 2$ degrees of freedom.

We've learned to use this model to test the hypothesis that there's no linear association between two quantitative variables, $H_0: \beta_1 = 0$.

And we've learned to construct and interpret a confidence interval around $b_1$ for the true slope $\beta_1$.

## TERMS

**Conditions for inference in regression (and checks for some of them)**

- **Straight Enough Condition** for linearity. (Check that the scatterplot of y against x has linear form and that the scatterplot of residuals against predicted values has no obvious pattern.) (p. 722)
- **Independence Assumption.** (Think about the nature of the data.) (p. 722)
- **Randomization Condition** for data that come from a random (or at least representative) sample of the population. (p. 723)
- **Random Residuals Condition** for any evidence of patterns, trends, or clumping. (p. 723)
- **Does the Plot Thicken? Condition** for constant variance. (Check that the scatterplot shows consistent spread across the range of the x-variable, and that the residuals plot has constant variance, too. A common problem is increasing spread with increasing predicted values—the *plot thickens*!) (p. 723)
- **Nearly Normal Condition** for Normality of the residuals. (Check a histogram of the residuals.) (p. 723)

**Residual standard deviation**

The spread of the data around the regression line is measured with the residual standard deviation, $s_e$:

$$s_e = \sqrt{\frac{\sum(y - \hat{y})^2}{n - 2}} = \sqrt{\frac{\sum e^2}{n - 2}}. \text{ (p. 728)}$$

**Standard error for the slope**

The standard error of $b_1$ estimates the standard deviation of the sampling distribution model for slopes of regression lines:

$$SE(b_1) = \frac{s_e}{\sqrt{n - 1}\, s_x}. \text{ (p. 729)}$$

**t-test for the regression slope**

When the assumptions are satisfied, we can perform a test for the slope coefficient. We usually test the null hypothesis that the true value of the slope is zero against the alternative that it is not. A zero slope would indicate a complete absence of linear relationship between y and x.

To test $H_0: \beta_1 = 0$, we find the P-value from the Student's t-model with $n - 2$ degrees of freedom where

$$t = \frac{b_1 - 0}{SE(b_1)}. \quad (p.\ 730)$$

**Confidence interval for the regression slope β**

When the assumptions are satisfied, we can find a confidence interval for the slope parameter from $b_1 \pm t^*_{n-2} \times SE(b_1)$. The critical value, $t^*_{n-2}$, depends on the confidence level specified and on Student's t-model with $n - 2$ degrees of freedom. (p. 730)

## ON THE COMPUTER

### Regression Analysis

All statistics packages make a table of results for a regression analysis. These tables differ slightly from one package to another, but all are essentially the same. We've seen other examples of such tables already.

All packages offer analyses of the residuals. With some, you must request plots of the residuals as you request the regression. Others let you find the regression model first and then analyze the residuals afterward. Either way, your analysis is not complete if you don't check the residuals with a histogram or Normal probability plot and a scatterplot of the residuals against x or the predicted values.

You should, of course, always look at the scatterplot of your two variables before computing a regression analysis.

Regression models are almost always found with a computer or calculator. The calculations are too long to do conveniently by hand for datasets of any reasonable size. No matter how the regression is computed, the results are usually presented in a table that has a standard form. Here's a portion of a typical regression results table, along with annotations showing where the numbers come from:

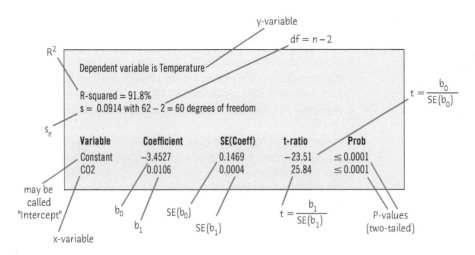

The regression table gives the coefficients (once you find them in the middle of all this other information), so we can see that the regression equation is

$$\widehat{Temp} = -3.4527 + 0.0106(CO_2)$$

and that the $R^2$ for the regression is 91.8%.

The column of t-ratios gives the test statistics for the respective null hypotheses that the true values of the coefficients are zero. The corresponding P-values are also usually reported.

# EXERCISES

1. **Graduate earnings** Does attending college pay back the investment? What factors predict higher earnings for graduates? *Money* magazine surveyed graduates, asking about their point of view of the colleges they had attended (*Money's Best Colleges*). One good predictor of early career earnings ($/year) turned out to be the average SAT score of entering students. Here are the regression model and associated plots. Write the regression model and explain what the slope coefficient means in this context.

   Response variable is: Earn
   R squared = 30.7%
   s = 5603 with 706 − 2 = 704 degrees of freedom

   | Variable | Coefficient | SE(Coeff) | t-ratio | P-value |
   |---|---|---|---|---|
   | Intercept | 14468.1 | 1777 | 8.14 | <0.0001 |
   | SAT | 27.2642 | 1.545 | 17.6 | <0.0001 |

   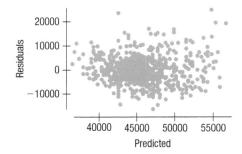
   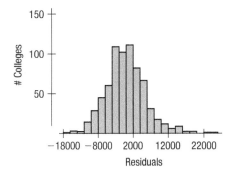

2. **Shoot to score 2016** A college hockey coach collected data from the 2016–2017 National Hockey League season. He hopes to convince his players that the number of shots taken has an effect on the number of goals scored. The coach performed a preliminary analysis, using the scoring statistics from 65 offensive players who had played at least 44 games by the middle of the season. (If you use the data file, note that it includes defensive players as well. Use the variable *Offense* to select the players in this analysis.) He predicts *Goals* from number of *Shots*.

   a) Discuss each of the conditions and assumptions required for him to proceed with the regression analysis.
   b) Write the regression model and explain what the slope coefficient means in this context.

   Response variable is Goals
   R-squared = 49.9%
   s = 2.983 with 65 − 2 = 63 degrees of freedom

   | Variable | Coefficient | SE(Coeff) | t-ratio | P-value |
   |---|---|---|---|---|
   | Intercept | 1.13495 | 1.231 | 0.922 | 0.3602 |
   | Shots | 0.099267 | 0.0125 | 7.93 | <0.0001 |

   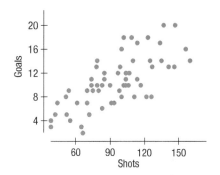
   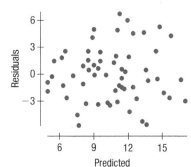
   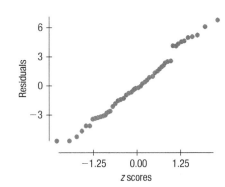

3. **Graduate earnings, part II** Using the regression output in Exercise 1, identify the standard deviation of the residuals and explain what it means in the context of the problem.

4. **Shoot to score another one** Using the regression output from Exercise 2, identify the standard deviation of the residuals and explain its meaning with a sentence in context.

5. **Graduate earnings, part III** Continuing with the regression of Exercise 1, write a sentence that explains the meaning of the standard error of the slope of the regression line, $SE(b_1) = 1.545$.

**6. Shoot to score, hat trick** Returning to the results of Exercise 2, write a sentence to explain the meaning of the standard error of the slope of the regression line, $SE(b_1) = 0.0125$.

**7. Earnings, part IV** Exercise 1 tests the hypotheses $H_0: \beta_1 = 0$ vs. $H_A: \beta_1 \neq 0$ and rejected the null hypothesis because the P-value was less than 0.0001. What can they conclude about the relationship between admission rates and graduation rates?

**8. Shoot to score, number four** What can the hockey coach in Exercise 2 conclude about shooting and scoring goals from the fact that the P-value <0.0001 for the slope of the regression line? Write a sentence in context.

**9. Earnings, part V** The college administrators in Exercise 1 constructed a 95% confidence interval for the slope of their regression line. Interpret the meaning of their interval (24.23, 30.32) within the context of the problem.

**10. Shoot to score, overtime** The coach in Exercise 2 found a 95% confidence interval for the slope of his regression line. Recall that he is trying to predict total goals scored based on shots taken. Interpret with a sentence the meaning of the interval $0.099 \pm 0.025$.

**T 11. Tracking hurricanes 2018** In Chapter 6, we looked at data from the National Oceanic and Atmospheric Administration about their success in predicting hurricane tracks. Here is a scatterplot of the error (in nautical miles) for predicting hurricane locations 24 hours in the future vs. the year in which the prediction (and the hurricane) occurred.

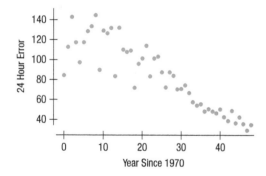

In Chapter 6, we could describe this relationship only in general terms. Now we can learn more. Here is the regression analysis:

Response variable is: Error_24h
R squared = 79.8%
s = 15.14 with 49 − 2 = 47 degrees of freedom

| Variable | Coefficient | SE(Coeff) | t-ratio | P-value |
|---|---|---|---|---|
| Intercept | 133.343 | 4.261 | 31.29 | <0.0001 |
| Year Since 1970 | −2.081 | 0.153 | −13.60 | <0.0001 |

a) Explain in words and numbers what the regression says.
b) State the hypothesis about the slope (both numerically and in words) that describes how hurricane prediction quality has changed.
c) Assuming that the assumptions for inference are satisfied, perform the hypothesis test and state your conclusion. Be sure to state it in terms of prediction errors and years.
d) Explain what the $R$-squared means in terms of this regression.

**12. Drug use 2013** The *2013 World Drug Report* investigated the prevalence of drug use as a percentage of the population aged 15 to 64. Data from 32 European countries are shown in the following scatterplot and regression analysis. (*World Drug Report*, 2013. www.unodc.org/wdr2013/)

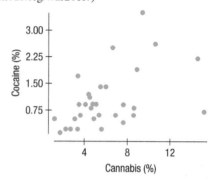

Response variable is Cocaine
R-squared = 25.8%
s = 0.7076 with 32 − 2 = 30 degrees of freedom

| Variable | Coefficient | SE(Coeff) | t-ratio | P-value |
|---|---|---|---|---|
| Intercept | 0.296899 | 0.2627 | 1.13 | 0.2673 |
| Cannabis | 0.123638 | 0.0382 | 3.23 | 0.0030 |

a) Explain in context what the regression says.
b) State the hypothesis about the slope (both numerically and in words) that describes how use of cannabis is associated with other drugs.
c) Assuming that the assumptions for inference are satisfied, perform the hypothesis test and state your conclusion in context.
d) Explain what $R$-squared means in context.
e) Do these results indicate that cannabis use leads to the use of harder drugs? Explain.

**T 13. Sea ice 2020** Climate scientists have been observing the extent of sea ice in the northern Arctic using satellite observations. Many have expressed concern because in recent decades the extent of sea ice has declined precipitously—possibly due to global climate change. Here is an analysis relating the minimum (September) *Extent* of sea ice ($km^2$) to the *Mean Global Temperature* (°C) for the years 1979–2020:

Response variable is: Extent
R squared = 69.3%
s = 0.626 with 41 − 2 = 39 degrees of freedom

| Variable | Coefficient | SE(Coeff) | t-ratio | P-value |
|---|---|---|---|---|
| Intercept | 64.470 | 6.227 | 10.35 | <0.0001 |
| Mean global temp | −3.817 | 0.407 | −9.38 | <0.0001 |

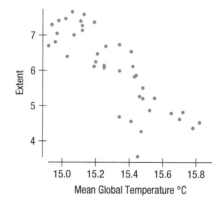

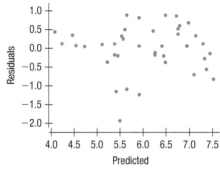
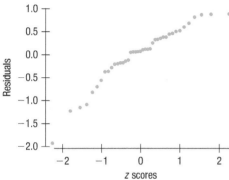

a) Explain in context what the regression says.
b) Check the assumptions and conditions for regression inference.
c) The output reports $s = 0.626$. Explain what that means in this context.
d) What's the value of the standard error of the slope of the regression line?
e) Explain what that means in this context.
f) Does this analysis prove that global temperature changes are causing sea ice to melt? Explain.

**14. Saratoga house prices** How does the price of a house depend on its size? Data from Saratoga, New York, on 1063 randomly selected houses that had been sold include data on price ($1000's) and size (1000 ft²), producing the following graphs and computer output:

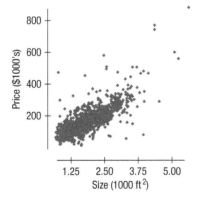

Dependent variable is Price
R-squared = 59.5%
s = 53.79 with 1063 − 2 = 1061 degrees of freedom

| Variable | Coefficient | SE(Coeff) | t-ratio | P-value |
|---|---|---|---|---|
| Intercept | −3.11686 | 4.688 | −0.665 | 0.5063 |
| Size | 94.4539 | 2.395 | 39.5 | ≤0.0001 |

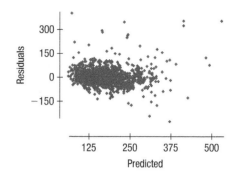
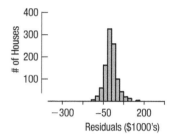

a) Explain in context what the regression says.
b) The intercept is negative. What does this mean? (Hint: Notice the P-value.)
c) The output reports $s = 53.79$. Explain what that means in this context.
d) What's the value of the standard error of the slope of the regression line?
e) Explain what that means in this context.

**15. More sea ice** Exercise 13 shows computer output examining the association between Arctic sea ice extent and global mean temperature. Find a 95% confidence interval for the slope and interpret it in context.

**16. Second home** Exercise 14 shows computer output examining the association between the sizes of houses and their sale prices.

a) Check the assumptions and conditions for inference.
b) Find a 95% confidence interval for the slope and interpret it in context.

**17. Hot dogs** Healthy eating probably doesn't include hot dogs, but if you are going to have one, you'd probably hope it's low in both calories and sodium. *Consumer Reports* listed the number of calories and sodium content (in milligrams) for 13 brands of all-beef hot dogs it tested. Examine the association, assuming that the data satisfy the conditions for inference.

Dependent variable is: Sodium
R-squared = 60.5%
s = 59.66 with 13 − 2 = 11 degrees of freedom

| Variable | Coefficient | SE(Coeff) | t-ratio | P-value |
|---|---|---|---|---|
| Constant | 90.9783 | 77.69 | 1.17 | 0.2663 |
| Calories | 2.29959 | 0.5607 | 4.10 | 0.0018 |

a) State the appropriate hypotheses about the slope.
b) Test your hypotheses and state your conclusion in the proper context.

**18. Cholesterol** Does a person's cholesterol level tend to change with age? Data collected from 1406 adults aged 45 to 62 produced the regression analysis shown. Assuming that the data satisfy the conditions for inference, examine the association between age and cholesterol level.

Dependent variable is: Chol
s = 46.16

| Variable | Coefficient | SE(Coeff) | t-ratio | P-value |
|---|---|---|---|---|
| Intercept | 194.232 | 13.55 | 14.3 | ≤0.0001 |
| Age | 0.771639 | 0.2574 | 3.00 | 0.0027 |

a) State the appropriate hypothesis for the slope.
b) Test your hypothesis and state your conclusion in the proper context.

**19. Second frank** Look again at Exercise 17's regression output for the calorie and sodium content of hot dogs.

a) The output reports $s = 59.66$. Explain what that means in this context.
b) What's the value of the standard error of the slope of the regression line?
c) Explain what that means in this context.

**20. More cholesterol** Look again at Exercise 18's regression output for age and cholesterol level.

a) The output reports $s = 46.16$. Explain what that means in this context.
b) What's the value of the standard error of the slope of the regression line?
c) Explain what that means in this context.

**21. Last dog** Based on the regression output seen in Exercise 17, create a 95% confidence interval for the slope of the regression line and interpret your interval in context.

**22. Cholesterol, finis** Based on the regression output seen in Exercise 18, create a 95% confidence interval for the slope of the regression line and interpret it in context.

**23. Marriage age 2019** The graph below shows, for years between 1890 and 2020, the average age at first marriage for men and women.

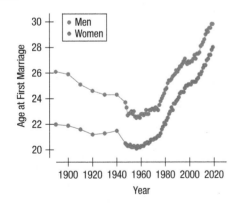

Generally, men are older at their first marriage than are women, but the gap seems to have been closing. Has the difference in age at first marriage between men and women really been declining?

Response variable is: Men – Women
R squared = 71.0%
s = 0.2457 with 79 − 2 = 77 degrees of freedom

| Variable | Coefficient | SE(Coeff) | t-ratio | P-value |
|---|---|---|---|---|
| Intercept | 29.672 | 1.987 | 14.93 | <0.0001 |
| Year | −0.0138 | 0.0010 | −13.75 | <0.0001 |

a) Write appropriate hypotheses.
b) Here are plots of the residuals. Comment on what they say about the regression. (Remember that some computer output shows residuals plotted against predicted values instead of against the explanatory variable. You're still looking for the the same thing: no pattern.)

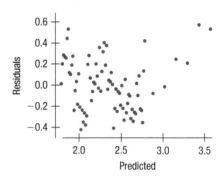

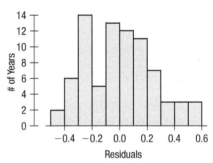

And here is a normal probability plot of the residuals:

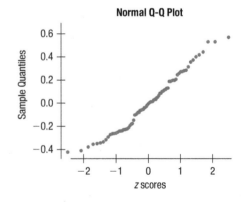

c) Test the hypothesis and state your conclusion about the trend in age at first marriage.

**24. Used Civics 2020** On October 3, 2020, autolist.com listed 55 used Honda Civics for sale by owner. Here's a scatterplot of the asking price vs. the number of miles on the odometer (in thousands):

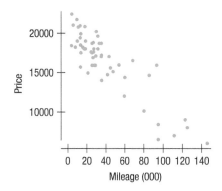

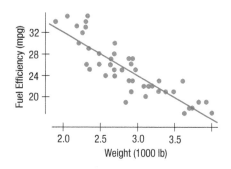

a) Do you think a linear model is appropriate? Explain.

Here is the regression model:

Response variable is: Price
R squared = 77.9%
s = 1850 with 55 − 2 = 53 degrees of freedom

| Variable | Coefficient | SE(Coeff) |
|---|---|---|
| Intercept | 20451.67 | 387.47 |
| Miles(000) | −101.91 | 7.451 |

b) State the null and alternative hypotheses under investigation.
c) Assuming that the assumptions for regression inference are reasonable, find the $t$- and P-values.
d) State your conclusion.

Here is a plot of the residuals.

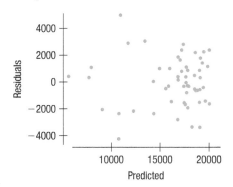

| Variable | Count | Mean | StdDev |
|---|---|---|---|
| MPG | 50 | 25.0200 | 4.83394 |
| wt/1000 | 50 | 2.88780 | 0.511656 |

Dependent variable is: MPG
R-squared = 75.6%
s = 2.413 with 50 − 2 = 48 df

| Variable | Coefficient | SE(Coeff) | t-ratio | P-value |
|---|---|---|---|---|
| Intercept | 48.7393 | 1.976 | 24.7 | ≤0.0001 |
| Weight | −8.21362 | 0.6738 | −12.2 | ≤0.0001 |

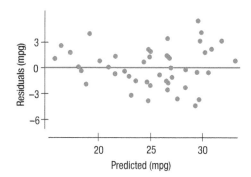

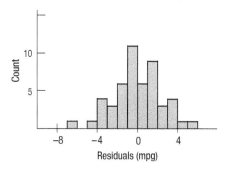

e) Do the assumptions and conditions for regression inference appear to be satisfied? If not, what would you suggest doing to improve the model?

**25. Marriage age 2019 again** Based on the analysis of marriage ages given in Exercise 23, find a 95% confidence interval for the rate at which the age gap is closing. Explain what your confidence interval means.

**26. Used Civics 2020 again** Based on the analysis of used car prices in Exercise 24, create a 95% confidence interval for the slope of the regression line and explain what your interval means in context.

**27. Fuel economy** A consumer organization has reported test data for 50 car models. We will examine the association between the weight of the car (in thousands of pounds) and the fuel efficiency (in miles per gallon). Here are the scatterplot, summary statistics, and regression analysis:

a) Is there strong evidence of an association between the weight of a car and its gas mileage? Write an appropriate hypothesis.
b) Are the assumptions for regression satisfied?
c) Test your hypothesis and state your conclusion.

**28. Roller Coasters** In Chapter 7 we saw that for roller coasters in the United States classified as "Thrill" rides, the association between their speeds and their heights is roughly linear. Following are a scatterplot showing that relationship, along with some regression output from a computer.

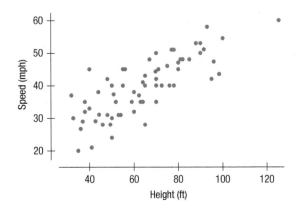

Response variable is: Speed
R squared = 64.65%
s = 5.151 with 91 − 2 = 89 degrees of freedom

| Variable | Coefficient | SE(Coeff) | t-ratio | P-value |
|---|---|---|---|---|
| Intercept | 15.586 | 1.806 | 8.63 | <0.001 |
| Height | 0.370 | 0.029 | 12.76 | <0.001 |

a) Write the equation of the regression line.
b) Explain the meaning of the number 12.76 in the "t-ratio" column.
c) Determine a 90% confidence interval for the slope of the regression line, and interpret it in this context.

**29. Fuel economy, part II** Consider again the data in Exercise 27 about the gas mileage and weights of cars.

a) Create a 95% confidence interval for the slope of the regression line.
b) Explain in this context what your confidence interval means.

**30. Climate change, part II** Consider the $CO_2$ and global temperature data of Exercise 28.

a) Find a 90% confidence interval for the slope of the true line describing the association between *Temp* and $CO_2$.
b) Explain in this context what your confidence interval means.

**31. Cereal** A healthy cereal should be low in both calories and sodium. Data for 77 cereals were examined and judged acceptable for inference. The 77 cereals had between 50 and 160 calories per serving and between 0 and 320 mg of sodium per serving. Here's the regression analysis:

Dependent variable is: Sodium
R-squared = 9.0%
s = 80.49 with 77 − 2 = 75 degrees of freedom

| Variable | Coefficient | SE(Coeff) | t-ratio | P-value |
|---|---|---|---|---|
| Intercept | 21.4143 | 51.47 | 0.416 | 0.6786 |
| Calories | 1.29357 | 0.4738 | 2.73 | 0.0079 |

a) Is there an association between the number of calories and the sodium content of cereals? Explain.
b) Do you think this association is strong enough to be useful? Explain.

**32. Brain size** Does your IQ depend on the size of your brain? A group of female college students took a test that measured their verbal IQs and also underwent an MRI scan to measure the size of their brains (in 1000's of pixels). The scatterplot and regression analysis are shown, and the assumptions for inference were satisfied.

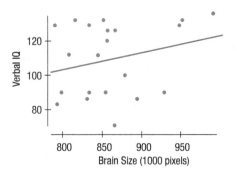

Dependent variable is: IQ_Verbal
R-squared = 6.5%   s = 21.5291   df = 18

| Variable | Coefficient | SE(Coeff) |
|---|---|---|
| Intercept | 24.1835 | 76.38 |
| Size | 0.098842 | 0.0884 |

a) Test an appropriate hypothesis about the association between brain size and IQ.
b) State your conclusion about the strength of this association.

**33. Another bowl** Further analysis of the data for the breakfast cereals in Exercise 31 looked for an association between *Fiber* content and *Calories* by attempting to construct a linear model. Below and on the next page are several graphs. Which of the assumptions for inference are violated? Explain.

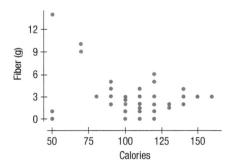

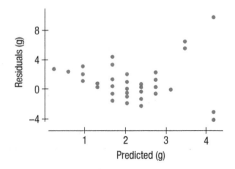

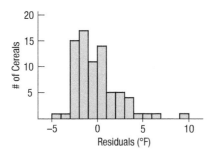

**34. Winter** The output shows an attempt to model the association between average *January Temperature* (in degrees Fahrenheit) and *Latitude* (in degrees north of the equator) for 59 U.S. cities. Which of the assumptions for inference do you think are violated? Explain.

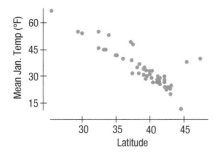

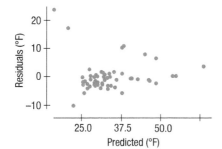

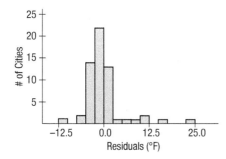

**35. Acid rain** Biologists studying the effects of acid rain on wildlife collected data from 163 streams in the Adirondack Mountains. They recorded the *pH* (acidity) of the water and the *BCI*, a measure of biological diversity. Here's a scatterplot of *BCI* against *pH*:

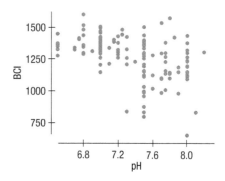

And here is part of the regression analysis:

Dependent variable is: BCI
R-squared = 27.1%
s = 140.4 with 163 − 2 = 161 degrees of freedom

| Variable | Coefficient | SE(Coeff) |
|---|---|---|
| Intercept | 2733.37 | 187.9 |
| pH | −197.694 | 25.57 |

a) State the null and alternative hypotheses under investigation.
b) Assuming that the assumptions for regression inference are reasonable, find the *t*- and P-values.
c) State your conclusion.

**36. SAT scores** How strong was the association between student scores on the Math and Verbal sections of the old SAT? Scores on each ranged from 200 to 800 and were widely used by college admissions offices. Here are summaries and plots of the scores for a graduating class at Ithaca High School:

| Variable | Count | Mean | Median | StdDev | Range | IntQRange |
|---|---|---|---|---|---|---|
| Verbal | 162 | 596.296 | 610 | 99.5199 | 490 | 140 |
| Math | 162 | 612.099 | 630 | 98.1343 | 440 | 150 |

Dependent variable is: Math
R-squared = 46.9%
s = 71.75 with 162 − 2 = 160 df

| Variable | Coefficient | SE(Coeff) | t-ratio | P-value |
|---|---|---|---|---|
| Intercept | 209.554 | 34.35 | 6.10 | ≤0.0001 |
| Verbal | 0.675075 | 0.0568 | 11.9 | ≤0.0001 |

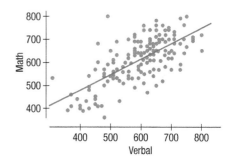

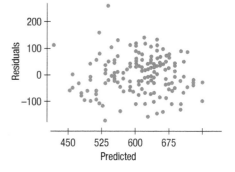

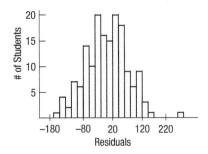

a) Is there evidence of an association between Math and Verbal scores? Write an appropriate hypothesis.
b) Discuss the assumptions for inference.
c) Test your hypothesis and state an appropriate conclusion.

**37. Ozone** The Environmental Protection Agency is examining the relationship between the ozone level (in parts per million) and the population (in millions) of U.S. cities. Part of the regression analysis is shown.

Dependent variable is: Ozone
R-squared = 84.4%
s = 5.454 with 16 − 2 = 14 df

| Variable | Coefficient | SE(Coeff) |
|---|---|---|
| Intercept | 18.892 | 2.395 |
| Population | 6.650 | 1.910 |

a) We suspect that the greater the population of a city, the higher its ozone level. Is the relationship significant? Assuming the conditions for inference are satisfied, test an appropriate hypothesis and state your conclusion in context.
b) Do you think that the population of a city is a useful predictor of ozone level? Use the values of both $R^2$ and $s$ in your explanation.

**38. Sales and profits** A business analyst was interested in the relationship between a company's sales and its profits. They collected data (in millions of dollars) from a random sample of *Fortune* 500 companies and created the regression analysis and summary statistics shown. The assumptions for regression inference appeared to be satisfied.

|  | Profits | Sales |
|---|---|---|
| Count | 79 | 79 |
| Mean | 209.839 | 4178.29 |
| Variance | 635,172 | 49,163,000 |
| Std Dev | 796.977 | 7011.63 |

Dependent variable is: Profits
R-squared = 66.2%  s = 466.2

| Variable | Coefficient | SE(Coeff) |
|---|---|---|
| Intercept | −176.644 | 61.16 |
| Sales | 0.092498 | 0.0075 |

a) Is there a significant association between sales and profits? Test an appropriate hypothesis and state your conclusion in context.
b) Do you think that a company's sales serve as a useful predictor of its profits? Use the values of both $R^2$ and $s$ in your explanation.

**39. Ozone, again** Consider again the relationship between the population and ozone level of U.S. cities that you analyzed in Exercise 37.

Give a 90% confidence interval for the approximate increase in ozone level associated with each additional million city inhabitants.

**40. More sales and profits** Consider again the relationship between the sales and profits of *Fortune* 500 companies that you analyzed in Exercise 38.

Find a 95% confidence interval for the slope of the regression line. Interpret your interval in context.

**41. Tablet computers** CNET.com tests tablet computers and continuously updates its list. As of October 2020, the list included the battery life (in hours) and luminous intensity (i.e., screen brightness, in cd/m$^2$). We want to know if *Battery life* is related to the maximum *Screen Brightness*. (www.cnet.com/news/cnet-tablet-battery-life-results/)

Response variable is Battery life (hrs)
R-squared = 11.2%
s = 1.191 with 39 − 2 = 37 degrees of freedom

| Variable | Coefficient | SE(Coeff) | t-ratio | P-value |
|---|---|---|---|---|
| Intercept | 5.75061 | 1.523 | 3.78 | 0.0006 |
| Max Brightness | 0.00874 | 0.0040 | 2.16 | 0.0372 |

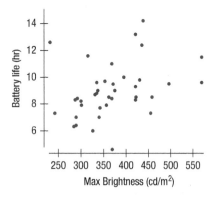

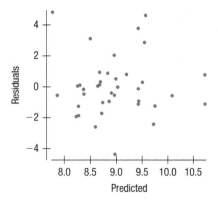

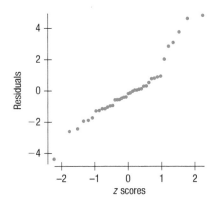

a) How many tablet computers were tested?
b) Are the conditions for inference satisfied? Explain.
c) Is there evidence of an association between maximum brightness of the screen and battery life? Test an appropriate hypothesis and state your conclusion.
d) Is the association strong? Explain.
e) What is the equation of the regression line?
f) Create a 90% confidence interval for the slope of the true line.
g) Interpret your interval in this context.

**42. Crawling** Researchers at the University of Denver Infant Study Center wondered whether temperature might influence the age at which babies learn to crawl. Perhaps the extra clothing that babies wear in cold weather would restrict movement and delay the age at which they started crawling. Data were collected on 208 boys and 206 girls. Parents reported the month of the baby's birth and the age (in weeks) at which their child first crawled. The table gives the average *Temperature* (°F) when the babies were 6 months old and average *Crawling Age* (in weeks) for each month of the year. Make the plots and compute the analyses necessary to answer the following questions.

| Birth Month | 6-Month Temperature | Average Crawling Age |
|---|---|---|
| Jan. | 66 | 29.84 |
| Feb. | 73 | 30.52 |
| Mar. | 72 | 29.70 |
| April | 63 | 31.84 |
| May | 52 | 28.58 |
| June | 39 | 31.44 |
| July | 33 | 33.64 |
| Aug. | 30 | 32.82 |
| Sept. | 33 | 33.83 |
| Oct. | 37 | 33.35 |
| Nov. | 48 | 33.38 |
| Dec. | 57 | 32.32 |

a) Would this association appear to be weaker, stronger, or the same if data had been plotted for individual babies instead of using monthly averages? Explain.

b) Is there evidence of an association between *Temperature* and *Crawling Age*? Test an appropriate hypothesis and state your conclusion. Don't forget to check the assumptions.
c) Create and interpret a 95% confidence interval for the slope of the true relationship.

**43. Body fat** Do the data shown in the table below indicate an association between *Waist Size* and *% Body Fat*? Test an appropriate hypothesis and state your conclusion.

| Waist Size (in.) | Weight (lb) | Body Fat (%) | Waist Size (in.) | Weight (lb) | Body Fat (%) |
|---|---|---|---|---|---|
| 32 | 175 | 6 | 33 | 188 | 10 |
| 36 | 181 | 21 | 40 | 240 | 20 |
| 38 | 200 | 15 | 36 | 175 | 22 |
| 33 | 159 | 6 | 32 | 168 | 9 |
| 39 | 196 | 22 | 44 | 246 | 38 |
| 40 | 192 | 31 | 33 | 160 | 10 |
| 41 | 205 | 32 | 41 | 215 | 27 |
| 35 | 173 | 21 | 34 | 159 | 12 |
| 38 | 187 | 25 | 34 | 146 | 10 |
| 38 | 188 | 30 | 44 | 219 | 28 |

**44. Body fat, again** Use the data from Exercise 45 to examine the association between *Weight Size* and *% Body Fat*.
a) Find a 90% confidence interval for the slope of the regression line of *% Body Fat* on *Weight Size*.
b) Interpret your interval in context.

**45. More roller coasters** Some roller coasters that are unusually fast or have special features are classified as "extreme" by aficionados. There were 190 such coasters operating in the Unites States at the end of 2020, according to the web site rcdb.com. A scatterplot of their lengths and their durations shows that the relationship between length and duration is linear with no outliers or unusual features, so a linear regression model is appropriate. Following is some computer output based on this regression model. (Lengths are measured in feet, durations in seconds.)

```
Response variable is: Duration
R-squared = 42.03%
s = 29.414 with 190 − 2 = 188 degrees of freedom
Variable    Coefficient   SE(Coeff)   t-ratio   P-value
Intercept   64.2128       5.2892      12.14     <0.0001
Length      0.0189        0.0016      11.67     <0.0001
```

a) Interpret the meaning of R-squared in this context.
b) Interpret the meaning of the slope in this context.
c) If we test the null hypothesis that the slope of the line is zero, what is the P-value? What conclusion would you draw from this hypothesis test?

**46. Still more roller coasters** Of the 91 "Thrill" roller coasters that were described in Exercise 28, data were available on the durations of 65 of them. A scatterplot of their durations against their heights showed no unusual features that would suggest that a linear model is inappropriate. Computer output for the linear regression of Duration on Height is shown below.

Response variable is: Duration
R-squared = 2.40%
s = 29.329 with 65 − 2 = 63 degrees of freedom

| Variable | Coefficient | SE(Coeff) | t-ratio | P-value |
|---|---|---|---|---|
| Intercept | 91.3283 | 11.8319 | 7.72 | <0.0001 |
| Height | 0.2329 | 0.1869 | 1.25 | 0.2175 |

a) How strong is the evidence of an association between *Height* and *Duration*? Explain how you know.
b) Describe how the value of R-squared is either consistent or inconsistent with your answer to part a).
c) Using only the computer output above, determine a reasonable estimate of the standard deviation of these 65 roller coasters' durations.

**47. Strike two** Remember the Little League instructional video discussed in Chapter 24? Ads claimed it would improve the performances of Little League pitchers. To test this claim, 20 Little Leaguers threw 50 pitches each, and we recorded the number of strikes. After the players participated in the training program, we repeated the test. The table shows the number of strikes each player threw before and after the training. A test of paired differences failed to show that this training improves ability to throw strikes. Is there any evidence that the effectiveness of the video (*After − Before*) depends on the player's initial ability to throw strikes (*Before*)? Test an appropriate hypothesis and state your conclusion. Propose an explanation for what you find.

**Number of Strikes (out of 50)**

| Before | After | Before | After |
|---|---|---|---|
| 28 | 35 | 33 | 33 |
| 29 | 36 | 33 | 35 |
| 30 | 32 | 34 | 32 |
| 32 | 28 | 34 | 30 |
| 32 | 30 | 34 | 33 |
| 32 | 31 | 35 | 34 |
| 32 | 32 | 36 | 37 |
| 32 | 34 | 36 | 33 |
| 32 | 35 | 37 | 35 |
| 33 | 36 | 37 | 32 |

**48. All the efficiency money can buy 2011** A sample of 84 model-2011 cars from an online information service was examined to see how fuel efficiency (as highway mpg) relates to the cost (Manufacturer's Suggested Retail Price in dollars) of cars. Here are displays and computer output:

Dependent variable is MPG
R-squared = 0.0216%
s = 3.54

| Variable | Coefficient | SE(Coeff) | t-ratio | P-value |
|---|---|---|---|---|
| Intercept | 36.514 | 1.496 | 24.406 | <0.0001 |
| Slope | −8.089E −6 | 6.439E −5 | −0.1256 | 0.900 |

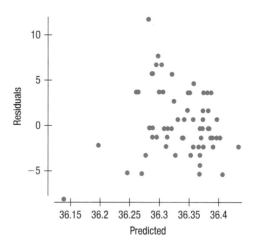

a) State what you want to know, identify the variables, and give the appropriate hypotheses.
b) Check the assumptions and conditions.
c) If the conditions are met, complete the analysis.

**49. Education and mortality** The software output below is based on the mortality rate (deaths per 100,000 people) and the education level (average number of years in school) for 58 U.S. cities.

| Variable | Count | Mean | StdDev |
|---|---|---|---|
| Mortality | 58 | 942.501 | 61.8490 |
| Education | 58 | 11.0328 | 0.793480 |

Dependent variable is: Mortality
R-squared = 41.0%
s = 47.92 with 58 − 2 = 56 degrees of freedom

| Variable | Coefficient | SE(Coeff) |
|---|---|---|
| Intercept | 1493.26 | 88.48 |
| Education | −49.9202 | 8.000 |

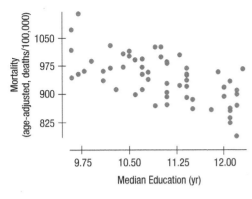
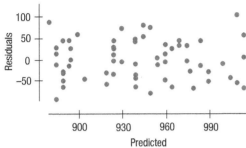
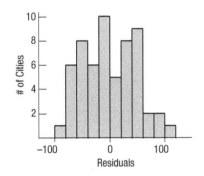

a) Comment on the assumptions for inference.
b) Is there evidence of a strong association between the level of *Education* in a city and the *Mortality* rate? Test an appropriate hypothesis and state your conclusion.
c) Can we conclude that getting more education is likely (on average) to prolong your life? Why or why not?
d) Find a 95% confidence interval for the slope of the true relationship.
e) Explain what your interval means.

**50.** Property assessments The software outputs below provide information about the *Size* (in square feet) of 18 homes in Ithaca, New York, and the city's assessed *Value* of those homes.

| Variable | Count | Mean | StdDev | Range |
|---|---|---|---|---|
| Size | 18 | 2003.39 | 264.727 | 890 |
| Value | 18 | 60946.7 | 5527.62 | 19710 |

Dependent variable is: Value
R-squared = 32.5%
s = 4682 with 18 − 2 = 16 degrees of freedom

| Variable | Coefficient | SE(Coeff) |
|---|---|---|
| Intercept | 37108.8 | 8664 |
| Size | 11.8987 | 4.290 |

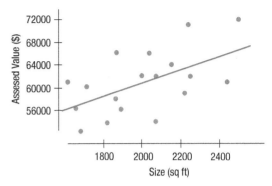
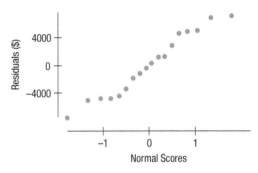
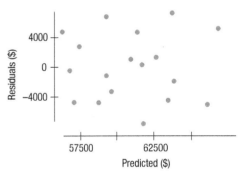

a) Explain why inference for linear regression is appropriate with these data.
b) Is there a significant association between the *Size* of a home and its assessed *Value*? Test an appropriate hypothesis and state your conclusion.
c) What percentage of the variability in assessed *Value* is explained by this regression model?
d) Give a 90% confidence interval for the slope of the true regression line, and explain its meaning in the proper context.
e) From this analysis, can we conclude that adding a room to your house will increase its assessed *Value*? Why or why not?

## JUST CHECKING

### Answers

1. A high $t$-ratio of 3.27 indicates that the slope is different from zero—that is, that there is a linear relationship between height and mouth size. The small P-value says that a slope this large would be very unlikely to occur by chance if, in fact, there was no linear relationship between the variables.

2. Not really. The $R^2$ for this regression model only 15.3%, so height doesn't account for very much of the variability in mouth size.

3. The value of $s$ tells the standard deviation of the residuals. Mouth sizes have a mean of 60.3 cubic centimeters. A standard deviation of 15.7 in the residuals indicates that the errors made by this regression model can be quite large relative to what we are estimating. Errors of 15 to 30 cubic centimeters would be common.

4. In repeated sampling, we estimate slopes of regression lines predicting mouth volume would vary with a standard deviation of 18.77 cubic centimeters per meter of height.

# Review of Part VII

## INFERENCE WHEN VARIABLES ARE RELATED

### Quick Review

With these last two chapters, you have added important analytical tools to your ways of looking at data. Here's a brief summary of those key concepts and skills, as well as an overview of statistical inference:

- Inferences about distributions of counts use chi-square models.
  - To see if an observed distribution is consistent with a proposed distribution model, use a goodness-of-fit test.
  - To see if two or more observed distributions could have arisen from populations with the same distribution model, use a test of homogeneity.
- Inference about association between two variables tests the hypothesis that it is plausible to consider the variables independent.
  - If the variables are categorical, display the data in a contingency table and use a chi-square test of independence.
  - If the variables are quantitative, display them with a scatterplot. You may use a linear regression $t$-test if there appears to be a linear association for which the residuals are random, consistent in terms of spread, and approximately Normal.
- You can now use statistical inference to answer questions about means, proportions, distributions, and associations.
  - No inference procedure is valid unless the underlying assumptions are true. Always check the conditions before proceeding. Many of those checks should be made by examining a graph.
  - You can make inferences about a single proportion or the difference of two proportions using Normal models.
- You can make inferences about one mean, the difference of two independent means, or the mean of paired differences using $t$-models.
- You can make inferences about distributions using chi-square models.
- You can make inferences about associations between categorical variables using chi-square models.
- You can make inferences about linear associations between quantitative variables using $t$-models.

If you look back at where we've been in this book, you'll see that statistical inference relies on almost everything we've seen. In Chapters 11 and 12 we learned techniques of collecting data using randomization—that's what makes inference possible at all. In Chapters 2, 3, and 6 we learned to plot our data and to look for the patterns and relationships we use to check the conditions that allow inference. In Chapters 2, 4, and 7 we learned about the summary statistics we use to do the mechanics of inference. We use our knowledge of randomness and probability from Chapters 10, 13, and 14 to help us think clearly about uncertainty, and the probability models of Chapters 5, 15, and 16 to measure our uncertainty precisely. Ultimately, the Central Limit Theorem of Chapter 17 makes all of inference possible.

Remember (have we said this often enough yet?): Never use any inference procedure without first checking the assumptions and conditions. Below we summarize the new types of inference procedures, the corresponding formulas, and the assumptions and conditions. You'll find complete summaries of all our inference procedures inside the back cover of the book. Have a look. Then you'll be ready for more opportunities to practice using these concepts and skills.

| Quick Guide to Inference | | | | | | | |
|---|---|---|---|---|---|---|---|
| **Think** | | | **Show** | | | | **Tell?** |
| Inference About? | One Group or Two? | Procedure | Model | Parameter | Estimate | SE | Chapter |
| Distributions (one categorical variable) | One sample | Goodness-of-Fit | $\chi^2$ df = cells − 1 | | | $\sum \dfrac{(\text{obs} - \text{exp})^2}{\text{exp}}$ | 25 |
| | Many independent groups | Homogeneity $\chi^2$ Test | | | | | |
| Independence (two categorical variables) | One sample | Independence $\chi^2$ Test | $\chi^2$ df = $(r-1)(c-1)$ | | | | |
| Association (two quantitative variables) | One sample | Linear Regression $t$-Test or Confidence Interval for $\beta$ | $t$ df = $n - 2$ | $\beta_1$ | $b_1$ | $\dfrac{s_e}{s_x\sqrt{n-1}}$ (compute with technology) | 26 |

| Assumptions for Inference | And the Conditions That Support or Override Them |
|---|---|

**Distributions/Association ($\chi^2$)**

- **Goodness-of-fit** [df = # of cells − 1; one variable, one sample compared with population model]

  1. Data are counts.
  2. Data in sample are independent.
  3. Sample is sufficiently large.

  1. (Are they?)
  2. SRS and $n < 10\%$ of the population.
  3. All expected counts $> 5$.

- **Homogeneity** [df = $(r − 1)(c − 1)$; samples from many populations compared on one variable]

  1. Data are counts.
  2. Data in groups are independent.
  3. Groups are sufficiently large.

  1. (Are they?)
  2. SRSs and $n < 10\%$ OR random allocation.
  3. All expected counts $> 5$.

- **Independence** [df = $(r − 1)(c − 1)$; sample from one population classified on two variables]

  1. Data are counts.
  2. Data are independent.
  3. Sample is sufficiently large.

  1. (Are they?)
  2. SRSs and $n < 10\%$ of the population.
  3. All expected counts $> 5$.

**Regression ($t$, df = $n − 2$)**

- **Association** between two quantitative variables ($\beta = 0$?)

  1. Form of relationship is linear.
  2. Errors are independent.
  3. Variability of errors is constant.
  4. Errors have a Normal model.

  1. Scatterplot looks approximately linear.
  2. No apparent pattern in residuals plot.
  3. Residuals plot has consistent spread.
  4. Histogram of residuals is approximately unimodal and symmetric or Normal probability plot reasonably straight (less critical as $n$ increases).

# REVIEW EXERCISES

**1. Genetics** Two human traits controlled by a single gene are the ability to roll one's tongue and whether one's ear lobes are free or attached to the neck. Genetic theory says that people will have neither, one, or both of these traits in the ratio 1:3:3:9 (1 attached, noncurling; 3 attached, curling; 3 free, noncurling; 9 free, curling). An Introductory Biology class of 122 students collected the data shown. Are they consistent with the genetic theory? Test an appropriate hypothesis and state your conclusion.

| | \multicolumn{4}{c}{Trait} | | | |
|---|---|---|---|---|
| | Attached, Noncurling | Attached, Curling | Free, Noncurling | Free, Curling |
| Count | 10 | 22 | 31 | 59 |

**2. Tableware** Nambe Mills manufactures plates, bowls, and other tableware made from an alloy of several metals. Each item must go through several steps, including polishing. To better understand the production process and its impact on pricing, the company checked the polishing time (in minutes) and the retail price (in US$) of these items. The regression analysis is shown below. The scatterplot showed a linear pattern, and residuals were deemed suitable for inference.

Dependent variable is: Price
R-squared = 84.5%
s = 20.50 with 59 − 2 = 57 degrees of freedom

| Variable | Coefficient | SE(Coeff) |
|---|---|---|
| Intercept | −2.89054 | 5.730 |
| Time | 2.49244 | 0.1416 |

a) How many different products were included in this analysis?
b) What fraction of the variation in retail price is explained by the polishing time?
c) Create a 95% confidence interval for the slope of this relationship.
d) Interpret your interval in this context.

**3. Hard water** In an investigation of environmental causes of disease, data were collected on the annual mortality rate (deaths per 100,000) for males in 61 large towns in England and Wales. In addition, the water hardness was recorded as the calcium concentration (parts per million, or ppm) in the drinking water. Here are the scatterplot and regression analysis of the relationship between mortality and calcium concentration.

Dependent variable is: mortality
R-squared = 43%
s = 143.0 with 61 − 2 = 59 degrees of freedom

| Variable  | Coefficient | SE(Coeff) |
|-----------|-------------|-----------|
| Intercept | 1676        | 29.30     |
| Calcium   | −3.23       | 0.48      |

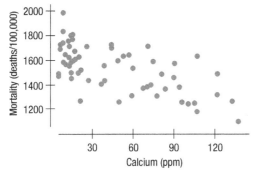

a) Is there an association between the hardness of the water and the mortality rate? Write the appropriate hypothesis.
b) Assuming the assumptions for regression inference are met, what do you conclude?
c) Create a 95% confidence interval for the slope of the true line relating calcium concentration and mortality.
d) Interpret your interval in context.

**4. Mutual funds** In July 2011, the *Wall Street Journal Online* reported the rate of return for the top 20 large-cap mutual funds over the last 10 years. ("Large cap" refers to companies worth over $10 billion.) Among other results, the *Journal* listed the 3-year and 5-year returns. (online.wsj.com)

a) Create a 95% confidence interval for the difference in rate of return for the 3- and 5-year periods covered by these data. Clearly explain what your interval means.
b) It's common for advertisements to carry the disclaimer "Past returns may not be indicative of future performance," but do these data indicate that there was an association between 3-year and 5-year rates of return?

| Annualized Returns (%) | | |
|---|---|---|
| Fund Name | 3-year | 5-year |
| Yact man Focused | 18.48 | 11.63 |
| Yackt man | 17.5 | 10.64 |
| CGM Focus | −19.78 | 2.3 |
| Fairholme | 4.76 | 5.89 |
| Mass Mutual | 10.66 | 6.87 |
| Amana Trust Income | 5.26 | 7.56 |
| Amana Trust Growth | 5.01 | 7.34 |
| Columbia Strategic | 2.37 | 4.69 |
| Columbia Masico | −0.62 | 2.54 |

| Annualized Returns (%) | | |
|---|---|---|
| Fund Name | 3-year | 5-year |
| Marsico 21st Century | −0.87 | 2.03 |
| Wasatch | 1.17 | 4.47 |
| Fidelity Contrafund | 2.06 | 5.25 |
| Gabelli Asset | 6.04 | 6.6 |
| Parnassus Equity | 4.93 | 7.2 |
| Balck Rock Equity | 1.72 | 4.46 |
| CGM Mutual | −5.32 | 4.63 |
| Eaton Vance | −5.46 | 3.66 |
| Gabelli Equity | 4.79 | 5.17 |
| Auxier Focus | 7.18 | 5.87 |
| Oppenheimer Equity | 9.15 | 5.82 |

**5. Resume fraud** In 2002 the Veritas Software company found out that its chief financial officer did not actually have the MBA he had listed on his resume. They fired him, and the value of the company's stock dropped 19%. Kroll, Inc., a firm that specializes in investigating such matters, said that they believe as many as 25% of background checks might reveal false information. How many such random checks would they have to do to estimate the true percentage of people who misrepresent their backgrounds to within ±5% with 98% confidence?

**6. Paper airplanes** In preparation for a regional paper airplane competition, a student tried out her latest design. The distances her plane traveled (in feet) in 11 trial flights are given here. (The world record is an astounding 193.01 feet!) The data were 62, 52, 68, 23, 34, 45, 27, 42, 83, 56, and 40 feet. Here are some summaries:

| Count | 11 |
|---|---|
| Mean | 48.3636 |
| Median | 45 |
| StdDev | 18.0846 |
| StdErr | 5.45273 |
| IntQRange | 25 |
| 25th %tile | 35.5000 |
| 75th %tile | 60.5000 |

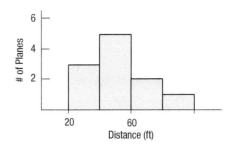

a) Construct a 95% confidence interval for the true distance.
b) Based on your confidence interval, is it plausible that the mean distance is 40 ft? Explain.
c) How would a 99% confidence interval for the true distance differ from your answer in part a? Explain briefly, without actually calculating a new interval.
d) How large a sample size would the student need to get a confidence interval half as wide as the one you got in part a, at the same confidence level?

**7. Economic inequity 2020** The following table is based on a Pew Research Poll of 6593 U.S. adults in January 2020. Respondents were classified as high income (over $120,400), middle income ($40,100–$120,400), or low income (less than $40,100). Those polled were asked for their views on choosing one option in the sentence "Reducing economic inequality should be ____ for the federal government to address." The data are summarized in the table below.

|  | A Top Priority | Important | Not Too Important | Should Not Be Done |
|---|---|---|---|---|
| High Income | 581 | 613 | 274 | 145 |
| Middle Income | 1307 | 1273 | 536 | 235 |
| Low Income | 855 | 576 | 132 | 66 |

Is there any evidence that income level is associated with feelings toward economic inequity in the United States? Test an appropriate hypothesis about this table, and state your conclusions.

**8. Wild horses** Large herds of wild horses can become a problem on some federal lands in the West. Researchers hoping to improve the management of these herds collected data to see if they could predict the number of foals that would be born based on the size of the current herd. Their attempt to model this herd growth is summarized in the graphs and output shown.

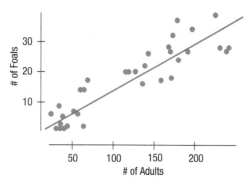

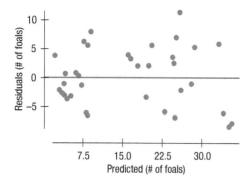

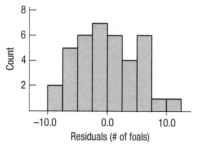

| Variable | Count | Mean | StdDev |
|---|---|---|---|
| Adults | 38 | 110.237 | 71.1809 |
| Foals | 38 | 15.3947 | 11.9945 |

Dependent variable is: Foals
R-squared = 83.5%
s = 4.941 with 38 − 2 = 36 degrees of freedom

| Variable | Coefficient | SE(Coeff) | t-ratio | P-value |
|---|---|---|---|---|
| Intercept | −1.57835 | 1.492 | −1.06 | 0.2970 |
| Adults | 0.153969 | 0.0114 | 13.5 | ≤0.0001 |

a) How many herds of wild horses were studied?
b) Are the conditions necessary for inference satisfied? Explain.
c) Create a 95% confidence interval for the slope of this relationship.
d) Explain in this context what that slope means.

**9. Lefties and music** In an experiment to see if left- and right-handed people have different abilities in music, subjects heard a tone and were then asked to identify which of several other tones matched the first. Of 76 right-handed subjects, 38 were successful in completing this test, compared with 33 of 53 lefties. Is this strong evidence of a difference in musical abilities based on handedness?

**10. AP Statistics scores 2016** In 2016, about 200,000 Statistics students nationwide took the Advanced Placement Examination in Statistics. The national distribution of scores and the results at Ithaca High School are shown in the table.

| Score | National Distribution | Ithaca High School Counts |
|---|---|---|
| 5 | 13.9% | 27 |
| 4 | 21.7% | 18 |
| 3 | 24.7% | 11 |
| 2 | 15.7% | 11 |
| 1 | 24.0% | 3 |

Is the distribution of scores at this high school significantly different from the national results? Explain.

**11. Polling** How accurate are pollsters in predicting the outcomes of congressional elections? The table shows the actual number of Democratic party seats in the House of Representatives and the number predicted by the Gallup organization for nonpresidential election years between World War II and 1998.

| Democratic Party Congressional Representatives | | |
|---|---|---|
| Year | Predicted | Actual |
| 1946 | 190 | 188 |
| 1950 | 235 | 234 |
| 1954 | 232 | 232 |
| 1958 | 272 | 283 |
| 1962 | 259 | 258 |
| 1966 | 247 | 248 |
| 1970 | 260 | 255 |
| 1974 | 292 | 291 |
| 1978 | 277 | 277 |
| 1982 | 275 | 269 |
| 1986 | 264 | 258 |
| 1990 | 260 | 267 |
| 1994 | 201 | 204 |
| 1998 | 211 | 211 |

a) Is there a significant difference between the number of seats predicted for the Democrats and the number they actually held? Test an appropriate hypothesis and state your conclusions.

b) Is there a strong association between the pollsters' predictions and the outcomes of the elections? Test an appropriate hypothesis and state your conclusions.

**12. Twins** In 2000 the *Journal of the American Medical Association* published a study that examined a sample of pregnancies that resulted in the birth of twins. Births were classified as preterm with intervention (induced labor or cesarean), preterm without such procedures, or term or post-term. Researchers also classified the pregnancies by the level of prenatal medical care the mother received (inadequate, adequate, or intensive). The data, from the years 1995–1997, are summarized in the table below. Figures are in thousands of births. (*JAMA* 284 [2000]: 335–341)

| | Twin Births, 1995–1997 (in thousands) | | | |
|---|---|---|---|---|
| | Preterm (induced or Cesarean) | Preterm (without procedures) | Term or Post-Term | Total |
| **Intensive** | 18 | 15 | 28 | 61 |
| **Adequate** | 46 | 43 | 65 | 154 |
| **Inadequate** | 12 | 13 | 38 | 63 |
| **Total** | 76 | 71 | 131 | 278 |

Level of Prenatal Care

Is there evidence of an association between the duration of the pregnancy and the level of care received by the mother?

**13. Twins, again** After reading of the *JAMA* study in Exercise 12, a large city hospital examined its records of twin births for several years and found the data summarized in the table below. Is there evidence that the way the hospital deals with pregnancies involving twins may have changed?

| | | 1990 | 1995 | 2000 |
|---|---|---|---|---|
| Outcome of Pregnancy | Preterm (induced or cesarean) | 11 | 13 | 19 |
| | Preterm (without procedures) | 13 | 14 | 18 |
| | Term or Post-Term | 27 | 26 | 32 |

**14. Preemies** Do the effects of being born prematurely linger into adulthood? Researchers examined 242 Cleveland-area children born prematurely between 1977 and 1979, and compared them with 233 children of normal birth weight; 24 of the "preemies" and 12 of the other children were described as being of "subnormal height" as adults. Is this evidence that babies born with a very low birth weight are more likely to be smaller than normal adults? ("Outcomes in Young Adulthood for Very-Low-Birth-Weight Infants," *New England Journal of Medicine*, 346, no. 3 [January 2002])

**15. LA rainfall** The Los Angeles Almanac website reports recent annual rainfall (in inches), as shown in the table.

| Year | Rain (in.) | Year | Rain (in.) |
|---|---|---|---|
| 1980 | 8.96 | 1991 | 21.00 |
| 1981 | 10.71 | 1992 | 27.36 |
| 1982 | 31.28 | 1993 | 8.14 |
| 1983 | 10.43 | 1994 | 24.35 |
| 1984 | 12.82 | 1995 | 12.46 |
| 1985 | 17.86 | 1996 | 12.40 |
| 1986 | 7.66 | 1997 | 31.01 |
| 1987 | 12.48 | 1998 | 9.09 |
| 1988 | 8.08 | 1999 | 11.57 |
| 1989 | 7.35 | 2000 | 17.94 |
| 1990 | 11.99 | 2001 | 4.42 |

a) Create a 90% confidence interval for the mean annual rainfall in LA.

b) Do these data suggest any change in annual rainfall as time passes? Check for an association between rainfall and year.

**16. Age and party 2019** The Pew Research Center conducted representative telephone surveys in 2018 and 2019. The following table shows the preferred political party affiliation of respondents and their ages for a sample of 1003 voters. Is there evidence of age-based differences in party affiliation in the United States for all voters?

| | Leaning Republican | Leaning Democrat | Neither/No Answer |
|---|---|---|---|
| **Millenial (1981–1996)** | 95 | 185 | 20 |
| **Gen X (1965–1980)** | 113 | 120 | 18 |
| **Boomer (1946–1964)** | 118 | 115 | 18 |
| **Silent (1923–1945)** | 123 | 120 | 8 |

a) Will you conduct a test of homogeneity or independence? Why?

b) Test an appropriate hypothesis.

c) State your conclusion, including an analysis of differences you find (if any).

**17. Birth days** During a 2-month period, 72 babies were born at the Tompkins Community Hospital in upstate New York. The table shows how many babies were born on each day of the week.

| Day | Births |
|---|---|
| Mon. | 7 |
| Tues. | 17 |
| Wed. | 8 |
| Thurs. | 12 |
| Fri. | 9 |
| Sat. | 10 |
| Sun. | 9 |

a) If births are uniformly distributed across all days of the week, how many would you expect on each day?
b) Test the hypothesis that babies are equally likely to be born on any of the days of the week.
c) Given the results of part b, do you think that the 7 births on Monday and 17 births on Tuesday indicate that women might be less likely to give birth on Monday, or more likely to give birth on Tuesday?
d) Can you think of any reasons why births may not occur completely at random?

**18. Wealth redistribution 2015** The following table is based on a Gallup Poll of 3045 U.S. adults on April 9–12, 2015. Respondents were stratified as high income (over $75,000), middle income ($30,000–$75,000), or low income (less than $30,000). Those polled were asked for their views on redistributing U.S. wealth by heavily taxing the rich. The data are summarized in the table below.

|  | Should Redistribute Wealth | Should Not | No Opinion |
|---|---|---|---|
| High Income | 426 | 579 | 10 |
| Middle Income | 558 | 447 | 10 |
| Low Income | 619 | 355 | 41 |

Is there any evidence that income level is associated with feelings toward wealth distribution in the United States? Test an appropriate hypothesis about this table, and state your conclusions.

**19. Eye and hair color** A survey of 1021 school-age children was conducted by randomly selecting children from several large urban elementary schools. Two of the questions concerned eye and hair color. In the survey, the following codes were used:

| Hair Color | Eye Color |
|---|---|
| 1 = Blond | 1 = Blue |
| 2 = Brown | 2 = Green |
| 3 = Black | 3 = Brown |
| 4 = Red | 4 = Grey |
| 5 = Other | 5 = Other |

The Statistics students analyzing the data were asked to study the relationship between eye and hair color.

a) One group of students produced the output shown below. What kind of analysis is this? What are the null and alternative hypotheses? Is the analysis appropriate? If so, summarize the findings, being sure to include any assumptions you've made and/or limitations to the analysis. If it's not an appropriate analysis, state explicitly why not.

Dependent variable is: Eyes
R-squared = 3.7%
s = 1.112 with 1021 − 2 = 1019 degrees of freedom

| Variable | Coefficient | SE(Coeff) | t-ratio | P-value |
|---|---|---|---|---|
| Intercept | 1.99541 | 0.08346 | 23.9 | ≤0.0001 |
| Hair | 0.211809 | 0.03372 | 6.28 | ≤0.0001 |

b) A second group of students used the same data to produce the output shown below. The table displays counts and standardized residuals in each cell. What kind of analysis is this? What are the null and alternative hypotheses? Is the analysis appropriate? If so, summarize the findings, being sure to include any assumptions you've made and/or limitations to the analysis. If it's not an appropriate analysis, state explicitly why not.

|  |  | Eye Color |  |  |  |  |
|---|---|---|---|---|---|---|
|  |  | 1 | 2 | 3 | 4 | 5 |
| Hair Color | 1 | 143 | 30 | 58 | 15 | 12 |
|  |  | 76.1 | 27.8 | 123.3 | 17.7 | 13.1 |
|  | 2 | 90 | 45 | 215 | 30 | 20 |
|  |  | 117.9 | 43.1 | 191.2 | 27.4 | 20.4 |
|  | 3 | 28 | 15 | 190 | 10 | 10 |
|  |  | 74.6 | 27.3 | 120.9 | 17.3 | 12.9 |
|  | 4 | 30 | 15 | 10 | 10 | 5 |
|  |  | 20.6 | 7.5 | 33.5 | 4.8 | 3.6 |
|  | 5 | 10 | 5 | 15 | 5 | 5 |
|  |  | 11.8 | 4.4 | 19.1 | 2.7 | 2.0 |

$$\sum \frac{(Observed - Expected)^2}{Expected} = 223.6 \quad \text{P-value} < 0.00001$$

**20. Depression and the Internet** The September 1998 issue of the *American Psychologist* published an article reporting on an experiment examining "the social and psychological impact of the Internet on 162 people in 73 households during their first 1 to 2 years online." In the experiment, a sample of households was offered free Internet access for one or two years in return for allowing their time and activity online to be tracked. The members of the households who participated in the study were also given a battery of tests at the beginning and again at the end of the study. One of the tests measured the subjects' levels of depression on a 4-point scale, with higher numbers meaning the person was more depressed. Internet usage was measured in average number of hours per week. The regression analysis examines the association between the subjects' depression levels and the amounts of Internet use. The conditions for inference were satisfied.

Dependent variable is: Depression After
R-squared = 4.6%
s = 0.4563 with 162 − 2 = 160 degrees of freedom

| Variable | Coefficient | SE(coeff) | t-ratio | Prob |
|---|---|---|---|---|
| Constant | 0.565485 | 0.0399 | 14.2 | ≤0.0001 |
| Intr_use | 0.019948 | 0.0072 | 2.76 | 0.0064 |

a) Do these data indicate that there is an association between Internet use and depression? Test an appropriate hypothesis and state your conclusion clearly.
b) One conclusion of the study was that those who spent more time online tended to be more depressed at the end of the experiment. News headlines said that too much time on the Internet can lead to depression. Does the study support this conclusion? Explain.

c) As noted, the subjects' depression levels were tested at both the beginning and the end of this study; higher scores indicated the person was more depressed. Results are summarized in the computer output shown below. Is there evidence that the depression level of the subjects changed during this study?

Depression Level
162 subjects

| Variable | Mean | StdDev |
|---|---|---|
| DeprBfore | 0.730370 | 0.487817 |
| DeprAfter | 0.611914 | 0.461932 |
| Difference | −0.118457 | 0.552417 |

**21. Pregnancy** In 1998 a San Diego reproductive clinic reported 42 live births to 157 women under the age of 38, but only 7 live births for 89 clients aged 38 and older. Is this evidence of a difference in the effectiveness of the clinic's methods for older women?

a) Test the appropriate hypotheses using the two-proportion $z$-procedure.
b) Repeat the analysis using an appropriate chi-square procedure.
c) Explain how the two results are equivalent.

**22. Eating in front of the TV** Roper Reports asked a random sample of people in 30 countries whether they agreed with the statement "I like to nibble while reading or watching TV." Allowable responses were "Agree Completely," "Agree Somewhat," "Neither Disagree Nor Agree," "Disagree Somewhat," "Disagree Completely," and "I Don't Know/No Response." Does a person's age influence their response? The table summarizes data from 3792 respondents in the 2006 sample of five countries (China, India, France, United Kingdom, and United States) for three age groups (Teens, 30's (30–39), and Over 60):

| | Agree Completely | Agree Somewhat | Neither Disagree Nor Agree | Disagree Somewhat | Disagree Completely |
|---|---|---|---|---|---|
| Teen | 369 | 540 | 299 | 175 | 106 |
| 30's | 272 | 522 | 325 | 229 | 170 |
| 60+ | 93 | 207 | 153 | 154 | 178 |

a) Make an appropriate display of these data.
b) Does a person's age seem to affect their response to the question about nibbling?

**T 23. Old Faithful** As you saw in an earlier chapter, Old Faithful isn't all that faithful. Eruptions do not occur at uniform intervals and may vary greatly. Can we improve our chances of predicting the time of the next eruption if we know how long the previous eruption lasted?

a) Describe what you see in this scatterplot.

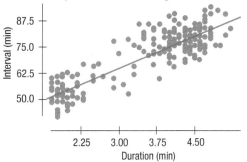

b) Write an appropriate hypothesis.
c) Here are a histogram of the residuals and the residuals plot. Do you think the assumptions for inference are met? Explain.

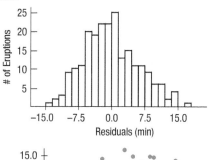

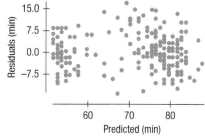

d) State a conclusion based on this regression analysis:

Dependent variable is: Interval
R-squared = 77.0%
s = 6.159 with 222 − 2 = 220 degrees of freedom

| Variable | Coefficient | SE(Coeff) | t-ratio | P-value |
|---|---|---|---|---|
| Intercept | 33.9668 | 1.428 | 23.8 | ≤0.0001 |
| Duration | 10.3582 | 0.3822 | 27.1 | ≤0.0001 |

| Variable | Mean | StdDev |
|---|---|---|
| Duration | 3.57613 | 1.08395 |
| Interval | 71.0090 | 12.7992 |

**24. Togetherness** Are good grades in high school associated with family togetherness? A simple random sample of 142 high-school students was asked how many meals per week their families ate together. Their responses produced a mean of 3.78 meals per week, with a standard deviation of 2.2. Researchers then matched these responses against the students' grade point averages. The scatterplot appeared to be reasonably linear, so they went ahead with the regression analysis, seen below. No apparent pattern emerged in the residuals plot.

Dependent variable is: GPA
R-squared = 11.0%
s = 0.6682 with 142 − 2 = 140 df

| Variable | Coefficient | SE(Coeff) |
|---|---|---|
| Intercept | 2.7288 | 0.1148 |
| Meals/wk | 0.1093 | 0.0263 |

a) Is there evidence of an association? Test an appropriate hypothesis and state your conclusion.
b) Do you think this association would be useful in predicting a student's grade point average? Explain.
c) Are your answers to parts a and b contradictory? Explain.

**25. Learning math** Developers of a new math curriculum called "Accelerated Math" compared performances of students taught by their system with control groups of students in the same schools who were taught using traditional instructional methods and materials. Statistics about pretest and posttest scores are shown in the table. (J. Ysseldyke and S. Tardrew, *Differentiating Math Instruction*, Renaissance Learning, 2002)

a) Did the groups differ in average math score at the start of this study?
b) Did the group taught using the Accelerated Math program show a significant improvement in test scores?
c) Did the control group show a significant improvement in test scores?
d) Were gains significantly higher for the Accelerated Math group than for the control group?

|  |  | Instructional Method | |
|---|---|---|---|
|  |  | Acc. Math | Control |
| Number of Students | | 231 | 245 |
| Pretest | Mean | 560.01 | 549.65 |
|  | St. Dev | 84.29 | 74.68 |
| Posttest | Mean | 637.55 | 588.76 |
|  | St. Dev | 82.9 | 83.24 |
| Individual gain | Mean | 77.53 | 39.11 |
|  | St. Dev. | 78.01 | 66.25 |

**26. Pesticides** A study published in 2002 in the journal *Environmental Health Perspectives* examined the gender ratios of children born to workers exposed to dioxin in Russian pesticide factories. The data covered the years from 1961 to 1988 in the city of Ufa, Bashkortostan, Russia. Of 227 children born to workers exposed to dioxin, 40% were male. Overall in the city of Ufa, the proportion of males was 51.2% (which is typical of human births, in general). Is this evidence that human exposure to dioxin may result in the birth of more girls?

**27. Dairy sales** Peninsula Creameries sells both cottage cheese and ice cream. The CEO recently noticed that in months when the company sells more cottage cheese, it seems to sell more ice cream as well. Two of his aides were assigned to test whether this is true or not. The first aide's plot and analysis of sales data for the past 12 months (in millions of pounds for cottage cheese and for ice cream) appear below.

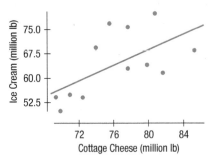

Dependent variable is: Ice Cream
R-squared = 36.9%
s = 8.320 with 12 − 2 = 10 degrees of freedom

| Variable | Coefficient | SE(Coeff) | t-ratio | P-value |
|---|---|---|---|---|
| Constant | −26.5306 | 37.68 | −0.704 | 0.4975 |
| Cottage cheese | 1.19334 | 0.4936 | 2.42 | 0.0362 |

The other aide looked at the differences in sales of ice cream and cottage cheese for each month and created the following output:

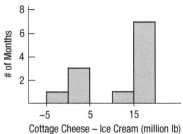

Cottage Cheese − Ice Cream
Count       12
Mean        11.8000
Median      15.3500
StdDev      7.99386
IntQRange   14.3000
25th %tile  3.20000
75th %tile  17.5000

Test $H_0$: $\mu(CC - IC) = 0$ vs. $H_a$: $\mu(CC - IC) \neq 0$
Sample Mean = 11.800000
t-Statistic = 5.113 w/11 df
Prob = 0.0003
Lower 95% bound = 6.7209429
Upper 95% bound = 16.879057

a) Which analysis would you use to answer the CEO's question? Why?
b) What would you tell the CEO?
c) Which analysis would you use to test whether the company sells more cottage cheese or ice cream in a typical month? Why?
d) What would you tell the CEO about this other result?
e) What assumptions are you making in the analysis you chose in part a? What assumptions are you making in the analysis in part c?
f) Next month's cottage cheese sales are 82 million pounds. Ice cream sales are not yet available. How much ice cream do you predict Peninsula Creameries will sell?
g) Give a 95% confidence interval for the true slope of the regression equation of ice cream sales by cottage cheese sales.
h) Explain what your interval means.

**28. Infliximab** In an article appearing in the journal *The Lancet* in 2002, medical researchers reported on the experimental use of the arthritis drug infliximab in treating Crohn's disease. In a trial, 573 patients were given initial 5-mg injections of the drug. Two weeks later, 335 had responded positively. These patients were then randomly assigned to three groups. Group I received continued injections of a placebo, Group II continued with 5 mg of infliximab, and Group III received 10 mg of the drug. After 30 weeks, 23 of 110 Group I patients were in remission, compared with 44 of 113 Group II and 50 of 112 Group III patients. Do these data indicate that continued treatment with infliximab is of value for Crohn's disease patients who exhibit a positive initial response to the drug?

**29. Weight loss** A weight loss clinic advertises that its program of diet and exercise will allow clients to lose 10 pounds in one month. A local reporter investigating weight reduction gets permission to interview a randomly selected sample of clients who report the given weight losses during their first month in this program. Create a confidence interval to test the clinic's claim that the typical weight loss is 10 pounds.

| Pounds Lost | |
|---|---|
| 9.5 | 9.5 |
| 13 | 9 |
| 9 | 8 |
| 10 | 7.5 |
| 11 | 10 |
| 9 | 7 |
| 5 | 8 |
| 9 | 10.5 |
| 12.5 | 10.5 |
| 6 | 9 |

**30. Education vs. income** The following displays examine the median income and education level (years in school) for several U.S. cities:

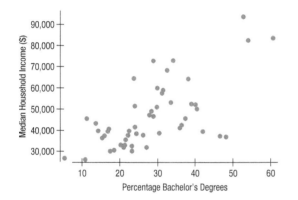

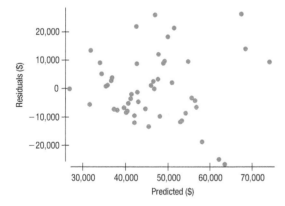

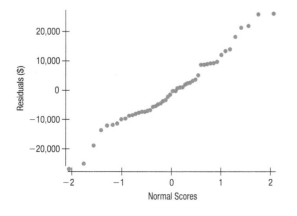

| Variable | Count | Mean | StdDev |
|---|---|---|---|
| Education | 57 | 10.9509 | 0.848344 |
| Income | 57 | 32742.6 | 3618.01 |

Dependent variable is: Income
R-squared = 32.9%
s = 2991 with 57 − 2 = 55 degrees of freedom

| Variable | Coefficient | SE(Coeff) | t-ratio | P-value |
|---|---|---|---|---|
| Intercept | 5970.05 | 5175 | 1.15 | 0.2537 |
| Education | 2444.79 | 471.2 | 5.19 | ≤0.0001 |

a) Do you think the assumptions for inference are met? Explain.
b) Does there appear to be an association between education and income levels in these cities?
c) Would this association appear to be weaker, stronger, or the same if data were plotted for individual people rather than for cities in aggregate? Explain.
d) Create and interpret a 95% confidence interval for the slope of the true line that describes the association between income and education.

**31. Diet** Thirteen overweight women volunteered for a study to determine whether eating specially prepared crackers before a meal could help them lose weight. The subjects were randomly assigned to eat crackers with different types of fiber (bran fiber, gum fiber, both, and a control cracker). Unfortunately, some of the women developed uncomfortable bloating and upset stomachs. Researchers suspected that some of the crackers might be at fault. The contingency table of "Cracker" versus "Bloat" shows the relationship between the four different types of crackers and the reported bloating. The study was paid for by the manufacturers of the gum fiber. What would you recommend to them about the prospects for marketing their new diet cracker?

|  |  | Bloat | |
|---|---|---|---|
|  |  | Little/None | Moderate/Severe |
| Cracker | Bran | 11 | 2 |
|  | Gum | 4 | 9 |
|  | Combo | 7 | 6 |
|  | Control | 8 | 4 |

**32. Cramming** Students in two basic Spanish classes were required to learn 50 new vocabulary words. One group of 45 students received the list on Monday and studied the words all week. Statistics summarizing this group's scores on Friday's quiz are given. The other group of 25 students did not get the vocabulary list until Thursday. They also took the quiz on Friday, after "cramming" Thursday night. Then, when they returned to class the following Monday, they were retested—without advance warning. Both sets of test scores for these students are shown.

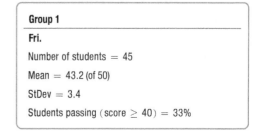

Group 1
Fri.
Number of students = 45
Mean = 43.2 (of 50)
StDev = 3.4
Students passing (score ≥ 40) = 33%

| Group 2 | | | |
|---|---|---|---|
| Fri. | Mon. | Fri. | Mon. |
| 42 | 36 | 50 | 47 |
| 44 | 44 | 34 | 34 |
| 45 | 46 | 38 | 31 |
| 48 | 38 | 43 | 40 |
| 44 | 40 | 39 | 41 |
| 43 | 38 | 46 | 32 |
| 41 | 37 | 37 | 36 |
| 35 | 31 | 40 | 31 |
| 43 | 32 | 41 | 32 |
| 48 | 37 | 48 | 39 |
| 43 | 41 | 37 | 31 |
| 45 | 32 | 36 | 41 |
| 47 | 44 | | |

a) On Friday, did the week-long study group have a mean score significantly higher than that of the overnight crammers?
b) Was there a significant difference in the percentages of students who passed the quiz on Friday?
c) Is there any evidence that when students cram for a test, their "learning" does not last for 3 days?
d) Use a 95% confidence interval to estimate the mean number of words that might be forgotten by crammers.
e) Is there any evidence that how much students forget depends on how much they "learned" to begin with?

**33.** Free throws 2017 At the middle of the 2016–2017 NBA season, James Hardin led the league by making 468 of 544 free throws, for a success rate of 86%. But Russell Westbrook was close behind with 425 of 517 (82.2%).
a) Find a 95% confidence interval for the difference in their free throw percentages.
b) Based on your confidence interval, is it certain that Hardin is better than Westbrook at making free throws?

**34.** More errors A corporation with a fleet of vehicles wanted to test the cost-effectiveness of using Motor Silk oil additive. For the study, 6100 delivery and passenger vehicles were tested for the same 3-month period in one year and then again in the subsequent year. In the initial year, the fleet was driven without Motor Silk. Then in the second year, Motor Silk was used according to the standard instructions. The average fuel economy increased from 18.97 mpg to 21.72 mpg.
a) What kind of a study is this?
b) Will the corporation do a one-tailed or a two-tailed test?
c) Explain in this context what a Type I error would be.
d) Explain in this context what a Type II error would be.
e) Which type of error would the additive manufacturer consider more serious?
f) If the vehicles with the additive are indeed statistically significantly better, can the company conclude it is an effect of the additive? Can it generalize this result and recommend the additive for all cars? Explain.

**35.** Seat belts 2015 The National Highway Traffic Safety Administration reported seat belt use and fatalities in car accidents. (*Seat Belt Use in 2015—Use Rates in the States and Territories.*

Report no. DOT HS-812-274) Is the rate of seat belt use different in New England compared to the Mountain states? We hope to create a 95% confidence interval for the difference in seat belt use proportions. Are these data appropriate for inference? If so, create the interval and make a concluding remark. Data are percentages.

| New England | 85 | 86 | 74 | 70 | 87 | 86 | |
|---|---|---|---|---|---|---|---|
| Mountain | 87 | 85 | 81 | 77 | 92 | 93 | 87 | 80 |

**36.** Family planning A 1954 study of 1438 pregnant women examined the association between the woman's education level and the occurrence of unplanned pregnancies, producing these data:

| | Education Level | | |
|---|---|---|---|
| | <3 Yr HS | 3+ Yr HS | Some College |
| Number of Pregnancies | 591 | 608 | 239 |
| % Unplanned | 66.2% | 55.4% | 42.7% |

Do these data provide evidence of an association between family planning and education in this era? (*Fertility Planning and Fertility Rates by Socio-Economic Status*, Social and Psychological Factors Affecting Fertility, 1954)

**37.** Sleepy An AP U.S. History teacher is worried that their first period class is sleepy when they take their exams. They decide to compare the scores from their first period to their second period class. Here are their scores:

| 1st Period | 2nd Period |
|---|---|
| 100 | 98 |
| 97 | 91 |
| 86 | 72 |
| 76 | 83 |
| 98 | 87 |
| 99 | 95 |
| 78 | 76 |
| 67 | 78 |
| 45 | 92 |
| Mean = 82.9 | Mean = 85.8 |

a) Which inference procedure is appropriate?
b) The teacher calculates the difference of the two means (1st period − 2nd period). What is their statistic?
c) As the sample size is rather small, a simulation is the chosen method. The scores are randomly sorted into two groups and the difference in means is calculated. This procedure is repeated 100 times. Given the dotplot below, find a P-value and write your conclusion.

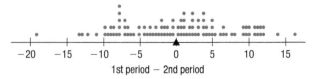
1st period − 2nd period

**38. Eat kale?** The Vegan Club on campus is curious about using documentaries to inform students about healthy foods. To begin, members of the club poll a few classmates and find that 3 out of 15 students are willing to eat a kale salad. Then they show the documentary *Forks over Knives* to a different group of students who are invited to a movie night. They poll these students, and 6 out of 17 say they would eat a kale salad.

a) What inference procedure is appropriate?
b) If we calculate our sample statistic by subtracting (#movie − #control), what is the value of our statistic? If we measure this difference in counts instead of in percentages, what is the difference?
c) As we've seen before, small sample sizes like these are best dealt with by running a simulation. Given the simulation run shown here (100 times), what is your conclusion about the effectiveness of students watching this documentary?

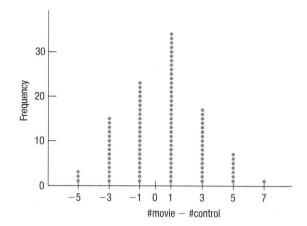

# PRACTICE EXAM

## I. MULTIPLE CHOICE

1. In a large university, 56% of the undergraduate students are female, as are 42% of the graduate students. How many females would we expect there to be in a random sample of 50 undergraduate and 30 graduate students?

   A) 18.8  B) 19.6  C) 37.8
   D) 39.2  E) 40.6

2. What's the standard deviation of the number of women found in samples of college students like the one described in Question 1 above?

   A) 2.49  B) 3.16  C) 4.43
   D) 4.47  E) 6.21

3. Researchers are investigating the association between the temperature in degrees and sales of hot dogs (in hundreds) at an outdoor sports stadium for a random sample of days. They constructed a 95% confidence interval for the slope of the regression line for predicting *Sales* from *Temperature*, obtaining $(-0.24, 1.68)$. Which statement is correct?

   A) There is not a significant association because the interval contains 0.
   B) The association is probably positive, because most of the interval is above 0.
   C) The association must be positive, because the interval contains 1.
   D) The researchers cannot make any conclusion, because the margin of error is too large.
   E) The researchers made an error, because the upper endpoint is greater than 1.

4. The price of video games often starts high when initial demand is high and then decreases as time goes on. A scatterplot of data collected for a sample of games suggests a linear model is reasonable, and an analysis produced a correlation coefficient of $-0.78$. What percent of the variation in price remains unexplained by this linear model?

   A) 4.8%  B) 22%  C) 39.2%
   D) 60.8%  E) 78%

5. A school district will use grade 1–5 students at one elementary school to study whether a new reading program can produce better reading comprehension scores. Half of the students will continue with the current program, and half will receive instruction with the new program. Which is the most appropriate design?

   A) Completely randomized, randomly assigning students to reading programs
   B) Randomized blocks, blocked by reading program
   C) Randomized blocks, blocked by gender
   D) Randomized blocks, blocked by grade level
   E) Randomized blocks, blocked by both grade level and gender

6. In Las Vegas $1 slot machines average a 95% payout; in other words, the expected value of a $1 bet is $0.95. Given that the machines' outcomes are random, which of these is true?

   A) A gambler who spends $100 will win $95 back.
   B) If a gambler plays long enough, they'll get all but 5% of his money back.
   C) There is a 95% chance that a gambler will lose money.
   D) In the long run, the casinos' profits should be about 5% of what the gamblers bet.
   E) A gambler who has lost many times in a row is more likely to win on the next bet.

7. A consumer group doing marketing research on consumer preferences for soft drinks will conduct a taste test comparing cola C and cola P. Which of the following is not a good suggestion for an effective study?

   A) Include people of different ages, genders, and other characteristics.
   B) Randomly select which cola is served first.
   C) Serve the colas in unmarked cups, rather than cans or bottles with a label.
   D) Ask a specific question, such as "Did you prefer cola C?"

E) Use statistical analysis to check for a significant difference in preference.

8. What is the smallest number of babies that must be randomly selected and tested in order to estimate the rate of jaundice among newborns to within ±3% with 95% confidence?

A) 250  B) 500  C) 800
D) 1200  E) 1500

9. The table shows the probability distribution for the number of points (X) a player can win by spinning a child's game spinner. If every point won moves you 5 spaces closer to winning, what is the average number of spaces moved per spin?

| X    | 1   | 2   | 3   | 4   | 5   |
|------|-----|-----|-----|-----|-----|
| P(X) | 0.4 | 0.2 | 0.2 | 0.1 | 0.1 |

A) 1.35  B) 2.3  C) 6.75
D) 7.3  E) 11.5

10. Which of the following statements is true?

A) A P-value of 0.011 is weaker evidence against the null hypothesis than a P-value of 0.033.
B) If we reject the null hypothesis at $\alpha = 0.05$, we'd also reject it at $\alpha = 0.01$.
C) We can increase the power of a test by decreasing the significance level from 5% to 1%.
D) We can decrease the risk of a Type I error by increasing the significance level from 1% to 5%.
E) The farther the true parameter is from the value we hypothesized, the lower the risk of a Type II error.

11. Soda companies have begun marketing 10-calorie sodas. The companies believe men prefer the taste and name of these drinks over traditional diet sodas. The table below summarizes preferences expressed by a random sample of 50 men and 50 women and also shows expected cell counts in parentheses.

|       | Diet   | 10-calorie |
|-------|--------|------------|
| Men   | 12     | 38         |
|       | (26.5) | (23.5)     |
| Women | 41     | 9          |
|       | (26.5) | (23.5)     |

In calculating the $\chi^2$ test statistic, what value is contributed by the cell for men who prefer 10-calorie drinks?

A) 0.62  B) 5.53  C) 8.95
D) 14.50  E) 210.25

12. Among the seniors at a certain high school, 65% attended the prom and 45% went on the senior trip, but 25% of the seniors did not participate in either the prom or the trip. What's the probability that a senior who went on the trip attended the prom?

A) 0.20  B) 0.35  C) 0.54
D) 0.69  E) 0.78

13. In June 2012 the Gallup organization conducted a poll using a random sample of 1004 adults, aged 18 and older, living in all 50 U.S. states and the District of Columbia. When asked, "On the whole, do you think immigration is a good thing or a bad thing for this country?" 66% responded "a good thing." The news report adds, "For results based on the total sample of national adults, one can say with 95% confidence that the maximum margin of sampling error is ±4 percentage points." (Source: www.gallup.com/poll/155210/Americans-Positive-Immigration.aspx) Based on this report, we can conclude that:

A) The percentage of all adult Americans who think that immigration is a good thing for this country is between 62% and 70%.
B) There's a 5% chance that the proportion of all adults who think immigration is a good thing is not between 62% and 70%.
C) In only 5% of all samples like this would the results differ by more than 4% from the proportion of all adults who think immigration is a good thing.
D) In 95% of all samples like this, between 62% and 70% of the respondents would say immigration is a good thing.
E) Between 62% and 70% of Americans think immigration is a good thing for 95% of all immigrants.

14. An analysis of laboratory data collected with the goal of modeling the weight (in grams) of a bacterial culture after several hours of growth produced the least squares regression line $\overline{\log(Weight)} = 0.25 + 0.61\, Hours$. Estimate the weight of the culture after 3 hours.

A) 0.32 g  B) 2.08 g  C) 8.0 g
D) 67.9 g  E) 120.2 g

15. Which is the appropriate interpretation of the slope of the bacterial growth model in Question 14 above?

A) For every additional 0.61 hours of growth, the culture is predicted to weigh 1 more gram.
B) For every additional hour of growth, the culture is predicted to weigh 0.61 more grams.
C) For every additional hour of growth, the logarithm of the culture's weight in grams is predicted to increase by 0.61.
D) At the beginning of the experiment, the logarithm of the culture's weight in grams was approximately 0.61 ounces.
E) 61% of variability in weight is explained by the number of hours the culture has been growing.

16. The PSA is a screening test for prostate cancer often recommended for men over 50. Unfortunately, it's not very reliable. Further testing reveals that only 30% of the men whose PSA comes back positive actually do have prostate cancer. Suppose a lab processing several PSAs reports 8 positive results one day. What's the probability that no more than 2 of those men really have prostate cancer?

A) 0.093  B) 0.296  C) 0.379
D) 0.552  E) 0.774

17. Could listening to rock music raise your blood pressure? To find out, researchers randomly divided 50 subjects into two groups. For 10 minutes, people in one group listened to loud rock music and those in the other group listened to soft jazz. The researchers recorded each person's change in blood pressure, and plan to run a hypothesis test to see if there's a significant difference between the mean changes for the two groups. Which condition does NOT need to be checked before proceeding?

A) Treatment should be randomly assigned.
B) The two treatment groups should be independent and not interfere with each other.
C) Both samples should be approximately normal.
D) Both samples have a big enough size to satisfy the Central Limit Theorem.
E) All of these are needed.

**18.** Some pet shop owners will soon begin selling a new variety of turtle, but first they want to determine the best environment for raising them. They want to try three aquarium temperatures and two different mixtures of food. They will randomly assign the different combinations of temperature and food to 18 aquariums, each of which holds 4 turtles. Which of the following statements is correct?

A) This experiment is poorly designed because there is no control group.
B) This experiment is poorly designed because there is no replication.
C) The factor is temperature and the treatments are the 2 food mixtures.
D) The 5 treatments are the 2 food mixtures and the 3 temperatures.
E) The factors are food and temperature, and there are 6 treatments.

**19.** A researcher discovered a data error that led her to eliminate the outlier seen in the lower left of her original scatterplot below.

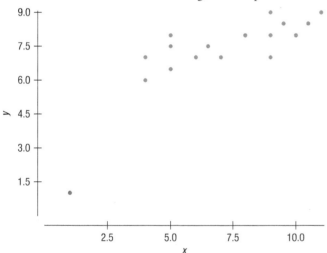

What effect did removing that point have on the slope of the regression line and the value of $R^2$ she had previously calculated?

A) The slope decreased and $R^2$ increased.
B) The slope increased and $R^2$ decreased.
C) The slope decreased, but $R^2$ remained the same.
D) Both the slope and $R^2$ increased.
E) Both the slope and $R^2$ decreased.

**20.** Maria has data from a random sample of 16 subjects and is constructing a 95% confidence interval for the population mean. Which value should she use for $t^*$?

A) 1.746  B) 1.753  C) 1.960
D) 2.120  E) 2.131

**21.** A company selling home furnaces claims that installing its new high efficiency furnace will "cut your heating bill in half." A consumer agency collects data for a random sample of homes and plots this year's heating costs with the new furnaces against previous costs with the old furnaces. Which are the most appropriate hypotheses to test the company's claim?

A) $H_0: \beta_1 = 0$ vs. $H_A: \beta_1 \neq 0$
B) $H_0: \beta_1 = 0$ vs. $H_A: \beta_1 > 0$
C) $H_0: \beta_1 = 0$ vs. $H_A: \beta_1 < 0$
D) $H_0: \beta = 0$ vs. $H_A: \beta < 0.5$
E) $H_0: \beta = 0.5$ vs. $H_A: \beta \neq 0.5$

**22.** In 2003 the U.S. Department of Education conducted a National Assessment of Adult Literacy by interviewing a random sample of 18,102 adults (aged 16+). Researchers assessed the respondents' literacy levels and also asked them how often they engaged in volunteer activities. The data are summarized in the table below.

|  |  | Literacy Level | | | |
| --- | --- | --- | --- | --- | --- |
|  |  | Below Basic | Basic | Intermediate | Proficient |
| Volunteer Activity | Once per week or more | 263 | 818 | 1654 | 611 |
|  | Less than once per week | 184 | 872 | 2068 | 782 |
|  | Never | 2185 | 3762 | 4550 | 1051 |

*Source:* U.S. Department of Education, National Center for Education Statistics.

If we wish to test whether *Volunteer Activity* and *Literacy Level* are independent, how many degrees of freedom are there?

A) 4  B) 6  C) 11
D) 12  E) 18,801

**23.** A citrus grower offers gourmet gift boxes of 6 Ruby Red grapefruits. The grapefruits used in these boxes weigh an average of 18 ounces with a standard deviation of 1.2 ounces. Assuming the weights are approximately normally distributed, what is the probability that the average weight of the Ruby Reds in one of these boxes is between 17.5 and 18.5 ounces?

A) 0.323  B) 0.383  C) 0.693
D) 0.986  E) 0.997

**24.** A zoologist studying a certain species of rats investigated the relationship between the weights of adult rats (in grams) and the lengths of their tails (in centimeters). The regression analysis for his data is shown below.

Response variable: Weight

| Variable | Coef | Std Error | t-ratio | P-value |
| --- | --- | --- | --- | --- |
| Constant | 23.5 | 3.89 | 6.04 | 0.0001 |
| Length | 2.1 | 0.24 | 8.79 | 0.0001 |

s = 14.31   R-squared = 75.4%

By how much do weight estimates produced by this model typically differ from the actual weights of the rats?

A) 0.24 g  B) 3.89 g  C) 14.31 g
D) 323.5 g  E) 24.6%

**25.** One advantage of using a control group with a placebo in an experiment is that the experiment will be

A) better because of the larger number of treatment groups.
B) better able to account for variability arising from extraneous factors.

C) better able to avoid bias.
D) better able to establish which factor is cause and which is effect.
E) better able to detect changes in the response variable.

26. Which sample result would provide evidence that there is an association between two quantitative variables based on a hypothesis test for the true population slope $\beta_1$?
A) Any sample with $b_1 = 0$
B) Any sample with $b_1 \neq 0$
C) Any sample with $b_1 > 0$
D) Any sample with $b_1$ significantly close to 0
E) Any sample with $b_1$ significantly different from 0

27. The Pew Research Center asked a random sample of 1047 adult (aged 18+) social media users and 623 teen (aged 12–17) social media users the following question: "How often do you witness online cruelty and meanness?" The table below summarizes the responses Pew obtained.

| Witness Cruelty and Meanness | Teens (12–17) | Adults (18+) | Total |
| --- | --- | --- | --- |
| Frequently | 75 | 73 | 148 |
| Sometimes | 181 | 188 | 369 |
| Only once in a while | 293 | 461 | 754 |
| Never | 68 | 304 | 372 |
| Don't know | 6 | 21 | 27 |
| Total | 623 | 1047 | 1670 |

To test for differences in the two age group distributions, what is the expected count for the cell that counts adults who reply "Sometimes"?

A) 167.0   B) 184.5   C) 188.0
D) 209.4   E) 231.34

28. Logs to be sawed into boards are first rough-cut to be a bit over 8 feet long. Measurements found the actual mean length to be 98.5 inches and 20% of the logs were longer than 100 inches. Assuming the logs' lengths are normally distributed, approximately what percentage of such logs should be between 97 and 100 inches long?

A) 20%   B) 40%   C) 60%
D) 68%   E) 80%

29. An investigation of water plant growth compares the heights (in inches) of a species of plant growing in two different lakes. The null hypothesis is $H_0: \mu_1 - \mu_2$. Here are the summaries of the data collected:

|  | Mean | StdDev | n |
| --- | --- | --- | --- |
| Lake 1 | 8.6 | 3.8 | 25 |
| Lake 2 | 7.6 | 2.3 | 25 |

What are the mean and standard error of the sampling distribution for the difference in sample means to be used by the hypothesis test?

A) 0 and 0.888   B) 0 and 1.5   C) 0 and 4.44
D) 1 and 0.888   E) 1 and 4.44

30. The regression analysis below models the relationship between the fuel efficiency (in miles per gallon) and horsepower (in 100's) for a random sample of 15 cars.

Dependent variable is: MPG
R-squared = 82.6%   R-squared (adjusted) = 81.3%
s = 2.435 with 15 − 2 = 13 degrees of freedom

| Variable | Coefficient | s.e. of (Coeff) | t-ratio | prob |
| --- | --- | --- | --- | --- |
| Constant | 43.4518 | 2.057 | 21.1 | ≤0.0001 |
| HP100 | −7.0166 | 0.89 | −7.86 | ≤0.0001 |

The highlighted value of 0.89 estimates the variability in
A) horsepower for this sample of cars.
B) fuel economy for this sample of cars.
C) slopes among this sample of cars.
D) slopes among all samples from this population of cars.
E) errors for predictions made by this model.

31. In a symmetric distribution, which of the following should be true?
  I. Maximum value − Median ≈ Median − Minimum value
  II. Q3 − Median ≈ Median − Q1
  III. Maximum value − Q3 ≈ Q1 − Minimum value
A) I only   B) II only   C) III only
D) II and III only   E) I, II, and III

32. For a Statistics class project at a large high school, a senior asked a random sample of students how often they use social networking sites. The relative frequencies of responses by male and female students are displayed in the bar chart below.

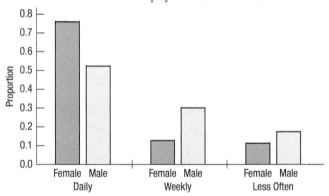

Which of the following statements can be supported by the display?

A) More females use social networking sites than males.
B) In general, females use the sites more often than males.
C) Approximately the same number of males and females were surveyed.
D) About 43% of all students surveyed said they use social networking sites weekly.
E) There appears to be no association between gender and frequency of use of social networking sites.

33. The salmon filets a seafood shop buys from a supplier weigh an average of 5.6 ounces, but 12% of them weigh less than 4 ounces. The shop owner believes the weights are normally distributed. What's the standard deviation of the filet weights?

A) 0.73 oz   B) 1.03 oz   C) 1.17 oz
D) 1.36 oz   E) 3.40 oz

**34.** Factory supervisors conducting a quality control study sampled twenty production batches and recorded the number of items in each sample that failed to meet standards. The results are summarized in the cumulative relative frequency histogram below.

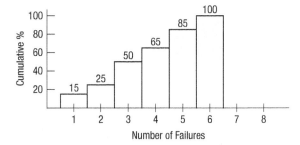

What is the probability that a sample had exactly five items that didn't meet standards?

A) 0.167 B) 0.15 C) 0.20
D) 0.65 E) 0.85

**35.** A study conducted in 2007 by the Baylor Department of Sociology asked the following question: "Please indicate your level of agreement with the following statement about science: Humans evolved from other primates over millions of years." The researcher summarized the data in a table categorizing the responses by level of *Belief in Evolution* and the *Age* group of the respondent. A chi-square test of independence between these two variables yielded a P-value of 0.025. Which statement correctly interprets this P-value?

A) The probability that the variables *Age* and *Belief in Evolution* are independent is 0.025.

B) The probability that the variables *Age* and *Belief in Evolution* are independent is 0.975.

C) Given the data that were observed, the probability that the variables *Age* and *Belief in Evolution* are independent is 0.025.

D) If *Belief in Evolution* is independent of *Age*, the probability of seeing results at least this extreme is only 0.025.

E) If there is an association between *Belief in Evolution* and *Age*, the probability of seeing results at least this extreme is only 0.025.

**36.** A researcher looks for a linear relationship between increasing the dosage of a vaccine the presence of antibodies. A scatterplot shows a weak linear relationship and the P-value for the slope is 0.349. Which of these is an appropriate conclusion?

A) With statistically significant evidence, more research is the next step.

B) The large P-value proves that as dosage increases, so do the antibodies.

C) Given that the results are not statistically significant, there is a linear relationship.

D) With such a large P-value, it is unlikely that increasing dosage has a linear relationship with increasing antibodies.

E) No conclusion can be reached without a control group.

**37.** To study the effectiveness of an interactive e-book, a researcher gave 20 students a pretest, had them study from the e-book, and then gave them a posttest. The table below summarizes the data they collected.

|        | Pretest | Posttest | Posttest − Pretest |
|--------|---------|----------|--------------------|
| Mean   | 78.9    | 81.4     | 2.5                |
| StdDev | 6.1     | 5.5      | 8.5                |

The researcher conducted a hypothesis test to see if using the e-book would produce a significant increase in mean score. What P-value did they find?

A) 0.028 B) 0.087 C) 0.091
D) 0.094 E) 0.102

**38.** A supplier tells a large hospital stockroom that the facility should buy 30% small, 45% medium, and 25% large surgical gloves. When the stockroom manager randomly surveys 110 workers about their size preferences, 23 request small, 57 medium, and 30 large gloves. The manager wonders whether this constitutes evidence that the supplier's recommendation won't meet the hospital's needs. What test should they run?

A) Chi-square goodness-of-fit test
B) Chi-square test of homogeneity
C) 1-sample *t*-test
D) 1-proportion *z*-test
E) 2-proportion *z*-test

**39.** To make a decision about closing time, a fast-food restaurant has recorded data on the number of customers using its drive-through window between 10:00 and 11:00 p.m. on Sunday through Thursday. A summary is shown in the table.

| Number of Customers | Relative Frequency |
|---------------------|--------------------|
| 1 | 0.08 |
| 2 | 0.10 |
| 3 | 0.14 |
| 4 | 0.12 |
| 5 | 0.15 |
| 6 | 0.10 |
| 7 | 0.20 |
| 8 | 0.05 |
| 9 | 0.06 |

What is the interquartile range for these data?

A) 3 B) 4 C) 5 D) 6 E) 7

**40.** Researchers are conducting an experiment at a zoo to compare the effect of two food supplements on weight gain among meerkats. The zoo has a large population of meerkats that live in 8 groups, each with separate living areas. Researchers will randomly assign four of the living areas to receive each food supplement. What are the experimental units in this study?

A) The two supplements
B) The individual meerkats
C) The 8 groups of meerkats
D) The weight gain of each meerkat
E) The mean weight gain for each of the 8 groups

## II. FREE RESPONSE

1. Origami USA hosts a conference in New York City every summer. Classes are offered on beginner, intermediate, and advanced levels of folding. Simple paper models use only a few folds, while the more complex models require more folds. The boxplot and statistics summarize the number of folds required for each of the models taught in a random sample of 15 classes.

   | Variable | Folds |
   |---|---|
   | N | 15 |
   | Mean | 72 |
   | StdDev | 24.07429 |
   | Median | 78 |
   | Range | 88 |
   | Min | 15 |
   | Max | 103 |
   | Q1 | 61 |
   | Q3 | 90 |

   a) Verify that minimum number of folds is an outlier. Show your work and justify your result.
   b) Do these data indicate that the conference is best suited for beginners or for more advanced folders? Use the graph and the summary statistics to justify your answer.
   c) Would it be appropriate to use these data to create a confidence interval for the number of folds in all the models taught in this conference's classes? Check the condition(s) for inference. Describe any concern(s) you have about creating such an interval. (Do not find the confidence interval.)

2. A small town has a Farmer's Market each Thursday night in a downtown park. Local growers, as well as craft makers and food vendors, set up booths for three hours. The Chamber of Commerce believes that the Market helps other downtown businesses because it attracts more visitors to the area. However, the Chamber has recently received complaints from some of the downtown businesses that the Farmer's Market hurts their normal sales on Thursdays. The Chamber wishes to survey more merchants in the area to determine the impact of the Market on their businesses.

   a) The Chamber of Commerce first considers inviting local merchants to fill out a poll on the Chamber's website. Describe a problem you see with this method.
   b) The downtown business district encompasses stores along 20 blocks of street frontage. The Chamber can improve its design by interviewing merchants in 3 of those blocks. Describe the procedure for this method.
   c) The park where the Farmer's Market is located is in one corner of the downtown business district. The Chamber might also consider a sample of businesses stratified by distance from the Market. Explain an advantage such a stratified sample may have over the cluster sampling method.
   d) Identify another way to stratify the downtown businesses, and explain why the Chamber should consider doing that.

3. Jane's Java serves both coffee and tea, each in a small or large size. History shows that 70% of the customers who order just one drink order coffee. Of those customers, 80% order a large; 60% of the customers who order tea order a large.

   a) Compute the probability that a random customer who orders one drink will get a small tea.
   b) Compute the probability that a random customer who orders one large drink gets coffee.
   c) Jane's prices are $3.00 for large coffee, $2.50 for large tea, $2.00 for small coffee, and $1.50 for small tea. What is the expected price for a drink sold by Jane?

4. A study published in 2005 examined health and diet among a representative sample of Canadian women. Researchers reported that only 6.8% of women who were vegetarians had high cholesterol, compared to 11.3% of nonvegetarians.

   a) Explain two reasons why this study would not justify a newspaper headline proclaiming, "Want to Lower Your Cholesterol? Become a Vegetarian!" Include an alternative explanation for the reported difference.
   b) If these results were based on data collected from 162 vegetarians and 177 nonvegetarians, do they provide statistical evidence that there's a difference in the rates of high cholesterol in the two populations?

5. Some states are moving to eliminate the sale of gas-powered cars and only allowing the sales of electric cars. While some citizens are excited at the prospect of a world with only electric cars, other people have an attachment to the type of cars they grew up with. Here is a survey, conducted using random dialing, asking people if they are likely to buy an electric car for their next purchase.

   | | | Buy Electric? | |
   |---|---|---|---|
   | | | Yes | No |
   | Age | 18–35 | 254 | 205 |
   | | 36–50 | 365 | 304 |
   | | 51–65 | 314 | 211 |
   | | 60+ | 287 | 113 |
   | | Total | 1220 | 833 |

   Is there evidence that there is an association between age and a person's likelihood of buying an electric car?

6. Although the event may seem startling to their students, 119 Statistics teachers participated in a fun run. Before the run, each person was asked to guess what their time would be (in seconds). After the run, the actual times were plotted against the teachers' guesses, and the assumptions for inference appeared to be reasonable. Here is the regression output:

   Dependent variable: actual time
   R-squared = 75.01%
   s = 278.1940

   | Variable | Coefficient | SE(C) | t-ratio | P-value |
   |---|---|---|---|---|
   | Intercept | 156.3777 | 79.3023 | 1.9719 | 0.051 |
   | Guess | 0.887006 | 0.04734 | 18.739 | <0.0001 |

   a) Does this output provide evidence of an association between guessed time and actual time? Explain.
   b) Assuming these 119 teachers represent a representative sample from a larger population who might participate in such a run, construct and interpret a 95% confidence interval for the slope of the regression line for that population.
   c) Does the interval you constructed in part b support your conclusion in part a? Explain.

# APPENDIX A: Selected Formulas

$Range = Max - Min$

$IQR = Q3 - Q1$

Outlier Rule-of-Thumb: $y < Q1 - 1.5 \times IQR$ or $y > Q3 + 1.5 \times IQR$

$$\bar{y} = \frac{\sum y}{n}$$

$$s = \sqrt{\frac{\sum(y - \bar{y})^2}{n - 1}}$$

$$z = \frac{y - \mu}{\sigma} \text{ (model based)}$$

$$z = \frac{y - \bar{y}}{s} \text{ (data based)}$$

$$r = \frac{\sum z_x z_y}{n - 1}$$

$\hat{y} = b_0 + b_1 x$ where $b_1 = \dfrac{r s_y}{s_x}$ and $b_0 = \bar{y} - b_1 \bar{x}$

$P(\mathbf{A}) = 1 - P(\mathbf{A}^C)$

$P(\mathbf{A} \cup \mathbf{B}) = P(\mathbf{A}) + P(\mathbf{B}) - P(\mathbf{A} \cap \mathbf{B})$

$P(\mathbf{A} \cap \mathbf{B}) = P(\mathbf{A}) \times P(\mathbf{B}|\mathbf{A})$

$$P(\mathbf{B}|\mathbf{A}) = \frac{P(\mathbf{A} \cap \mathbf{B})}{P(\mathbf{A})}$$

If **A** and **B** are independent, $P(\mathbf{B}|\mathbf{A}) = P(\mathbf{B})$

$E(X) = \mu = \sum x \cdot P(x)$      $Var(X) = \sigma^2 = \sum(x - \mu)^2 P(x)$

$E(X \pm c) = E(X) \pm c$      $Var(X \pm c) = Var(X)$

$E(aX) = aE(X)$      $Var(aX) = a^2 Var(X)$

$E(X \pm Y) = E(X) \pm E(Y)$      $Var(X \pm Y) = Var(X) + Var(Y)$, if $X$ and $Y$ are independent

Geometric: $P(x) = q^{x-1} p$      $\mu = \dfrac{1}{p}$      $\sigma = \sqrt{\dfrac{q}{p^2}}$

Binomial: $P(x) = \binom{n}{x} p^x q^{n-x}$      $\mu = np$      $\sigma = \sqrt{npq}$

$\hat{p} = \dfrac{x}{n}$      $\mu(\hat{p}) = p$      $SD(\hat{p}) = \sqrt{\dfrac{pq}{n}}$

Sampling distribution of $\bar{y}$:

(CLT) As $n$ grows, the sampling distribution approaches the Normal model with

$$\mu(\bar{y}) = \mu_y \quad SD(\bar{y}) = \frac{\sigma}{\sqrt{n}}$$

## Inference:

Confidence interval for parameter = ***statistic ± critical value × SD(statistic)***

Test statistic = $\dfrac{Statistic - Parameter}{SD(statistic)}$

| Parameter | Statistic | SD(statistic) | SE(statistic) |
|---|---|---|---|
| $p$ | $\hat{p}$ | $\sqrt{\dfrac{pq}{n}}$ | $\sqrt{\dfrac{\hat{p}\hat{q}}{n}}$ |
| $p_1 - p_2$ | $\hat{p}_1 - \hat{p}_2$ | $\sqrt{\dfrac{p_1 q_1}{n_1} + \dfrac{p_2 q_2}{n_2}}$ | $\sqrt{\dfrac{\hat{p}_1 \hat{q}_1}{n_1} + \dfrac{\hat{p}_2 \hat{q}_2}{n_2}}$ |
| $\mu$ | $\bar{y}$ | $\dfrac{\sigma}{\sqrt{n}}$ | $\dfrac{s}{\sqrt{n}}$ |
| $\mu_1 - \mu_2$ | $\bar{y}_1 - \bar{y}_2$ | $\sqrt{\dfrac{\sigma_1^2}{n_1} + \dfrac{\sigma_2^2}{n_2}}$ | $\sqrt{\dfrac{s_1^2}{n_1} + \dfrac{s_2^2}{n_2}}$ |
| $\mu_d$ | $\bar{d}$ | $\dfrac{\sigma_d}{\sqrt{n}}$ | $\dfrac{s_d}{\sqrt{n}}$ |
| $\sigma_\varepsilon$ | $s_e = \sqrt{\dfrac{\sum(y - \hat{y})^2}{n - 2}}$ | | |
| $\beta_1$ | $b_1$ | | $\dfrac{s_e}{s_x \sqrt{n - 1}}$ |

Pooling: For testing difference between proportions: $\hat{p}_{pooled} = \dfrac{y_1 + y_2}{n_1 + n_2}$

*For testing difference between means: $s_p = \sqrt{\dfrac{(n_1 - 1)s_1^2 + (n_2 - 1)s_2^2}{n_1 + n_2 - 2}}$

Substitute these pooled estimates in the respective SE formulas for both groups when assumptions and conditions are met.

Chi-square: $\chi^2 = \sum \dfrac{(obs - exp)^2}{exp}$

# APPENDIX B: Guide to Statistical Software

## Chapter 2: Displaying and Describing Categorical Data

### DATA DESK

To make a bar chart or pie chart:
- In the Toolbar, click on the display you want to make. Data desk will open a blank window and invite you to drag in a variable.
- Drag the variable to plot into the display widow.

Data Desk cannot make stacked bar charts.

### EXCEL

To make a bar chart:
- First make a pivot table (Excel's name for a frequency table). From the **Data** menu, choose **Pivot Table** and **Pivot Chart Report**.
- When you reach the Layout window, drag your variable to the row area and drag your variable again to the data area. This tells Excel to count the occurrences of each category. Once you have an Excel pivot table, you can construct bar charts and pie charts.
- Click inside the Pivot Table.
- Click the Pivot Table Chart Wizard button. Excel creates a bar chart.
- A longer path leads to a pie chart; see your Excel documentation.

**COMMENTS**

Excel uses the pivot table to specify the category names and find counts within each category. If you already have that information, you can proceed directly to the Chart Wizard.

### JMP

JMP makes a bar chart and frequency table together:
- From the **Analyze** menu, choose **Distribution**.
- In the Distribution dialog, drag the name of the variable into the empty variable window beside the label "Y, Columns"; click **OK**.

To make a pie chart:
- Choose **Chart** from the **Graph** menu.
- In the Chart dialog, select the variable name from the Columns list.
- Click on the button labeled "Statistics," and select "N" from the drop-down menu.
- Click the "**Categories, X, Levels**" button to assign the same variable name to the $x$-axis.
- Under Options, click on the **second** button—labeled "**Bar Chart**"—and select "Pie" from the drop-down menu.

### MINITAB

To make a bar chart:
- Choose **Bar Chart** from the **Graph** menu.
- Select "Counts of unique values" in the first menu, and select "Simple" for the type of graph. Click **OK**.
- In the Chart dialog, enter the name of the variable that you wish to display in the box labeled "Categorical variables."
- Click **OK**.

### STATCRUNCH

To make a bar chart or pie chart
- Click on **Graph**.
- Choose the type of plot » **With Data** or » **With Summary**.
- Choose the variable name from the list of **Columns**; if using summaries, also choose the counts.
- Click on **Next**.
- Choose **Frequency/Counts** or (usually) **Relative frequency/Percents**. Note that you may elect to group categories under a specified percentage as "Other."
- Click on **Compute!**

## TI-NSPIRE

To create a bar chart or pie chart on the same page as the spreadsheet:

- On a Lists & Spreadsheets page, name a list by typing a variable name in the top row of the column. Variable names can only contain letters and numbers and are not case sensitive.
- Enter categories into the named list; enter frequencies (or relative frequencies) into a second named list.
- Press £ to get to the name of the category list. Press £ again so that the entire list is highlighted. Press b→Data →Quick Graph. This creates a dot plot with one dot per category. Press b→Plot Properties→Add Y Summary List and select the frequency list to complete the Bar Chart with frequencies.

- To change a graph to a pie chart, press b→Plot Type→ Pie Chart.

To create the plot on a full page:

- Enter frequency table as above.
- Add a Data & Statistics page to the problem. Insert the name of the category list for "Click to add variable." This creates a dot plot with one dot per category.
- Continue as above to add frequency list and create the graphs.

## Chapter 3: Displaying and Summarizing Quantitative Data

### DATA DESK

To make a histogram:

- Select the variable to display.
- In the **Plot** menu, choose **Histogram**.

To calculate summaries:

- In the **Calc** menu, open the summaries submenu. Options offer separate tables, a single unified table, and other formats.

### EXCEL

To make a histogram:

- First make a bar chart. A longer path leads to a histogram. You must determine the bar boundaries yourself and tell Excel to eliminate the spaces between the bars.

In Excel, there is another way to find some of the standard summary statistics. For example, to compute the mean:

- Click on an empty cell.
- Go to the Formulas tab in the Ribbon. Click on the drop down arrow next to "AutoSum" and choose "**Average**."
- Enter the data range in the formula displayed in the empty box you selected earlier.
- Press **Enter**. This computes the mean for the values in that range.

To compute the standard deviation:

- Click on an empty cell.
- Go to the Formulas tab in the Ribbon and click the drop down arrow next to "AutoSum" and select "**More functions . . .**"
- In the dialog window that opens, select "**STDEV**" from the list of functions and click OK. A new dialog window opens. Enter a range of fields into the text fields and click **OK**. Excel computes the standard deviation for the values in that range and places it in the specified cell of the spreadsheet.

### JMP

To make a histogram and find summary statistics:

- Choose **Distribution** from the **Analyze** menu.
- In the **Distribution** dialog, drag the name of the variable that you wish to analyze into the empty window beside the label "**Y, Columns**."

- Click **OK**. JMP computes standard summary statistics along with displays of the variables.

## MINITAB

To make a histogram:
- Choose **Histogram** from the **Graph** menu.
- Select "Simple" for the type of graph and click **OK**.
- Enter the name of the quantitative variable you wish to display in the box labeled "Graph variables." Click **OK**.

To calculate summary statistics:
- Choose **Basic statistics** from the **Stat** menu. From the **Basic Statistics** submenu, choose **Display Descriptive Statistics**.
- Assign variables from the variable list box to the Variables box. MINITAB makes a Descriptive Statistics table.

## STATCRUNCH

To make a histogram, dotplot, or stem-and-leaf plot:
- Click on **Graphics**.
- Choose the type of plot.
- Choose the variable name from the list of Columns.
- Click on **Next**.
- (For a histogram) Choose **Frequency** or (usually) **Relative frequency**, and (if desired) set the axis scale by entering the Start value and Bin width.
- Click on **Create Graph**.

To calculate summaries:
- Click on **Stat**.
- Choose **Summary Stats » Columns**.
- Choose the variable name from the list of Columns.
- Click on **Calculate**.

**COMMENTS**

You may need to hold down the Ctrl or Command key to choose more than one variable to summarize.

Before calculating, you can click on **Next** to choose additional summary statistics.

## TI-NSPIRE

To create a dot plot or histogram on the same page as the spreadsheet:
- On a Lists & Spreadsheets page, name a list by typing a variable name in the top row of the column. Variable names can only contain letters and numbers and are not case sensitive.
- Enter the data into the named list.
- Press £ to get to the name of the data list. Press £ again so that the entire list is highlighted. Press b→Data→ Quick Graph. This creates a dot plot.
- To change the type of plot, press b→Plot Type→ Histogram.

To create the plot on a full page:
- Enter the data as above.
- Add a Data & Statistics page to the problem. Use **a** to select the name of the data list for "Click to add variable." This creates a dot plot.
- Continue as above.

To change the width of the histogram bars:
- Press b→Plot Properties→Histogram Properties→Bin Settings→Equal Bin Width. Fill in window with desired bin width and alignment (cut point). Use **e** to get between cells. Click OK.
- If needed, resize the graph to fit the window by pressing b→Window/Zoom→Zoom – Data.

To compute summary statistics:
- On a Lists & Spreadsheets page, enter the data into a named list.
- With the cursor in the first cell of the data list, press b→ Statistics→Stat Calculations→One-Variable Statistics. Select the number of lists (1) in the One-Variable Statistics window (you can find summary statistics on multiple lists at the same time). Click OK.
- In the second window, use **a** to click in the box to show the list name(s) and select the desired list. Leave Frequency List as 1 unless you have entered a frequency list. e down and click OK.
- The summary statistics will start in the column to the right of your data list.

Note: You can also calculate summary statistics on a Calculator page; however, on a Calculator page, the summary statistics will not change if you modify the data.

# Chapter 4: Understanding and Comparing Distributions

There are two ways to organize data when we want to compare groups. Each group can be in its own variable (or list, on a calculator). In this form, the experiment comparing cups would have four variables, one for each type of cup:

| CUPPS | SIGG | Nissan | Starbucks |
|---|---|---|---|
| 6 | 2 | 12 | 13 |
| 6 | 1.5 | 16 | 7 |
| 6 | 2 | 9 | 7 |
| 18.5 | 3 | 23 | 17.5 |
| 10 | 0 | 11 | 10 |
| 17.5 | 7 | 20.5 | 15.5 |
| 11 | 0.5 | 12.5 | 6 |
| 6.5 | 6 | 24.5 | 6 |

But there's another way to think about and organize the data. What is the variable of interest (the *What*) in this experiment? It's the number of degrees lost by the water in each cup. And the *Who* is each time she tested a cup. We could gather all of the temperature values into one variable and put the names of the cups in a second variable listing the individual results, one on each row. Now, the *Who* is clearer—it's an experimental run, one row of the table. Most statistics packages prefer data on groups organized in this way.

| Cup | Temperature Difference | Cup | Temperature Difference |
|---|---|---|---|
| CUPPS | 6 | SIGG | 12 |
| CUPPS | 6 | SIGG | 16 |
| CUPPS | 6 | SIGG | 9 |
| CUPPS | 18.5 | SIGG | 23 |
| CUPPS | 10 | SIGG | 11 |
| CUPPS | 17.5 | SIGG | 20.5 |
| CUPPS | 11 | SIGG | 12.5 |
| CUPPS | 6.5 | SIGG | 24.5 |
| Nissan | 2 | Starbucks | 13 |
| Nissan | 1.5 | Starbucks | 7 |
| Nissan | 2 | Starbucks | 7 |
| Nissan | 3 | Starbucks | 17.5 |
| Nissan | 0 | Starbucks | 10 |
| Nissan | 7 | Starbucks | 15.5 |
| Nissan | 0.5 | Starbucks | 6 |
| Nissan | 6 | Starbucks | 6 |

That's actually the way we've thought about the wind speed data in this chapter, treating wind speeds as one variable and the groups (whether seasons, months, or days) as a second variable.

## DATA DESK

If the data are in separate variables:
- Select the variables and choose **Boxplot side by side** from the **Plot** menu. The boxes will appear in the order in which the variables were selected.

If the data are a single quantitative variable and a second variable holding group names:
- Select the quantitative variable as Y and the group variable as X.

- Then choose **Boxplot y by x** from the **Plot** menu. The boxes will appear in alphabetical order by group name.

### COMMENT

Data Desk offers options for assessing whether any pair of medians differ. Check the Boxplot Options under the menu in the upper left corner of the boxplot window.

APPENDIX B  Guide to Statistical Software   A-7

### EXCEL

Excel cannot make boxplots.

### JMP

- Choose **Fit y by x**.
- Assign a continuous response variable to **Y, Response** and a nominal group variable holding the group names to **X, Factor**, and click **OK**. JMP will offer (among other things) dotplots of the data.
- Click the red triangle and, under **Display Options**, select Boxplots.

Note: If the variables are of the wrong type, the display options might not offer boxplots.

### MINITAB

- Choose **Boxplot . . .** from the **Graph** menu.

If your data are in the form of one quantitative variable and one group variable:

- Choose **One Y** and **with Groups**.

If your data are in separate columns of the worksheet:

- Choose **Multiple Y's**.

### STATCRUNCH

To make a boxplot
- Click on **Graph**.
- Choose **Boxplot**.
- Choose the variable name from the list of **Columns**.
- Indicate that you want to **identify outliers**.
- Click on **Compute!**

To make side-by-side boxplots
- Click on **Graph.**
- Choose **Boxplot**.
- Choose the variable name from the list of **Columns**.
- Choose the column that holds the categories to **Group by**.
- Indicate that you want to **Plot groups for each column**.
- Indicate that you want to **identify outliers**.
- Click on **Compute!**

### TI-NSPIRE

To create a box plot using a named list:
- On a Lists & Spreadsheets page, enter the data into a named list.
- Add a Data & Statistics page to the problem. Use **a** to select the name of the data list for "Click to add variable." This creates a dot plot.
- To change the type of plot, press b→Plot Type→Box Plot.

To create multiple box plots on the same page:
- On a Lists & Spreadsheets page, enter the data into multiple named lists.
- Add a Data & Statistics page to the problem. Use **a** to select the name of the first data list for "Click to add variable." Change the dot plot to a box plot.
- Press b→Plot Properties→Add X Variable and select the name of the second variable to graph. Repeat for additional variables.

## Chapter 5: The Standard Deviation as a Ruler and the Normal Model

### DATA DESK

To make a "Normal Probability Plot" in Data Desk:
- Select the Variable.
- Choose **Normal Prob Plot** from the **Plot** menu.

**COMMENTS**

Data Desk places the ordered data values on the vertical axis and the Normal scores on the horizontal axis.

You can also find Normal Probability Plot commands in the HyperView menus of variable names on the axis of a histogram and in other plots and analyses we'll see in later chapters.

## EXCEL

Excel offers a "Normal probability plot" as part of the Regression command in the Data Analysis extension, but (as of this writing) it is not a correct Normal probability plot and should not be used.

## JMP

To make a "Normal Quantile Plot" in JMP:
- Make a histogram using **Distributions** from the **Analyze** menu.
- Click on the drop-down menu next to the variable name.
- Choose **Normal Quantile Plot** from the drop-down menu.
- JMP opens the plot next to the histogram.

**COMMENTS**

JMP places the ordered data on the vertical axis and the Normal scores on the horizontal axis. The vertical axis aligns with the histogram's axis, a useful feature.

## MINITAB

To make a "Normal Probability Plot" in MINITAB:
- Choose **Probability Plot** from the **Graph** menu.
- Select "Single" for the type of plot. Click **OK**.
- Enter the name of the variable in the "Graph variables" box. Click **OK**.

**COMMENTS**

MINITAB places the ordered data on the horizontal axis and the Normal scores on the vertical axis.

## STATCRUNCH

To make a Normal probability plot:
- Click on **Graph**.
- Choose **QQ Plot**.
- Choose the variable name from the list of **Columns**.
- Click on **Create Graph**.

To work with Normal percentiles:
- Click on **Stat**.
- Choose **Calculators » Normal**.
- Choose a lower tail (≤) or upper tail (≥) region.
- Enter the z-score cutoff, and then click on **Compute** to find the probability.

OR

Enter the desired probability, and then click on **Compute** to find the z-score cutoff.

## TI-NSPIRE

To find normal percentages:
- Make a sketch of a normal curve on paper, labeled with the mean and standard deviation. Mark the boundaries and shade the area of interest.
- On a Calculator page, press b→Probability→ Distributions→Normal Cdf.
- Fill in the lower and upper bounds of the region of interest (use **e** to move between entries).
- Fill in the mean and standard deviation of the distribution. Click OK.

To find normal cutpoints:
- Make a sketch of a normal curve on paper, labeled with mean and standard deviation. Draw and shade the given area.

- On a Calculator page, press b→Probability→ Distributions→Inverse Normal.
- Enter the known area (percentage) below the cutpoint as a decimal. Enter the mean and standard deviation of the distribution. Click OK.

To create a Normal Probability Plot:
- On a Lists & Spreadsheets page, enter the data into a named list.
- Add a Data & Statistics page to the problem. Use **a** to select the name of the data list for "Click to add variable." This creates a dot plot.
- Press b→Plot Type→Normal Probability Plot to change the type of plot.

# Chapter 6: Scatterplots, Association, and Correlation

## DATA DESK

To make a scatterplot of two variables:
- Click the scatterplot tool in the tool bar or choose **Scatterplot** from the Calc menu. Data desk opens a plot window.
- Drag variables to the Y and X axes and drop them there.
- You can find the correlation by choosing **Correlation** from the scatterplot's HyperView menu.

Alternatively, click the **Correlation tool** in the toolbar and drag variables into the Correlation window it opens.

**COMMENTS**

We prefer that you look at the scatterplot first and then find the correlation. But if you've found the correlation first, click on the correlation value to drop down a menu that offers to make the scatterplot.

## EXCEL

To make a scatterplot in Excel:
- Select the columns of data to use in the scatterplot. You can select more than one column by holding down the control key while clicking.
- In the Insert tab, click on the **Scatter** button and select the **Scatter with only Markers** chart from the menu.

Unfortunately, the plot this creates is often statistically useless. To make the plot useful, we need to change the display:
- With the chart selected click on the Gridlines button in the Layout tab to cause the Chart Tools tab to appear.
- Within Primary Horizontal Gridlines, select None. This will remove the gridlines from the scatterplot.
- To change the axis scaling, click on the numbers of each axis of the chart, and click on the **Format Selection** button in the Layout tab.
- Select the Fixed option instead of the Auto option, and type a value more suited for the scatterplot. You can use the pop-up dialog window as a straightedge to approximate the appropriate values.

Excel automatically places the leftmost of the two columns you select on the *x*-axis, and the rightmost one on the *y*-axis. If that's not what you'd prefer for your plot, you'll want to switch them.

- To switch the X- and Y-variables:
- Click the chart to access the **Chart Tools** tabs.
- Click on the **Select Data** button in the Design tab.
- In the pop-up window's Legend Entries box, click on **Edit**.
- Highlight and delete everything in the Series X Values line, and select new data from the spreadsheet. (Note that selecting the column would inadvertently select the title of the column, which would not work well here.)
- Do the same with the Series Y Values line.
- Press **OK**, then press **OK** again.

**COMMENTS**

The CORREL(array1, array2) function computes the correlation coefficient in a cell.

You can also find correlations with the Analysis Toolpack.

## JMP

To make a scatterplot and compute correlation:
- Choose **Fit Y by X** from the **Analyze** menu.
- In the Fit Y by X dialog, drag the Y variable into the "**Y, Response**" box, and drag the X variable into the "**X, Factor**" box.
- Click the **OK** button.

Once JMP has made the scatterplot, click on the red triangle next to the plot title to reveal a menu of options:
- Select **Density Ellipse** and select 0.95. JMP draws an ellipse around the data and reveals the **Correlation** tab.
- Click the blue triangle next to Correlation to reveal a table containing the correlation coefficient.

## MINITAB

To make a scatterplot:
- Choose **Scatterplot** from the **Graph** menu.
- Choose "Simple" for the type of graph. Click **OK**.
- Enter variable names for the Y variable and X variable into the table. Click **OK**.

To compute a correlation coefficient:
- Choose **Basic Statistics** from the **Stat** menu.
- From the Basic Statistics submenu, choose **Correlation**. Specify the names of at least two quantitative variables in the "Variables" box.
- Click **OK** to compute the correlation table.

## STATCRUNCH

To make a scatterplot:
- Click on **Graphics**.
- Choose **Scatter Plot**.
- Choose **X** and **Y** variable names from the list of **Columns**.
- Click on **Create Graph**.

To find a correlation:
- Click on **Stat**.
- Choose **Summary Stats » Correlation**.
- Choose two variable names from the list of **Columns**. (You may need to hold down the ctrl or command key to choose the second one.)
- Click on **Calculate!**

## TI-NSPIRE

To make a scatterplot of two variables:
- On a Lists & Spreadsheets page, enter the data into two named lists.
- Add a Data & Statistics page to the problem. Use **a** to select the name of the explanatory variable list for "Click to add variable" on the X-axis. Use **a** to select the name of the response variable list for "Click to add variable" on the Y-axis.
- Plot the least-squares regression line on the graph by pressing b→Analyze→Regression→Show Linear (a+bx).

To find the correlation coefficient, the coefficient of determination, and the equation of the line:
- With the cursor in the first cell of the second data list, press b→Statistics→Stat Calculations→Linear Regression (a+bx).
- Complete the dialog window with the names of the X List and Y List (use **a** to click in the box to show the list names). Leave Frequency List as 1 unless you have entered a frequency list. e down and click OK.
- The regression analysis will start in the column to the right of your second data list.

Note: You can run a regression analysis on a Calculator page; however, on a Calculator page, the statistics will not change if you modify the data.

# Chapter 7: Linear Regression

## DATA DESK

- Select the **Regression** button in the control bar. Data desk will make a Regression window.
- Drag y and x variables into the appropriate parts of the table.
- Or in a scatterplot HyperView menu, choose **Regression** to compute the regression.
- To display the line, from a scatterplot HyperView menu, choose **Add Regression Line**.
- To plot the residuals, click on the **HyperView** menu in the upper left corner of the **Regression** output table and select the residual plot you want from the menu that drops down.

**COMMENTS**

Alternatively,
- Select the *y*-**variable** and the *x*-**variable** icons.
- In the Calc menu, choose **Regression**.

## EXCEL

- Click on a blank cell in the spreadsheet.
- Go to the **Formulas** tab in the Ribbon and click **More Functions » Statistical**.
- The data analysis add-in includes a Regression command.
- In the dialog that pops up, enter the range of one of the variables in the space provided.
- Enter the range of the other variable in the space provided.
- Click **OK**.
- Excel offers alternatives. The LINEST function performs regressions. It is an "array function" in Excel. Consult the Excel documentation for help on array functions and on LINEST.
- The SLOPE(y-range, x-range) and INTERCEPT(y-range, x-range) functions compute the slope and intercept, but don't provide the rest of the usual regression results.

### COMMENTS

The correlation is computed in the selected cell. Correlations computed this way will update if any of the data values are changed.

Before you interpret a correlation coefficient, always make a scatterplot to check for nonlinearity and outliers. If the variables are not linearly related, the correlation coefficient cannot be interpreted.

## JMP

- Choose **Fit Y by X** from the **Analyze** menu.
- Specify the *y*-variable in the Select Columns box and click the **y, Response** button.
- Specify the *x*-variable and click the **X, Factor** button.
- Click **OK** to make a scatterplot.
- In the scatterplot window, click on the red triangle beside the heading labeled "**Bivariate Fit . . .**" and choose **Fit Line**. JMP draws the least squares regression line on the scatterplot and displays the results of the regression in tables below the plot.

## MINITAB

- Choose **Regression** from the **Stat** menu.
- From the Regression submenu, choose **Fitted Line Plot**.
- In the Fitted Line Plot dialog, click in the **Response Y** box, and assign the *y*-variable from the variable list.
- Click in the **Predictor X** box, and assign the *x*-variable from the Variable list. Make sure that the Type of Regression Model is set to Linear. Click the **OK** button.

## STATCRUNCH

- Click on **Stat**.
- Choose **Regression » Simple Linear**.
- Choose X and Y variable names from the list of columns.
- Click on **Next** (twice) to **Plot the fitted line** on the scatterplot.
- Click on **Calculate** to see the regression analysis.
- Click on **Next** to see the scatterplot.

### COMMENTS

Remember to check the scatterplot to be sure a linear model is appropriate.

Note that before you **Calculate** you can:
- enter an X-value for which you want to find the predicted Y-value;
- save all the fitted values;
- save the residuals;
- ask for a residuals plot.

## TI-NSPIRE

To make a scatterplot of two variables:

- On a Lists & Spreadsheets page, enter the data into two named lists.
- Add a Data & Statistics page to the problem. Use **a** to select the name of the explanatory variable list for "Click to add variable" on the X-axis. Use **a** to select the name of the response variable list for "Click to add variable" on the Y-axis.
- Plot the least-squares regression line on the graph by pressing b→Analyze→Regression→Show Linear (a+bx).

To find the correlation coefficient, the coefficient of determination, and the equation of the line:

- With the cursor in the first cell of the second data list, press b→Statistics→Stat Calculations→Linear Regression (a+bx).
- Complete the dialog window with the names of the X List and Y List (use **a** to click in the box to show the list names). Leave Frequency List as 1 unless you have entered a frequency list. e down and click OK.
- The regression analysis will start in the column to the right of your second data list.

Note: You can run a regression analysis on a Calculator page; however, on a Calculator page, the statistics will not change if you modify the data.

## Chapter 8: Regression Wisdom

### DATA DESK

- Click on the **HyperView** menu on the **Regression** output table. A menu drops down to offer scatterplots of residuals against predicted values, Normal probability plots of residuals, or just the ability to save the residuals and predicted values.
- Click on the name of a predictor in the regression table to be offered a scatterplot of the residuals against that predictor.

**COMMENTS**

If you change any of the variables in the regression analysis, Data Desk will offer to update the plots of residuals.

### EXCEL

The Data Analysis add-in for Excel includes a Regression command. The dialog box it shows offers to make plots of residuals.

**COMMENTS**

Do not use the Normal probability plot offered in the regression dialog. It is not what it claims to be and is wrong.

### JMP

- From the **Analyze** menu, choose **Fit Y by X**. Select **Fit Line**.
- Under Linear Fit, select **Plot Residuals**. You can also choose to **Save Residuals**.

- Subsequently, from the **Distribution menu**, choose **Normal quantile plot** or **histogram** for the residuals.

### MINITAB

- From the **Stat** menu, choose **Regression**.
- From the **Regression** submenu, select Regression again.
- In the Regression dialog, enter the response variable name in the "Response" box and the predictor variable name in the "Predictor" box.
- To specify saved results, in the Regression dialog, click **Storage**.
- Check "Residuals" and "Fits." Click **OK**.

- To specify displays, in the Regression dialog, click **Graphs**.
- Under "Residual Plots," select "Individual plots" and check "Residuals versus fits."
- Click **OK**. Now back in the Regression dialog, click **OK**. Minitab computes the regression and the requested saved values and graphs.

## STATCRUNCH

To create a residuals plot:
- Click on **Stat**.
- Choose **Regression » Simple Linear** and choose **X** and **Y**.
- Click on **Next** and click on **Next** again.
- Indicate which type of residuals plot you want.
- Click on **Calculate**.

### COMMENTS

Note that before you click on **Next** for the second time you may indicate that you want to save the values of the residuals. Residuals becomes a new column, and you may use that variable to create a histogram or residuals plot.

## TI-NSPIRE

To graph a residuals plot:
- On a Lists & Spreadsheets page, enter the data into two named lists.
- Add a Data & Statistics page to the problem. Use **a** to select the name of the explanatory variable list for "Click to add variable" on the X-axis. Use **a** to select the name of the response variable list for "Click to add variable" on the Y-axis.
- Plot the least-squares regression line on the graph by pressing b→Analyze→Regression→Show Linear (a+bx).
- Plot the residuals plot underneath the scatterplot by pressing b→Analyze→Residuals→Show Residual Plot.

To graph a residuals plot on a separate page:
- Find the regression equation by either calculating Linear Regression or by creating the scatterplot and showing the regression line on the graph.
- Add a new Data & Statistics page.
- Use **a** to select the name of the explanatory variable for "Click to add variable" on the X-axis.
- Move the cursor to "Click to add variable" on the Y-axis, press **h**, and select **stat.resid**.

### COMMENTS

Note: You cannot do graph residuals without first plotting the least-squares regression line.

# Chapter 9: Re-expressing Data: Get It Straight!

## DATA DESK

To re-express a variable in Data Desk, select the variable and Choose the function to re-express it from the **Manip > Transform** menu. Square root, log, reciprocal, and reciprocal root are immediately available. For others, make a derived variable and type the function. Data Desk makes a new derived variable that holds the re-expressed values. Any value changed in the original variable will immediately be re-expressed in the derived variable.

### COMMENTS

Or choose **Manip > Transform > Dynamic > Box-Cox** to generate a continuously changeable variable and a slider that specifies the power. Set plots to **Automatic Update** in their HyperView menus and watch them change dynamically as you drag the slider.

## EXCEL

To re-express a variable in Excel, use Excel's built-in functions as you would for any calculation. Changing a value in the original column will change the re-expressed value.

## JMP

To re-express a variable in JMP, double-click to the right of the last column of data to create a new column. Name the new column and select it. Choose **Formula** from the **Cols** menu. In the Formula dialog, choose the transformation and variable that you wish to assign to the new column. Click the **OK** button. JMP places the re-expressed data in the new column.

**COMMENTS**

The log and square root re-expressions are found in the **Transcendental** menu of functions in the formula dialog.

As of version 11, JMP allows re-expression "on the fly" in most dialogs. For example, in the Fit Y by X platform, right click on either variable to bring up a Transform option that offers most re-expressions.

## MINITAB

To re-express a variable in MINITAB:
▶ Choose **Calculator** from the **Calc** menu.
▶ In the Calculator dialog, specify a name for the new re-expressed variable.
▶ Use the **Functions List,** the calculator buttons, and the **Variables list** box to build the expression.
▶ Click **OK.**

## STATCRUNCH

To re-express data:
▶ From the Data menu, select **Compute Expression**.
▶ In the Compute Expression box, type the expression and name the new variable.
▶ Click the **Compute!** button to make the new variable.

## TI-NSPIRE

To re-express data:
▶ On a Lists & Spreadsheets page, enter the data in a named list. Add a Data & Statistics page and draw a scatterplot for the data.
▶ Go back to the Lists & Spreadsheets page, create and name a new list. For example, if the original list is weight, the new list could be logweight.
▶ Arrow down to the cell below the name. Press · and insert the re-expression formula. Press · again to populate the list. Remember to use parentheses in the formula.
▶ Add a new Data & Statistics page, and create a scatterplot.

Note: You can re-express the X-variable, the Y-variable, or both X- and Y-variables. Plot the scatterplots of all combinations to see which re-expression does the best job of straightening the line.

# Chapter 10: Understanding Randomness

## DATA DESK

Generate random numbers in Data Desk with the **Generate Random Numbers . . .** command in the **Manip** menu. A dialog guides you in specifying the number of variables to fill, the number of cases, and details about the values. For most simulations, generate random uniform values.

**COMMENTS**

**Bernoulli Trials** generate random values that are 0 or 1, with a specified chance of a 1.

**Binomial Experiments** automatically generate a specified number of Bernoulli trials and count the number of 1s.

### EXCEL

The **RAND** function generates a random value between 0 and 1. You can multiply to scale it up to any range you like and use the INT function to turn the result into an integer.

**COMMENTS**

Published tests of Excel's random-number generation have declared it to be inadequate. However, for simple simulations, it should be OK. Don't trust it for important large simulations.

### JMP

- In a new column, in the **Cols** menu choose **Column Info** . . . .
- In the dialog, click the **New Property** button, and choose **Formula** from the drop-down menu.

- Click the **Edit Formula** button, and in the **Functions(grouped)** window click on Random. **Random Integer (10)**, for example, will generate a random integer between 1 and 10.

### MINITAB

- In the **Calc** menu, choose **Random Data** . . . .
- In the Random Data submenu, choose **Uniform** . . . .

A dialog guides you in specifying details of range and number of columns to generate.

### STATCRUNCH

To generate a list of random numbers:
- Click on **Data**.
- Choose **Simulate data » Uniform**.
- Enter the number of rows and columns of random numbers you want. (Often you'll specify the desired number of random values as **Rows** and just 1 **Column**.)
- Enter the interval of possible values for the random numbers ($a \leq x < b$).
- Click on **Compute!** The random numbers will appear in the data table.

**COMMENTS**

To get random *integers* from 0 to 99, set **a** = 0 and **b** = 100, and then simply ignore the decimal places in the numbers generated.

OR

- Click **Build**.
- Add Function **Floor**.
- Add Column.
- Click on **Okay**.
- Click on **Compute!**

### TI-NSPIRE

To generate random integers on a Calculator page:
- Press b→Probability→Random→Integer. Type the desired range within the parentheses: RandInt(1,6) will generate a single random integer between 1 and 6.
- Adding a third value will generate a list of that length: RandInt(1,6,10) will generate a list of ten random integers between 1 and 6.
- You can also type randint(1,6,10) instead of using the menus. Typing is not case-sensitive.

To generate random integers on a Lists & Spreadsheets page:
- Name the list and press ..
- In the formula cell (directly below the name), press .. Press b→Data→Random→Integer and enter the values or type in the formula.

## Chapter 15: Random Variables

### STATCRUNCH

To compute mean and SD for a random variable:
- Enter the values in one column and the probabilities in another.
- Choose **Stat >> Calculators >> Custom**.

- Select the **Values** and the probabilities as the **Weights**.
- Click on **Compute!**

## TI-NSPIRE

To compute the mean and standard deviation for a discrete random variable:

- On a Lists & Spreadsheet page, name two lists—one for the data values and one for the probabilities. Enter the data values in the first named list and the probabilities in the second.
- With the cursor in the first cell of the second data list, press b→Statistics→Stat Calculations→One-Variable Statistics. Select the number of lists (1) in the One-Variable Statistics window (you only have one list of values). Click OK.
- In the second window, use **a** to click in the box to show the list name(s) and select the list of values.
- Select the probabilities list for the Frequency List. e down and click OK.
- The summary statistics will start in the column to the right of your frequency list.

Note that there is no sample standard deviation.

## Chapter 16: Probability Models

Most statistics packages offer functions that compute probabilities for various probability models. The only important differences among these functions are in what they are named and the order of their arguments. In these functions, pdf stands for "probability density function"—what we've been calling a probability model. The letters cdf stand for "cumulative distribution function," the technical term when we want to accumulate probabilities over a range of values. These technical terms show up in many of the function names.

### DATA DESK

**BinomDistr**(*x, n, prob*) (pdf)
**CumBinomDistr**(*x, n, prob*) (cdf)

**COMMENTS**
These functions work in derived variables or in scratchpads.

### EXCEL

**Binomdist**(*x, n, prob, cumulative*)

**COMMENTS**
Set cumulative = *true* for cdf, *false* for pdf.
Excel's function fails when *x* or *n* is large.

### JMP

**Binomial Probability** (*prob, n, x*) (pdf)
**Binomial Distribution** (*prob, n, x*) (cdf)

### MINITAB

- Choose **Probability Distributions** from the **Calc** menu.
- Choose **Binomial** from the Probability Distributions submenu.
- To calculate the probability of getting *x* successes in *n* trials, choose **Probability**.
- To calculate the probability of getting *x* or fewer successes among *n* trials, choose **Cumulative Probability**.
- For Geometric, choose **Geometric** from the Probability Distribution submenu.

### STATCRUNCH

To calculate binomial probabilities:

- Click on **Stat**.
- Choose **Calculators » Binomial**.
- Enter the parameters, **n** and **p**.
- Choose a specific outcome (=) or a lower tail (≤ or <) or upper tail (≥ or >) sum.
- Enter the number of successes **x**.
- Click on **Compute**.

## TI-NSPIRE

To compute binomial probabilities:
- On a calculator page, select b→Probability→Distributions→Binomial Pdf (probability distribution function)
- OR→Binomial Cdf (cumulative probabilities for a selected range).
- Binomial Pdf—enter the Number of Trials, n; the Probability of Success, p; and X value, the value of interest.
- Binomial Cdf—enter the Number of Trials; the Probability of Success; and the Lower Bound and Upper Bound.
- Binomial distributions can also be accessed by starting with b→Statistics→Distributions.

Geometric probabilities are computed in a similar fashion.

## Chapter 18: Confidence Intervals for Proportions

### DATA DESK

Data Desk does not offer built-in methods for inference with proportions.

**COMMENTS**

For summarized data, open a Scratchpad to compute the standard deviation and margin of error by typing the calculation. Then use **z-interval for individual μs.**

### EXCEL

Inference methods for proportions are not part of the standard Excel tool set.

**COMMENTS**

For summarized data, type the calculation into any cell and evaluate it.

### JMP

For a **categorical** variable that holds category labels, the **Distribution** platform includes tests and intervals for proportions. For summarized data, put the category names in one variable and the frequencies in an adjacent variable. Designate the frequency column to have the **role** of **frequency**. Then use the **Distribution** platform.

**COMMENTS**

JMP uses slightly different methods for proportion inferences than those discussed in this text. Your answers are likely to be slightly different, especially for small samples.

### MINITAB

- Choose **Basic Statistics** from the **Stat** menu.
- Choose **1Proportion** from the Basic Statistics submenu.
- If the data are category names in a variable, assign the variable from the variable list box to the **Samples in columns** box. If you have summarized data, click the **Summarized Data** button and fill in the number of trials and the number of successes.
- Click the **Options** button and specify the remaining details.
- If you have a large sample, check **Use test and interval based on normal distribution**.
Click the **OK** button.

**COMMENTS**

When working from a variable that names categories, MINITAB treats the last category as the "success" category. You can specify how the categories should be ordered.

## STATCRUNCH

To create a confidence interval for a proportion using summaries:

▶ Click on **Stat**.

▶ Choose **Proportion Statistics » One sample » With Summary**.

▶ Enter the **Number of successes** (x) and **Number of observations** (n).

▶ Indicate **Confidence Interval for p** (Standard-Wald), and then enter the **Level** of confidence.

▶ Click on **Compute!**

To create a confidence interval for a proportion using data:

▶ Click on **Stat**.

▶ Choose **Proportion Statistics » One sample » With Data**.

▶ Choose the variable **Column** listing the **Outcomes**.

▶ Enter the outcome to be considered a **Success**.

▶ Indicate **Confidence Interval for p** (Standard-Wald), and then enter the **Level** of confidence.

▶ Click on **Compute!**

## TI-NSPIRE

To compute a confidence interval for a population proportion:

▶ On a Calculator page, press b→Statistics→Confidence Intervals→1-Prop z Interval.

▶ Enter the number of Successes, x, as a whole number. Use standard mathematical rounding if needed.

▶ Enter the sample size, n, and the desired Confidence Level. Click OK.

Note: Confidence level can be entered with or without the decimal: .95 or 95 for 95% confident.

# Chapter 19: Testing Hypotheses About Proportions

## DATA DESK

Data Desk does not offer built-in methods for inference with proportions. The Replicate Y by X command in the Manip menu will "reconstruct" summarized count data so that you can display it.

For means:

▶ Select variables.

▶ From the Calc menu, choose **Test** for hypothesis tests.

▶ Select the test from the drop-down menu and make other choices in the dialog.

**COMMENTS**

For summarized data, open a Scratchpad to compute the standard deviation and margin of error by typing the calculation. Then perform the test with the **z-test for individual μs** found in the Test command.

## EXCEL

Inference methods for proportions are not part of the standard Excel tool set.

For means, specify formulas. Find $t^*$ with the TINV(alpha, df) function.

**COMMENTS**

For summarized data, type the calculation into any cell and evaluate it.

Hypothesis tests are not really automatic. There's no easy way to find P-values in Excel. For the examples in this chapter, substitute 0.05 for "alpha" in the TINV command.

## JMP

For a **categorical** variable that holds category labels, the **Distribution** platform includes tests and intervals of proportions. For summarized data:

▶ Put the category names in one variable and the frequencies in an adjacent variable.

▶ Designate the frequency column to have the **role of frequency**. Then use the **Distribution** platform.

**COMMENTS**

JMP uses slightly different methods for proportion inferences than those discussed in this text. Your answers are likely to be slightly different.

## MINITAB

Choose **Basic Statistics** from the **Stat** menu:
- Choose **1Proportion** from the Basic Statistics submenu.
- If the data are category names in a variable, assign the variable from the variable list box to the **Samples in columns** box.
- If you have summarized **data**, click the **Summarized Data** button and fill in the number of trials and the number of successes.
- Click the **Options** button and specify the remaining details.
- If you have a large sample, check **Use test and interval based on Normal distribution**.
- Click the **OK** button.

**COMMENTS**

When working from a variable that names categories, Minitab treats the last category as the "success" category. You can specify how the categories should be ordered.

## STATCRUNCH

To test a hypothesis for a proportion using summaries:
- Click on **Stat**.
- Choose **Proportions » One sample » with summary**.
- Enter the **Number of successes** (x) and **Number of observations** (n).
- Click on **Next**.
- Indicate **Hypothesis Test**, then enter the hypothesized Null proportion, and choose the **Alternative** hypothesis.
- Click on **Calculate**.

To test a hypothesis for a proportion using data:
- Click on **Stat**.
- Choose **Proportions » One sample » with data**.
- Choose the variable **Column** listing the Outcomes.
- Enter the outcome to be considered a Success.
- Click on **Next**.
- Indicate **Hypothesis Test**, then enter the hypothesized Null proportion, and choose the **Alternative** hypothesis.
- Click on **Calculate**.

## TI-NSPIRE

To test a hypothesis about a population proportion:
- On a Calculator page, press b→Statistics→Stat Tests→ 1-Prop z Test.
- Enter the null hypothesized value, P0, as a decimal.
- Enter the number of Successes, x, as a whole number. Use standard mathematical rounding if needed. Enter sample size, n.
- Choose the Alternative Hypothesis: Ha: prop > p0, Ha: prop ≠ p0, or Ha: prop < p0.
- Click OK.

Note: Hypothesis tests can also be run from a Lists & Spreadsheets page with the option to draw a plot of the test statistic with the P Value shaded. Press a to check the draw box.

# Chapter 21: Comparing Two Proportions

## DATA DESK

Data Desk does not offer built-in methods for inference with proportions. Use Replicate Y by X to construct data corresponding to given proportions and totals. The Test and Estimate commands in the Calc menu offer two-sample t methods to test or find confidence intervals for means of two independent groups. Select the two variables first, then go to the menus.

**COMMENTS**

For inference on proportions with summarized data, open a Scratchpad to compute the standard deviations and margin of error by typing the calculation.

## EXCEL

Inference methods for proportions are not part of the standard Excel tool set.

**COMMENTS**

For summarized data, type the calculation into any cell and evaluate it.

## JMP

For a **categorical** variable that holds category labels, the **Distribution** platform includes tests and intervals of proportions:

- For summarized data, put the category names in one variable and the frequencies in an adjacent variable.
- Designate the frequency column to have the **role of frequency**. Then use the **Distribution platform**.

**COMMENTS**

JMP uses slightly different methods for proportion inferences than those discussed in this text. Your answers are likely to be slightly different.

## MINITAB

To find a hypothesis test for a proportion:

- Choose **Basic Statistics** from the **Stat** menu.
- Choose **2Proportions**... from the Basic Statistics submenu. If the data are organized as category names in one column and case IDs in another, assign the variables from the variable list box to the **Samples in one column** box.
- If the data are organized as two separate columns of responses, click on **Samples in different columns**: and assign the variables from the variable list box. If you have summarized data, click the **Summarized Data** button and fill in the number of trials and the number of successes for each group.
- Click the **Options** button and specify the remaining details. Remember to click the **Use pooled estimate** of $p$ **for test** box when testing the null hypothesis of no difference between proportions.
- Click the **OK** button.

**COMMENTS**

When working from a variable that names categories, MINITAB treats the last category as the "success" category. You can specify how the categories should be ordered.

## STATCRUNCH

To do inference for the difference between two proportions using summaries:

- Click on **Stat**.
- Choose **Proportion Statistics » Two sample » With Summary**.
- Enter the **Number of successes** (x) and **Number of observations** (n) in each group.
- Click on **Next**.
- Indicate **Hypothesis Test**, then enter the hypothesized Null proportion difference (usually 0), and choose the Alternative hypothesis.

OR

Indicate **Confidence Interval**, and then enter the **Level** of confidence.

- Click on **Compute!**

To do inference for the difference between two proportions using data:

- Click on **Stat**.
- Choose **Proportion Statistics » Two sample » With Data**.
- For each group, choose the variable **Column** listing the **Outcomes**, and enter the outcome to be considered a **Success**.
- Indicate **Hypothesis Test**, then enter the hypothesized Null proportion difference (usually 0), and choose the Alternative hypothesis.

OR

Indicate **Confidence Interval**, and then enter the **Level** of confidence.

- Click on **Compute!**

### TI-NSPIRE

To compute a confidence interval for the difference between two population proportions:

- On a Calculator page, press b→Statistics→Confidence Intervals→2-Prop z Interval.
- Enter the number of Successes, x1 and x2, as whole numbers. Use standard mathematical rounding if needed.
- Enter the sample sizes, n1 and n2, and the desired Confidence Level (level can be entered with or without the decimal: .95 or 95). Click OK.

To test a hypothesis about the difference between two population proportions:

- On a Calculator page, press b→Statistics→Stat Tests→ 2-Prop z Test.
- Enter the number of Successes, x1 and x2, as whole numbers. Enter the sample sizes, n1 and n2.
- Choose the Alternative Hypothesis: Ha: p1 > p2, Ha: p1 ≠ p2, or Ha: p1 < p2. Press OK.

Note: Hypothesis tests can also be run from a Lists & Spreadsheets page with the option to Draw a plot of the test statistic with the P Value shaded. Press a to check the draw box.

## Chapter 22: Inferences About Means

### DATA DESK

- Select variables.
- From the **Calc** menu, choose **Estimate** for confidence intervals or **Test** for hypothesis tests.
- Select the interval or test from the drop-down menu and make other choices in the dialog.

### EXCEL

Specify formulas. Find t* with the TINV(alpha, df) function.

**COMMENTS**

Not really automatic. There's no easy way to find P-values in Excel. For the examples in this chapter, substitute 0.05 for "alpha" in the TINV command.

### JMP

- From the **Analyze** menu, select **Distribution**.
- For a confidence interval, scroll down to the "Moments" section to find the interval limits.
- For a hypothesis test, click the red triangle next to the variable's name and choose **Test Mean** from the menu.
- Then fill in the resulting dialog.

**COMMENTS**

"Moment" is a fancy statistical term for means, standard deviations, and other related statistics.

### MINITAB

- From the **Stat** menu, choose the **Basic Statistics** submenu.
- From that menu, choose **1-sample t . . . .**
- Then fill in the dialog.

**COMMENTS**

The dialog offers a clear choice between confidence interval and test.

## STATCRUNCH

To do inference for a mean using summaries:
- Click on **Stat**.
- Choose **T Statistics » One sample » With Summary**.
- Enter the **Sample mean, Sample std dev**, and **Sample size**.
- Click on **Next**.
- Indicate **Hypothesis Test**, then enter the hypothesized Null mean, and choose the Alternative hypothesis.

OR

- Indicate **Confidence Interval**, and then enter the **Level** of confidence.
- Click on **Calculate!**

To do inference for a mean using data:
- Click on **Stat**.
- Choose **T Statistics » One sample » With Data**.
- Choose the variable **Column**.
- Click on **Next**.
- Indicate **Hypothesis Test**, then enter the hypothesized Null mean, and choose the Alternative hypothesis.

OR

- Indicate **Confidence Interval**, and then enter the **Level** of confidence.
- Click on **Calculate!**

## TI-NSPIRE

To compute a confidence interval for a population mean:
- On a Calculator page, press b→Statistics→Confidence Intervals→t Interval.
- Select Data Input Method as Data or Stats (summary statistics).
- Complete the dialog window that follows. You must have entered the data values into a named list to use the Data option.
- Enter the desired Confidence Level (confidence level can be entered with or without the decimal: .95 or 95). Click OK.

Note: Using a Lists & Spreadsheets page is recommended when using data. If you change the data, the analysis will change automatically.

To test a hypothesis about a population mean:
- On a Calculator page, press b→Statistics→Stat Tests→ t Test.
- Select Data Input Method as Data or Stats (summary statistics).
- Complete the dialog window that follows. You must have entered the data values into a named list to use the Data option.
- Choose the Alternative Hypothesis: Ha: $\mu > \mu_0$, Ha: $\mu \neq \mu_0$, Ha: $\mu < \mu_0$. Click OK.

Note: Running a Hypothesis test from a Lists & Spreadsheets page also gives the option to Draw a plot of the test statistic with the P Value shaded. Press a to check the draw box.

## Chapter 23: Comparing Means

There are two ways to organize data when we want to compare two independent groups. The data can be in two lists, as in the table at the start of this chapter. Each list can be thought of as a variable. In this method, the variables in the batteries example would be *Brand Name* and *Generic*. Graphing calculators usually prefer this form, and some computer programs can use it as well.

There's another way to think about the data. What is the response variable for the battery life experiment? It's the *Time* until the music stopped. But the values of this variable are in both columns, and actually there's an experiment factor here, too—namely, the *Brand* of the battery. So, we could put the data into two different columns, one with the *Times* in it and one with the *Brand*. Then the data would look as shown in the table to the right.

This way of organizing the data makes sense as well. Now the factor and the response variables are clearly visible. You'll have to see which method your program requires. Some packages even allow you to structure the data either way.

The commands to do inference for two independent groups on common statistics technology are not always found in obvious places. Here are some starting guidelines.

| Time | Brand |
|---|---|
| 194.0 | Brand name |
| 205.5 | Brand name |
| 199.2 | Brand name |
| 172.4 | Brand name |
| 184.0 | Brand name |
| 169.5 | Brand name |
| 190.7 | Generic |
| 203.5 | Generic |
| 203.5 | Generic |
| 206.5 | Generic |
| 222.5 | Generic |
| 209.4 | Generic |

## DATA DESK

- Select variables.
- From the **Calc** menu, choose **Estimate** for confidence intervals or **Test** for hypothesis tests.
- Select the interval or test from the drop-down menu and make other choices in the dialog.

**COMMENTS**

Data Desk expects the two groups to be in separate variables.

## EXCEL

- From the Data Tab, Analysis Group, choose **Data Analysis.**
- Alternatively (if the Data Analysis Tool Pack is not installed), in the Formulas Tab, choose More functions > Statistical > TTEST, and specify Type = 3 in the resulting dialog.
- Fill in the cell ranges for the two groups, the hypothesized difference, and the alpha level.

**COMMENTS**

Excel expects the two groups to be in separate cell ranges. Notice that, contrary to Excel's wording, we do not need to assume that the variances are *not* equal; we simply choose not to assume that they *are* equal.

## JMP

- From the **Analyze** menu, select **Fit y by x.**
- Select variables: a **Y, Response** variable that holds the data and an **X, Factor** variable that holds the group names. JMP will make a dotplot.
- Click the **red triangle** in the dotplot title, and choose **Unequal variances.**
  The *t*-test is at the bottom of the resulting table.
- Find the P-value from the Prob > F section of the table (they are the same).

**COMMENTS**

JMP expects data in one variable and category names in the other. Don't be misled: There is no need for the variances to be unequal to use two-sample *t* methods.

## MINITAB

- From the **Stat** menu, choose the **Basic Statistics** submenu.
- From that menu, choose **2-sample t . . . .** Then fill in the dialog.

**COMMENTS**

The dialog offers a choice of data in two variables, or data in one variable and category names in the other.

## STATCRUNCH

To do inference for the difference between two means using summaries:

- Click on **Stat**.
- Choose **T Statistics » Two sample » With Summary.**
- Enter the **Sample mean, Standard deviation**, and sample **Size** for each group.
- De-select **Pool variances**.
- Indicate **Hypothesis Test**, then enter the hypothesized Null mean difference (usually 0), and choose the Alternative hypothesis.

OR

Indicate **Confidence Interval**, and then enter the **Level** of confidence.

- Click on **Compute!**

To do inference for the difference between two means using data:

- Click on **Stat**.
- Choose **T Statistics » Two sample » With Data.**
- Choose the variable Column for each group.
- De-select **Pool variances**.
- Indicate **Hypothesis Test**, then enter the hypothesized Null mean difference (usually 0), and choose the Alternative hypothesis.

OR

Indicate **Confidence Interval**, and then enter the **Level** of confidence.

- Click on **Compute!**

## TI-NSPIRE

To compute a confidence interval for the difference between two population means:

▶ On a Calculator page, press b→Statistics→Confidence Intervals→2-Sample t Interval.

▶ Select Data Input Method as Data or Stats (summary statistics).

▶ Complete the dialog window that follows. You must have entered the data values into named lists to use the Data option.

▶ Enter the desired Confidence Level (confidence level can be entered with or without the decimal: .95 or 95). Click OK.

Note: Using a Lists & Spreadsheets page is recommended when using data instead of Stats. If you change the data, the analysis will change automatically.

To test a hypothesis about the difference between two population means:

▶ On a Calculator page, press b→Statistics→Stat Tests→ 2-Sample t Test.

▶ Select Data Input Method as Data or Stats (summary statistics).

▶ Complete the dialog window that follows. You must have entered the data values into named lists to use the Data option.

▶ Choose the Alternative Hypothesis: Ha: $\mu1 > \mu2$, Ha: $\mu1 \neq \mu2$, or Ha: $\mu1 < \mu2$. Click OK.

Note: Running a Hypothesis test from a Lists & Spreadsheets page also gives the option to Draw a plot of the test statistic with the P Value shaded. Press a to check the draw box.

## Chapter 24: Paired Samples and Blocks

### DATA DESK

▶ Select variables.

▶ From the **Calc** menu, choose **Estimate** for confidence intervals or **Test** for hypothesis tests.

▶ **Select** the interval or test from the drop-down menu, and make other choices in the dialog.

**COMMENTS**

Data Desk expects the two groups to be in separate variables and in the same "Relation"—that is, about the same cases.

### EXCEL

▶ In Excel 2003 and earlier, select **Data Analysis** from the **Tools menu**.

▶ In Excel 2007, select **Data Analysis** from the **Analysis** Group on the **Data Tab**.

▶ From the **Data Analysis** menu, choose **t-test: paired two-sample for Means**.
Fill in the cell ranges for the two groups, the hypothesized difference, and the alpha level.

**COMMENTS**

Excel expects the two groups to be in separate cell ranges. **Warning:** Do not compute this test in Excel without checking for missing values. If there are any missing values (empty cells), Excel will usually give a wrong answer. Excel compacts each list, pushing values up to cover the missing cells, and then checks only that it has the same number of values in each list. The result is mismatched pairs and an entirely wrong analysis.

### JMP

▶ From the **Analyze** menu, select **Matched Pairs**.

▶ Specify the columns holding the two groups in the **Y Paired Response** dialog.

▶ Click **OK**.

### MINITAB

▶ From the **Stat** menu, choose the **Basic Statistics** submenu.

▶ From that menu, choose **Paired t** . . .

▶ Then fill in the dialog.

**COMMENTS**

Minitab takes "First sample" minus "Second sample."

### STATCRUNCH

To do inference for the mean of paired differences:
- Click on **Stat**.
- Choose **T Statistics » Paired**.
- Choose the **Column** for each variable.
- Check **Save differences** so you can look at a histogram to be sure the Nearly Normal condition is satisfied.

- Indicate **Hypothesis Test**, then enter the hypothesized Null mean difference (usually 0), and choose the Alternative hypothesis.

OR

- Indicate **Confidence Interval**, and then enter the **Level** of confidence.
- Click on **Compute!**

### TI-NSPIRE

For inference on a matched-pair design, compute a list of differences:
- In a Lists & Spreadsheets page, enter values for the two datasets into named lists, i.e., time1 and time2.
- Name a third list, i.e., diff. Type · to get to the formula line and · again to enter the difference formula, i.e., diff:=time2−time1. Press · to generate the list of differences.

- Construct a confidence interval using a one-variable t Interval with the list of differences as your data.
- Test a hypothesis using a one-variable t Test with the list of differences as your data.

## Chapter 25: Comparing Counts

### DATA DESK

- Select the contingency table icon in the tool bar.
- Drag variables into the table specifying which is the row and which is the column variable.
- From the **Calc** menu, choose **Contingency Table**.
- From the table's HyperView menu choose **Table Options**. (Or Choose **Calc** > **Calculation Options** > **Table Options**.)
- In the dialog, check the boxes for **Chi Square** and for **Standardized Residuals**. Data Desk will display the chi-square and its P-value below the table, and the standardized residuals within the table.

**COMMENTS**

Data Desk automatically treats variables selected for this command as categorical variables even if their elements are numerals. The **Compute** Counts command in the table's HyperView menu will make variables that hold the table contents (as selected in the Table Options dialog), including the standardized residuals.

### EXCEL

Excel offers the function **CHITEST(actual_range, expected_range)**, which computes a chi-square value for homogeneity. Both ranges are of the form UpperLeftcell:LowerRightCell, specifying two rectangular tables that must hold counts (although Excel will not check for integer values). The two tables must be of the same size and shape.

**COMMENTS**

Excel's documentation claims this is a test for independence and labels the input ranges accordingly, but Excel offers no way to find expected counts, so the function is not particularly useful for testing independence. You can use this function only if you already know both tables of counts or are willing to program additional calculations.

## JMP

- From the **Analyze** menu, select **Fit Y by X**.
- Choose one variable as the Y, response variable, and the other as the X, factor variable. Both selected variables must be Nominal or Ordinal.
- **JMP** will make a plot and a contingency table. Below the contingency table, **JMP** offers a **Tests** panel. In that panel, the Chi Square for independence is called a **Pearson ChiSquare**. The table also offers the P-value.
- Click on the **Contingency Table** title bar to drop down a menu that offers to include a **Deviation** and **Cell Chi square** in each cell of the table.

**COMMENTS**

JMP will choose a chi-square analysis for a **Fit Y by X** if both variables are nominal or ordinal (marked with an N or O), but not otherwise. Be sure the variables have the right type.

Deviations are the observed–expected differences in counts. Cell chi-squares are the squares of the standardized residuals. Refer to the deviations for the sign of the difference.

Look under **Distributions** in the **Analyze** menu to find a chi-square test for goodness-of-fit.

## MINITAB

- From the **Stat** menu, choose the **Tables** submenu.
- From that menu, choose **Chi Square Test** . . . .
- In the dialog, identify the columns that make up the table. Minitab will display the table and print the chi-square value and its P-value.

**COMMENTS**

Alternatively, select the **Cross Tabulation** . . . command to see more options for the table, including expected counts and standardized residuals.

## STATCRUNCH

To perform a Goodness-of-Fit test:
- Enter the observed counts in one column of a data table, and the expected counts in another.
- Click on **Stat**.
- Choose **Goodness-of-fit » Chi-Square test**.
- Choose the **Observed Column** and the **Expected Column**.
- Click on **Compute!**

**COMMENTS**

These Chi-square tests may also be performed using the actual data table instead of summary counts. See the StatCrunch **Help page** for details.

To perform a Test of Homogeneity or Independence:
- Create a table (without totals):
- Name the first column as one variable, enter the categories underneath.
- Name the adjacent columns as the categories of the other variable, entering the observed counts underneath.
- Click on **Stat**.
- Choose **Tables » Contingency » With Summary**.
- Choose the **Columns** holding counts.
- Choose the **Row labels column**.
- Enter the **Column label** name.
- Choose **Expected Count** (and, optionally, Contributions to Chi-Square).
- Click on **Compute!**

### TI-NSPIRE

To conduct a $X^2$ Goodness-of-Fit test:

- On a Lists & Spreadsheets page, enter the observed and expected counts into two named lists. (To name a list, type a variable name in the top row of the column. Variable names can only contain letters and numbers and are not case-sensitive.)
- With the cursor in the first cell of the data list, press b→ Statistics→Stat Tests→$X^2$ GOF.
- In the dialogue window, select the names for the Observed List and Expected List. Enter the Degrees of Freedom.
- Check the box if you want to Draw a plot of the test statistic with the P Value shaded. Click OK.

To conduct a $X^2$ Test of Independence or Homogeneity:

- Enter the contingency table data into a matrix: on a Calculator page, press t and select the three-by-three matrix icon. Complete the Create a Matrix dialogue box with the matrix dimensions. Click OK.
- Enter the data into the matrix, using **e** to move between the cells of the matrix.
- Press **e** again to exit the matrix.
- Name the matrix by pressing **/h** to store (sto) a name for the matrix. Press ·.
- Press b→Statistics→Stat Tests→$X^2$ 2-way Test. Select or type the name of the matrix in the box as the Observed Matrix. Click OK.
- To access the expected matrix, press **h** and select stat. ExpMatrix.

Note: A $X^2$ Test can also be run from a Lists & Spreadsheets page.

## Chapter 26: Inferences for Regression

### DATA DESK

- In the tool ribbon or from the Calc menu, choose Regression. Data desk makes a regression table.
- Drag variables into the appropriate parts of the table. Data desk recomputes the regression immediately.
- Select plots of residuals from the Regression table's HyperView menu.
- Click on any coefficient to drop down a HyperView that offers the partial regression plot for that coefficient.

**COMMENTS**

You can replace any variable in the regression table by dragging the icon of another variable over the variable name in the table and dropping it there. You can add predictors to the model by dragging them into the predictor part of the table and dropping them there. You can remove any variable by click and hold over the variable's name.

### EXCEL

- Select **Data Analysis** from the Analysis Group on the Data Tab.
- Select **Regression** from the Analysis Tools list.
- Click the **OK** button.
- Enter the data range holding the Y-variable in the box labeled Y-range.
- Enter the range of cells holding the X-variable in the box labeled X-range.
- Select the **New Worksheet Ply** option.
- Select Residuals options. Click the OK button.

Alternatively, The LINEST function can compute a multiple regression. LINEST is an array function, producing a table of results that is not identical to, but contains the same results as the regression table shown above. Consult your Excel documentation for details on array functions and the LINEST function.

**COMMENTS**

The Y and X ranges do not need to be in the same rows of the spreadsheet, although they must cover the same number of cells. But it is a good idea to arrange your data in parallel columns as in a data table.

Excel calls the standard deviation of the residuals the Standard Error. This is a common error. Don't be confused; it is not $SE(y)$, but rather $s_e$.

## JMP

- From the **Analyze** menu, select **Fit Y by X**.
- Select variables: a Y, Response variable, and an X, Factor variable. Both must be continuous (quantitative).
- JMP makes a scatterplot.
- Click on the red triangle beside the heading labeled **Bivariate Fit** ... and choose **Fit Line**. JMP draws the least squares regression line on the scatterplot and displays the results of the regression in tables below the plot.
- The portion of the table labeled "Parameter Estimates" gives the coefficients and their standard errors, *t*-ratios, and P-values.

**COMMENTS**

JMP chooses a regression analysis when both variables are "Continuous." If you get a different analysis, check the variable types.

The Parameter table does not include the residual standard deviation $s_e$. You can find that as Root Mean Square Error in the Summary of Fit panel of the output.

## MINITAB

- Choose **Regression** from the **Stat** menu.
- Choose **Regression** ... from the **Regression** submenu.
- In the Regression dialog, assign the Y-variable to the Response box and assign the X-variable to the Predictors box.
- Click the **Graphs** button.
- In the Regression-Graphs dialog, select **Standardized residuals**, and check **Normal plot of residuals** and **Residuals versus fits**.
- Click the **OK** button to return to the Regression dialog.
- Click the **OK** button to compute the regression.

**COMMENTS**

You can also start by choosing a Fitted Line plot from the **Regression** submenu to see the scatterplot first—usually good practice.

In Minitab Express, regression and multiple regression are found in the Regression menu of the toolbar. Minitab offers an output option that flags individual values that may be unusual due to large residual or leverage.

## STATCRUNCH

- Click on **Stat**.
- Choose **Regression » Simple Linear**.
- Choose X and Y variable names from the list of columns.
- Indicate that you want to see a residuals plot and a histogram of the residuals.
- Indicate **Hypothesis tests**, then enter the hypothesized null slope (usually 0) and choose the alternative hypothesis.

OR

Indicate **Confidence Interval**, then enter the **Level** of confidence.

- Click on **Compute!**
- Click on **Next** to see any plots you chose.

**COMMENTS**

Be sure to check the conditions for regression inference by looking at both the residuals plot and a histogram of the residuals.

## TI-NSPIRE

To compute a confidence interval for the population slope:

- Enter the data into two named lists.
- On a Calculator page, press b→Statistics→Confidence Intervals→Linear Reg t Intervals.
- In the first box, select Slope. Press OK.
- Complete the dialogue box with the X List, the Y List, Frequency List if used, and Confidence Level. Click OK.

To test a hypothesis about a population slope:

- Enter the data into two named lists.
- Press b→Statistics→Stat Tests→Linear Reg t Test.
- Complete the dialogue box with the X List, the Y List, and Frequency List if used. Select the Alternate Hypothesis from Ha:$\beta$ & $\rho > 0$, Ha:$\beta$ & $\rho \neq 0$, or Ha:$\beta$ & $\rho < 0$. Click OK.

# APPENDIX C Answers

## IMPORTANT NOTE

Here are the "answers" to the exercises for the chapters and the unit reviews. **Note that these answers are only outlines of the complete solution.** *Your* solution should follow the model of the Step-by-Step examples, where appropriate. You should explain the context, show your reasoning and calculations, and write complete interpretations and conclusions (in context!). For some problems, what you decide to include in an argument may differ somewhat from the answers here. But, of course, the numerical part of your answer should match the numbers in the answers shown.

### Chapter 1

1. Retailers, and suppliers to whom they sell the information, will use the information about what products consumers buy to target their advertisements to customers more likely to buy their products.

3. Owners can advertise about the availability of parking. They can also communicate with businesses about hours when more spots are available and when they should encourage more business.

5. The individual games.

7. The *Who* is the selected subjects; the *What* includes medical, dental, and physiological measurements and laboratory test results.

9. Categorical.

11. Quantitative.

13. Answers will vary.

15. *Who*—Pedestrians and bicyclists killed or severely injured by turning vehicles between 2010 and 2014 in New York City; *What*—Cause of death or injury; *Population*—All pedestrians and bicyclists killed or severely injured in New York City between 2010 and 2014.

17. *Who*—Middle school, high school, and college students in 12 states that were part of the study; *What*—Their ability to distinguish between real and fake news stories; *Population*—All middle school, high school, and college students in these 12 states.

19. *Who*—24 blind patients; *What*—Response to embryonic stem cell treatment; *Population*—All patients with one of these two forms of blindness.

21. *Who*—54 bears; *Cases*—Each bear is a case; *What*—Weight, neck size, length, and sex; *When*—Not specified; *Where*—Not specified; *Why*—To estimate weight from easier-to-measure variables; *How*—Researchers collected data on 54 bears they were able to catch; *Variable*—Weight; *Type*—Quantitative; *Units*—Not specified; *Variable*—Neck size; *Type*—Quantitative; *Units*—Not specified; *Variable*—Length; *Type*—Quantitative; *Units*—Not specified; *Variable*—Sex; *Type*—Categorical.

23. *Who*—sandwiches on Arby's menu; *Cases*—Each sandwich is a case; *What*—Type of meat, number of calories, and serving size; *When*—Not specified; *Where*—Arby's restaurants; *Why*—To assess nutritional value of sandwiches; *How*—Report by Arby's restaurants; *Variable*—Type of meat; *Type*—Categorical; *Variable*—Number of calories; *Type*—Quantitative; *Units*—Calories; *Variable*—Serving size; *Type*—Quantitative; *Units*—Ounces.

25. *Who*—882 births; *Cases*—Each of the 882 births is a case; *What*—Mother's age, length of pregnancy, type of birth, level of prenatal care, birth weight of baby, sex of baby, and baby's health problems; *When*—1998–2000; *Where*—Large city hospital; *Why*—Researchers were investigating the impact of prenatal care on newborn health; *How*—Not specified exactly, but probably from hospital records; *Variable*—Mother's age; *Type*—Quantitative; *Units*—Not specified (probably years); *Variable*—Length of pregnancy; *Type*—Quantitative; *Units*—Weeks; *Variable*—Birth weight of baby; *Type*—Quantitative; *Units*—Not specified (probably pounds and ounces); *Variable*—Type of birth; *Type*—Categorical; *Variable*—Level of prenatal care; *Type*—Categorical; *Variable*—Sex; *Type*—Categorical; *Variable*—Baby's health problems; *Type*—Categorical.

27. *Who*—Experiment subjects; *Cases*—Each subject is a case; *What*—Treatment (herbal cold remedy or sugar solution) and cold severity; *When*—Not specified; *Where*—Not specified; *Why*—To test efficacy of herbal remedy on common cold; *How*—The scientists set up an experiment; *Variable*—Treatment; *Type*—Categorical; *Variable*—Cold severity rating; *Type*—Quantitative; *Units*—Scale from 0 to 5.

29. *Who*—Streams; *Cases*—Each stream is a case; *What*—Name of stream, substrate of the stream, acidity of the water, temperature, BCI; *When*—Not specified; *Where*—Upstate New York; *Why*—To study ecology of streams; *How*—Not specified; *Variable*—Stream name; *Type*—Identifier; *Variable*—Substrate; *Type*—Categorical; *Variable*—Acidity of water; *Type*—Quantitative; *Units*—pH; *Variable*—Temperature; *Type*—Quantitative; *Units*—Degrees Celsius; *Variable*—BCI; *Type*—Quantitative; *Units*—Not specified.

31. *Who*—Dogs trained to identity odors; *What*—Whether dogs can be trained to identify COVID-19 by odor; *When*—2020; *Where*—Helsinki; *Why*—Possible way to identity COVID patients; *Variables*—Whether dogs could identify the scent—Categorical.

33. *Who*—Kentucky Derby races; *What*—Date, winner, jockey, trainer, owner, and time; *When*—1875 to 2020; *Where*—Churchill Downs, Louisville, Kentucky; *Why*—Not specified

(To see trends in horse racing?); *How*—Official statistics collected at race; *Variable*—Year; *Type*—Identifier and Quantitative; *Units*—Year; *Variable*—Winner; *Type*—Identifier; *Variable*—Jockey; *Type*—Categorical; *Variable*—Trainer; *Type*—Categorical; *Variable*—Owner; *Type*—Categorical; *Variable*—Time; *Type*—Quantitative; *Units*—Minutes and seconds.

## Chapter 2 (See note, page A-29)

1. Answers will vary.

3. Answers will vary.

5. a) Yes; each movie is categorized in a single genre.
   b) Documentary.

7. Comedy is the most popular genre, followed closely by Adventure, Action, Drama, and Thriller/Suspense. Horror was much less frequent. Documentaries and Concert/Musicals are a very, very small percentage of movies.

9. a) Omnivores are more liberal and less conservative than Carnivores. Vegetarians are the most liberal. Other comments are appropriate.
   b) The differences are very large. It does appear there is a strong association between diet preference and political alignment.

11. 1755 students applied for admission to the magnet schools program. 53% were accepted, 17% were wait-listed, and the other 30% were turned away.

13. a) Yes. We can add because these categories do not overlap. (Each person is assigned only one cause of death.)
    b) 36.8%.
    c) Either a bar chart or pie chart with "other" added would be appropriate. A bar chart is shown.

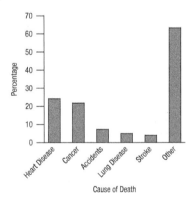

Cause of Death

15. a) The bar chart shows that grounding and collision are the most frequent causes of oil spills. Very few have unknown causes.
    b) A pie chart seems appropriate as well.

17. The percentages total 141%, and the three-dimensional display distorts the sizes of the regions, violating the area principle.

19. In both the South and West, about 58% of the eighth-grade smokers preferred Marlboro. Newport was the next most popular brand, but was far more popular in the South than in the West, where Camel was cited nearly 3 times as often as in the South. Nearly twice the proportion of in the West as in the South indicated that they had no usual brand (12.9% to 6.7%).

21. a) 79.4%.
    b) 8.1%.
    c)

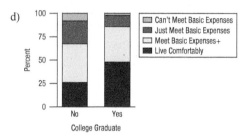

| Completed college? | Can't meet basic expenses | Just meet basic expenses | Meet basic expenses+ | Live comfortably |
|---|---|---|---|---|
| No | 8.1% | 24.4% | 42.0% | 25.6% |
| Yes | 2.4% | 11.8% | 38.1% | 47.7% |

d)

No, household financial situation is not independent of college completion status. The stacked barchart above shows, for example, college grads are about twice as likely to live comfortably as non-college grads.

23. a) 82.5%.   b) 12.9%.   c) 11.1%.   d) 13.4%.   e) 85.7%.

25. a) 73.9% 4-year college, 13.4% 2-year college, 1.5% military, 5.2% employment, 6.0% other.
    b) 77.2% 4-year college, 10.5% 2-year college, 1.8% military, 5.3% employment, 5.3% other.
    c) Many charts are possible. Here is a side-by-side bar chart.

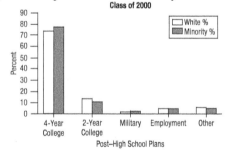

d) The white and minority students' plans are very similar. The small differences should be interpreted with caution because the total number of minority students is small. There is little evidence of an association between race and plans.

27. a) 16.6%.   b) 11.8%.   c) 37.7%.   d) 53.0%.

29. 1755 students applied for admission to the magnet schools program: 53% were accepted, 17% were wait-listed, and the other 30% were turned away. While the overall acceptance rate was 53%, 93.8% of blacks and Hispanics were accepted, compared to only 37.7% of Asians and 35.5% of whites. Overall, 29.5% of applicants were black or Hispanic, but only 6% of those turned away were. Asians accounted for 16.6% of all applicants, but 25.4% of those turned away. Whites were 54% of the applicants and 68.5% of those who were turned away. It appears that the admissions decisions were not independent of the applicant's ethnicity.

31. a) Men: 9/178 = 5.06%
    Women: 1/111 = 0.9%
    Men are much more likely than women to be conservative carnivores.

b) Women. Of the 17 liberal vegetarians more than 70% are women.

**33.** a) The most frequently chosen superpower (by about a third of those in the sample) was Flying. The next most appealing superpowers were Freezing Time and Telepathy, followed by Invisibility. The least frequently chosen was Super Strength.
b) All of the graphs look about the same. This suggests that the proportions of students choosing these superpowers in these samples are pretty to the proportions of students in the whole population who selected them. The differences in proportions across superpowers aren't just quirks of one or two samples—they reflect real differences in the population.

**35.** a) About 10%.
b) There are more men who didn't have cancer who never seldom ate fish.

**37.** a) Men. The male columns are wider than the female columns.
b) Yes, females are more likely to be liberal, and males are more likely to be conservative.
c) Yes, liberals are more likely to be vegetarians, and conservatives are more likely to be carnivores.
d) Yes, the differences in vegetarians are more pronounced in females than in males. The differences in carnivores are more pronounced in males than in females.
e) It has hard to compare the areas, but it appears the number of females may slightly outnumber the males in this category.
f) Yes. The percentage of females in the category is much larger than the percentage of males.

**39.** a) Low 20.0%, Normal 48.9%, High 31.0%.
b)

|  |  | Age |  |  |
|---|---|---|---|---|
|  |  | Under 30 | 30–49 | Over 50 |
| Blood Pressure | Low | 27.6% | 20.7% | 15.7% |
|  | Normal | 49.0% | 50.8% | 47.2% |
|  | High | 23.5% | 28.5% | 37.1% |

c)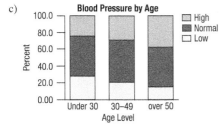

d) As age increases, the percentage of adults with high blood pressure increases. By contrast, the percentage of adults with low blood pressure decreases.
e) No, but it gives an indication that it might. We would need to follow the same individuals over time.

**41.** No, there's no evidence that Prozac is effective. The relapse rates were nearly identical: 28.6% among the people treated with Prozac, compared to 27.3% among those who took the placebo.

**43.** a) 4%.
b) 49.5%.
c) Only a slightly higher percentage of younger drivers are male than in the overall population. This difference decreases as drivers are older. But in the oldest categories, the drivers are increasingly women.
d) No, the answer to part c describes a relationship that is not independent.

**45.** a) 160 of 1300, or 12.3%.
b) Yes. Major surgery: 15.3% vs. minor surgery: 6.7%.
c) Large hospital: 13%; small hospital: 10%.
d) Large hospital: Major 15% vs. minor 5%; small hospital: Major 20% vs. minor 8%.
e) No. Smaller hospitals have a higher rate for both kinds of surgery, even though it's lower "overall."
f) The small hospital has a larger percentage of minor surgeries (83.3%) than the large hospital (20%). Minor surgeries have a lower delay rate, so the small hospital looks better "overall."

**47.** a) 42.6%.
b) A higher percentage of males than females were admitted: Males: 47.2% to females: 30.9%.
c) Program 1: Males 61.9%, females 82.4%.
Program 2: Males 62.9%, females 68.0%.
Program 3: Males 33.7%, females 35.2%.
Program 4: Males 5.9%, females 7.0%.
d) The comparison in part c is better, because it takes into account the differences in admission rates among programs; those differences are lost in part b.

## Chapter 3 (See note, page A-29)

**1.** Answers will vary.

**3.** Answers will vary.

**5.** a) Unimodal (near 0) and skewed to the right. Many seniors will have 0 or 1 speeding ticket. Some may have several, and a few may have more than that.
b) Probably unimodal and slightly skewed to the right. It is easier to score 15 strokes over the mean than 15 strokes under the mean.
c) Probably unimodal and symmetric. Weights may be equally likely to be over or under the average.
d) Probably bimodal. Men's and women's distributions may have different modes. It may also be skewed to the right, because it is possible to have very long hair, but hair length can't be negative.

**7.** a) The distribution of cereal carbohydrates is bimodal with a center at around 14 grams and a second center at 21 grams. The spread is from 5 grams to 23 grams, with most cereals between 10 and 18 grams.
b) Cereals made for kids can be high in sugar and carbs, so that probably explains the mode at 21 carbs.

**9.** a) About 80%.
b) Unimodal and skewed to the right with at least one high outlier. Most of the vineyards are between 0 and 90 acres with a few larger ones. The mode is between 0 and 30 acres.

**11.** a) Because the distribution is skewed to the right, we expect the mean to be larger.
b) Bimodal and skewed to the right. Center mode near 8 days. Another mode at 1 day (may represent patients who didn't survive). Most of the patients stay between 1 and 15 days. There are some extremely high values above 25 days.
c) The median and IQR, because the distribution is strongly skewed.

**13.** a) 46.5 points.
b) Q1 = 37, Q3 = 56.
c)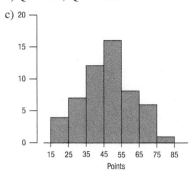

The distribution is fairly symmetric. It is centered at 46.5 with an IQR of 19 points. The minimum is 16 and the maximum is 75.

**15.** a) The boxplots do not show clusters and gaps, nor locations of multiple modes.
b) Boxplots can give only general ideas about overall shape and should not be used when more detail is needed.

**17.** a) Shape is difficult to discern from a boxplot, but there is a suggestion of a possible skew to the right with four potential outliers and this is confirmed in the histogram.
b) The actual shape is discernable in the histogram. You can see that there are no gaps until the potential outliers and that the distribution is unimodal with the peak near 0 to 500 adoptions.
c) The median would be the most appropriate measure of center because of the skew and the outliers.
d) The IQR would be the most appropriate measure of spread because of the skew and the extreme outliers.

**19.** a) The distribution is strongly skewed to the right, so use the median and IQR.
b) The IQR is 50, so the upper fence is the upper quartile +1.5 IQRs; that is, 78 + 75 = 153. There appear to be 3 to 5 parks that should be considered as outliers with more than 153 camp sites.
c)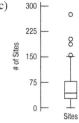
d) The distribution is unimodal with a strong skew to the right. There are several outliers past the 1.5 × IQR upper fence of 153 camp sites. The median number of camp sites is 43.5 sites. The mean is 62.8 sites. The mean is larger than the median because it has been influenced by the strong skew and the outliers.

**21.** a) The typical amount by which the cereals differ in their carb content from the mean is 3.8 grams.
b) Given the bimodal nature of the graph, spread is best communicated by describing the variation within each group. Separating the kids' cereals from the regular cereals would be the best option.

**23.** The mean and standard deviation because the distribution is unimodal and symmetric.

**25.** a) The mean is closest to $2.60 because that's the balancing point of the histogram.
b) The standard deviation is closest to $0.15 because that's a typical distance from the mean. There are no prices as far as $0.50 or $1.00 from the mean.

**27.** a) About 105 minutes.
b) Yes, only 2 of these 150 movies run that long.
c) They'd probably be about the same because the distribution is reasonably symmetric.

**29.** a) i. The middle 50% of movies ran between 97 and 115 minutes.
ii. Typically, movie lengths varied from the mean run time by 17.3 minutes.
b) Because the distribution is reasonably symmetric, either measure of spread would be appropriate, although the presence of outliers may inflate the SD somewhat.

**31.** a) The IQR might change a small amount but will not change drastically. Removing the largest point will shift measures of position down by one-half of a data point—this will not affect the IQR greatly. However, the standard deviation measures the distance of the data from the mean and now the largest of those distances has been removed, thus making the standard deviation much smaller.
b) Similarly to part (a), the median will only be shifted slightly lower and may not even change. However, the mean will now have its total sum reduced by a large amount and will decrease.

**33.** a) The median will be unaffected. The mean will be larger.
b) The range and standard deviation will increase; the IQR will be unaffected.

**35.** The publication is using the median; the watchdog group is using the mean, pulled higher by the several very successful movies in the long right tail.

**37.** a) Mean $525; median $450.
b) 2 employees earn more than the mean.
c) The median, because of the outlier.
d) The IQR will be least sensitive to the outlier of $1200, so it would be the best to report.

**39.** a) Price
```
24 | 5
24 |
23 | 5
23 | 0 1 2 3
22 | 5 5 5 8 9 9
22 | 1 2 3
21 | 9      21|9 = $2.19 per gallon
```
b) The distribution of gas prices is roughly symmetric with a median of $2.285 per gallon. Q1 = $2.24 and Q3 = $2.315 for an IQR of $0.075 with one outlier at $2.45.
c) The one outlier at $2.45 is an unusual feature.

**41.** a) Because the distribution is strongly skewed, the median and IQR are most appropriate.
b) Because the distribution is skewed to the higher values, the mean will be greater than the median.

c) The median is 4.5 million. The IQR is 5 million (Q1 = 2 million; Q3 = 7 million).
d) The distribution of populations is unimodal and skewed to the right. The median population is 4.5 million. Two states are clear outliers, with populations of 37 million and 25 million. The two states with populations of 19 million are also outliers by the 1.5 IQR rule-of-thumb.

**43.** The distribution is skewed low. Except for a few seasons with fewer than 20 home runs, the number of homers was typically in the 30's or 40's, with a high of 57. The distribution appears to be bimodal, with most seasons having more than 30 home runs and a few seasons having fewer than 25 home runs.

**45.** a)

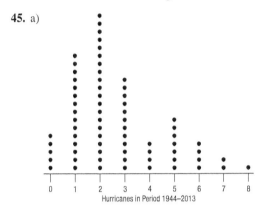

b) Slightly skewed to the right. Unimodal, mode near 2 hurricanes. Counts of 0 to 5 hurricanes are all typical. Possibly a second mode near 5. No outliers.

**47.** a) This is not a histogram. The horizontal axis should split the number of home runs hit in each year into bins. The vertical axis should show the number of years in each bin.

b)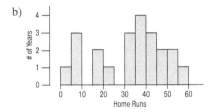

**49.** Skewed to higher values, with one fairly symmetric cluster near 4.4, another small cluster at 5.6. Two stray values in middle seem not to belong to either group. pH values between 4.12 and 4.64 are typical.

**51.** Histogram bins are too wide to be useful.

**53.** a) Asking price is skewed right with a center around $10,000–$15,000. Most cars have an asking price between $5,000 and $20,000. A small number of cars have asking prices between $40,000 and about $75,000, with the highest priced cars certainly being outliers.
b) Between $10,000–$15,000.
c) Around 2% (this is a *rough* estimate!).
d) Because the distribution is skewed right, the mean asking price would be more than the median asking price.

**55.** a) Median 280.5, IQR 10, mean 279.9, SD 10.14.
b) Because there is a low outlier (you *did* make a histogram first, didn't you?) it is better to report median and IQR.
c) We should set the outlier aside, discuss it separately, and report the mean and SD because the rest of the data are unimodal and symmetric. Eighth graders scored 281, on average, with a standard deviation of 6. Puerto Rico has a lower score.

**57.** The histogram shows that the distribution of Percent Change is unimodal and skewed to the right. The states vary from a minimum of −0.6 (Michigan) to 35.1% (Nevada) growth in the decade. The median was 7.8% and half of the states had growth between 4.3% and 14.1%. Not surprisingly, the top three states in terms of growth were all from the West: Nevada (35.1%), Arizona (24.6%), and Utah (23.8%).

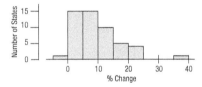

**59.** a) About 36 mph.
b) Q1 about 35 mph and Q3 about 37 mph.
c) The range appears to be about 7 mph, because the data extend from about 31 to 38 mph. The IQR is about 2 mph.
d) We can't know exactly, but the boxplot may look something like this:

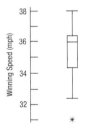

e) The median winning speed has been about 36 mph, with a max of about 38 and a min of about 31 mph. Half have run between about 35 and 37 mph, for an IQR of 2 mph.

## Chapter 4 (See note, page A-29)

**1.** Answers will vary.

**3.** Answers will vary.

**5.** a) Prices appear to be both higher on average and more variable in Baltimore than in the other three cities. Prices in Chicago may be slightly higher than in Dallas and Denver, but the difference is very small.

b) There are outliers on the low end in Baltimore and Chicago and one high outlier in Dallas, but these do not affect the overall conclusions reached in part a).

7. The distributions are both unimodal and skewed to the right, with 2011 being more strongly skewed. Though there are potential outliers on the high end for both sets, none seem to be departures from the pattern. The median for 2011 is somewhat lower and the IQR is slightly larger than for 2016.

9. a) About 59%.
   b) Bimodal.
   c) Some cereals are very sugary; others are healthier low-sugar brands.
   d) Yes.
   e) Although the ranges appear to be comparable for both groups (about 28%), the IQR is larger for the adults' cereals, indicating that there's more variability in the sugar content of the middle 50% of adults' cereals.

11. a)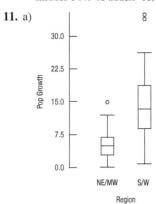

    b) Growth rates in NE/MW states are tightly clustered near 5%. S/W states are more variable, and bimodal with modes near 14 and 22. The S/W states have an outlier as well. Around all the modes, the distributions are fairly symmetric.

13. a) They should be put on the same scale, from 0 to 20 days.
    b) Lengths of men's stays appear to vary more than for women. Men have a mode at 1 day and then taper off from there. Women have a mode near 5 days, with a sharp drop afterward.
    c) A possible reason is childbirth.

15. a) Both girls have a median score of about 17 points per game, but Scyrine is much more consistent. Her IQR is about 2 points, while Alexandra's is over 10.
    b) If the coach wants a consistent performer, she should take Scyrine. She'll almost certainly deliver somewhere between 15 and 20 points. But if she wants to take a chance and needs a "big game," she should take Alexandra. Alexandra scores over 24 points about a quarter of the time. (On the other hand, she scores under 11 points as often.)

17. a) On average, eggs are the most expensive of the three.
    b) No, the prices of all three items overlap.

19. Compact cars generally get the best gas mileage. The first quartile for cars is above all the values for trucks and above the third quartile for SUVs. The trucks have the lowest median mileage and the least variability. There are a few low outliers in compact cars, and one low outlier in trucks.

21. Load factors are generally highest and least variable in the summer months (June–August). They are lower and more variable in the winter and spring.

23. Class A is 1, class B is 2, and class C is 3.

25. a) Probably slightly left skewed. The mean is slightly below the median, and the 25th percentile is farther from the median than the 75th percentile.
    b) No, all data are within the fences.
    c)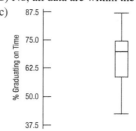

    d) The 48 universities graduate, on average, about 68% of freshmen "on time," with percents ranging from 43% to 87%. The middle 50% of these universities graduate between 59% and 75% of their freshmen in 4 years.

27. a) *Who:* Student volunteers
    *What:* Memory test
    *Where, when:* Not specified
    *How:* Students took memory test 2 hours after drinking caffeine-free, half-dose caffeine, or high-caffeine soda.
    *Why:* To see if caffeine makes you more alert and aids memory retention.
    b) Drink: categorical; test score: quantitative.
    c)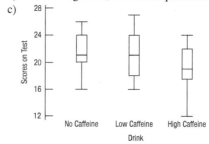

    d) The participants scored about the same with no caffeine and low caffeine. The medians for both were 21 points, with slightly more variation for the low-caffeine group. The high-caffeine group generally scored lower than the other two groups on all measures of the 5-number summary: min, lower quartile, median, upper quartile, and max.

29. a) The median of the heights is about the same for both steel and wooden coasters (just over 100 feet), but the wooden coasters have a much smaller spread (an IQR of about 44 compared to 88). Both distributions are slightly skewed right. And both steel and wooden coasters have at least one high outlier that is much taller than the other coasters.
    b) The upper fence for steel coasters is 295 ft and all (eight) outliers are above that value. For wooden coasters the outlier is above 188.75 ft.
    c) The mean is greater than the median for both distributions because both are slightly skewed right and have high outliers. The mean is nonresistant and is pulled higher by the skewness and outliers.

**31.** a) Boys.
b) Boys.
c) Girls.
d) The boys appeared to have more skew, as their scores were less symmetric between quartiles. The girls' quartiles are the same distance from the median, although the left tail stretches a bit farther to the left.
e) Girls. Their median and upper quartiles are larger. The lower quartile is slightly lower, but close.
f) $[14(4.2) + 11(4.6)]/25 = 4.38$.

**33.**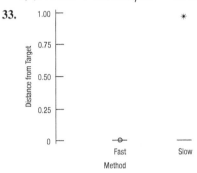

There appears to be an outlier! This point should be investigated. We'll proceed by redoing the plots with the outlier omitted:

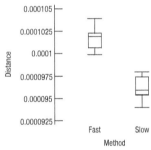

It appears that slow speed provides much greater accuracy. But the outlier should be investigated. It is possible that slow speed can induce an infrequent very large distance.

**35.** a)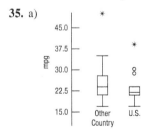

b) Mileage for U.S. models is typically lower, and less variable than for cars made elsewhere. The median for U.S. models is around 22 mpg, compared to 24 for the others. Both groups have some high outliers—most likely hybrid vehicles. (Other answers possible.)

**37.** a) Males have a longer median travel time and a much larger spread. Both male and female travel times are skewed to the right.
b) All five distributions are skewed to the right (which makes sense, as we expect long commutes to be less common). However, the spreads in the populations are probably about the same. Sometimes the IQRs are about the same, sometimes the male IQR is larger, and sometimes the female IQR is larger. The medians are probably about the same in the population as they are roughly the same in the five samples.

**39.** a) Day 16 (but any estimate near 20 is okay).
b) Day 65 (but anything around 60 is okay).
c) Around day 50.

**41.** a) Most of the data are found in the far left of this histogram. The distribution is very skewed to the right.
b) Re-expressing the data by, for example, logs or square roots might help make the distribution more nearly symmetric.

**43.** a) The logarithm makes the histogram more symmetric. It is easy to see that the center is around 3.5 in log assets.
b) Around 2500 million dollars.
c) Around 1000 million dollars.

**45.** a) Fusion time and group.
b) Fusion time is quantitative (units = seconds). Group is categorical.
c) Both distributions are skewed to the right with high outliers. The boxplot indicates that visual information may reduce fusion time. The median for the VV group seems to be about the same as the lower quartile of the NV group.

## Chapter 5 (See note, page A-29)

**1.** 1.

**3.** McGwire's performance was more impressive from a comparison of $z$-scores (3.87 versus 3.16). But you'll have to decide if the steroid use makes it less impressive anyway.

**5.** Lowest score = 910. Mean = 1230. SD = 120. Q3 = 1350. Median = 1270. IQR = 240.

**7.** Your score was 2.2 standard deviations higher than the mean score in the class.

**9.** 65.

**11.** In January, a high of 55 is not quite 2 standard deviations above the mean, whereas in July a high of 55 is more than 2 standard deviations lower than the mean. So it's less likely to happen in July.

**13.** The $z$-scores, which account for the difference in the distributions of the two tests, are 1.5 and 0 for Derrick and 0.5 and 2 for Julie. Derrick's total is 1.5, which is less than Julie's 2.5.

**15.** a) Megan did (her average was 86); Anna did not (her average was 83).
b) Anna had a higher combined z-score, 1.0 compared to 0.6.

**17.** a) About 1.33 standard deviations below the mean.
b) $6(z = -1.33)$ is more unusual than $9(z = 0.667)$.

**19.** a) Mean = 0 pounds; SD is unchanged at 1.5 pounds.
b) Mean = 0.50(8) = \$4; SD = 0.50(1.5) = \$0.75.

**21.** Min = 16(5) + 30 = 110 oz; median = 16(7.5) + 30 = 150 oz; SD = 16(1.5) = 24 oz; IQR = 16(2) = 32 oz.

**23.** College professors can have between 0 and maybe 40 (or possibly 50) years' experience. A standard deviation of 1/2 year is impossible, because many professors would be 10 or 20 SDs away from the mean, whatever it is. An SD of 16 years would mean that 2 SDs on either side of the mean is plus or minus 32, for a range of 64 years. That's too high. So, the SD must be 6 years.

**25.** a)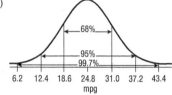

b) 18.6 to 31.0 mpg.  c) 16%.
d) 13.5%.  e) Less than 12.4 mpg.

**27.** Any weight more than 2 standard deviations below the mean, or less than $8 - 2(1.5) = 5$ pounds, is unusually low. We expect to see a quokka below $8 - 3(1.5) = 3.5$ pounds only rarely.

**29.** a)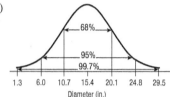

b) Between 6.0 and 24.8 inches.  c) 2.5%.
d) 34%.  e) 16%.

**31.** Because the histogram is not unimodal and symmetric, it is not wise to have faith in numbers from the Normal model.

**33.** a) The histogram is bimodal; and there are too many observations within one SD of the mean (many more than the 68% we'd expect in a Normal model).
b) The distribution for all schools is not approximately Normally distributed. The distribution with Milwaukee schools removed has a slight skewness, but is reasonably symmetric so a Normal model may be reasonable for those schools. With the lower Milwaukee schools removed, the mean would increase and the standard deviation would decrease.

**35.** a) 3.94% ($z$-score is 1.76).  b) $5/53 = 9.4\%$.
c) Because the Normal model doesn't fit.

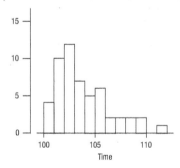

d) The distribution is skewed to the high end.

**37.** a) 2.5%.
b) 2.5% of the 508 receivers, or about 13 of them, should gain more than $253.76 + 2 \times 312.36 = 878.5$ yards. (In fact, 34 receivers exceeded this value.)
c) The distribution is strongly skewed to the right, not symmetric.

**39.** a) 6.3%.  b) 76.8%.  c) 16.9%.
**41.** a) 9.92 lb.  b) 6.74 lb.  c) 7.21 lb to 8.79 lb.
**43.** a) 7.62 lb.  b) 11.49 lb.  c) 2.02 lb.

**45.** a)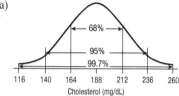

b) 30.85%.  c) 17.00%.
d) 32.38 points.  e) 212.9 points.

**47.** a) 11.1%.  b) (35.9, 40.5) inches.
c) 40.5 inches.

**49.** a) 5.3 grams.  b) 6.4 grams.
c) Younger because SD is smaller.

**51.** The distribution for all schools is not approximately Normally distributed. The distribution with Milwaukee schools removed has a slight skewness, but is reasonably straight so a Normal model may be reasonable for those schools.

**53.** a) Yes. The Normal probability plot looks reasonably straight, though there is some variability around that. It may be reasonable to attribute this variability to random variation. The dotplot looks reasonably symmetric with no outliers.
b) Because a Normal model was deemed appropriate, we can create such an interval. It is about (17.76, 81.84).

**55.** a) 2.5%.
b) No, the distribution is skewed.
c) No, the distribution is skewed.
d) The rule would work fairly well even though there is a slight asymmetry to the distribution.
e) Yes. Now the distribution of the means is nearly Normal. The 68–95–99.7 Rule works well.

## Part I Review (See note, page A-29)

**1.** a)

[histogram: # of Bananas vs Price (cents), bars at 40.0, 42.5, 45.0, 47.5, 50.0, 52.5, 55.0]

b) Median = 49 cents; IQR = 7 cents.
c) The distribution is unimodal and left skewed. The center is near 50 cents; values extend from 42 cents to 53 cents.

**3.** a) The newer cars had a higher median asking price and a larger range and IQR. Both distributions are skewed to the right and both have a number of high outlier asking prices.
b) Original price, miles driven, condition, etc. . . . (Answers will vary).
c) All summary statistics will be multiplied by this factor. This will increase the measures of center and the boxplots will become more spread out. The max and min will become more extreme.

**5.** a) It means their heights are also more variable.
b) The $z$-score for women to qualify is 2.40, greater than the 1.75 for men, so it is harder for women to qualify.

**7.** a) *Who*—850 people who live near State University; *What*—Age, attended college? Favorable opinion of State U?; *When*—Not stated; *Where*—Region around State U; *Why*—To report to the university's directors; *How*—Sampled and phoned 850 local residents.
  b) Age—Quantitative (years); attended college?—categorical; favorable opinion?—categorical.
  c) The fact that the respondents know they are being interviewed by the university's staff may influence answers.

**9.** a) These are categorical data, so mean and standard deviation are meaningless.
  b) Not appropriate. Even if it fits well, the Normal model is meaningless for categorical data.

**11.** a)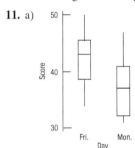

  b) The scores on Friday were higher by about 5 points on average. This is a drop of more than 10% off the average score and shows that students fared worse on Monday after preparing for the test on Friday. The spreads are about the same, but the scores on Monday are a bit skewed to the right.
  c)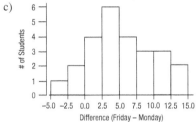

  d) The changes (Friday − Monday) are unimodal and centered near 4 points, with a spread of about 5 (SD). They are fairly symmetric, but slightly skewed to the right. Only 3 students did better on Monday (had a negative difference).

**13.** a) Quantitative.
  b) Categorical.
  c) You: boxplot, histogram, dotplot; Friend: pie or bar graph.
  d) You: yes; mean, median, IQR, SD, etc . . . ; Friend: no; just (relative) frequencies of the groups.

**15.** a) Annual mortality rate for males (quantitative) in deaths per 100,000 and water hardness (quantitative) in parts per million.
  b) Calcium is skewed right, possibly bimodal. There looks to be a mode down near 12 ppm that is the center of a fairly tight symmetric distribution of values between 0 ppm and 25 ppm and another mode near 62.5 ppm that is the center of a much more spread out, symmetric (almost uniform) distribution of values between 25 ppm and 150 ppm. Mortality, however, appears unimodal and symmetric with the mode near 1500 deaths per 100,000, and with a range of 1000 deaths per 100,000.

**17.** a) They are on different scales.
  b) January's values are lower and more spread out.
  c) Roughly symmetric but slightly skewed to the left. There are more low outliers than high ones. Center is around 40 degrees with an IQR of around 7.5 degrees.

**19.** a) Bimodal with modes near 2 and 4.5 minutes; fairly symmetric around each mode.
  b) Because there are two modes, which probably correspond to two different groups of eruptions, an average might not make sense.
  c) The intervals between eruptions are longer for long eruptions. There is very little overlap. More than 75% of the short eruptions had intervals less than about an hour (62.5 minutes), while more than 75% of the long eruptions had intervals longer than about 75 minutes. Perhaps the interval could even be used to predict whether the next eruption will be long or short.

**21.** a)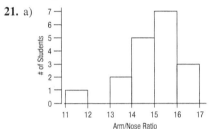

  The distribution is left skewed with a center of about 15. It has an outlier between 11 and 12. Except for the outlier, all of the arm-to-nose ratios are between 13 and 17.
  b) Even though the distribution is somewhat skewed, the mean and median are close. The mean is 15.0 and the SD is 1.25.
  c) Yes. 11.8 is already an outlier. 9.3 is more than 4.5 SDs below the mean. It is a very low outlier.

**23.** a)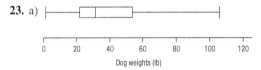

  b) Dog weights have a median of 29 lb, an IQR of 37 lb, a range of 98.5 lb, and are skewed to the right.
  c) None of the summary statistics will change, because both 80 and 60 are between Q3 and the maximum.

**25.** a)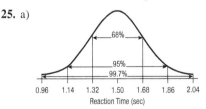

  b) According to the model, reaction times are symmetric with center at 1.5 seconds. About 95% of all reaction times are between 1.14 and 1.86 seconds.
  c) 8.2%.
  d) 24.1%.
  e) Quartiles are 1.38 and 1.62 seconds, so the IQR is 0.24 second.
  f) The slowest 1/3 of all drivers have reaction times of 1.58 seconds or more.

**27.** a)

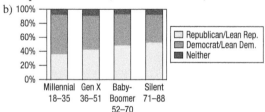

b) Mean = 100.25; SD = 25.54 pieces of mail.
c) The distribution is somewhat symmetric and unimodal, but the center is rather flat, almost uniform. The range is a little more than 100 pieces of mail.
d) 64%. The Normal model seems to work reasonably well, because it predicts 68%.

**29.** a) *Who*—100 health food store customers; *What*—Have you taken a cold remedy?, and Effectiveness (scale 1 to 10); *When*—Not stated; *Where*—Not stated; *Why*—Promotion of herbal medicine; *How*—In-person interviews.
b) Have you taken a cold remedy?—categorical; Effectiveness—quantitative or ordinal.
c) No. Customers are not necessarily representative, and the Council had an interest in promoting the herbal remedy.

**31.** a) 38 cars.
b) Possibly because the distribution is skewed to the right.
c) Center—median is 148.5 cubic inches. Spread—IQR is 126 cubic inches.
d) No. It's bigger than average, but smaller than more than 25% of cars. The upper quartile is at 231 cubic inches.
e) No. 1.5 IQR is 189, and $105 - 189$ is negative, so there can't be any low outliers. $231 + 189 = 420$. There aren't any cars with engines bigger than this, because the maximum has to be at most 105 (the lower quartile) + 275 (the range) = 380.
f) Because the distribution is skewed to the right, this is probably not a good approximation.
g) Mean, median, range, quartiles, IQR, and SD all get multiplied by 16.4.

**33.** a) 45.4%.
b) If this were a random sample of all voters, yes.
c) 34.9%.
d) 1.4%.
e) 20%.
f) 7%.

**35.** a) Republican—3603; Democrat—3765; Neither—562. Or, Republican—45.4%; Democrat—47.5%; Neither—7.1%.
b) 

c) It appears that the older the voter, the less likely they are to lean Democratic and the more likely to lean Republican.
d) No. There seems to be an association between age and affiliation. Younger voters tend to be more Democratic and less Republican.

**37.** a) 0.43 hour.
b) 1.4 hours.
c) 0.89 hour (or 53.4 minutes).
d) Survey results vary, and the mean and the SD may have changed.

**39.** Answers will vary.

## Practice Exam Answers

**I. 1.** D.   **3.** D.   **5.** A.   **7.** D.   **9.** E.

**II.**

**1.** a) The NBA players have a higher median average points per game than the WNBA players ($\approx 11$ vs. $\approx 6$). The NBA average points per game is slightly more variable than the WNBA; both the IQR and the range are slightly larger. Both leagues are skewed right with a few outliers for the best (highest scoring) players in both leagues.
b) $10 + 1.5(10 - 3.3) = 20.05$. Della Donne is not an outlier as her average is slightly below the upper fence; she is (probably) the end of the whisker. Griner's average of 20.7 is above the upper fence, making her average an outlier.
c) $(157 \times 6.97 + 259 \times 11.998)/(157 + 259) = 10.1$ pts/game on average.
d) Because the NBA distribution is skewed to the right and has several large outliers, the mean is larger than the median. Having a small number of players who score an unusually large amount contributes to overall sum but only increases the denominator of the mean by one, which has a large affect on the mean.

## Chapter 6 (See note, page A-29)

**1.** a) Weight in ounces: explanatory; Weight in grams: response. (Could be other way around.) To predict the weight in grams based on ounces. Scatterplot: positive, straight, strong (perfectly linear relationship).
b) Ice cream cone sales: explanatory. Air conditioner sales: response—although the other direction would work as well. To predict one from the other. Scatterplot: positive, straight, moderate.
c) Shoe size: explanatory; GPA: response. To try to predict GPA from shoe size. Scatterplot: no direction, no form, very weak.
d) Miles driven: explanatory; Gallons remaining: response. To predict the gallons remaining in the tank based on the miles driven since filling up. Scatterplot: negative, straight, moderate.

**3.** a) None.   b) 3 and 4.   c) 2, 3, and 4.
d) 1 and 2.   e) 3 and possibly 1.

5. a)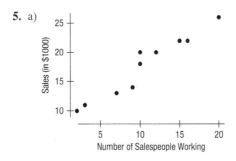

b) It is positive. Days with more salespeople working tend to have greater sales.
c) It has a linear shape.
d) The relationship between number of salespeople working and sales is strong.
e) There are no outliers.

7. There seems to be a very weak—or possibly no—accociation between brain size and performance IQ.

9. a) True.
b) False. It will not change the correlation.
c) False. Correlation has no units.

11. Correlation does not demonstrate causation. The analyst's argument is that sales staff cause sales. However, the data may reflect the store hiring more people as sales increase, so any causation may run the other way.

13. a)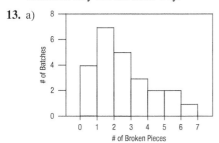

b) Unimodal, skewed to the right. The skew.
c) The positive, somewhat linear relation between batch number and broken pieces.

15. a) 0.006.   b) 0.777.   c) −0.923.   d) −0.487.

17. There may be an association, but not a correlation unless the variables are quantitative. There could be a correlation between average number of hours of TV watched per week per person and number of crimes committed per year. Even if there is a relationship, it doesn't mean one causes the other.

19. a) $r = 0.965$.
b) Because the relationship looks linear, it's appropriate to calculate the correlation. The value for the correlation is positive, indicating the positive trend. The value for the correlation is around 0.965, which confirms the strong association between book sales and number of salespeople working.

21. The scatterplot is not linear; correlation is not appropriate.

23. The correlation may be near 0. We expect nighttime temperatures to be low in January, increase through spring and into the summer months, then decrease again in the fall and winter. The relationship is not linear.

25. The correlation coefficient won't change, because it's based on z-scores. The z-scores of the prediction errors are the same whether they are expressed in nautical miles or miles.

27. a) Assuming the relation is linear, a correlation of −0.772 shows a moderately strong relation in a negative direction.
b) Continent is a categorical variable. Correlation does not apply.

29. a) Actually, yes, taller children will tend to have higher reading scores, but this doesn't imply causation.
b) Older children are generally both taller and are better readers. Age is the lurking variable.

31. a) No. We don't know this from the correlation alone. There may be a nonlinear relationship or outliers.
b) No. We can't tell from the correlation what the form of the relationship is.
c) No. We don't know from the correlation coefficient.
d) Yes, the correlation doesn't depend on the units used to measure the variables.

33. The researchers reported an association between eating away from home and the death rate, but the author assumed eating out caused the higher death rate.

35. This is categorical data even though it is represented by numbers. The correlation is meaningless.

37. a) The association is positive, moderately strong, and roughly straight, with several states whose HCI seems high for their median income and one state whose HCI appears low given its median income.
b) The correlation would still be 0.65.
c) The correlation wouldn't change.
d) DC would be a moderate outlier whose HCI is high for its median income. It would lower the correlation slightly.
e) No. We can only say that higher median incomes are associated with higher housing costs, but we don't know why. There may be other economic variables at work.

39. a) There is a moderately strong, negative relationship between gas mileage and engine size. Cars with larger engines tend to have lower gas mileage. The shape seems almost linear, but the points for the largest two engine sizes seem to indicate a bit of a curve.
b) Since the relationship is not quite linear, the correlation isn't appropriate, but it would indicate a negative, moderately strong relationship.

41.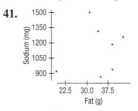

(Plot could have explanatory and predictor variables swapped.) Correlation is 0.199. There does not appear to be a relation between sodium and fat content in burgers, especially without the low-fat, low-sodium item. The correlation of 0.199 shows a weak relationship, even with the outlier excluded.

43. a) Yes, the form does seem to be somewhat linear.
b) Positive. Teams with more runs scored tend to have higher attendance.

c) There is a positive association, but even if it were stronger it does not *prove* that more fans would come if the number of runs increased. Association does not indicate causality.

45. a) The relationship between position number and distance is nonlinear, with a positive direction. There is very little scatter from the trend.
    b) The relationship is not linear.

## Chapter 7 (See note, page A-29)

1. 281 milligrams.

3. The potassium content is actually lower than the model predicts for a cereal with that much fiber. A residual of $-22$ means that particular cereal has 22 mg less potassium than the model predicts.

5. The model predicts that cereals will have approximately 27 more milligrams of potassium, on average, for every additional gram of fiber.

7. 81.5%.

9. The true potassium contents of cereals vary from the predicted amounts with a standard deviation of 30.77 milligrams.

11. a) Model is appropriate.
    b) Model is not appropriate. Relationship is nonlinear.
    c) Model may not be appropriate. Spread is changing.

13. a) People spend about the same average amount on food regardless of how many people are attending the game.
    b) 5.00. The slope is the average number of dollars each person spends on food. $5 is the most reasonable amount.
    c) Possibilities include age of the attendees, time of day, duration of the game, and temperature.

15. a) False. The line usually touches none of the points. We minimize the sum of the squared errors.
    b) True.
    c) False. It is the sum of the squares of all the residuals that is minimized.

17. a) $-4.358 \frac{MPG}{liter}$.
    b) $\widehat{Fuel\ economy} = 37.492 - 4.358 \times Engine\ size$.

19. a) $\widehat{Sales} = 8.1006 + 0.9134\ (Salespeople)$.
    b) A day with one more salesperson than another tends to have $913.40 more in sales.
    c) The model predicts that a day with 0 salespeople will have $8101 in sales. Since they would not sell books without salespeople, this value is not meaningful.
    d) $15,407.80.
    e) $R^2 = 93.2\%$. 93.2% of the variation in *Sales* is explained by the regression model using *Number of Salespeople* as a predictor.

21. a) Thousands of dollars.
    b) The day with ten salespeople working and $20,000 in sales
    c) The day with two salespeople working.

23. Samples will vary, so it's not surprising that the second sample gives a slope that's a little different from the first one. It would be surprising if the slopes were very different, but 0.9134 and 0.8815 are pretty similar.

25. a) Linearity Assumption.
    b) Outlier Condition.
    c) Equal Spread Condition.

27. a) *Price* (in thousands of dollars) is *y* and *Size* (in square feet) is *x*.
    b) Slope is thousands of $ per square foot.
    c) Positive. Larger homes should cost more.

29. A linear model on *Size* accounts for 71.4% of the variation in home *Price* for these data.

31. a) 0.845; + because larger homes cost more.
    b) Price should be 0.845 SDs above the mean in price.
    c) Price should be 1.690 SDs below the mean in price.

33. a) Predicted *Price* increases by about $0.061 \times 1000$, or $61.00 per additional sq ft.
    b) 230.82 thousand, or $230,820.
    c) $115,020; $-$6000 is the residual.

35. a) $R^2$ does not tell whether the model is appropriate, but measures the strength of the linear relationship. High $R^2$ could also be due to an outlier.
    b) Predictions based on a regression line are for average values of *y* for a given *x*. The actual wingspan will vary around the prediction.

37. a) Probably not. Your score is better than about 97.5% of people, assuming scores follow the Normal model. Your next score is likely to be closer to the mean.
    b) The friend should probably should retake the test. His score is better than only about 16% of people (again, assuming scores follow a Normal model). His score is likely to be closer to the mean.

39. a) Yes. The residuals are a bit larger for cigarettes with more tar, but it is not extreme and there is no curvature.
    b) The linear model on *Tar* content accounts for 81.4% of the variability in *Nicotine*.

41. a) $r = 0.902$.
    b) Nicotine should be 1.804 SDs below average.
    c) Tar should be 0.902 SDs above average.

43. a) $\widehat{Nicotine} = 0.148305 + 0.062163\ Tar$.
    b) 0.708 mg.
    c) A cigarette with one more milligram of tar than another is predicted to contain 0.062 mg more nicotine.
    d) We'd predict that a cigarette with no tar has about 0.148 mg of nicotine.
    e) 0.533 mg.

45. a) Yes. The relationship is straight enough, with a few outliers. The spread increases a bit for states with large median incomes, but we can still fit a regression line.
    b) From summary statistics: $\widehat{HCI} = -156.50 + 0.0107\ MFI$; From original data: $\widehat{HCI} = -157.64 + 0.0107\ MFI$
    c) From summary Statistics: Predicted $HCI = 324.93$; from original data: 324.87.
    d) 223.09.
    e) $\widehat{z_{HCI}} = 0.65\ z_{MFI}$.
    f) $\widehat{z_{MFI}} = 0.65\ z_{HCI}$.

47. a) $\widehat{Total} = 539.803 + 1.103\ Age$.
    b) Yes. Both variables are quantitative; the plot is straight (although nearly flat); there are no apparent outliers; the plot does not appear to change spread throughout the range of *Age*.
    c) $559.65; $594.94.

d) 0.14%.
e) No. The plot is nearly flat. The model explains almost none of the variation in *Total Yearly Purchases*.

**49.** a) Moderately strong, fairly straight, and positive; possibly some outliers (higher-than-expected math scores).
b) The student with 500 verbal and 800 math.
c) $r = 0.68$, indicating a positive, fairly strong linear relationship. between math scores and verbal scores.
d) $\widehat{Math} = 209.6 + 0.675 \times Verbal$.
e) Every point of verbal score adds 0.675 points to the predicted average math score.
f) 547.1 points.
g) 50.4 points.

**51.** a) 0.685.
b) $\widehat{Verbal} = 171.3 + 0.694 \times Math$.
c) The observed verbal score is higher than predicted from the math score.
d) 518.5 points.
e) 559.6 points.
f) This regression equation cannot make predictions in the other direction.

**53.** a) The relationship is straight enough, but very weak. In fact, there may be no relationship at all between these variables.
b) The predicted number of wildfires has been decreasing by about 222 per year.
c) Yes, the intercept estimates the number of wildfires in 1985 as about 78,792.
d) The residuals are distributed around zero with a standard deviation of 12,397 fires. Compared to the observed values, most of which are between 60,000 and 90,000 fires, this residual standard deviation in our model's predictions is quite large so the model isn't very effective for prediction.
e) Only 2.7% of the variation in the number of wildfires can be accounted for by the linear model on *Years since 1985*. This confirms the impression from the scatterplot that there is very little association between these variables—that is, that any change in the number of wildfires during this period is not linear over time. The average number of fires might provide as good a prediction as the model.

**55.** a)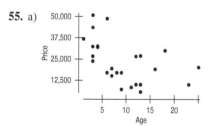

b) There is a moderately strong negative linear association between *Price* and *Age*.
c) Yes.
d) −0.49.
e) *Age* accounts for 24.1% of the variation in *Advertised Price*.
f) Other factors contribute—options, condition, mileage, etc.

**57.** a) $\widehat{Price} = 32{,}791.1 - 984.5\ Years$.
b) Predicted prices decline at the rate of about $984.50 per year.
c) The model predicts that a typical new convertible costs $32,791.10.

d) $22,946.
e) Negative residual. Its price is below the predicted value for its age.
f) The model predicts a much lower price. The Mclaren is an extraordinary car and would be an outlier in this dataset.
g) $1554.
h) No. This is extrapolating beyond the oldest car in the dataset.

**59.** a)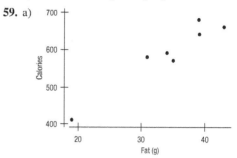

b) 92.3% of the variation in calories can be accounted for by the fat content.
c) $\widehat{Calories} = 211.0 + 11.06 \times Fat$.
d)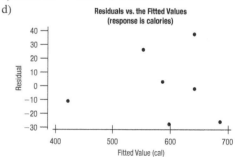

Residuals show no clear pattern, so the model seems appropriate.
e) You could say a fat-free burger still has 211.0 calories, but this is extrapolation (no data close to 0).
f) Every gram of fat adds 11.06 calories, on average.
g) 553.5 calories.

**61.** a) The regression line was for predicting calories from fat, not the other way around.
b) $\widehat{Fat} = -15.0 + 0.083 \times Calories$. Predict 35.1 grams of fat.

**63.** a) 0.958.
b) $CO_2$ levels account for 91.8% of the variation in mean temperature.
c) $\widehat{Mean\ temp\ anomaly} = -3.4527 + 0.0106\ CO_2$.
d) The predicted mean temperature has been increasing at an average rate of 0.011 degrees (C)/ppm of $CO_2$.
e) It makes no sense to interpret the intercept as being about the atmosphere with no $CO_2$, so we treat this just as a starting value for prediction.
f) No.
g) 1.34 degrees C.
h) No. This is an extrapolation. The model can't say what will happen under other circumstances. At best it can give a general idea of what to expect.

**65.** a) $\widehat{\%\ Body\ Fat} = -27.4 + 0.25 \times Weight$.
b) Residuals look randomly scattered around 0, so conditions are satisfied.
c) *% Body Fat* increases, on average, by 0.25 percent per pound of *Weight*.

d) Reliable is relative. $R^2$ is 48.5%, but residuals have a standard deviation of 7%, so variation around the line is large.
e) 0.9 percent.

67. a) $\widehat{LongJump} = 7.595 - 0.01(800mTime)$. The model suggests that long jump distances average about 1 cm less for every additional second it took competitors to run 800 m.
b) 9.4%.
c) Yes, the slope is negative. Faster runners tend to jump higher as well.
d) There is an extraordinary point (Akela Jones) that is influential. This model is not appropriate. Except for 2 points, there seems to be little evidence of a relationship.
e) No. The residual standard deviation is 0.23 meters, almost exactly the same as the SD of all long jumps. The influential point dominates the model making it unsafe to use.

69. The sum of the squared vertical distances to any other line would be greater than 1790.

## Chapter 8 (See note, page A-29)

1. a) The trend appears to be somewhat linear but fairly flat up to about 1940; from 1940 to about 1960 the trend appears to be nonlinear with a large drop from 1940 to 1950. From 1960 or so to the present, the trend appears to be increasing and somewhat linear.
b) Relatively strong for certain periods.
c) No, as a whole the graph is clearly nonlinear. Within certain periods (e.g., 1975 to the present) the correlation is high.
d) Overall, no. You could fit a linear model to the period from 1975 to 2003, but why? You don't need to interpolate, since every year is reported, and extrapolation seems dangerous.

3. a) The relationship is not straight.
b) It will be curved downward.

5. a) No. We need to see the scatterplot first to see if the conditions are satisfied, and models are always wrong.
b) No, the linear model might not fit the data everywhere.

7. a) Millions of dollars per minute of run time.
b) Costs for movies increase at approximately the same rate per minute.
c) On average dramas cost about $20 million less for the same run time.

9. This observation was influential. After it was removed, the $R^2$ value and the slope of the regression line both changed by a large amount.

11. No; we cannot infer causation. In warm weather, more children go outside and play.

13. Individual student scores will vary greatly. The class averages will have much less variability and may disguise important trends.

15. a) There are several features to comment on in this plot. There is a strong monthly pattern around the general trend. From 1997 to 2008, passengers increased fairly steadily with a notable exception of Sept. 2001, probably due to the attack on the Twin Towers. Then sometime in late 2008, departures dropped dramatically, possibly due to the economic crisis. Recently, they have been recovering, but not at the same rate as their previous increase.

b) The trend was fairly linear until late 2008; then passengers dropped suddenly.
c) The trend since 2009 has been linear (overall, ignoring monthly oscillations). If the increase continues to be linear, the predictions should be reasonable for the short term.
d) With the travel restrictions due to COVID-19, we should see a big drop in 2020–2021. We can't know what's going to happen after that.

17. a) 1) High leverage, small residual.
2) No. not influential for the slope.
3) Correlation would decrease because outlier has large $z_x$ and $z_y$, increasing correlation.
4) Slope wouldn't change much because the outlier is in line with other points.
b) 1) High leverage, probably small residual.
2) Yes, influential.
3) Correlation would weaken, increasing toward zero.
4) Slope would increase toward 0 because the outlier makes it negative.
c) 1) Some leverage, large residual.
2) Yes. somewhat influential.
3) Correlation would increase because scatter would decrease.
4) Slope would increase slightly.
d) 1) Little leverage, large residual.
2) No. not influential.
3) Correlation would become stronger and become more negative because scatter would decrease.
4) Slope would change very little.

19. 1) e.  2) d.  3) c.  4) b.  5) a.

21. Perhaps high blood pressure causes high body fat, high body fat causes high blood pressure, or both could be caused by a lurking variable such as a genetic or lifestyle issue.

23. a) The graph shows that, on average, students progress at about one reading level per year. This graph shows averages for each grade. The linear trend has been enhanced by using averages.
b) Very close to 1.
c) The individual data points would show much more scatter, and the correlation would be lower.
d) A slope of 1 would indicate that for each 1-year grade level increase, the average reading level is increasing by 1 year.

25. a) *Cost* decreases by $2.13 per degree of average daily *Temp*. So warmer temperatures indicate lower costs.
b) For an average monthly temperature of 0°F, the cost is predicted to be $133.
c) Too high; the residuals (observed − predicted) around 32°F are negative, showing that the model overestimates the costs.
d) $111.70.
e) About $105.70.
f) No, the residuals show a definite curved pattern. The data are probably not linear.
g) No, there would be no difference. The relationship does not depend on the units.

27. a) 0.881.
b) Treasury bill rates during this period grew at about 0.25% per year, starting from an interest rate of about 0.61%.
c) Substituting 80 in the model yields a predicted rate of about 20.4%.
d) Not really. Extrapolating 50 years beyond the end of these data would be dangerous and likely to be inaccurate.

**29.** a) Interest rates peaked around 1980 and decreased afterward. This regression model has a negative slope and a high intercept.
b) This model predicts −3.43%! Much lower than the prediction with the other model.
c) Even though we separated the data, there is no way of knowing if this trend will continue. And the rate cannot become negative, so we have clearly extrapolated far beyond what the data can support.
d) It is clear from the scatterplot that we can't count on TBill rates to change in a linear way over many years, so it would not be wise to use any regression model to predict rates.

**31.** a) Stronger. Both slope and correlation would increase.
b) Restricting the study to nonhuman animals would justify it.
c) Moderately strong.
d) For every year increase in life expectancy, the gestation period increases by about 15.5 days, on average.
e) About 270.5 days.

**33.** a) Removing hippos would make the association stronger because hippos are more of a departure from the pattern.
b) Increase.
c) No, there must be a good reason for removing data points.
d) Yes, removing it lowered the slope from 15.5 to 11.6 days per year.

**35.** a) The prediction is 26.13 years old.
b) Not much, because the data are not truly linear and 2025 is 10 years from the last data point (extrapolating is risky).
c) No, that extrapolation of more than 50 years would be absurd. There's no reason to believe the trend from 1955 to 2015 will continue.
d) 28.97 years old.
e) Not very much. We would have to assume that the current trend would continue until 2025. That's a strong assumption, and probably not true.
f) No. That would simply be an unreasonable extrapolation.

**37.** a) The residual plot is reasonably scattered with no evidence of nonlinearity, so we can fit the regression model. But there seems to be high leverage point, which could be affecting the regression model.
b) Niger is an outlier with a higher life expectancy than typical for its large family size.
c) 74.2% of the variation in life expectancy is accounted for by the regression on birthrate.
d) Although there is an association, there is no reason to expect causality. Lurking variables are likely to affect both *Fertility* and *Life Expectancy*.

**39.** a)

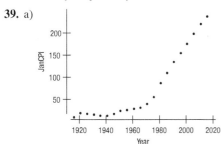

The scatterplot is clearly nonlinear; however, the last few years—say, from 1970 on—do appear to be linear.
b) Using the data from 1970 to 2020 gives $r = 0.998$ and $\widehat{CPI} = -8761.06 + 4.4653\, Year$

Predicted CPI in 2030 = 303.45 (This is an extrapolation of doubtful accuracy. There are techniques for making such predictions, but they are beyond the scope of this course.)

**41.** a) The slope means that for each year one bridge is newer than another, its condition is predicted to be 0.0196 points higher. The *y*-intercept means that a bridge built in year 0 would be predicted to have a rating of −33.4. This is, of course, meaningless.
b) About 40% of the variability in the condition ratings of bridges in Tompkins County can be explained by the regression line with *year built* as a predictor.
c) The predictions of condition rating made by our model are typically off by about 0.69 points.
d) No, that would require us to extrapolate outside the range of our data.
e) These points would pull up on the left end of the regression line, making it less steep. The slope would decrease.
f) Because these points do not fit the pattern of the rest of the data, these points make the relationship weaker. $R^2$ would likely decrease.
g) Then that point would be right in the middle of the rest of the points, so it would not be remarkable at all.

## Chapter 9 (See note, page A-29)

**1.** a) No re-expression needed.
b) Re-express to straighten the relationship.
c) Re-express to equalize spread.

**3.** a) There's an annual pattern in when people fly, so the residuals cycle up and down. There was also a sudden decrease in passenger traffic after 2008.
b) No, this kind of pattern can't be helped by re-expression.

**5.** a) 16.44.  b) 7.84.  c) 0.36.  d) 1.75.  e) 27.59.

**7.** a) Fairly linear, negative, moderately strong.
b) The model predicts that gas mileage decreases an average of 7.652 mpg for each thousand pounds of weight.
c) No. Residuals show a curved pattern.

**9.** a) Residuals are more randomly spread around 0, with some low outliers.
b) $\widehat{Fuel\ Consumption} = 0.625 + 1.178 \times Weight$.
c) For each additional 1000 pounds of *Weight*, the model estimates an additional 1.178 gallons will be needed to drive 100 miles.
d) 23.89 mpg.

**11.** a) Although nearly 97% of the variation in GDP can be accounted for by this model, we should examine a scatterplot of the residuals to see if it's appropriate.
b) No. The residuals show clear curvature.

**13.** No. The residual pattern shows that this model is still not appropriate. Because the residuals plot now bends the other way, try a square root.

**15.** a) There is a curve in the data, more easily seen in the residual plot.
b) The data are now more linear.
c) 193 ft.
d) 32 ft.
e) 10 mph is outside the scope of our data, so this extrapolation is risky.

**17.** Improve homoscedasticity (more nearly equal variances between groups).

**19.** a) There is a strong, positive, curved relationship between length of year and distance from the sun. Bodies that are farther from the sun tend to have longer years.
b) log(*Length of Year*) vs. log(*Distance*).
c) The scatterplot appears linear and there is not a clear pattern in the residual plot, so a linear model is appropriate.

**21.** This *adjusted position* results in a residual plot with more of a random scatter, so it is preferable.

**23.** Some models have a high $R^2$, but the Exercise 19 model has $R^2 = 1$. That indicates that perhaps we have found a physical law. If we find another system with the same pattern, it would add evidence for the Titius-Bode "law"; otherwise, we would discount the law.

**25.** This relationship is not strictly increasing or strictly decreasing and cannot be made straight by the methods of this chapter.

**27.** One possible model of transformed data is

$$\widehat{\frac{1}{Weight}} = 1.603 + 0.00112(Oranges).$$ However, the oscillating pattern in the residuals warns that these data may not be suitable for re-expression.

**29.**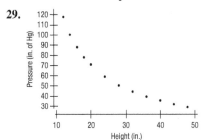

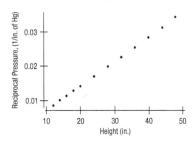

$1/Pressure$ is an exact relationship. $R^2$ is 100%.
$\widehat{Reciprocal\ Pressure} = -0.000077 + 0.000713\,Height$

| Predictor | Coeff | SE Coeff | T | P |
|---|---|---|---|---|
| Intercept | −0.00007670 | 0.00007813 | −0.98 | 0.349 |
| Height | 0.00071307 | 0.00000260 | 274.30 | 0.000 |

s = 0.0001057   R − Sq = 100.0%   R − Sq(adj) = 100.0%
or $\widehat{\log(Pressure)} = 3.15 - 1.001\log(Height)$.

## Part II Review (See note, page A-29)

**1.** % over 50, 0.69.
% under 20, −0.71.
% Full-time faculty, 0.09.
% Graduating on time, −0.51.

**3.** a) There does not appear to be a linear relationship.

b) Nothing, there is no reason to believe that the results for the Finger Lakes region are representative of the vineyards of the world.
c) $\widehat{CasePrice} = 92.77 + 0.567\,Years$.
d) Only 2.7% of the variation in case price is accounted for by the ages of vineyards. Most of that is due to two outliers. We are better off using the mean price rather than this model.

**5.** a) $\widehat{TwinBirthsRate} = 17.77 + 0.551\,Years\,Since\,1980$.
b) Twin births have increased, on average, by 0.55 births per year.
c) 38.7 births predicted; 32.6 observed in fact; the prediction is too great by about 6 births.
d) $\widehat{TwinBirths} = 18.21 + 0.508\,Years\,Since\,1980$. The fit is very good. $R^2$ is 97%. However, the residual plot reveals a fluctuating pattern, so there may be more to understand about twin births than is available from these data.

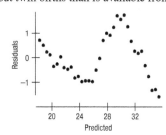

**7.** a) −0.520.
b) Negative, not strong, somewhat linear, but with more variation as *pH* increases.
c) The *BCI* would also be average.
d) The predicted *BCI* will be 1.56 SDs of *BCI* below the mean *BCI*.

**9.** a) $\widehat{BCI} = 2733.37 - 197.694\,pH$.
b) $\widehat{BCI}$ decreases by about 197.69 points on average per point of *pH*.
c) $\widehat{BCI} = 1112.3$.

**11.** a) 73 points.   b) 7.
c) $r = 0.75$.   d) 100 points.
e) The regression equation is designed to predict final exam scores based on midterm exam scores. You would need to find the regression equation to predict midterm scores based on final exam scores.
f) −85 points.
g) Increase. The point is unusual and has a high negative residual that would decrease the correlation; removing it would increase the correlation and the $R^2$ value.
h) Slope will increase.

**13.** a) 0.473.
b) Weak, not much structure.
c) She did better on Monday than her performance on Friday would predict.
d) Their Monday score will be about 0.473 standard deviations below average.
e) $\widehat{Monday} = 14.592 + 0.536\,Friday$.
f) 36.02.

**15.** *Sex* and *Colorblindness* are both categorical variables, not quantitative. Correlation is meaningless for them, but we can say that the variables are associated.

**17.** a) Yes, the $R^2$ values indicate that 97.2% of the variation in the Indian crocodile length and 98.1% of the variation in the Australian crocodile length are accounted for by the variation in the head size.

b) The slopes of the regression equations are similar, as are the $R^2$ values.
c) The two models have different y-intercepts. That means that the Indian crocodile is smaller.
d) Predicted body length for the Indian crocodile is 389.4 cm but is 463.969 cm for the Australian croc. The skeleton was probably from an Indian crocodile.

**19.** a) Yes. The scatterplot looks linear and there is no pattern in the residual plot.
b) 38 years.
c) $s = 5.598$ so predictions are typically off by about 5.6 years.
d) 26 inches is outside the range of diameters in our data. Extrapolating is risky.

**21.** a) There is a moderately strong, curved, positive association between *Happiness* and *Income per Capita*. People in countries with more income per capita tend to be happier.
b) No, the scatterplot shows a clear curve.
c) The scatterplot looks linear and there is no clear pattern in the residual plot, so the transformation appears to have been effective.
d) The predictions of happiness level made from this model are typically off by about 0.7 points on the ten point scale.
e) 60.7%.
f) Health, levels of crime, safety, income inequality, and worklife balance are some examples.

**23.** a) The very low correlation suggests there is little or no linear relationship between *Quality of Service* and *Tip Size*.
b) $R^2 = 1.21\%$, indicating that the variation in *Quality of Service* accounts for only about 1% of the variation in *Tip Size* even if the relationship is straight.

**25.** a) 71.9%.
b) At higher latitudes the average January temperature is lower.
c) $\widehat{January\ Temperature} = 108.80 - 2.111\ Latitude$.
d) Average January temperature is lower by 2.11°F per degree of latitude.
e) The intercept predicts an average January temperature of 108.8° at the equator (where latitude $= 0$). This is an extrapolation and probably not trustworthy.
f) 24.4 degrees.
g) The equation underestimates the average January temperature.

**27.** a) The scatterplot shows a strong, linear, positive association.
b) There is an association, but it isn't causal. Athletes who perform well in the high jump are likely to also perform well in the long jump.
c) Neither; the change in units does not affect the correlation.
d) I would predict the winning long jump to be 0.910 SDs above the mean long jump.

**29.** a) No relation; the correlation would probably be close to 0.
b) The relation would have a positive direction and the correlation would be strong, assuming that students were studying French in each grade level. Otherwise, no correlation.
c) No relation; correlation close to 0.
d) The relation would have a positive direction and the correlation would be strong, because vocabulary would increase with each grade level.

**31.** $\widehat{Calories} = 560.7 - 3.08\ Time$.
Toddlers who spend an additional minute at the table consume, on average, 3.08 fewer calories.

**33.** There seems to be a strong, positive, linear relationship with one high-leverage point (Northern Ireland) that makes the overall $R^2$ quite low. Without that point, the $R^2$ increases to 61.5%. Of course, these data are averaged across thousands of households, so the correlation appears to be higher than it would be for individuals. Any conclusions about individuals would be suspect.

**35.** a) 3.842.   b) 501.187.   c) 4.0.

**37.** a) 30,818 pounds.
b) 1302 pounds.
c) 31,187.6 pounds.
d) Negative residuals will be more of a problem, as the predicted weight would overestimate the weight of the truck; trucking companies might be inclined to take the ticket to court.

**39.** The original data are nonlinear, with a significant curvature. Using reciprocal square root of drain time gave a scatterplot that is nearly linear:

$$\widehat{1/\sqrt{Drain\ Time}} = 0.0024 + 0.219\ Diameter.$$

## Practice Exam Answers

**I. 1.** D.   **3.** B.   **5.** C.   **7.** D.
**9.** C.   **11.** E.   **13.** B.   **15.** E.

**II. 1.** a) $\widehat{Length} = 17.047 - 1.914\ Time$.
b) 92.3% of the variability of the length of the pencil can be explained by the number of hours it has been used.
c) The model suggests that new pencils are about 17.047 cm long, and then get about 1.914 cm shorter during each additional hour of use.
d) 6.597 cm.
e) No. There are other variables that may affect this model, most important, the type of pencil and the way each person works. She may write slower, press harder, sharpen her pencil more often, etc.

**3.** a) Answers will vary. Two sample displays are shown here. Students should *not* stack bars or use any other display that has them adding the percentages, nor should they use pie charts.

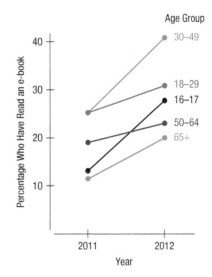

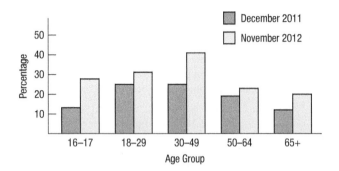

b) The percentage of book readers who had read at least one e-book in the past year increased from 2011 to 2012 in every age category. In both years the 18–29 and 30–49 age groups had a larger proportion of e-book readers than the other age groups.

c) Students may look at the growth of e-book readership in terms of absolute differences in percentages, or in terms of percent difference.

|  | Age Group | | | | |
| --- | --- | --- | --- | --- | --- |
|  | 16–17 | 18–29 | 30–49 | 50–64 | 65+ |
| Absolute Difference | 15 | 6 | 16 | 4 | 8 |
| Percent Difference | 115% | 24% | 64% | 21% | 67% |

Either way, the growth in e-book readership among book readers is different for different age groups, so there seems to be an association. For example, e-book reader growth in the 16–17 age group is far greater than in the over-50 age groups.

## Chapter 10 (See note, page A-29)

1. a) Yes, who takes out the trash cannot be predicted before the flip of a coin.
   b) No, it is not random, because you will probably name a favorite team.
   c) Yes, your new roommate cannot be predicted before names are drawn.

3. A machine pops up numbered balls. If it is truly random, the outcome cannot be predicted and all possible outcomes would be equally likely. It is random only if the balls generate numbers in equal frequencies in the long run.

5. Use two-digit numbers 00–99; let 00–16 = pool/spa, 03–99 = no pool/spa. Or, if you use technology, use 1–100; 1–17 = pool/spa, 18–100 = no pool/spa.

7. a) 45, 10.  b) 17, 22.

9. If the lottery is random, it doesn't matter which number you play; all are equally likely to win.

11. a) The outcomes are not equally likely; for example, tossing 5 heads does not have the same probability as tossing 0 or 9 heads, but the simulation assumes they are equally likely.
    b) The even-odd assignment assumes that a Canadian is equally likely to have or not have the antibody. We cannot assume that this is a 50-50 probability.
    c) The likelihood for the first ace in the hand is not the same as for the second or third or fourth. But with this simulation, the likelihood is the same for each. (And it allows you to get 5 aces, which could get you in trouble in a real poker game!)

13. The conclusion should indicate that the simulation *suggests* that the average length of the line would be about 3.2 people. Future results might not match the simulated results exactly.

15. a) The component is one voter voting. An outcome is a vote for our candidate or not. Use two random digits, giving 00–54 a vote for your candidate and 55–99 for the underdog.
    b) A trial is 100 votes. Examine 100 two-digit random numbers, and count how many people voted for each candidate. Whoever gets the majority of votes wins that trial.
    c) The response variable is whether the underdog wins or not.

17. Answers will vary, but average answer will be about 51%.

19. Answers will vary, but average answer will be about 26%.

21. a) Answers will vary, but you should win about 10% of the time.
    b) You should win at the same rate with any number.

23. Answers will vary, but you should win about 10% of the time.

25. a) Sample response: Assign digit pairs 00–33 to represent a candidate passing on the first trial, 34–99 to represent failing the first time. Pairs 00–71 represent a pass on subsequent trials, and 72–99 represent a failure. In a random digit table, select pairs of digits, allowing repeated pairs, until a "pass" is reached. Record the number of pairs needed to get a pass. Repeat many times.
    b) About 2.8 attempts.

27. Answers will vary, but average answer will be about 18%.

29. Do the simulation in two steps. First simulate the payoffs. Then count until $500 is reached. Answers will vary, but average should be near 10.8 customers.

31. Answers will vary, but average answer will be about 3 children.

33. Answers will vary, but average answer will be about 7.5 rolls.

35. No, it will happen about 40% of the time.

37. Answers will vary, but average answer will be about 37.5%.

39. Answers will vary, but simulations should have 1 or more positive result about 15% of the time.

41. a) Answers will vary. One option: Look at digits in groups of 4, letting each digit represent one ferret. The first two digits in each group will represent the ferrets that got the nasal spray and the last two will represent those that got the placebo. Let the digits 0, 1, 2, 3 and 4 represent "gets infected" and the digits 5, 6, 7, 8, and 9 represent "doesn't get infected." Record whether the outcome of interest occurred with those four digits. Repeat this many times with different sets of four digits. The proportion of times that the outcome of interest occurred among all your trials is the estimated chance of the observed experimental outcome, assuming the nasal spray is ineffective.
    b) Answers will vary, but the average answer will be around 5%.
    c) My probability estimate is low, meaning that if the nasal spray is ineffective, then the observed outcome would have been pretty unlikely. That provides some evidence that the nasal spray is effective.

43. a) Repeat the simulation three times. All three simulations would need to match the results of the experiment.

b) The new probability will be much lower. In fact, so low that it now appears very unlikely that random chance can explain the results that the researcher found. COVID nasal spray has strong evidence of working! (in ferrets, anyway . . . ).

## Chapter 11 (See note, page A-29)

1. a) No. It would be nearly impossible to get exactly 500 men and 500 women from every country by random chance.
   b) A stratified sample, stratified by whether the respondent is male or female.

3. a) Voluntary response.
   b) The percentage of people who prefer vanilla ice cream (and the other choices).
   c) We have no confidence at all in estimates from such methods.

5. a) The population of interest is all adults in the United States aged 18 and older.
   b) The sampling frame is U.S. adults with telephones.

7. a) Population—All U.S. adults.
   b) Parameter—Proportion who have used and benefited from alternative medicine.
   c) Sampling Frame—All Consumers Union subscribers.
   d) Sample—Those who responded.
   e) Method—The list of subscribers makes a convenience sample.
   f) Left Out—Those who are not Consumers Union subscribers and those who didn't respond.
   g) Bias—Undercoverage: The convenience sample leaves out all nonsubscribers. Nonresponse: Those who do not respond may have different responses from those that did.

9. a) Population—All adults from that particular city.
   b) Parameter—Proportion who think drinking and driving is a serious problem.
   c) Sampling Frame—Patrons of this particular bar.
   d) Sample—Every 10th person leaving the bar.
   e) Method—Systematic sampling with a random start.
   f) People who don't frequent that particular bar.
   g) Bias—Those interviewed had just left a bar. They may think drinking and driving is less of a problem than do other adults. Also, a particular bar would tend to cater to a certain demographic, which would not be representative of the population.

11. a) Simple random sample.
    b) No.
    c) They could stratify by distance from the dump site.

13. a) If the police used a list of registered automobiles, they would have to track down each of those owners, and unregistered vehicles would be unlikely to be on the list.
    b) They could use a systematic sample, checking every 20th vehicle that passes a checkpoint. They would randomly select a car out of the first 20, then check every 20th vehicle after that. They could also use a 20-sided die or a random number generator and roll for each car. If they roll a 1, the car gets checked.

15. Bias. Only people watching the news who care enough will respond, and their preference may differ from that of other voters. The sampling method may systematically produce samples that don't represent the population of interest.

17. a) Voluntary response. Only those who see the ad, have Internet access, *and* feel strongly enough will respond.
    b) Cluster sampling. One school may not be typical of all.
    c) Attempted census. Will have nonresponse bias.
    d) Stratified sampling with follow-up. Should be unbiased.

19. a) This is a multistage design, with a cluster sample at the first stage and a simple random sample for each cluster.
    b) Begin by generating 3 unique random numbers between 1 and 17, inclusive. From each of those 3 churches, number the list of members from 1 to $n$. Randomly select 100 random numbers between 1 and $n$ for each church and survey those members.
    c) Answers may vary. Church members who are not active will probably be more difficult to reach as they have moved, resulting in the parameter being underestimated.

21. a) This is a systematic sample.
    b) The sampling frame is patrons willing to wait for the roller coaster on that day at that time. It should be representative of the people in line, but not of all people at the amusement park.
    c) It is likely to be representative of those waiting for the roller coaster. Indeed, it may do quite well if those at the front of the line respond differently (after their long wait) than those at the back of the line.
    d) The patrons who are not willing to wait in line for the ride.

23. a) Answers will definitely differ. Question 1 will probably get many "No" answers, while Question 2 will get many "Yes" answers. This is response bias.
    b) "Do you think standardized tests are appropriate for deciding whether a student should be promoted to the next grade?" (Other answers will vary.)

25. a) Biased toward yes because of "pollute." "Should companies be responsible for any costs of environmental cleanup?"
    b) Biased toward no because of "old enough to serve in the military." "Do you think the drinking age should be lowered from 21?"

27. a) Not everyone has an equal chance. Misses people with unlisted numbers, or without landline phones, or at work.
    b) Generate random numbers and call at random times.
    c) Under the original plan, those families in which one person stays home are more likely to be included. Under the second plan, many more are included. People without landline phones are still excluded.
    d) It improves the chance of selected households being included.
    e) This takes care of phone numbers. Time of day may be an issue. People without landline phones are still excluded.

29. a) Answers will vary.
    b) Your own arm length. Parameter is your own arm length; population is all possible measurements of it.
    c) Population is now the arm lengths of you and your friends. The average estimates the mean of these lengths.
    d) Probably not. Friends are likely to be of the same age and not very diverse or representative of the larger population.

31. a) Assign unique numbers 001 to 120 to each order. Use random numbers to select 10 transactions to examine.
    b) Sample proportionately within each type. (Do a stratified random sample.)

**33.** a) Using a computer, generate a random number from 1 to 200. Select that case of salsa. Inspect all 12 jars in that case.
b) Divide the day into (for example) four time periods. Randomly select a case from each time period. Inspect a randomly chosen jar from each case. Note: This is multistage sampling, technically, but the first step stratifies by time of day, as requested.
c) Weight (quantitative) and taste (categorical).

**35.** a) We would be concerned that some doctors are not listed in Yelp and thus would not be undercovered by this sampling method. Those businesses might be more (or less) likely to accept Medicaid than the ones in our sample.
b) Not appropriate. This convenience sample will almost certainly be biased. For example, the first 40 businesses are probably successful and more likely to be hiring than the businesses later in the list or not on Yelp at all.

**37.** a) Let a consecutive pair of digits in a random digits table represent one student. If the first digit is a 1, 2, or 3 then the student participates in teams sports; otherwise they don't. The second digit indicates whether the student likes pep rallies, but its meaning depends on the first digit. If the first digit indicates that the student participates in team sports, then let all digits except for 9 represent "likes pep rallies, and let 9 represent "doesn't like pep rallies." If the first digit indicates that the student *doesn't* participate in team sports, then let digits 0 and 1 represent "likes pep rallies" and let digits 2 through 9 represent "doesn't like pep rallies." For each pair of digits, record whether the student likes pep rallies or not. Repeat for 50 pairs of digits and compute the proportion of simulated "yes" responses there are. That constitutes one trial. Repeat this for many trials to see the distribution of the simulated sample proportion.
b) We want to sample 15 team sports participants (30% of 50) and 35 non–team sportsparticipants. Look at 15 digits to represent the 15 team sport participants in the sample. For them, digits 0–8 represent a "yes" and 9 is a "no." Then the next 35 digits represent nonparticipants. For those, 0 or 1 is a "yes." 2–9 is a "no." Record the proportion of digits that represent "yes" responses among all 50.
c) Answers will vary.
d) The two dotplots are both centered in about the same place, and both are mound-shaped and reasonably symmetric. The difference is that the distribution of results from the SRS has more variability than the distribution of results from the stratified sample.
e) The rightmost dot in the SRS picture corresponds to a very high estimate of the proportion of students who like the pep rallies. This could easily be because that particular sample had more than the 15 participants in it than you'd expect for a sample of 50 students. In the other dotplot, the highest estimate isn't as different from the others. That's because all of the samples in this simulation had exactly 15 participants and exactly 35 nonparticipants, so there is no variability in the dotplot that results from an under- or overrepresentation of the generally pep-rally-loving team sports participants.

**39.** a) Several terms are poorly defined. The survey needs to specify the meaning of "family" for this purpose and the meaning of "higher education." The term "seek" may also be poorly defined (for example, would applying to college but not being admitted qualify for seeking more education?).
b) i) Cluster sample.
ii) Stratified sample.
iii) Systematic sample.
c) This is not an SRS. Although each student may have an equal chance of being in the survey, groups of friends who choose to sit together will either all be in or out of the sample, so the selection is not independent.
d) i) This would suffer from voluntary response bias.
ii) This would be a convenience sample.
e) The proportion in the sample is a statistic. The proportion of all students is the parameter of interest. The statistic estimates that parameter, but is not likely to be exactly the same.

## Chapter 12 (See note, page A-29)

**1.** a) No, there are no manipulated factors. It is an observational study.
b) There may be lurking variables that are associated with both parental income and performance on the SAT.

**3.** a) This is a retrospective observational study.
b) That's appropriate because MS is a relatively rare disease.
c) The subjects were U.S. military personnel, some of whom had developed MS.
d) The variables were the vitamin D blood levels and whether or not the subject developed MS.

**5.** a) This was a randomized, double-blind experiment.
b) Yes, such an experiment is the right way to determine whether maggots were as effective as surgery.
c) 100 men with wounds on their lower limbs.
d) The treatments were sterile maggots and a traditional surgical procedure. The response was the percentage of dead tissue in the wounds.

**7.** a) Experiment.
b) 130 patients with eligible lacerations.
c) Type of treatment; two levels.
d) 2 treatments.
e) Scarring, pain level, and speed of healing.
f) Completely randomized.
g) The evaluators were blind. You cannot blind the patients!
h) On these subjects, the two treatments worked equally well regarding scarring, but the adhesive was less painful and worked faster.

**9.** a) Observational study.
b) Prospective.
c) Men and women with moderately high blood pressure and normal blood pressure; unknown selection process.
d) Memory and reaction time.
e) As there is no random assignment, there is no way to know that high blood pressure *caused* subjects to do worse on memory and reaction-time tests. A lurking variable may also be the cause.

**11.** a) Experiment.
b) Postmenopausal women.
c) Alcohol—2 levels; blocking variable—estrogen supplements (2 levels).
d) 1 factor (alcohol) at 2 levels = 2 treatments.

e) Increase in estrogen levels.
f) Blocked.
g) Not blind unless beverages look, smell, and taste the same
h) Indicates that alcohol consumption *for those taking estrogen supplements* may increase estrogen levels.

**13.** a) Observational study.
b) Retrospective.
c) Women in Finland; unknown selection process with data from church records.
d) Women's life spans.
e) As there is no random assignment, there is no way to know that having sons or daughters shortens or lengthens the life span of mothers.

**15.** a) Observational study.
b) Prospective.
c) People with or without depression; unknown selection process.
d) Frequency of crying in response to sad situations.
e) There is no apparent difference in crying response (to sad movies) for depressed and nondepressed groups.

**17.** a) This is an experiment because the treatments were randomly assigned to the subjects.
b) The treatments were having at least one 15-minute neuromuscular warm-up exercise session per week and having no such warm-up session. The response variable was the number of ACL tears in each group.
c) Less variation in ACL tears would be expected among players of the same gender in the same sport, which would allow a difference between the treatments to be detected more easily.
d) The conclusions of the study cannot be generalized to males, females in other countries, or to players of other sports.

**19.** a) The factors are type of pill (levels: standard pain reliever and placebo) and water (levels: ice water or no ice water).
b) The four treatments are standard pain reliever with ice water, standard pain reliever with no ice water, placebo with ice water, and placebo with no ice water. The response variable is the self-reported pain level.
c) Yes, the placebo is to prevent participants from knowing which pill they get. They cannot be blinded with respect to the ice water treatment.
d) There are several possibilities. Gender, age, frequency of migraines. But the justification given must be that there is reason to think that the variable is associated with the response variable of self-reported pain level.

**21.** a) The subjects for the experiment are the individuals; the vitamin supplement/placebo is explanatory, the change in the condition of the bipolar disorder is the response variable, both categorical.
b) They need to compare omega-3 results to something. Perhaps bipolarity is seasonal and would have improved during the experiment anyway.

**23.** a) Subjects' responses might be related to many other factors (diet, exercise, genetics, etc.). Randomization should equalize the two groups with respect to unknown factors.
b) More subjects would minimize the impact of individual variability in the responses, but the experiment would become more costly and time consuming.

**25.** People who engage in regular exercise might differ from others with respect to bipolar disorder, and that additional variability could obscure the effectiveness of this treatment.

**27.** Answers may vary. Line up the 24 plants. Put 8 red, 8 white, and 8 blue poker chips in a bag. Mix thoroughly. Draw chips at random, placing one by each plant in sequence. Assign all the plants with red chips to get a full dose of fertilizer, blue to get a half-dose, and white to get none.

**29.** a) The placebo provides a baseline for comparison. Without the placebo the researchers would not know the current rate of infection.
b) The difference in infection rate between the two groups was extremely unlikely to happen by chance.
c) We can conclude that there is very strong evidence that the vaccine causes a reduction in the infection rate. But only for these ethnic groups and for people 18 years or older.

**31.** a) Allowing athletes to self-select treatments could confound the results. Other issues such as severity of injury, diet, age, etc., could also affect time to heal; randomization should equalize the treatment groups with respect to any such variables.
b) A control group could have revealed whether either exercise program was better (or worse) than just letting the injury heal.
c) Doctors who evaluated the athletes to approve their return to sports should not know which treatment the subject had.
d) It's hard to tell. The difference of 15 days seems large, but the standard deviations indicate that there was a great deal of variability in the times.

**33.** a) The differences among the Mozart and quiet groups were more than would have been expected from sampling variation.
b)
c) The Mozart group seems to have the smallest median difference and thus the *least* improvement, but there does not appear to be a significant difference.
d) No, if anything, there is less improvement, but the difference does not seem significant compared with the usual variation.

**35.** a) Observational, prospective study.
b) The supposed relation between health and wine consumption might be explained by the confounding variables of income and education.
c) No. While the variables have a relation, there is no causality indicated for the relation.

**37.** a) Number the ferrets and randomly choose half to receive the medication and half to receive the placebo.
b) Yes, this result seems to be a bigger difference than would happen by chance.
c) Yes, because this is a controlled experiment, the researcher can conclude that he has evidence that the treatment is the cause of the difference.

**39.** a) Randomly assign half the teachers in the district to use each method. Students should be randomly assigned to teachers as well, if possible. Construct an appropriate reading test to be used at the end of the year, and compare scores.

b) If teachers are allowed to self-select their teaching method, they will pick the method they are most familiar with. It is very possible that some of the better teachers all prefer the same method and then that method will have higher scores (due to the teacher, not the method). Note: Different sample sizes are not a problem. Later in the course you will learn formulas that adjust for different size treatment groups.

**41.** a) They mean that the difference is higher than they would expect from normal sampling variability.

b) An observational study.

c) No. Perhaps the differences are attributable to some confounding variable (e.g., people are more likely to engage in riskier behaviors on the weekend, routine procedures are rarely scheduled for weekends, etc.) rather than the day of admission.

d) Perhaps people have more serious accidents and traumas on weekends and are thus more likely to die as a result.

**43.** Answers may vary. This experiment has 1 factor (pesticide), at 3 levels (pesticide A, pesticide B, no pesticide), resulting in 3 treatments. The response variable is the number of beetle larvae found on each plant. Randomly select a third of the plots to be sprayed with pesticide A, a third with pesticide B, and a third with no pesticide (because the researcher also wants to know whether the pesticides even work at all). To control the experiment, the plots of land should be as similar as possible with regard to amount of sunlight, water, proximity to other plants, etc. If not, plots with similar characteristics should be blocked together. If possible, use some inert substance as a placebo pesticide on the control group, and do not tell the counters of the beetle larvae which plants have been treated with pesticides. After a given period of time, count the number of beetle larvae on each plant and compare the results.

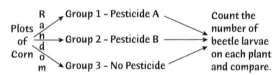

**45.** Answers may vary. Find a group of volunteers. Each volunteer will be required to shut off the machine with their dominant hand and nondominant hand. Randomly assign the dominant or nondominant hand to be used first. Complete the first attempt for the whole group. Now repeat the experiment with the alternate hand. Check the differences in time for the dominant and nondominant hands.

**47.** a) Jumping with or without a working parachute.
b) Volunteer skydivers (the dimwitted ones).
c) A parachute that looks real but doesn't work.
d) A good parachute and a placebo parachute.
e) Whether parachutist survives the jump (or extent of injuries).
f) All should jump from the same altitude in similar weather conditions and land on similar surfaces.
g) Randomly assign people the parachutes.
h) The skydivers (and the people involved in distributing the parachute packs) shouldn't know who got a working chute.

And the people evaluating the subjects after the jumps should not be told who had a real parachute either!

**49.** a) This is an observational study because the student is not randomly assigning companies to use or not use the Internet for business transactions.

b) If profitability did increase in the 2000s, it could have been due to a number of factors, not specifically the Internet as a means for conducting business.

## Part III Review (See note, page A-29)

**1.** Observational prospective study. Indications of behavior differences can be seen in the two groups. May show a link between premature birth and behavior, but there may be lurking variables involved.

**3.** Experiment, matched by gender and weight, randomization within blocks of two pups of same gender and weight. Explanatory variable: type of diet. Treatments: low-calorie diet and allowing the dog to eat all it wants. Response variable: length of life. Can conclude that, on average, dogs (of this breed) with a lower-calorie diet live longer.

**5.** Observational prospective study. Indications of differences in mortality among those who drink a lot of coffee compared to those who drink none. May show that heavy coffee drinkers tend to live longer but a causal relationship cannot be established.

**7.** Sampling. Probably a simple random sample, although may be stratified by type of firework. Population is all fireworks produced each day. Parameter is proportion of duds. Can determine if the day's production is ready for sale.

**9.** Observational prospective study. Researcher can conclude that for anyone's lunch, even when packed with ice, food temperatures are rising to unsafe levels.

**11.** Experiment, with a control group being the genetically engineered mice who received no antidepressant and the treatment group being the mice who received the drug. The response variable is the amount of plaque in their brains after one dose and after four months. There is no mention of blinding or matching. Conclusions can be drawn to the general population of mice and we should assume treatments were randomized. To conclude the same for humans would be risky, but researchers might propose an experiment on humans based on this study.

**13.** Experiment. Explanatory variable is gene therapy. Hamsters were randomized to treatments. Treatments were gene therapy or not. Response variable is heart muscle condition. Can conclude that gene therapy is beneficial (at least in hamsters).

**15.** Sampling. Population is all oranges on the truck. Parameter is proportion of unsuitable oranges. Procedure is probably stratified random sampling with regions inside the truck being the strata. Can conclude whether or not to accept the truckload.

**17.** Observational retrospective study performed as a telephone-based randomized survey. Based on the excerpt, it seems reasonable to conclude that more education is associated with a higher Emotional Health Index score, but the insistence on causality would be faulty reasoning.

19. Answers will vary. This is a simulation problem. Using a random digits table or software, call 0–4 a loss and 5–9 a win for the gambler on a game. Use blocks of 5 digits to simulate a week's pick.

21. Answers will vary.

23. a) Experiment. Actively manipulated candy giving, diners were randomly assigned treatments, control group was those with no candy, lots of dining parties.
    b) It depends on when the decision was made. If early in the meal, the server may give better treatment to those who will receive candy—biasing the results.
    c) A difference in response so large it cannot be attributed to natural sampling variability.

25. a) 100 obese persons.
    b) Restrictive diet, alternate fasting diet, control (no special diet).
    c) Weight loss (reported as % of initial weight).
    d) No, it could not be blind because participants must know what they are eating.
    e) No, it is not necessary that participants be a random sample from the population. All that is needed is that they be randomly assigned to treatment groups.

27. a) Water (quality and temperature) and material can vary. Results may be influenced by these confounding variables.
    b) Unrealistic conditions. This won't say how SparkleKleen works in normal situations.
    c) Might work, but if all the swatches were stained at the same time, the stains on the later swatches will have more time to "set in," causing bias against SparkleKleen. Other variables (changes in water temperature or pressure) won't be randomized.
    d) No guarantee that conditions are comparable. And we should question the fairness of data supplied by SparkleKleen itself.

29. a) 10.6% chance to break at 900 pounds or less.
    b) Simulation results will vary. Use groups of three digits to simulate each rivet. For every one that is less than "106," denote as a failed rivet. Count how many rivets you need until 3 failures are reached.

31. a) Eyes (open or closed), Music (On or Off), Moving (sitting or moving).
    b) It is a blocking variable.
    c) Each of 4 subjects did 8 runs, so 32 in all.
    d) Randomizing completely is better to reduce the possibility of confounding with factors that weren't controlled for.

33. a) A difference in response so large it cannot be attributed to natural sampling variability.
    b) More likely, younger-looking individuals are more sexually active than older ones. We have no means of comparison (different levels of sexual activity in people of the same age, for example).

35. a) Prospective observational study (it feels like we are administering a treatment, but we are only observing a behavior).
    b) No, but it does limit the scope of the conclusion. Any observed difference can only be generalized to other people like these students.
    c) 100% vs. 34% appears very statistically significant. That is, a difference this large seems unlikely to have happened by chance.

37. a) *Who*—19 medical students; *What*—observation/diagnosis skill; *Why*—to see if art observation improves medical observation; *When*—not stated; *Where*—University of Texas; *How*—experiment (lacking control!).
    b) An experiment, the treatment (the course with VTS) was applied to the students.
    c) Medical students similar to those at University of Texas.
    d) Yes, because a treatment was applied to the students.
    e) The experiment lacks a control group. It would be helpful to use a control group too if VTS can be taught only as a medical procedure, or if evaluating art in addition to medical images causes an improvment.

39. a) Nonresponse bias by conservatives.
    b) U.S. adults.
    c) Results do not total 100%.
    d) Observed differences are of the same size one might expect from natural sampling variability.

41. Since players vary in their ability to hit the ball, I will have each batter hit with both types of bats several times in a randomly chosen order. For each batter, calculate the average difference in distance, with metal or wood as the response variable.

43. a) Factors: Tire Pressure (levels 28 and 32 psi) and Gasoline Type (levels regular and premium).
    b) 8 runs. 2 each at each of the 4 treatment combinations.
    c) Conditions (especially weather) may vary and may confound the effect of the gasoline types.

45. a) Use stratified sampling to select 2 first-class passengers and 12 from coach.
    b) Number passengers alphabetically, 01 = Bergman to 20 = Testut. Read in blocks of two, ignoring any numbers more than 20. This gives 65, 43, 67, 11 (selects Fontana), 27, 04 (selects Castillo).
    c) Number passengers alphabetically from 001 to 120. Use the random-number table to find three-digit numbers in this range until 12 different values have been selected.

47. Simulation results will vary. (Use integers 00 to 99 as a basis. Use integers 00 to 69 to represent a tee shot on the fairway. If on the fairway, use digits 00 to 79 to represent on the green. If off the fairway, use 00 to 39 to represent getting on the green. If not on the green, use digits 00 to 89 to represent landing on the green. For the first putt, use digits 00 to 19 to represent making the shot. For subsequent putts, use digits 00 to 89 to represent making the shot.)

49. a) Factors—Computer (Mac or PC), Library (Main or Science), Time of Week (Weekday or Weekend), Time of Day (Before or After 5 p.m.), and Load (Running Other Jobs or Not Running Other Jobs).
    b) 32.
    c) Random order.

## Practice Exam Answers

I.  1. C.    3. B.    5. D.    7. D.    9. E.
    11. A.   13. C.   15. B.   17. D.   19. B.

**II. 1.** a)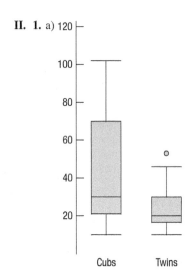

b) Both distributions of runs scored are skewed to the right. The Twins have an outlier at 53 runs scored. The median number of runs is a bit higher for the Cubs (30) than for the Twins (20). The Cubs have much greater variability in the number of runs scored than do the Twins.

c) Both distributions are skewed to the right, but since the variability of numbers of runs scored for the Cubs is so much greater, the mean number of runs for the Cubs is pulled toward the larger values more than that of the Twins.

**3.** a) Assign each student a number label from 1 to 1200. Use a random number generator to select 120 different numbers.
b) Randomly select 30 students from each grade level.
c) Assign each homeroom a number label from 1 to 40. Use a random number generator to select 4 unique numbers. Survey all 30 students in each of those 4 homerooms.
d) Stratifying by grade would reduce the variation arising from differing computer usage among grade levels. Cluster sampling will make it easier to access students, but may not represent all the grade levels.

## Chapter 13 (See note, page A-29)

**1.** a) S = {HH, HT, TH, TT}, equally likely.
b) S = {0, 1, 2, 3}, not equally likely.
c) S = {H, TH, TTH, TTT}, not equally likely.
d) S = {1, 2, 3, 4, 5, 6}, not equally likely.

**3.** In this context, "truly random" should mean that every number is equally likely to occur.

**5.** There is no "Law of Averages." She would be wrong to think that they are "due" for a harsh winter.

**7.** There is no "Law of Averages." If at bats are independent, his chance for a hit does not change based on recent successes or failures.

**9.** a) There is some chance you would have to pay out much more than the $1500.
b) Many customers pay for insurance. The small risk for any one customer is spread among all.

**11.** a) 0.30. b) 0.80.

**13.** a) 0.0204. b) 0.9796. c) 0.9537.

**15.** a) Legitimate. b) Legitimate.
c) Not legitimate (sum more than 1).
d) Legitimate.
e) Not legitimate (can't have negatives or values more than 1).

**17.** A family may own both a computer and an HD TV. The events are not disjoint, so the Addition Rule does not apply.

**19.** When cars are traveling close together, their speeds are not independent, so the Multiplication Rule does not apply.

**21.** a) He has multiplied the two probabilities.
b) He assumes that being accepted at the colleges are independent events.
c) No. Colleges use similar criteria for acceptance, so the decisions are not independent.

**23.** a) 0.72. b) 0.89. c) 0.28.

**25.** a) 0.5184. b) 0.0784. c) 0.4816.

**27.** a) Repair needs for the two cars must be independent.
b) Maybe not. An owner may treat the two cars similarly, taking good (or poor) care of both. This may decrease (or increase) the likelihood that each needs to be repaired.

**29.** a) Yes. Every 20th person should be independent from each other.
b) No. Groups of people will probably have a similar reaction.
c) No. Whether or not it rains one day will probably make it more (or less) likely that it will rain on the next day.
d) Yes. The cities should be independent from each other because a random sample was taken.

**31.** a) This is probably just random variation.
b) It might be that the player got "hot." But it could just be a lucky streak!
c) There is a strong indication here that the restaurant has changed its procedures. You might need to go elsewhere.

**33.** a) 662/1010 = 0.655.
b) 182/1010 + 2/1010 = 184/1010 = 0.182.

**35.** a) 0.281. b) 0.551.
c) Responses are independent.
d) People were polled at random.

**37.** a) 0.4324. b) 0.9724.
c) (1 − 0.46) + 0.46(1 − 0.06) or 1 − (0.46)(0.06).

**39.** a) 1) 0.124. 2) 0.34. 3) 0.802. 4) 0.0.
b) 1) 0.0021. 2) 0.0989. 3) 0.6472. 4) 0.4842.

**41.** a) Disjoint (can't be both red and orange).
b) Independent (unless you're drawing from a small bag).
c) No. Once you know that one of a pair of disjoint events has occurred, the other is impossible.

**43.** a) 0.0046. b) 0.125. c) 0.296.
d) 0.421. e) 0.995.

**45.** a) 0.027. b) 0.063. c) 0.973.
d) 0.014.

**47.** a) 0.024. b) 0.250. c) 0.543.

**49.** 0.078.

**51.** a) For any day with a valid three-digit date, the chance is 0.001, or 1 in 1000. For many dates in October through December, the probability is 0. (No three digits will make 10/15, for example.)

b) There are 65 days when the chance to match is 0 (Oct. 10–31, Nov. 10–30, and Dec. 10–31). The chance for no matches on the remaining 300 days is 0.741.
c) 0.259.    d) 0.049.

## Chapter 14 (See note, page A-29)

1. 0.42.    3. 0.675.
5. 0.135.
7. No, $P(\text{Basketball}) = 0.4$, but $P(\text{Basketball}|\text{Football}) = 0.415$. These are not the same. (But they're so close that for all practical purposes, the events are essentially independent.)
9. 

|  | Snapchat | No Snapchat | Total |
|---|---|---|---|
| **13–34** | 0.17 | 0.11 | **0.28** |
| **Not 13–34** | 0.03 | 0.69 | **0.72** |
| **Total** | **0.20** | **0.80** | **1.00** |

(Age labels rows)

11. a) 99%.    b) 1%.    c) 53%.
13. a) $P(\text{TV} \cap \text{Fridge}^c)$, 0.31.
    b) $P(\text{TV} \cup \text{Fridge}) \cap (\text{TV} \cup \text{Fridge}^c)$, 0.48.
    c) $P(\text{TV}^c \cap \text{Fridge}^c)$, 0.31.
15. a) 0.202.5.    b) 0.6965.    c) 0.2404.
    d) 0.0402.
17. a) 0.50.    b) 1.00.    c) 0.077.
    d) 0.333.
19. a) 0.11.    b) 0.27.    c) 0.407.
    d) 0.344.
21. a) 0.011.    b) 0.222.    c) 0.054.
    d) 0.337, $P(\text{Primary}|\text{China})$.    e) 0.436, $P(\text{China}|\text{Primary})$.
23. 0.21.
25. a) 0.145.    b) 0.118.    c) 0.414.
    d) 0.217.
27. a) 0.318.    b) 0.955.    c) 0.071.
    d) 0.009.
29. a) 32%.    b) 0.135.
    c) No, 7% of juniors have taken both.
    d) No, the probability that a junior has taken a computer course is 0.23. The probability that a junior has taken a computer course *given* they have taken a Statistics course is 0.135.
31. a) $42/46 = 91.3\%$.
    b) They might be. If you have a landline, there is a 91.3% chance you have a cell phone, while cell phone use overall is at 95%. So the chances are fairly similar.
    c) No, because 42% have both.
33. Yes, $P(\text{Ace}) = 4/52 = 1/13$. $P(\text{Ace}|\text{any suit}) = 1/13$.
35. a) 0.33.
    b) No, 9% of the chickens had both contaminants.
    c) Perhaps. Although $P(C|S) = 0.64 \neq P(C)$, the chickens contaminated with *salmonella* were only slightly more likely also to have *campylobacter*.
37. No, only 32% of all men have high cholesterol, but 40.7% of those with high blood pressure do.
39. a) 55%.
    b) Yes. Overall, 37% use social media for work-related purposes, but of those who find social media distracting, 55% are on social media for work.
41. No. For example, 81.3% of elementary teachers polled favor the proposal, but only 60% of all teachers do.
43. 

| | | Interest in Cooking Club | | |
|---|---|---|---|---|
| | | Yes | No | Total |
| **Year** | Junior | 4 | 8 | 12 |
| | Senior | 5 | 10 | 15 |
| | Total | 9 | 18 | 27 |

45. No. If they were independent, the horizontal divisions in the bars would line up in the "Yes" and "No" columns. Since they do not, *Prostate Cancer* and *Fish Consumption* are not independent. (But they're *very* close; they may be independent for all practical purposes.)

47. a)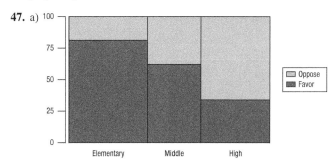

b) 

| | Elementary | Middle | High |
|---|---|---|---|
| **Favor** | 45 | 30 | 39 |
| **Oppose** | 30 | 20 | 26 |

c)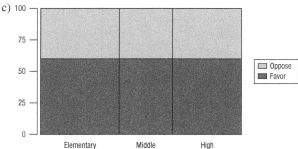

49. a) No, the probability that the luggage arrives on time depends on whether the flight is on time. The probability is 95% if the flight is on time and only 65% if not.
    b) 0.695.
51. 0.975.
53. a) No, the probability of missing work for day-shift employees is 0.01. It is 0.02 for night-shift employees. The probability depends on whether they work the day or night shift.
    b) 1.4%.
55. 57.1%.
57. a) 0.20.    b) 0.272.    c) 0.353.
    d) 0.033.
59. 0.563.
61. Over 0.999.

**63.** 32.4%.

**65.** a) 6.25%.  b) 16.7%.

**67.** 1.19%.

## Chapter 15 (See note, page A-29)

**1.** a) 19.  b) 4.2.

**3.** a)

[histogram of Swings probability distribution, x-axis 0 to 6, peak near 3]

b) 2.22 swings.
c) 2 swings.
d) The mean is a bit larger than the median. That is what you should generally expect from a distribution that is skewed right.

**5.** 1.38 swings.

**7.** a)

| Number of Points | +5 | +2 | −1 |
|---|---|---|---|
| Probability | 0.125 | 0.65625 | 0.21875 |

b) 1.72 points.
c) Depends! While the average is positive, you have roughly a 1 in 5 chance of losing a point. Are you desperate?

**9.** a)

| Number of Students Called On | 1 | 2 | 3 | 4 |
|---|---|---|---|---|
| Probability | 0.85 | 0.1275 | 0.019 | 0.0034 |

b) 1.18 students.
c) 0.176 wrong answers.

| Number Wrong | 0 | 1 | 2 | 3 | 4 |
|---|---|---|---|---|---|
| P(Number Wrong) | 0.85 | 0.1275 | 0.019 | 0.002869 | 0.000506 |

**11.** $30,000.

**13.** a) 7.  b) 1.89.

**15.** 1.74 points.

**17.** 0.45 students called on.

**19.** a) 1.7.  b) 0.9.

**21.** $\mu = 0.64$, $\sigma = 0.93$.

**23.** a) $50.  b) $100.

**25.** a) No, the probability of winning the second depends on the outcome of the first.
b) 0.42.  c) 0.08.
d)

| Games Won | 0 | 1 | 2 |
|---|---|---|---|
| P(Games Won) | 0.42 | 0.50 | 0.08 |

e) $\mu = 0.66$, $\sigma = 0.62$.

**27.** a)

| Number Good | 0 | 1 | 2 |
|---|---|---|---|
| P(Number Good) | 0.067 | 0.467 | 0.467 |

b) 1.40.  c) 0.61.

**29.** a) $\mu = 30$, $\sigma = 6$.  b) $\mu = 26$, $\sigma = 5$.
c) $\mu = 30$, $\sigma = 5.39$.  d) $\mu = -10$, $\sigma = 5.39$.
e) $\mu = 20$, $\sigma = 2.83$.

**31.** a) $\mu = 240$, $\sigma = 12.80$.  b) $\mu = 240$, $\sigma = 24$.
c) $\mu = 720$, $\sigma = 34.18$.  d) $\mu = 60$, $\sigma = 39.40$.
e) $\mu = 600$, $\sigma = 22.63$.

**33.** a)

| Value | 1 | 2 | 3 | 4 | 5 | 6 |
|---|---|---|---|---|---|---|
| Probability | $\frac{1}{6}$ | $\frac{1}{6}$ | $\frac{1}{6}$ | $\frac{1}{6}$ | $\frac{1}{6}$ | $\frac{1}{6}$ |

b) $E(X) = 3.5$, $\sigma(X) = 1.71$.
c) $E(X) = 8.5$, $\sigma(X) = 1.71$.
d)

| Value | 6 | 7 | 8 | 9 | 10 | 11 |
|---|---|---|---|---|---|---|
| Probability | $\frac{1}{6}$ | $\frac{1}{6}$ | $\frac{1}{6}$ | $\frac{1}{6}$ | $\frac{1}{6}$ | $\frac{1}{6}$ |

$E(X) = 8.5$, $\sigma(X) = 1.71$.
e) They are the same.

**35.** a,b) See the answers for 33a and b.
c) $E(X) = 7$, $\sigma(X) = 2.42$.
d)

| Value | 2 | 3 | 4 | 5 | 6 | 7 | 8 | 9 | 10 | 11 | 12 |
|---|---|---|---|---|---|---|---|---|---|---|---|
| Probability | $\frac{1}{36}$ | $\frac{2}{36}$ | $\frac{3}{36}$ | $\frac{4}{36}$ | $\frac{5}{36}$ | $\frac{6}{36}$ | $\frac{5}{36}$ | $\frac{4}{36}$ | $\frac{3}{36}$ | $\frac{2}{36}$ | $\frac{1}{36}$ |

$E(X) = 7$, $\sigma(X) = 2.42$.
e) They are the same.

**37.** $1700, $141.

**39.** a) 1.8.  b) 0.87.
c) Cartons are independent of each other.

**41.** a) 34.2 eggs.  b) 0.87.
c) The sum of the eggs is always 36, so the amount of variation in the good eggs is the same as the amount of variation in the bad eggs.

**43.** a) 1310.  b) 200.

**45.** a) The average mean and median can be quite different, depending on the shape of the distribution.
b) The distribution may be strongly skewed to the right.
c) It would be approximately 0.5.
d) It would be much less than 0.5.

**47.** a) $\mu = 13.6$, $\sigma = 2.55$.
b) Assuming the hours are independent of each other.
c) A typical 8-hour day will have about 11 to 16 repair calls.
d) 19 or more repair calls would be a lot! That's more than two standard deviations above average.

**49.** a) $\mu = 23.4$, $\sigma = 2.97$.
b) We assume each truck gets tickets independently.

**51.** a) There will be many gains of $150 with a few large losses.
b) $\mu = \$300$, $\sigma = \$8485.28$.
c) $\mu = \$1,500,000$, $\sigma = \$600,000$.
d) Yes, $0 is 2.5 SDs below the mean for 10,000 policies.
e) Losses are independent of each other. A major catastrophe with many policies in an area would violate the assumption.

**53.** a) $E(X) = 1.7$, $\sigma(X) = 0.9$.

b)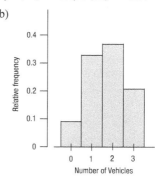

The distribution of the number of vehicles looks fairly symmetric. The expected value is 1.7 vehicles and the standard deviation is 0.9 vehicles.

c) Some of the threes would be fours, fives, and so on. Higher numbers of vehicles would probably have lower frequencies. This distribution would probably look skewed to the right.

d) Since some of the largest values would be even larger, both the expected value and the standard deviation would probably increase.

**55.** a) 1 oz.     b) 0.5 oz.
c) 0.02.     d) $\mu = 4$ oz, $\sigma = 0.5$ oz.
e) 0.159.     f) $\mu = 12.3$ oz, $\sigma = 0.54$ oz.

**57.** a) 12.2 oz.     b) 0.51 oz.     c) 0.058.

**59.** a) The plan that compares each runner to their own time is a better plan. Individual runners may vary, causing confusion in isolating the effect of the backpack in the first plan. But if each volunteer runs with and without the backpack, we will isolate the effect that the backpack is making.

b) We would expect that the standard deviation of the difference in the second plan would become smaller. This is because we would expect faster runners to deal with the extra weight better than slower runners (probably) so the standard deviation of the difference will be smaller than it would be if we compared two independent groups.

**61.** a) $\mu = 200.57$ sec, $\sigma = 0.46$ sec.

b) No, $z = \dfrac{199.48 - 200.57}{0.461} = -2.36$. There is only 0.009 probability of swimming that fast or faster.

**63.** a) $A =$ price of a pound of apples; $P =$ price of a pound of potatoes; $Profit = 100A + 50P - 2$.

b) $63.00.     c) $20.62.

d) Mean, no; SD, yes (independent sales prices).

e) Probably not. Prices tend to go up and down together, to some extent.

**65.** a) $\mu = 1920$, $\sigma = 48.99$; $P(C > 2000) = 0.051$.

b) $\mu = \$220$, $\sigma = 11.09$; No—$300 is more than 7 SDs above the mean.

c) $P(D - C/2 > 0) \approx 0.26$.

## Chapter 16 (See note, page A-29)

**1.** a) No. More than two outcomes are possible on each roll.

b) Yes, assuming the people are equivalent to a random sample of the population.

c) No. The chance of a heart changes based on which cards have been dealt so the trials are not independent.

d) No, because they are sampling without replacement the probability of selecting a voter in favor of the proposed budget changes based on previous selections. And the sample size (500) is more than 10% of the population (3000) so we cannot even use the approximation.

e) If packages in a case are independent of each other, yes.

**3.** a) Use single random digits. Let 0, 1 = Biles. Count the number of random numbers until a 0 or 1 occurs.

b) Results will vary.

c) Results will vary.

d) 

| x | 1 | 2 | 3 | 4 | 5 | 6 | 7 | 8 | ≥9 |
|---|---|---|---|---|---|---|---|---|---|
| P(x) | 0.2 | 0.16 | 0.128 | 0.102 | 0.082 | 0.066 | 0.052 | 0.042 | 0.168 |

**5.** a) Use single random digits. Let 0, 1 = Biles. Examine random digits in groups of five, counting the number of 0's and 1's.

b) Results will vary.

c) Results will vary.

d)

| x | 0 | 1 | 2 | 3 | 4 | 5 |
|---|---|---|---|---|---|---|
| P(x) | 0.33 | 0.41 | 0.20 | 0.05 | 0.01 | 0.0 |

**7.** No. Departures from the same airport during a 2-hour interval may not be independent. All could be delayed by weather, for example.

**9.** a) 0.0819.     b) 0.00128.     c) 0.992.

**11.** a) 5.     b) 8 shots.     c) 1.26 shots.

**13.** 20 calls.

**15.** a) 25.     b) 0.185.     c) 0.217.     d) 0.693.

**17.** a) 0.00344.     b) 0.99999957.     c) 0.1559.
d) 0.2518.     e) 0.8774.     f) 0.3744.

**19.** a) 6.72.     b) 1.037.     c) 1.19 people.

**21.** a) $\mu = 3.2$, $\sigma = 1.64$.
b) i. 0.969.    ii. 0.206.    iii. 0.0000732.    iv. 0.000427.

**23.** $\mu = 20.28$, $\sigma = 4.22$.

**25.** a) 0.118.     b) 0.324.     c) 0.744.     d) 0.580.

**27.** a) Success or failure: serve in or fault. 70% probability needs to stay constant and each serve independent. Counting the number of good serves out of 6 attempts.

b) These assumptions are reasonable. (Although we might wonder if their probability of success changes under certain conditions. For example, they become tired, nervous, more confident, etc.)

**29.** a) $\mu = 56$, $\sigma = 4.10$.

b) Yes, $np = 56 \geq 10$, $nq = 24 \geq 10$; serves are independent.

c) In a match with 80 serves, approximately 68% of the time they will have between 51.9 and 60.1 good serves, approximately 95% of the time they will have between 47.8 and 64.2 good serves, and approximately 99.7% of the time they will have between 43.7 and 68.3 good serves.

d) Normal, approx.: 0.014; Binomial, exact: 0.016.

**31.** a) Assuming apples fall and become blemished independently of each other, $Binom(300, 0.06)$ is appropriate. Because $np \geq 10$ and $nq \geq 10$, $N(18, 4.11)$ is also appropriate.

b) Normal, approx.: 0.072; Binomial, exact: 0.085.

c) No, 50 is 7.8 SDs above the mean.

33. Normal, approx.: 0.053; Binomial, exact: 0.061.

35. The mean number of sales should be 24 with SD 4.60. Ten sales is more than 3.0 SDs below the mean. They were probably misled.

37. a) 5.   b) 0.066.   c) 0.107.   d) $\mu = 24$, $\sigma = 2.19$.
    e) Normal, approx.: 0.819; Binomial, exact: 0.848.

39. $\mu = 20$, $\sigma = 4$. I'd want *at least* 32 (3 SDs above the mean). (Answers will vary.)

41. Probably not. There's a more than 9% chance that they could hit 4 shots in a row, so they can expect this to happen nearly once in every 10 sets of 4 shots they take. That does not seem unusual.

43. Yes. We'd expect them to make 22 shots, with a standard deviation of 3.15 shots. 32 shots is more than 3 standard deviations above the expected value, an unusually high rate of success.

## Part IV Review (See note, page A-29)

1. a) 0.34.   b) 0.27.   c) 0.069.
   d) No, 2% of cars have both types of defects.
   e) Of all cars with cosmetic defects, 6.9% have functional defects. Overall, 7.0% of cars have functional defects. The probabilities here are estimates, so these are probably close enough to say the defects are independent.

3. a) $C$ = Price to China; $F$ = Price to France; Total $= 3C + 5F$.
   b) $\mu = \$5500$, $\sigma = \$672.68$.
   c) $\mu = \$500$, $\sigma = \$180.28$.
   d) Means—no. Standard deviations—yes; ticket prices must be independent of each other for different countries, but all tickets to the same country are at the same price.

5. a) $\mu = -\$0.20$, $\sigma = \$1.89$.   b) $\mu = -\$0.40$, $\sigma = \$2.67$.

7. a) 0.99998.   b) 0.9267.

9. Expected (extra) cost of the cheaper policy with the deductible is $2.50, much less than the $12 for the no-deductible surcharge, so on average, she will save money by going with the deductible. But the standard deviation ($35.27) is evidence of risk. Value of the car shouldn't influence the decision.

11. a) $\mu = 6.775$, $\sigma = 2.60$.   b) No, because $np = 6.78 < 10$.

13. a) 
| Spaces | 5 | 10 | 20 |
|---|---|---|---|
| P(Spaces) | 0.5 | 0.25 | 0.25 |

   b) $\mu = 10$, $\sigma = 6.12$.
   c) 
| Spaces | 0 | 1 | 2 | 3 | 4 |
|---|---|---|---|---|---|
| P(Spaces) | 1/3 | 1/6 | 1/6 | 1/6 | 1/6 |

   d) $\mu = 1.67$, $\sigma = 1.49$.
   e) $\mu = 11.67$, $\sigma = 6.30$.

15. $\mu = 8.33$, $\sigma = 6.30$.

17. a) 0.389.   b) 0.284.   c) 0.793.   d) 0.896.

19. a) No, this does not confirm the advice. If you follow the advice, it seems there's only an 80% chance it goes up in the third year.
    b) It actually has risen in 72% of all years. Not much difference from their strategy.

21. a) 0.716.   b) 0.053.   c) 0.960.

23. a) 12.   b) 3.32.
    c) Because both $np = 12 \geq 10$ and $nq = 138 \geq 10$.
    d) There is a 68% chance between 8.68 and 15.32 (9 and 15 students); 95% chance between 5.35 and 18.65 (6 and 18 students); 99.7% chance between 2.03 and 21.97 (3 and 21 students).

25. a) $\mu = 100$, $\sigma = 8$.   b) $\mu = 1000$, $\sigma = 60$.
    c) $\mu = 100$, $\sigma = 8.54$.   d) $\mu = 50$, $\sigma = 10$.
    e) $\mu = 100$, $\sigma = 11.31$.

27. a) Many do both, so the two categories can total more than 100%.
    b) No. They can't be disjoint. If they were, the total would be 100% or less.
    c) No. Probabilities are different for boys and girls.
    d) 0.0524.

29. a) 21 days.   b) 1649.73 som.
    c) 3300 som extra; about 157-som "cushion" each day.

31. Yes. You'd expect 539.6 homeowners, with an SD of 13.58 homeowners. 498 is just over 3 SDs below the mean; this is unusually low. Whether that's the mayor's fault is another question!

33. a) 0.0079.   b) 0.4117.   c) 0.199 (approximately).

35. a) 6.   b) 15.   c) 0.402.

37. a) 10%.   b) 42%.   c) 37.3%.
    d) 48% of departments used statistics software, 37.3% of departments that used calculators used software. Because these are not the same, these events are not independent.

39. a) 1/11.   b) 7/22.   c) 5/11.   d) 0.   e) 19/69.

41. a) $990.   b) $200.

43. a) Expected number of stars with planets.
    b) Expected number of planets with life.
    c) Probability of a planet with a suitable environment having intelligent life.
    d) $f_l$: If a planet has a suitable environment, the probability that life develops.
    $f_i$: If a planet develops life, the probability that the life evolves intelligence.
    $f_c$: If a planet has intelligent life, the probability that it develops radio communication.

45. 0.991.

## Practice Exam Answers

I. 1. B.   3. C.   5. A.   7. E.   9. E.
   11. D.   13. E.   15. D.   17. A.   19. C.
   21. B.   23. C.   25. E.

II. 1. a) 72.2%.   b) 43.3%.
    c)
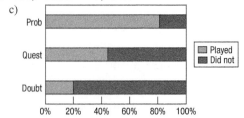

d) No. Players listed as "Probable" are more likely to play than those listed as "Questionable," and players listed as "Doubtful" are far less likely to play.

3. a) The patients are the units; the response variable is the pain rating.
   b) The treatments are the two dosage regimens.
   c) Number the patients 1–50. Generate a series of random numbers between 1 and 50. Ignoring repeats, assign the first 25 patients whose numbers are found to take 400 mg every 6 hours. Assign the remaining 25 patients to take 200 mg every 3 hours.
   d) The control group would allow us to see how much improvement in pain levels might occur in 8 days even without the medication, making it clearer how much effect the medication may be having with either dosing regimen.

5. a) Perhaps not. Three friends might have similar spending habits, so the trials may not be independent.
   b) 0.512.  c) $6.00.  d) $6.633.
   e) Yes. When sold at full price, the expected profit on 100 cards is $600. If sold at a $500 discount, the expected profit would be only $100. The standard deviation of that profit on 100 cards is $66.33, making the possibility of a loss only about 1.5 standard deviations below the mean, not an unlikely outcome.

## Chapter 17 (See note, page A-29)

1. All the histograms are centered near 0.05. As $n$ gets larger, the histograms approach the Normal shape, and the variability in the sample proportions decreases.

3. a) Mean = 3.4; SD = 1.517.
   b) 5, 4; 5, 4; 5, 3; 5, 1; 4, 4; 4, 3; 4, 1; 4, 3; 4, 1; 3, 1.
   c) Means: 4.5, 4.5, 4, 3, 4, 3.5, 2.5, 3.5, 2.5, 2.
   d) The mean is the same as that of the population. The standard deviation is smaller. The mean is an unbiased estimate of the population mean.

5. a) 0.536.
   b) 0.420. There is a difference of about 0.115.
   c) 0.125.
   d) 0.117. There is a difference of about 0.008.
   e) The Normal model does not give a close approximation when $np$ and $nq$ are too small. (Even when they are large, it's still an approximation!).

7. a) 

| $n$ | Observed Mean | Theoretical Mean | Observed SD | Theoretical SD |
|---|---|---|---|---|
| 20 | 0.0497 | 0.05 | 0.0479 | 0.0487 |
| 50 | 0.0516 | 0.05 | 0.0309 | 0.0308 |
| 100 | 0.0497 | 0.05 | 0.0215 | 0.0218 |
| 200 | 0.0501 | 0.05 | 0.0152 | 0.0154 |

   b) They are all quite close to what we expect from the theory.
   c) The histogram is unimodal and symmetric for $n = 200$.
   d) The Success/Failure Condition says that $np$ and $nq$ should both be at least 10, which is not satisfied until $n = 200$ for $p = 0.05$. The theory predicted my choice.

9. a) No; highly skewed.
   b) No; Normal does not give a good approximation.

11. a) Unimodal and symmetric, because probability of heads and tails is equal.
    b) 0.5.  c) 0.125.  d) $np = 8 < 10$.

13. a) About 68% should have proportions between 0.4 and 0.6, about 95% between 0.3 and 0.7, and about 99.7% between 0.2 and 0.8.
    b) $np = 12.5$, $nq = 12.5$; both are $\geq 10$.
    c)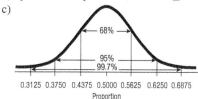
    
    $np = nq = 32$; both are $\geq 10$. Also, coin tosses are fair (unbiased) and independent, by their nature.
    d) The center remains at 0.5, the shape becomes more normal, and the spread becomes smaller as the sample size increases.

15. This is a fairly unusual result: about 2.26 SDs below the mean. The probability of that is about 0.012. So, in a class of 100 this is certainly a reasonable possibility.

17. a)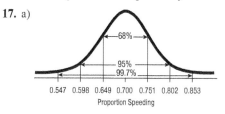

    b) Both $np = 56$ and $nq = 24 \geq 10$. However, this is not a random sample. And the drivers *may* not be independent of each other. For example, drivers might slow down because they see other drivers slow down after they see the officer. Or it might be an unusually fast (or slow) time of day. These problems could *completely* invalidate the model in part a.

19. a) Assume that these children are typical of the population. (*Note:* While we prefer random samples, sometimes the real world intrudes with its messiness and we are forced to work with the sample we have.) They represent fewer than 10% of all children. We expect 20.4 nearsighted and 149.6 not; both are at least 10.
    b)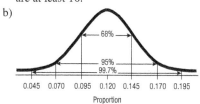
    c) Probably between 12 and 29.

21. a) Mean = 7%; SD = 1.8%.
    b) We have a random sample that we can reasonably assume represents 200 clients who are independent from each other. $np = 14$ and $nq = 186$; both are $\geq 10$.
    c) 0.048.

**23.**

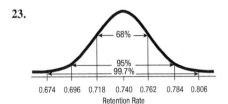

We have a random sample (assuming that students who take the ACT are typical of college students) and certainly 400 < 10% of college first-years; 296 and 104 are both ≥10.

**25.** Yes; if their students were typical, a retention rate of 522/603 = 86.6% would be over 7 standard deviations above the expected rate of 74%.

**27.** 0.212. Reasonable that those polled are independent of each other and represent less than 10% of all potential voters. The sample was selected at random. Success/Failure Condition met: $np = 208$, $nq = 192$; both ≥10.

**29.** 0.0008 using $N(0.65, 0.048)$ model.

**31.** Answers will vary. Using $\mu + 3\sigma$ for "very sure," the restaurant should have 52 with-kids seats. Assumes customers at any time are independent of each other, a random sample, and represent less than 10% of all potential customers. $np = 36$, $nq = 84$, so Normal model is reasonable ($\mu = 0.30$, $\sigma = 0.042$).

**33.** a) Normal, center at $\mu$, standard deviation $\sigma/\sqrt{n}$.
b) Standard deviation will be smaller. Center will remain the same. Shape will be closer to Normal.

**35.** a) The histogram is unimodal and slightly skewed to the right, centered at 36 inches with a standard deviation near 4 inches.
b) All the histograms are centered near 36 inches. As $n$ gets larger, the histograms approach the Normal shape and the variability in the sample means decreases. The histograms are fairly Normal by the time the sample reaches size 5.

**37.** a)

| n | Observed Mean | Theoretical Mean | Observed SD | Theoretical SD |
|---|---|---|---|---|
| 2 | 36.314 | 36.33 | 2.855 | 2.842 |
| 5 | 36.314 | 36.33 | 1.805 | 1.797 |
| 10 | 36.341 | 36.33 | 1.276 | 1.271 |
| 20 | 36.339 | 36.33 | 0.895 | 0.899 |

b) They are all very close to what we would expect.
c) For samples as small as 5, the sampling distribution of sample means is unimodal and very symmetric.
d) The distribution of the original data is nearly unimodal and symmetric, so it doesn't take a very large sample size for the distribution of sample means to be approximately Normal.

**39.**

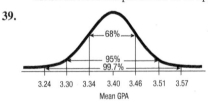

Approximately Normal, $\mu = 3.4$, $\sigma = 0.055$. We assume that the students are randomly assigned to the seminars, less than 10% of all possible students, and that individual's GPAs are independent of one another.

**41.** a) As the CLT predicts, there is more variability in the smaller outlets.
b) If the lottery is random, all outlets are equally likely to sell winning tickets.

**43.** a) 21.1%.   b) 276.8 days or more.
c) $N(266, 2.07)$.   d) 0.002.

**45.** a) There are more premature births than very long pregnancies. Modern practice of medicine stops pregnancies at about 2 weeks past normal due date.
b) Parts a and b—yes—we can't use Normal model if it's very skewed. Part c—no—CLT guarantees a Normal model for this large sample size.

**47.** a) $\mu = \$2.00$, $\sigma = \$3.61$.
b) $\mu = \$4.00$, $\sigma = \$5.10$.
c) 0.190. Model is $N(80, 22.80)$.

**49.** a) $\mu = 2.865$, $\sigma = 1.337$.
b) No. The score distribution in the sample should resemble that in the population, somewhat uniform for scores 1–4 slightly more 3's and fewer 5's.
c) Approximately $N\left(2.865, \dfrac{1.337}{\sqrt{40}}\right)$.

**51.** About 21%, based on $N(2.865, 0.168)$.

**53.** a) $N(2.9, 0.045)$.   b) 0.0131.
c) 2.97 gm/mi.

**55.** a) Can't use a Normal model to estimate probabilities. The distribution is skewed right—not Normal.
b) 4 is probably not a large enough sample to say the average follows the Normal model.
c) No. This is 3.16 SDs above the mean.

**57.** a) 0.0003. Model is $N(384, 34.15)$.
b) $427.77 or more.

**59.** a) 0.734.
b) 0.652. Model is $N(10, 12.81)$.
c) 0.193. Model is $N(120, 5.774)$.
d) 0.751. Model is $N(10, 7.394)$.

**61.** a) Binomial (25, 0.58); $P(x = 14) = 15.6\%$.
b) Center = 0.58, shape is approximately normal, standard deviation = 0.099.
c) $0.58 \pm 2(0.99) \approx$ anything below 38% or above 78% would be unusual. Because $n = 25$ is fairly small, the sample proportions will vary a lot!

## Chapter 18 (See note, page A-29)

**1.** She believes the true proportion is within 4% of her estimate, with some (probably 95%) degree of confidence.

**3.** a) Population—all cars; sample—those actually stopped at the checkpoint; $p$—proportion of all cars with safety problems; $\hat{p}$—proportion actually seen with safety problems (10.4%); if sample (a cluster sample) is representative, then the methods of this chapter will apply.
b) Population—general public; sample—those who logged onto the website; $p$—population proportion of those who favor prayer in school; $\hat{p}$—proportion of those who voted in the poll who favored prayer in school (81.1%); can't use methods of this chapter—sample is biased and nonrandom.

c) Population—parents at the school; sample—those who returned the questionnaire; $p$—proportion of all parents who favor uniforms; $\hat{p}$—proportion of respondents who favor uniforms (60%); should not use methods of this chapter because not SRS (possible nonresponse bias).

d) Population—students at the college; sample—the 1632 students who entered that year; $p$—proportion of all students who will graduate on time; $\hat{p}$—proportion of that year's students who graduate on time (85.0%); can use methods of this chapter if that year's students (a cluster sample) are viewed as a representative sample of all possible students at the school.

5. a) Not correct. This implies certainty.
b) Not correct. Different samples will give different results. Many fewer than 95% will have 88% on-time orders.
c) Not correct. The interval is about the population proportion, not the sample proportion in different samples.
d) Not correct. In this sample, we *know* 88% arrived on time.
e) Not correct. The interval is about the parameter, not the days.

7. a) False.  b) True.  c) True.  d) False.

9. a) This means that 45% of the 1058 teens reported that they were online "almost constantly." This is our best estimate of $p$, the proportion of *all* U.S. teens who would say that.
b) $SE(\hat{p}) = \sqrt{\frac{(0.45)(0.55)}{1058}} \approx 0.0153$.
c) Because we don't know $p$, we use $\hat{p}$ to estimate the standard deviation of the sampling distribution. So the standard error is our estimate of the sample to sample variation in the sample proportion when we ask 1058 teens how often they are online.

11. a) We are 95% confident that, if we were to ask all U.S. teens whether they are online almost constantly, between 42% and 48% of them would say they are.
b) If we were to collect many random samples of 1058 teens, about 95% of the confidence intervals we construct would contain the proportion of all U.S. teens who say they go online several times a day.

13. a) $SE(\hat{p}) = \sqrt{\frac{\hat{p}\hat{q}}{n}} = \sqrt{\frac{0.75 \times 0.25}{1997}} = 0.00969$
$ME = 1.645 \times 0.00969 = 0.01594$.
b) We are 90% confident that the observed proportion responding "wrong track" is within 0.016 of the population proportion.

15. a) Yes. Random sample and sufficiently large sample.
b) Larger. Higher confidence requires a wider confidence interval.

17. On the basis of this sample, we are 90% confident that the proportion of Japanese cars is between 29.9% and 47.0%.

19. a) It's not clear how the sample was chosen, but coming from many stores, it may be representative of all seafood sold. The sample was far less than 10% of the population. There were 12 successes and 10 failures, both at least 10.
b) (0.34, 0.75).
c) We're 95% confident that between 34% and 75% of all "red snapper" sold in food stores and restaurants in these three states is not actually red snapper, if this sample can be considered a random sample of all red snapper sold in these three states.

21. a) This means that 56% of the 1060 teens in the sample said they go online several times a day. This is our best estimate of $p$, the proportion of *all* U.S. teens who would say they do so.
b) $SE(\hat{p}) = \sqrt{\frac{(0.56)(0.44)}{1060}} \approx 0.0152$.
c) Because we don't know $p$, we use $\hat{p}$ to estimate the standard deviation of the sampling distribution. So the standard error is our estimate of the amount of variation in the sample proportion we expect to see from sample to sample when we ask 1060 teens whether they go online several times a day.

23. a) (0.0223, 0.0250). The sample is random and 50,000 is assumed to be less than 10% of the cardholders. 1184 & 48,816 ≥ 10.
b) Because 0.02 is below the confidence interval, it isn't one of the plausible values. Proceed with the plan.

25. a) (12.7%, 18.6%) Random sample. 582 < 10% of all U.S. accidents. 91 & 491 ≥ 10.
b) We are 95% confident, based on this sample, that the proportion of all auto accidents that involve American teenage drivers is between 12.7% and 18.6%.
c) About 95% of all random samples will produce confidence intervals that contain the true population proportion.
d) Contradicts. The interval is completely below 20%.

27. Probably nothing. Those who bothered to fill out the survey may be a biased sample.

29. a) Response bias (wording).
b) (53%, 59%).
c) Smaller—the sample size was larger.

31. a) (18.2%, 21.8%).
b) We are 98% confident, based on the sample, that between 18.2% and 21.8% of English children are deficient in vitamin D.
c) About 98% of all random samples will produce a confidence interval that contains the true proportion of children deficient in vitamin D.

33. a) $0.68 \pm 1.96 \times \sqrt{0.68 \times \frac{1 - 0.68}{97}} = (0.587, 0.773)$.
We assume the sample is representative. 66 & 31 ≥ 10.
b) We are 95% confident that between 58.7% and 77.3% of all college presidents were concerned about faculty readiness to conduct online learning or hybrid learning in June 2020.
c) No, the confidence interval includes 60%; it is not entirely above 60%, and we do not have evidence that *more than* 60% of all college presidents were concerned about faculty readiness to conduct online learning or hybrid learning.

35. a) (15.5%, 26.3%).   b) 612.
c) Sample may not be random or representative. Deer that are legally hunted may not represent all sexes and ages.

37. a) 141.   b) 318.   c) 564.

39. 1801.

41. 384 total, using $p = 0.15$.

43. 90%.

45. a) The Success/Failure Condition is not met (5 successes is less than 10).
b) (0.095, 0.572).

c) No. The interval seems reasonable, although too wide to be useful.
d) (−0.039, 0.305). Yes, one endpoint of the interval is negative, which is not reasonable for a proportion.
e) Intervals created by his process will have a capture rate less than 95%.

## Chapter 19 (See note, page A-29)

1. a) $H_0: p = 0.30$; $H_A: p < 0.30$.
   b) $H_0: p = 0.50$; $H_A: p \neq 0.50$.
   c) $H_0: p = 0.20$; $H_A: p > 0.20$.

3. Statement d is correct.

5. No. We can say only that there is a 27% chance of seeing the observed effectiveness just from natural sampling variation. There is no *evidence* that the new formula is more effective, but we can't conclude that they are equally effective.

7. a) No. There's a 25% chance of losing twice in a row. That's not unusual.
   b) 0.125.
   c) No. We expect that to happen 1 time in 8.
   d) Maybe 5? The chance of 5 losses in a row is only 1 in 32, which seems unusual.

9. a) The new drug is not more effective than aspirin.
   b) The new drug is more effective than aspirin.
   c) There is not sufficient evidence to conclude that the new drug is better than aspirin.
   d) There is evidence that the new drug is more effective than aspirin.

11. 1) Use $p$, not $\hat{p}$ in hypotheses.
    2) The question was about failing to meet the goal, so $H_A$ should be $p < 0.96$.
    3) Said SRS, but there's no mention of randomization.
    4) Did not check of the 10% Condition.
    5) Did not check $0.04(200) = 8$. Because $nq < 10$, the Success/Failure Condition is violated.
    6) $188/200 = 0.94$; $SD(\hat{p}) = \sqrt{\frac{(0.96)(0.04)}{200}} = 0.014$.
    7) $z$ is incorrect; it should be $z = \frac{0.94 - 0.96}{0.014} = -1.43$.
    8) $P = P(z < -1.43) = 0.076$
    9) We lack evidence that the new instructions do not work.

13. a) $H_0: p = 0.30$; $H_A: p > 0.30$.
    b) Random sample. $80 < 10\%$ of 1000. $(0.3)(80) \geq 10$ and $(0.7)(80) \geq 10$. Wells are independent only if customers don't have farms on the same underground springs.
    c) $z = 0.73$; P-value $= 0.232$.
    d) If his dowsing is no different from standard methods, there is more than a 23% chance of seeing results as good as those of the dowser's, or better, by natural sampling variation.
    e) These data provide no evidence that the dowser's chance of finding water is any better than normal drilling.

15. a) $H_0: p_{2000} = 0.34$; $H_A: p_{2000} \neq 0.34$.
    b) Students were randomly sampled and should be independent. 34% and 66% of 8302 are greater than 10. 8302 students is less than 10% of the entire student population of the United States.

c) $z = -1.92$; P-value $= 0.055$.
d) With such a P-value greater than 5%, I fail to reject $H_0$. There is insufficient evidence to say there has been a change in the proportion of students who have no absences.
e) No. A difference this small, which is not statistically significant, is also not practically meaningful. We might look at new data in a few years.

17. a) $H_0: p = 0.05$ vs. $H_A: p < 0.05$.
    b) This is a random sample, 100,000 is less than 10% of millions, and we expect 5000 successes and 95,000 failures, both greater than 10.
    c) $z = -3.178$; P-value $= 0.00074$, so we reject $H_0$; there is strong evidence that the donation rate would be below 5%.

19. a) $H_0: p = 0.66$; $H_A: p > 0.66$.
    b) The sample is representative. $240 < 10\%$ of all law school applicants. We expect $240(0.66) = 158.4$ to be admitted and $240(0.34) = 81.6$ not to be; both at least 10. $z = 0.63$; P-value $= 0.26$.
    c) Because the P-value is high, there is not sufficient evidence to claim that LSATisfaction demonstrates improvement over the national average.

21. 22 is more than 10% of the population of 150 and $(0.20)(22) < 10$. Do not proceed with a test. The standard deviation would not be correct and the sampling distribution of the sample proportion would not be normal.

23. $H_0: p = 0.03$; $\hat{p} = 0.015$. One mother having twins will not affect another, so observations are independent; sample is less than 10% of all such births. We are assuming that this observational study is providing us with a representational sample; however, the mothers at this hospital may not be representative of all teenagers; $(0.03)(469) = 14.07 \geq 10$; $(0.97)(469) \geq 10$. $z = -1.91$; P-value $= 0.0556$. The P-value is not low enough to reject the null hypothesis at a significance level of $\alpha = 0.05$, but there is some weak evidence that the rate of twins born to teenage girls at this hospital is less than the national rate of 3%. It is not clear whether this can be generalized to all teenagers.

25. $H_0: p = 0.25$; $H_A: p > 0.25$. Random sample; sample is less than 10% of all potential subscribers; $(0.25)(500) \geq 10$; $(0.75)(500) \geq 10$. $z = 1.24$; P-value $= 0.1076$. The P-value is high, so do not reject $H_0$. These data do not show that more than 25% of current readers would subscribe; the company should not go ahead with the WebZine on the basis of these data.

27. $H_0: p = 34\%$; $H_A: p > 34\%$ $p = $ the true percentage of searchers that click through. The sample is random; given the large number of searches, it is very reasonable that we have sample $< 10\%$; 192.8 and 374.2 $> 10$. $\hat{p} = 0.4497$; $z = 5.516$ With a P-value that is essentially zero, we reject the null hypothesis. We found very strong evidence that the percentage of searchers have a click-through rate of more than 34%.

29. $H_0: p = 0.053$; $H_A: p > 0.053$. We are assuming that this sample is representative of the current dropout rate. 94.4 and 1687.6 $\geq 10$. $\hat{p} = 0.073$; $z = 3.76$; P-value $= 0.0001$. Because the P-value is very, very low, we reject $H_0$. These data do provide very convincing evidence that the dropout rate was unusually high in this district last year.

31. $H_0: p = 0.90$; $H_A: p < 0.90$. Random sample; $122 < 10$ of travelers who lost luggage; $122(0.9)$, $122(0.1) \geq 10$. $\hat{p} = 0.844$; $z = -2.05$; P-value $= 0.0201$. Because the P-value is so low, we reject $H_0$. There is strong evidence that the actual rate at which passengers with lost luggage are reunited with it within 24 hours is less than the 90% claimed by the airline.

33. a) Yes; assuming this sample to be a typical group of people, $P = 0.0008$. This cancer rate is very unusual.
    b) No, this group of people may be atypical for reasons that have nothing to do with the radiation.

35. a) 0.017.
    b) Because $0.017 < \alpha = 0.05$, reject $H_0$. The researchers have evidence that the new trap attracts more than 70% of cockroaches that pass within 20 cm of the trap.

37. Because confidence interval is completely below the value of 45%, the mayor will conclude that there is evidence that the percentage of people who will come to a full and complete stop is below 45%.

## Chapter 20 (See note, page A-29)

1. a) Let $p =$ probability of winning on the slot machine. $H_0: p = 0.01$ vs. $H_A: p \neq 0.01$.
   b) Let $p =$ proportion of patients cured by the new drug. $H_0: p = 0.3$ vs. $H_A: p \neq 0.3$.
   c) Let $p =$ proportion of clients now using the website. $H_0: p = 0.4$ vs. $H_A: p \neq 0.4$.

3. a) False. It provides evidence against it but does not show it is false.
   b) False. The P-value is not the probability that the null hypothesis is true.
   c) True.
   d) False. Whether a P-value provides enough evidence to reject the null hypothesis depends on the risk of a Type I error that one is willing to assume (the $\alpha$ level).

5. a) False. A high P-value shows that the data are consistent with the null hypothesis but provides no evidence for rejecting the null hypothesis.
   b) False. It results in rejecting the null hypothesis but does not prove that it is false.
   c) False. A high P-value shows that the data are consistent with the null hypothesis but does not prove that the null hypothesis is true.
   d) False. Whether a P-value provides enough evidence to reject the null hypothesis depends on the risk of a Type I error that one is willing to assume (the $\alpha$ level).

7. a) $H_0: p = 0.40$ vs. $H_A: p \neq 0.40$. Two-sided.
   b) $H_0: p = 0.42$ vs. $H_A: p > 0.42$. One-sided.
   c) $H_0: p = 0.50$ vs. $H_A: p > 0.50$. One-sided.

9. a) True.
   b) False. The alpha level is set independently and does not depend on the sample size.
   c) False. The P-value would have to be less than 0.01 to reject the null hypothesis.
   d) False. It simply means we do not have enough evidence at that alpha level to reject the null hypothesis.

11. a) Type I error. The actual value is not greater than 0.3 but it rejected the null hypothesis.
    b) No error. The actual value is 0.50, which was not rejected.
    c) Type II error. The null hypothesis was not rejected, but it was false. The true relief rate was greater than 0.25.

13. a) Two-sided. Let $p$ be the percentage of students who prefer Diet Pepsi. $H_0: p = 0.5$ vs. $H_A: p \neq 0.5$.
    b) One-sided. Let $p$ be the percentage of teenagers who prefer the new formulation. $H_0: p = 0.5$ vs. $H_A: p > 0.5$.
    c) One-sided. Let $p$ be the percentage of people who intend to vote for the override. $H_0: p = 2/3$ vs. $H_A: p > 2/3$.
    d) Two-sided. Let $p$ be the percentage of days that the market goes up. $H_0: p = 0.5$ vs. $H_A: p \neq 0.5$.

15. If there is no difference in effectiveness, the chance of seeing an observed difference this large or larger is 4.7% by natural sampling variation.

17. At $\alpha = 0.10$:, yes. The P-value is less than 0.05, so it's less than 0.10. But to reject $H_0$ at $\alpha = 0.01$, the P-value must be below 0.01, which isn't necessarily the case.

19. a) There is only a 1.1% chance of seeing a sample proportion as low as 89.4% vaccinated by natural sampling variation if 90% have really been vaccinated.
    b) We conclude that $p$ is below 0.9, but a 95% confidence interval would suggest that the true proportion is between (0.889, 0.899). Most likely, a decrease from 90% to 89.9% would not be considered important. On the other hand, with 1,000,000 children a year vaccinated, even 0.1% represents about 1000 kids—so this may very well be important.

21. a) Random sample; $4000 < 10\%$ of all gamers, 1520 and 2480 $> 10$; (0.362, 0.398).
    b) Because 35% is not in the interval we have strong evidence that more than 35% of gamers are 18 to 34.
    c) $\alpha = 0.01$; it's an upper tail test based on a 98% confidence interval.

23. a) (0.375, 0.425).
    b) Because 52.7% is not in the confidence interval, we can reject the hypothesis that $p = 0.527$. Because the entire interval is lower than 0.527, it appears that Trump's level of support is lower.

25. a) The Success/Failure Condition is violated: only 5 pups had dysplasia.
    b) Dogs whose owners take them to a vaccination clinic may have different overall health and care than the general population of dogs.

27. a) Type II error.
    b) Type I error.
    c) By making it easier to get the loan, the bank has reduced the alpha level.
    d) The risk of a Type I error is decreased and the risk of a Type II error is increased.

29. a) Power is the probability that the bank denies a loan that would not have been repaid.
    b) Raise the cutoff score.
    c) A larger number of trustworthy people would be denied credit, and the bank would miss the opportunity to collect interest on those loans.

**31.** a) The null is that the level of homeownership remains the same. The alternative is that it rises.
b) The city concludes that homeownership is on the rise, but in fact the tax breaks don't help.
c) The city abandons the tax breaks, but they were helping.
d) A Type I error causes the city to forgo tax revenue, while a Type II error withdraws help from those who might have otherwise been able to buy a home.
e) The power of the test is the city's ability to detect an actual increase in homeownership.

**33.** a) It is decided that the shop is not meeting standards when it is.
b) The shop is certified as meeting standards when it is not.
c) Type I.
d) Type II.

**35.** a) The probability of detecting a shop that is not meeting standards.
b) 40 cars; larger $n$.
c) 10%; more chance to reject $H_0$.
d) A lot. Larger differences are easier to detect.

**37.** a) One-tailed. The seniors' complaint is that they are underrepresented.
b) Deciding the seniors were underrepresented, when it was just random variation.
c) Deciding the difference was just random variation, when in fact more fewer seniors were selected.
d) The probability of correctly detecting that seniors were underselected.
e) Increases power.
f) Lower, because $n$ is smaller.

**39.** a) One-tailed. Software is supposed to decrease the dropout rate.
b) $H_0: p = 0.13$; $H_A: p < 0.13$.
c) He buys the software when it doesn't help students.
d) He doesn't buy the software when it does help students.
e) The probability of correctly deciding the software is helpful.

**41.** a) $z = -3.21$, $p = 0.0007$. The change is statistically significant. A 95% confidence interval is (2.3%, 8.5%). This is clearly lower than 13%. If the cost of the software justifies it, the professor should consider buying the software.
b) The chance of observing 11 or fewer dropouts in a class of 203 is only 0.07% if the dropout rate is really 13%.

**43.** a) $H_A: p = 0.30$, where $p$ is the probability of heads
b) Reject the null hypothesis if the coin comes up tails—otherwise fail to reject.
c) P(tails given the null hypothesis) = $0.1 = \alpha$.
d) P(tails given the alternative hypothesis) = power = 0.70.
e) Spin the coin more than once and base the decision on the sample proportion of heads.

**45.** a) 0.0464.   b) Type I.   c) 37.6%.
d) Increase the number of shots. Or keep the number of shots at 10, but increase alpha by declaring that 8, 9, or 10 will be deemed as having improved.

**47.** a) $H_0: p = 0.4$, where $p$ is the probability a person is cured of the rash within one month; $H_A: p > 0.4$
b) $\mu_{\hat{p}} = p_0 = 0.4$; $\sigma_{\hat{p}} = \sqrt{\dfrac{p_0(1 - p_0)}{n}} = \sqrt{\dfrac{0.4(1 - 0.4)}{50}} = 0.069$.

c) $0.4 \pm 1.96 \times 0.069$, so between 26% and 54%.
d) No; yes; no error; $(200 - 6)/200 = 97\%$.
e) Yes; no; Type I error; $6/200 = 3\%$.

## Chapter 21 (See note, page A-29)

**1.** 0.0161.

**3.** We are 95% confident that, based on these data, the proportion of foreign-born Canadians is between 3.24% and 9.56% more than the proportion of foreign-born Americans.

**5.** If we were to take repeated samples of these sizes of Canadians and Americans, and compute two-proportion confidence intervals, we would expect 95% of the intervals to contain the true difference in the proportions of foreign-born citizens.

**7.** We must assume the data were collected randomly and that the Americans selected are independent of the Canadians selected. Both assumptions should be met. Also, for both groups, we have at least 10 national-born and foreign-born citizens and the sample sizes are less than 10% of the population sizes. All conditions for inference are met.

**9.** a) 0.064.   b) 4.03.
c) P-value is $< 0.001$. Very strong evidence, so reject the null hypothesis that the proportions are the same in the two countries.

**11.** It's very unlikely that samples would show an observed difference this large if in fact there is no real difference in political knowledge between those who get their news from social media and those who watch cable news.

**13.** The ads may be working. If there had been no real change in name recognition, there'd be only about a 3% chance the percentage of sampled voters who heard of this candidate would be at least this much higher.

**15.** The responses are not from two independent groups, but are from the same individuals.

**17.** a) Stratified.   b) 6% higher among males.   c) 4%.
d)

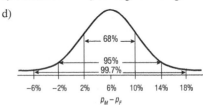

e) Yes; a poll result showing little difference is only 1–2 standard deviations below the expected outcome.

**19.** a) Yes; random sample; less than 10% of the population; samples are independent; more than 10 successes and failures in each sample.
b) $(0.055, 0.140)$.
c) We are 95% confident, based on these samples, that the proportion of American women age 65 and older who suffer from arthritis is between 5.5% and 14.0% more than the proportion of American men of the same age who suffer from arthritis.
d) Yes; the entire interval lies above 0.

**21.** a) 0.035.
b) Randomly selected homes; 827 and 130 are less than 10% of the population; 473, 354, 19, and 111 are more than 10; (0.356, 0.495).
c) We are 95% confident, based on these data, that the proportion of pets with a malignant lymphoma in homes where herbicides are used is between 35.6% and 49.5% higher than the proportion of pets with lymphoma in homes where no pesticides are used.

**23.** a) Experiment; men were randomly assigned to have surgery or not.
b) (0.006, 0.080).
c) Because the entire interval lies above 0, there is evidence that surgery may be effective in preventing death from prostate cancer.

**25.** a) Yes; subjects were randomly divided into independent groups, and more than 10 successes and failures were observed in each group.
b) (4.7%, 8.9%).
c) Yes; we're 95% confident that the rate of infection is 5–9 percentage points lower. That's a meaningful reduction, considering the 20% infection rate among the unvaccinated kids.

**27.** a) $H_0: p_V - p_{NV} = 0$, $H_A: p_V - p_{NV} < 0$.
b) Because 0 is not in the confidence interval, reject the null. There's evidence that the vaccine reduces the rate of ear infections.
c) 2.5%.
d) Type I.
e) Babies would be given ineffective vaccinations.

**29.** a) Prospective study.
b) $H_0: p_1 - p_2 = 0$, $H_A: p_1 - p_2 \neq 0$, where $p_1$ is the proportion of students whose parents disapproved of smoking who became smokers and $p_2$ is the proportion of students whose parents are lenient about smoking who became smokers.
c) Yes; we assume the students are a representative sample; they are less than 10% of the population; samples are independent; 57.7, 226.3, 10.4, 40.6 are all $> 10$.
d) $z = -1.38$, P-value $= 0.1678$. Because the P-value is greater than any reasonable significance level such as 0.05, these samples do not show evidence that parental attitudes influence teens' decisions to smoke.
e) If there is no difference in the proportions, there is about a 17% chance of seeing the observed difference or larger by natural sampling variation.
f) Type II.

**31.** a) (−0.046, 0.225).
b) We are 95% confident that the proportion of teens whose parents disapprove of smoking who will eventually smoke is between 22.5% less and 4.6% more than for teens with parents who are lenient about smoking.
c) 95% of all random samples will produce intervals that contain the true difference in smoking rates between teens who have lenient vs. disapproving parents in the population.

**33.** a) No; subjects weren't assigned to treatment groups. It's an observational study.
b) $H_0: p_Y - p_O = 0$, $H_A: p_Y - p_O \neq 0$. $z = 3.56$, P-value $= 0.0004$. With a P-value this low, we reject $H_0$. There is evidence of a difference in the clinic's effectiveness. We have evidence that younger mothers have a higher birth rate than older mothers.
c) We are 95% confident, based on these data, that the proportion of successful live births at the clinic is between 10.0% and 27.8% higher for mothers under 38 than in those 38 and older.

**35.** a) The polls are independent random samples; 328.4, 301.6, 526.6, 483.4 are all $> 10$. $H_0: p_1 - p_2 = 0$, $H_A: p_1 - p_2 > 0$. $z = 1.18$, P-value $= 0.120$. With a P-value this high, we fail to reject $H_0$. These data do not show evidence of a decrease in the voter support for the candidate.
b) Type II.

**37.** a) $H_0: p_1 - p_2 = 0$; $H_A: p_1 - p_2 > 0$, where $p_1$ is the proportion of women who never had mammograms who die of breast cancer and $p_2$ is the proportion of women who undergo screening who die of breast cancer. $z = 2.17$, P-value $= 0.0148$. With a P-value this low, we reject $H_0$. These data do suggest that mammograms are associated with fewer breast cancer deaths.
b) Type I.

**39.** a) We are 95% confident that between 67.0% and 83.0% of patients with joint pain will find medication A effective.
b) We are 95% confident that between 51.9% and 70.3% of patients with joint pain will find medication B effective.
c) Yes; they overlap. This might indicate no difference in the effectiveness of the medications. (Not a proper test.)
d) We are 95% confident that the proportion of patients with joint pain who will find medication A effective is between 1.7% and 26.1% higher than the proportion who will find medication B effective.
e) No; the evidence suggests there is a difference in the effectiveness of the medications.
f) To estimate the variability in the difference of proportions, we must add variances. The two one-sample intervals do not. The two-sample method is the correct approach.

**41.** Independent random samples; 400, 246.3, 95.3, and 58.7 are all $> 10$. $H_0: p_{urban} - p_{rural} = 0$ vs. $H_A: p_{urban} - p_{rural} \neq 0$. Yes; $z = 3.19$. With a low P-value of 0.0013, reject the null hypothesis of no difference. There is strong evidence to suggest that the percentages are different for the two groups: It appears that people from urban areas are more likely to agree with the statement than those from rural areas.

**43.** Yes; $z = 3.44$. With a low P-value of 0.0003, reject the null hypothesis of no difference. There's evidence of an increase in the proportion of parents checking the websites visited by their teens.

**45.** A 95% confidence interval is (−1.6%, 5.6%); 0%—or no bounce—is a plausible value. They should have said that there was no evidence of a bounce from their poll, however, because they can't prove there was none at all. (OR $z = -1.09$; P-value $= 0.136$).

**47.** We need to assume their data is taken from 2 random and independent samples; 73.4, 55.6, 104.6, and 79.4 are all $> 10$. $H_0: p_{young} - p_{old} = 0$ against $H_A: p_{young} - p_{old} > 0$. $z = 1.54$. The one-sided P-value is 0.062, so we may not reject the null hypothesis at $\alpha = 0.05$, but would at $\alpha = 0.10$. Although the evidence is not strong, *Time* may be justified in saying that younger men are more comfortable discussing personal problems.

**49.** a) 20%.

b) Because a difference of four or more can happen just by chance 20% of the time (which is greater than any reasonable significance level), we are not convinced that children with their parents are more polite than those without their parents. The difference we observed could just be due to random chance.

## Part V Review (See note, page A-29)

**1.** Call the omega-3 subjects Group 2. $H_0$: There is no difference in relapse rates, $p_1 - p_2 = 0$. $H_A$: The relapse rate in those who use omega-3 fatty acids is lower, $p_1 - p_2 > 0$.

**3.** a) 10.29.

b) Not really. The $z$-score is $-1.11$. Not any evidence to suggest that the proportion for Monday is low.

c) It seems high. The $z$-score is 2.26 with a P-value of 0.024 (two-sided). But note that it may not be that unusual for *some* day of the week (if not necessarily Tuesdays) to have a count this high. This issue will be addressed in Chapter 25!

d) Some births are scheduled for the convenience of the doctor and/or the mother.

**5.** a) $H_0$: $p = 0.77$; $H_A$: $p < 0.77$.

b) Random sample; less than 10% of all California gas stations, $0.77 \times 47 = 36.2$, $0.23 \times 47 = 10.8$. Both are greater than 10. Assumptions and conditions are met.

c) $z = -1.106$, P-value $= 0.134$.

d) With a P-value this high, we fail to reject $H_0$. These data do not provide convincing evidence that the proportion of leaking gas tanks in California is less than 77%.

e) Yes, Type II.

f) Increase $\alpha$; increase the sample size.

g) Increasing $\alpha$—increases power, lowers chance of Type II error, but increases chance of Type I error. Increasing sample size—increases power, costs more time and money.

**7.** a) The researcher believes that the true proportion of "A's" is within 10% of the estimated 54%, namely, between 44% and 64%.

b) Small sample or high confidence level.

c) No, 63% is contained in the interval.

**9.** a) We can be 95% confident that the true proportion is within 1.4 percentage points of 38%—that is, that it is between 36.6% and 39.4%.

b) The mail group would have the larger ME because its sample size is smaller.

c) CI $= (78.3\%, 81.7\%)$.

d) The ME is 0.017, which is larger than the 1.4% ME in part a, largely because of the smaller sample size.

**11.** a) Bimodal (Read the question carefully!).

b) $\mu$, the population mean. Sample size does not matter.

c) $\sigma/\sqrt{n}$; sample size does matter.

d) It becomes closer to a Normal model and narrower as the sample size increases.

**13.** a) For $\hat{p}$, $\mu = 0.80$, $\sigma = 0.028$.

b) Yes. $0.8(200) = 160$, $0.2(200) = 40$. Both $\geq 10$. Assume shots are independent.

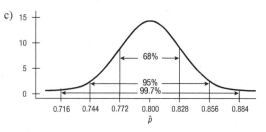

d) 0.039.

**15.** Let Group 1 be 1990 and Group 2 be 2000. $H_0$: There is no difference, $p_1 - p_2 = 0$. $H_A$: Early births have increased, $p_1 - p_2 < 0$. $z = -0.729$, P-value $= 0.2329$. Because the P-value is so high, we do not reject $H_0$. These data do not show evidence of an increase in the incidence of early birth of twins.

**17.** a) Let 1 represent the magnesium sulfate group. $H_0$: There is no difference, $p_1 - p_2 = 0$. $H_A$: Treatment prevents deaths from eclampsia, $p_1 - p_2 < 0$.

b) Samples are random and independent; less than 10% of all pregnancies (or eclampsia cases); more than 10 successes and failures in each group.

c) 0.8008.

d) There is insufficient evidence to conclude that magnesium sulfate is effective in preventing eclampsia deaths.

e) Type II.

f) Increase the sample size; increase $\alpha$.

g) Increasing sample size: decreases variation in the sampling distribution, is costly. Increasing $\alpha$: Increases likelihood of rejecting $H_0$, increases chance of Type I error.

**19.** a) $1.96 \times \sqrt{\dfrac{(0.81)(0.19)}{1100}} \approx 0.02318$, or about 2.3 percentage points.

b) It would be greater. Smaller groups yield wider margins of error, all other things being equal.

c) Different response rates by different demographic groups can lead to sampling bias even when random digit dialing is used. Demographics that are under-represented in responses relative to the Census (e.g., young people, who tend to respond in very low rates) are "weighted" more heavily—that is, their responses count more in the estimate of $p$. (And a special formula for the margin of error is used, accounting for the different weights.)

**21.** a) $H_0$: $p = 0.36$, where $p$ is the proportion of all drivers age 15 to 20 that speed. $H_A$: $p < 0.36$.
$n\hat{p} = 99$, $n(1 - \hat{p}) = 726$ are both at least 10. 825 is fewer than 10% all drivers age 15 to 20. Conditions are met for a one-proportion $z$-test.

$$z = \dfrac{\hat{p} - p_0}{\sqrt{\dfrac{p_0(1-p_0)}{n}}} = \dfrac{0.12 - 0.36}{\sqrt{\dfrac{0.36(0.64)}{825}}} = -14.4$$

P-value $=$ almost 0.

Because the *P*-value of almost 0 is so low, we reject the null hypothesis. We have convincing evidence that the proportion of teens that speed is less than 0.36.

b) If 36% of all teens actually speed, the probability of getting a random sample of 825 teens in which 99 or fewer of them speed is almost 0.

**23.** a) One would expect many small fish, with a few large ones.
b) We don't know the exact distribution, but we know it's not Normal.
c) Probably not. With a skewed distribution, a sample size of five is not a large enough sample to say the sampling model for the mean is approximately Normal.
d) 0.961.

**25.** a) Yes. $0.8(60) = 48$, $0.2(60) = 12$. Both are $\geq 10$.
b) 0.834.
c) Higher. Bigger sample has smaller standard deviation for $\hat{p}$.
d) Answers will vary. For $n = 500$, the probability is 0.997.

**27.** a) Assume this is a representative group of less than 10% of patients; 335, $238 \geq 10$. 54.4 to 62.5%.
b) Based on this study, with 95% confidence the proportion of Crohn's disease patients who will respond favorable to infliximab is between 54.4% and 62.5%.
c) 95% of all such random samples will produce confidence intervals that contain the true proportion of patients who respond favorably.

**29.** At least 423, assuming that $p$ is near 50%.

**31.** a) Assume it's a representative sample, certainly less than 10% of all preemies and normal babies; more than 10 failures and successes in each group. 1.5% to 16.2% greater for normal-birth-weight children.
b) Because 0 is not in the interval, there is evidence that preemies have a lower high school graduation rate than children of normal birth weight.
c) Type I, because we rejected the null hypothesis.

**33.** a) $H_0$: The computer is undamaged. $H_A$: The computer is damaged.
b) 20% of good PCs will be classified as damaged (bad), while all damaged PCs will be detected (good).
c) 3 or more.
d) 20%.
e) By switching to two or more as the rejection criterion, 7% of the good PCs will be misclassified, but only 10% of the bad ones will, increasing the power from 20% to 90%.

**35.** The null hypothesis is that Obama's approval proportion was 66%—the Clinton benchmark. The one-tailed test has a $z$-value of $-2.00$, so the P-value is 0.023. It looks like there is reasonable evidence to suggest that Obama's January 2017 ratings were lower than the Clinton benchmark high.

**37.** a) The company is interested only in confirming that the athlete is well known.
b) Type I: the company concludes that the athlete is well known, but that's not true. It offers an endorsement contract to someone who lacks name recognition. Type II: the company overlooks a well-known athlete, missing the opportunity to sign a potentially effective spokesperson.
c) Type I would be more likely, Type II less likely.

**39.** I am 95% confident that the proportion of U.S. adults who favor nuclear energy is between 7 and 19 percentage points higher than the proportion who would accept a nuclear plant near their area.

**41.** a) We are 95% confident that between 45.36% and 61.22% of patients with metastatic melanoma will have at least a partial response to vemurafenib based on this study.
b) 266 or 267 patients (using 0.5 or 0.53 as the proportion).

## Practice Exam Answers

**I.**
| | | | | |
|---|---|---|---|---|
| **1.** B. | **3.** C. | **5.** D. | **7.** B. | **9.** C. |
| **11.** D. | **13.** D. | **15.** E. | **17.** C. | **19.** E. |
| **21.** B. | **23.** D. | **25.** D. | **27.** A. | **29.** E. |

**II. 1.** a) Slightly skewed to the right. The mean is a bit larger than the median and the right tail is longer than the left tail.
b) Yes. Any point above $Q3 + 1.5IQR = 21 + 1.5(4) = 27$ is an outlier, and there is at least one value at 28. Any point below $Q1 - 1.5IQR = 17 - 1.5(4) = 11$ would be an outlier, but the minimum is 14 so there are no low outliers.
c) At least 25% of the cars get 17 mpg or less, so that's not "exceptionally gas-thirsty."

**3.** (*Answers may vary.*)
a) The wording of the question implies people are unhappy, which may influence employees to also say that they are unhappy, causing managers to overestimate employee dissatisfaction.
b) The fact that e-mail responses would not be anonymous might lead employees to say they're happier than they really are, causing managers to underestimate employee dissatisfaction.
c) Those who chose not to reply may feel differently from those who did. If respondents tend to be (or at least claim to be) happier, this would lead managers to underestimate employee dissatisfaction.

**5.** a) $H_0$: $p = 0.0435$; $H_A$: $p > 0.0435$.
We assume these 841 birds are representative of all Chernobyl area barn swallows; also that they comprise fewer than 10% of such barn swallows; $841(0.0435) = 36.6 \geq 10$ and $841(0.9565) = 804.4 \geq 10$. OK to do a 1-proportion $z$-test.
$$\hat{p} = \frac{112}{841} = 0.1332 \Rightarrow$$
$$z = \frac{0.1332}{\sqrt{\frac{(0.0435)(0.9565)}{841}}} = 12.75; P \approx 0$$
Because the P-value is extremely small, we reject $H_0$. There is very strong evidence that the rate of albinism in barn swallows is abnormally high in the Chernobyl area.
b) Nothing; there may be other causes.

## Chapter 22 (See note, page A-29)

**1.** a) SDs are 1 lb, 0.5 lb, and 0.2 lb, respectively.
b) The distribution of pallets. The CLT tells us that the Normal model is approached in the limit regardless of the underlying distribution. As samples get larger, the approximation gets better.

**3.** a) 1.74. b) 2.37. c) 0.0524. d) 0.0889.

**5.** Shape becomes closer to Normal; center does not change; spread becomes narrower.

**7.** a) The confidence interval is for the population mean, not the individual cows in the study.
b) The confidence interval is not for individual cows.
c) We *know* the average gain in this study was 56 pounds!
d) The average weight gain of all cows does not vary. It's what we're trying to estimate.
e) No. There is not a 95% chance for another sample to have an average weight gain between 45 and 67 pounds. There is a 95% chance that another sample will have its average weight gain within two standard errors of the population mean.

**9.** a) No. A confidence interval is not about individuals in the population.
b) No. It's not about individuals in the sample, either.
c) No. We know the mean cost for students in the sample was $1467.
d) No. A confidence interval is not about other sample means.
e) Yes. A confidence interval estimates a population parameter.

**11.** a) Based on this sample, we can say, with 95% confidence, that the mean pulse rate of adults is between 70.9 and 74.5 beats per minute.
b) 1.8 beats per minute.
c) Larger.

**13.** a) Houses are independent; randomly sampled. We should check the Nearly Normal Condition by making a histogram. Sample size: 36 should be big enough.
b) ($9,052.50, $10,067.50).
c) We are 95% confident that the interval ($9052.50, $10,067.50) contains the mean gain in home value for all homes in the community.

**15.** The assumptions and conditions for a *t*-interval are not met. The distribution is highly skewed to the right and there is a large outlier.

**17.** a) Yes. Randomly selected group; less than 10% of the population; the histogram is not unimodal and symmetric, but it is not highly skewed and there are no outliers, so with a sample size of 52, the CLT says $\bar{y}$ is approximately Normal.
b) (98.06, 98.51) degrees F.
c) We are 98% confident, based on the data, that the average body temperature for an adult is between 98.06°F and 98.51°F.
d) 98% of all such random samples will produce intervals containing the mean temperature of all adults in the population.
e) These data suggest that the mean normal temperature is somewhat less than 98.6°F.

**19.** We are 95% confident that the interval ($9052.50, $10,067.50) contains the mean gain in home value for all houses in the community. That is, 95% of all random samples of bids will produce intervals that will contain this population mean.

**21.** The interval is a range of possible values for the mean shoe size. The average is not a value that any individual in the population will have, but an average of all the individuals.

**23.** a) Narrower. A smaller margin of error, so less confident.
b) Advantage: more chance of including the true value. Disadvantage: wider interval.
c) Narrower; due to the larger sample, the SE will be smaller.
d) About 252.

**25.** a) The distribution is a bit skewed to the right, so with a sample size of 12 the "nearly normal" condition is not met.
b) If anything, this distribution looks more skewed. But with $n = 32 > 30$, the sampling distribution of the sample mean should be approximately normal. The "nearly normal" condition is now met.

**27.** a) (709.90, 802.54).
b) With 95% confidence, based on these data, the speed of light is between 299,709.9 and 299,802.5 km/sec.
c) Normal model for the distribution, independent measurements. These seem reasonable here, but it would be nice to see if the Nearly Normal Condition held for the data.

**29.** a) Given no time trend, the monthly on-time departure rates should be independent. Though not a random sample, these months should be representative. The histogram looks unimodal, possibly left-skewed; not a concern with this large sample.
b) 77.84% < $\mu$(OT Departure%) < 78.75%.
c) We can be 90% confident that the interval from 77.84% to 78.75% holds the mean monthly percentage all flight departures that are on time.

**31.** $t = 2.2$ on 35 df, $P = 0.034$. We reject the null hypothesis because 0.034 is small, and conclude that the gain of home values in this community does appear to be unusual.

**33.** The 95% confidence interval lies entirely above the 0.08 ppm limit, evidence that mirex contamination is too high and consistent with rejecting the null. We used an upper-tail test, so the P-value should therefore be smaller than $(1 - (\frac{1}{2}) 0.95) = 0.025$, and it was.

**35.** If in fact the mean cholesterol of pizza eaters does not indicate a health risk, then 7 of every 100 samples would have mean cholesterol levels as high (or higher) as observed in this sample.

**37.** a) Upper-tail. We want to show it will hold 500 pounds (or more) easily.
b) They will decide the stands are safe when they're not.
c) They will decide the stands are unsafe when they are in fact safe.

**39.** a) Decrease $\alpha$. This means a smaller chance of declaring the stands safe if they are not.
b) The probability of correctly detecting that the stands are capable of holding more than 500 pounds.
c) Decrease the standard deviation—probably costly. Increase the sample size—takes more time for testing and is costly. Increase $\alpha$—more Type I errors. Increase the "design load" to be well above 500 pounds—again, costly.

**41.** a) $H_0$: $\mu = 23.3$; $H_A$: $\mu > 23.3$.
b) We have a random sample of the population. Population may not be normally distributed, as it would be easier to have a few much older men at their first marriage than some very young men. However, with a sample size of 40, $\bar{y}$ should be approximately Normal. We should check the histogram for severity of skewness and possible outliers.
c) $(\bar{y} - 23.3)/(s/\sqrt{40}) \sim t_{39}$.
d) 0.1447.
e) If the average age at first marriage is still 23.3 years, there is a 14.5% chance of getting a sample mean of 24.2 years or older simply from natural sampling variation.

f) We lack evidence that the average age at first marriage has increased from the mean of 23.3 years.

43. a) Random sample; the Nearly Normal Condition seems reasonable. (Show a Normal probability plot or histogram.) The histogram is nearly uniform, with no outliers or skewness.
b) $\bar{y} = 28.78$, $s = 0.40$.
c) (28.36, 29.21) grams.
d) Based on this sample, we are 95% confident the average weight of the content of Ruffles bags is between 28.36 and 29.21 grams.

45. a) Type I; he mistakenly rejected the null hypothesis that $p = 0.10$ (or worse).
b) Yes. These are a random sample of bags and the Nearly Normal Condition is met (show a Normal probability plot or histogram); $t = -2.51$ with 7 df for a one-sided P-value of 0.0203.

47. a) Random sample; the Nearly Normal Condition seems reasonable from a Normal probability plot. The histogram is roughly unimodal and symmetric with no outliers. (Show your plot.)
b) (1187.9, 1288.4) chips.
c) Based on this sample, the mean number of chips in an 18-ounce bag is between 1187.9 and 1288.4, with 95% confidence. The *mean* number of chips is clearly greater than 1000. However, if the claim is about individual bags, then it's not necessarily true. If the mean is 1188 and the SD deviation is near 94, then we'd expect about 2.5% of the bags to have fewer than 1000 chips, based on a Normal model. If in fact the mean is 1288, we'd expect about 0.1% to be below 1000.

49. 27 using *z*, 31 using *t*.

51. a) The Normal probability plot is relatively straight, with one outlier at 93.8 sec. Without the outlier, the conditions seem to be met. The histogram is roughly unimodal and symmetric with no other outliers. (Show your plot.)
b) $t = -2.63$, P-value $= 0.0160$. With the outlier included, we might conclude that the mean completion time for the maze is not 60 seconds; in fact, it is less.
c) $t = -4.46$, P-value $= 0.0003$. Because the P-value is so small, we reject $H_0$. Without the outlier, we see strong evidence that the average completion time for the maze is less than 60 seconds. The outlier here did not change the conclusion.
d) The maze does not meet the "one-minute average" requirement. Both tests rejected a null hypothesis of a mean of 60 seconds.

53. a) The study is well designed and the data show no outliers. We are 95% confident that the average wooden tip penetration is between 199 mm and 213.3 mm.
b) The experiment is well designed and the data show no outliers.
We are 95% confident that the average stone tip penetration is between 212.3 mm and 236.8 mm.

55. a) We have a large sample size. We are 90% confident that the mean number of lawsuits will be between 4220 and 7249.

b) Because these are aggregate data taken over 8 years from 50 different states, the overall mean may not prove useful, especially to individual states. It is possible that this interval could be used on a federal level to see if the number of lawsuits is increasing.

57. a) Random sample; the Nearly Normal Condition seems reasonable. (Show a Normal probability plot or histogram.) The histogram is nearly uniform, with no outliers or skewness.
b) $\bar{y} = 28.78$, $s = 0.40$.
c) $t = 2.94$, $df = 5$, P-value $= 0.032$. Because the P-value is low, we reject $H_0$. We have convincing evidence that the mean weight of bags of Ruffles potato chips is not 28.3 grams. It appears to be higher.

59. a) We are 95% confident that the interval from 22.1 mpg to 26.5 mpg contains the mean gas mileage for all 2019 vehicles.
b) Since this was a random sample of 2019 vehicles, we can be 95% confident that this interval captures the mean mileage for all 2019 vehicles.

61. a) $294.6 < \mu(\text{Drive Distance}) < 296.96$.
b) These data are not a random sample of golfers. The top professionals are (unfortunately) not representative and were not selected at random. We might consider the 2021 data to represent the population of all professional golfers, past, present, and future but, as performances tend to change through the years, we have no reason to believe this is a representative sample of those, either.
c) The data are means for each golfer, so they are less variable than if we looked at all the separate drives.

## Chapter 23 (See note, page A-29)

1. Yes. The high P-value means that we lack evidence of a difference, so 0 is a possible value for $\mu_{\text{Meat}} - \mu_{\text{Beef}}$.

3. a) Plausible values of $\mu_{\text{Meat}} - \mu_{\text{Beef}}$ are all negative, so the mean fat content is probably higher for beef hot dogs.
b) The difference is significant.
c) 10%.

5. a) False. The confidence interval is about means, not about individual hot dogs.
b) False. The confidence interval is about means, not about individual hot dogs.
c) True.
d) False. CIs based on other samples will also try to estimate the true difference in population means; there's no reason to expect other samples to conform to this result.
e) True.

7. a) 2.927.
b) Larger.
c) Based on this sample, we are 95% confident that students who learn Math using the CPMP method will score, on average, between 5.57 and 11.43 points better on a test solving applied Algebra problems with a calculator than students who learn by traditional methods.
d) Yes; 0 is not in the interval; plausible differences are positive.

9. a) $H_0$: $\mu_C - \mu_T = 0$ vs. $H_A$: $\mu_C - \mu_T \neq 0$.

b) Yes. Groups are independent, though we don't know if students were randomly assigned to the programs. Sample sizes are large, so CLT applies.

c) If the means for the two programs are really equal, there is less than a 1 in 10,000 chance of seeing a difference as large as or larger than the observed difference just from natural sampling variation.

d) On average, students who learn with the CPMP method do significantly worse on Algebra tests that do not allow them to use calculators than students who learn by traditional methods.

11. We must assume the samples were random or otherwise independent of each other. We also assume that the distributions are roughly normal, but the groups are large enough to proceed.

13. We are 95% confident that the mean purchase amount at Walmart is between $1.85 and $14.15 less than the mean purchase amount at Target.

15. a) (1.36, 4.64).
b) No; 5 minutes is beyond the high end of the interval.

17.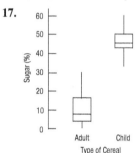

Random sample—given, independent samples, less than 10% of all cereals; boxplot shows no outliers—not exactly symmetric, but these are reasonable sample sizes. Based on these samples, with 95% confidence, sugar makes up a greater percentage of the weight of children's cereals than of adults' cereals by an average of between 32 and 41 percentage points.

19. Random assignment; both distributions are unimodal and symmetric. $H_0: \mu_N - \mu_C = 0$ vs. $H_A: \mu_N - \mu_C > 0$. $t = 2.207$; P-value $= 0.0168$; df $= 37.3$. Because of the small P-value, we reject $H_0$. These data do suggest that new activities are better. The mean reading comprehension score for the group with new activities is significantly (at $\alpha = 0.05$) higher than the mean score for the control group.

21. a)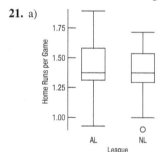

Both are reasonably symmetric. There is one possible low outlier among National League teams.

b) Based on these data, the average number of home runs hit per game in an American League stadium is between 1.27 and 1.59 with 95% confidence.

c) No. The boxplot indicates that 1.38 is in the middle of the NL values.

d) We want to work directly with the average difference. The two separate confidence intervals do not answer questions about the difference. The difference has a different standard deviation, found by adding variances.

e) $(-0.119, 0.269)$ (AL $-$ NL).

f) Based on these data, with 95% confidence, American League stadiums average between 0.119 fewer and 0.269 more home runs per game than National League stadiums.

g) No, 0 is in the interval.

23. a) We want to work directly with the average difference. The two separate confidence intervals do not answer questions about the difference. The difference has a different standard deviation, found by adding variances. Using the individual intervals is the equivalent of adding standard deviations.

b) $(-0.1171, 0.2704)$(AL $-$ NL).

c) Based on these data, with 95% confidence, American League stadiums average between 0.1171 fewer and 0.2704 more home runs per game than National League stadiums.

d) No; 0 is in the interval.

25. These are not two independent samples. These are before and after scores for the same individuals (paired data).

27. a) These data do not provide evidence of a difference in ad recall between shows with sexual content and violent content.

b) Random assignment; large groups. $H_0: \mu_S - \mu_N = 0$ vs. $H_A: \mu_S - \mu_N \neq 0$. $t = -6.08$; df $= 213.99$; P-value $= 5.5 \times 10^{-9}$. Because the P-value is low, we reject $H_0$. These data suggest that ad recall between shows with sexual and neutral content is different; those who saw shows with neutral content had higher average recall.

29. a) Groups are large and randomly assigned. $H_0: \mu_V - \mu_N = 0$ vs. $H_A: \mu_V - \mu_N \neq 0$. $t = -7.21$; df $= 201.96$; P-value $= 1.1 \times 10^{-11}$. Because of the very small P-value, we reject $H_0$. There appears to be a difference in mean ad recall between shows with violent content and neutral content; viewers of shows with neutral content remember more brand names, on average.

b) With 95% confidence, the average number of brand names remembered 24 hours later is between 1.45 and 2.41 higher for viewers of neutral content shows than for viewers of sexual content shows, based on these data (df $= 204.8$).

31. These were random samples, both less than 10% of properties sold. Prices of houses should be independent, and random sampling makes the two groups independent. The boxplots make the price distributions appear to be reasonably symmetric, and with the large sample sizes the few outliers don't affect the means much. Based on this sample, we're 95% confident that, in New York, having a waterfront is worth, on average, about $59,121 to $140,898 more in sale price.

33. $H_0: \mu_W = \mu_S$; $H_A: \mu_W < \mu_S$; assume these arrows are representative, samples are independent and reasonably symmetric. $t = -3.2$; df $= 9.66$; P-value $= 0.005$. With a P-value $< 0.05$, we reject $H_0$. We found statistically

significant evidence that the stone-tipped arrows penetrated deeper than the wooden-tipped arrows.

35. $H_0$: $\mu_P = \mu_l$; $H_A$: $\mu_P > \mu_l$. The data were carefully collected from 2 independent groups. There are outliers in the lay individual's observations, but both datasets are symmetric. $t = 4.33$; df = 11.6; P-value = 0.0005, < 0.05 and we reject $H_0$. We found statistically significant evidence that the mean % collagen is higher in the priests' bones than the lay individuals.

37. $H_0$: $\mu_{big} - \mu_{small} = 0$ vs. $H_A$: $\mu_{big} - \mu_{small} \neq 0$; bowl size was assigned randomly; amount scooped by individuals and by the two groups should be independent. $t = 2.104$; df = 34.3; P-value = 0.0428. The low P-value leads us to reject the null hypothesis. There is evidence of a difference in the average amount of ice cream that people scoop when given a bigger bowl.

39. a) The 95% confidence interval for the difference is (0.61, 5.39). 0 is not in the interval, so scores in 1996 were significantly higher. (Or the $t$, with more than 7500 df, is 2.459 for a P-value of 0.0070.)
    b) Because both samples were very large, there shouldn't be a difference in how certain you are, assuming conditions are met.

41. Independent Groups Assumption: The runners are different women, so the groups are independent. The Randomization Condition is satisfied because the runners are selected at random for these heats.

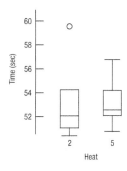

Nearly Normal Condition: The boxplots show an outlier, but we will proceed and then redo the analysis with the outlier deleted. When we include the outlier, $t = 0.0484$ with a two-sided P-value of 0.963. With the outlier deleted, $t = -1.10$ with $P = 0.300$. Either P-value is so large that we fail to reject the null hypothesis of equal means and conclude that there is no evidence of a difference in the mean times for randomly assigned heats.

43. With $t = -4.57$ (df = 7.03) and a very low P-value of 0.0013, we reject the null hypothesis of equal mean velocities. There is strong evidence that golf balls hit off Stinger tees will have a higher mean initial velocity.

45. The distributions of penetration depths for both types of arrow tips are reasonably symmetric with no outliers. This is an experiment where the arrow types are the treatments. We don't know if they were shot in a randomized order so the randomization condition may or may not be met.

This is a $t$-interval for the difference in means. The 95% confidence interval (stone − wood) is (5.5, 31.4). We are 95% confident that the stone-tipped arrows penetrate 5.5 mm to 31.4 mm deeper than the wood-tipped arrows on average.

47. a) $H_0$: $\mu_M - \mu_R = 0$ vs. $H_A$: $\mu_M - \mu_R > 0$. $t = -0.70$; df = 45.88; P-value = 0.7563. Because the P-value is so large, we do not reject $H_0$. These data provide no evidence that listening to Mozart while studying is better than listening to rap. (With only the summary statistics and group sizes less than 30, we must simply assume that the responses in each group are distributed approximately normally.)
    b) With 90% confidence, those who listen to no music while studying remember an average of between 0.189 and 5.351 objects more than those who listen to Mozart.

49.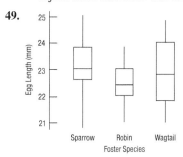

$H_0$: $\mu_S - \mu_R = 0$ vs. $H_A$: $\mu_S - \mu_R \neq 0$. $t = 1.641$; df = 21.60; P-value = 0.115. Because $P > 0.05$, fail to reject $H_0$. There is no evidence of a difference in mean cuckoo egg length between robin and sparrow foster parents.
$H_0$: $\mu_S - \mu_W = 0$ vs. $H_A$: $\mu_S - \mu_W \neq 0$. $t = 0.549$; df = 26.86; P-value = 0.587. Because $P > 0.05$, fail to reject $H_0$. There is no evidence of a difference in mean cuckoo egg length between sparrow and wagtail foster parents.
$H_0$: $\mu_R - \mu_W = 0$ vs. $H_A$: $\mu_R - \mu_W \neq 0$. $t = -1.012$; df = 23.60; P-value = 0.322. Because $P > 0.05$, fail to reject $H_0$. There is no evidence of a difference in mean cuckoo egg length between robin and wagtail foster parents. In general, we should be wary of doing three $t$-tests on the same data. Our Type I error rate is not the same for doing three tests as it is for doing one test. However, because none of the tests showed significant differences, this is less of a concern here.

51. a)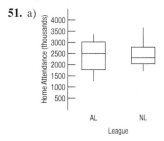

    b) A difference of 60,000 is very near the center of the distribution of simulated differences. The vast majority of the simulated differences are more than 60,000 in absolute value, so the difference is not statistically significant.

**53.** [Answers will vary.] Using the randomization test, 500 simulated differences are plotted below. The actual difference of 0.07833 is almost exactly in the middle of the distribution of simulated differences. The simulated 2-tailed P-value is 0.952, which means that 95.2% of the simulated differences were at least 0.7833 in absolute value; there is no evidence of a difference in mean times between heats.

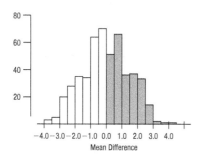

## Chapter 24 (See note, page A-29)

**1.** a) Paired.   b) Not paired.   c) Paired.

**3.** a) Randomly assign 50 hens to each of the two kinds of feed. Compare production at the end of the month.
b) Give all 100 hens the new feed for 2 weeks and the old feed for 2 weeks, randomly selecting which feed the hens get first. Analyze the differences in production for all 100 hens.
c) Matched pairs. Because hens vary in egg production, the matched-pairs design will control for that.

**5.** a) Show the same people ads with and without sexual images, and record how many products they remember in each group. Randomly decide which ads a person sees first. Examine the differences for each person.
b) Randomly divide volunteers into two groups. Show one group ads with sexual images and the other group ads without. Compare how many products each group remembers.

**7.** a) Matched pairs—same cities in different periods.
b) There is a significant difference (P-value = 0.0244) in the labor force participation rate for women in these cities; women's participation seems to have increased between 1968 and 1972.

**9.** a) Use the paired $t$-test because we have pairs of Fridays in 5 different months. Data from adjacent Fridays within a month may be more similar than data from randomly chosen Fridays.
b) We conclude that there is evidence (P-value = 0.0212) that the mean number of cars found on the M25 motorway on Friday the 13th is less than on the previous Friday.
c) We don't know if these Friday pairs were selected at random. If these are the Fridays with the largest differences, this will affect our conclusion. The Nearly Normal Condition appears to be met by the differences, but the sample size is small.

**11.** Adding variances requires that the variables be independent. These price quotes are for the same cars, so they are paired. Drivers quoted high insurance premiums by the local company will be likely to get a high rate from the online company, too.

**13.** a) The histogram—we care about differences in price.
b) Insurance cost is based on risk, so drivers are likely to see similar quotes from each company, making the differences relatively smaller.
c) The price quotes are paired; they were for a random sample of fewer than 10% of the agent's customers; the histogram of differences looks approximately Normal.

**15.** $H_0$: $\mu(Local - Online) = 0$ vs. $H_A$: $\mu(Local - Online) > 0$. $t = 0.83$; df = 9. With a high P-value of 0.215, we don't reject the null hypothesis. These data don't provide evidence that online premiums are lower, on average.

**17.** a) No. The vehicles have no natural pairing.
b) Possibly. The data are quantitative and paired by vehicle.
c) The sample size is large, but there is at least one extreme outlier that should be investigated before applying these methods.

**19.** (7.17, 7.57). We are 95% confident that the interval from 7.17 mpg to 7.57 mpg captures the true improvement in highway gas mileage compared to city gas mileage.

**21.** The difference between fuel efficiency of cars and that of trucks can be large, but isn't relevant to the question asked about highway vs. city driving. Pairing places each vehicle in its own block to remove that variation from consideration.

**23.** a) The order of treatments was randomly assigned. A group of $n = 52$ differences is more than 30, so conditions are met.
b) With $df = 51$, the $t = \pm 2.57$, depending on the direction of subtraction.
c) Because $p = 0.013 < 0.05$, we reject the null hypothesis. We have convincing evidence that listening to Korean traditional music reduced the reported pain levels of the patients.

**25.** Based on these data, we are 90% confident that boys, on average, can do between 1.6 and 13.0 more push-ups than girls (independent samples—not paired).

**27.** a) Paired sample test. Data are before/after for the same workers; workers randomly selected; assume fewer than 10% of all this company's workers; boxplot of differences shows them to be symmetric, with no outliers.

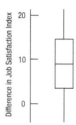

b) $H_0$: $\mu_D = 0$ vs. $H_A$: $\mu_D > 0$. $t = 3.60$; P-value = 0.0029. Because $P < 0.01$, reject $H_0$. These data show evidence that average job satisfaction has increased after implementation of the program.
c) Type I.

**29.** $H_0: \mu_D = 0$ vs. $H_A: \mu_D \neq 0$. Data are paired by brand; brands are independent of each other; fewer than 10% of all yogurts (questionable); boxplot of differences shows an outlier (100) for Great Value:

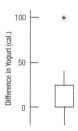

With the outlier included, the mean difference (Strawberry – Vanilla) is 12.5 calories with a *t*-stat of 1.332, with 11 df, for a P-value of 0.2098. Deleting the outlier, the difference is even smaller, 4.55 calories with a *t*-stat of only 0.833 and a P-value of 0.4241. With P-values so large, we do not reject $H_0$. We conclude that the data do not provide evidence of a difference in mean calories.

**31.** a) Cars were probably not a simple random sample, but may be representative in terms of stopping distance; boxplot does not show outliers, but does indicate right skewness. A 95% confidence interval for the mean stopping distance on dry pavement is (131.8, 145.6) feet.
b) Data are paired by car; cars were probably not randomly chosen, but representative; boxplot shows an outlier (car 4) with a difference of 12. With deletion of that car, a Normal probability plot of the differences is relatively straight.

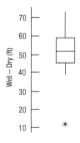

Retaining the outlier, we estimate with 95% confidence that the average braking distance is between 38.8 and 62.6 feet more on wet pavement than on dry, based on this sample. (Without the outlier, the confidence interval is 47.2 to 62.8 feet.)

**33.** a) Paired Data Condition: Data are paired by college. Randomization Condition: This was a random sample of public colleges and universities.

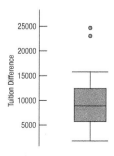

Normal Population Assumption: UC Irvine and New College of Florida appear to be outliers. We may want to consider setting them aside, as they may not represent the typical difference.
b) Having deleted the outliers, we are 90% confident, based on the remaining data, that nonresidents pay, on average, between $7424 and $11,149 more than residents. If we retain the outliers, the interval is ($8339, $13,578).
c) The assertion is not reasonable; With or without the outliers, $7000 is lower than the values in the interval.

**35.** a) 60% is 30 strikes; $H_0: \mu = 30$ vs. $H_A: \mu > 30$. $t = 6.07$; P-value $= 3.92 \times 10^{-6}$. With a very small P-value, we reject $H_0$. There is very strong evidence that players can throw more than 60% strikes after training, based on this sample.
b) $H_0: \mu_D = 0$ vs. $H_A: \mu_D > 0$. $t = 0.135$; P-value $= 0.4472$. With such a high P-value, we do not reject $H_0$. These data provide no evidence that the program has improved pitching in these Little League players.

**37.** a) We would use a paired *t*-test here since the data are paired by job title in the two years. The observed average increase in pay is $12,670.86. This corresponds to a *t*-value of 3.4422, and with 13 degrees of freedom, we find that a one-tailed P-value is 0.0022. At a significance level of 0.01, we conclude that average White House pay grew significantly higher (but see part c).
b) ($4718.42, $20623.29).
c) Adjusting for inflation, the difference is now only $770.70 and in fact the adjusted 2016 salaries are higher. Statistically there is no difference in the means with a *t*-value of −0.1968 and a one sided P-value of 0. 577. The confidence interval is (−$9229.5, $7688.17).
d) In that case, the data would not be paired, so you would instead use a two-sample *t*-test.

## Part VI Review (See note, page A-29)

**1.** a) $H_0: \mu_{Jan} - \mu_{Jul} = 0$; $H_A: \mu_{Jan} - \mu_{Jul} \neq 0$. $t = -1.94$; df $= 43.68$; P-value $= 0.0590$. Because P $< 0.10$, reject the null. These data show a significant difference in mean age to crawl between January and July babies.
b) $H_0: \mu_{Apr} - \mu_{Oct} = 0$; $H_A: \mu_{Apr} - \mu_{Oct} \neq 0$. $t = -0.92$; df $= 59.40$; P-value $= 0.3610$. Because P $> 0.10$, do not reject the null; these data do not show a significant difference between April and October with regard to the mean age at which crawling begins.
c) These results are not consistent with the claim.

**3.** $H_0: p = 0.26$; $H_A: p \neq 0.26$. $z = 0.946$; P-value $= 0.3443$. Because the P-value is high, we do not reject $H_0$. These data do not show that the Denver-area rate is different from the national rate in the proportion of businesses with women owners.

**5.** Based on these data, we are 95% confident that the mean difference in aluminum oxide content is between $-3.37$ and $1.65$. Because the interval contains 0, the means in aluminum oxide content of the pottery made at the two sites could reasonably be the same.

**7.** a) $H_0: p_{ALS} - p_{Other} = 0$; $H_A: p_{ALS} - p_{Other} > 0$. $z = 2.52$; P-value $= 0.0058$. With such a low P-value, we reject $H_0$. This is strong evidence that there is a higher proportion of varsity athletes among ALS patients than those with other disorders.

b) Observational retrospective study. To make the inference, one must assume the patients studied are representative.

9. $H_0: \mu = 7.41$; $H_A: \mu \neq 7.41$. $t = 2.18$; df $= 111$; P-value $= 0.0313$. With such a low P-value, we reject $H_0$. Assuming that Missouri babies fairly represent the United States, these data suggest that American babies are different from Australian babies in birth weight; it appears American babies are heavier, on average.

11. a) If there is no difference in the average fish sizes, the chance of seeing an observed difference this large just by natural sampling variation is less than 0.1%.
    b) If cost justified, feed them a natural diet.
    c) Type I.

13. a) Assuming the conditions are met, from these data we are 95% confident that patients with cardiac disease average between 3.39 and 5.01 years older than those without cardiac disease.
    b) Older patients are at greater risk from a variety of other health issues, and perhaps more depressed.

15. a) Stratified sample survey.
    b) We are 95% confident that the proportion of 12- to 15-year-old gamers who use in-game chat features is between 13.9 and 26.1 percentage points higher than that of 8- to 11-year-old gamers.
    c) Yes. The entire interval lies above 0.

17. a) Bimodal.
    b) $\mu$, the population mean. Sample size does not matter.
    c) $\sigma/\sqrt{n}$; sample size does matter.
    d) It becomes closer to a Normal model and less variable as the sample size increases.

19. Based on the data, we are 95% confident that the mean difference in words misunderstood is between $-3.76$ and $3.10$. Because 0 is in the confidence interval, we would conclude that the two tapes could be equivalent.

21. a) If dogs are equally likely to face any direction while pooping, getting a test statistic as large or larger than they got has a probability of $6.2 \times 10^{-8}$.
    b) Because the P-value is so low, we reject the null hypothesis. We have convincing evidence that dogs have a preference for aligning themselves with the magnetic field when pooping.

23. Based on the survey, we are 95% confident that the proportion of all American adults who felt at that time that the federal government was responsible for ensuring health coverage for all Americans was between 58.1% and 61.9%.

25. $H_0$: There is no difference in cancer rates, $p_1 - p_2 = 0$.
    $H_A$: The cancer rate in those who use the herb is higher, $p_1 - p_2 > 0$.

27. a) We are 95% confident that the mean difference in rainfall between Victorville and 29 Palms is $(-57.6 \text{ cm}, 43.2 \text{ cm})$.
    b) We are 95% confident that the mean difference in rainfall between Victorville and Mitchell's Cavern is $(-37.2 \text{ cm}, 66.0 \text{ cm})$.
    c) We are 95% confident that the mean difference in rainfall between 29 Palms and Mitchell's Cavern is $(-15.2 \text{ cm}, 58.5 \text{ cm})$.

d) Zero is in the intervals. It appears that none of these regions is statistically significantly different in average rainfall from the others.

29. Data are matched pairs (before and after for the same rooms); less than 10% of all rooms in a large hotel; uncertain how these rooms were selected (are they representative?). Histogram shows that differences are roughly unimodal and symmetric, with no outliers. A 95% confidence interval for the difference, before − after, is $(0.58, 2.65)$ counts. Because the entire interval is above 0, these data suggest that the new air-conditioning system was effective in reducing average bacteria counts.

31. a) We are 95% confident that between 4.27% and 10.26% of e-cigarette users would quit smoking.
    b) 1151 people estimating $p$ using 21/289; 4269 people using 0.5.

33. a) From this histogram, about 115 loaves or more. (Not Normal.) This assumes the last 100 days are typical.
    b) Large sample size; CLT says $\bar{y}$ will be approximately Normal.
    c) From the data, we are 95% confident that on average the bakery will sell between 101.2 and 104.8 loaves of bread a day.
    d) 25.
    e) Yes, 100 loaves per day is too low—the entire confidence interval is above that.

35. a) $H_0: p_{\text{High}} - p_{\text{Low}} = 0$; $H_A: p_{\text{High}} - p_{\text{Low}} \neq 0$. $z = -3.57$; P-value $= 0.0004$. Because the P-value is so low, we reject $H_0$. These data suggest the IRS risk is different in the two groups; it appears people who consume dairy products often have a lower risk, on average.
    b) Doesn't indicate causality; this is not an experiment.

37. Based on these data, we are 95% confident that seeded clouds will produce an average of between $-4.76$ and $559.56$ more acre-feet of rain than unseeded clouds. Because the interval contains negative values, it may be that seeding is unproductive.

39. a) Randomizing order of the tasks helps avoid bias and memory effects. Randomizing the cards helps avoid bias as well.
    b) $H_0: \mu_D = 0$; $H_A: \mu_D \neq 0$.

Boxplot of the differences looks symmetric with no outliers.

$t = -1.70$; P-value $= 0.0999$. Do not reject $H_0$, because $P > 0.05$. The data do not provide evidence that the color or written word dominates.

41. a) Different samples give different means; this is a fairly small sample. The difference may be due to natural sampling variation.
    b) $H_0: \mu = 100$; $H_A: \mu < 100$.
    c) Batteries selected are a SRS (representative); fewer than 10% of the company's batteries; lifetimes are approximately Normal.
    d) $t = -1.0$; P-value $= 0.1666$. Do not reject $H_0$. This sample does not show that the average life of the batteries is significantly less than 100 hours.
    e) Type II.

**43.** a) $H_0: \mu_1 - \mu_2 = 0$; $H_A: \mu_1 - \mu_2 > 0$. $t = 0.90$; P-value $= 0.1864$. Because $P > 0.05$, we do not reject $H_0$. Weeklong study scores were not significantly higher.
b) $H_0: p_1 - p_2 = 0$; $H_A: p_1 - p_2 \neq 0$. $z = -3.10$; P-value $= 0.0019$. With such a small P-value, we reject $H_0$. There is evidence of a difference in proportion for passing on Friday; it appears that cramming may be more effective.
c) $H_0: \mu_D = 0$; $H_A: \mu_D > 0$. $t = 5.17$; P-value $< 0.0001$. These data show evidence that learning does not last for 3 days because mean score declined.

**45.** a) We are 99% confident that the interval 5.9% to 16.5% contains the true difference in violent felony rates when comparing individual therapy to MST.
b) Because the entire interval is above 0, we can conclude that MST is successful in reducing the proportion of juvenile offenders who commit violent felonies. The population of interest is adolescents with mental health problems.

## Practice Exam Answers

**I.**
| | | | | |
|---|---|---|---|---|
| 1. C. | 3. B. | 5. E. | 7. E. | 9. D. |
| 11. D. | 13. D. | 15. C. | 17. E. | 19. E. |
| 21. B. | 23. C. | 25. E. | 27. A. | 29. E. |
| 31. D. | 33. D. | 35. B. | | |

**II.** **1.** a) $\widehat{Low} = 108.798 - 2.111 Lat$.
b) The model suggest that average January low temperatures decrease about 2.111°F for every 1° increase in north latitude.
c) Probably not. It suggests that the average January low temperature at the equator would be over 108°F.
d) This city's average January low temperature is 4.2°F lower than the model predicts.

**3.** a) $(0.5)(0.2) + (0.5)(0.75) = 0.475$.
b) $\frac{0.1}{0.475} \approx 0.21$.
c) $100(0.525) + 1000(0.475) = \$527.50$.

**5.** a) To see if the data suggest the two time populations are Normal.
b) Random allocation makes the groups independent; assume normality of population distributions. OK to use 2-sample $t$-interval: $(-4.60, 1.50)$. We can be 95% confident that the mean commuting time for the carbon frame bike is between 4.6 minutes shorter and 1.5 minutes longer than for the steel frame bike.
c) If the doctor ran this experiment repeatedly, we'd expect about 95% of the resulting intervals to capture the true difference in mean commuting times.
d) No; the interval offers the possibilities that the bike may be faster or slower.

## Chapter 25 (See note, page A-29)

**1.** a) $(30, 30, 30, 30)$, 30 for each season.
b) 1.933. c) 3.

**3.** a) 3.
b) No. It's smaller than the mean.

c) It would say that there is not enough evidence to reject the null hypothesis that births are distributed uniformly across the seasons.

**5.** a) The age distributions of customers at the two branches are the same.
b) Chi-square test of homogeneity.
c)

| | Age | | | |
|---|---|---|---|---|
| | Less Than 30 | 30–55 | 56 or Older | Total |
| In-Town Branch | 25 | 45 | 30 | 100 |
| Mall Branch | 25 | 45 | 30 | 100 |
| Total | 50 | 90 | 60 | 200 |

d) 9.778. e) 2. f) 0.0075.
g) Reject $H_0$ and conclude we have evidence that the age distributions at the two branches are not the same.

**7.** Random variation happens all the time. It would actually be quite strange if 120 birthdays were perfectly distributed with 30 in each season. The amount of variation that we see in this data is not more than would happen by random chance. And chi-square is the tool that measures this variation.

**9.** a)

| | Fatal | Not Fatal |
|---|---|---|
| Body | 54.15 | 12.85 |
| Head/neck | 36.37 | 8.63 |
| Limb | 27.48 | 6.52 |

b) 52.65. c) 2. d) $P < 0.0001$.
e) Reject $H_0$. There is evidence that site of injury and whether or not it was lethal are not independent.

**11.** a) Chi-square test of independence. We have one sample and two variables. We want to see if the variable *Account Type* is independent of the variable *Trade Type*.
b) Other test. *Account Size* is quantitative, not counts.
c) Chi-square test of homogeneity. We want to see if the distribution of one variable, *Courses*, is the same for two groups (resident and nonresident students).

**13.** a) 10. b) Goodness-of-fit.
c) $H_0$: The die is fair (all faces have $p = 1/6$). $H_A$: The die is not fair.
d) Count data; rolls are random and independent; expected frequencies are all bigger than 5.
e) 5. f) $\chi^2 = 5.600$; P-value $= 0.3471$.
g) Because the P-value is high, do not reject $H_0$. The data show no evidence that the die is unfair.

**15.** a) 45.3333 for each topping.
b) 2.485. c) 2.

**17.** a) Weights are quantitative, not counts.
b) Count the number of each kind of nut, assuming the company's percentages are based on counts rather than weights.

**19.** $H_0$: The police force represents the population (29.2% white, 28.2% black, etc.). $H_A$: The police force is not representative of the population. $\chi^2 = 16516.88$; df $= 4$; P-value $= 0.0000$. Because the P-value is so low, we reject $H_0$. These data show evidence that the police force is not representative of the population. In particular, there are too many white officers in relationship to their membership in the community.

21. a) $\chi^2 = 5.671$; df = 3; P-value = 0.1288. With a P-value this high, we fail to reject $H_0$. Yes, these data are consistent with those predicted by genetic theory.
b) $\chi^2 = 11.342$; df = 3; P-value = 0.0100. Because of the low P-value, we reject $H_0$. These data provide evidence that the distribution is not as specified by genetictheory.
c) All other things being equal, a statistical test has greater power when the sample size is larger. So if the null hypothesis is false, a larger dataset is more likely to provide evidence against it than would a smaller dataset.

23. a) 101/17 = 5.94.   b) Goodness-of-fit.
c) $H_0$: The number of large hurricanes remains constant over decades.
$H_A$: The number of large hurricanes has changed.
d) 15.   e) P-value = 0.676.
f) The very high P-value means these data offer no evidence that the number of large hurricanes has changed.

25. a) Homogeneity.
b) $H_0$: There is no difference in the proportion of converted sales for the two different versions of the Amazon page.
$H_A$: There is a difference in the proportion of converted sales for the two different versions of the Amazon page.

27. a) 1.   b) 38.73.
c) We assume that Version A and Version B are independent trials conducted over similar sales windows. The expected counts, 38.7, 82,3, 364.3, 773.7 are all greater than 5. Note that the 10% condition does not apply, given that this an experiment, not random sampling.

29. a) 6.02.   b) 0.00175.
c) With a P-value of 0.002 < α, we reject Ho. We found strong evidence that the proportion of converted sales is not the same for the two versions of the Amazon page.

31. This is a potential indication that the two time windows for the different versions were not, in fact, independence. For example, if Version B was used closer to a busy holiday season, this may have increased the total views of the site.

33. We assume that the rolls are a representative sample; 20 is greater than 5. (Note that the 10% condition is off-topic, due to this being an experiment.) $H_0$: The die is uniform. $H_A$: At least one side of the die is not uniform. $\chi^2 = 7.1$, P-value = 0.791, df = 11. With a very, very large P-value, we fail to reject the null. We do not have evidence that the die is not uniform.

35. a)

| | Did You Use the Internet Yesterday? | |
|---|---|---|
| | Yes | No |
| White | 2506.38 | 895.62 |
| Black | 338.90 | 121.10 |
| Hispanic/Other | 445.73 | 159.27 |

b) 11.176.   c) 2.   d) 0.0037.
e) Reject the null hypothesis; there is evidence that Race and Internet Use are not independent.

37. a) 40.2%.   b) 8.1%.   c) 62.2%.   d) 285.48.
e) $H_0$: Survival was independent of status on the ship.
$H_A$: Survival depended on the status.
f) 3.
g) We reject the null hypothesis. There is evidence that survival depended on status. It appears that first-class passengers were more likely to survive than passengers of any other class.

39. First-class passengers were most likely to survive, while third-class passengers and crew were underrepresented among the survivors.

41. a) Experiment—actively imposed treatments (different drinks).
b) Homogeneity.
c) $H_0$: The rate of urinary tract infection is the same for all three groups.
$H_A$: The rate of urinary tract infection is different among the groups.
d) Count data; random assignment to treatments; all expected frequencies larger than 5.
e) 2.
f) $\chi^2 = 7.776$; P-value = 0.020.
g) With a P-value this low, we reject $H_0$. These data provide reasonably strong evidence that there is a difference in urinary tract infection rates between cranberry juice drinkers, lactobacillus drinkers, and the control group.

43. a) Independence.
b) $H_0$: *Political Affiliation* is independent of *Sex*. $H_A$: There is a relationship between *Political Affiliation* and *Sex*.
c) Counted data; probably a random sample, but can't extend results to other states; all expected frequencies greater than 5.
d) $\chi^2 = 4.851$; df = 2; P-value = 0.0884.
e) Because P > 0.05, we do not reject $H_0$. These data do not provide evidence of a relationship between *Political Affiliation* and *Sex*.

45. $H_0$: *Political Affiliation* is independent of *Region*. $H_A$: There is a relationship between *Political Affiliation* and *Region*. $\chi^2 = 13.849$; df = 4; P-value = 0.0078. With a P-value this low, we reject $H_0$. *Political Affiliation* and *Region* seem to be related. Examination of the residuals suggests that those in the West are more likely to be Democrat than Republican; those in the Northeast are more likely to be Republican than Democrat.

47. $H_0$: All digits 0–9 are equally likely to appear. $H_A$: Some numbers are more likely than others. Draws should be independent; these weeks should be representative of all draws, and are fewer than 10% of all possible Pick-3 lotteries. All expected values are at least 5. Goodness-of-fit test: $\chi^2 = 6.46$; df = 9; P-value = 0.693. These data fail to provide evidence that all the digits are not equally likely.

49. a) Homogeneity.
b) $H_0$: The grade distribution is the same for both professors.
$H_A$: The grade distributions are different.
c)

| | Dr. Alpha | Dr. Beta |
|---|---|---|
| A | 6.667 | 5.333 |
| B | 12.778 | 10.222 |
| C | 12.222 | 9.778 |
| D | 6.111 | 4.889 |
| F | 2.222 | 1.778 |

Three cells have expected frequencies less than 5.

**51.** a)

| | Dr. Alpha | Dr. Beta |
|---|---|---|
| A | 6.667 | 5.333 |
| B | 12.778 | 10.222 |
| C | 12.222 | 9.778 |
| Below C | 8.333 | 6.667 |

All expected frequencies are now larger than 5.
b) Decreased from 4 to 3.
c) $\chi^2 = 9.306$; P-value $= 0.0255$. Because the P-value is so low, we reject $H_0$. The grade distributions for the two professors seem to be different. Dr. Alpha gives fewer A's and more grades below C than Dr. Beta.

**53.** $\chi^2 = 14.058$; df $= 1$; P-value $= 0.0002$. With a P-value this low, we reject $H_0$. There is evidence of racial steering. Blacks are much less likely to rent in Section A than Section B.

**55.** a) $z = 3.74936$; $z^2 = 14.058$.
b) P-value $(z) = 0.0002$ (same as in Exercise 53).

**57.** $\chi^2 = 2.699$; df $= 3$; $P = 0.4404$. Because the P-value is $> 0.05$, these data do not show enough evidence of an association between the mother's age group and the outcome of the pregnancy.

**59.** $\chi^2 = 13.67$; df $= 4$; P-value $= 0.008$. With a P-value this low, we reject $H_0$. We have strong evidence that there is a difference in superpower preferences. (Note: While 2 of the expected counts are less than 5, we may proceed cautiously; see p. 705. This P-value is very small, making it safe in this case to reject the null hypothesis.)

## Chapter 26 (See note, page A-29)

**1.** $\widehat{Earn} = 14{,}468 + 27.264(SAT)$; graduates earn, on average $27.26/year per point of SAT score.

**3.** $s = 5603$. Because the range of *Earnings* is about 15,000, this number is reasonably large and we observe a fair amount of spread about the regression model.

**5.** The standard error for the slope tells us how much the slope of the regression equation would vary from sample to sample.

**7.** We conclude that there is evidence of a relationship between *Earnings* and *SAT Scores* for these schools.

**9.** We can conclude, with 95% confidence, that the increase in earnings, on average is between about $24 to $30 per point on the SAT.

**11.** a) $\widehat{Error\_24th} = 133.3 - 2.08\ Years\ Since\ 1970$; according to the model, the error made in predicting a hurricane's path has been declining at a rate of about 2.08 nautical miles per year starting from about 133 nautical miles in 1970.
b) $H_0: \beta_1 = 0$; there has been no change in prediction accuracy. $H_A: \beta_1 \neq 0$; there has been a change in prediction accuracy.
c) With a P-value $< 0.0001$, I reject the null hypothesis and conclude that prediction accuracies have in fact been changing during this period.
d) 79.8% of the variation in hurricane prediction accuracy is accounted for by this linear model on time.

**13.** a) $\widehat{Extent} = 64.47 - 3.82(Mean\ global\ temp)$; starting from an extent of $64.47\ km^2$ in 1979, the model predicts a decrease in extent of $3.82\ km^2$ per degree Celsius increase in mean global temperature.
b) The scatterplot shows a moderate linear relationship. The residual plot shows a possible bend and slightly greater variation on the left than on the right. There are no striking outliers. The normal probability plot looks reasonably straight. We should proceed with caution because the conditions are almost satisfied.
c) $s$ is the standard deviation of the residuals.
d) 0.407.
e) The standard error is the estimated standard deviation of the sampling distribution of the slope coefficient. Over many random samples from this population we'd expect to see slopes of the samples varying by this much.
f) No. We can see an association, but we cannot establish causation from this study.

**15.** We are 95% confident that the slope is between $-4.64$ and $-3.00$.

**17.** a) $H_0: \beta_1 = 0$; there's no association between calories and sodium content in all-beef hot dogs. $H_A: \beta_1 \neq 0$: there is an association.
b) Based on the low P-value (0.0018), I reject the null. There is evidence of an association between the number of calories in all-beef hot dogs and their sodium contents.

**19.** a) Among all-beef hot dogs with the same number of calories, the sodium content varies, with a standard deviation of about 60 mg.
b) 0.561 mg/cal.
c) If we tested many other samples of all-beef hot dogs, the slopes of the resulting regression lines would be expected to vary, with a standard deviation of about 0.56 mg of sodium per calorie.

**21.** I'm 95% confident that for every additional calorie, all-beef hot dogs have, on average, between 1.07 and 3.53 mg more sodium.

**23.** a) $H_0$: Difference in age at first marriage has not been changing, $\beta_1 = 0$. $H_A$: Difference at first marriage has been changing, $\beta_1 \neq 0$.
b) Plot of residuals vs. predicted values shows a "wave" shape, oscillating. The histogram of the residuals is somewhat bimodal. The normal probability plot suggests that the residuals are nearly Normal, but we should be cautious in interpreting the regression model.
c) $t = -13.8$, $P < 0.0001$. We can be confident that the slope is not zero. It looks like the gap in age between men and women at first marriage has been shrinking.

**25.** We are 95% confident that the slope of *Men − Women* vs. *Year* is between $-0.016$ and $-0.012$.

**27.** a) $H_0$: *Fuel Economy* and *Weight* are not (linearly) related, $\beta_1 = 0$. $H_A$: *Fuel Economy* changes with *Weight*, $\beta_1 \neq 0$. P-value $< 0.0001$, indicating strong evidence of an association.
b) Yes, the conditions seem satisfied. Histogram of residuals is unimodal and symmetric; residuals plot looks OK, but some "thickening" of the plot with increasing values.
c) $t = -12.2$; P-value $< 0.0001$. These data show evidence that *Fuel Economy* decreases with the *Weight* of the car.

**29.** a) $(-9.57, -6.86)$ mpg per 1000 pounds.

b) Based on these data, we are 95% confident that *Fuel Efficiency* decreases between 6.86 and 9.57 miles per gallon, on average, for each additional 1000 pounds of *Weight*.

**31.** a) Yes. $t = 2.73$; P-value $= 0.0079$. With a P-value so low, we reject $H_0$. There is evidence of a positive relationship between *Calories* and *Sodium* content.
b) No. $R^2 = 9\%$ and $s$ appears to be large, although without seeing the data, it is a bit hard to tell.

**33.** Plot of *Calories* against *Fiber* does not look linear; the residuals plot also shows increasing variance as predicted values get large. The histogram of residuals is right skewed.

**35.** a) $H_0$: No (linear) relationship between *BCI* and *pH*, $\beta_1 = 0$. $H_A$: There is a relationship, $\beta_1 \neq 0$.
b) $t = -7.73$ with 161 df; P-value $< 0.0001$.
c) There seems to be a negative relationship; *BCI* decreases as *pH* increases at an average of 197.7 *BCI* units per increase of 1 *pH*.

**37.** a) $H_0$: No linear relationship between *Population* and *Ozone*, $\beta_1 = 0$. $H_A$: *Ozone* increases with *Population*, $\beta_1 > 0$. $t = 3.48$; P-value $= 0.0018$. With a P-value so low, we reject $H_0$. These data show evidence that *Ozone* increases with *Population*.
b) Yes, *Population* accounts for 84% of the variability in *Ozone* level, and $s$ is just over 5 parts per million.

**39.** Based on this regression analysis, each additional million residents corresponds to an increase in average ozone level of between 3.29 and 10.01 ppm, with 90% confidence.

**41.** a) 39.
b) The scatterplot appears "straight enough," there are no extreme outliers; the residuals show no pattern and a roughly constant spread, and their distribution appears approximately normal. So yes, regression is reasonable.
c) The P-value associated with a null hypothesis of no slope is 0.0372. This is smaller than a significance level of 0.05, so yes—there is evidence of an association between *Battery Life* and *Maximum Brightness*.
d) The association is not very strong. The scatterplot shows a lot of variability around the line, and R-squared is only 11.2%.
e) $\overline{Battery\ Life} = 5.751 + 0.0087\ (Max\ Brightness)$.
f) $0.00874 \pm (1.687)0.0040$, or about $(0.002, 0.015)$.
g) We can be 90% confident that for every additional unit cd/m$^2$ of *Max Brightness*, the battery life is estimated to be between 0.002 hours and 0.015 hours longer, on averge.

**43.** $H_0$: No linear relationship between *Waist Size* and *% Body Fat*, $\beta_1 = 0$. $H_A$: *% Body Fat* changes with *Waist Size*, $\beta_1 \neq 0$. $t = 8.14$; P-value $< 0.0001$. There's evidence that *% Body Fat* seems to increase with *Waist Size*.

**45.** a) About 42% of the variability in extreme coasters' durations can be explained by this linear model based on the coasters' lengths.
b) On average, extreme coasters gain about 0.0189 seconds for every additional foot of length.
c) The P-value is less than 0.0001. Because the P-value is so small, we reject the null hypothesis of the slope being zero and conclude that there is an association between duration and length.

**47.** $H_0$: Slope of *Effectiveness* vs. *Initial Ability* $= 0$; $H_A$: Slope $\neq 0$

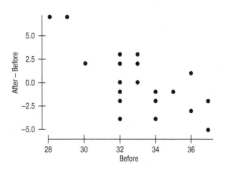

Scatterplot is straight enough. Regression conditions appear to be met. $t = -4.34$; df $= 18$; P-value $= 0.0004$. With a P-value this small, we reject the null hypothesis. There is strong evidence that the effectiveness of the video depends on the player's initial ability. The negative slope observed that the method is more effective for those whose initial performance was poorest and less so for those whose initial performance was better. This looks like a case of regression to the mean. Those who were above average initially tended to be worse after training. Those who were below average initially tended to improve.

**49.** a) Scatterplot looks linear; no overt pattern in residuals; histogram of residuals roughly symmetric and unimodal.
b) $H_0$: No linear relationship between *Education* and *Mortality*, $\beta_1 = 0$. $H_A$: $\beta_1 \neq 0$. $t = -6.24$; P-value $< 0.001$. There is evidence that cities in which the mean education level is higher also tend to have a lower mortality rate.
c) No. Data are on cities, not individuals. Also, these are observational data. We cannot predict causal consequences from them.
d) $(-65.95, -33.89)$ deaths per 100,000 people.
e) *Mortality* decreases, on average, between 33.89 and 65.95 deaths per 100,000 for each extra year of average *Education*.

## Part VII Review (See note, page A-29)

**1.** $H_0$: The proportions are as specified by the ratio 1:3:3:9; $H_A$: The proportions are not as stated. $\chi^2 = 5.01$; df $= 3$; P-value $= 0.1711$. Because P $> 0.05$, we fail to reject $H_0$. These data do not provide evidence to indicate that the proportions are other than 1:3:3:9.

**3.** a) $H_0$: *Mortality* and *Calcium Concentration* in water are not linearly related, $\beta_1 = 0$; $H_A$: They are linearly related, $\beta_1 \neq 0$.
b) $t = -6.73$; P-value $< 0.0001$. There is strong evidence of a negative association between calcium in drinking water and mortality.
c) $(-4.19, -2.27)$ deaths per 100,000 for each ppm calcium
d) Based on the regression analysis, we are 95% confident that mortality (deaths per 100,000) decreases, on average, between 2.27 and 4.19 for each part per million of calcium in drinking water.

5. 406 checks.

7. $H_0$: Income and opinion are independent.
   $H_A$: Income and opinion are not independent.
   $\chi^2 = 154.69$; df $= 6$; P-value $< 0.0001$. With such a small P-value, we reject $H_0$. These data show evidence that income level and opinion about economic inequality are not independent. Examination of the components shows that the low-income respondents are more likely to approve of redistribution compared to the high-income respondents.

9. $H_0$: $p_L - p_R = 0$; $H_A$: $p_L - p_R \neq 0$. $z = 1.38$; P-value $= 0.1683$. Because P $> 0.05$, we do not reject $H_0$. These data do not provide evidence of a difference in musical abilities between right- and left-handed people.

11. a) $H_0$: $\mu_D = 0$; $H_A$: $\mu_D \neq 0$. Boxplot of the differences indicates a strong outlier (1958). With the outlier kept in, the $t$-stat is 0, with a P-value of 1.00 (two-sided). There is no evidence of a difference (on average) of actual and that predicted by Gallup. With the outlier taken out, the $t$-stat is still only $-0.8525$ with a P-value of 0.4106, so the conclusion is the same.
    b) $H_0$: There is no (linear) relationship between predicted and actual number of Democratic seats won ($\beta_1 = 0$). $H_A$: There is a relationship ($\beta_1 \neq 0$). Reject $H_0$. The relationship is very strong, with an $R^2$ of 97.7%. The $t$-stat is 22.56. Even with only 12 df, this is clearly significant (P-value $< 0.0001$). There is an outlying residual (1958), but without it, the relationship is even stronger.

13. Conditions are met; df $= 4$; $\chi^2 = 0.69$; P-value $= 0.9526$. Because P $> 0.05$, we do not reject $H_0$. We do not have evidence that the way the hospital deals with twin pregnancies has changed.

15. a) Based on these data, the average annual rainfall in LA is between 11.65 and 17.39 inches, with 90% confidence.
    b) No. The regression equation is $Rain = -51.684 + 0.033$ $Year$. $R^2 = 0.1\%$. A test of $H_0$: $\beta = 0$ yields $t = 0.12$ with P-value $= 0.9029$.

17. a) 10.29.
    b) $\chi^2 = 6.56$; 6 df; P-value $= 0.3639$. No evidence to suggest that births are not distributed equally across days of the week.
    c) The chi-square components for Monday and Tuesday are 1.05 and 4.4, respectively. Monday's value is not unusual at all. Tuesday's is borderline high, but we concluded that there is not evidence that births are not uniform. With 7 categories, it is not surprising that one is large.
    d) Some births are scheduled for the convenience of the doctor and/or the mother.

19. a) Linear regression is meaningless—the data are categorical.
    b) This is a two-way table that is appropriate. $H_0$: *Eye* and *Hair* color are independent. $H_A$: *Eye* and *Hair* color are not independent. Five cells have expected counts less than 5, so the $\chi^2$ analysis is not valid unless cells are merged. However, with a $\chi^2$ value of 223.6 with 16 df and a P-value $< 0.0001$, the results are not likely to change if we merge appropriate eye colors.

21. a) $H_0$: $p_Y - p_O = 0$; $H_A$: $p_Y - p_O \neq 0$. $z = 3.56$; P-value $= 0.0004$. With such a small P-value, we reject $H_0$. We conclude there is evidence of a difference in effectiveness; it appears the methods are not as good for older women.
    b) $\chi^2 = 12.70$; P-value $= 0.0004$. Same conclusion.
    c) The P-values are the same; $z^2 = (3.563944)^2 = 12.70 = \chi^2$.

23. a) Positive direction; generally linear trend; moderate scatter
    b) $H_0$: There is no linear relationship between *Interval* and *Duration*. $\beta_1 = 0$. $H_A$: There is a linear relationship, $\beta_1 \neq 0$.
    c) Yes; histogram is unimodal and roughly symmetric; residuals plot shows random scatter.
    d) $t = 27.1$; P-value $\leq 0.001$. With such a small P-value, we reject $H_0$. There is evidence of a positive linear relationship between duration and time to next eruption of Old Faithful.

25. a) $t = 1.42$; df $= 459.3$; P-value $= 0.1574$. Because $P > 0.05$, we do not reject $H_0$. We lack evidence that the two groups differed in ability at the start of the study.
    b) $t = 15.11$; P-value $< 0.0001$. The group taught using the Accelerated Math program showed a significant improvement.
    c) $t = 9.24$; P-value $< 0.0001$. The control group showed a significant improvement in test scores.
    d) $t = 5.78$; P-value $< 0.0001$. The Accelerated Math group had significantly higher gains than the control group.

27. a) The regression—he wanted to know about association.
    b) There is a moderate relationship between cottage cheese and ice cream sales; for every million pounds of cottage cheese, 1.19 million pounds of ice cream are sold, on average.
    c) Testing if the mean difference is 0 (matched $t$-test). Regression won't answer this question.
    d) The company sells more cottage cheese than ice cream, on average.
    e) part (a)—linear relationship; residuals have a Normal distribution; residuals are independent with equal variation about the line. (c)—Observations are independent; differences are approximately Normal; less than 10% of all possible months' data.
    f) About 71.32 million pounds.
    g) (0.09, 2.29).
    h) From this regression model, every million pounds of cottage cheese sold is associated with an average increase in ice cream sales of between 0.09 and 2.29 million pounds.

29. Based on these data, the average weight loss for the clinic is between 8.24 and 10.06 pounds, with 95% confidence. The clinic's claim is plausible.

31. $\chi^2 = 8.23$; P-value $= 0.0414$. There is evidence of an association between *Cracker Type* and *Bloating*. The large amount of moderate/severe bloating with gum fiber is alarming. Prospects for marketing this cracker are not good.

33. a) $(-0.60\%, 8.2\%)$.
    b) No. The interval for the difference includes 0.

35. With the concern that 2 low outliers in the New England data cast doubt on any inferences, we cautiously suggest that the percentage of New England people who wear seat belts is between 12 percentage points lower and 4.2 percentage points higher than the percentage of Mountain people who do.

**37.** a) 2-mean *t*-test.
b) $-2.9$.
c) About 35%. Because this P-value is so large, we fail to reject the null. We failed to find evidence that the 1st period has a lower mean score than the 2nd period.

## Practice Exam Answers

**I.** 
1. E.  3. A.  5. D.  7. D.  9. E.
11. C.  13. C.  15. C.  17. C.  19. A.
21. E.  23. C.  25. B.  27. E.  29. A.
31. E.  33. D.  35. D.  37. E.  39. B.

**II. 1.** a) $Q1 - 1.5(IQR) = 61 - 1.5(90 - 61) = 17.5$; $15 < 17.5$.
b) Advanced; 50% of the classes use at least 78 folds and fewer than 25% involve simpler models with 40–60 folds.
c) No; the sample size is small and there is an outlier.

**3.** a) 0.12.  b) 0.757.  c) $2.59.

**5.** Chi-square test of homogeneity. $H_0$: There is no association between age and likelihood of buying an electric car, $H_A$: The age groups are di!erent. We have counts of people in each category. We assume the age groups are independent. Random sample and less than 10% USA. Expected counts (smallest is 162.3) are all at least 5. The conditions seem to be met, so we can use a chi-square model with $(4 - 1)(2 - 1) = 3$ degrees of freedom to conduct a chi-square test of homogeneity. $\chi^2 = 34.98$ with 3 df; $P = 0.001$. With $P < 0.05$ we reject $H_0$; there is strong evidence that the age groups do not have similar likelihood of buying an electric car.

# APPENDIX D: Photo and Text Acknowledgments

**Chapter 1 Photo Credits:**
**001 Shutterstock:** Rawpixel.com/Shutterstock; **003 Alamy Stock Photo:** David Cheskin/AP Images/Alamy Stock Photo; **004 Shutterstock:** Makistock/Shutterstock; **007b Shutterstock:** NicoElNino/Shutterstock; **007t Alamy Stock Photo:** Stefano Rellandini/Reuters/Alamy Stock Photo; **008 Texas Instruments Incorporated:** Texas Instruments Incorporated; **009 Shutterstock:** Rawpixel.com/Shutterstock.

**Chapter 1 Text Credits:**
**001 Carroll, Lewis:** Lewis Carroll, Alice's Adventures in Wonderland; **010 Picasso, Pablo:** Pablo Picasso; **012 Churchill Downs Incorporated:** Source: excerpt from Horsehats.com. Published by thoroughbred promotion.

**Chapter 2 Photo Credits:**
**014 Alamy Stock Photo:** AF archive/Alamy Stock Photo; **019 Munroe, Randall:** 2013 Randall Munroe. Reprinted with permission. All rights reserved; **020 Shutterstock:** Alastair Grant/AP/Shutterstock; **026 123RF GB Ltd:** Pieter De Pauw/123RF; **033 Alamy Stock Photo:** AF archive/Alamy Stock Photo.

**Chapter 2 Text Credits:**
**015bt Kingofthedead:** Source: https://upload.wikimedia.org/wikipedia/commons/thumb/0/06/ElectoralCollege2020_with_results.svg/1200px-ElectoralCollege2020_with_results.svg.png; **017 Gallup, Inc:** Source: https://news.gallup.com/poll/1600/congress-public.aspx; **035 SAGE Publications:** Source: Data in Student survey; **036 Center for Disease Control:** Source: www.cdc.gov/nchs/fastats/deaths.htm; **036 PlaneCrashInfo.com:** (www.planecrashinfo.com/cause.htm); **036 ITOPF: Source:** (www.itopf.com); **037 Elsevier B.V.:** Source: Scott F. Nadler, Michael Prybicien, Gerard A. Malanga, and Dan Sicher, "Complications from Therapeutic Modalities: Results of a National Survey of Athletic Trainers" Archives of Physical Medical Rehabilation 84 (june 2003); **038 Pew Research Center:** Source: pewsocialtrends.org/2019/12/11/most-americans-saythecurrent-economy-is-helping-the-rich-hurting-the-poor-andmiddle-class/; **039 MedRxiv:** Source: https://www.medrxiv.org/content/10.1101/2020.03.11.20031096v2; **039 Pearson Education:** Source: https://ww2.amstat.org/censusatschool/; **040 Oxford University Press:** Source: Data from Fish diet; **041 Pew Research Center:** Data from www.pewsocialtrends.org/files/2012/04/Women-in-the-Workplace.pdf); **042 U.S. Department of Transportation:** Source: https://www.fhwa.dot.gov/policyinformation/statistics.cfm.

**Chapter 3 Photo Credits:**
**044 Alamy Stock Photo:** Toru Hanai/REUTERS/Alamy Stock Photo; **046 Texas Instruments Incorporated:** Texas Instruments Incorporated; **047 123RF GB Ltd:** Iurii Sokolov/123RF; **051b Alamy stock photo:** ZUMA Press Inc/Alamy Stock Photo; **051t Shutterstock:** Michael Brochstein/SOPA Images/Shutterstock; **053 Shutterstock:** Olivier Juneau/Shutterstock; **054 Alamy Stock Photo:** AF archive/Alamy Stock Photo; **060 Shutterstock:** CGi Heart/Shutterstock; **061 Harrison and sons.:** Harrison and sons; **062 Texas Instruments Incorporated:** Texas Instruments Incorporated; **068 Alamy Stock Photo:** Car Collection/Alamy Stock Photo; **069 Texas Instruments Incorporated:** Texas Instruments Incorporated; **070 Munroe, Randall:** 2013 Randall Munroe. Reprinted with permission. All rights reserved; **072 Alamy Stock Photo:** Toru Hanai/REUTERS/Alamy Stock Photo.

**Chapter 3 Text Credits:**
**083 National Center for Education Statistics:** nces.ed.gov/nationsreportcard/); **084 United States Census Bureau:** Source: www.census.gov/compendia/statab/rankings.html; **084 TJC Media Ventures, Inc:** data in kentucky derby 2020.

### Chapter 4 Photo Credits:
**086 Shutterstock:** Ondrej Prosicky/Shutterstock; **089 Shutterstock:** Orlok/Shutterstock; **090 Shutterstock:** BGStock72/Shutterstock; **091 Texas Instruments Incorporated:** Texas Instruments Incorporated; **092 Alamy Stock Photo:** Erin Paul Donovan/Alamy Stock Photo; **092 Texas Instruments Incorporated:** Texas Instruments Incorporated; **095 Shutterstock:** James Atoa/UPI/Shutterstock; **097 Ithaca Times:** Ithaca Times; **098c Munroe, Randall:** Randall Munroe. Reprinted with permission. All rights reserved; **098t Shutterstock:** Ondrej Prosicky/Shutterstock.

### Chapter 4 Text Credits:
**100 U.S. Department of Transportation:** Source: www.TranStats.bts.gov/Data_Elements.aspx?Data=2) reports; **101 Elsevier B.V.:** Mulcahey, Lutz, Kozen, Betz, Prospective Evaluation of Biceps to Triceps and Deltoid to Triceps for Elbow Extension in Tetraplegia, Journal of Hand Surgery, 28, 6, 2003; **102 U.S. Department of Health & Human Services:** A National Vital Statistics Report www.cdc.gov/nchs/; **103 U.S. Department of Transportation:** The Research and Innovative Technology Administration of the Bureau of Transportation Statistics (www.TranStats.bts.gov); **103 U.S. Department of Energy:** Source: www.fueleconomy.gov; **108 Alcohol Alert:** www.alcoholalert.com.

### Chapter 5 Photo Credits:
**110 Shutterstock:** Diego Azubel/EPA/Shutterstock; **111 Alamy Stock Photo:** Pcn Black/Pcn Photography/Alamy Stock Photo; **115 Alamy Stock Photo:** Pcn Black/Pcn Photography/Alamy Stock Photo; **116 123RF GB Ltd:** Serezniy/123RF; **121 Alamy Stock Photo:** Charlie Newham/Alamy Stock Photo; **122 Texas Instruments Incorporated:** Texas Instruments Incorporated; **123 Texas Instruments Incorporated:** Texas Instruments Incorporated; **123 Shutterstock:** Frank Franklin II/AP/Shutterstock; **125 Alamy Stock Photo:** Enigma/Alamy Stock Photo; **125 Texas Instruments Incorporated:** Texas Instruments Incorporated; **126 Shutterstock:** Olga Nayashkova/Shutterstock; **130 Texas Instruments Incorporated:** Texas Instruments Incorporated; **133 Shutterstock:** Diego Azubel/EPA/ Shutterstock; **134 Munroe, Randall:** © 2013 Randall Munroe. Reprinted with permission. All rights reserved; **134 Texas Instruments Incorporated:** Texas Instruments Incorporated; **136 Shutterstock:** Earth Trotter Photography/Shutterstock.

### Chapter 5 Text Credits:
**118 Box, George:** George Box, famous statistician; **138 MLB Advanced Media, LP:** Source: https://baseballsavant.mlb.com.

### Chapter 6 Photo Credits:
**155 Alamy Stock Photo:** Nature and Science/Alamy Stock Photo; **157 Alamy Stock Photo:** Iain Masterton/Alamy Stock Photo; **159 Munroe, Randall:** © 2013 Randall Munroe. Reprinted with permission. All rights reserved; **159 Texas Instruments Incorporated:** Texas Instruments Incorporated; **160 Texas Instruments Incorporated:** Texas Instruments Incorporated; **160 Shutterstock:** Andrii Vodolazhskyi/Shutterstock; **162 Shutterstock:** Andrey Popov/Shutterstock; **163 Texas Instruments Incorporated:** Texas Instruments Incorporated; **164 Texas Instruments Incorporated:** Texas Instruments Incorporated; **168 Getty Images:** Colin Anderson/Photographer's Choice RF/Getty images; **168 Getty Images:** Colin Anderson/Photographer's Choice RF/Getty images; **170t 123RF GB Ltd:** Ewastudio/123RF; **170b Alamy Stock Photo:** Nature and Science/Alamy Stock Photo; **171 Munroe, Randall:** © 2013 Randall Munroe. Reprinted with permission. All rights reserved.

### Chapter 6 Text Credits:
**168 The University of Adelaide Library, eBooks @ Adelaide:** Sir Ronald Aylmer Fisher (1974). Collected Papers of R. A. Fisher: 1948-62; **168 Ornithologische Monatsberichte:** (Ornithologishe Monatsberichte, 44, no. 2); **177 ESPN Internet Ventures:** "MLB Attendance Report - 2017" www.espn.com/mlb/attendance; **178 U.S. Department of Transportation:** "Bureau Of Transportation Statistics" www.transtats.bts.gov/homepage.asp.

### Chapter 7 Photo Credits:
**179 Getty Images:** Peter Cade/Photodisc/Getty Images; **184 Shutterstock:** Gemphoto/Shutterstock; **186 NASA:** Jeff Schmaltz/NASA; **187 Alamy Stock Photo:** US Navy Photo/Alamy Stock Photo; **197 Shutterstock:** MaraZe/Shutterstock; **199 Texas Instruments:** Texas Instruments Incorporated; **201b Getty Images:** Peter Cade/Photodisc/Getty Images; **201t Munroe, Randall:** 2013 Randall Munroe. Reprinted with permission. All rights reserved.

### Chapter 7 Text Credits:

**180 Box, George:** George Box, famous statistician; **211 U.S. Department of Health & Human Services:** National Center for Health Statistics, www.cdc.gov/nchs/; **212 ProOxygen:** Data in $CO_2$ monthly 2020; **212 FastFoodNutrition.org:** Data from tests on 15 different sandwiches randomly selected from thewebsite http://fast-food-nutrition.findthebest.com/d/a/Chicke Sandwich produced the Calories vs. Fat scatterplot.

### Chapter 8 Photo Credits:

**215 Alamy Stock Photo:** Doug Allan/Bluegreen Pictures /Alamy Stock Photo; **217 Munroe, Randall:** © 2013 Randall Munroe. Reprinted with permission. All rights reserved; **220 Alamy Stock Photo:** Scott J. Ferrell/Congressional Quarterly/Alamy Stock Photo; **220t Shutterstock:** Rawpixel.com/Shutterstock; **220c Alamy Stock Photo:** Scott J. Ferrell/Congressional Quarterly/Alamy Stock Photo; **220b Associated Press:** Jim Cole/AP Images; **221 Alamy Stock Photo:** World History Archive/Alamy Stock Photo; **221 Alamy Stock Photo:** Marc Serota/Reuters/Alamy Stock Photo; **226 123RF GB Ltd:** William Perugini/123RF; **229 Alamy Stock Photo:** Doug Allan/Bluegreen Pictures /Alamy Stock Photo.

### Chapter 8 Text Credits:

**216 Jessica U. Meir:** Excerpt from Research Note on Emperor Penguins, Scripps Institution of Oceanography's Center for Marine Biotechnology and Biomedicine at the University of California at San Diego by Jessica Meir. Published by Meir, Jessica; **217 Bohr, Niels:** Niels Bohr, Danish physicist; **221 ARCHIMEDES:** Archimedes (c. 287 BC–c. 212 BC); **230 U.S. Department of Commerce:** www.census.gov; **231 Port of Oakland:** www.oaklandairport.com; **232 U.S. Department of Commerce:** www.nhc.noaa.gov; **233 Federal Reserve Bank of St. Louis:** (https://fred.stlouisfed.org/series/TB3MS); **234 U.S. Department of Commerce:** WWW.Census.com.

### Chapter 9 Photo Credits:

**239 Shutterstock:** Drozdin Vladimir/Shutterstock; **239 Getty Images:** Electravk/iStock/Getty Images; **247 Alamy Stock Photo:** Dmitry Deshevykh/Alamy Stock Photo; **250 Texas Instruments:** Texas Instruments Incorporated; **251 Getty Images:** Electravk/iStock/Getty Images; **251 Texas Instruments:** Texas Instruments Incorporated; **252 Texas Instruments:** Texas Instruments Incorporated; **254 Shutterstock:** Drozdin Vladimir/Shutterstock.

### Chapter 9 Text Credits:

**240 Pearson Education:** DE VEAUX, RICHARD D.; VELLEMAN, PAUL F.; BOCK, DAVID E., INTRO STATS, 5th Ed., © 2018; **241 Pearson Education:** DE VEAUX, RICHARD D.; VELLEMAN, PAUL F.; BOCK, DAVID E., INTRO STATS, 5th Ed., © 2018; **256 Gapminder:** www.gapminder.org.

### Chapter 10 Photo Credits:

**276 Shutterstock:** Suzanne Tucker/Shutterstock; **277 xkcd:** Source: https://xkcd.com/1210/; **278 Shutterstock:** Richard Goldberg/Shutterstock; **278 Alamy Images:** Imageplotter/Alamy Stock Photo; **278 Shutterstock:** Jose Breton- Pics Action/Shutterstock; **278 Shutterstock:** Leonard Zhukovsky/shutterstock; **280 Shutterstock:** Jonathan D. Wilson/Shutterstock; **281 Shutterstock:** Wavebreakmedia/Shutterstock; **282 Texas Instruments:** Texas Instruments Incorporated; **283 Texas Instruments:** Texas Instruments Incorporated; **285 Shutterstock:** Suzanne Tucker/Shutterstock.

### Chapter 10 Text Credits:

**276 Cambridge University Press:** Ian Hocking, The Taming of Chance; **277 Coveyou, Robert R.:** Robert R. Coveyou, Oak Ridge National Laboratory; **278 Australian Broadcasting Corporation:** Dr Karl Kruszelnicki "Dr Karl: How Russian criminals cheated the pokies" http://www.abc.net.au/news/science/2017-06-13/dr-karl-how-russian-cheats-beat-the-pokies/8607598; **278 Greene, Brian:** Brian Greene, The Fabric of the Cosmos: Space, Time, and the Texture of Reality (p. 91).

### Chapter 11 Photo Credits:

**290 Shutterstock:** AnnaStills/Shutterstock; **291 Literary Digest:** Literary Digest; **292 Alamy Stock Photo:** Everett Collection Historical/Alamy Stock Photo; **296 Shutterstock:** Shutterstock; **297 Shutterstock:** StockphotoVideo/Shutterstock; **298 Shutterstock:** Daisy Daisy/Shutterstock; **299 123RF:** Mark Bowden/123RF; **300 Shutterstock:** Maxfeld/Shutterstock; **301 Alamy Stock Photo:** Richard G. Bingham II /Alamy Stock Photo; **307 Shutterstock:** AnnaStills/Shutterstock.

### Chapter 11 Text Credits:

**293 Pew Research Center:** The Pew Research Center; **303 The New York Times Company:** "The New York Times/CBS News Poll, January 20–25, 2006." Published in The New York Times, © 2006; **305 MediaNews Group, Inc.:** Source: Jennifer Hieger, "Portrait of Homebuyer Household: 2 Kids and a PC," Orange County Register, 27 July 2001.

### Chapter 12 Photo Credits:

**315 Alamy Stock Photo:** Hill Street Studios/Tetra Images, LLC/Alamy Stock Photo; **316 Getty Images:** Shimmo/E+/Getty Images; **317 Alamy Stock Photo:** Photo Researchers/Science History Images/Alamy Stock Photo; **318 Alamy Stock Photo:** The History Collection/Alamy Stock Photo; **318 Shutterstock:** Jaromir Chalabala/Shutterstock; **319 Shutterstock:** Dave Newman/Shutterstock; **320 Shutterstock:** Iurii Davydov/Shutterstock; **321 123RF GB Ltd:** Kostic Dusan/123RF; **322 Munroe, Randall:** © 2013 Randall Munroe. Reprinted with permission. All rights reserved; **324 Shutterstock:** Pictrider/Shutterstock; **325 Getty Images:** Laurence Mouton/PhotoAlto/Getty Images; **326 Shutterstock:** Galinka/Shutterstock; **327 Munroe, Randall:** © 2013 Randall Munroe. Reprinted with permission. All rights reserved; **328 Shutterstock:** Orkhan Aslanov/Shutterstock; **328 Shutterstock:** Nvuk/Shutterstock; **329 Shutterstock:** Fotokostic/Sutterstock; **330 Getty Images:** Joss/iStock/Getty Images; **330 Shutterstock:** Katrina Leigh/Shutterstock; **334 Alamy Stock Photo:** Hill Street Studios/Blend Images/Alamy Stock Photo.

### Chapter 12 Text Credits:

**316 Lord Halifax:** Lord Halifax; **324 Condé Nast:** Experiments on Trial," by Hannah Fry, The New Yorker, 2 March 2020; **340 Elsevier B.V.:** R.G. Janssen, D.A. Schwartz, and P.F. Velleman, "A Randomized Controlled study of contrast Baths on patients with carpel Tunnel syndrome," Journal of Hand Therapy, 2009.

### Chapter 13 Photo Credits:

**354 Shutterstock:** Jessica Bethke/Shutterstock; **357 Getty Images:** SpxChrome/iStock/Getty Images; **357 123RF GB Ltd:** James Steidl/123RF; **358 Alamy Stock Photo:** OJO Images Ltd/Alamy Stock Photo; **358 123RF GB Ltd:** Zentilia/123RF; **359 Shutterstock:** Bakounine/Shutterstock; **361 Getty Images:** Chris Hepburn/iStock/Getty Images; **362 Shutterstock:** Sovastock/Shutterstock; **365 Alamy Stock Photo:** Carolyn Jenkins/Alamy Stock Photo; **367 Shutterstock:** Jessica Bethke/Shutterstock; **372 Getty Images:** Mario Tama/Gettty Images.

### Chapter 13 Text Credits:

**356 Bernoulli, Jacob:** Jacob Bernoulli, 1713, discoverer of the LLN; **357 Yogi Berra:** Yogi Berra; **362 Yogi Berra:** Yogi Berra; **368 Associated Press:** Associated Press, fall 1992, quoted by Schaeffer et al; **369 U.S. Department of Health and Human Services:** http://www.cdc.gov/nchs/data/nhis/earlyrelease/wireless201512.pdf; **370 The New York Times Company:** The New york Times, "A Great year for Ivy League schools, but Not so Good for Applicants to Them," April 4, 2007.

### Chapter 14 Photo Credits:

**374b U.S. Department of the Treasury:** U. S. Department of the Treasury; **374t Shutterstock:** Worradirek/Shutterstock; **375 Shutterstock:** CandyBox Images/Shutterstock; **376 Getty Images:** Apomares/E+/Getty Images; **377 Shutterstock:** Dean Drobot/Shutterstock; **380 Shutterstock:** CandyBox Images/Shutterstock; **381 123RF:** Graham Oliver/123RF; **383 Alamy Stock Photo:** Randy Miramontez/Alamy Stock Photo; **384 Shutterstock:** WAYHOME studio/Shutterstock; **386 Shutterstock:** Andrey_Popov/Shutterstock; **387 Shutterstock:** Simev/Shutterstock; **391 Alamy Stock Photo:** ZUMA Press, Inc/Alamy Stock Photo; **392 Munroe, Randall:** 2013 Randall Munroe. Reprinted with permission; **394 Shutterstock:** Worradirek/Shutterstock.

### Chapter 14 Text Credits:

**399 Pew Research Center:** LEE RAINIE, KATHRYN ZICKUHR, A Snapshot of Reading in America in 2013. Pew Research Center http://www.pewinternet.org/2014/01/16/a-snapshot-of-reading-in-america-in-2013/.

APPENDIX D  Photo and Text Acknowledgments   **A-83**

### Chapter 15 Photo Credits:
**402 Getty Images:** DigitalVision/Getty Images; **404 123RF GB Ltd:** Michael Brown/123RF; **405 123RF:** Hemant Mehta/123RF; **406b Shutterstock:** Shutterstock; **406t Shutterstock:** Mythja/Shutterstock; **407 Texas Instruments Incorporated:** Texas Instruments Incorporated; **408 Texas Instruments Incorporated:** Texas Instruments Incorporated; **409 Shutterstock:** Merzzie/Shutterstock; **410 123RF:** Esther Derksen/123RF; **412 Alamy Stock Photo:** Design Pics/Alamy Stock Photo; **416b Central Intelligence Agency:** Central Intelligence Agency; **416t Alamy Stock Photo:** Karl Mondon/Contra Costa Times/ZUMA Press; **418 Shutterstock:** Rade Kovac/Shutterstock; **421 Getty Images:** DigitalVision/Getty Images.

### Chapter 15 Text Credits:
**408 U.S. Department of Commerce:** U.S. Department of Commerce.

### Chapter 16 Photo Credits:
**429b Shutterstock:** Salty View/Shutterstock; **429t Shutterstock:** Stockphotofan1/Shutterstock; **430 Alamy Stock Photo:** Photo 12/Alamy Stock Photo; **431 Getty Images:** Tpopova/iStock/Getty Images; **432 Shutterstock:** 963 Creation/Shutterstock; **433 Texas Instruments Incorporated:** Texas Instruments Incorporated; **436b Getty Images:** Vladm/iStock/Getty Images; **436t Getty Images:** Roman Okopny/iStock/Getty Images; **437 Texas Instruments Incorporated:** Texas Instruments Incorporated; **438 Texas Instruments Incorporated:** Texas Instruments Incorporated; **438t Getty Images:** Ichaka/E+/Getty images; **440 Getty Images:** Ene/iStock/Getty Images; **441 Getty Images:** Marilyn Nieves/Vetta/Getty images; **442 Getty Images:** Myhrcat/iStock/Getty Images; **442 123RF GB Ltd:** Paras Soni/123RF; **444 Shutterstock:** Stockphotofan1/Shutterstock.

### Chapter 17 Photo Credits:
**460 Shutterstock:** Danie Nel Photography/Shutterstock; **466 Shutterstock:** Pavel L. Photo and Video/Shutterstock; **470 Shutterstock:** Javier Galeano/AP/Shutterstock; **472 Getty Images:** Keith Brofsky/Photodisc/Getty Images; **476 Shutterstock:** Erika Cross/Shhutterstock; **478 Shutterstock:** Danie Nel Photography/Shutterstock.

### Chapter 17 Text Credits:
**465 U.S. Department of Health and Human Services:** Cynthia L. Ogden, Cheryl D. Fryar, Margaret D. Carroll, and Katherine M. Flegal, Mean Body Weight, Height, and Body Mass Index, United States 1960–2002, Advance Data from Vital and Health Statistics Number 347, October 27, 2004. www.cdc.gov/nchs; **468 Laplace, Pierre-Simon:** Pierre-Simon Laplace, 1812; **476 The Seattle Times:** Lynn Thompson, School size: Is smaller really better? October 26, 2005.

### Chapter 18 Photo Credits:
**488b Shutterstock:** Rich Carey/Shutterstock; **488t Shutterstock:** unterwegs/Shutterstock; **491 Shutterstock:** Alexey Boldin/Shutterstock; **494 Andrews McMeel Syndication:** GARFIELD © 1999 Paws, Inc. Distributed by Andrews McMeel Syndication. Reprinted with permission. All rights reserved; **496 Shutterstock:** Pictrider/Shutterstock; **498 Texas Instruments Incorporated:** Texas Instruments Incorporated; **498 123RF:** sdecoret/123RF; **499 Shutterstock:** Florida Stock/Shutterstock; **504b Shutterstock:** Rich Carey/Shutterstock; **504t Shutterstock:** unterwegs/Shutterstock.

### Chapter 18 Text Credits:
**490 Institute of Mathematical Statistics:** Tukey, John W. The Future of Data Analysis. Ann. Math. Statist. 33 (1962), no. 1, 1–67; **498 Pew Research Center:** COURTNEY KENNEDY AND ARNOLD LAU, "Most Americans believe in intelligent life beyond Earth; few see UFOs as a major national security threat," JUNE 30, 2021, https://www.pewresearch.org/fact-tank/2021/06/30/most-americans-believe-in-intelligent-life-beyond-earth-few-see-ufos-as-a-major-national-security-threat/.

### Chapter 19 Photo Credits:
**511 Shutterstock:** Lotus_studio/Shutterstock; **518 Shutterstock:** Tommaso79/Shutterstock; **520 Shutterstock:** Teamdaddy/Shutterstock; **522 Texas Instruments Incorporated:** Texas Instruments Incorporated; **524 Shutterstock:** CJ GUNTHER/EPA-EFE/Shutterstock; **527 Munroe, Randall:** 2013 Randall Munroe. Reprinted with permission. All rights reserved; **529 Shutterstock:** Lotus_studio/Shutterstock.

### Chapter 19 Text Credits:
**512 Merriam-Webster, Inc.:** Webster's Unabridged Dictionary, 1913; **514 New York State Office of Court Administration:** NY state jury instructions; **515 Oliver and Boyd:** Sir Ronald Fisher, The Design of Experiments; **519 Russell Lowell, James:** James Russell Lowell, Credidimus Jovem Regnare; **523 The Zetetic Scholar:** From Marcello Truzzi, "On the Extraordinary: An Attempt At Clarification." The Zetetic Scholar, Volume 1, Number 1, 1978, p.11; **533 Pew Research Center:** Pew Research Center, http://www.pewinternet.org/2017/07/11/online-harassment-2017/, Online Harassment 2017.

### Chapter 20 Photo Credits:
**535 Getty Images:** Karl Weatherly/Photographer's Choice RF/Getty Images; **536 Munroe, Randall:** Randall Munroe. Reprinted with permission. All rights reserved; **537 Shutterstock:** Nikkolia/Shutterstock; **540 Velleman, Paul:** Paul Velleman; **541 Shutterstock:** Limpido/Shutterstock; **547 Munroe, Randall:** 2013 Randall Munroe. Reprinted with permission. All rights reserved; **549 123RF:** Dean Drobot/123RF; **553 Getty Images:** Karl Weatherly/Photographer's Choice RF/Getty Images.

### Chapter 20 Text Credits:
**539 Hume, David:** David Hume, "Enquiry Concerning Human Understanding," 1748; **542 Hunter, Stu:** Stu Hunter; **546 Houghton Mifflin Harcourt:** 11J. R. R. Tolkien, The Hobbit. 75th anniversary edition, Houghton Mifflin Harcourt (September 18, 2012), p. 7; **557 CareerBuilder, LLC:** Career Builder, www.careerbuilder.com.

### Chapter 21 Photo Credits:
**562 123RF:** Milkos/123RF; **563 Alamy Stock Photo:** Marka/Press/Alamy Stock Photo; **566 Shutterstock:** Aslysun/Shutterstock; **567 Shutterstock:** Shutterstock; **568 Texas Instruments:** Texas Instruments; **569 Texas Instruments:** Texas Instruments; **569b Shutterstock:** Syda Productions/Shutterstock; **569t Shutterstock:** Sunshine Seeds/Shutterstock; **573 Texas Instruments:** Texas Instruments; **573 Texas Instruments:** Texas Instruments; **574 Shutterstock:** Amble Design/Shutterstock; **576 123RF:** Milkos/123RF.

### Chapter 21 Text Credits:
**578 Gallup, Inc.:** Evolution, Creationism, Intelligent Design; © 2021, Gallup, Inc.

### Chapter 22 Photo Credits:
**593 123RF:** Dean Drobot/123RF; **594 Shutterstock:** Robert Crow/Shutterstock; **599 Alamy Stock Photo:** John Angerson/Alamy Stock Photo; **600 Texas Instruments:** Texas Instruments; **603 Shutterstock:** AshTproductions/Shutterstock; **605 Texas Instruments:** Texas Instruments; **608 Shutterstock:** Kzenon/Shutterstock; **609 Texas Instruments:** Texas Instruments; **610 Texas Instruments:** Texas Instruments; **614 123RF:** Dean Drobot/123RF.

### Chapter 22 Text Credits:
**593 AuthorHouse:** Jim Maas and Rebecca Robbins, in their book Sleep for Success!; **593 AuthorHouse:** B. Maas, Dr. James, RSleep for Success! Everything You Must Know; © 2011, AuthorHouse; **602 United States Census Bureau:** United States Census Bureau; **611 United States Census Bureau:** Census Bureau; **619 U.S. Department of Transportation:** www.transtats.bts.gov/HomeDrillchart.asp; **621 Cambridge University Press:** Waguespack, Nicole, et. al., Antiquity, 2009; **622 Business Horizons:** Business Horizons, 2012, 55; **622 PGA TOUR:** www.pgatour.com.

### Chapter 23 Photo Credits:
**624 Shutterstock:** Thanasit Tippawan/Shutterstock; **626 Getty Images:** Juanmonino/E+/Getty Images; **631 Texas Instruments:** Texas Instruments; **632 Shutterstock:** Jenifoto1/Shutterstock; **633 Getty Images:** Aldomurillo/E+/Getty Images; **635 Texas Instruments:** Texas Instruments; **637 Getty Images:** Caroline Woodham/Photodisc/Getty Images; **641 Shutterstock:** Thanasit Tippawan/Shutterstock.

### Chapter 23 Text Credits:

**643 National Council of Teachers of Mathematics:** Journal for Research in Mathematics Education, 31, no. 3[2000]); **644 Yahoo:** Data from www.sports.yahoo.com; **645 Baseball Info Solutions:** Data in Baseball 2016; **648 Sage Publications, Ltd. (UK):** Journal of Applied Behavioral Science, March 2008.

### Chapter 24 Photo Credits:

**652 DORIT H. AARON:** Dorit H. Aaron; **656 Shutterstock:** Vladislav Gajic/Shutterstock; **659 Texas Instruments:** Texas Instruments; **660 Texas Instruments:** Texas Instruments; **661 DORIT H. AARON:** Dorit H. Aaron; **662 Texas Instruments:** Texas Instruments; **665 DORIT H. AARON:** Dorit H. Aaron.

### Chapter 24 Text Credits:

**667 Routledge, Garland Publishing, Taylor & Francis BOOKS:** Technometrics 1975; **668 British Medical Journal (BMJ Publishing Group):** The British Medical Journal, Friday the 13th Bad for Your Health?; **670 American Association for the Advancement of Science:** UENOYAMA REIKO, The characteristic response of domestic cats to plant iridoids allows them to gain chemical defense against mosquitoes; © 2021 American Association for the Advancement of Science; **672 College Board:** Data from www.collegeboard.com.

### Chapter 25 Photo Credits:

**689 123RF GB Ltd:** Baloncici/123RF; **690 123RF GB Ltd:** Vladimir Mucibabic/123RF; **691 Shutterstock:** Photo Works/Shutterstock; **692l Shutterstock:** ALEKSANDR S. KHACHUNTS/Shutterstock; **692r Shutterstock:** B-D-S Piotr Marcinski/Shutterstock; **693 123RF:** Deagreez/123RF; **696 Texas Instruments:** Texas Instruments; 696 Texas Instruments: Texas Instruments; **697 Texas Instruments:** Texas Instruments; **699 Shutterstock:** Tomas del amo/Shutterstock; **700 Shutterstock:** Iofoto/Shutterstock; **702 United States Department of Agriculture:** Peggy Greb/United States Department of Agriculture; **703 123RF:** Kenny Kierna/123RF; **704 123RF GB Ltd:** Semmick photo/123RF; **706 Texas Instruments:** Texas Instruments; **709 123RF GB Ltd:** Baloncici/123RF.

### Chapter 25 Text Credits:

**713 National Hurricane Center:** www.nhc.noaa.gov/dcmi.shtml; **714 The National Lottery:** http://www.national-lottery.co.za/lotto/statistics.aspx; **714 Jungle scout:** Connolly Brian, Amazon A/B Testing: Step-by-Step Guide Brian Connolly, 2021.

### Chapter 26 Photo Credits:

**720 Shutterstock**: VLADJ55/Shutterstock; **724 Getty Images:** Stocktrek/Photodisc/Getty Images; **726 123RF:** Dipressionist/123RF; **730 Texas Instruments:** Texas Instruments; **731 123RF:** Jatesada Natayo/123RF; **735 Texas Instruments:** Texas Instruments; **736 Texas Instruments:** Texas Instruments; **739 Shutterstock:** VLADJ55/Shutterstock.

### Chapter 26 Text Credits:

**725 Bacon, Francis:** Francis Bacon (1561–1626); **742 United Nations Office on Drugs and Crime (UNODC):** World Drug REPORT, 2013. www.unodc.org/unodc/en/data-and-analysis/WDR-2013.html; **748 CBS Interactive, Inc.:** www.cnet.com/news/cnet-tablet-battery-life-results/.

### Review Part 1 Text Credits:

**153 Classy:** https://www.classy.org/blog/infographic-social-media-demographics-numbers/.

### Review Part 2 Text Credits:

**264 Sophocles:** Sophocles (495–406 bce); **265 U.S. Department of Health & Human Services:** www.cdc.gov/nchs/births.htm); **266 David E. Bock:** www.uscgboating.org/library/accident-statistics/Recreational-Boating-Statistics-2015.pdf); **271 American Medical Association:** Source: M D Kogan, "Trends in twin birth outcomes and prenatal care utilization in the United States, 1981–1997," JAMA, 284 [2000]: 335–341.

### Review Part 3 Text Credits:

**345 V. H. Winston & Son, Inc:** Source: "Sweetening the Till: The Use of Candy to Increase Restaurant Tipping," Journal of Applied Social Psychology 32, no. 2 [2002]: 300–309; **347 The Weekly Standard LLC:** Source: The Weekly Standard, March 10, 1997); **351 Board of Regents of the University of Wisconsin System:** Source: http://www.stat.wisc.edu/~st571-1/15-regression-4.pdf).

### Review Part 4 Text Credits:

**440 Gallup, Inc.:** http://www.gallup.com/poll/9916/Homosexuality.aspx#5; **451 Science News:** Science News, 161 no. 24 [2002]; **453 Catholic News Service (CNS):** Homeownership rate in the United States from 1990 to 2020, Statista. © Statista 2021. https://www.statista.com/statistics/184902/homeownership-rate-in-the-us-since-2003/; **455 The New York Times Company:** Jane E. Brody, A Host Of Ills When Iron Out Of Balance, August 2012 New York Times Article; **457 American Educator:** Sahlberg, pasi, "A model Lesson: Finland shows Us What Equal Opportunity Looks Like," American Educator Spring 2012: 25; **458 U.S. Department of Health & Human Services:** Receipt of Services for Substance Use and Mental Health Issues among Adults: Results from the 2016 National Survey on Drug Use and Health, NSDUH Data Review, September 2017. SAMHSA. https://www.samhsa.gov/data/sites/default/files/NSDUH-DR-FFR2-2016/NSDUH-DR-FFR2-2016.htm.

### Review Part 5 Text Credits:

**585 Elsevier B.V.:** Data from Lancet, June 1, 2002; **586 Society for Science & the Public:** Science News, 161, no. 24 [2002]; **587 Massachusetts Medical Society:** Outcomes in Young Adulthood for Very-Low-Birth-Weight Infants, New England Journal of Medicine, 346, no. 3 [2002]; **588 Gallup, Inc.:** The Gallup Poll; **590 U.S. Department of Justice:** D. Levine and P. Mangan, "Recidivism of Prisoners Released in 1994"; NCJ193427; Bureau of Justice; File name rpr94bxl.csv http://bjs.ojp.usdoj.gov/index.cfm?ty=pbdetail&iid=1134; **592 VGChartz Ltd:** Modified from https://www.vgchartz.com/tools/hw_date.php

### Review Part 6 Text Credits:

**676 Elsevier B.V.:** Benson and Janette, Infant Behavior and Development [1993]; **676 Smell and Taste Treatment and Research Foundation Ltd.:** A. R. Hirsch and L. H. Johnston, Odors and Learning Chicago: Smell and Taste Treatment and Research Foundation; **677 John Wiley & Sons, Inc:** A. Tubb, A. J. Parker, and G. Nickless, The Analysis of Romano British Pottery by Atomic Absorption Spectrophotometry. Archaeometry, 22[1980]:153:171; **677 Society for Science & the Public:** Science News, Sept. 28 [2002]); **677 World Health Organization (US):** Health and Health Behavior Among Young People. Copenhagen: World Health Organization, 2000; **678 Wesson, R.L.:** Ross and Wenzel, An Analysis of Traffic Deaths by Vehicle Type and Model, March 2002; **679 University of Iowa:** Faith Loven, A Study of the Interlist Equivalency of the CID W-22 Word List Presented in Quiet and in Noise. University of Iowa [1981]); **680 Education Policy Analysis Archives:** The Effectiveness of Teach for America and Other Under-certified Teachers on Student Academic Achievement: A Case of Harmful Public Policy. Education Policy Analysis Archives [2002]); **687 British Medical Journal (BMJ Publishing Group):** Jeremy Groves, "Bicycle Weight and Commuting Time: A Randomized Trial," Significance: Statistics Making Sense (June 2011), vol. 8, no. 2: 95–97.

### Review Part 7 Text Credits:

**757 American Medical Association Journal of Ethics:** (JAMA 284 [2000]: 335–341); **757 American Medical Association:** JAMA 284 [2000]: 335–341; **758 American Psychologist:** A Social Technology That Reduces Social Involvement and Psychological Well-Being Copyright 1998 by the American Psychological Association, Inc. 0003-066X/98/$2.00 Vol. 53, No. 9, 1017-1031; **762 The Milbank Memorial Fund Quarterly:** Fertility Planning and Fertility Rates by Socio-Economic Status, Social and Psychological Factors Affecting Fertility, 1954; **764 Gallup, Inc.:** Gallup, (Source: www.gallup.com/poll/155210/Americans-PositiveImmigration.aspx); **765 U.S. Department of Education:** U.S. Department of Education, National Center for education statistics.

# APPENDIX E: Index

*Note:* Page numbers in **boldface** indicate chapter-level topics; FE indicates For Example or Step-by-Step references; n indicates footnotes; TI indicates TI Tips instructions.

## Numbers

5-number summary, 57
95% confidence, 491–493, 493FE
68–95–99.7 Rule, 118–120, 119FE, 121FE
10% Condition
    Bernoulli trials and, 430–431, 433FE, 437FE, 443FE
    Central Limit Theorem and, 469, 471FE
    for chi-square test of homogeneity, 699, 705FE
    for comparing counts, 691, 692FE
    for comparing means, 627, 629FE
    for comparing proportions, 565
    confidence intervals for proportions and, 496, 497FE
    hypothesis testing and, 518FE, 520FE
    inferences about means and, 601, 603FE
    for Normal model, 463, 464
    one-proportion $z$-tests and, 538FE
    one-sample $t$-test for the mean and, 608FE
    for paired data, 654, 655FE
    sampling distribution model for a mean and, 472FE
    sampling distribution model for proportions and, 465FE, 466FE
    for two-proportion $z$-intervals, 567FE

## A

Accidental correlation, 165–167
Accuracy, 67
Actuaries, 402
Addition Rule, probability and, 361–362, 361FE
Addition Rule for Variances, 413
A.G. Edwards, 88FE
Ages of U.S. senators, 50–51
Airline flight cancellations, 60FE–61FE
Airline on-time arrivals, 91, 362–363
Alpha levels, 541–542
Alternative hypotheses, 512, 515–516
    one-sided, 516
    two- and one-tailed, 516n
    two-sided, 515–516
Amazon, 3, 4, 5
Amazon Standard Identification Number (ASIN), 5
Anecdotal evidence, 319
Angus cows' weight, 594FE–595FE
Archimedes, 221
Area principle, 16
Armstrong, Lance, 7
Associations, 156–158, 162FE–163FE
    direction of, 156
    negative, 156
    positive, 156
    strength of, 157
    using statistical software, A-9–A-10
Avandia, 536FE, 539FE, 541FE
Average, 63. *See also* Mean(s)

## B

Bacon, Francis, 725
Ballplayers' birth months, 690, 693FE
Bar charts, 17–18, 17FE–18FE
    segmented, 22, 23FE, 24FE
    side-by-side, 28
Base $e$ logarithms, 245n
Base 10 logarithms, 245n
Battery life, 624–625, 629FE–631FE
Battery production, 386FE, 390FE
Bayesian reasoning, 392–394
Bernoulli, Daniel, 430
Bernoulli, Jacob, 356
Bernoulli trials, 429–431, 435
    10% rule and, 430–431, 433FE, 437FE, 443FE
Berra, Yogi, 156, 362
Bias
    nonresponse, 293, 306
    reducing, 306
    response, 306
    in sampling, 291–292
    voluntary response, 304
Biden, Joe, 15
Bill and Melinda Gates Foundation, 475
Bimodal histograms, 52
Binge drinking, 388–389
Binomial model, 434–438, 436FE–437FE, 437TI–438TI, 445TI
Binomial probability, 434–435
    Normal approximation to, 440FE–441FE
Birth months of ballplayers, 690, 693FE
Blinding, 325–326, 326FE
Blocking, 327–329, 328FE, 663–664
    in experiments, 319–320
    using statistical software, A-24–A-25
Blood types, 432FE–433FE, 436FE–437FE, 438–439, 442FE–443FE
Bohr, Niels, 217
Boomerang CEOs, 184
Box, George, 118, 180
Boxplots, 59–62, 60FE–61FE
    comparing groups with, 88–92, 89FE, 90FE–92FE
    making, 62TI
Bridges, 191–193
Buchanan, Pat, 220, 221
Bush, George W., 220, 222

## C

Campus meal programs, 375FE, 382
Campus room assignments, 386–388
Cancer and smoking, 168
Candy colors, 365FE–366FE, 387FE
Carpal tunnel surgery, 632
Carroll, Lewis, 1
Cases, 4
Categorical data, **14–33**
    displaying, 34TI
    displaying and describing using statistical software, A-3–A-4
Categorical Data Condition, 19, 26FE
Categorical variables, 5, 6
Causation
    chi-square tests and, 707–708
    correlation and, 167–168
Cawthorn, Madison, 51
Ceci, Stephen, 330
Cells, 690–691
Censuses, 293–294
Centers
    of distributions, 52, 60FE–61FE, 62–65, 64FE
    moving, 113–114
    of random variables, 403–405, 404FE, 407TI–408TI
Central Limit Theorem (CLT), 468–469
    assumptions and conditions for, 469
    inferences about means and, 594–595, 594FE–595FE
    for means, 471FE–472FE
    when population is very skewed, 473–475
Cereal, 126FE–128FE, 224, 563
Chi-square models, 692
Chi-square statistic, 692

Chi-square tests, 710TI
    causation and, 707–708
    of goodness-of-fit, 693–697,
        693FE–695FE, 696TI–697TI
        calculation for, 695
    of homogeneity, 698–702, 700FE–
        701FE, 706TI
        assumptions and conditions for,
            698–699
        calculations for, assumptions and
            conditions for, 699
    of independence, 702–707, 702FE–
        704FE, 704FE–705FE, 706FE–
        707FE, 706TI
Chief executive officers' earnings, 94–95,
    119FE
Cluster sampling, 299–300, 299FE
Coffee containers, 90FE–91FE
Coin flips, 357FE, 442
Common cold, 332
Company assets, 241–242
Complement(s), 360
Complement Rule, probability and, 360,
    361FE
Completely randomized design,
    321FE–322FE
Completely randomized two-factor
    experiments, 329
Computer shipments, 406FE–407FE
Computer tips. *See also* Statistical software
    for binomial model, 445TI
    for calculating statistics, 69TI
    for chi-square tests, 710TI
    for comparing distributions, 99TI
    for confidence intervals for proportions,
        498TI, 505TI
    for correlation, 171TI
    for creating a normal probability plot,
        130TI
    for creating scatterplots, 159TI–160TI
    for data, 8TI
    for displaying and summarizing
        quantitative variables, 75TI
    for displaying categorical data, 34TI
    for experiments, 336TI
    for finding confidence intervals,
        568TI–569TI
    for finding correlations, 163TI–164TI
    for finding geometric probabilities,
        433TI–434TI
    for generating random numbers,
        282TI–283TI
    for hypothesis testing, 521TI–522TI,
        554TI, 573TI
    for inference for difference of means,
        642TI
    for inferences about means, 615TI
    for logarithmic re-expressions,
        251TI–252TI
    for making boxplots, 62TI
    for making histograms, 46TI

for Normal model, 134TI
for paired $t$-tests, 666TI
for random variables, 422TI
for re-expressing to achieve linearity,
    250TI
for re-expression, 254TI
for regression analysis, 203TI, 230TI,
    740TI
for regression lines, 199TI
for residuals plots, 199TI
for sampling, 309TI
for scatterplots, 171TI
for simulations or generating random
    numbers, 285TI
Conclusions, hypothesis testing and, 519,
    519FE
Conditional distributions, 21–27, 22FE–
    23FE, 24FE–27FE
Conditional probability, 378–380, 380FE,
    518, 537. *See also* P-values
    reversing the conditioning and, 390–392,
        391FE–392FE, 393FE–394FE
Confidence intervals, 543–545
    for comparing proportions, 565–569,
        566FE, 567FE–568FE
    for counts, 498
    for difference in sample means, 627FE–
        628FE, 631TI–632TI
    finding, 568TI–569TI
    for means, 506FE–606FE, 598–599,
        599FE
        hypothesis testing and, 610–611
        interpreting, 606
    95%, for small samples, 544–545
    pooled $t$-test for the difference between
        means and, 638
    regression inference and, 730
Confidence intervals for proportions,
    **488–510**, 505TI
    assumptions and conditions for, 495–498,
        496FE–497FE
    critical values and, 494–495, 495FE
    finding, 498TI
    margin of error and, 493–494, 494FE
    95%, meaning of, 491–493, 493FE
    proof of, 500–502
    sample size and, 498–500, 499FE, 500FE
    using statistical software, A-17–A-18
Confounding, 330–332, 330FE
    lurking variables vs., 331
Congress, confidence in, 17FE–18FE
Constants
    adding and subtracting, standard devia-
        tion and, 411–412, 412FE
    multiplying by, mean and variance and,
        412, 412FE
Contaminants in salmon, 599TE
Context, 3
Contingency tables, 19–21, 21FE, 702
Continuous random variables, 403,
    417–420, 418FE–420FE, 441

Control, in experiments, 318, 320FE
Control groups, 324
Control treatments, 324
Convenience sampling, 305, 305FE
Corals, 488–490
Correlation, 160–168, 162FE–163FE,
    171TI
    accidental, 165–167
    causation and, 167–168
    conditions and, 161–164
    finding, 163TI–164TI
    least squares line and, 181–183
    properties of, 164–165
    for scatterplot patterns, 162FE
    strength of, 165
    using statistical software, A-9–A-10
Correlation coefficient, 161
Cotinine in blood, 96–97
Count(s)
    comparing, **689–719**
        chi-square test of independence and,
            702–707, 703FE–705FE, 706FE–
            707FE, 706TI
        goodness-of-fit tests and. *See*
            Goodness-of-fit tests
        homogeneity and. *See* Homogeneity
        using statistical software,
            A-25–A-27
    confidence intervals for, 498
Counted Data Condition, 698, 700FE,
    705FE
    for comparing counts, 691, 691FE,
        694FE
Coveyou, Robert R., 277
Craters on earth, 724FE–725FE, 731FE
Credit card use, 54FE, 64FE, 66FE, 67FE,
    331
Critical values, 494–495, 495FE
    inferences about means and, 600TI
Cumulative distribution graphs, 49–51
Cumulative probability distributions, 409

# D

Data, **1–13**
    area principle and, 16
    bar charts and, 17–18, 17FE–18FE
    calculator tips for, 8TI
    categorical. *See* Categorical data
    on computer, 10TI
    conditional distributions and, 21–27,
        22FE–23FE, 24FE–27FE
    contingency tables and, 19–21, 21FE
    defined, 1
    frequency tables and, 16–17
    on Internet, 7
    mosaic plots and, 28–29
    pie charts and, 19
    plotting, 625
    re-expressing. *See* Re-expressing data
    rescaling, 114–115, 115FE
    shifting, 113–114

side-by-side bar charts and, 28
simulations and, 29–30
technique for tips for, 8TI, 10TI
transforming. *See* Re-expressing data
Data analysis, rules of, 15, 61
Data collection, 4–5
Data Desk. *See* Statistical software
Data-gathering devices, 2
Data tables, definition of, 3
De Moivre, Abraham, 118n
Death and disability insurance, 402–404, 405–406, 411–413
Degrees of freedom, 597
$t$-table and, 604
Dependent variable, 158n, 159
*The Design of Experiments* (Fisher), 541
Diagramming, of experiments, 320–321
Dice, rolling, 431
Dice game simulation, 280
Direction, of associations, 156
Discrete random variables, 403
Disjoint events, 361
independence vs., 382–383
probability and, 382–383, 384FE–385FE
Distributions, 17, 44, **86–109**
center of, 52, 60FE–61FE, 62–65, 64FE
comparing, 99TI
using statistical software, A-6–A-7
conditional, 21–27, 22FE–23FE, 24FE–27FE
cumulative probability, 409
marginal, 20–21, 21FE
sampling. *See* Sampling distribution models; Sampling distributions
shapes of, 52–55, 54FE, 60FE–61FE
spread of, 52, 60FE–61FE, 65–66, 66FE
Standard Normal, 117
symmetric
mean of, 62–65
median of, 63–65
Does the Plot Thicken? Condition, 195, 196, 197FE, 723, 723FE, 725FE, 726FE, 733FE
Dorsey, Jack, 95n
Dotplots, 49
Double-blind experiments, 326
Driving test pass rate, 517, 518FE–519FE
*Dungeons and Dragons*, 410, 411

## E

Ears and tongue rolling, 690–691, 692, 697
Earthquake magnitudes, 57, 59
Effect size, 519, 542, 548–550
paired data and, 662–663, 663FE
Electronic gadget use, 18FE
Empirical Rule, 118n
Employee honesty, 636FE–637FE
Equal Variance Assumption, for pooled $t$-test for the difference between means, 637–638

Error
margin of, 493–494, 494FE
sampling, 296, 463
standard, 190–191, 489, 512
of difference between two proportions, 564FE
of difference in independent sample means, 626
of the slope, 729–730
Types I and II, 545–547, 546FE, 552FE
Events, 355
disjoint (mutually exclusive), 361
independence vs., 382–383
probability and, 382–383, 384FE–385FE
independent, 356
Excel. *See* Statistical software
Executives' zodiac signs, 693FE–695FE, 696TI–697TI
Expected Cell Frequency Condition
for chi-square test of homogeneity, 699, 700FE, 705FE
for comparing counts, 691, 692FE, 694FE
Expected value, 403–405, 404FE
Experiment(s), **316–342**, 336TI
blinding and, 325–326, 326FE
blocking and, 327–329, 328FE
confounding and, 330–332, 330FE
control treatments and, 324
definition of, 316
diagramming, 320–321
double-blind, 326
experimental units and, 317
factors and, 317, 329
placebos and, 326–327
random assignment and, 316
response variables and, 317, 318FE
samples vs., 324
single-blind, 326
statistical significance and, 322–323, 332–333
subjects (participants) in, 317
treatments and, 317, 318FE
two-factor, completely randomized, 329
Experimental design, 317, 318–320, 320FE, 321FE–322FE
principles of, 318–320
Experimental units, 4, 317
Explanatory variable, 158–159
Extrapolation, 217–220, 219FE–220FE, 240

## F

Facebook, 1–2, 5, 377FE–378FE, 383–385, 384FE–385FE
Factors
completely randomized two-factor experiments and, 329
confounding and, 331
experiments and, 329, 317
Failures, 464

False negatives, 545–547, 546FE
False positives, 545–547, 546FE
Farr, William, 61
Fechner, Gustav, 317
Fish in diet, 26FE–27FE
Fisher, Ronald Aylmer, 168, 515, 541, 549, 601
Fishing line strength, 247FE–249FE, 251
5-number summary, 58
Flexible workweek and miles driven, 653FE, 658FE–659FE, 659TI–660TI, 663FE
Flight cancellations, 60FE–61FE
Flight on-time arrivals, 91, 362–363
Food and Drug Administration (FDA), 317
Form
linear, 156
of scatterplots, 156
Formal probability, 360–362, 362FE
Formulas, A-1–A-2
Frequency tables, 16–17
Fry, Hanna, 324
Fuel efficiency, 68FE, 226FE–227FE, 240–241
Fundamental Theorem of Statistics, 468–469

## G

Gallup, George, 292
General Addition Rule, probability and, 374–378, 375FE, 376FE, 377FE–378FE
General Multiplication Rule, probability and, 386, 386FE
Geometric model, 431–434, 431FE–433FE, 433TI–434TI
Global warming, 319, 499FE, 720–721, 723, 726FE–727FE, 734–735
Goodness-of-fit tests, chi-square, 689–697, 693FE–695FE, 696TI–697TI
assumptions and conditions for, 691–692, 691FE–692FE
calculations for, 692–693, 693FE, 695
finding expected counts and, 690FE
Google, 5
Gore, Al, 220–221
Gosset, William S., 595–597, 598
Grading on a curve, 110
Groups
comparing with boxplots, 88–92, 90FE–92FE
comparing with histograms, 87–88, 88FE

## H

Halifax, Lord, 316
Harvell, Drew, 488
Heart rates of penguins, 216, 244FE, 246, 249FE
Height/age correlation of baseball players, 165–167
Heights of pitchers, 123FE–124FE

Hess, David, 95n
Hirscher, Marcel, 111, 112
Histograms, 44–46
   bimodal, 52
   comparing groups with, 87–88, 88FE
   designing, 45
   making, 46TI
   relative frequency, 45
   skewed, 53
   symmetric, 53
   uniform, 52
   unimodal, 52
Hocking, Ian, 276
Homogeneity, chi-square test of, 698–702, 700FE–701FE, 706TI
   assumptions and conditions for, 698–699
   calculations for, 699
Hoosic River, 92
Household sizes, 408–409
Hume, David, 539
Hunter, Stu, 542
Hurricane Agnes, 157
Hurricane Irene, 92
Hurricane Maria, 155
Hurricane prediction, 155–157, 159
Hypotheses, 512–513
   alternative, 512, 515–516
      one- and two-tailed, 516n
      one-sided, 516
      two-sided, 515–516
   framing, 513FE
   null, 512, 535–536
      failing to accept, 515
   writing, 517FE, 526FE
Hypothesis testing, **535–561**, 554TI
   about differences in means, 635TI
   about proportions, **511–534**, 520FE–521FE, 521TI–522TI
      alternative hypotheses and, 512, 515–516
      checking conditions for, 518FE
      conclusions of, 519, 519FE
      costs and, 520
      jury trials as, 513–515
      P-values and, 514–515, 514FE–515FE, 518–519, 518FE–519FE
      reasoning of, 516–522
      using statistical software, A-18–A-19
   alpha levels and, 541–542
   confidence intervals and, 543–545, 544FE–545FE
   for means, 610–611
   for mean, 606–610, 607FE, 608FE–609FE, 609TI–610TI
   P-values and, 537–541, 541FE
      small, 537, 539
   power and, 548–552, 552FE
      effect size and, 548–550
      sample size and, 550–551, 552FE
      significance level and, 548
      unexplained variability and, 551
   practical vs. statistical significance and, 542
   Type I and Type II errors and, 545–547, 546FE, 552FE

## I

Ice breakup dates, 732, 732FE–734FE
Identifier variables, 5
Independence, 356, 381–382
   checking for, 362–364, 363FE–364FE, 382FE, 383
   chi-square test of, 702–707, 702FE–704FE, 704FE–705FE, 706FE–707FE, 706TI
   disjoint events vs., 382–383
   probability and, 362–364, 363FE–364FE
Independence Assumption
   Central Limit Theorem and, 469
   for chi-square test of homogeneity, 699, 700FE, 705FE
   for comparing counts, 691, 691FE, 694FE
   for comparing means, 627, 629FE, 634FE, 636FE
   for comparing proportions, 565
   confidence intervals for proportions and, 495
   hypothesis testing and, 520FE
   inferences about means and, 601, 602FE
   for Normal model, 463
   one-proportion $z$-tests and, 538FE
   for paired data, 654, 655FE, 657FE, 661FE
   probability and, 362, 365FE, 366FE
   random variables and, 416FE, 419FE
   for regression, 722, 725FE, 726FE, 733FE
   sampling distribution model for means and, 472FE
   sampling distribution model for proportions and, 466FE
   for two-proportion $z$-intervals, 567FE
Independent Groups Assumption
   for comparing means, 627, 628FE, 629FE, 634FE, 636FE
   for comparing proportions, 564–565
   for two-proportion $z$-intervals, 567FE
Independent variable, 158n, 159
Inferences
   for comparing proportions, 577TI
   regression. *See* Regression inference
Inferences about means, **593–623**, 615TI
   assumptions and conditions for, 601–604, 602FE, 603FE–604FE
   Central Limit Theorem and, 594–595, 594FE–595FE
   confidence intervals and, 598–599, 599FE, 610–611
      finding, 605TI–606TI
      interpreting, 606
   finding $t$-model probabilities and critical values and, 600TI
   finding $t$-values using table and, 604–605
   Gosset's $t$ and, 595–598
   hypothesis test for, 606–610, 607FE, 608FE–609FE, 609TI–610TI
   sample size and, 611–612, 612FE
   Student's $t$ and, 597–598
   using statistical software, A-21–A-22
Influence, 222
Ingot cracking rate, 511
Intercepts, 184–185
Internet
   data on, 7
   privacy and, 5
Internet surveys, 305
Interquartile range (IQR), 56–58

## J

Jastrow, J., 318
Jewelry gauges, 251TI–252TI
JMP. *See* Statistical software
Johnson-Thompson, Katarina, 110, 111, 112
Jolie, Angelina, 359
Jury trials as hypothesis testing, 513–515

## K

Keno, 357
Kentucky Derby, 49, 52, 58

## L

Ladder of Powers, 245–250
Landers, Ann, 304
LaPlace, Pierre-Simon, 468, 469
Large Enough Sample Condition
   Central Limit Theorem and, 469, 472FE
   sampling distribution model for a mean and, 473FE
Law of Averages, 356–358, 357FE
Law of Large Numbers (LLN), 356, 357
Least squares line, 181, 199TI
   correlation and, 181–183
   different samples and, 191–193
   in real units, 184–188, 186FE, 187FE–188FE
   variance and, 193–195, 194FE
Legitimate assignment of probabilities, 361–362
Leverage, 221–222
Life expectancy, 223–224, 225
Line of best fit. *See* Least squares line
Linear form, 156
Linear models, 180
Linear regression, **179–214**, 197FE–198FE, 203TI. *See also* Least squares line; *entries beginning with term* Regression
   assumptions and conditions for, 195–196
   calculating regression equations and, 187FE–188FE
   correlation and, 181–183
   reasonableness of, 200
   regression to the mean and, 183–184
   residuals and, 180–181, 188–191, 189FE, 199TI
   nonlinearity and, 215–217

squared, as measure of fit, 181
  typical size of, 190–191
  standard error and, 190–191
  using statistical software, A-10–A-13
  variation and, 193–195, 194FE
Linearity, for re-expressing to achieve, 250TI
Linearity Assumption, for regression, 722, 725FE
Logarithms, 96, 250–252
  base 10 vs. natural (base $e$), 245n
  re-expressions and, 251TI–252TI
Lowell, James Russell, 519
Lower quartile, 57
Lucky Lovers discount, 404FE, 406FE, 412FE, 413FE, 415FE–416FE
Lurking variables, 168, 223–224
  confounding vs., 331

## M

Maas, Jim, 593
Male births, 520FE–521FE
Manual dexterity, 243, 652, 661FE–662FE
Marathon runners and temperature, 735TI–736TI
Margin of error (*ME*), 493–494, 494FE
Marginal distributions, 20–21, 21FE
Matching, 328
Mean(s)
  comparing, **624–651**
    assumptions and conditions for, 627–631, 628FE
    confidence interval for, 628FE–629FE, 631FE–632FE
    inference for, 642TI
    permutation test for, 638–639
    pooled *t*-test for, 637–639
    standard error of the difference in independent sample means and, 626FE
    test for, 632–637, 633FE–635FE, 635TI
    two-sample *t*-interval and, 629FE–631FE
    two-sample *t*-test for, 636FE–637FE
    using statistical software, A-22–A-24
  confidence intervals for, 506FE–606FE, 598–599, 599FE
    hypothesis testing and, 610–611
    interpreting, 606
  hypothesis testing for, 606–610, 607FE, 608FE–609FE, 609TI–610TI
  inferences about. *See* Inferences about means
  multiplying by a constant and, 412, 412FE
  sample, sampling distribution model for, 470
  sampling distribution of, simulating, 467–468
Mean Absolute Deviation (MAD), 65n
Median, 55–56
Meir, Jessica, 216
Minitab. *See* Statistical software

Mode, 52
Model(s), 180
  binomial, 434–438, 436FE–437FE, 437TI–438TI, 445TI
  chi-square, 692
  hypothesis testing and, 517–518
  linear, 180
  Normal. *See* Normal models
  probability. *See* Probability models
  regression, interpreting, 731FE
  sampling distribution, for porportions, 464–467, 465FE, 466FE–467FE
Model world, real world vs., 476
Mosaic plots, 28–29
Motorcycle helmets, 535, 537FE–538FE
Mouth volume, 731
Multiplication Rule, probability and, 362, 363FE–364FE
Multistage sampling, 300
Mutually exclusive events, 361
  independence vs., 382–383
  probability and, 382–383, 384FE–385FE

## N

Nader, Ralph, 220
National Council of Teachers of Mathematics (NCTM), 297
National Health and Nutrition Examination Survey (NHANES) 2001–2002, 113–115
National Hurricane Center (NHC), 155
National Oceanic and Atmospheric Administration (NOAA), 155
Natural logarithms, 245n
Nearly Normal Condition, 118, 120
  for comparing means, 628FE, 629FE, 634FE, 636FE
  inferences about means and, 601, 602FE, 603FE
  Normal models and, 124FE, 125FE, 126FE, 127FE, 128FE
  one-sample *t*-test for the mean and, 608FE
  for paired data, 655, 655FE, 657FE, 661FE
  for regression, 723–724, 723FE, 725FE, 727FE, 733FE
Negative associations, 156
Nest Egg Index, 88FE
Newspaper readership, 493FE, 495FE
Nightingale, Florence, 61
95% confidence, 491–493, 493FE
No Outliers Condition
  for correlation, 161
  for regression, 187FE
Nonlinear regression models, 252
Nonresponse bias, 293, 306
Normal curve, areas under, table of, A-97–A-98
Normal model(s), **117–142**, 134TI, 438–441, 440FE–441FE
  assessing for population, 130–131
  assumptions and conditions for, 463–464

determining *z*-scores from percentiles and, 124–128, 125FE–128FE, 125TI
finding normal percentiles and, 122–124, 122TI–123TI, 123FE–124FE
normal probability plots and, 129–130, 130FE
rules for working with, 120–122, 121FE
as sampling distribution model, 463–464
sampling distribution of a proportion and, 462–463
68–95–99.7 Rule and, 118–120, 119FE
using statistical software, A-7–A-8
Normal Model Assumption, random variables and, 419FE
Normal percentiles, determining, 122–124, 122TI–123TI
Normal probability plots, 129–130, 130FE
Normality Assumption, 118
Notation
  *, 494
  $\alpha$, 541
  $b$, 295
  $\beta$, 295, 546, 721
  $e$, 188
  $E$, 692
  $E(X)$, 403
  $H$, 512
  hat over variables, 182
  $\mu$, 117, 295
  $n$, 56, 63
  $n!$, 435
  $O$, 692
  $p$, 431, 462
  $P$, 358
  $P(\mathbf{B}|\mathbf{A})$, 380
  $\hat{p}$, 295, 462
  $q$, 431, 462
  Q1, Q2, Q3, 58
  $\hat{q}$, 462
  $r$, 161, 295
  $R$, 699
  $\rho$, 295
  $s$, 65, 295
  $s^2$, 65
  $SD(\hat{p})$, 471
  $SD(\bar{y})$, 471
  $s_e$, 729
  $SE(\hat{p})$, 489
  $SE(b_1)$, 729
  $\sigma$, 117, 295
  $\Sigma$, 62
  $t$, 598
  $t^*$, 598
  $x$, 158
  $y$, 158
  $\bar{y}$, 295
  $\hat{y}$, 462
  $z$, 111
  $\Delta_0$, 633
  $\mu$, 403
  $X$, 692

Null hypothesis, 512, 535–536
　failing to accept, 515
Numeric results, differences in, 69

# O

Observational studies, 315–316, 316FE
Occam's razor, 253
Ogives, 49–51
Oil price predictions, 217–219
Olympics, 110, 111, 111FE–112FE, 112, 112FE–113FE
One-proportion $z$-interval, 496
One-proportion $z$-tests, 517–518, 520FE, 537FE–538FE
One-sample $t$-interval for the mean, 603FE–604FE
One-sample $t$-test for the mean, 606–610, 607FE, 608FE–609FE
One-sided alternatives, 516
One-tail probability, table of, A-99
One-tailed alternatives, 516n
Ordinal variables, 6, 6FE
Origin, scatterplots and, 156
Outcomes, 355
Outlier(s), 53, 92, 92FE, 157
Outlier Condition
　for correlation, 163FE
　for regression, 186, 195, 196, 197FE, 723–724, 727FE, 733FE

# P

P-values, 514–515, 514FE–515FE, 518–519, 518FE–519FE, 537–541, 541FE
　high, 540–541
　small, 537, 539
Page, Larry, 95n
Paired data, **652–663**
　assumptions and conditions for, 654–656
　confidence intervals for matched pairs and, 660–662, 661FE–662FE, 662TI
　effect size and, 662–663, 663FE
　hypothesis testing with, 659TI–660TI
　identifying, 653FE
　independence vs., 653
　paired $t$-test and, 656FE–659FE, 666TI
　using statistical software, A-24–A-25
Paired Data Condition, 654, 655FE, 657FE, 661FE
Paired $t$-interval, 660, 661FE–662FE
Paired $t$-test, 654, 656, 656FE–659FE
Parameters, 117
Participants, 4, 317
Penguin heart rates, 216, 244FE, 246, 249FE
Percentages, 21
Percentiles, 57
　determining $z$-scores from, 124–128, 125FE–128FE, 125TI
　Normal, determining, 122–124, 122TI–123TI
Permutation test for two means, 638–639
Pet food safety, 316FE, 318FE, 320FE, 326FE, 328FE, 330FE

Picasso, Pablo, 10
Pie charts, 19
Pierce, C. S., 318
Pilots, 303
Placebo(s), 326–327
Placebo effect, 326–327
Plant fertilizer, 321FE–322FE
Playing cards, 277
Polls, confidence intervals and, 493FE
Ponganis, Paul, 216
Pooled $t$-test for the difference between means, 637–639
Pooling, 570–571
Population(s), 4
Population density, 239, 291
Population parameters, 294–295
Portion size judgments, 626FE, 628FE–629FE
Positive associations, 156
Post-graduation plans, 698, 699, 700FE–701FE
Potato flea beetles, 702
Power, hypothesis testing and, 548–552, 552FE
　effect size and, 548–550
　sample size and, 550–551, 552FE
　significance level and, 548
　unexplained variability and, 551
Practical significance, statistical significance vs., 542
Predicted value, 180
Presidential elections, 15, 220–222, 225
Price comparisons, 157FE–158FE
Privacy, Internet and, 5
Probability, 356, **358–401**, 365FE–366FE
　Addition Rule and, 361–362, 361FE
　Bayesian reasoning and, 392–394, 393FE–394FE
　binomial, 434–435
　Complement Rule and, 360, 361FE
　conditional, 378–380, 380FE, 518, 537. See also P-values
　disjoint events and, 382–383, 384FE–385FE
　drawing without replacement and, 386–388, 387FE
　formal, 360–362, 362FE
　General Addition Rule and, 374–378, 375FE, 376FE, 377FE–378FE
　General Multiplication Rule and, 386, 386FE
　independence and, 362–364, 363FE–364FE, 381–383, 382FE
　legitimate assignment of, 361–362
　Multiplication Rule and, 362, 363FE–364FE
　one-tail, table of, A-99
　Probability Assignment Rule and, 360
　proportion vs., 460
　reversing conditioning and, 390–392, 391FE–392FE, 393FE–394FE
　right-tail, table of, A-100–A-101

rules for working with, 359
　subjective, 359
　$t$-model, 600TI
　tables and, 383–384
　theoretical, 358–359
　tree diagrams and, 388–390, 390FE
　two-tail, table of, A-99
　Venn diagrams and, 376, 376FE, 383–384
Probability Assignment Rule, 360
Probability distributions, cumulative, 409
Probability models, 403, **429–449**
　Bernoulli trials and, 429–431
　　10% rule and, 430–431, 433FE, 437FE, 443FE
　binomial, 434–438, 436FE–437FE, 437TI–438TI, 445TI
　geometric, 431–434, 431FE–433FE, 433TI–434TI
　Normal. See Normal model(s)
　statistical significance and, 442–443, 442FE–443FE
　using statistical software, A-16–A-17
Proportions
　comparing, **562–582**, 573TI
　　assumptions and conditions for, 564–565
　　confidence intervals and, 565–569, 566FE, 567FE–569FE
　　inferences for, 577TI
　　permutation test for, 574–575
　　pooling and, 570–571
　　standard deviation of difference and, 563–564
　　two-proportion $z$-test for, 571–574, 571FE–573FE
　　using statistical software, A-19–A-21
　confidence intervals for. See Confidence intervals for proportions
　hypothesis testing about. See Hypothesis testing, about proportions
　probability vs., 460
　sampling distribution model for, 464–467, 465FE, 466FE–467FE
　sampling distribution of, 460–463
Prospective studies, 316
Prostate cancer, 26FE–27FE
Pseudorandomness, 277, 278
Pulse rates of women, 47
Pythagorean Theorem of Statistics, 414

# Q

Quantiles, 125n
Quantitative data, **44–85**
　boxplots and, 59–62, 60FE–61FE
　cumulative distribution graphs and, 49–51
　displaying, 75TI
　displaying and describing using statistical software, A-4–A-5
　5-number summary and, 58
　histograms and, 44–46, 46TI
　mean and, 62–65, 64FE

median and, 55–56, 63–65, 64FE
mode and, 52
shape of distribution and, 52–55, 54FE
spread and, 56–58
    interquartile range and, 56–58
    range and, 56
    standard deviation and, 65–66, 66FE
stem-and-leaf displays and, 47–49
summarizing, 58, 62–65, 64FE, 67FE, 68FE, 75TI
what to *tell* about, 67–68, 67FE, 68FE
Quantitative Data Condition, 51, 68FE, 116FE
Quantitative variables, 5–6, 6FE
Quantitative Variables Condition
    for correlation, 161, 163FE
    for regressions, 186, 187FE, 195, 196, 197FE
Quartiles, 57–58

## R

Random assignment, 316
Random numbers, generating, 277, 280, 282TI–283TI
Random phenomena, 354–356
Random Residuals Condition, for regression, 723, 725FE, 726FE
Random variables, **402–428**, 422TI
    continuous, 403, 417–420, 418FE–420FE, 441
    discrete, 403
    expected value of, 403–405, 404FE, 407TI–408TI
    means and variances of, 411–417, 412FE, 413FE, 415FE–417FE
    representations of, 408–411
    spread of, 405–408, 406FE–407FE, 407TI–408TI
    using statistical software, A-15–A-16
Randomization
    in experiments, 319, 320FE
    sampling and, 292
Randomization Condition
    Central Limit Theorem and, 469, 471FE
    for chi-square test of homogeneity, 699, 705FE
    for comparing counts, 691, 692FE, 694FE
    for comparing means, 627, 628FE, 630FE, 634FE, 636FE
    for comparing proportions, 565
    confidence intervals for proportions and, 496, 497FE
    hypothesis testing and, 518FE, 520FE
    inferences about means and, 601, 603FE
    for Normal model, 463
    one-proportion $z$-tests and, 538FE
    one-sample $t$-test for the mean and, 608FE
    for paired data, 654, 655FE, 657FE
    for regression, 723, 725FE, 726FE, 733FE
    sampling distribution model for means and, 472FE
    sampling distribution model for proportions and, 465FE, 466FE
    for two-proportion $z$-intervals, 567FE
Randomized block design, 327
Randomness, **276–289**
    generating numbers and, 277, 280, 282TI–283TI
    random phenomena and, 354–356
    using statistical software, A-14–A-15
Range, 56
Re-expressing data, 94–97, **239–263**, 246FE, 247FE–249FE, 254TI
    to achieve linearity, 250TI
    comparison of, 249FE
    curves vs., 252
    to equalize spread across groups, 96–97
    goals of, 241–244
    to improve symmetry, 94–95
    Ladder of Powers and, 245–250
    logarithmic, 250–252, 251TI–252TI
    recognizing uses of, 244FE
    using statistical software, A-13–A-14
Real world, model world vs., 476
Records, 4
Regression. *See also* Linear regression
    extrapolation and, 217–220, 219FE–220FE, 240
    lurking variables and causation and, 223–224
    nonlinearity and, 215–217, 252
    outliers, leverage, and influence and, 220–223
    summary variables and, 225–227, 226FE–227FE
Regression analysis, 230TI, 740TI
Regression inference, **720–752**, 726FE–727FE, 731FE, 732FE–734FE, 735TI–736TI
    assumptions and conditions for, 722–727, 724FE–725FE
    differing samples and slopes and, 736–738
    intuition about, 727–729
    model and data for, 721–722
    standard error for the slope and, 729–730
    using statistical software, A-27–A-28
Regression line, 181, 199TI
    correlation and, 181–183
    different samples and, 191–193
    in real units, 184–188, 186FE, 187FE–188FE
    variance and, 193–195, 194FE
Regression models, interpreting, 731FE
Regression to the mean, 183
Relative frequency histograms, 45
Relative frequency tables, 17
Replication, of experiments, 319, 320FE
Representativeness, of samples, 292, 292FE, 295
Rescaling data, 114–115, 115FE
Residual(s), 180–181, 188–191, 189FE, 199TI
    nonlinearity and, 215–217
    squared, as measure of fit, 181
    typical size of, 190–191
Residual standard deviation, 728
Respondents, 4
Response bias, 306
Response variables, 158–159, 279, 317, 318FE
Retrospective studies, 315–316
Right-tail probability, table of, A-100–A-101
Robbins, Rebecca, 593
Rock, Paper, Scissors, 549–551
Roller coasters, 89FE, 91FE–92FE, 160–161, 179–183, 184–185, 189, 191, 193, 196

## S

Sale of used SUV, 416FE–417FE
Salmon contamination, 599FE
Sample(s), 4, 4FE, 291, 324
Sample mean, sampling distribution model for, 470
Sample size, 292, 293
    confidence intervals for proportions and, 498–500, 499FE, 500FE
    hypothesis testing and, 550–551, 552FE
    inferences about means and, 611–612, 612FE
Sample Size Assumption
    Central Limit Theorem and, 469
    for Normal model, 464
Sample space, 355
Sample statistics, 294
Sample surveys, **290–314**. *See also* Sampling
    pilots and, 303
    validity of, 302–303
Sampling, 301FE–302FE, 309TI
    bad, 304–306
    bias and, 291–292
    censuses and, 293–294
    cluster, 299–300, 299FE
    convenience, 305, 305FE
    incomplete sampling frames and, 305
    multistage, 300
    nonresponse bias and, 293, 306
    population parameters and, 294–295
    random, simple, 295–296, 296FE
    randomization and, 292
    representativeness of sample and, 292, 292FE, 295
    response bias and, 306
    sample size and, 293
    stratified, 296–299, 298FE–299FE
    systematic, 300–301
    undercoverage and, 305
    voluntary response, 304, 304FE

Sampling distribution models, **460–487**
   assumptions and conditions of, 469
   Central Limit Theorem and, 468–469, 473–475
   for difference between two independent sample proportions, 566
   of a mean, simulating, 467–468
   Normal model and, 462–464, 470–473, 471FE–473FE
   of other statistics, 467
   for a proportion, 464–467, 465FE, 466FE–467FE
   real world vs. model world and, 476
   variation and, 475–476
Sampling distributions
   for difference between two sample means, 627
   of a mean, simulation of, 467–468
   of a proportion, 460–463
   for regression slopes, 729
Sampling error, 296, 463
Sampling frames, 296
Sampling variability, 296, 463
Scale, adjusting, 114–115
Scatterplots, 156–158, 157FE–158FE, 171TI
   creating, 159TI–160TI
   form of, 156
   re-expression to straighten, 247FE–249FE
   using statistical software, A-9–A-10
Seat belt use, 562
Segmented bar charts, 22, 23FE, 24FE
Shifting data, 113–114
Side-by-side bar charts, 28
Significance level, 548
Similar Spreads Condition, for pooled $t$-test for the difference between means, 638
Simple random samples (SRSs), 295–296, 296FE
Simpson's paradox, 32
Simulation(s), 29–30, 278–284, 281FE–282FE, 285TI
   building, 278–283, 280FE
   of sampling distribution of a mean, 467–468
   trials and, 278–284
   number of, 283–284
Simulation components, 279
Single-blind experiments, 326
68–95–99.7 Rule, 118–120, 119FE, 121FE
Skewed histograms, 53
Skewness, 65
*Sleep for Success!* (Maas and Robbins), 593
Slope, 184–185
   regression, $t$-tests for, 730–731, 732FE–734FE
   regression inference and, 736–738
   standard error of, 729–730
Smoking and cancer, 168
Social media use, 377FE–378FE, 383–385, 384FE–385FE
Software. *See* Statistical software

Soup, 292, 293
Spam e-mail, 431FE–432FE, 436FE, 440FE–441FE
Spouses' ages, 660, 662, 663–664
Spread
   across groups, re-expressing data to equalize, 96–97
   of random variables, 405–408, 406FE–408FE, 407TI–408TI
Standard deviation, 65–66, 512
   adding and subtracting a constant and, 411–412, 412FE
   of difference between two proportions, 563–564
   of random variables, 405
   as ruler, 110–111
   using statistical software, A-7–A-8
Standard error, 190–191, 489, 512
   of difference between two proportions, 564FE
   of difference in independent sample means, 626
   of the slope, 729–730
Standard Normal distribution, 117
Standard Normal model, 117
Standardization, with $z$-scores, 111–113, 111FE–113FE
Standardized values, 111
StatCrunch. *See* Statistical software
Statistical significance, 322–323, 332–333, 442–443, 541–542
   practical significance vs., 542
Statistical software. *See also* Computer tips
   associations using, A-9–A-10
   blocks using, A-24–A-25
   categorical data display and description using, A-3–A-4
   comparing counts using, A-25–A-27
   comparing means using, A-22–A-24
   comparing proportions using, A-19–A-21
   confidence intervals for proportions using, A-17–A-18
   correlation using, A-9–A-10
   distributions using, A-6–A-7
   hypothesis testing about proportions using, A-18–A-19
   inferences about means using, A-21–A-22
   linear regression using, A-10–A-13
   Normal model using, A-7–A-8
   paired samples using, A-24–A-25
   probability models using, A-16–A-17
   quantitative data display and description using, A-4–A-5
   random variables using, A-15–A-16
   randomness using, A-14–A-15
   re-expressing data using, A-13–A-14
   regression inference using, A-27–A-28
   scatterplots using, A-9–A-10

   standard deviation as ruler using, A-7–A-8
Statistics, 117
   calculating, 69TI
   definition of, 294
   sample, 294
Stem-and-leaf displays (stemplots), 47–49
   comparing groups with, 88FE
Straight Enough Condition
   for correlation, 161, 163FE
   for regression, 186, 187FE, 195, 196, 197FE, 722, 726FE, 733FE
Stratified random sampling, 296–297
Stratified sampling, 297–299, 298FE–299FE
Strength
   of associations, 157
   of correlations, 165
Students' hours of sleep, 593–594
Student's $t$, 597–598
Su, Lisa, 95
Subject(s), 4, 317
Subjective probability, 359
Success(es), 464
Success/Failure Condition
   for comparing proportions, 565
   confidence intervals for proportions and, 496, 497FE
   hypothesis testing and, 518FE, 520FE, 544FE
   for Normal model, 440, 443FE, 464
   one-proportion $z$-tests and, 538FE
   sampling distribution model for proportions and, 465FE, 466FE
   for two-proportion $z$-intervals, 567FE
Super Bowl, watching, 21FE, 22FE–23FE, 24FE–25FE
Superpowers, 703FE–704FE, 706FE–707FE
Symmetric distributions
   mean of, 62–65
   median of, 63–65
Symmetric histograms, 53
Symmetry, re-expressing data to improve, 94–95
Systematic sampling, 300–302

# T

$t$-intervals
   for the mean, one-sample, 603FE–604FE
   paired, 660, 661FE–662FE
   two-sample, 626
     for difference between means, 627, 629FE–631FE
$t$-tests
   for the mean, one-sample, 606–610, 607FE, 608FE–609FE
   paired, 654, 656, 656FE–659FE
   pooled, for the difference between means, 637–639

for the regression slope, 730–731, 732FE–734FE
Student's *t*, 597–598
two-sample, 626
    for the difference between means, 633, 633FE–635FE, 636FE–637FE
Tails, 53
Tattoos and alcohol consumption, 381
Tattoos and hepatitis C, 704FE–705FE
TB testing, 391FE–392FE
Technology. *See* Computer tips; Statistical software
10% Condition
    Bernoulli trials and, 430–431, 433FE, 437FE, 443FE
    Central Limit Theorem and, 469, 471FE
    for chi-square test of homogeneity, 699, 705FE
    for comparing counts, 691, 692FE
    for comparing means, 627, 629FE
    for comparing proportions, 565
    confidence intervals for proportions and, 496, 497FE
    hypothesis testing and, 518FE, 520FE
    inferences about means and, 601, 603FE
    for Normal model, 463, 464
    one-proportion $z$-tests and, 538FE
    one-sample $t$-test for the mean and, 608FE
    for paired data, 654, 655FE
    sampling distribution model for a mean and, 472FE
    sampling distribution model for proportions and, 465FE, 466FE
    for two-proportion $z$-intervals, 567FE
Terrestrial craters, 724FE–725FE
Texting while driving, 2
Theoretical probability, 358–359
Therapeutic touch (TT), 540
Thiam, Nafissatou, 110, 111, 112
TI-Nspire. *See* Statistical software
Timeplots, 93–94, 156
*Titanic* sinking, 14, 16–17, 19–22, 23–24, 63–64, 378–380
Tongue rolling and ears, 690–691, 692, 697
Traffic lights, 354–355, 361FE, 363FE–364FE
Transforming data. *See* Re-expressing data
Treatments, 317, 318FE
Tree diagrams, 388–390, 390FE

Trials, 355
    Bernoulli, 429–431, 435
        10% rule and, 430–431, 433FE, 437FE, 443FE
    response variables of, 279
    simulations and, 278–284
Trump, Donald, 15, 224
Tsunamis, 44, 45, 55–56
Tukey, John W., 47, 490
Twitter, 377FE–378FE, 383–385, 384FE–385FE
Two-proportion $z$-intervals, 566, 566FE, 567FE–568FE
Two-proportion $z$-test, 571–574, 571FE–573FE
Two-sample $t$-interval, 626
    for difference between means, 627, 629FE–631FE
Two-sample $t$-tests, 626
    for the difference between means, 633, 633FE–635FE, 636FE–637FE
Two-sided alternatives, 515–516
Two-tail probability, table of, A-99
Two-tailed alternatives, 516n
Type I and II errors, 545–547, 546FE, 552FE

# U

Undercoverage, 305
Uniform histograms, 52
Unimodal histograms, 52
Unit(s), 5
U.S. senators' ages, 50–51
Upper quartile, 57
Used camera purchase, 632, 633FE–635FE, 635TI

# V

Validity of surveys, 302–303
Variability
    hypothesis testing and, 551
    sampling, 463
Variables, 4, 158–160
    categorical, 5, 6
    dependent, 158n, 159
    explanatory, 158–159
    identifier, 5
    independent, 158n, 159
    independent and dependent, 158n, 159
    lurking, 168, 223–224, 331

    ordinal, 6, 6FE
    quantitative, 5–6, 6FE
    random. *See* Random variables
    response, 158–159, 279, 317, 318FE
Variance, 65
    addition of, 563
    multiplying by a constant and, 412, 412FE
    of random variables, 405
Variation
    regression and, 193–195, 194FE
    sampling distribution models and, 475–476
Venn diagrams, 376, 376FE, 383–384
Voluntary response bias, 304
Voluntary response samples, 304, 304FE

# W

Wainer, Howard, 475
Waiting in line, 66
Website and sales, 513FE, 514FE–515FE
William of Occam, 253
Wind speed, 86–89, 92–94, 186FE, 189FE
Women's Health Initiative, 317
Women's pulse rates, 47
Wrist circumference, 119FE
Wrist surgery, 544FE–545FE

# X

*x*-variable, 158n, 159

# Y

*y*-intercept, 184–185
*y*-variable, 158n, 159
Young, Don, 51

# Z

$z$-intervals
    one-proportion, 496
    two-proportion, 566, 566FE, 567FE–568FE
$z$-scores, 111–113, 116–118, 116FE–117FE
    combining, 112FE–113FE
    determining from percentiles, 124–128, 125FE–128FE, 125TI
$z$-tests
    one-proportion, 517–518, 520FE, 537FE–538FE
    two-proportion, 571–574, 571FE–573FE
Zwerling, Harris, 475

# APPENDIX F: Tables

**Table Z**
Areas under the standard Normal curve

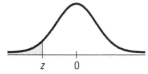

An interactive version of this table is at www.astools.datadesk.com/normal_table.html

| 0.09 | 0.08 | 0.07 | 0.06 | 0.05 | 0.04 | 0.03 | 0.02 | 0.01 | 0.00 | z |
|---|---|---|---|---|---|---|---|---|---|---|
| | | | | | | | | | 0.0000† | −3.9 |
| 0.0001 | 0.0001 | 0.0001 | 0.0001 | 0.0001 | 0.0001 | 0.0001 | 0.0001 | 0.0001 | 0.0001 | −3.8 |
| 0.0001 | 0.0001 | 0.0001 | 0.0001 | 0.0001 | 0.0001 | 0.0001 | 0.0001 | 0.0001 | 0.0001 | −3.7 |
| 0.0001 | 0.0001 | 0.0001 | 0.0001 | 0.0001 | 0.0001 | 0.0001 | 0.0001 | 0.0002 | 0.0002 | −3.6 |
| 0.0002 | 0.0002 | 0.0002 | 0.0002 | 0.0002 | 0.0002 | 0.0002 | 0.0002 | 0.0002 | 0.0002 | −3.5 |
| 0.0002 | 0.0003 | 0.0003 | 0.0003 | 0.0003 | 0.0003 | 0.0003 | 0.0003 | 0.0003 | 0.0003 | −3.4 |
| 0.0003 | 0.0004 | 0.0004 | 0.0004 | 0.0004 | 0.0004 | 0.0004 | 0.0005 | 0.0005 | 0.0005 | −3.3 |
| 0.0005 | 0.0005 | 0.0005 | 0.0006 | 0.0006 | 0.0006 | 0.0006 | 0.0006 | 0.0007 | 0.0007 | −3.2 |
| 0.0007 | 0.0007 | 0.0008 | 0.0008 | 0.0008 | 0.0008 | 0.0009 | 0.0009 | 0.0009 | 0.0010 | −3.1 |
| 0.0010 | 0.0010 | 0.0011 | 0.0011 | 0.0011 | 0.0012 | 0.0012 | 0.0013 | 0.0013 | 0.0013 | −3.0 |
| 0.0014 | 0.0014 | 0.0015 | 0.0015 | 0.0016 | 0.0016 | 0.0017 | 0.0018 | 0.0018 | 0.0019 | −2.9 |
| 0.0019 | 0.0020 | 0.0021 | 0.0021 | 0.0022 | 0.0023 | 0.0023 | 0.0024 | 0.0025 | 0.0026 | −2.8 |
| 0.0026 | 0.0027 | 0.0028 | 0.0029 | 0.0030 | 0.0031 | 0.0032 | 0.0033 | 0.0034 | 0.0035 | −2.7 |
| 0.0036 | 0.0037 | 0.0038 | 0.0039 | 0.0040 | 0.0041 | 0.0043 | 0.0044 | 0.0045 | 0.0047 | −2.6 |
| 0.0048 | 0.0049 | 0.0051 | 0.0052 | 0.0054 | 0.0055 | 0.0057 | 0.0059 | 0.0060 | 0.0062 | −2.5 |
| 0.0064 | 0.0066 | 0.0068 | 0.0069 | 0.0071 | 0.0073 | 0.0075 | 0.0078 | 0.0080 | 0.0082 | −2.4 |
| 0.0084 | 0.0087 | 0.0089 | 0.0091 | 0.0094 | 0.0096 | 0.0099 | 0.0102 | 0.0104 | 0.0107 | −2.3 |
| 0.0110 | 0.0113 | 0.0116 | 0.0119 | 0.0122 | 0.0125 | 0.0129 | 0.0132 | 0.0136 | 0.0139 | −2.2 |
| 0.0143 | 0.0146 | 0.0150 | 0.0154 | 0.0158 | 0.0162 | 0.0166 | 0.0170 | 0.0174 | 0.0179 | −2.1 |
| 0.0183 | 0.0188 | 0.0192 | 0.0197 | 0.0202 | 0.0207 | 0.0212 | 0.0217 | 0.0222 | 0.0228 | −2.0 |
| 0.0233 | 0.0239 | 0.0244 | 0.0250 | 0.0256 | 0.0262 | 0.0268 | 0.0274 | 0.0281 | 0.0287 | −1.9 |
| 0.0294 | 0.0301 | 0.0307 | 0.0314 | 0.0322 | 0.0329 | 0.0336 | 0.0344 | 0.0351 | 0.0359 | −1.8 |
| 0.0367 | 0.0375 | 0.0384 | 0.0392 | 0.0401 | 0.0409 | 0.0418 | 0.0427 | 0.0436 | 0.0446 | −1.7 |
| 0.0455 | 0.0465 | 0.0475 | 0.0485 | 0.0495 | 0.0505 | 0.0516 | 0.0526 | 0.0537 | 0.0548 | −1.6 |
| 0.0559 | 0.0571 | 0.0582 | 0.0594 | 0.0606 | 0.0618 | 0.0630 | 0.0643 | 0.0655 | 0.0668 | −1.5 |
| 0.0681 | 0.0694 | 0.0708 | 0.0721 | 0.0735 | 0.0749 | 0.0764 | 0.0778 | 0.0793 | 0.0808 | −1.4 |
| 0.0823 | 0.0838 | 0.0853 | 0.0869 | 0.0885 | 0.0901 | 0.0918 | 0.0934 | 0.0951 | 0.0968 | −1.3 |
| 0.0985 | 0.1003 | 0.1020 | 0.1038 | 0.1056 | 0.1075 | 0.1093 | 0.1112 | 0.1131 | 0.1151 | −1.2 |
| 0.1170 | 0.1190 | 0.1210 | 0.1230 | 0.1251 | 0.1271 | 0.1292 | 0.1314 | 0.1335 | 0.1357 | −1.1 |
| 0.1379 | 0.1401 | 0.1423 | 0.1446 | 0.1469 | 0.1492 | 0.1515 | 0.1539 | 0.1562 | 0.1587 | −1.0 |
| 0.1611 | 0.1635 | 0.1660 | 0.1685 | 0.1711 | 0.1736 | 0.1762 | 0.1788 | 0.1814 | 0.1841 | −0.9 |
| 0.1867 | 0.1894 | 0.1922 | 0.1949 | 0.1977 | 0.2005 | 0.2033 | 0.2061 | 0.2090 | 0.2119 | −0.8 |
| 0.2148 | 0.2177 | 0.2206 | 0.2236 | 0.2266 | 0.2296 | 0.2327 | 0.2358 | 0.2389 | 0.2420 | −0.7 |
| 0.2451 | 0.2483 | 0.2514 | 0.2546 | 0.2578 | 0.2611 | 0.2643 | 0.2676 | 0.2709 | 0.2743 | −0.6 |
| 0.2776 | 0.2810 | 0.2843 | 0.2877 | 0.2912 | 0.2946 | 0.2981 | 0.3015 | 0.3050 | 0.3085 | −0.5 |
| 0.3121 | 0.3156 | 0.3192 | 0.3228 | 0.3264 | 0.3300 | 0.3336 | 0.3372 | 0.3409 | 0.3446 | −0.4 |
| 0.3483 | 0.3520 | 0.3557 | 0.3594 | 0.3632 | 0.3669 | 0.3707 | 0.3745 | 0.3783 | 0.3821 | −0.3 |
| 0.3859 | 0.3897 | 0.3936 | 0.3974 | 0.4013 | 0.4052 | 0.4090 | 0.4129 | 0.4168 | 0.4207 | −0.2 |
| 0.4247 | 0.4286 | 0.4325 | 0.4364 | 0.4404 | 0.4443 | 0.4483 | 0.4522 | 0.4562 | 0.4602 | −0.1 |
| 0.4641 | 0.4681 | 0.4721 | 0.4761 | 0.4801 | 0.4840 | 0.4880 | 0.4920 | 0.4960 | 0.5000 | −0.0 |

Second decimal place in z

†For z ≤ −3.90, the areas are 0.0000 to four decimal places.

**Table Z** (cont.)
Areas under the standard Normal curve

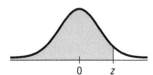

| z | 0.00 | 0.01 | 0.02 | 0.03 | 0.04 | 0.05 | 0.06 | 0.07 | 0.08 | 0.09 |
|---|---|---|---|---|---|---|---|---|---|---|
| 0.0 | 0.5000 | 0.5040 | 0.5080 | 0.5120 | 0.5160 | 0.5199 | 0.5239 | 0.5279 | 0.5319 | 0.5359 |
| 0.1 | 0.5398 | 0.5438 | 0.5478 | 0.5517 | 0.5557 | 0.5596 | 0.5636 | 0.5675 | 0.5714 | 0.5753 |
| 0.2 | 0.5793 | 0.5832 | 0.5871 | 0.5910 | 0.5948 | 0.5987 | 0.6026 | 0.6064 | 0.6103 | 0.6141 |
| 0.3 | 0.6179 | 0.6217 | 0.6255 | 0.6293 | 0.6331 | 0.6368 | 0.6406 | 0.6443 | 0.6480 | 0.6517 |
| 0.4 | 0.6554 | 0.6591 | 0.6628 | 0.6664 | 0.6700 | 0.6736 | 0.6772 | 0.6808 | 0.6844 | 0.6879 |
| 0.5 | 0.6915 | 0.6950 | 0.6985 | 0.7019 | 0.7054 | 0.7088 | 0.7123 | 0.7157 | 0.7190 | 0.7224 |
| 0.6 | 0.7257 | 0.7291 | 0.7324 | 0.7357 | 0.7389 | 0.7422 | 0.7454 | 0.7486 | 0.7517 | 0.7549 |
| 0.7 | 0.7580 | 0.7611 | 0.7642 | 0.7673 | 0.7704 | 0.7734 | 0.7764 | 0.7794 | 0.7823 | 0.7852 |
| 0.8 | 0.7881 | 0.7910 | 0.7939 | 0.7967 | 0.7995 | 0.8023 | 0.8051 | 0.8078 | 0.8106 | 0.8133 |
| 0.9 | 0.8159 | 0.8186 | 0.8212 | 0.8238 | 0.8264 | 0.8289 | 0.8315 | 0.8340 | 0.8365 | 0.8389 |
| 1.0 | 0.8413 | 0.8438 | 0.8461 | 0.8485 | 0.8508 | 0.8531 | 0.8554 | 0.8577 | 0.8599 | 0.8621 |
| 1.1 | 0.8643 | 0.8665 | 0.8686 | 0.8708 | 0.8729 | 0.8749 | 0.8770 | 0.8790 | 0.8810 | 0.8830 |
| 1.2 | 0.8849 | 0.8869 | 0.8888 | 0.8907 | 0.8925 | 0.8944 | 0.8962 | 0.8980 | 0.8997 | 0.9015 |
| 1.3 | 0.9032 | 0.9049 | 0.9066 | 0.9082 | 0.9099 | 0.9115 | 0.9131 | 0.9147 | 0.9162 | 0.9177 |
| 1.4 | 0.9192 | 0.9207 | 0.9222 | 0.9236 | 0.9251 | 0.9265 | 0.9279 | 0.9292 | 0.9306 | 0.9319 |
| 1.5 | 0.9332 | 0.9345 | 0.9357 | 0.9370 | 0.9382 | 0.9394 | 0.9406 | 0.9418 | 0.9429 | 0.9441 |
| 1.6 | 0.9452 | 0.9463 | 0.9474 | 0.9484 | 0.9495 | 0.9505 | 0.9515 | 0.9525 | 0.9535 | 0.9545 |
| 1.7 | 0.9554 | 0.9564 | 0.9573 | 0.9582 | 0.9591 | 0.9599 | 0.9608 | 0.9616 | 0.9625 | 0.9633 |
| 1.8 | 0.9641 | 0.9649 | 0.9656 | 0.9664 | 0.9671 | 0.9678 | 0.9686 | 0.9693 | 0.9699 | 0.9706 |
| 1.9 | 0.9713 | 0.9719 | 0.9726 | 0.9732 | 0.9738 | 0.9744 | 0.9750 | 0.9756 | 0.9761 | 0.9767 |
| 2.0 | 0.9772 | 0.9778 | 0.9783 | 0.9788 | 0.9793 | 0.9798 | 0.9803 | 0.9808 | 0.9812 | 0.9817 |
| 2.1 | 0.9821 | 0.9826 | 0.9830 | 0.9834 | 0.9838 | 0.9842 | 0.9846 | 0.9850 | 0.9854 | 0.9857 |
| 2.2 | 0.9861 | 0.9864 | 0.9868 | 0.9871 | 0.9875 | 0.9878 | 0.9881 | 0.9884 | 0.9887 | 0.9890 |
| 2.3 | 0.9893 | 0.9896 | 0.9898 | 0.9901 | 0.9904 | 0.9906 | 0.9909 | 0.9911 | 0.9913 | 0.9916 |
| 2.4 | 0.9918 | 0.9920 | 0.9922 | 0.9925 | 0.9927 | 0.9929 | 0.9931 | 0.9932 | 0.9934 | 0.9936 |
| 2.5 | 0.9938 | 0.9940 | 0.9941 | 0.9943 | 0.9945 | 0.9946 | 0.9948 | 0.9949 | 0.9951 | 0.9952 |
| 2.6 | 0.9953 | 0.9955 | 0.9956 | 0.9957 | 0.9959 | 0.9960 | 0.9961 | 0.9962 | 0.9963 | 0.9964 |
| 2.7 | 0.9965 | 0.9966 | 0.9967 | 0.9968 | 0.9969 | 0.9970 | 0.9971 | 0.9972 | 0.9973 | 0.9974 |
| 2.8 | 0.9974 | 0.9975 | 0.9976 | 0.9977 | 0.9977 | 0.9978 | 0.9979 | 0.9979 | 0.9980 | 0.9981 |
| 2.9 | 0.9981 | 0.9982 | 0.9982 | 0.9983 | 0.9984 | 0.9984 | 0.9985 | 0.9985 | 0.9986 | 0.9986 |
| 3.0 | 0.9987 | 0.9987 | 0.9987 | 0.9988 | 0.9988 | 0.9989 | 0.9989 | 0.9989 | 0.9990 | 0.9990 |
| 3.1 | 0.9990 | 0.9991 | 0.9991 | 0.9991 | 0.9992 | 0.9992 | 0.9992 | 0.9992 | 0.9993 | 0.9993 |
| 3.2 | 0.9993 | 0.9993 | 0.9994 | 0.9994 | 0.9994 | 0.9994 | 0.9994 | 0.9995 | 0.9995 | 0.9995 |
| 3.3 | 0.9995 | 0.9995 | 0.9995 | 0.9996 | 0.9996 | 0.9996 | 0.9996 | 0.9996 | 0.9996 | 0.9997 |
| 3.4 | 0.9997 | 0.9997 | 0.9997 | 0.9997 | 0.9997 | 0.9997 | 0.9997 | 0.9997 | 0.9997 | 0.9998 |
| 3.5 | 0.9998 | 0.9998 | 0.9998 | 0.9998 | 0.9998 | 0.9998 | 0.9998 | 0.9998 | 0.9998 | 0.9998 |
| 3.6 | 0.9998 | 0.9998 | 0.9999 | 0.9999 | 0.9999 | 0.9999 | 0.9999 | 0.9999 | 0.9999 | 0.9999 |
| 3.7 | 0.9999 | 0.9999 | 0.9999 | 0.9999 | 0.9999 | 0.9999 | 0.9999 | 0.9999 | 0.9999 | 0.9999 |
| 3.8 | 0.9999 | 0.9999 | 0.9999 | 0.9999 | 0.9999 | 0.9999 | 0.9999 | 0.9999 | 0.9999 | 0.9999 |
| 3.9 | 1.0000† | | | | | | | | | |

†For $z \geq 3.90$, the areas are 1.0000 to four decimal places.

| | Two-tail probability | 0.20 | 0.10 | 0.05 | 0.02 | 0.01 | |
|---|---|---|---|---|---|---|---|
| | One-tail probability | 0.10 | 0.05 | 0.025 | 0.01 | 0.005 | |
| **Table T** Values of $t_\alpha$ | df | | | | | | df |
| | 1 | 3.078 | 6.314 | 12.706 | 31.821 | 63.657 | 1 |
| | 2 | 1.886 | 2.920 | 4.303 | 6.965 | 9.925 | 2 |
| | 3 | 1.638 | 2.353 | 3.182 | 4.541 | 5.841 | 3 |
| | 4 | 1.533 | 2.132 | 2.776 | 3.747 | 4.604 | 4 |
| | 5 | 1.476 | 2.015 | 2.571 | 3.365 | 4.032 | 5 |
| | 6 | 1.440 | 1.943 | 2.447 | 3.143 | 3.707 | 6 |
| | 7 | 1.415 | 1.895 | 2.365 | 2.998 | 3.499 | 7 |
| | 8 | 1.397 | 1.860 | 2.306 | 2.896 | 3.355 | 8 |
| | 9 | 1.383 | 1.833 | 2.262 | 2.821 | 3.250 | 9 |
| | 10 | 1.372 | 1.812 | 2.228 | 2.764 | 3.169 | 10 |
| | 11 | 1.363 | 1.796 | 2.201 | 2.718 | 3.106 | 11 |
| | 12 | 1.356 | 1.782 | 2.179 | 2.681 | 3.055 | 12 |
| | 13 | 1.350 | 1.771 | 2.160 | 2.650 | 3.012 | 13 |
| | 14 | 1.345 | 1.761 | 2.145 | 2.624 | 2.977 | 14 |
| | 15 | 1.341 | 1.753 | 2.131 | 2.602 | 2.947 | 15 |
| | 16 | 1.337 | 1.746 | 2.120 | 2.583 | 2.921 | 16 |
| | 17 | 1.333 | 1.740 | 2.110 | 2.567 | 2.898 | 17 |
| | 18 | 1.330 | 1.734 | 2.101 | 2.552 | 2.878 | 18 |
| | 19 | 1.328 | 1.729 | 2.093 | 2.539 | 2.861 | 19 |
| | 20 | 1.325 | 1.725 | 2.086 | 2.528 | 2.845 | 20 |
| | 21 | 1.323 | 1.721 | 2.080 | 2.518 | 2.831 | 21 |
| | 22 | 1.321 | 1.717 | 2.074 | 2.508 | 2.819 | 22 |
| | 23 | 1.319 | 1.714 | 2.069 | 2.500 | 2.807 | 23 |
| | 24 | 1.318 | 1.711 | 2.064 | 2.492 | 2.797 | 24 |
| | 25 | 1.316 | 1.708 | 2.060 | 2.485 | 2.787 | 25 |
| | 26 | 1.315 | 1.706 | 2.056 | 2.479 | 2.779 | 26 |
| | 27 | 1.314 | 1.703 | 2.052 | 2.473 | 2.771 | 27 |
| | 28 | 1.313 | 1.701 | 2.048 | 2.467 | 2.763 | 28 |
| | 29 | 1.311 | 1.699 | 2.045 | 2.462 | 2.756 | 29 |
| | 30 | 1.310 | 1.697 | 2.042 | 2.457 | 2.750 | 30 |
| | 32 | 1.309 | 1.694 | 2.037 | 2.449 | 2.738 | 32 |
| | 35 | 1.306 | 1.690 | 2.030 | 2.438 | 2.725 | 35 |
| | 40 | 1.303 | 1.684 | 2.021 | 2.423 | 2.704 | 40 |
| | 45 | 1.301 | 1.679 | 2.014 | 2.412 | 2.690 | 45 |
| | 50 | 1.299 | 1.676 | 2.009 | 2.403 | 2.678 | 50 |
| | 60 | 1.296 | 1.671 | 2.000 | 2.390 | 2.660 | 60 |
| | 75 | 1.293 | 1.665 | 1.992 | 2.377 | 2.643 | 75 |
| | 100 | 1.290 | 1.660 | 1.984 | 2.364 | 2.626 | 100 |
| | 120 | 1.289 | 1.658 | 1.980 | 2.358 | 2.617 | 120 |
| | 140 | 1.288 | 1.656 | 1.977 | 2.353 | 2.611 | 140 |
| | 180 | 1.286 | 1.653 | 1.973 | 2.347 | 2.603 | 180 |
| | 250 | 1.285 | 1.651 | 1.969 | 2.341 | 2.596 | 250 |
| | 400 | 1.284 | 1.649 | 1.966 | 2.336 | 2.588 | 400 |
| | 1000 | 1.282 | 1.646 | 1.962 | 2.330 | 2.581 | 1000 |
| | $\infty$ | 1.282 | 1.645 | 1.960 | 2.326 | 2.576 | $\infty$ |
| **Confidence levels** | | 80% | 90% | 95% | 98% | 99% | |

An interactive version of this table is at www.astools.datadesk.com/tdist_table.html

**Table X**

Values of $\chi_\alpha^2$

An interactive version of this table is at www.astools.datadesk.com/chi_table.html

| Right-tail probability | 0.10 | 0.05 | 0.025 | 0.01 | 0.005 |
|---|---|---|---|---|---|
| df | | | | | |
| 1 | 2.706 | 3.841 | 5.024 | 6.635 | 7.879 |
| 2 | 4.605 | 5.991 | 7.378 | 9.210 | 10.597 |
| 3 | 6.251 | 7.815 | 9.348 | 11.345 | 12.838 |
| 4 | 7.779 | 9.488 | 11.143 | 13.277 | 14.860 |
| 5 | 9.236 | 11.070 | 12.833 | 15.086 | 16.750 |
| 6 | 10.645 | 12.592 | 14.449 | 16.812 | 18.548 |
| 7 | 12.017 | 14.067 | 16.013 | 18.475 | 20.278 |
| 8 | 13.362 | 15.507 | 17.535 | 20.090 | 21.955 |
| 9 | 14.684 | 16.919 | 19.023 | 21.666 | 23.589 |
| 10 | 15.987 | 18.307 | 20.483 | 23.209 | 25.188 |
| 11 | 17.275 | 19.675 | 21.920 | 24.725 | 26.757 |
| 12 | 18.549 | 21.026 | 23.337 | 26.217 | 28.300 |
| 13 | 19.812 | 22.362 | 24.736 | 27.688 | 29.819 |
| 14 | 21.064 | 23.685 | 26.119 | 29.141 | 31.319 |
| 15 | 22.307 | 24.996 | 27.488 | 30.578 | 32.801 |
| 16 | 23.542 | 26.296 | 28.845 | 32.000 | 34.267 |
| 17 | 24.769 | 27.587 | 30.191 | 33.409 | 35.718 |
| 18 | 25.989 | 28.869 | 31.526 | 34.805 | 37.156 |
| 19 | 27.204 | 30.143 | 32.852 | 36.191 | 38.582 |
| 20 | 28.412 | 31.410 | 34.170 | 37.566 | 39.997 |
| 21 | 29.615 | 32.671 | 35.479 | 38.932 | 41.401 |
| 22 | 30.813 | 33.924 | 36.781 | 40.290 | 42.796 |
| 23 | 32.007 | 35.172 | 38.076 | 41.638 | 44.181 |
| 24 | 33.196 | 36.415 | 39.364 | 42.980 | 45.559 |
| 25 | 34.382 | 37.653 | 40.647 | 44.314 | 46.928 |
| 26 | 35.563 | 38.885 | 41.923 | 45.642 | 48.290 |
| 27 | 36.741 | 40.113 | 43.195 | 46.963 | 49.645 |
| 28 | 37.916 | 41.337 | 44.461 | 48.278 | 50.994 |
| 29 | 39.087 | 42.557 | 45.722 | 59.588 | 52.336 |
| 30 | 40.256 | 43.773 | 46.979 | 50.892 | 53.672 |
| 40 | 51.805 | 55.759 | 59.342 | 63.691 | 66.767 |
| 50 | 63.167 | 67.505 | 71.420 | 76.154 | 79.490 |
| 60 | 74.397 | 79.082 | 83.298 | 88.381 | 91.955 |
| 70 | 85.527 | 90.531 | 95.023 | 100.424 | 104.213 |
| 80 | 96.578 | 101.879 | 106.628 | 112.328 | 116.320 |
| 90 | 107.565 | 113.145 | 118.135 | 124.115 | 128.296 |
| 100 | 118.499 | 124.343 | 129.563 | 135.811 | 140.177 |

## Table of random digits

| Row | | | | | | | | | | |
|---|---|---|---|---|---|---|---|---|---|---|
| 1 | 96299 | 07196 | 98642 | 20639 | 23185 | 56282 | 69929 | 14125 | 38872 | 94168 |
| 2 | 71622 | 35940 | 81807 | 59225 | 18192 | 08710 | 80777 | 84395 | 69563 | 86280 |
| 3 | 03272 | 41230 | 81739 | 74797 | 70406 | 18564 | 69273 | 72532 | 78340 | 36699 |
| 4 | 46376 | 58596 | 14365 | 63685 | 56555 | 42974 | 72944 | 96463 | 63533 | 24152 |
| 5 | 47352 | 42853 | 42903 | 97504 | 56655 | 70355 | 88606 | 61406 | 38757 | 70657 |
| 6 | 20064 | 04266 | 74017 | 79319 | 70170 | 96572 | 08523 | 56025 | 89077 | 57678 |
| 7 | 73184 | 95907 | 05179 | 51002 | 83374 | 52297 | 07769 | 99792 | 78365 | 93487 |
| 8 | 72753 | 36216 | 07230 | 35793 | 71907 | 65571 | 66784 | 25548 | 91861 | 15725 |
| 9 | 03939 | 30763 | 06138 | 80062 | 02537 | 23561 | 93136 | 61260 | 77935 | 93159 |
| 10 | 75998 | 37203 | 07959 | 38264 | 78120 | 77525 | 86481 | 54986 | 33042 | 70648 |
| 11 | 94435 | 97441 | 90998 | 25104 | 49761 | 14967 | 70724 | 67030 | 53887 | 81293 |
| 12 | 04362 | 40989 | 69167 | 38894 | 00172 | 02999 | 97377 | 33305 | 60782 | 29810 |
| 13 | 89059 | 43528 | 10547 | 40115 | 82234 | 86902 | 04121 | 83889 | 76208 | 31076 |
| 14 | 87736 | 04666 | 75145 | 49175 | 76754 | 07884 | 92564 | 80793 | 22573 | 67902 |
| 15 | 76488 | 88899 | 15860 | 07370 | 13431 | 84041 | 69202 | 18912 | 83173 | 11983 |
| 16 | 36460 | 53772 | 66634 | 25045 | 79007 | 78518 | 73580 | 14191 | 50353 | 32064 |
| 17 | 13205 | 69237 | 21820 | 20952 | 16635 | 58867 | 97650 | 82983 | 64865 | 93298 |
| 18 | 51242 | 12215 | 90739 | 36812 | 00436 | 31609 | 80333 | 96606 | 30430 | 31803 |
| 19 | 67819 | 00354 | 91439 | 91073 | 49258 | 15992 | 41277 | 75111 | 67496 | 68430 |
| 20 | 09875 | 08990 | 27656 | 15871 | 23637 | 00952 | 97818 | 64234 | 50199 | 05715 |
| 21 | 18192 | 95308 | 72975 | 01191 | 29958 | 09275 | 89141 | 19558 | 50524 | 32041 |
| 22 | 02763 | 33701 | 66188 | 50226 | 35813 | 72951 | 11638 | 01876 | 93664 | 37001 |
| 23 | 13349 | 46328 | 01856 | 29935 | 80563 | 03742 | 49470 | 67749 | 08578 | 21956 |
| 24 | 69238 | 92878 | 80067 | 80807 | 45096 | 22936 | 64325 | 19265 | 37755 | 69794 |
| 25 | 92207 | 63527 | 59398 | 29818 | 24789 | 94309 | 88380 | 57000 | 50171 | 17891 |
| 26 | 66679 | 99100 | 37072 | 30593 | 29665 | 84286 | 44458 | 60180 | 81451 | 58273 |
| 27 | 31087 | 42430 | 60322 | 34765 | 15757 | 53300 | 97392 | 98035 | 05228 | 68970 |
| 28 | 84432 | 04916 | 52949 | 78533 | 31666 | 62350 | 20584 | 56367 | 19701 | 60584 |
| 29 | 72042 | 12287 | 21081 | 48426 | 44321 | 58765 | 41760 | 43304 | 13399 | 02043 |
| 30 | 94534 | 73559 | 82135 | 70260 | 87936 | 85162 | 11937 | 18263 | 54138 | 69564 |
| 31 | 63971 | 97198 | 40974 | 45301 | 60177 | 35604 | 21580 | 68107 | 25184 | 42810 |
| 32 | 11227 | 58474 | 17272 | 37619 | 69517 | 62964 | 67962 | 34510 | 12607 | 52255 |
| 33 | 28541 | 02029 | 08068 | 96656 | 17795 | 21484 | 57722 | 76511 | 27849 | 61738 |
| 34 | 11282 | 43632 | 49531 | 78981 | 81980 | 08530 | 08629 | 32279 | 29478 | 50228 |
| 35 | 42907 | 15137 | 21918 | 13248 | 39129 | 49559 | 94540 | 24070 | 88151 | 36782 |
| 36 | 47119 | 76651 | 21732 | 32364 | 58545 | 50277 | 57558 | 30390 | 18771 | 72703 |
| 37 | 11232 | 99884 | 05087 | 76839 | 65142 | 19994 | 91397 | 29350 | 83852 | 04905 |
| 38 | 64725 | 06719 | 86262 | 53356 | 57999 | 50193 | 79936 | 97230 | 52073 | 94467 |
| 39 | 77007 | 26962 | 55466 | 12521 | 48125 | 12280 | 54985 | 26239 | 76044 | 54398 |
| 40 | 18375 | 19310 | 59796 | 89832 | 59417 | 18553 | 17238 | 05474 | 33259 | 50595 |

| Assumptions for Inference | And the Conditions That Support or Override Them |
|---|---|
| **Proportions ($z$)** | |
| ◆ **One sample** | |
| 1. Individuals are independent. | 1. SRS and $n < 10\%$ of the population. |
| 2. Sample is sufficiently large. | 2. Successes and failures each $\geq 10$. |
| ◆ **Two groups** | |
| 1. Groups are independent. | 1. (Think about how the data were collected.) |
| 2. Data in each group are independent. | 2. Both are SRSs and $n < 10\%$ of populations OR random allocation. |
| 3. Both groups are sufficiently large. | 3. Successes and failures each $\geq 10$ for both groups. |
| **Means ($t$)** | |
| ◆ **One sample** (df $= n - 1$) | |
| 1. Individuals are independent. | 1. SRS and $n < 10\%$ of the population. |
| 2. Population has a Normal model. | 2. Histogram is unimodal and symmetric.* |
| ◆ **Matched pairs** (df $= n - 1$) | |
| 1. Data are matched. | 1. (Think about the design.) |
| 2. Individuals are independent. | 2. SRS and $n < 10\%$ OR random allocation. |
| 3. Population of differences is Normal. | 3. Histogram of differences is unimodal and symmetric.* |
| ◆ **Two independent groups** (df from technology) | |
| 1. Groups are independent. | 1. (Think about the design.) |
| 2. Data in each group are independent. | 2. SRSs and $n < 10\%$ OR random allocation. |
| 3. Both populations are Normal. | 3. Both histograms are unimodal and symmetric.* |
| **Distributions/Association ($\chi^2$)** | |
| ◆ **Goodness of fit** [df $=$ # of cells $-1$; one variable, one sample compared with population model] | |
| 1. Data are counts. | 1. (Are they?) |
| 2. Data in sample are independent. | 2. SRS and $n < 10\%$ of the population. |
| 3. Sample is sufficiently large. | 3. All expected counts $\geq 5$. |
| ◆ **Homogeneity** [df $= (r-1)(c-1)$; many groups compared on one variable] | |
| 1. Data are counts. | 1. (Are they?) |
| 2. Data in groups are independent. | 2. SRSs and $n < 10\%$ OR random allocation. |
| 3. Groups are sufficiently large. | 3. All expected counts $\geq 5$. |
| ◆ **Independence** [df $= (r-1)(c-1)$; sample from one population classified on two variables] | |
| 1. Data are counts. | 1. (Are they?) |
| 2. Data are independent. | 2. SRSs and $n < 10\%$ of the population. |
| 3. Sample is sufficiently large. | 3. All expected counts $\geq 5$. |
| **Regression ($t$, df $= n - 2$)** | |
| ◆ **Association** between two quantitative variables ($\beta = 0$?) | |
| 1. Form of relationship is linear. | 1. Scatterplot looks approximately linear. |
| 2. Errors are independent. | 2. No apparent pattern in residuals plot. |
| 3. Variability of errors is constant. | 3. Residuals plot has consistent spread. |
| 4. Errors have a Normal model. | 4. Histogram of residuals is approximately unimodal and symmetric, or normal probability plot reasonably straight.* |
| | (*less critical as $n$ increases) |